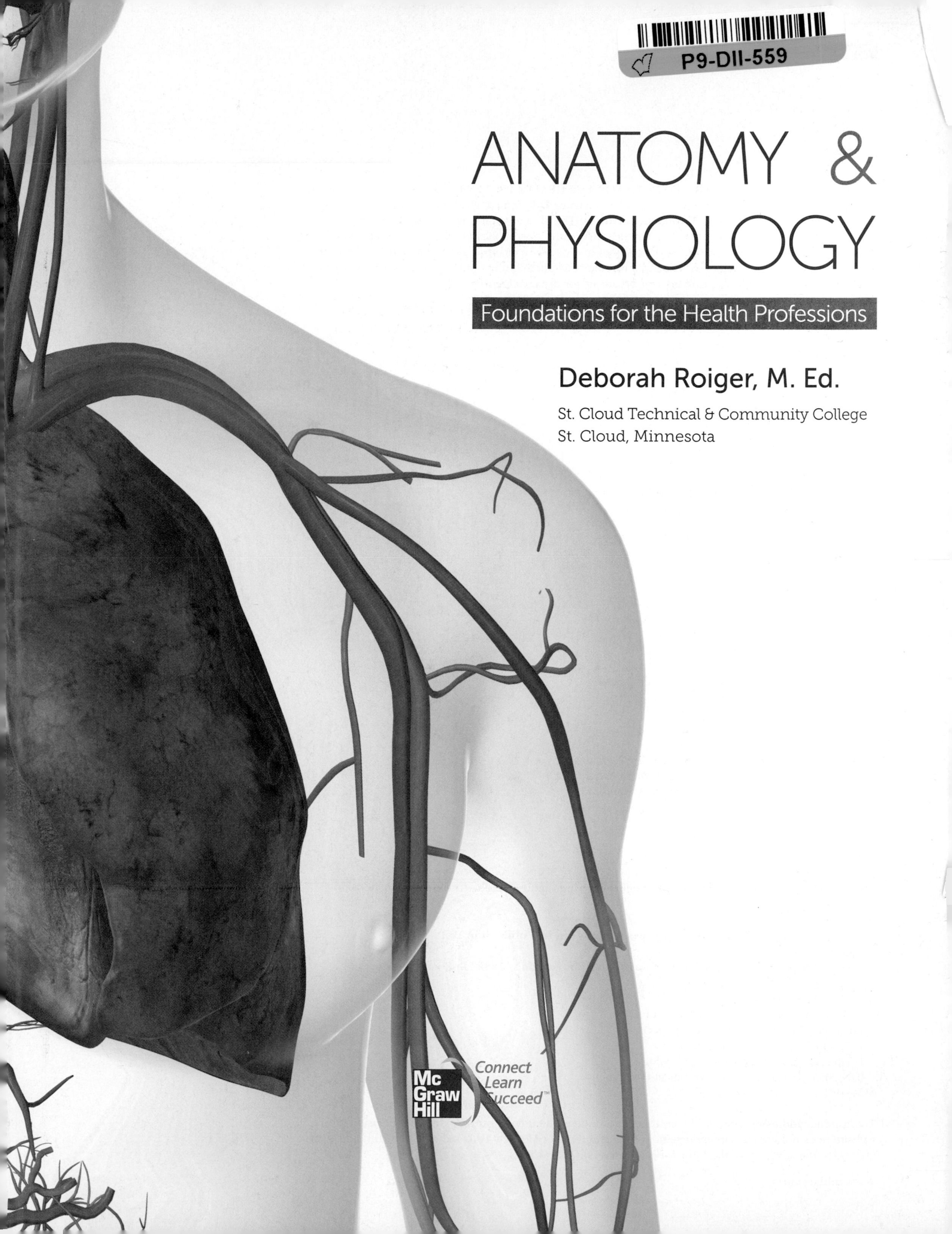

ANATOMY & PHYSIOLOGY

Foundations for the Health Professions

Deborah Roiger, M. Ed.

St. Cloud Technical & Community College
St. Cloud, Minnesota

Mc Graw Hill
Connect Learn Succeed™

ANATOMY & PHYSIOLOGY: FOUNDATIONS FOR THE HEALTH PROFESSIONS
Published by McGraw-Hill, a business unit of The McGraw-Hill Companies, Inc., 1221 Avenue of the Americas, New York, NY, 10020.

Some ancillaries, including electronic and print components, may not be available to customers outside the United States.

This book is printed on acid-free paper.

1 2 3 4 5 6 7 8 9 0 DOW/DOW 1 0 9 8 7 6 5 4 3 2

ISBN 978-0-07-340212-3
MHID 0-07-340212-5

Editorial director: *Michael S. Ledbetter*
Vice president/Director of marketing: *Alice Harra*
Publisher: *Kenneth S. Kasee Jr.*
Director, digital products: *Crystal Szewczyk*
Managing development editor: *Christine Scheid*
Development editor: *Edward Helmold*
Marketing manager: *Mary B. Haran*
Marketing specialist: *Ada Bjorklund-Moore*
Digital development editor: *Katherine Ward*
Director, Editing/Design/Production: *Jess Ann Kosic*
Project manager: *Marlena Pechan*
Senior buyer: *Sandy Ludovissy*
Senior designer: *Marianna Kinigakis*
Senior photo research coordinator: *John C. Leland*
Photo researcher: *Danny Meldung, Photo Affairs*
Manager, digital production: *Janean A. Utley*
Media project manager: *Brent dela Cruz*
Media project manager: *Cathy L. Tepper*
Outside development house: *TripleSSS/Andrea Edwards*
Outside development house: *Laserwords/Jodie Bernard*
Cover design: *Cody B. Wallis and Nathan Kirkman*
Interior design: *Maureen McCutcheon Design*
Typeface: *10/12 ITC Garamond Std Light*
Compositor: *Laserwords Private Limited*
Printer: *R. R. Donnelley*
Cover credit: Illustration © Nucleus Medical Media, all rights reserved
Credits: The credits section for this book begins on page Credits-1 and is considered an extension of the copyright page.

Library of Congress Cataloging-in-Publication Data

Roiger, Deborah.
Anatomy & physiology : foundations for the health professions / Deborah Roiger.
p. ; cm.
Anatomy and physiology
Includes bibliographical references and index.
ISBN-13: 978-0-07-340212-3 (alk. paper)
ISBN-10: 0-07-340212-5 (alk. paper)
I. Title. II. Title: Anatomy and physiology.
[DNLM: 1. Anatomy. 2. Physiological Phenomena. QS 4]
612—dc23

2011040248

The Internet addresses listed in the text were accurate at the time of publication. The inclusion of a website does not indicate an endorsement by the authors or McGraw-Hill, and McGraw-Hill does not guarantee the accuracy of the information presented at these sites.

www.mhhe.com

McGraw-Hill Higher Education and Blackboard have teamed up. What does this mean for you?

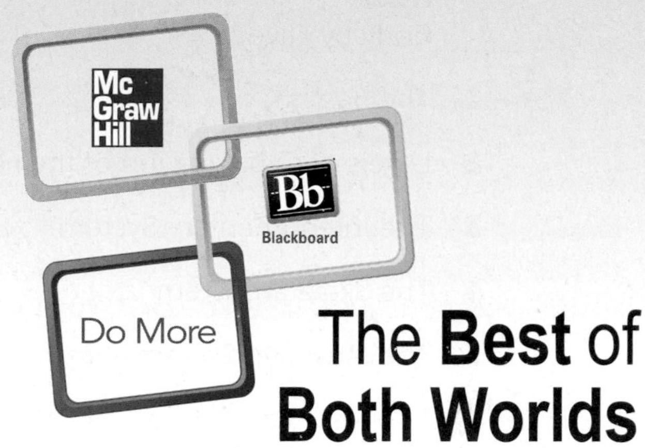

1. **Your life, simplified.** Now you and your students can access McGraw-Hill's *Connect Plus* and Create right from within your Blackboard course—all with one single sign-on. Say goodbye to the days of logging in to multiple applications.

2. **Deep integration of content and tools.** Not only do you get single sign-on with *Connect Plus* and Create, you also get deep integration of McGraw-Hill content and content engines right in Blackboard. Whether you're choosing a book for your course or building *Connect Plus* assignments, all the tools you need are right where you want them—inside Blackboard.

3. **Seamless gradebooks.** Are you tired of keeping multiple gradebooks and manually synchronizing grades into Blackboard? We thought so. When a student completes an integrated *Connect Plus* assignment, the grade for that assignment authomatically (and instantly) feeds your Blackboard grade center.

4. **A solution for everyone.** Whether your institution is already using Blackboard, or you just want to try Blackboard on your own, we have a solution for you. McGraw-Hill and Blackboard can now offer you easy access to industry-leading technology and content, whether your campus hosts it, or we do. Be sure to ask your local McGraw-Hill representative for details.

brief contents

about the **author**

Deborah Roiger

M.A. Education, St. Mary's
University of Minnesota

B.S. Biology and Earth Science,
St. Cloud State University

Deborah Roiger had originally wanted to pursue a career as a doctor, but an undergraduate education in child psychology took her in a different direction. After graduation, she became a therapist at a residential treatment center for boys. There she learned that clear expectations, specific objectives, and high standards are crucial for the success of all individuals. Her experience at the St. Cloud Children's Home prepared her well in the basic values of teaching.

After starting her family, Deborah went back to school for a teaching degree in life science. She also took the place of her high-school mentor when he retired. She earned her master's degree in education and began teaching anatomy and physiology at St. Cloud State University. Four years later, Deborah was recruited by St. Cloud Technical College (SCTC) to develop the A&P curriculum, design and furnish an A&P lab, and teach anatomy and physiology to their many health professions students.

Deborah has taught for 12 years and has experienced success in the Minnesota State Colleges and University system (MNSCU) of 32 state-sponsored technical colleges, community colleges, and universities. She was voted instructor of the year in 2007 by the students at SCTC, received five awards for excellence by MNSCU for projects she developed, and was named educator of the year in 2009 by the MNSCU board of trustees. Deborah's development of digital resources for teaching anatomy and physiology online also earned an Innovation of the Year award in 2009 from the League for Innovation in the Community College.

Deborah has presented her work in the development of anatomy and physiology resources at three national conferences. She is a member of Human Anatomy and Physiology Society (HAPS), American Academy for the Advancement of Science (AAAS), and National Science Teachers' Association.

preface

Not only is the structure of the human body fascinating, but no subject is more relevant to students' lives than how their own bodies function. Students seeking careers as health care professionals can be doubly motivated when taking an anatomy and physiology course, as the content is relevant to their lives and prepares them for their chosen careers.

As instructors, we look for a text that is, first and foremost, accurate; is written at an appropriate level for our students—neither too high nor too low; can be customized to accommodate the organization of our individual courses; and will be interesting and appealing to our students. This textbook and accompanying suite of products went through a development process that included:

- No less than five levels of rigorous market review and accuracy checking, from first-draft manuscript through the final round of page proofs.
- Reading-level assessments completed throughout the development of the textbook to ensure readability.
- The mapping of every piece of content to specific learning outcomes, giving instructors the option to choose which learning outcomes they want to cover and easily assign only the content that addresses those learning outcomes.
- The creation and incorporation of a brand-new art program, with enlarged figures to accommodate visual learners.

Throughout the course of a development process that is first and foremost market-driven, many instructors reviewed this entire product suite multiple times and agreed that this is what they've been waiting for. Reviewer comments include:

Great job. I felt that you have a lot of passion.
—Adrian Rios, *Newbridge College*

The writing style is easy to read and the topics are understandable. The author doesn't lose the student in too much detail.
—Tim Feltmeyer, *Erie Business Center*

It is well diagramed, it contains a great deal of questions to stimulate thought about the subject, it allows entry level students to pick this text up and learn from it.
—Gerald Heins, *Northeast Wisconsin Technical College*

Physiology is explained very well, and quite understandable.
—Kathleen Metzger, *Valdosta Technical College*

A good introductory level textbook that students might actually like to read!
—Tina Squire, *Northland Community & Technical College*

Three Key Principles

The pedagogical approach of this product suite is founded on three key principles:

1. **Tell our students what we are going to teach them.** Each chapter in this text and workbook begins with a list of specific learning outcomes.
2. **Teach our students what we told them we would.** Everything in this text, workbook, and all ancillary materials directly relates to the learning outcomes stated at the beginning of every chapter. Real-life situations, analogies, and a commonsense, direct approach are used to teach anatomy and physiology concepts.
3. **Test our students on what we taught.** All assignments, activities, discussion questions, review questions, and test questions designed for this text and workbook directly assess only the learning outcomes stated in the beginning of each chapter.

Following these principles means there will be no surprises for the student. If the learning outcomes are stressed through lecture and assignments, students quickly find comfort in having a guideline to follow in tackling this content.

Walkthrough

Pedagogical Features

As educators, we need tools to help with our instruction. The following elements are built into this text to facilitate student learning.

I. Specific Learning Outcomes

Specific learning outcomes give clear expectations and direct student learning. Every piece of content, including the text, figures, and tables in each chapter, directly relates to the learning outcomes.

Outcome tabs on the edge of the text page allow students to easily access the content corresponding to a specific learning outcome.

The learning outcomes are very helpful to students. These give a target or goal set of what one is about to accomplish by studying this text.

—Ryan Morris
Pierce College

Anatomy of a Neuron

6.3 learning **outcome**
Describe the anatomy of a neuron.

A single neuron can be very long (measuring a meter or more), as it may start at the tip of your finger and end at your spinal cord. A neuron has three basic parts: one or more **dendrites** that receive information; a **body** containing the nucleus and organelles for protein synthesis; and an **axon** that carries the nerve impulse along

Chapter summaries and review questions are located at the end of each chapter. As with the rest of the content, all summaries and review questions directly relate to the chapter learning outcomes.

Chapter Mapping at the end of each chapter quickly directs the student to the location of all text, figures, tables, and review questions relevant to each of the learning outcomes. This feature also allows instructors to easily customize the content to meet their needs.

7 **chapter mapping**

This section of the chapter is designed to help you find where each outcome is covered in this text.

	Outcomes	Readings, figures, and tables	Assessments
7.1	Use medical terminology related to the senses of the nervous system.	Word roots: p. 263	Word Building: 1–5
7.2	Classify the senses in terms of what is sensed and where the receptors are located.	Overview: p. 264	Multiple Select: 3
7.3	Describe the sensory receptors for the general senses in the skin.	General senses: pp. 264–265 Figure 7.2 Table 7.1	Spot Check: 1, 2 Matching: 7
7.4	Explain the types of information transmitted by sensory receptors in the skin.	Physiology of the general senses of the skin: pp. 265–267 Figure 7.3	
7.5	Describe the pathway for pain.	Pathway for pain: pp. 267–268 Figure 7.4	Multiple Select: 2
7.6	Describe the sensory receptors for taste.	Taste: pp. 268–269 Figure 7.5	Matching: 6

II. Anatomy and Physiology in Context

Physiology concepts are emphasized and put in context with real-world examples. The functions of each system are explained in the context of a specific individual, and interconnections are made between new concepts and content in previous chapters.

The author has very good balance of physiology and anatomy, and integrates them well consistently throughout the chapter. Each concept is completely developed in context, giving much greater understanding of utilization of the physiology with anatomy.

—Pamela McNamara
Beckfield College

A homeostasis theme carries throughout the text, indicated by an icon that shows the interdependent relationship between homeostasis and acid-base balance, fluid and electrolytes, and nutrition. These concepts are covered in the context of the body systems, not isolated in separate chapters. As each topic is relevant to homeostasis in the text, the appropriate section of the icon is highlighted in the margin.

Putting the Pieces Together ties the system covered in each system chapter to the other 10 systems of the human body.

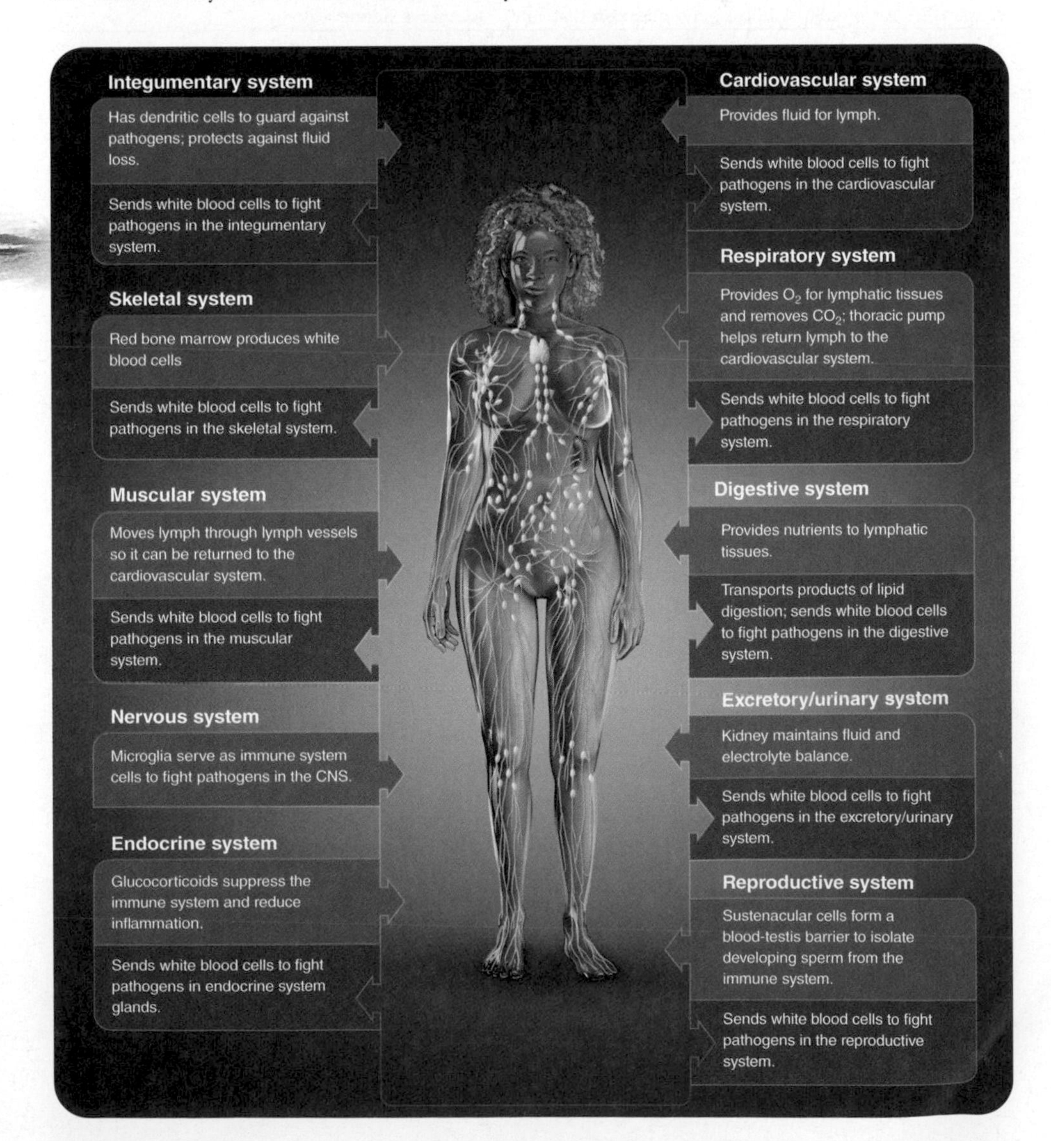

Cadaver photos are paired with **illustrations.** Students will ultimately be working with real bodies, not models and illustrations; therefore, they need to see how a real body looks.

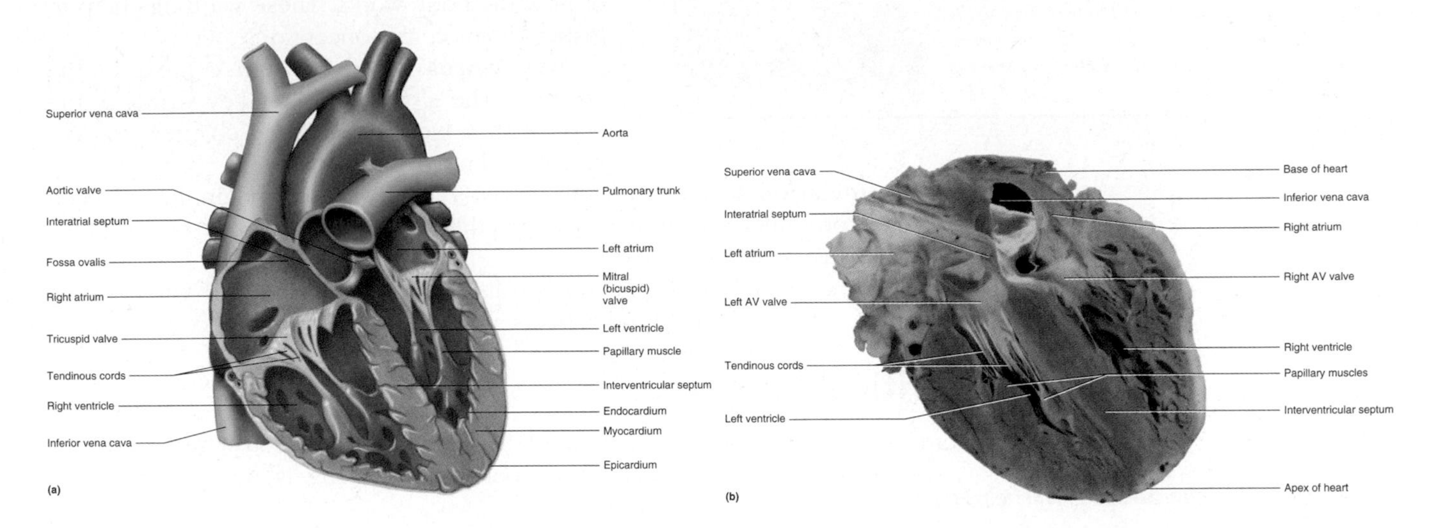

Clinical Point boxes make the connection between anatomy and physiology concepts and applications in health professions.

Applied Genetics boxes focus on genetic applications to anatomy and physiology.

clinical point

A **murmur** is an abnormal heart sound. It may be a **functional murmur** (not a problem) or a **pathological murmur** (possible leaky valve). The murmur often makes a *ssh* sound. If the heart sound is lubbssh-dupp, lubbssh-dupp, the AV valves may be suspected of leaking because the abnormal sound is occurring with the first sound in the cardiac cycle.

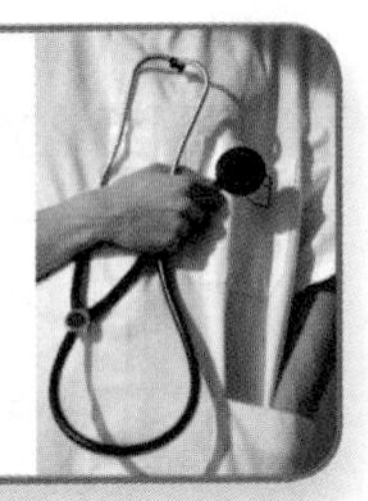

applied genetics

Women with mutations in their BRCA1 or BRCA2 gene have a substantially greater risk of developing breast and ovarian cancer. It is possible to genetically test for these changes. Once the presence of the mutations in either the BRCA1 or BRCA2 gene has been determined, some women choose to have double mastectomies and **oophorectomies** (surgical removal of the ovaries) before any cancer has been detected. The National Cancer Institute reports that this prophylactic removal of the breasts and ovaries is an effective way to reduce the risk of breast and ovarian cancer.[1]

An effects of aging section is included in each system chapter to help students understand the ever-increasing geriatric patient population.

Disorders common to each system are included in each system chapter to help students understand the relationship of abnormal anatomy and physiology to pathology.

Spot Checks are interspersed throughout each chapter to give the student an opportunity to check for understanding before moving on. The answers to these questions are found in Appendix C.

I like the "spot checks", "clinical points", marginal outcomes, and the variety of the end of chapter review activities. Each of these will help with different learning styles.

—Marilyn Turner
Ogeechee Technical College

spot check **3** What is common in the naming of hormones produced by the hypothalamus that target the anterior pituitary?

warning

Whether skin is "thick" or "thin" does not have to do with the depth of the skin or the number of cell layers. It has to do with the presence of the stratum lucidum, which is present only in thick skin. Thick skin is found on the lips, on the palms of the hands, and on the plantar surface of the feet. Thick skin has no hair, while thin skin does.

Warning boxes alert the student to possible mistakes and misconceptions. Most students come to the course with preconceived notions of how the body works. These warnings help to dispel common misconceptions.

Key words are defined in context in the chapters. The glossary defines key words in the context they were used in the chapters to activate prior knowledge.

A pronunciation key for difficult terms is located at the beginning of each chapter. Each pronunciation is also located in the margin next to the first occurrence of the vocabulary word in the text.

Word roots specific to the system are listed at the beginning of each system chapter, and common prefixes and suffixes are listed inside the back cover.

III. Critical Thinking

Anatomy and physiology is so much more than memorization of structures and their functions. A deeper understanding is needed by health professionals.

The inclusion of clinical cases will help Health Science students see how this is applicable to what they are learning or will learn in their program, and that this information is indeed what they need to know to be in Health Sciences.

—Karen Dunbar
Ivy Tech Community College

Case studies appear throughout the chapters to help the students make connections in real-life situations.

Critical-thinking questions are included as part of the chapter review questions in every chapter.

Critical Thinking

1. The effects of smoking on Carol's respiratory system function were discussed in this chapter. What other environmental factors and lifestyle choices would adversely affect the respiratory system? Explain.
2. What can be done to minimize the effects of aging on the respiratory system? Explain.
3. Bob participated in an A&P lab during the respiratory system unit. He recorded his respiratory rate and his tidal volume at rest and then again after running around the parking lot at his school. His results were as shown to the right:

	At rest	Immediately after the run
Respiratory rate:	12 breaths/min	20 breaths/min
Tidal volume:	400 mL	650 mL

How much more air per minute was Bob's respiratory system able to move after the run than before it? What caused his values to change? Explain your response in terms of the physiology of the respiratory system.

Instructor Resources

The Online Learning Center (**www.mhhe.com/roigerap**) that is included with this text contains a host of teaching resources at your disposal, including an Instructor's Manual written by the author; PowerPoint presentations with talking points for each chapter and an Image Library containing all figures—both with and without labels; nearly 1,500 EZTest Online test questions; a correlation grid of all content tied to CAAHEP and ABHES competencies; a Lesson Plan with hot links to all the content and resources tied to each learning outcome; and a student Review Guide.

Workbook

A **full-color workbook**, written by the author, is available for purchase to reinforce the lessons that students learn in each chapter of the text. Workbook features include a coloring book section, lab exercises and activities, key-word concept maps, and review questions.

acknowledgments

I would like to thank my editors at McGraw-Hill—Ken Kasee and Edward Helmold—and Andrea Edwards for their guidance on the writing process; Jodie Bernard for her creative eye in planning the art; and Marlena Pechan, who helped me through the production process. I would especially like to thank Nadine Holland, Nancy Neuwirth, and Dr. Nia Bullock—Nadine for her careful eye as a reader of my rough drafts and her advice concerning pedagogy; Nancy, my lab assistant, for her attention to detail in proofing the manuscript and the art; and Nia for always being there as a sounding board with good advice and support. Lastly, I would like to thank my husband David, whose patience and encouragement make all things seem possible.

contributors

Test Bank Author
Nia Bullock, Ph.D.
Miller-Motte College

PowerPoint Presentation Author
Jason Lapres, M.H.S.
Lone Star College

reviewers

Fazal Aasi
El Camino College

Meghan Andrikanich
Lorain County Community College

Anna Avola
Hodges University

Christina Bain
The Salter School, Tewksbury

Katherine Baus
Southwest Florida College

Wendi Bennet
Antonelli College

Stephanie Bernard
Sanford Brown College, Jacksonville

Daniel Bickerton
Ogeechee Technical College

Gerry Brasin
Premier Education Group

Kim Bricker
Harrison College

Linda Ciarleglio
Stone Academy

Teresa Cowan
Baker College of Clinton Township

Tammy Denesha
Branford Hall Career Institute

Karen Dunbar
Ivy Tech Community College

Lori Ebert
Brown Mackie College

Rhonda Epps, CMA, RMA
National College of Business and Technology, Knoxville

Todd Farney
Wichita Technical Institute

Tim Feltmeyer
Erie Business Center

Paul Fierimonte
Middlesex Community College

Ruby Fogg
Manchester Community College

Karen Frederick
Terra State Community College

Daniel Graetzer
Northwest University

Brittaney Harp
Beckfield College

Gerald Heins
Northwest Wisconsin Technical College

Brian Scott Hobson
University of Arkansas Community College

Linda Iavarone
Rhode Island Hospital School of Diagnostic Imaging

Vanessa Ingrassia
Seacoast Career School

Helena Kronick
The Salter School, Tewksbury

Jason LaPres
Lone Star College

Rosanne Magarelli
Mesa Community College

Nancy Maisonet
Branford Hall Career Institute

Sundeep Majumdar
Texas A&M University

Philip A. Mayo
Premier Education Group

Pamela McNamara
Beckfield College, Tri-County Campus

Kathleen Metzger
Valdosta Technical College

Narine Mirijanian
Baker College of Auburn Hills

Ryan Morris
Pierce College

Judy Pento
Seacoast Career School

Paula C. Perkins
Stone Academy

Wanda Ragland
Macomb Community College

Adrian Rios
Newbridge College

Matthew Robinson
Seacoast Career School

Janette Rodriguez
Wood Tobe-Coburn School

Dan Ropek
Columbia Gorge Community College

Robert Roth
Miller Motte Technical College

Gwen Ruff
Gwinnett College

Carolyn Santiago
Santa Barbara Business College

Lisa Schaffer
Newbridge College

Melody Schweitzer
Brown Mackie College, Louisville

Ramon Selove
The Ohio State University

Ann Simpson
Lord Fairfax Community College

Deborah Smith
Southeastern Technical College

Tina Squire
Northland College

Marilyn Turner
Ogeechee Technical College

Hildreth Diane Youmans
Ogeechee Technical College

focus group attendees

Linda Ciarleglio
Stone Academy

Estelle Coffino
The College of Westchester

Todd Farney
Wichita Technical Institute

Grant Iannelli
Kaplan University

Robin Jones
Brown Mackie College

Katin Keirstead
Seacoast Career School

Heather Kies
Goodwin College

career education summit attendees

Khaliff Ali
Fortis College

Scott Eide
Minnesota School of Business

Linda Hitt
Pima Medical Institute

Penny Lee
Med Tech College

Deborah McCray
Florida Career College

Deward Reece
Sanford Brown College

Cathy Teel-Borowski
McCann School of Business and Technology

Jae Sands
Southern Technical College

Jenn Shulte
Academy of Healing Arts

dedication

I would like to dedicate this book to all of my anatomy and physiology students who have taught me so much.

contents

4 The Skeletal System

5 The Muscular System

6 The Nervous System

7 The Nervous System—Senses

8 The Endocrine System

9 The Cardiovascular System—Blood

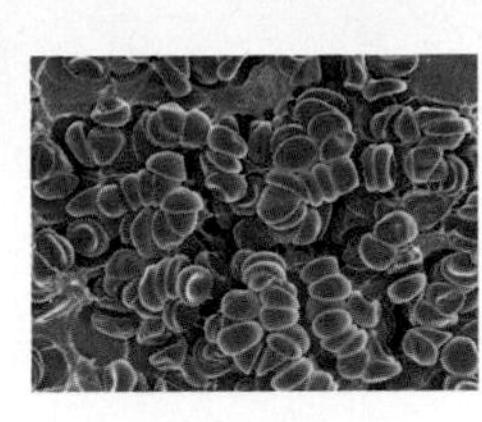

10 The Cardiovascular System—Heart and Vessels

11 The Lymphatic System

12 The Respiratory System

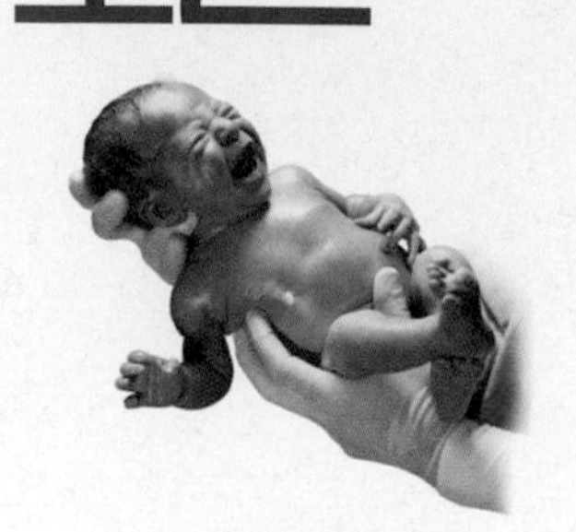

13 The Digestive System

14 The Excretory/Urinary System

15 The Male Reproductive System

16 The Female Reproductive System

1 The Basics

Welcome to the start of your education on the structures and functions of the human body. Whether you are pursuing a health career or reading this book for personal interest, no subject will be more relevant to you than your own body and how it functions.

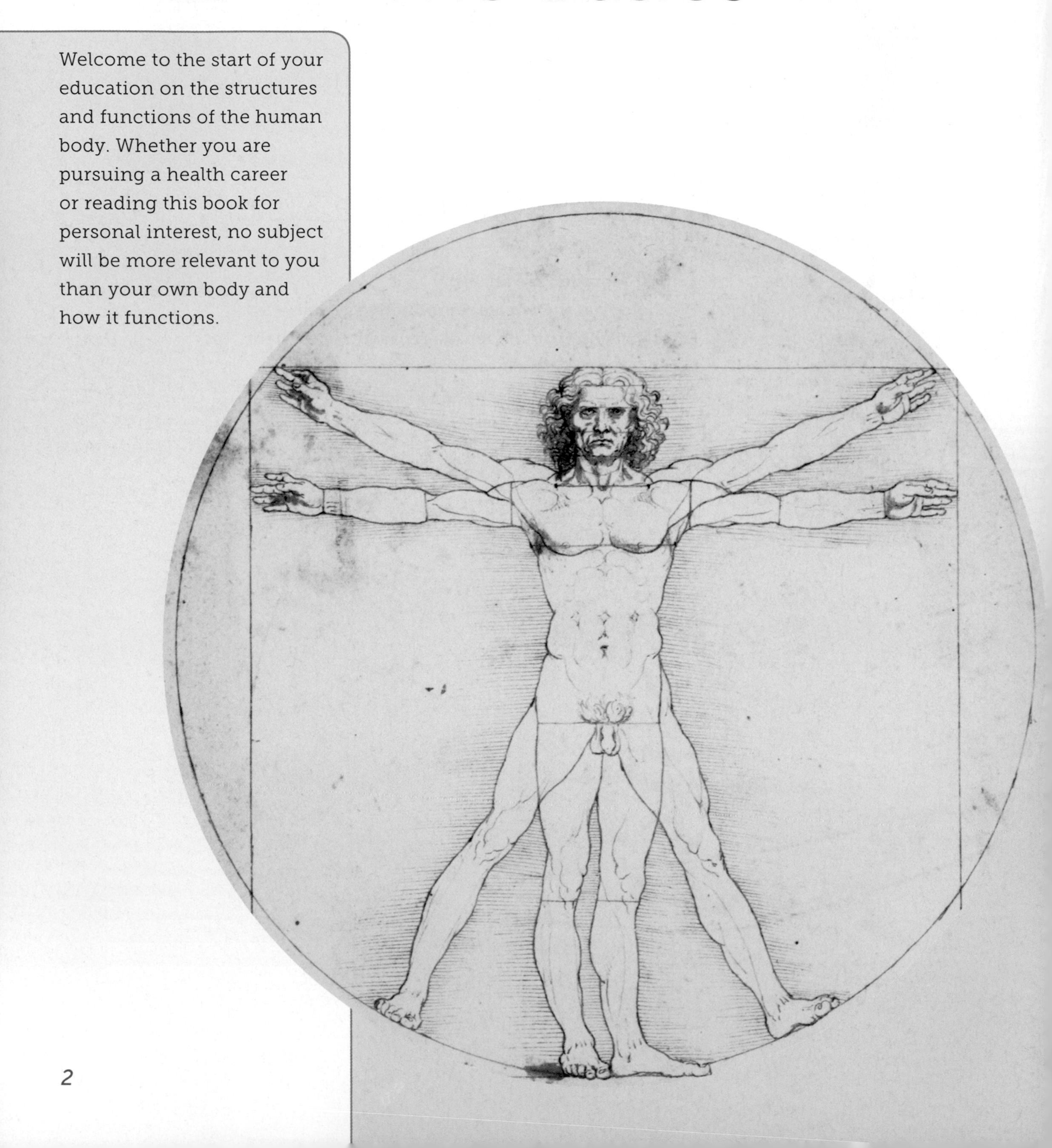

learning outcomes

After completing this chapter, you should be able to:

1.1 Define *anatomy* and *physiology*.

1.2 Describe the location of structures in the human body using anatomical terms of direction, regions, planes, positions, and cavities.

1.3 Locate serous membranes by their individual names and relative location to organs.

1.4 Define *homeostasis* and explain why it is so important in human physiology.

1.5 Define *negative feedback* and *positive feedback* and explain their importance to homeostasis.

pronunciation **key**

homeostasis: ho-mee-oh-STAY-sis

mediastinum: MEE-dee-ass-TIE-num

meninges: meh-NIN-jeez

mesenteries: MESS-en-ter-reez

parietal: pah-RYE-eh-tal

pleura: PLUR-ah

sagittal: SAJ-ih-tal

serous: SEER-us

visceral: VISS-er-al

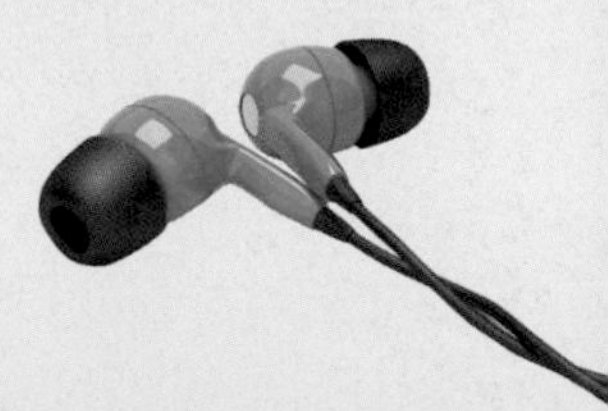

1.1 learning **outcome**

Define *anatomy* and *physiology*.

Overview

Anatomy is the study of body structures, including structures of all sizes, from microscopic red blood cells to the heart, which is the size of a fist. **Physiology** is the study of exactly how all of these structures function. For example, why are red blood cells thinner in the middle than they are along the edge? What do red blood cells actually do, and how does their shape aid in the way they work? What causes your heart to speed up or slow down, and why does it have four chambers?

These important questions of anatomy and physiology give us an idea of how interconnected structure and function are. You will need a fundamental knowledge of the human body and how it works under normal circumstances before you can begin to approach, study, and understand abnormal functioning.

The foundation of that knowledge is the basic terms of anatomy.

1.2 learning **outcome**

Describe the location of structures in the human body using anatomical terms of direction, regions, planes, positions, and cavities.

Anatomical Terms

As a student pursuing a career in allied health, part of the essential knowledge you need to learn at the start is the vocabulary of anatomy. To describe the location of various structures, you need to be able to effectively use anatomical terms of direction, regions, planes, positions, and cavities.

Picture the body in the **standard anatomical position** before you begin to identify and describe the location of a structure. A body in the standard anatomical position looks like this: The body is upright, the legs are close together, the feet are flat on the floor, the arms are close to the sides, and the head, toes, and palms of the hands are facing forward. See **Figure 1.1**.

Anatomical Terms of Direction

Once the standard anatomical position has been considered, anatomical terms of direction are used to describe the following: the location of a particular structure in the body, the location of a structure relative to another structure, or the location of something within a structure. See Table 1.1 and Figures 1.2 to 1.4.

FIGURE 1.1 Standard anatomical position. Head, toes, and palms of the hands are facing forward.

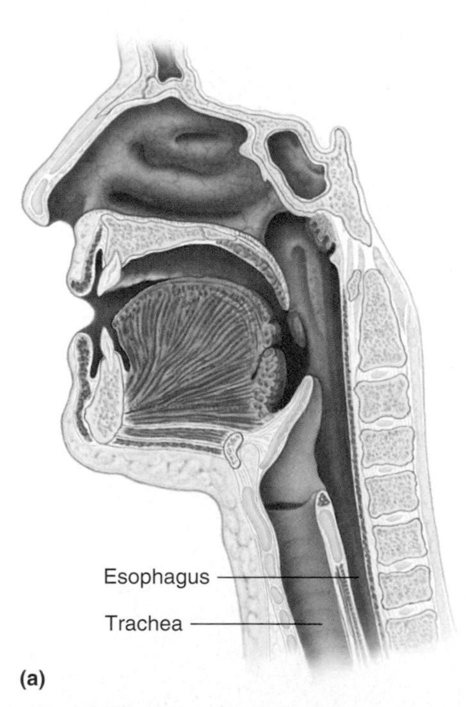

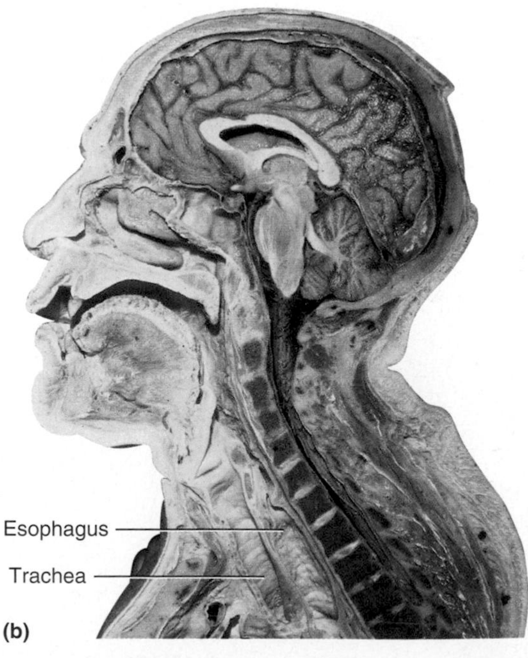

FIGURE 1.2 Sagittal view of the head: (a) the trachea anterior to the esophagus, (b) a cadaver head.

TABLE 1.1 Anatomical terms of direction

Term of direction	Definition	Example
Anterior or **ventral**	Front or belly side	The trachea is anterior to the esophagus.
Posterior or **dorsal**	Back side	The esophagus is posterior to the trachea.
Superior	Closer to the top of the head (used for head, neck, and trunk)	The lungs are superior to the diaphragm.
Inferior	Farther from the top of the head (used for head, neck, and trunk)	The stomach is inferior to the diaphragm.
Medial	Toward the midline of the body	The heart is medial to the lungs.
Lateral	Away from the midline of the body	The lungs are lateral to the heart.
Bilateral	Relating to or affecting two sides	She had bilateral ovarian tumors.
Proximal	Closer to the connection to the body	The elbow is proximal to the wrist.
Distal	Farther from the connection to the body	Fingernails are at the distal end of the fingers.
Superficial	Closer to the surface (used for layered structures)	The epidermis is superficial to the dermis of the skin.
Deep	Farther from the surface (used for layered structures)	The hypodermis is deep to the dermis of the skin.
Right	On the body's right side (not the viewer's right side)	The liver is on the right side of the body.
Left	On the body's left side (not the viewer's left side)	The stomach is on the left side of the body.

spot check 1 How would you describe the location of your nose using at least three anatomical terms?

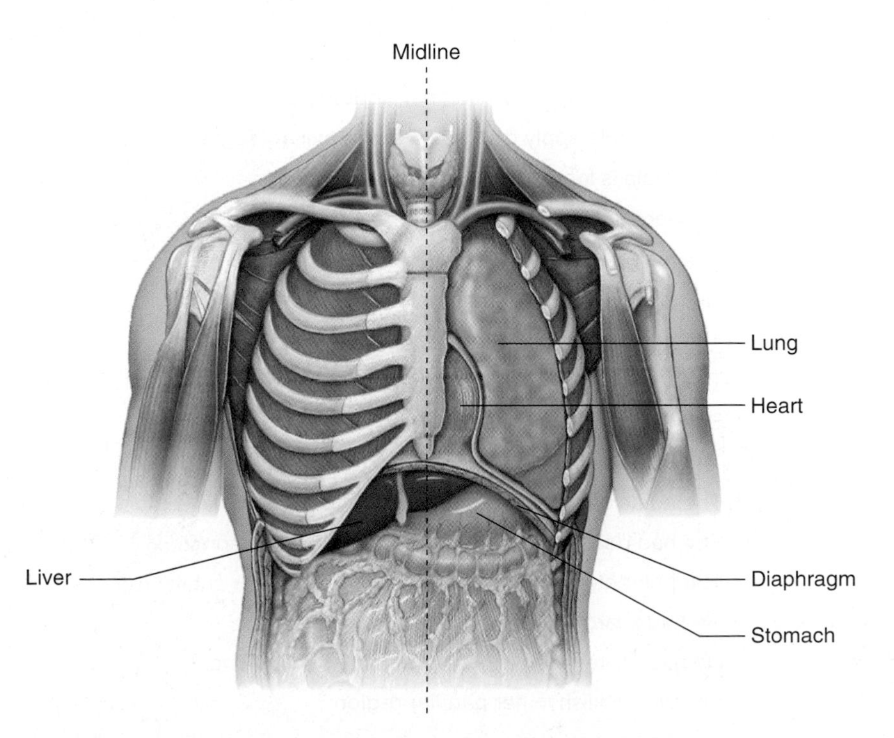

FIGURE 1.3 Chest and abdominal organs. The lungs are lateral to the heart. The stomach is inferior to the diaphragm.

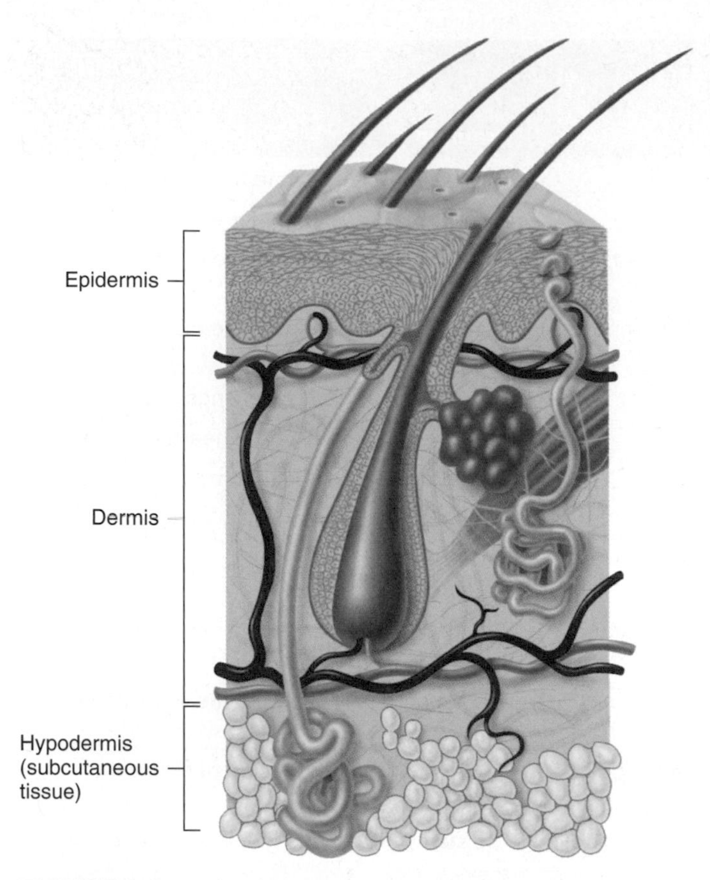

FIGURE 1.4 Layers of the skin. The epidermis is superficial to the dermis. The hypodermis is deep to the dermis.

Anatomical Regions

In addition to being described by its location within the body, a structure may also be described by identifying the specific region in which it is located. The two major regions of the body are the **axial** region (head, neck, and trunk) and the **appendicular** region (arms and legs). These regions are further subdivided. See Table 1.2 and **Figure 1.5**.

The abdominal region can be divided in either of two ways: into four quadrants or into nine regions similar to a tic-tac-toe grid. See **Figure 1.6**. By mentally dividing this region, you can communicate about it and the structures within it more quickly and accurately. For example, when dealing with abdominal pain, you can use either method to describe the location of the pain. The four quadrants are the right upper quadrant, left upper quadrant, right lower quadrant, and left lower quadrant. The nine regions of the abdomen are the right hypochondriac region, epigastric region, left hypochondriac region, right lumbar region, umbilical region, left lumbar region, right inguinal region, hypogastric region, and left inguinal region.

TABLE 1.2 Anatomical regions

Region	Definition	Example
Axial	Head, neck, and trunk	The ribs are in the axial region.
Abdominal	Belly	She had a measles rash on her abdomen.
Axillary	Armpit	Many people apply deodorant in the axillary region.
Cephalic or **cranial**	Head	The brain is located in the cranial region.
Cervical	Neck	He wore a cervical collar after the car accident.
Facial	Face	The maxilla is a facial bone.
Inguinal	Groin	He had an inguinal hernia.
Pelvic	Lower end of the trunk	The urinary bladder is located in the pelvic region.
Thoracic	Chest	The breasts are in the thoracic region.
Umbilical	Navel	She developed an umbilical hernia during her pregnancy.
Appendicular	Arms and legs	The elbow is in the appendicular region.
Brachial	Arm	The nurse administered the vaccine to his brachial region.
Carpal	Wrist	She had pain in the carpal region from a wrist sprain.
Cubital	Elbow	The phlebotomist drew blood from the anterior cubital region.
Femoral	Thigh	He had a large bruise on his femoral region.
Palmar	Palms of the hands	He had blisters on his palmar surface from raking.
Patellar	Knee	She was ticklish in her patellar region.
Plantar	Soles of the feet	She had her plantar warts removed.
Tarsal	Ankle	He had pain in his tarsal region from an ankle sprain

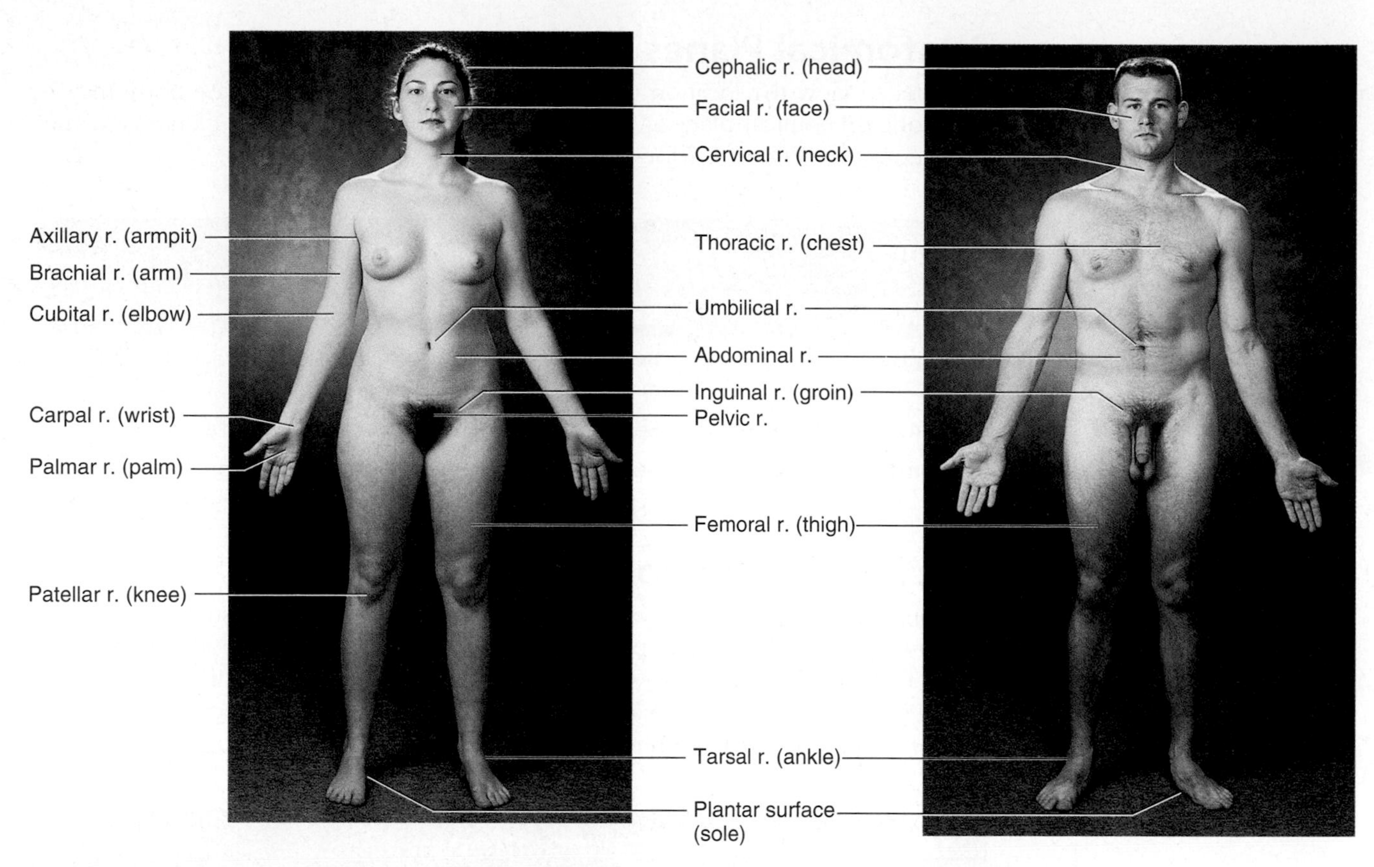

FIGURE 1.5 Anatomical regions: anterior (ventral) view.

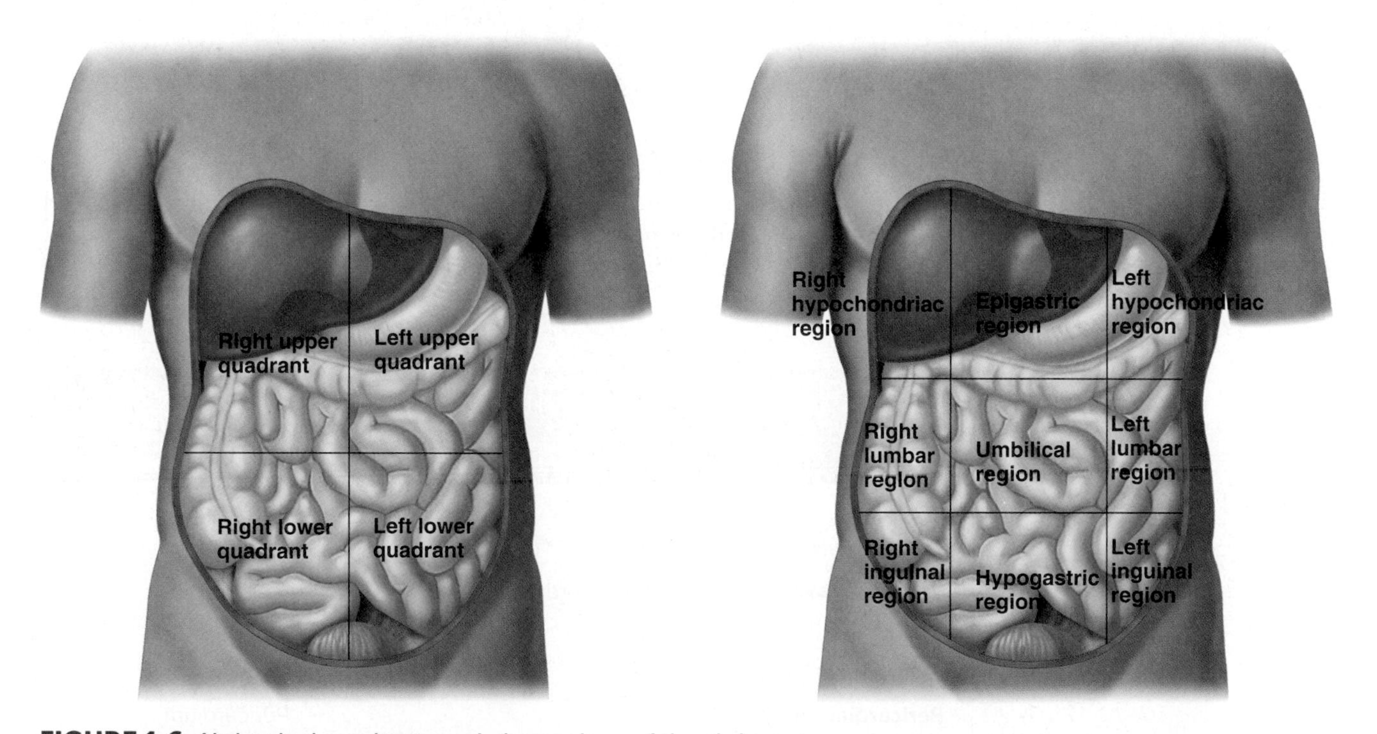

FIGURE 1.6 Abdominal quadrants and nine regions of the abdomen.

spot check ❷ Describe the location of liver pain by specific abdominal region in two different ways.

Anatomical Planes

In order to view the location of structures from different angles, the body may be cut along anatomical planes. Table 1.3 lists the three anatomical planes and their definitions. See **Figure 1.7** for examples.

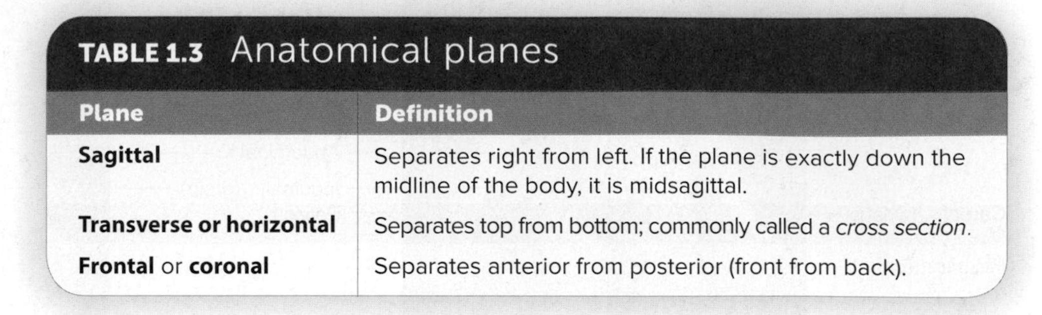

TABLE 1.3 Anatomical planes

Plane	Definition
Sagittal	Separates right from left. If the plane is exactly down the midline of the body, it is midsagittal.
Transverse or horizontal	Separates top from bottom; commonly called a *cross section*.
Frontal or **coronal**	Separates anterior from posterior (front from back).

sagittal: SAJ-ih-tal

Anatomical Positions

Anatomical terms may also be used to describe body position, such as *standard anatomical position,* which appears earlier in this chapter. **Prone** and **supine** are terms that may be used to describe the position of either the entire body or individual parts of the body such as the hands. For example, your hands would be in a supine position if you held them out in front of you, palms facing up. See Table 1.4.

TABLE 1.4 Anatomical positions

Position	Definition
Supine	Anterior surface facing up
Prone	Anterior surface facing down

Anatomical Cavities

Most of the body's organs are located in cavities, which are pocketlike spaces of various sizes. There are three general cavities—the dorsal cavity, the thoracic cavity, and the abdominopelvic cavity—that can be further subdivided. See Table 1.5 for a specific breakdown of cavities, their associated organs, and the lining membranes.

TABLE 1.5 Anatomical cavities

Cavity	Associated Organs	Lining Membranes
Dorsal:		
Cranial cavity	Brain	Meninges
Vertebral cavity	Spinal cord	Meninges
Thoracic:		
Pleural cavities (2)	Lungs	Pleurae
Pericardial cavity	Heart	Pericardium
Abdominopelvic:		
Abdominal cavity	Digestive organs, spleen	Peritoneum
Pelvic cavity	Urinary bladder, rectum, reproductive organs	Peritoneum

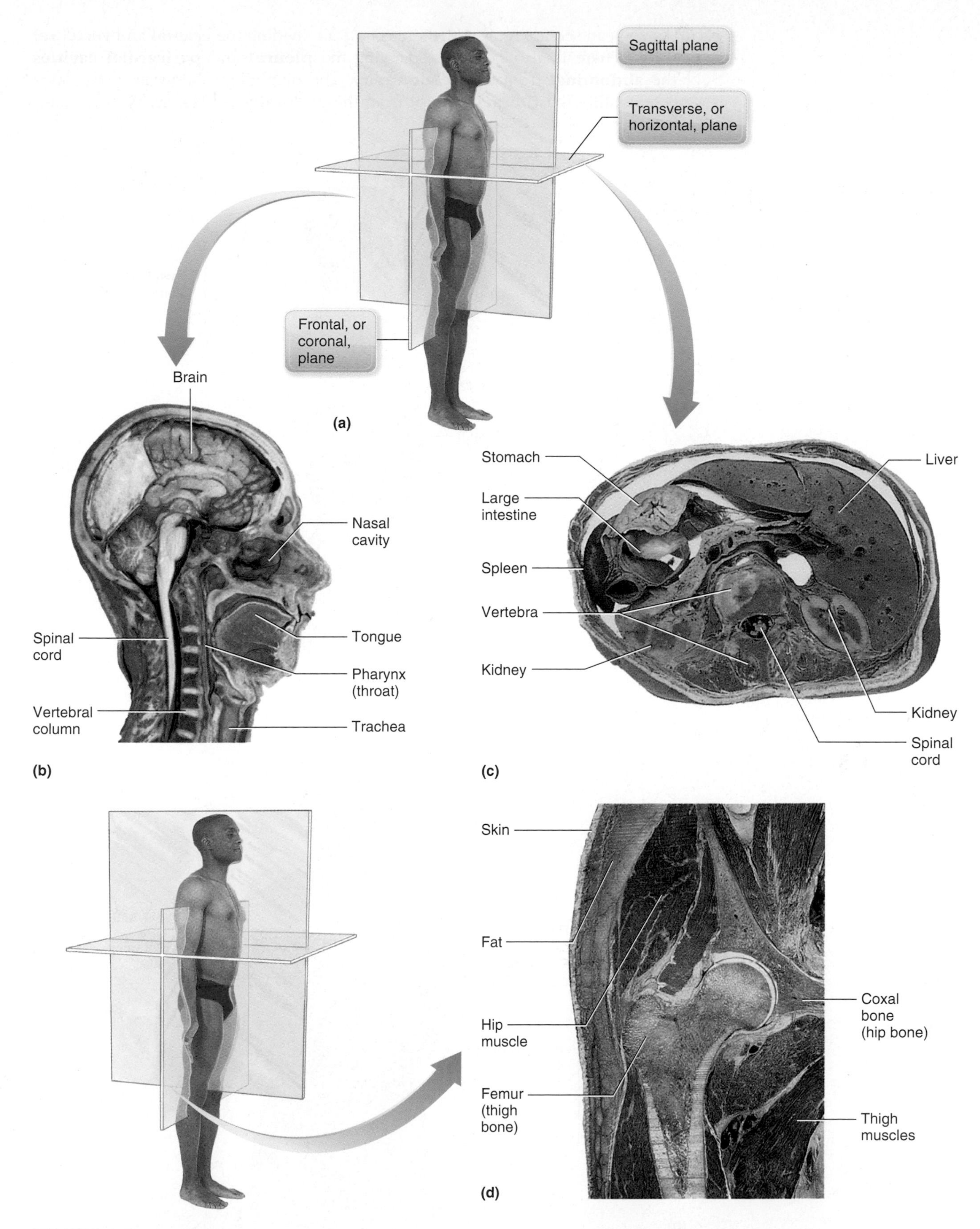

FIGURE 1.7 Anatomical planes: (a) full figure showing planes, (b) sagittal section of the head, (c) transverse section through the abdomen, (d) frontal section through the right hip.

As you can see in **Figure 1.8**, there is no wall dividing the **cranial** and **vertebral cavities.** There is also no wall separating the **pleural** and **pericardial cavities** or the **abdominal** from the **pelvic cavity.** The diaphragm, however, serves as a wall separating the **thoracic cavity** from the **abdominopelvic cavity.** The space

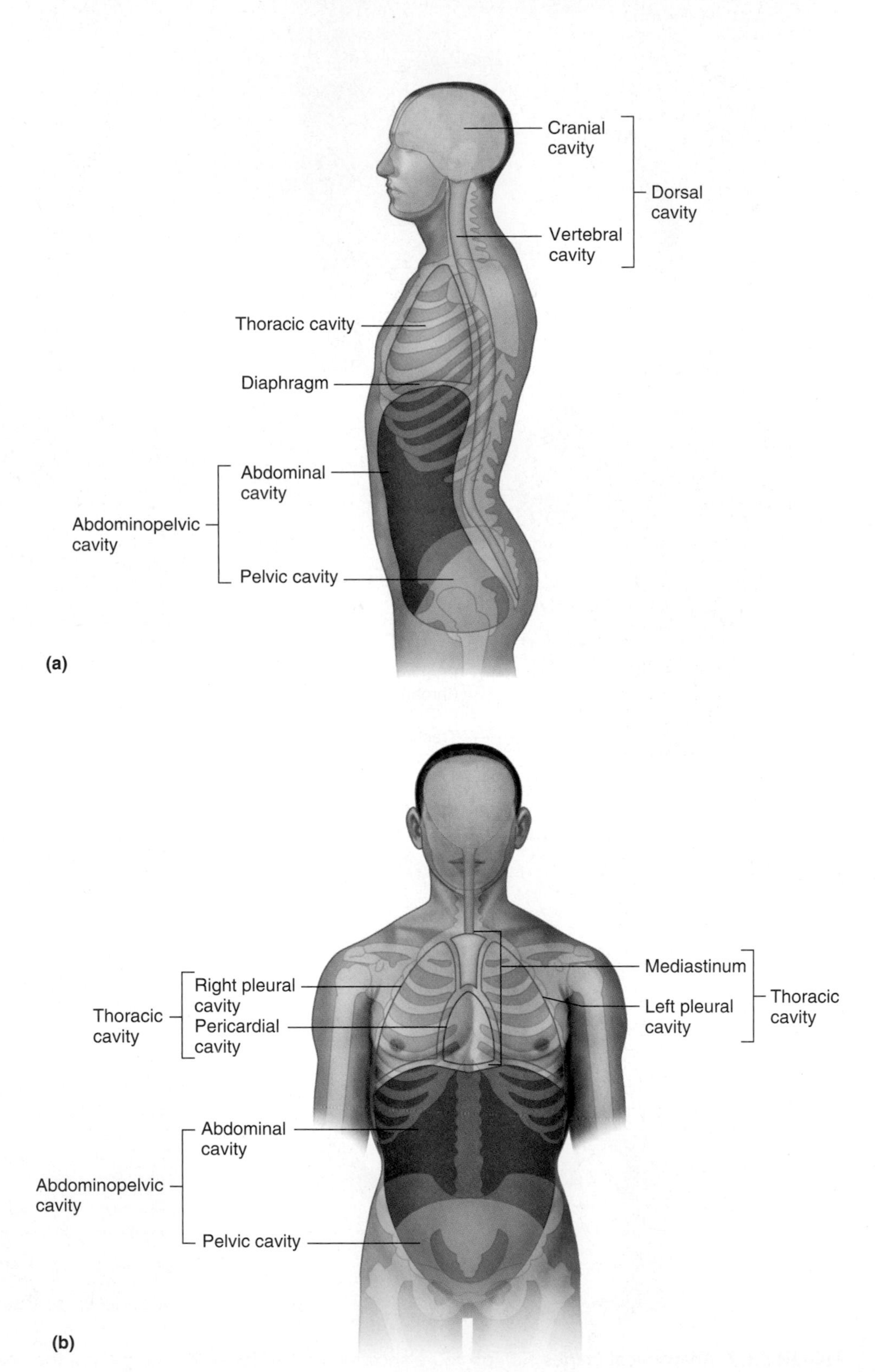

FIGURE 1.8 Body cavities. (a) lateral view, (b) anterior view.

between the pleural cavities that contains the heart, esophagus, trachea, thymus, and major vessels is called the **mediastinum.** Again, see **Figure 1.8** for the locations of these cavities in the body.

mediastinum: MEE-dee-ass-TIE-num

Now that you know where the cavities are in relation to the organs they house, it is important to note that these cavities are lined with membranes. The cranial and vertebral cavities are lined by the **meninges**—three layers of membrane surrounding the brain and spinal cord. You will learn more about the meninges in the nervous system chapter. The thoracic and abdominopelvic cavities contain fluid-filled **serous membranes,** which line the cavities and surround the organs.

meninges: meh-NIN-jeez

serous: SEER-us

Serous Membranes

1.3 learning **outcome**

Locate serous membranes by their individual names and relative location to organs.

Serous membranes are double-layered membranes that contain fluid between the two layers. A good analogy for a serous membrane is a very soft water balloon. You slowly push your fist into the balloon so that it does not break. Part of the water balloon will be in direct contact with your fist, and part of the balloon will not. Water exists between the two layers. Your fist represents the individual organ, while the layers of the water balloon represent the layers of the serous membrane. See **Figure 1.9**.

A serous membrane called the *pleura* surrounds each of the lungs in the **thoracic cavity** just as the water balloon surrounds your fist. The part of the pleural membrane in direct contact with the lung is called the **visceral pleura.** The part of the pleural membrane not in direct contact with the lung is the **parietal pleura.** Pleural fluid fills the space between the two layers. The heart is surrounded by a similar serous membrane called the *pericardium.* The pericardium has two layers: the **visceral pericardium** (in contact with the heart) and the **parietal pericardium** (not in contact with the heart). Pericardial fluid fills the space between these two layers. See **Figure 1.10**.

visceral: VISS-er-al

pleura: PLUR-ah

parietal: pah-RYE-eh-tal

The **abdominopelvic cavity** contains another serous membrane called the **peritoneum.** This double-layered serous membrane has many abdominal organs pushed into it from the posterior wall of the abdominopelvic cavity. The parietal portion of the peritoneum lines the anterior wall of the abdominopelvic cavity, while the visceral portion of the peritoneum covers several, but not all, of the organs in the abdominal cavity. The kidneys and most of the pancreas are **retroperitoneal,** meaning they are between the parietal peritoneum and the posterior abdominal wall. See **Figure 1.11**.

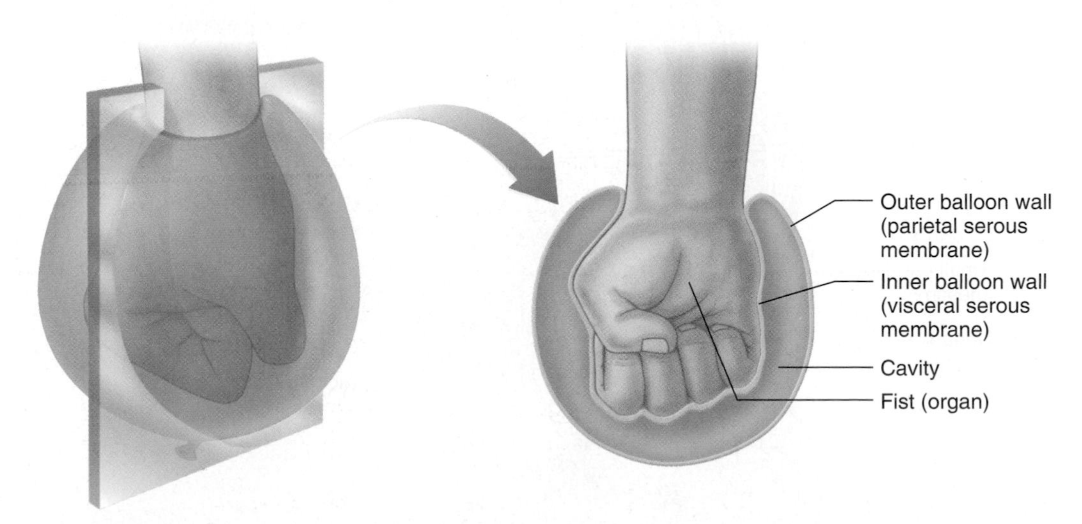

FIGURE 1.9 Water balloon. A fist is slowly pushed into the water balloon.

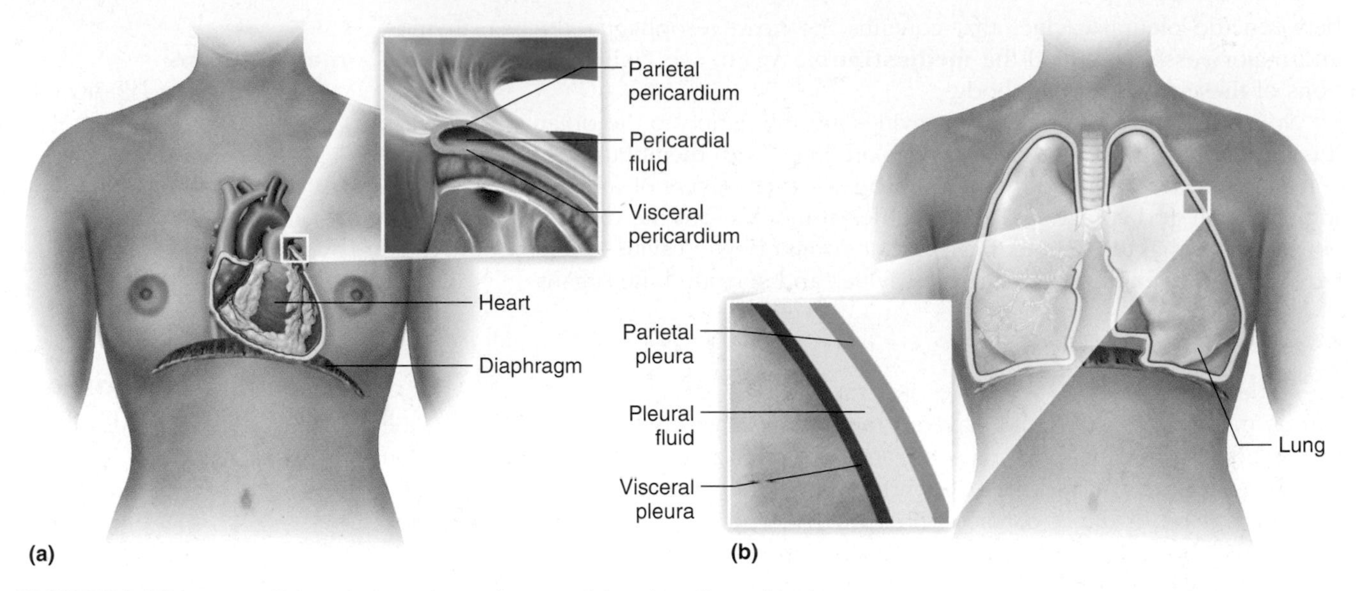

FIGURE 1.10 Pericardial and pleural membranes (a) pericardium, (b) pleurae.

warning

The diagram in Figure 1.11 may lead to a misconception. Although there appears to be significant space between the organs in this drawing, there is actually no extra space in the abdominopelvic cavity. The diagram has been drawn with extra space to highlight the peritoneal membrane.

FIGURE 1.11 Peritoneal membrane.

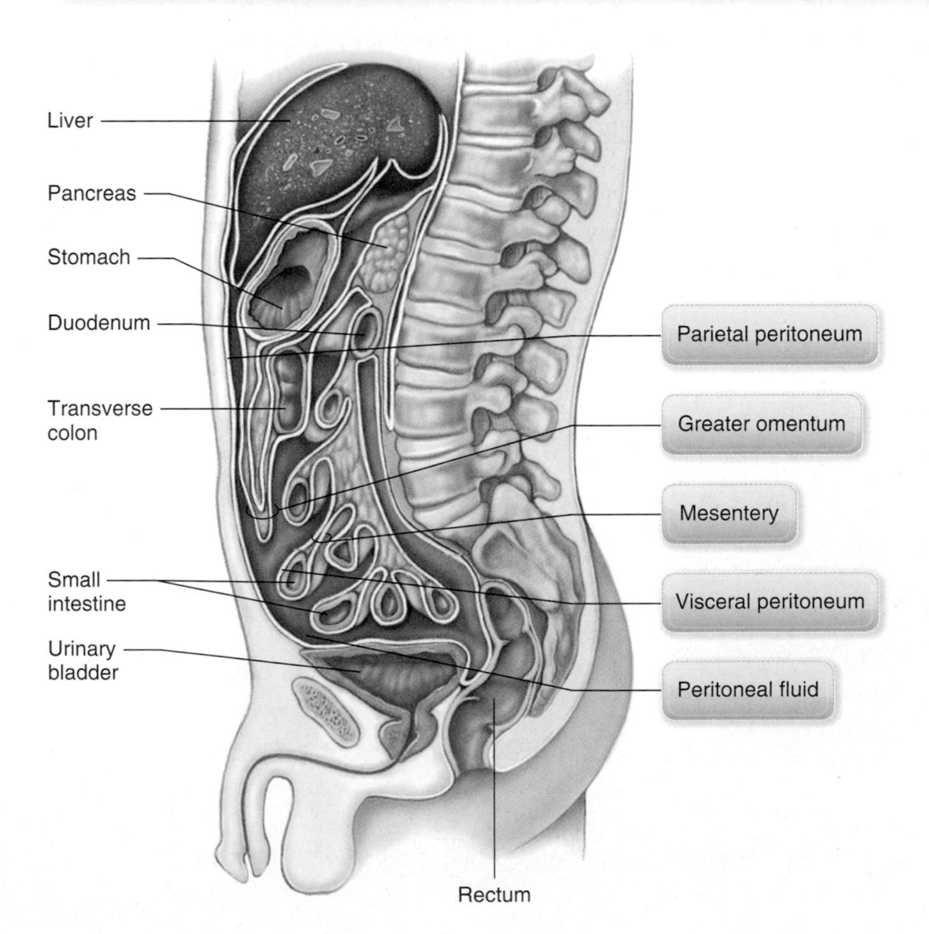

In **Figure 1.11**, look closely at the peritoneum surrounding a section of the small intestine. The membrane starts at the posterior abdominal wall, extends out and surrounds the section of small intestine (this part of the membrane is the **visceral peritoneum**), and then returns parallel to itself to the posterior wall before extending out again to surround the next organ. This membrane repeatedly extends from the posterior wall to surround individual organs. The membrane then returns parallel to itself before finally becoming the parietal peritoneum that lines the anterior abdominal wall. The peritoneal fluid is found between the parietal peritoneum (lining the anterior wall) and the visceral peritoneum (surrounding each organ). The sections of the membrane where the peritoneum comes back parallel to itself are called the **mesenteries.** The mesenteries do not involve any fluid because they are outside the water balloon–like serous membrane. Blood vessels and nerves, going out to individual organs, can be found neatly arranged in the mesenteries. See **Figure 1.12**.

mesenteries:
MESS-en-ter-reez

The visceral peritoneum (shown in **Figure 1.11**) has two extensions. See **Figure 1.13**. The **greater omentum** looks like a fatty apron lying over all the abdominal viscera; it extends from the inferior margin of the stomach. The **lesser omentum** is smaller and extends from the superior edge of the stomach to the liver.

spot check 3 What specific serous membrane is attached to the inferior surface of the diaphragm? What specific serous membrane(s) is (are) attached to the superior surface of the diaphragm?

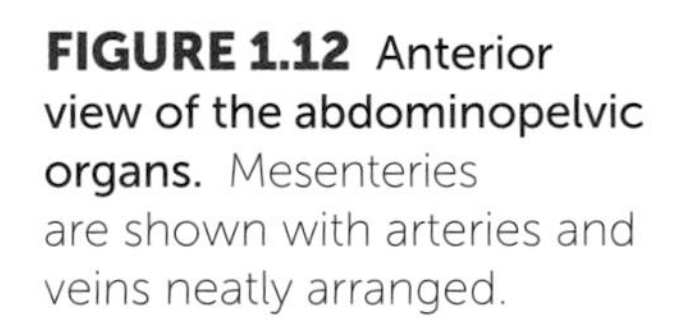

FIGURE 1.12 Anterior view of the abdominopelvic organs. Mesenteries are shown with arteries and veins neatly arranged.

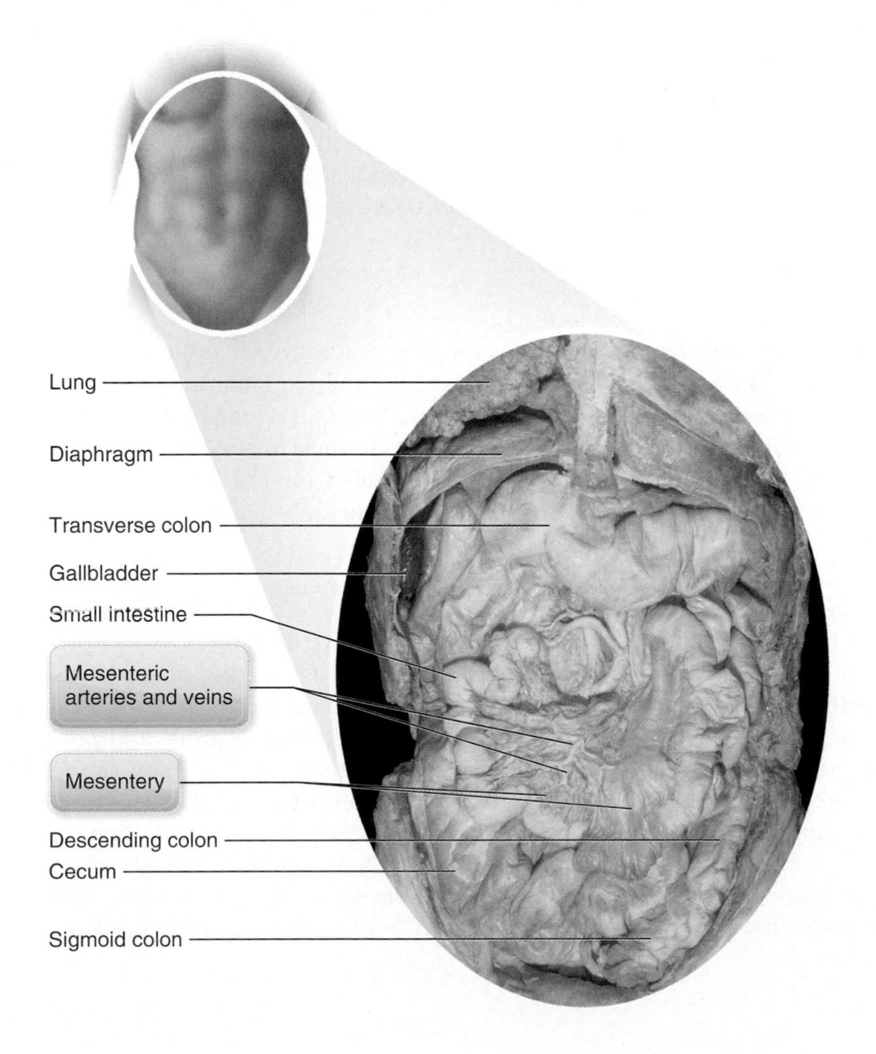

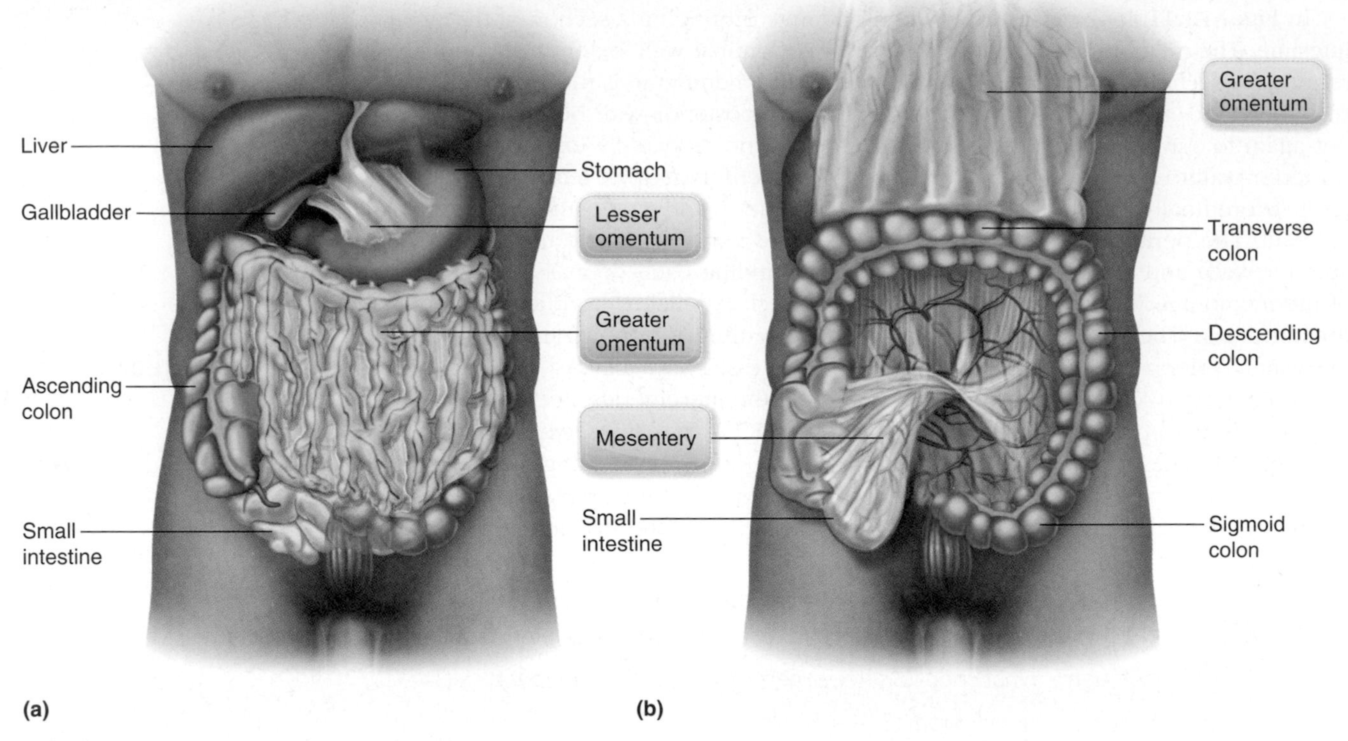

FIGURE 1.13 Anterior view of omentums and mesenteries: (a) the greater omentum covering the abdominopelvic organs, and the lesser omentum; (b) the greater omentum and small intestines retracted to show the mesenteries and the mesenteric blood vessels neatly arranged within.

Physiological Terms

Now that you have become familiar with basic anatomical terms, let's look at the basic terms of physiology that will be used throughout this book. These terms include *homeostasis, negative feedback,* and *positive feedback.*

1.4 learning **outcome**

Define *homeostasis* and explain why it is so important in human physiology.

Homeostasis

Homeostasis is an important unifying concept in physiology. All structures function together in the human body to maintain a steady internal environment. Think of homeostasis as a series of ranges—a range of blood pressure, temperature, blood oxygen levels, and blood calcium levels that the body must maintain for normal functioning. If these levels move outside the homeostasis range (too high or too low), the body will usually attempt to bring the levels back to the optimal range.

1.5 learning **outcome**

Define *negative feedback* and *positive feedback* and explain their importance to homeostasis.

homeostasis: ho-mee-oh-STAY-sis

If the body detects a movement outside the homeostasis range, it may use either of two feedback mechanisms in response. **Negative feedback** is the process the body uses to reverse the direction of movement away from homeostasis. **Positive feedback** is the process the body uses to increase the movement away from homeostasis. It is important to understand that the terms *negative* and *positive* in this context are not value judgments indicating good or bad. They simply describe the direction of movement. If the body detects a change outside its normal homeostasis range (either too high or too low) and it works to reach its homeostasis range by reversing the direction of movement, that is negative feedback. If the body detects a change and works to make the levels move even farther away from homeostasis, that is positive feedback. Examples of each type of feedback are presented below.

Negative Feedback Jen takes a break from studying and goes to the vending machines for a soda and a candy bar. Her digestive system digests the soda and the candy bar and absorbs the digested sugar into her bloodstream, causing her blood sugar level to rise above normal. Her pancreas recognizes the increased blood sugar level and, in response to that stimulus, releases the hormone insulin. The insulin then travels to most cells in her body, telling these cells to take in sugar from the blood. Through this process, her blood sugar level is lowered back to homeostasis. This scenario is an example of negative feedback. The blood sugar level became too high from the soda and candy, and the body brought the level back down to normal.

Now consider another example of negative feedback: Paul has not eaten for a long time, so his blood sugar level has fallen below normal. His pancreas detects the decreased blood sugar level and, in response to this stimulus, releases the hormone glucagon. Glucagon then travels to his liver, telling it to release glucose (sugar) into his bloodstream. This raises his blood sugar level to homeostasis. The blood sugar level became too low, and his body brought the level back up to normal. See **Figure 1.14**.

Positive Feedback The next scenario is an example of positive feedback as it relates to the concept of homeostasis. A woman is at the end of her pregnancy. In homeostasis, the uterus does not contract during pregnancy. As the fetus reaches full term, its head pushes on the neck of the uterus (the cervix). The increased pressure on the cervix causes the cervix to release chemicals (prostaglandins) that cause the uterus to contract, moving away from homeostasis. The contractions cause the fetal head to push harder on the cervix, and this increases the pressure. The cervix responds by making more prostaglandins, leading to more contractions

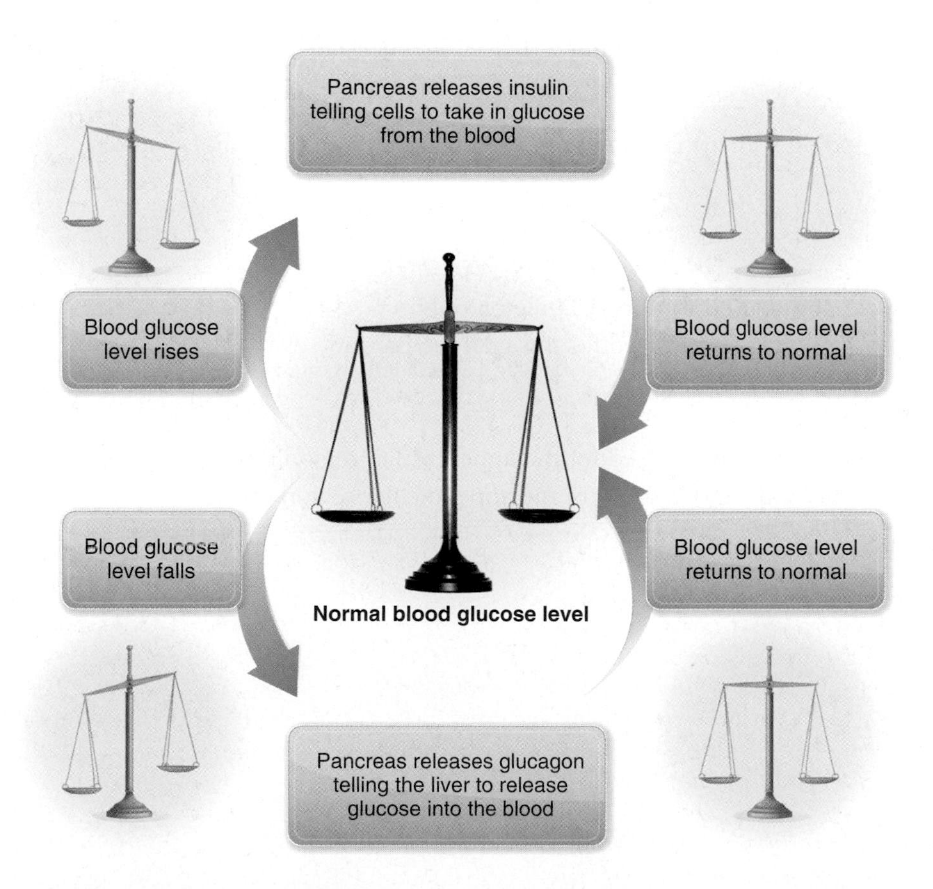

FIGURE 1.14 Negative feedback mechanisms for blood sugar regulation.

and further increasing the fetal head's pressure on the cervix. As you can see, this has a "runaway-train effect," creating more and more contractions until the baby is finally born.

spot check ❹ It is 90 degrees and sunny. You are wearing your cap and gown for graduation. You want to look your best for the outdoor graduation photos, but you are sweating profusely. Is the production of sweat a negative or positive feedback mechanism?

Homeostasis as a unifying concept is covered in all the system chapters in this book. An icon in the margin indicates the introduction of homeostasis concepts. The icon highlights topics important to homeostasis: acid-base balance, fluid and electrolyte balance, and nutrition. These topics are explained in later chapters.

summary

Overview

- Anatomy is the study of structures, and physiology is the study of how structures function.

Anatomical Terms

- The standard anatomical position is standing with arms at the sides and palms and head facing forward.

Anatomical Terms of Direction

- Anterior or ventral means front or belly side.
- Posterior or dorsal means back side.
- Superior is closer to the top of the head and is used for the axial region only.
- Inferior is farther away from the top of the head and is used for the axial region only.
- Medial is toward the midline of the body.
- Lateral is away from the midline of the body.
- Proximal is closer to the connection to the body and is used for the appendicular region only.
- Distal is farther from the connection to the body and is used for the appendicular region only.
- Superficial is closer to the surface.
- Deep is farther from the surface.
- Right is on the body's right side.
- Left is on the body's left side.

Anatomical Regions

- Axial is head, neck, and trunk.
- Abdominal is belly.
- Axillary is armpit.
- Cranial or cephalic is head.
- Cervical is neck.

- Facial is face.
- Inguinal is groin.
- Pelvic is lower trunk.
- Thoracic is chest.
- Umbilical is navel.
- Appendicular is arms and legs.
- Brachial is arm.
- Carpal is wrist.
- Cubital is elbow.
- Femoral is thigh.
- Palmar is palms of the hands.
- Patellar is knee.
- Plantar is soles of the feet.
- Tarsal is ankle.

Anatomical Planes

- The sagittal plane separates right from left.
- The transverse or horizontal plane separates top from bottom.
- The frontal or coronal plane separates front from back.

Anatomical Positions

- Supine is anterior surface facing up.
- Prone is anterior surface facing down.

Anatomical Cavities

- The dorsal cavity contains the cranial and vertebral cavities, which are lined by the meninges.
- The thoracic cavity contains two pleural cavities lined by pleural membranes, the pericardial cavity lined by the pericardial membranes, and the mediastinum.
- The abdominopelvic cavity contains the abdominal cavity and the pelvic cavity, together lined by the peritoneum.

Serous Membranes

- Serous membranes form two layers when surrounding an organ, similar to a water balloon surrounding a fist.
- The pericardial membranes surround the heart.
- The pleural membranes surround the lungs.
- The peritoneal membranes surround many of the abdominopelvic organs.
- The portion of the serous membrane in contact with the organ is the visceral pericardium, visceral pleura, or visceral peritoneum.
- The portion of the serous membrane not in contact with the organ is the parietal pericardium, parietal pleura, or parietal peritoneum.
- The mesenteries are sections of the peritoneum that neatly arrange blood vessels and nerves to organs.
- The greater and lesser omentums are extensions of the peritoneal membrane.

Physiological Terms

Homeostasis

- Homeostasis is a steady internal environment in which the body works best. If the body detects a change away from homeostasis, it will use either of two feedback mechanisms.
- Negative feedback is the process the body uses to reverse the direction of movement away from homeostasis.
- Positive feedback is the process the body uses to increase the movement away from homeostasis.

key words for review

The following terms are defined in the glossary.

abdominal
anatomy
appendicular
axial
cranial
greater omentum
homeostasis
meninges
mesentery
negative feedback
parietal
pelvic
peritoneum
physiology
pleura
positive feedback
proximal
serous membrane
thoracic
visceral

chapter review questions

Multiple Select: *Select the correct choices for each statement. The choices may be all correct, all incorrect, or a combination of correct and incorrect.*

1. What will happen if I eat two candy bars and my blood sugar level rises above normal?
   a. The body will try to bring the blood sugar level back to homeostasis.
   b. The body will use a positive-feedback mechanism to correct it.
   c. There is no need for my body to do anything.
   d. Homeostasis must be achieved to keep the body working optimally.
   e. The body will use a negative-feedback mechanism to correct it.
2. Which of the following statements use(s) anatomical terms of direction correctly?
   a. The esophagus is superior to the trachea.
   b. The ventral side of the arm has more hair than the dorsal side.
   c. The knee is superior to the ankle.
   d. The diaphragm is superior to the stomach.
   e. The thumb is medial to the ring finger.
3. A bullet enters the body in the left axillary region and lodges in the right lung. Which of the following statements accurately describe(s) the bullet's path?
   a. The bullet passes through the parietal peritoneum before it passes through the visceral peritoneum.
   b. The bullet passes through six layers of serous membrane before lodging in the lung.
   c. The bullet passes through a layer of the pericardial membrane four times before lodging in the lung.
   d. The bullet passes through the visceral peritoneum before lodging in the lung.
   e. The bullet passes through three layers of visceral pleura before lodging in the right lung.
4. Which of the following statements is (are) accurate for the abdominal region?
   a. The liver is primarily in the upper right quadrant.
   b. The stomach is in the upper left quadrant.
   c. The epigastric region is superior to the umbilical region.
   d. The hypogastric region is superior to the umbilical region.
   e. The hypochondriac regions are lateral to the epigastric region.

Matching: *Match each subregion with the major region to which it belongs.*

_____ 1. Carpal region
_____ 2. Tarsal region
_____ 3. Pelvic region
_____ 4. Cervical region
_____ 5. Cubital region

a. Axial region
b. Appendicular region

Matching: *Match each cavity to the membranes it contains. Some questions may have more than one answer.*

_____ 6. Dorsal cavity
_____ 7. Thoracic cavity
_____ 8. Vertebral cavity
_____ 9. Cranial cavity
_____ 10. Abdominopelvic cavity

a. Pleural membranes
b. Pericardial membranes
c. Peritoneal membranes
d. Meninges

Critical Thinking:

1. What anatomical plane of the body would be used to illustrate the relative position of the liver, stomach, and spinal column?
2. What anatomical plane of the body would be used to illustrate the relative position of the urinary bladder, the stomach, and the liver?

1 chapter mapping

This section of the chapter is designed to help you find where each outcome is covered in this text.

	Outcomes	Readings, figures, and tables	Assessments
1.1	Define *anatomy and physiology.*	Overview: p. 4	
1.2	Describe the location of structures in the human body using anatomical terms of direction, regions, planes, positions, and cavities.	Anatomical terms: pp. 4–11 Tables 1.1–1.5 Figures 1.1–1.8	Spot Check: 1, 2 Multiple Select: 2, 4 Matching: 1–5, 6–10 Critical Thinking: 1–2
1.3	Locate serous membranes by their individual names and relative location to organs.	Serous membranes: pp. 11–14 Figures 1.9–1.13	Spot Check: 3 Multiple Select: 3
1.4	Define *homeostasis* and explain why it is so important in human physiology.	Homeostasis: p. 14	Multiple Select: 1
1.5	Define *negative feedback* and *positive feedback* and explain their importance to homeostasis.	Homeostasis: pp. 14–16 Figure 1.14	Spot Check: 4 Multiple Select: 1

2 Levels of Organization of the Human Body

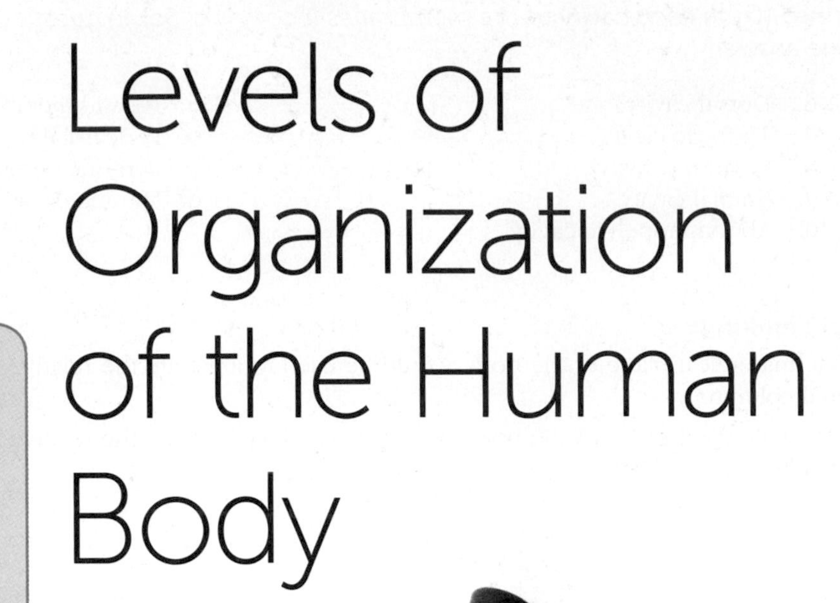

Now that you have completed Chapter 1, you have started to see the connection between anatomy and physiology and to understand how the structures of the body work and relate to the bigger picture of normal functioning. In this chapter, you will learn how the body's structures and functions are organized, and this will further clarify precisely how human anatomy and physiology are so interrelated.

learning outcomes

After completing this chapter, you should be able to:

2.1 List the levels of organization of the human body from simplest to most complex.

2.2 Define the terms *matter, element, atom,* and *isotope*.

2.3 Define *molecule* and describe two methods of bonding that may form molecules.

2.4 Summarize the five functions of water in the human body and give an explanation or example of each.

2.5 Compare solutions based on tonicity.

2.6 Determine whether a substance is an acid or a base and its relative strength if given its pH.

2.7 Describe the four types of organic molecules in the body by giving the elements present in each, their building blocks, an example of each, the location of each example in the body, and the function of each example.

2.8 Explain three factors governing the speed of chemical reactions.

2.9 Write the equation for cellular respiration using chemical symbols and describe it in words.

2.10 Explain the importance of ATP in terms of energy use in the cell.

2.11 Describe cell organelles and explain their functions.

2.12 Compare four methods of passive transport and active transport across a cell membrane in terms of materials moved, direction of movement, and the amount of energy required.

2.13 Describe bulk transport, including endocytosis and exocytosis.

2.14 Describe the processes of transcription and translation in protein synthesis in terms of location and the relevant nucleic acids involved.

2.15 Describe what happens to a protein after translation.

2.16 Explain the possible consequences of mistakes in protein synthesis.

2.17 Describe the process of mitosis, including a comparison of the chromosomes in a parent cell to the chromosomes in the daughter cells.

2.18 Explain the possible consequences of mistakes in replication.

2.19 Describe the effects of aging on cell division.

2.20 Describe the four classifications of tissues in the human body.

2.21 Describe the modes of tissue growth, change, shrinkage, and death.

2.22 Describe the possible effects of uncontrolled growth of abnormal cells in cancer.

2.23 Explain how genetic and environmental factors can cause cancer.

2.24 Identify the human body systems and their major organs.

pronunciation **key**

apoptosis: AP-op-TOE-sis

atrophy: AT-roh-fee

chromatin: KROH-ma-tin

cilia: SILL-ee-ah

Golgi: GOAL-jee

hyaline: HIGH-ah-lin

hyperplasia: high-per-PLAY-zee-ah

meiosis: my-OH-sis

mitosis: my-TOE-sis

monosaccharide: MON-oh-SACK-ah-ride

necrosis: neh-KROH-sis

osmosis: oz-MO-sis

phagocytize: FAG-oh-sit-ize

squamous: SKWAY-mus

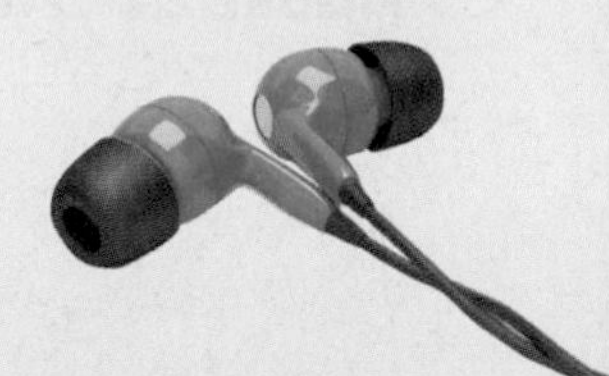

2.1 learning outcome

List the levels of organization of the human body from simplest to most complex.

Overview

As you can see in **Figure 2.1**, our bodies are organized in a hierarchy, from simplest to most complex.

The **chemical level**—the simplest level—deals with the body's chemistry and involves individual atoms and molecules. At the **organelle level,** molecules work together as organelles to perform specific functions. Mitochondria, ribosomes, and Golgi complexes are examples of organelles. At the **cellular level,** the organelles work together to perform specific functions for a **cell** (the basic unit of life). The body is composed of trillions of cells that work together to complete specific functions at the **tissue level.** Different tissues work together to perform particular functions at the **organ level.** The stomach, liver, and lungs are all organs. The organs work together at the **systems level** to perform functions like digestion and respiration. All the body systems must work together to accomplish the entire organism's functions. In this case, the organism is a human. The **organism level** is the most complex.

This chapter covers the essential information about each level of this hierarchy up to an introduction to the systems. The individual systems are covered later in their own chapters.

FIGURE 2.1 Levels of organization in the human body.

Levels of Organization

In this section, we take a detailed look at each level of the human body's hierarchy of organization, starting with the most basic—the chemical level—as all other levels are based on chemistry.

Chemical Level

2.2 learning outcome

Define the terms *matter, element, atom,* and *isotope.*

The chemical level of organization includes many chemistry concepts: atoms and how they bond to form molecules, water and how it is the basis for solutions in the body, acids and bases and how they are measured by pH, organic molecules and their importance in the body, and chemical reactions, such as cellular respiration. We begin our exploration with the basics.

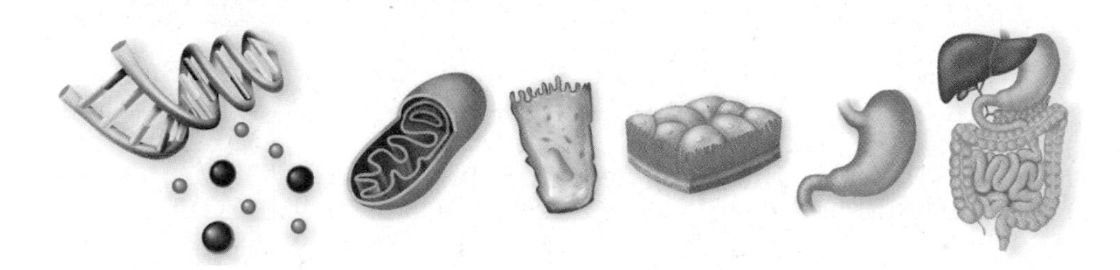

All solids, liquids, and gases are composed of **matter.** Matter is defined as anything that takes up space and has mass. The chair you are sitting on as you read this, the beverage you may have at your side, and the air you breathe are all composed of matter. The purest form of matter is an **element,** which has a unique set of chemical properties.

The elements are listed in the Periodic Table of the Elements by their composition and by their chemical properties. See **Figure 2.2**. The most common elements in the human body are listed in Table 2.1.

Atoms and Isotopes The smallest piece of an element still exhibiting the element's unique set of chemical properties is an **atom.** Atoms are composed of **protons, electrons,** and **neutrons.** For an example, see the diagram of a carbon atom in **Figure 2.3**.

The number of protons for each element is fixed and is indicated by the **atomic number.** For carbon (C), the number of protons in each atom is six. The number of electrons equals the number of protons. The **atomic mass** is the combined number of protons and neutrons. The atomic mass of carbon is 12.0112. Therefore, one

TABLE 2.1 Major elements of the human body

Element, by weight	Symbol	Percentage of body weight
Oxygen	O	65.00%
Carbon	C	18.00
Hydrogen	H	10.00
Nitrogen	N	3.00
Calcium	Ca	1.50
Phosphorus	P	1.00
Sulfur	S	0.25
Potassium	K	0.20

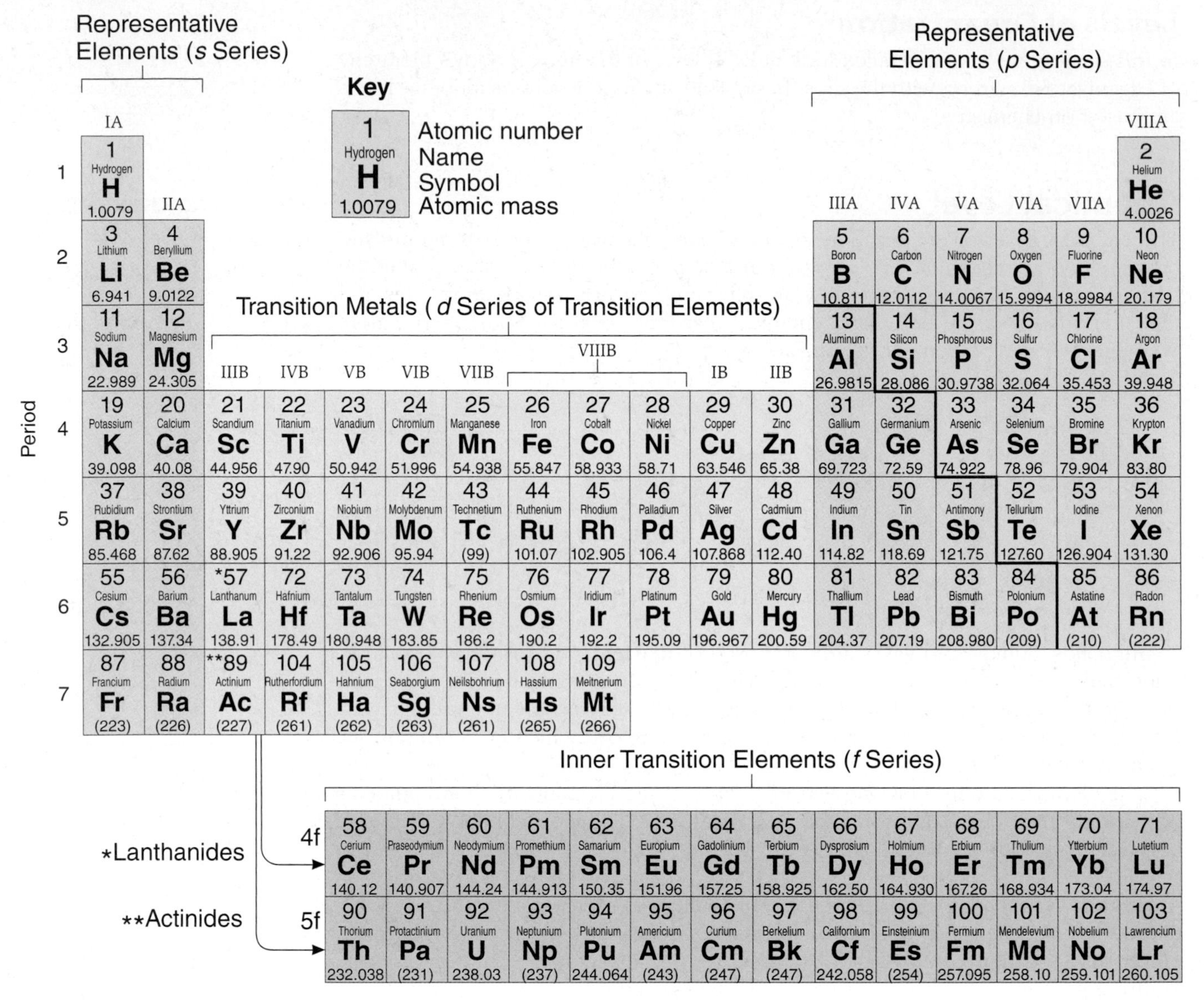

Period	IA	IIA	IIIB	IVB	VB	VIB	VIIB	VIIIB	VIIIB	VIIIB	IB	IIB	IIIA	IVA	VA	VIA	VIIA	VIIIA
1	1 Hydrogen H 1.0079																	2 Helium He 4.0026
2	3 Lithium Li 6.941	4 Beryllium Be 9.0122											5 Boron B 10.811	6 Carbon C 12.0112	7 Nitrogen N 14.0067	8 Oxygen O 15.9994	9 Fluorine F 18.9984	10 Neon Ne 20.179
3	11 Sodium Na 22.989	12 Magnesium Mg 24.305											13 Aluminum Al 26.9815	14 Silicon Si 28.086	15 Phosphorous P 30.9738	16 Sulfur S 32.064	17 Chlorine Cl 35.453	18 Argon Ar 39.948
4	19 Potassium K 39.098	20 Calcium Ca 40.08	21 Scandium Sc 44.956	22 Titanium Ti 47.90	23 Vanadium V 50.942	24 Chromium Cr 51.996	25 Manganese Mn 54.938	26 Iron Fe 55.847	27 Cobalt Co 58.933	28 Nickel Ni 58.71	29 Copper Cu 63.546	30 Zinc Zn 65.38	31 Gallium Ga 69.723	32 Germanium Ge 72.59	33 Arsenic As 74.922	34 Selenium Se 78.96	35 Bromine Br 79.904	36 Krypton Kr 83.80
5	37 Rubidium Rb 85.468	38 Strontium Sr 87.62	39 Yttrium Y 88.905	40 Zirconium Zr 91.22	41 Niobium Nb 92.906	42 Molybdenum Mo 95.94	43 Technetium Tc (99)	44 Ruthenium Ru 101.07	45 Rhodium Rh 102.905	46 Palladium Pd 106.4	47 Silver Ag 107.868	48 Cadmium Cd 112.40	49 Indium In 114.82	50 Tin Sn 118.69	51 Antimony Sb 121.75	52 Tellurium Te 127.60	53 Iodine I 126.904	54 Xenon Xe 131.30
6	55 Cesium Cs 132.905	56 Barium Ba 137.34	*57 Lanthanum La 138.91	72 Hafnium Hf 178.49	73 Tantalum Ta 180.948	74 Tungsten W 183.85	75 Rhenium Re 186.2	76 Osmium Os 190.2	77 Iridium Ir 192.2	78 Platinum Pt 195.09	79 Gold Au 196.967	80 Mercury Hg 200.59	81 Thallium Tl 204.37	82 Lead Pb 207.19	83 Bismuth Bi 208.980	84 Polonium Po (209)	85 Astatine At (210)	86 Radon Rn (222)
7	87 Francium Fr (223)	88 Radium Ra (226)	**89 Actinium Ac (227)	104 Rutherfordium Rf (261)	105 Hahnium Ha (262)	106 Seaborgium Sg (263)	107 Neilsbohrium Ns (261)	108 Hassium Hs (265)	109 Meitnerium Mt (266)									

*Lanthanides 4f	58 Cerium Ce 140.12	59 Praseodymium Pr 140.907	60 Neodymium Nd 144.24	61 Promethium Pm 144.913	62 Samarium Sm 150.35	63 Europium Eu 151.96	64 Gadolinium Gd 157.25	65 Terbium Tb 158.925	66 Dysprosium Dy 162.50	67 Holmium Ho 164.930	68 Erbium Er 167.26	69 Thulium Tm 168.934	70 Ytterbium Yb 173.04	71 Lutetium Lu 174.97
**Actinides 5f	90 Thorium Th 232.038	91 Protactinium Pa (231)	92 Uranium U 238.03	93 Neptunium Np (237)	94 Plutonium Pu 244.064	95 Americium Am (243)	96 Curium Cm (247)	97 Berkelium Bk (247)	98 Californium Cf 242.058	99 Einsteinium Es (254)	100 Fermium Fm 257.095	101 Mendelevium Md 258.10	102 Nobelium No 259.101	103 Lawrencium Lr 260.105

FIGURE 2.2 Periodic Table of the Elements.

FIGURE 2.3 Carbon atom diagram: Periodic Table information for a carbon atom and a model of a carbon atom.

atom of carbon contains six protons and six neutrons. Notice that the atomic mass listed in **Figure 2.3** is not a whole number. The number of neutrons may vary for atoms of the same element. The atomic mass listed is the average mass for an atom of carbon.

Isotopes of an element are atoms that have the same number of protons as every other atom of that element but have a different number of neutrons. The atomic mass listed in the Periodic Table is an average taken of all the atoms and isotopes for that element.

The protons in an atom are positively charged, and the electrons are negatively charged. Neutrons have no charge—they are electrically neutral. Because the number of positive protons is equal to the number of negative electrons, atoms are also electrically neutral. Electrons travel in orbits or

clinical point

Some isotopes are unstable and freely emit particles to get to a more stable form. If they do, they are called **radioisotopes** and their decay, called **radioactivity,** can be very useful in medicine for diagnosis and treatment. For example, the thyroid gland (located in the cervical region) is soft tissue and therefore does not show up well on an x-ray. It is also one of the few parts of the body that requires the element iodine, which it uses to make thyroid hormone. (You will learn more about this in the chapter on the endocrine system.) A physician can inject a small amount of the iodine radioisotope into a patient. The patient's thyroid gland quickly takes in the radioisotope of iodine, and the radiation given off by the isotope causes the thyroid to be visible on an x-ray. This gives the physician more diagnostic information. For treatment, other radioisotopes in different strengths may be administered as radioactive seeds to tumor sites. These seeds can deliver radiation directly to the tumor site to kill cancerous cells.

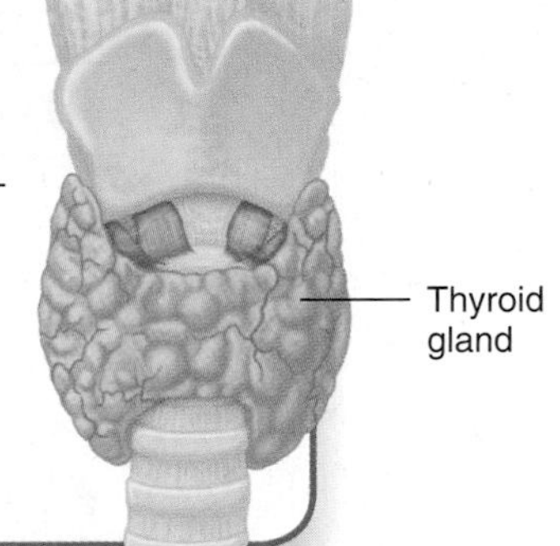

shells around the centrally located protons and neutrons. See **Figure 2.3**. Each shell has a set limit as to how many electrons it can hold.

spot check 1 How many protons, electrons, and neutrons are in a typical potassium (K) atom? Use the Periodic Table in Figure 2.2 to derive your answer.

2.3 learning outcome

Define *molecule* and describe two methods of bonding that may form molecules.

Bonding to Form Molecules Atoms will bind with other atoms to form **molecules** in order to fill their outer shells with electrons. This makes them more stable. The definition of a molecule is two or more atoms bonded together. A very stable bond called a **covalent bond,** which is often formed by carbon atoms, occurs when two or more atoms *share* electrons to fill their outer shells. Another type of bond is an **ionic bond.** Here, two or more atoms bind to form a molecule by *giving up or receiving electrons* from each other to fill their outer shells. The transfer of electrons means each of the individual atoms is no longer electrically neutral. Sodium (Na) and chlorine (Cl) atoms typically form an ionic bond. The sodium atom, which gives up an electron, becomes positively charged, while the chlorine atom, which receives an electron, becomes negatively charged. Atoms with a charge are called **ions.** Ions are attracted to each other because of their opposite charges, so they form a bond resulting in an electrically neutral molecule. The molecule, sodium chloride (NaCl), is table salt. See **Figure 2.4**.

When placed in water, ionically bonded molecules separate into their individual ions. These ions in solution, called **electrolytes,** are capable of conducting electricity. You will study electrolytes more thoroughly in the muscle and nervous system chapters.

Some examples of electrolytes are sodium (Na^+), potassium (K^+), calcium (Ca^{2+}), and chloride ions (Cl^-). Electrolyte balance is very important, as it can mean life or death. For example, diarrhea is a major concern in infants because of the loss of electrolytes in the runny stool. For this reason, a physician may prescribe a commercial electrolyte solution to restore electrolyte balance.

FIGURE 2.4 Bonding: (a) covalent bonding for carbon dioxide (CO_2), in which atoms share electrons; (b) ionic bonding for sodium chloride (NaCl), in which atoms gain and lose electrons to form ions.

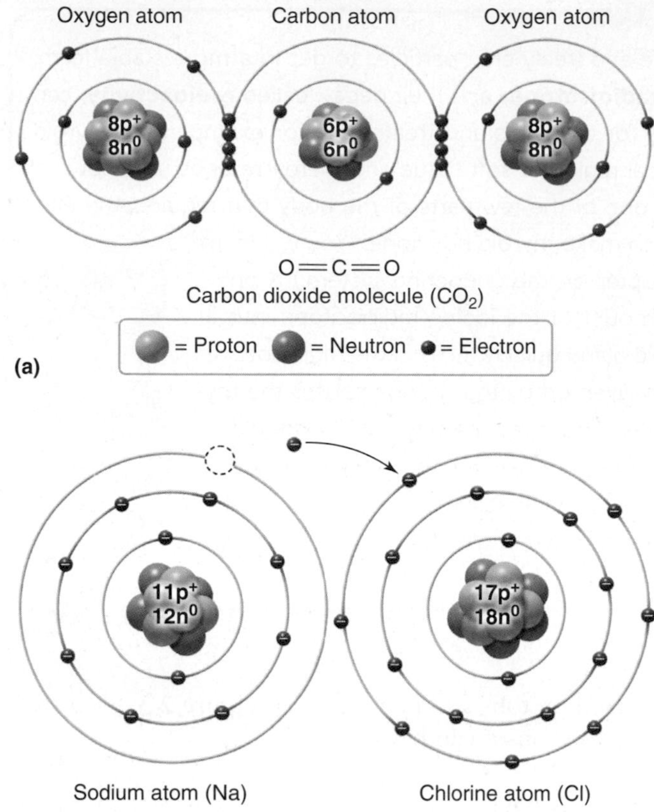

Separate atoms
If a sodium atom loses an electron to a chlorine atom, the sodium atom becomes a sodium ion (Na^+) and the chlorine atom becomes a chlorine ion (Cl^-).

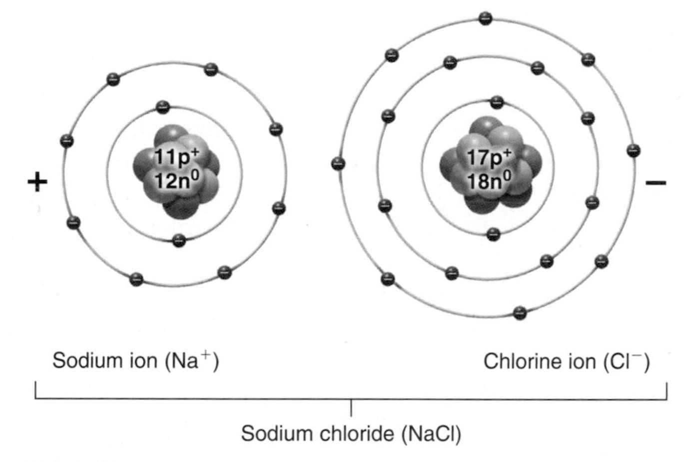

Bonded ions
These oppositely charged particles attract electrically and join by an ionic bond.

(b)

Water Water serves key functions in the body. You have just seen one of them—it can separate ionically bonded atoms into ions to create electrolytes. The human body is approximately 50 to 75 percent water (H_2O). As the main component of many body fluids, water carries out the following five functions for the body:

- Water chemically separates ionically bonded molecules into individual ions called electrolytes.
- Water works as a lubricant in tears and the fluid of joints.
- Water aids in chemical reactions, as in saliva during digestion.
- Water is used to transport nutrients and wastes in blood plasma.
- Water is used for temperature regulation. It has a high heat capacity to maintain body temperature and can also be used to cool the body when it evaporates from the surface as sweat.

2.4 learning **outcome**

Summarize the five functions of water in the human body and give an explanation or example of each.

Solutions Body fluids, like tears, sweat, saliva, and plasma, are not pure water. They are solutions. Every **solution** is composed of two basic parts: one or more **solutes** and a **solvent.**

Consider, for example, a salt solution: If you add a tablespoon of salt to a beaker of water, the salt settles to the bottom of the beaker. If you stir the contents of the beaker, the salt disappears as it dissolves. You now have a salt solution. This solution is composed of a solute (the salt) dissolved in a solvent (the water). Solutes may be solids, liquids, or gases. Examples are salt, alcohol, and carbon dioxide, respectively. Water is a common solvent in the body. **Concentration** refers to the amount of solute present in a solution relative to the amount of solvent. To create a more concentrated salt solution than the one in our example, you would add more salt to the same amount of water.

The term **tonicity** is used when comparing solutions. A solution may be **hypertonic, isotonic,** or **hypotonic** when compared to another solution. If a solution is hypertonic, it is more concentrated with solutes than the other solution. If a solution is isotonic, it has the same concentration of solutes as the other solution. If a solution is hypotonic, it is less concentrated with solutes than the other solution. This basic information on solutions will be important for our discussion of transport across cell membranes later in this chapter. See **Figure 2.5**.

2.5 learning **outcome**

Compare solutions based on tonicity.

FIGURE 2.5 Tonicity. The beaker with the hypotonic solution has less solutes than solution A. The beaker with the isotonic solution has the same amount of solutes as solution A. The beaker with the hypertonic solution has more solutes than solution A.

2.6 learning **outcome**

Determine whether a substance is an acid or a base and its relative strength if given its pH.

Acids, Bases, and pH At first glance, the terms *acids, bases,* and *pH* may look familiar. You take an antacid to relieve excess stomach acid, many household cleaners you use are bases, and your shampoo is often pH-balanced. However, you will need to know these terms chemically to understand their importance in human physiology.

An **acid** is a molecule that releases a **hydrogen ion (H^+)** when added to water. A **base** is a molecule that will accept the hydrogen ion, often by releasing a **hydroxide ion (OH^-)** when added to water. You can measure the strength of acids and bases by using a **pH** (potential of hydrogen) scale. This scale is a number range from 0 to 14. See **Figure 2.6**. A molecule with a pH of 7 is considered neutral because it is composed of equal amounts of hydrogen and hydroxide ions (H^+ and OH^-). Water (H_2O or HOH) is a good example of a molecule with a pH of 7. Acids have a pH less than 7. Bases have a pH greater than 7. Each one-number difference in pH indicates 10 times more ions released.

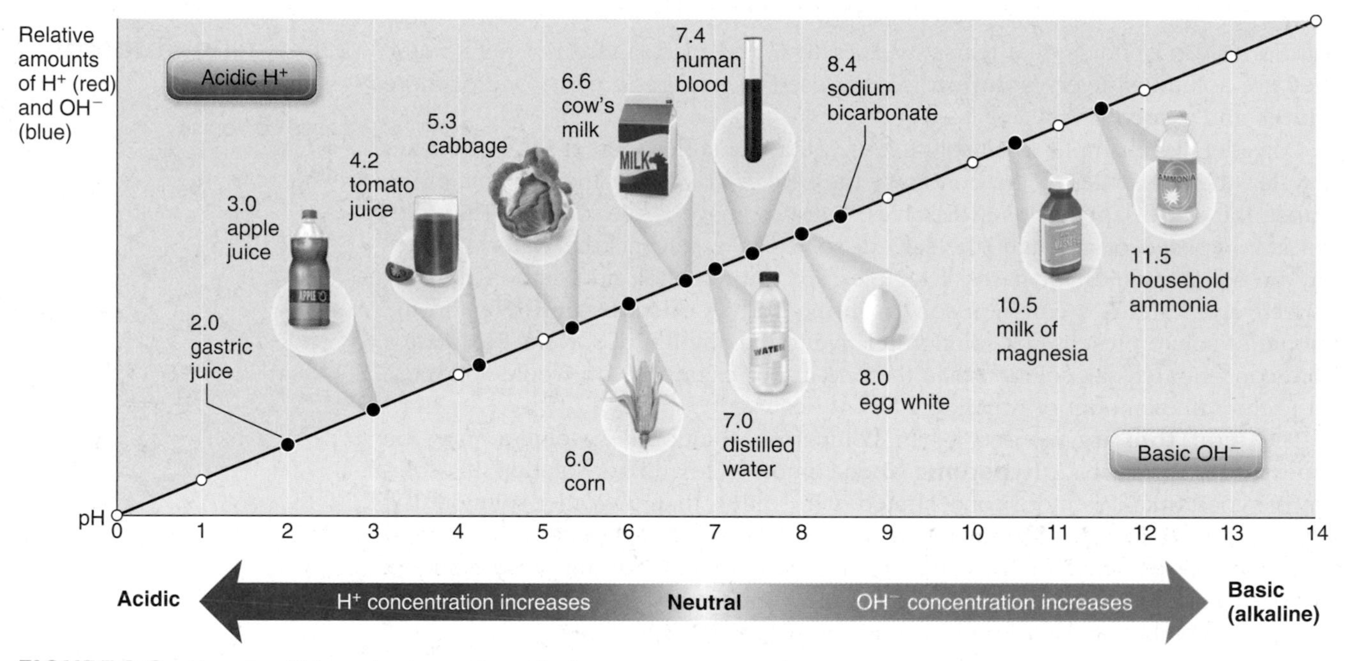

FIGURE 2.6 pH scale. This scale shows the pH of many common substances.

Consider three substances, A, B, and C. Substance A has a pH of 3. Substance B has a pH of 4. Both substances are acids because their pHs are less than 7. Substance A is a stronger acid because its pH is further away from neutral than is substance B's pH. Both substances release H^+ when placed in water because they are acids. Substance A releases 10 times more H^+ than substance B because A's pH is one number lower than B's. If substance C has a pH of 2, it too is an acid and it releases H^+ when placed in water. Substance C would release 10 times more H^+ than substance A because its pH is one number lower than A's. Substance C would release 100 times more H^+ than substance B because substance C's pH is two numbers lower than B's pH (10 times 10). The lower than 7 the pH is, the more hydrogen ions the substance has to release; this makes it more **acidic.** The same relationship exists for bases. The higher than 7 the pH is, the more hydroxide ions (OH^-) the substance has to release; this makes it more basic **(alkaline).** See **Figure 2.7**.

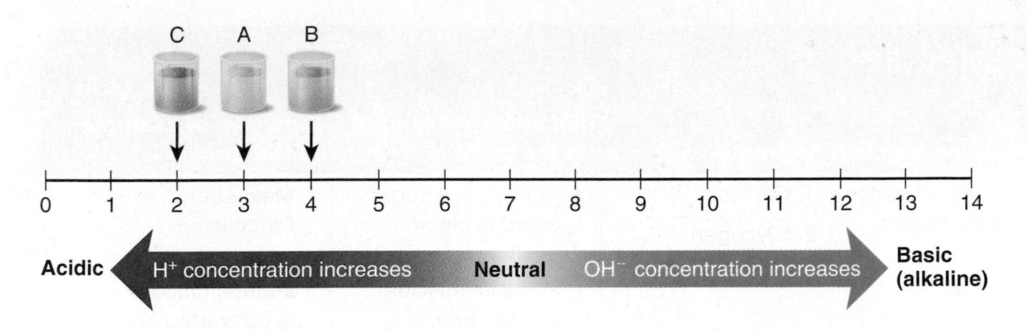

FIGURE 2.7 pH comparison. Substance A has a pH of 3. Substance B has a pH of 4. Substance C has a pH of 2.

The acid-base balance is very important to the body's homeostasis. For example, the pH of human blood needs to be within the very narrow pH range of 7.35 to 7.45. A blood pH lower than 7.35 is considered **acidosis,** and a blood pH higher than 7.45 is considered **alkalosis.** Either condition can be fatal. In the system chapters of this book, you will investigate how the body's acid-base balance is maintained in many body fluids. This investigation will include how the body regulates blood pH, what may cause an acid-base imbalance in the blood, and what disorders may result from that imbalance.

spot check 2 Liquid X has a pH of 8. Liquid Y has a pH of 11. Are these liquids acids or bases? Which ion will they release (H^+ or OH^-) when placed in water? Which liquid is stronger? How many times more ion will be released in the stronger liquid than in the other liquid?

2.7 learning outcome

Describe the four types of organic molecules in the body by giving the elements present in each, their building blocks, an example of each, the location of each example in the body, and the function of each example.

Organic Molecules In common usage, the term *organic* refers to something that is healthy, free of pesticides, and naturally grown. The meaning in chemistry is very different. In chemistry, **organic molecules** come from life and must contain atoms of the elements carbon *and* hydrogen. You may know that you are 50 to 75 percent water (H_2O) and that you exhale carbon dioxide (CO_2). These molecules are certainly involved in life, but CO_2 contains carbon but not hydrogen and H_2O contains hydrogen but not carbon. Therefore, these molecules are not organic molecules. They are inorganic molecules. The four types of organic molecules are **carbohydrates, lipids, proteins,** and **nucleic acids.** See Table 2.2. These organic molecules are discussed in the paragraphs that follow.

As shown in Table 2.2, building blocks are subunits of complex molecules, and each organic molecule has a unique building block. Below, we take a closer look at the building blocks for each of these molecules.

Carbohydrates. These organic molecules contain atoms of carbon, hydrogen, and oxygen in a ratio of 1:2:1. **Monosaccharides** (simple sugars)—the simplest form of a carbohydrate—are the building blocks of carbohydrates. Glucose is an example of a monosaccharide. If you combine two monosaccharides to form a single molecule, you have a disaccharide. If you string many monosaccharides together like beads in a necklace to form a single molecule, you have a polysaccharide. See **Figure 2.8.** Glycogen (a starch) is an example of a polysaccharide. Carbohydrates are an energy source for the cell. You will learn how they are used for energy later in this chapter when you explore cellular respiration.

monosaccharide:
MON-oh-SACK-ah-ride

TABLE 2.2 Organic molecules

Organic molecule	Elements	Building blocks	Example	Location of example in the body	Function of example in the body
Carbohydrate	C, H, O in a ratio of 1:2:1	Monosaccharides (simple sugars)	1. Glucose 2. Glycogen	1. Glucose is a sugar found in blood. 2. Glycogen is a starch found in muscles and the liver.	1. Main energy source for cells 2. Stored energy source (glucose is converted to glycogen for storage)
Lipid	C, H, O not in a ratio of 1:2:1	Fatty acids and glycerol	1. Fats 2. Steroids 3. Phospholipids	1. Adipose tissue 2. Hormones found in blood 3. Cell membranes	1. Stored energy 2. Regulate the body 3. Give structure to cell and regulate what goes in and out of the cell
Protein	C, H, O, N	Amino acids (20 different amino acids)	1. Keratin and collagen 2. Hormones 3. Transport proteins 4. Enzymes 5. Antibodies 6. Muscle proteins 7. Binding and receptor proteins	1. Skin 2. Blood 3. Blood 4. Everywhere 5. Blood 6. Muscles 7. Cell membranes	1. Give strength 2. Regulate the body 3. Transport other molecules 4. Aid in chemical reactions 5. Fight foreign invaders 6. Allow for contraction of muscles 7. Hold cells together
Nucleic acid	C, H, O, N, P	Nucleotides	1. Deoxyribonucleic acid (DNA) 2. Ribonucleic acid (RNA)	1. Nucleus of a cell 2. Many places in a cell	1. DNA is the genetic information. 2. RNA processes the genetic information.

(a) Monosaccharide (glucose)

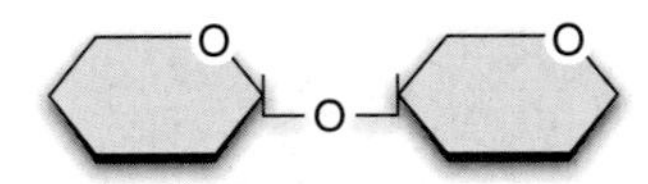

(b) Disaccharide

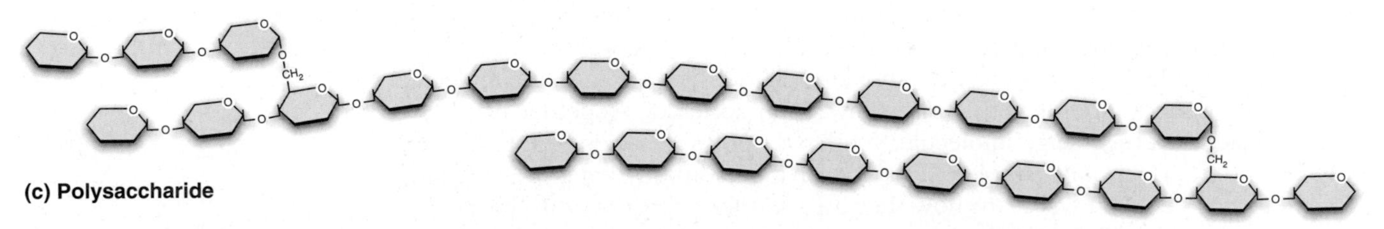

(c) Polysaccharide

FIGURE 2.8 Carbohydrates: (a) a monosaccharide, (b) two monosaccharides combined to form a disaccharide, (c) monosaccharides strung together to form a polysaccharide like glycogen.

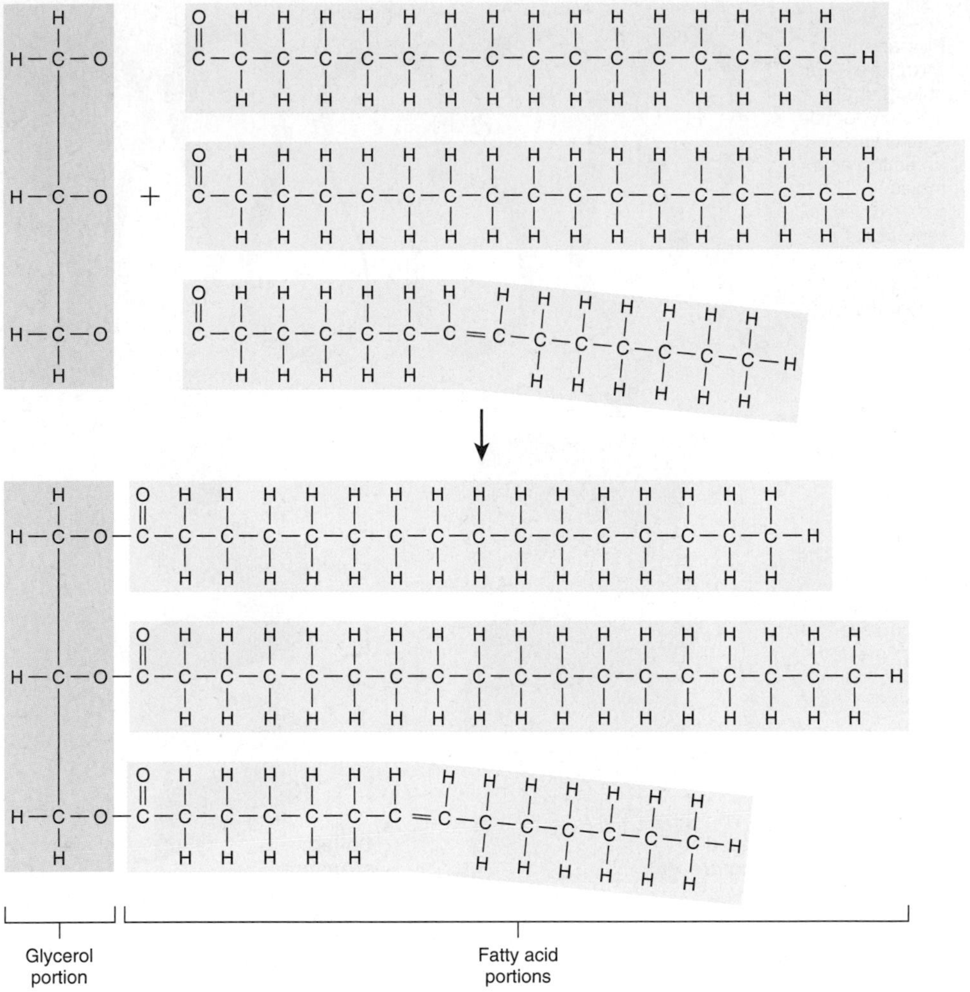

Glycerol portion

Fatty acid portions

FIGURE 2.9 **Triglyceride (fat) synthesis.** A glycerol molecule combines with three fatty acid molecules to form a triglyceride (fat).

Lipids. These organic molecules contain atoms of the same elements as those in carbohydrates (carbon, hydrogen, and oxygen) but not in the 1:2:1 ratio. The building blocks of lipids are **fatty acids** and **glycerol.** See **Figure 2.9**. The three types of lipids you will focus on in later chapters are fats, steroids, and phospholipids.

Proteins. These organic molecules contain nitrogen in addition to carbon, hydrogen, and oxygen. The building blocks for proteins are the 20 different **amino acids.** It is very important that you understand that proteins function in many ways according to their shape. This shape is determined by the order of the protein's amino acids. Parts of amino acids may be attracted to segments of other amino acids. This attraction causes the amino acid chain to bend, fold, and pleat, which results in a very unique shape.

Consider this comparison: Think about the keys on your key ring. One key opens the front door to your home, another opens the lock for your bike, a third key unlocks your car. Although your keys are all basically made from the same metal, they do not open the same locks. A tiny variation in shape makes a huge difference as to whether or not the key works. The same is true for proteins. You will investigate how proteins are made later in this chapter and how proteins are used in the body in every following chapter. See **Figure 2.10** for an example of protein structure.

FIGURE 2.10 Protein structure. This diagram shows the four levels of protein structure. Each protein is a string of amino acids that coils, sheets, and folds and may assemble with additional chains to form the protein's unique shape.

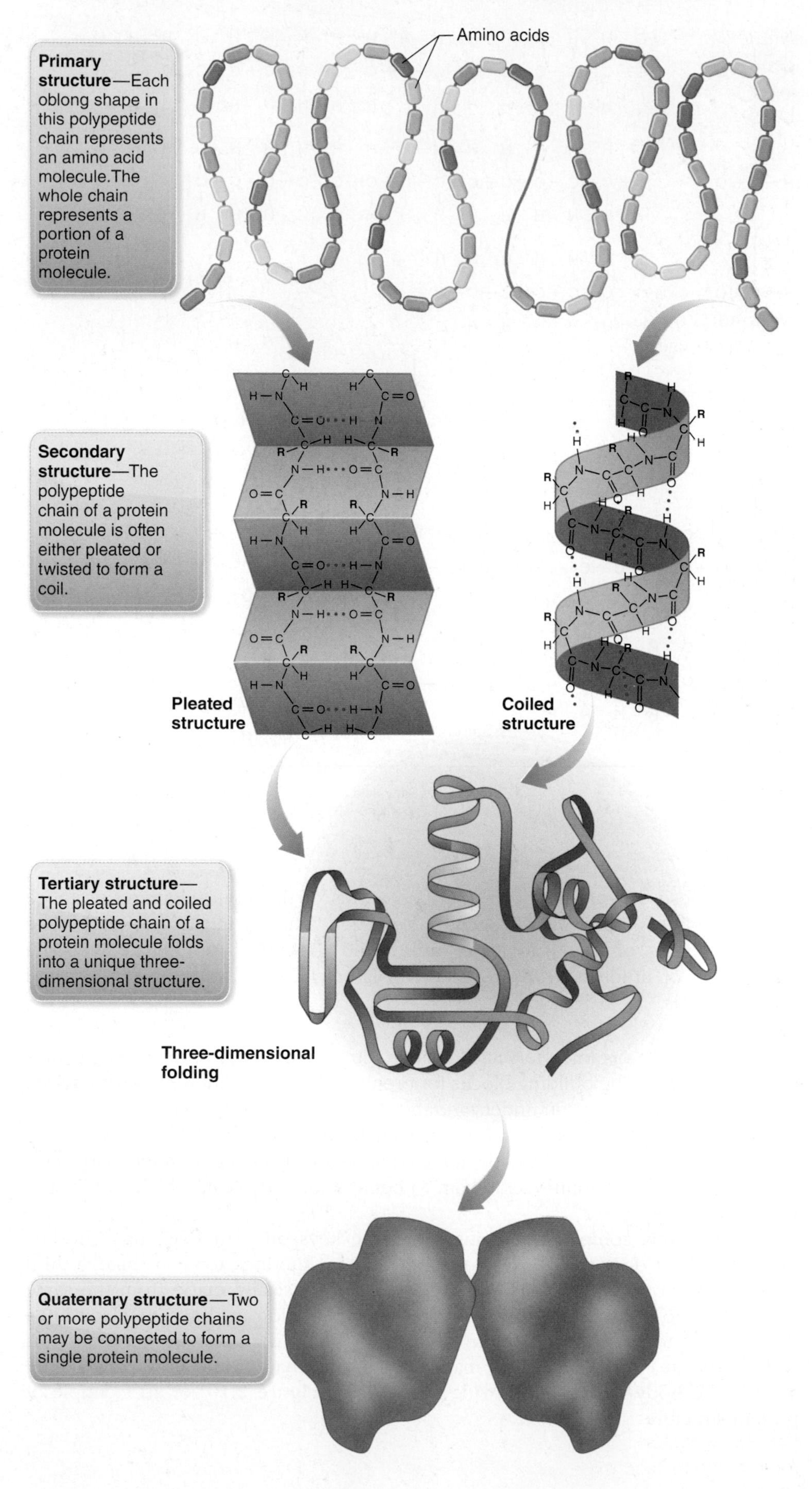

Nucleic acids. These organic molecules are composed of carbon, hydrogen, oxygen, nitrogen, and phosphorus. The building blocks for nucleic acids are nucleotides strung together in a twisted, double strand (double helix), as in deoxyribonucleic acid (DNA), or in a single strand as in ribonucleic acid (RNA). The nucleotides in DNA are guanine (G), cytosine (C), adenine (A), and thymine (T). In RNA, uracil (U) is substituted for thymine. The double helix of DNA resembles a twisted ladder with pairs of nucleotides forming base pairs, the rungs of the ladder. Guanine always pairs with cytosine (G-C) and adenine always pairs with thymine (A-T). See **Figure 2.11**.

Your DNA contains all of the genetic information that is you. It is not written in English; it is written in DNA language called the **genetic code.**

English uses 26 letters combined in many different ways to form words of various lengths. The words are separated by spaces, and thoughts are separated by punctuation to give them meaning. You must read the letters in correct order to make sense of what is written. For example, *tar* has a very different meaning than *rat*.

DNA language is much simpler. The genetic code has just four letters, the nucleotides (G, C, A, T), that are always read three nucleotides at a time (called a **triplet**). If you know always to read three letters at a time, there is no need for spaces. Punctuation is not necessary. Code triplets tell you where to begin and where to stop reading the DNA. Your DNA has approximately 3 billion base pairs of nucleotides to record all the information and directions on how your body works. A **gene** is the amount of DNA that must be read to give you the directions to make one specific protein. Your DNA contains about 35,000 genes. RNA is used to process the DNA genetic information, as you will learn later in this chapter.

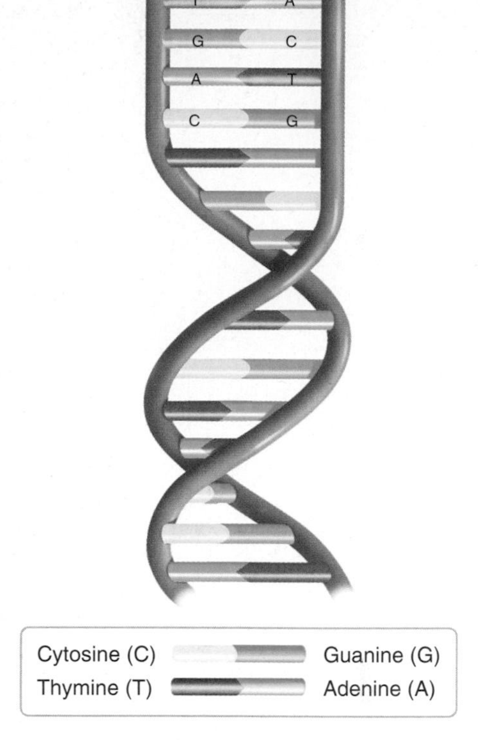

FIGURE 2.11 DNA structure: nucleotides pairing to form the DNA molecule.

applied **genetics**

Up to this point of the chapter, approximately 4,000 words involving the use of 16,000 letters (A to Z) were used to convey the information on the pages. The gene that contains the directions on how to make a single chloride channel protein on a cell membrane uses approximately 250,000 base pairs of nucleotides (G, C, T, A) to convey the information. Cystic fibrosis, a genetic disorder that reduces life expectancy to approximately 35 years, results when just 3 of the 250,000 base pairs of nucleotides are missing. This results in the loss of 1 amino acid from the chain of 1,480 that are needed to make this chloride channel protein.[1] As you will see later in this chapter, this small mistake in directions results in an amino acid sequence that is slightly off, causing the shape of the resulting protein to be slightly off too. Because the shape is not exactly what it is supposed to be, the chloride channel protein does not function properly. Cystic fibrosis is the result.

spot check **3** What type of organic molecule is $C_6H_{12}O_6$? Use Table 2.2 to derive your answer.

2.8 learning **outcome**

Explain three factors governing the speed of chemical reactions.

Chemical Reactions Now that we have discussed atoms and molecules, we can look at what happens when they are combined. Some molecules will react with others in what is called a **chemical reaction.** The sum total of all the chemical reactions that take place in the human body is called **metabolism.** In any reaction you start with **reactants** and end with **products:**

Reactants → Products *or* Products ← Reactants

Read the arrow as "yield," and know that it always points to the products. It does not matter in what order the reaction is written, as long as you remember that you always end with the products.

Speed of reactions. You must bring molecules together for them to react with one another. Imagine a playland ball pit where children are playing. The children represent reactants/molecules. You need to have children collide for them to interact. So what can you do to get them to come together more quickly?

1. **Increase the concentration of the reactants.** Add more kids to the ball pit. More children playing increases the likelihood that two or more will come together.
2. **Increase the speed of the reactants.** Giving all the kids caffeine and sugar makes them move faster and therefore more likely to come together. Just as caffeine and sugar speeds up kids, heat speeds up the motion of molecules.
3. **Use a catalyst.** Have an adult (the catalyst) step into the pit, grab one child with one hand and another child with the other hand, and then bring them together. A catalyst is any molecule that speeds up a reaction without becoming chemically involved. Enzymes (one of the protein examples) often act as catalysts to speed up chemical reactions. They facilitate the reaction without becoming chemically involved.

spot check 4 How does putting leftovers in the refrigerator relate to bacteria metabolism?

2.9 learning outcome

Write the equation for cellular respiration using chemical symbols and describe it in words.

Cellular respiration. Some reactions require that energy be added for them to occur. Other reactions release the energy held in the chemical bonds between the atoms in a molecule. **Cellular respiration** is such a reaction, and it is one of the most important chemical reactions in the body. Chemical reactions are written using a shorthand of chemical symbols and numbers. Here is the cellular respiration reaction, followed by the meaning of the symbols and numbers:

$$C_6H_{12}O_6 + O_2 \rightarrow CO_2 + H_2O$$

Glucose + Oxygen *yields* Carbon dioxide and Water

A law in nature states that matter can neither be created nor destroyed. So there must be an equal number of atoms of each element on both sides of the sample reaction above. A subscript number indicates the number of atoms present for the element just before it. No subscript is understood as 1. For example, the molecule CO_2 (carbon dioxide) contains 1 atom of carbon and 2 atoms of oxygen.

Count the atoms on each side of the reaction:

$C_6H_{12}O_6 + O_2$		→	$CO_2 + H_2O$	
Reactant side		→	**Product side**	
Carbon	6		Carbon	1
Hydrogen	12		Hydrogen	2
Oxygen	8		Oxygen	3

As you can see, an unequal number of atoms for each element exists on each side of this reaction. To balance the equation, numbers are placed in front of the molecules:

- If you have $\mathbf{H_2O}$, you have 1 molecule of water having 2 hydrogen atoms and 1 oxygen atom.
- If you have $\mathbf{3H_2O}$, you have 3 molecules of water, each molecule having 2 hydrogen atoms and 1 oxygen atom. Three molecules of water have a total of 6 hydrogen atoms and 3 oxygen atoms.

Now count the number of atoms for each element on each side of the reaction:

$\mathbf{C_6H_{12}O_6 + 6O_2}$		$\longrightarrow$	$\mathbf{6CO_2 + 6H_2O}$	
Reactant side		$\longrightarrow$	**Product side**	
Carbon	6		Carbon	6
Hydrogen	12		Hydrogen	12
Oxygen	18		Oxygen	18

So the correct, balanced formula for this equation is:

$$C_6H_{12}O_6 + 6O_2 \rightarrow 6CO_2 + 6H_2O$$

Glucose + Oxygen *yields* Carbon dioxide and Water

But, hold on, you are not finished just yet. Cellular respiration is performed in cells. It is the reason you eat—to deliver glucose to cells. It is one of the reasons you breathe—to deliver oxygen to your cells. But why is cellular respiration important? So that you can produce carbon dioxide and water? Surely this cannot be the reason. Carbon dioxide is a waste product you get rid of by exhaling. And producing water is not the reason either, as you can easily add water to your body by drinking. The reason our cells perform cellular respiration is to release the energy within the bonds of the glucose molecule. You will learn much more about this in the muscular system chapter.

The final, complete version of this chemical reaction is the one you must remember:

$$C_6H_{12}O_6 + 6O_2 \rightarrow 6CO_2 + 6H_2O + \text{Energy}$$

Glucose + Oxygen *yields* Carbon dioxide + Water + Energy

2.10 learning **outcome**

Explain the importance of ATP in terms of energy use in the cell.

ATP The energy released from the glucose molecule in cellular respiration must be converted to a usable form. A helpful analogy is that energy is like money. You may have a job for which you earn money (or energy, in this example), which may be issued as a paper check. There is monetary value in the check just as there is chemical energy within the bonds of the glucose molecule. However, you cannot take your paycheck to the vending machine to buy a soda. Although the check has monetary value, it is not in a usable form. So the check must first be converted to cash before it can be used. The same idea can be applied to your body cells. These cells cannot use the energy contained in the chemical bonds of a glucose molecule until this energy has been released through cellular respiration and converted to a usable form. This involves another chemical reaction that requires the addition of energy:

$$\text{Energy} + \text{ADP} + \text{P} \rightarrow \text{ATP}$$

Energy + Adenosine diphosphate + Phosphate → Adenosine triphosphate

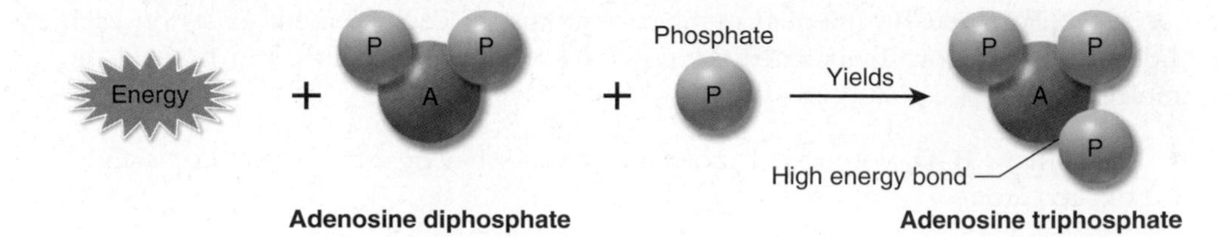

FIGURE 2.12 Formation of ATP. ADP combines with P to form ATP, a high-energy molecule.

Adenosine triphosphate (ATP) contains the usable form of energy for the cell. This energy is released from a glucose molecule's chemical bonds via cellular respiration, and then it helps to form a chemical bond between adenosine diphosphate and a third phosphate. See **Figure 2.12**. Cells can easily break the bond between the third phosphate and adenosine, releasing the energy as needed. The usable form of energy for the cell contained in the ATP molecule is like the cash in your pocket for the soda in that vending machine.

You have now finished the basic concepts of chemistry in the chemical level of the human body's organization. You will need this understanding for the upcoming levels, including the system levels in chapters that follow. It is vital that you see each level not as an end point but, instead, as part of a building process in which the levels build upon and support each other. In the section below, for example, you will see how molecules function together as organelles in the next level of organization.

2.11 learning **outcome**

Describe cell organelles and explain their functions.

Organelle Level

Molecules of different types come together to form **organelles.** Organelles are specific structural components of cells. Each organelle has its own function, and together they carry out the necessary cellular functions. Descriptions of organelles and their functions are listed in Table 2.3. Not all cells have the same organelles. The type and number of organelles in each cell depend upon the cell's function.

The organelles are suspended within the cell in a fluid called **cytoplasm,** which is a solution. Cytoplasm contains electrolytes, nutrients, wastes, and gases (such as oxygen and carbon dioxide) as the solutes and water as the solvent. A **cytoskeleton** of protein fibers organizes the organelles within the cytoplasm. See **Figure 2.13**.

Here are some specific examples of cell types and the organelles they need to perform their functions. Cells of the pancreas produce the protein insulin for export out of the cell. On the basis of the organelle functions listed in Table 2.3, you might predict that pancreas cells need the following:

- Well-developed, rough endoplasmic reticulum (site of protein production).
- Many ribosomes (to assemble proteins).

TABLE 2.3 Organelles

Organelle	Description and location	Function
Cell membrane (plasma membrane)	Phospholipid bilayer Found in all cells	Gives structure to cell, defining what is intracellular (inside the cell) and what is extracellular (outside the cell); regulates what may enter or leave the cell
Cilia	Hairlike extensions of cell membrane Found in cells needing to move materials outside themselves (cells lining trachea, moving inhaled dust out of the trachea)	Move in wavelike motion to move materials past the cell
Microvilli	Hairlike extensions of cell membrane Found in cells requiring extra surface area (cells lining the intestines, absorbing nutrients)	Provide extra surface area for the cell
Nucleus	Enclosed by a membrane Found in all cells except red blood cells	Houses DNA
Mitochondria	Rod shaped; enclosed by a membrane Found in large numbers in cells with high energy demands	Carry out cellular respiration and process the energy released to form ATP
Ribosomes	Large and small subunits Found in large numbers in cells that produce proteins	Assemble amino acids into proteins
Endoplasmic reticulum (ER) **1. Rough ER** **2. Smooth ER**	Sheets of membrane extending from nuclear membrane 1. Has ribosomes on its surface Extensive in cells producing proteins 2. Does not have ribosomes on its surface Extensive in cells producing lipids	1. Site of protein production 2. Site of lipid production
Golgi complex	Membrane-enclosed folds usually close to the ER Extensive in cells involved in protein and lipid production	Inspects and modifies proteins and lipids produced in the cell
Secretory vesicles	Membrane packages bubbled off the Golgi complex that contain the inspected and modified products of the Golgi complex Found in large numbers in cells that produce proteins for export out of the cell	Carry materials from the Golgi complex to the cell membrane for export outside the cell
Lysosomes	Membrane-bound packages of enzymes Found in large numbers in cells required to destroy materials (white blood cells destroy bacteria)	Store and isolate enzymes often used for intracellular digestion until they are needed

- Many Golgi complexes (to inspect and modify proteins produced).
- Many secretory vesicles (to serve as packages for export out of the cell).

On the other hand, white blood cells that **phagocytize** (eat and destroy) bacteria do not need the same relative amounts of organelles. White blood cells need many lysosomes containing digestive enzymes to destroy the bacteria.

cilia: SILL-ee-ah

Golgi: GOAL-jee

phagocytize: FAG-oh-sit-ize

spot check 5 Predict the relative amounts of organelles needed for a cell in a testicle that produces the steroid hormone testosterone. Use Table 2.3 to derive your answer.

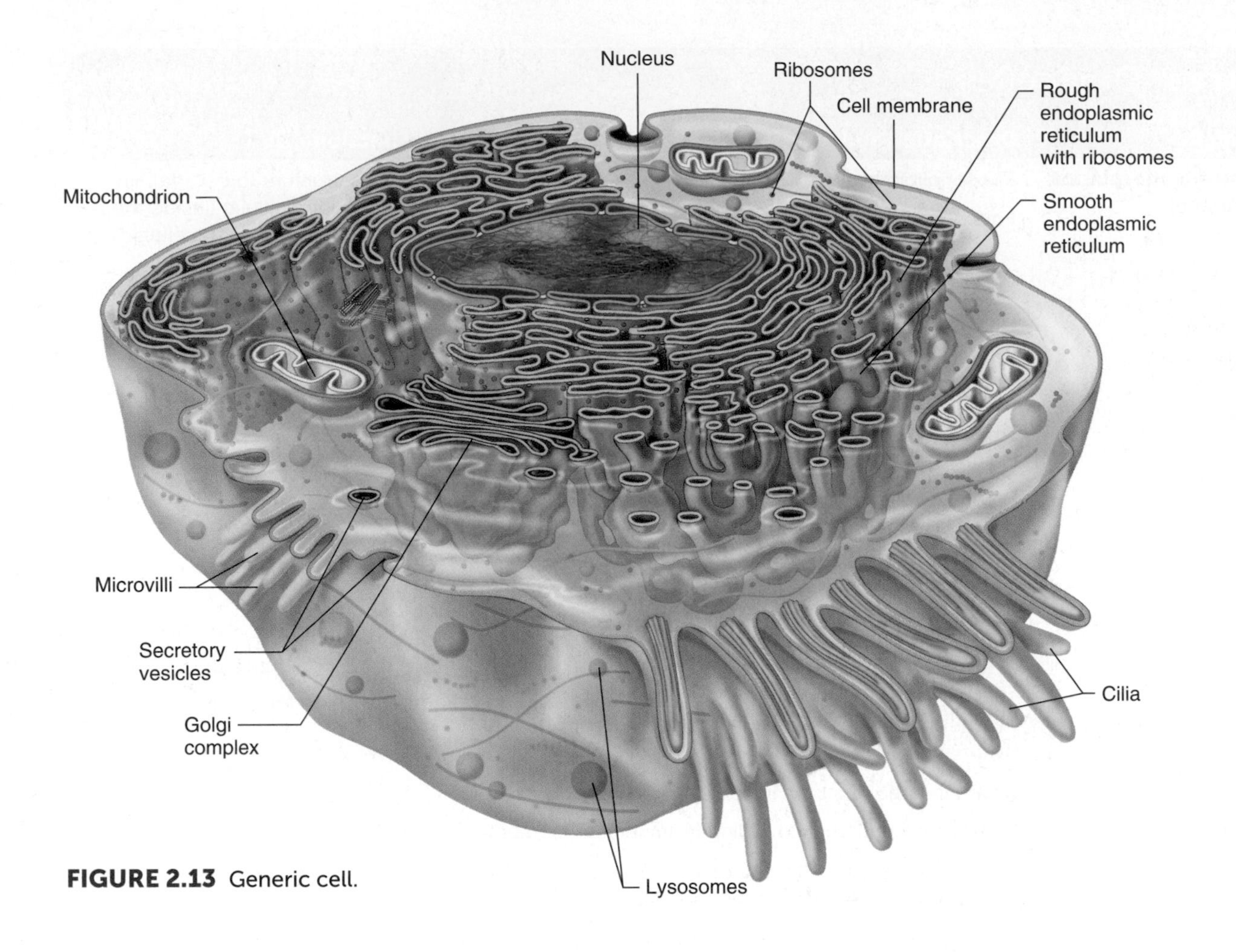

FIGURE 2.13 Generic cell.

> **warning**
>
> Figure 2.13 may lead to a misconception, as it depicts a *generic* cell showing the different possible types of organelles. Each type of cell is unique on the basis of its structure and function. Each cell has its own unique set and number of organelles to carry out that function. See Table 2.3.

Cell Membrane By studying the **cell membrane,** you will see how different organic molecules work together to function as an organelle.

> **warning**
>
> Be careful of the terminology for this organelle, as it is a cell membrane, *not* a cell wall. Cell walls are not found in human (animal) cells; they are found in bacteria and plant cells. Cell walls in plants are made of a carbohydrate called *cellulose,* which is indigestible for humans. In nutrition, cellulose is commonly called *fiber.* Nutrition will be covered in many of the system chapters. But we return here to our discussion of the cell membrane.

The cell membrane is designed to define the cell by separating the **extracellular fluid** (fluid outside the cell) from the **intracellular fluid** (fluid inside the cell). Both fluids are primarily water. Phospholipids, the primary component of a cell membrane, are composed of a **hydrophilic** (water-loving) glycerol head and two fatty acid **hydrophobic** (water-fearing) chains. The phospholipids arrange themselves in a **bilayer.** This means that the glycerol heads face the extracellular and intracellular fluids and that the fatty acid chains face toward each other (away from the fluids). The phospholipids are not rigidly connected; they float side by side in what is called a *fluid mosaic*. **Cholesterol** molecules (an example of a steroid lipid) may be found between the fatty acid chains of the phospholipids. Their presence may cause a stiffening of the cell membrane. See **Figure 2.14**.

In addition to containing phospholipids and steroids, the cell membrane may also contain a variety of embedded proteins. Some of these proteins serve as **receptors.** Their unique shape on the extracellular side of the cell membrane can make them sensitive to specific chemicals, such as hormones. Other proteins serve as **channels** to allow materials (like those unable to travel through the phospholipid bilayer) to enter the cell. You will read more about membrane transport shortly. Still other proteins have a carbohydrate on their extracellular surface. These **glycoproteins** serve as an identifying tag that enables the immune system to tell what is *self* from what is *foreign*. You will explore glycoproteins more fully in the cardiovascular and lymphatic systems chapter.

All of these organic molecules come together to carry out the functions of the cell membrane. These functions are the following:

- To give structure to the cell.
- To define what is intracellular (inside the cell) and what is extracellular (outside the cell).
- To regulate what may enter or leave the cell by a process called *membrane transport*.

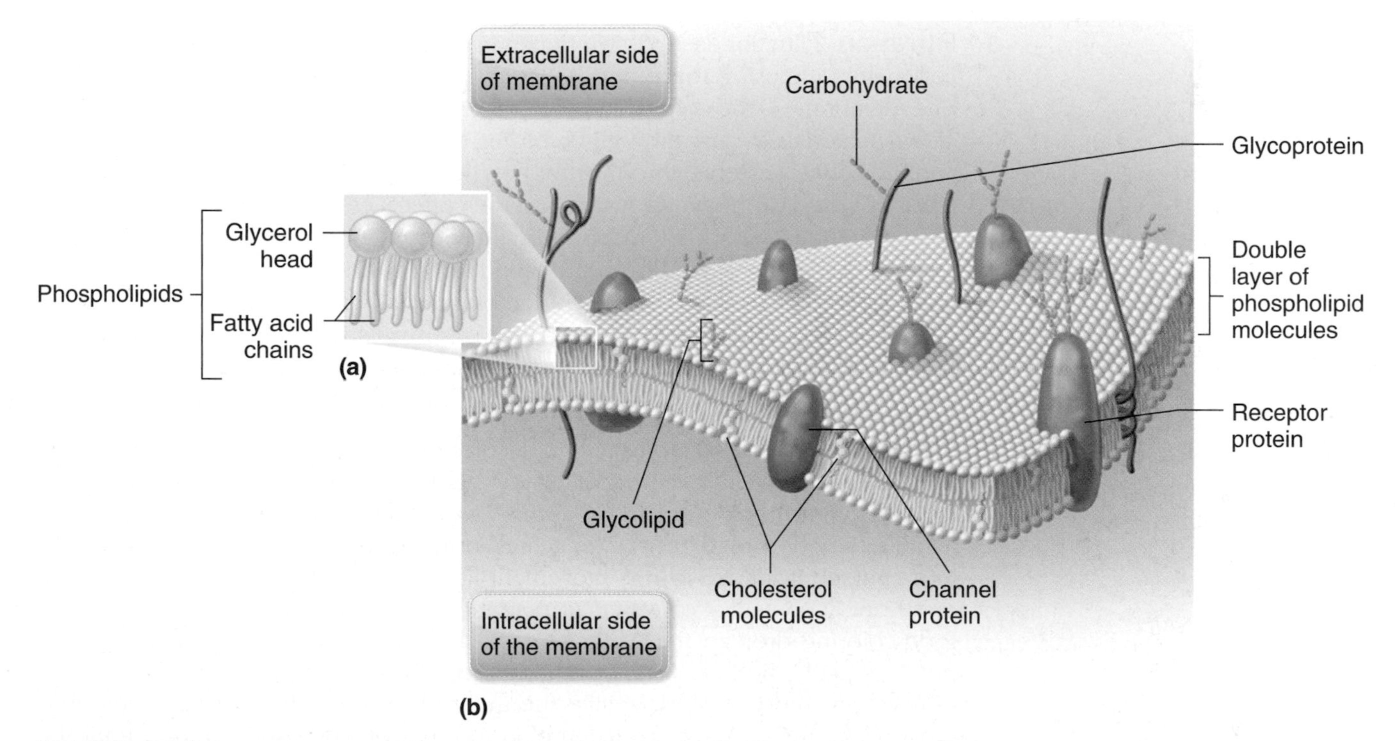

FIGURE 2.14 **Cell membrane (plasma membrane):** (a) phospholipids, (b) cell membrane.

Now that you understand the organelle level, you are ready to explore how the organelles (listed in Table 2.3) work together to complete the functions of the next level of your body's organization.

Cellular Level

By now you should be familiar with the anatomy of cells, which are defined by a cell membrane and contain various organelles depending upon their function. Protein fibers suspend the organelles in a solution called *cytoplasm,* and together the organelles carry out the cell's functions. We discuss below the cellular functions of membrane transport, protein synthesis, and cell division.

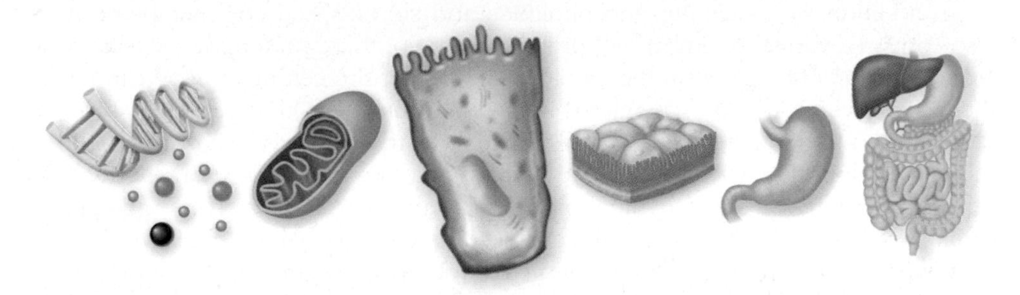

2.12 learning **outcome**

Compare four methods of passive transport and active transport across a cell membrane in terms of materials moved, direction of movement, and the amount of energy required.

Membrane Transport Movement of ions and molecules across the cell membrane can happen in either of two general ways: passively (requiring no energy) or actively (requiring energy). Materials can also be moved across the cell membrane in bulk.

Passive transport. Passive transport can move materials across the cell membrane in either direction—into or out of the cell. It is important to remember that with each of the four following methods, no energy is required for this type of movement to happen.

1. **Filtration.** Filtration is a passive-transport method that moves materials across a cell membrane using **force** but no energy. An analogy for this is a coffee maker. The membrane is the coffee filter. You place coffee grounds in the filter. The filter allows materials to pass through according to the size of its pores. It allows water and the coffee's essence through, but not the grounds. The coffee maker uses the force of *gravity* to move the water and the coffee's essence through the filter. An example of this method of transport in the body occurs in the kidneys. Water and some wastes are removed from the blood in the kidney to form urine. The filter in this case is the cell membranes of the blood vessel walls. It is not gravity, however, that allows the wastes to cross the vessel walls in the kidney (you can still produce urine standing on your head). Instead, the *blood pressure* forces wastes out of the bloodstream to form urine in the kidney. No energy is required.

2. **Simple diffusion.** All atoms and molecules maintain a constant state of motion. Even the molecules in the paper of this book are moving while you are reading. Although they do not move very far from each other because the paper is a solid, they do in fact move. Molecules in liquids and gases move more freely. You can see evidence of this movement if you add a drop of food coloring to a stationary container of water. At first you can see exactly the position of the food-coloring drop in the water. But, as time goes on, the food coloring disperses equally throughout the container's contents. In simple diffusion, as in all diffusion methods, materials (like the molecules in the drop of food coloring) move from areas of high concentration to areas of low concentration until the concentrations become equal. See **Figure 2.15**.

   Of course, not all materials can pass through the cell membrane by simple diffusion. In fact, the cell membrane is selectively permeable, which means that only select molecules can pass through it by simple diffusion. Oxygen is an example of a molecule that can pass through the phospholipid bilayer on its own.

Factors that govern the speed of simple diffusion are:

FIGURE 2.15 Simple diffusion.

- **Temperature.** Heat causes molecules to move faster; increased temperature increases the speed of simple diffusion.
- **Molecular weight.** Heavy proteins move slower than lighter, smaller molecules like electrolytes, and gases diffuse faster.
- **Concentration gradient.** This is the amount of difference in concentration on either side of the membrane; the greater the difference, the faster the diffusion. Materials are said to move *down a concentration gradient* (from areas of high concentration to areas of low concentration) until they are equal on both sides of the membrane.
- **Membrane surface area.** The speed of diffusion is increased with greater surface area, so there is more membrane for this to occur.

3. **Facilitated diffusion.** This passive-transport method is used for molecules that cannot diffuse through the selectively permeable membrane on their own (like glucose), so they need help getting through a channel protein. Consider again the negative feedback example in Chapter 1. If Jen consumes the soda and candy bar she bought from the vending machine on her study break, her digestive system digests the soda and candy bar and absorbs the sugar into her bloodstream. This causes her blood sugar (glucose) level to rise above normal. Her pancreas recognizes the increased glucose level and responds by releasing the hormone insulin, which travels to most cells. Insulin binds to a protein receptor (shaped specifically for insulin) on the cell's surface. When insulin is bound to this receptor, a gated channel protein opens. This allows glucose to diffuse into the cell from an area of high concentration to an area of low concentration until the concentrations on the outside and inside of the cell are equal. The transport of glucose across the cell membrane fits the description of diffusion, but it requires insulin fitting into the receptor before it can happen. Insulin facilitated this diffusion. The cell's glucose uptake lowers the blood glucose level, restoring homeostasis. This is negative feedback. See **Figure 2.16**.

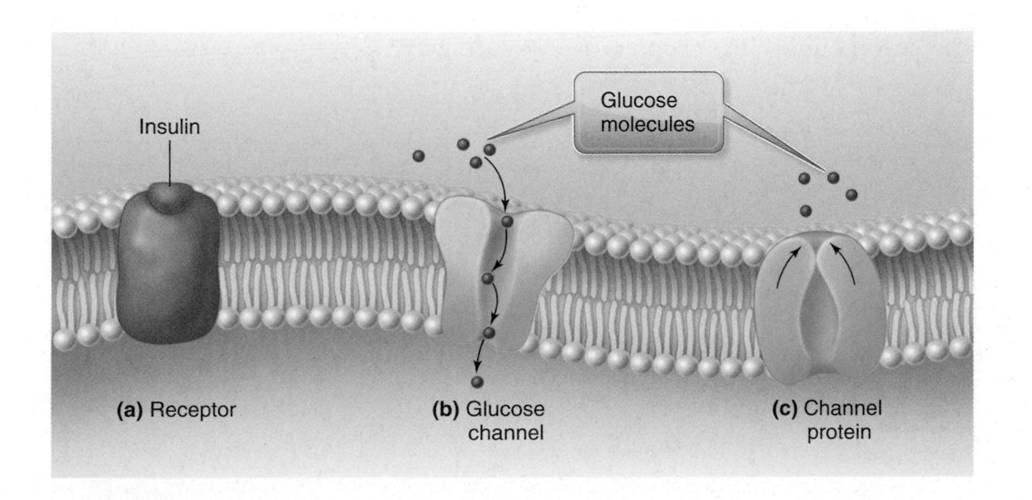

FIGURE 2.16 Examples of membrane proteins during transport. In facilitated diffusion, insulin binds to a protein receptor (a) based on shape. This causes a gated channel for glucose (b) to open, allowing glucose to diffuse into the cell. A channel protein (c) that is not for glucose remains closed.

osmosis: oz-MO-sis

4. **Osmosis.** Cell membranes are selectively permeable. As you have seen, some molecules can move across the membrane by filtration, some by simple diffusion, and others by facilitated diffusion. Some molecules, however, cannot move across the cell membrane to equalize their concentration by any passive means. In the human body, the intracellular fluid (cytoplasm) and the extracellular fluid are both solutions. If the solutes cannot move across the membrane, water will move across the cell membrane by a process called **osmosis** to equalize the concentration of both solutions. Water moves to the more concentrated solution to dilute it. See **Figure 2.17**.

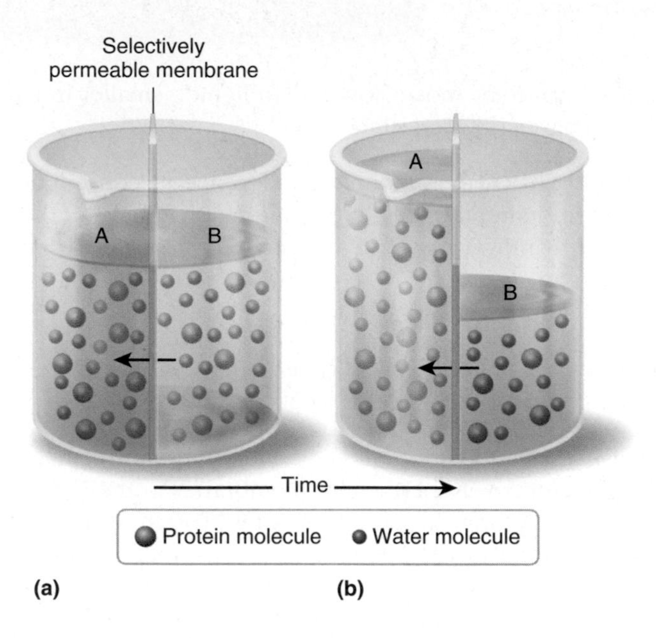

FIGURE 2.17 Osmosis. A selectively permeable membrane separates this container into two compartments. At the start, shown in (a), side A contains water and many large protein molecules. Side B contains water and fewer large protein molecules (hypotonic). The protein molecules cannot move across the selectively permeable membrane. So, in (b), water moves from side B to side A to try to equalize the concentrations.

An example of osmosis can be seen in the blood. Blood is composed of blood cells traveling in **plasma.** The cytoplasm inside the blood cells and the plasma on the outside are both solutions separated by the selectively permeable cell membrane. If the plasma is isotonic (has the same concentration of solutes relative to the cytoplasm of the blood cells), no movement of water by osmosis takes place. If the plasma is hypotonic (less concentrated with solutes than the cytoplasm of the blood cells), water from the plasma will move across the cell membrane by osmosis to reduce the concentration of solutes in the cytoplasm of the blood cells. The blood cells swell to accommodate the extra water. If the concentration gradient is too great, the blood cells may **lyse** (burst). If the plasma is hypertonic (more concentrated with solutes then the cytoplasm of the blood cells), water from the cytoplasm will move across the membrane by osmosis to reduce the concentration of solutes in the plasma. The blood cells may **crenate** (shrivel and appear spiky) from the water loss. Fluid balance is an important concept in several upcoming chapters. See **Figure 2.18**.

spot check 6 It is possible to chemically remove just the shell from a raw egg. You are then left with a membrane enclosing a highly concentrated solution: the egg white. The egg's membrane is selectively permeable and will not let the solutes inside the egg cross this membrane. Predict what would happen to the egg's weight if the egg is placed in a beaker of water. What membrane transport process would be responsible for the change, if any?

(a)

(b)

(c)

FIGURE 2.18 Red blood cells in three solutions: (a) hypotonic, (b) isotonic, (c) hypertonic.

Active transport. Active transport moves materials across the cell membrane from areas of low concentration to areas of high concentration. This involves moving materials *up a concentration gradient* against the natural trend of diffusion. It requires the cell's usable form of energy contained in ATP molecules. An analogy for this is a flood. The ceiling, walls, windows, and doors of a basement apartment are like a cell membrane, regulating what can enter or exit the apartment. In a flood, water enters the apartment through the somewhat leaky walls, windows, and doors from an area of high concentration (outside the apartment) to an area of low concentration (inside the apartment) until the levels are equal. This is simple diffusion, a passive process. The apartment resident, however, will want to make the water move in the opposite direction, from an area of low concentration (the apartment) to an area of high concentration (outside) to keep the apartment dry. This requires the work of a pump that needs energy to run. This is active transport.

An example of this type of membrane transport in the body is a **sodium-potassium pump.** This series of channels on the cell membrane regulates the concentration of sodium and potassium ions on both sides of the cell membrane. This is vital for neurons (cells of the nervous system that carry electrical messages). Ions are charged atoms. Electricity is the flow of charged particles. The sodium-potassium pump controls the flow of electricity by controlling the concentration of these ions and their movement across the cell membrane. About half of the energy you use each day is used to run the sodium-potassium pump. You will learn more about how this works in the nervous system chapter.

spot check ❼ What high-energy molecule will the sodium-potassium pump need?

2.13 learning outcome

Describe bulk transport, including endocytosis and exocytosis.

Bulk transport. This form of membrane transport moves large quantities of materials—not individual ions and molecules—across a cell membrane at one time. The two forms of bulk transport—depending upon the direction materials are transported—are explained below:

1. **Endocytosis** moves materials into the cell in bulk. For example, a white blood cell moves close to a foreign particle such as a bacterium, engulfs it by surrounding it with its cell membrane, and then pinches the membrane off, creating a

membrane-enclosed vesicle (packet) of the bacterium inside the cell. Here, in essence, the white blood cell has phagocytized (eaten) the bacterium. A lysosome, containing digesting enzymes, can then merge its membrane with the membrane enclosing the bacterium. The enzymes from the lysosome digest the bacterium, making it harmless.

2. **Exocytosis** moves materials out of the cell in bulk. Continuing the previous example, the membrane-enclosed packet of the digested bacterium moves to the inside of the cell membrane, merges with it, and then opens to the outside, expelling its contents (the digested bacterium) in bulk. See **Figure 2.19**.

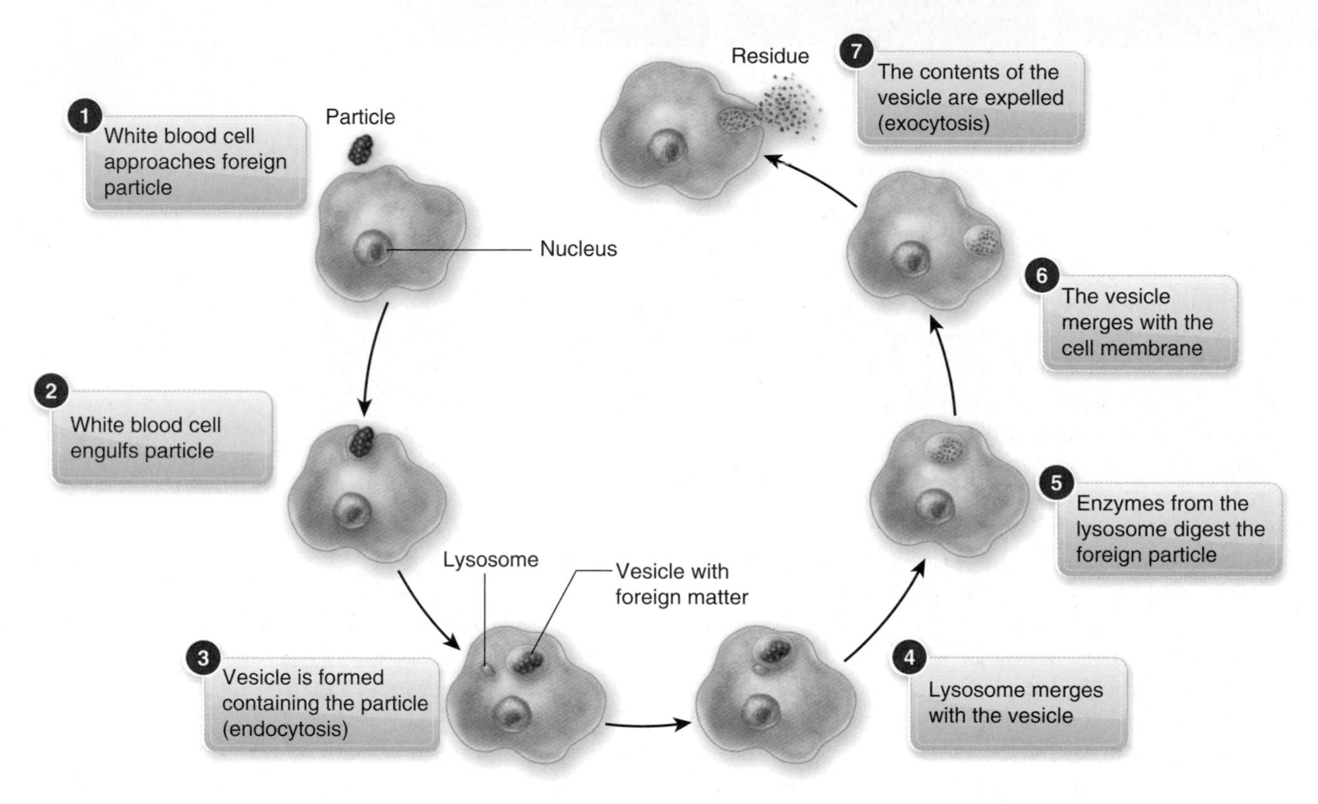

FIGURE 2.19 Endocytosis and exocytosis. Steps 1 through 3 show endocytosis. Steps 4 through 7 show exocytosis.

2.14 learning outcome

Describe the processes of transcription and translation in protein synthesis in terms of location and the relevant nucleic acids involved.

Protein Synthesis Making proteins, a process called **protein synthesis,** is another major function for many different cells. The production process from cell to cell is the same. However, the directions and the assembly of the amino acid building blocks can result in very different proteins. The directions for assembling the amino acids into a protein are written in the DNA, housed in the nucleus. It is the ribosome (floating free in the cytoplasm or located on the rough ER) that actually assembles the amino acid building blocks in the correct order to produce a functioning protein. Protein synthesis is a two-stage process involving **transcription** and **translation.**

Transcription. Transcription must happen first, and it occurs in the nucleus. The DNA contains all the information on creating a protein, but it cannot leave the nucleus. Somehow, the information contained in the DNA must be converted to a form that can be transported to where it needs to be used. This is similar to taking notes from a book on reserve in a library. You need the information where you will write your paper, but you cannot take the book from the library. In transcription the double-stranded DNA is opened at the section containing the directions for making the protein. Free nucleotides bind to one side of the open DNA (C's bind to G's and U's bind to A's) to form a single-stranded **messenger RNA (mRNA)** molecule. Uracil (U) replaces thymine (T) in RNA language. The mRNA, carrying the DNA directions in a usable form, leaves the nucleus and travels to the ribosome. See **Figure 2.20**.

Translation. Translation happens at the ribosome. The ribosome, free or on the rough ER, must use the directions contained in the mRNA to assemble amino acids into a functioning protein. There are 20 individual and unique amino acids. The specific amino acids used and their sequence are vital to the protein's eventual shape. These amino acids must be transported to the ribosome on the rough ER for assembly.

The process of translation is similar to automobile production. A car is made of auto parts. Each type of auto part is unique just as amino acids differ from one to another. Trucking companies transport and deliver the auto parts, which eventually end up in an assembly line at the assembly plant for the autoworkers. Different trucking companies deliver different parts. One delivers spark plugs; another delivers windshields. The placement of these auto parts—carburetors, tires, seat belts—must be specific if the car is to work properly. For example, a carburetor cannot be installed in place of a muffler, and a tire will not work in place of a steering wheel.

In translation, the ribosome reads three nucleotides (a **codon**) at a time on the mRNA. A **transfer RNA (tRNA)** molecule delivers a specific amino acid to the

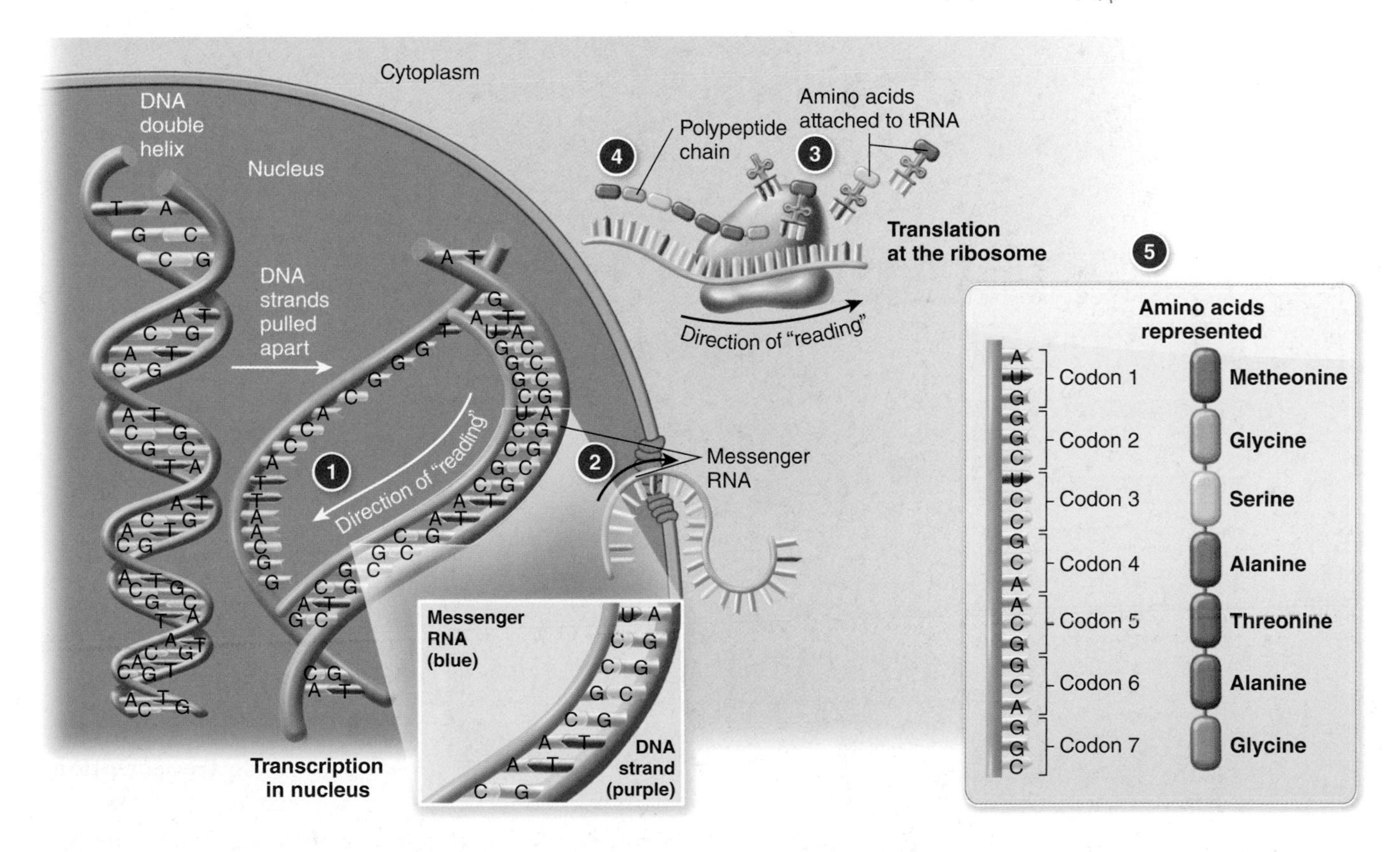

FIGURE 2.20 **Protein synthesis:** Steps 1 and 2 show transcription, steps 3 and 4 show translation, and step 5 shows the assembled protein and the mRNA that coded for it. (1) the DNA molecule is opened, and the DNA information is copied, or transcribed, into mRNA, (2) mRNA leaves the nucleus and attaches to a ribosome, (3) translation begins as tRNA molecules with anticodons to match mRNA codons bring specific amino acids to the ribsome, (4) translation continues as the ribosome moves along the mRNA and more amino acids are added, (5) the ribosome releases the assembled protein shown here with the mRNA that was used for its assembly.

FIGURE 2.21 Close-up look at translation at a ribosome.

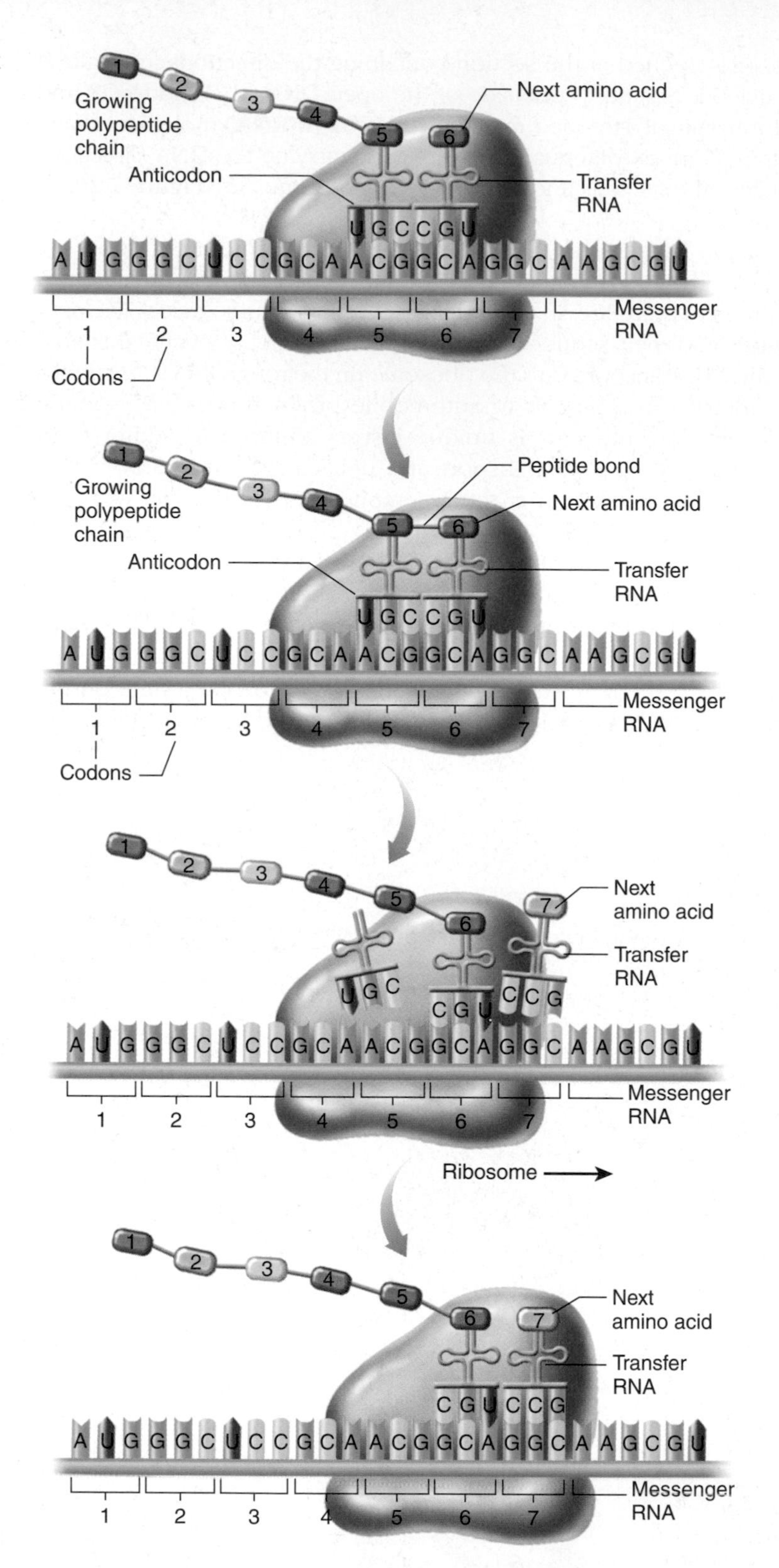

ribosome by matching its three nucleotides **(anticodon)** to the codon on the mRNA (C's match to G's and U's match to A's). Transfer RNA is the truck delivering a specific auto part (amino acid) to the plant for assembly. The ribosome reads the next codon on the mRNA. The tRNA that has the correct anticodon to match with it brings its specific amino acid to the ribosome. The amino acids are joined together at the ribosome, and the first tRNA falls away to pick up another of its specific amino acids from the cytoplasm (for later use). This process continues until the complete amino acid chain (like a complete car) is assembled. See **Figures 2.20** and **2.21**.

spot check ❽ If the third triplet on the DNA strand in the nucleus coding for a particular protein was GCC, what corresponding codon would be formed for the mRNA during transcription? What would have to be the anticodon of the tRNA used to match this mRNA codon during translation?

2.15 learning outcome

Describe what happens to a protein after translation.

Once the amino acid chain has been formed, the Golgi complex inspects and possibly modifies it. The cell may use the inspected protein directly from the Golgi complex, or the Golgi complex can package the protein in a secretory vesicle that carries the protein to the cell membrane to be exocytosed from the cell. See **Figure 2.22**.

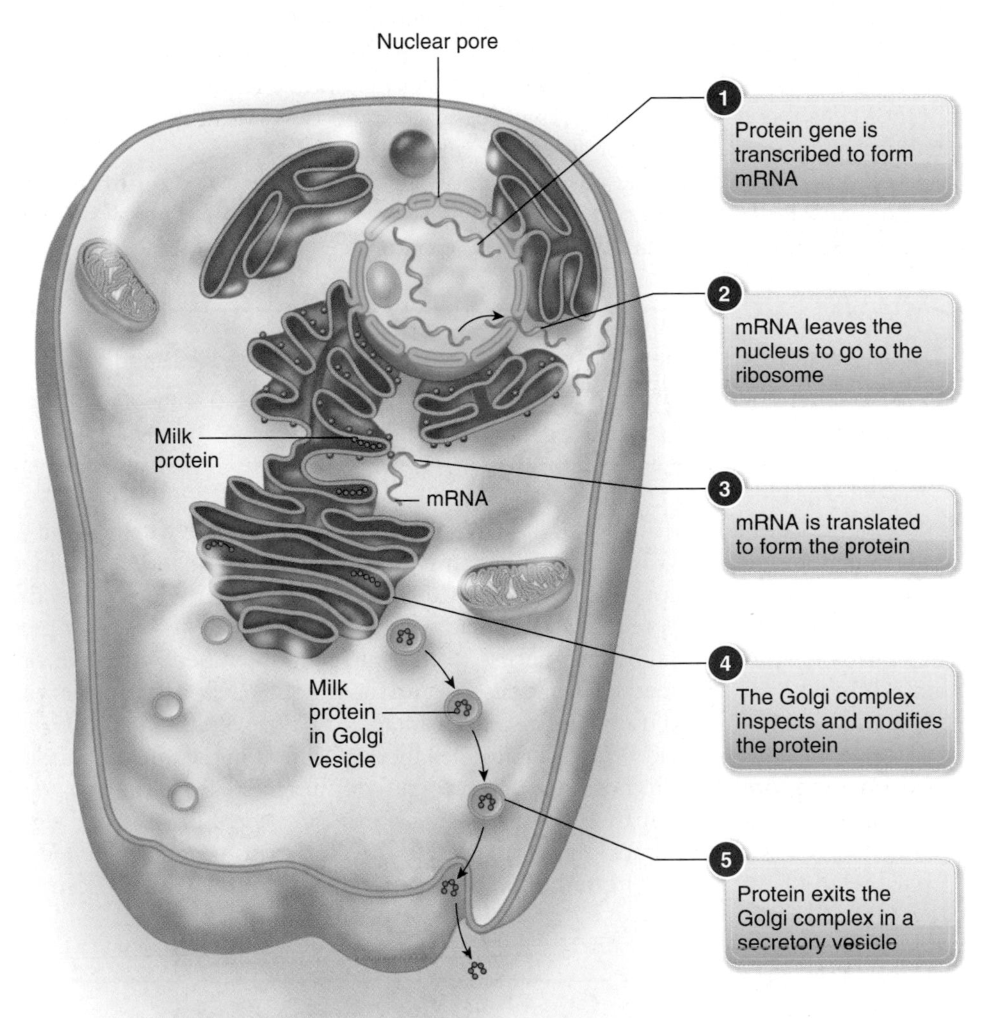

FIGURE 2.22 What happens after translation.

2.16 learning outcome

Explain the possible consequences of mistakes in protein synthesis.

Mistakes in protein synthesis. Although the examples above may give the impression that protein synthesis is always a smooth and successful process, mistakes can happen. Such mistakes may be catastrophic or of little consequence. Transfer RNA has a three-nucleotide anticodon on one end and uses the other end to transport one of 20 different amino acids. Four possible nucleotides can be combined to form the three-nucleotide anticodon on a tRNA molecule. There are more possible combinations for anticodons than there are different amino acids. As a result, some amino acids may be transported by more than one tRNA. If a mistake in transcription results in a faulty mRNA molecule, either of two outcomes may result:

1. The ribosome reading the faulty mRNA calls for a tRNA that brings an amino acid that is wrong.
2. The ribosome reading the faulty mRNA calls for a tRNA that just happens to be a tRNA that brings the correct amino acid.

In the first case, a different protein is made, with possible drastic consequences. In the second case, the correct protein is made and there are no consequences.

2.17 learning outcome

Describe the process of mitosis, including a comparison of the chromosomes in a parent cell to the chromosomes in the daughter cells.

meiosis: my-OH-sis

mitosis: my-TOE-sis

Cell Division Cell division is another key cellular function. Meiosis and mitosis are the two types of cell division. **Meiosis** is involved only in sperm and egg production; it is discussed in the reproductive system chapters. **Mitosis** is the process all other cells use to divide, and it is necessary for the development of the human anatomy, which is composed of 40 trillion cells.

In mitosis, a single cell, the **parent cell,** divides to become two **daughter cells.** Once the division has taken place, the parent cell no longer exists. The two daughter cells are identical to each other and to the parent cell that came before them.

warning

Do not be confused by the terminology of mitosis. The terms *parent cell* and *daughter cells* have nothing to do with sexual reproduction or gender; they are simply used to refer to generations, as in parents come before daughters.

chromatin: KROH-ma-tin

During most of a cell's life cycle, the DNA in the nucleus is loosely spread out in an arrangement called **chromatin.** This arrangement allows information contained in the DNA to be used more easily. Think of the way you arrange your books and notes when studying. By spreading these materials out in front of you—books open, notes visible—you have immediate access to the information. When you have finished studying, you will need to pack these materials in your backpack to take with you. At that point, it would no longer be an advantage to have everything spread out.

Back to the cell: To divide, the cell packages the DNA into tight, compact bundles called **chromosomes.** Chromosomes are present only for cell division. There are 46 chromosomes in all human cells other than sperm and eggs (sperm and eggs each contains 23 chromosomes). In a typical cell, the chromosomes are arranged in 23 pairs: Half of each pair is from the individual's mother; the other half is from the individual's father.

clinical point

The exception to this is red blood cells, which do not have nuclei and, hence, have no DNA. Forensic scientists use the DNA from white blood cells to test and establish identity or paternity.

All cells in an individual have the same DNA. Skin cells producing the skin pigment melanin have the same DNA as pancreas cells producing insulin. Salivary gland cells producing saliva have the same DNA as stomach cells producing stomach acid. Muscle cells have the same DNA as bone cells. They all contain the complete set of information, all 3 billion base pairs. Each type of cell uses only the part of the DNA that is relevant to its specific function.

The reason that every cell has the same complete set of human DNA is further explained here: A fully developed human is composed of approximately 40 trillion cells. These cells begin from one fertilized egg (a zygote). The zygote received half of its DNA from the father's sperm and the other half from the mother's egg. The zygote has 46 chromosomes when it begins its first division by mitosis. It becomes two identical daughter cells when mitosis is complete. It seems reasonable to conclude that when a cell with 46 chromosomes divides into two equal parts, the resulting two cells will each have 23 chromosomes. But this is not the case. The parent cell, the zygote in this case, makes an identical copy of the DNA before dividing so that each daughter cell gets a complete set that is identical to the DNA set of the other daughter cell and identical to the DNA set of the parent cell. This copying, called **replication,** ensures that every cell (other than the exceptions mentioned earlier) has a full identical set of DNA. The process of mitosis continues over and over until the fetus is fully developed. Even after birth, mitosis is important for the growth of organs and tissues, the repair of damaged cells, and the replacement of dead cells. Therefore, mitosis is a multistage process, as you can see in **Figure 2.23**, and its basic concept is relatively simple, as shown in **Figure 2.24**.

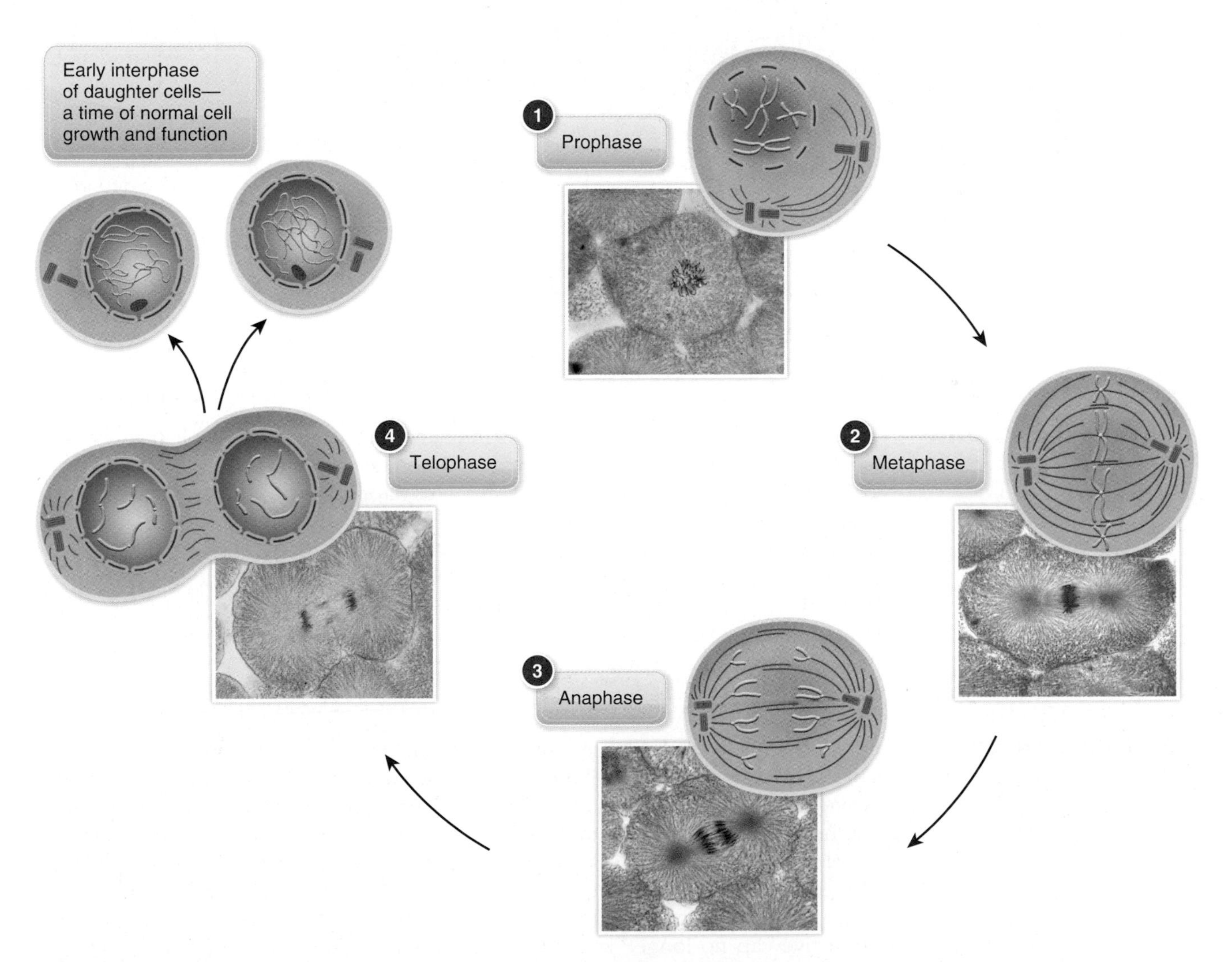

FIGURE 2.23 **Mitosis.** The series of micrographs shows the stages of mitosis in whitefish eggs. The diagram depicts a hypothetical cell with two chromosome pairs. Human cells contain 23 pairs of chromosomes. The stages are (1) prophase: chromatin condenses to form chromosomes, and the nuclear membrane breaks down; (2) metaphase: chromosomes align along the midline of the cell; (3) anaphase: chromosomes move to opposite poles; (4) telophase: chromosomes decondense to form chromatin, the nuclear membrane forms, and the cell finishes division.

FIGURE 2.24 **Mitosis simplified.** One parent cell with 46 chromosomes replicates and divides to form two daughter cells each having 46 chromosomes. The daughter cells' DNA sets are identical to each other and to the parent cell's DNA, which no longer exists once mitosis has taken place.

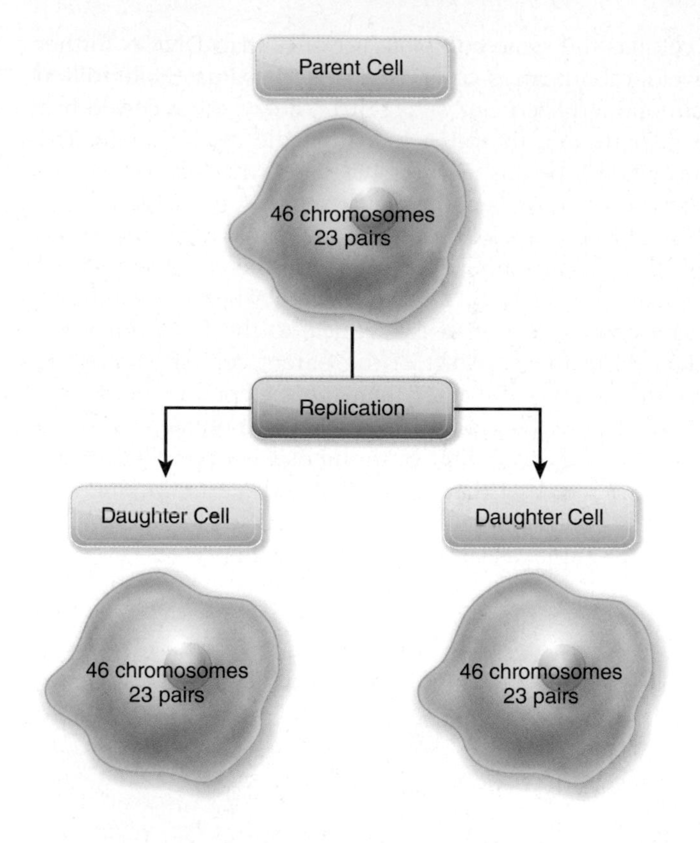

spot check ⑨ How does the DNA of a brain cell compare to the DNA of a bone cell? How does the DNA of a brain cell differ from the DNA of a sperm cell?

2.18 learning outcome

Explain the possible consequences of mistakes in replication.

Replicating 3 billion base pairs of human DNA each time a cell divides is an amazing feat. Each cell has special enzymes to check for accuracy during the replication process, but nothing is perfect. Mistakes happen. Even if the enzymes mistake only a few base pairs of nucleotides, there may be consequences. If the mistake in replication occurs in the part of the DNA that the cell does not use, there are no consequences. If it occurs in a section of DNA the cell does use, the mistake will be obvious. This **mutation** may or may not be of benefit to the cell. Regardless of the benefit or harm the mistake causes for the cell, the mutation will be carried on to the next daughter cells when the cell divides.

2.19 learning outcome

Describe the effects of aging on cell division.

Effect of Aging on Cells A human cell can divide for only so long. One possible explanation involves telomeres. **Telomeres** are sequences of nucleotides that provide a protective cap on the ends of chromosomes. Although they do not code for the production of proteins, they are believed to stabilize the chromosome by keeping it from unraveling and preventing it from sticking to other chromosomes. The cell cannot replicate the very ends of the chromosome. So each time the cell divides, a fraction of the telomeres is lost. Once all of the telomeres are lost, the protective cap no longer exists and vital protein-coding nucleotide sequences are then at the ends of the chromosome. The cell still cannot replicate the ends of the chromosomes, and with vital nucleotides at these ends, more and more mistakes in replication are bound to happen. As a result, the cell's chances of producing more and more dysfunctional proteins increase with age. Eventually the cell becomes nonfunctional and dies. See **Figure 2.25**.

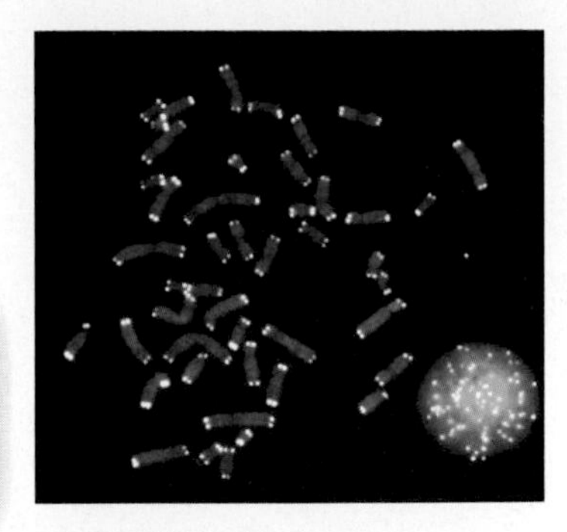

FIGURE 2.25 Telomeres: micrograph of human chromosomes with telomeres highlighted.

> **clinical point**
>
> Cancer cells have an active enzyme that is absent in normal cells. This enzyme, **telomerase,** repairs the telomere damage during replication, making the cancer cell immortal.

You now have a better understanding of cellular functions and the effects of aging on cells. In the next section, you will see how cells function together at the tissue level.

Tissue Level

2.20 learning outcome

Describe the four classifications of tissues in the human body.

Histology is the study of tissues. The four basic classifications of tissues are epithelial, connective, muscle, and nervous. Below you will find a basic explanation of each of these tissue classes. Specific tissues are covered in the relevant system chapters.

Epithelial Tissues These tissues cover and line all body surfaces. Epithelial tissues cover organs, vessels, and ducts and line hollow organs, vessels, and ducts. As a result, all epithelial cells have a free edge that borders an open area on the outside surface or as a lining of an inside surface. Epithelial tissues are named, first, for the shape of their cells and, second, for the amount of layering of the cells.

It is important to know that cells are three-dimensional objects—with a width, height, and depth—that can have different shapes. Epithelial cells can be **squamous** (flat and thin), **cuboidal** (cube-shaped), or **columnar** (tall column-shaped). A **basement membrane** separates epithelial tissue from other tissues.

squamous: SKWAY-mus

Tissue layering is described in three ways: **Simple epithelial tissue** has a single layer of epithelial cells; **stratified epithelial tissue** is composed of stacked layers of epithelial cells; and **pseudostratified epithelial tissue** appears to be layered, but all cells have contact with the basement membrane, so it is a false *(pseudo)* layering. See **Figure 2.26**.

Some examples of epithelial tissue are:

- Simple squamous epithelial tissue lining the alveoli (air sacs) of the lung. See **Figure 2.27**.
- Simple cuboidal epithelial tissue that lines the tubules in the kidneys. See **Figure 2.28**.

- Stratified squamous epithelial tissue lining the mouth and esophagus. See **Figure 2.29**.
- Simple columnar epithelial tissue that lines the small intestines. See **Figure 2.30**.
- Pseudostratified ciliated columnar epithelial tissue that lines much of the respiratory tract. In addition to its ciliated columnar cells that move debris in the respiratory tract, this tissue contains goblet cells that function to produce mucus. See **Figure 2.31**.

An exception to the naming of epithelial tissue is **transitional epithelial tissue.** This epithelium is stratified (layered), but its cell shape is difficult to describe because it is so changeable. Transitional epithelial tissue is designed to stretch, and it lines structures like the urinary bladder. If stretched, the cells appear to be more squamous. If not stretched, they appear to be more cuboidal. See **Figure 2.32**.

FIGURE 2.26 Epithelial cell shapes and layering.

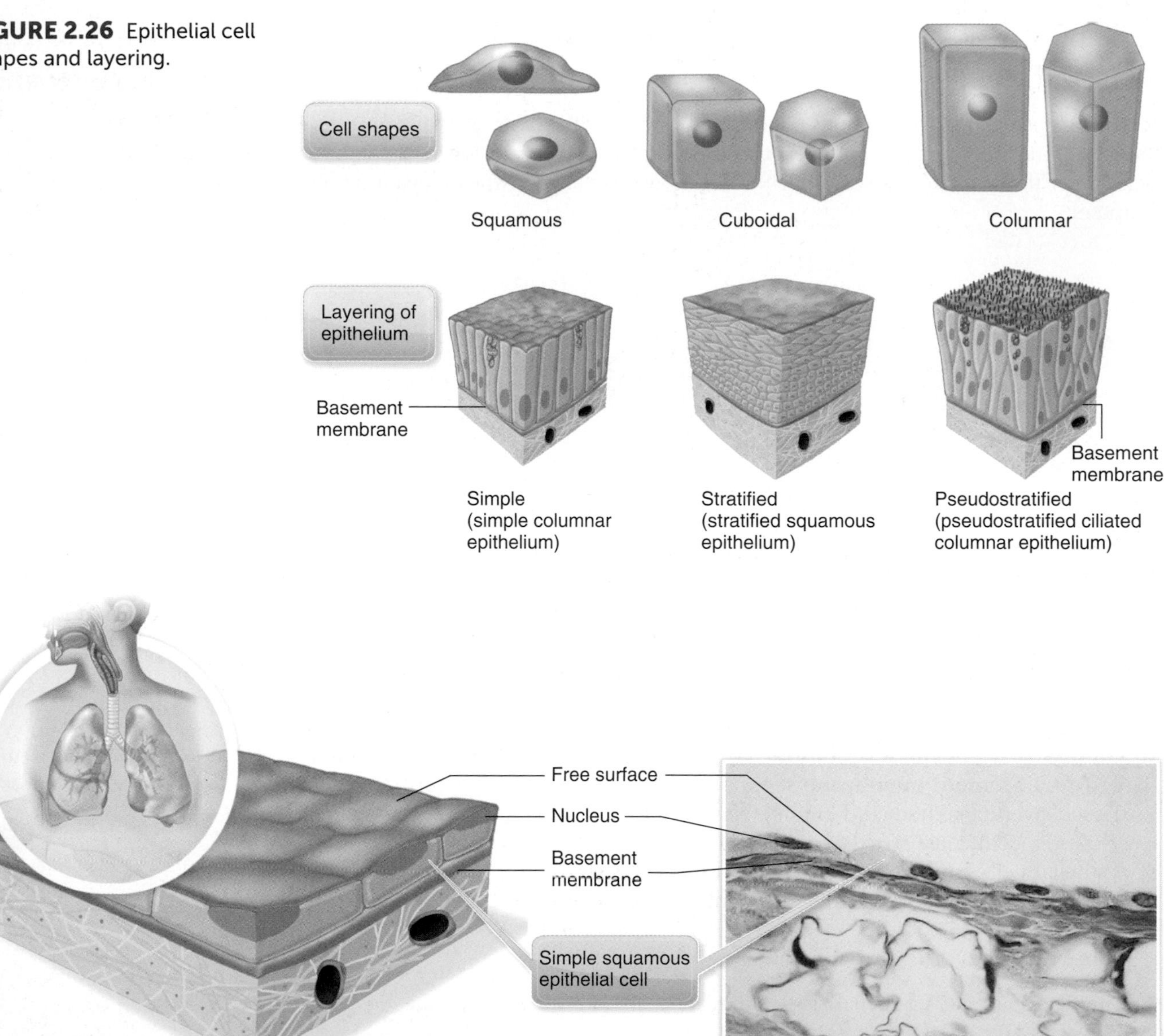

FIGURE 2.27 Simple squamous epithelial tissue, showing a single layer of squamous cells (nuclei of epithelial cells appear as dark circles), the free edge (free surface) bordering the lumen (open space), and the basement membrane separating the epithelial tissue from other tissue. *Micrograph:* simple squamous epithelial tissue lining the alveoli (air sacs) of the lung.

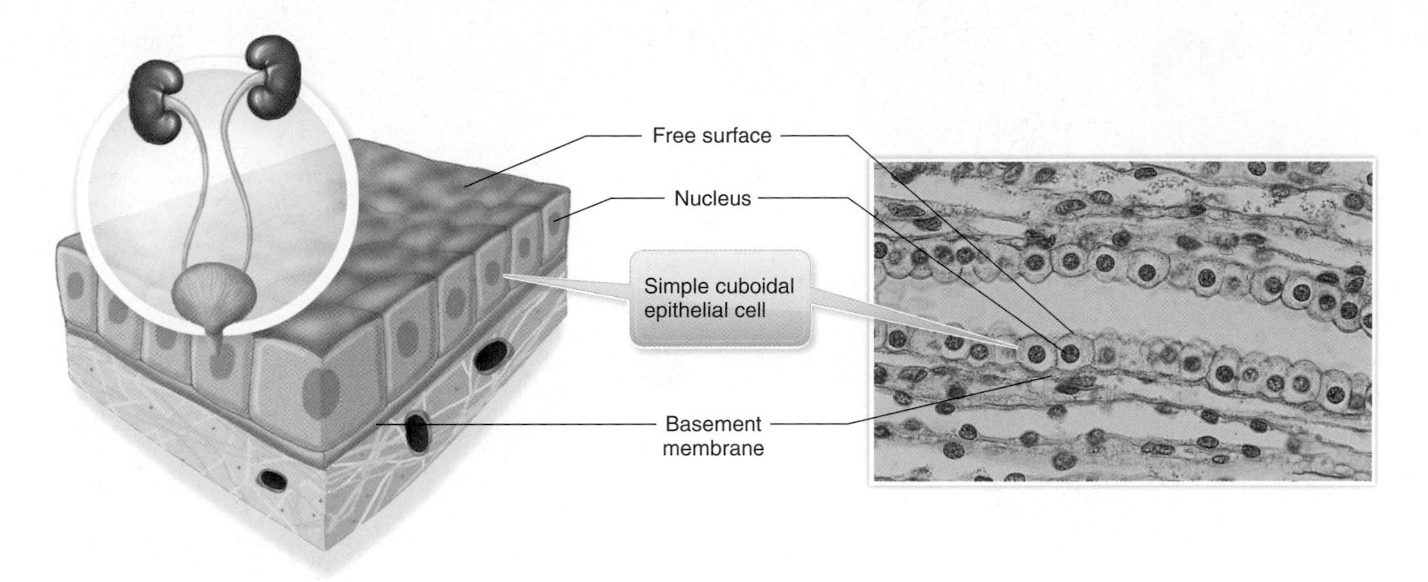

FIGURE 2.28 Simple cuboidal epithelial tissue, showing a single layer of cuboidal (cube-shaped) cells (nuclei of epithelial cells appear as dark circles), the free edge bordering the lumen (open space), and the basement membrane separating the epithelial tissue from other tissue. *Micrograph:* simple cuboidal tissue lining kidney tubules.

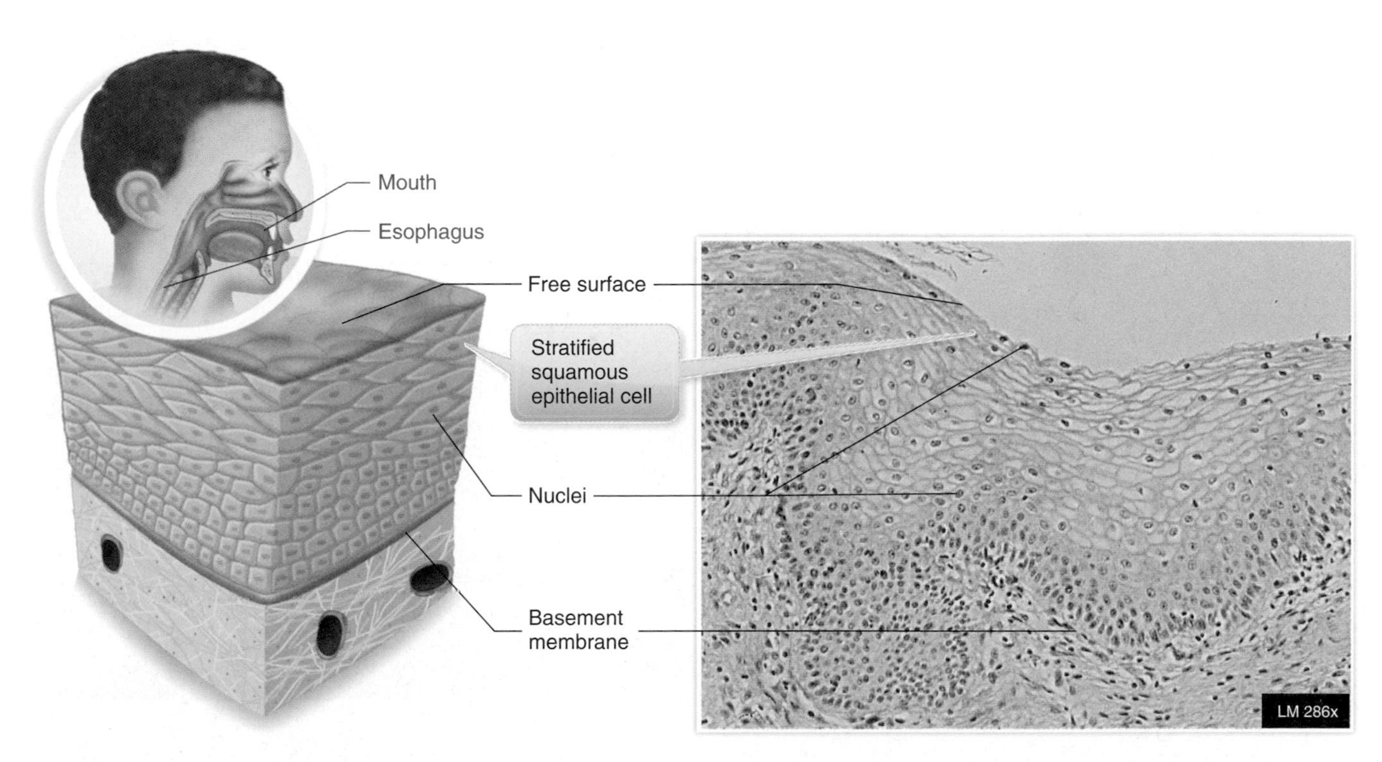

FIGURE 2.29 Stratified squamous epithelial tissue, showing stratified squamous cells, the free edge bordering an open area, and a basement membrane separating the epithelial tissue from other tissue. *Micrograph:* stratified squamous epithelial tissue lining the mouth and esophagus.

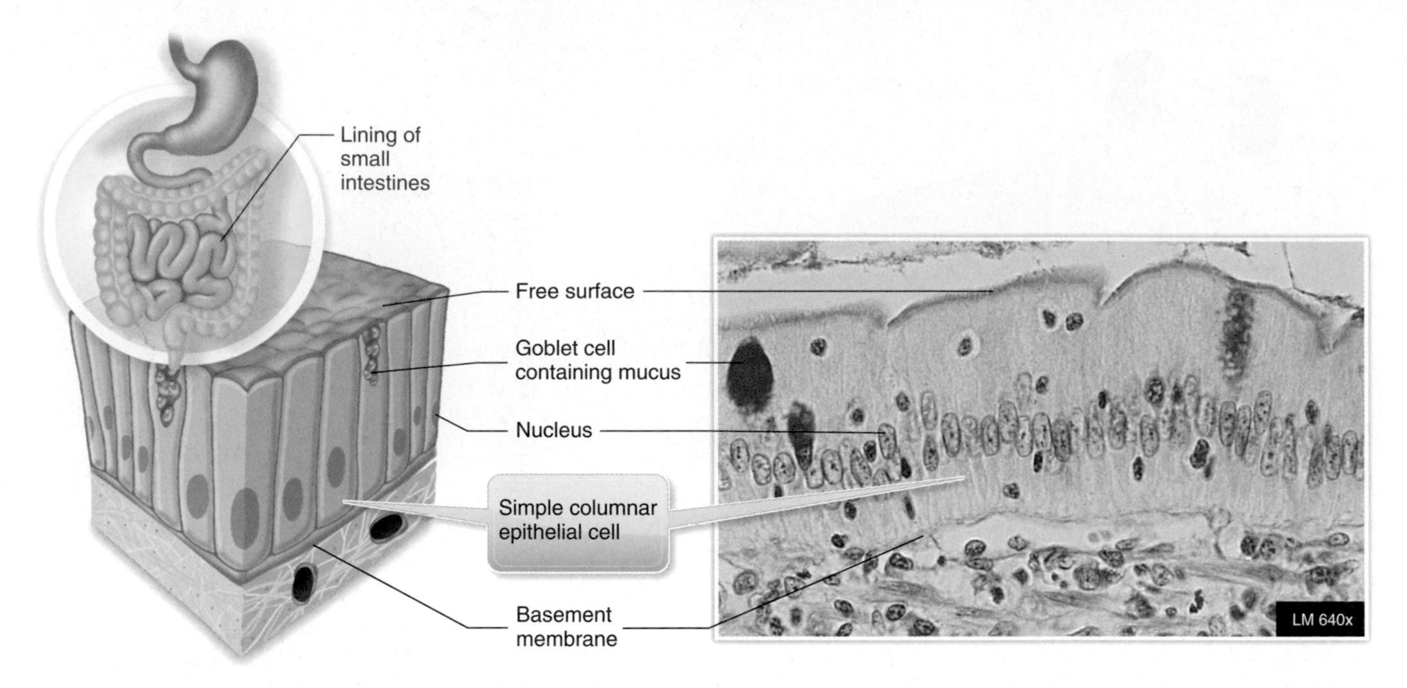

FIGURE 2.30 **Simple columnar epithelial tissue,** showing a single layer of columnar-shaped cells, the free edge bordering the lumen, goblet cells that produce mucus, and a basement membrane separating the epithelial tissue from the connective tissue. *Micrograph:* simple columnar epithelial tissue lining the small intestines.

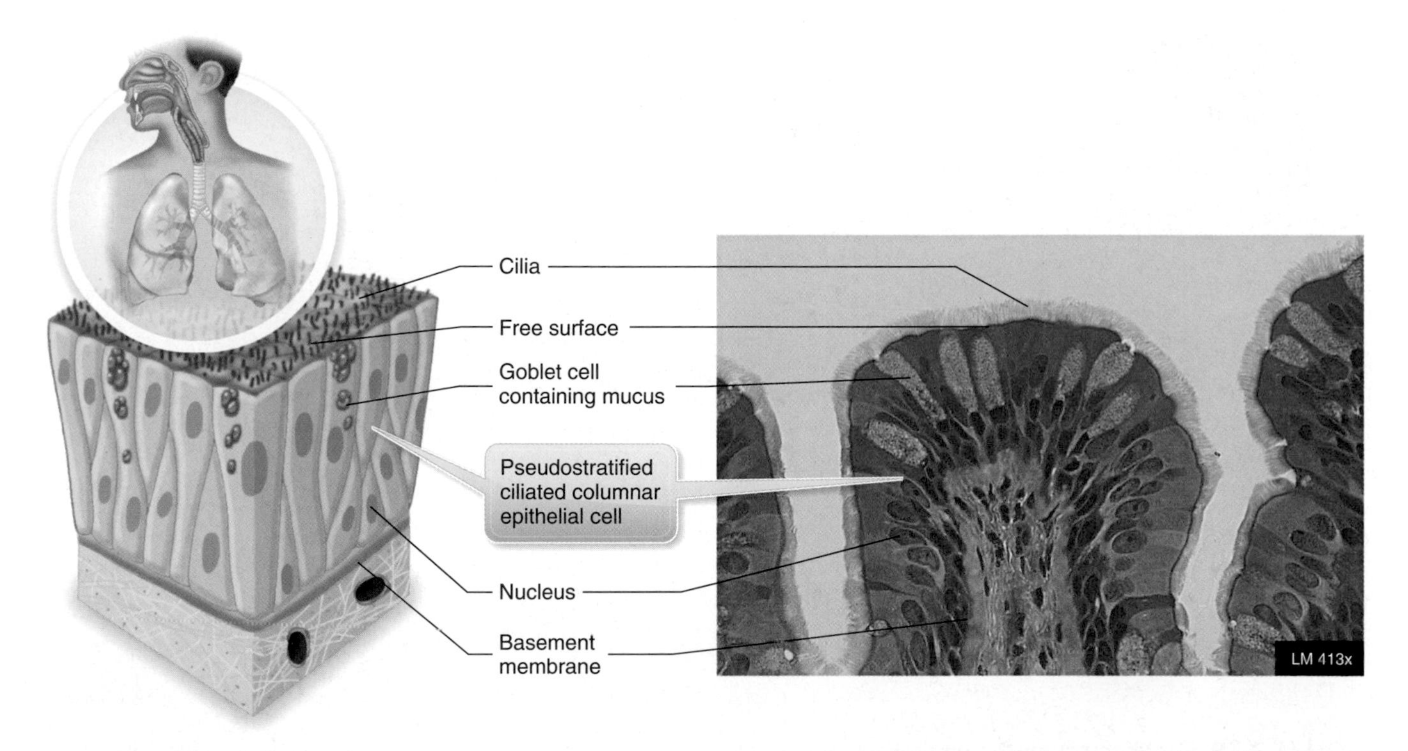

FIGURE 2.31 **Pseudostratified ciliated columnar epithelial tissue,** showing a pseudostratified layer of ciliated columnar-shaped cells, the free edge bordering the lumen, goblet cells that produce mucus, and a basement membrane separating the epithelial tissue from the connective tissue. *Micrograph:* ciliated pseudostratified columnar epithelial tissue lining much of the respiratory tract.

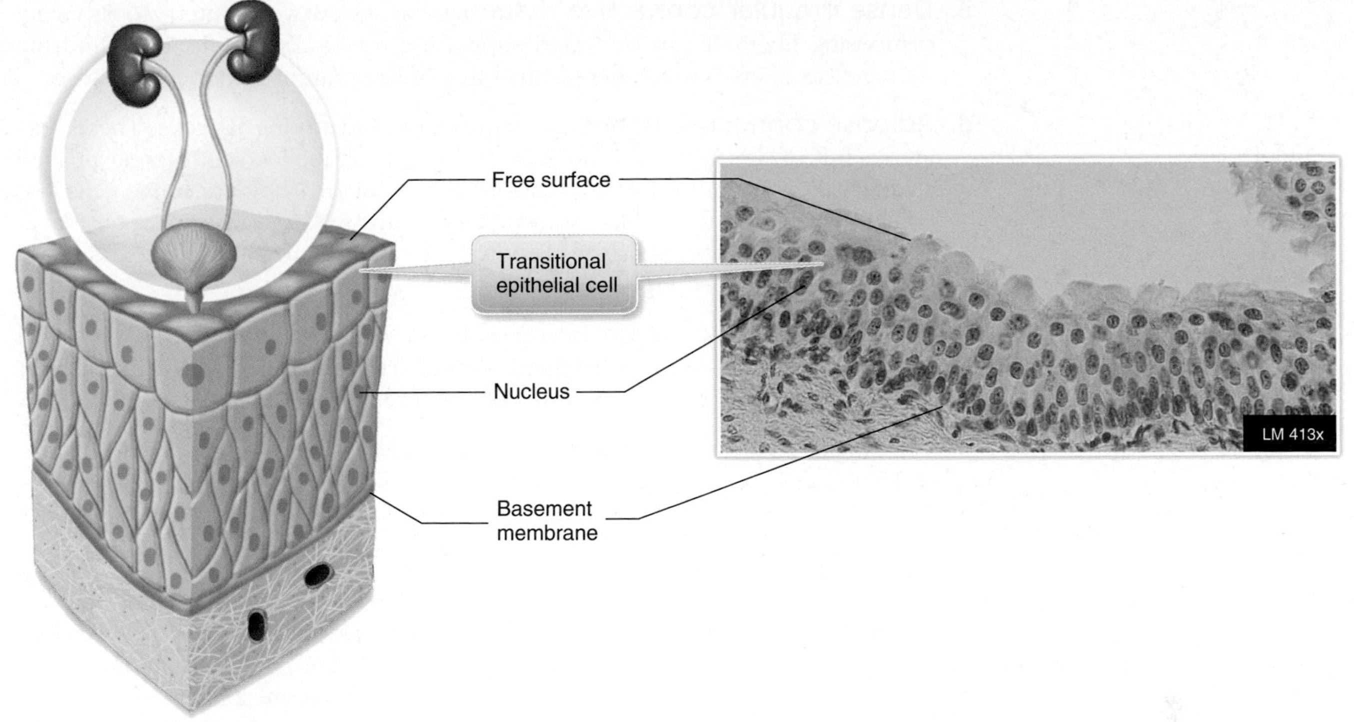

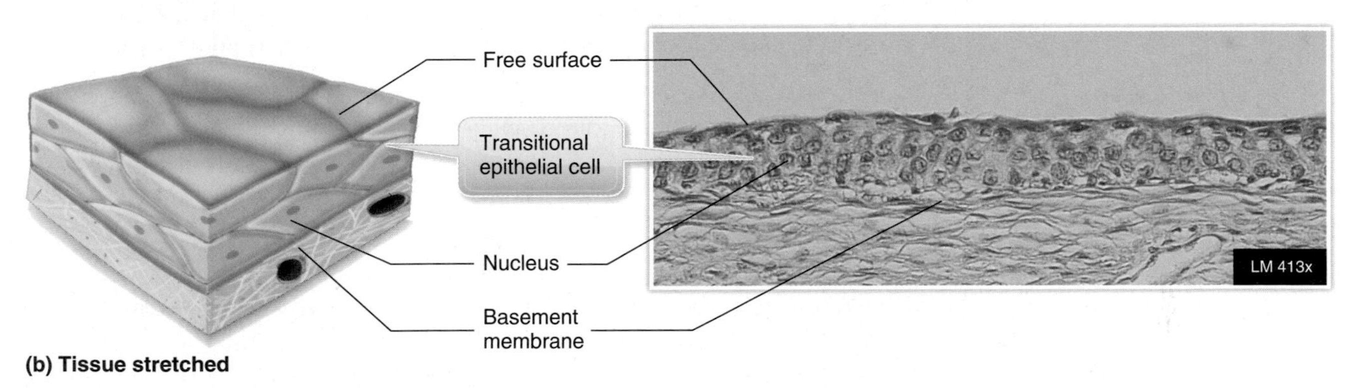

FIGURE 2.32 **Transitional epithelial tissue,** showing the free edge, the transitional epithelial cells, and the basement membrane separating the epithelial tissue from other tissue. Notice that (a) shows this tissue when not stretched and (b) shows this tissue stretched. *Micrographs:* transitional epithelial tissue lining the urinary bladder.

Connective Tissues Connective tissues all have cells and fibers in a **matrix** (background substance). The density of the matrix and the type of cells and fibers determine the type of connective tissue. The density of the matrix is highly variable; it can be very fluid, as in blood, or as dense and hard as concrete, as in bone.

Types of connective tissue include the following:

1. **Loose/areolar connective tissue** has a loose arrangement of fibers in a matrix with a thick fluid consistency. A variety of cells are able to move through the matrix. It is found, for example, in the middle layer of the skin (dermis) and between the serous layers of the mesenteries. See **Figure 2.33**.

2. **Dense regular connective tissue** has mostly dense bundles of collagen (protein) fibers that run parallel to each other. Fiber-making cells (fibroblasts) are occasionally interspersed between fibers. The cells in this tissue are not able to move (immobile). This arrangement of fibers gives strength and resistance to pulling forces for the tendons and ligaments composed of this tissue. See **Figure 2.34**.

3. **Dense irregular connective tissue** has an interwoven pattern to its many composing fibers. It can be found supporting the skin's middle layer, and the weave of its fibers is much denser than that of loose/areolar connective tissue.
4. **Adipose connective tissue** is composed of lipid-storing fat cells. These cells are so full of lipids that the nucleus and other organelles seem to be pushed aside to allow room for the lipid droplet they contain. They are active cells that convert carbohydrates to fats. Adipose tissue can be found in the deepest layer of the skin, where it serves as insulation; in the breast; around organs; and in the greater omentum. See **Figure 2.35**.
5. **Blood connective tissue** is composed of red and white blood cells and platelets in a very fluid matrix called *plasma*. This tissue is covered extensively in the cardiovascular system chapter. See **Figure 2.36**.
6. **Cartilage connective tissue** is of three types: hyaline, elastic, and fibrocartilage. The fibers involved determine their type. All three types of cartilage have cells surrounded by a very durable gel-like matrix. These tissues are covered extensively in the skeletal system chapter.

hyaline: HIGH-ah-lin

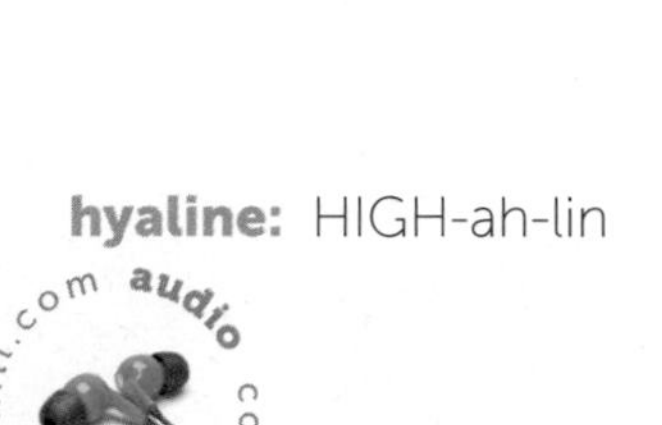

   - **Hyaline cartilage connective tissue** has a very smooth, glassy appearance. Its collagen fibers are so fine that they are virtually invisible. This cartilage is found at the ends of long bones, the larynx, the nose, bronchi, and the cartilages between the ribs and sternum. See **Figure 2.37**.
   - **Elastic cartilage connective tissue** has elastic fibers running in all directions. These fibers allow this cartilage to snap back to shape if bent. Elastic cartilage can be found in the ear and the epiglottis. See **Figure 2.38**.
   - **Fibrocartilage connective tissue** has dense bundles of collagen fibers all running in the same direction. These fibers allow this cartilage to function as a shock absorber. Fibrocartilage connective tissue can be found in the disks between vertebrae and in the meniscus of the knee. See **Figure 2.39**.
7. **Bone connective tissue** has bone cells isolated by a dense, concretelike matrix that makes bone very hard. Collagen fibers in the matrix allow a little bit of flex so that the bone is not brittle. This tissue is covered extensively in the skeletal system chapter. See **Figure 2.40**.

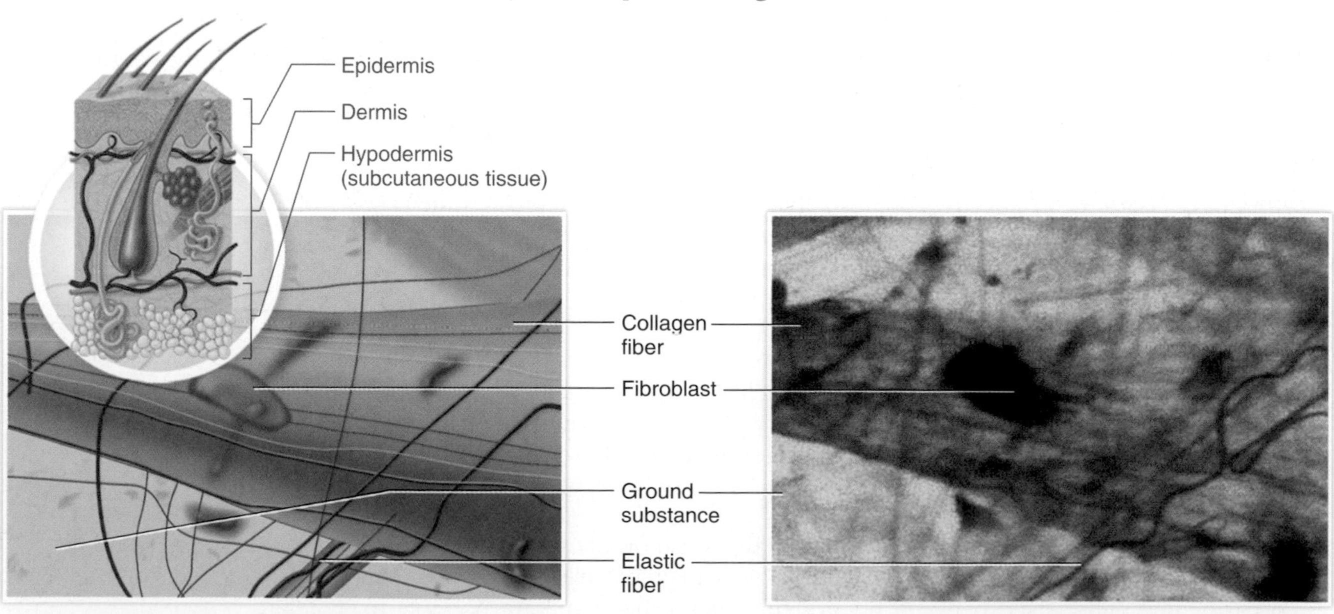

FIGURE 2.33 Loose/areolar connective tissue, showing cells that produce fibers (fibroblasts) in a loose weave of collagen and elastic fibers suspended in the ground substance (matrix). *Micrograph:* loose/areolar connective tissue of the dermis.

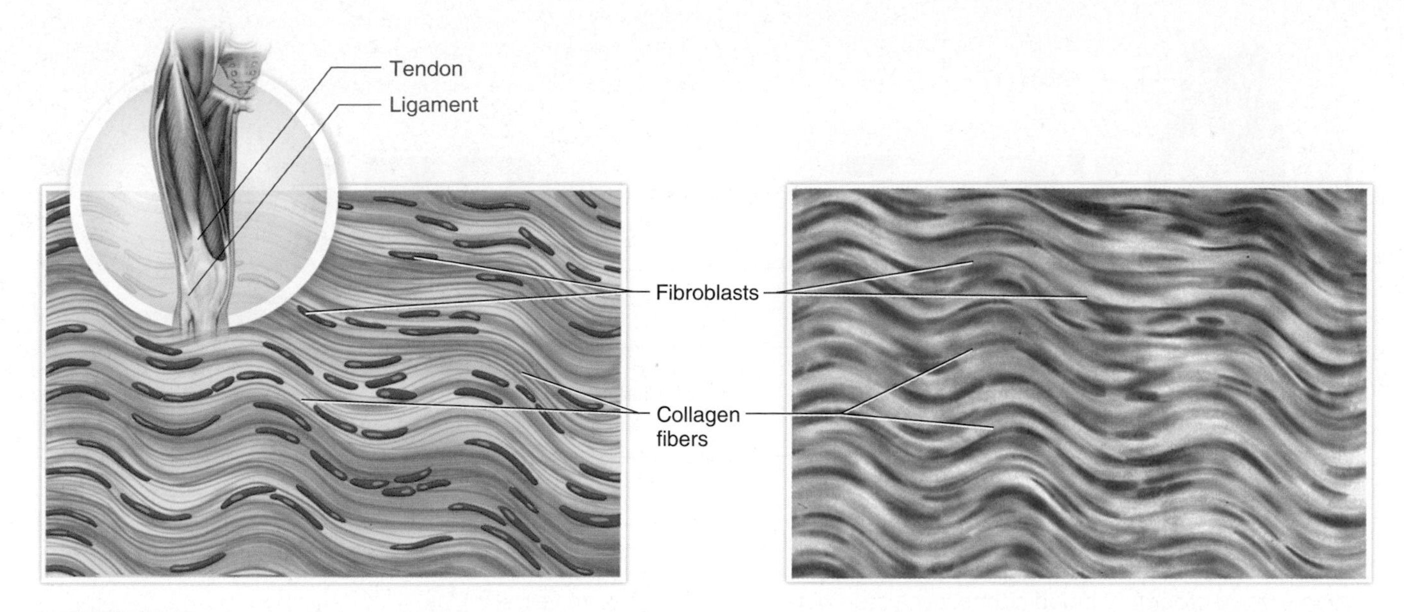

FIGURE 2.34 **Dense regular connective tissue,** showing cells that produce fibers (fibroblasts) and a dense arrangement of parallel collagen fibers in a ground substance (matrix). *Micrograph:* dense regular connective tissue in a tendon.

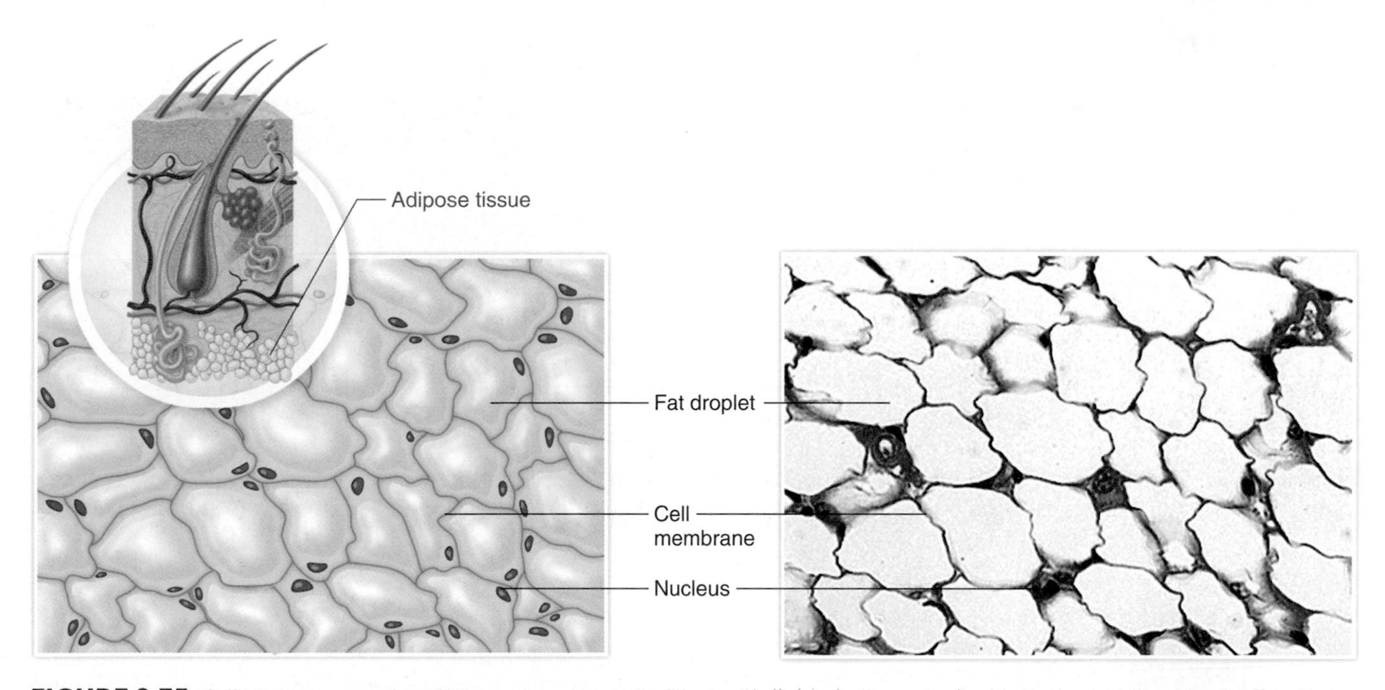

FIGURE 2.35 **Adipose connective tissue,** showing cells filled with lipids (adipocytes) with their nuclei pushed off to the side. *Micrograph:* adipose connective tissue in the deepest layer of the skin.

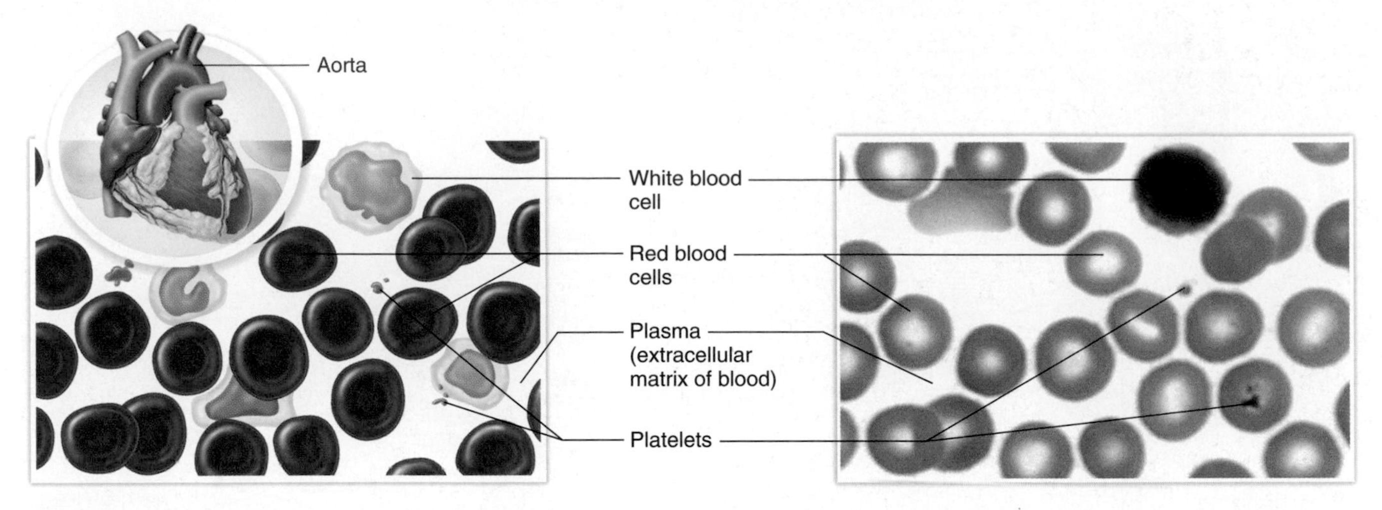

FIGURE 2.36 **Blood connective tissue,** showing blood cells and platelets in a fluid matrix called *plasma*. The white blood cell (dark purple) is stained to make it more visible. Red blood cells are the most numerous. Platelets are cell fragments. *Micrograph:* blood connective tissue obtained from any blood vessel.

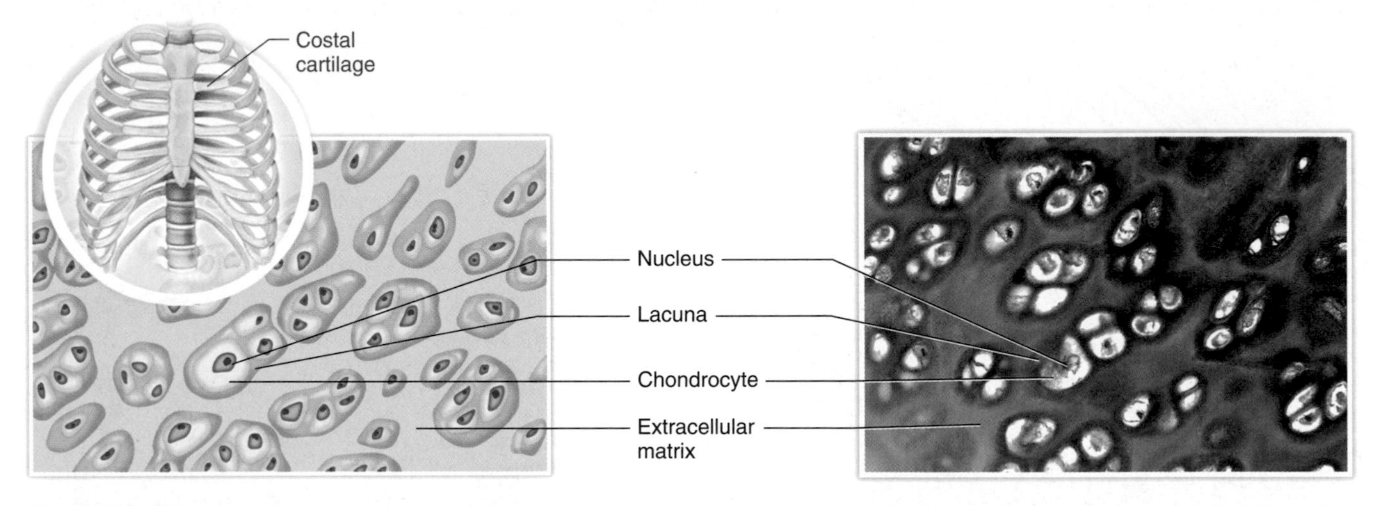

FIGURE 2.37 **Hyaline cartilage connective tissue,** showing cartilage cells (chondrocytes) in spaces (lacunae) within the matrix. *Micrograph:* hyaline cartilage connective tissue in the costal cartilages that connect the ribs to the sternum.

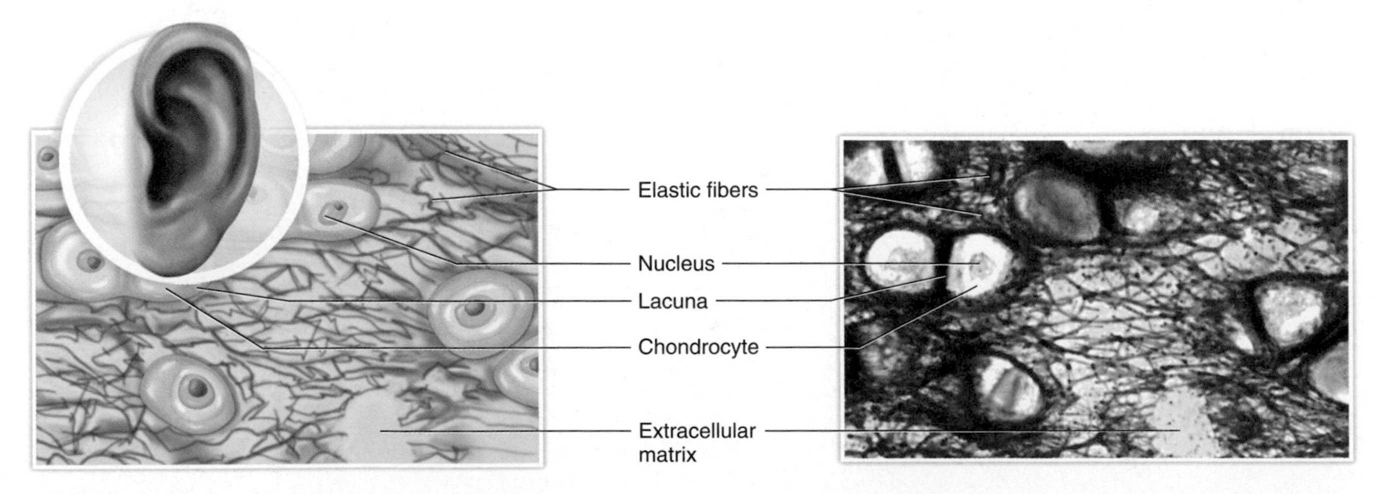

FIGURE 2.38 **Elastic cartilage connective tissue,** showing cartilage cells (chondrocytes) in spaces (lacunae) within the matrix and elastic fibers running in all directions. Micrograph: elastic cartilage connective tissue of the ear.

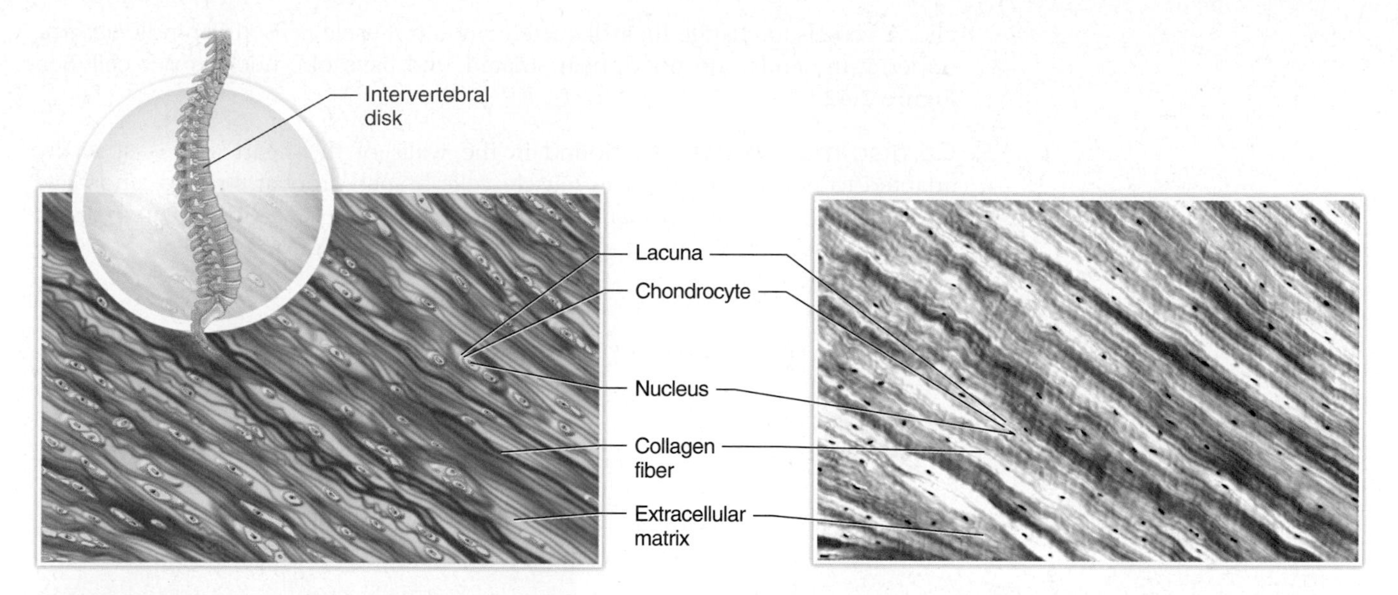

FIGURE 2.39 **Fibrocartilage connective tissue,** showing cartilage cells (chondrocytes) in spaces (lacunae) within the matrix and bands of parallel collagen fibers running in one direction. Micrograph: fibrocartilage connective tissue in an intervertebral disk.

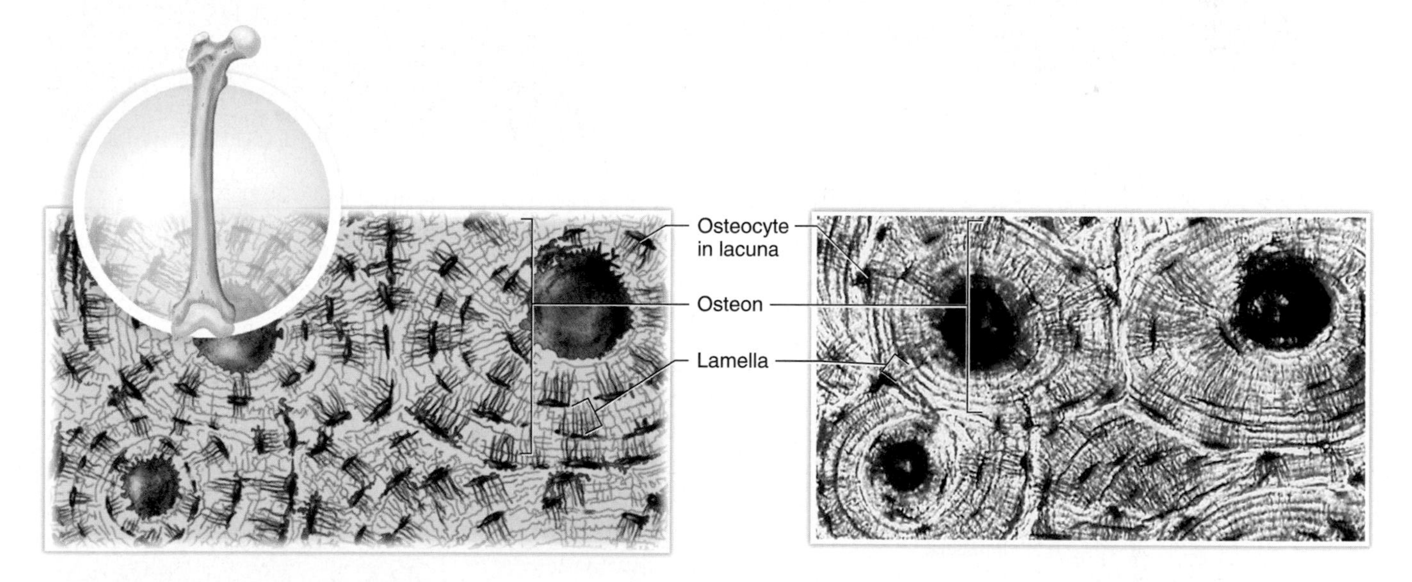

FIGURE 2.40 **Bone connective tissue,** showing an organized target-like arrangement (osteon), lacunae that house bone cells (osteocytes), and layers of hard matrix (lamellae) in bone. Micrograph: bone connective tissue found in the compact bone of a bone shaft.

Muscle Tissues The three types of muscle tissue are skeletal muscle, smooth muscle, and cardiac muscle. All of these are composed of cells with a high concentration of proteins. The proteins and their arrangement allow muscle cells to contract. These tissues are covered extensively in the muscular system chapter.

1. **Skeletal muscle tissue** makes up the skeletal muscles that move the body and control body openings. Skeletal muscle cells are cylindrical, appear striated (striped), and have multiple nuclei pushed off to the side. See **Figure 2.41**.
2. **Smooth muscle tissue** can be found in the walls of hollow organs, veins, and arteries. This tissue allows hollow organs to move materials through them and

allows vessels to change their diameter. Smooth muscle cells are spindle-shaped (taper at the ends), do not appear striated, and have one nucleus per cell. See **Figure 2.42**.

3. **Cardiac muscle tissue** is found in the walls of the heart and is specially adapted to not fatigue. Cardiac muscle cells branch, appear striated, and have one nucleus per cell. Specialized junctions between cells (intercalated disks) allow for fast transmission of electrical impulses. See **Figure 2.43**.

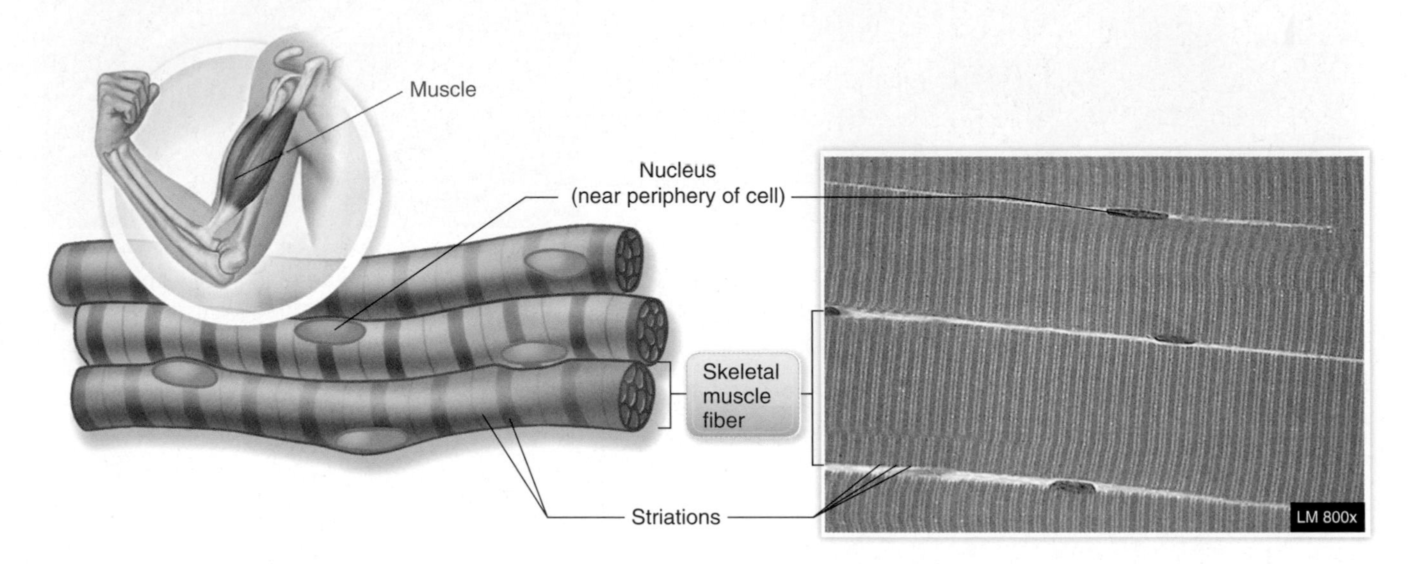

FIGURE 2.41 **Skeletal muscle tissue,** showing parallel, cylindrical muscle cells (muscle fibers) that have striations and multiple nuclei pushed off to the side. Micrograph: skeletal muscle tissue of the biceps brachii muscle in the arm.

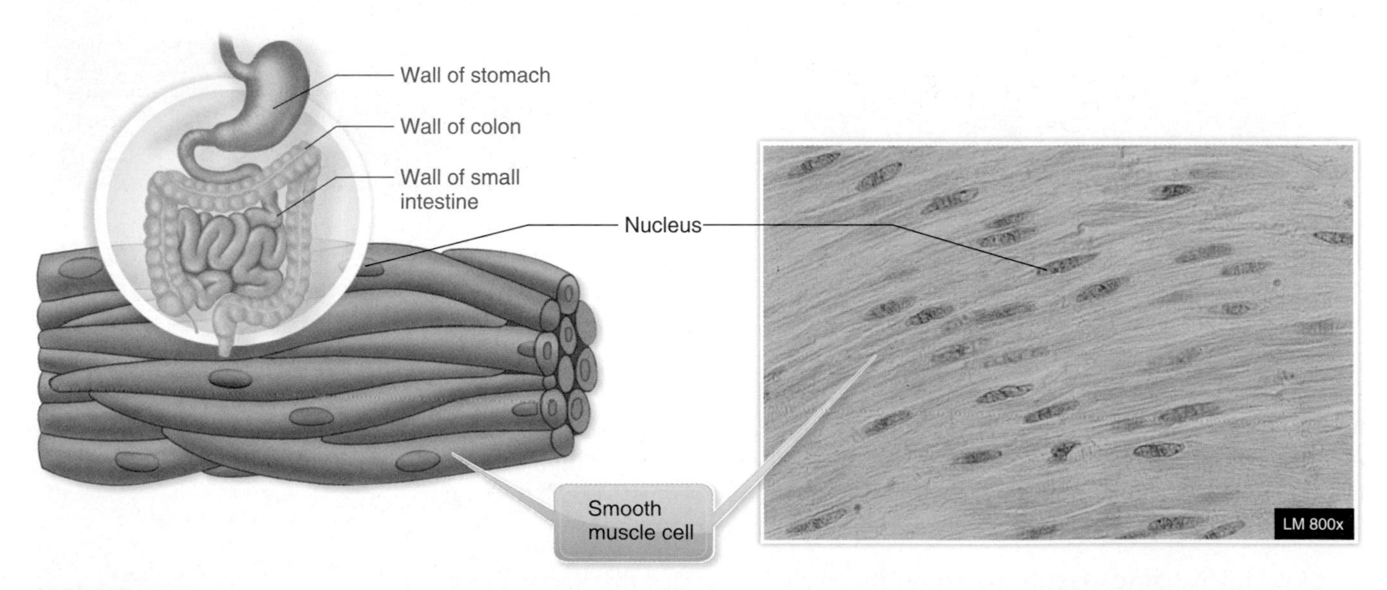

FIGURE 2.42 **Smooth muscle tissue,** showing tapered, nonstriated muscle cells with one nucleus per cell. The micrograph shows smooth muscle tissue in the wall of the stomach.

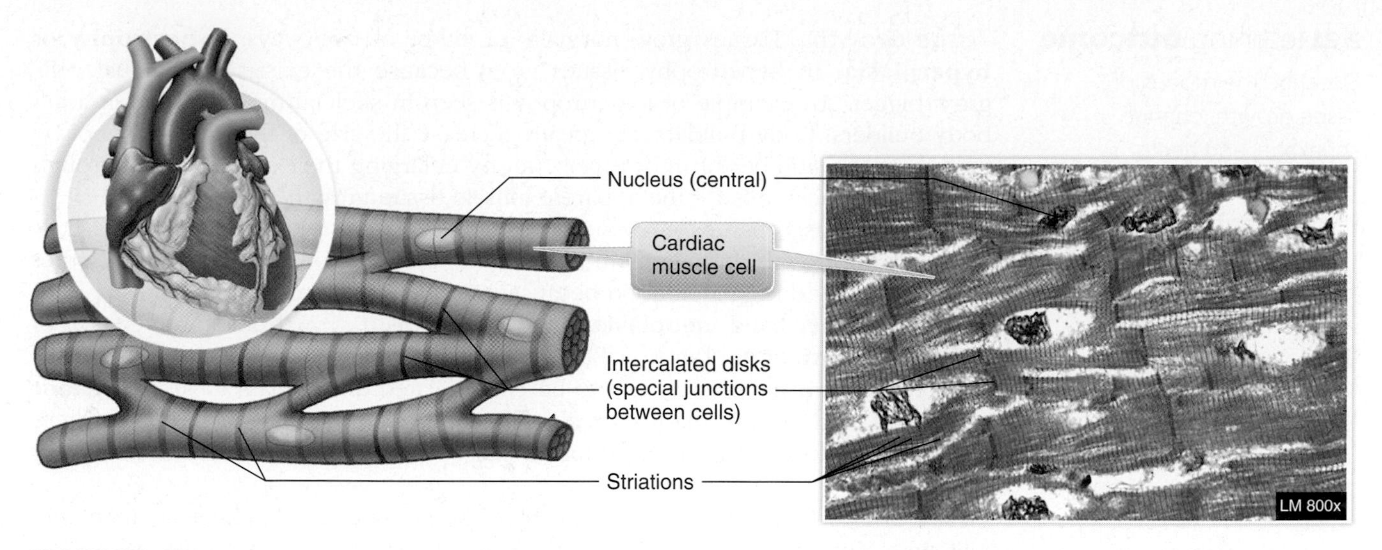

FIGURE 2.43 Cardiac muscle tissue, showing branching, striated muscle cells with one nucleus per cell, and intercalated disks between cells. Micrograph: cardiac muscle of the heart.

Nervous Tissue The body uses nervous tissue for communication through electrical and chemical signals. This tissue is composed of nerve cells called **neurons** and many more support cells called **neuroglia** that protect and assist neurons in their function. Neurons can vary greatly in size and shape. Nervous tissue is covered extensively in the nervous system chapter. See **Figure 2.44**.

The tissues you just covered are not stagnant. In fact, they can grow, change, shrink, and die. Let's look at what can happen to tissues.

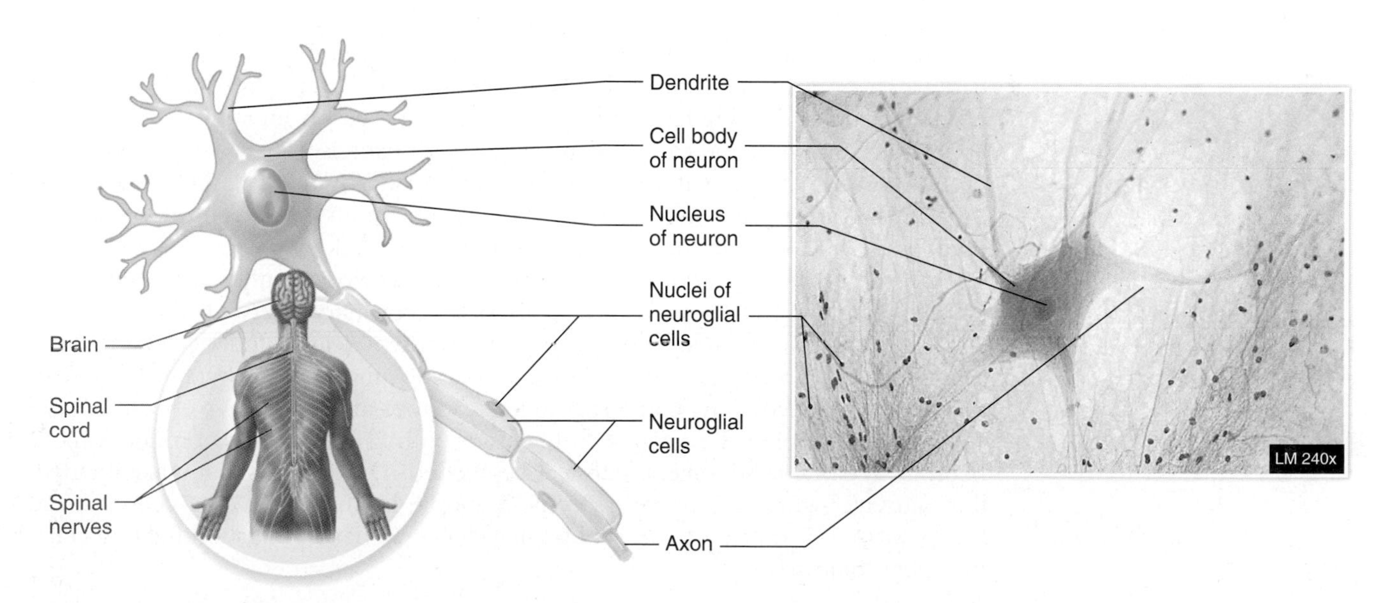

FIGURE 2.44 A neuron and surrounding neuroglial cells of nervous tissue. The neuron has a long axon and many dendrites radiating from the cell body. *Micrograph:* a neuron surrounded by neuroglial support cells.

2.21 learning outcome

Describe the modes of tissue growth, change, shrinkage, and death.

hyperplasia: high-per-PLAY-zee-ah

Tissue Growth Tissues grow normally in either of two ways: **hypertrophy** or **hyperplasia.** In hypertrophy, tissues grow because the existing individual cells grow bigger. An example of hypertrophy is seen in skeletal muscle tissue in adult body builders. Body builders can greatly increase the size of their muscles, not by increasing the number of muscle cells but by enlarging their existing cells through training. This accounts for the apparent muscle tissue growth.

In hyperplasia, tissues grow because more cells are produced. Hyperplasia is the mode of growth during childhood. Skeletal muscles, organs, vessels, and other structures grow during childhood because more cells are produced.

On the other hand, **neoplasia** is the uncontrolled growth and proliferation of cells of abnormal or nonfunctional tissue. This kind of growth results in a **neoplasm** (tumor). **Benign** neoplasms tend to be encapsulated and remain local. **Malignant** neoplasms tend to have cells that break off and travel to other parts of the body where they continue to produce more abnormal cells. This migration is called **metastasis.**

Tissue Change Tissue type is not absolute. Some types of tissue may change over a lifetime. The change of a tissue from one type to another is called **metaplasia.** An example of this can be seen in the normal development of the lining of the vagina. As a child, a girl's vagina is lined by simple cuboidal epithelial tissue. At puberty, due to the influence of hormones, the lining changes from simple cuboidal epithelial tissue to stratified squamous epithelial tissue. The stratified squamous epithelium is a more durable lining, better adapted for intercourse and childbirth. Environmental factors may also cause metaplasia. An example is seen in the ciliated pseudostratified columnar epithelium lining the bronchi of a heavy smoker. The ciliated cells function to move debris (such as dust and smoke particles) out of the bronchi to the throat to be swallowed. Due to the constant irritation to the respiratory lining caused by the heavy smoking, the lining changes from ciliated pseudostratified columnar epithelium to stratified squamous epithelium, a much stronger tissue. The loss of ciliated cells reduces the lining's capacity to remove debris normally. The heavy smoker then must cough up what the lining can no longer remove. This is commonly called "smoker's hack."

atrophy: AT-roh-fee

Tissue Shrinkage and Death **Atrophy** is the shrinkage of tissue due to a decrease in cell size or number. It can be caused by aging or lack of use. Anyone who has had a cast on a broken arm or leg has experienced muscle atrophy. The muscles of the broken arm or leg have atrophied beneath the cast from lack of use. Normal appearance and function can be achieved through exercise.

necrosis: neh-KROH-sis

Necrosis is the premature death of tissue, caused by disease, infection, toxins, or trauma. **Gangrene** is tissue necrosis resulting from an insufficient blood supply, often associated with an infection. **Infarction** is the sudden death of tissue, which often results from a loss of blood supply. An example is a myocardial infarction (sudden death of heart muscle) due to a blocked coronary artery.

apoptosis: AP-op-TOE-sis

Apoptosis is programmed cell death. This mode of death removes cells that have fulfilled their function and are no longer needed. Examples of this type of death can be seen in the developing fetus. The fingers and toes are originally webbed, but the cells of the webbing die by apoptosis before birth. The fetus also produces more neurons in early development than will be needed. The neurons that make connections survive, while any unnecessary neurons die by apoptosis. The cells that die by apoptosis are quickly consumed by macrophages (large, infection-fighting cells) that clear them away.

2.22 learning outcome

Describe the possible effects of uncontrolled growth of abnormal cells in cancer.

Cancer Cancer is the uncontrolled growth of tissue-forming neoplasms. The cancer cells function solely to grow and divide, nothing else. The tumor cells compete with healthy tissue for nutrients, often causing **angiogenesis** (blood vessel growth) to feed the cancer cells. The tumors can block passageways; break through organ and vessel walls, causing hemorrhaging; and reduce the body's ability to fight infection.

There are many causes of cancer. Most cancers are the result of mutations. For example, mutations can stem from mistakes made in replication or from environmental factors called **carcinogens** that affected the DNA and caused the mutation. Radiation from ultraviolet light, chemicals like nitrites in bacon and lunch meat, and viruses like the human papilloma virus (HPV) are all carcinogens. The mutations may turn on **oncogenes,** which are genes that code for uncontrolled production of cellular growth factors stimulating mitosis or the receptors for the growth factors. Either way, the cells grow and divide uncontrollably. The mutations may also damage tumor suppressor genes. Healthy tumor suppressor genes inhibit oncogenes. If the tumor suppressor gene is damaged through mutation, there is little to keep the oncogene from expressing itself.

2.23 learning **outcome**

Explain how genetic and environmental factors can cause cancer.

Cancers are named for their tissue of origin:

- Carcinomas originate in epithelial tissues.
- Sarcomas originate in connective tissues or muscle.
- Lymphomas originate in lymphoid tissue, discussed in the lymphatic system chapter.
- Leukemias originate in blood-forming tissues in the red bone marrow, discussed in the cardiovascular system chapter.

spot check 10 Where in a cell would oncogenes be located? What other carcinogens do you know of that might stimulate an oncogene?

Organ Level

2.24 learning **outcome**

Identify the human body systems and their major organs.

Different tissue types work together in organs, the next level in the human body's hierarchy of organization. For example, the heart is an organ composed of cardiac muscle tissue that allows its chambers to forcefully contract to expel blood. It also has a fibrous skeleton of elastic fibers. This allows the heart chambers to return to shape after a contraction so that they can fill with blood again. Each of the chambers is lined with epithelial tissue. The epithelial tissue extends between chambers to cover the heart valves, which regulate blood flow. Coronary blood vessels with smooth muscle in their walls are located in the heart to feed the cardiac muscle tissue. All of these different types of tissue work together to form the heart, a functioning pump for blood. See **Figure 2.45**.

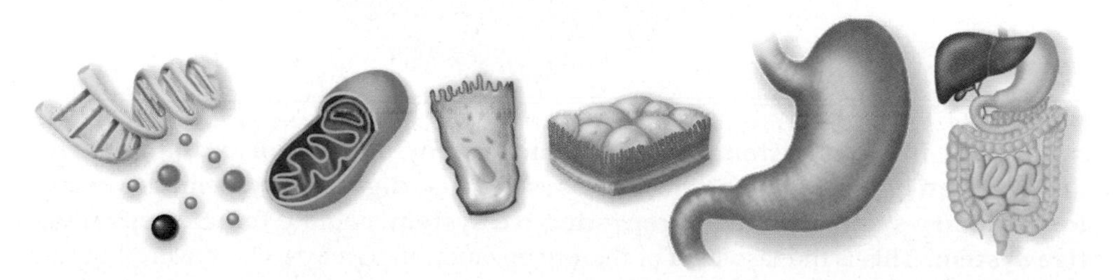

Because the major function of the heart is to pump and circulate blood, the heart belongs to the cardiovascular system. Like many other organs, however, the heart plays a minor role in other systems. For example, the heart produces a hormone that helps regulate urine production in the excretory system. You will cover all of the organs in their relevant system chapters. See Table 2.4.

System Level

Organs and other structures work together to carry out the system function. The body systems are the **integumentary system,** the **skeletal system,** the **muscular**

FIGURE 2.45 Heart: (a) diagram of heart chambers (atria and ventricles) and valves, (b) cadaver heart and valves.

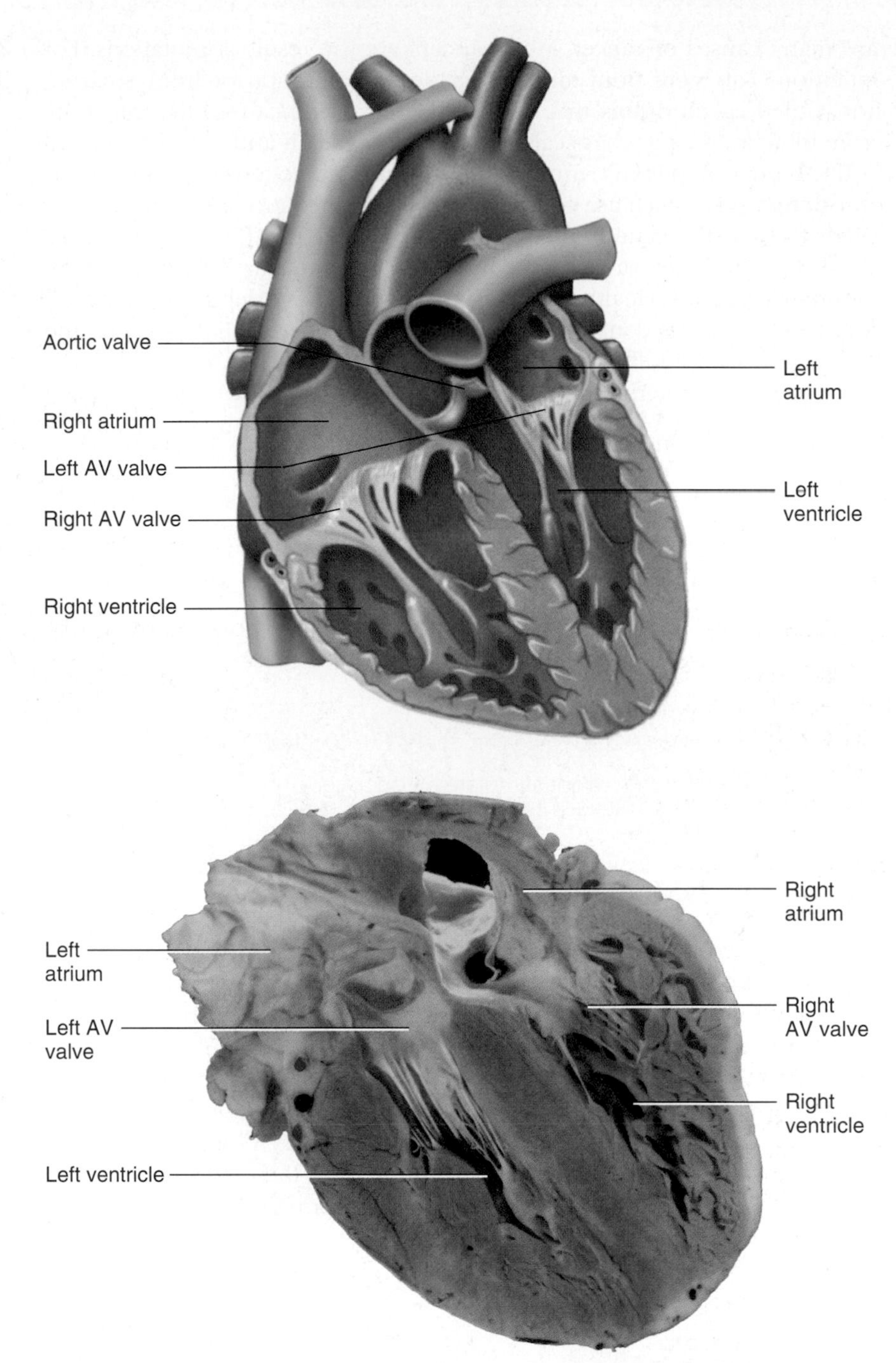

system, the **nervous system,** the **endocrine system,** the **cardiovascular system.** the **lymphatic system,** the **respiratory system,** the **digestive system,** the **excretory/urinary system,** the **male reproductive system,** and the **female reproductive system.** This is the last level of the organization hierarchy to be covered in this text. Each of the remaining chapters covers a system of the human body and begins with the relevant section of Table 2.4. See Table 2.4 and **Figure 2.46**.

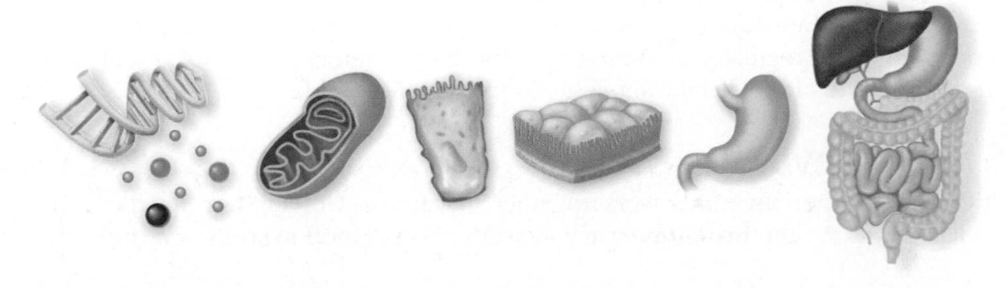

BODY BY LAYERS

Overview

The human body is three-dimensional. In anatomy, it is important that you understand the relationships of structures with depth. This section of the text allows you to peel back the layers of the body so that you can see what lies beneath. The torsos of two bodies are represented—a female and a male. An anterior view of each body is shown in seven layers with structures numbered on each layer. A key is provided on each page for the structures drawn on that layer. A master key for all of the structures shown in both bodies is located before the layered view of each body.

warning

Each human body is unique. Body by Layers represents a typical female and male human being. Individuals may and often do vary from one to another.

study **hint**

In each system chapter, structures are often introduced isolated from other structures so that they can be examined more closely. As you work through the system chapters, it will be helpful to keep coming back to this section to identify structures for each system in the context of the whole Body by Layers. You can easily cover the key on each layer and quiz yourself on structures for the system. Then reveal the key to check your answers. Remember that structures may be shown on more than one layer.

Master Key

This master key lists by number all of the labeled structures in the female and male Body by Layers, and the numbering is consistent. For example, the number "1" indicates the aorta in both the female and male Body by Layers. Indentations in the key indicate that the indented structures are part of the structure listed previously. For example, "abdominal aorta," "arch of aorta," and "descending (thoracic) aorta" are all part of the aorta. Structures are grouped by category. The following abbreviations are used: a.= artery, v.= vein, and m.= muscle.

Blood Vessels

1. Aorta
2. Abdominal aorta
3. Arch of aorta
4. Descending (thoracic) aorta
5. Brachiocephalic a.
6. Brachiocephalic v.
7. Common carotid a.
8. Common iliac a.
9. External jugular v.
10. Femoral a.
11. Femoral v.
12. Great saphenous v.
13. Inferior mesenteric a.
14. Inferior vena cava
15. Internal jugular v.
16. Pulmonary a.
17. Pulmonary trunk
18. Pulmonary v.
19. Subclavian a.
20. Subclavian v.
21. Superior mesenteric a.
22. Superior mesenteric v.
23. Superior vena cava

Bones, Cartilage, and Joints

24. Clavicle
25. Costal cartilage
26. Femur
27. Fifth lumbar vertebra
28. Humerus
29. Intervertebral disk
30. Larynx
31. Pubic bone
32. Pubic symphysis
33. Rib
34. Sacrum
35. Scapula
36. Sternum
37. Manubrium
38. Xiphoid process

Cavities

39. Abdominal cavity
40. Pelvic cavity
41. Pleural cavity
42. Thoracic cavity

Glands

43. Adrenal gland
44. Thymus gland
45. Thyroid gland

Lymph Vessels

46. Lymph vessels
47. Right lymphatic duct
48. Thoracic duct
49. Thoracic lymph nodes

Membranes

50. Greater omentum
51. Parietal pleura
52. Pericardial sac
53. Lesser omentum
54. Mesentery

Muscles

55. Biceps brachii m.
56. Deltoid m.
57. Diaphragm
58. External abdominal oblique m.
59. External intercostal m.
60. Gluteus medius m.
61. Adductor longus m.
62. Gracilis m.
63. Iliacus m.
64. Internal abdominal oblique m.
65. Internal intercostal m.
66. Latissimus dorsi m.
67. Pectoralis major m.
68. Pectoralis minor m.
69. Psoas major m.
70. Rectus abdominus m.
71. Rectus femoris m.
72. Sartorius m.
73. Serratus anterior m.
74. Sternocleidomastoid m.
75. Tensor fasciae latae m.
76. Transverse abdominal m.
77. Trapezius m.
78. Vastus intermedius m.
79. Vastus lateralis m.
80. Vastus medialis m.

Organs

81. Gallbladder
82. Heart
83. Left atrium
84. Left ventricle
85. Right atrium
86. Right auricle
87. Right ventricle
88. Kidney
89. Large intestine
90. Appendix
91. Ascending colon
92. Cecum
93. Descending colon
94. Rectum
95. Sigmoid colon
96. Transverse colon
97. Liver
98. Falciform ligament
99. Left lobe of liver
100. Right lobe of liver
101. Lung
102. Inferior lobe of right lung
103. Middle lobe of right lung
104. Superior lobe of right lung
105. Pancreas
106. Small intestine
107. Duodenum
108. Ileum
109. Jejunum
110. Spleen
111. Stomach
112. Urinary bladder

Reproductive Organs and Structures

113. Breast
114. Areola
115. Mammary gland
116. Nipple
117. Epididymis
118. Inguinal canal
119. Mons pubis
120. Ovary
121. Penis
122. Corpus cavernosum
123. Corpus spongiosum
124. Round ligament of uterus
125. Scrotum
126. Spermatic cord
127. Testis
128. Uterine tube
129. Uterus
130. Vagina

Tubes and Ducts

131. Cystic duct
132. Esophagus
133. Right bronchus
134. Trachea
135. Ureter
136. Urethra

Other Structures

137. Inguinal ligament
138. Linea alba
139. Umbilicus

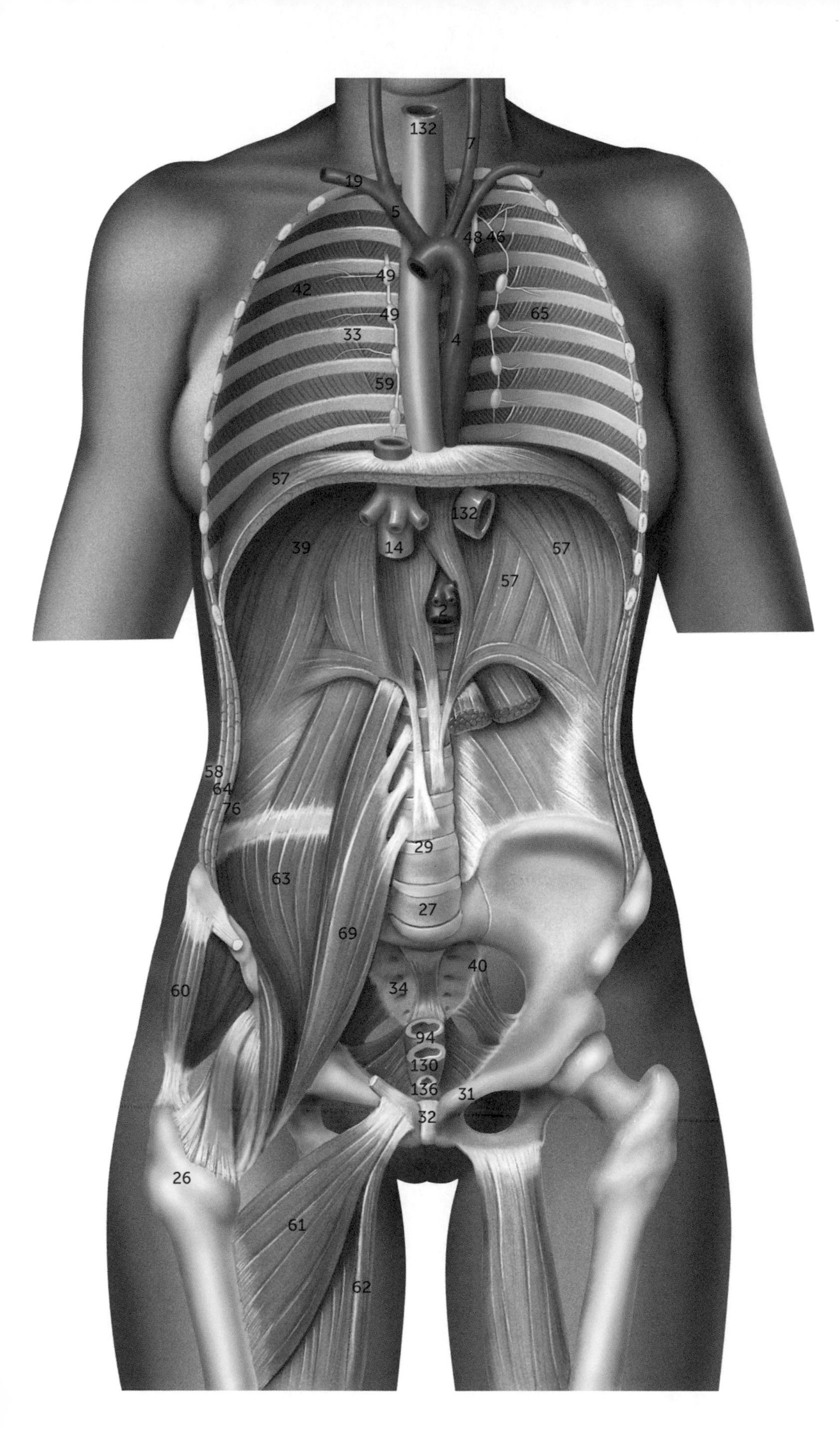

Female **Plate 7**

2 Abdominal aorta
4 Descending (thoracic) aorta
5 Brachiocephalic a.
7 Common carotid a.
14 Inferior vena cava
19 Subclavian a.
26 Femur
27 Fifth lumbar vertebra
29 Intervertebral disc
31 Pubic bone
32 Pubic symphysis
33 Rib
34 Sacrum
39 Abdominal cavity
40 Pelvic cavity
42 Thoracic cavity
46 Lymph vessels
48 Thoracic duct
49 Thoracic lymph nodes
57 Diaphragm
58 External abdominal oblique m.
59 External intercostal m.
60 Gluteus medius m.
61 Adductor longus m.
62 Gracilis m.
63 Iliacus m.
64 Internal abdominal oblique m.
65 Internal intercostal m.
69 Psoas major m.
76 Transverse abdominal m.
94 Rectum
130 Vagina
132 Esophagus
136 Urethra

Master Key

This master key lists by number all of the labeled structures in the female and male Body by Layers, and the numbering is consistent. For example, the number "1" indicates the aorta in both the female and the male Body by Layers. Indentations in the key indicate that the indented structures are part of the structure listed previously. For example, "abdominal aorta," "arch of aorta," and "descending (thoracic) aorta" are all part of the aorta. Structures are grouped by category. The following abbreviations are used: a.= artery, v.= vein, and m.= muscle.

Blood Vessels

1. Aorta
2. Abdominal aorta
3. Arch of aorta
4. Descending (thoracic) aorta
5. Brachiocephalic a.
6. Brachiocephalic v.
7. Common carotid a.
8. Common iliac a.
9. External jugular v.
10. Femoral a.
11. Femoral v.
12. Great saphenous v.
13. Inferior mesenteric a.
14. Inferior vena cava
15. Internal jugular v.
16. Pulmonary a.
17. Pulmonary trunk
18. Pulmonary v.
19. Subclavian a.
20. Subclavian v.
21. Superior mesenteric a.
22. Superior mesenteric v.
23. Superior vena cava

Bones, Cartilage, and Joints

24. Clavicle
25. Costal cartilage
26. Femur
27. Fifth lumbar vertebra
28. Humerus
29. Intervertebral disk
30. Larynx
31. Pubic bone
32. Pubic symphysis
33. Rib
34. Sacrum
35. Scapula
36. Sternum
37. Manubrium
38. Xiphoid process

Cavities

39. Abdominal cavity
40. Pelvic cavity
41. Pleural cavity
42. Thoracic cavity

Glands

43. Adrenal gland
44. Thymus gland
45. Thyroid gland

Lymph Vessels

46. Lymph vessels
47. Right lymphatic duct
48. Thoracic duct
49. Thoracic lymph nodes

Membranes

50. Greater omentum
51. Parietal pleura
52. Pericardial sac
53. Lesser omentum
54. Mesentery

Muscles

55. Biceps brachii m.
56. Deltoid m.
57. Diaphragm
58. External abdominal oblique m.
59. External intercostal m.
60. Gluteus medius m.
61. Adductor longus m.
62. Gracilis m.
63. Iliacus m.
64. Internal abdominal oblique m.
65. Internal intercostal m.
66. Latissimus dorsi m.
67. Pectoralis major m.
68. Pectoralis minor m.
69. Psoas major m.
70. Rectus abdominus m.
71. Rectus femoris m.
72. Sartorius m.
73. Serratus anterior m.
74. Sternocleidomastoid m.
75. Tensor fasciae latae m.
76. Transverse abdominal m.
77. Trapezius m.
78. Vastus intermedius m.
79. Vastus lateralis m.
80. Vastus medialis m.

Organs

81. Gallbladder
82. Heart
83. Left atrium
84. Left ventricle
85. Right atrium
86. Right auricle
87. Right ventricle
88. Kidney
89. Large intestine
90. Appendix
91. Ascending colon
92. Cecum
93. Descending colon
94. Rectum
95. Sigmoid colon
96. Transverse colon
97. Liver
98. Falciform ligament
99. Left lobe of liver
100. Right lobe of liver
101. Lung
102. Inferior lobe of right lung
103. Middle lobe of right lung
104. Superior lobe of right lung
105. Pancreas
106. Small intestine
107. Duodenum
108. Ileum
109. Jejunum
110. Spleen
111. Stomach
112. Urinary bladder

Reproductive Organs and Structures

113. Breast
114. Areola
115. Mammary gland
116. Nipple
117. Epididymis
118. Inguinal canal
119. Mons pubis
120. Ovary
121. Penis
122. Corpus cavernosum
123. Corpus spongiosum
124. Round ligament of uterus
125. Scrotum
126. Spermatic cord
127. Testis
128. Uterine tube
129. Uterus
130. Vagina

Tubes and Ducts

131. Cystic duct
132. Esophagus
133. Right bronchus
134. Trachea
135. Ureter
136. Urethra

Other Structures

137. Inguinal ligament
138. Linea alba
139. Umbilicus

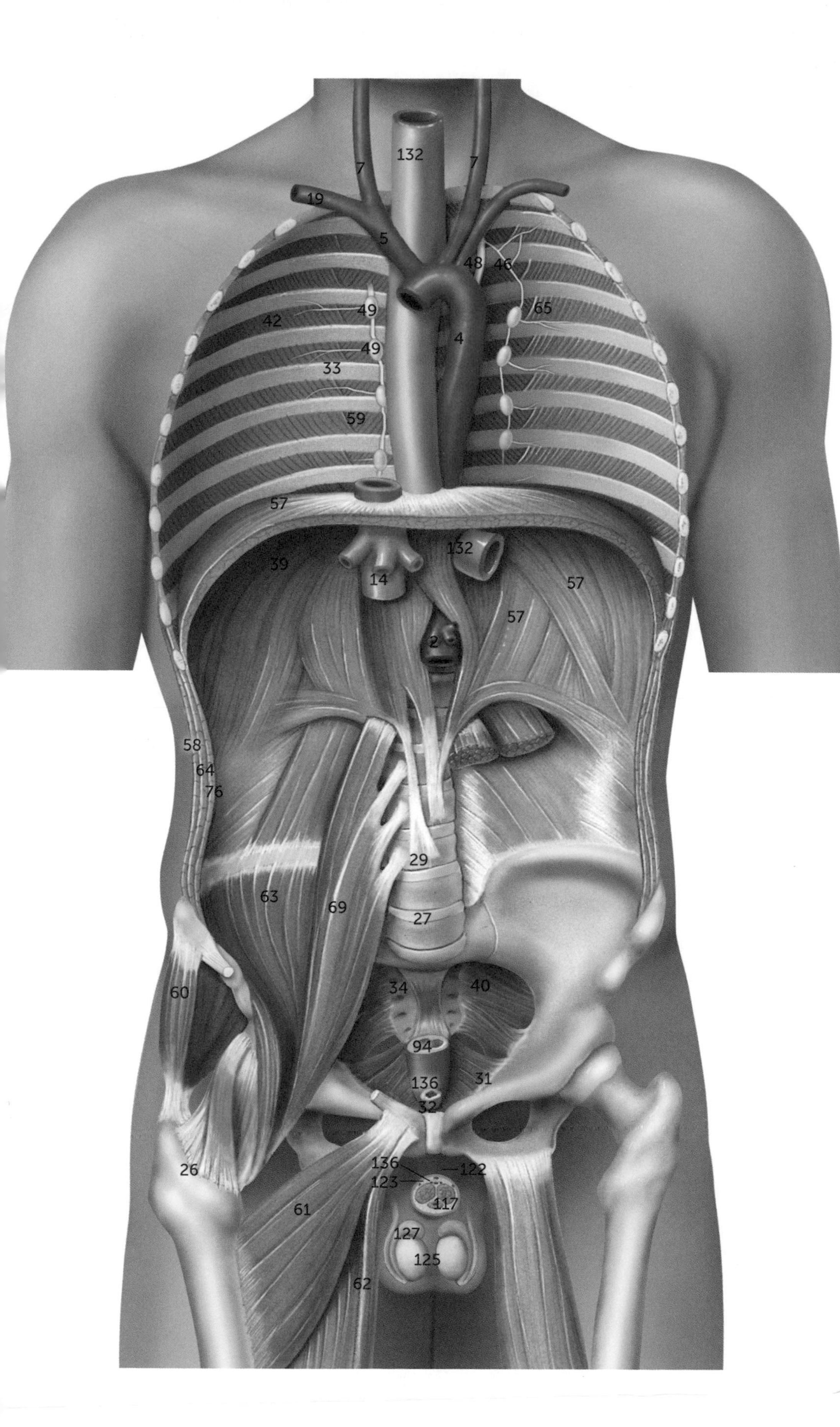

Male **Plate 7**

- 2 Abdominal aorta
- 4 Descending (thoracic) aorta
- 5 Brachiocephalic a.
- 7 Common carotid a.
- 14 Inferior vena cava
- 19 Subclavian a.
- 26 Femur
- 27 Fifth lumbar vertebra
- 29 Intervertebral disc
- 31 Pubic bone
- 32 Pubic symphysis
- 33 Rib
- 34 Sacrum
- 39 Abdominal cavity
- 40 Pelvic cavity
- 42 Thoracic cavity
- 46 Lymph vessels
- 48 Thoracic duct
- 49 Thoracic lymph nodes
- 57 Diaphragm
- 58 External abdominal oblique m.
- 59 External intercostal m.
- 60 Gluteus medius m.
- 61 Adductor longus m.
- 62 Gracilis m.
- 63 Iliacus m.
- 64 Internal abdominal oblique m.
- 65 Internal intercostal m.
- 69 Psoas major m.
- 76 Transverse abdominal m.
- 94 Rectum
- 117 Epididymis
- 122 Corpus cavernosum
- 123 Corpus spongiosum
- 125 Scrotum
- 127 Testis
- 132 Esophagus
- 136 Urethra

TABLE 2.4 Human body systems

System	Major organs and structures	Accessory structures	Function
Integumentary	Skin, hair, nails, cutaneous glands		Protection, vitamin D production, temperature regulation, water retention, sensation, nonverbal communication
Skeletal	Bones	Ligaments, cartilages	Support, movement, protection, acid-base balance, electrolyte balance, blood formation
Muscular	Muscles	Tendons	Movement, stability, control of body openings and passages, communication, heat production
Nervous	Brain, spinal cord, nerves	Meninges, sympathetic chain of ganglia	Communication, motor control, sensation
Endocrine	Pineal gland, hypothalamus, pituitary gland, thyroid gland, adrenal glands, pancreas, testes, ovaries		Communication, hormone production
Cardiovascular	Heart, aorta, superior and inferior venae cavae	Arteries, veins, capillaries	Transportation, protection by fighting foreign invaders and clotting to prevent its own loss, acid-base balance, fluid and electrolyte balance, temperature regulation
Lymphatic	Thymus gland, spleen, tonsils	Thoracic duct, right lymphatic duct, lymph nodes, lymph vessels, MALT, Peyer's patches	Fluid balance, immunity, lipid absorption, defense against disease
Respiratory	Nose, pharynx, larynx, trachea, bronchi, lungs	Diaphragm, sinuses, nasal cavity	Gas exchange, acid-base balance, speech, sense of smell, creation of pressure gradients necessary to circulate blood and lymph
Digestive	Esophagus, stomach, small intestine, large intestine	Liver, pancreas, gallbladder, cecum, teeth, salivary glands	Ingestion, digestion, absorption, defecation
Excretory/urinary	Kidney, ureters, urinary bladder, urethra	Lungs, skin, liver	Removal of metabolic wastes, fluid and electrolyte balance, acid-base balance, blood pressure regulation
Male reproductive	Testes	Scrotum, spermatic ducts (epididymis, ductus deferens), accessory glands (seminal vesicles, prostate gland, bulbourethral glands), penis	Production and delivery of sperm, secretion of sex hormones
Female reproductive	Ovaries	Uterus, uterine tubes, vagina, vulva, breasts	Production of an egg, housing of the fetus, birth, lactation, secretion of sex hormones

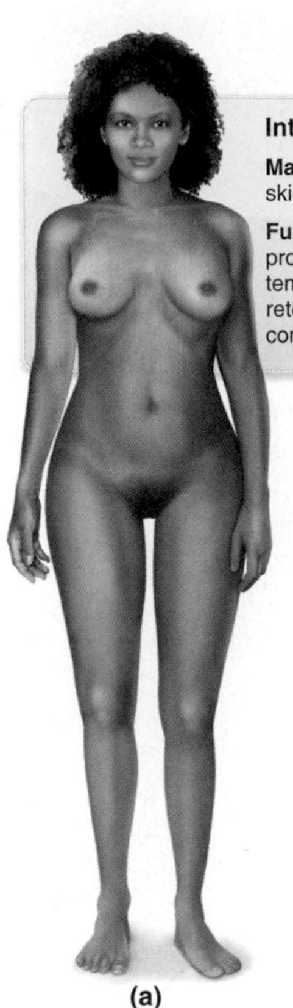

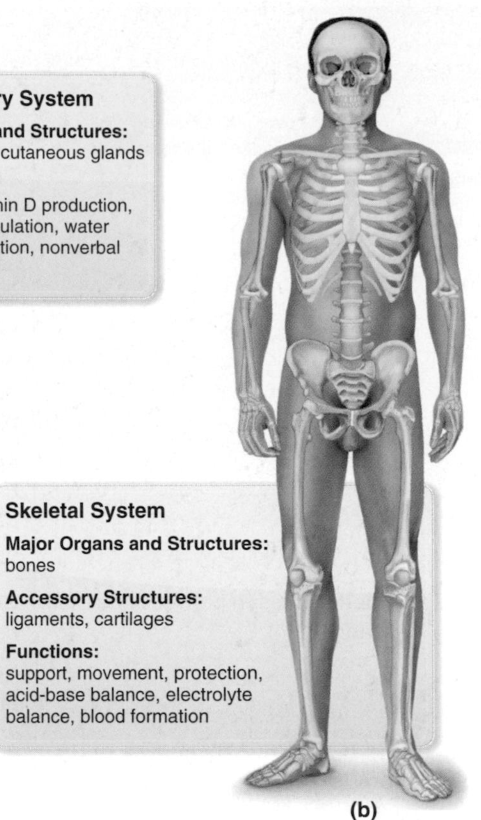

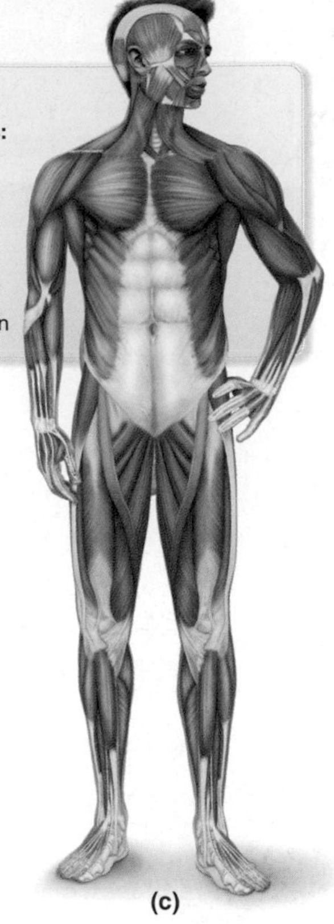

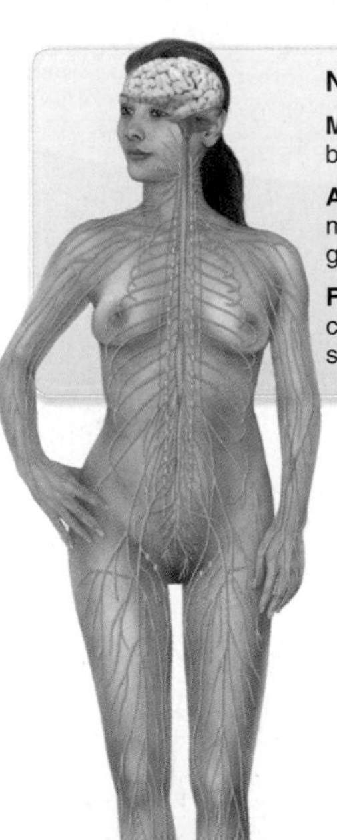

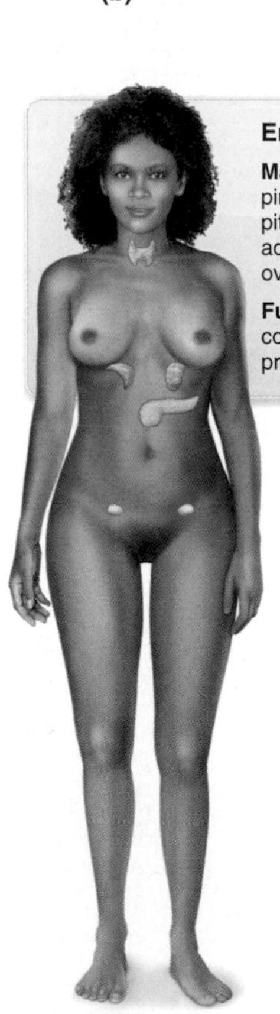

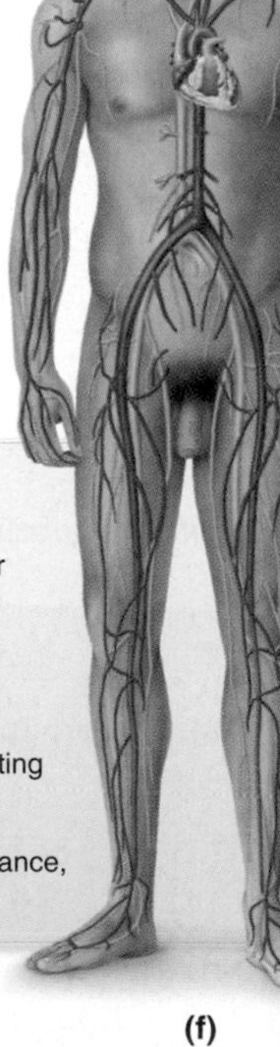
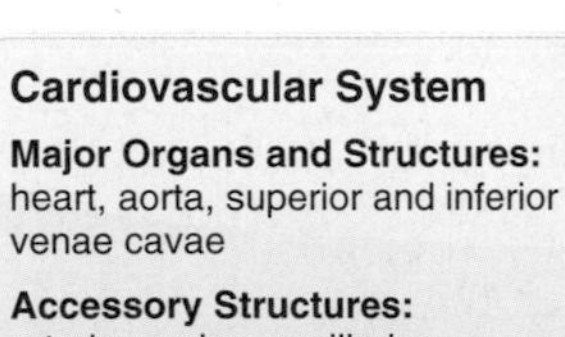

FIGURE 2.46 Human body systems: (a) integumentary system, (b) skeletal system, (c) muscular system, (d) nervous system, (e) endocrine system, (f) cardiovascular system, (g) lymphatic system, (h) respiratory system, (i) digestive system, (j) excretory/urinary system, (k) male reproductive system, (l) female reproductive system.

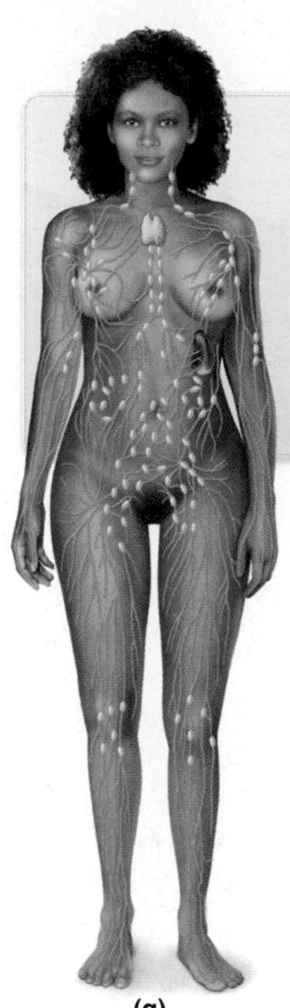

Lymphatic System

Major Organs and Structures:
thymus gland, spleen, tonsils

Accessory Structures:
thoracic duct, right lymphatic duct, lymph nodes, lymph vessels, MALT, Peyer's patches

Functions:
fluid balance, immunity, lipid absorption, defense against disease

(g)

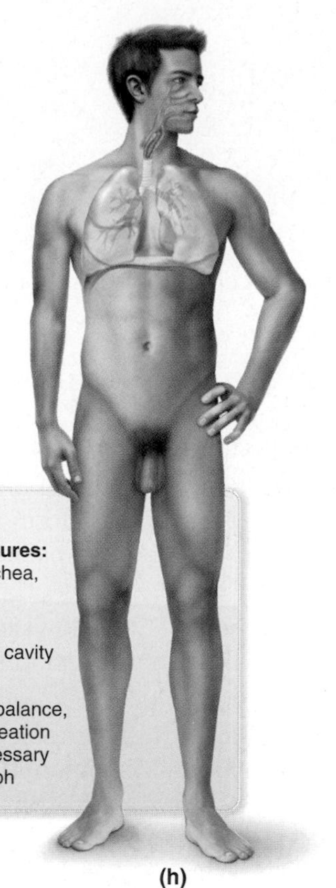

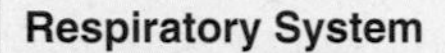

Respiratory System

Major Organs and Structures:
nose, pharynx, larynx, trachea, bronchi, lungs

Accessory Structures:
diaphragm, sinuses, nasal cavity

Functions:
gas exchange, acid-base balance, speech, sense of smell, creation of pressure gradients necessary to circulate blood and lymph

(h)

Digestive System

Major Organs and Structures:
esophagus, stomach, small intestine, large intestine

Accessory Structures:
liver, pancreas, gallbladder, cecum, teeth, salivary glands

Functions:
ingestion, digestion, absorption, defecation

(i)

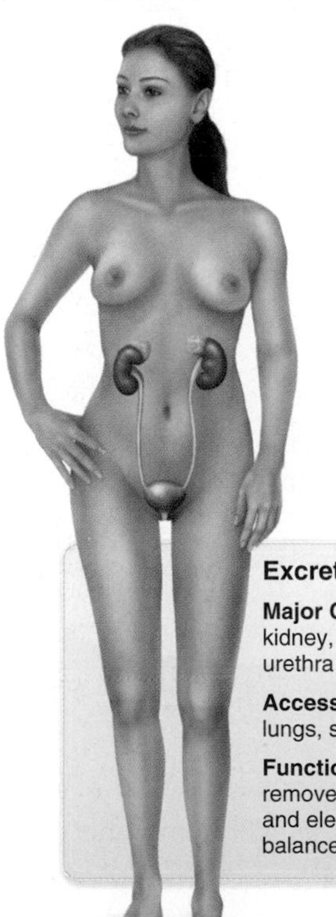

Excretory/Urinary System

Major Organs and Structures:
kidney, ureters, urinary bladder, urethra

Accessory Structures:
lungs, skin, liver

Functions:
remove metabolic wastes, fluid and electrolyte balance, acid-base balance, blood pressure regulation

(j)

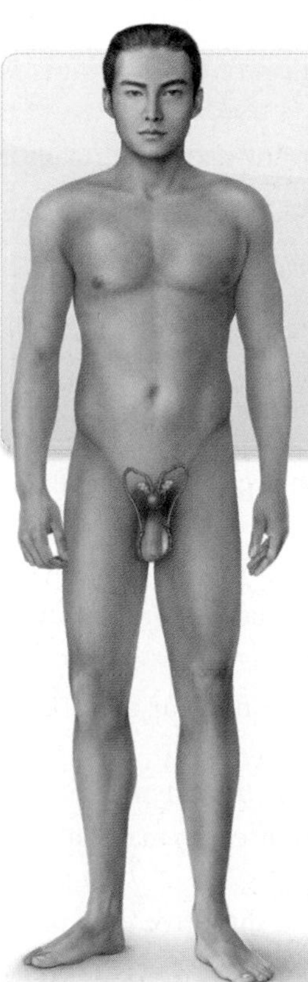

Male Reproductive System

Major Organs and Structures:
testes

Accessory Structures:
scrotum, spermatic ducts (epididymis, ductus deferens), accessory glands (seminal vesicles, prostate gland, bulbourethral glands), penis

Functions:
production and delivery of sperm, secretion of sex hormones

(k)

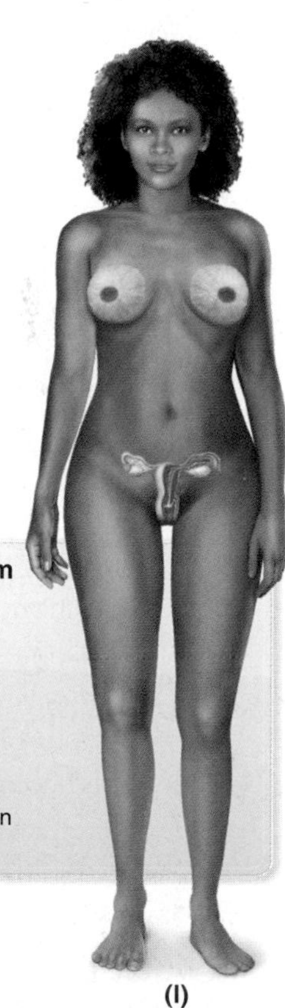

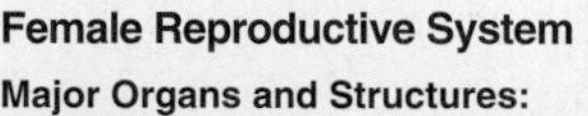

Female Reproductive System

Major Organs and Structures:
ovaries

Accessory Structures:
uterus, uterine tubes, vagina, vulva, breasts

Functions:
production of an egg, housing of the fetus, birth, lactation, secretion of sex hormones

(l)

summary

Overview

- There is a hierarchy of organization to the body:

chemical → organelles → cells → tissues → organs → systems → organism

Levels of Organization

Chemical Level

- All matter is composed of atoms containing protons, neutrons, and electrons. Isotopes are atoms that have additional neutrons. If an isotope freely gives off particles, it is called a *radioisotope.*
- Atoms bond to form molecules. Atoms share electrons to form a covalent bond. Atoms give up and receive electrons in an ionic bond. Ionically bonded molecules separate to become ions in water. Ions in solution are electrolytes.
- Water carries out five vital functions in the body: (1) It allows for ions in solution. (2) It works as a lubricant. (3) It aids in chemical reactions. (4) It helps with transportation. (5) It is used for temperature regulation.
- A solution is composed of a solute dissolved in a solvent. *Concentration* refers to the amount of solute relative to the amount of solvent. In comparing solutions, a hypertonic solution is more concentrated, an isotonic solution is the same concentration, and a hypotonic solution is less concentrated.
- Acids are hydrogen ion donors. Bases accept hydrogen ions. The pH scale is used to measure acidity and alkalinity; every one-number difference is a tenfold difference in the amount of H^+ or OH^-.
- The four types of organic molecules are carbohydrates, lipids, proteins, and nucleic acids. All organic molecules must contain carbon and hydrogen. Organic molecules are composed of building blocks.
- Molecules must come together to react. Metabolism is the total of all the chemical reactions in the body. Reactions happen faster if (1) the concentration of the reactants is increased, (2) the speed of the reactants is increased by adding heat, and (3) a catalyst is used.
- Cellular respiration is one of the most important chemical reactions in the body.

$$C_6H_{12}O_6 + 6O_2 \rightarrow 6CO_2 + 6H_2O + \text{Energy}$$

Glucose + Oxygen *yields* Carbon dioxide + Water + Energy

- The molecule ATP contains the usable form of energy for the cell.

Organelle Level

- Organelles are suspended within the cell in a fluid called *cytoplasm.* The organelles are:
  1. Cell membrane: a phospholipid bilayer that gives structure to the cell and regulates what may enter and leave the cell.
  2. Nucleus: the part of the cell that houses the DNA.
  3. Mitochondria: organelles that perform cellular respiration and process the energy to ATP.
  4. Ribosomes: organelles that assemble proteins.
  5. Endoplasmic reticulum: an extension of the nuclear membrane. Rough ER is the site of protein synthesis. Smooth ER is the site of lipid synthesis.
  6. Golgi complexes: membrane-enclosed folds that inspect and modify proteins and lipids produced in the cell.
  7. Secretory vesicles: membrane packages that carry materials from the Golgi complex to the cell membrane for export from the cell.
  8. Lysosomes: membrane-bound packages of digestive enzymes.

Cellular Level

- Materials can be moved across the selectively permeable cell membrane through passive, active, or bulk transport.
- Passive transport moves materials from areas of high concentration to areas of low concentration with no energy required. Methods of passive transport include filtration, simple diffusion, facilitated diffusion, and osmosis.

- Active transport moves materials across a cell membrane from areas of low concentration to areas of high concentration (against a concentration gradient). Active transport requires energy.
- Bulk transport moves large quantities of materials, not individual molecules, across a cell membrane. Materials are moved out of the cell through exocytosis. Materials are moved into the cell through endocytosis.
- Protein synthesis is a two-step process. Transcription happens in the nucleus. It produces mRNA, which carries the instructions written in the DNA to the ribosome on the rough endoplasmic reticulum. Translation happens at the ribosome. The tRNAs transport specific amino acids to the ribosome for assembly based on the information in the mRNA.
- The amino acid sequence that was assembled through translation is inspected and modified by the Golgi complex.
- Mistakes in protein synthesis may or may not have any consequences.
- Human DNA is spread out for use as chromatin. Before dividing, the DNA is tightly packaged into 46 chromosomes.
- Mitosis is the method of cell division used by all cells other than those producing sperm and eggs.
- A parent cell replicates its 46 chromosomes before dividing to become two daughter cells. Each daughter has a complete set of 46 chromosomes that is identical to the other daughter's set and to the parent cell's set, which no longer exists once cell division is complete.
- Mistakes in replication may or may not have any consequences.
- Telomeres are noncoding strings of nucleotides at the ends of chromosomes. Their purpose is believed to be protection of the chromosome ends. Part of the telomere is lost with each cell division. When all of the telomeres are lost, more and more mistakes in replication may occur. Therefore, more mistakes potentially affecting cellular function occur with age.

Tissue Level

- There are four basic classifications of tissues:
  1. Epithelial tissue covers or lines all of the body's surfaces. The various types are named for the cell shape and amount of layering.
  2. Connective tissue has cells and fibers in a matrix. The matrix may be very fluid, as in blood, or very hard and dense as in bone.
  3. Muscle tissue has a high concentration of proteins; this arrangement allows the tissues to contract.
  4. Nervous tissue is composed of neurons and support cells called neuroglia. The function of nervous tissue is communication through electrical and chemical signals.
- Tissues grow normally through hyperplasia (making more cells) or hypertrophy (making existing cells bigger).
- Tissues grow abnormally to form neoplasms (tumors). Neoplasms may be benign or malignant.
- Metaplasia is the change of tissue from one type to another.
- Atrophy is the shrinkage of tissue due to age or disuse.
- Necrosis is the premature death of tissue. If the death is sudden, it is called an infarction. Gangrene is the death of tissue due to an insufficient blood supply, usually associated with an infection.
- Apoptosis is programmed cell death.
- Cancer is the uncontrollable growth of cells that function solely for the purpose of growing and dividing. Cancer cells compete with healthy tissue for space and nutrients, invade passageways, break through organ and vessel walls to cause hemorrhage, and reduce the body's ability to fight infection.
- Most cancers are caused by mutations through mistakes in replication or through environmental factors called carcinogens.

Organ Level

- Many different types of tissue function collectively as organs to carry out functions in the various systems.

System Level

- There are 11 systems in the human body: integumentary, skeletal, muscular, nervous, endocrine, cardiovascular, lymphatic, respiratory, digestive, excretory/urinary, and reproductive.

key words for review

The following terms are defined in the glossary.

acid
atom
base
cellular respiration
chemical reaction
epithelial tissues
histology
integumentary system
membrane transport
metabolism
mitosis
molecule
mutation
organelles
organic molecules
osmosis
protein synthesis
reactants
replication
solution

chapter review questions

Multiple Select: *Select the correct choices for each statement. The choices may be all correct, all incorrect, or a combination of correct and incorrect.*

1. Which of the following statements is (are) accurate for gastric juice produced in the stomach that has a pH of 0.8?
   a. It is a strong base.
   b. It will have more hydroxide ions than hydrogen ions.
   c. It contains 20 times more hydrogen ions than a liquid with a pH of 2.8.
   d. It is a strong acid.
   e. It is more alkaline than human blood.
2. Which of the following statements would be true for testosterone given that it is a steroid?
   a. It has a carbon-hydrogen-oxygen ratio of 1:2:1.
   b. It is a lipid.
   c. It is composed of fatty acids.
   d. It has monosaccharide building blocks.
   e. It contains carbon, hydrogen, and oxygen.
3. Which of the following statements is (are) true concerning transport across a cell membrane?
   a. Molecules move from high concentration to low concentration in facilitated diffusion.
   b. Molecules move because of a force in filtration.
   c. Water moves during osmosis because the cell membrane is selectively permeable.
   d. ATP is needed for active transport.
   e. Water would flow into a cell placed in a hypertonic solution.
4. Which of the following statements is (are) true if a cell has a moderate number of mitochondria, well-developed rough endoplasmic reticulum, many ribosomes, a well-developed Golgi complex, and no secretory vesicles?
   a. This cell's function is to produce steroids.
   b. There would need to be fatty acids and glycerol in the cytoplasm.
   c. The product made in this cell would be exported from the cell.
   d. There are moderate energy demands in this cell.
   e. There would need to be amino acids in the cytoplasm of this cell.
5. What happens in mitosis?
   a. Two daughter cells have identical DNA to the parent cell.
   b. The mRNA is replicated.
   c. Errors in replication may have no consequences.
   d. Errors in replication may result in mutations.
   e. The parent cell's DNA is formed in 23 chromosomes before division.

6. Which of the following statements is (are) true for tissues?
   a. Tissues change from one type to another in hyperplasia.
   b. Tissues grow by making existing cells larger through apoptosis.
   c. Tissues grow uncontrollably, ceasing normal function in metaplasia.
   d. Tissues shrink from lack of use in atrophy.
   e. Tissues suffer from premature death in necrosis.
7. What happens in cancer?
   a. An oncogene may cause an increase in growth factor receptors on a cell.
   b. A neoplasm may metastasize.
   c. Tumors may block passages or poke holes in them.
   d. Tumors encourage blood vessel growth to themselves so they may be fed.
   e. Benign tumors compete for space.
8. What does water do in the body?
   a. It separates ionically bonded molecules to form electrolytes.
   b. It aids in chemical reactions.
   c. It helps regulate body temperature.
   d. It is used for transportation of substances within the body.
   e. It lubricates joints.
9. Which of the following statements is (are) true about the chemical reaction $C_6H_{12}O_6 + 6O_2 \longrightarrow 6CO_2 + 6H_2O$?
   a. This reaction is called *cellular respiration*.
   b. The reactants in this reaction are glucose and oxygen.
   c. Decreasing the concentration of reactants would speed up this reaction.
   d. Energy is a product in this reaction.
   e. There are six glucose molecules represented in this reaction.
10. Which of the following statements is (are) true concerning protein synthesis?
   a. Translation happens first.
   b. Translation occurs in the nucleus.
   c. Transcription happens at the ribosome.
   d. Secretory vesicles carry proteins to the cell membrane for export from the cell.
   e. Messenger RNA delivers amino acids to the Golgi complex.

Matching: *Match the organelles to their function.*

_____	1. Assembles amino acids to form proteins	a. Lysosome
_____	2. Inspects and modifies proteins produced	b. Cell membrane
_____	3. Contains digestive enzymes	c. Mitochondrion
_____	4. Regulates what enters or leaves the cell	d. Golgi complex
_____	5. Performs cellular respiration	e. Ribosome

Matching: *Match the tissue to the classification. Choices may be used more than once.*

_____	6. Adipose	a. Epithelial tissue
_____	7. Blood	b. Connective tissue
_____	8. Transitional	c. Muscle tissue
_____	9. Neurons	d. Nervous tissue
_____	10. Hyaline cartilage	

Critical Thinking

1. Paramedics arrive on the scene of a car accident. They assess the scene and call the emergency room with the victim's condition. The ER doctor recommends starting an IV, not to treat the patient at this time but to establish an intravenous line should drugs need to be quickly administered later, on the way to the hospital. Should the IV fluids be hypotonic, isotonic, or hypertonic to blood plasma? Explain.
2. Henri is a gardener. He works outside and is exposed to the sun's ultraviolet rays. Skin is fast-growing tissue that grows by hyperplasia. What will be the effects for Henri if the gene that codes for insulin (made by cells in the pancreas) is damaged by ultraviolet rays in some of the skin cells on his arm?

2 chapter mapping

This section of the chapter is designed to help you find where each outcome is covered in this text.

	Outcomes	Readings, figures, and tables	Assessments
2.1	List the levels of organization of the human body from simplest to most complex.	Overview: pp. 22–23 Figure 2.1	
2.2	Define the terms *matter, element, atom,* and *isotope.*	Chemical level: pp. 23–25 Figures 2.2, 2.3 Table 2.1	Spot Check: 1
2.3	Define *molecule* and describe two methods of bonding that may form molecules.	Bonding to form molecules: pp. 25–26 Figure 2.4	
2.4	Summarize the five functions of water in the human body and give an explanation or example of each.	Water: p. 27	Multiple Select: 8
2.5	Compare solutions based on tonicity.	Solutions: p. 27 Figure 2.5	Critical Thinking: 1
2.6	Determine whether a substance is an acid or a base and its relative strength if given its pH.	Acids, bases, and pH: pp. 28–29 Figures 2.6, 2.7	Spot Check: 2 Multiple Select: 1
2.7	Describe the four types of organic molecules in the body by giving the elements present in each, their building blocks, an example of each, the location of each example in the body, and the function of each example.	Organic molecules: pp. 29–33 Figures 2.8–2.11 Table 2.2	Spot Check: 3 Multiple Select: 2
2.8	Explain three factors governing the speed of chemical reactions.	Chemical reactions: pp. 33–34	Spot Check: 4 Multiple Select: 9
2.9	Write the equation for cellular respiration using chemical symbols and describe it in words.	Cellular respiration: pp. 34–35	Multiple Select: 9
2.10	Explain the importance of ATP in terms of energy use in the cell.	ATP: pp. 35–36 Figure 2.12	Spot Check: 7
2.11	Describe cell organelles and explain their functions.	Organelle level: pp. 36–40 Figures 2.13, 2.14 Table 2.3	Spot Check: 5 Multiple Select: 4 Matching: 1–5
2.12	Compare four methods of passive transport and active transport across a cell membrane in terms of materials moved, direction of movement, and the amount of energy required.	Membrane transport: pp. 40–43 Figures 2.15–2.18	Spot Check: 6 Multiple Select: 3 Critical Thinking: 1
2.13	Describe bulk transport, including endocytosis and exocytosis.	Bulk transport: pp. 43–44 Figure 2.19	
2.14	Describe the processes of transcription and translation in protein synthesis in terms of location and the relevant nucleic acids involved.	Protein synthesis: pp. 44–47 Figures 2.20, 2.21	Spot Check: 8 Multiple Select: 10
2.15	Describe what happens to a protein after translation.	Protein synthesis: p. 47 Figure 2.22	Multiple Select: 10
2.16	Explain the possible consequences of mistakes in protein synthesis.	Mistakes in protein synthesis: pp. 47–48	

	Outcomes	Readings, figures, and tables	Assessments
2.17	Describe the process of mitosis, including a comparison of the chromosomes in a parent cell to the chromosomes in the daughter cells.	Cell division: pp. 48–50 Figures 2.23, 2.24	Spot Check: 9 Multiple Select: 5
2.18	Explain the possible consequences of mistakes in replication.	Cell division: p. 50	Critical Thinking: 2
2.19	Describe the effects of aging on cell division.	Effects of aging on cells: pp. 50–51 Figure 2.25	
2.20	Describe the four classifications of tissues in the human body.	Tissue level: pp. 51–61 Figures 2.26–2.44	Matching: 6–10
2.21	Describe the modes of tissue growth, change, shrinkage, and death.	Tissue growth: p. 62 Tissue change: p. 62 Tissue shrinkage and death: p. 62	Multiple Select: 6
2.22	Describe the possible effects of uncontrolled growth of abnormal cells in cancer.	Cancer: p. 62	Multiple Select: 7
2.23	Explain how genetic and environmental factors can cause cancer.	Cancer: p. 63	Spot Check: 10
2.24	Identify the human body systems and their major organs.	Organ level: p. 63 Figure 2.45 Table 2.4 System level: pp. 63–67 Figure 2.46	

footnote

1. Kerem, B., Rommens, J. M., Buchanan, J. A., Markiewicz, D., Cox, T. K., Chakravarti, A., Buchwald, M., et al. (1989, September 8). Identification of the cystic fibrosis gene: Genetic analysis [Electronic version]. *Science Magazine, 245,* 1073–1080.

references and resources

Kerem, B., Rommens, J. M., Buchanan, J. A., Markiewicz, D., Cox, T. K., Chakravarti, A., Buchwald, M., et al. (1989, September 8). Identification of the cystic fibrosis gene: Genetic analysis [Electronic version]. *Science Magazine, 245,* 1073–1080.

National Institutes of Health & National Institute on Aging. (2006, June). *Aging under the microscope: A biological quest.* Bethesda, MD: Department of Health and Human Services.

Porter, R. S., & Kaplan, J. L. (Eds.). (n.d.). Selected physiologic age-related changes. *Merck manual for healthcare professionals.* Retrieved March 19, 2011, from http://www.merckmanuals.com/media/professional/pdf/Table_337-1.pdf?qt=selected physiologic age-related changes&alt=sh.

Rommens, J. M., Iannuzzi, M. C., Kerem, B., Drumm, M. L., Melmer, G., Rozmahel, M. D. R., et al. (1989, September 8). Identification of the cystic fibrosis gene: Chromosome walking and jumping [Electronic version]. *Science Magazine, 245,* 1059–1065.

Saladin, Kenneth S. (2010). *Anatomy & physiology: The unity of form and function* (5th ed.). New York: McGraw-Hill.

Seeley, R. R., Stephens, T. D., & Tate, P. (2006). *Anatomy & physiology* (7th ed.). New York: McGraw-Hill.

Shier, D., Butler, J., & Lewis, R. (2010). *Hole's human anatomy & physiology* (12th ed.). New York: McGraw-Hill.

3 The Integumentary System

Although you might associate the word *organ* with the heart or one of the lungs or kidneys, the skin is in fact the largest organ in the human body. The skin plays an enormous role in the integumentary system, which is made up of skin, hair, nails, and cutaneous glands. See Figure 3.1. By its very nature and location, this system is the most visible of all the body systems, and its structures are constantly changing and growing.

learning outcomes

After completing this chapter, you should be able to:

3.1 Use medical terminology related to the integumentary system.

3.2 Describe the histology of the epidermis, dermis, and hypodermis.

3.3 Describe the cells of the epidermis and their function.

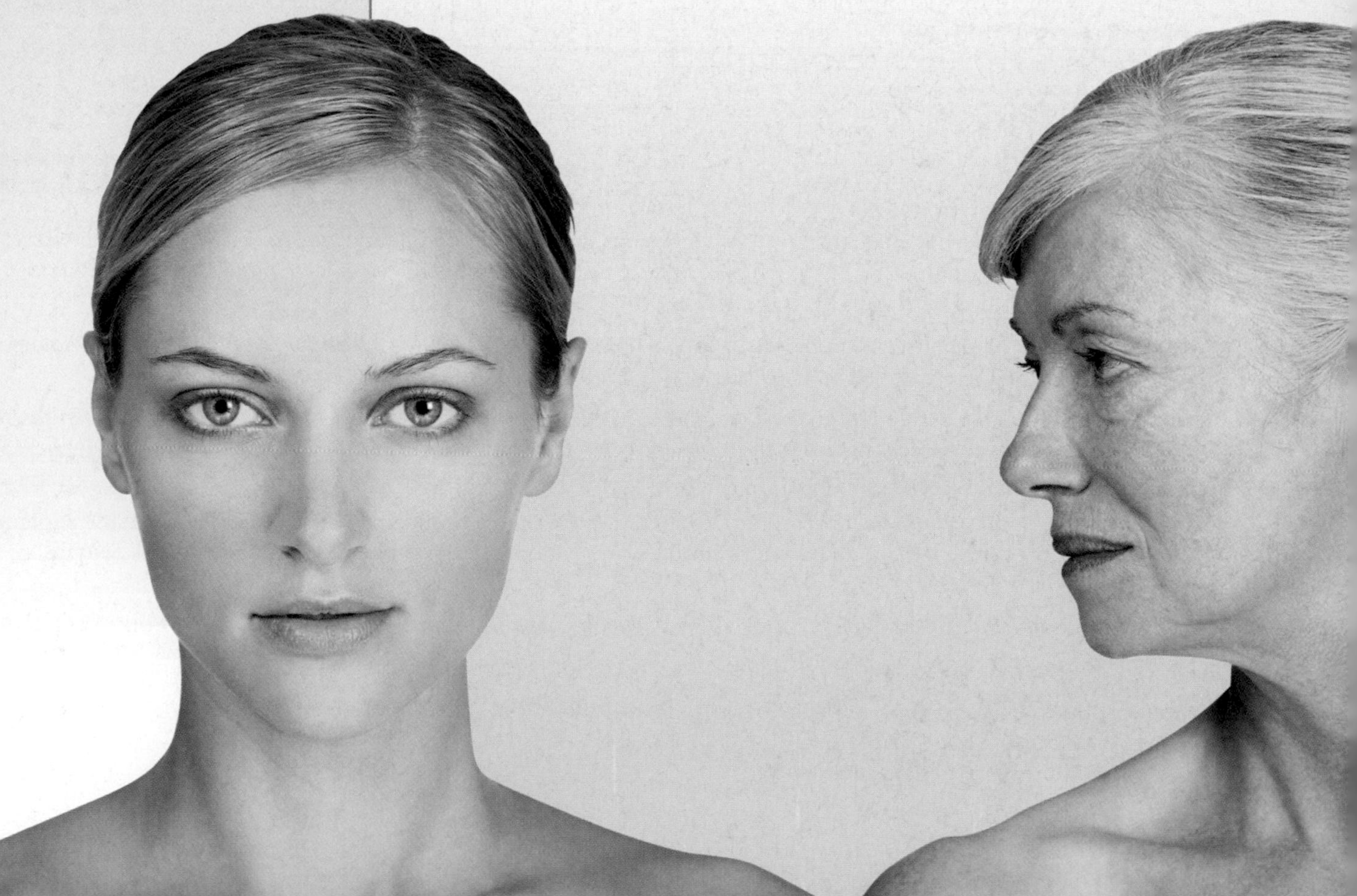

3.1 learning **outcome**

Use medical terminology related to the integumentary system.

3.4 Describe the structures of the dermis and their functions.

3.5 Compare and contrast the glands of the skin in terms of their structure, products, and functions.

3.6 Describe the histology of a hair and hair follicle.

3.7 Explain how a hair grows and is lost.

3.8 Describe the structure and function of a nail.

3.9 Explain how the layers and structures of the skin work together to carry out the functions of the system.

3.10 Explain how the skin responds to injury and repairs itself.

3.11 Describe the symptoms of inflammation and explain their cause in terms of the structure and function of the skin.

3.12 Compare and contrast three degrees of burns in terms of symptoms, layers of the skin affected, and method used by the body for healing.

3.13 Describe the extent of a burn using the rule of nines.

3.14 Summarize the effects of aging on the integumentary system.

3.15 Describe three forms of skin cancer in terms of the body area most affected, appearance, and ability to metastasize.

3.16 Describe an example of a bacterial, a viral, and a fungal infection of the skin.

word **roots** & combining **forms**

cutane/o: skin

cyan/o: blue

derm/o: skin

dermat/o: skin

kerat/o: hard

melan/o: black

onych/o: nail

seb/o: oil

pronunciation **key**

arrector pili: a-REK-tor PYE-lye

basale: ba-SAL-eh

ceruminous: seh-ROO-min-us

comedo: KOM-ee-doh

eponychium: ep-uh-NIK-ee-um

exocrine: EK-soh-krin

keratinocytes: ke-RAT-in-oh-sites

lucidum: loo-SEE-dum

lunule: LOON-yule

sebaceous: se-BAY-shus

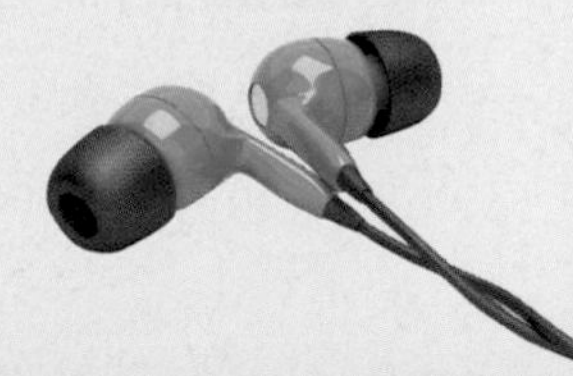

Overview

In this chapter, you will become familiar with the anatomy and physiology of the integumentary system. Note that the skin's physiology is explained in the context of going out for a morning run. You will also learn how skin responds to injury, inflammation, burns, the effects of aging, and skin disorders such as skin cancers and infections, as well as how skin repairs itself.

3.2 learning **outcome**

Describe the histology of the epidermis, dermis, and hypodermis.

Anatomy of the Skin, Hair, and Nails

As the body's largest organ, the skin makes up approximately 15 percent of the body's total weight. It consists of two layers, the **epidermis** and the **dermis.** The epidermis is the skin's most superficial layer, composed of stratified squamous epithelial tissue. Deep to the epidermis is the dermis, which is composed of loose/areolar connective tissue over dense irregular connective tissue. The **cutaneous** glands, the hair follicles, and most of the skin's nerve endings can be found in the dermis.

Deep to the dermis is the **hypodermis** or **subcutaneous layer** (subcutaneous tissue). Although it is technically not part of the skin, this layer attaches the skin to the rest of the body. Mainly composed of adipose connective tissue, the hypodermis serves as an insulating layer, a cushioning layer, and an energy source. This layer is generally thicker in women than in men. See Figure 3.2.

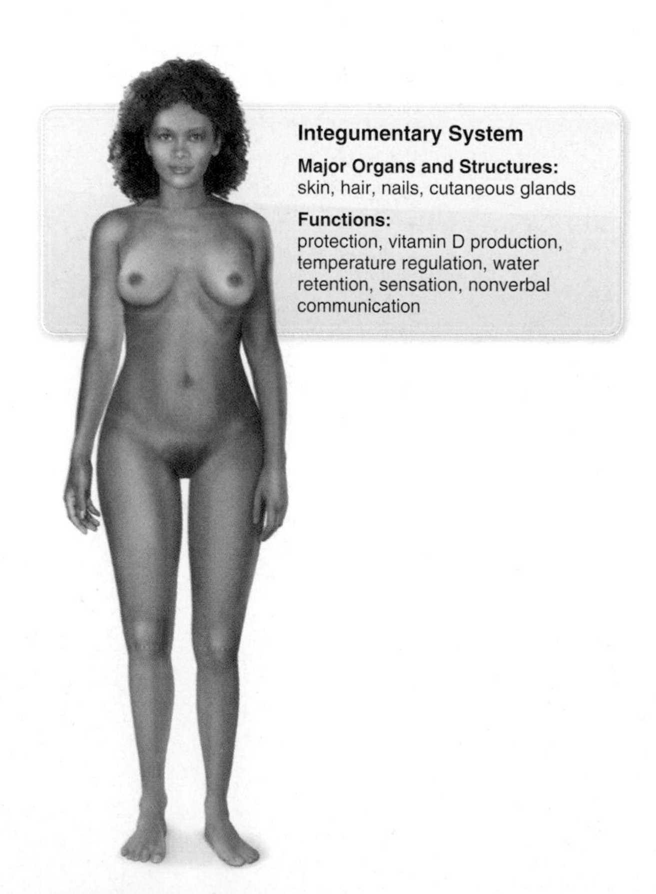

FIGURE 3.1 Integumentary system.

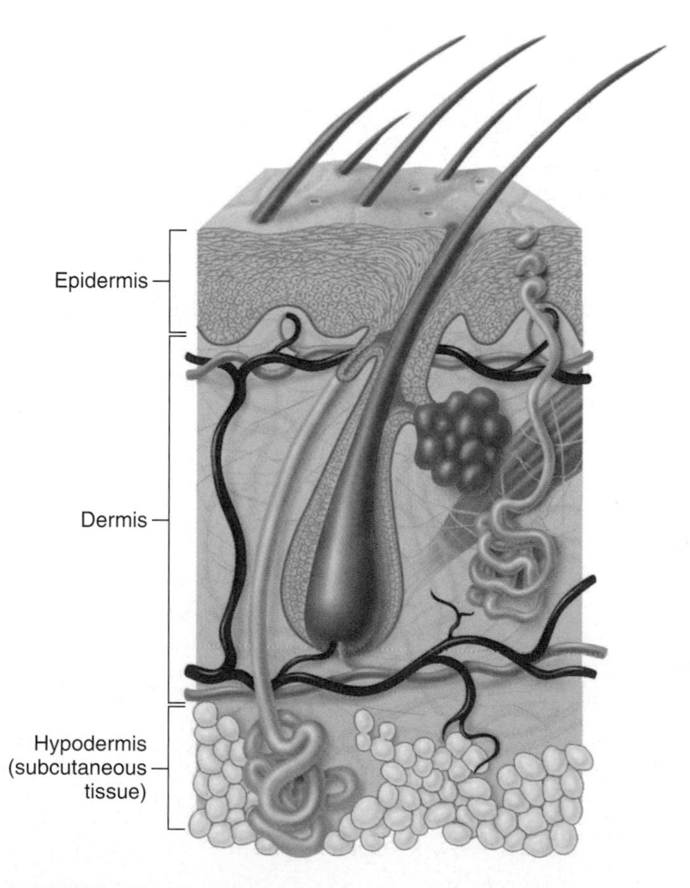

FIGURE 3.2 The skin.

clinical point

The hypodermis is an ideal site for subcutaneous drug injections (which are done with a hypodermic needle) because it has numerous blood vessels and can absorb the drugs easily. See Figure 3.3.

FIGURE 3.3 Hypodermic needle.

Epidermis

The epidermis (the superficial layer of the skin), is subdivided into four or five general layers called *strata*. These strata can be seen in **Figure 3.4**. The deepest layer is the **stratum basale,** which contains a single layer of cuboidal cells. It is the only stratum of the epidermis with cells that actively grow and divide to produce new epidermis. The stratum basale dips into the dermis of the skin to form the hair follicles. Superficial to the stratum basale are the stratum spinosum and the stratum granulosum. The **stratum lucidum** is found only in **thick skin.** It is not found

basale: ba-SAL-eh

lucidum: loo-SEE-dum

warning

Whether skin is "thick" or "thin" does not have to do with the depth of the skin or the number of cell layers. It has to do with the presence of the stratum lucidum, which is present only in thick skin. Thick skin is found on the lips, on the palms of the hands, and on the plantar surface of the feet. Thick skin has no hair, while thin skin does.

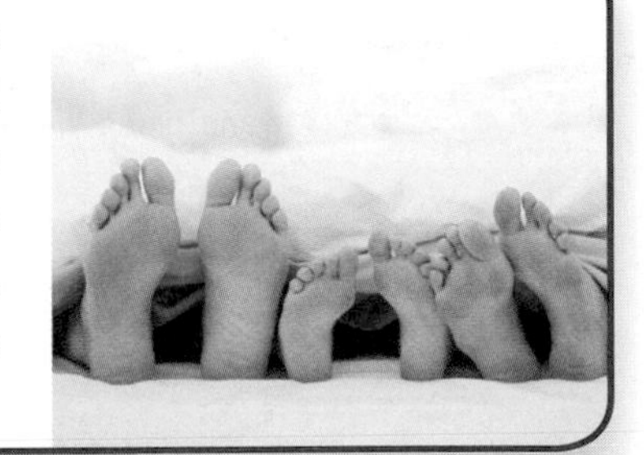

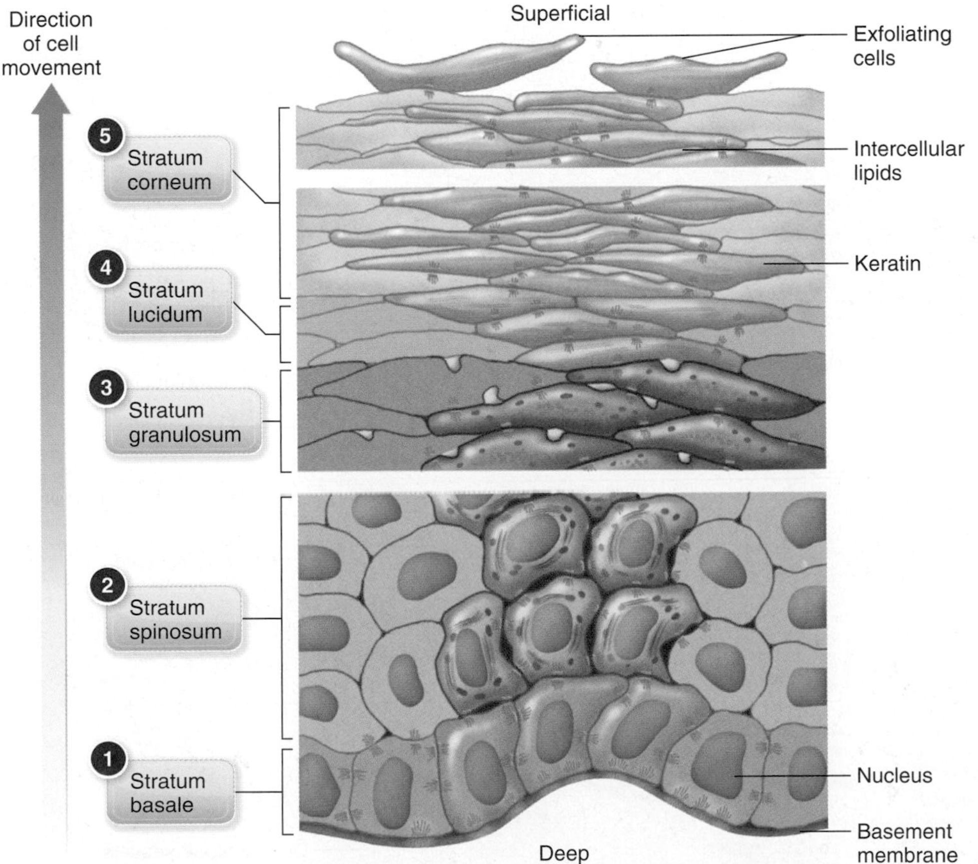

FIGURE 3.4 Strata of the epidermis. This is epidermis of thick skin. Stratum lucidum is present, and there is no hair.

in **thin skin.** The **stratum corneum** is the most superficial stratum. It may have as many as 20 layers of cells. It is composed of tough, waterproof dead cells that eventually flake off **(exfoliate).** The human body sheds thousands of these dead stratum corneum cells every day. The exfoliated cells make up the majority of all the dust you find on your floor and furniture when cleaning.

spot check 1 How does the anatomy of the epidermis of your forearm compare to the anatomy of the epidermis of your lips?

3.3 learning outcome

Describe the cells of the epidermis and their function.

Cells of the Epidermis Several different types of cells that serve specific functions can be found in the epidermis. These cells are defined in the list below. See **Figure 3.5**.

- **Keratinocytes** begin in the stratum basale and make up the majority of epidermal cells. Their purpose is to grow and divide. As they divide, they

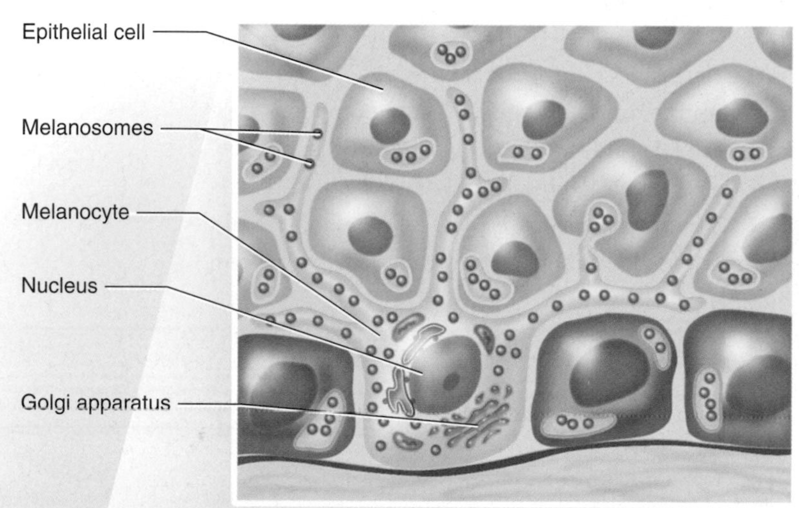

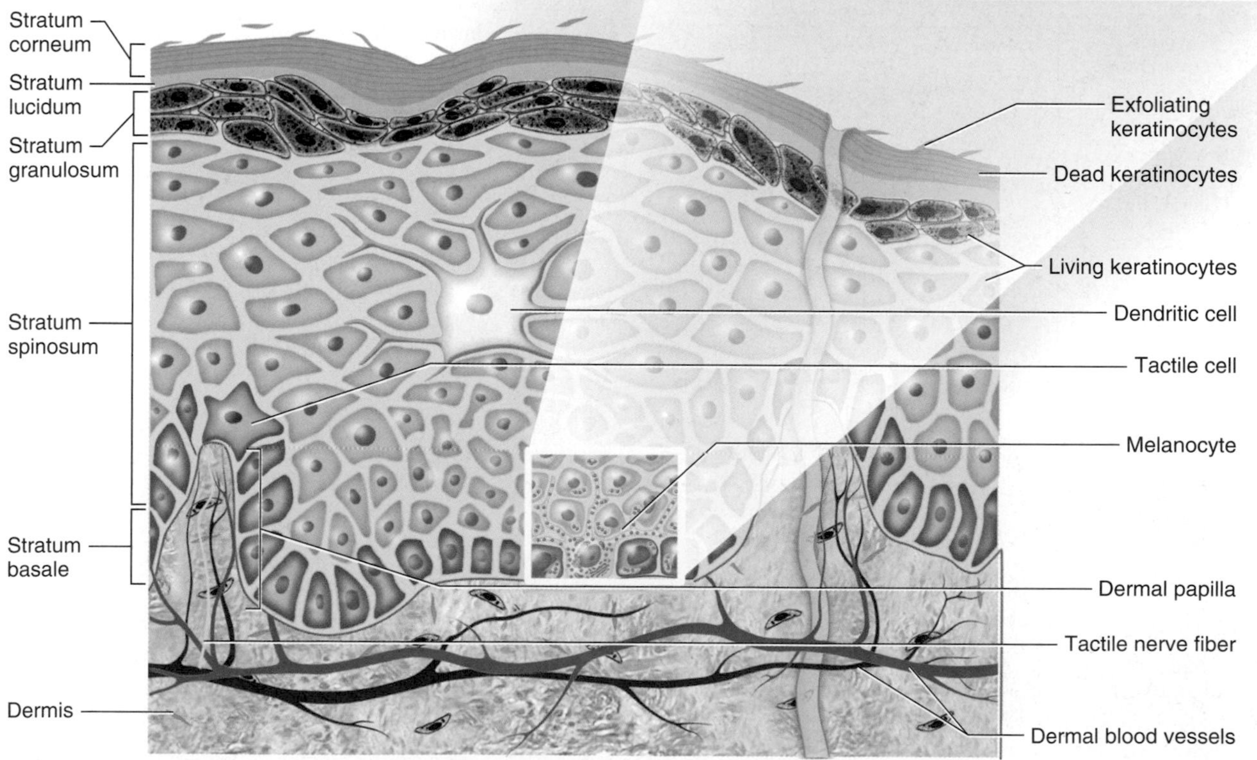

FIGURE 3.5 Cells of the epidermis.

push the older cells up toward the surface. The keratinocytes produce and fill themselves with **keratin** (a hard, waterproof protein) as they move toward the surface. By the time they reach the **stratum corneum,** the keratinocytes have completely filled with keratin and died. This process is called **cornification.** These cells form a very durable stratum corneum at the surface. See **Figures 3.4** and **3.5**.

keratinocytes: ke-RAT-in-oh-sites

- **Melanocytes** produce skin pigments called *melanin*. These cells stay in the stratum basale, but they have projections to more superficial layers. The keratinocytes take in the melanin produced by pinching off bits of the melanocyte extensions that contain melanin-filled vesicles called *melanosomes*. This is an example of endocytosis (covered in Chapter 2). See the inset in **Figure 3.5**. Melanocytes may not be evenly distributed across the skin, and denser patches of these cells account for freckles and moles.

warning

The idea that people with dark skin have more melanocytes than people with light skin is a misconception, as both types of skin contain the same number. Although the melanocytes function the same, the difference is the amount of melanin they produce. This amount is genetically determined. See Figure 3.6.

- **Tactile cells** serve as receptors for fine touch only. Although they are found in the stratum basale, they are associated with nerve cells in the underlying dermis.
- **Dendritic cells** are immune system cells found in the stratum spinosum and the stratum granulosum. They alert the body's immune system to the invasion of **pathogens** (disease-causing foreign invaders) that could make it through the stratum corneum.

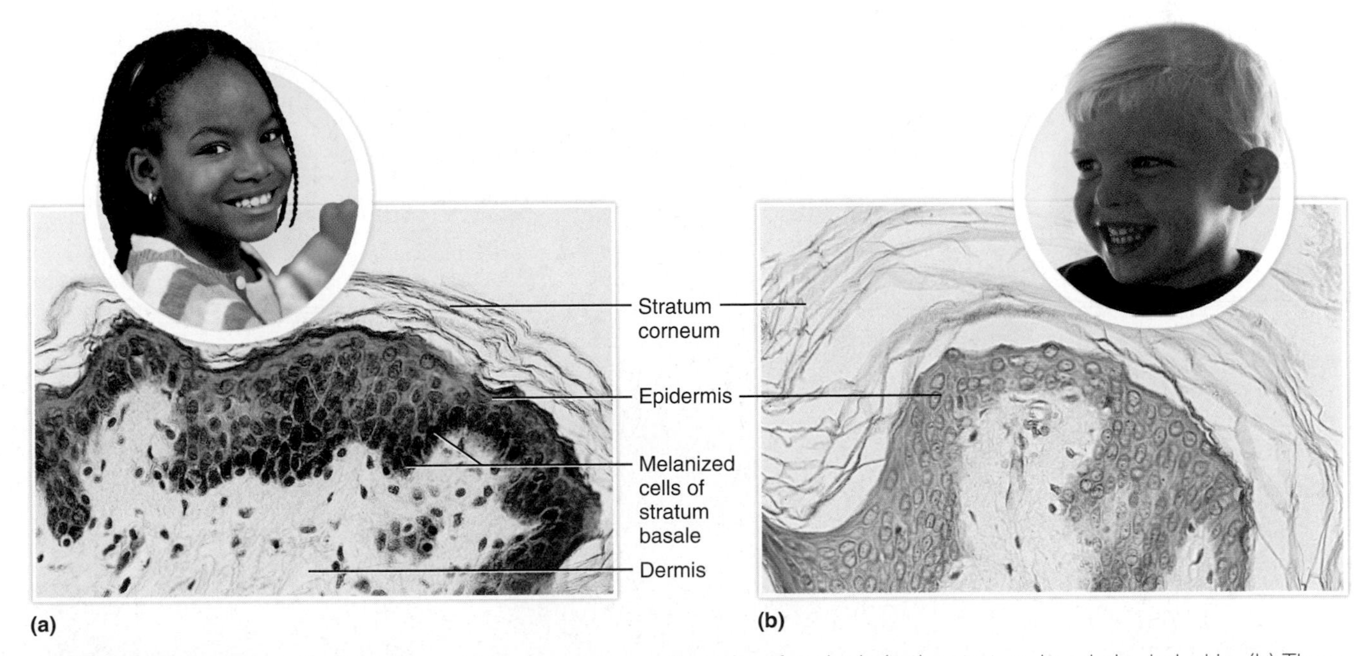

FIGURE 3.6 Variations in skin color. (a) There are heavy deposits of melanin in the stratum basale in dark skin. (b) There is little to no visible melanin in the stratum basale of light skin.

3.4 learning **outcome**

Describe the structures of the dermis and their functions.

Dermis

The dermis is sometimes called the "true skin." Blood vessels, fibers, nerve endings, hair follicles, and cutaneous glands are found in the dermis. See **Figure 3.7**. We explain this layer and its related structures in the paragraphs that follow.

Papillae The superficial edge of the dermis has bumps called **papillae,** which are in direct contact with the epidermis. These papillae, arranged in a random pattern over most of the body, form individual-specific patterns of ridges on the palmar and planter surfaces. This unique arrangement creates fingerprints. The papillae's numerous small blood vessels provide nearby stratum basale cells with nutrients they need to grow and divide. The stratum basale cells are fed by diffusion. Because blood vessels do not extend into the epidermis, the risk of blood loss if the epidermis becomes injured is reduced.

spot check ❷ How deep into layers of the skin would a cut have to be in order to cause bleeding?

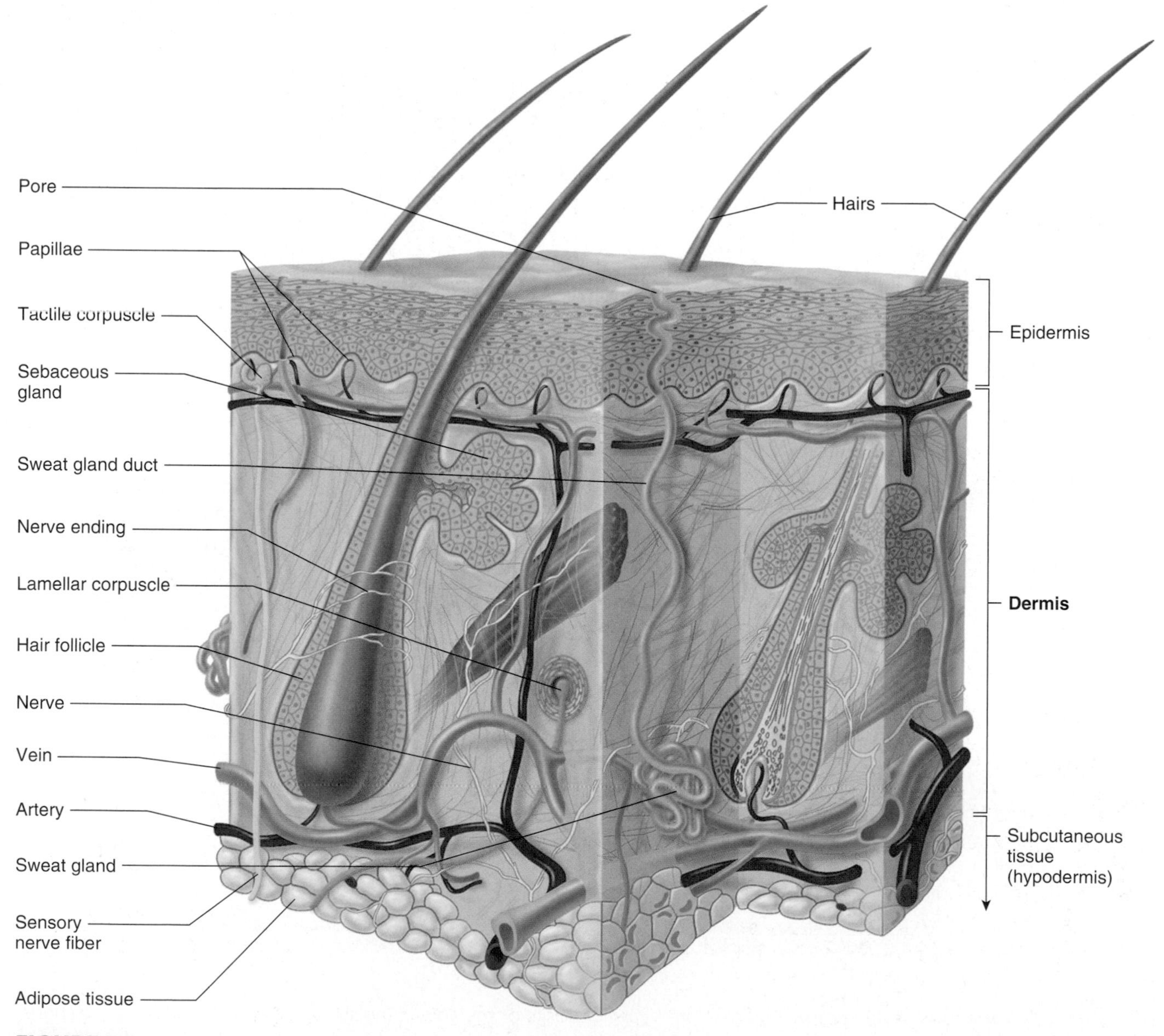

FIGURE 3.7 The dermis and its structures.

Fibers The dermis changes from loose/areolar tissue to dense irregular connective tissue with depth. Both of these tissue types are characterized by cells and fibers in a matrix, but they lack a neat dividing line between them. The cells **(fibroblasts)** of both tissues produce two types of protein fibers: collagen and elastin. **Collagen** fibers give the skin strength. **Elastin** fibers provide elasticity, which allows the skin to come back to shape if stretched. The number of fibers increases with depth in the dermis.

warning

Be careful of the word *elastic* as it does not mean the ability to be stretched. (That is the definition of *extensibility,* which is covered in the muscular system chapter.) **Elasticity** is the ability to come back to shape if something has first been stretched, like a rubber band. For example, think of an elastic waistband on an old pair of sweatpants. This waistband is no longer elastic if you stretch out the waistband, let go, and the pants fall off.

Vitamin A and vitamin C are important for healthy skin because they are necessary for collagen production. You can get vitamin A (which is also important in the maintenance of epithelial tissues) by eating green and yellow vegetables, dairy products, and liver. You can boost your vitamin C intake by consuming plenty of fruits and green vegetables.

Nerve Endings Nerve cells have endings in the dermis that serve as receptors (receiving devices). These receptors include warm receptors, cold receptors, pain receptors, and pressure receptors. For example, the lamellar and tactile corpuscles shown in **Figure 3.7** are nerve receptors. There are also receptors associated with hair follicles located in the dermis that can detect the movement of a single hair. The distribution of receptors varies by region. All of the skin's nerve endings are explained in more detail in Chapter 7, "The Nervous System—Senses."

3.5 learning **outcome**

Compare and contrast the glands of the skin in terms of their structure, products, and functions.

Cutaneous Glands The skin's cutaneous glands—which are considered **exocrine glands**—are located throughout the dermis. Exocrine glands produce and secrete products that are delivered to the appropriate locations through ducts. There are two basic types of cutaneous glands: **sebaceous glands** and **sweat glands.** See **Figure 3.7**

exocrine: EK-soh-krin

sebaceous: se-BAY-shus

Sebaceous glands. As you can see in **Figure 3.7**, sebaceous glands are associated with a hair follicle. The sebaceous gland duct leads to the hair follicle surrounding the hair root. **Sebum,** a very oily, lipid-rich substance, is produced by the sebaceous gland to moisturize the skin and hair. Sebum exits the hair follicle to the skin's surface with the emerging hair. You can remove sebum from your skin by washing with soap. Once your skin is clean, you can then replenish its moisture balance by replacing the removed sebum with a lanolin-containing lotion. Lanolin is sebum produced by sheep.

The hormones estrogen and testosterone increase the amount of sebum produced. The levels of these hormones greatly increase at puberty, which causes the sebaceous glands to become much more active, producing more and more sebum. In this case, the short ducts delivering sebum to the hair follicle can become plugged with the excess sebum and cells shed from the sebaceous gland. The gland continues to produce more sebum even though the duct is plugged. The plug prevents the sebum from reaching the surface, which results in a **comedo** (a whitehead or blackhead). It appears as a blackhead if the plug reaches the surface. In addition, because *P. acnes* bacteria normally live on the skin, oil and excess cells of the plug allow these

comedo: KOM-ee-doh

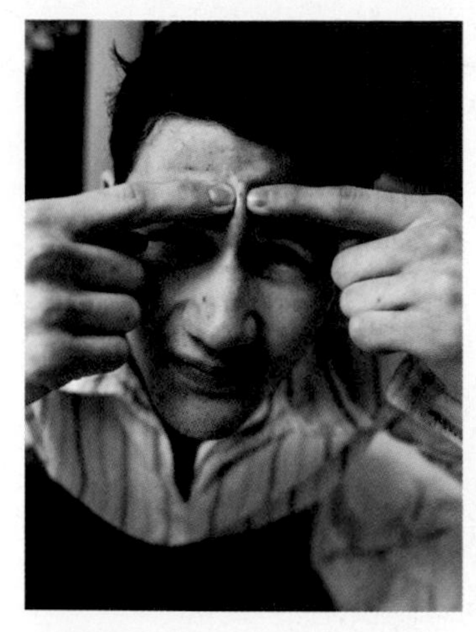

FIGURE 3.8 Male teen with acne.

bacteria to grow in the plugged follicles. This causes inflammation that may result in the breakdown in the hair follicle wall. If the inflammation continues, pus may form, resulting in a pimple. This condition is called **acne.** See **Figure 3.8**. Estrogen levels fluctuate for a woman both during her menstrual cycle and during pregnancy, making acne more prevalent for her on certain days of her cycle and during pregnancy.

clinical point

Methods of treating acne are based on the disorder's specific anatomy and physiology. Antibiotics kill the bacteria. Other drug therapies include the following treatments:

- Salicylic acid helps to reduce the shedding of cells that line the sebaceous gland and hair follicles.
- Benzoyl peroxide destroys *P. acnes* and may reduce sebum production.
- Vitamin A derivatives called *retinoids* unplug comedones (plural of *comedo*) to allow other topical medicines to enter the follicle.

Sweat glands. There are several types of sweat glands present in the dermis, and these glands differ in location, product, and function. See Table 3.1 and **Figure 3.7**.

For an example of how these glands work, consider the merocrine sweat glands—the most numerous sweat glands in your body—described in Table 3.1. The presence of lactic acid in merocrine sweat (secreted from the merocrine gland) causes the pH range of merocrine sweat to be 4 to 6. This forms an **acid mantle** on the skin that reduces the growth of bacteria.

TABLE 3.1 Sweat glands

Gland	Location	Product	Function
Merocrine sweat gland	This is the most numerous type of sweat gland. Merocrine glands deliver their sweat to the surface of the skin through a duct whose opening is called a *pore*. They are located all over the body and are highly concentrated in the palmar and plantar regions.	Watery sweat composed of 99% water, lactic acid, nitrogenous waste called *urea,* and some salt.	Helps cool the body through evaporation
Apocrine sweat gland	This type of sweat gland delivers its sweat through a duct leading to a hair follicle. Apocrine glands are associated with axillary hair, pubic hair, and the beard. They begin to produce their sweat at puberty.	Lipid-rich sweat that bacteria feed on. It is not the sweat itself but the waste from the bacteria feeding on the sweat that creates body odor.	Serves as scent to influence the behavior of others
Ceruminous gland	This is a modified sweat gland. Ceruminous glands are found only in the ear canal.	Cerumen (earwax).	Keeps the eardrum flexible, waterproofs the ear canal, kills bacteria, and protects the ear canal from foreign debris
Mammary gland	This is a modified sweat gland found in the breast. Mammary glands begin to develop at puberty and fully develop during pregnancy. They deliver their product to ducts that end at the nipple.	Breast milk, which is composed of water, carbohydrates, lipids, proteins, and minerals.	Nourishes an infant

Hair Follicles As you have already seen, hair follicles play an important role in the healthy functioning of the integumentary system. Let's take a closer look at precisely how the many hair follicles do their job. See **Figure 3.9**. The stratum basale dips down to form the hair follicles in the thin skin's dermis. Each hair follicle is positioned at an angle in the dermis with a dermal papilla at its base. The dermal papilla has a blood vessel, which feeds the keratinocytes and melanocytes contained in the hair follicle. The keratinocytes produce the hair, while the melanocytes produce the hair's pigment (color). Active keratinocytes form the **hair matrix** just above the dermal papilla. The **dermal papilla** is the hair's growth center, and it contains important nerves and blood vessels that provide amino acids for keratin production.

Associated with the hair follicle is a smooth muscle called an **arrector pili muscle.** This muscle attaches the hair follicle's base to the epidermis at an angle. See **Figure 3.9**. When this muscle contracts, it pulls on the hair follicle's base, making the hair stand perpendicular (in an upright position) to the skin's surface. For furry animals, this makes a thicker layer of fur insulation to help keep the skin warm. For humans, the visible effect is "goose bumps."

3.6 learning outcome

Describe the histology of a hair and hair follicle.

ceruminous: seh-ROO-min-us

arrector pili: a-REK-tor PYE-lye

connect.mcgraw-hill.com audio

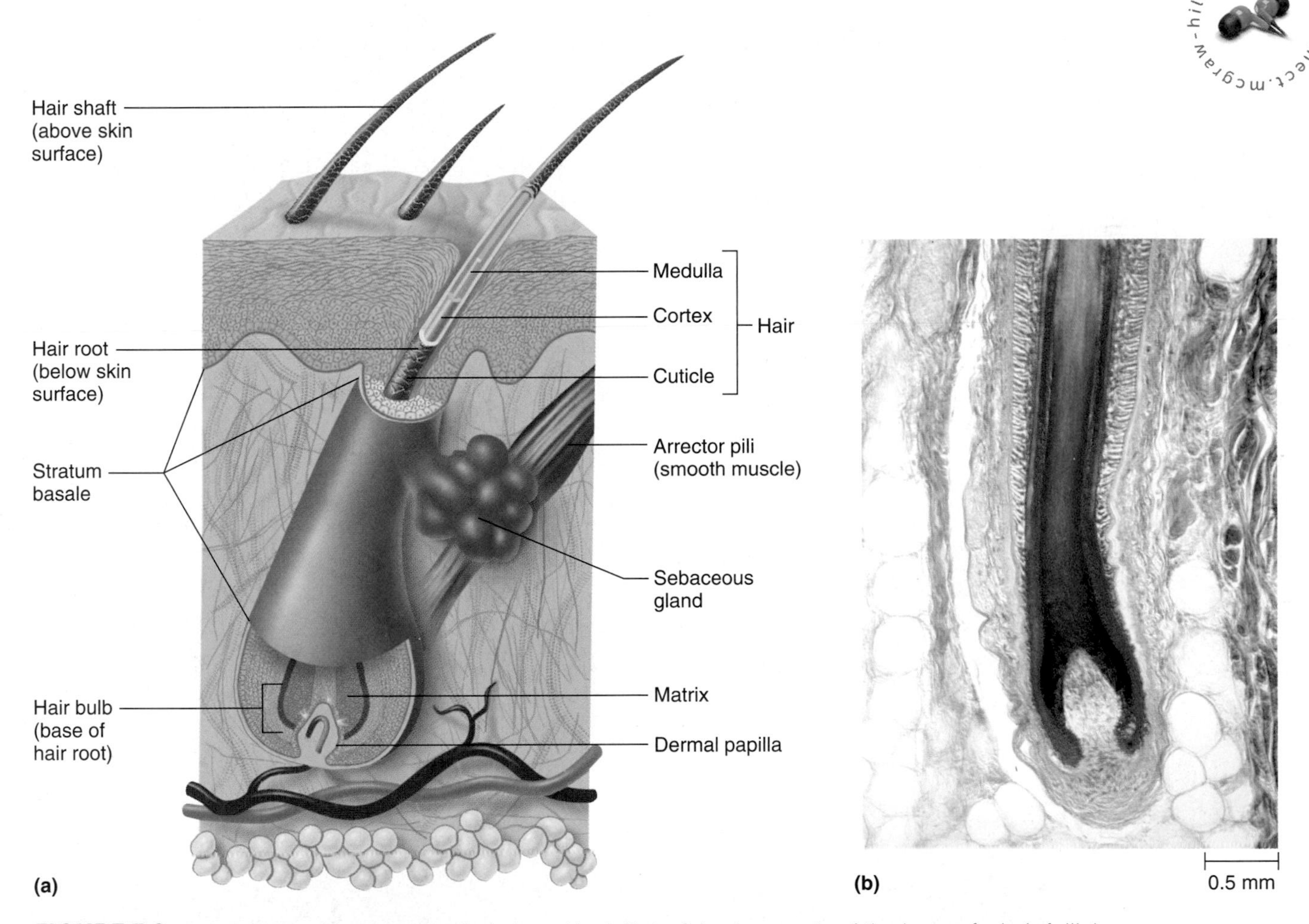

FIGURE 3.9 Hair follicle: (a) structure of a hair and its follicle, (b) micrograph of the base of a hair follicle.

Hair

Under normal circumstances, hair is present wherever there is thin skin on the body. There are three types of hair on every human body. They are the following:

- **Lanugo** hair, which is very fine and unpigmented (colorless), forms on a fetus during the last three months of its development. This hair is usually replaced by birth.

- **Vellus** hair, which is also unpigmented and very fine, replaces lanugo hair around the time of birth. An example of vellus hair is the body hair on most women and children.
- **Terminal** hair, which is thick, coarse, and heavily pigmented, forms the eyebrows, eyelashes, and hair on the scalp. At puberty, terminal hair forms in the axillary and pubic regions of both sexes. It also forms on the face and possibly on the trunk and limbs of men.

All types of human hair can be divided into three sections (shown in **Figure 3.9**):

- The **bulb** is a thickening of the hair at the end of the hair follicle.
- The **root** extends from the bulb to the skin's surface.
- The **shaft** is the section of the hair extending out from the skin's surface.

A cross section of a hair shows three layers: an inner medulla composed of soft keratin, a middle cortex composed of hard keratin, and an outer cuticle that appears as interlocking scaly plates of dead keratinocytes. The direction of the cuticle plates resists the follicle's cells, so the hair cannot be easily pulled from the follicle. The hair is easy to manage if the plates of the cuticle are lying smooth on the hair and not sticking out along the length of the shaft. Conditioners can be used to encourage the cuticle to lay flat.

The hair texture depends on the shaft's cross-sectional shape. Straight hair has a round shaft, wavy hair has an oval shaft, and curly hair has a flatter shaft. Rates of growth within the follicle matter too. Straight hair grows evenly within the follicle. Wavy and curly hair may grow alternately faster on one side of the follicle than on the other.

Hair color depends on the amount and type of melanin produced by the hair follicle's melanocytes. Gray to white hair results from a lack of melanin in the hair's cortex and from the possible presence of air in the medulla. See **Figure 3.10**.

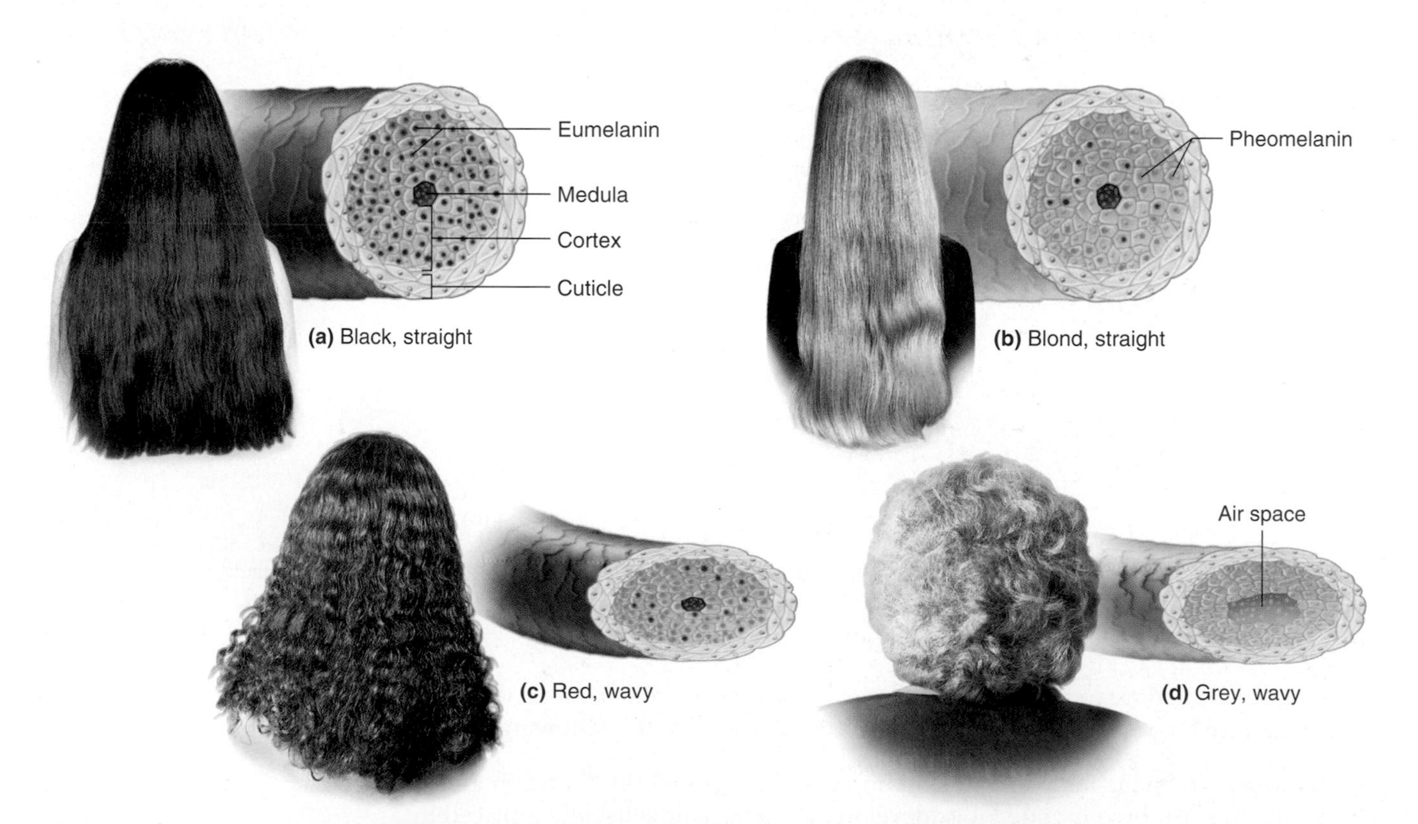

FIGURE 3.10 Hair color and texture. Hair colors are shown with the corresponding type of melanin; textures, with the corresponding shape of hair-shaft cross section. All three layers of the hair shaft are shown.

As with most living things, hair has a life cycle. There is a growing stage, a resting stage, and then a dying stage, when the hair finally falls out. However, not all of the hairs on the head cycle on the same schedule. Each hair grows about a half inch per month, and its growth stage lasts for approximately three years. Each hair then goes into a one- to two-year resting stage, and then it falls out. If all the hairs on your head were on the same schedule, you would go bald every four to five years. Instead, roughly 90 percent of the hairs on your head are somewhere in the growing stage at any given time. In fact, it is perfectly normal to lose about 100 hairs from your scalp every day. You probably notice this routine hair loss all the time, as your fallen hair accumulates in the tub or sink drain, on the bathroom floor, and in your brush when you wash, style, and blow-dry your hair. Eyelashes go through a similar cycle. They grow for about 30 days, rest for 105 days, and then fall out.

3.7 learning outcome

Explain how a hair grows and is lost.

spot check 3 Describe what eyelashes would look like if they had the same growth cycle as scalp hair.

Nails

Nails form on the distal end (away from the attachment point) of the fingers and toes. As you can see in **Figure 3.11**, the structure of the nail includes the skin-covered **nail root** and the visible **nail plate.** The free edge of the nail and the **nail body** make up the nail plate. The nail body lies over the **nail bed.** The nail bed appears pink because of the dermis's numerous blood vessels. Laterally, the skin rises to form a **nail fold** over the nail's lateral edge. Here the nail fits into a **nail groove.** An **eponychium,** or **cuticle,** at the distal edge of the nail body, is composed of stratum corneum cells extending onto the nail bed. The **nail matrix** (the nail's growth center) at the root of the nail is composed of active keratinocytes in the stratum basale. The nail is translucent, although a **lunula** or **lunule** (white crescent) may be visible under the nail especially on the thumbs. This is where the nail matrix is thick enough to hide the blood vessels of the dermis deep to it. Again, see **Figure 3.11**.

Composed of stratum corneum cells with hard keratin, nails grow distally throughout life and, unlike hair, do not have a resting stage. Fingernails grow faster than toenails. Nails protect the ends of the fingers and toes, aid in grasping small objects, and are used for scratching.

3.8 learning outcome

Describe the structure and function of a nail.

eponychium: ep-uh-NIK-ee-um

lunule: LOON-yule

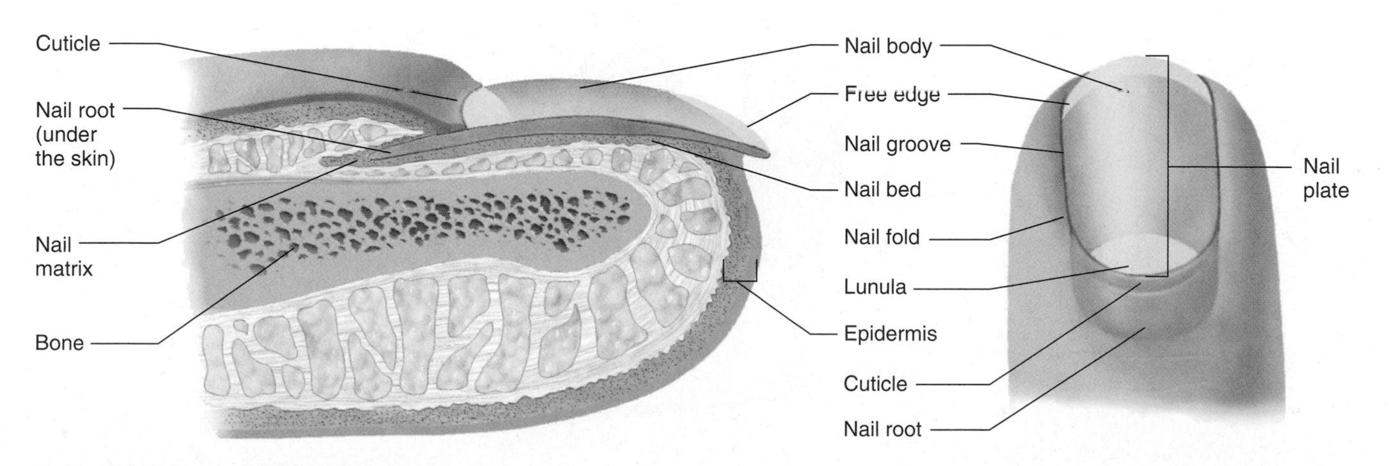

FIGURE 3.11 Anatomy of a nail.

clinical point

It is important to note that nails can be a good indicator of health. Respiratory diseases that result in chronic low blood oxygen levels can cause clubbed fingertips and cyanosis. With clubbed fingertips, the fingertips widen and the nails become more convex. In cyanosis (a bluish skin condition), the nail beds may appear blue if there is low blood oxygen circulating to the nail bed. White spots, pitting, ridges, and scaling of the nails may indicate **psoriasis.** Psoriasis is a condition characterized by an overgrowth and improper keratinization of the epidermis and nails. Observation of the nails is important in a physical examination.

3.9 learning **outcome**

Explain how the layers and structures of the skin work together to carry out the functions of the system.

Physiology of the Integumentary System

Now find a pen and draw four dots on the posterior surface of your hand to represent the corners of a square inch. This 1 square inch of skin you have on your own hand is typically composed of:

- 20 blood vessels
- 65 hairs and arrector pili muscles
- 78 nerve endings for heat
- 13 nerve endings for cold
- 160 to 165 sensors for pressure
- 100 sebaceous glands
- 650 sweat glands
- 19,500,000 cells[1]

All of these structures work together to accomplish the functions outlined in the following paragraphs. To see how they do this, let's look at Nick and Kate as they go for a run on a cool, sunny morning. See **Figure 3.12**.

FIGURE 3.12 Nick and Kate on a morning run.

Functions of Skin

Your skin has several necessary and important functions. These functions include:

- Protection
- Vitamin D production
- Temperature regulation
- Water retention
- Sensation
- Nonverbal communication

Protection Skin is the body's first line of defense against the foreign pathogens Nick and Kate will encounter on their run. Anything they come in contact with may be a source of potential pathogens. Luckily, the dead, cornified stratum corneum of their skin makes entry for bacteria difficult. The skin's surface is also dry, with an acidic pH, making it an unfriendly place for bacteria to grow. Dendritic cells of the epidermis stand guard if any pathogens try to make it past the stratum corneum.

spot check ❹ How might the histology of the dermis hamper the movement of bacteria as they go deeper into the skin?

Remember that the skin protects the body in other ways too. The sun's ultraviolet (UV) rays can damage the skin cells' DNA. Some stratum basale cells are actively going through mitosis every day to make new epidermis. Any damage to the DNA of those cells would be carried to future generations of cells. The stratum basale's melanocytes react to UV exposure by producing more melanin. This melanin is then carried toward the skin's surface by the keratinocytes. Melanin protects underlying cells from the potential damage of UV light. The skin exposed to the sun will begin to darken.

Vitamin D Production Despite the potential dangers of overexposure to the sun's UV rays, minimal exposure to these rays does have a benefit. UV light encourages skin to produce vitamin D, as it activates a precursor molecule in the skin. The liver and kidney then modify this molecule to become active vitamin D. If you live in a four-season climate, your exposure to UV light may be greatly reduced during the winter months. Not only is the sun less intense, but you are more likely to cover most of your skin with clothing to stay warm, therefore reducing your UV exposure. It is for this reason that vitamin D is added to most dairy products.

Vitamin D is important for the absorption of calcium from the diet into the bloodstream. Calcium is needed for bone development and maintenance. You will revisit this topic in the skeletal system chapter.

Temperature Regulation Let's get back to Nick and Kate's workout. They are outside running on a cool, sunny morning. If it is very cool out, their skin may get paler as the dermal blood vessels constrict to preserve heat for the body's core. As they continue to run, their muscles generate heat and they will notice that their skin is becoming redder. The dermal blood vessels are now dilating and increasing blood flow to the skin, which gives it more color. The increased blood flow radiates off some of the excess heat to keep the body's core from getting too warm. If this effort is not sufficient to remove the excess heat, the brain will trigger sweating. In sweating, fluid (not blood cells) leaks from the dilated blood vessels. Sweat glands take in this fluid, process it, and deliver it to the skin's surface as sweat to start evaporative cooling. This further reduces body temperature.

The scenario just described is an example of negative-feedback temperature regulation. The body becomes too warm, so blood vessels dilate and sweat glands produce sweat to cool the body down to a normal temperature. Another example of negative feedback concerning core body temperature occurs in freezing conditions. In this case, you may see the skin first redden as blood vessels dilate in the dermis to warm the skin and prevent it from freezing. If, however, this process draws too much heat from the body's core (heart, lungs, and other organs), the dermal blood vessels will constrict to preserve the heat for the core and thus preserve life. Frostbite of the skin may then occur.

Sensation During the run, the nerve endings in the stratum basale and dermis of the skin constantly send the brain messages. Nick and Kate will sense the cool temperature from the cold receptors in their dermis. They will feel the sun on their faces from the heat receptors and feel the presence of their clothing and shoes from the pressure sensors. Nick will even detect the ant crawling on his leg as it disturbs his leg hair, activating the sensors wrapped around his hair follicles. If Kate stumbles and falls, pain receptors will alert her to any skin injury.

Nonverbal Communication Aside from the skin's fluctuating temperature or any minor injury sustained from a fall while out running, it is crucial to note that the condition of their skin and their hair sends nonverbal messages to anyone seeing them. Nick may blush if he is embarrassed by something he has done, or Kate's face may become pale if she is frightened. Even the color, texture, silkiness, and other qualities of the skin and hair can all be indicative of overall health. Billion-dollar industries in hair care, skin care, and cosmetics have been developed to help convey positive messages to one's self and others to ensure social acceptance. So it is quite likely that Nick and Kate will want to freshen up their appearance after their run.

Water Retention Taking a bath after the run brings us to the last of the skin's functions: water retention. In this case, let's assume Kate is going to have a bath. As you know, her body is made up of trillions of cells. You learned in Chapter 2 that a cell will swell through osmosis if placed in a hypotonic solution. Yet nothing happens to Kate's cells when she immerses her body in a tub of bathwater (a hypotonic solution). As she relaxes in the tub, her body does not swell and her cells do not gain weight. The skin's epidermis waterproofs the body by keeping water from the environment out and body fluids in. (One of the major dangers in severe burns is fluid loss when the epidermis is completely destroyed.) If Kate stays in the bathwater long enough, she may notice that the skin on her fingers and toes becomes wrinkly, like a raisin. Why? Her fingers and toes are the areas of the skin most likely to have abrasions to the epidermis through everyday activities. The abrasions allow some water to pass through the epidermis, causing swelling in the underlying tissues. The wrinkling she sees while bathing is the swelling of these underlying cells going through osmosis.

Injuries to the Skin

3.10 learning outcome

Explain how the skin responds to injury and repairs itself.

Skin can be damaged by all sorts of injuries, including cuts, punctures, and burns. These injuries can happen in everyday life and are not necessarily from a disease process. Let's look at how the physiology of the skin works so the skin can repair itself.

Regeneration versus Fibrosis

Wounds to the skin can heal by **regeneration** or **fibrosis.** In regeneration the wound is healed with the same tissue that was damaged, and normal function is returned. In fibrosis, the wound is healed with scar tissue, and normal function is not returned. Consider a cut to the skin that goes into the dermis, as shown in **Figure 3.13**. What happens to heal the wound?

1. Cutting into the dermis means severing blood vessels, so the wound bleeds and then clots.

2. A scab forms and pulls the edges of the wound closer together as it dries. This is called **wound contracture.**

3. A race begins between the dermis's fibroblasts and the stratum basale's keratinocytes. The fibroblasts' collagen fibers produce **granulation tissue** to fill in the wound's clot. In the meantime, stratum basale cells are actively dividing. They normally divide and push the older cells toward the surface to form epidermis. But if they are not in contact with other stratum basale cells, the keratinocytes will divide and push cells laterally toward other stratum basale cells until the edges of stratum basale come in contact with each other again. This is called **contact inhibition.** Once these edges are joined, the keratinocytes resume the cornification process described earlier to form new epidermis. The type of healing is determined by who wins the race.

4. It all depends on whether the keratinocytes reach contact inhibition before the fibroblasts can fill the wound with so much collagen that keratinocytes are prevented from meeting. If the stratum basale wins, the wound heals by regeneration and normal function results from the new epidermis. If the fibroblasts win, the wound heals by fibrosis. The scar tissue cannot accomplish the same functions as normal epidermis. A key factor in who wins the race is how far apart the edges are. See **Figure 3.13**.

spot check ❺ How would the anatomy and physiology of scar tissue formed during fibrosis differ from the anatomy and physiology of normal epidermis formed through regeneration?

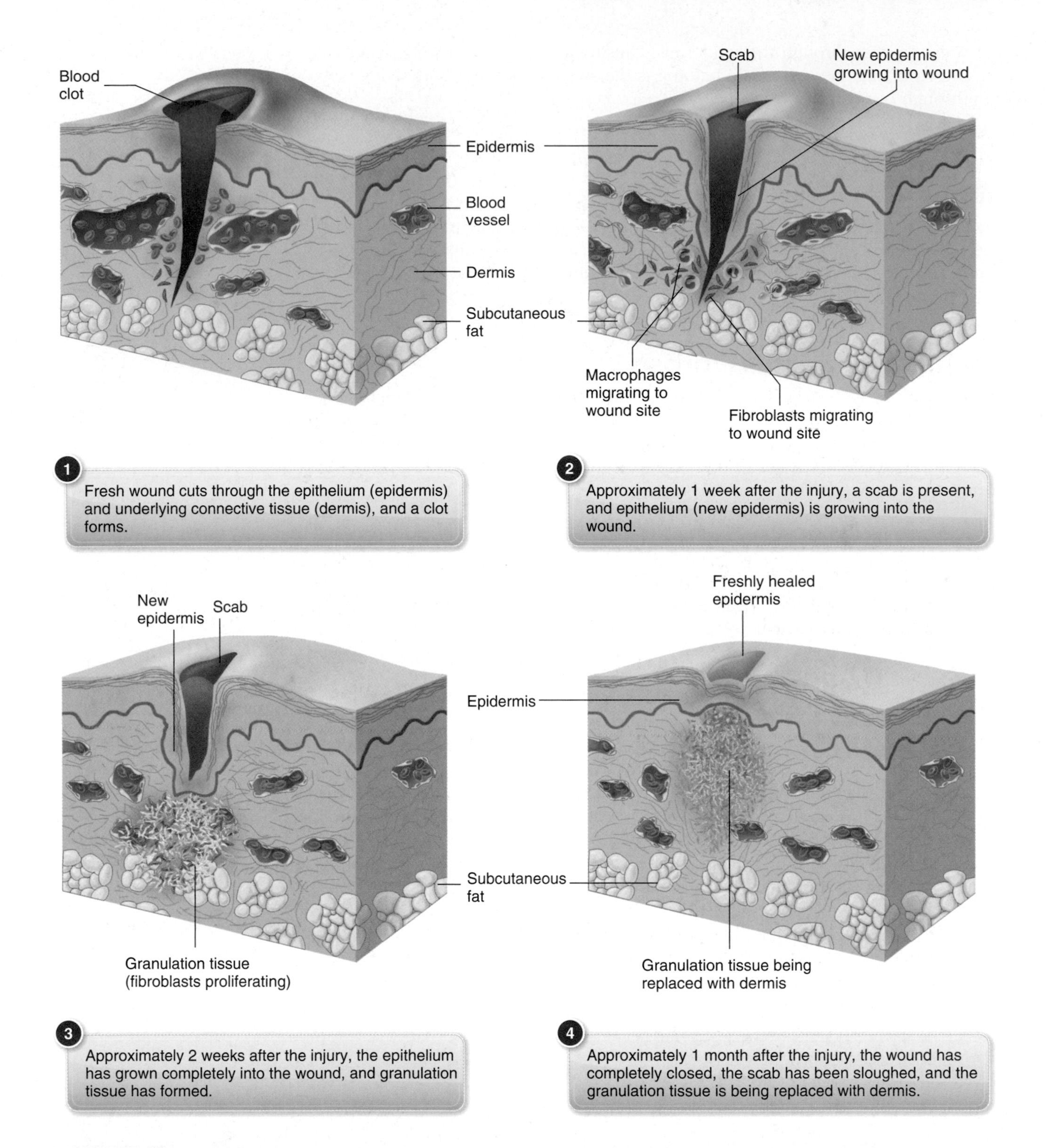

FIGURE 3.13 Wound healing showing regeneration. Stratum basale cells reached contact inhibition before fibroblasts filled the area with a fibrosis (scar).

Inflammation

Sometimes injury to the skin allows bacteria and other pathogens to enter underlying tissues. The body's response to the invasion is inflammation. Redness, heat, swelling, and pain are signs of inflammation. Think of how the body responds when a splinter cuts through the skin. See **Figure 3.14**. The splinter damages tissue of the dermis and pushes bacteria into the wound. The damaged tissues produce chemicals called **mediators of inflammation,** which diffuse away from the damaged area and cause any blood vessels they meet to dilate. This brings more blood flow to the area. The increased blood flow accounts for the redness and heat. Blood vessels become more permeable (leaky) when they dilate. Fluid from the blood leaks out into the surrounding damaged tissue, causing it to swell. This extra fluid increases pressure on the nerve endings, creating the pain. You will explore inflammation more fully in the lymphatic system chapter.

3.11 learning outcome

Describe the symptoms of inflammation and explain their cause in terms of the structure and function of the skin.

warning

It is important to understand that fluid, *not* red blood cells, can leave dilated blood vessels. Redness is not caused by red blood cells leaking out into the tissues. It is simply the result of small vessels becoming bigger in diameter, enabling more blood to flow to the area through the vessels.

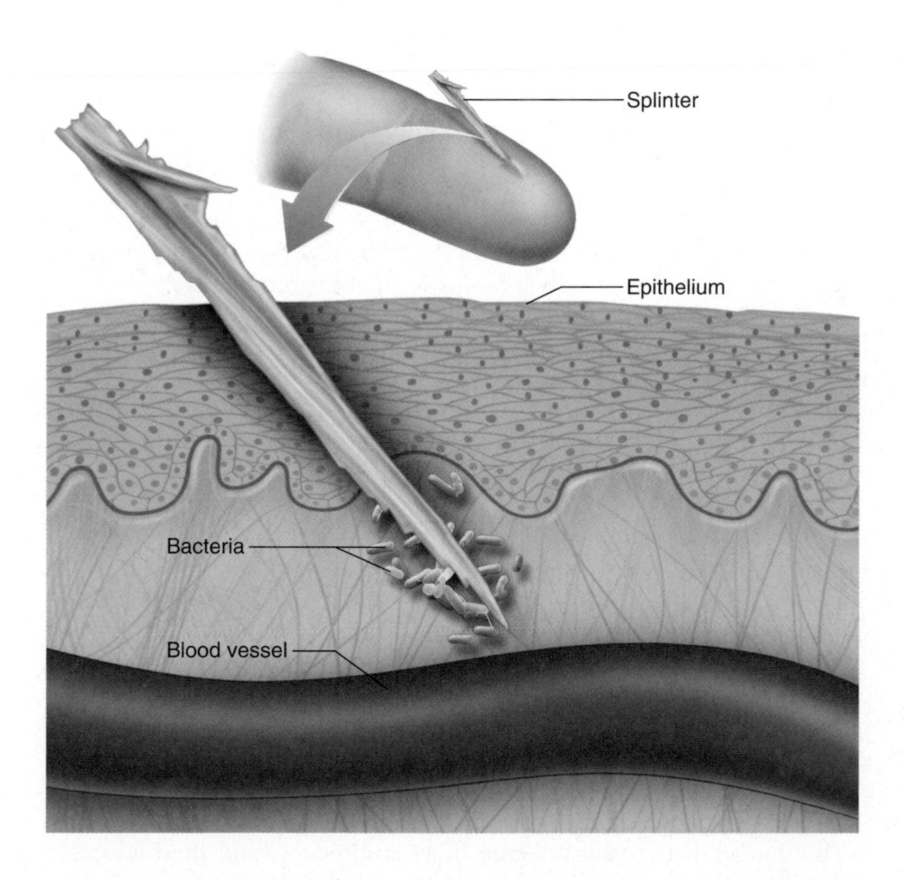

FIGURE 3.14 Splinter in the skin damages tissues and promotes an inflammatory response.

Pain and swelling are not desirable, so why does the body have an inflammatory response? Certainly, the pain will cause you to notice and pay attention to the inflammation. The increased blood brings more nutrients to the area and carries more wastes away. This speeds the healing process. Also, the increased heat discourages bacteria growth. So, in essence, inflammation is the body's way of fixing a problem.

clinical point

Some medications are anti-inflammatory agents, which help to reduce pain and swelling. Common mediators of inflammation produced by damaged tissues are histamines and prostaglandins. Antihistamines block the effects of histamine, and aspirin blocks the production of prostaglandins.

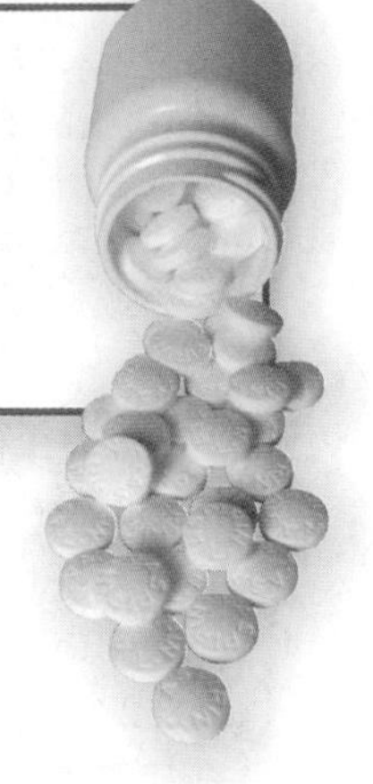

3.12 learning outcome

Compare and contrast three degrees of burns in terms of symptoms, layers of the skin affected, and method used by the body for healing.

Burns

These are a common skin injury. Just think how easy it is to burn a finger while lighting a candle or how quickly you can lose track of time while out in the summer sun without sunscreen. Burns can be categorized in three degrees by the skin layers involved. These categories are described below and shown in **Figure 3.15**.

- **First-degree burns** are the most common burns. Sunburns are often first-degree burns. They involve only the skin's epidermis. Symptoms are redness, pain, and swelling. New epidermis will be made by the stratum basale.
- **Second-degree burns,** sometimes called *partial-thickness burns,* involve the epidermis and dermis. Symptoms are redness, pain, swelling, and blisters. New epidermis will be made from the stratum basale cells at the burn's edges and the stratum basale of the hair follicles growing toward each other until contact inhibition is achieved. Normal epidermis production then proceeds.
- **Third-degree burns,** sometimes called *full-thickness burns,* are the most serious burns. They involve the epidermis, dermis, and hypodermis. The symptoms are charring and no pain. There is no pain at the site of a third-degree burn because all of the nerve endings have been destroyed in the dermis. However, there is pain in the second- and first-degree burns that typically surround the third-degree burn. The size of this type of burn is crucial to healing. Because the dermis has been completely destroyed, the third-degree burn site has no hair follicles. Therefore, there is no additional stratum basale of the hair follicles available to help heal the burn. The only stratum basale left is at the burn's edges. Skin grafting may be necessary if the stratum basale's edges are too far apart. The danger with a third-degree burn is infection and fluid loss. See **Figure 3.15**.

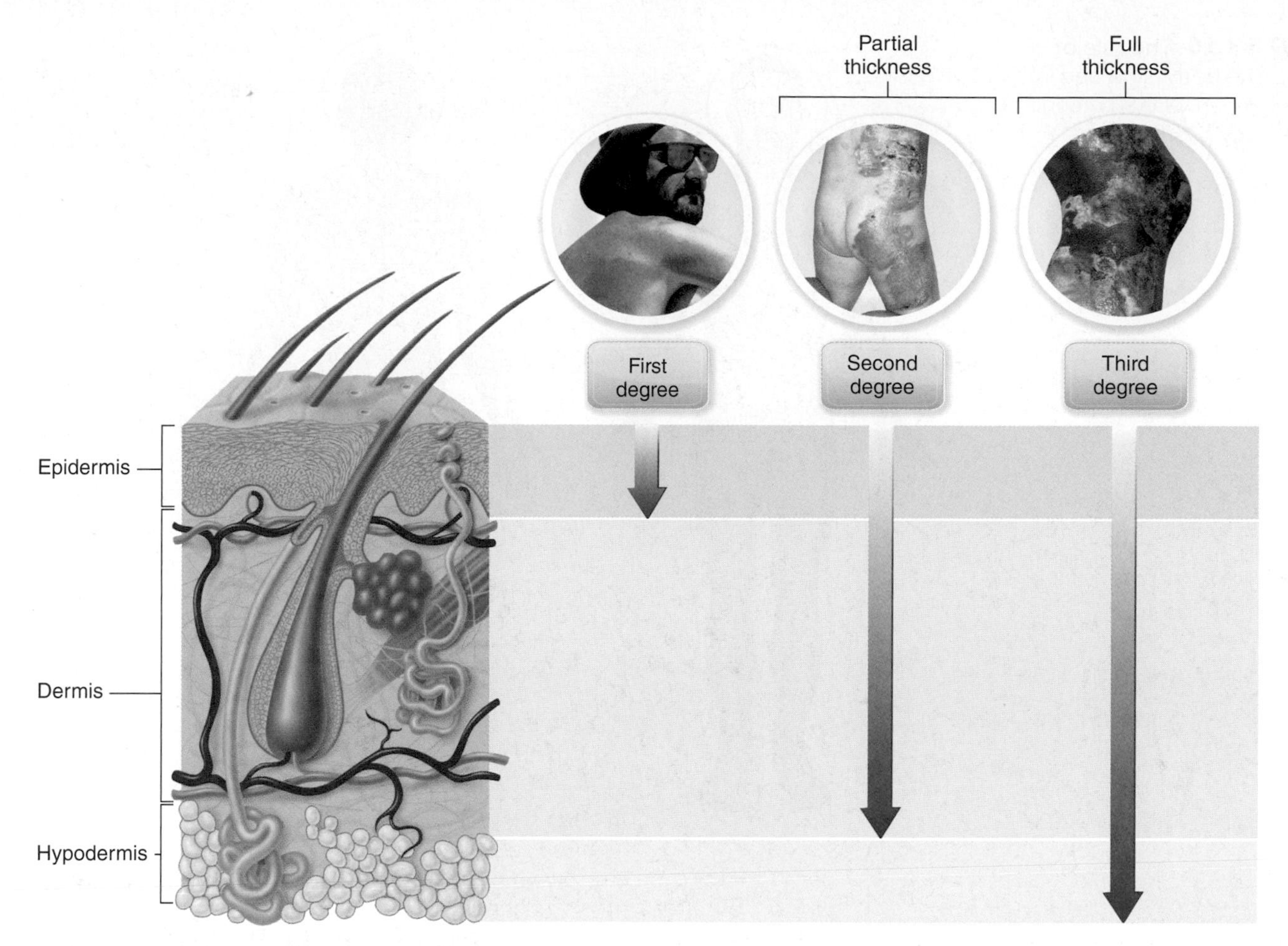

FIGURE 3.15 Three degrees of burns.

3.13 learning **outcome**

Describe the extent of a burn using the rule of nines.

clinical point

The treatment options and chance for recovery may be determined by the severity and extent of a burn. You have just read about the severity of burns categorized by degree. The extent of a burn can be determined by applying the "rule of nines": The body is divided into 11 areas, each of which represents approximately 9 percent of the body's surface area. See Figure 3.16.

spot check 6 Hailey went snow skiing in February on a bright, sunny day. It was cold and she was adequately dressed. She got a sunburn on her face, which was exposed. She did not have any blisters. Classify the severity and extent of her burn.

FIGURE 3.16 **The rule of nines.** The body is divided into 11 areas, each with 9 percent of the body's surface area.

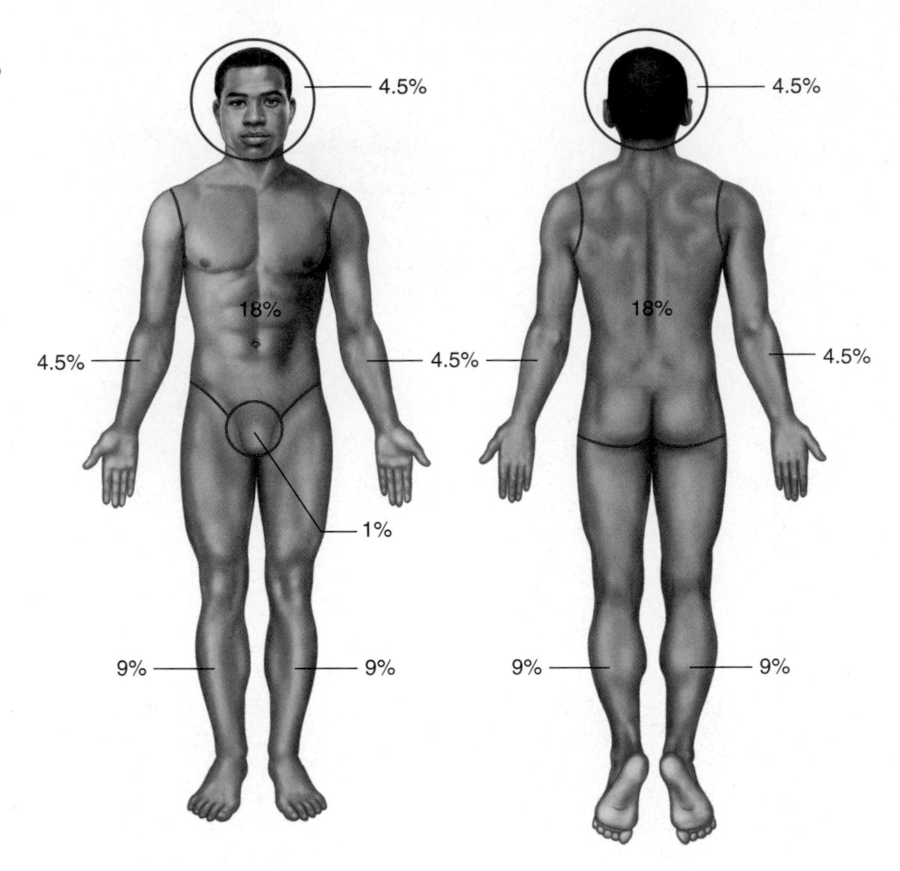

3.14 learning **outcome**

Summarize the effects of aging on the integumentary system.

Effects of Aging on the Integumentary System

Skin changes with age. The epidermis becomes dry due to the loss of some of the sebaceous and sweat glands in the dermis. Also, reductions in estrogen and testosterone slow the production of sebum from the existing sebaceous glands. Some melanocytes in the epidermis are also lost, while others begin to overproduce melanin. The result is uneven tanning and age spots. Age spots are flat, brown spots produced by melanocytes, sometimes called "liver spots." They have nothing to do with the liver, but they are colored like the liver. They typically occur on sun-exposed skin surfaces.

Over time, the dermis thins and the number of collagen and elastic fibers is reduced. This, together with gravity, causes the skin to sag and wrinkle. Blood vessels become more fragile and respond less quickly to temperature regulation. Combined with the decrease in dermis sweat glands and the reduction in hypodermis adipose tissue, all of this makes temperature regulation much more difficult.

The thinning of the hypodermis also reduces the cushioning layer between the skin and underlying tissues. Bumps resulting in bruises become more frequent and apparent. In the elderly, the skin over bony prominences may become necrotic (the tissue dies) if they remain in the same position for too long. This can cause pressure ulcers or bedsores.

Nails are also affected by aging. The nail plate and nail matrix thin with age. The thinning of the nail matrix is visible with the loss of the lunule. Nail fractures become more common as the nail body thins. Toenails may thicken and yellow due to fungal infections, which the body is less prepared to fight.

Hair, too, is affected by aging. Terminal hairs on the ears, nose, and chin become more coarse. The loss of hair follicle melanocytes causes hair to go gray. Because hair does not replace itself as rapidly in later years, the hair on the scalp becomes thinner. It also becomes drier, as the hair follicle has a reduced number of sebaceous glands.

spot check 7 What could be done to minimize specific effects of aging on the skin, hair, or nails?

Integumentary System Disorders

Skin disorders may or may not have anything to do with aging. There are many types of skin disorders, and in the following paragraphs you will learn about two: skin cancers and skin infections.

Skin Cancer

3.15 learning outcome

Describe three forms of skin cancer in terms of the body area most affected, appearance, and ability to metastasize.

Skin cancer is the most common type of cancer. Although cancer has many causes, skin cancer is usually associated with sun exposure and it is more common in older, light-skinned people. The amount of exposure is cumulative with age. Light-skinned people have less melanin present to protect them from the sun's UV rays. The list below explains the three forms of skin cancer and will help you see how they differ.

- **Basal cell carcinoma** is the most common form of skin cancer. It starts in stratum basale cells and first appears as a small, shiny bump on the face, hands, ears, and neck. As it progresses, it develops a central depression and a pearly edge. It rarely metastasizes. See **Figure 3.17**.
- **Squamous cell carcinoma** starts from keratinocytes in the stratum spinosum. First appearing as a red, scaly patch that develops a central crust, it commonly forms on the face, hands, ears, and neck. If detected early, it can be surgically removed. Unlike basal cell carcinoma, this form of skin cancer can metastasize. See **Figure 3.17**.
- **Malignant melanoma** is the rarest but deadliest form of skin cancer. It metastasizes easily and can occur anywhere on the body. It starts with melanocytes, usually in a preexisting mole. Malignant melanoma may be seen as a dark spot under a nail or as a mole that is asymmetrical and has uneven color and scalloped borders. See **Figure 3.17**.

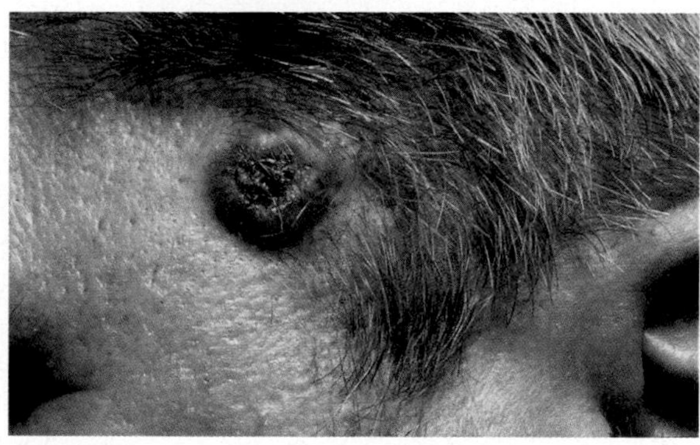

(a)

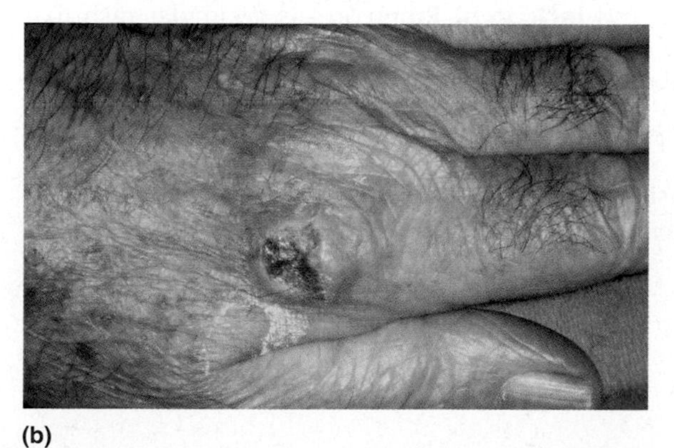

(b)

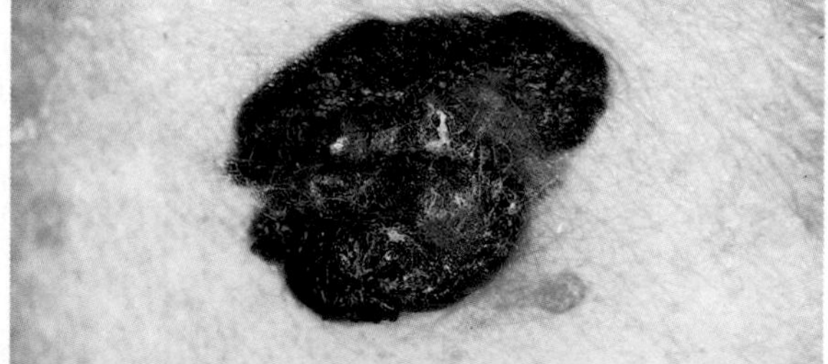

(c)

FIGURE 3.17 Skin cancer: (a) basal cell carcinoma, (b) squamous cell carcinoma, (c) malignant melanoma.

clinical point

A change in the number, size, or shape of moles or other highly pigmented areas of the skin can be evaluated through the ABCDs of skin cancer:

- *Asymmetry.* One half does not match the other half.
- *Border* irregularity. The edges are ragged or blurred.
- *Color.* More than one color is present, and the color is uneven.
- *Diameter.* The growth of a mole is in excess of 6 millimeters.

3.16 learning **outcome**

Describe an example of a bacterial, a viral, and a fungal infection of the skin.

Skin Infections

Unlike skin cancers, skin infections stem from bacterial, viral, and fungal infections. In the paragraphs below, we discuss a representative of each type of skin infection.

Cellulitis This infection of the skin's dermis or hypodermis is frequently caused by *Streptococcus* or *Staphylococcus* bacteria. Cellulitis commonly occurs on the face and lower legs; it is characterized by redness and swelling of an area of the skin that increases in size rapidly. As you can see in **Figure 3.18**, the infected area has a tight, glossy appearance, and it is tender or painful. Cellulitis is also accompanied by other signs of infection, such as fever, chills, and muscle aches. Cellulitis can be serious and even deadly. It is treated with antibiotics.

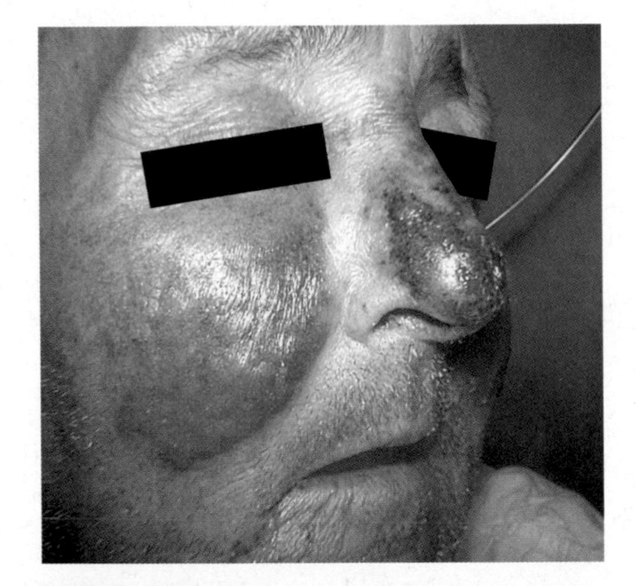

FIGURE 3.18 Cellulitis caused by streptococcal bacteria.

Warts These growths are caused by an infection of the skin produced by the human papilloma virus (HPV). Types of warts include common warts (found most often on the fingers); plantar warts (found on the soles of the feet); genital warts (which are sexually transmitted); and flat warts (found on frequently shaved surfaces).

Warts that appear in children often go away as the immune system develops to fight them, but they may also be removed by freezing.

Tinea Infections These infections are caused by a fungus. Contact with an infected person, damp surfaces like shower floors or pool decks, or even pets can transmit the fungus. Examples of tinea fungal infections include **ringworm, athlete's foot,** and **jock itch.** Ringworm is a circular rash that clears from the center, giving it a ringlike appearance. See **Figure 3.19**. It is a fungal infection and has nothing to do with worms. Athlete's foot causes burning, itching, and cracking of the skin between the toes. Jock itch causes an itchy, burning rash in the groin region. Tinea infections are treated with fungicides.

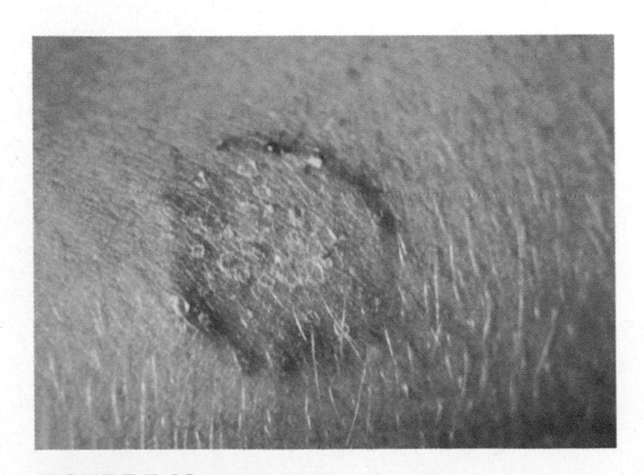

FIGURE 3.19 Ringworm.

putting the pieces **together**

The Integumentary System

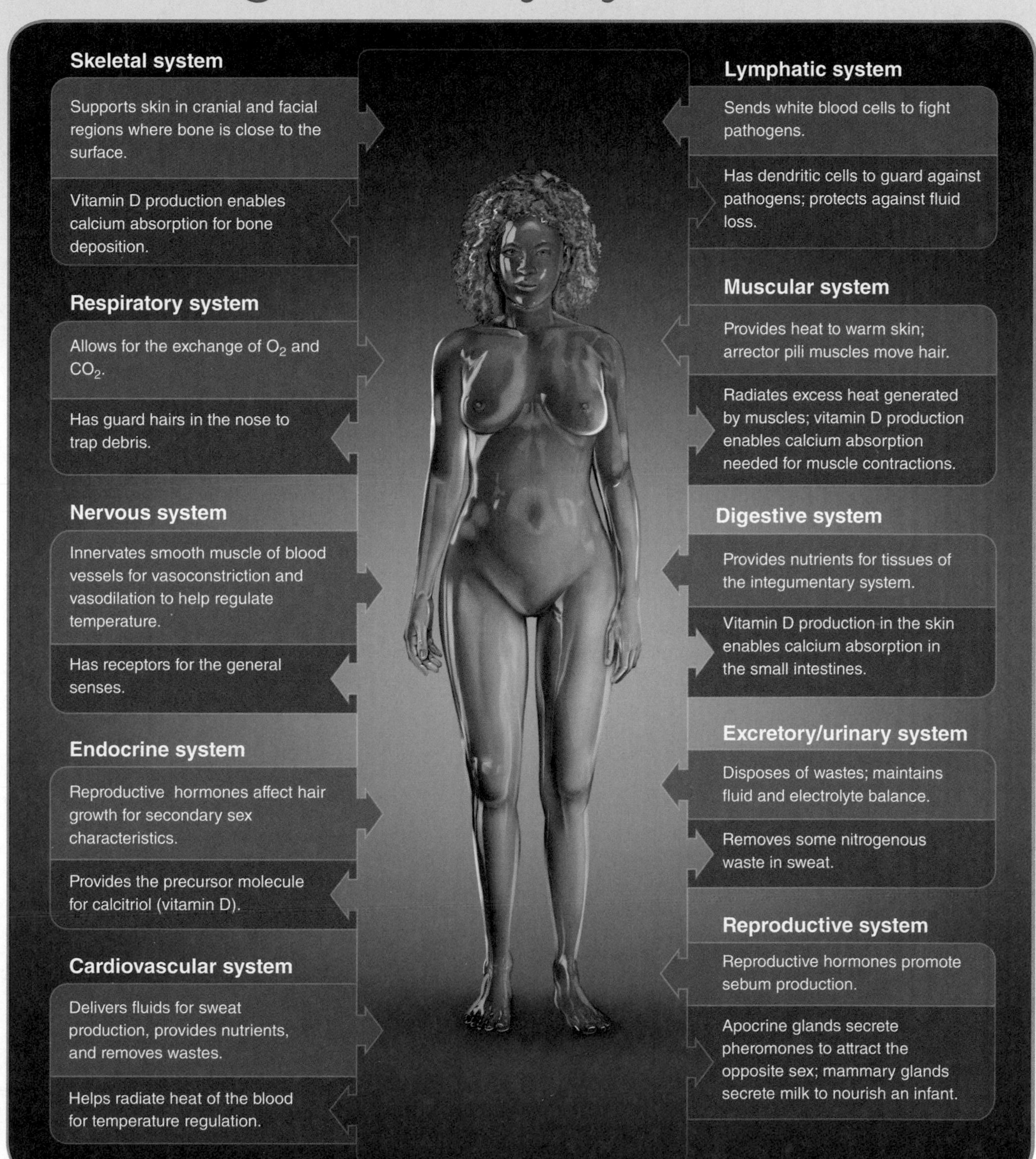

FIGURE 3.20 **Putting the Pieces Together—The Integumentary System:** connections between the integumentary system and the body's other systems.

summary

Overview

- The integumentary system is composed of the skin, hair, nails, and cutaneous glands.

Anatomy of the Skin, Hair, and Nails

- The skin is the largest organ of the body.
- It is composed of two layers: the epidermis and the dermis.
- The epidermis is stratified squamous epithelial tissue.
- The dermis is loose/areolar connective tissue over dense irregular connective tissue.
- The hypodermis is adipose connective tissue; it is not part of the skin, but it attaches the skin to the body.

Epidermis

- The stratum basale is composed of keratinocytes and melanocytes. This stratum actively divides to make new epidermis.
- The stratum corneum is composed of dead, keratin-filled cells.
- The stratum lucidum is found only in thick skin.
- Other cells of the epidermis include tactile cells and dendritic cells.

Dermis

- The dermis contains papillae, fibers, nerve endings, cutaneous glands, and hair follicles.
- There are two main types of cutaneous glands: sebaceous glands and sweat glands.
- Sebaceous glands produce sebum and are associated with a hair follicle.
- Sweat glands include merocrine glands, apocrine glands, ceruminous glands, and mammary glands.
- Hair follicles are formed by the stratum basale, which is the center for hair growth.

Hair

- The three types of hair are lanugo, vellus, and terminal.
- A hair can be divided into three sections: the bulb, the root, and the shaft.
- There are three layers to a hair: the inner medulla, the cortex, and the cuticle.
- Hair goes through a growing stage and a resting stage, and then it falls out.

Nails

- Nails protect the ends of the fingers and toes, aid in grasping objects, and are used for scratching.
- They are composed of hard keratin.
- They are formed by stratum basale cells in the nail matrix.

Physiology of the Integumentary System

Functions of the Skin

- The functions of the integumentary system include protection from pathogens and UV light, vitamin D production, temperature regulation, water retention, sensation, and nonverbal communication.

Injuries to the Skin

Regeneration versus Fibrosis

- Skin can heal by regeneration or fibrosis.
- In regeneration, normal function returns.
- In fibrosis, normal functioning tissue is replaced by scar tissue.

Inflammation

- Inflammation is the body's response to a foreign invader.
- Signs of inflammation are redness, heat, swelling, and pain.

Burns

- Burns can be classified by degree.
- First-degree burns involve only the epidermis. Symptoms are redness, pain, and swelling.
- Second-degree burns involve the epidermis and dermis. Symptoms include redness, pain, swelling, and blisters.
- Third-degree burns involve the epidermis, dermis, and hypodermis. Symptoms include charring and no pain at the burn site.
- The rule of nines is used to determine the extent of a burn.

Effects of Aging on the Integumentary System

- All parts of the integumentary system are affected by aging.
- The epidermis becomes drier, with uneven tanning and age spots.
- The dermis thins and produces less collagen and elastic fibers; this, along with gravity, causes sagging and wrinkling of the skin.
- The hypodermis thins, providing less cushioning and less insulation.
- Nails become thinner and more susceptible to fracture.
- Hair thins and turns gray.

Integumentary System Disorders

Skin Cancer

- Skin cancer is the most common cancer, and it is associated with sun exposure.
- Basal cell carcinoma is the most common skin cancer, and it tends not to metastasize.
- Squamous cell carcinoma can metastasize.
- Malignant melanoma is the rarest form of skin cancer, but it is also the most deadly because it metastasizes easily.

Skin Infections

- Skin can be infected by a type of bacteria, a virus, or a fungus.
- Cellulitis is an example of a bacterial skin infection.
- Warts are an example of a viral skin infection.
- Tinea infections of the skin are caused by a fungus.

key words for review

The following terms are defined in the glossary.

acne
contact inhibition
cornification
cutaneous
epidermis
exocrine glands
fibrosis
first-degree burn
keratin
mediators of inflammation
melanocytes
papillae
pathogens
regeneration
sebum
stratum basale
subcutaneous layer
sweat glands
thin skin
wound contracture

chapter review questions

Word Building: *Use the word roots listed at the beginning of the chapter and the prefixes and suffixes inside the back cover to build words with the following meanings:*

1. Condition of red skin: ______________________________
2. Skin inflammation: ______________________________
3. Condition of too much keratin: ______________________________
4. Tumor of the dermis: ______________________________
5. Pain in the nails: ______________________________

Multiple Select: *Select the correct choices for each statement. The choices may be all correct, all incorrect, or a combination of correct and incorrect.*

1. Which of the following statements is (are) true during inflammation?
   a. The symptoms are pain, redness, heat, and wrinkling.
   b. Redness is caused by red blood cells leaking out of the vessels.
   c. Blood vessels dilate and become more permeable.
   d. Mediators are released that cause blood vessels to constrict.
   e. Blood flow increases to the area.
2. Which of the following statements is (are) true about skin?
   a. The skin is involved in vitamin C production.
   b. The skin becomes thinner and less elastic with age.
   c. The skin excretes some wastes.
   d. The skin regulates temperature by producing sebum.
   e. The skin contains collagen and keratin fibers in the epidermis.
3. Which of the following statements is (are) true about hair?
   a. Terminal hair undergoes a growth stage and then a resting stage before it falls out.
   b. The hair follicle contains stratum basale cells.
   c. A hair is composed of hard and soft keratin.
   d. Lanugo hair forms during fetal development.
   e. Terminal hairs are formed in the axillary regions during puberty.
4. Which of the following statements is (are) true about the stratum basale?
   a. The stratum basale contains nerve endings for fine touch.
   b. The stratum basale contains more melanocytes in people with dark skin.
   c. The stratum basale cells are fully cornified.
   d. The stratum basale is a single layer of cells.
   e. The stratum basale is present in thick skin but not in thin skin.
5. Which of the following statements is (are) true about cutaneous glands?
   a. Cutaneous glands are endocrine glands because they have ducts to deliver their products.
   b. Sebaceous and sweat glands are found in the hypodermis.
   c. Mammary glands are modified sebaceous glands.
   d. Merocrine glands produce a lipid-rich sweat that bacteria like to feed on.
   e. Apocrine glands can be found on the palms of the hands.
6. Which cells are found in the stratum basale?
   a. Tactile cells
   b. Fibroblasts
   c. Dendritic cells
   d. Melanocytes
   e. Keratinocytes

7. Which of the following statements accurately describe(s) a structure of a nail?
   a. The eponychium is the cuticle.
   b. The nail bed is the growth center for the nail.
   c. The nail plate is composed of hard collagen.
   d. The nail matrix is mostly composed of dendritic cells.
   e. The nail folds are at the proximal edge of the nail.
8. Which of the following is (are) true about skin cancer?
   a. Malignant melanoma is usually found on the hands, face, neck, and ears.
   b. Squamous cell carcinoma is rare, but deadly.
   c. Basal cell carcinoma typically does not metastasize.
   d. Basal cell carcinoma starts as a shiny bump on the face, ears, neck, or hands.
   e. Malignant melanoma may be seen under a nail plate.
9. Which of the following statements is (are) true about burns?
   a. First-degree burns are the most serious.
   b. Second-degree burns involve the dermis and epidermis.
   c. First-degree burns may require skin grafting.
   d. The stratum basale of the hair follicles will help generate new epidermis in a second-degree burn.
   e. Blisters are present at the site of a third-degree burn.
10. Which of the following statements accurately describe(s) the function of structures found in the dermis?
   a. Fibroblasts form elastic fibers, which allow the skin to come back to shape if stretched.
   b. Apocrine glands produce sweat to provide a scent.
   c. Merocrine glands produce sweat used to cool the body.
   d. Cerumen from ceruminous glands is used to keep the eardrum flexible.
   e. Sebaceous glands deliver their product to the hair follicle to keep the hair moisturized.

Matching: *Match the type of tissue to the layer at which it may be found. Some answers may be used more than once. Some questions may have more than one correct answer.*

______	1. Epidermis	a. Simple cuboidal tissue
______	2. Dermis	b. Stratified squamous epithelial tissue
______	3. Hypodermis	c. Loose/areolar connective tissue
______	4. Stratum corneum	d. Adipose connective tissue
______	5. Stratum basale	e. Dense irregular connective tissue

Matching: *Match the pathogen to the infection. Some answers may be used more than once.*

______	6. Athlete's foot	a. Bacteria
______	7. Cellulitis	b. Worm
______	8. Ringworm	c. Virus
______	9. Warts	d. Fungus
______	10. Jock itch	

Critical Thinking

1. Daniel cut his arm. His mother dressed the wound and took him to the doctor. By the time the doctor looked at the wound, it had stopped bleeding. However, the doctor still recommended that the wound be stitched. Would you agree with her recommendation? Justify your answer in terms of the anatomy and physiology involved in wound healing.
2. Older people are more likely to develop skin cancer than younger people because the damage to cells caused by UV light is cumulative. Explain in terms of anatomy and physiology why the damage is cumulative.
3. Paul is a 30-year-old bald man who went on a cruise. On his first day at sea, he fell asleep on a lounge chair in the sun next to the pool. While sleeping, he was lying in a prone position with his arms extended over his head. He was wearing only a swim suit. By the time he woke up, Paul suffered from a second-degree burn. Describe the burn, and determine the extent using the rule of nines.

3

chapter mapping

This section of the chapter is designed to help you find where each outcome is covered in this text.

	Outcomes	Readings, figures, and tables	Assessments
3.1	Use medical terminology related to the integumentary system.	Word roots: p. 75	Word Building: 1–5
3.2	Describe the histology of the epidermis, dermis, and hypodermis.	Anatomy of the skin: pp. 76–78 Figures 3.2–3.4	Spot Check: 1 Matching: 1–5
3.3	Describe the cells of the epidermis and their function.	Cells of the epidermis: pp. 78–79 Figures 3.4–3.6	Multiple Select: 4, 6 Critical Thinking: 2
3.4	Describe the structures of the dermis and their functions.	Dermis: pp. 80–81 Figure 3.7	Spot Check: 2 Multiple Select: 10
3.5	Compare and contrast the glands of the skin in terms of their structure, products, and functions.	Cutaneous glands: pp. 81–82 Table 3.1 Figures 3.7, 3.8	Multiple Select: 5, 10
3.6	Describe the histology of a hair and hair follicle.	Hair follicles: pp. 83–84 Figures 3.9, 3.10	Multiple Select: 3
3.7	Explain how a hair grows and is lost.	Hair: p. 85	Spot Check: 3 Multiple Select: 3
3.8	Describe the structure and function of a nail.	Nails: pp. 85–86 Figure 3.11	Multiple Select: 7
3.9	Explain how the layers and structures of the skin work together to carry out the functions of the system.	Physiology of the skin: pp. 86–89 Figure 3.12	Spot Check: 4 Multiple Select: 2
3.10	Explain how the skin responds to injury and repairs itself.	Regeneration vs. fibrosis: pp. 89–90 Figure 3.13	Spot Check: 5 Multiple Select: 9 Critical Thinking: 1
3.11	Describe the symptoms of inflammation and explain their cause in terms of the structure and function of the skin.	Inflammation: pp. 91–92 Figure 3.14	Multiple Select: 1
3.12	Compare and contrast three degrees of burns in terms of symptoms, layers of the skin affected, and method used by the body for healing.	Burns: pp. 92–93 Figure 3.15	Spot Check: 6 Multiple Select: 9
3.13	Describe the extent of a burn using the rule of nines.	Burns: pp. 93–94 Figure 3.16	Spot Check: 6 Critical Thinking: 3
3.14	Summarize the effects of aging on the integumentary system.	Effects of aging: pp. 94–95	Spot Check: 7 Multiple Select: 2
3.15	Describe three forms of skin cancer in terms of the body area most affected, appearance, and ability to metastasize.	Skin cancer: pp. 95–96 Figure 3.17	Multiple Select: 8
3.16	Describe an example of a bacterial, a viral, and a fungal infection of the skin.	Skin infections: p. 96 Figures 3.18, 3.19	Matching: 6–10

footnote

1. Smith, R. R., & Kennedy, J. (1991). *Instructor's resource manual and test bank to accompany essentials of anatomy and physiology.* St. Louis, MO: Mosby-Year Book.

references and resources

Beers, M. H., Porter, R. S., Jones, T. V., Kaplan, J. L., & Berkwits, M. (Eds.). (2006). *Merck manual of diagnosis and therapy* (18th ed.). Whitehouse Station, NJ: Merck Research Laboratories.

Dermatology of Seattle. (2011). *Skin cancer.* Retrieved January 28, 2011, from http://www.dermatologyseattle.com/medical/skin-cancer/.

National Cancer Institute. (n.d.). *Skin cancer.* Retrieved March 17, 2011, from http://www.cancer.gov/cancertopics/types/skin.

National Institute of Arthritis and Musculoskeletal and Skin Diseases. (2010, October). *Acne.* Retrieved March 16, 2011, from http://www.niams.nih.gov/Health_Info/Acne/default.asp.

Saladin, Kenneth S. (2010). *Anatomy & physiology: The unity of form and function* (5th ed.). New York: McGraw-Hill.

Seeley, R. R., Stephens, T. D., & Tate, P. (2006). *Anatomy & physiology* (7th ed.). New York: McGraw-Hill.

Shier, D., Butler, J., & Lewis, R. (2010). *Hole's human anatomy & physiology* (12th ed.). New York: McGraw-Hill.

Smith, R. R., & Kennedy, J. (1991). *Instructor's resource manual and test bank to accompany essentials of anatomy and physiology.* St. Louis, MO: Mosby-Year Book.

4 The Skeletal System

Imagine life without bones. How would you move your feet, point your finger, practice a yoga posture, or chew an apple? How would you communicate with your friends? Would you be able to hold a phone to your ear or text a message? Your skeletal system makes it possible for you to do all of these things, as it allows you to function as a human being. See Figure 4.1.

4.1 learning **outcome**

Use medical terminology related to the skeletal system.

learning outcomes

After completing this chapter, you should be able to:

4.1	Use medical terminology related to the skeletal system.
4.2	Distinguish between the axial skeleton and the appendicular skeleton.
4.3	Describe five types of bones classified by shape.
4.4	Identify bones, markings, and structures of the axial skeleton and appendicular skeleton.
4.5	Describe the cells, fibers, and matrix of bone tissue.
4.6	Compare and contrast the histology of compact and cancellous bone.
4.7	Compare and contrast the histology of hyaline, elastic, and fibrocartilage connective tissues.
4.8	Describe the anatomy of a long bone.
4.9	Distinguish between two types of bone marrow in terms of location and function.
4.10	Describe three major structural classes of joints and the types of joints in each class.
4.11	Differentiate between rheumatoid arthritis and osteoarthritis.
4.12	Explain how minerals are deposited in bone.
4.13	Compare and contrast endochondral and intramembranous ossification.
4.14	Compare and contrast endochondral and appositional bone growth.
4.15	Explain how bone is remodeled by reabsorption.
4.16	Explain the nutritional requirements of the skeletal system.
4.17	Describe the negative-feedback mechanisms affecting bone deposition and reabsorption by identifying the relevant glands, hormones, target tissues, and hormone functions.
4.18	Summarize the six functions of the skeletal system and give an example or explanation of each.
4.19	Summarize the effects of aging on the skeletal system.
4.20	Classify fractures using descriptive terms.
4.21	Explain how a fracture heals.
4.22	Describe bone disorders and relate abnormal function to the pathology.

word roots & combining forms

ankyl/o: bent, crooked
arthr/o: joint
burs/o: sac
carp/o: wrist
chondr/o: cartilage
condyl/o: condyle
cost/o: rib
crani/o: head, skull
femor/o: femur, bone of the thigh
fibul/o: fibula, lateral bone of the lower leg
humer/o: humerus, bone of the upper arm
ili/o: ilium, bone of the hip
ischi/o: ischium, bone of the hip
lumb/o: lower back
maxill/o: maxilla, upper jaw
myel/o: bone marrow, spinal cord
orth/o: straight
oste/o: bone
patell/o: patella, kneecap
phalang/o: phalanges, bones of the fingers and toes
pub/o: pubis, bone of the hip
stern/o: sternum, breastbone
synov/i: synovial fluid, joint, or membrane
tars/o: tarsals, foot
tibi/o: tibia, medial bone of the lower leg

pronunciation key

acetabulum: as-eh-TAB-you-lum
acromion: ah-CROW-mee-on
calcaneus: kal-KAY-knee-us
calcitriol: kal-sih-TRY-ol
canaliculi: kan-ah-LIK-you-lie
capitulum: ka-PIT-you-lum
conchae: KON-kee

continued on next page

pronunciation **key** *continued*

condyle: KON-dile
coracoid: KOR-ah-koyd
cuneiform: KYU-ni-form
diaphysis: die-AF-ih-sis
epiphyseal: eh-PIF-ih-see-al
foramen: fo-RAY-men
gomphosis: gom-FOE-sis
ilium: ILL-ee-um
ischium: ISS-kee-um
malleolus: mal-LEE-oh-lus
olecranon: oh-LEK-kra-nun
ossa coxae: OS-sah COCK-see
osteoporosis: OS-tee-oh-poh-ROE-sis
phalanges: fah-LAN-jeez
phalanx: FAY-lanks
pisiform: PIS-ih-form
sacroiliac: say-kroh-ILL-ee-ak
sacrum: SAY-crum
sella turcica: SELL-ah TUR-sih-kah
sphenoid: SFEE-noyd
suture: SOO-chur
symphysis: SIM-feh-sis
synchondrosis: sin-kon-DROH-sis
syndesmosis: sin-dez-MOH-sis
synovial: si-NOH-vee-al
trabeculae: tra-BECK-you-lee
triquetrum: tri-KWE-trum
trochanter: troh-KAN-ter
trochlea: TROHK-lee-ah
vertebral: VER-teh-bral
xiphoid: ZIE-foyd

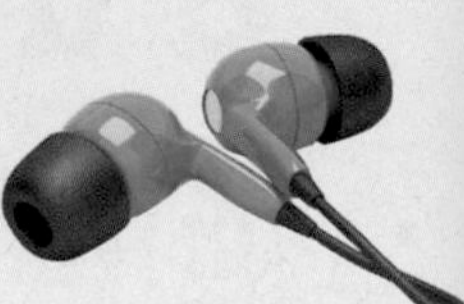

Overview

No one could deny that your skeletal system makes it possible for you to move your body from one place to another and to communicate with friends using a cell phone or a friendly wave of the hand. But your skeletal system does so much more than just give your body support and allow for movement. It protects your organs, produces blood cells, and helps maintain your electrolyte and acid-base balance. This chapter covers these functions in the context of everyday activities.

In this chapter, you will also learn about the anatomy of the skeletal system, which includes the bones, cartilage, and ligaments of the body. See **Figure 4.1**.

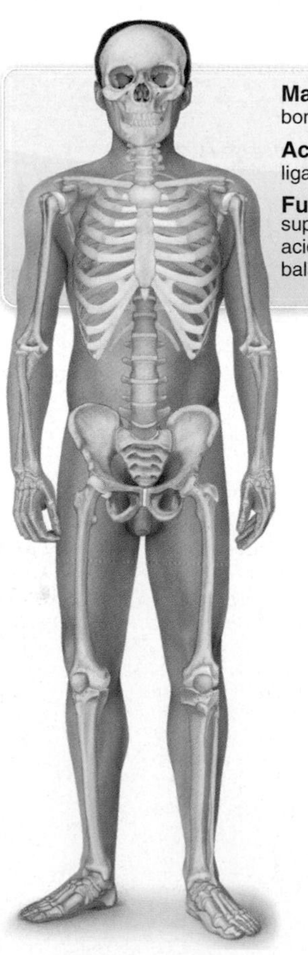

FIGURE 4.1 Skeletal system.

You will find that bone is dynamic tissue, changing daily, and that what you do can affect the changes. You will need to start at the cellular and tissue levels to understand how the structures of this system change. But first, you should become familiar with the different bones of the body.

Anatomy of the Skeletal System

4.2 learning outcome

Distinguish between the axial skeleton and the appendicular skeleton.

The human skeleton is divided into two major divisions: the axial skeleton and the appendicular skeleton. The **axial skeleton** consists of the bones of the head, neck, and trunk. The **appendicular skeleton** includes the bones of the arms, legs, and girdles (the bones that attach the arms and legs to the trunk). See **Figure 4.2**.

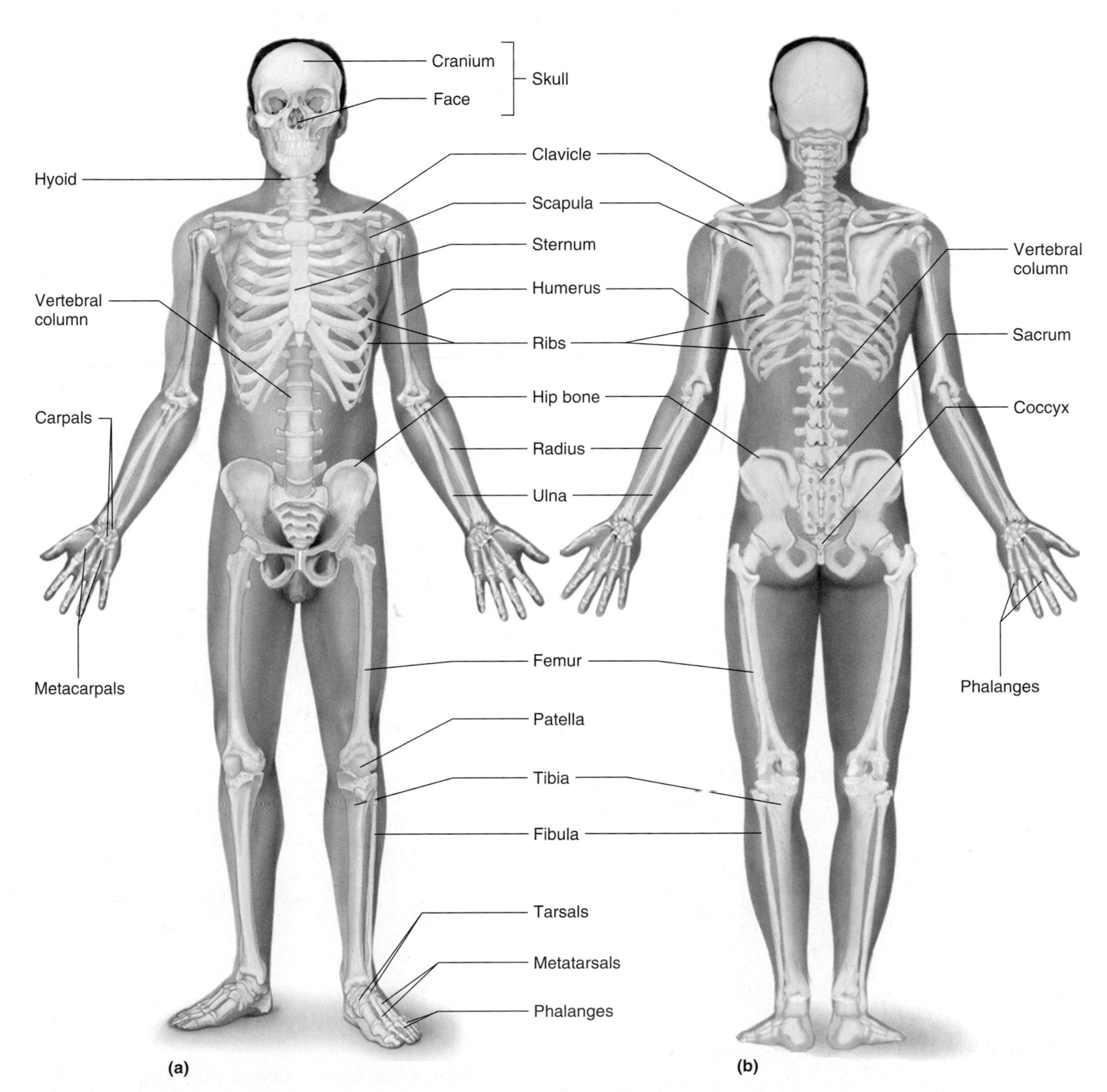

FIGURE 4.2 The axial and appendicular skeletons. (a) anterior, (b) posterior. The axial skeleton is shown in blue, and the appendicular skeleton is shown in gold.

4.3 learning **outcome**

Describe five types of bones classified by shape.

Classification of Bones

Bones are classified by their shape, and each bone fits into one of five classes:

- Long bones
- Short bones
- Flat bones
- Irregular bones
- Sesamoid bones

Long Bones A bone is considered to be a **long bone** if it is longer than it is wide and it has clubby ends. The bones of the arms, legs, fingers and toes are long bones. The most distal finger bone is not very long, but it fits the definition of a long bone because it is longer than it is wide and it has clubby ends. See **Figure 4.3**. You will learn about the anatomy of a long bone and how it develops and changes later in this chapter.

Short Bones Short bones are not longer than they are wide. These bones, which include wrist bones and proximal foot bones, are more cube-shaped. See **Figure 4.3**.

Flat Bones Flat bones are just that—flat. They look like a sheet of modeling clay that has been molded over an object. The sternum (breastbone), the cranial bones of the skull, and the ribs are all flat bones. See **Figure 4.4**. We discuss the anatomy and development of a flat bone later in this chapter.

Irregular Bones Irregular bones are also self-explanatory: They do not fit into any of the other categories. Irregular bones have processes, spines, and ridges that stick out and serve as attachment

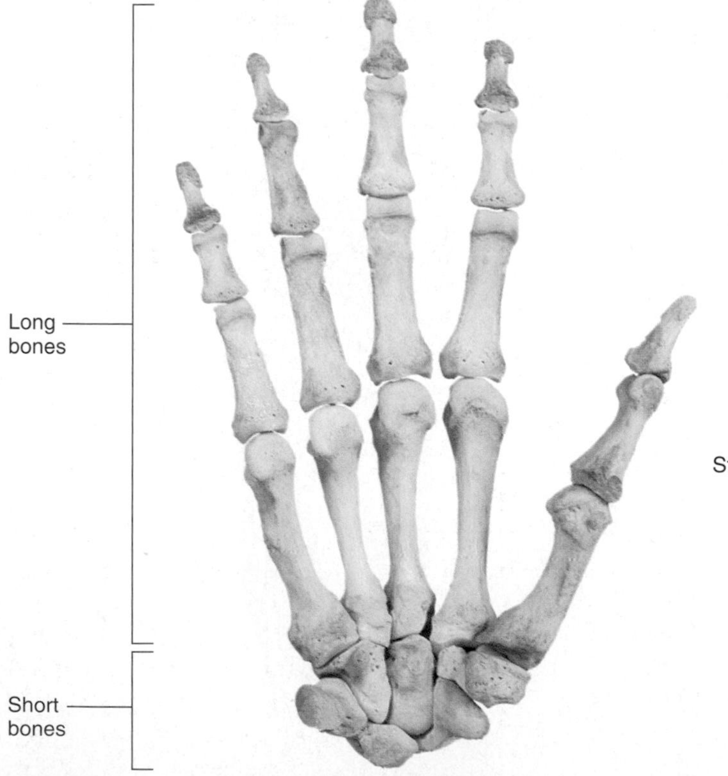

FIGURE 4.3 Long bones of the hand and short bones of the wrist.

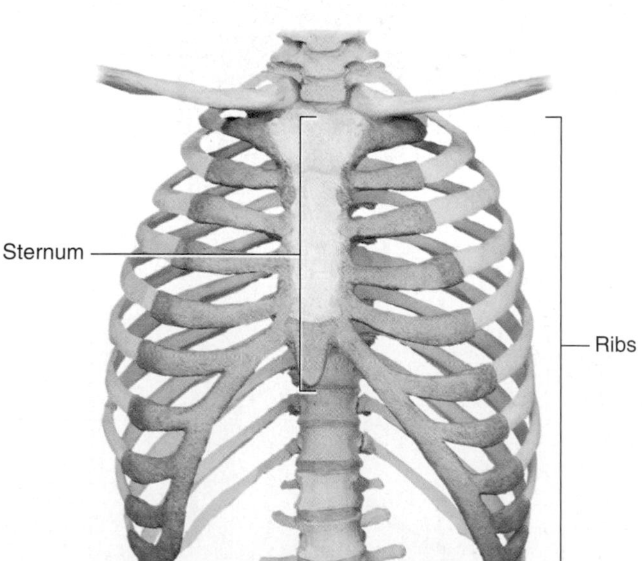

FIGURE 4.4 Flat bones—sternum and ribs.

warning

It is very easy to mistakenly think ribs are long bones because they are longer than they are wide. However, ribs do not have clubby ends, so they are considered flat bones.

points for tendons and ligaments. Tendons attach muscle to bone, and ligaments attach bone to bone. The vertebrae are good examples of this classification. See **Figure 4.5**.

Sesamoid Bones Sesamoid bones look like sesame seeds. They grow in tendons where there is a lot of friction. Their presence helps protect the tendon from wear and tear as the tendon slides over a bony prominence. The body has at least two sesamoid bones, which are the patellas, otherwise known as the *kneecaps*. Granted, these bones would be very huge sesame seeds, but they are seed-shaped nonetheless. Each patella protects the patellar tendon that goes across the anterior surface of the knee. See **Figure 4.6**.

The human body contains about 206 bones. This is an approximate figure because each human body may have a slightly different number of sesamoid bones. The two patellas are part of the 206 bones. Many people, however, may develop additional, smaller sesamoid bones in the tendons of their fingers and in the joints of their toes. You would need an x-ray of a body to know exactly how many bones it contains.

You have now viewed the skeleton as a whole, and you know that the bones of both parts—axial and appendicular—can be classified by their shape. In the next sections, you will look more closely at the individual bones and structures of the axial and appendicular skeletons.

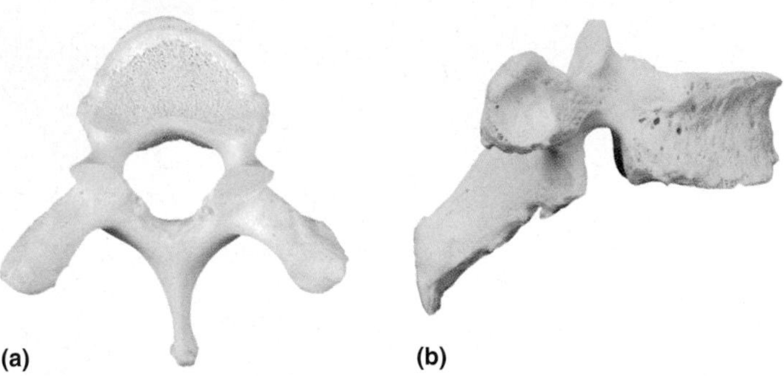

FIGURE 4.5 Irregular bone—typical vertebra: (a) superior view, (b) lateral view.

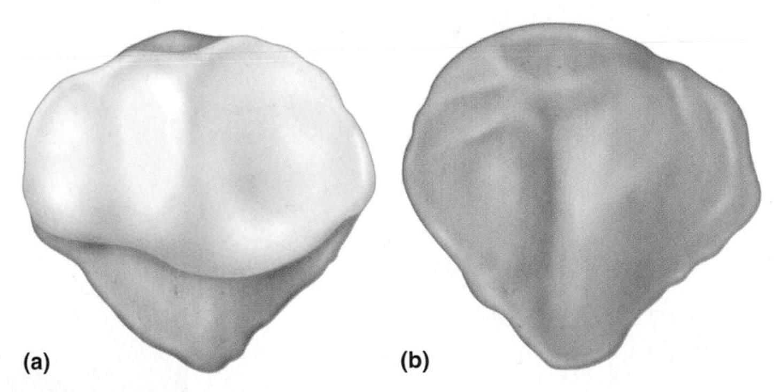

FIGURE 4.6 Sesamoid bone—the patella: (a) posterior, (b) anterior.

Axial Skeleton

4.4 learning outcome

Identify bones, markings, and structures of the axial skeleton and appendicular skeleton.

The axial skeleton includes the bones of the head, neck, and trunk. You can further break it down into the following specific bone types: cranial bones, facial bones, spinal column, sternum and ribs, and the hyoid bone. We start with the cranial and facial bones of the head.

The skull is composed of 8 cranial bones and 14 facial bones. The cranial bones form a cavity to house the brain. The facial bones provide structure, and they are the attachments for the facial expression muscles. See **Figures 4.7** to **4.12**.

Cranial Bones Cranial bones include the **frontal bone, occipital bone,** two **temporal bones,** and two **parietal bones,** which happen to be flat bones. Learning the location of these bones is crucial for any student pursuing a career in allied health. When you study the brain in the nervous system chapter, you will find that each lobe of the cerebrum is named for the bone it lies under. See **Figures 4.7**, **4.8**, and **4.12**.

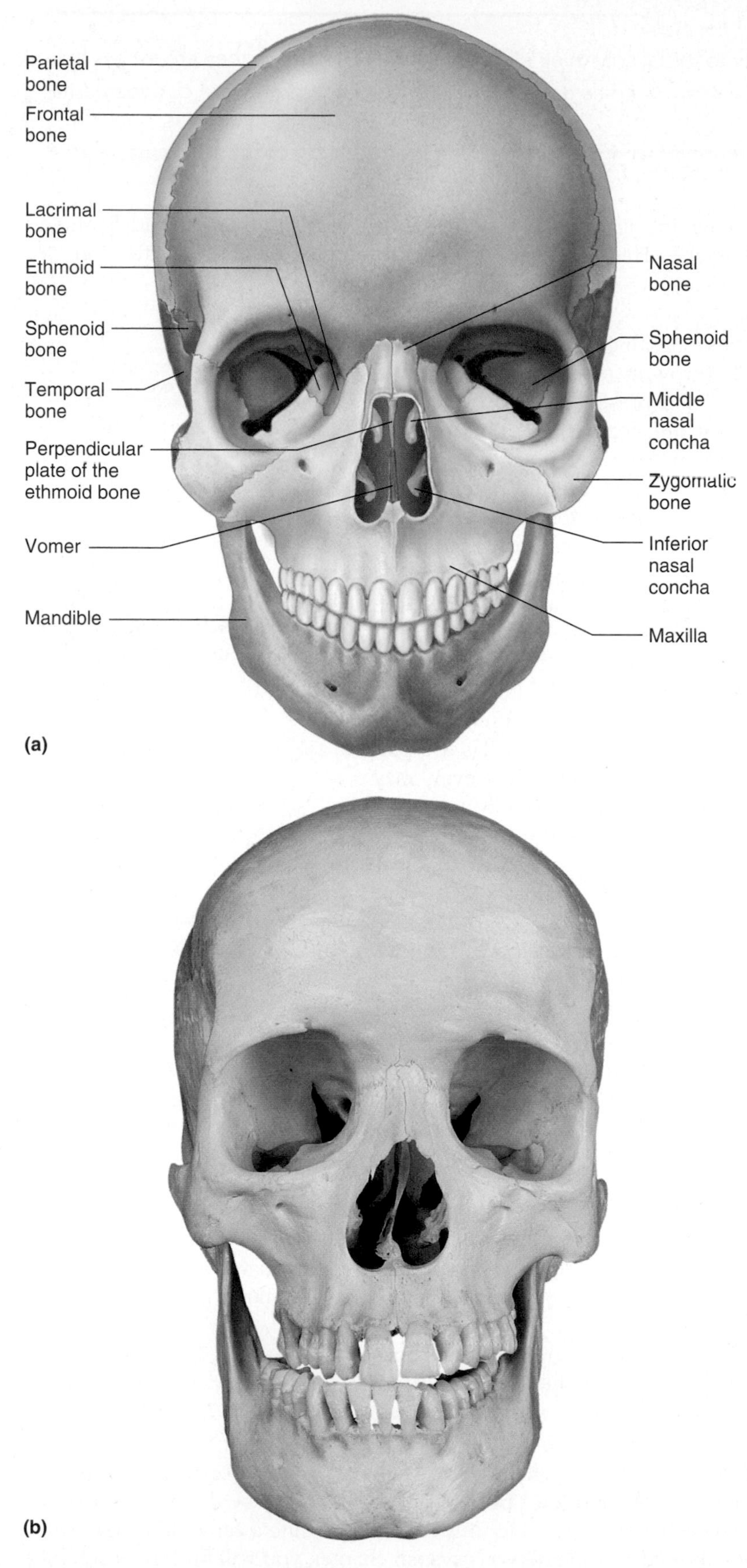

FIGURE 4.7 Anterior view of the skull: (a) colored image, (b) natural skull.

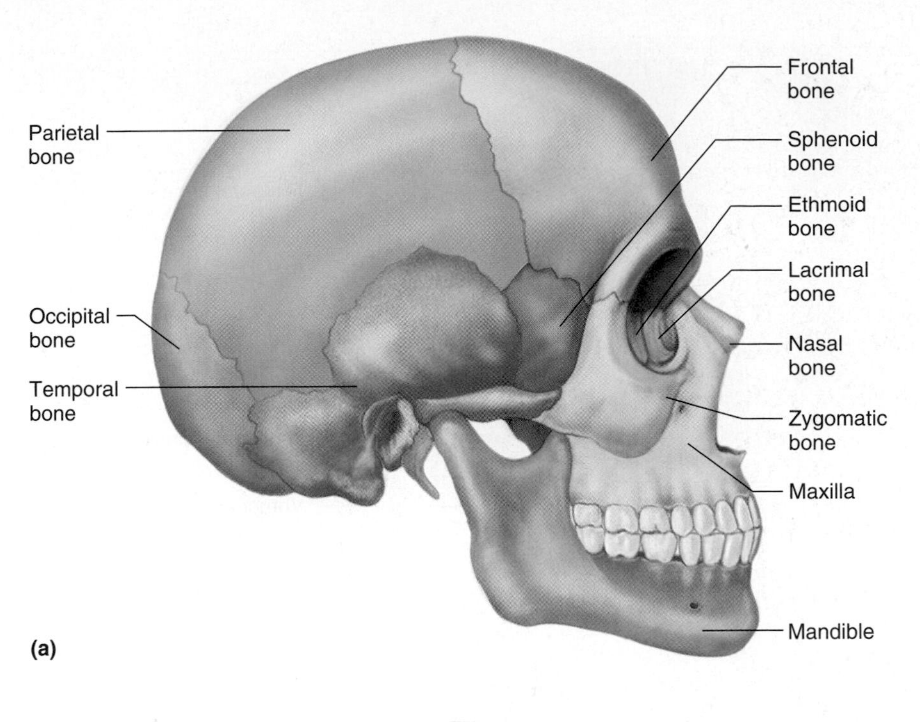

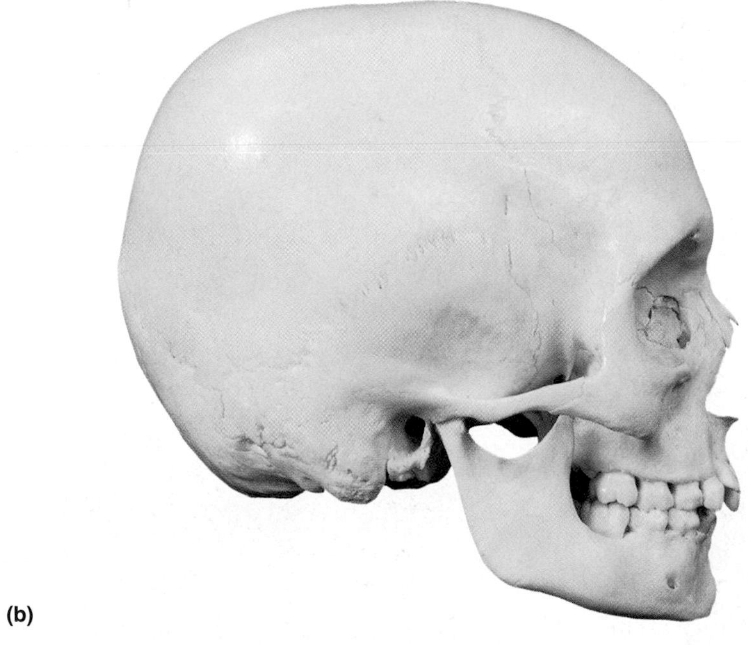

FIGURE 4.8 Lateral view of the skull: (a) colored image, (b) natural skull.

If you look closely at **Figure 4.10**, you will see many holes or openings in the skull. These openings are called **foramina** (sing.,/), **foramen,** and they allow the passage of blood vessels and nerves. The occipital bone contains a large opening called the **foramen magnum,** which allows the spinal cord to exit the cranial cavity.

foramen: fo-RAY-men

The **external occipital protuberance** is located on the occipital bone's posterior surface. See **Figure 4.10**. This protuberance is typically larger in males, and—like all other structures that stick out from a bone—it is the attachment point for a tendon to connect a muscle to the bone. More information about this particular muscle appears in the muscular system chapter.

sphenoid: SFEE-noyd

The **ethmoid** and **sphenoid** bones are irregular bones, and they form the majority of the cranial cavity floor. The ethmoid bone includes a structure called the

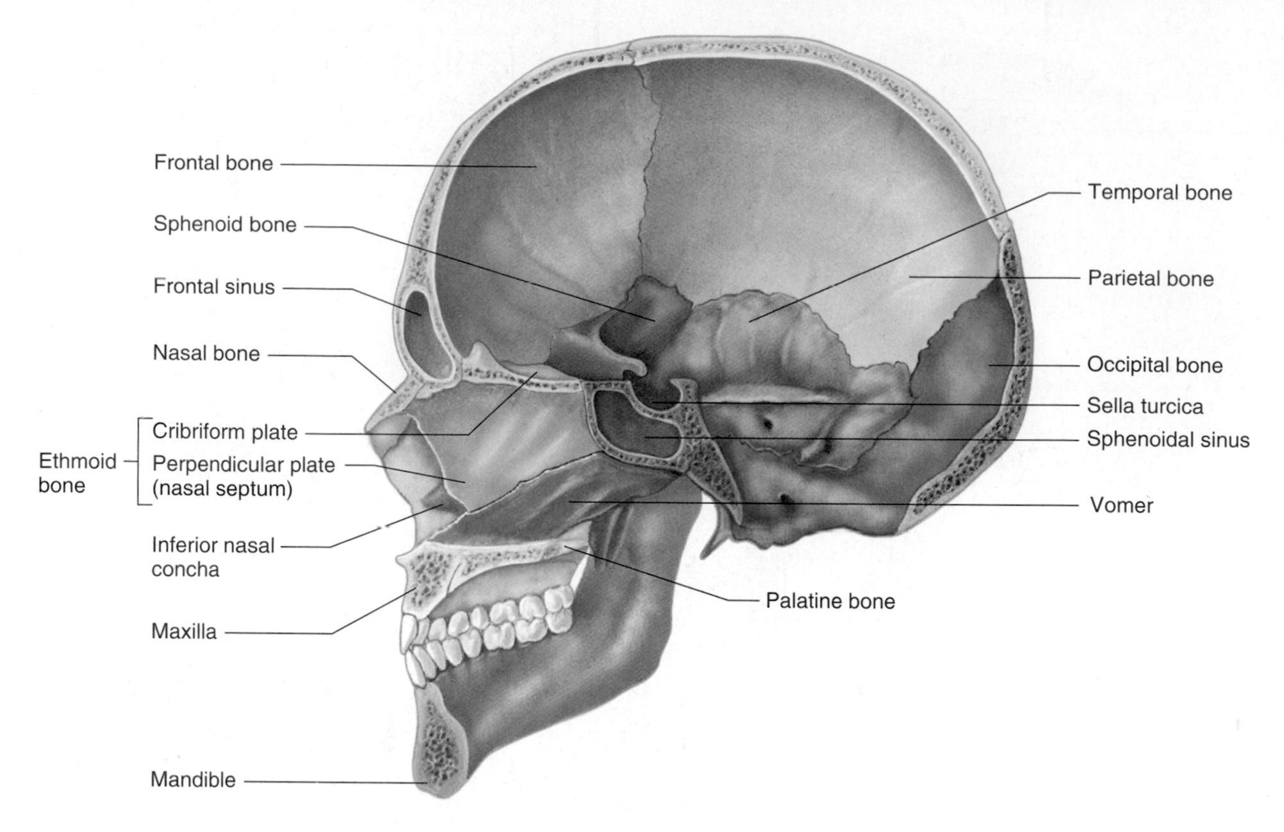

FIGURE 4.9 Medial view of the skull.

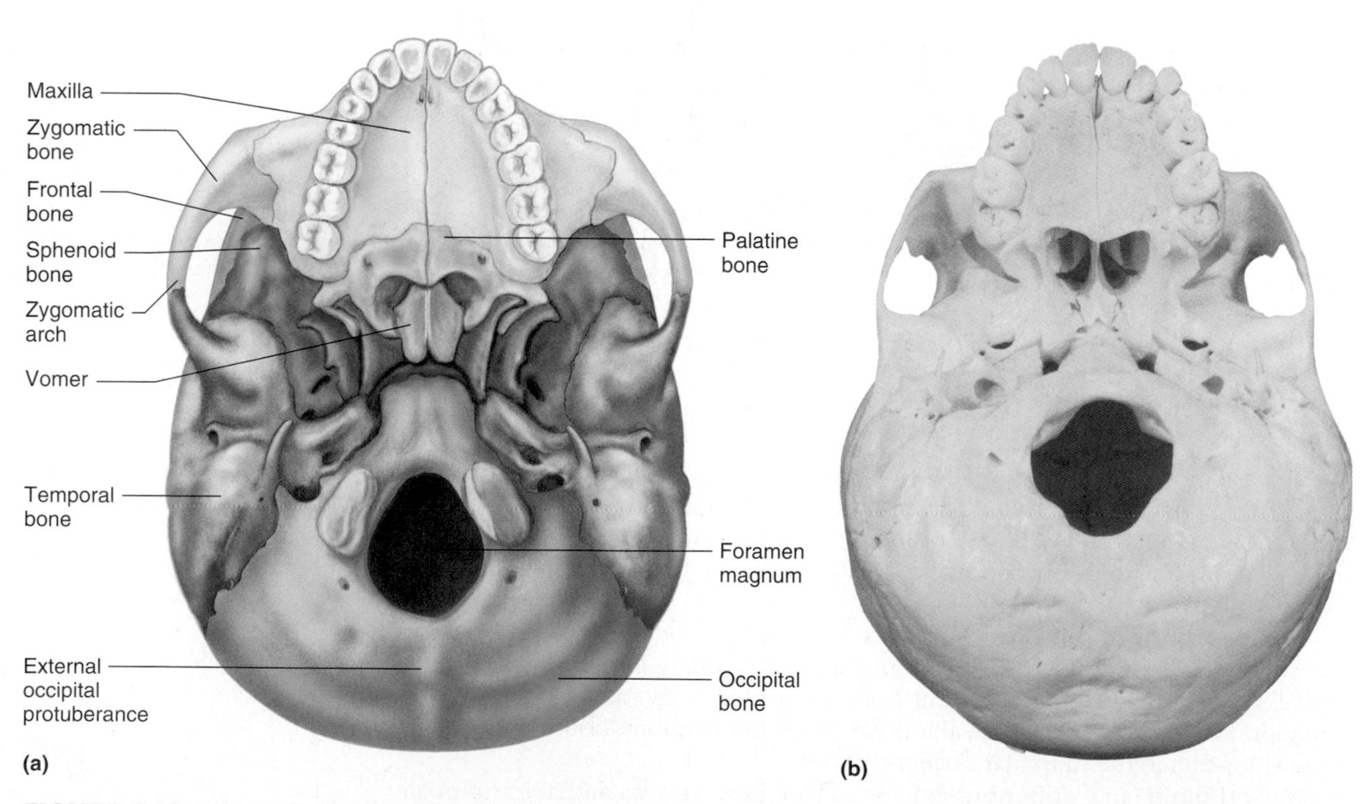

FIGURE 4.10 Inferior view of the skull: (a) colored image, (b) natural skull.

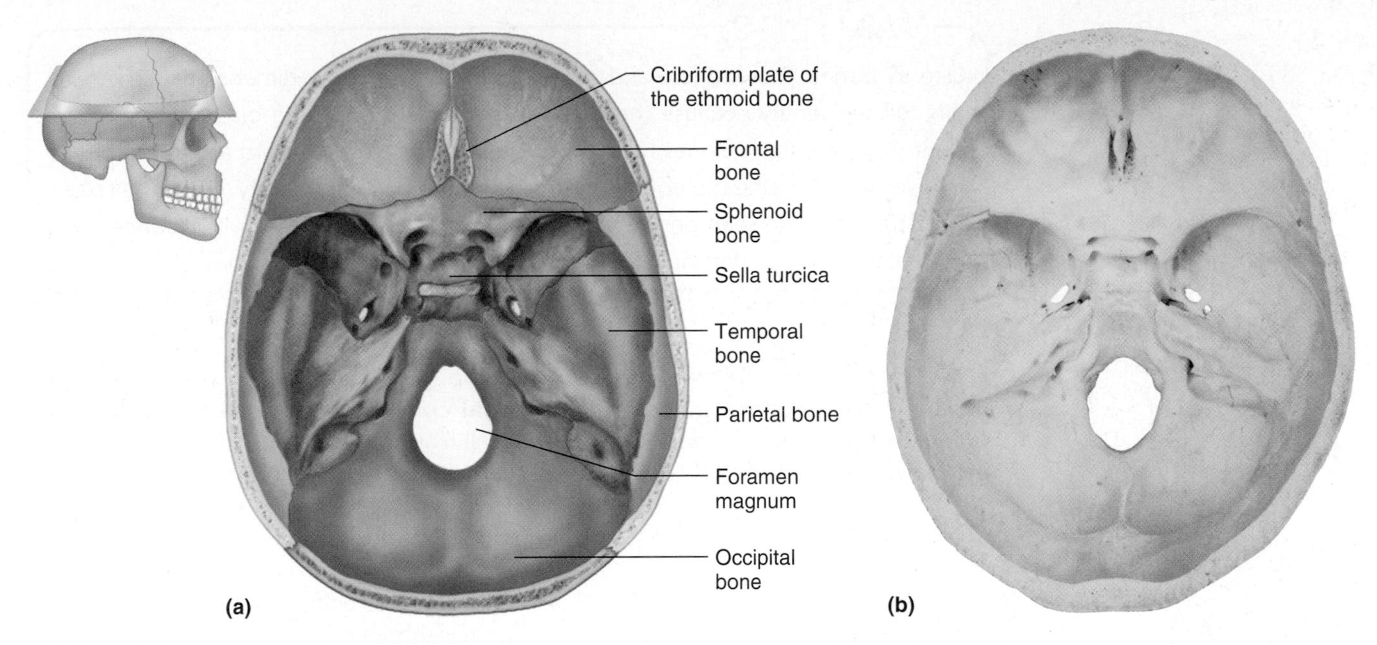

FIGURE 4.11 Cranial floor of the skull: (a) colored image, (b) natural skull.

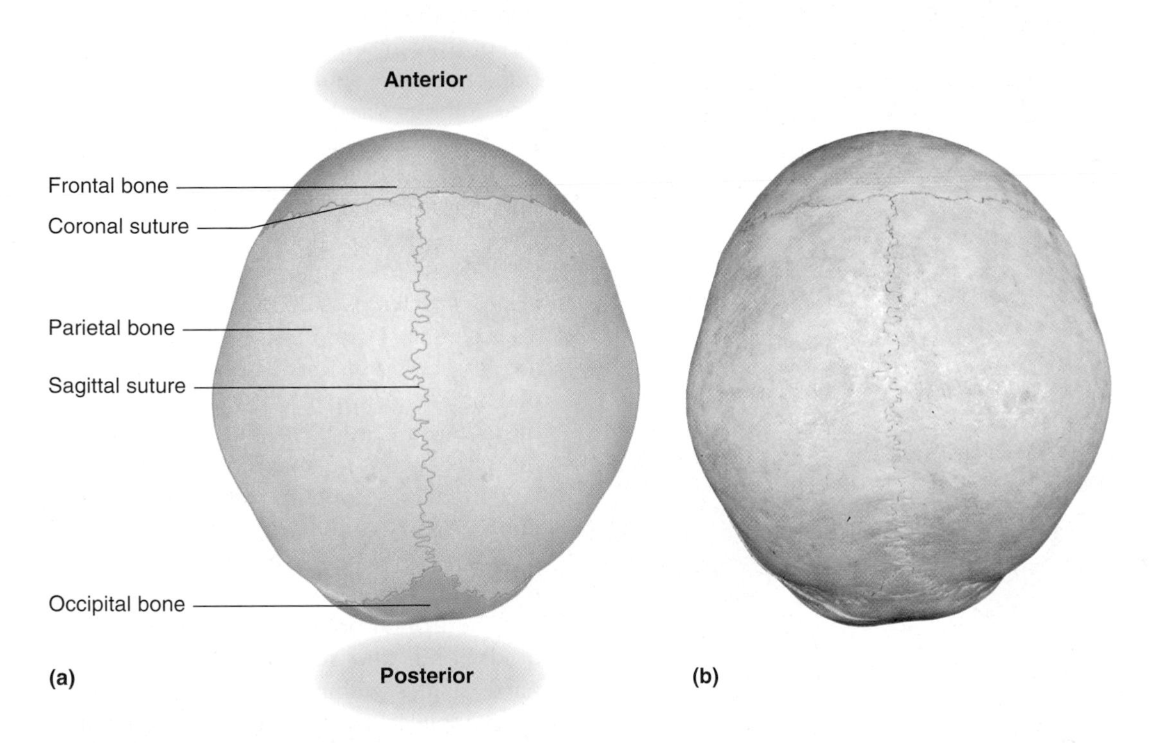

FIGURE 4.12 Superior view of the skull: (a) colored image, (b) natural skull.

cribriform plate. See **Figures 4.11** and **4.13a**. This plate consists of two depressions in the cranial cavity's anterior floor. The cribriform plate is perforated with many holes. These holes allow nerve endings from the first cranial nerve to have access to the nasal cavity for the sense of smell. The ethmoid bone also forms lateral bony ridges called **conchae** in the nasal cavity. The sphenoid bone appears butterfly-shaped, and it is visible from outside the skull next to the temporal bone. Inside the skull, the sphenoid bone forms another important structure called the **sella turcica** ("Turkish saddle"). The pituitary gland sits in this bony saddle, and the saddle's broad bar helps protect the pituitary gland by surrounding it in bone. See **Figures 4.11** and **4.13b**.

concha: KON-kee

sella turcica: SELL-ah TUR-sih-kah

> **warning**
>
> Like all bones, the ethmoid and sphenoid bones are three-dimensional. They are difficult to identify because they appear differently depending on the view of the skull. Together, both bones form the cranial vault floor, and both are found in the eye socket. The ethmoid bone forms structures in the nasal cavity, while the sphenoid bone can be seen on the exterior skull. It is important to look at bones from several angles to understand their location with respect to other bones.

sacrum: SAY-crum

connect.mcgraw-hill.com audio

Facial Bones The following bones make up your facial bones: two **nasal,** two **lacrimal,** two **zygomatic,** two **inferior nasal concha,** two **maxilla,** two **palatine,** one **mandible,** and one **vomer.** As you can tell from so many different kinds of bones, the facial part of the skull is quite complicated. Notice that in **Figure 4.7** it takes seven bones (palatine bone is obscured by the maxilla) to form the eye socket. If you feel the bridge of your nose, you will notice a change about halfway down. That is where the nasal bones end and plates of nasal cartilage begin. The vomer forms the inferior part of the nasal septum, which divides the nostrils into right and left. The ethmoid bone forms the superior part of the septum. The inferior nasal concha forms the inferior lateral ridge in the nasal cavity, while the ethmoid bone forms the superior and middle lateral ridges. The zygomatic bone is the cheekbone. Notice how it arches out to perfectly form an arch with the temporal bone. Muscles pass deep to this arch.

Again, it is important to realize these are three-dimensional bones. The maxilla forms the upper jaw, but it also forms the anterior hard palate in the roof of the mouth cavity. The mandible is the lower jaw. In fact, the only movable joint in your entire skull—the temporal mandibular joint—is where the mandible meets the temporal bone.

The frontal bone, ethmoid bone, sphenoid bone, and **maxilla** (a facial bone) have cavities within the bones themselves. These spaces are the sinuses. Each **sinus** is named for the bone that contains it. The sinuses are lined by mucous membranes and filled with air, and they help to warm and moisten inspired air and give resonance to the voice. See **Figure 4.14**.

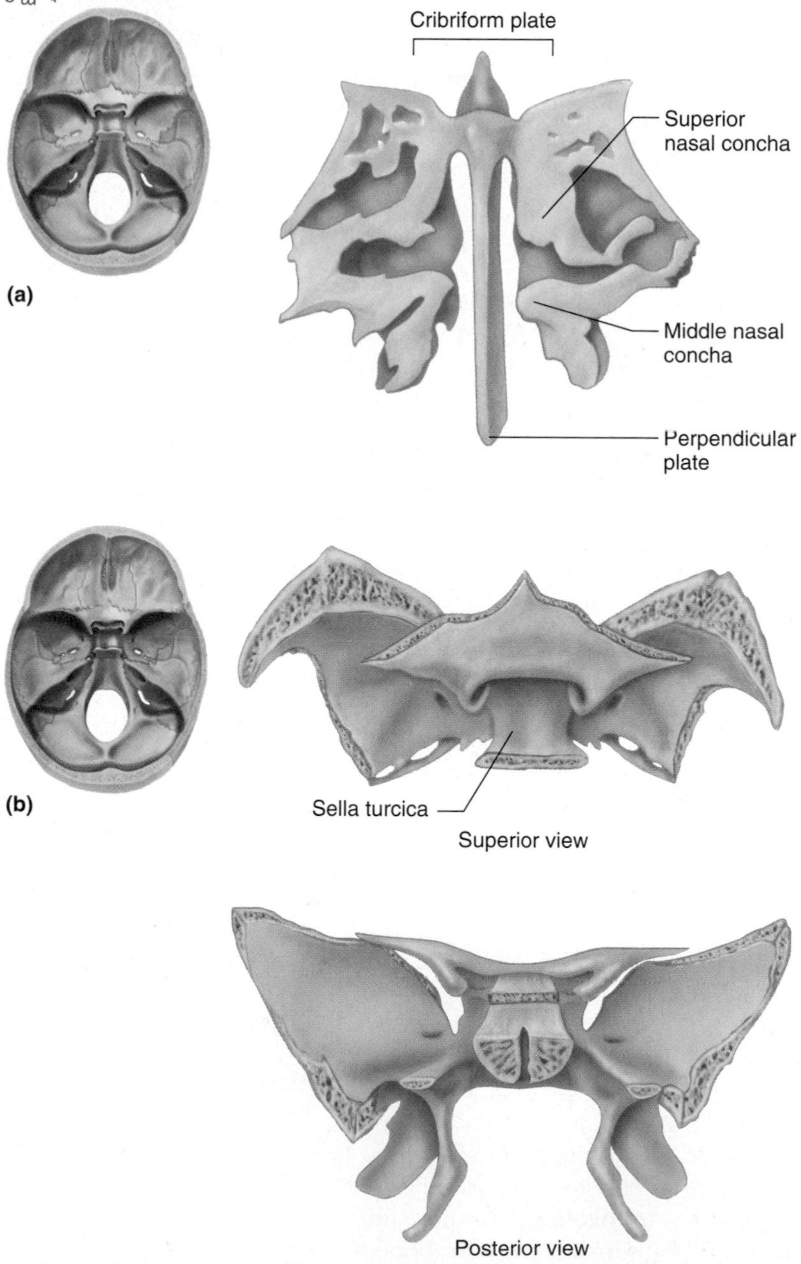

FIGURE 4.13 The ethmoid and sphenoid bones: (a) ethmoid, (b) sphenoid.

Spinal Column In an adult, the spinal column is composed of 26 bones. It contains three types of vertebrae, the **sacrum,** and the **coccyx.** There are 7 cervical vertebrae, 12 thoracic vertebrae, 5 lumbar vertebrae, 1 sacrum, and 1 coccyx. See **Figure 4.15**.

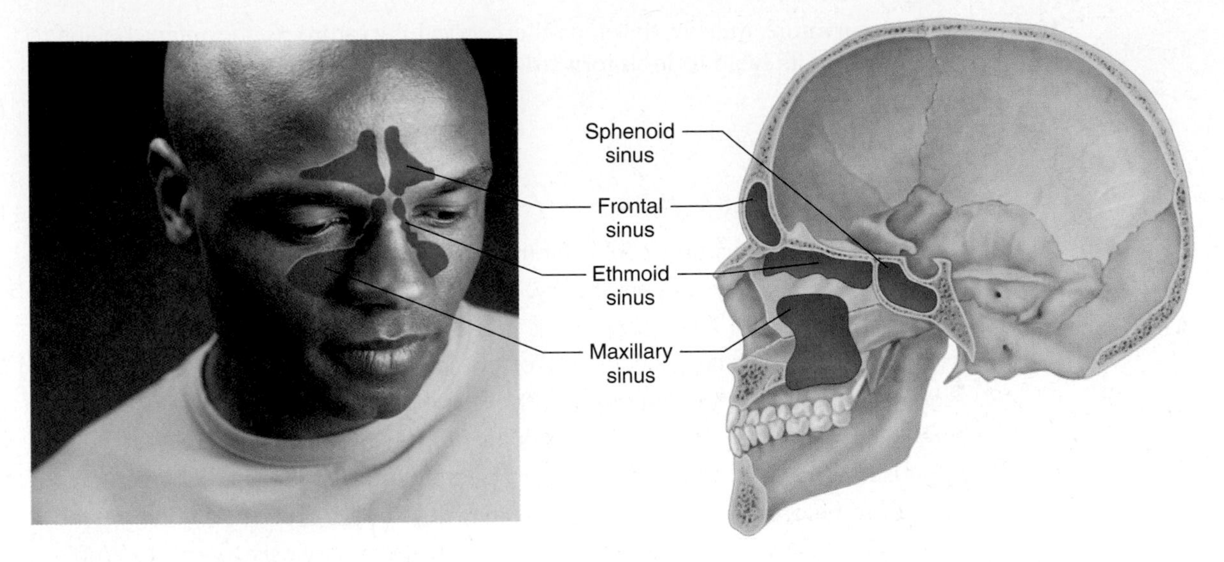

FIGURE 4.14 Sinuses.

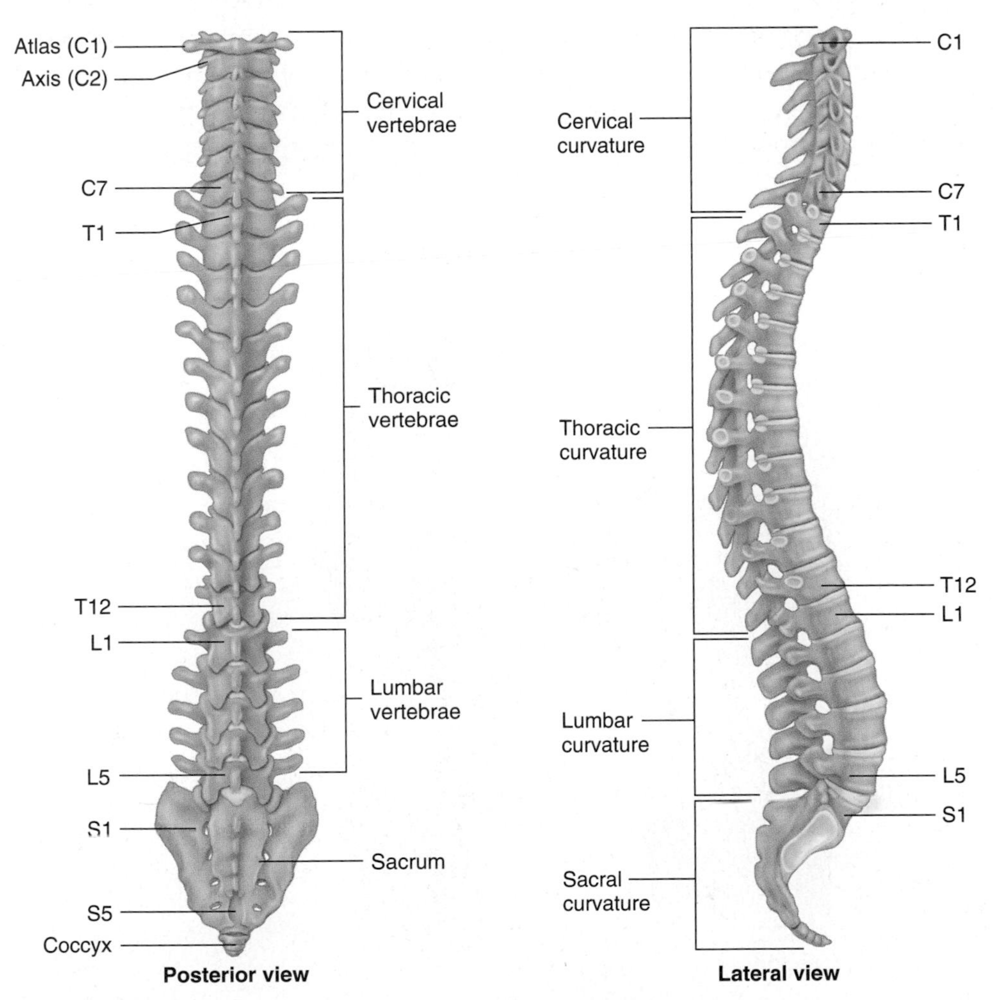

FIGURE 4.15 Spinal column.

FIGURE 4.16 Adult spinal column curvatures.

In **Figure 4.15**, you will notice that the spinal column forms a straight line when viewed from an anterior or posterior perspective. But when viewed laterally, as in **Figure 4.16**, the spinal column has an elongated S-shaped curve. The S curvature develops over time. The spinal column of a newborn has a C-shaped curvature. This

C-shaped curvature quickly develops the cervical curvature as the infant begins to crawl and raise its head to look forward. The lumbar curvature starts as the toddler begins to walk.

clinical point

Abnormal spinal curvatures can result from **congenital defects** (present at birth), disease, aging, obesity, and pregnancy. The most common abnormal curvature is **scoliosis.** See Figure 4.17a. Scoliosis is most often diagnosed in teenage girls. It is characterized by a lateral curvature of the spinal column, often in the thoracic region. During the screening process, the individual bends over at the waist. The screener looks to see if the shoulders appear to be level and the spine is straight. If this is not the case, a follow-up x-ray is taken to look for the lateral curvature. If diagnosed before the spinal column has fully developed, scoliosis can be treated with a back brace or surgery. It is important to treat scoliosis to ensure ample room for organs.

Kyphosis, commonly called "hunchback," is an exaggerated abnormal curvature of the thoracic vertebrae. It is associated with aging and osteoporosis. **Lordosis,** commonly called "swayback," is an exaggerated curvature of the lumbar vertebrae, often associated with obesity and pregnancy. See Figure 4.17b and c .

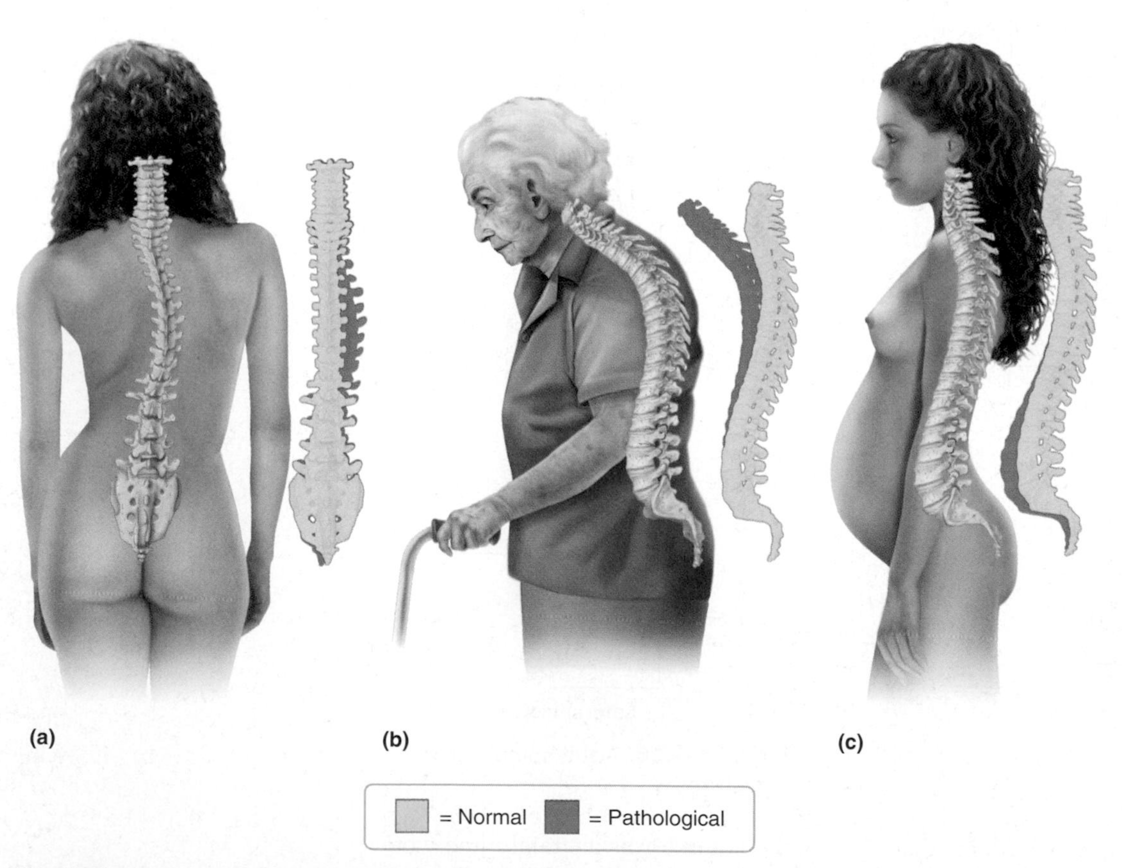

FIGURE 4.17 Abnormal curvatures of the spinal column: (a) scoliosis, (b) kyphosis, (c) lordosis.

Each vertebra is an irregular bone. Look closely at the lumbar vertebra shown in **Figure 4.18a**. It has a body, a vertebral foramen, a **spinous process,** and two **transverse processes.** The **body** of the vertebra supports the weight of the body. The spinous and transverse processes are attachment points for tendons and ligaments. The **vertebral foramen** allows the spinal cord to pass through the vertebra. Spinal nerves exit the spinal cord between vertebrae.

vertebral: VER-teh-bral

Between each of the vertebrae is an **intervertebral disk** of softer matrix surrounded by fibrocartilage. The disks support the body weight and act as shock absorbers, cushioning the vertebrae from the impact of each footstep. See **Figure 4.18b**.

clinical point

Improper heavy lifting compresses the intervertebral disks. The pressure of extra weight may cause one of these disks to bulge out laterally. This bulge may allow the softer matrix to ooze out. This condition is called a **herniated disk**. In common usage, this is often referred to as a *ruptured* or *slipped disk*. Depending on the direction of the bulge, a herniated disk can put pressure on the spinal cord or spinal nerves, causing severe pain. See Figure 4.18c.

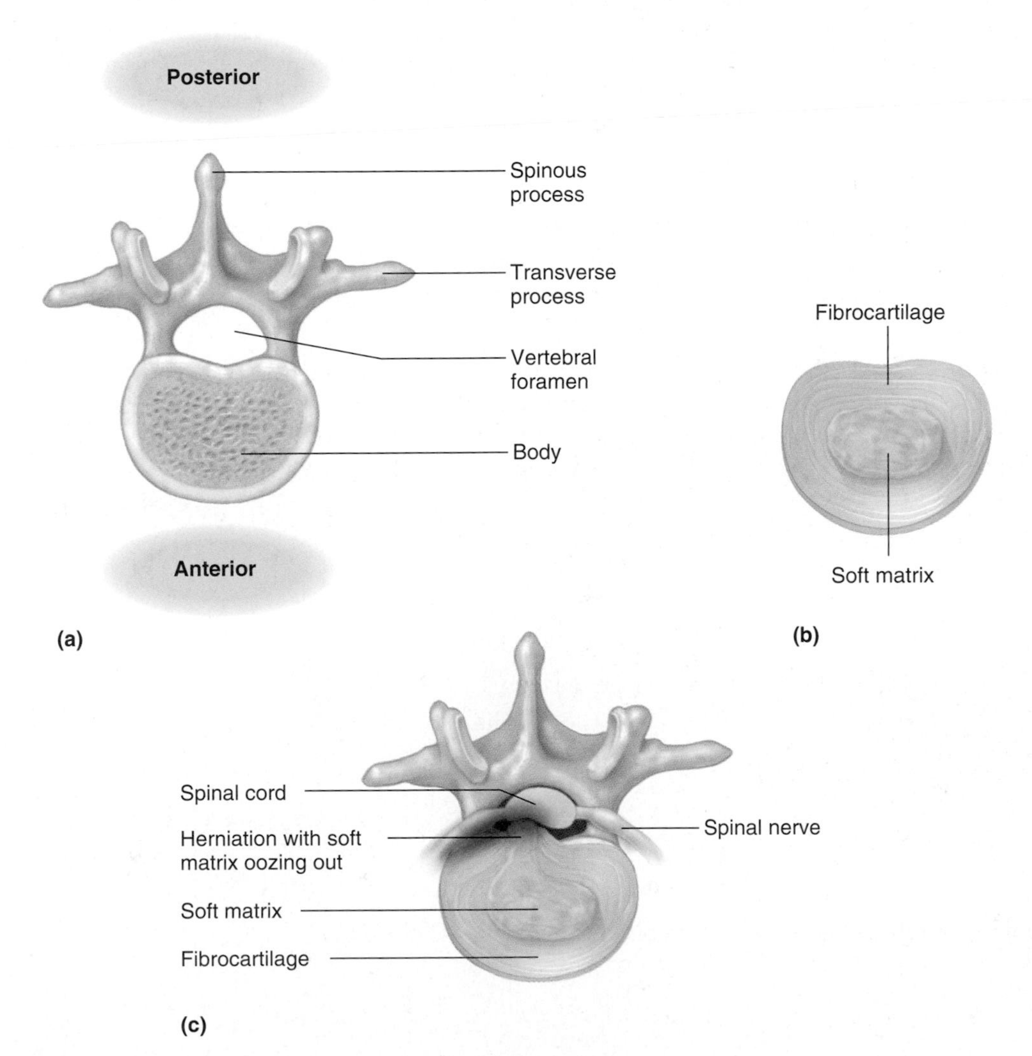

FIGURE 4.18 A vertebra: (a) second lumbar vertebra (L2), (b) intervertebral disk, (c) herniated disk.

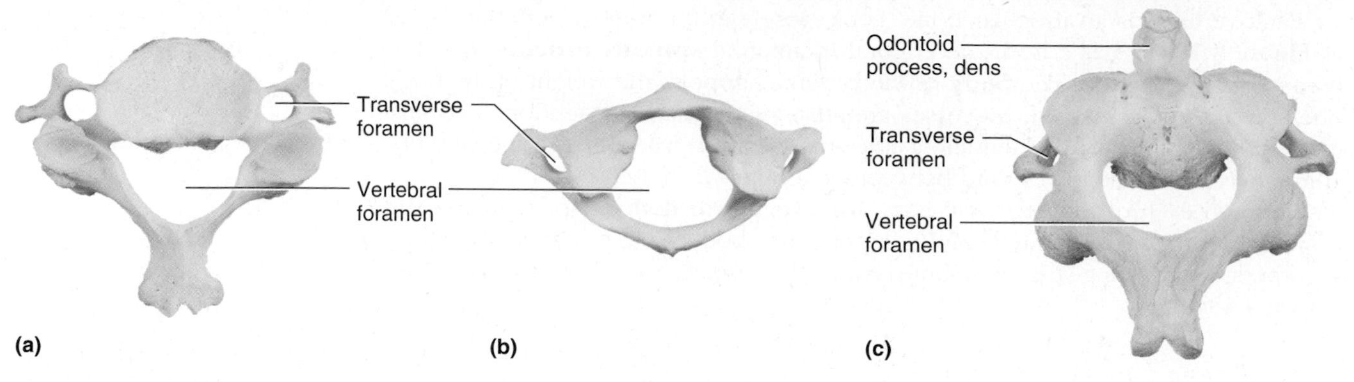

FIGURE 4.19 Cervical vertebrae: (a) typical cervical vertebra, (b) atlas (C1), (c) axis (C2).

Cervical vertebrae. Look closely at the typical cervical vertebra in **Figure 4.19a**. Notice that the seven cervical vertebrae have foramen in the transverse processes. The vertebral arteries pass through these openings on their way to the head. You will learn more about these in the cardiovascular system chapter. The cervical vertebrae are the only vertebrae with transverse foramen. The first two cervical vertebrae—the atlas and the axis—are unique. The first is the **atlas** (**Figure 4.19b**), which has a very little body and a very large vertebral foramen for the spinal cord. Second comes the **axis** (**Figure 4.19c**), which has a peglike structure called the **odontoid process** or **dens.** It sticks through the large vertebral foramen of the atlas and provides a pivot point so that the atlas can rotate on the axis. This allows you to turn your head to the right or left.

study **hint**

Here is an easy hint for remembering which of these vertebrae comes first: They are in alphabetical order—*atlas* comes before *axis*.

Thoracic vertebrae. The 12 thoracic vertebrae are distinctive because they are the only vertebrae in the body that have smooth surfaces called **costal facets.** Ribs attach to the facets on the bodies and transverse processes of these vertebrae. They are numbered T1 through T12. T1 is the most superior. See **Figure 4.20**.

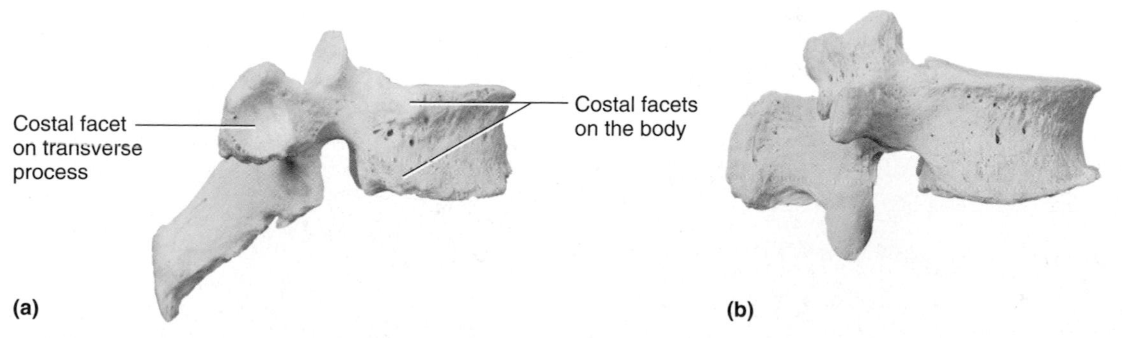

FIGURE 4.20 Thoracic and lumbar vertebrae: (a) thoracic vertebra with costal facets, (b) lumbar vertebra with no costal facets.

Lumbar vertebrae. The five lumbar vertebrae are the most massive because they support the weight of the body. There may be little difference in mass between T12 and L1, which are located right next to each other. T12, however, has facets for ribs to attach, whereas L1 does not. See **Figure 4.20**.

Sacrum and coccyx. The sacrum and the coccyx, which are still part of the axial skeleton, complete the inferior end of the spinal column. The sacrum is composed of five separate bones in a fetus that fuse to become one bone in an adult. The coccyx is composed of four to five bones in a fetus that fuse to become one bone in an adult. Therefore, there are 33 to 34 bones in a fetal spinal column and just 26 bones in an adult spinal column. See **Figure 4.21**.

Sternum The sternum is a flat bone composed of three parts: the **manubrium,** the **body,** and the **xiphoid process.** Together, they serve as a protective plate for the heart and as an attachment site for the ribs encasing the thorax. The manubrium is the attachment point for the **pectoral girdle.** You will read more about this shortly in regard to the appendicular skeleton. The xiphoid process is the most inferior part of the sternum. See **Figure 4.22**.

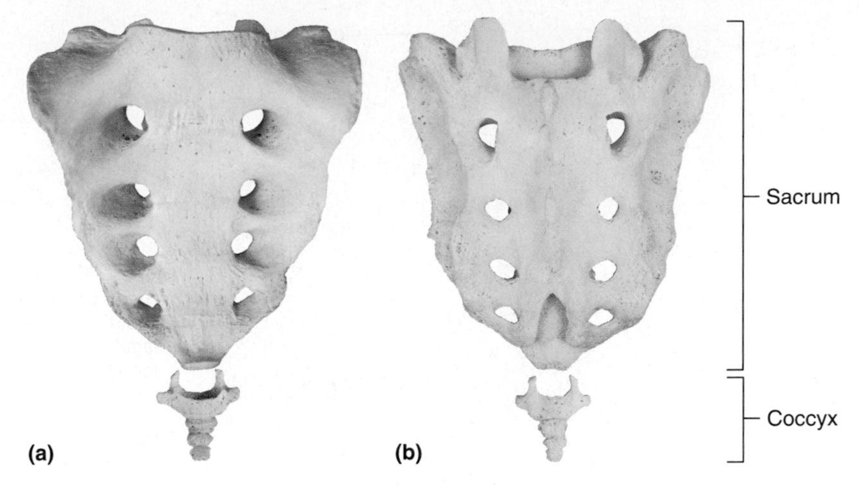

FIGURE 4.21 **Sacrum and coccyx:** (a) anterior view, (b) posterior view.

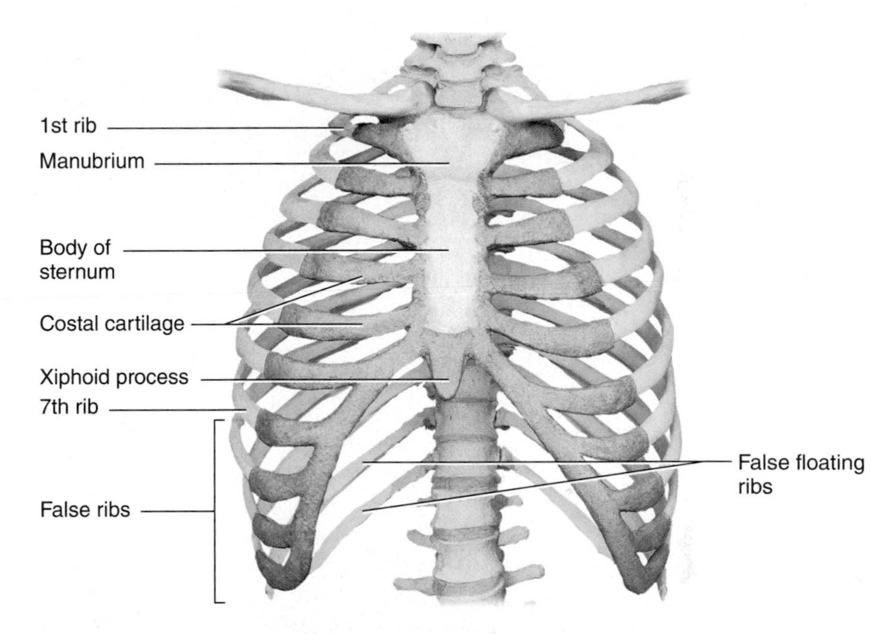

FIGURE 4.22 **The sternum and ribs:** anterior view.

clinical point

Before you administer CPR, care must be taken to locate the xiphoid process. During CPR, you must apply pressure to the sternum superior to the xiphoid process to ensure the xiphoid process does not break.

xiphoid: ZIE-foyd

audio connect.mcgraw-hill.com

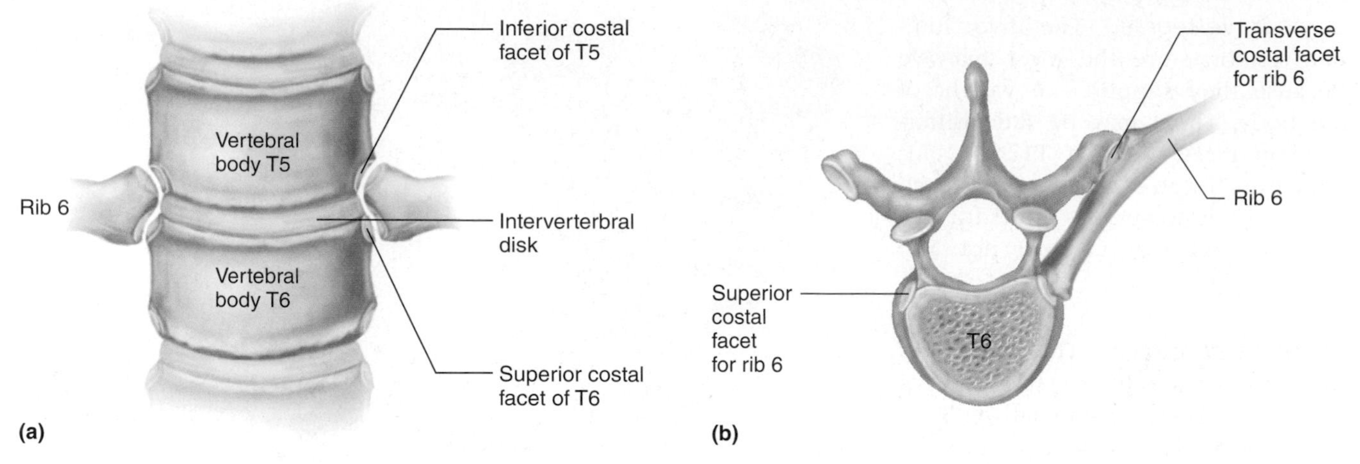

FIGURE 4.23 Rib 6 attachment to T5 and T6: (a) anterior view, (b) superior view.

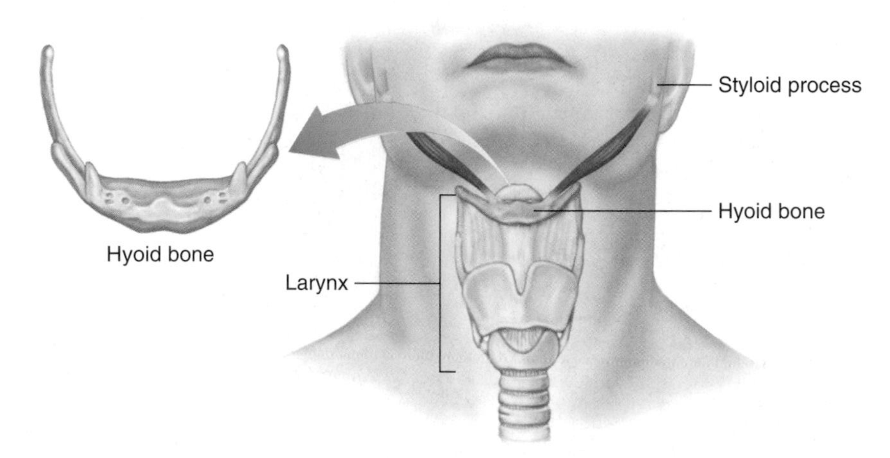

FIGURE 4.24 Hyoid bone.

Ribs As you can see in **Figure 4.22**, there are 12 pairs of ribs in the human body. They provide protection for the lungs in the thoracic cavity. The superior seven pairs of ribs are connected to the sternum by their individual **costal cartilages.** These pairs are considered **true ribs.** Pairs 8 through 12 are considered **false ribs** because they do not have individual costal cartilages connecting them to the sternum. Of the false ribs, pairs 8 through 10 share a costal cartilage to connect to the sternum. Pairs 11 and 12 are considered to be false **floating ribs** because they are not connected to the sternum. The costal cartilages are composed of hyaline cartilage connective tissue. The ribs attach posteriorly to the thoracic vertebrae at the vertebral bodies and transverse processes. See **Figure 4.23**.

Hyoid Bone The hyoid bone is a U-shaped bone found in the body's anterior cervical region between the mandible and the larynx. This bone is unique because it is not attached to another bone. Muscles attach to the hyoid bone to form the angle between the chin and the neck. See **Figure 4.24**.

clinical point

Forensic pathologists look for fractures of the hyoid bone as an indication of strangulation.

spot check 1 Name three flat bones of the axial skeleton.

spot check 2 Name two irregular bones of the axial skeleton.

Appendicular Skeleton

The appendicular skeleton is composed of the bones of the limbs and the bones (called *girdles*) that attach each limb to the axial skeleton. The **pectoral girdle** bones attach the arm bones to the axial skeleton. The **pelvic girdle** bones attach the leg bones to the axial skeleton. We begin our discussion with the bones of the pectoral girdle and the upper limb.

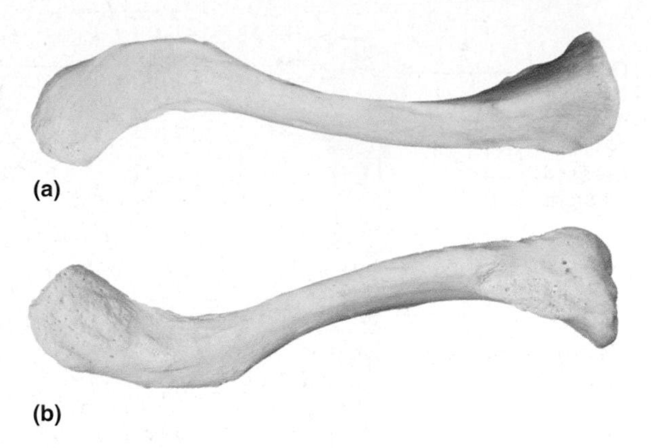

FIGURE 4.25 Right clavicle. (a) superior view, (b) inferior view.

Pectoral Girdle In the pectoral girdle, the **clavicle** and **scapula** connect the arm to the axial skeleton. The clavicle, commonly called the *collarbone,* is slightly S-shaped. It holds the shoulder out from the body laterally, preventing the chest muscles from pulling the shoulder medially. See Figure 4.25.

The scapula is the shoulder blade. It has a smooth anterior surface that allows it to slide over the ribs, while its posterior surface has a prominent spine for the attachment of muscles by tendons. The scapula contains the following three prominent lateral features: the **acromion process** articulates (joins together) with the clavicle; the **coracoid process** is an attachment point for muscles (see the next chapter) by tendons; and the **glenoid cavity** is a smooth surface that articulates with the upper arm bone called the **humerus.** See Figure 4.26.

acromion: ah-CROW-mee-on

coracoid: KOR-ah-koyd

clinical **point**

The only bone-to-bone connection of the arm and scapula to the axial skeleton is at the point where the clavicle and the manubrium of the sternum meet. The clavicle is a commonly fractured bone because of the stress placed on it when people reach out with an arm to break a fall. If a clavicle is broken, it is very important to immobilize the arm to the body. This will relieve the stress on the fracture from the weight of the arm.

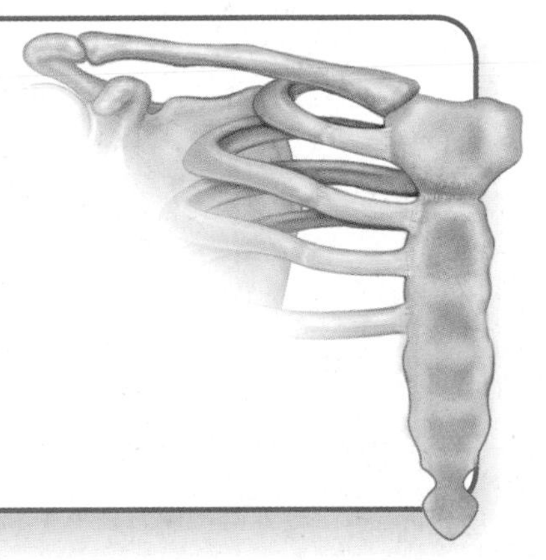

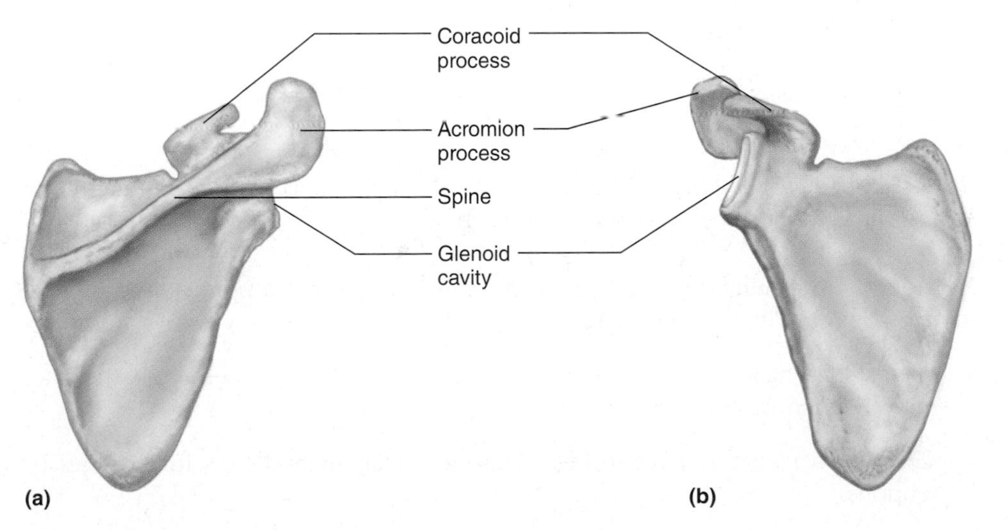

FIGURE 4.26 Scapula: (a) posterior view, (b) anterior view.

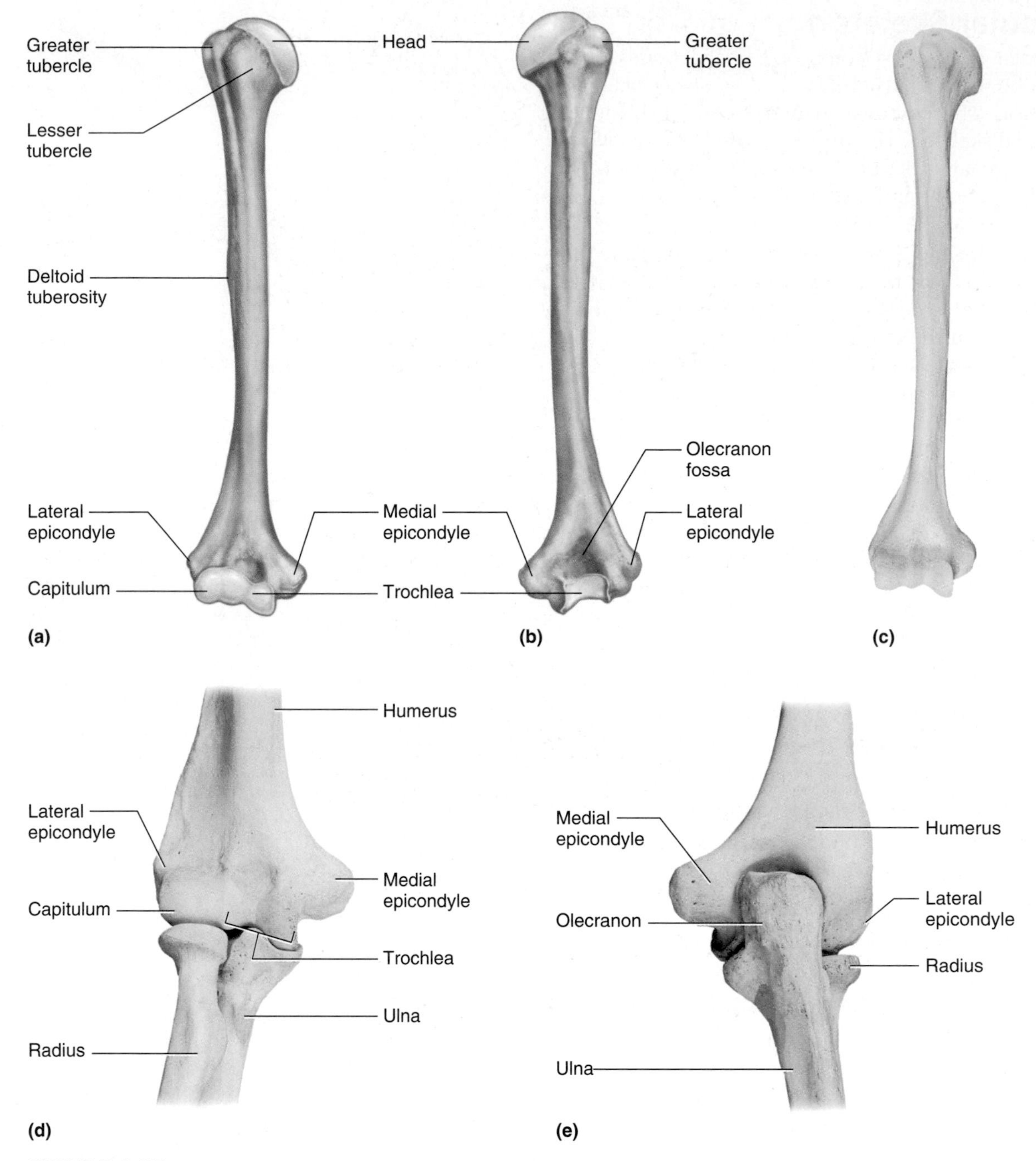

FIGURE 4.27 Humerus: (a) anterior view, (b) posterior view, (c) natural humerus, anterior view, (d) anterior elbow, (e) posterior elbow.

Bones of the Upper Limb The bones of the upper limb include the humerus, radius, ulna, carpals, metacarpals, and phalanges.

Humerus. The humerus is the proximal long bone of the arm, and its many features (shown in **Figure 4.27**) include the following:

- **Head.** The head is located at the proximal end of the humerus. It articulates with the glenoid cavity of the scapula.
- **Greater and lesser tubercles.** These are attachment points for muscles by tendons.

- **Deltoid tuberosity.** **Tuberosities** are rough areas on a bone that serve as attachment points of muscles by tendons. This tuberosity is an attachment point for the deltoid muscle.
- **Capitulum.** This is a rounded smooth surface on the distal end of the bone. It articulates with the radius, which you will read about shortly.
- **Trochlea.** This is a pulley-shaped smooth surface on the distal end of the humerus. It articulates with the ulna, which you will also read about shortly.
- **Lateral and medial epicondyles.** **Condyles** are smooth bone surfaces that articulate with another bone at a joint. *Head, capitulum,* and *trochlea* are specific names for the humerus's condyles. **Epicondyles** are rough bumps usually to the side of the condyles that work as attachment points for muscles by tendons. You will need to be able to determine medial from lateral epicondyles, as this will be very relevant when you study muscles in the next chapter.
- **Olecranon fossa.** A fossa is a depression in a bone. The olecranon fossa is a depression on the posterior surface of the humerus's distal end. The olecranon, a feature of the ulna, fits into this depression. You will learn about the ulna shortly. Again, see **Figure 4.27**.

Radius. The radius is a long bone of the forearm. See **Figure 4.28**. Special features of this bone include:

- **Head.** The head is at the proximal end of the radius. It articulates with the capitulum of the humerus.
- **Styloid process.** **Styloid** means pointy, while a **process** is something that sticks out. So the styloid process is a pointed process at the radius's distal end. It helps frame the lateral carpal bones of the wrist (thumb side). See **Figure 4.28**.

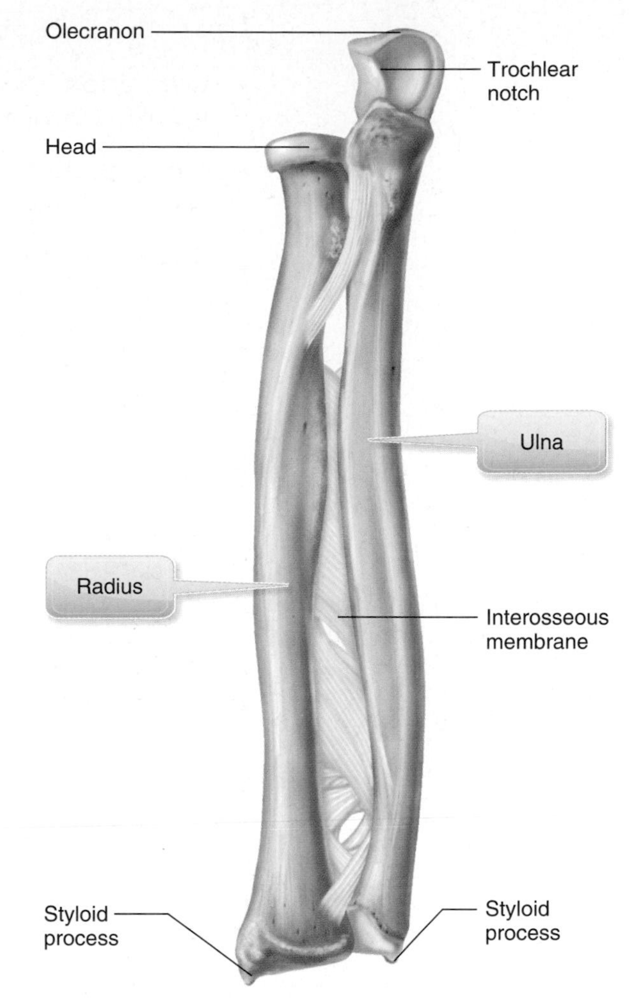

FIGURE 4.28 Radius and ulna: anterior view.

capitulum: ka-PIT-you-lum

condyles: KON-diles

trochlea: TROHK-lee-ah

olecranon: oh-LEK-kra-nun

study **hint**

If someone were to show you a humerus, would you be able to tell if it belonged to the right or left side of the body? You will have a very good understanding of the features of this bone when you can accomplish that task. Think about what you already know that might be helpful. You know that the head articulates with the glenoid cavity of the scapula. So the head has to face medially—it cannot face laterally away from the body. You also know that the olecranon fossa is on the posterior surface of the bone. Knowing the head is medial and the olecranon fossa is posterior is enough to allow you to determine right from left. It also is enough to help you determine the medial epicondyle from the lateral epicondyle, as long as you keep in mind the standard anatomical position.

Ulna. The ulna is another long bone of the forearm. Its special features include:

- **Olecranon.** The olecranon is a hook at the proximal end of the ulna; it fits into the olecranon fossa of the humerus.
- **Trochlear notch.** The trochlear notch is a smooth surface at the ulna's proximal end. It articulates with the trochlea of the humerus.
- **Styloid process.** The styloid process of the ulna frames the medial carpal bones of the wrist.

There is also an **interosseus membrane** that connects the radius and ulna along the length of the two bones. This is actually a ligament connecting bone to bone. See **Figure 4.28**. It helps distribute the pressure put on either bone equally to both bones and thus helps prevent wear and tear at the elbow joint. You will learn more about this membrane later in this chapter when you investigate joints.

triquetrum: tri-KWE-trum

pisiform: PIS-ih-form

Carpal bones. There are eight short bones in the wrist that are collectively called the *carpal bones*. They form two rows of cubelike bones that allow movement from side to side and front to back. The proximal row of bones (starting from the thumb side) is made up of the **scaphoid, lunate, triquetrum,** and **pisiform.** The distal row of bones (starting on the thumb side) is made up of the **trapezium, trapezoid, capitate,** and **hamate.** See **Figure 4.29**.

study **hint**

There is a mnemonic device that will help you remember the names of the carpal bones. The saying uses the first letter of each bone in order, and it goes like this: "Small little trains pull tiny train cars home."

Metacarpals. The five metacarpal bones are long bones. They make up the palm of the hand. The metacarpals are numbered, with 1 being proximal to the thumb and 5 being proximal to the little finger. See **Figure 4.29**.

phalanges: fah-LAN-jeez

phalanx: FAY-lanks

Phalanges. The 14 phalanges are long bones that make up the fingers. They are named by their position as proximal, middle, or distal and by the finger in which they reside. The thumb is 1 and the little finger is 5. People often wear a wedding ring on the left proximal phalanx 4. See **Figure 4.29**.

warning

On an x-ray, the metacarpal bones appear to be part of the fingers. For this reason, it is easy to confuse metacarpal bone 1 with proximal phalanx 1. Take a look at your thumb. There are only two phalanges in the thumb. All other fingers contain three phalanges. See Figure 4.30.

ilium: ILL-ee-um

ischium: ISS-kee-um

ossa coxae: OS-sah COCK-see

Now that you have covered the bones of the pectoral girdle and upper limb, it is time to examine the bones of the pelvic girdle and lower limb.

Pelvic Girdle The right and left pelvic girdles attach the lower limbs to the axial skeleton at the sacrum. Each of the pelvic girdles is composed of three bones: the **ilium,** the **ischium,** and the **pubis.** These three bones are fused together, and they

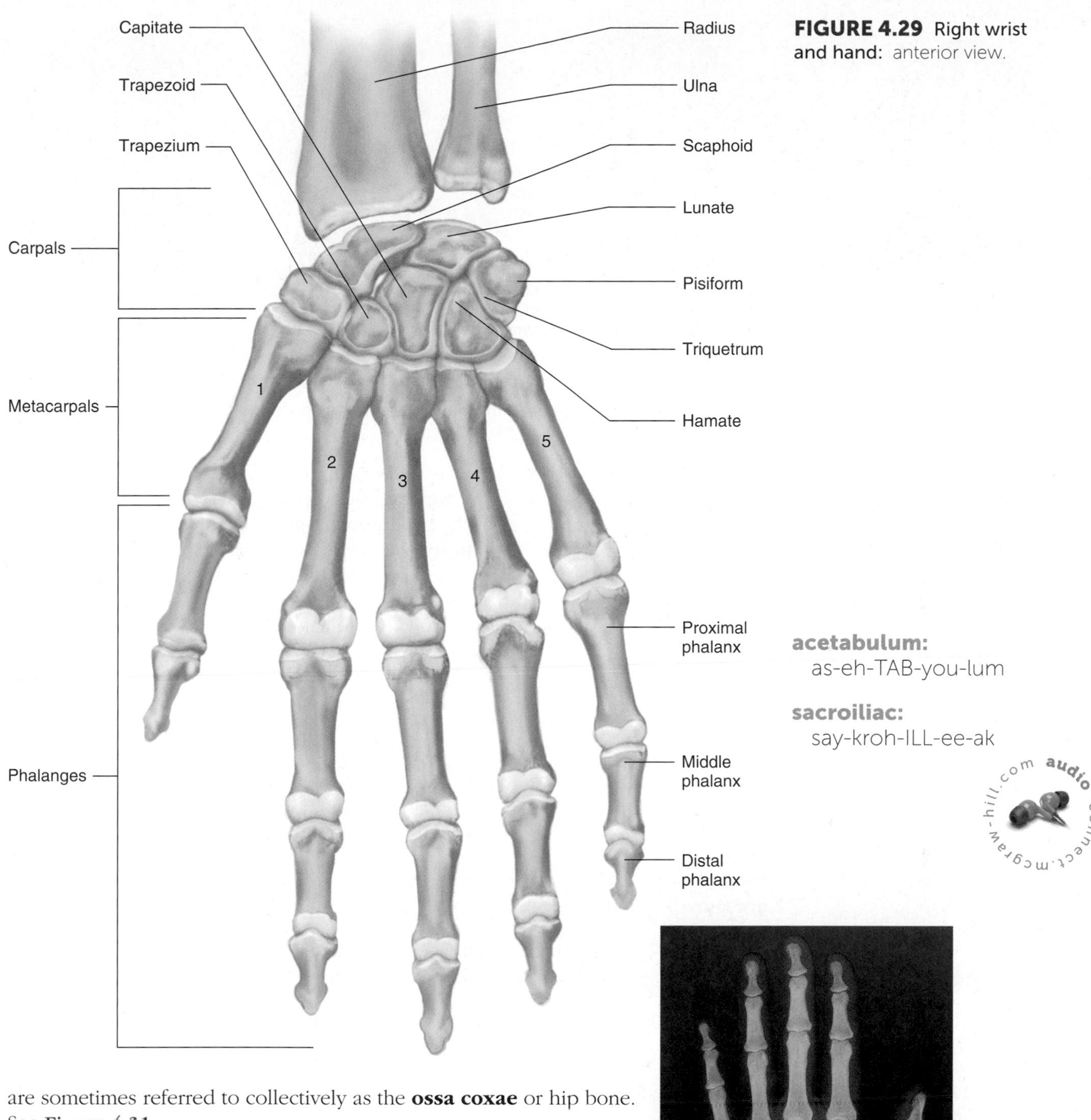

FIGURE 4.29 Right wrist and hand: anterior view.

acetabulum: as-eh-TAB-you-lum

sacroiliac: say-kroh-ILL-ee-ak

connect.mcgraw-hill.com audio

are sometimes referred to collectively as the **ossa coxae** or hip bone. See **Figure 4.31**.

The ilium is the most superior bone of a pelvic girdle, and the ischium is the most inferior. The third bone of the pelvic girdle, the pubis, is the most anterior bone. Together these three bones form a lateral feature called the **acetabulum.** This is the smooth hip socket that articulates with the thigh bone (femur). It is remarkable to think that three bones developed together to form a socket that would perfectly fit a fourth bone.

The right and left pelvic girdles along with the sacrum and coccyx form the **pelvis.** The joint between the ilium and the sacrum is called the **sacroiliac** joint. This knowledge will help you to determine the right and left pelvic girdles. The ilium has a rough area that must be

Sesamoid bone

FIGURE 4.30 X-ray of a hand, including a sesamoid bone.

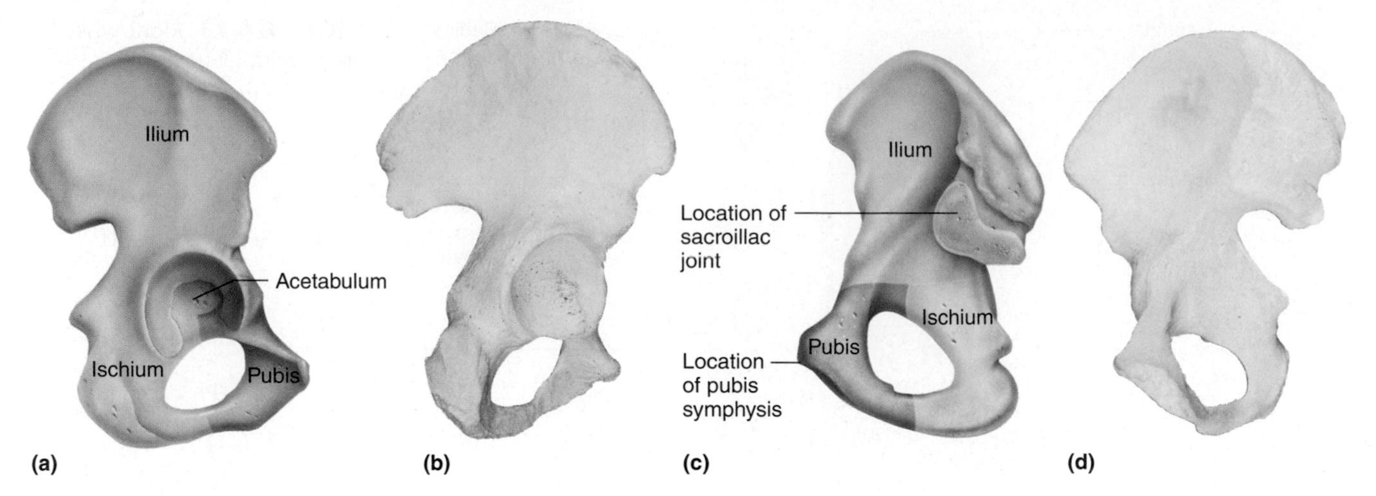

FIGURE 4.31 **Pelvic girdle:** (a) lateral view, (b) natural bone, lateral view, (c) medial view, (d) natural bone, medial view.

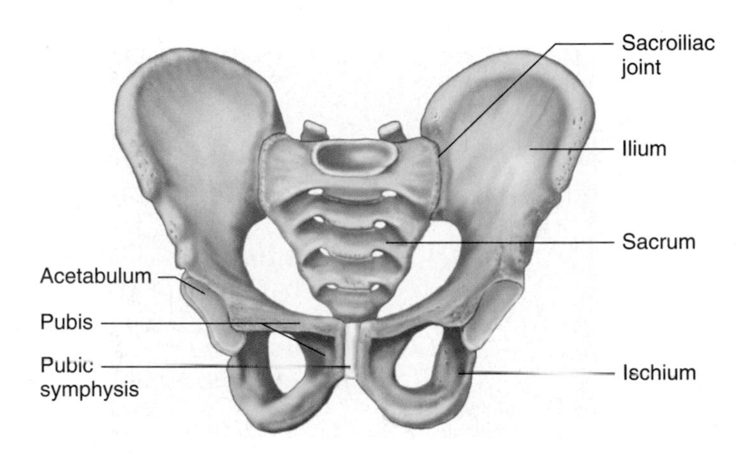

FIGURE 4.32 **Pelvis:** anterosuperior view.

medial to form the sacroiliac joint with the sacrum. The right and left pelvic girdles are connected anteriorly by fibrocartilage at a joint called the **pubic symphysis.** See Figure 4.32.

The male and female pelvises have specific differences, but, mainly, it is important for you to know the following: The female pelvis is wider and shallower, has a more rounded pelvic brim, and has a larger pelvic inlet or opening than the male pelvis. All of these differences are accommodations for pregnancy and birth. See Figure 4.33.

Bones of the Lower Limb The bones of the lower limb include the femur, tibia, and fibula and the tarsals, metatarsals, and phalanges of the foot.

symphysis: SIM-feh-sis

trochanter: troh-KAN-ter

Femur. The femur is the proximal long bone of the leg. Its special features (shown in Figure 4.34) include:

- **Head.** The head of the femur is a round ball-like structure at the proximal end of the femur. It articulates with the acetabulum of the pelvic girdle.
- **Neck.** The neck of the femur connects the head to the bone's shaft.

- **Greater and lesser trochanters.** The greater and lesser trochanters are similar to the tubercles found on the humerus of the arm. The trochanters are more

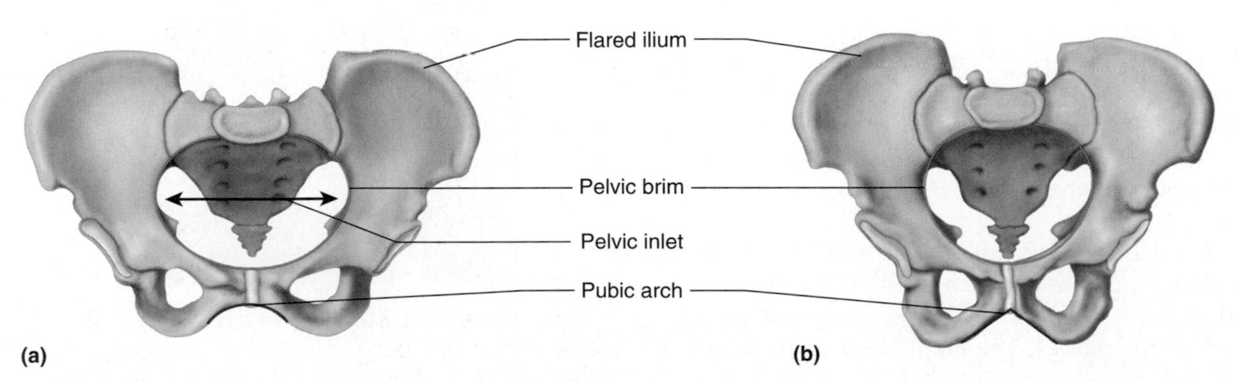

FIGURE 4.33 **Female and male pelvises:** (a) female pelvis, (b) male pelvis.

massive than the tubercles. Like the tubercles, the trochanters are attachment points for muscles by tendons.

- **Medial and lateral condyles.** The condyles are smooth surfaces at the distal end of the femur. They articulate with the tibia, which you will read about shortly. The condyles bulge and are more prominent posteriorly.
- **Medial and lateral epicondyles.** The epicondyles at the distal end of the femur are similar to the medial and lateral epicondyles of the humerus. They are rough prominences at the side of the condyles and serve as attachment sites for muscles by tendons. See **Figure 4.34**.

Patella. The patella is a sesamoid bone in the patellar ligament. It protects the patellar ligament from wear and tear as it glides over the femur and tibia at the knee joint. See **Figure 4.34**.

Tibia. The tibia is the more massive long bone of the lower leg. It articulates with the femur at the knee. Special features of the tibia (shown in **Figure 4.35**) include:

- **Medial and lateral condyles.** The medial and lateral condyles of the tibia articulate with the medial and lateral condyles of the femur at the knee.
- **Tibial tuberosity.** The tibial tuberosity is similar to the deltoid tuberosity of the humerus. It is a roughened area for the attachment of muscles by tendons.

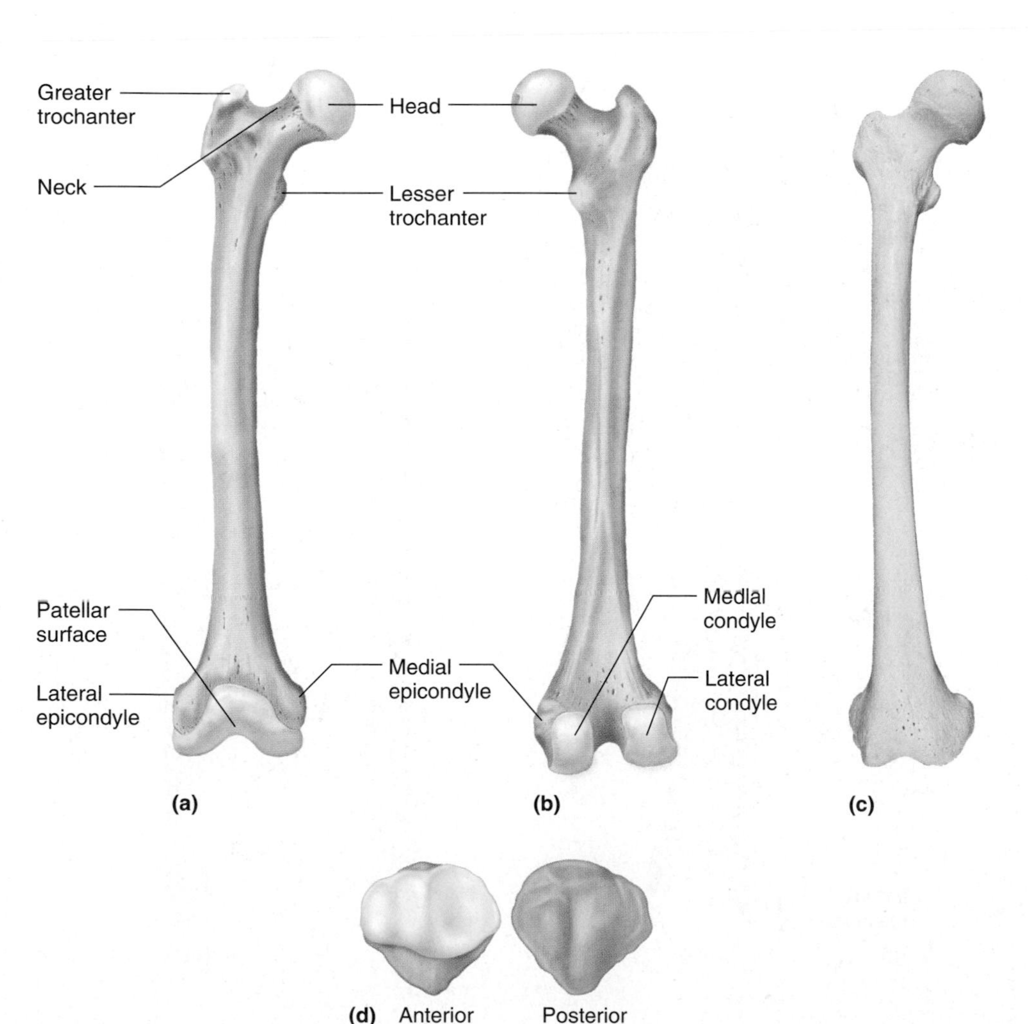

FIGURE 4.34 Femur and patella: (a) anterior view, (b) posterior view, (c) natural femur, anterior view, (d) patella.

- **Anterior crest.** The anterior crest of the tibia is a sharp ridge running along the anterior shaft of the bone. It is very close to the surface and is commonly referred to as the shin.

malleolus:
mal-LEE-oh-lus

- **Medial malleolus.** The medial malleolus is a specific name for a prominent medial epicondyle. It forms the knob of the medial ankle and is an attachment site for muscles by tendons. See **Figure 4.35**.

Fibula. The fibula is the less massive long bone of the lower leg. See **Figure 4.35**. It is lateral to the tibia, and it does not articulate with the femur at the knee. The head at the proximal end of the fibula articulates with the tibia. The lateral malleolus at the fibula's distal end forms the lateral knob of the ankle. It is an attachment point for muscles by tendons. It also anchors the ankle to prevent it from turning laterally.

Similar to the membrane located between the radius and the ulna, there is another interosseous membrane that runs between the shafts of the tibia and fibula. You will examine this further when you learn about joints.

calcaneus:
kal-KAY-knee-us

audio connect.mcgraw-hill.com

Tarsal bones. The tarsal bones are short bones of the ankle and foot. There are seven tarsal bones, which are described below and shown in **Figure 4.36**.

- **Talus.** The talus articulates with the distal end of the tibia.
- **Calcaneus.** The calcaneus is the heel bone. The **calcaneal (Achilles) tendon** attaches the calf muscles to the calcaneus.

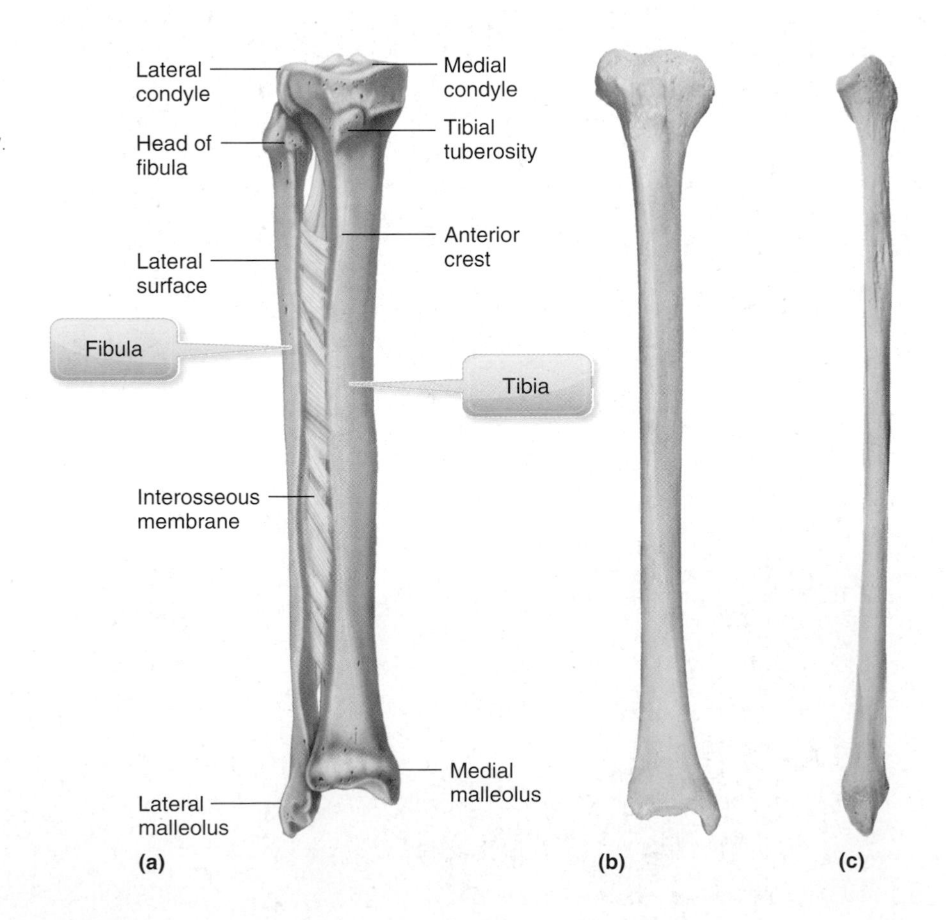

FIGURE 4.35 Right tibia and fibula: (a) anterior view, (b) natural tibia, anterior view, (c) natural fibula, anterior view.

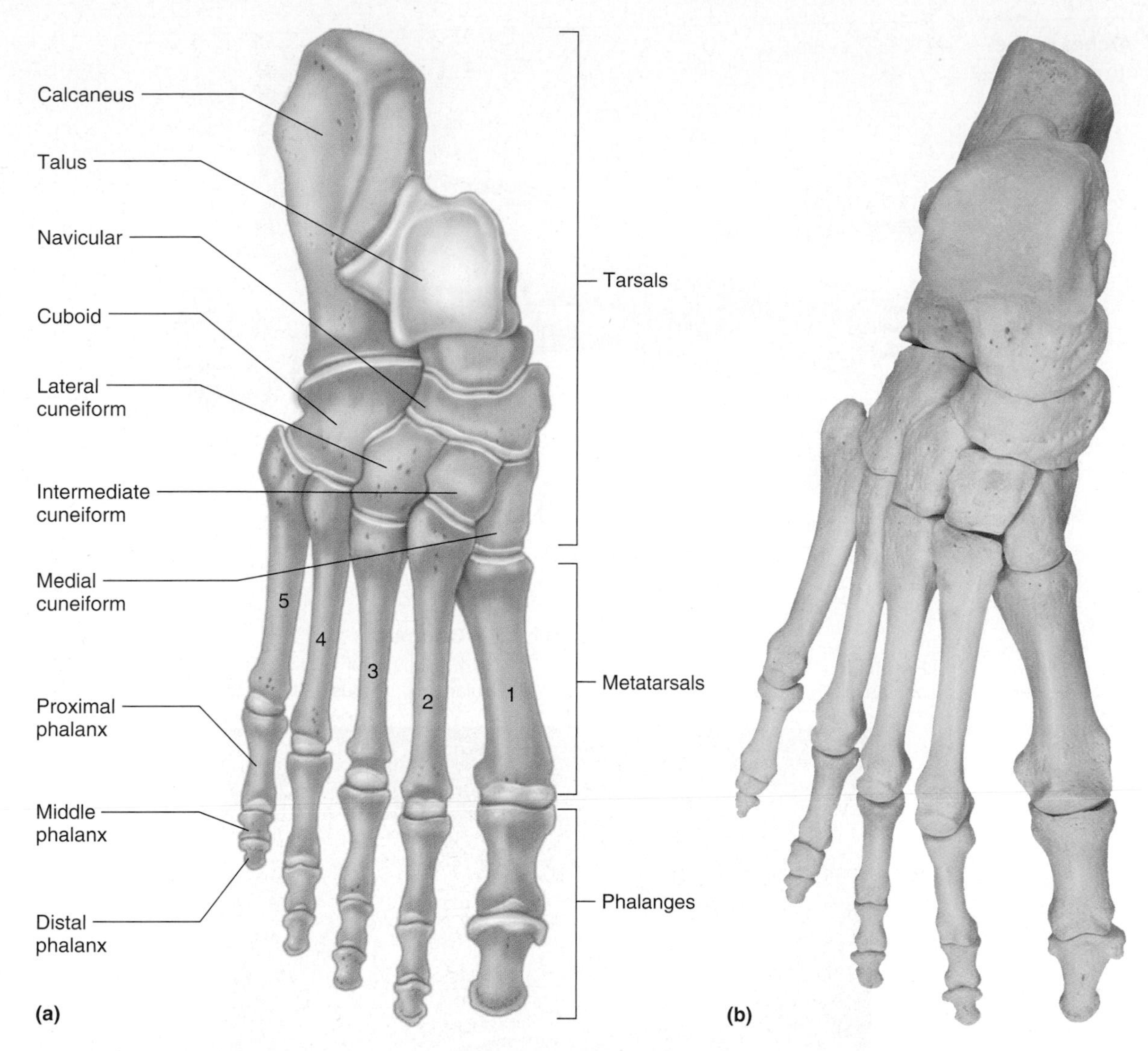

FIGURE 4.36 The foot: (a) superior (dorsal) view of right foot, (b) natural foot, superior view of right foot.

- **Navicular.** The navicular is a large, wedge-shaped bone.
- **Cuneiforms** and **cuboid.** The three cuneiforms and the cuboid bone make up the distal row of tarsals in the foot. The cuboid is the most lateral. See **Figure 4.36**.

cuneiform: KYU-ni-form

Metatarsals. There are five metatarsal bones in the foot, which are proximal to the toes and distal to the tarsals. Like the metacarpals of the hand, they are numbered 1 through 5. The metatarsal proximal to the great (big) toe is metatarsal 1. See **Figure 4.36**.

Phalanges. The 14 phalanges are long bones that make up the toes. They are named by their position as proximal, middle, or distal and by the toe in which they reside. The great toe is 1 and the little toe is 5. See **Figure 4.36**.

The tarsals and metatarsals of the foot form a longitudinal arch and a transverse arch to the sole of the foot. You can see this when viewing a footprint. Not all of the plantar surface comes in contact with the ground, as the arches are supported by strong ligaments that attach the bones together. See **Figure 4.37**.

FIGURE 4.37 Arches of the foot: (a) inferior (medial) view, (b) x-ray of right foot, lateral view.

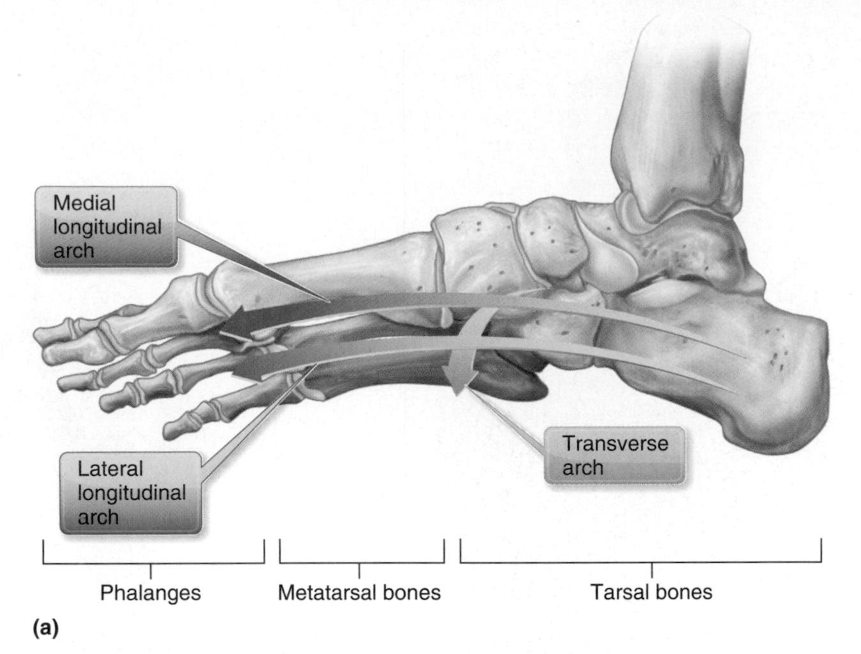

(a)

Cuboid
Cuneiform
Calcaneus
Proximal phalanx I
Metatarsal I
Navicular
Talus
Tibia
Fibula

(b)

clinical point

Congenital weakness, obesity, or repetitive stress can cause the foot ligaments to stretch, leading to a condition called *flat feet* or *fallen arches*. Here, the foot's entire plantar surface comes in contact with the ground when standing. People with this condition may be less tolerant to prolonged standing and walking.

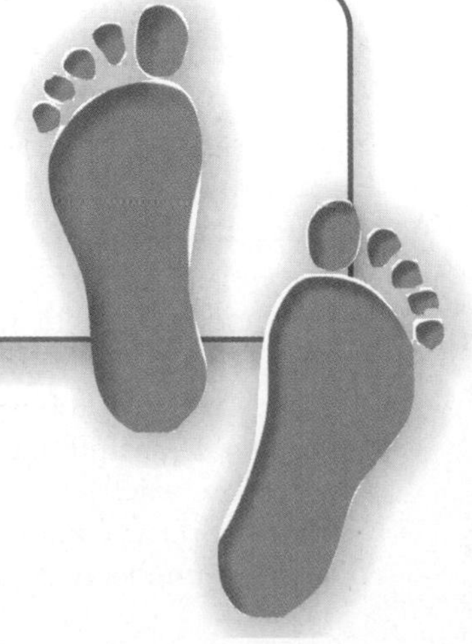

spot check 3 What bones make up the acetabulum?

spot check  Where does the pelvic girdle join the axial skeleton?

Histology of the Skeletal System

Now that you have a better understanding of the bones of the skeletal system, you are ready to explore the skeletal system at the microscopic level. At this level, you will need to look at bone, hyaline cartilage, elastic cartilage, and fibrocartilage connective tissues. You have already identified these tissues in Chapter 2. In this chapter, you will study the cells, fibers, and matrix of these tissues more completely to see how they function in the skeletal system.

4.5 learning outcome

Describe the cells, fibers, and matrix of bone tissue.

Bone Connective Tissue. Bone is dynamic tissue. It changes daily. You will learn why and how it changes shortly, but first you need to understand the cells, fibers, and matrix involved.

Osteoblasts and **osteoclasts** are types of **osteocytes** (bone cells). Osteoblasts build bone tissue, while osteoclasts destroy it. Both types of cells are necessary for the skeletal system to function properly. Osteoblasts build new bone by forming a soft matrix of protein and carbohydrate molecules with collagen fibers. The osteoblasts then allow hard mineral crystals to be deposited in the matrix. The mineral crystals are mostly **hydroxyapatite,** a calcium phosphate mineral salt. The hydroxyapatite crystals make the matrix hard. The collagen fibers give the matrix some flexibility. This process is somewhat similar to that in a sidewalk or bridge decking: Cement makes the sidewalk or decking hard, but steel rods embedded in the sidewalk or bridge decking give it flexibility that reinforces it. So, without sufficient hydroxyapatite, bones become soft. Without collagen fibers, bones become brittle. You can see this at home if you soak a chicken bone in vinegar for a few days. The acid of the vinegar will dissolve the hard mineral crystals in the chicken bone, leaving the flexible collagen fibers. You might even be able to tie the bone in a knot if enough mineral crystals have been dissolved.

clinical point

Osteogenesis imperfecta, commonly called *brittle bones,* is a congenital defect in which the bones lack collagen fibers. With this defect, the bones are very brittle and break easily. Until this condition is accurately diagnosed, affected children may appear to be victims of abuse because of their large number of broken bones.

Rickets is a childhood disorder in which an inadequate amount of mineral crystals is deposited in the bone. The bones are therefore too soft. The leg bones may not be able to completely support the weight of the body. As a result, the legs become bowed and deformed as they develop.

4.6 learning outcome

Compare and contrast the histology of compact and cancellous bone.

There are two types of bone tissue. **Compact bone** is very dense and highly organized. **Cancellous bone** is spongy in appearance, characterized by delicate slivers and plates of bone with spaces between.

Compact bone. This type of bone is found in the shafts of long bones and the surfaces of flat bones. Compact bone tissue is arranged in a series of **osteons (Haversian systems)** that appear as targets. See **Figure 4.38b** and **c**. The central

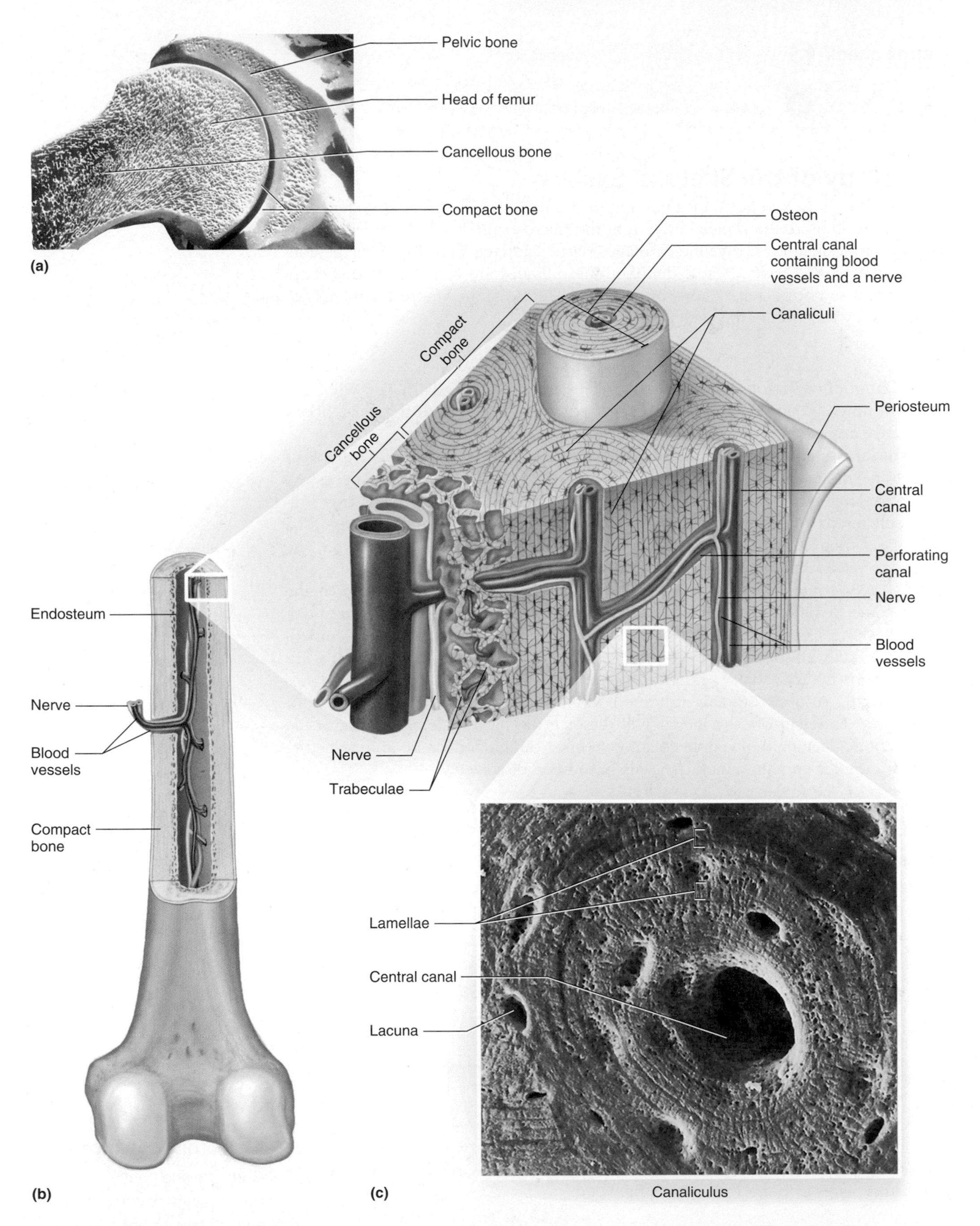

FIGURE 4.38 **Histology of bone tissue:** (a) compact and cancellous bone in a frontal section of the hip joint, (b) three-dimensional illustration of an osteon, (c) microscopic view of an osteon. © Dr. Richard Kessel & Dr. Randy Kardon/ Visuals Unlimited, Inc.

osteonic or **Haversian canal** contains blood vessels and a nerve. Matrix is formed around the canal in concentric layers called **lamellae.** The osteocytes are found in spaces called **lacunae** arranged in circles around the central canal. They have boxed themselves in the lacunae by depositing matrix around themselves. Tiny cracks called **canaliculi** in the matrix allow the osteocytes to reach out to each other and to the central canal for nutrients. It is important to realize that canaliculi are microscopic. The one osteon you see in **Figure 4.38c** has been highly magnified. It represents a very small sliver of bone on the microscope slide.

canaliculi: kan-ah-LIK-you-lie

It is also important to realize that bone cells have a very good supply of nutrients through the blood supply to the central (Haversian) canals and the canaliculi. Bone cells need these features because the matrix is very dense. Diffusion does not work (to supply osteocytes with the nutrients they need for survival or to remove their wastes) across this dense mineral crystal matrix.

Cancellous bone. This type of bone is found in the end of long bones and in the middle of flat and irregular bones. If you cut through a flat bone, it would look like a cancellous bone sandwich: It would have a thin layer of compact bone on the surfaces as the bread, and the filling would be cancellous bone. Cancellous bone is not as organized as compact bone, and it does not have Haversian systems. Here, the matrix is arranged in delicate slivers and plates called **trabeculae.** The spaces between the interlocking trabeculae give cancellous bone a spongy appearance. See **Figure 4.38a**.

trabeculae: tra-BECK-you-lee

You have now explored the cells, fibers, and matrix of bone tissue and compared the histology of two types of bone: compact bone and cancellous bone. Cartilage—previously mentioned in regard to the bones and structures of the axial and appendicular skeletons—also plays an important role in this system. We explain the histology of cartilage connective tissue in detail below.

4.7 learning outcome

Compare and contrast the histology of hyaline, elastic, and fibrocartilage connective tissues.

Cartilage Connective Tissue There are three types of cartilage in this system: hyaline, elastic, and fibrocartilage. All three types of cartilage have the same cells, fibers, and matrix. How the fibers are arranged is what makes the difference.

In cartilage connective tissue, the cells are called **chondrocytes.** They produce a matrix of **proteoglycans** and water. A proteoglycan is basically a protein molecule with a carbohydrate added to it. When viewed through a microscope, the matrix looks like clear gelatin containing bubbles. That is not a bad analogy, as gelatin is basically protein, sugar (a carbohydrate), and water, just like the matrix. The bubbles are lacunae—spaces similar to the lacunae of compact bone. Chondrocytes have deposited the matrix in every direction, trapping themselves in the lacunae. The fibers of cartilage are made of collagen. See **Figure 4.39**.

It is important to notice that this type of tissue lacks a blood supply. Unlike bone, cartilage does not have blood vessels feeding the chondrocytes. The reason for the difference is seen in the matrices: The matrix of cartilage is like gelatin, whereas the matrix of bone is like cement. If you put a drop of ink on the surface of a cube of each matrix, the ink would be able to diffuse across the cartilage matrix but it would not be able to diffuse across the mineral matrix of bone. Chondrocytes are therefore fed and have their wastes removed by diffusion across the matrix. They do not need a direct blood supply. Diffusion through the matrix is a slower process than having a direct blood supply.

Hyaline cartilage connective tissue. This type of cartilage is found covering the ends of long bones, in the costal cartilages of the ribs, and in the nasal cartilages of the nose. The matrix looks like gelatin, smooth and clear. The lacunae are prominent, with chondrocytes visible within them. Whatever collagen fibers are present are so fine that they are barely visible. See **Figure 4.39a**.

Elastic cartilage connective tissue. This cartilage is found in the pinna of the ear (outer ear flap) and in the epiglottis in the throat. It has the same chondrocytes, matrix, and lacunae as the other cartilages. The difference is the direction of the fibers. In elastic cartilage the fibers run in all directions, giving the cartilage elasticity. A good example is the outer ear flap. You can bend it over and let go, and it immediately goes back to shape. See **Figure 4.39b**.

Fibrocartilage connective tissue. This type of cartilage is found in the intervertebral disks, the pubic symphysis, and the meniscus of the knee. Fibrocartilage has the same matrix and chondrocytes in lacunae as the other types of cartilage. The difference is that the collagen fibers are very visible and run in only one direction in fibrocartilage. This arrangement of fibers allows fibrocartilage to serve as a shock absorber. The knee's **meniscus** is a prime example. Each footstep causes the femur to pound against the tibia. The meniscus (made of fibrocartilage) lies between the femur and the tibia in the knee joint. It absorbs the shock of the impact and prevents wear and tear on the joint. See **Figure 4.39c**.

spot check **5** How does the matrix of compact bone differ from that of cancellous bone? How are the matrices of the three types of cartilage similar?

spot check **6** How do the matrices of bone and cartilage differ in terms of how cells are fed and have their wastes removed?

4.8 learning **outcome**

Describe the anatomy of a long bone.

Anatomy of a Long Bone

Now that you have looked at the bones and the histology of the skeletal system, it is time to look at the anatomy of a long bone. This will bring the gross anatomy and the histology together. See **Figure 4.40**.

The **epiphyses** are the clubby ends of a long bone. Each epiphysis is covered by **articular cartilage,** which is composed of hyaline cartilage connective tissue. The articular cartilage provides a smooth surface for the end of the long bone to articulate with another bone. Articular cartilage is firmly attached to the bone. Cancellous bone is found in the epiphyses. Red bone marrow fills the spaces between the cancellous bone's trabeculae.

warning

Many students mistake articular cartilage for bone when dissecting a specimen. It appears to be the hard, white, smooth end of the bone. But articular cartilage can be sliced with a scalpel, and bone cannot.

diaphysis: die-AF-ih-sis

The **diaphysis** is the shaft of the long bone. It is composed of compact bone, but it is not solid bone. The diaphysis is a hollow tube of compact bone filled with yellow bone marrow in what is called a **marrow (medullary) cavity.** Why is the diaphysis hollow? After all, wouldn't a solid rod of bone be stronger than a pipe of bone? Yes, it would, but it would also be much heavier. The presence of the marrow cavity reduces the bone's weight.

The diaphysis has a fibrous covering called the **periosteum.** It starts where the articular cartilage ends (at the proximal epiphysis) and goes to where the articular cartilage begins (at the distal epiphysis). The periosteum encircles the diaphysis,

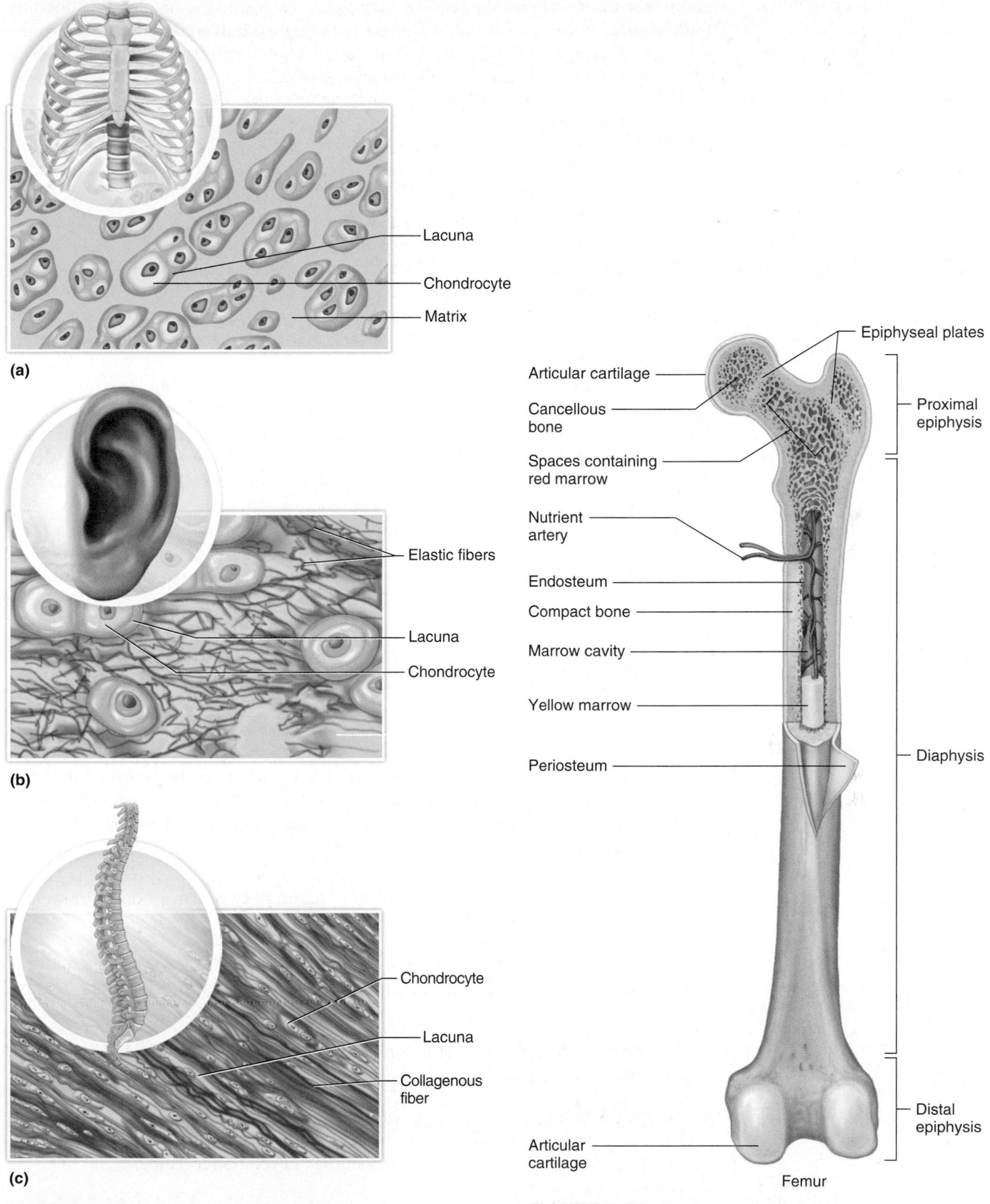

FIGURE 4.39 Cartilage connective tissue: (a) hyaline cartilage, (b) elastic cartilage, (c) fibrocartilage.

FIGURE 4.40 Anatomy of a long bone.

and it is a source of osteoblasts. The marrow cavity inside the diaphysis is lined by **endosteum,** which is a source of osteoclasts. A **nutrient artery** enters the bone through a foramen in the diaphysis. This artery branches to form the blood vessels in the many Haversian canals in the compact bone.

spot check  If the shaft of the humerus were broken, would the bone bleed? Explain.

4.9 learning **outcome**
Distinguish between two types of bone marrow in terms of location and function.

Red bone marrow is found in the spaces of cancellous bone. This includes flat bones like the sternum, irregular bones like the vertebrae, and the epiphyses of long bones. Red bone marrow is composed of stem cells, which produce both red and white blood cells and platelets. You will learn more about blood cell production in Chapter 9.

Yellow bone marrow is found in the marrow cavity of mature long bones. The marrow cavity in a developing long bone originally contains red marrow. By the time the bone matures, the marrow has become yellow marrow composed mostly of fatty tissue. Yellow marrow reduces the bone's weight because fat is less dense than bone. Yellow bone marrow does not produce blood cells, but it can convert back to red bone marrow in cases of extreme anemia. You can see yellow bone marrow in the meat department of a grocery store. The round bone seen in a thick slice of ham or a thin beef roast is a transverse section of the diaphysis of a long bone. The bone is compact bone. The fatty substance in the center of the bone is yellow bone marrow.

Now you understand the anatomy of a long bone, including the articular cartilage that covers the ends of the bone. This cartilage provides a smooth surface where two or more long bones meet. Below, you will learn about other ways bones meet.

4.10 learning **outcome**
Describe three major structural classes of joints and the types of joints in each class.

Joints

When two or more bones meet, they form a **joint** or an **articulation.** Joints can be classified by their anatomy or by the amount of motion they allow. By their anatomy, joints can be classified as fibrous, cartilaginous, or synovial. There are subdivisions called *types* within each class.

Fibrous Joints Fibrous joints have fibrous tissue between the bones. There are three types of joints in this class (see Table 4.1):

suture: SOO-chur

gomphosis: gom-FOE-sis

syndesmosis: sin-dez-MOH-sis

- **Sutures.** A suture has a fibrous membrane between bones until the suture is completely closed. It can be found between cranial bones of the skull. Sutures are immovable joints.
- **Gomphoses.** A gomphosis is formed by fibrous ligaments holding a tooth in its socket. Gomphoses are immovable joints.
- **Syndesmoses.** A syndesmosis is formed by an interosseous membrane. It can be found between the radius and the ulna and between the tibia and the fibula. Syndesmoses may allow some movement, as in rotating a hand.

Cartilaginous Joints Cartilaginous joints have cartilage between the bones. There are two types of joints in this class (see Table 4.1):

- **Symphyses.** The pubic symphysis has fibrocartilage between the two pubic bones. This joint becomes more elastic and slightly movable during the birth process.

TABLE 4.1 Joints

Class	Type	Description	Movement	Location
Fibrous		Fibrous tissue connecting bones together		
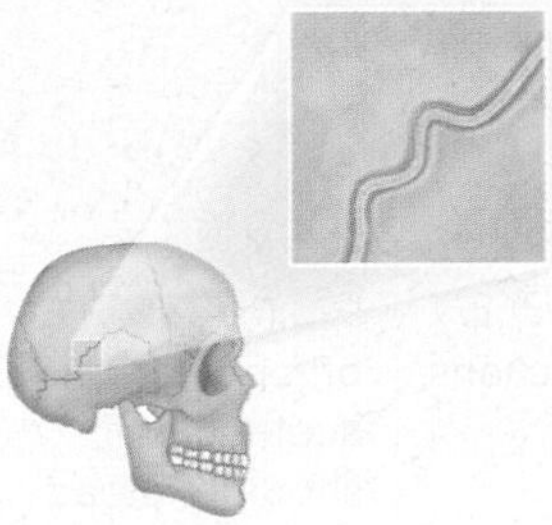	Suture	Membrane of the skull.	Immovable	Between flat bones of the skull
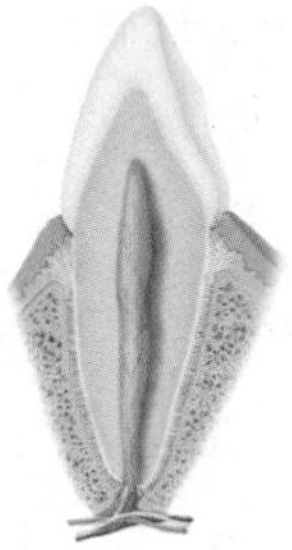	Gomphosis	Ligament.	Immovable	Tooth in socket
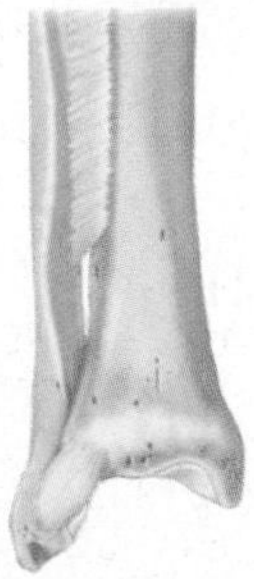	Syndesmosis	Formed by interosseus ligament.	Partly movable	Between tibia and fibula; between radius and ulna
Cartilaginous		Cartilage connecting bones together.		
	Symphysis	Fibrocartilage connecting bones.	Very little movement	Between vertebrae of the spinal column; between pubic bones
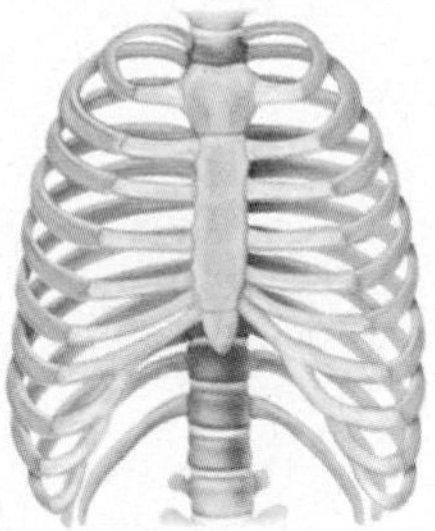	Synchondrosis	Hyaline cartilage connecting bones.	Partly movable	Between the ribs and sternum
			Immovable	Between epiphysis and diaphysis of a developing long bone

Class	Type	Description	Movement	Location
Synovial		Bones capped with cartilage articulate within a fluid-filled cavity.	Freely movable	
	Hinge	C-shaped surface of one bone swings about the rounded surface of another bone.	Very movable in one direction, like a door hinge	Elbow, knee, interphalangeal joints
	Ball and socket	Ball of one bone fits into a socket of another.	Very movable in all directions	Hip (head of femur to acetabulum); shoulder (head of humerus in glenoid cavity)
	Saddle	Concave surfaces of two bones articulate with one another.	All movements possible, but rotation limited	Carpometacarpal joint of the thumb
	Gliding	Two opposed flat surfaces of bone glide past one another.	Up-down wave of the hand at the wrist	Carpal bones
	Ellipsoid	Reduced ball and socket.	All movements, but rotation severely limited: side-to-side wave of the hand at the wrist; abduction and adduction of the fingers	Carpal bones; joint between metacarpals and phalanges
	Pivot	Ring of bone articulates with a post of bone.	Rotation	Atlas on the odontoid process/dens of the axis

- **Synchondroses.** Synchondroses can be found in the long bones of children. A synchondrosis is the cartilage joint between an epiphysis and the diaphysis of a long bone. This is an immovable joint.

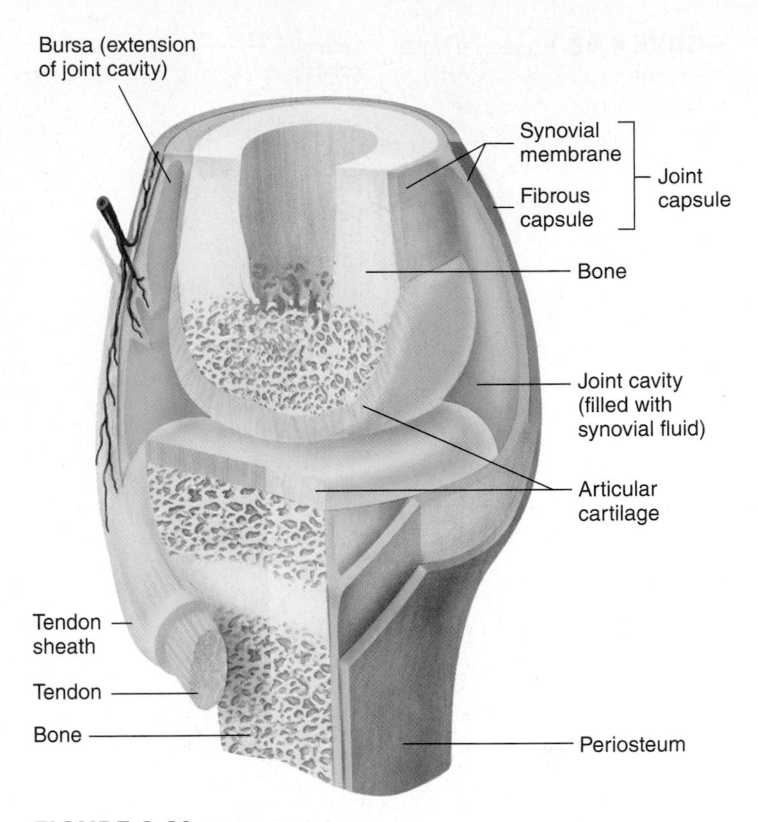

FIGURE 4.41 Synovial joint.

Synovial Joints Look at the anatomy of a synovial joint in Figure 4.41. Synovial joints have a joint cavity. This joint space is formed by a **joint capsule** that surrounds and seals the joint space. The capsule is composed of fibrous connective tissue continuous with the periosteum of the articulating bones. The joint space is lined by a **synovial membrane,** which produces a very slippery **synovial fluid.** This fluid lubricates the joint, reducing the heat of friction as the bones articulate. The synovial fluid also feeds and removes wastes from the articular cartilage that covers the bones' surfaces in the joint space. The synovial membrane may extend at the joint to form a pocket called a **bursa.** The bursa acts as a cushion for tendons rubbing against the bone.

The knee is a prime example of a synovial joint. The femur, tibia, and patella form the knee joint. See Figure 4.42. The knee is a relatively unstable joint because the distal end of the femur articulates with the rather flat-surfaced tibia at the knee. In comparison, the hip is a more stable joint because the head of the femur fits into a deep, bony socket called the *acetabulum*. Five ligaments connect the bones to help support the knee:

synchondrosis: sin-kon-DROH-sis

synovial: si-NOH-vee-al

- Medial and lateral **collateral ligaments** attach the epicondyles of the femur to the epicondyles of the tibia and fibula. They prevent side-to-side movement at the knee.

- Anterior and posterior **cruciate ligaments** attach the femur to the tibia. They cross to form an X between the femur's condyles, and they are named for their attachment relative to the tibia: The anterior cruciate ligament attaches to the tibia's anterior, and the posterior cruciate ligament attaches to the tibia's posterior side. These ligaments prevent the femur from sliding forward or backward relative to the tibia.

- The **patellar ligament,** sometimes called the *patellar tendon,* attaches the patella to the tibia. It also attaches the quadriceps muscles of the anterior thigh to the tibia at the tibial tuberosity. It qualifies as a tendon and a ligament because it attaches muscle to bone and bone to bone.

There are two C-shaped fibrocartilage pads—the menisci—between the femur and the tibia. Each meniscus acts as a shock absorber for a femoral condyle and prevents it from sliding from side to side. See **Figure 4.42d**. You can also see in **Figure 4.42c** the many bursae that cushion the knee joint.

The human body contains six types of synovial joints (shown in Table 4.1):

- **Hinge.** A hinge joint is very movable in one direction, like a door hinge. Example: elbow.

- **Ball and socket.** The ball of one bone fits into a socket of another bone. This type of synovial joint is very movable in all directions. Example: hip.

FIGURE 4.42 Knee: (a) right knee, anterior view, (b) right knee, posterior view, (c) left knee, medial view, (d) left tibia and menisci, superior view.

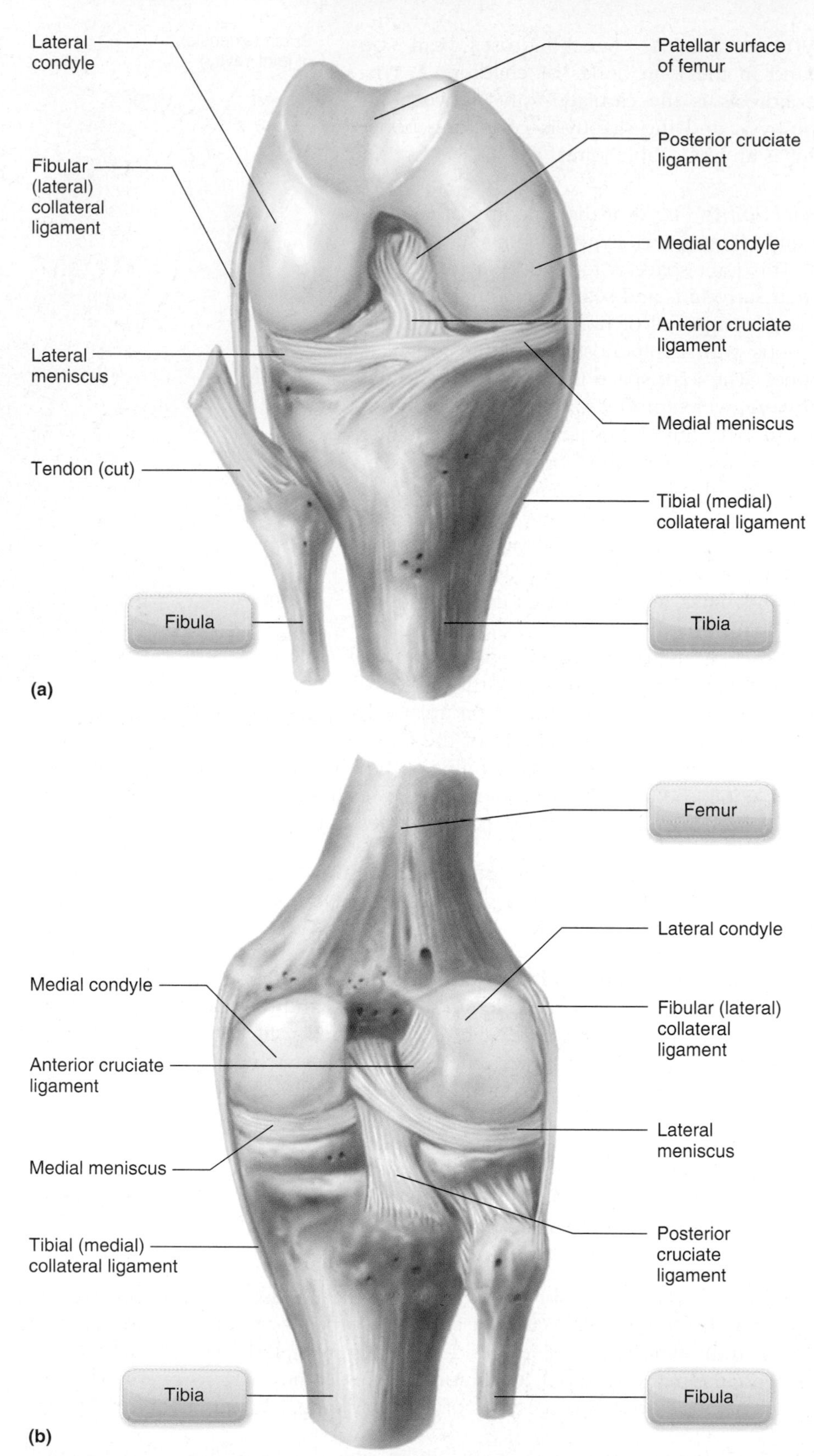

FIGURE 4.42 concluded

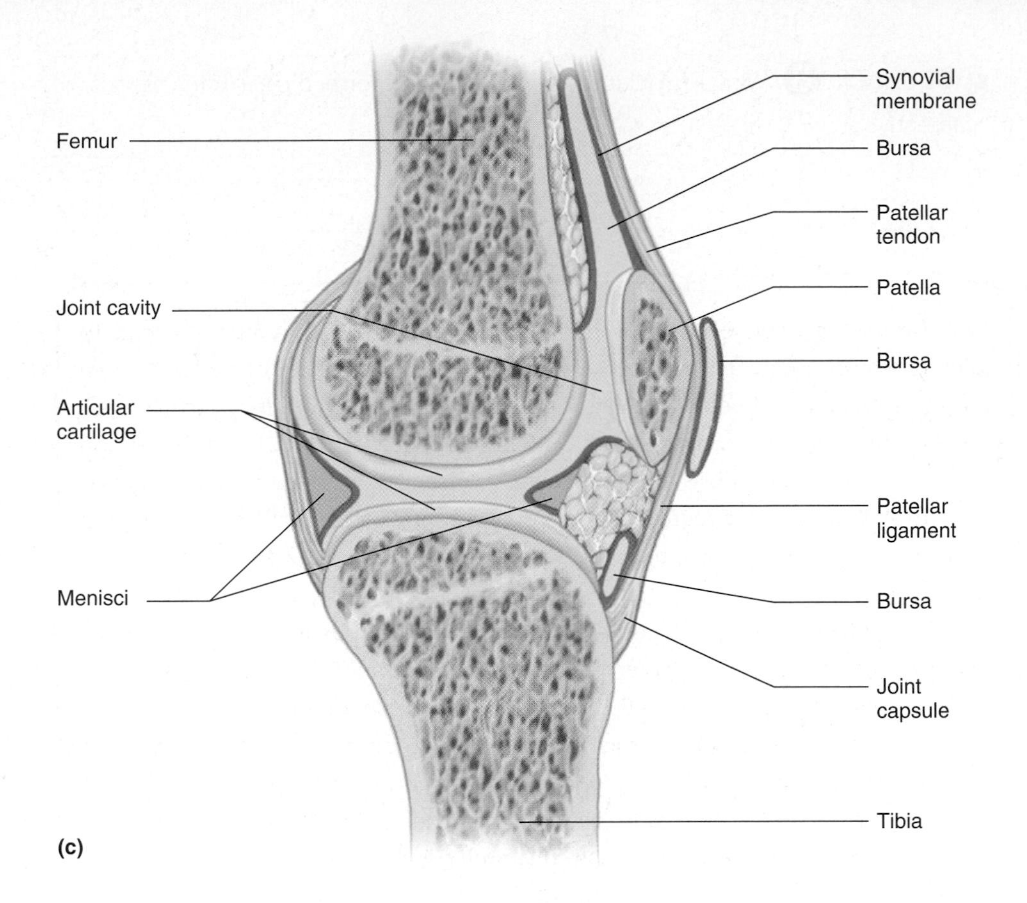

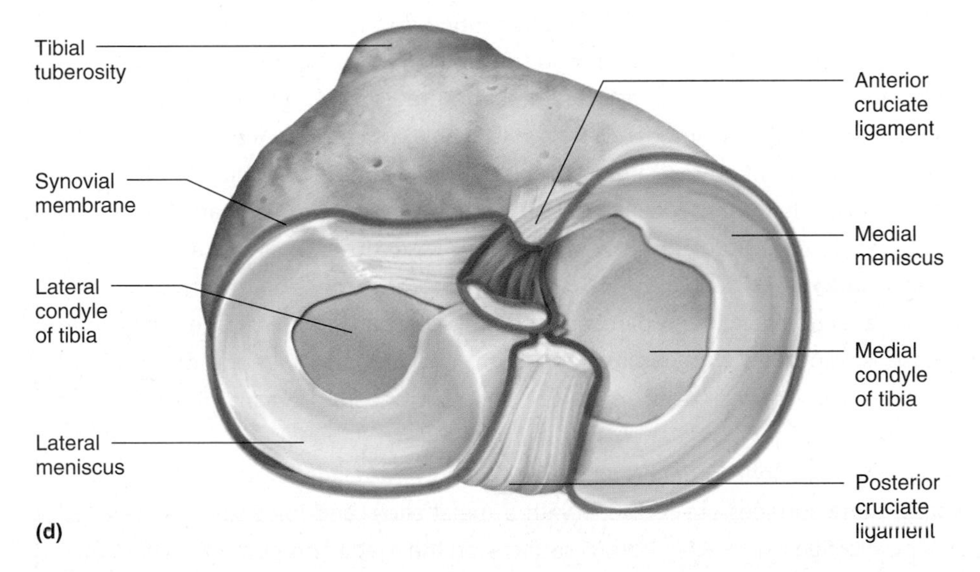

- **Saddle.** Concave surfaces of the bones articulate with one another. All movements are possible, but rotation is limited. Example: carpal-metacarpal joint of the thumb.
- **Gliding.** Flat surfaces of the bones glide past one another. Example: carpal bones, where this joint produces an up-and-down wave of the hand.
- **Ellipsoid.** This is a reduced ball and socket. Example: carpal bones, where this joint produces a side-to-side wave of the hand. Another example: metacarpophalangeal joints, which enable the fingers to be spread apart.
- **Pivot.** A ring of bone articulates with a post of bone. Example: atlas on the odontoid process of the axis, which enables rotation of the head.

spot check 8 What kind (class and type) of joint is formed by an interosseous ligament?

4.11 learning outcome

Differentiate between rheumatoid arthritis and osteoarthritis.

clinical point

Because we use our joints daily, normal wear and tear can eventually lead to joint problems, such as **arthritis**. Arthritis is an inflammation of a joint. There are more than 100 types of arthritis, and some of them can affect other tissues beyond the joints. According to the Centers for Disease Control and Prevention (CDC), arthritis is the most common cause of disability in the United States, with over 19 million adults affected.[1] All forms of arthritis involve stiff or painful joint movement. With this constant pain comes fatigue. There are two different forms of the disorder.

Osteoarthritis is the most common form of arthritis. It usually occurs in people over the age of 40, and 85 percent of people over the age of 70 show some signs of this condition. It is caused by the normal wear and tear of a joint or injury to the joint as the articular cartilage wears with age and becomes rough. **Crepitus** is the creaking sound that may be heard during the movement of osteoarthritic joints. This form of arthritis usually occurs in joints of the fingers, hips, knees, and vertebrae. Osteoarthritis is often treated with anti-inflammatory drugs to relieve the symptoms and physical therapy to improve function.

Rheumatoid arthritis (RA) is an autoimmune disease that can happen to anyone at any age. Children may develop juvenile RA. In rheumatoid arthritis, the body's own immune system attacks the structures of the joint. Antibodies (produced by the body's immune system) mistakenly attack a joint's synovial membranes, causing inflammation. As the synovial membranes thicken, enzymes produced by inflammatory cells erode the articular cartilage. The cartilage can be eaten away to the point that the articulating bones fuse with one another. This is called **ankylosis.** See Figure 4.43. RA tends to have periods of remission, and then it flares up again. Joint damage, however, is progressive. Management of the disease is important to minimize long-term joint damage. A range of treatment options exists, from anti-inflammatory and immunosuppressant drugs to joint replacement surgery.

Joint replacement surgery is an option for badly damaged joints. Here, the articulating bone surfaces are replaced with a metal alloy, and joint sockets are lined with plastic. See Figure 4.44. Porous surfaces on the metal components (where they attach to the bone) allow osteoblasts to deposit bone into the component surface to ensure a tight bond. Joint replacement can be very successful in restoring mobility. The majority of artificial knees last more than 20 years.

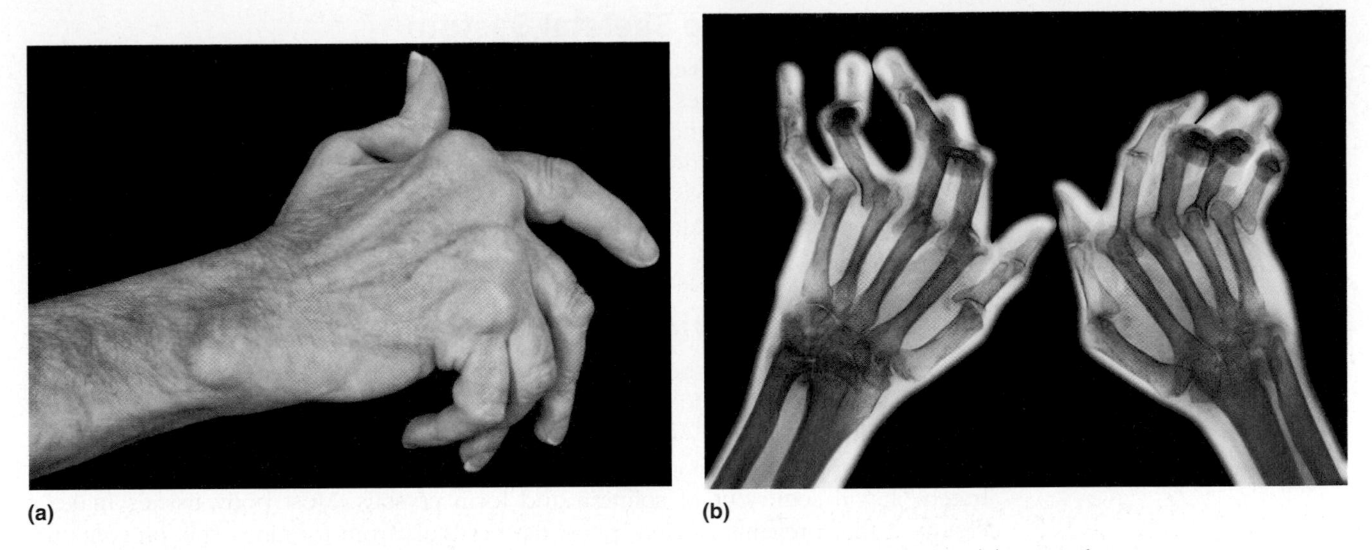

FIGURE 4.43 Rheumatoid arthritis: (a) a severe case showing ankylosis of the joints, (b) x-ray of a severe case.

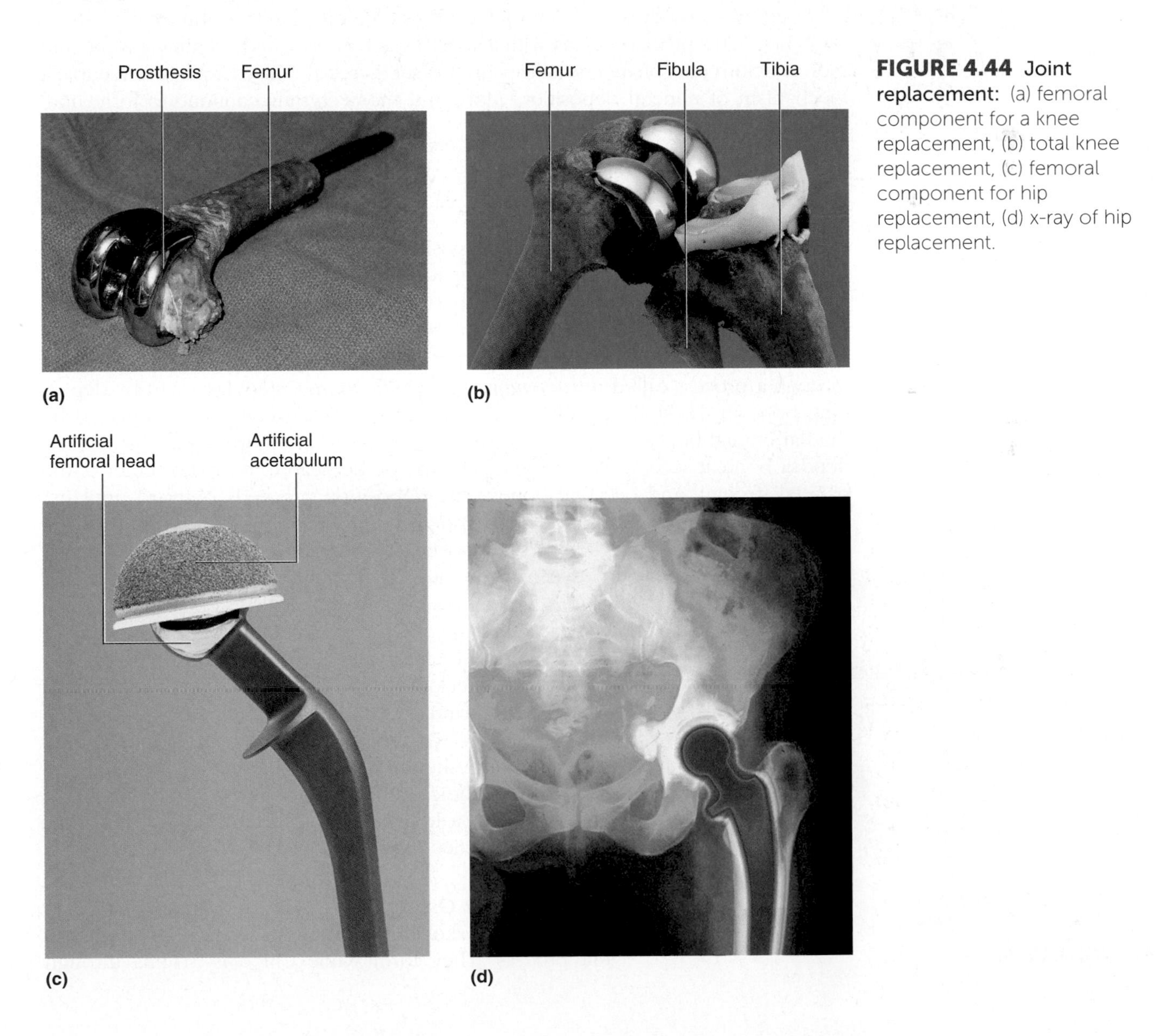

FIGURE 4.44 Joint replacement: (a) femoral component for a knee replacement, (b) total knee replacement, (c) femoral component for hip replacement, (d) x-ray of hip replacement.

Physiology of the Skeletal System

Up to this point, you have covered all of the skeletal system's anatomy. You have looked at bones, cartilages, and joints. Now it is time to understand how all of the anatomy functions. As you know, bone is dynamic tissue that changes daily. To show you how the structures of this system develop, grow, and continue to change throughout life, we discuss their physiology below in the context of Maria and her soon-to-be-born baby, Chloe. See **Figure 4.45**.

4.12 learning **outcome**

Explain how minerals are deposited in bone.

Mineral Deposition

To understand how your bones develop, first you need to learn how bone matrix is formed. Osteoblasts produce the collagen fibers of bone, but they do not produce bone's hydroxyapatite crystals. They simply allow hydroxyapatite to be deposited. Calcium phosphate is dissolved in body fluids and blood. If it is highly concentrated, it will settle out of solution and form crystals. Most body tissues make a chemical that prevents calcium phosphate crystals from forming. This prevents tissues like those in the muscles, liver, and eyes from calcifying.

Osteoblasts produce a chemical that allows the dissolved calcium phosphate to crystallize. This process begins with a single "seed" crystal, and eventually more and more calcium phosphate crystallizes on this seed crystal. This is a positive-feedback mechanism of mineral deposition. More and more crystals continue to form until they are out of the osteoblasts' range.

4.13 learning **outcome**

Compare and contrast endochondral and intramembranous ossification.

Bone Development

Now that you know how osteoblasts deposit matrix in general, you are ready to see how flat bones and long bones are formed in a fetus. The processes involved are explained below and summarized in Table 4.2.

Intramembranous Ossification The flat bones of baby Chloe's skull are forming through a process called *intramembranous ossification*. Osteoblasts start by depositing bone in the skull membrane. The membrane will eventually be replaced by the flat cranial bones. Look at the top of Chloe's skull in **Figure 4.46a**. Osteoblasts deposit bone at several sites in the membrane of the head to form the center of the parietal, frontal, and occipital bones. The deposition proceeds outward until the bones eventually fuse. This will not happen, however, by the time Chloe is born. At birth, there will be **fontanelles** (membranous areas) between the bones of the skull. Chloe's mom may refer to the larger fontanelle between the frontal and parietal bones as the baby's "soft spot." Once the bones meet, they form a joint called a *suture*. The flat bones will cease to grow any bigger once the sutures have completely closed in adulthood. Until that time, the membranes continue to grow as the osteoblasts continue to deposit more bone. This allows the skull to continue to grow larger. Imagine if this were not the case, and the sutures closed prematurely. How would you look with a head the size of a newborn on your body? See **Figure 4.46b**.

FIGURE 4.45 Maria.

Endochondral Ossification The long bones of Chloe's appendicular skeleton are formed through a different process. They form while Chloe is a fetus through

TABLE 4.2 Types of ossification

Intramembranous ossification	Endochondral ossification
1. This process forms flat bones of the skull of developing fetus.	1. This process forms long bones, vertebrae, ribs, sternum, scapula, pelvis, and bones of the limbs of developing fetus.
2. Osteoblasts deposit bones in membranes.	2. Osteoblasts deposit bone in hyaline cartilage models.
3. Site of ossification is the center of the future bone.	3. Primary site of ossification is the diaphysis, with deposition toward the epiphyses. Secondary sites of ossification are in each epiphysis, depositing toward the diaphysis.
4. Membranes continue to grow as bone is being deposited.	4. Cartilage of the model continues to grow as bone is being deposited.
5. Existing membrane, not yet bone, present at birth is called a fontanelle.	5. Existing cartilage between the epiphyses and the diaphysis is called an epiphyseal plate.
6. Continued membrane growth allows bones of the skull to get bigger as the child grows.	6. Epiphyseal plate continues to grow, allowing bones to increase in length.
7. Development is finished when sutures are completely closed.	7. Development is finished when the epiphyseal plates are completely closed.

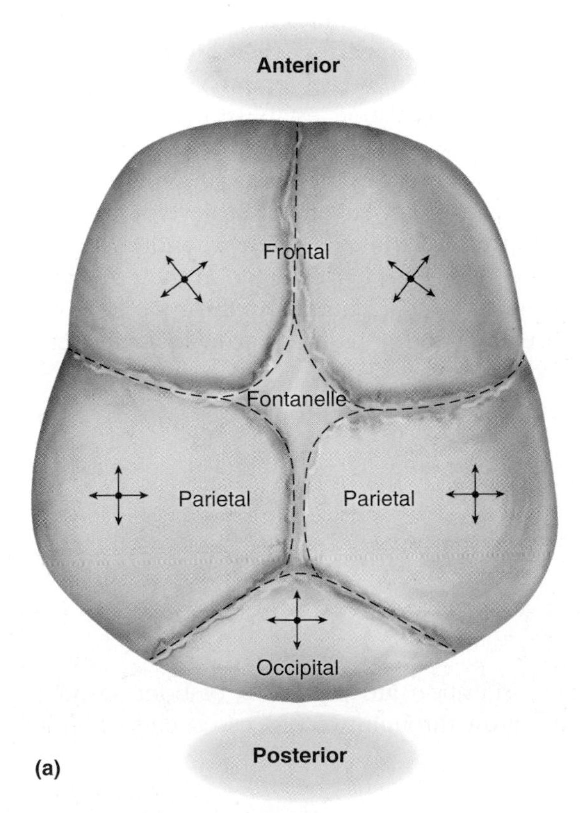

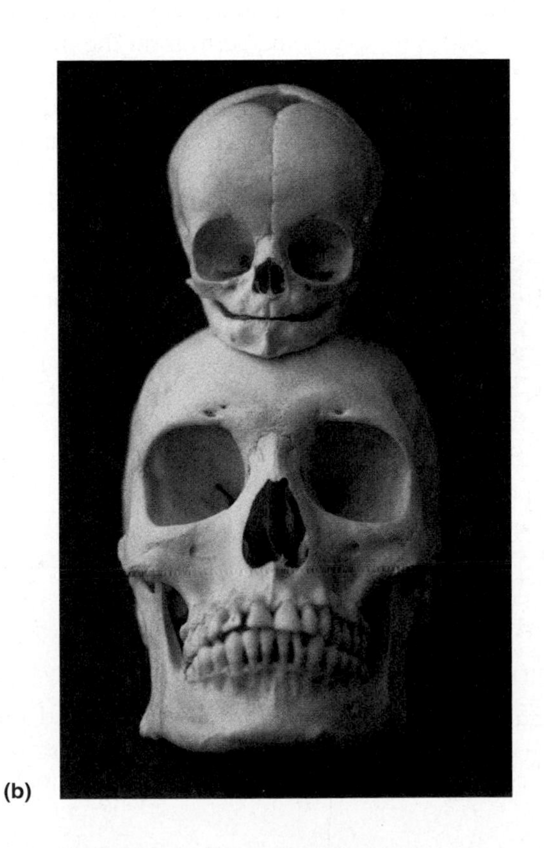

FIGURE 4.46 Intramembranous ossification of the skull: (a) fetal skull, superior view, (b) baby skull (showing a fontanelle) atop an adult skull.

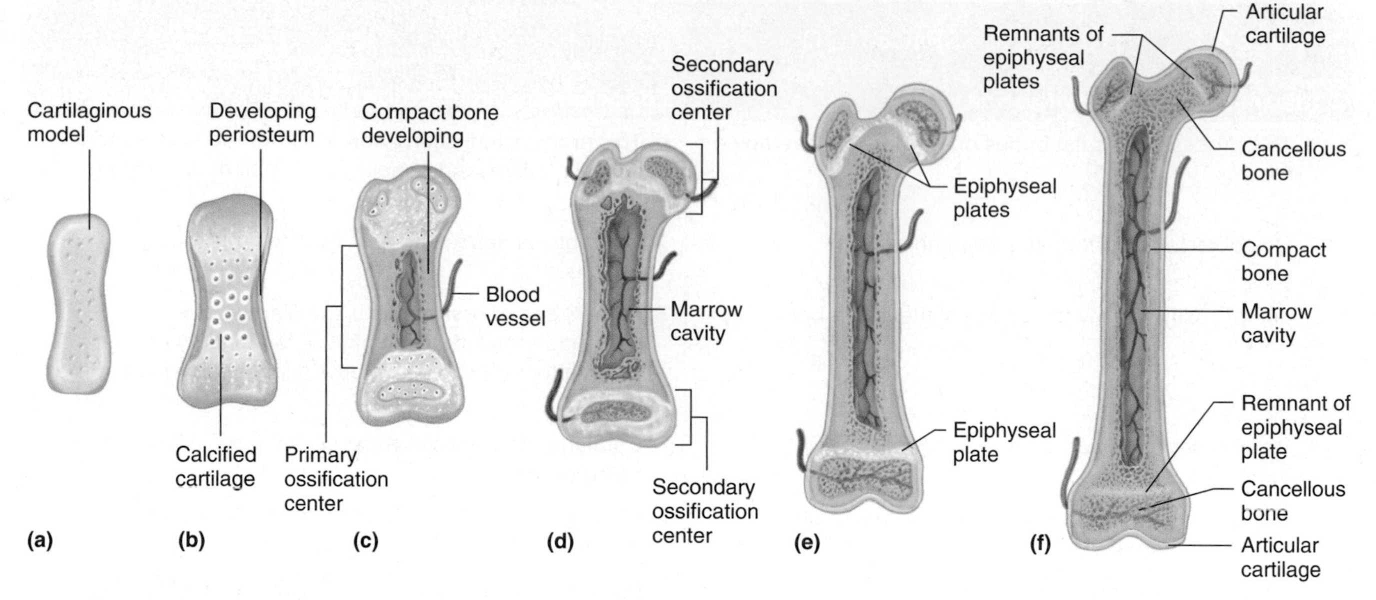

FIGURE 4.47 Endochondral ossification of a long bone.

endochondral ossification. From the term *endochondral,* you should understand that this form of bone development happens inside cartilage. It starts with a small hyaline cartilage model of the long bone. See **Figure 4.47**. Osteoblasts begin depositing bone around the diaphysis of the model, forming a bony collar. This blocks the chondrocytes' blood supply in the diaphysis of the model, so they die. The lacunae merge to form the marrow cavity. A blood vessel penetrates through the beginning bone of the fetus and establishes stem cells in the marrow cavity. This creates red bone marrow. Secondary ossification centers begin in each epiphysis. A blood vessel penetrates through the beginning bone here also to establish the stem cells of red bone marrow between the trabeculae. Osteoblasts deposit bone in all directions. In the meantime, the chondrocytes between the epiphyses and the diaphysis continue to produce more cartilage, extending the length of the developing bone. This zone of cartilage is called the **epiphyseal plate,** and the chondrocytes will continue to expand it. Osteoblasts will continue to deposit bone in the epiphyseal plate until all long bone growth is finished. You will learn about this type of growth shortly. Other bones formed through endochondral ossification include the vertebrae, ribs, sternum, scapula, and pelvic bones.

epiphyseal:
eh-PIF-ih-see-al

audio connect.mcgraw-hill.com

spot check 9 What type of ossification would form a humerus in Chloe as a fetus?

4.14 learning outcome
Compare and contrast endochondral and appositional bone growth.

Bone Growth

Intramembranous and endochondral ossification are processes of bone development in the fetus. After birth, the bones grow through two processes: endochondral growth and appositional growth.

Endochondral Growth As you can see in **Figure 4.48**, Chloe is a happy, healthy baby girl. Her skeletal system is continuing to grow. The osteoblasts are continuing to deposit bone in the epiphyseal plates of her long bones while the chondrocytes continue to expand the plates with cartilage. See **Figure 4.49**. It is a race between the two

FIGURE 4.48 Chloe.

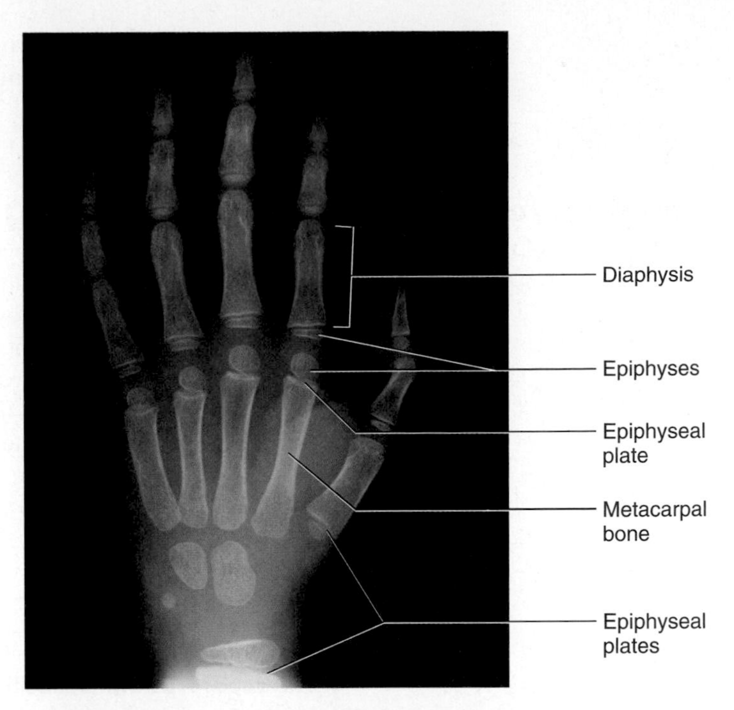

FIGURE 4.49 X-ray of a child's hand. The long bones of the hand have only one epiphyseal plate. The epiphyseal plates are still evident in the long bones of this hand. When endochondral growth is complete, they will no longer be visible.

types of cells, which will be in the race until Chloe reaches puberty. At puberty, the hormone estrogen will speed up the osteoblasts, which will deposit more bone than the chondrocytes can produce cartilage. The osteocytes will eventually close the epiphyseal plates, and endochondral growth will have stopped. Chloe's long bones will not get any longer. All that will remain of the growth plate is an **epiphyseal line** indicating where the epiphyseal plate was located. If Chloe had been a boy, it would have been the hormone testosterone at puberty that would have caused the osteoblasts to deposit bone faster.

Appositional Bone Growth **Appositional bone growth** occurs in all types of bone. In long bones, it does not make the bone longer, but it makes it more massive. Osteoblasts of the periosteum deposit more bone on the bone's shaft. Osteoblasts of the cancellous bone's trabeculae in the epiphyses deposit more bone along the bone's lines of stress. This makes the bone stronger to handle the stress. See **Figure 4.50**.

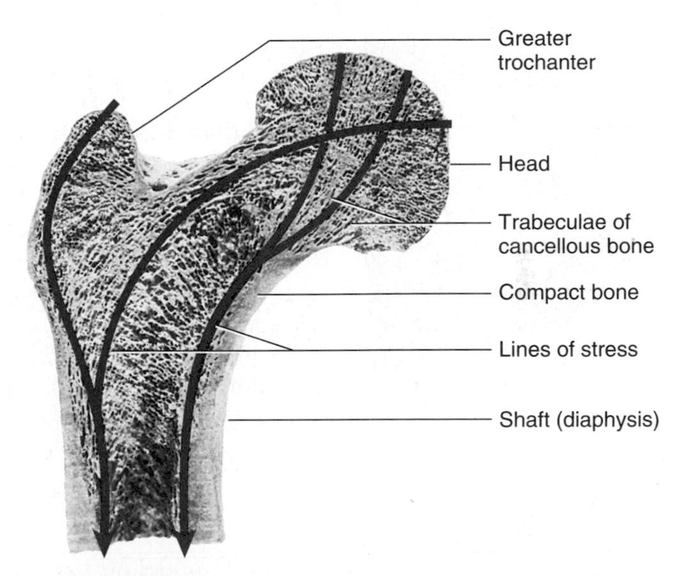

FIGURE 4.50 Appositional bone growth. Bone deposition increases along lines of stress.

Chloe is now already able to sit. She will continue to develop her muscles as she learns to crawl and walk. This will cause more stress on her bones at the epicondyles where the muscles are attached by tendons. The tubercles of the humerus and trochanters of the femur will continue to enlarge through appositional bone growth as more stress is applied. This is also true of her other epicondyles and the spines and processes of other bones like the vertebrae.

Maria, Chloe's mother, is also experiencing appositional growth as long as she supplies her osteoblasts with sufficient calcium and continues to put stress on her bones with exercise. It is important to note that Maria is relatively young. Appositional growth becomes more difficult with age. You will learn more about this shortly.

Again, bone is dynamic tissue. You have seen how it develops and how it continues to grow. But bone can also be lost. The next section explains why and how this happens.

4.15 learning outcome

Explain how bone is remodeled by reabsorption.

Bone Remodeling

Calcium is used by many systems in your body. Muscles require calcium to contract. Blood requires calcium to clot. The skeletal, muscular, and cardiovascular systems take the calcium they need from the blood. If the blood contains sufficient calcium, bone will be maintained. If there is more calcium than necessary in the blood, appositional bone growth will take place. If there is insufficient calcium in the blood, calcium will be taken from the bone to bring blood calcium levels back to normal. It is the job of osteoclasts to reabsorb calcium by producing hydrochloric acid. The acid dissolves the calcium phosphate crystals into solution, which then allows the calcium and phosphate ions to return to the blood.

warning

The following three terms are often misused by students: **deposition, absorption,** and **reabsorption.** Deposition is the process of putting calcium phosphate crystals into the bone. Absorption is the process of putting calcium into the blood for the first time. This occurs through the diet. Reabsorption is the process of putting calcium into the blood again, not the first time. To clarify, reabsorption involves dissolving calcium phosphate crystals from the bone and putting the calcium back into the blood again.

Bone therefore serves as a reservoir for calcium. While Chloe was a fetus, her osteoblasts took calcium from Maria's blood. This reduced Maria's blood calcium levels. If Maria did not absorb sufficient calcium from her diet, her osteoclasts would have reabsorbed calcium from her bones to keep her blood calcium levels normal.

clinical point

Pregnant women often need to take calcium supplements to ensure sufficient calcium levels in their blood. The condition **osteomalacia** is a softening of the bones due to reabsorption of calcium to meet the needs of pregnancy. See Figure 4.51.

FIGURE 4.51 Calcium supplements.

Bone remodeling happens daily depending on the calcium levels in the blood and the stress applied to the bone. Both sufficient calcium and stress on the bone are required to ensure healthy bone. This can be seen in astronauts returning from extended stays in a weightless environment. The lack of gravity reduces the bone stress. They may also have bone loss through remodeling even though their diet may be rich in calcium and vitamin D.

4.16 learning outcome

Explain the nutritional requirements of the skeletal system.

Nutritional Requirements of the Skeletal System

As you can see in **Figure 4.52**, dairy products are a major source of calcium in the diet. Calcium is not in the fat portion of the dairy product, so removing some of the fat does not affect the calcium content. Active vitamin D, also called **calcitriol,** is required for the small intestines to absorb calcium from the diet. As you may remember from the integumentary system chapter, vitamin D is produced in the skin and modified by the liver and kidney to become calcitriol. The calcium of the diet moves through the digestive system without ever being absorbed into the blood if calcitriol is not present. For that reason, vitamin D is added to most dairy products. Calcium can also be found in green leafy vegetables, such as broccoli, collards, kale, turnip greens, and bok choy. Again, see **Figure 4.52**. The phosphorus for phosphates is found in dairy products and meats. It is readily absorbed by the small intestines and does not require vitamin D.

calcitriol:
kal-sih-TRY-ol

spot check **10** What specifically can be done to ensure strong, healthy bones in adulthood? Explain.

Hormonal Regulation of Bone Deposition and Reabsorption

4.17 learning outcome

Describe the negative-feedback mechanisms affecting bone deposition and reabsorption by identifying the relevant glands, hormones, target tissues, and hormone functions.

How do the osteoblasts and osteoclasts know when to deposit or reabsorb bone? They do so because their work is regulated by hormones. You have already covered the hormones estrogen and testosterone. These hormones speed up osteoblasts at puberty to encourage the epiphyseal plates to begin closing. Testosterone has less of an effect than estrogen in speeding up the osteoblasts. As a result, it takes the epiphyseal plates in men longer to completely close. Men have a longer growth period and are therefore typically taller than women. Estrogen is produced by the ovaries in women. Testosterone is primarily produced by the testes in men. Both estrogen and testosterone also serve as a lock on calcium in the bone, making it more difficult for osteoclasts to reabsorb bone.

Two other hormones—from the thyroid gland and the parathyroid gland—regulate bone deposition and reabsorption based on the level of calcium in the blood through negative-feedback mechanisms. The thyroid gland produces the hormone **calcitonin** when blood calcium levels are high. Calcitonin travels through the blood. Although it comes in contact with many cells, it works only on osteoblasts and osteoclasts because they have receptors for the hormone. Calcitonin tells osteoblasts to deposit calcium in the bone, and it prevents osteoclasts from reabsorbing calcium. The net effect of calcitonin's action is bone deposition to reduce blood calcium levels.

The **parathyroid gland** looks like small buttons on the posterior side of the thyroid gland. See **Figure 4.53**. It produces **parathyroid hormone (PTH)** when blood

FIGURE 4.52 Sources of calcium in the diet.

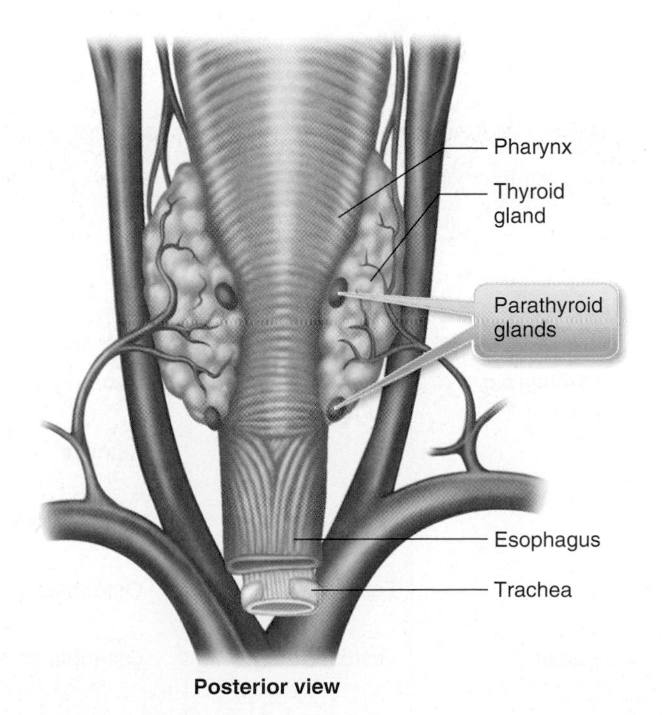

FIGURE 4.53 Parathyroid gland.

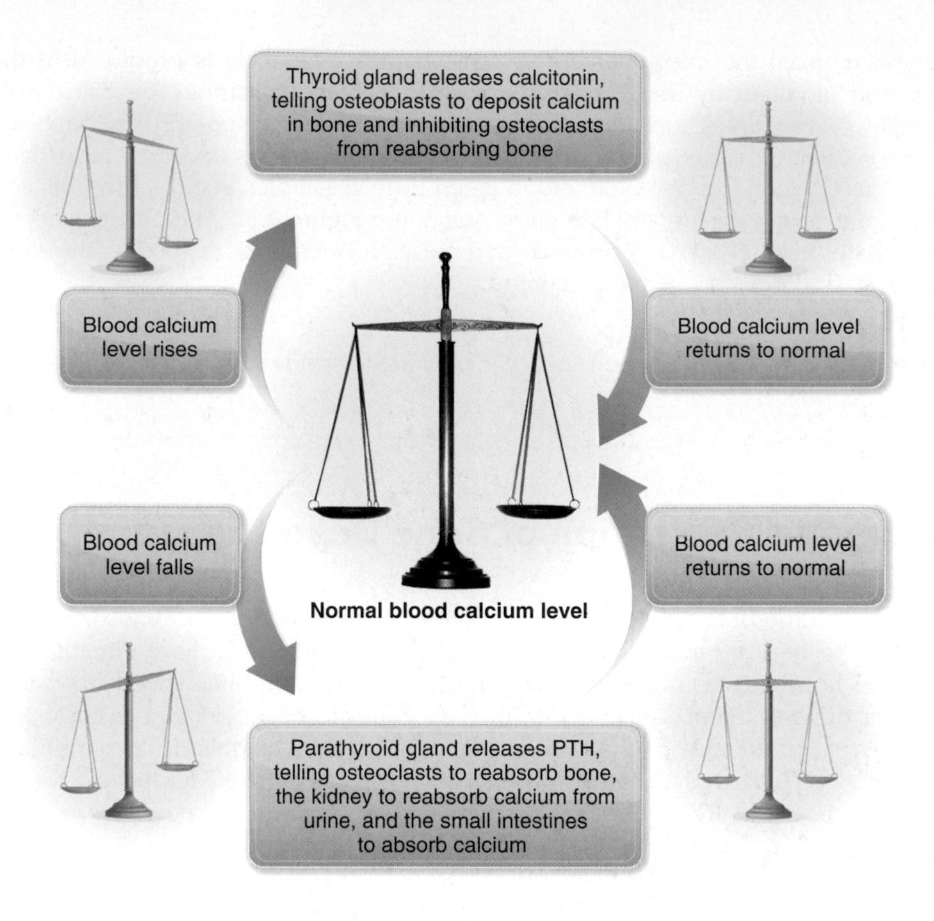

FIGURE 4.54 Homeostasis of calcium.

calcium levels are low. PTH travels through the blood and has three target tissues. It tells osteoclasts to reabsorb bone, increasing blood calcium levels. It tells the kidney to reabsorb any calcium that may be contained in the urine (this does not increase blood calcium, but it maintains the current levels by preventing the loss of calcium in the urine). Lastly, PTH targets the small intestine, telling it to absorb any calcium present in the small intestine from digestion. See **Figures 4.53** and **4.54** and Table 4.3.

TABLE 4.3 Hormonal regulation of bone deposition and reabsorption

Gland	Hormone	Target tissue	Function
Thyroid gland	Calcitonin	1. Osteoblasts	1. Tells osteoblasts to deposit bone, reducing blood calcium levels
		2. Osteoclasts	2. Inhibits osteoclasts from reabsorbing bone
Parathyroid gland	Parathyroid hormone	1. Osteoclasts	1. Tells osteoclasts to reabsorb bone, increasing blood calcium levels
		2. Kidneys	2. Tells kidneys to reabsorb calcium, maintaining blood calcium levels
		3. Small intestine	3. Tells small intestine to absorb calcium, increasing blood calcium levels (Vitamin D is required for this to work.)
Ovaries	Estrogen	Osteoblasts	Tells osteoblasts to work faster; serves as a lock on bone by inhibiting osteoclasts
Testes	Testosterone	Osteoblasts	Tells osteoblasts to work faster; serves as a lock on bone by inhibiting osteoclasts (Testosterone has less of an effect than does estrogen.)

You have now covered the anatomy and the physiology of your skeletal system. Next, you will see how they work together to carry out the system's functions.

Functions of the Skeletal System

4.18 learning **outcome**

Summarize the six functions of the skeletal system and give an example or explanation of each.

The skeletal system serves six important functions for the human body. It provides support, allows for movement, provides protection for organs, aids in acid-base balance, aids in electrolyte balance, and is responsible for blood formation. We discuss each of these functions below.

Support Chloe's spinal column was C-shaped at birth. As she developed, she learned to crawl and then walk. Her spinal column developed an S-shape to support her weight as she became more upright. The lumbar vertebrae became the most massive to support the majority of her weight. Her femurs and tibias also became more massive through appositional growth to support her weight.

Movement The arrangement of bones is very important in this function. Each of Chloe's little hands at birth contained 14 phalanges, so she could grasp Maria's finger. She would not be able to grasp anything if her fingers did not have multiple bones and joints. Chloe's spine would be nice and straight if it were a single bone, but her adult spinal column will have 26 bones. The joint between each bone allows slight motion. Added together, the motions of all the joints between vertebrae will allow Chloe to touch her toes when standing. If you held your arm straight out to the side, would you be able to then touch the tip of your nose with your index finger if you did not have a shoulder, an elbow, a wrist, and finger joints?

Protection Chloe's flat bones underwent intramembranous ossification to form her skull's cranial bones, which provide protection for her brain. Her ribs and sternum developed to protect her lungs and heart. Her vertebrae developed the central vertebral foramen to protect her spinal cord. Even within the skull, the sella turcica developed to give added protection to her pituitary gland.

Acid-Base Balance Maintaining normal blood pH (7.35 to 7.45) is very important. Acidosis results if the blood pH is too low. Acidosis is characterized by too many free hydrogen ions in the blood. Calcium phosphate is an ionically bonded molecule that forms calcium ions and phosphate ions in solution. If the pH of the blood is too low, the phosphate ions will bind to excess free hydrogen ions, causing the pH to rise. So phosphate ions act as a **buffer,** resisting a change in normal blood pH.

Electrolyte Balance Bone serves as a reservoir for the electrolyte calcium. Suppose Maria has a glass of milk for breakfast, a grilled cheese sandwich for lunch, and vegetables with a chicken breast for dinner. During the course of this particular day, Maria has consumed calcium in her diet. The calcium will be absorbed into her bloodstream as long as she has sufficient calcitriol. If this process raises her blood calcium levels above normal, the calcium will be deposited in her bones. If she had not consumed the calcium and her blood calcium levels fell below normal, calcium would have been reabsorbed from the bone to bring her blood calcium levels to homeostasis.

Blood Formation Red blood cells, white blood cells, and platelets are produced by stem cells in the red bone marrow. You will learn about this in detail when you study the cardiovascular system.

You have looked at how the skeletal system functions as a whole in the human body. But what are the effects of aging on this system? Does it continue to function normally?

clinical point

Red bone marrow may be harvested to give to another person who is a close match. It can even be harvested to give back to the donor. Some cancer treatments include chemotherapy so strong that it kills all fast-growing tissues like bone marrow along with the cancer cells. In this case, bone marrow can be harvested before treatment and the harvested bone marrow can be given back to the patient after the chemotherapy is over.

4.19 learning outcome

Summarize the effects of aging on the skeletal system.

Effects of Aging on the Skeletal System

The ratio of deposition and reabsorption changes as we age. In general:

- Deposition > Reabsorption Birth to age 25 Increasing bone mass and density
- Deposition = Reabsorption Age 25 to 45 Maintaining bone mass and density
- Deposition < Reabsorption Age 45 and over Decreasing bone mass and density

A major reason for this is that the levels of estrogen and testosterone decrease with age. Both of these hormones serve as a lock on calcium in the bone. It is much easier for osteoclasts to reabsorb bone when the levels of these hormones are decreased. This has a greater impact on women, as their production of estrogen ceases after menopause. Men's production of testosterone diminishes as they age, but not to the extent of women's production of estrogen.

The effects of the decreased bone mass and density with age are many:

- Each vertebra becomes thinner. The spinal column becomes more curved and compressed, resulting in a shorter trunk and a more stooped posture. The neck may also become tilted. Compression fractures of the vertebrae become more common.
- The change in posture affects the gait and balance. This makes the elderly more prone to falls. The decreased bone mass and density result in fractures from falls.
- Long bones lose mass, but not length. The arms and legs appear longer in relation to the trunk.
- Scapulae thin and become more porous. On an x-ray, they may appear to have holes.
- Joints stiffen and become less flexible as osteoarthritis sets in. The amount of synovial fluid may decrease.
- Minerals may deposit in the joints, especially the shoulder.
- Phalangeal joints lose cartilage, and the bones may thicken slightly.

Age-related changes occur in the system's cartilage too. You have already read that articular cartilage erodes in osteoarthritis. The intervertebral disks become thinner from the effect of gravity over the years. This, along with the vertebrae compressing, accounts for the way we seem to get shorter as we age.

The effects of aging can be minimized, but not eliminated. The best way to ensure good bone health for life is to build strong bones while deposition exceeds reabsorption. Proper nutrition with ample calcium and vitamin D, along with exercise, throughout life can reduce the effects of aging.

Now that you are familiar with what normally happens to structures in this system over the years, it is time to consider what can happen when things are not quite normal, as in the case of fractures and bone disorders.

Fractures

4.20 learning outcome

Classify fractures using descriptive terms.

A fracture is a break in a bone. It can result from injury or trauma, like a fall, or it can result from a disease process that has weakened the bone. Fractures may produce a simple crack, a dent, or a bone that is shattered in many pieces. See **Figure 4.55**.

Types of Fractures

There are several descriptive terms used to classify fractures, and they include the following:

- **Closed.** A closed fracture (formerly called a *simple fracture*) does not cause a break in the skin. A shattered bone may not break through the skin, but it hardly seems appropriate to refer to it as simple.
- **Open.** An open fracture (formerly called a *compound fracture*) breaks through the skin.
- **Complete.** The bone is in two or more pieces.
- **Displaced.** The bone is no longer in proper alignment.
- **Nondisplaced.** The bone is in proper alignment.
- **Hairline.** There is a crack in the bone.
- **Greenstick.** The bone has broken through one side but not completely through the other side.
- **Depressed.** The bone has been dented. This fracture is found where there is cancellous bone, as in skull fractures.
- **Transverse.** The bone is broken perpendicular to its length.
- **Oblique.** The break in the bone is at an angle.
- **Spiral.** The break in the bone spirals up the bone. This type of break often results from twisting the bone. This may occur when children fall while kneeling on a chair at the table with their feet sticking through the chair's spokes.
- **Epiphyseal.** The break occurs at the epiphyseal plate in a child.
- **Comminuted.** The bone is broken into three or more pieces (commonly referred to as *shattered*).
- **Compression.** Cancellous bone has been compressed. This type of fracture may occur in the vertebrae.

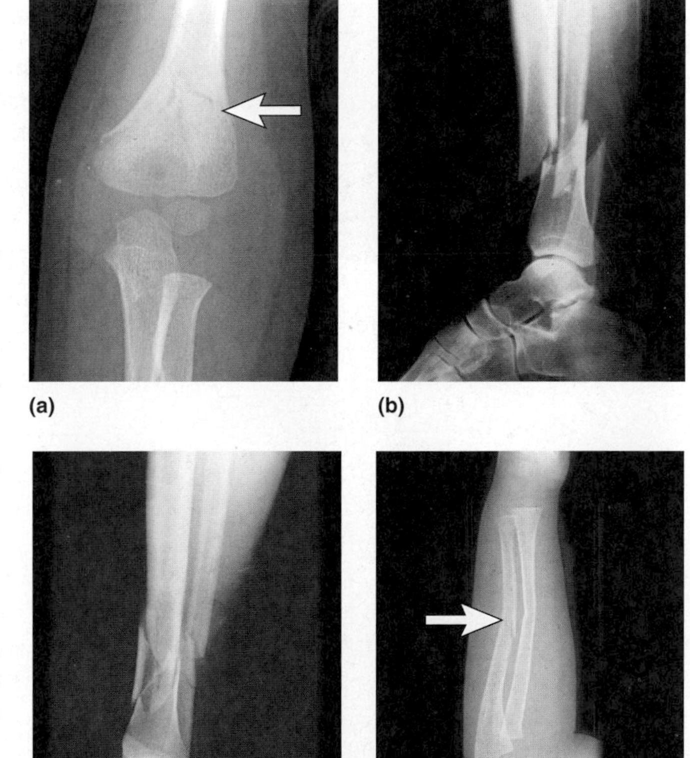

FIGURE 4.55 X-rays of fractures: (a) nondisplaced, (b) displaced, (c) comminuted, (d) greenstick.

spot check 11 What terms would be used to describe a fracture that broke the shaft of a humerus at an angle into two separate pieces, with one of the pieces sticking out of the skin?

4.21 learning outcome

Explain how a fracture heals.

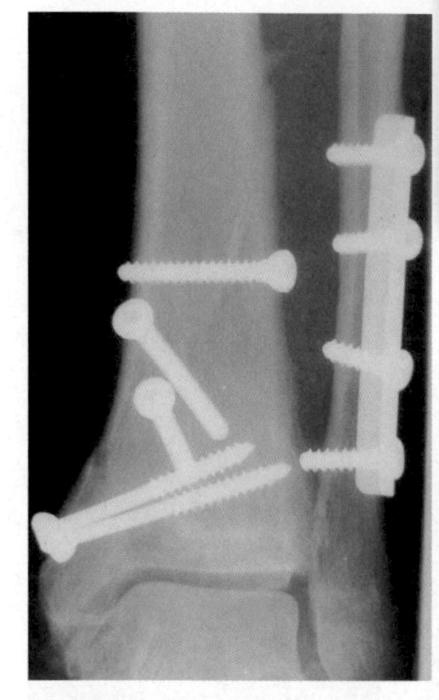

FIGURE 4.56 Open reduction of an ankle fracture.

clinical point

Chapter 3 covered injuries to the skin. As you read in that chapter, it is important to bring the edges of a cut together as much as possible for proper healing or a primary union. It is just as important to bring the edges of a fractured bone together as close as possible with the proper alignment to ensure the best healing. A **closed reduction** sets the edges of the fracture in proper alignment by manipulating the bone without surgery. An **open reduction** sets the bones in proper alignment through surgery. Plates, pins, and screws may be used in an open reduction to ensure that the bone fragments stay in proper alignment. It is important to immobilize the break once either type of reduction is performed to give the body the best chance of healing the fracture properly. See Figure 4.56.

Fracture Healing

The body uses several processes, in steps, to heal a complete transverse fracture of a long bone (see **Figure 4.57**):

1. A bone bleeds when broken. Not only is the bone broken, but so are the blood vessels in all of the Haversian canals. Blood clots once it is out of the vessels, and a hematoma is formed.
2. Stem cells from the periosteum deposit collagen and fibrocartilage in the break to form a soft **callus.** The callus is thicker than the original bone.
3. Osteoblasts deposit bone in the soft callus, forming a hard callus. Bone deposition extends into the marrow cavity.

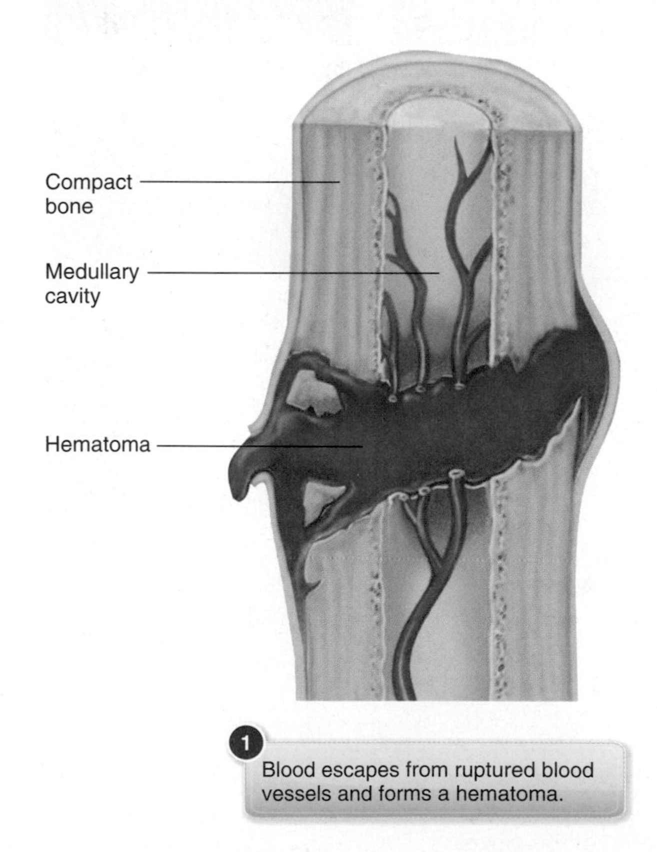

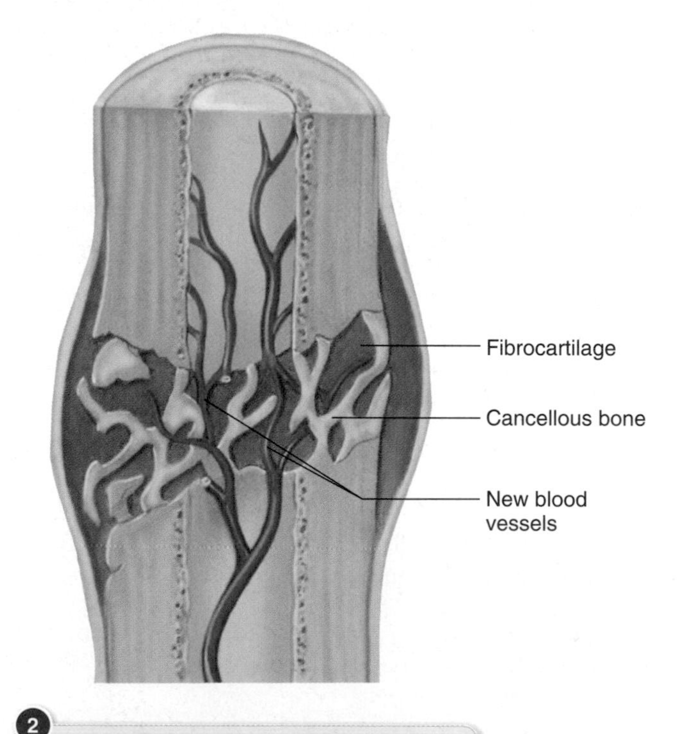

FIGURE 4.57 The healing of a bone fracture.

FIGURE 4.57 concluded

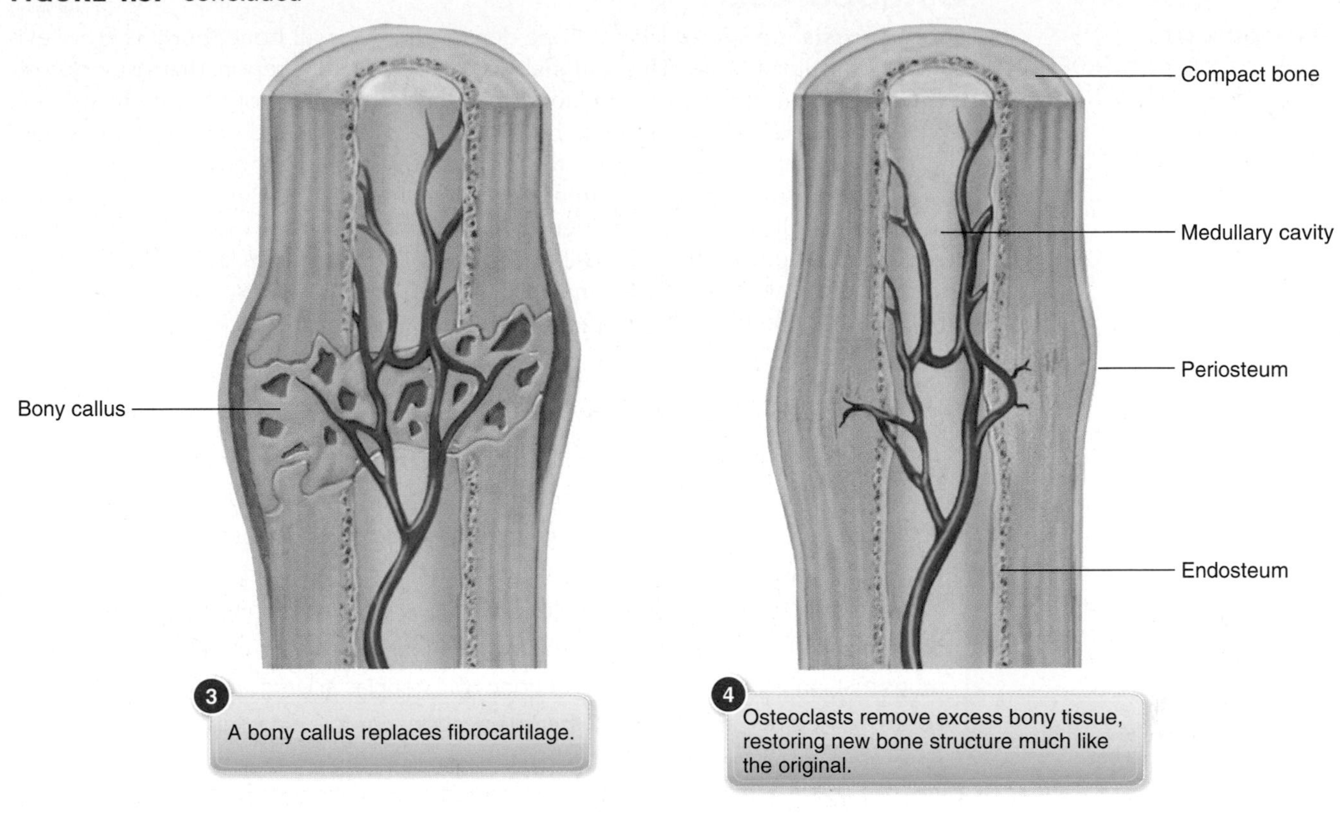

4. Osteoblasts continue forming compact bone in the break while osteoclasts from the endosteum remodel the bone to reestablish the marrow cavity. The remnant of the hard callus can be seen as a thickening of the bone after the bone has healed.

Skeletal System Disorders

4.22 learning **outcome**

Describe bone disorders and relate abnormal function to the pathology.

You have already studied several bone disorders in the context of the anatomy and physiology of the skeletal system. Below, you will learn about two more. Table 4.4 summarizes all the bone disorders discussed in this chapter.

TABLE 4.4 Bone disorders

Type of disorder	Disorder	Description
Bone softening	Osteoporosis	Severe lack of bone density
	Rickets	Lack of bone deposition in children
	Osteomalacia	Bone loss due to meeting the demands of pregnancy
Brittle bones	Osteogenesis imperfecta	Lack of collagen fibers in bone
Abnormal spinal curvatures	Scoliosis	Lateral curvature of vertebral column
	Kyphosis	Exaggerated thoracic curvature
	Lordosis	Exaggerated lumbar curvature
Joint inflammation	Osteoarthritis	Wear and tear on a joint
	Rheumatoid arthritis	Autoimmune disease
Infection	Osteomyelitis	Acute or chronic infection usually from bacteria or fungi

Osteoporosis

osteoporosis: OS-tee-oh-poh-ROE-sis

Osteoporosis is a severe lack of bone density. It affects all bone, but it is more evident in cancellous bone. The National Institutes of Health report that osteoporosis is a major health threat for 44 million Americans, 68 percent of whom are women. A diet deficient in calcium and vitamin D, lack of exercise, and diminished estrogen and testosterone due to aging are the major causes of osteoporosis. The presence of this disorder may not be apparent until a fracture occurs. Half of all women and one-fourth of all men will experience a bone fracture from osteoporosis in their lifetime. "Osteoporosis is responsible for 1.5 million fractures annually, including approximately 300,000 hip fractures, 700,000 vertebral fractures, 250,000 wrist fractures, and 300,000 fractures at other sites."[2] See **Figure 4.58**.

clinical point

A DEXA (dual-energy x-ray absorptiometry) scan uses low-dose radiation to measure bone density in the hip and vertebrae. It can be performed on a yearly basis to detect a decrease in bone density before osteoporosis progresses to the point of causing fractures.

Treatment for osteoporosis includes improving nutrition, increasing exercise, preventing falls, and using therapeutic medications. Bisphosphonates are a class of drugs that inhibit the reabsorption of bone by osteoclasts. Drugs in this class include Fosamax, Actonel, and Boniva. Other medications for this disorder include calcitonin and estrogen therapy.

Osteomyelitis

Osteomyelitis is a bone infection that can reach the bone from the blood, from surrounding tissues, or from trauma that exposes the bone to a pathogen (such as a bacteria or fungus). This type of trauma is typical of an open fracture or a break in the shin's skin because the anterior surface of the tibia is very superficial. The infection may be acute (lasting several months or less), or it may be chronic (lasting several months to years). Treatment may include antibiotics and/or surgery to drain the area and remove damaged bone.

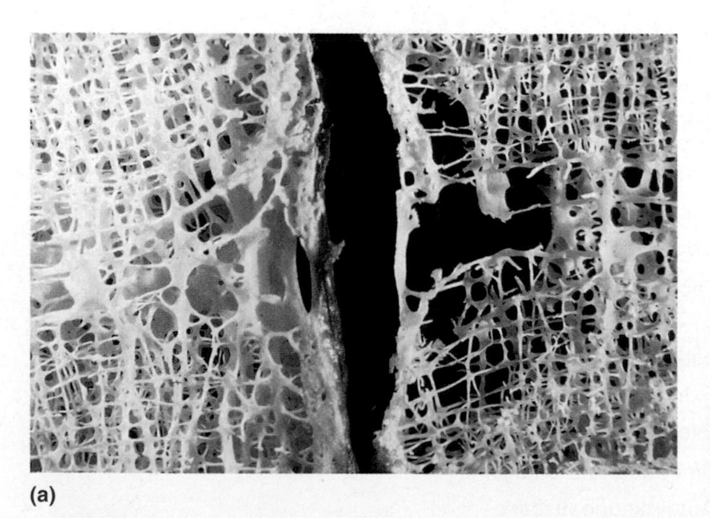
(a)

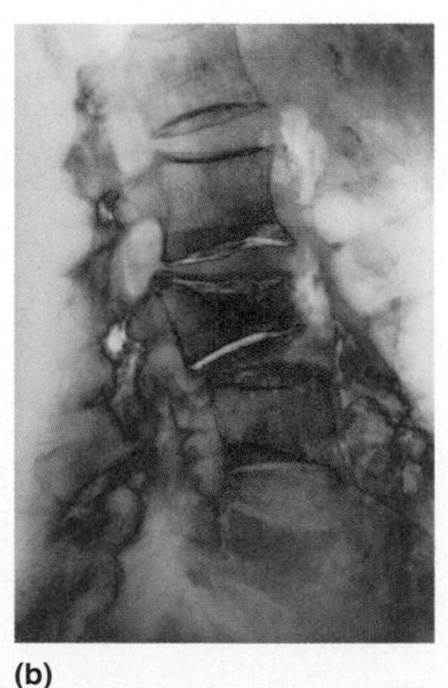
(b)

(c)

FIGURE 4.58 Osteoporosis: (a) comparison of healthy cancellous bone and bone showing osteoporosis, (b) x-ray of lumbar vertebrae showing osteoporosis, (c) kyphosis due to osteoporosis.

putting the pieces **together**

The Skeletal System

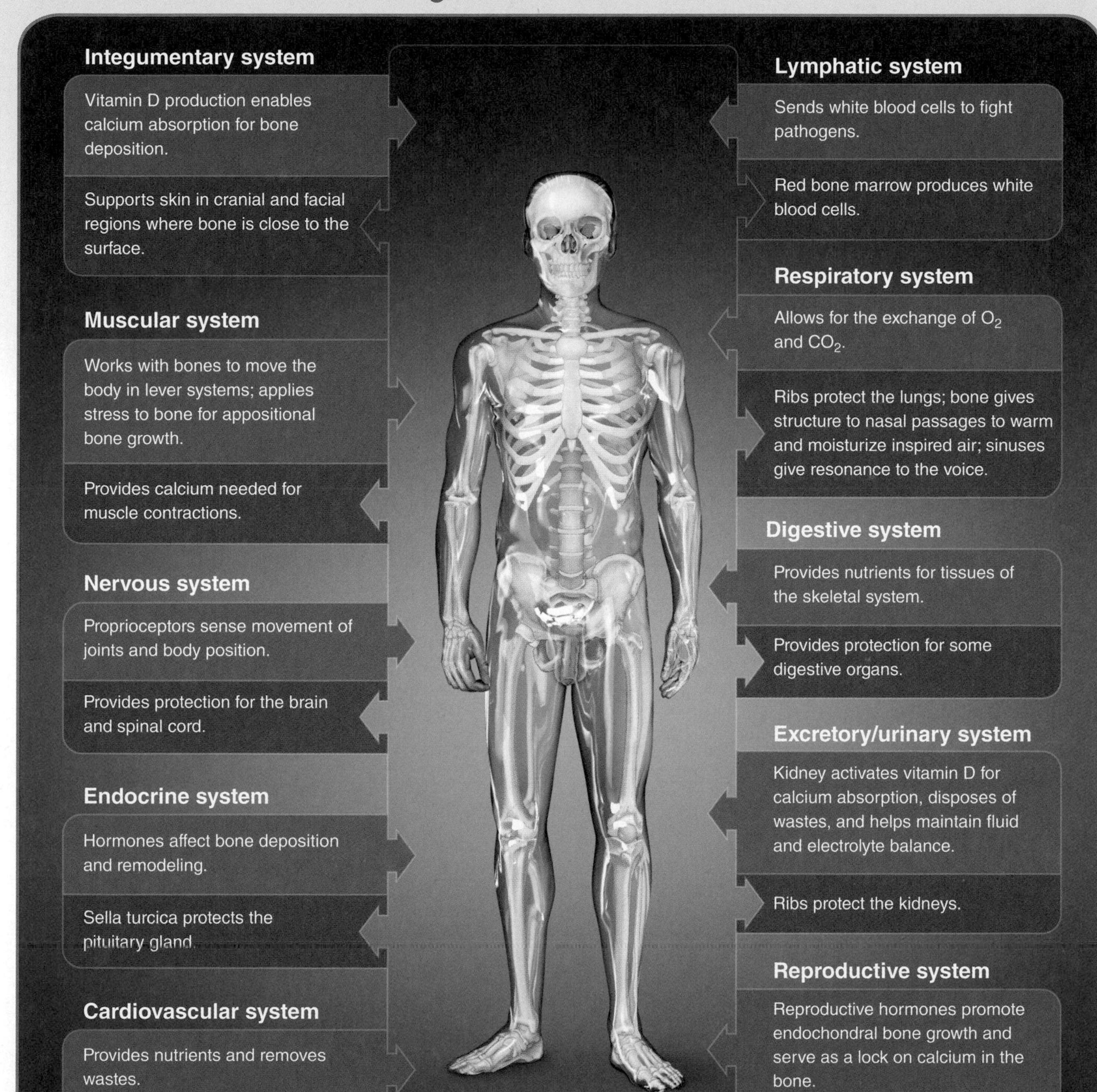

FIGURE 4.59 Putting the Pieces Together—The Skeletal System: connections between the skeletal system and the body's other systems.

summary

Overview

- The skeletal system is composed of bones, cartilages, and ligaments. It is a dynamic system that changes daily.

Anatomy of the Skeletal System

- The skeleton can be divided into two parts: the axial skeleton and the appendicular skeleton.

Classification of Bones

Bones can be classified by shape:

- Long bones are longer than they are wide and have clubby ends. Example: tibia.
- Short bones are cube-like. Example: carpal bones.
- Flat bones look like they are a sheet of clay that has been molded. Example: parietal bone.
- Irregular bones have many projections and spines. Example: vertebrae.
- Sesamoid bones grow in tendons where there is a lot of friction. Example: patella.

Axial Skeleton

The axial skeleton contains the following bones:

- Cranial bones: frontal, occipital, temporal, parietal.
- Ethmoid and sphenoid.
- Facial bones: nasal, lacrimal, zygomatic, inferior nasal concha, maxilla, palatine, mandible, vomer.
- Spinal column: 7 cervical vertebrae, 12 thoracic vertebrae, 5 lumbar vertebrae, sacrum, coccyx.
- Sternum.
- Ribs: 12 pairs. Of these, 7 pairs are true ribs with individual costal cartilages and 5 pairs are false ribs. Of the 5 pairs of false ribs, 2 pairs are floating because they do not have or share a costal cartilage.
- Hyoid bone.

Appendicular Skeleton

The appendicular skeleton is composed of the bones of the limbs and the bones of the girdles that connect the limbs to the axial skeleton.

- Pectoral girdle: clavicle and scapula.
- Bones of the upper limb: humerus, radius, ulna, carpal bones, metacarpals, phalanges.
- Pelvic girdle: ilium, ischium, pubis.
- Bones of the lower limb: femur, patella, tibia, fibula, tarsal bones, metatarsals, phalanges.

Histology of the Skeletal System

The skeletal system contains bone connective tissue and cartilage connective tissue.

Bone connective tissue:

- Osteoblasts build bone tissue by making matrix with collagen fibers and allowing hydroxyapatite (calcium phosphate crystals) to deposit.
- Collagen fibers give bone some flexibility.
- Calcium phosphate crystals make bone hard.
- Osteoclasts destroy bone.
- Compact bone is well organized into osteons (Haversian systems).
- Cancellous bone is loosely organized as trabeculae.

Cartilage connective tissue:

- Chondrocytes produce a matrix composed of proteoglycans and water.
- Hyaline cartilage matrix is smooth and clear. It is found as nasal cartilages, costal cartilages, and articular cartilages covering the ends of long bones.

- Fibrocartilage matrix has fibers going in one direction to act as a shock absorber. It is found in the intervertebral disks, the menisci of the knee, and the pubic symphysis.
- Elastic cartilage matrix has fibers going in all directions so as to be elastic. It is found in the pinna of the ear and the epiglottis.

Anatomy of a Long Bone

- Epiphyses are the clubby ends of the bone. They are composed of cancellous bone.
- The diaphysis is the shaft of the bone. It is composed of compact bone.
- The periosteum covers the diaphysis of the bone. The endosteum lines the marrow cavity.
- Red bone marrow is found in the epiphyses.
- The marrow cavity in the diaphysis is filled with yellow marrow.

Joints

Joints can be classified on the basis of their anatomy:

- Fibrous joints have fibrous tissue between bones.
  - Suture. This is formed by the membranes of intramembranous ossification.
  - Gomphosis. This is formed by ligaments holding the tooth in its socket.
  - Syndesmosis. This is formed by an interosseous membrane.
- Cartilaginous joints have cartilage between the bones.
  - Symphysis. This is formed by fibrocartilage between the pubic bones.
  - Synchondrosis. This is formed by hyaline cartilage between the diaphysis and the epiphyses of bones in children.
- Synovial joints are lined by a synovial membrane and have synovial fluid in the joint space.
- The knee is a relatively unstable joint held together by five ligaments: the medial and lateral collateral ligaments, the anterior and posterior cruciate ligaments, and the patellar ligament. It also contains fibrocartilage pads called *menisci* that act as a shock absorber.
- There are six types of synovial joints: hinge, ball and socket, saddle, gliding, ellipsoid, pivot.
- Osteoarthritis is wear and tear on a joint. Most people develop osteoarthritis as they age.
- Rheumatoid arthritis is an autoimmune disease. Anyone, including children, can develop rheumatoid arthritis.

Physiology of the Skeletal System

Mineral Deposition

- Osteoblasts produce a chemical that allows calcium phosphate crystals to be deposited. This is a positive-feedback mechanism starting with a seed crystal.

Bone Development

- Flat bones are formed through intramembranous ossification.
- Long bones are formed through endochondral ossification.

Bone Growth

- Long bones continue to grow longer after birth through endochondral growth until the epiphyseal plates are closed.
- Appositional growth makes bones more massive. It occurs along lines of stress.

Bone Remodeling

- Bone remodeling is done by osteoclasts. They remove bone by producing hydrochloric acid, which dissolves the calcium phosphate crystals.
- Bones act as a reservoir for calcium.

Nutritional Requirements of the Skeletal System

- Calcium can be found in dairy products and in green leafy vegetables such as broccoli, collards, kale, turnip greens, and bok choy.

- Phosphorus for phosphates is found in dairy products and meats.
- Vitamin D is needed for calcium absorption.

Hormonal Regulation of Bone Deposition and Reabsorption

- Bone deposition and reabsorption are regulated by hormones on the basis of blood calcium levels.
- If blood levels of calcium are too high, calcitonin tells osteoblasts to deposit bone.
- If blood levels of calcium are too low, PTH tells osteoclasts to reabsorb bone.
- Estrogen and testosterone speed up deposition at puberty and serve as a lock on calcium in the bone.

Functions of the Skeletal System

- Support. Vertebral column allows the body to be erect.
- Movement. The arrangement of bones and joints allows a range of movements.
- Protection. The cranial bones protect the brain. The sternum and rib cage protect the lungs and heart.
- Acid-base balance. Phosphate ions can bind to excess hydrogen ions to buffer the pH of the blood.
- Electrolyte balance. Bones serve as a reservoir for calcium.
- Blood formation. Red blood cells, white blood cells, and platelets are produced in the red bone marrow.

Effects of Aging on the Skeletal System

- The ratio of deposition to reabsorption changes as we age.
- Decreases in estrogen and testosterone levels are responsible for the change.
- More reabsorption decreases the bone mass and density.
- Vertebrae thin, and the spinal column becomes more curved and compressed.
- The elderly are more prone to falls, resulting in fractures.
- Joints stiffen and become less flexible.
- Minerals may deposit in joints.
- The best way to ensure good bone health in later life is to build strong bones when deposition exceeds reabsorption.
- Exercise and good nutrition, including calcium and vitamin D, can minimize the effects of aging.

Fractures

Types of Fractures

- Fractures can be classified by using descriptive terms.
- Fractured bones can be set back into proper alignment by closed or open reduction.

Fracture Healing

- The healing of a fracture starts with stem cells forming a soft callus in a hematoma.
- Osteoblasts deposit bone in the soft callus to form a hard callus.
- Osteoclasts finish the healing of the fracture by remodeling the hard callus to reestablish the marrow cavity.

Skeletal System Disorders

- Bone-softening disorders include osteoporosis, rickets, and osteomalacia.
- *Brittle bones* is another name for osteogenesis imperfecta.
- Abnormal spinal curvatures include scoliosis, kyphosis, and lordosis.
- Joint inflammations include osteoarthritis and rheumatoid arthritis.

Osteoporosis

- Osteoporosis is a severe lack of bone density.

Osteomyelitis

- Osteomyelitis is a bone infection.

key words for review

The following terms are defined in the glossary.

absorption
appendicular skeleton
appositional bone growth
axial skeleton
cancellous bone
chondrocyte
comminuted
compact bone
deposition
diaphysis
endochondral ossification
epiphyseal plate
fontanelle
foramen magnum
hydroxyapatite
meniscus
osteon (Haversian system)
reabsorption
synovial membrane
trabeculae

chapter review questions

Word Building: *Use the word roots listed in the beginning of the chapter and the prefixes and suffixes inside the back cover to build words with the following meanings:*

1. Inflammation of the extension of the synovial membrane that acts as a cushion: ______________________
2. Cartilage cell: ______________________
3. Joint pain: ______________________
4. Pertaining to the bones of the wrist: ______________________
5. Pertaining to the articulating surface of a long bone: ______________________

Multiple Select: *Select the correct choices for each statement. The choices may be all correct, all incorrect, or a combination of correct and incorrect.*

1. Which of the following groups of bones belong(s) to the axial skeleton?
   a. Parietal, pelvic, and sesamoid bones.
   b. Ribs, vertebrae, and radius.
   c. Facial bones, sternum, and vertebral column.
   d. Ilium, ischium, and sacrum.
   e. Sternum, clavicle, and sacrum.
2. Which of the following statements is (are) true about the marrow cavity of a bone?
   a. It is the same as the Haversian canal.
   b. It contains yellow marrow.
   c. It contains red marrow in a mature bone.
   d. It contains fat.
   e. It is lined by endosteum.
3. Which of the following statements is (are) true concerning the normal development of a long bone?
   a. The cartilage is replaced by bone.
   b. The marrow cavity disappears.
   c. The epiphyseal plate disappears.
   d. The epiphysis fuses with the diaphysis.
   e. The epiphysis fuses to form a suture.

4. What will happen if the calcium level in the blood is above normal?
   a. Osteoclasts will deposit calcium in the bone.
   b. Osteoblasts will reabsorb bone.
   c. The parathyroid gland will produce PTH.
   d. A negative-feedback mechanism will correct it.
   e. The thyroid will produce calcitriol.
5. What will happen if blood calcium levels fall below normal?
   a. A positive-feedback mechanism will correct it.
   b. The parathyroid gland will produce calcitonin.
   c. Osteoclasts will reabsorb bone.
   d. PTH will target osteoclasts, the kidney, and the small intestines.
   e. Osteoclasts will be inhibited from depositing calcium.
6. Which of the following statements is (are) true for chondrocytes?
   a. A chondrocyte Golgi complex adds a carbohydrate to a protein to form matrix.
   b. A chondrocyte produces keratin fibers to form hyaline cartilage.
   c. A chondrocyte in articular cartilage has its wastes removed by synovial fluid.
   d. A chondrocyte undergoes protein synthesis to form hydroxyapatite.
   e. A chondrocyte is nourished through simple diffusion through the matrix.
7. Which of the following statements is (are) true concerning osteoblasts or osteoclasts?
   a. Osteocytes tend to deposit more calcium than they reabsorb as a person ages.
   b. Osteocytes will work faster to deposit calcium if testosterone is present.
   c. Osteocytes will reabsorb calcium more efficiently in a female if estrogen is absent.
   d. Osteocytes will reabsorb calcium if the bone is under stress.
   e. Osteocytes undergo protein synthesis to form collagen fibers.
8. Which of the following statements would be good advice to a 15-year-old female gymnast?
   a. "You need plenty of calcium and vitamin D now because this is the prime time of your life for depositing calcium in bone."
   b. "Your testosterone will make your bone cells work faster to close your growth plates."
   c. "You need plenty of vitamin D to absorb phosphates."
   d. "It is good that you are training because stress on the bone causes appositional bone growth."
   e. "Dairy products and meats are good sources of calcium."
9. Which of the following statements is (are) true concerning calcium?
   a. Calcium is an electrolyte whose balance in the blood is maintained by the skeletal system through deposition and reabsorption.
   b. Calcium is reabsorbed by the production of an acid by osteoclasts.
   c. Calcium is found in trabeculae.
   d. Calcium is deposited in bone through a positive-feedback mechanism of mineral crystal deposition.
   e. PTH causes calcium to be absorbed by the small intestines if there is calcitriol.
10. Which of the following statements is (are) true concerning bone disorders?
   a. Osteomyelitis is the depletion of calcium from the bone to meet the demands of pregnancy.
   b. Kyphosis is an exaggerated curvature of the cervical vertebrae.
   c. Osteoarthritis is an autoimmune disease.
   d. Osteomalacia, rickets, osteoporosis, and osteogenesis imperfecta result in brittle bones.
   e. Rheumatoid arthritis can happen to people of any age.

Matching: *Match the structural class of joint to the type of joint in that class. Some answers may be used more than once.*

_____	1. Syndesmosis	a. Synovial
_____	2. Saddle	b. Fibrous
_____	3. Gomphosis	c. Cartilaginous
_____	4. Symphysis	
_____	5. Pivot	

Matching: *Match the fracture to the description. Some questions may have more than one correct answer.*

_____ 6. A shattered bone with pieces poking through the skin
_____ 7. A fracture caused by twisting the bone
_____ 8. A fracture that causes a dent in the bone
_____ 9. A diagonal fracture of the tibia that does not break the skin
_____ 10. A crack in the femur

a. Complete
b. Spiral
c. Open
d. Comminuted
e. Oblique
f. Closed
g. Compression
h. Hairline
i. Greenstick

Critical Thinking

1. Chloe, at 16, is a very good basketball player. She is 5 feet 5 inches tall. She is an average student academically. She is sure she will be recruited by Division 1 schools for basketball because of her skills and because she will continue to grow taller. What information do you need to know to advise her about her potential height? What tests or procedures could you perform to get that information? Explain.
2. Female athletes who train very hard and restrict their fat intake may lose their periods due to the lack of estrogen (estrogen is a modified cholesterol, a fat often found in the diet). What effect might this have on the athletes' bones? Explain.
3. Consider your new knowledge of bone and cartilage histology, the blood supply to both tissues, and your understanding of bone growth. Why would a fracture of the distal epiphyseal plate of the femur possibly result in a leg that is shorter than the other one? Explain.

4 chapter mapping

This section of the chapter is designed to help you find where each outcome is covered in this text.

	Outcomes	Readings, figures, and tables	Assessments
4.1	Use medical terminology related to the skeletal system.	Word roots: pp. 105–106	Word Building: 1–5
4.2	Distinguish between the axial skeleton and the appendicular skeleton.	Anatomy of the skeletal system: p. 107 Figure 4.2	Multiple Select: 1
4.3	Describe five types of bones classified by shape.	Classification of bones: pp. 108–109 Figures 4.3–4.6	
4.4	Identify bones, markings, and structures of the axial skeleton and appendicular skeleton.	Axial skeleton: pp. 109–120 Figures 4.7–4.24 Appendicular skeleton: pp. 121–131 Figures 4.25–4.37	Spot Check: 1–4
4.5	Describe the cells, fibers, and matrix of bone tissue.	Bone connective tissue: p. 131	Spot Check: 5, 6 Multiple Select: 7 Critical Thinking: 3

	Outcomes	Readings, figures, and tables	Assessments
4.6	Compare and contrast the histology of compact and cancellous bone.	Bone connective tissue: pp. 131–133 Figure 4.38	Multiple Select: 9
4.7	Compare and contrast the histology of hyaline, elastic, and fibrocartilage connective tissues.	Cartilage connective tissue: pp. 133–134 Figure 4.39	Spot Check: 5, 6 Multiple Select: 6 Critical Thinking: 3
4.8	Describe the anatomy of a long bone.	Anatomy of a long bone: pp. 134–136 Figure 4.40	Spot Check: 7 Multiple Select: 2
4.9	Distinguish between two types of bone marrow in terms of location and function.	Red bone marrow: p. 136	Multiple Select: 2
4.10	Describe three major structural classes of joints and the types of joints in each class.	Joints: pp. 136–142 Figures 4.41, 4.42 Table 4.1	Spot Check: 8 Matching: 1–5
4.11	Differentiate between rheumatoid arthritis and osteoarthritis.	Joints: pp. 142–143 Figures 4.43, 4.44	Multiple Select: 10
4.12	Explain how minerals are deposited in bone.	Mineral deposition: p. 144	Multiple Select: 7, 8, 9
4.13	Compare and contrast endochondral and intramembranous ossification.	Bone development: pp. 144–146 Figures 4.46, 4.47 Table 4.2	Spot Check: 9 Multiple Select: 3 Critical Thinking: 1
4.14	Compare and contrast endochondral and appositional bone growth.	Bone growth: pp. 146–148 Figures 4.48–4.50	Spot Check: 10 Multiple Select: 3, 8
4.15	Explain how bone is remodeled by reabsorption.	Bone remodeling: p. 148 Figure 4.51	Multiple Select: 5, 7, 9
4.16	Explain the nutritional requirements of the skeletal system.	Nutritional requirements of the skeletal system: pp. 148–149 Figure 4.52	Spot Check:10 Multiple Select: 8
4.17	Describe the negative-feedback mechanisms affecting bone deposition and reabsorption by identifying the relevant glands, hormones, target tissues, and hormone functions.	Hormonal regulation of bone deposition and reabsorption: pp. 149–151 Figures 4.53, 4.54 Table 4.3	Multiple Select: 4, 5, 9 Critical Thinking: 2
4.18	Summarize the six functions of the skeletal system and give an example or explanation of each.	Functions of the skeletal system: pp. 151–152	Multiple Select: 9
4.19	Summarize the effects of aging on the skeletal system.	Effects of aging on the skeletal system: p. 152	Multiple Select: 7, 8
4.20	Classify fractures using descriptive terms.	Fractures: p. 153 Figure 4.55	Spot Check: 11 Matching: 6–10
4.21	Explain how a fracture heals.	Fractures: pp. 154–155 Figures 4.56, 4.57	Critical Thinking: 3
4.22	Describe bone disorders and relate abnormal function to the pathology.	Skeletal system disorders: pp. 155–156 Figure 4.58 Table 4.4	Multiple Select: 10

footnotes

1. Centers for Disease Control and Prevention. (2010, April 15). *Arthritis*. Retrieved April 29, 2010, from http://www.cdc.gov/arthritis/.
2. National Institutes of Health, Osteoporosis and Related Bone Diseases National Resource Center. (2010). *Osteoporosis overview.* Retrieved March 7, 2010, from http://www.niams.nih.gov/Health_Info/Bone/Osteoporosis/overview.asp.

references and resources

Beers, M. H., Porter, R. S., Jones, T. V., Kaplan, J. L., & Berkwits, M. (Eds.). (2006). *Merck manual of diagnosis and therapy* (18th ed.). Whitehouse Station, NJ: Merck Research Laboratories.

Centers for Disease Control and Prevention. (2010, April 15). *Arthritis*. Retrieved April 29, 2010, from http://www.cdc.gov/arthritis/.

Frassetto, L. A., Morris, R. C., Sellmeyer, D. E., & Sebastian, A. (2008, February). Adverse effects of sodium chloride on bone in the aging human population resulting from habitual consumption of typical American diets [Electronic version]. *Journal of Nutrition, 138,* 419s–422s.

Gebhardt, S. E., & Thomas, R. G. (2002). Nutritive value of foods. *United States Department of Agriculture, Agricultural Research Services, Home and Garden Bulletin, 72.*

National Institutes of Health, Osteoporosis and Related Bone Diseases National Resource Center. (2010). *Osteoporosis overview.* Retrieved March 7, 2010, from http://www.niams.nih.gov/Health_Info/Bone/Osteoporosis/overview.asp.

Porter, R. S., & Kaplan, J. L. (Eds.). (n.d.). Selected physiologic age-related changes. *Merck manual for healthcare professionals*. Retrieved March 19, 2011, from http://www.merckmanuals.com/media/professional/pdf/Table_337-1.pdf?qt=selected physiologic age-related changes&alt=sh.

Raisz, L. G. (2008, February). Osteoporosis: Musculoskeletal and connective tissue disorders. *Merck manual of diagnosis and therapy.* Retrieved March 17, 2011, from http://www.merckmanuals.com/professional/sec04/ch036/ch036a.html.

Saladin, Kenneth S. (2010). *Anatomy & physiology: The unity of form and function* (5th ed.). New York: McGraw-Hill.

Seeley, R. R., Stephens, T. D., & Tate, P. (2006). *Anatomy & physiology* (7th ed.). New York: McGraw-Hill.

Shier, D., Butler, J., & Lewis, R. (2010). *Hole's human anatomy & physiology* (12th ed.). New York: McGraw-Hill.

University of Maryland Medical Center. (2008, August 10). *Aging changes in the bones—muscles—joints—Overview.* Retrieved April 8, 2010, from http://www.umm.edu/ency/article/004015.htm.

5 The Muscular System

When you open your text to Chapter 5 or start your computer to access this chapter online, muscle action is required. Even the expression on your face as you read the learning outcomes puts your muscles to use. While you are reading, you are breathing and your heart is pumping. Your muscles contract to give you control over areas that, on first thought, you might not associate with muscles, such as your bladder and bowels. In fact, muscles have many functions. See Figure 5.1. They are even responsible for much of your body heat.

5.1 learning outcome

Use medical terminology related to the muscular system.

learning outcomes

After completing this chapter, you should be able to:

5.1 Use medical terminology related to the muscular system.

5.2 Define terms concerning muscle attachments and the ways muscles work in groups to aid, oppose, or modify each other's actions.

5.3 Demonstrate actions caused by muscles.

5.4 Identify muscles, giving the origin, insertion, and action.

5.5 Describe the structural components of a muscle, including the connective tissues.

5.6 Describe the structural components of a skeletal muscle fiber, including the major proteins.

5.7 Explain the five physiological characteristics of all muscle tissue.

5.8 Explain how a nerve stimulates a muscle cell at a neuromuscular junction.

5.9 Describe a muscle contraction at the molecular level.

5.10 Compare and contrast a muscle twitch and tetany with regard to the steps of a muscle contraction at the molecular level.

5.11 Define a motor unit and explain the effect of recruitment.

5.12 Compare and contrast isotonic and isometric contractions.

5.13 Describe an example of a lever system in the human body, giving the resistance, effort, and fulcrum.

5.14 Compare aerobic and anaerobic respiration in terms of amount of ATP produced, speed, and duration.

5.15 Explain the basis of muscle fatigue and soreness.

5.16 Compare and contrast skeletal, cardiac, and smooth muscle tissue in terms of appearance, structure, type of nerve stimulation, type of respiration, and location.

5.17 Explain the nutritional requirements of the muscular system.

5.18 Summarize the five functions of the muscular system and give an example or explanation of each.

5.19 Summarize the effects of aging on the muscular system.

5.20 Describe muscle disorders and relate abnormal function to pathology.

word roots & combining forms

muscul/o: muscle

my/o: muscle

sarco: flesh

sthen/o: strength

pronunciation key

acetylcholine: AS-eh-til-KOH-leen

biceps brachii: BYE-sepz BRAY-kee-eye

brachialis: BRAY-kee-al-is

brachioradialis: BRAY-kee-oh-RAY-dee-al-is

buccinator: BUCK-sih-NAY-tor

digitorum: DIJ-ih-TOR-um

endomysium: EN-doh-MISS-ee-um

erector spinae: ee-REK-tor SPY-nee

fascia: FASH-ee-ah

fascicle: FAS-ih-kull

fibularis: FIB-you-LAH-ris

flexor carpi radialis: FLEK-sor KAR-pee RAY-dee-al-is

flexor carpi ulnaris: FLEK-sor KAR-pee ul-NAR-is

frontalis: fron-TAY-lis

gastrocnemius: gas-trok-NEE-me-us

gracilis: GRASS-ih-lis

iliacus: ill-EE-ah-kuss

latissimus dorsi: lah-TISS-ih-muss DOOR-sigh

continued on next page

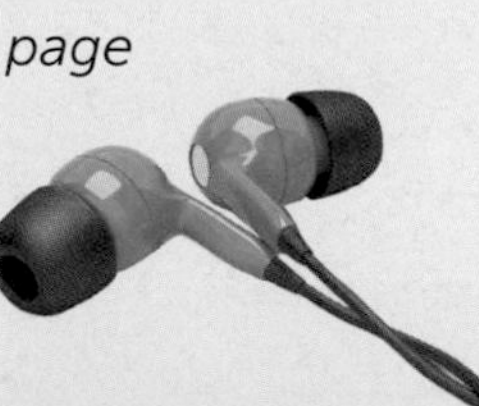

pronunciation **key** *continued*

masseter: MASS-eh-ter

occipitalis: ock-SIP-ih-TAY-lis

orbicularis oculi: or-BIK-you-LAIR-is OCK-you-lie

orbicularis oris: or-BIK-you-LAIR-is OR-is

palmaris: pahlm-AIR-is

pectineus: pek-TIN-ee-us

pectoralis: pek-tor-AL-is

platysma: plah-TIZ-ma

psoas: SO-ahz

quadriceps femoris: KWAD-rih-seps FEM-or-is

semimembranosus: sem-ee-MEM-bran-OH-sis

semitendinosus: sem-ee-TEN-dih-NOH-sis

serratus: seh-RAY-tus

soleus: SO-lee-us

sternocleidomastoid: STIR-noh-KLIDE-oh-MASS-toyd

temporalis: tem-poh-RAHL-is

tensor fasciae latae: TEN-sor FASH-ee-ee LAY-tee

tibialis: tih-bee-AH-lis

trapezius: trah-PEE-zee-us

triceps brachii: TRY-sepz BRAY-kee-eye

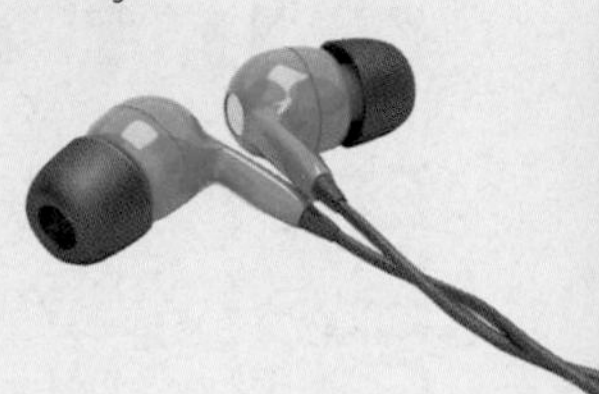

Overview

Skeletal muscles are the primary structures in the muscular system. Each skeletal muscle contains muscle fibers that contract, connective tissues that focus the force of the contraction, and nerves that control the contraction. These muscles contract to move most of the bones you learned about in Chapter 4 in a variety of motions.

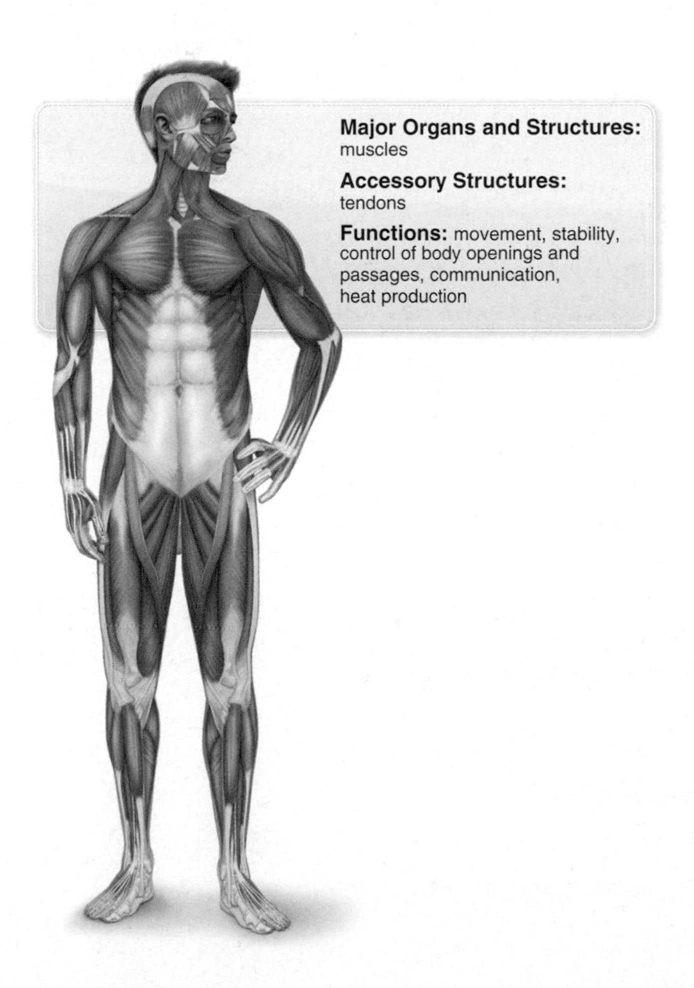

FIGURE 5.1 The muscular system.

This chapter covers prominent skeletal muscles by region and the motions they cause in the body. It also covers skeletal muscle cell anatomy and metabolism to help you learn precisely how a muscle cell contracts at the molecular level and how it produces the energy needed for the contractions to happen.

Although skeletal muscles are the primary structures in the muscular system, there are actually three types of muscle tissue in this system: skeletal, cardiac, and smooth muscle tissue. This chapter compares the different types of muscle tissue in terms of appearance, location, and the kind of metabolism used by each to process energy.

Once you have become familiar with the different types of muscle tissue, you can then combine this knowledge to see how the functions of this system are accomplished. This chapter ends with information about the effects of aging on the muscular system and muscular system disorders.

Anatomy of the Muscular System

To explore the skeletal muscles of the human body, you must first understand the definitions of muscle terms and of the different muscle actions. You will need to effectively use these terms when describing the muscles of the body and their actions.

Anatomical Terms

Defined below are terms of muscle attachment and terms that indicate the interrelated actions of muscles.

5.2 learning outcome

Define terms concerning muscle attachments and the ways muscles work in groups to aid, oppose, or modify each other's actions.

Terms of Muscle Attachment

- **Origin:** the attachment of a muscle to a bone or structure that *does not move* when the muscle contracts. A muscle has an attachment to a bone or another structure at each end. One attachment must be anchored for the muscle to be able to pull at the other end. The origin is the site of the anchored end.
- **Insertion:** the attachment of a muscle to a bone or structure that *does move* when the muscle contracts.
- **Intrinsic muscle:** a muscle that has its origin and insertion located in the same body region. Example: The temporalis muscle is intrinsic to the head because its origin and insertion are both in the head. (To see this, look ahead at **Figure 5.10**.)
- **Extrinsic muscle:** a muscle that has its origin located in a body region different from that of its insertion. Example: The sternocleidomastoid muscle is extrinsic to the head because its origin is in the head but its insertion is in the thorax. (Again, turn to **Figure 5.10**.)

Terms That Indicate the Interrelated Actions of Muscles The terms below classify muscles according to how they work together to aid, oppose, or modify each other's actions. Examples of how these terms are used are provided in the descriptions of the muscles later in the chapter.

- **Fixator:** a muscle that holds an origin stable for another muscle.
- **Synergists:** muscles that have the same action.
- **Prime mover:** the main muscle of the synergists that performs the action.
- **Antagonist:** a muscle that has an opposing action.

5.3 learning outcome

Demonstrate actions caused by muscles.

Muscle Actions

Muscle actions are the motions caused by the muscle contraction. **Figures 5.2** to **5.9** show the following common actions:

- **Flexion:** action that bends a part of the body anteriorly, such as flexing the elbow. To flex the arm is to bring the entire arm anteriorly, as in using your entire arm to point at something ahead of you. The exception is the knee. Flexion of the knee moves the lower leg posteriorly. See **Figure 5.2a** and **b**.
- **Extension:** action that bends a part of the body posteriorly, such as straightening the arm at the elbow. As with flexion, the exception is the knee. Extending the knee straightens the lower leg. See **Figure 5.2a** and **b**.
- **Abduction:** movement of a part of the body away from the midline. See **Figure 5.3a**.
- **Adduction:** movement of a part of the body toward the midline. See **Figure 5.3b**. Doing "jumping jacks" is to alternately abduct and adduct the arms and legs.
- **Protraction:** movement that brings part of the body forward. See **Figure 5.4a**.
- **Retraction:** movement that brings part of the body backward. For example, a chicken protracts and retracts its head as it walks. See **Figure 5.4b**.
- **Lateral excursion:** movement of the jaw laterally to either side. See **Figure 5.4c**.
- **Medial excursion:** movement of the jaw back to the midline. See **Figure 5.4d**.
- **Dorsiflexion:** position of standing on the heels with the toes pointing up off the floor. See **Figure 5.5a**.
- **Plantar flexion:** position of standing on tiptoes with the heels off the floor. See **Figure 5.5a**.
- **Inversion:** position in which the soles of the feet are together, facing each other. See **Figure 5.5b**.
- **Eversion:** position in which the soles of the feet point away from each other. See **Figure 5.5c**.

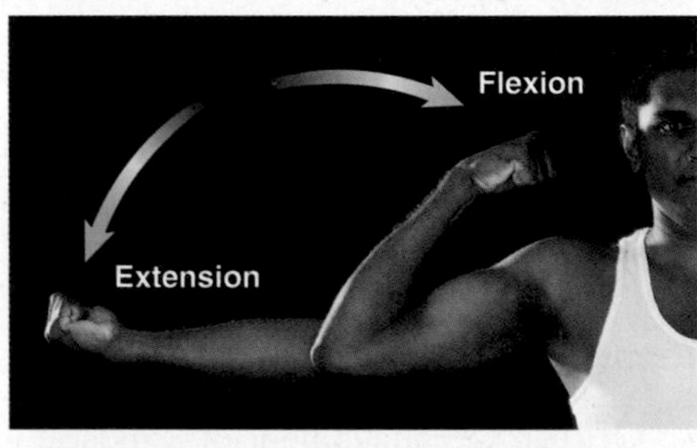

(a)

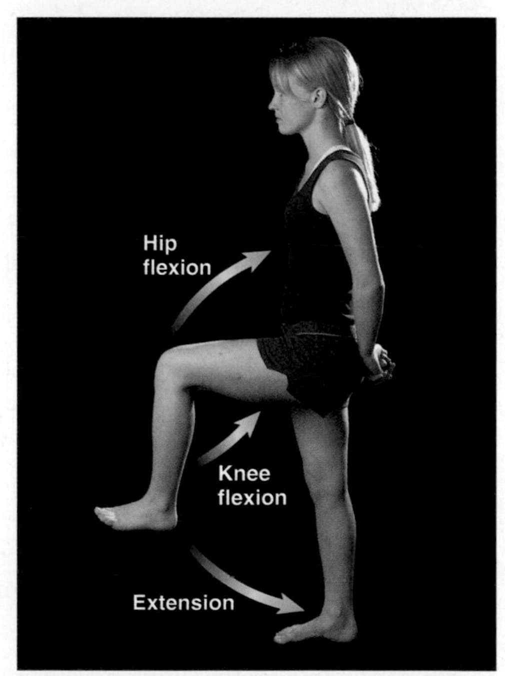

(b)

FIGURE 5.2 Flexion and extension: (a) flexion and extension of the elbow, (b) flexion and extension of the knee and hip.

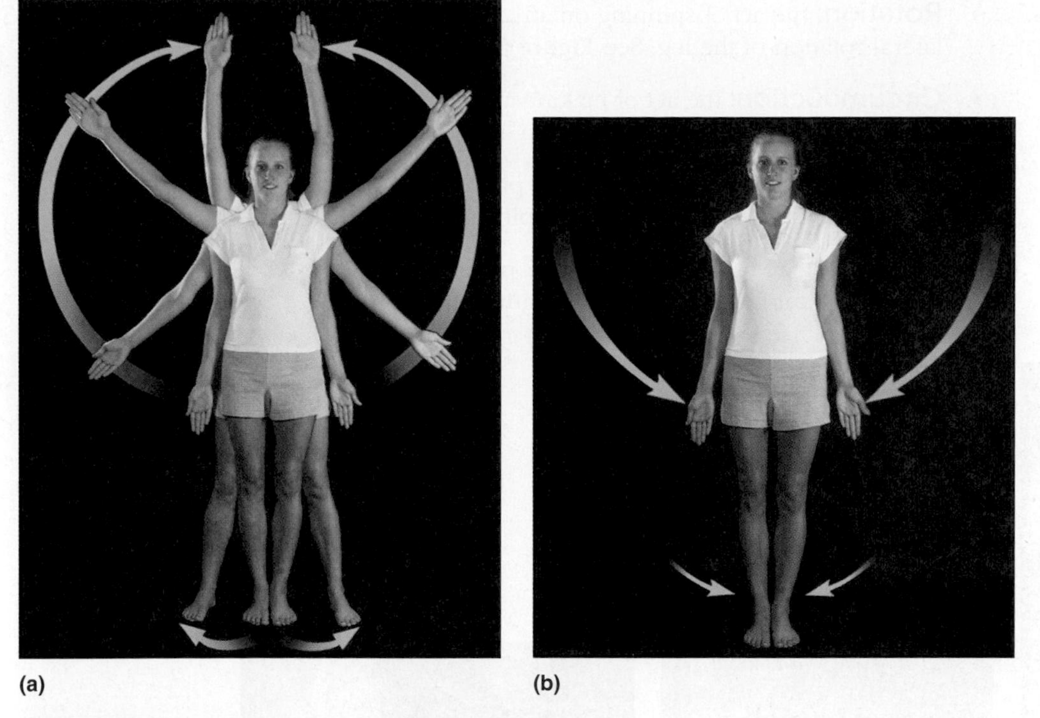

(a) (b)

FIGURE 5.3 Abduction and adduction: (a) abduction, (b) adduction.

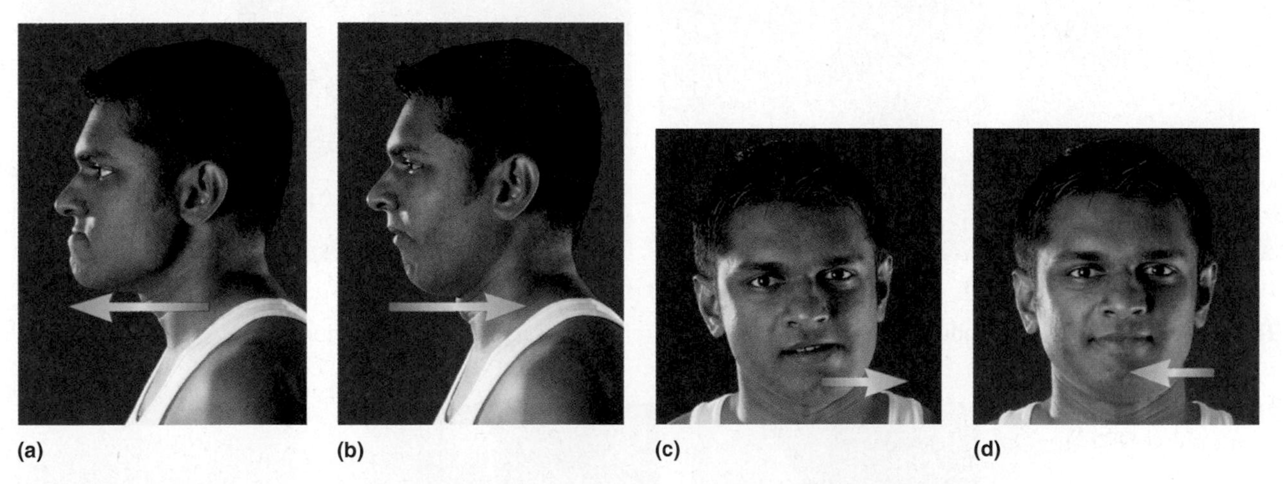

(a) (b) (c) (d)

FIGURE 5.4 Protraction, retraction, lateral excursion, and medial excursion: (a) protraction, (b) retraction, (c) lateral excursion, (d) medial excursion.

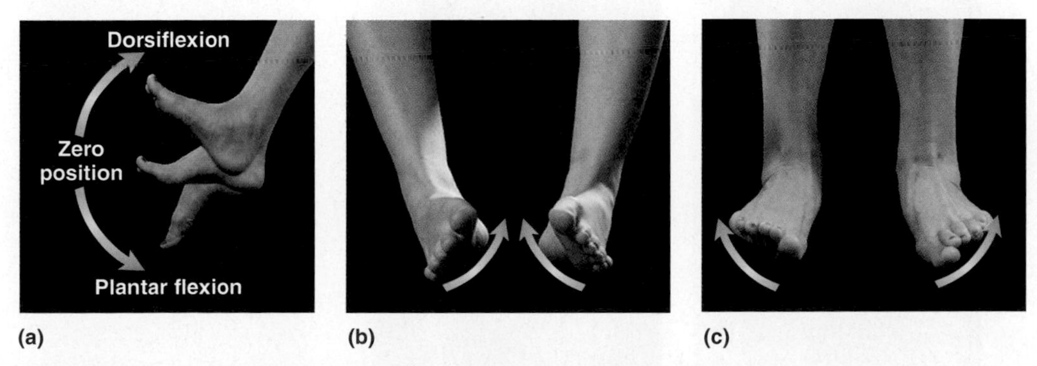

(a) (b) (c)

FIGURE 5.5 Dorsiflexion, plantar flexion, inversion, and eversion: (a) dorsiflexion and plantar flexion, (b) inversion, (c) eversion.

- **Rotation:** the act of spinning on an axis. Pointing your toes to the side involves lateral rotation of the leg. See **Figure 5.6a** and **b**.
- **Circumduction:** the act of making a circle with part of the body. A baseball or softball pitch involves circumduction at the shoulder. See **Figure 5.6c**.
- **Supination:** rotation that turns the palms up. You could hold soup in the palm of your hand when your palm is supinated. See **Figure 5.7a**.
- **Pronation:** rotation that turns the palms down. You would pour soup from your hand during pronation. See **Figure 5.7b**.

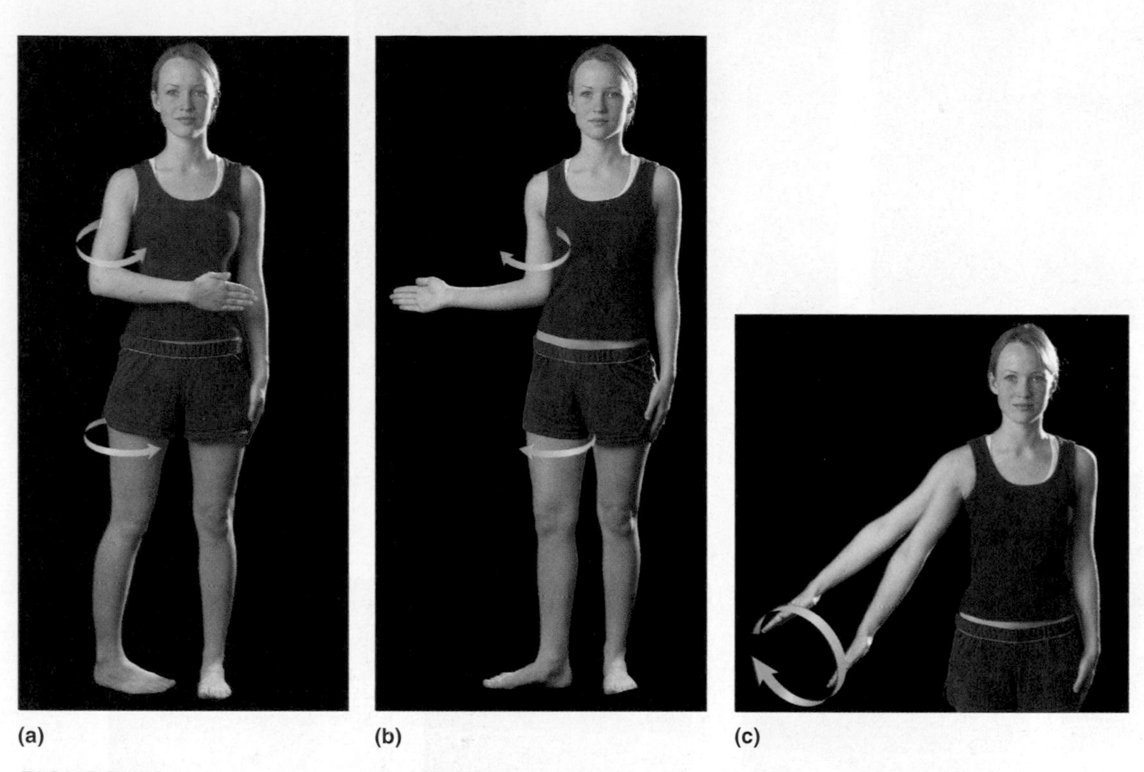

FIGURE 5.6 Rotation and circumduction: (a) medial rotation, (b) lateral rotation, (c) circumduction.

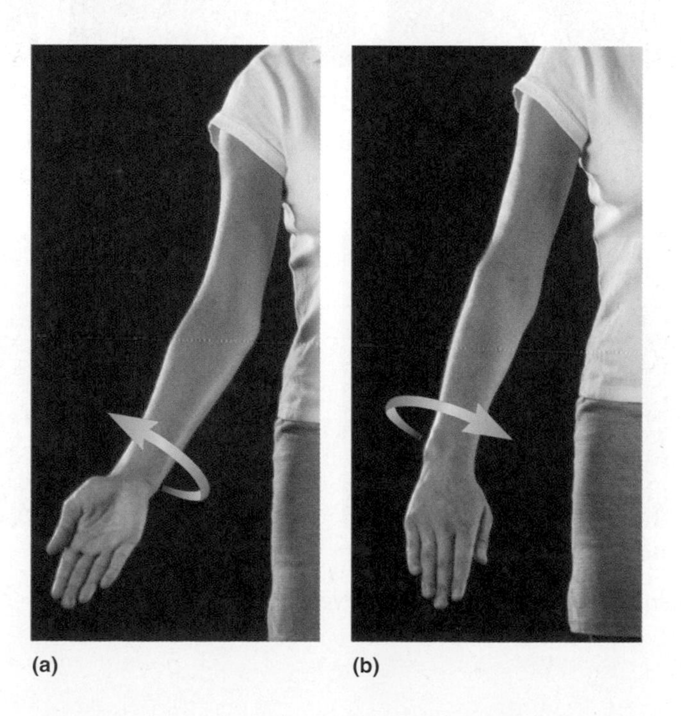

FIGURE 5.7 Supination and pronation: (a) supination, (b) pronation.

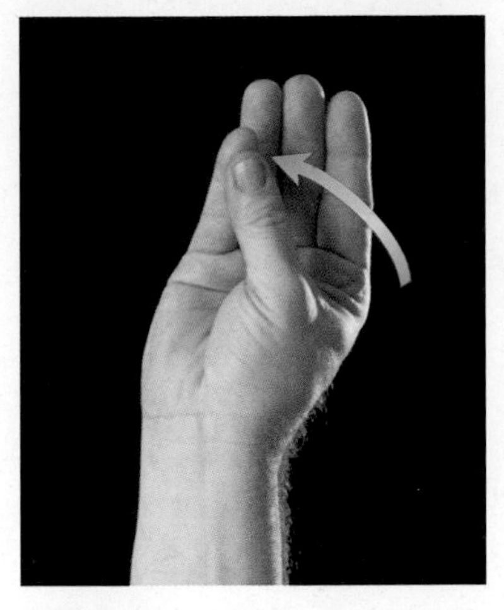

FIGURE 5.8 Opposition.

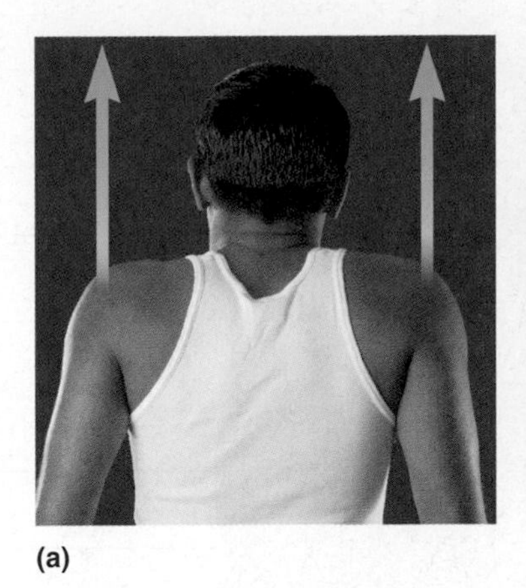

(a)

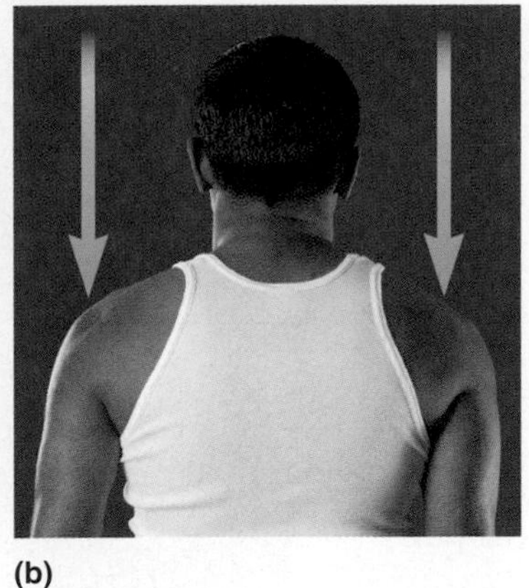

(b)

FIGURE 5.9 Elevation and depression: (a) elevation, (b) depression.

- **Opposition:** the act of bringing the thumb to the palm. See **Figure 5.8**.
- **Reposition:** the act of taking the thumb away from the palm.
- **Elevation:** the act of closing the jaw or raising the shoulders. See **Figure 5.9a**.
- **Depression:** the act of opening the jaw or lowering the shoulders. See **Figure 5.9b**.

study hint

Studying this system can be difficult because there are so many muscles with names that are difficult to pronounce. Remember, you have a complete set of muscles with you at all times in your own body.

A good study technique is to find a muscle on your body and contract it. Point to the muscle's origin and insertion and feel the action while saying the name aloud. If you can hear yourself say the name, you will have an easier time spelling it.

5.4 learning outcome

Identify muscles, giving the origin, insertion, and action.

Figures 5.10 and **5.11** show the superficial muscles of the entire body. The next section of the chapter focuses on the muscles by region. Closer views of muscles are shown in each body region.

Muscles by Region

Muscles of the Head and Neck While larger, more obviously visible muscles might come to mind when you hear the word *muscles,* even the head and neck muscles, no matter their size, are crucial to the body's functioning. See **Figure 5.12**. The names, origins, insertions, and functions of the muscles of the head and neck are outlined in Table 5.1 and explained in the following list:

- The **orbicularis oris** is an example of a muscle that does not have a bone as an origin or insertion. The insertion is the lips, and the origin is a complex cord at the corner of the mouth.
- The **frontalis** muscle originates on the epicranial aponeurosis (a membranous sheath that rides and can slide back and forth slightly over the top of the skull).

frontalis: fron-TAY-lis

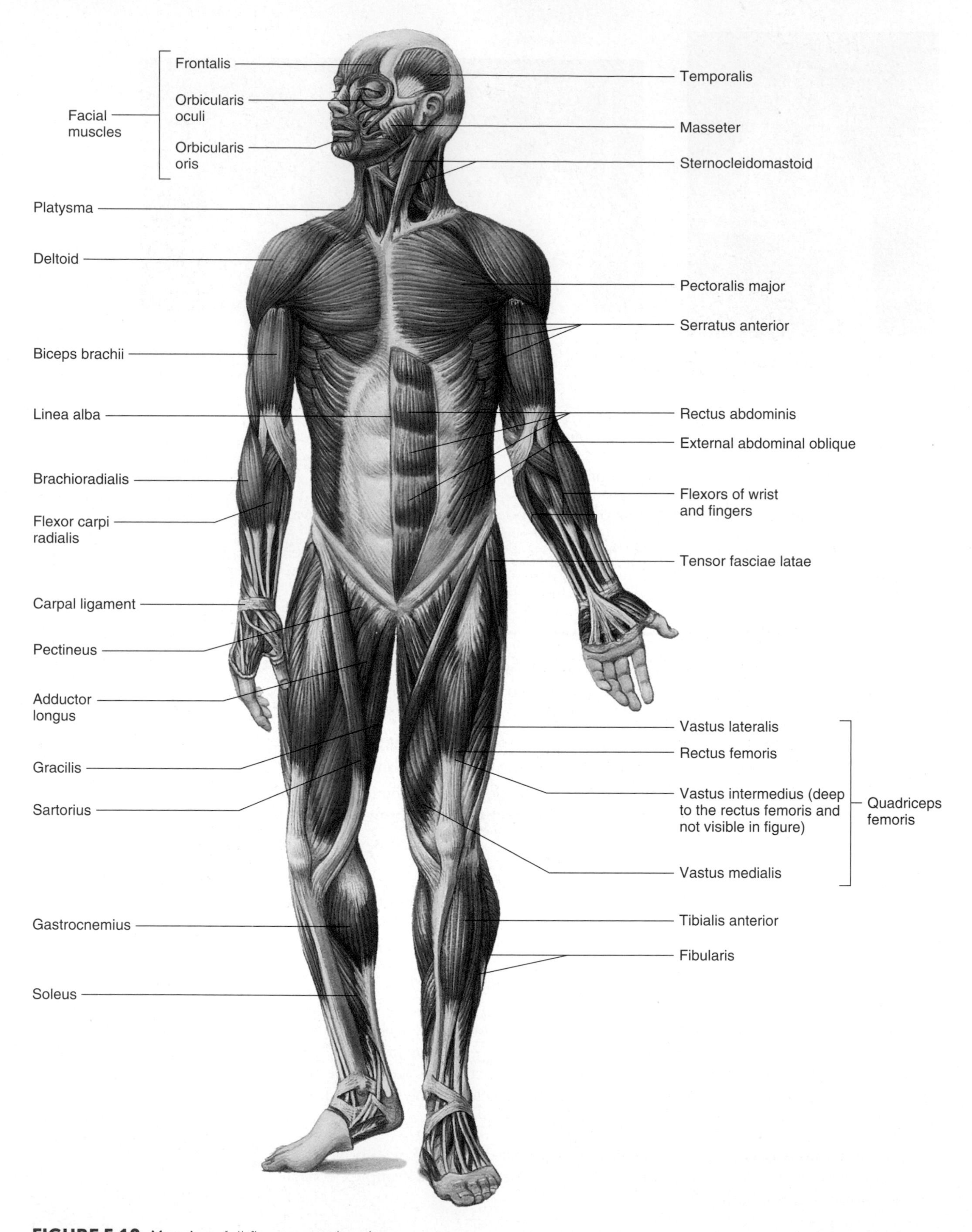

FIGURE 5.10 Muscles: full figure, anterior view.

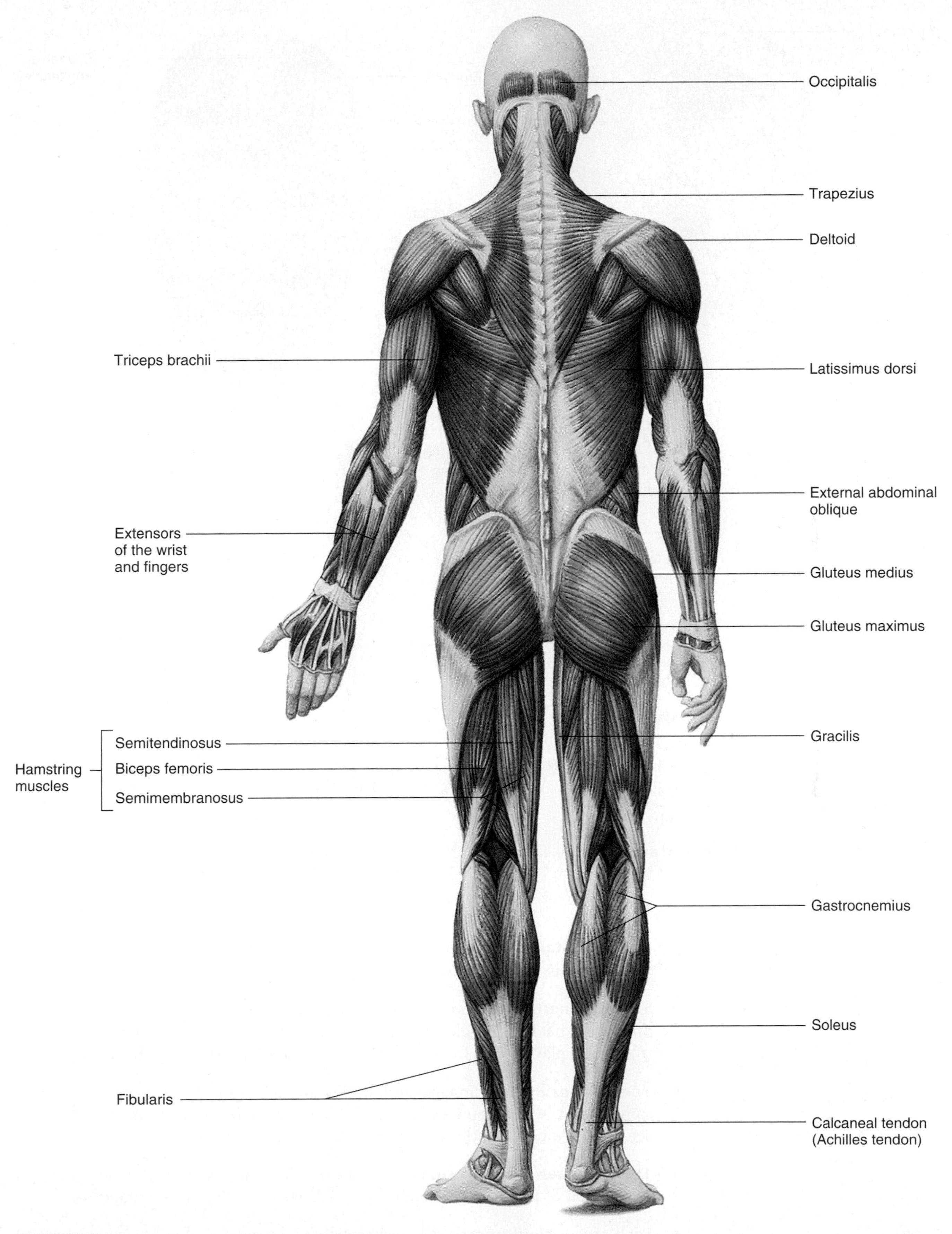

FIGURE 5.11 Muscles: full figure, posterior view.

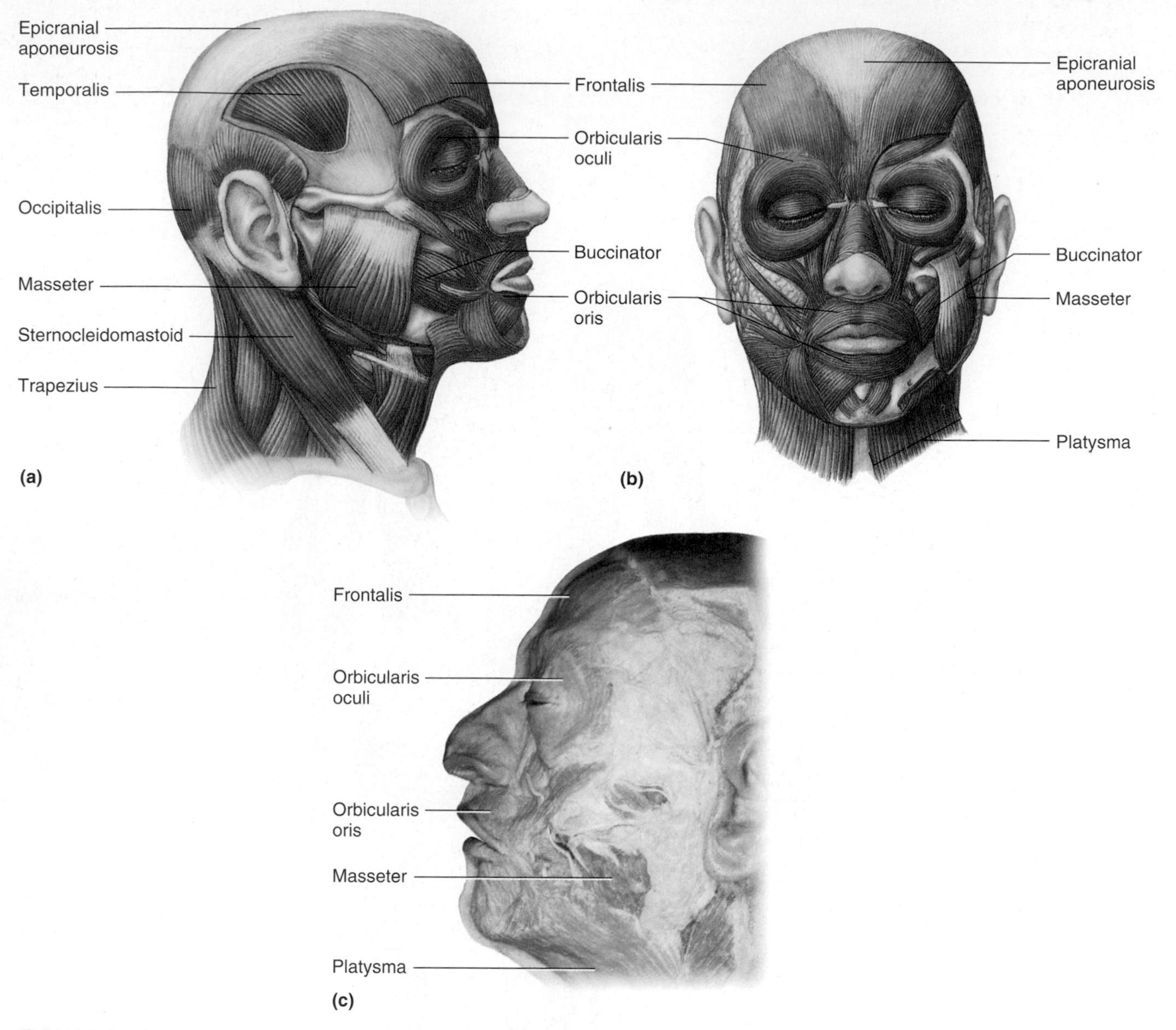

FIGURE 5.12 Muscles of the head and neck: (a) lateral view, (b) anterior view, (c) lateral view of cadaver.

buccinator: BUCK-sih-NAY-tor

occipitalis: ock-SIP-ih-TAY-lis

masseter: MASS-eh-ter

platysma: plah-TIZ-ma

sternocleidomastoid: STIR-noh-KLIDE-oh-MASS-toyd

The **occipitalis** contracts when the frontalis raises the eyebrows. In this case, the occipitalis acts as a fixator holding the origin for the frontalis steady.

- The **buccinator** is often called the "trumpeter's muscle," but do not be confused and think that this muscle causes a kissing motion of the lips. That is not how a trumpet is played. The buccinator flattens and compresses the cheeks.
- The **temporalis** and **masseter** muscles are synergists because they have the same action. When you look at the insertion for each on the mandible, the masseter has a much broader connection. It is the prime mover.
- The **masseter** and **platysma** have opposite functions as far as the elevation and depression of the mandible are concerned. They are therefore antagonists.
- Each **sternocleidomastoid** muscle rotates the head.

TABLE 5.1 Muscles of the head and neck

Muscle	Origin	Insertion	Function
Orbicularis oris	Complex cord at the corner of the mouth	Lips	Closes and protrudes lips, as in kissing.
Orbicularis oculi	Medial eye orbit	Eyelids	Closes eye.
Frontalis	Sheath of fibrous tissue over the top of the skull, called the **epicranial aponeurosis**	Skin of forehead	Raises eyebrows and wrinkles the skin of the forehead.
Occipitalis	Temporal bone	Epicranial aponeurosis	Fixes epicranial aponeurosis as an origin for the frontalis muscle.
Temporalis	Temporal bone	Mandible	Elevates, retracts, and causes medial and lateral excursion of the mandible.
Buccinator	Maxilla and mandible	Orbicularis oris	Compresses cheeks.
Masseter	Zygomatic arch	Mandible	Elevates mandible.
Platysma	Fascia of the deltoid and pectoralis major muscles	Mandible	Depresses mandible and draws the corner of the mouth and lower lip downward.
Sternocleidomastoid	Manubrium of the sternum and medial portion of the clavicle	Mastoid process of the temporal bone	Individually, each muscle rotates the head. Together, the muscles bring the head forward and down.

spot check 1 If you put one finger on the origin of the right sternocleidomastoid (the mastoid process posterior to your right earlobe) and another finger on the insertion of the right sternocleidomastoid (the manubrium at the base of the neck), you can determine which direction the right sternocleidomastoid rotates the head. Muscles can only shorten when they contract. First turn your head right, and then turn your head left. Which direction caused your fingers to come closer together, indicating that the muscle shortened?

Muscles of the Thorax and Abdomen The muscles of the thorax and abdomen are shown in **Figure 5.13**. The names, origins, insertions, and functions of these muscles are outlined in Table 5.2 and explained in the following list:

- The **pectoralis minor** is deep to the **pectoralis major.**
- The **serratus anterior** has a serrated or jagged edge like that of a serrated knife.
- The **diaphragm** is a dome-shaped skeletal muscle composed of skeletal muscle tissue. It divides the thorax from the abdomen. A central tendon at the center of the dome serves as the insertion. The diaphragm is the prime mover for breathing and flattens as it contracts, increasing the size of the thoracic cavity.
- The **rectus abdominis** is the "six-pack muscle" you see on abdomens of people who are very physically fit. Whenever you see the word *rectus,* think straight up and down. A fibrous sheath—formed by the tendons of the internal and external abdominal oblique muscles—encloses the rectus abdominis.

orbicularis oculi: or-BIK-you-LAIR-is OCK-you-lie

orbicularis oris: or-BIK-you-LAIR-is OR-is

pectoralis: pek-tor-AL-is

serratus: seh-RAY-tus

temporalis: tem-poh-RAHL-is

- The abdominal wall on either side of the rectus abdominis has three layers. The most superficial layer is the **external abdominal oblique.** *Oblique* means that it runs at an angle. Deep to this layer is the **internal abdominal oblique.** The deepest layer is the **transverse abdominal.** Its cells run on the transverse plane.
- The **rectus abdominis** and the **external and internal obliques** are synergists for flexing the spine, but only the external and internal abdominal obliques are synergists with the **transverse abdominal** for compressing the abdomen.

FIGURE 5.13 Muscles of the thorax and abdomen: (a) superficial and deep muscles, anterior view, (b) deep muscles, anterior view, (c) abdominal muscles of a cadaver.

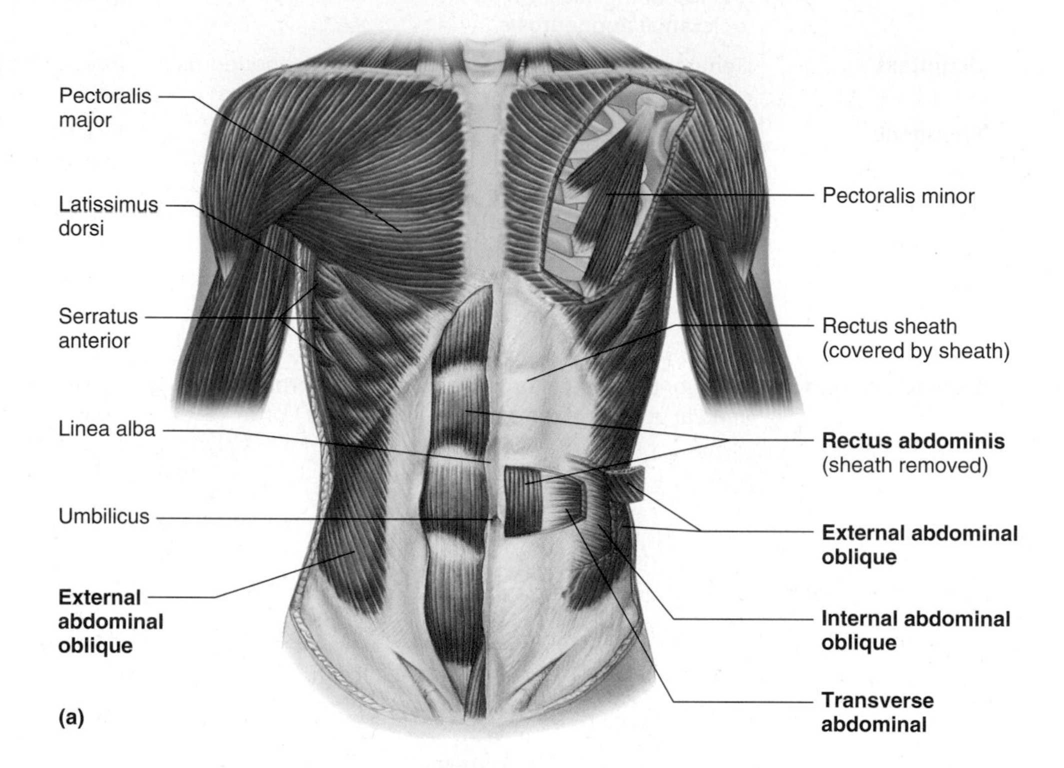

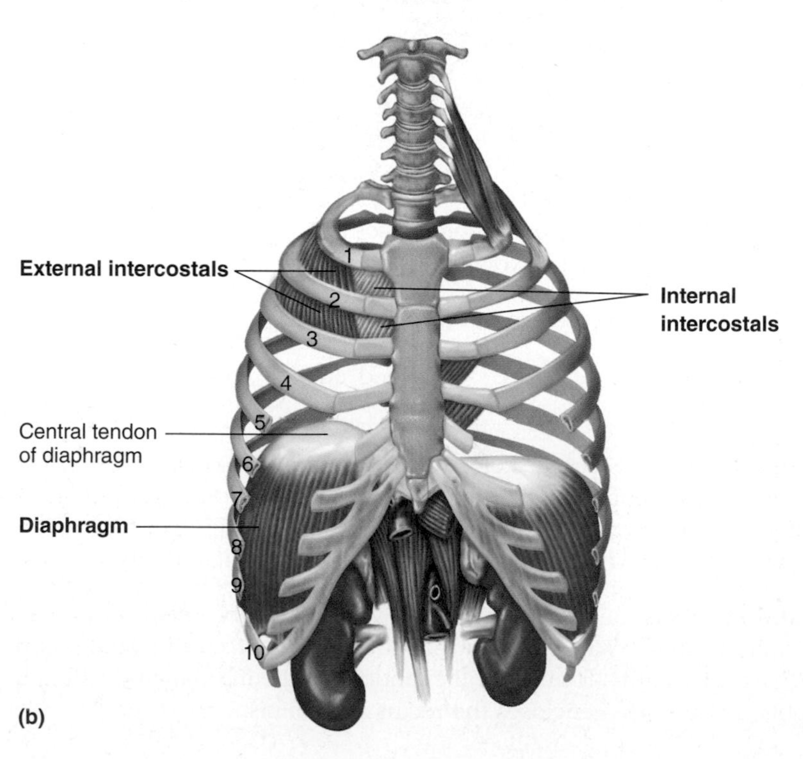

FIGURE 5.13 concluded

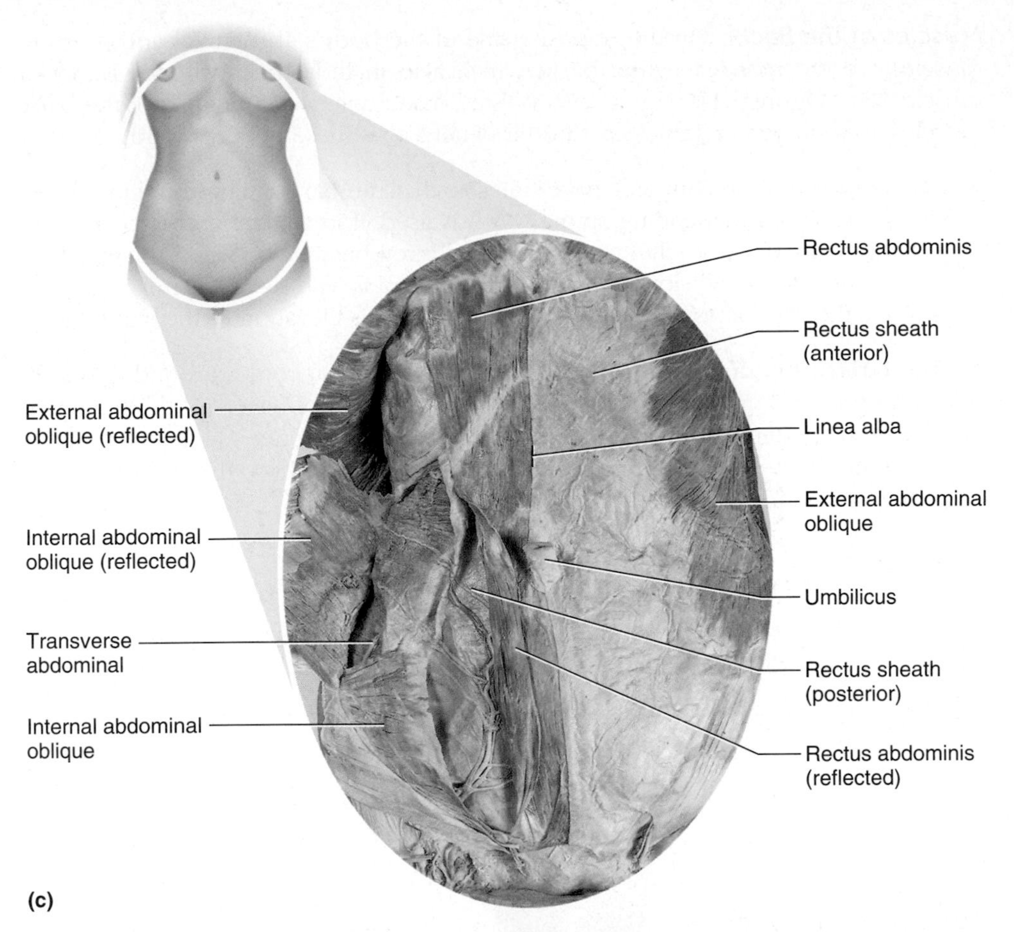

(c)

TABLE 5.2 Muscles of the thorax and abdomen

Muscle	Origin	Insertion	Function
Pectoralis major	Clavicle and costal cartilages	Humerus	Flexes and adducts humerus
Pectoralis minor	Ribs	Coracoid process of scapula	Depresses and protracts shoulder
Serratus anterior	Anterior ribs	Scapula	Protracts shoulder
Diaphragm	Xyphoid process, costal cartilages, and lumbar vertebrae	Central tendon of diaphragm	Prime mover for breathing
External intercostals	Inferior margins of ribs	Superior margin of next-lower rib	Expand the thoracic cavity during inspiration
Internal intercostals	Superior margins of ribs	Inferior margin of next-higher rib	Compress the thoracic cavity during forced expiration
External abdominal oblique	Ribs	Iliac crest	Compresses abdomen, flexes spine, and allows rotation at the waist
Internal abdominal oblique	Iliac crest	Ribs	Compresses abdomen, flexes spine, and allows rotation at the waist
Rectus abdominis	Pubis	Xyphoid process and costal cartilages	Flexes spine
Transverse abdominal	Inguinal ligament, iliac crest, and costal cartilages	Linea alba and pubis	Compresses abdomen

Muscles of the Back On the opposite side of the body's abdominal and thoracic muscles are the muscles of the back, which also include the neck and buttocks muscles. See **Figure 5.14**. The names, origins, insertions, and functions of the back muscles are outlined in Table 5.3. Further detail about these muscles follows:

trapezius:
trah-PEE-zee-us

latissimus dorsi:
lah-TISS-ih-muss
DOOR-sigh

- The **trapezius** is a diamond-shaped muscle that can have multiple origins, insertions, and actions depending on the way it is used. If its attachment to the scapula is held steady (fixated), the external occipital protuberance becomes the insertion and the action is extension of the head. If the head is held steady as the origin, the scapula then becomes the insertion and the action is elevation of the shoulders.
- The **latissimus dorsi** is a very broad back muscle that comes around the body to insert on the groove between the greater and lesser tubercles of the humerus. Its job is to extend, adduct, and medially rotate the humerus. The arm motion of the butterfly stroke in swimming is a good example of how these three movements are performed together by this muscle.

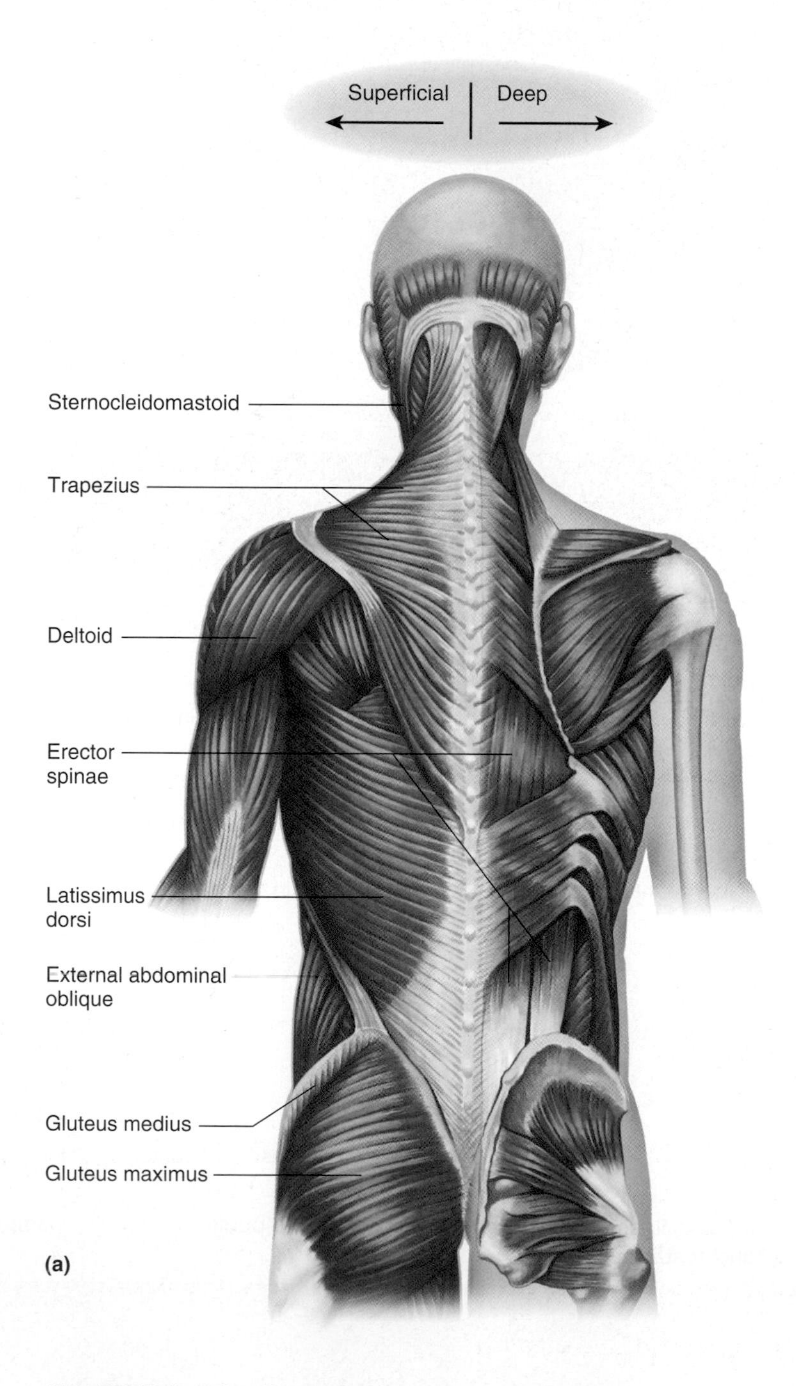

FIGURE 5.14 **Muscles of the neck, back and buttocks:** (a) superficial muscles *(left)* and deep muscles *(right)*, (b) back muscles of a cadaver, with superficial muscles *(right)* and deep muscles *(left)*.

FIGURE 5.14 concluded

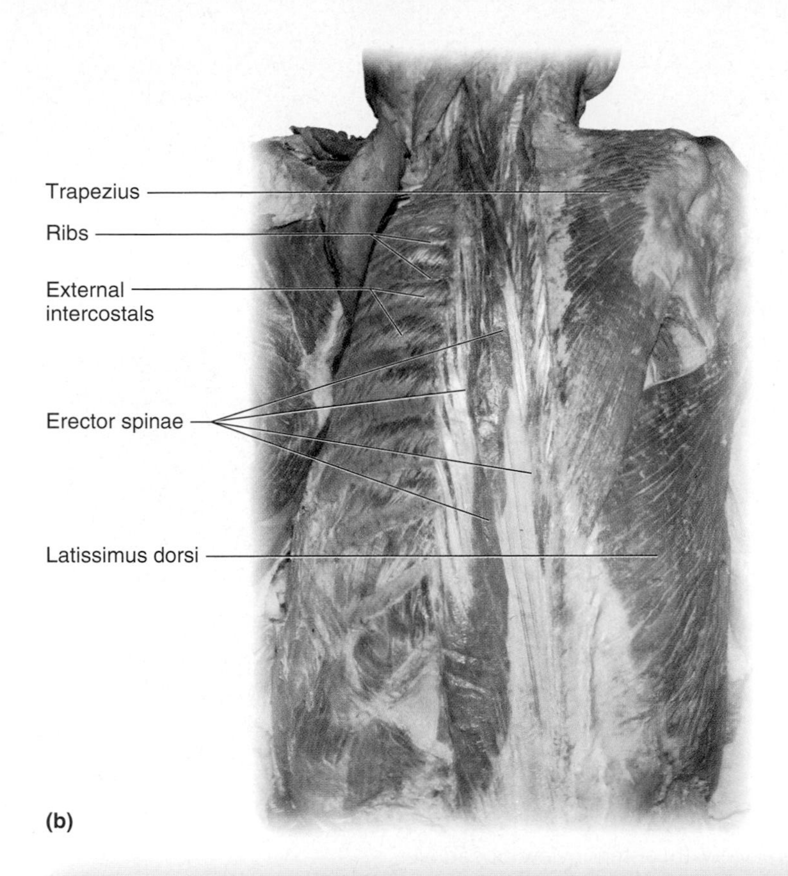

TABLE 5.3 Muscles of the neck, back, and buttocks

Muscle	Origin	Insertion	Function
Trapezius	Scapula	External occipital protuberance	Extends head
Latissimus dorsi	Lumbar vertebrae	Intertubercular groove of humerus	Extends, adducts, and medially rotates the humerus
Erector spinae	Ribs, vertebrae, and ilium	Ribs and vertebrae	Holds spine erect for posture and extends spine
Gluteus maximus	Ilium, sacrum, and coccyx	Femur	Extends and laterally rotates hip
Gluteus medius	Ilium	Greater trochanter of femur	Abducts and medially rotates hip

- The **erector spinae** is a whole group of muscles that runs straight up and down, deep in the back. Its function is to hold the spine erect for posture and further extension of the spine.

erector spinae:
ee-REK-tor SPY-nee

- The **gluteus maximus** is the largest muscle of the buttocks. It extends the leg and laterally rotates the hip.
- The **gluteus medius** is also located in the buttocks. It is an antagonist to the gluteus maximus in regard to hip rotation, as it medially rotates the hip.

Muscles of the Arm The muscles of the arm are shown in **Figure 5.15**. Their names, origins, insertions, and functions are outlined in Table 5.4, and the following item provides additional details:

- The **biceps brachii, brachialis,** and **brachioradialis** are synergists for flexing the elbow. The brachialis muscle is the prime mover. The antagonist to this group is the **triceps brachii.**

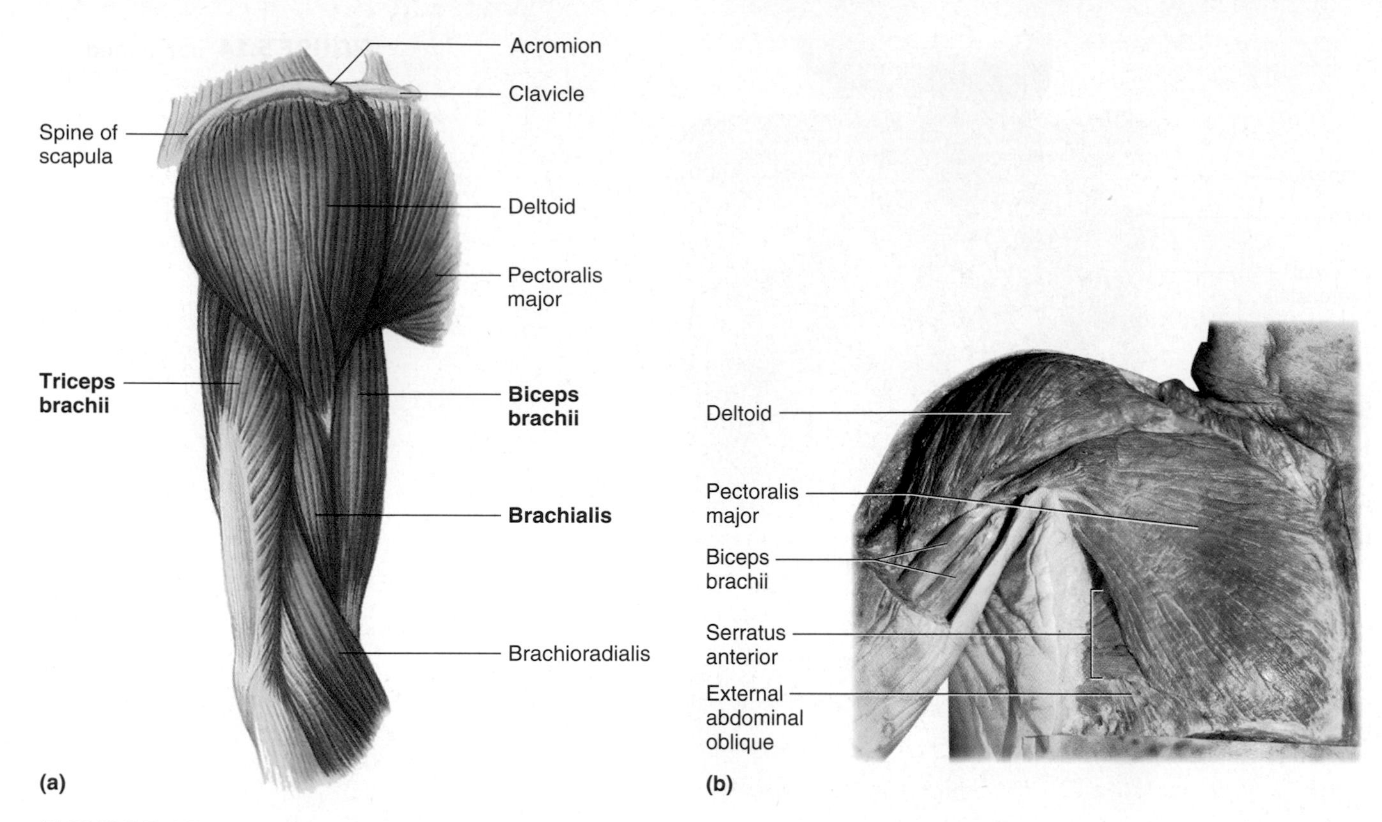

FIGURE 5.15 Pectoral and brachial muscles: (a) lateral view, (b) anterior view of cadaver.

biceps brachii: BYE-sepz BRAY-kee-eye

triceps brachii: TRY-sepz BRAY-kee-eye

brachialis: BRAY-kee-al-is

brachioradialis: BRAY-kee-oh-RAY-dee-al-is

TABLE 5.4 Muscles of the arm

Muscle	Origin	Insertion	Function
Deltoid	Clavicle and scapula	Deltoid tuberosity of humerus	Abducts humerus
Biceps brachii	Scapula	Radius	Flexes elbow
Triceps brachii	Humerus and scapula	Olecranon of ulna	Extends elbow
Brachialis	Humerus	Ulna	Flexes elbow
Brachioradialis	Humerus	Radius	Flexes elbow

Muscles of the Forearm Connected to the muscles of the arm are the muscles of the forearm. See **Figure 5.16**. The names, origins, insertions, and functions of the muscles of the forearm are outlined in Table 5.5. Further detail about these muscles is provided below:

- The forearm has many muscles. It is important to pay attention to their origins because the origins help with understanding the actions of these muscles. For example, all flexors originate at the medial epicondyle of the humerus, while all extensors originate at the lateral epicondyle of the humerus.
- Carpi radialis and carpi ulnaris muscles have long and short versions called *longus* and *brevis* components. The longus or brevis components may also be called **carpi radialis** or **carpi ulnaris.**
- **Flexor carpi ulnaris** and **extensor carpi ulnaris** muscles have different origins, but both run along the ulna. They are antagonists to each other for wrist flexion and extension, but they are synergists for wrist adduction.

flexor carpi ulnaris: FLEK-sor KAR-pee ul-NAR-is

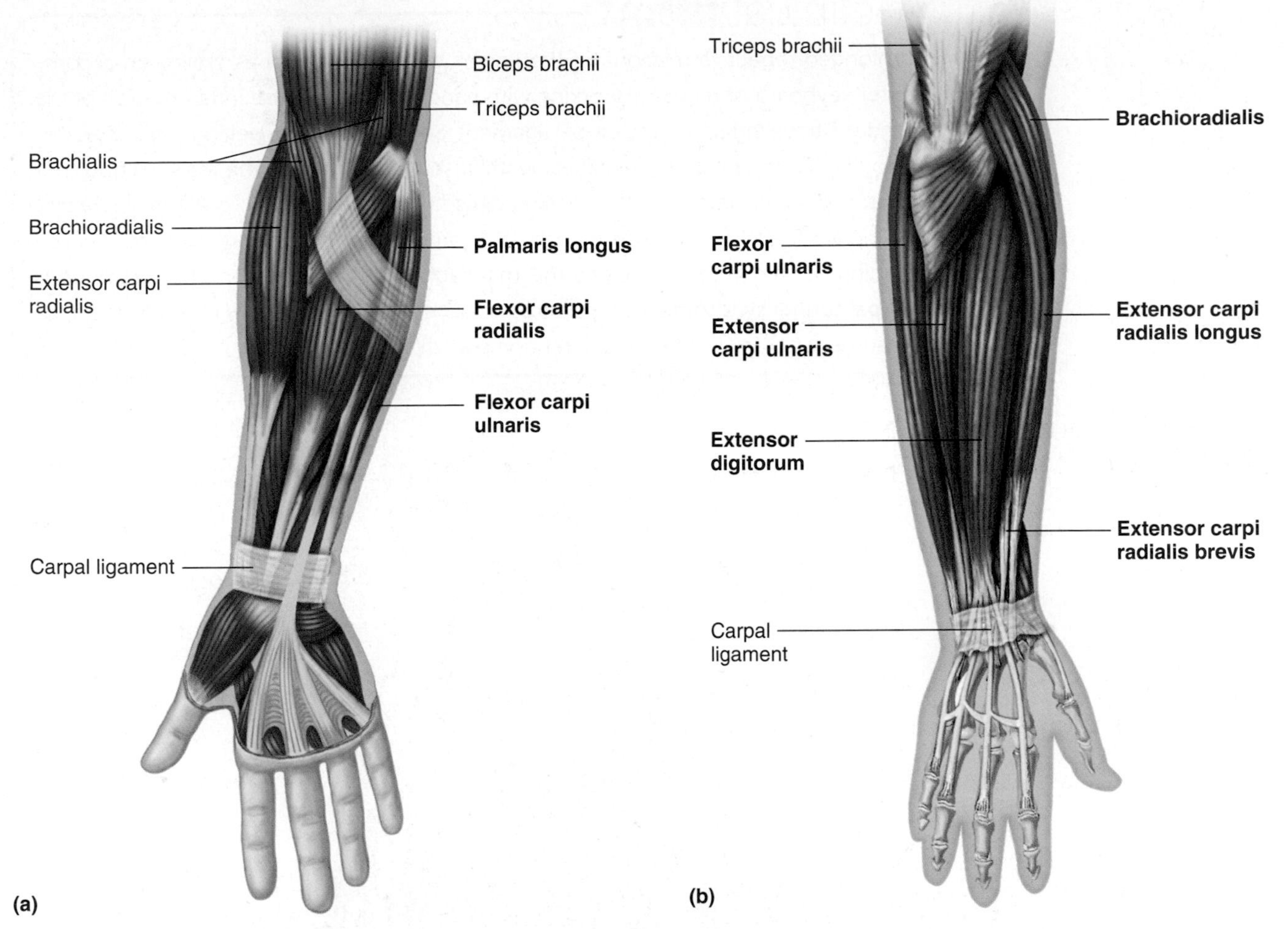

FIGURE 5.16 Muscles of the forearm: (a) superficial flexors, anterior view, (b) superficial extensors, posterior view.

TABLE 5.5 Muscles of the forearm

Muscle	Origin	Insertion	Function
Extensor carpi radialis	Lateral epicondyle of humerus	Metacarpals	Extends and abducts wrist
Extensor carpi ulnaris	Lateral epicondyle of humerus	Carpals and metacarpals	Extends and adducts wrist
Palmaris longus	Medial epicondyle of humerus	Palmar aponeurosis	Flexes wrist
Flexor carpi radialis	Medial epicondyle of humerus	Metacarpals	Flexes and abducts wrist
Flexor carpi ulnaris	Medial epicondyle of humerus	Metacarpals	Flexes and adducts wrist
Extensor digitorum	Lateral epicondyle of humerus	Posterior phalanges	Extends fingers
Flexor digitorum	Ulna	Phalanges	Flexes fingers

- **Flexor carpi radialis** and **extensor carpi radialis** muscles have different origins, but both run along the radius. They are antagonists to each other for wrist flexion and extension, but they are synergists for wrist abduction.

- **Palmaris longus** is the only muscle of the forearm with a tendon that is superficial to the carpal ligament (bracelet-like ligament at the wrist). All other tendons, nerves, and vessels travel deep to this ligament.

flexor carpi radialis: FLEK-sor KAR-pee RAY-dee-al-is

palmaris: pahlm-AIR-is

digitorum: DIJ-ih-TOR-um

connect.mcgraw-hill.com audio

clinical point

Prolonged repetitive motions of the fingers and hands, such as typing on a computer keyboard or regularly working with hand tools, can cause inflammation of the tendons traveling under the carpal ligament. See Figure 5.17. Swelling is a characteristic of inflammation. Because the sheath (fascia) superficial to the tendons does not stretch to accommodate the swelling caused by inflammation, pressure increases on nerves traveling deep to the carpal ligament. The added pressure may cause tingling or pain in the palm, and this may radiate to the shoulder. This condition is **carpal tunnel syndrome**. It can be treated with anti-inflammatory medications and/or surgery to remove part of the tendon sheath.

FIGURE 5.17 Carpal tunnel syndrome: (a) anterior view, (b) cross section.

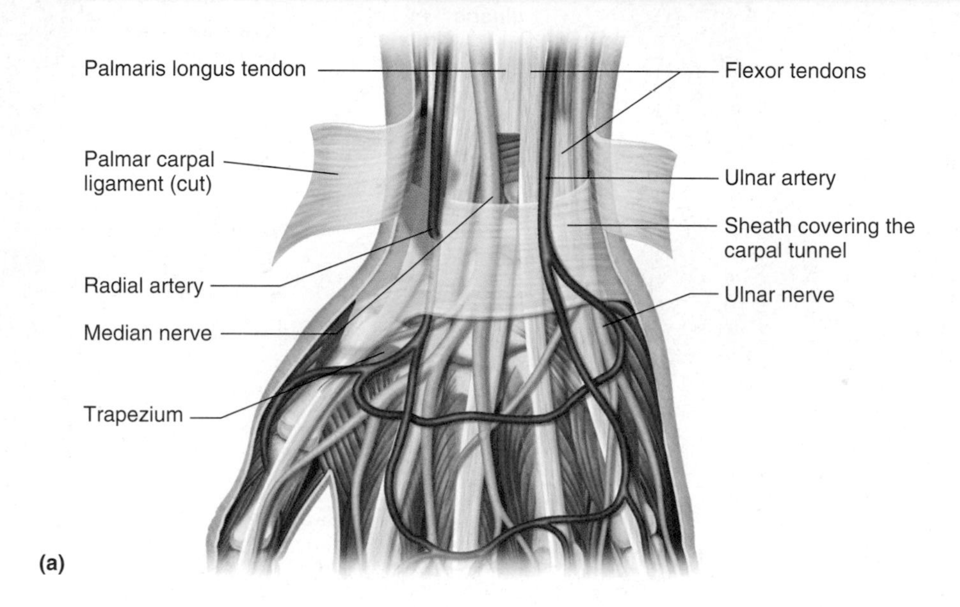

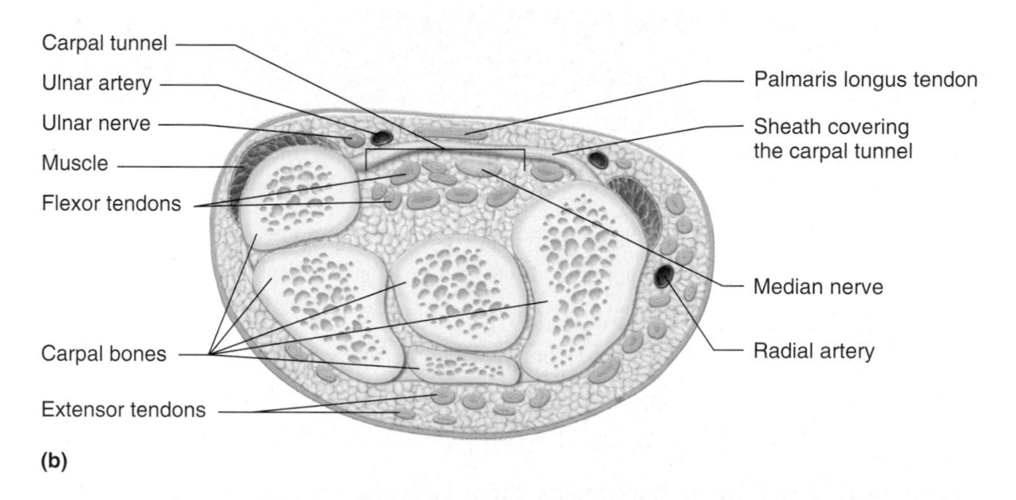

iliacus: ill-EE-ah-kuss

psoas: SO-ahz

Muscles of the Thigh Now that you have reviewed the muscles of the arm and upper body, you are ready to study the body's lower limbs, focusing first on the thigh muscles. See **Figures 5.18** and **5.19**. Table 5.6 outlines the names, origins, insertions, and functions of the thigh muscles, and the following list provides additional details:

- The **psoas major** originates high on the posterior abdominal wall (T12–L5), while the **iliacus** originates on and appears to line the ilium. Both muscles

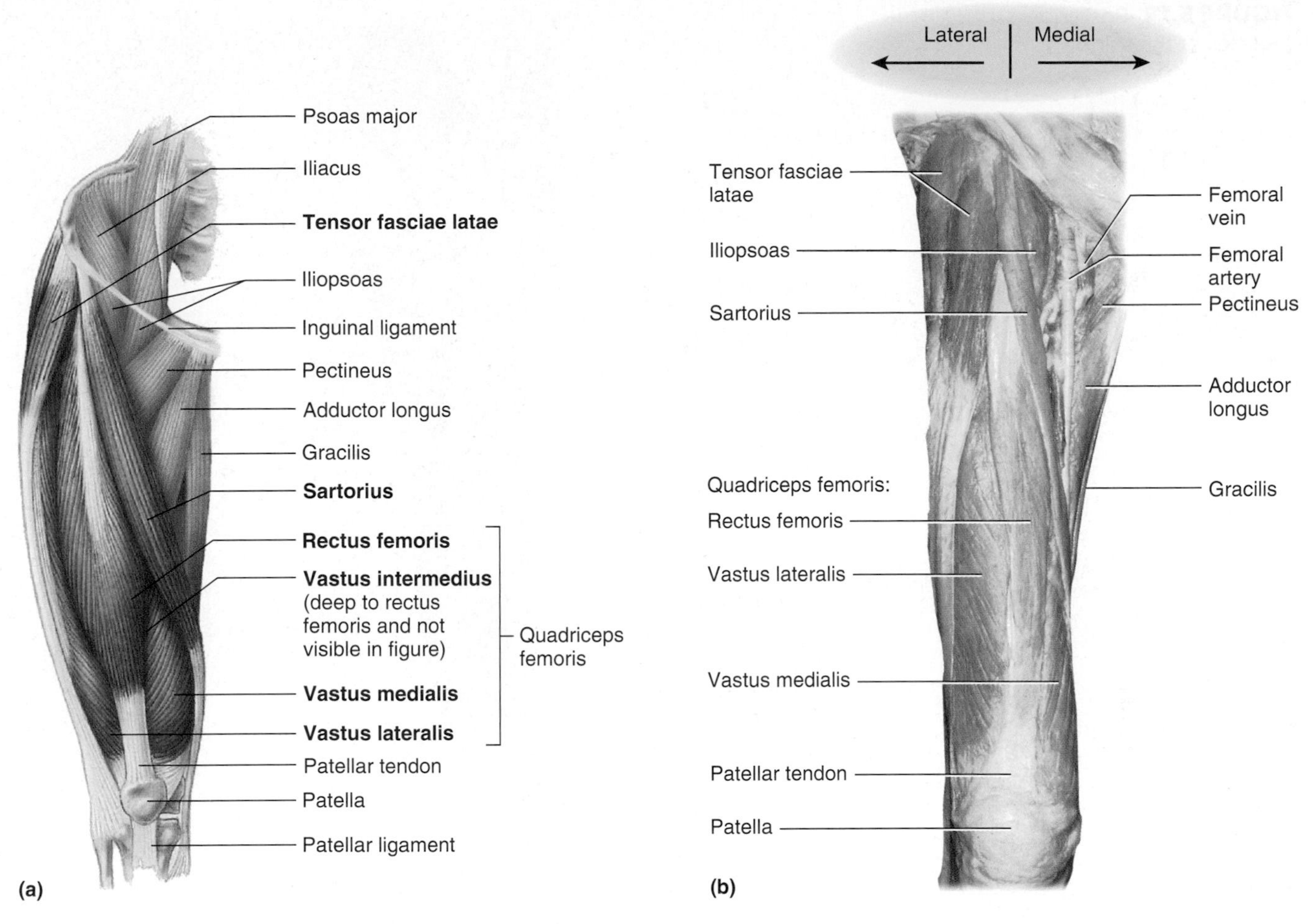

FIGURE 5.18 Anterior muscles of the thigh: (a) anterior view, (b) superficial muscles of a cadaver.

run deep to the **inguinal ligament** and insert on the lesser trochanter of the femur. When they pass deep to the inguinal ligament, they merge to form the **iliopsoas.**

- The **rectus femoris, vastus lateralis, vastus medialis,** and **vastus intermedius** make up the group of muscles called the **quadriceps femoris.** They are synergists for knee extension. Note that they all have the same origin except for the rectus femoris, which originates above the hip at the ilium. Due to its origin, the rectus femoris is solely responsible for hip flexion.
- The **biceps femoris, semitendinosus,** and **semimembranosus** make up the group called the **hamstrings,** which are synergists for knee flexion. The hamstrings are antagonists to the quadriceps. The biceps femoris has its origin above the hip at the ischium, so it is also responsible for hip extension.
- The **gracilis, adductor longus,** and **pectineus** are synergists for hip adduction.

quadriceps femoris: KWAD-rih-seps FEM-or-is

gracilis: GRASS-ih-lis

pectineus: pek TIN-ee-us

semimembranosus: sem-ee-MEM-bran-OH-sis

semitendinosus: sem-ee-TEN-dih-NOH-sis

study **hint**

A study tip for remembering the gracilis, adductor longus, and pectineus muscles involves the first letter of each muscle. Going medial to lateral with the first letter of each, you have GAP. These muscles are in that order, and they close the *gap* between the legs (adduction).

FIGURE 5.19 Posterior muscles of the thigh.

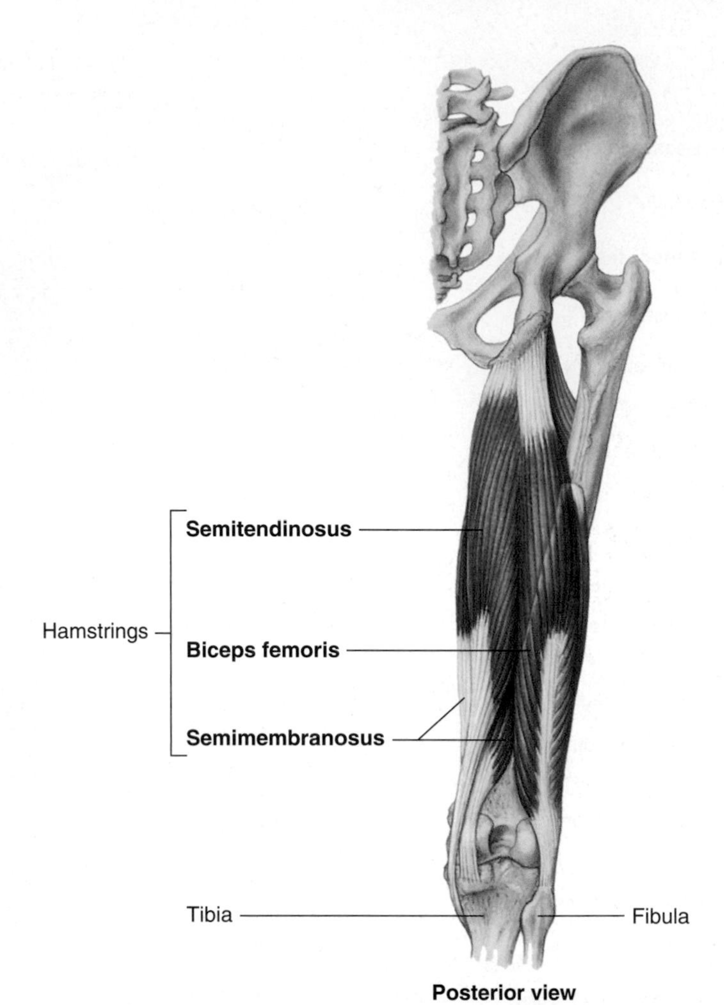

tensor fasciae latae:
TEN-sor FASH-ee-ee LAY-tee

TABLE 5.6 Muscles of the thigh

Muscle	Origin	Insertion	Function
Tensor fasciae latae	Ilium	Tibia	Extends knee and medially rotates hip
Gracilis	Pubis	Tibia	Flexes knee and adducts hip
Adductor longus	Pubis	Femur	Adducts hip
Pectineus	Pubis	Femur	Adducts and flexes hip
Iliacus	Ilium	Lesser trochanter of femur	Flexes hip
Iliopsoas	Ilium and vertebrae	Lesser trochanter of femur	Flexes hip
Psoas major	Vertebrae	Lesser trochanter of femur	Flexes hip
Sartorius	Ilium	Tibia	Flexes knee and flexes hip
Rectus femoris	Ilium	Tibial tuberosity	Extends knee and flexes hip
Vastus lateralis	Greater trochanter of femur	Tibial tuberosity	Extends knee
Vastus medialis	Femur	Tibial tuberosity	Extends knee
Vastus intermedius	Femur	Tibial tuberosity	Extends knee
Biceps femoris	Ischium	Fibula	Flexes knee and extends hip
Semitendinosus	Ischium	Tibia	Flexes knee
Semimembranosus	Ischium	Tibia	Flexes knee

spot check ❷ Which muscles are synergists for hip flexion? Which muscles are antagonists to hip flexors?

Muscles of the Leg **Figure 5.20** shows the muscles of the leg, and Table 5.7 outlines their names, origins, insertions, and functions. Additional information is provided below:

- The **gastrocnemius** and **soleus** are synergists for plantar flexion. They are antagonists to the **tibialis anterior,** which is responsible for dorsiflexion. The gastrocnemius and soleus share the calcaneal (Achilles) tendon for their insertion.
- The **peroneus** muscles may also be called **fibularis** muscles.

gastrocnemius: gas-trok-NEE-me-us

soleus: SO-lee-us

fibularis: FIB-you-LAH-ris

tibialis: tih-bee-AH-lis

TABLE 5.7 Muscles of the leg

Muscle	Origin	Insertion	Function
Gastrocnemius	Femur	Calcaneus	Plantar-flexes foot
Soleus	Fibula and tibia	Calcaneus	Plantar-flexes foot
Peroneus, or fibularis	Fibula	Metatarsal	Everts foot
Tibialis anterior	Tibia	Metatarsal	Dorsiflexes and inverts foot

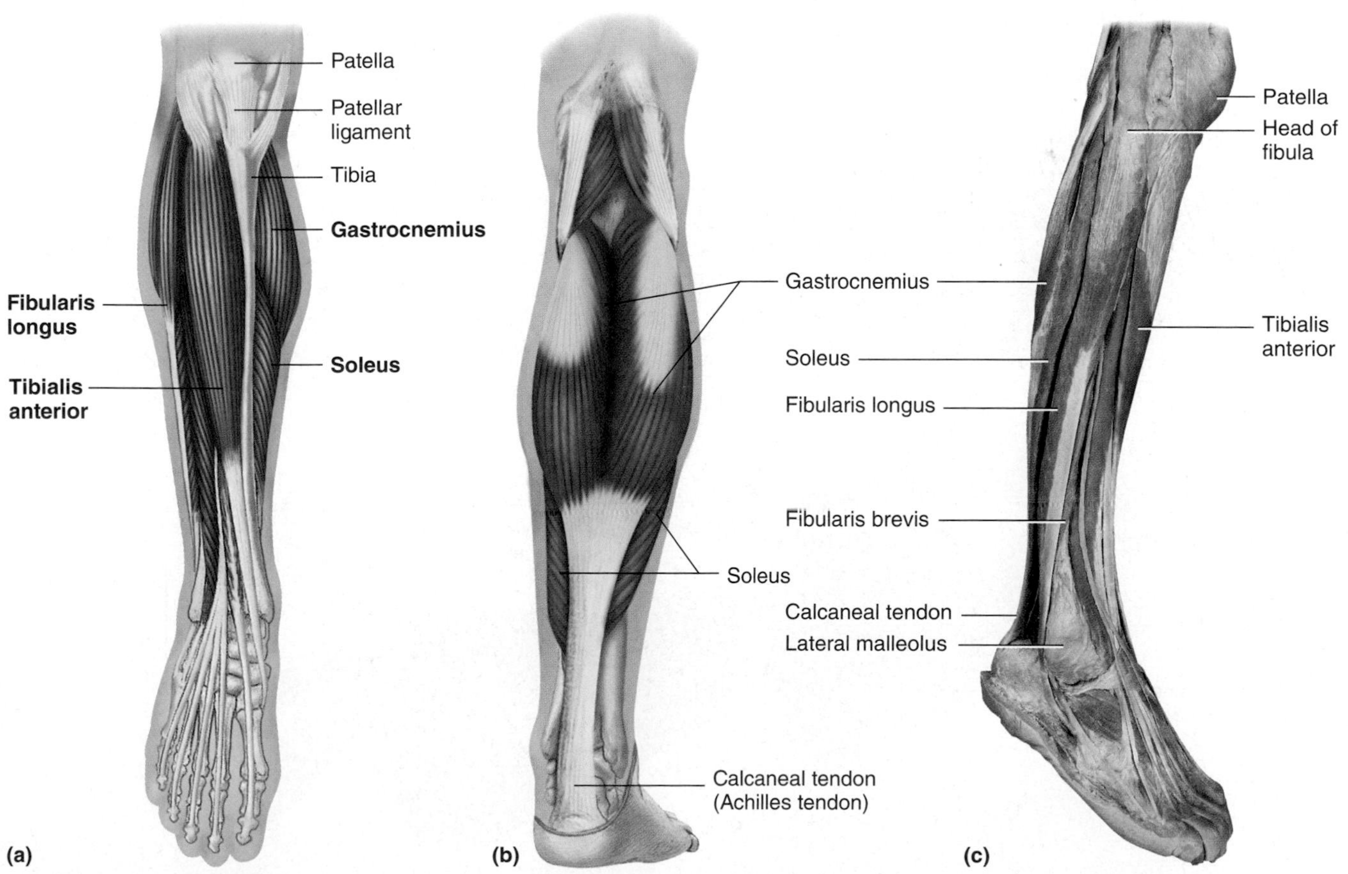

FIGURE 5.20 Muscles of the leg: (a) anterior view, (b) posterior view, (c) lateral view of cadaver.

5.5 learning outcome

Describe the structural components of a muscle, including the connective tissues.

endomysium: EN-doh-MISS-ee-um

fascicle: FAS-ih-kull

fascia: FASH-ee-ah

audio connect.mcgraw-hill.com

Anatomy of a Skeletal Muscle

Now that you have looked at the prominent muscles by region, it is time to consider the anatomy of a single muscle and its connective tissues, starting with the smallest structure in **Figure 5.21a**. (Follow along with the diagram while reading this section.)

An individual muscle cell **(muscle fiber)** is surrounded by connective tissue called the **endomysium.** Muscle fibers are grouped together to form a **fascicle,** which is surrounded by connective tissue called the **perimysium.** The fascicles are grouped together to form the muscle. The entire muscle is surrounded by a connective tissue called the **epimysium.** It is important to understand that all of these connective tissues run through the entire length of the muscle. The endomysium between muscle fibers, the perimysium surrounding each fascicle, and the epimysium surrounding the entire muscle merge at the ends of the muscle to form the tendon that attaches the muscle to the bone. A tendon is not just a piece of fibrous connective tissue attached to each end of the muscle. Its components are an integral part of the entire muscle, not just the ends.

A fourth connective tissue, called **fascia,** is shown in **Figure 5.21b**. Fascia is a tough, fibrous tissue that does not allow for expansion. It surrounds several muscles of an area, forming muscle compartments, and it separates muscle from the hypodermis.

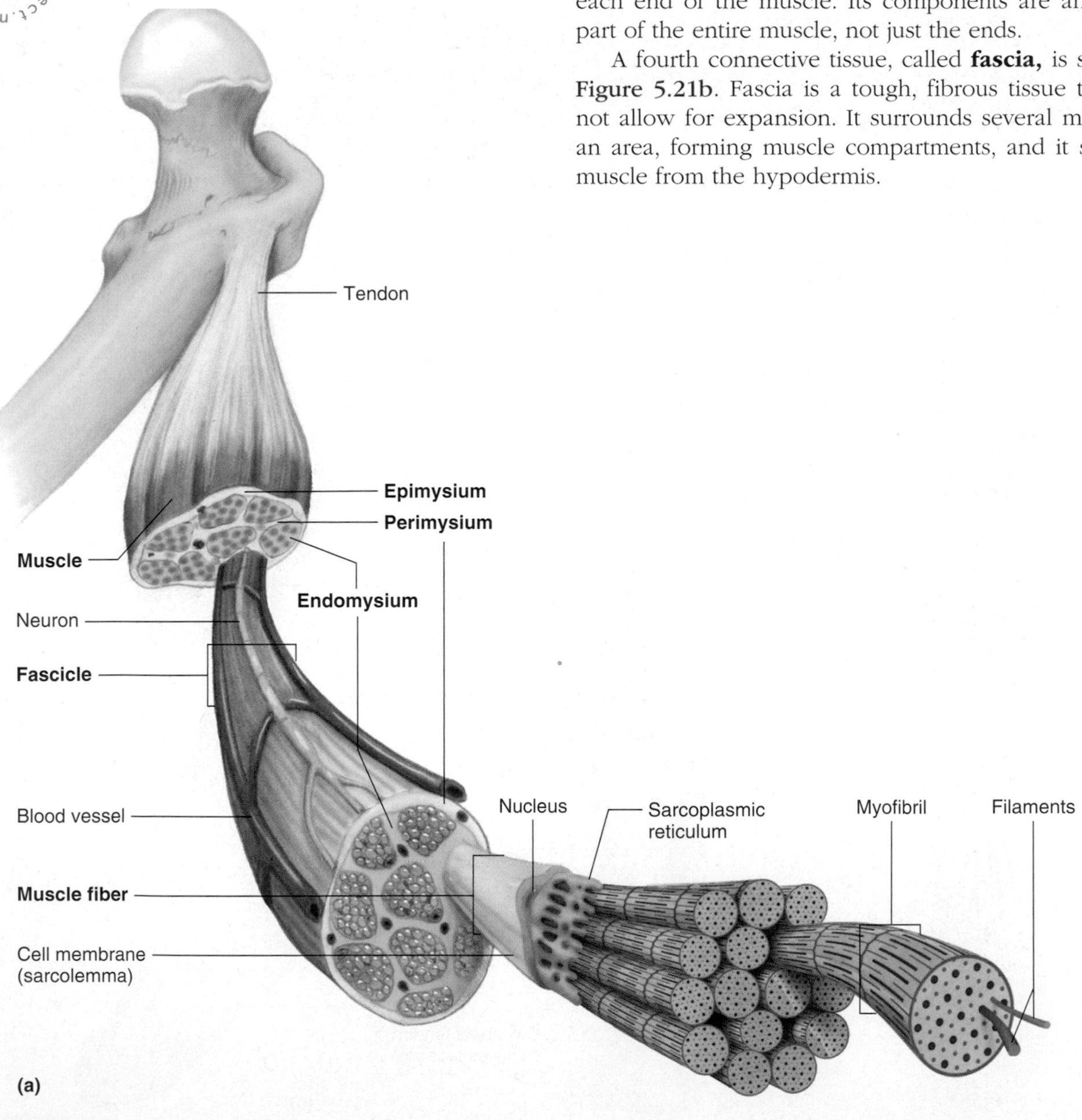

FIGURE 5.21 Connective tissues and structural components of a thigh muscle: (a) structural components of a muscle and attachment of muscle to bone, (b) cross section of the lower leg showing the relationship of individual muscles and fasciae.

FIGURE 5.21 concluded

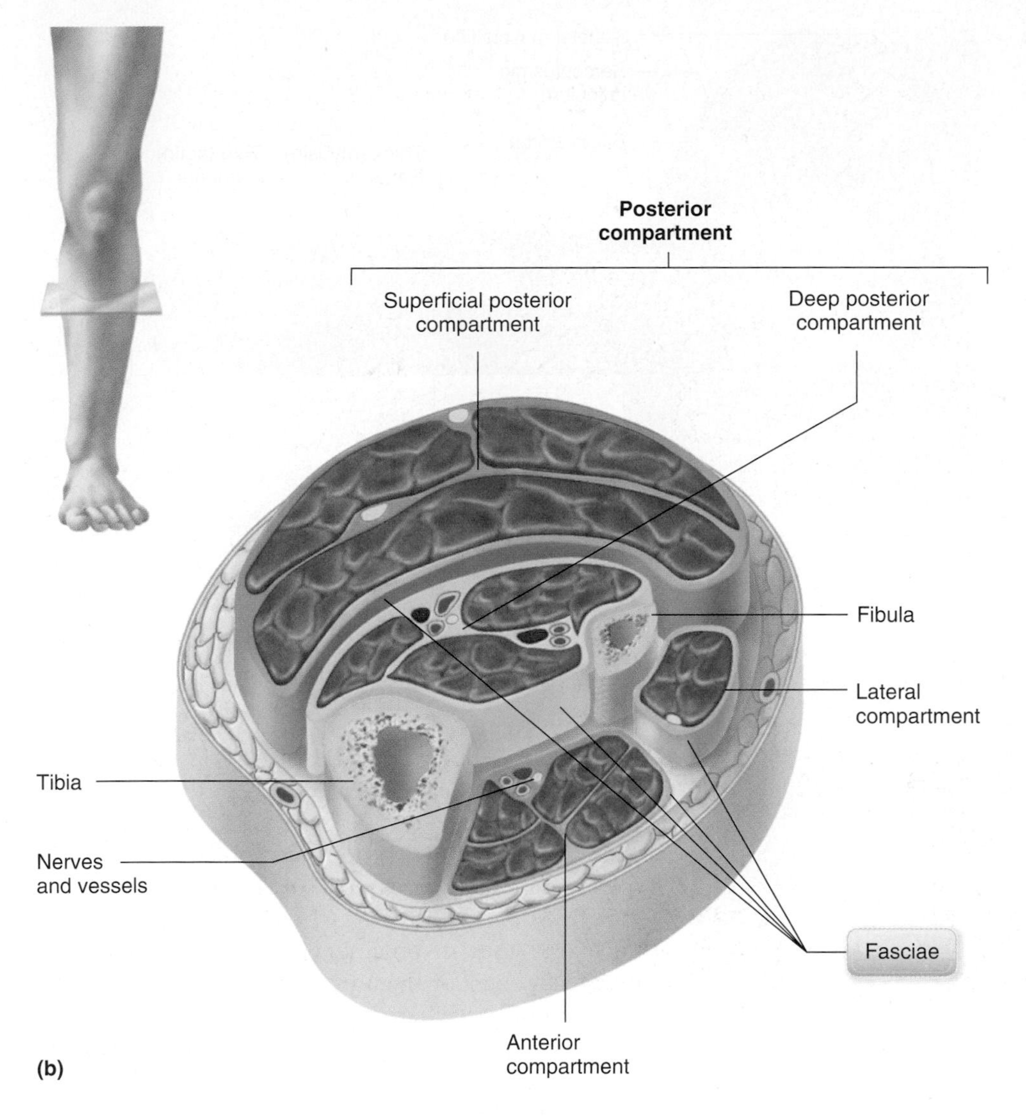

clinical point

Overactivity or trauma may cause the muscles of a compartment to become inflamed and swell. Because the fascia does not stretch enough to accommodate the swelling, pressure builds within the compartment. The increased pressure hampers blood flow and muscle activity. This is called **compartment syndrome**. It may require cutting the fascia of the compartment to relieve the pressure and reestablish blood flow.

Anatomy of a Skeletal Muscle Cell

5.6 learning outcome

Describe the structural components of a skeletal muscle fiber, including the major proteins.

Figure 5.22 shows the anatomy of a single muscle fiber (muscle cell). Muscle fibers have specialized names for many of their organelles. The cell membrane is called the **sarcolemma.** The smooth endoplasmic reticulum of the cell is called the **sarcoplasmic reticulum** (shown in yellow). Its function is to store calcium ions until they are needed. Notice that a muscle fiber is a bundle of **myofibrils,** but there is no connective tissue surrounding each of the myofibrils.

FIGURE 5.22 Structure of a muscle fiber: (a) the structures of a muscle fiber (cell), (b) two sarcomeres.

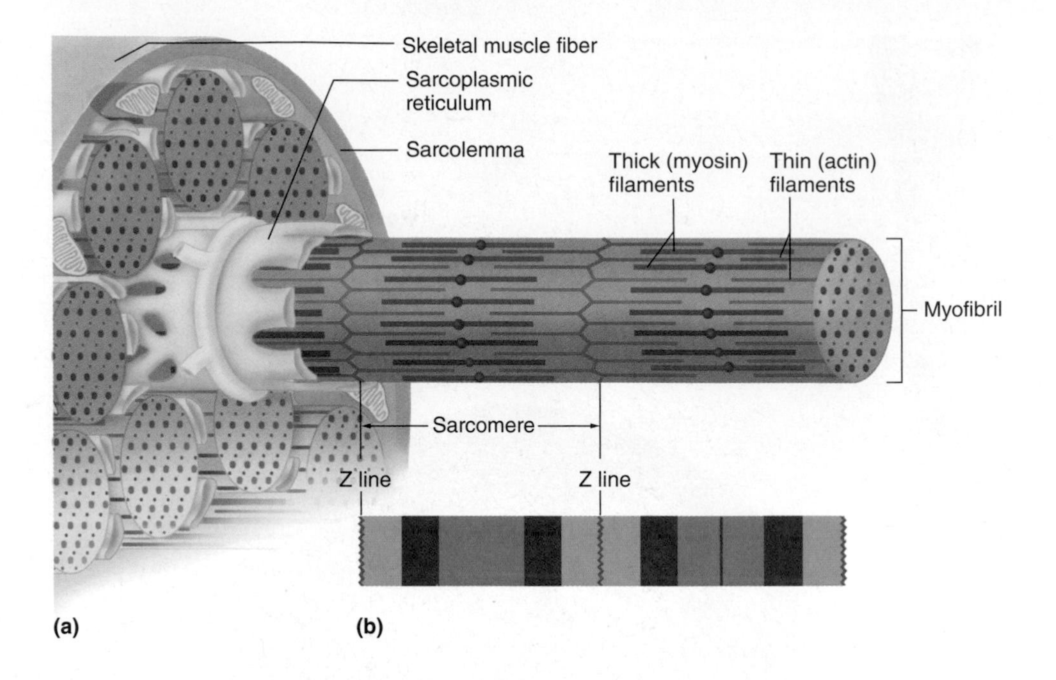

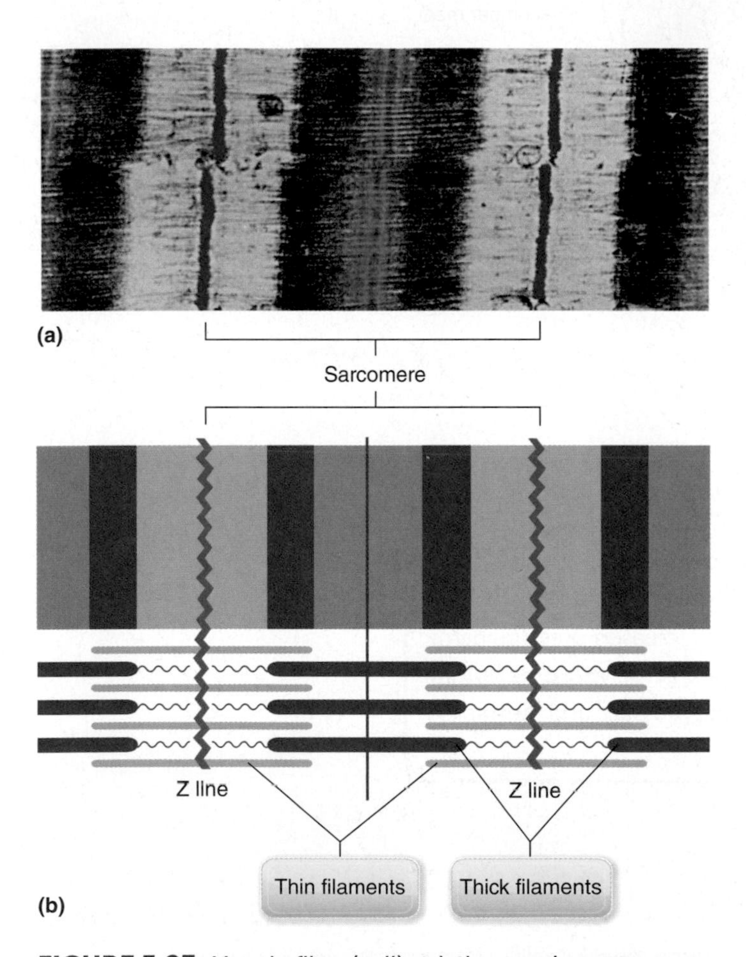

FIGURE 5.23 Muscle fiber (cell) striations and sarcomeres: (a) micrograph showing striations, (b) the overlapping of thick and thin myofilaments responsible for the striations.

Myofibrils are composed of a series of repeated functional units called **sarcomeres** running end to end. See **Figure 5.22**. **Z lines** form the ends of a sarcomere, which is composed of thick and thin filaments **(myofilaments)** made of protein. Only the thin myofilaments are attached at the Z lines. The thick and thin myofilaments overlap in two areas near the center of the sarcomere. These sections of the sarcomere (where thick and thin myofilaments overlap) appear darker. This accounts for the striated appearance of skeletal muscle tissue when seen on a microscope slide. See **Figure 5.23**.

Thick myofilaments are made of several hundred protein molecules called **myosins.** A myosin molecule looks like a golf club with a pincher-like head **(cross-bridge),** which can grab and bend. See **Figure 5.24**. As shown in the figure, the myosin molecules are grouped with the cross-bridges pointing toward the two ends of the thick myofilament.

Thin myofilaments are composed of three different protein molecules—actin, tropomyosin, and troponin. **Actin** looks like a double chain of beads twisted together. **Tropomyosin** resembles a thread running through the actin chain. It covers spots called **active sites** on actin where myosin molecules could grab hold. **Troponin** is a small protein attached to tropomyosin. See **Figure 5.24**.

This information about the proteins of the myofilaments will be used later in the chapter to explain how a muscle fiber contracts at a molecular level.

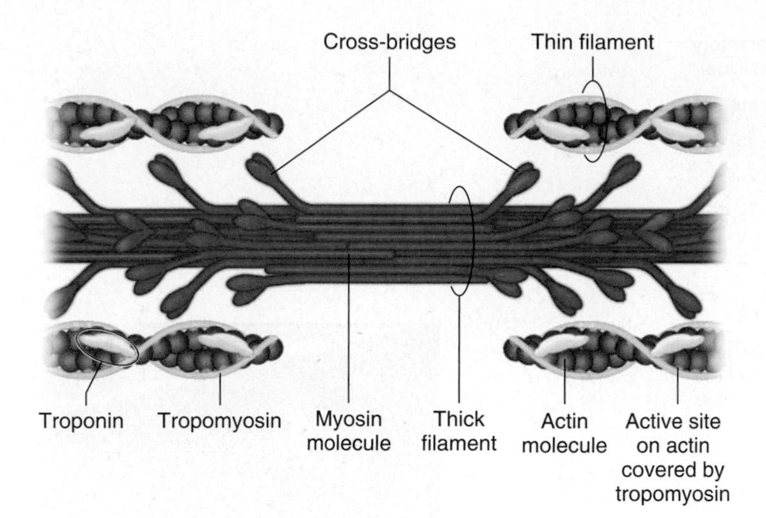

FIGURE 5.24 Protein structure of thick and thin myofilaments. Active sites are shown as blue dots on actin that are covered by tropomyosin.

Physiology of the Muscular System

You have become familiar with the prominent skeletal muscles of this system, the anatomy of a muscle, and the anatomy of a skeletal muscle cell down to its molecular level. How exactly does that anatomy work to carry out a muscle contraction? To answer that question, you first need to understand the physiological characteristics shown by all muscle tissues.

Physiological Characteristics of Muscle Tissue

5.7 learning outcome

Explain the five physiological characteristics of all muscle tissue.

All muscle tissues display five physiological characteristics. These characteristics and their meanings are as follows:

1. **Excitability.** A muscle cell can be stimulated by a nerve to contract.
2. **Conductivity.** The stimulation from the nerve moves quickly along the length of the muscle cell.
3. **Contractility.** A muscle cell can shorten with force. Muscles can only pull; they cannot push.
4. **Extensibility.** A muscle cell can be stretched. If the biceps brachii contracts to flex the arm, the triceps brachii needs to stretch to accommodate the motion. Muscles are stretched by the contraction of other muscles.
5. **Elasticity.** If a muscle cell is stretched, it will return to its original shape.

Look for excitability, conductivity, contractility, and elasticity as you study how a muscle contracts. Extensibility is not included in a description of a muscle contraction because it is not a characteristic shown by the contracting muscle. The ability to stretch would be characteristic of the muscle opposing the contraction.

Neuromuscular Junction

5.8 learning outcome

Explain how a nerve stimulates a muscle cell at a neuromuscular junction.

To see how a muscle cell shows excitability, you begin by studying a neuromuscular junction. See **Figure 5.25**. There is an indentation in the muscle cell that forms a gap **(synapse,** also called a **synaptic cleft)** where the nerve ending meets the muscle

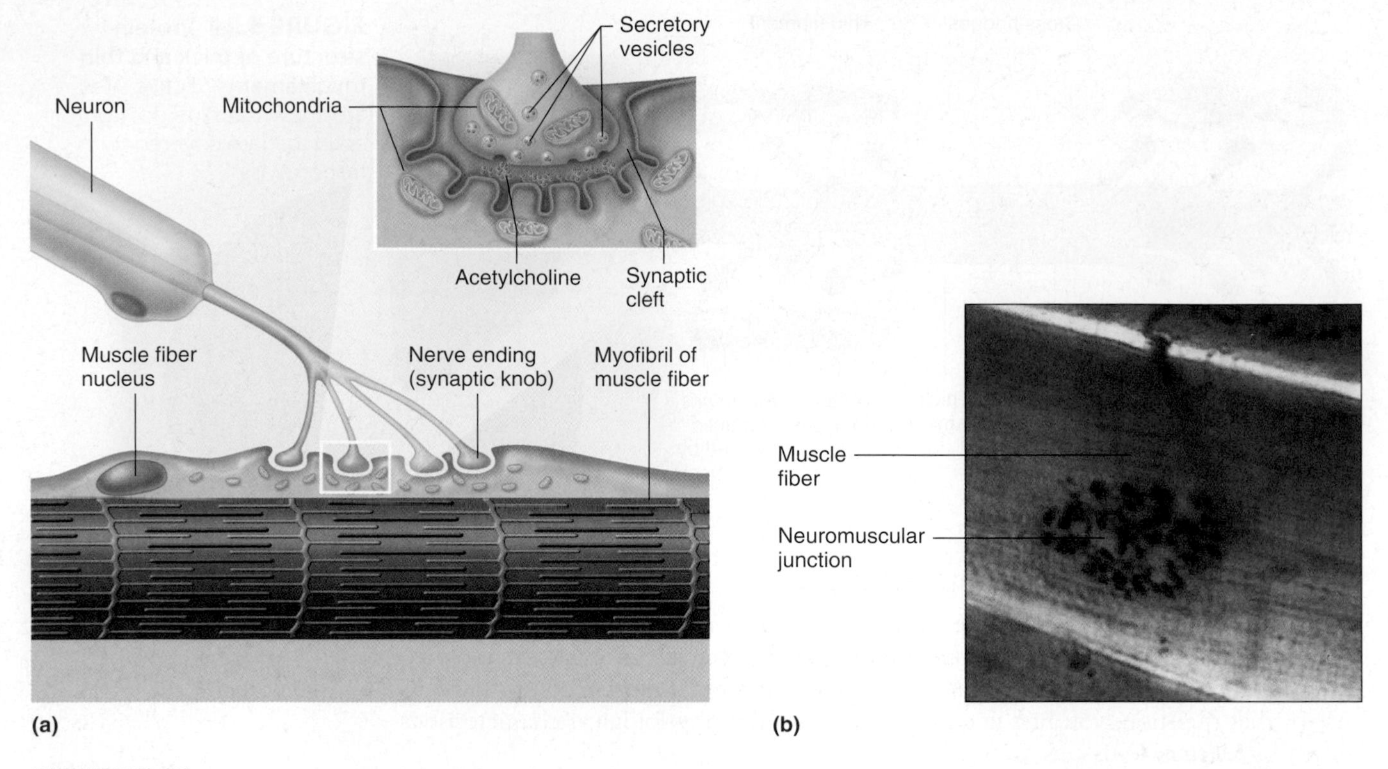

FIGURE 5.25 A neuromuscular junction: (a) anatomy of a neuromuscular junction, (b) micrograph of a neuromuscular junction.

cell. The nerve ending and muscle cell do not actually touch. Instead, receptors with a very specific shape are located on the muscle cell in the space formed by the neuromuscular junction. These receptors work to communicate information between the nerve ending and the muscle cell.

Stimulating the muscle cell begins with an electrical signal traveling down the nerve cell (neuron). This signal triggers the release of a chemical **(neurotransmitter)** from the **synaptic knob** at the nerve end. The neurotransmitter is designed to fit into the receptors on the muscle cell on the basis of its unique shape, like a key fitting into a lock. **Acetylcholine (ACh),** a protein, is the neurotransmitter released for skeletal muscle tissue. The presence of acetylcholine in the receptors is the signal for the muscle cell to contract. There is a minimum amount, called the **threshold,** of acetylcholine that is necessary in receptors for the muscle to react. The muscle cell will not respond if there is less acetylcholine fitting into receptors than the threshold amount. A muscle's response to nerve stimulation is **all or nothing,** based on a threshold amount of acetylcholine fitting into receptors. Adding more acetylcholine than the threshold will not give a bigger response. Once the threshold is reached, the stimulation is conducted down the length of the entire muscle cell (conductivity).

acetylcholine: AS-eh-til-KOH-leen

5.9 learning **outcome**

Describe a muscle contraction at the molecular level.

Muscle Contraction at the Molecular Level

A skeletal muscle cell also shows contractility. The **sliding filament theory** of muscle contraction involves thick myofilaments grabbing thin myofilaments and pulling them toward the center of the sarcomere. The thin myofilaments are attached to the Z lines, which are then drawn closer together as the thick myofilaments pull on the thin. As all of the sarcomeres are shortened, so too is the muscle cell.

To see how the proteins in the muscle cell work together to shorten the sarcomere, look at the steps of a contraction at the molecular level. Follow along with the diagram in **Figure 5.26** as you read through the steps below, which describe

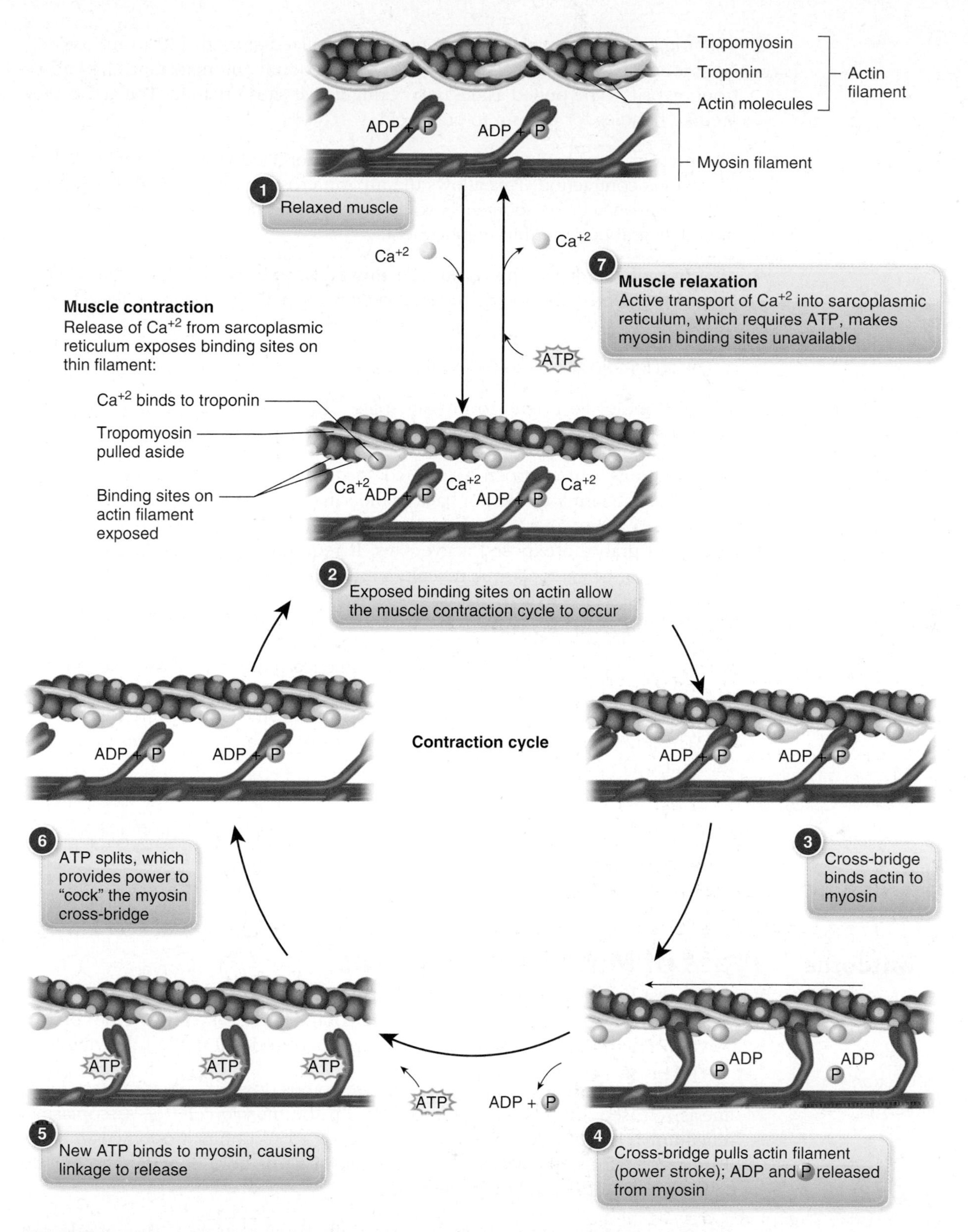

FIGURE 5.26 Sliding filament model.

this process. Pay close attention to where energy is required. Remember, the usable form of energy for the cell is contained in an ATP molecule.

1. An electrical impulse travels down the neuron. Acetylcholine is released from the synaptic knob of the neuron and fits into the receptors on the relaxed muscle cell (the neuromuscular junction is not shown on the figure).

2. This triggers the sarcoplasmic reticulum (specialized smooth ER) to release calcium ions. The released calcium ions bind to troponin automatically. This causes tropomyosin to be pulled aside, uncovering active sites on actin. The active sites are like handles for myosin to grab hold.
3. The myosin cross-bridges have already been energized ("cocked") with ATP in a previous contraction. This allows the myosin cross-bridges to grab hold of the active sites on actin as soon as they are exposed. At this point, the thick myofilament has grabbed the thin myofilament.
4. Myosin pulls on actin. This is called a **power stroke,** which draws the Z lines toward the center, shortening the sarcomere. An ADP and a P are released from myosin.
5. A new ATP molecule binds to myosin, causing it to let go of the active site on actin.
6. The ATP molecule bound to myosin splits. This provides the energy to "cock" the myosin cross-bridge for the next power stroke. (See steps 3 and 4.)
7. Several things must happen for the contraction to be over and the muscle to relax. The calcium must be put back in the sarcoplasmic reticulum so that tropomyosin can again cover the active sites on actin. If this does not occur, myosin will again grab the exposed active sites. It requires ATP to return the calcium ions back to the sarcoplasmic reticulum by *active transport.*
8. The acetylcholine must also be removed from the receptors at the neuromuscular junction to keep the calcium in the sarcoplasmic reticulum. As long as acetylcholine is in the receptors, the sarcoplasmic reticulum will allow calcium ions to diffuse out and myosin will grab hold of the active sites on actin. The muscle cell removes the acetylcholine by producing an enzyme called acetylcholinesterase. This enzyme removes the acetylcholine from the receptors. (This step is not shown on **Figure** 5.26).

spot check 3 How much acetylcholine was necessary for the contraction to happen?

5.10 learning outcome

Compare and contrast a muscle twitch and tetany with regard to the steps of a muscle contraction at the molecular level.

Types of Muscle Contractions

What has just been described is a **muscle twitch**—the contraction of one muscle cell due to one nerve impulse. A graph of the contraction broken down into phases is shown in **Figure** 5.27. These phases are explained in detail in the following list:

- During the **latent phase,** the nerve impulse comes down the neuron; acetylcholine is released and fits into the receptors on the muscle cell; the sarcoplasmic reticulum releases calcium; the calcium binds to troponin; tropomyosin shifts position to expose the active sites; and myosin grabs hold of actin. The muscle cell has not shortened during this phase.
- During the **contraction phase,** myosin pulls (power stroke). The muscle cell shortens.
- During the **relaxation phase,** myosin lets go. The muscle goes back to shape because it is elastic (elasticity).
- During the **refractory phase,** the calcium is actively transported back to the sarcoplasmic reticulum and the muscle produces acetylcholinesterase to remove the acetylcholine from the receptors. The muscle still appears to be relaxed.

clinical point

When muscle contractions are not carried out normally, the body may be dealing with an autoimmune disorder, such as **myasthenia gravis.** Myasthenia gravis usually affects women between the ages of 20 and 40. The immune system produces antibodies that normally fight foreign pathogens, but in people with myasthenia gravis the antibodies attack the acetylcholine receptors in the neuromuscular junctions. This leaves the muscle with fewer receptors, so it is less sensitive to acetylcholine. Muscle weakness results because the muscle cannot fully respond to the nerve impulses.

Each muscle twitch has the four phases shown in the graph. However, your hand is likely not going through a series of twitches (contractions and relaxations) as you hold a pencil to take notes. Your flexor digitorum muscles are in a sustained contraction to flex your fingers to hold your pencil. So how is this done? It is controlled by the *frequency* of the nerve impulses. If more and more nerve impulses come and complete their latent phases before the muscle cell can even begin to enter the relaxation phase from the first nerve impulse, the effect is a sustained contraction called **tetany.** It is the frequency of the nerve impulses that determines the occurrence either of a series of twitches with relaxation between nerve impulses or of sustained contractions with no relaxation between nerve impulses. See **Figure 5.28**.

warning

Be careful not to confuse tetany with the disorder **tetanus** (sometimes commonly referred to as *lockjaw*). Tetany is a purposeful, sustained contraction caused by nerve impulses coming in rapid succession. Tetanus is a disorder caused by a powerful neurotoxin produced by the bacterium *Clostridium tetani.* Tetanus produces painful, unwanted, sustained muscle contractions that often begin in the masseter muscle.

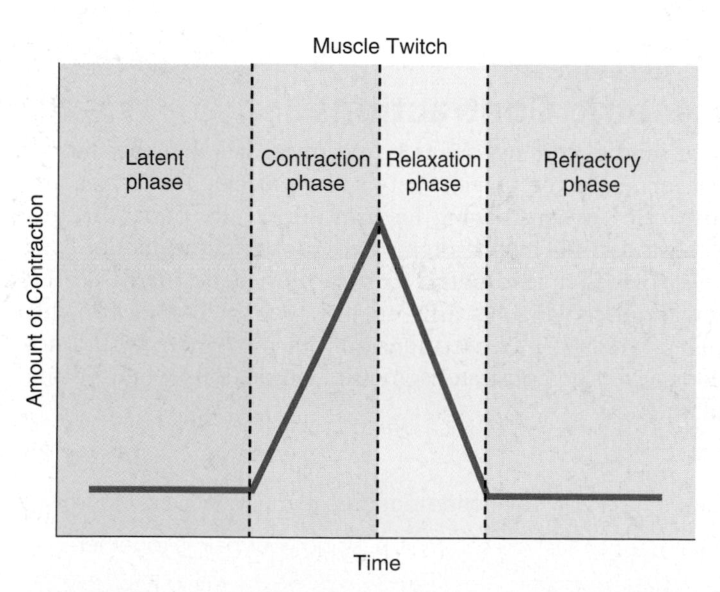

FIGURE 5.27 Graph of a muscle twitch: muscle contraction over time divided into latent, contraction, relaxation, and refractory phases.

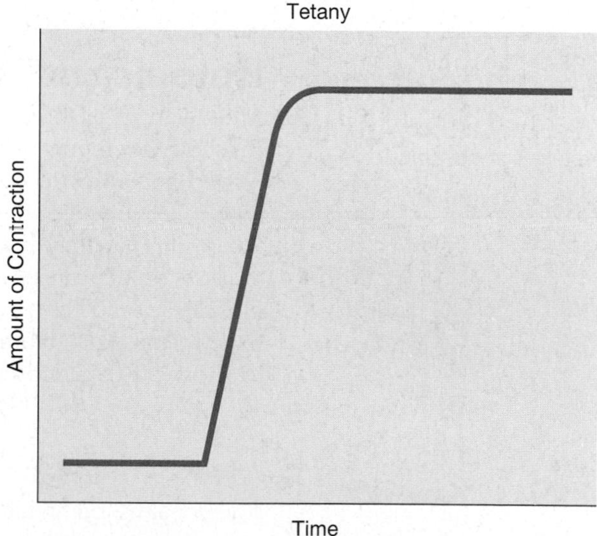

FIGURE 5.28 Graph showing tetany.

You now have a good idea of the difference (and its cause) between a single muscle twitch and tetany (a sustained contraction for a muscle cell). But what determines how much a muscle as a whole moves? Imagine a boxer's punch. His fist is clenched, involving flexion of the fingers in a sustained contraction (tetany). But his arm is first pulled back and then rapidly pushed forward to deliver the punch. There must be a difference in the use of his muscles, whether he is stopping halfway or going the full extent. To understand just how much an entire muscle actually moves, you need to look at motor units and recruitment.

5.11 learning outcome

Define a motor unit and explain the effect of recruitment.

Motor Units and Recruitment

A **motor unit** is defined as a single nerve cell and all the muscle cells it stimulates. A nerve cell can have more than one nerve ending. For example, it may have just a few nerve endings stimulating a few muscle cells, or it may have hundreds of nerve endings stimulating hundreds of muscle cells. A single nerve cell stimulates all of the muscle cells in its motor unit at the same time. If the size of the motor unit is small (a few muscle cells per nerve cell), there is a small motion. If the motor unit is large (a hundred muscle cells per nerve cell), there is a large motion. You need to be able to do fine, precise motions with your hands. Therefore, you need small motor units in the flexors and extensors of the fingers and wrist. However, you do not need to do fine, precise motions using your hamstrings and quadriceps muscles to flex and extend your knee, so the motor units are large in these muscles.

Consider again the boxer's punch, which involves getting more and more motor units involved during the punch. This is called **recruitment.** Even though motor units are small—for muscles moving the arm, wrist, and fingers—the boxer can get a bigger motion by stimulating more and more motor units. So there is rapid recruitment in muscles of the shoulder and arm during his punch.

spot check 4 What causes the difference between gently holding an egg in your fist and crushing an egg in your fist?

5.12 learning outcome

Compare and contrast isotonic and isometric contractions.

Isotonic and Isometric Contractions

Another way of looking at muscle contractions is based upon whether the muscle caused movement or just increased the tension within the muscle. The boxer displayed an **isotonic contraction** when moving his arm during the punch. In this type of contraction, the tension in his muscles remained constant, and motion was the result. He may have displayed an **isometric contraction** if he increased the tension of his muscles without moving his arm. In this type of contraction, his muscles would have bulged with the increased tension, but movement would not have resulted. Bodybuilders often demonstrate isometric contractions when posing to show off their musculature.

spot check 5 Put the palms of your hands together at your chest. Now push your palms together equally as hard as you can. Was pushing your palms together an isotonic or isometric contraction?

Levers

> **5.13 learning outcome**
>
> Describe an example of a lever system in the human body, giving the resistance, effort, and fulcrum.

Most isotonic contractions result in the movement of a bone by a muscle using a lever system. You may have studied lever systems in a physics course. A lever is a rigid object that can be used to lift something. For example, you open a can of soup with a can opener. You then use the tip of a knife to pry under the sharp edge of the lid to lift it so that it can be removed from the can without your cutting your finger. In this case, the knife was used as a lever. In the human body, bones act as levers. A basic lever system has three parts:

1. **Resistance** is a weight to be lifted.
2. **Effort** is the force applied to lift the weight. In a muscle system, the effort is the insertion of the muscle.
3. **Fulcrum** is a pivot point on the lever that does not move. In muscle lever systems, the fulcrum is a joint.

The order of the resistance (R), effort (E), and fulcrum (F) determines the class of lever used. **Figure 5.29** shows examples of the three classes of levers. Look closely at **Figure 5.29c**, the third-class lever, which is the most common lever system used in the body. The diagram shows the biceps brachii muscle flexing the elbow. Try this motion yourself. In this lever system, the radius is the lever. The

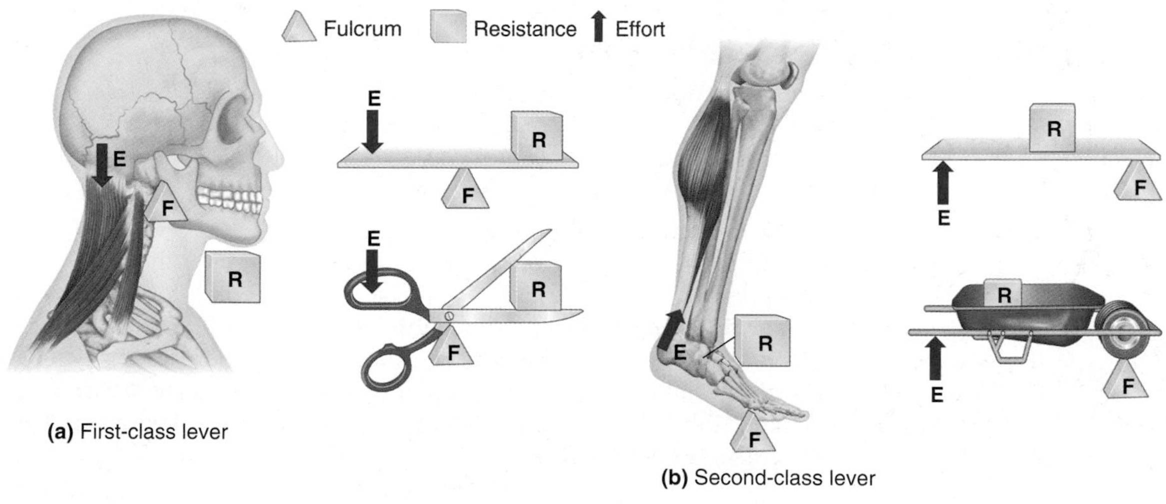

FIGURE 5.29 **Lever systems showing the resistance (R), the effort (E), and the fulcrum (F):** (a) first-class lever, the trapezius muscle extends the head to lift the chin, (b) second-class lever, the gastrocnemius muscle does plantar flexion to lift the body (centered over the base of the tibia), (c) third-class lever, the biceps brachii muscle flexes the elbow to lift the hand.

weight being lifted is the hand (R). The effort is the insertion of the muscle on the radius (E). Notice that it is very close to the fulcrum, the elbow joint (F). The elbow is the pivot point for the radius. The radius does not move at the elbow; it pivots there. All third-class lever systems have the effort between the resistance and the fulcrum.

Having the effort so close to the fulcrum is a disadvantage mechanically because it limits how much you can lift. For example, try holding down one end (F) of a meter stick (the lever) and placing a book (R) on the opposite end. Now try lifting the book by pulling on the lever close to the end you are holding down (F), and then try again close to the book (R). It is much easier to lift the weight if the effort (E) is closer to the resistance than to the fulcrum. Why then is the insertion for the biceps brachii so close to the elbow and not closer to the hand? Imagine what the arm would look like if the biceps brachii's insertion were closer to the hand. Would you be able to straighten your arm? If so, imagine how the triceps brachii, the antagonist, would look and work. Although having the insertion of a muscle so close to the fulcrum is a disadvantage mechanically, it does allow for a greater range of motion.

You have now covered how a muscle cell is stimulated and how it contracts at a molecular level. You have looked at different types of contractions and how they work. Energy was used during all of the contractions to move the thin myofilaments and to actively transport calcium ions back to the sarcoplasmic reticulum. Next, you will see how the muscle cell processes energy.

5.14 learning **outcome**

Compare aerobic and anaerobic respiration in terms of amount of ATP produced, speed, and duration.

Muscle Metabolism

Muscle metabolism can be defined as the chemical reactions a muscle cell uses to process energy. To better comprehend this process, let's first review what you studied in Chapter 2. Mitochondria produce energy for a cell through cellular respiration: $C_6H_{12}O_6 + 6O_2 \longrightarrow 6CO_2 + 6H_2O +$ Energy. The energy produced in this reaction is used to form an ATP molecule: Energy + ADP + P $\longrightarrow$ ATP. The third phosphate bond of the ATP molecule holds the usable energy for the cell.

The mitochondria of muscle cells can use two different forms of cellular respiration—**aerobic respiration** or **anaerobic respiration**—to obtain the energy needed to form ATP. The difference between the two is in the use of oxygen, the products produced, and the amount of energy made available to generate ATP. **Figure 5.30** and Table 5.8 compare the two methods of respiration, and each method is discussed in detail below.

Aerobic Respiration Aerobic respiration begins with a glucose molecule. The first step is to produce pyruvic acid from the glucose. This step, called **glycolysis,** produces enough energy to form two ATP molecules. The dashed line in **Figure 5.30a** indicates several further steps in aerobic respiration, beyond glycolysis, that release

TABLE 5.8 Comparison of aerobic and anaerobic respiration

	Aerobic respiration	Anaerobic respiration
Number of ATPs produced per glucose molecule	36	2
Oxygen required	Yes	No
Speed of the process	Slower (more steps)	Faster (fewer steps)
Products other than energy	Carbon dioxide and water	Lactic acid
Duration	Long duration (hours)	Few minutes

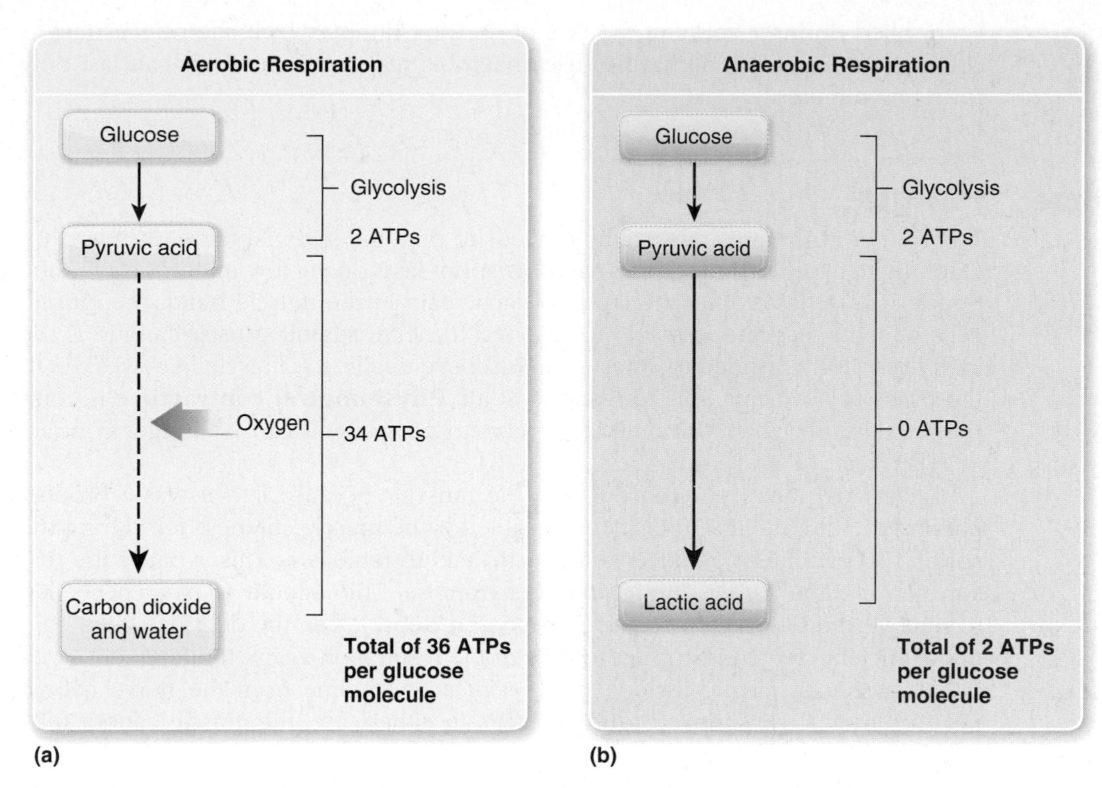

FIGURE 5.30 Muscle metabolism: (a) aerobic respiration, (b) anaerobic respiration.

additional energy. As you can see in the figure, aerobic respiration is a multistep process that requires the addition of oxygen, results in carbon dioxide and water, and produces enough energy to form 36 ATP molecules. As long as sufficient oxygen is available, aerobic respiration can continue for long periods of time. For example, you may be able to go on a leisurely walk for hours.

Mitochondria perform aerobic respiration whether or not there is an immediate need for energy. As you already know, ATP is the usable form of energy for a cell, but you also need to know that ATP is an unstable molecule. ATP can waste the energy it contains by releasing the energy when the energy is not needed. Therefore, mitochondria convert ATP to a more stable molecule called **creatine phosphate (CP)** when energy is not immediately needed. Mitochondria do this by adding creatine to ATP. Creatine phosphate is a storage molecule only. It must be converted back to ATP for the energy to be used.

Anaerobic Respiration As **Figure 5.30b** shows, anaerobic respiration also begins with a glucose molecule. The first step is glycolysis, which produces enough energy to form two ATPs. Notice that this step is the same in both aerobic and anaerobic respiration. In anaerobic respiration, however, there is insufficient oxygen to add to the process after glycolysis is completed. Pyruvic acid is then changed to lactic acid, a waste product that must be removed. Anaerobic respiration involves fewer steps than aerobic respiration, does not require oxygen, and produces enough energy to form just two ATP molecules. Why does the body go through the process of anaerobic respiration if it produces less energy and a waste product that must be removed? It does so because this process is faster (has fewer steps) and still provides energy to form ATP molecules when the supply of oxygen is insufficient to perform aerobic respiration. If sufficient oxygen is available, the body will perform aerobic respiration because it is more efficient. The buildup of lactic acid limits the length of time anaerobic respiration can be done. For example, a long-distance runner may run aerobically for a long time. But if insufficient oxygen is supplied during a sprint at the end of her run, during which she runs as fast as she can, her muscles will have to use anaerobic respiration, and this can last only for a few minutes.

5.15 learning **outcome**

Explain the basis of muscle fatigue and soreness.

Fatigue

The runner in the previous example needs to time her sprint at the end of her run carefully to avoid **fatigue.** The anaerobic sprint lasts only a few minutes due to the buildup of lactic acid. As the levels of lactic acid in the muscle build, the muscle cells become less and less able to respond to nerve stimuli. Muscle fatigue is the inability to fully respond to a nerve impulse; eventually, the muscle may weaken to the point of not being able to respond at all. **Physiological contracture** is complete fatigue in which the muscle appears to be stuck. It can no longer contract or relax.

Lactic acid must be removed from the muscles because it is a waste product bathing the muscle in acid. It is responsible for the muscle soreness felt during the sprint. Oxygen must be added to the lactic acid to remove it. This is partly the reason you breathe harder during and after exercise. The amount of oxygen needed to remove the lactic acid is called the **oxygen debt.** When the debt has been paid and sufficient oxygen is present to do aerobic respiration again, the fatigue is over. Other causes of fatigue include insufficient acetylcholine from the nerve cell or insufficient glucose supplied to the muscle. In either case, the muscle cannot fully respond.

On the other hand, some skeletal muscle fibers are specially adapted to stay aerobic, so they are less likely to fatigue. These muscle fibers are called **slow-twitch fibers** (aerobic respiration is a slower process). They have extra mitochondria, a better blood supply to deliver oxygen and glucose, glycogen stores within the muscle cell that can be converted to glucose, and a protein called myoglobin to store oxygen until it is needed. Other skeletal muscle fibers called **fast-twitch fibers** excel at anaerobic respiration. Each skeletal muscle in your body has cells of each type. The ratio of slow- to fast-twitch fibers is genetically determined. You can train to increase the efficiency of the fibers you have, but you cannot train to change their type.

spot check 6 Jessica and Jennifer are twins but not identical. They are both going out for track at their high school. They live together, have similar diets, and train together. Explain why Jessica is better than Jennifer at running the 100-meter dash and why Jennifer is better than Jessica at the 1,600-meter run. What other events or activities may show similar results?

5.16 learning **outcome**

Compare and contrast skeletal, cardiac, and smooth muscle tissue in terms of appearance, structure, type of nerve stimulation, type of respiration, and location.

Comparison of Muscle Tissues

You have looked at the anatomy and physiology of skeletal muscle tissue. Now you will see how skeletal muscle tissue compares to the two other types of muscle tissue in this system. See Table 5.9.

All three types of muscle tissue—skeletal, cardiac, and smooth—are primarily composed of protein molecules. As you saw with skeletal muscle, muscle cells use these proteins to contract. To maintain the proteins necessary for contractions, these muscle tissues must carry out protein synthesis. The next section, therefore, focuses on the nutritional requirements of muscle tissue.

TABLE 5.9 Comparison of skeletal, cardiac, and smooth muscle tissues

	Skeletal muscle tissue	Cardiac muscle tissue	Smooth muscle tissue
Appearance	Long, striated cells with many nuclei per cell pushed off to the side	Branched, striated cells with a single nucleus and junctions between cells called *intercalated disks* (covered in Chapter 10)	Spindle-shaped cells with no striations and a single nucleus
Type of nerve stimulus	Voluntary, under conscious control	Autorhythmic, i.e., self-stimulating (No nerve stimulus is needed for cardiac muscle cells to contract. Nerve stimuli may modify the frequency of contractions.)	Involuntary, not under conscious control
Type of respiration	Aerobic and anaerobic	Aerobic	Aerobic
Location	Associated with the bones and skin and with circular muscles called *sphincters* that control body openings	Heart	Hollow organs and blood vessel walls

Nutritional Requirements of Muscle Tissue

5.17 learning **outcome**

Explain the nutritional requirements of the muscular system.

As you read in Chapter 2, the building blocks of proteins are the 20 amino acids. All 20 amino acids must be present in the cell during protein synthesis to ensure the production of a functioning protein with exactly the right shape. A nonfunctional protein may be produced if just one amino acid is missing. The body can make 11 amino acids, which are called **nonessential amino acids.** See Table 5.10. The nine other amino acids—**essential amino acids**—must come from the diet.

TABLE 5.10 Amino acids

Nonessential	Essential
Alanine	Histidine
Arginine	Isoleucine
Asparagine	Leucine
Aspartic acid	Lysine
Cysteine	Methionine
Glutamic acid	Phenylalanine
Glutamine	Threonine
Glycine	Tryptophan
Proline	Valine
Serine	
Tyrosine	

The dietary recommended daily allowance (RDA) of protein for an adult of average weight is 46 to 56 grams (g). This protein may come from a variety of sources. Given that the nine essential amino acids must come directly from your diet, it is important to note that **complete proteins** have all the amino acids necessary for the human body. Meats, eggs, and dairy products are good sources of complete proteins. **Incomplete proteins** are those that are missing one or more of the needed amino acids. For example, legumes like beans are low in methionine, and cereals like rice are low in lysine. But having a vegetarian meal of beans and rice supplies all of the amino acids.

In addition to both nonessential and essential amino acids, the minerals calcium and potassium are also necessary for muscle cells to function properly. The RDA for calcium is 1,000 milligrams (mg). Dairy products, fish, shellfish, and green leafy vegetables are good sources of calcium. As mentioned in the integumentary and skeletal system chapters, vitamin D is necessary for calcium absorption. The RDA for potassium is 4,700 mg. Red meat, poultry, fish, cereals, spinach, and bananas are good sources of potassium. See Appendix B for dietary guidelines.

spot check 7 Which type of diet, vegetarian or nonvegetarian, requires more planning to meet the needs of the muscular system? Explain.

study **hint**

In the previous paragraphs, the amounts for the muscular system's required nutrients are expressed in grams. As a health care professional, you will need to effectively use the metric system for measurements of weight (grams), volume (liters), and length (meters). It may help you to associate known quantities to their equivalents in the metric system. For example, you already know that a large bottle of soda contains 2 liters (L) of liquid. The burger of the quarter-pound burger (4 ounces precooked) you may have had for lunch weighs approximately 113 grams. You can even use your body as a measurement tool. In Appendix A, you will find metric conversion tables and a ruler marked in inches and centimeters. Measure the length and width of your thumb and the maximum distance you can spread your thumb and index finger apart in both inches and centimeters. Knowing these measurements will help you visualize the metric lengths used in future chapters.

Functions of the Muscular System

5.18 learning outcome

Summarize the five functions of the muscular system and give an example or explanation of each.

You have now covered all the anatomy, physiology, and requirements of this system. It is time to put all of that information together to see how the functions of the muscular system are carried out.

The muscular system carries out five important functions in the human body: movement, stability, control of body openings and passages, communication, and heat production. Consider the role of each of these functions in Sam's body as he is drinking his tea (see **Figure 5.31**):

1. Movement. The flexor digitorum muscles of Sam's right arm are in tetany to hold his cup while his right biceps brachii, brachioradialis, and brachialis muscles are working as synergists to flex his elbow in an isotonic contraction. This allows Sam to bring his cup to his mouth. A gradual recruitment of additional motor units makes a smooth contraction going the full distance to his mouth.
2. Stability. Sam is sitting upright with his head stable. Some of the motor units in his trapezius muscle are taking turns in isometric contractions to maintain the stability of his head. This is called **muscle tone.** Your posture is the result of muscle tone.
3. Control of body openings and passages. As skeletal muscles, Sam's urinary and anal sphincters are under his voluntary control. He will decide when he wants them to relax so that he can pass urine and defecate.
4. Communication. It may be the sunny day, the tea, the company, or just a pleasant thought, but something has prompted Sam to use his facial muscles to communicate through his smile that he is pleased. Of course, Sam may also use his muscles in his throat, jaw, tongue, and diaphragm to communicate through speech.
5. Heat production. Sam is sitting outside in a short-sleeve shirt. He does not need to find an outside heat source to maintain his body temperature, as his muscles provide body heat. Sam's muscle cells are performing cellular respiration to supply the energy for his muscular system, but not all of the energy is used efficiently. Some energy is lost as heat in the process. This is similar to a car burning gasoline. The energy (in the bonds of the gasoline molecules) is released in the engine to move the drivetrain of a car, but the engine is not totally efficient. Some energy is lost as heat during the process. The heat can be felt on the hot hood of the car.

FIGURE 5.31 Sam.

In the photo, you can see that Sam is no longer a young man. We focus next on the effects of aging on this system.

Effects of Aging on the Muscular System

5.19 learning outcome

Summarize the effects of aging on the muscular system.

Lean muscle mass decreases with age due to atrophy. Fat is deposited in muscle, the muscle fibers shrink, and some of the muscle tissue is replaced by fibrous tissue. Muscle changes begin in the 20s for men and in the 40s for women. The rate and extent of muscle loss is genetically determined. The decrease of muscle mass in weight-bearing muscles is fiber-type specific. Fast-twitch fibers are more affected than slow-twitch fibers.

Changes in the muscle tissue, along with the effects of aging on the nervous system (covered in Chapter 6), have several effects:

- Strength is decreased.
- Fatigue occurs more quickly.
- Reduced muscle tone limits stability.
- Movement slows and becomes more limited.
- The gait shortens and is slower.
- Muscle tremors become more common.

Exercise is one of the best ways to limit the effects of aging on the muscular system. Resistance exercises, such as weight lifting, increase strength by increasing muscle mass through hypertrophy. Exercises that increase cardiovascular function, such as brisk walking or jogging, increase the supply of oxygen and other nutrients to the muscle tissue, so muscles work more efficiently.

spot check 8 What would be the effect of reduced muscle tone on posture?

5.20 learning **outcome**

Describe muscle disorders and relate abnormal function to pathology.

Muscular System Disorders

You have already read about carpal tunnel syndrome and myasthenia gravis in Clinical Points earlier in this chapter. In this section, you will learn about three other disorders of the muscular system—hernias, cramps, and muscular dystrophy:

- **Hernias.** A hernia occurs when any part of the viscera protrudes through the muscle of the abdominal wall. An *inguinal hernia* involves a loop of the intestine protruding through the inguinal canal (an opening in the muscle of the abdominal wall for blood vessels to reach the testes). This type of hernia can occur if a man improperly lifts heavy weights. A *hiatal hernia* involves the stomach protruding through the diaphragm. This type of hernia is most common in obese people over 40.
- **Cramps.** A cramp is a painful muscle spasm. Heavy exercise, dehydration, electrolyte imbalance, extreme cold, low blood glucose levels, or lack of blood flow can cause painful muscle spasms. A cramp should not be confused with a *Charley horse,* which is often an athletic injury involving a painful tear, stiffness, and blood clotting in a muscle.
- **Muscular dystrophy.** This term is used for a group of genetic disorders that result in progressive weakening, degeneration, and replacement of muscle tissue with fibrous scar tissue.

putting the pieces **together**

The Muscular System

Integumentary system

Radiates excess heat generated by muscles; vitamin D production enables calcium absorption needed for muscle contractions.

Provides heat to warm skin; arrector pili muscles move hair.

Skeletal system

Provides calcium needed for muscle contractions.

Works with bones to move the body in lever systems; applies stress to bone for appositional bone growth.

Nervous system

Stimulates muscle contractions.

Muscles carry out movements initiated in the central nervous system.

Endocrine system

Hormones regulate blood calcium and glucose levels needed for muscle contractions.

Skeletal muscles protect some endocrine glands, such as the adrenal glands.

Cardiovascular system

Provides nutrients and removes wastes.

Moves blood through veins so it can return to the heart.

Lymphatic system

Sends white blood cells to fight pathogens.

Moves lymph through lymph vessels so it can be returned to the cardiovascular system.

Respiratory system

Provides O_2 to muscle tissue and removes CO_2 produced through aerobic respiration.

Skeletal muscle contractions are responsible for inspiration and forced expiration.

Digestive system

Provides nutrients for tissues of the muscular system.

Skeletal muscles are used for chewing and swallowing; muscles provide protection for some digestive organs.

Excretory/urinary system

Kidney disposes of wastes and maintains electrolyte balance needed by muscles.

Skeletal muscles control the passing of urine.

Reproductive system

Reproductive hormones promote muscle growth and development.

Muscle contractions are involved in ejaculation and childbirth.

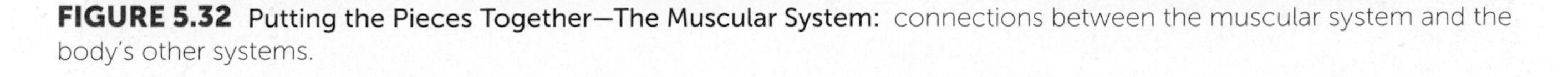

FIGURE 5.32 Putting the Pieces Together—The Muscular System: connections between the muscular system and the body's other systems.

summary

Overview

- Skeletal muscles are primary structures in the muscular system.

Anatomy of the Muscular System

Anatomical Terms

- Terms such as *origin* and *insertion* are used to indicate muscle attachments.

Muscle Actions

- Muscle actions are the motions produced by muscles.

Muscles by Region

- Muscles of the head and neck include orbicularis oris, orbicularis oculi, frontalis, occipitalis, temporalis, buccinator, masseter, platysma, and sternocleidomastoid.
- Muscles of the thorax and abdomen include pectoralis major, pectoralis minor, serratus anterior, diaphragm, external intercostals, internal intercostals, external abdominal obliques, internal abdominal obliques, rectus abdominis, and transverse abdominal.
- Muscles of the back and buttocks include trapezius, latissimus dorsi, erector spinae, gluteus maximus, and gluteus medius.
- Muscles of the arm include deltoid, biceps brachii, triceps brachii, brachialis, brachioradialis.
- Muscles of the forearm include extensor carpi radialis, extensor carpi ulnaris, palmaris longus, flexor carpi radialis, flexor carpi ulnaris, extensor digitorum, and flexor digitorum.
- Muscles of the thigh include tensor fasciae latae, gracilus, adductor longus, pectineus, iliacus, iliopsoas, psoas major, sartorius, rectus femoris, vastus lateralis, vastus medialis, vastus intermedius, biceps femoris, semitendinosus, and semimembranosus.
- Muscles of the leg include gastrocnemius, soleus, peroneus/fibularis, and tibialis anterior.

Anatomy of a Skeletal Muscle

- A muscle has a fibrous covering called the *epimysium*.
- A muscle is composed of a bundle of fascicles.
- Each fascicle is surrounded by perimysium.
- A fascicle is composed of muscle cells (muscle fibers) surrounded by endomysium.
- The connective tissues of the muscle come together at the end of the muscle cell, or fiber, to form a tendon.

Anatomy of a Skeletal Muscle Cell

- A muscle cell is composed of myofibrils.
- Each myofibril is composed of thick and thin myofilaments arranged in sarcomeres.
- Thick and thin myofilaments are composed of protein molecules.

Physiology of the Muscular System

Physiological Characteristics of Muscle Tissue

- All muscle tissues display five physiological characteristics: excitability, conductivity, contractility, extensibility, and elasticity.

Neuromuscular Junction

- Stimulation of a muscle cell by a nerve happens at a neuromuscular junction.
- An electrical stimulation along the nerve cell results in the release of acetylcholine.
- Acetylcholine fits into receptors on the muscle cell to stimulate it to contract.

- A minimal amount of stimulus called a *threshold* is needed for the muscle to respond.
- As long as the threshold is reached, the muscle cell will contract in an all-or-nothing manner.

Muscle Contraction at the Molecular Level

- The sliding filament theory of muscle contraction involves thick myofilaments grabbing thin myofilaments and pulling them toward the center of the sarcomere.
- As all of the sarcomeres are shortened, so too is the muscle cell.
- Energy contained in ATP is needed for the contraction to happen and to actively transport calcium ions back to the sarcoplasmic reticulum so that the muscle can relax.

Types of Muscle Contractions

- A twitch is a contraction of a muscle cell in response to a single nerve stimulus.
- A twitch has a latent phase, contraction phase, relaxation phase, and refractory phase.
- Tetany is a sustained contraction due to repetitive nerve signals.
- It is the frequency of the nerve impulses that determines whether the contraction will be a twitch or tetany.

Motor Units and Recruitment

- A motor unit is a single nerve cell and all of the muscle cells it stimulates.
- Small motor units are needed for fine, precise movements.
- Getting more motor units involved is recruitment.
- More and more motor units can be recruited to achieve a larger motion.

Isotonic and Isometric Contractions

- In an isotonic contraction, tension in the muscle remains constant as the muscle shortens.
- In an isometric contraction, tension in the muscle increases, but there is no shortening of the muscle.

Levers

- Muscles move bones in lever systems.
- There are three parts to a lever system: the resistance (the weight to be moved), the effort (the force applied at the insertion of the muscle), and the fulcrum (the pivot point for the lever, always a joint).
- Lever systems are classified as first, second, or third class on the basis of the location of the fulcrum, effort, and resistance.
- Most levers in the human body are third-class levers.

Muscle Metabolism

- Muscle cells can do either aerobic or anaerobic respiration to process energy.
- Aerobic respiration is a many-step process that produces enough energy to generate 36 ATP molecules for every glucose molecule but requires the addition of oxygen.
- Anaerobic respiration is a shorter process that produces enough energy to generate 2 ATP molecules per glucose molecule and does not require oxygen.
- Anaerobic respiration results in lactic acid, which must be removed by adding oxygen.
- Aerobic respiration can be done for long periods of time, while anaerobic respiration can be done only for short periods of time.

Fatigue

- Fatigue is the inability of a muscle to fully respond to a nerve stimulus.
- Fatigue can result from the buildup of lactic acid, the lack of acetylcholine, or the lack of glucose.
- Fast-twitch fibers are specialized for anaerobic respiration and therefore fatigue quickly.
- Slow-twitch fibers are specialized for aerobic respiration, so they do not fatigue quickly.

Comparison of Muscle Tissues

- Skeletal muscle tissue is composed of long, striated cells with multiple nuclei pushed off to the side. The cells are under voluntary control, rely on aerobic or anaerobic respiration for energy production, and are associated with bones, skin, and body openings.
- Cardiac muscle tissue is composed of branched, striated cells with a single nucleus and junctions between cells called *intercalated disks*. The cells are autorhythmic, rely on aerobic respiration for energy production, and are located in the heart.
- Smooth muscle tissue is composed of spindle-shaped cells with a single nucleus. The cells are not under voluntary control, rely on aerobic respiration for energy production, and are located in the walls of blood vessels and hollow organs.

Nutritional Requirements of Muscle Tissue

- Muscle tissue must maintain the proteins needed for contraction. Therefore, amino acids, the building blocks of proteins, must be included in the diet.
- The body can make nonessential amino acids.
- Essential amino acids must be supplied through the diet.
- The mineral potassium is also needed for proper muscle function.

Functions of the Muscular System

- The muscular system carries out five important functions in the human body: movement, stability, control of body openings and passages, communication, and heat production.

Effects of Aging on the Muscular System

- Lean muscle mass decreases with age.
- The amount of loss is genetically determined.
- Fast-twitch fibers are more affected than slow-twitch fibers.
- The effects of decreased muscle mass include the following: Strength is decreased, fatigue occurs more quickly, stability is reduced, movement slows and becomes more limited, and the gait shortens.
- Exercise is the best way to limit the effects of aging.

Muscular System Disorders

- Hernia is the protrusion of viscera through the muscle of the abdominal wall.
- Cramp is a painful muscle spasm that may have many causes.
- *Muscular dystrophy* is a term for a group of hereditary disorders that result in the progressive degeneration of muscle tissue.

key words for review

The following terms are defined in the glossary.

acetylcholine (ACh)
aerobic respiration
anaerobic respiration
antagonist
extension
fascicle
fatigue
flexion
insertion
isometric
isotonic
lever
motor unit
muscle twitch
origin
recruitment
sarcomere
sliding filament theory
synergists
tetany

chapter review questions

Word Building: *Use the word roots listed in the beginning of the chapter and the prefixes and suffixes inside the back cover to build words with the following meanings:*

1. The sudden death of heart muscle: ____________________
2. Muscle cell: ____________________
3. Cytoplasm of a muscle cell: ____________________
4. Pertaining to heart muscle: ____________________
5. Muscle tumor: ____________________

Multiple Select: *Select the correct choices for each statement. The choices may be all correct, all incorrect, or any combination of correct and incorrect.*

1. Which of the following statements is (are) accurate about aerobic and anaerobic respiration?
   a. Aerobic respiration is best for sprints.
   b. Anaerobic respiration is best for long distances because it does not need large amounts of oxygen.
   c. Aerobic respiration is faster.
   d. Aerobic respiration produces more energy per glucose molecule.
   e. Both aerobic respiration and anaerobic respiration start with glycolysis.
2. What happens in a muscle twitch?
   a. Myosin grabs a thick myofilament.
   b. Sarcomeres shorten during the contraction phase.
   c. ATP is put back in the sarcoplasmic reticulum during the refractory phase.
   d. A threshold stimulus must be reached before anything will happen.
   e. Thin myofilaments are pulled toward the center of a sarcomere.
3. Nutritional requirements of muscle tissue include which of the following?
   a. Essential amino acids
   b. Potassium
   c. Nucleic acids
   d. Balanced incomplete proteins
   e. Cholesterol
4. Which of the following statements is (are) accurate in regard to skeletal, cardiac, and smooth muscle tissues?
   a. Cardiac muscle is autorhythmic.
   b. Smooth muscle is voluntary.
   c. Skeletal muscle is arranged in sarcomeres.
   d. Skeletal, cardiac, and smooth muscle tissues are striated.
   e. Smooth muscle can be found in the walls of arteries and veins
5. Which of the following effects result(s) from aging on the muscular system?
   a. Muscle mass declines.
   b. Slow-twitch fibers are affected most by aging.
   c. Fatigue occurs more quickly.
   d. Increased muscle tone limits stability.
   e. The gait lengthens and is slower.
6. Which of the following accurately state(s) the action of a muscle?
   a. The orbicularis oris closes the eye.
   b. The deltoid abducts the arm.

c. The hamstring muscles extend the knee.
d. The biceps brachii flexes the elbow.
e. The triceps brachii extends the elbow.

7. Which of the following statements is (are) true about lever systems?
   a. The effort is the origin of the muscle.
   b. The fulcrum is a joint.
   c. The resistance is between the effort and the fulcrum in a third-class lever.
   d. First-class levers are the most common in the human body.
   e. The quadriceps muscles use a third-class lever to extend the knee.
8. Which of the following statements appropriately use(s) the underlined term?
   a. A synergist helps another muscle do its action.
   b. A fixator holds the insertion steady for another muscle.
   c. An antagonist opposes the action of another muscle.
   d. An origin is always a joint.
   e. The insertion is located on the bone that moves.
9. What happens at a neuromuscular junction?
   a. An electrical impulse causes the release of a chemical.
   b. Acetylcholine is released by the muscle cell.
   c. Acetylcholinesterase is released from the nerve cell.
   d. A neurotransmitter is released.
   e. Acetylcholine fits into receptors on the muscle cell.
10. Which of the following statements accurately describe(s) the action?
   a. Making a fist is flexing the fingers.
   b. Standing on tiptoes is dorsiflexion.
   c. Pitching a baseball is circumduction.
   d. Pointing with your arm straight in front of you is extending the arm.
   e. Squatting involves knee flexion.

Matching: *Match the structure to the description. Some answers may be used more than once.*

_____ 1. Bundle of muscle fibers
_____ 2. Connective tissue surrounding a muscle fiber
_____ 3. Connective tissue surrounding a fascicle
_____ 4. Z line to Z line
_____ 5. Where thick myofilaments attach

a. Perimysium
b. Sarcomere
c. Endomysium
d. Fascicle
e. Tendon
f. Myofibril
g. Active site

Matching: *Match the disorder to the description. Some answers may be used more than once.*

_____ 6. Muscle spasm due to the buildup of lactic acid
_____ 7. Compression of the nerves and flexor tendons of the fingers
_____ 8. Progressive degeneration and replacement of muscle fibers
_____ 9. Destruction of acetylcholine receptors
_____ 10. Blood flow to swollen muscles restricted by fascia

a. Muscular dystrophy
b. Myasthenia gravis
c. Carpal tunnel syndrome
d. Cramp
e. Compartment syndrome
f. Charley horse
g. Hernia

Critical Thinking

1. Asim was raking his yard all afternoon. After several hours, his hands became very tired, and he was unable to let go of the rake. Explain what was happening.
2. Ian swam the 1,500-meter freestyle event in 18 minutes 23.4 seconds. He paced himself at first but finished the event swimming as fast as he could. He continued to breathe heavily after the event. What type of respiration was he using (a) at the beginning of the event, (b) near the end of the event, and (c) after the event? Explain.
3. What type of contraction occurs at the molecular level if nerve impulses occur at a frequency so great that the calcium cannot be actively transported back to the sarcoplasmic reticulum between nerve impulses? Give an example of when this type of contraction is used.

5 chapter mapping

This section of the chapter is designed to help you find where each outcome is covered in this text.

	Outcomes	Readings, figures, and tables	Assessments
5.1	Use medical terminology related to the muscular system.	Word roots: pp. 167–168	Word Building: 1–5
5.2	Define terms concerning muscle attachments and the ways muscles work in groups to aid, oppose, or modify each other's actions.	Anatomical terms: p. 169 Figure 5.10	Multiple Select: 8
5.3	Demonstrate actions caused by muscles.	Muscle actions: pp. 170–173 Figures 5.2–5.9	Multiple Select: 10
5.4	Identify muscles, giving the origin, insertion, and action.	Muscles by region: pp. 173–187 Figures 5.10–5.20 Tables 5.1–5.7	Spot Check: 1, 2 Multiple Select: 6
5.5	Describe the structural components of a muscle, including the connective tissues.	Anatomy of a skeletal muscle: pp. 188–189 Figure 5.21	Matching: 1–5
5.6	Describe the structural components of a skeletal muscle fiber, including the major proteins.	Anatomy of a skeletal muscle cell: pp. 189–191 Figures 5.22–5.24	
5.7	Explain the five physiological characteristics of all muscle tissue.	Physiological characteristics of muscle tissue: p. 191	
5.8	Explain how a nerve stimulates a muscle cell at a neuromuscular junction.	Neuromuscular junction: pp. 191–192 Figure 5.25	Multiple Select: 9
5.9	Describe a muscle contraction at the molecular level.	Muscular contraction at the molecular level: pp. 192–194 Figure 5.26	Spot Check: 3 Multiple Select: 2

	Outcomes	Readings, figures, and tables	Assessments
5.10	Compare and contrast a muscle twitch and tetany with regard to the steps of a muscle contraction at the molecular level.	Types of muscle contractions: pp. 194–196 Figures 5.27, 5.28	Critical Thinking: 3
5.11	Define a motor unit and explain the effect of recruitment.	Motor units and recruitment: p. 196	Spot Check: 4
5.12	Compare and contrast isotonic and isometric contractions.	Isotonic and isometric contractions: p. 196	Spot Check: 5
5.13	Describe an example of a lever system in the human body, giving the resistance, effort, and fulcrum.	Levers: pp. 197–198 Figure 5.29	Multiple Select: 7
5.14	Compare aerobic and anaerobic respiration in terms of amount of ATP produced, speed, and duration.	Muscle metabolism: pp. 198–200 Figure 5.30 Table 5.8	Multiple Select: 1 Critical Thinking: 2
5.15	Explain the basis of muscle fatigue and soreness.	Fatigue: p. 200	Spot Check: 6 Critical Thinking: 1
5.16	Compare and contrast skeletal, cardiac, and smooth muscle tissue in terms of appearance, structure, type of nerve stimulation, type of respiration, and location.	Comparison of muscle tissues: pp. 200–201 Table 5.9	Multiple Select: 4
5.17	Explain the nutritional requirements of the muscular system.	Nutritional requirements of muscle tissue: pp. 201–202 Table 5.10	Spot Check: 7 Multiple Select: 3
5.18	Summarize the five functions of the muscular system and give an example or explanation of each.	Functions of the muscular system: p. 203 Figure 5.31	
5.19	Summarize the effects of aging on the muscular system.	Effects of aging on the muscular system: pp. 203–204	Spot Check: 8 Multiple Select: 5
5.20	Describe muscle disorders and relate abnormal function to pathology.	Muscular system disorders: pp. 204	Matching: 6–10

references and resources

Beers, M. H., Porter, R. S., Jones, T. V., Kaplan, J. L., & Berkwits, M. (Eds.). (2006). *Merck manual of diagnosis and therapy* (18th ed.). Whitehouse Station, NJ: Merck Research Laboratories.

Mooar, P. (2007, July). Effects of aging: Biology of the musculoskeletal system. *Merck manual home edition*. Retrieved April 8, 2010, from http://www.merckmanuals.com/home/sec05/ch058/ch058g.html.

Saladin, Kenneth S. (2010). *Anatomy & physiology: The unity of form and function* (5th ed.). New York: McGraw-Hill.

Seeley, R. R., Stephens, T. D., & Tate, P. (2006). *Anatomy & physiology* (7th ed.). New York: McGraw-Hill.

Shier, D., Butler, J., & Lewis, R. (2010). *Hole's human anatomy & physiology* (12th ed.). New York: McGraw-Hill.

Thompson, L. V. (1994, January). Effects of age and training on skeletal muscle physiology and performance [Electronic version]. *Physical Therapy, 74*(1), 71–81.

University of Maryland Medical Center. (2008, August 10). *Aging changes in the bones—muscles—joints—Overview*. Retrieved April 8, 2010, from http://www.umm.edu/ency/article/004015.htm.

6 The Nervous System

Whether it is smiling while daydreaming on a sunny day, speeding up the heart when frightened by a stranger, or squinting in response to a bright light, the human body's nervous system is all about communication, motor control, and sensation. See Figure 6.1. It is a fast, highly efficient way for one part of the body to communicate with another.

6.1 learning **outcome**

Use medical terminology related to the nervous system.

learning outcomes

After completing this chapter, you should be able to:

- **6.1** Use medical terminology related to the nervous system.
- **6.2** Describe the organization of the nervous system in regard to structure and function.
- **6.3** Describe the anatomy of a neuron.
- **6.4** Differentiate multipolar, bipolar, and unipolar neurons in terms of anatomy, location, and direction of nerve impulses.
- **6.5** Describe neuroglial cells and state their function.
- **6.6** Describe the meninges covering the brain and spinal cord.
- **6.7** Explain the importance of cerebrospinal fluid, including its production, circulation, and function.
- **6.8** Describe the major landmarks and subdivisions of the brain and state their functions.
- **6.9** Describe the spinal cord.
- **6.10** Describe the anatomy of a nerve and its connective tissues.
- **6.11** List the cranial nerves in order, stating their function and whether they are sensory, motor, or both.
- **6.12** Describe the attachment of nerves to the spinal cord.
- **6.13** Compare the parasympathetic and sympathetic divisions of the autonomic nervous system in terms of anatomy and function.
- **6.14** Describe a resting membrane potential.
- **6.15** Compare and contrast a local potential and an action potential.
- **6.16** Describe a specific reflex and list the components of its reflex arc.
- **6.17** Explain the difference between short-term and long-term memory.
- **6.18** Differentiate between Broca's area and Wernicke's area in regard to their location and function in speech.
- **6.19** Explain the function of the nervous system by writing a pathway for a sensory message sent to the brain to be processed for a motor response.
- **6.20** Explain the nutritional requirements of the nervous system.
- **6.21** Explain the effects of aging on the nervous system.
- **6.22** Describe nervous system disorders.

word roots & combining forms

cephal/o: head

cerebell/o: cerebellum

cerebr/o: cerebrum

dur/o: tough

encephal/o: brain

gangli/o: ganglion

gli/o: glue

medull/o: medulla

mening/o: meninges

myel/o: spinal cord

neur/o: nerve

poli/o: gray matter

pronunciation key

aphasia: ah-FAY-zee-ah

dura mater: DYU-rah MAY-ter

ependymal: ep-EN-dih-mahl

gyri: JI-rye

myelin: MY-eh-lin

neuroglia: nyu-roh-GLEE-ah

oligodendrocyte: OL-ih-goh-DEN-droh-site

Ranvier: rahn-vee-AY

Schwann: SHWANN

sulci: SUL-sye

Wernicke: WUR-ni-keh

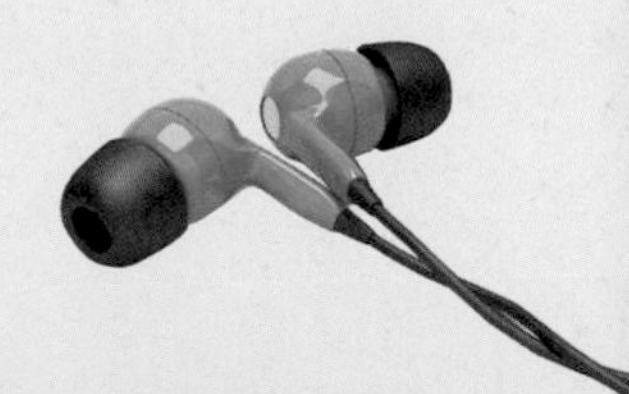

6.2 learning outcome

Describe the organization of the nervous system in regard to structure and function.

Overview

The nervous system has two main divisions. One of these divisions is the **central nervous system (CNS),** which is composed of the brain and spinal cord and serves as the central processing center. The other main division is the **peripheral nervous system (PNS),** a network of all the nerves in the body that sends messages to and from the central processing center. **Figure 6.2** shows how the nervous system is organized.

study hint

As you read through this chapter, you may find it helpful to refer to Figure 6.2 to refresh your mind on the organization of this complex system.

This chapter will follow the organization of the nervous system that is shown in **Figure 6.2**, starting with the **central nervous system (CNS),** shown on the left. You will begin your study of the CNS with the four subdivisions of the brain. These four subdivisions are the **cerebrum** and its lobes; the **diencephalon,** composed of the **hypothalamus** and **thalamus;** the **brainstem,** composed of the **medulla oblongata, pons, midbrain,** and **reticular formation;** and the **cerebellum.** You will also learn about the anatomy of the **spinal cord** and its functions. After you become familiar with the brain and the spinal cord, you will then move on to study the divisions within the **peripheral nervous system (PNS),** shown on the right in **Figure 6.2**.

As you can see in **Figure 6.2**, the PNS is composed of nerves carrying messages in two directions. **Sensory** neurons carry **afferent** (incoming) messages to the brain or spinal cord. **Motor** neurons carry **efferent** (outgoing) messages away from the brain and spinal cord.

Some of the efferent messages travel on motor neurons to stimulate skeletal muscles to move the body, so they belong to the **somatomotor** (*soma* means "body," *motor* means "movement") division. Other efferent messages go to glands, the cardiac muscle of the heart, or the smooth muscle of hollow organs and blood vessels. These messages make up the **autonomic** division. (Again, see **Figure 6.2**.)

The autonomic division has two subdivisions: **parasympathetic** and **sympathetic.** The parasympathetic division sends electrical messages to carry out functions for vegetative activities such as digestion, defecation, and urination. The sympathetic division sends electrical messages to prepare the body for physical activity, often referred to as *fight or flight*. You will become familiar with the differences in anatomy of these two subdivisions and learn more about their functions later in the chapter.

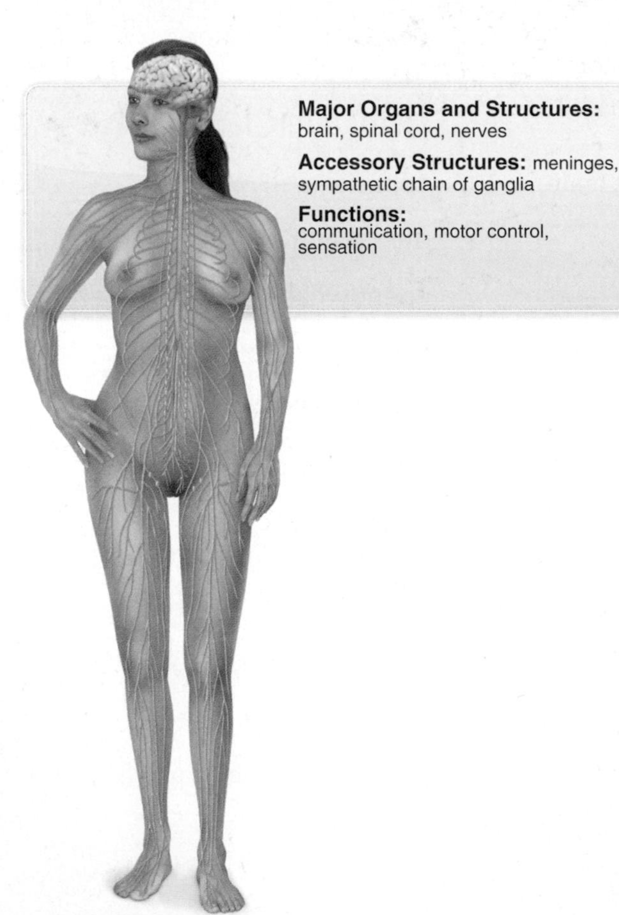

FIGURE 6.1 The nervous system.

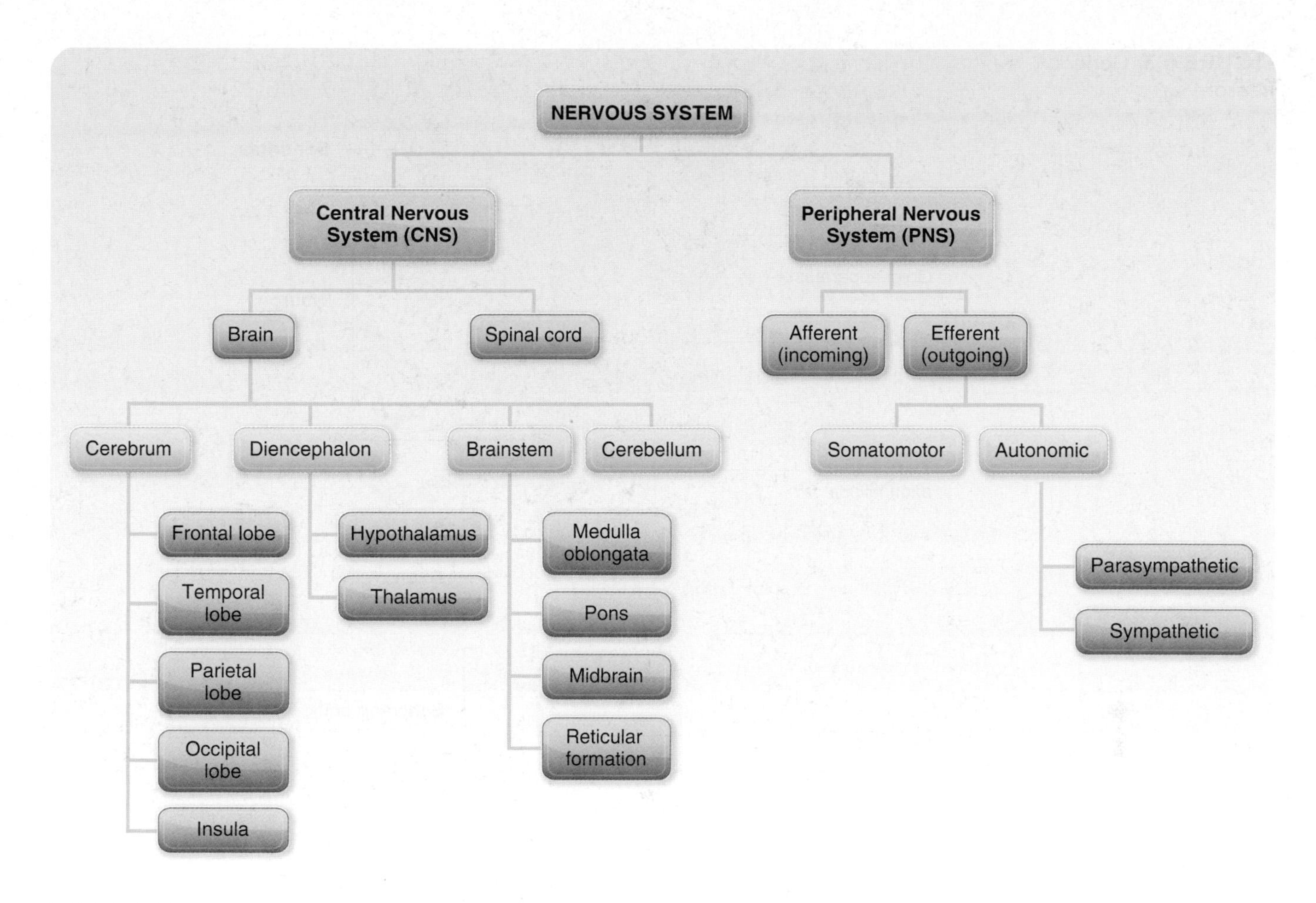

FIGURE 6.2 Organization of the nervous system.

Anatomy of the Nervous System

To understand the divisions of this system, you must first take a look at the anatomy of a **neuron** (nerve cell) and the **neuroglial** cells that aid neurons in their function.

neuroglia:
nyu-roh-GLEE-ah

Anatomy of a Neuron

6.3 learning **outcome**
Describe the anatomy of a neuron.

A single neuron can be very long (measuring a meter or more), as it may start at the tip of your finger and end at your spinal cord. A neuron has three basic parts: one or more **dendrites** that receive information; a **body** containing the nucleus and organelles for protein synthesis; and an **axon** that carries the nerve impulse along its length to the **synaptic knobs** at the end of the neuron. The basic parts of a neuron are shown in **Figure 6.3** and discussed below:

- **Dendrites.** A neuron may have anywhere from 1 to 1,000 dendrites. The more dendrites the neuron has, the more information it can process. The dendrites branch to make multiple connections and form precise pathways. Incoming messages travel from the dendrites toward the body.

FIGURE 6.3 Generic neuron.

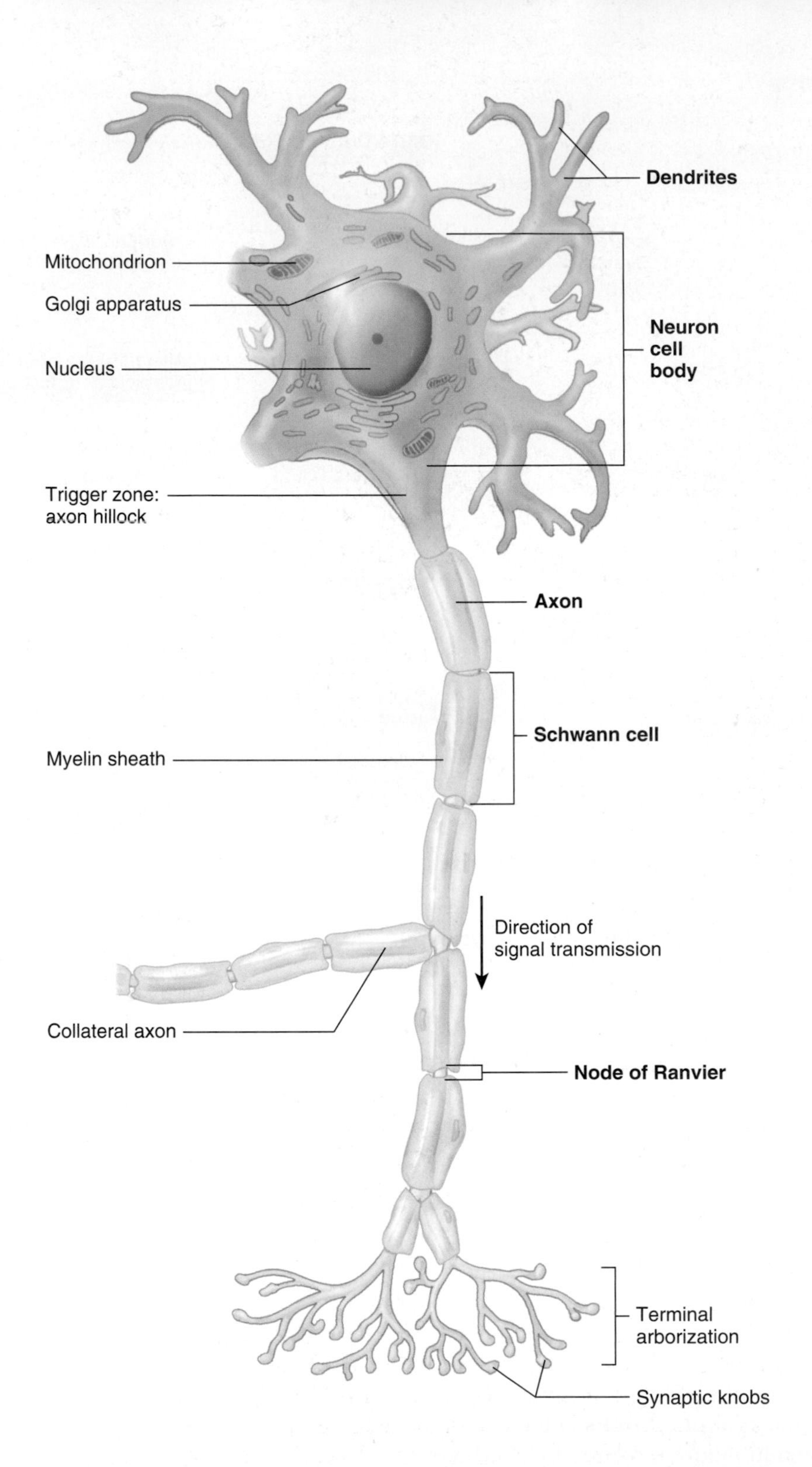

- **Body.** The body contains the nucleus and organelles for protein synthesis. In the previous chapter you learned that protein neurotransmitters were released from the knob at the end of a nerve cell to stimulate a muscle cell to contract. In the case of skeletal muscle, the neurotransmitter is acetylcholine. Neurotransmitters are made in the body of the neuron through protein synthesis. The Golgi complex then packages the neurotransmitter in secretory vesicles called **synaptic vesicles.** These vesicles carry the neurotransmitter down the next part of the neuron (the axon) to the synaptic knob at the end. This is called **axonal transport.**

clinical point

Axonal transport, in the direction from the body to the synaptic knob, is relatively fast. The empty synaptic vesicles are carried in the opposite direction (back to the body) for reuse in what is called **retrograde axonal transport.** Viruses, such as the rabies virus, can hitch a ride with either type of transport. A bite on the hand or arm from a rabies-infected animal may damage a neuron. When the neuron is damaged, the virus gains entry into the neuron and travels by axonal and/or retrograde axonal transport to move from one neuron to another and finally reach the brain.

- **Axon.** The axon leaves the neuron body at the **axon hillock,** also called the **trigger zone.** Electrical messages travel away from the body down to the axon's end, which branches in the **terminal arborization** (*arbor* means "tree"). A synaptic knob at the end of each branch forms a **synapse** (junction) with another cell. The synapse may be with a muscle cell, a gland cell, or the dendrite of another nerve cell. As you read in Chapter 5, a motor unit is defined as a single neuron and all of the muscle fibers it stimulates. By having the terminal arborization and multiple synaptic knobs, one neuron can potentially stimulate hundreds of muscle fibers in a large motor unit. Each synaptic knob meets a muscle cell at a neuromuscular junction (a synapse). Here, the synaptic vesicles release acetylcholine by exocytosis to stimulate the muscle cell.

It is important to note in **Figure 6.3** that the axon is only intermittently covered with a **myelin sheath.** Myelin does not cover the body or dendrites of a neuron, just the axons. Very short axons may not need this covering at all. Myelin is lipid-rich, and it insulates the axons much like the wire coating insulates the wires of electrical appliances in your home. But unlike the coating on wires, the myelin on an axon has gaps. These gaps, called **nodes of Ranvier,** are very important in the conduction of nerve impulses. Later in this chapter you will read more about their role. It is also important to note that myelin is white in color. So it makes sense that **white matter,** in the brain and spinal cord, is a concentration of myelinated axons. **Gray matter,** in the brain and spinal cord, is a concentration of dendrites, cell bodies, and unmyelinated axons. Peripheral nerves have myelinated axons.

myelin: MY-eh-lin

Ranvier: rahn-vee-AY

6.4 learning outcome

Differentiate multipolar, bipolar, and unipolar neurons in terms of anatomy, location, and direction of nerve impulses.

Types of Neurons There are three basic types of neurons: multipolar, bipolar, and unipolar. Table 6.1 compares their anatomy, their location, the types of messages they carry, and the direction these messages are sent. These neuron types and their functions are further described in the following list:

- **Multipolar neurons** look like the neuron shown in **Figure 6.3**. They have multiple dendrites and an axon that may or may not have a collateral branch. This is the most common type of neuron in the brain and spinal cord. Motor neurons are multipolar. They carry electrical messages away from (efferent) the brain and spinal cord.
- **Bipolar neurons** have one dendrite and one axon. They can be found in the nasal cavity, the retina of the eye, and the inner ear. They are sensory neurons, so the electrical messages they carry travel toward (afferent) the brain.
- **Unipolar neurons** seem to have one process that serves as dendrite and axon with the cell body pushed off to the side. They are sensory neurons in the body, located in areas such as the skin, organs, and other areas where bipolar neurons

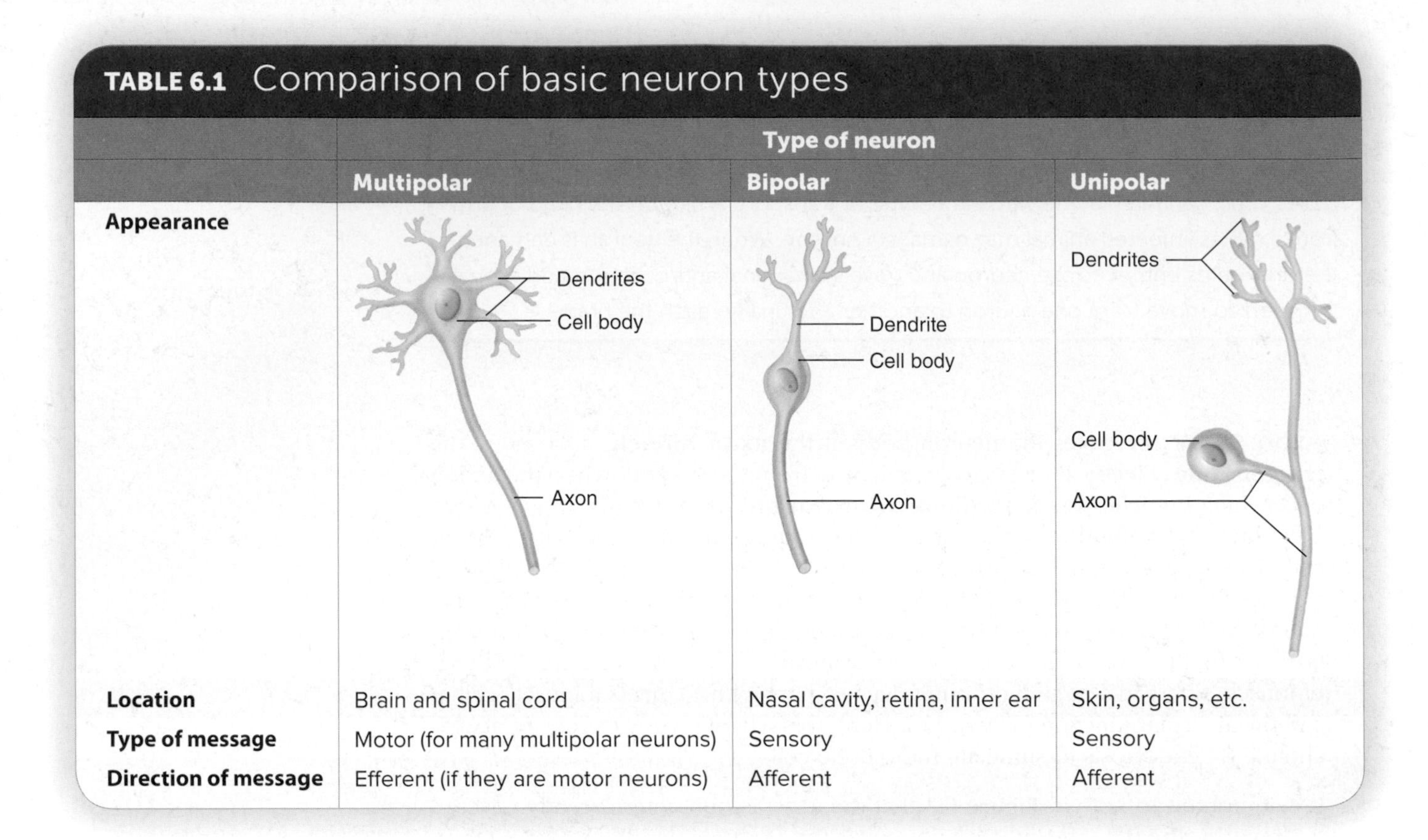

TABLE 6.1 Comparison of basic neuron types

	Type of neuron		
	Multipolar	**Bipolar**	**Unipolar**
Appearance			
Location	Brain and spinal cord	Nasal cavity, retina, inner ear	Skin, organs, etc.
Type of message	Motor (for many multipolar neurons)	Sensory	Sensory
Direction of message	Efferent (if they are motor neurons)	Afferent	Afferent

are not present. Since they are sensory, the electrical messages carried by these neurons are afferent. The cell bodies of many unipolar neurons are grouped together to form **ganglia.** You will learn more about this when you study how nerves connect to the spinal cord.

spot check 1 Which type of neuron would make up the somatomotor division of the PNS?

In total, the body contains over a trillion neurons of these three types. However, these neurons cannot function alone. For every one neuron, there are approximately 50 support cells, which are collectively called *neuroglia*. The term *neuroglia* literally means "nerve glue."

6.5 learning outcome

Describe neuroglial cells and state their function.

oligodendrocyte: OL-ih-goh-DEN-droh-site

ependymal: ep-EN-dih-mahl

Schwann: SHWANN

Neuroglia

There are six types of neuroglial cells in the human body: **oligodendrocytes, ependymal cells, astrocytes, microglia, Schwann cells,** and **satellite cells.** We discuss the appearance, location, and function of these cells below. See Table 6.2 for a comparison.

As you can see in **Figure 6.4**, four of the six types of neuroglial cells are located solely in the brain and spinal cord:

- **Oligodendrocytes** resemble octopi that reach out with tentacles to tightly wrap around the axons of neurons in the CNS. These cells form the myelin in the CNS. Imagine a piece of tape. It would have a top and a sticky bottom surface. If you wrapped the tape around your finger several times, you would have several layers of tape. The part of the oligodendrocyte that wraps around the axon is very thin (like the tape), but the top and bottom surfaces are cell

TABLE 6.2 Comparison of neuroglia		
Type of neuroglia	**Location**	**Function**
Oligodendrocytes	CNS	Form myelin in the CNS
Ependymal cells	CNS	Produce cerebrospinal fluid
Astrocytes	CNS	Form the blood-brain barrier, regulate composition of CSF, and form scar tissue
Microglia	CNS	Provide protection by seeking and removing damaged cells, debris, and pathogens
Schwann cells	PNS	Form myelin in the PNS and help damaged myelinated axons regenerate
Satellite cells	PNS	Regulate the chemical environment of ganglia in the PNS

membranes composed mostly of phospholipids. The oligodendrocytes wrap around a section of an axon several times, just as you imagined the tape around your finger. So there are many layers of phospholipids of the oligodendrocyte cell membrane wrapped around the axon. This explains why myelin is very lipid-rich. As you can see in **Figure 6.4**, an oligodendrocyte can reach out and form myelin on more than one axon at a time.

- Ependymal cells line fluid-filled cavities and spaces in the CNS. They produce **cerebrospinal fluid (CSF).** Cilia on the ependymal cells are responsible in part for the circulation of the fluid. You will learn more about cerebrospinal fluid shortly.

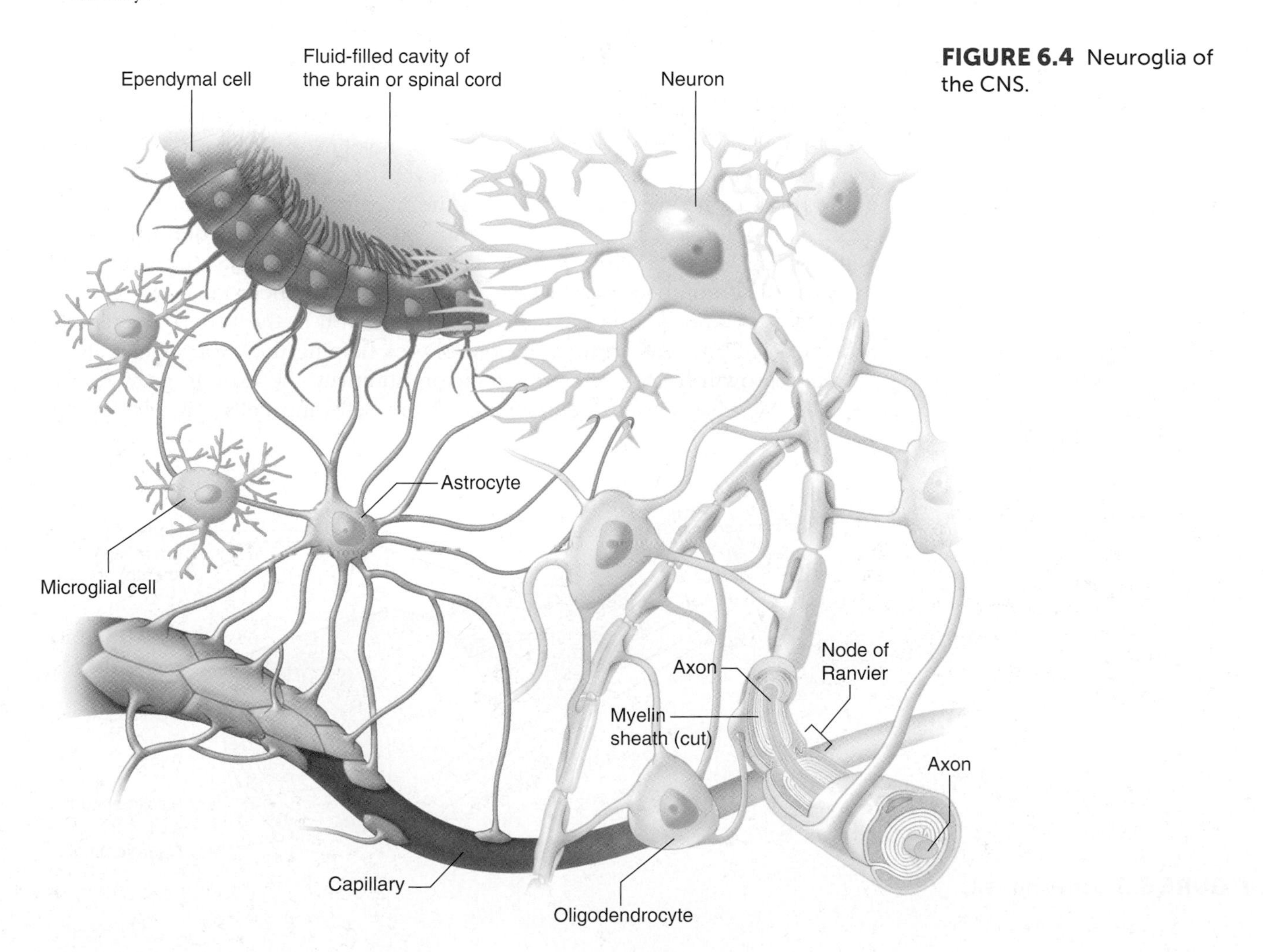

FIGURE 6.4 Neuroglia of the CNS.

- **Astrocytes** are the most numerous neuroglial cells in the CNS. They have a somewhat star-shaped appearance and have several functions. Astrocytes cover the nonmyelin portions of neurons and blood vessels in the CNS, forming a **blood-brain barrier.** This barrier allows astrocytes to regulate what can leave the bloodstream to enter the CNS, protecting the CNS from potentially toxic chemicals. Astrocytes also regulate the contents of the cerebrospinal fluid by absorbing excess neurotransmitters and potassium ions. If a neuron in the CNS becomes damaged, astrocytes fill the space with scar tissue. This process is called **sclerosis.**
- **Microglia** are small macrophages in the CNS that look for cell damage, debris, and pathogens. If these items are found, it is the microglia's responsibility to remove them. They constantly wander through the CNS as an important line of defense.

clinical point

Even though neurons have little capacity for mitosis past puberty, neoplasms (tumors) may develop in the CNS. It is often neuroglial cells that divide out of control to form the tumor. These tumors **(gliomas)** tend to be malignant and grow rapidly, competing with the brain for space in the cranial cavity. Because the blood-brain barrier formed by the astrocytes prevents many chemotherapies from reaching the tumor cells, these tumors often need to be treated with radiation and/or surgery.

The last two types of neuroglial cells are found only in the PNS:

- **Schwann cells** form the myelin in the PNS. Unlike oligodendrocytes of the CNS, each Schwann cell forms one piece of myelin on one axon of one neuron. See **Figure 6.5**. Like the oligodendrocytes, they wrap their cell membrane around the axon several times to form the lipid-rich myelin insulation for the axon. The outermost layer of a Schwann cell is called a **neurolemma.** Along with forming myelin, Schwann cells are important in the PNS if the axon of a myelinated neuron becomes severed. If the body of the neuron is damaged, the cell dies. However, if a myelinated axon of a neuron in the PNS is severed, the Schwann cell past the break is responsible for helping the axon regenerate by secreting growth factors. The growth factors stimulate the axon to grow toward the sleeve created by the neurolemma of the Schwann cells. An analogy will

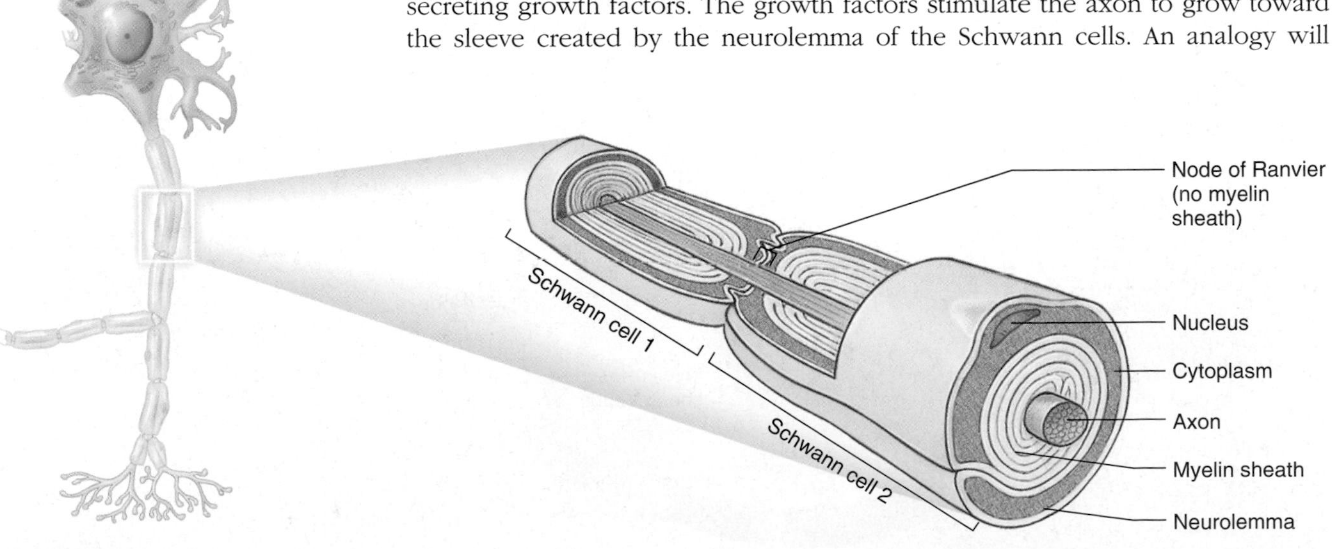

FIGURE 6.5 Schwann cell.

help you to understand this concept. You may have seen someone take a cutting from an ivy plant and place it in a jar of water so that the cutting can develop roots. After a couple of weeks, extensions can be seen branching in many directions from the cutting's severed end. Just like the ivy cutting, a severed axon will develop extensions growing out of the severed end, but only the growth in the direction of the neurolemma sleeve is stimulated to continue to grow by growth factors released by the Schwann cell. All of the other extensions stop growing. In this way, the axon regenerates in the proper location.

- **Satellite cells** surround neuron cell bodies in the ganglia of the PNS. They help regulate the chemical environment of the neurons.

spot check ❷ Encephalitis is often caused by an infection in the brain. It is difficult to treat with medication. Which neuroglial cell makes it difficult to treat encephalitis with drugs? Which neuroglial cell will fight the pathogen causing the infection?

Now that you have covered the primary and support cells of this system, you can focus on the anatomy of the CNS and PNS.

Anatomy of the Central Nervous System

6.6 learning outcome

Describe the meninges covering the brain and spinal cord.

Meninges In exploring the anatomy of the CNS, we begin with the role of the **meninges.** The brain and spinal cord are covered by meninges, which are three membranes that line the cranial and vertebral cavities. The most superficial layer of the meninges is the **dura mater** ("tough mother"). Deep to it is a very delicate, weblike layer called the **arachnoid mater** ("spider-like mother"). Tight to the brain and spinal cord is the **pia mater** ("affectionate mother"). See Figure 6.6. Each of the meninges forms a continuous covering over the brain and spinal cord. Between the vertebrae and the dura mater surrounding the spinal cord is the **epidural space,** which contains blood vessels, adipose tissue, and loose connective tissue. Anesthetics can be administered into this space during surgeries and childbirth. The space between the arachnoid mater and pia mater is the **subarachnoid space.** It contains cerebrospinal fluid.

dura mater: DYU-rah MAY-ter

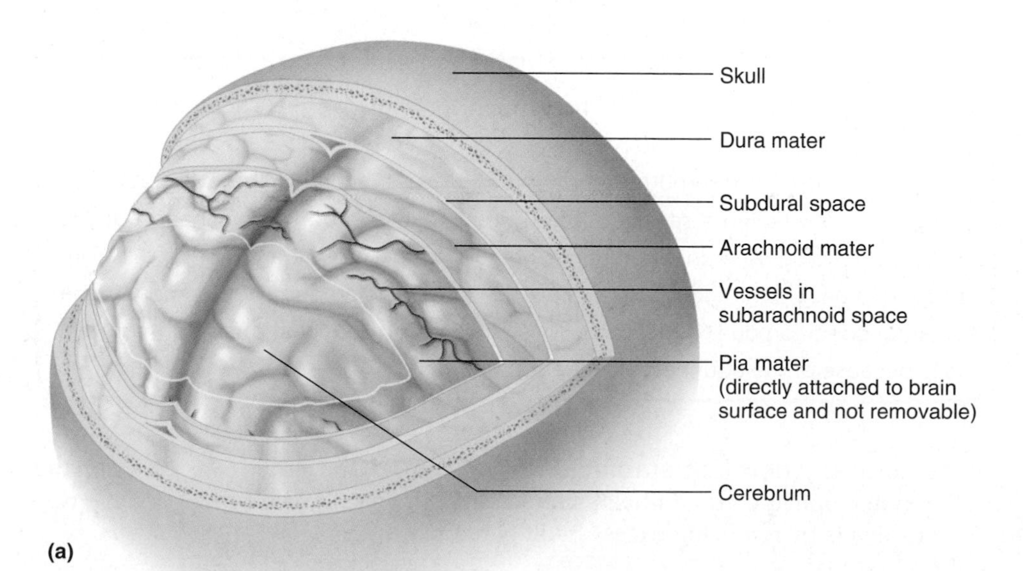

FIGURE 6.6 The meninges: (a) covering the brain, (b) covering the spinal cord, (c) transverse section of the spinal cord between vertebrae, showing the meninges and epidural space.

FIGURE 6.6 concluded

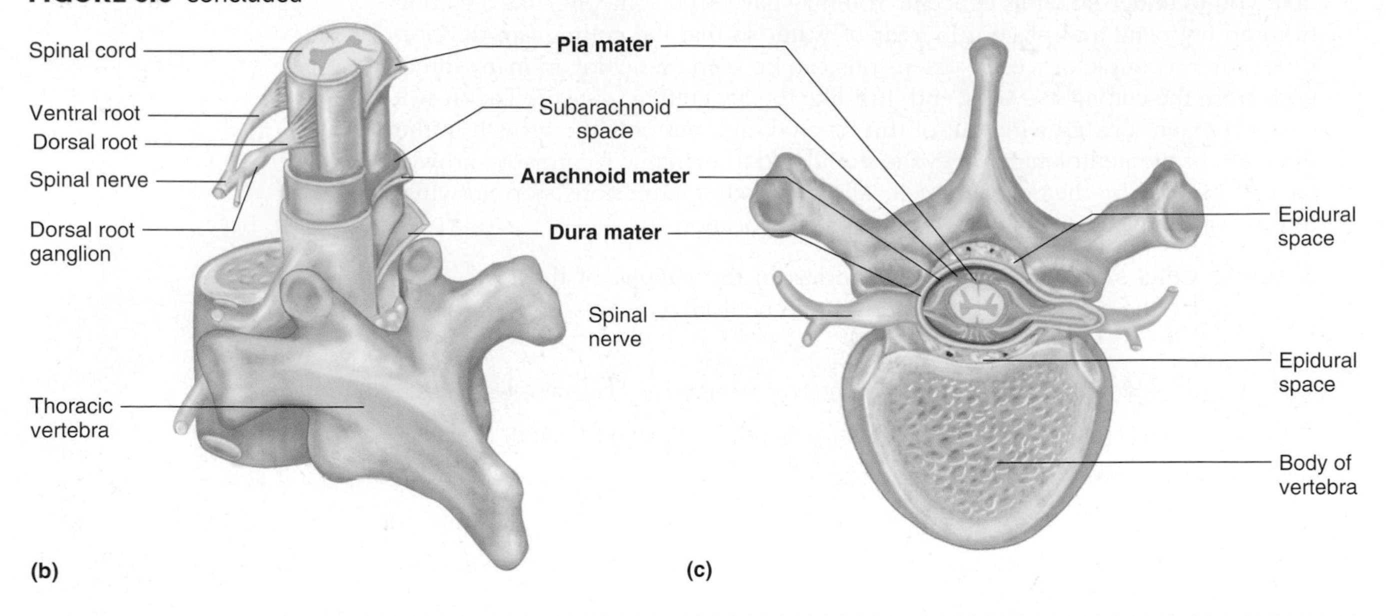

6.7 learning outcome

Explain the importance of cerebrospinal fluid, including its production, circulation, and function.

Cerebrospinal Fluid Cerebrospinal fluid is made by ependymal cells lining cavities in the brain called **ventricles.** See **Figure 6.7**. A bed of capillaries (small blood vessels), called a **choroid plexus,** exists in the walls of the ventricles. Ependymal cells cover the capillaries and take what they need to make CSF from capillary blood. CSF is a clear, colorless fluid that circulates between the ventricles and the subarachnoid space to bathe the brain and spinal cord. The cilia of the ependymal cells, gravity, and the pulsation of arteries in the brain are responsible for the CSF's circulation. See **Figure 6.7c**. Approximately 500 milliliters (mL) of CSF are produced daily. The same amount of CSF is absorbed through **arachnoid villi** back into the bloodstream every day. So there are only approximately 100 to 160 mL of CSF present at any one time. The CSF has several important functions:

1. **Provides buoyancy.** The floor of the cranial cavity is bone with various ridges. The CSF allows the brain to float in the cranial cavity. Without the CSF, nervous tissue would be damaged by the sheer weight of the brain against the bony floor.
2. **Provides protection.** The CSF cushions the brain from impact. The following Clinical Point puts this important job of the CSF into perspective.

clinical point

If you are traveling 50 miles per hour (mph) in your car, the car is traveling at 50 mph, you are traveling at 50 mph, and your brain is traveling at 50 mph. If the car hits a pole, it is now going 0 mph but you and your brain are still moving 50 mph. If you are not wearing a seat belt, the window will slow down your head but the brain is still going 50 mph. The brain is going to slam into the anterior skull before it bounces back to hit the posterior skull. CSF limits the amount of the impact. The resulting condition due to the impact is a **concussion.** Concussions often come in pairs because of the bounce-back effect.

3. **Facilitates chemical stability.** The CSF rinses metabolic wastes from the brain and spinal cord and helps regulate the chemical environment. One way it does this is by removing excess hydrogen ions.
4. **Provides nutrients.** The CSF provides CNS tissues with some nutrients—like glucose.

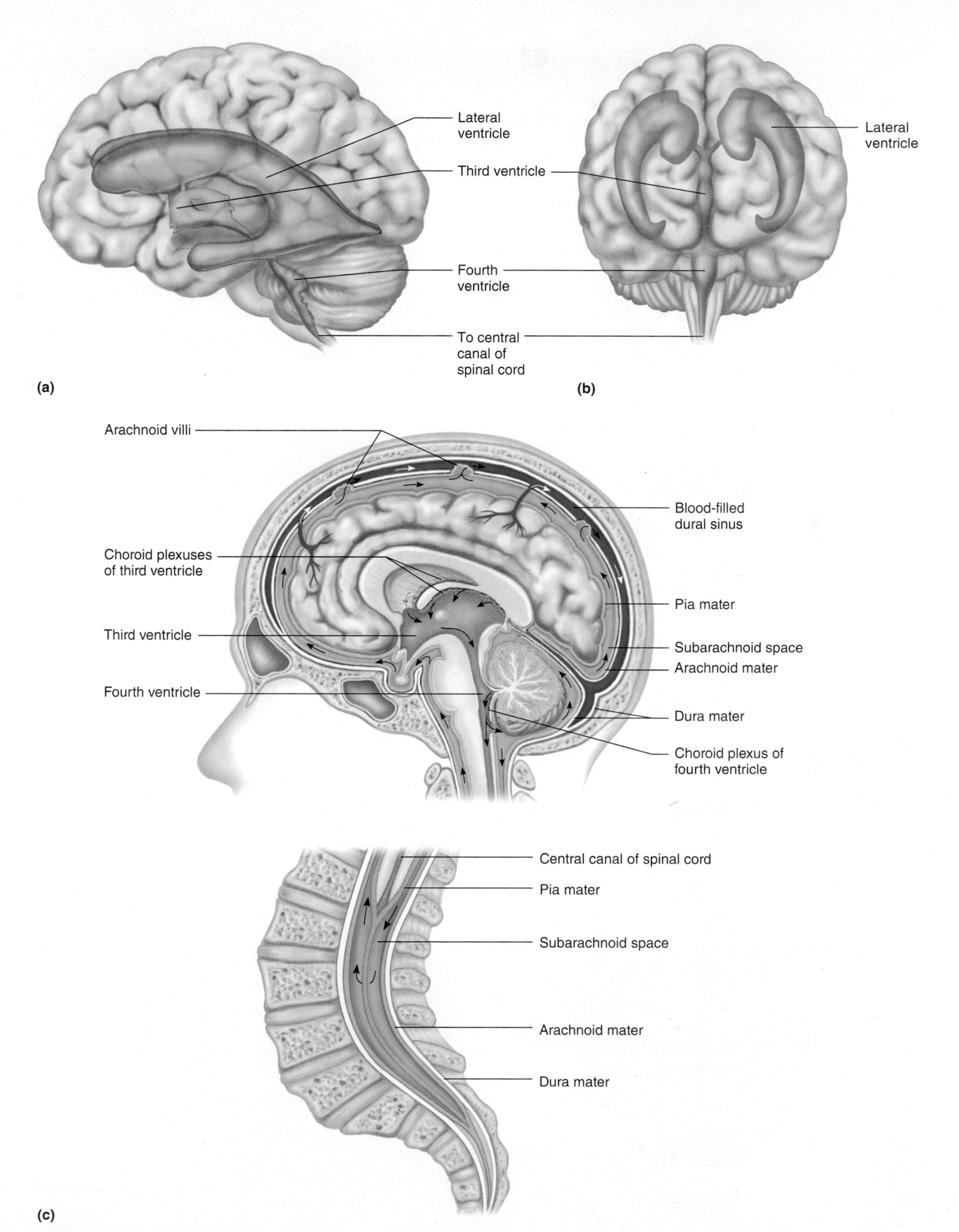

FIGURE 6.7 Ventricles of the brain: (a) lateral view, (b) anterior view, (c) circulation of CSF around the brain and spinal cord.

spot check 3 What will happen to pH if the CSF removes excess hydrogen ions?

You have studied the meninges and the cerebrospinal fluid surrounding the brain. You are now ready to study the anatomy of the brain itself.

6.8 learning outcome

Describe the major landmarks and subdivisions of the brain and state their functions.

sulci: SUL-sye

gyri: JI-rye

Brain As mentioned earlier in this chapter, the brain can be divided into four subdivisions: the cerebrum, the diencephalon, the brainstem, and the cerebellum. See **Figure 6.8** and Table 6.3.

Cerebrum The cerebrum is characterized by a series of grooves and folds on its surface. The grooves are called **sulci** (*sulcus,* singular). The folds are called **gyri** (*gyrus,* singular). The purpose of these structures is to give extra surface area. There is a **longitudinal fissure** that separates the cerebrum into right and left

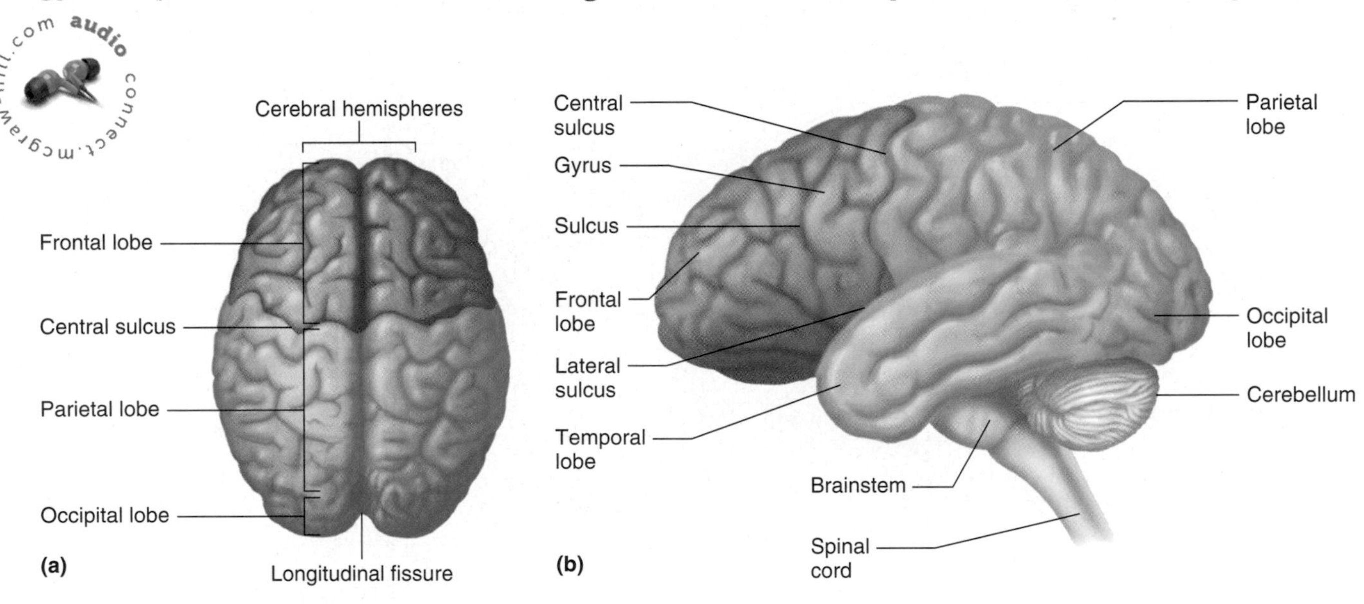

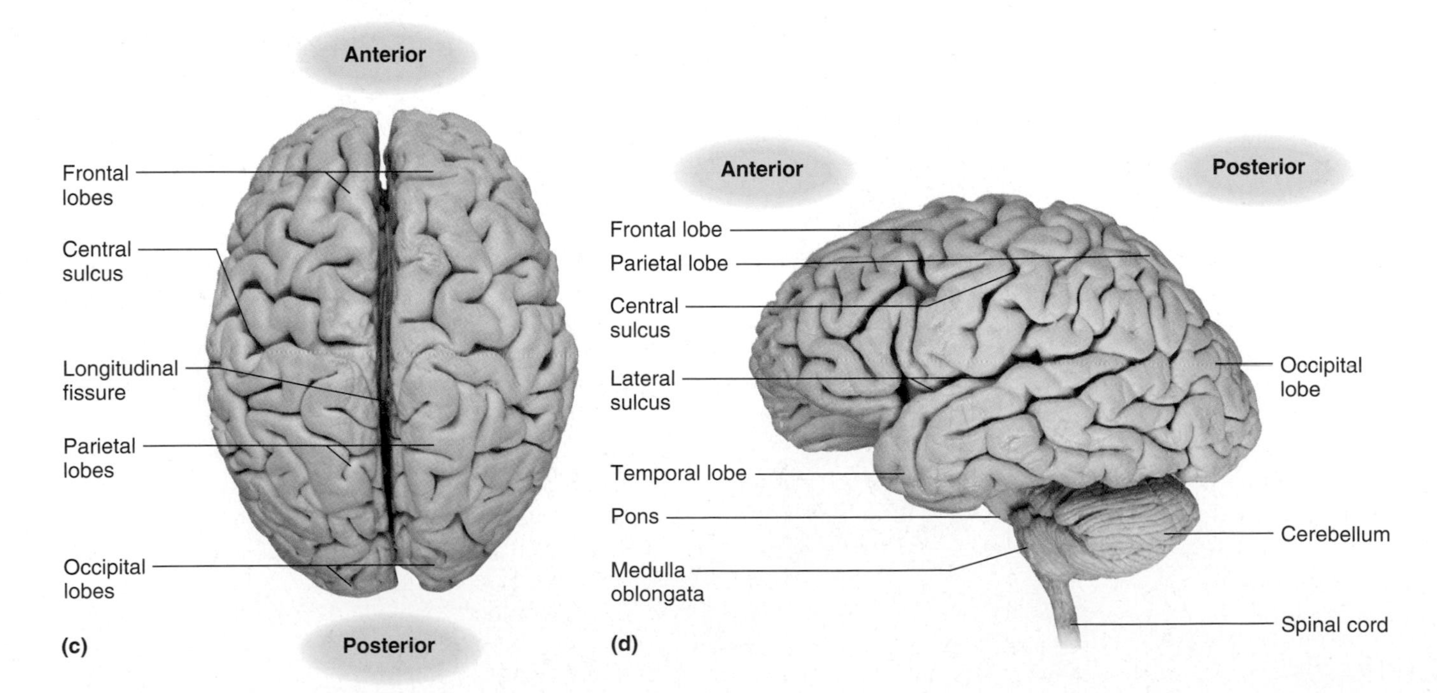

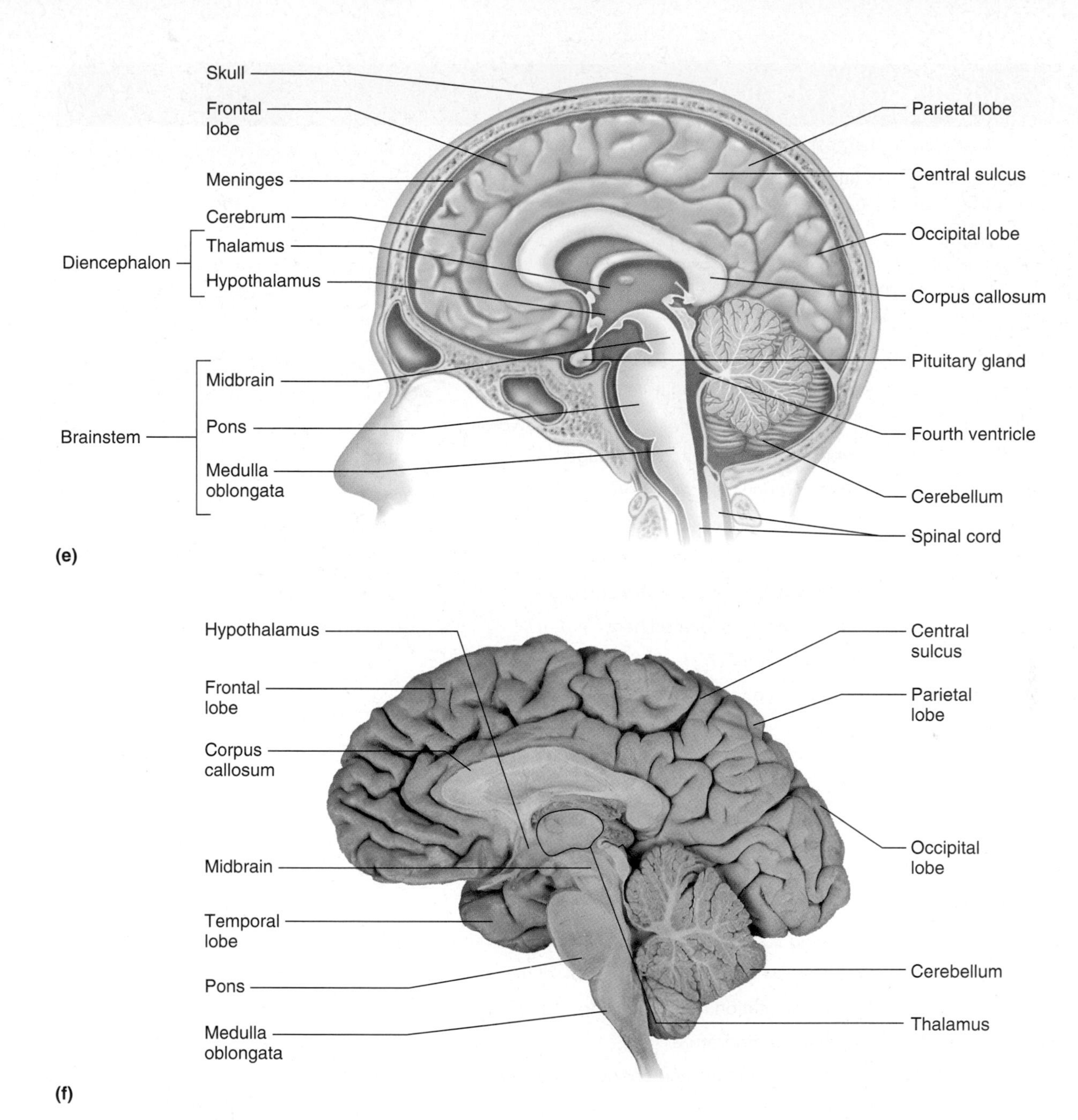

FIGURE 6.8 The brain: (a) superior view, (b) lateral view, (c) superior view of cadaver brain, (d) lateral view of cadaver brain, (e) midsagittal view, (f) midsagittal view of cadaver brain.

hemispheres. A white fibrous band called the **corpus callosum** is deep to the fissure and keeps the right and left hemispheres connected. The corpus callosum appears white because it is composed of myelinated axons carrying messages back and forth between the two hemispheres, allowing them to communicate with each other. The superficial part of the cerebrum, the **cortex,** is composed of gray matter. Most of the brain's dendrites and cell bodies are located here. Conscious thought and voluntary actions arise in the cortex. The rest of the cerebrum is composed of white matter: myelinated axons carrying messages. See Figure 6.8.

The cerebrum can be divided into four major lobes, named for the cranial bone they lie beneath, and a fifth lobe, called the **insula,** that lies deep to the lateral sulcus. Many of the lobes are responsible for sensory information. In each of these lobes, there is a **general sensory area** to identify sensory messages coming in and an **association area** to interpret the message by comparing it to what has come

TABLE 6.3 Subdivisions of the brain

Cerebrum 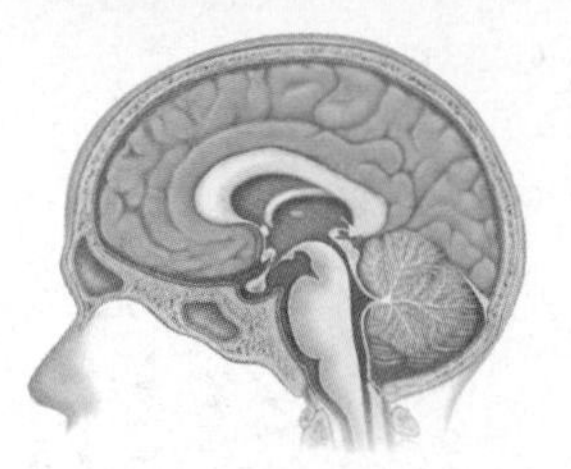	Largest subdivision, divided into two hemispheres. Characterized by gyri and sulci. Divided into lobes.
Frontal lobe	Contains premotor and primary motor areas. Motivation and aggression are located here. Contains Broca's area for language.
Parietal lobe	Sense of touch is located here. Higher-level processes for math and problem solving are also located here.
Temporal lobe	Sense of hearing is located here. Contains Wernicke's area for language.
Occipital lobe	Sense of vision is located here.
Insula	Not much is known.
Diencephalon 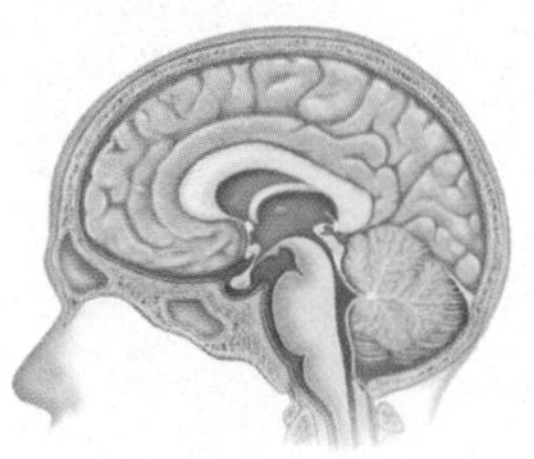	Located deep to the cerebrum.
Thalamus	Switching station for incoming sensory messages. Sends message to appropriate lobe of the cerebrum.
Hypothalamus	Monitoring station for maintaining homeostasis. Regulates temperature. Performs autonomic and endocrine functions.
Brainstem 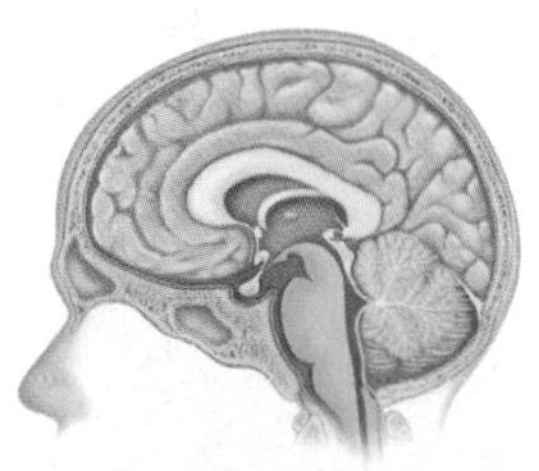	Located in the cranial cavity inferior to the diencephalon and anterior to the cerebellum. All parts include tracts of neurons traveling to and from the spinal cord.
Midbrain 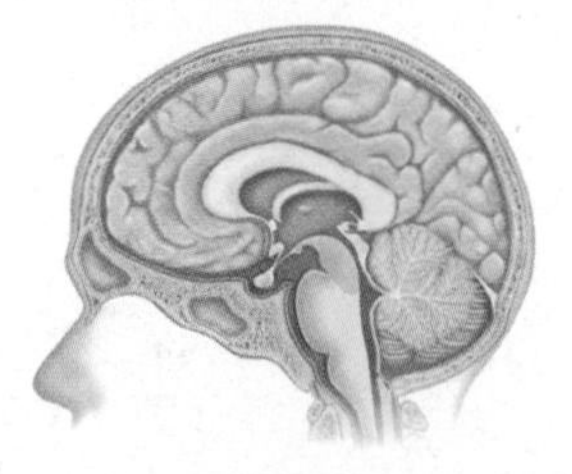	Appears as a hook. Has colliculi for vision and hearing.

TABLE 6.3 concluded

Pons 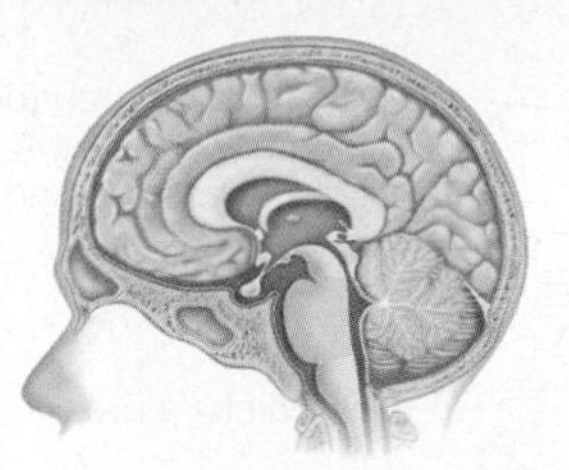	Appears as a bulge between the midbrain and the cerebellum. Serves as a bridge to the cerebellum for efferent motor messages.
Medulla oblongata 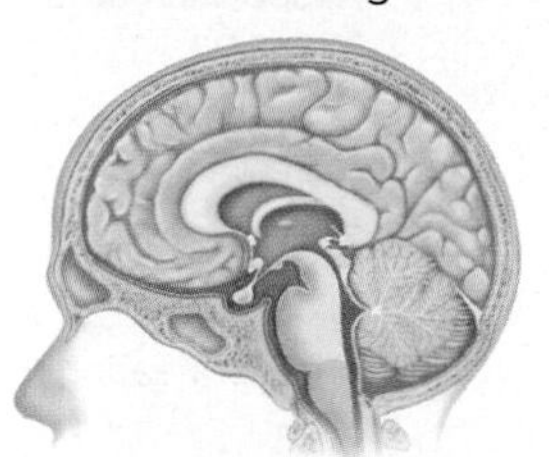	Most inferior part of the brainstem. Motor messages cross sides at the pyramids. Contains centers to regulate heart rate, blood pressure, respiratory rate, and blood vessel diameter.
Reticular formation 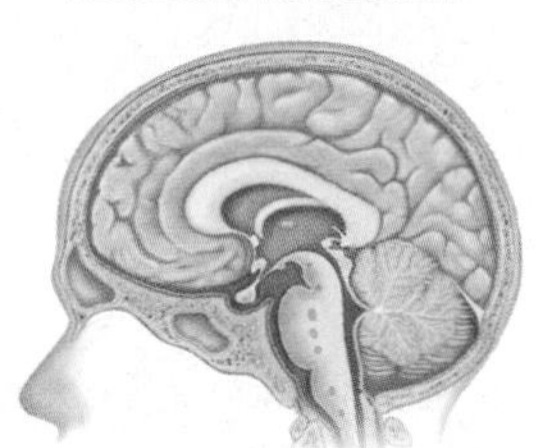	Groups of cell bodies located throughout the brainstem. Determines if sensory messages will be consciously perceived by the cerebrum. Responsible for sleep-wake cycle.
Cerebellum 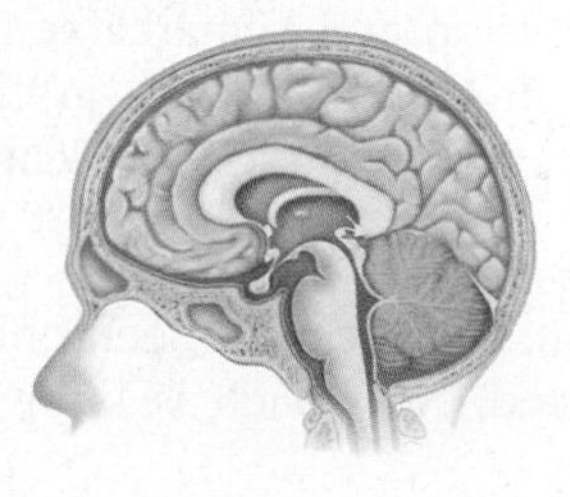	Receives input of body-part location. Uses that information in fine-tuning efferent motor messages to maintain coordination, balance, and smooth motions.

before. For detailed information about each of the lobes, see **Figure 6.9** and review the following list:

- Frontal lobe. The frontal lobe is responsible for the sense of smell. Olfactory messages have a complicated route but are ultimately interpreted in the frontal lobe. The frontal lobe also has two motor areas: a **premotor area** that plans efferent skeletal muscle messages and a **primary motor area** that then sends out the planned, voluntary skeletal muscle messages. This lobe is also responsible for motivation, judgment, and aggression. **Broca's area,** an important area concerning language, is also located in the frontal lobe. It is discussed later in the chapter.
- Parietal lobe. The parietal lobe is responsible for the general senses, like touch. (The general senses and other senses are discussed in Chapter 7.) The cortex of this lobe is also responsible for higher-level processes, such as math and problem solving.

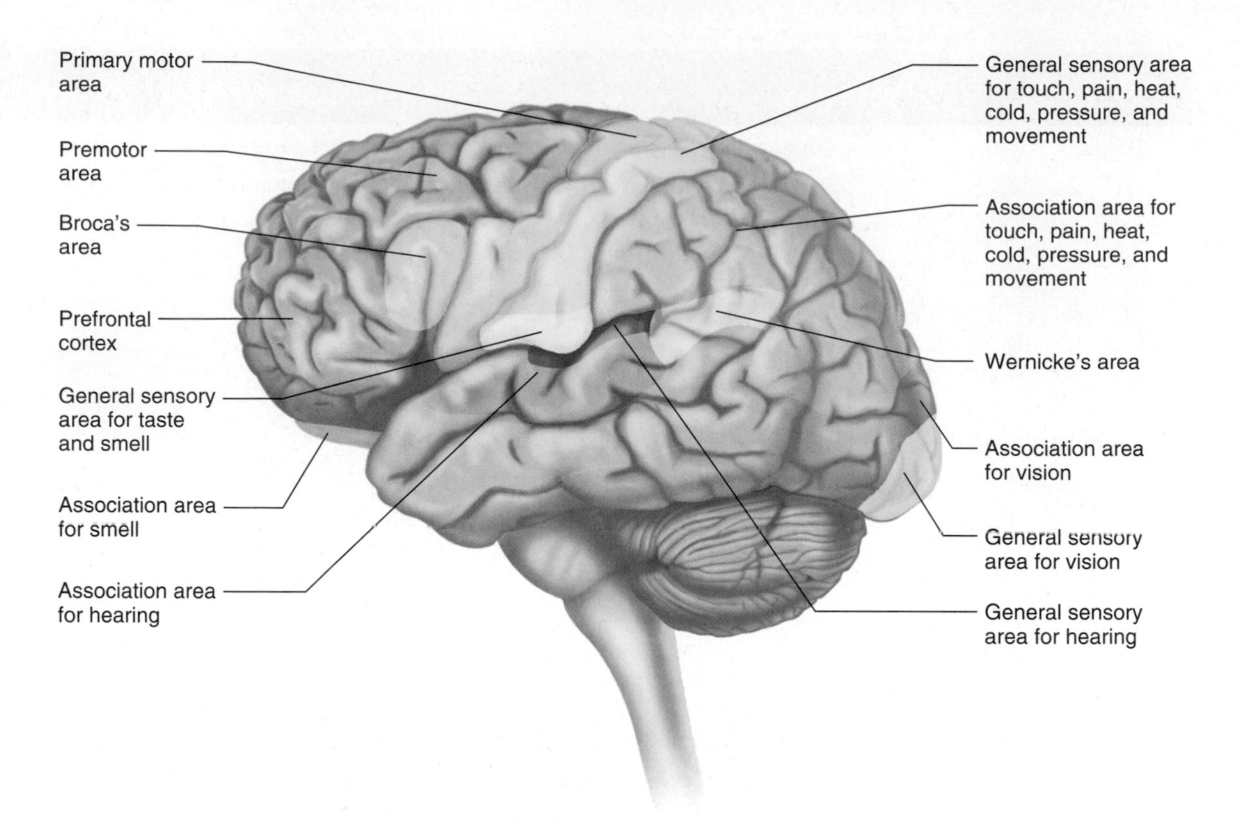

FIGURE 6.9 Functional regions of the cerebral cortex.

Wernicke: WUR-ni-keh

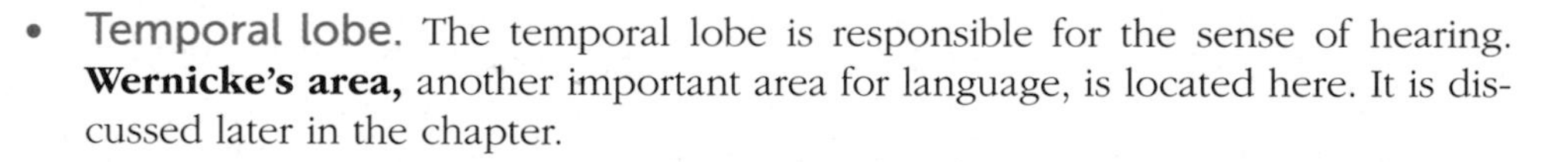

- **Temporal lobe.** The temporal lobe is responsible for the sense of hearing. **Wernicke's area,** another important area for language, is located here. It is discussed later in the chapter.
- **Occipital lobe.** The occipital lobe is responsible for vision. It is remarkable to consider that the receptors for vision are located at the front of the head (in the eye), but the input received is interpreted at the back of the brain. You will learn how this is done in the next chapter.
- **Insula.** This small lobe is located deep in the cerebrum, and it can be seen only when part of the cerebrum (temporal lobe) is retracted. Not much is known about the function of this lobe.

In addition to having five lobes, the cerebrum contains large parts of the **limbic system.** There is no defined anterior boundary for the ring of structures in the limbic system, as they make up parts of several of the lobes. The limbic system includes structures important for memory and learning **(hippocampus)** and others for emotions **(amygdala).** See **Figure 6.10**. Parts of the limbic system also involve the diencephalon, the subdivision of the brain you will study next.

Diencephalon The diencephalon has two major components, the thalamus and the hypothalamus. Each is described below and shown in **Figure 6.8e**.

study hint

When you are looking at a midsagittal view of the brain, such as Figure 6.8e, the diencephalon looks like the head of a duck. Upon closer inspection, seeing that the diencephalon has two components, imagine that the thalamus is the head of the duck and the hypothalamus is its beak.

- **Thalamus.** The thalamus serves as a switching station for incoming sensory messages except those for smell. It directs the sensory messages to the appropriate lobe of the cerebrum.

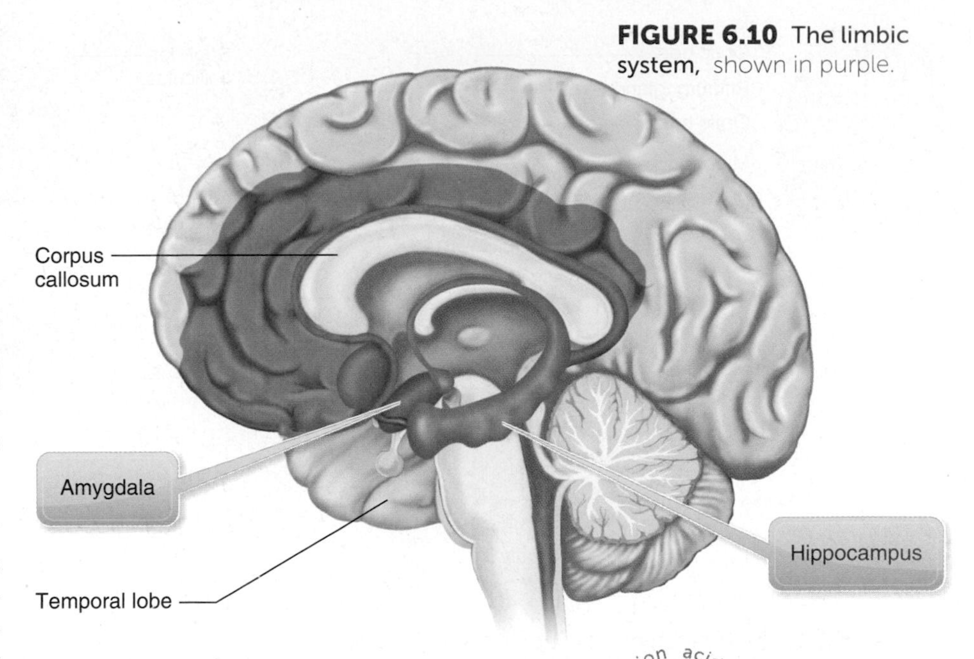

FIGURE 6.10 The limbic system, shown in purple.

- **Hypothalamus.** The hypothalamus is very important in maintaining homeostasis. It monitors the internal environment and has several functions, including the following:
    - **Temperature regulation.** The hypothalamus monitors body temperature, and it may stimulate an increase in sweat production to cool the body or promote muscle action (shivering) to produce heat to warm the body. These are two more examples of negative-feedback mechanisms.

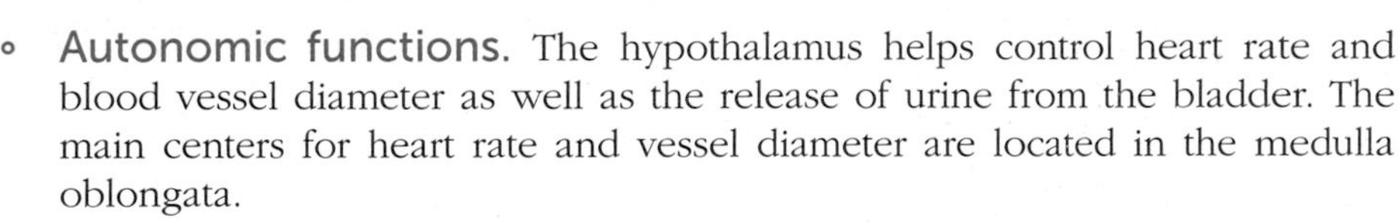

    - **Autonomic functions.** The hypothalamus helps control heart rate and blood vessel diameter as well as the release of urine from the bladder. The main centers for heart rate and vessel diameter are located in the medulla oblongata.
    - **Endocrine functions.** The hypothalamus produces the hormones ADH and oxytocin as well as several releasing hormones that regulate hormone production of the pituitary gland (covered in the endocrine system chapter).
    - **Food and water intake.** The hypothalamus monitors blood glucose and amino acid levels, and it is responsible for the sensation of hunger if the levels are below homeostasis. The hypothalamus also monitors the concentration of the blood, and it is responsible for the sensation of thirst in cases of dehydration.

    - **Sexual development.** The hypothalamus stimulates sexual development and arousal.

Neurons connect the hypothalamus to the posterior pituitary through the infundibulum, a stalk extending from the tip of the hypothalamus. You will learn more about this in the endocrine system chapter.

Brainstem The brainstem is composed of four parts: the medulla oblongata, pons, midbrain, and reticular formation. See **Figures 6.8e** and **6.11**.

- **Medulla oblongata.** The medulla oblongata is the most inferior section of the brainstem. All ascending sensory messages pass through the medulla oblongata on their way from the spinal cord to the thalamus. All descending motor messages also travel through the medulla oblongata on their way from the primary motor area in the frontal lobe of the cerebrum to the spinal cord and ultimately out to the skeletal muscles. The efferent motor messages pass through two anterior raised areas on the medulla oblongata, called the **pyramids.** The motor messages cross sides at the pyramids, so messages coming from the right frontal lobe go to muscles on the left side of the body and motor messages sent by

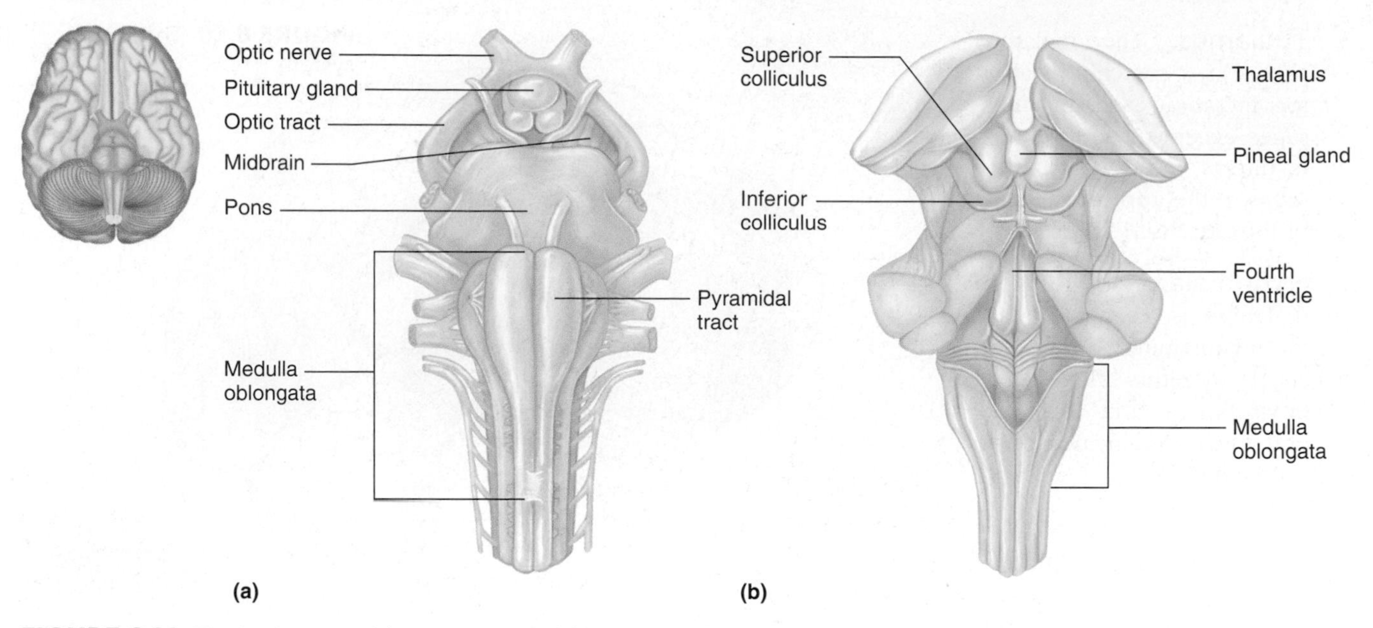

FIGURE 6.11 The brainstem: (a) anterior view, (b) posterior view.

the left frontal lobe go to muscles on the right side of the body. The medulla oblongata also contains centers to regulate heart rate, respiratory rate, and blood vessel diameter. Even the emetic center that controls vomiting is located in the medulla oblongata. You will continue to learn more about this section of the brainstem as it becomes relevant in other systems.

warning

It is important to realize that all of the brainstem is located in the cranial cavity; the brainstem does not extend beyond the cranial cavity into the neck.

clinical point

Encephalitis can cause swelling of the brain, which is located in a bony cavity with no room for expansion. If the swelling is severe enough, the medulla oblongata (the most inferior part of the brainstem) will be pressed toward the foramen magnum (an opening in the skull that cannot stretch). The medulla oblongata is wider than the foramen magnum. Given that the medulla oblongata contains the centers for heart rate, respiratory rate, and vessel diameter, you can see why this may be a lethal situation if these centers are damaged from the pressure caused by the swelling. Medication and/or removal of part of the skull may be necessary to reduce the swelling or give temporary room for expansion.

- **Pons.** The pons appears as a large bulge on the anterior surface of the brainstem between the medulla oblongata and the midbrain. Like the medulla oblongata, the pons has ascending tracts of neurons carrying sensory messages to the thalamus and descending motor tracts from the cerebrum. The pons serves as a bridge for motor tracts to the cerebellum, where they are fine-tuned before continuing on their way to the medulla oblongata and spinal cord.

- **Midbrain.** When viewed in a sagittal section, the midbrain appears as a hook at the superior end of the brainstem. See **Figure 6.8e**. The midbrain has four bulges, called **colliculi,** on its posterior surface. The two superior colliculi are important for visual reflexes, as in tracking the movement of an object or focusing on something seen off to the side. The two inferior colliculi are important for auditory reflexes, such as turning your head toward a sound or jumping at the sound of a loud noise. Both colliculi direct the sensory messages on to the thalamus.
- **Reticular formation.** The reticular formation is composed of groups of cell bodies (called **nuclei** in the CNS, **ganglia** in the PNS) scattered throughout the brainstem. The reticular formation is important for arousal, as it determines whether sensory messages will be consciously noticed by the cerebrum. It is also responsible for sleep-wake cycles. If the reticular formation is not working, a coma results.

clinical point

Many drugs such as barbiturates, benzodiazepines, and opiates depress the CNS by affecting the reticular formation. The effects can range from mild calming to sleep (sedation) to loss of sensory sensitivity (anesthesia). On the other hand, smelling salts stimulate the reticular formation.

Cerebellum The cerebellum is the last subdivision of the brain that you need to cover before moving on to the spinal cord. The cerebellum is located posteriorly in the brain, inferior to the occipital lobes of the cerebrum. As you can see in **Figure 6.12**, the cerebellum has tracts of white matter, called the **arbor vitae,** which branch like a tree. The cerebellum receives sensory messages concerning the position of limbs, muscles, and joints. It uses this information to fine-tune efferent skeletal muscle messages to coordinate position, balance, and movement. The effect is smooth, coordinated movement, such as touching the end of your nose with your finger while your eyes are closed. People who are proficient at keyboarding—either typing or playing piano—have **reflexive memory.** When they were first learning to keyboard, it may have been necessary for them to watch their fingers move to each key, but by now their cerebellum has fine-tuned messages so often that keyboarding seems effortless.

clinical point

The normally smooth, coordinated movements for which the cerebellum is responsible are affected by alcohol consumption. Alcohol can impair cerebellar function. A sobriety test may include walking a straight line, alternately touching the right and left index finger to the nose while the eyes are closed, or standing on one foot. All of these activities are difficult to perform smoothly if the cerebellum is impaired.

At this point, you have finished a very basic overview of the human brain. It is important to know that research is constantly being done to broaden understanding of this very complicated structure. We focus next on the other division of the central nervous system, the spinal cord.

FIGURE 6.12
The cerebellum.

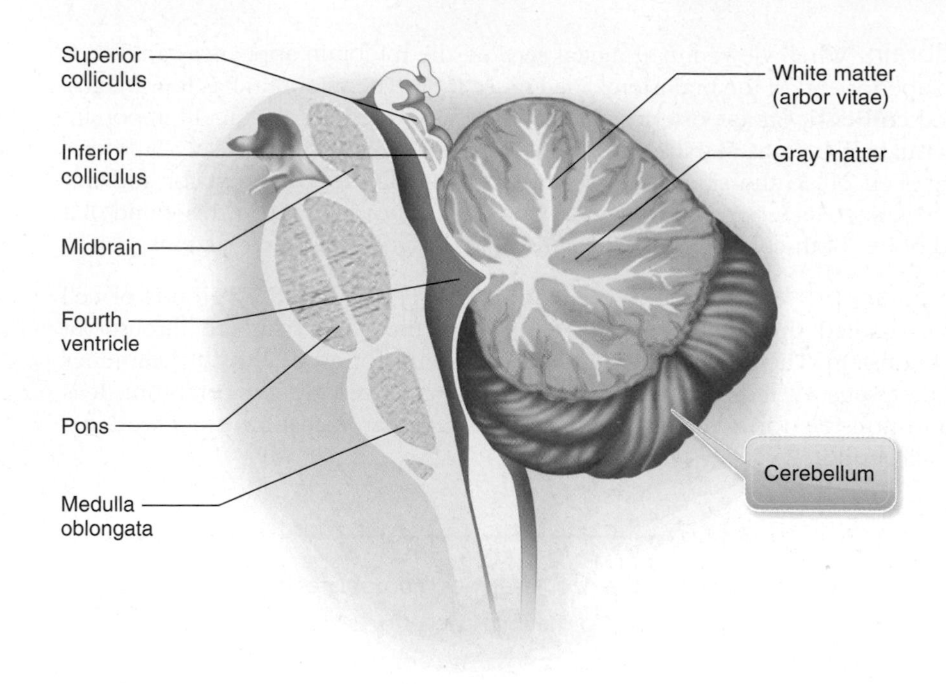

6.9 learning **outcome**

Describe the spinal cord.

Spinal Cord The spinal cord is a solid cylindrical structure in the vertebral cavity that extends from the foramen magnum to the inferior margin of the first lumbar vertebra. From there, a bundle of nerve roots, called the **cauda equina** ("horse's tail") extends from L1 to S5 in the vertebral cavity. See **Figure 6.13**. The spinal cord is enlarged in the cervical and lumbar regions to accommodate the number of nerve fibers going to and from the limbs. Thirty-one pairs of spinal nerves attach to the spinal cord between vertebrae. You will learn more about these nerves when you cover the peripheral nervous system.

clinical point

Meningitis is a serious inflammation of the meninges caused by viruses or bacteria often acquired through a respiratory, throat, or ear infection. A test, called a **lumbar puncture,** can be done to look for the presence of a pathogen in the CSF. This test involves inserting a needle through the dura mater and arachnoid mater to access the CSF in the subarachnoid space. The test is performed in the lumbar region below the end of the cord to reduce the risk of accidental damage to the cord. The strands of the cauda equina in this region are bathed in CSF. Imagine trying to stab a single piece of wet spaghetti with a fork. Because the spaghetti is wet and slippery, it tends to slide away as the fork approaches. The intent is not to stab the spinal cord or nerves in a lumbar puncture. However, if the needle does go too far, the strands of the cauda equina (bathed in CSF) tend to move out of the way, reducing the risk of injury.

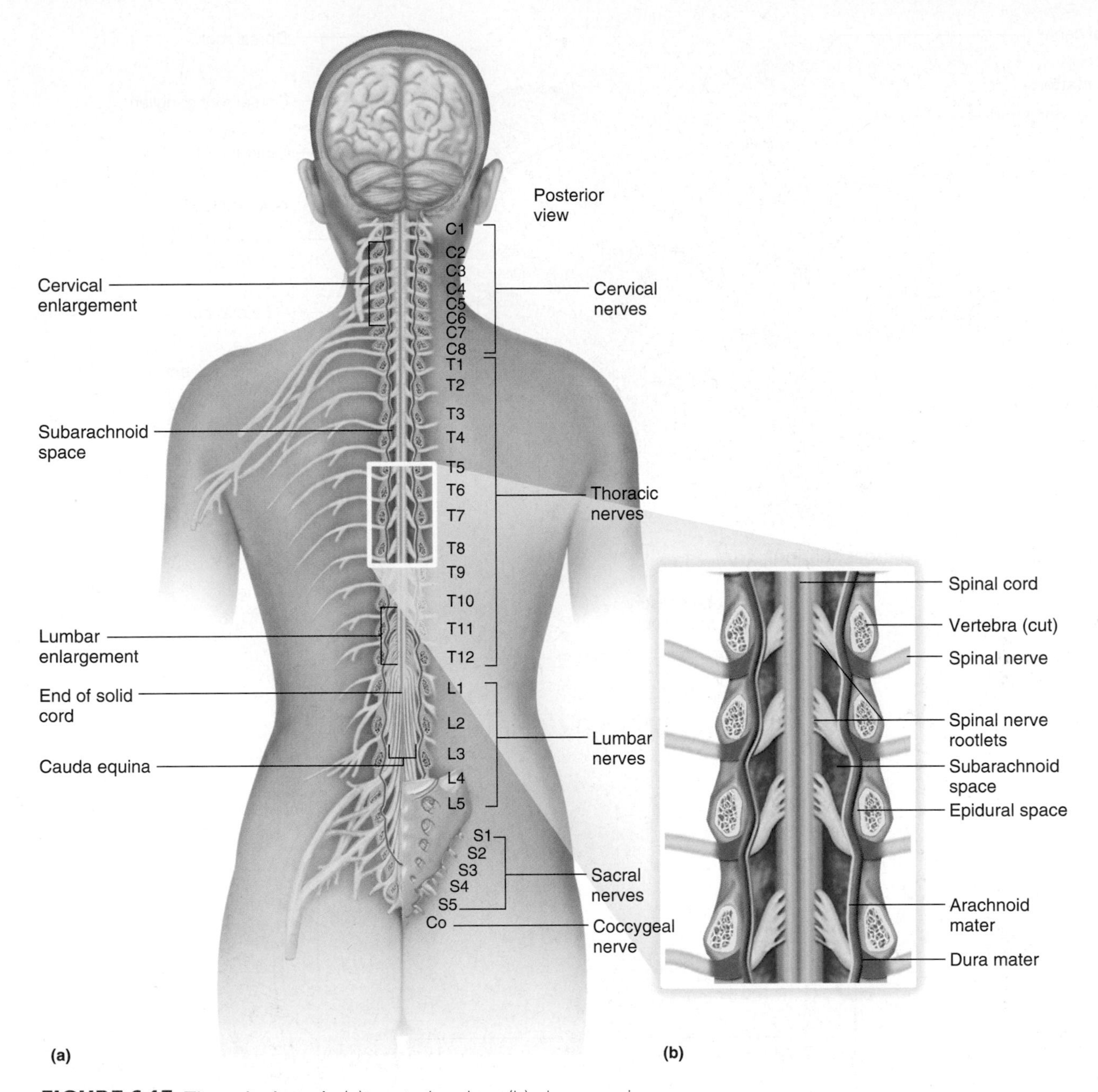

FIGURE 6.13 The spinal cord: (a) posterior view, (b) close-up view.

As you can see in **Figure 6.14**, a cross section of the cord shows both gray and white matter. The gray matter is in the center of the cord and arranged in an "H," the points of which are called **horns.** The gray matter is composed of dendrites, cell bodies, and short unmyelinated neurons **(interneurons).** These interneurons synapse with other neurons in the spinal cord to carry a message from the cord up to the brain or out to the body. You will see them again later when you study reflexes. The white matter of the spinal cord contains myelinated axons of neurons arranged in **columns.** Axons in ascending columns carry messages to the brain, while axons in descending columns carry messages away from the brain. The axons are grouped together in tracts with similar functions. For instance, **Figure 6.15a** shows how a sensory (primary) neuron from the skin goes to the spinal cord. There, it synapses with an interneuron, so its message can be eventually carried to the brain on myelinated axons in the lateral column on the left side of the cord. Descending tracts from the cerebral cortex carry motor messages on myelinated axons in the anterior and lateral columns of the cord. See **Figure 6.15b**.

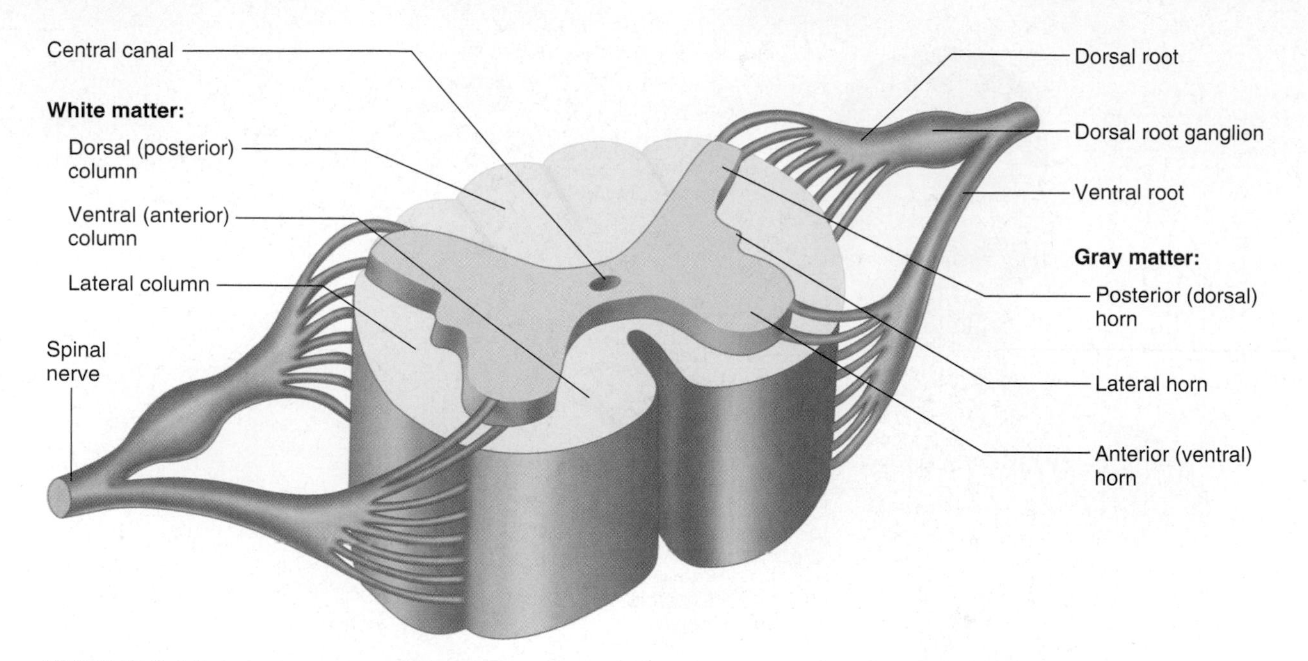

FIGURE 6.14 Spinal cord cross section.

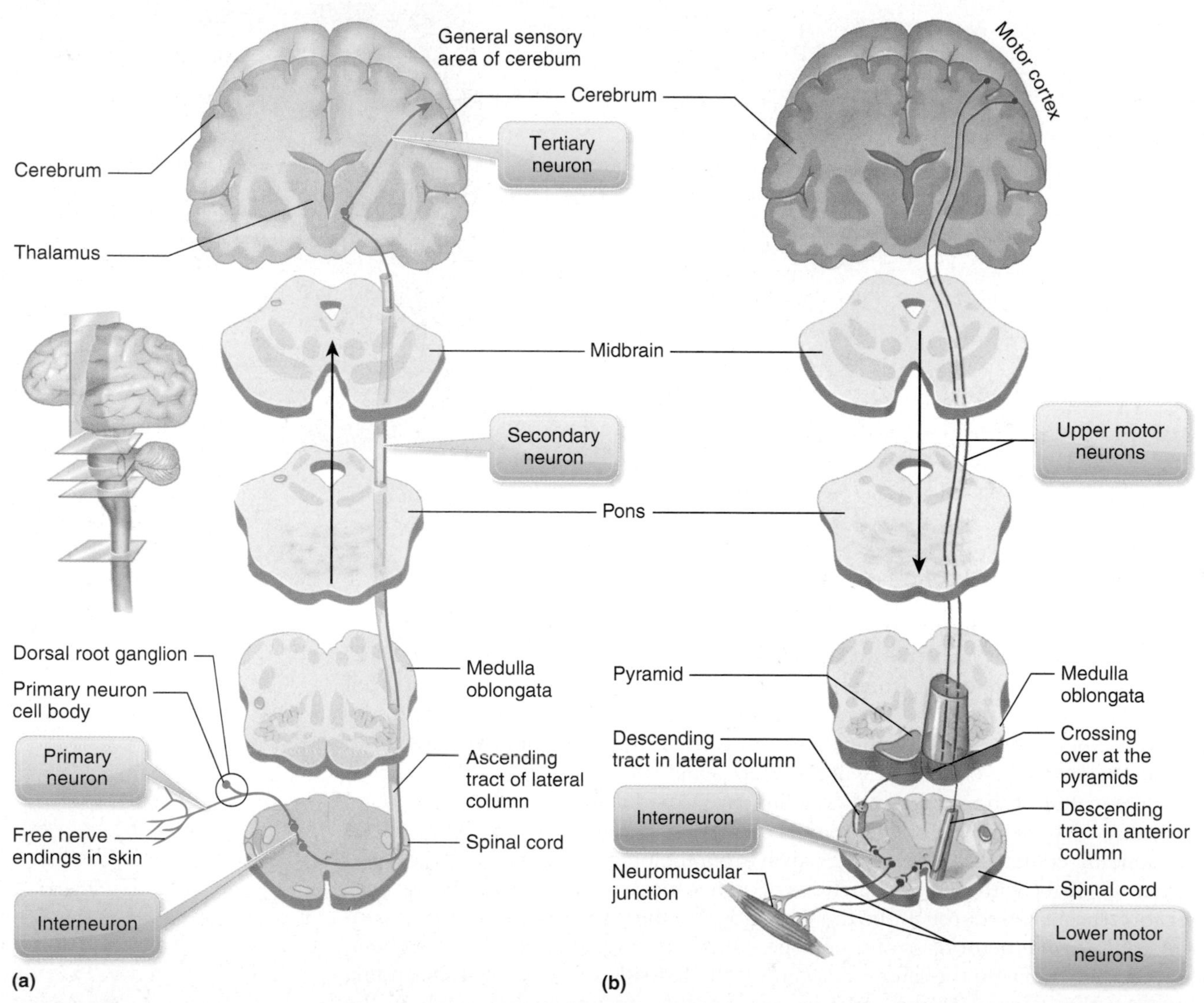

FIGURE 6.15 Ascending and descending tracts: (a) ascending tracts from skin receptors travel through the spinal cord to the cerebral cortex, (b) descending tracts from the cerebral cortex cross over at the medulla oblongata and travel to the spinal cord and out to muscles.

Now that you have studied the organization of the central nervous system, you are ready to explore the nerves of the peripheral nervous system. The PNS nerves carry messages to and from the CNS. To tackle this system, you can begin with the basic anatomy of a nerve.

Anatomy of the Peripheral Nervous System

6.10 learning **outcome**

Describe the anatomy of a nerve and its connective tissues.

Anatomy of a Nerve A nerve is a bundle of nerve fibers with an arrangement and connective tissues similar to those of the muscle fibers of a muscle. See **Figure 6.16**. **Endoneurium** is the connective tissue surrounding an axon of an individual neuron. Axons are arranged in bundles called *fascicles* that are surrounded by

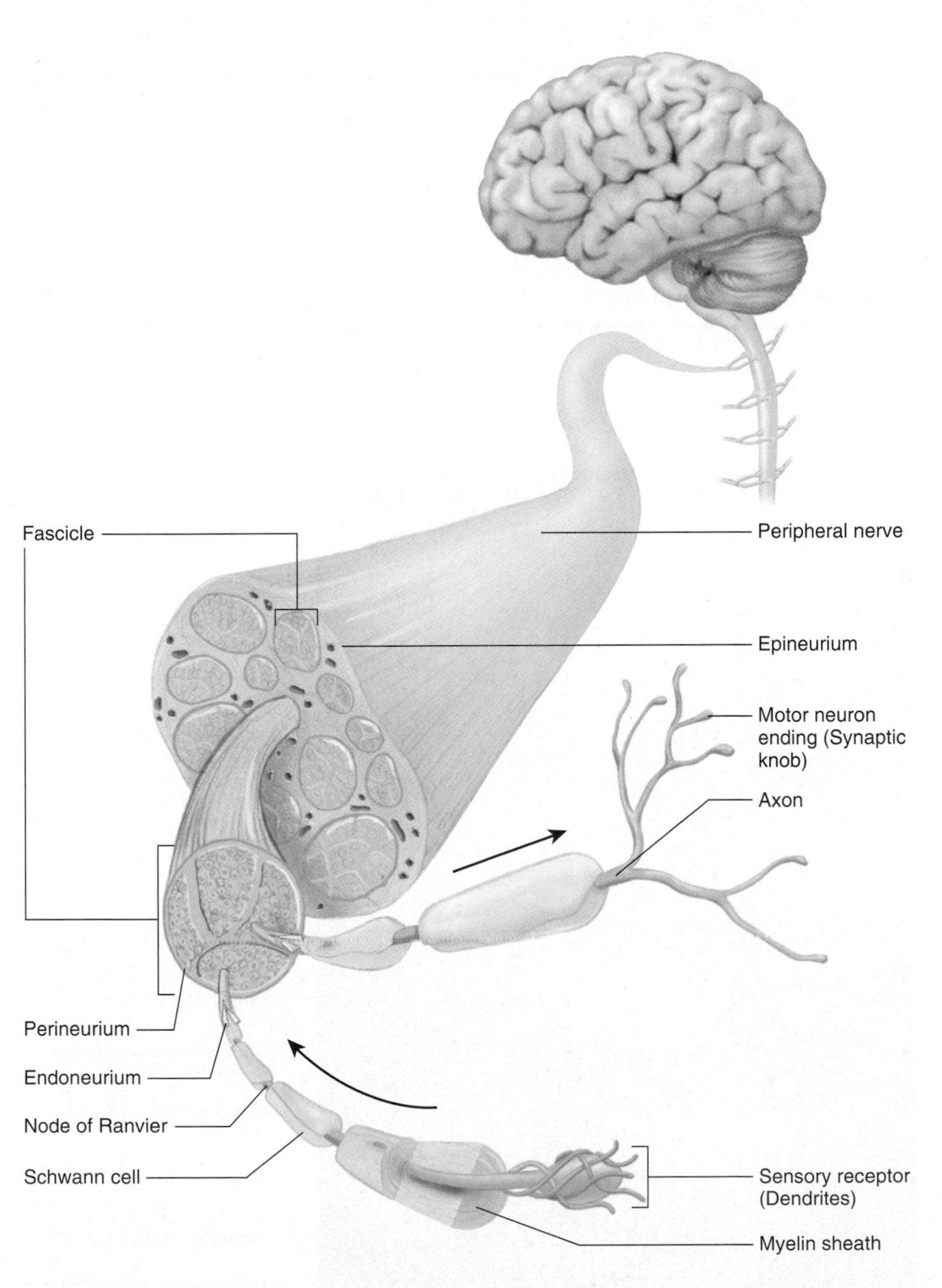

FIGURE 6.16 The anatomy of a nerve: a nerve with sensory and motor neurons. Arrows show the direction of nerve signals.

perineurium. The fascicles are bundled to form a nerve, which is surrounded by **epineurium.** An individual nerve may contain all afferent axons and be called *sensory,* all efferent axons and be called *motor,* or both sensory and motor neuron axons carrying messages in both directions. Two categories of nerves based upon their connection to the CNS are the cranial nerves and the spinal nerves.

6.11 learning outcome

List the cranial nerves in order, stating their function and whether they are sensory, motor, or both.

Cranial Nerves Cranial nerves connect directly to the brain. Their messages do not go through the spinal cord. The 12 pairs of cranial nerves are numbered in the order that they come off the inferior surface of the brain. See **Figure 6.17** and Table 6.4.

study **hint**

Mnemonic devices can help you remember the cranial nerves. Here is one for remembering the names in order: Look at the first letter of each name—OOOTTAFAGVAH. Imagine the three O's as pills. The rest of the letters then represent "Take Three Aspirin For A Giant, Very Awful, Headache." Here is one for the type of message carried by nerves, in order, given S = sensory, M = motor, and B = both: "Some Say Marry Money, But My Brother Says, Bad Business Marry Money."

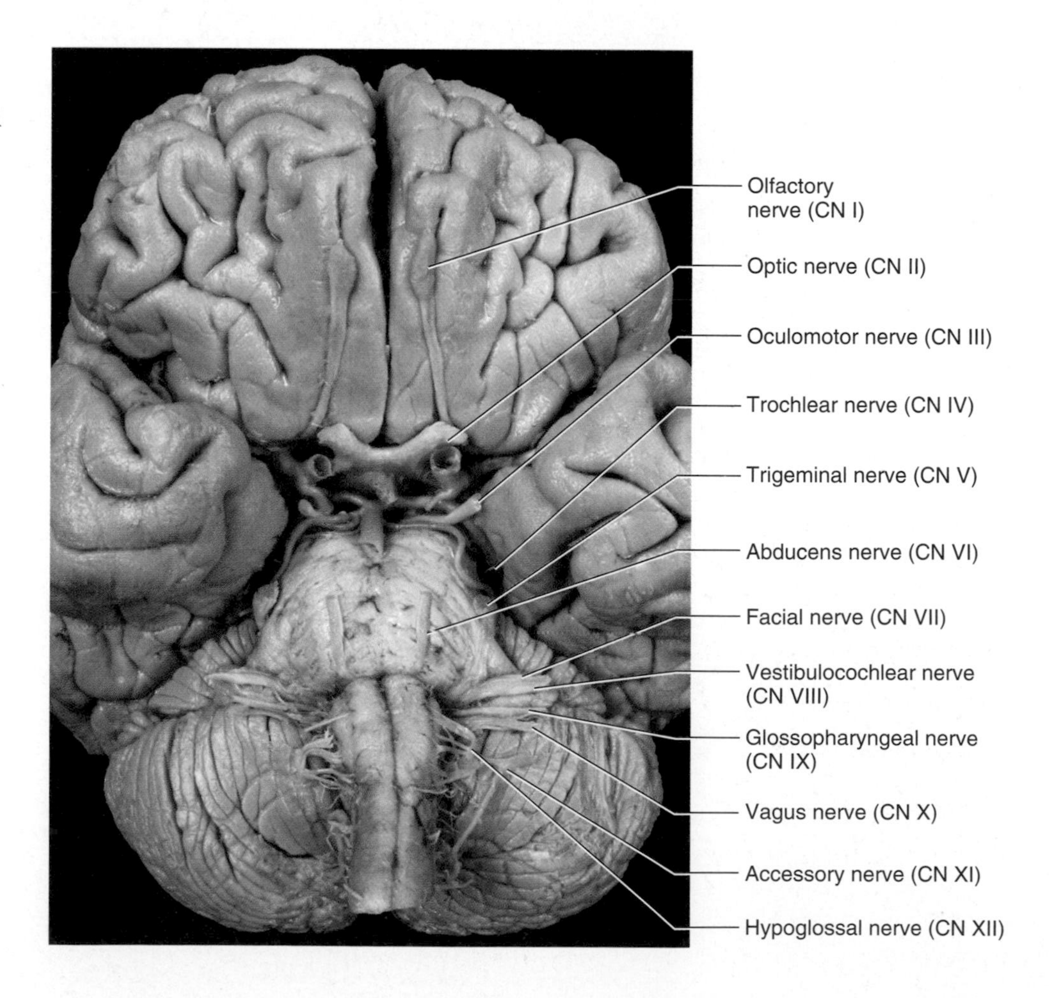

FIGURE 6.17 Cranial nerves: Inferior view of cadaver brain showing cranial nerves.

TABLE 6.4 Cranial nerves

Name	Type of messages: sensory, motor, or both	Function
I. Olfactory nerve	S	Sensory for smell
II. Optic nerve	S	Sensory for vision
III. Oculomotor nerve	M	Motor for eye movement
IV. Trochlear nerve	M	Motor for eye movement
V. Trigeminal nerve	B	Sensory for pain, touch, and temperature for the eye and lower and upper jaws Motor for muscles for chewing
VI. Abducens nerve	M	Motor for eye movement.
VII. Facial nerve	B	Sensory for taste Motor for facial expression
VIII. Auditory (vestibulocochlear) nerve	S	Sensory for hearing and equilibrium
IX. Glossopharyngeal nerve	B	Sensory for taste Motor for swallowing
X. Vagus nerve	B	Sensory and motor for organs in the thoracic and abdominal cavities Motor for larynx.
XI. Accessory nerve	M	Motor for the trapezius, sternocleidomastoid, and muscles of the larynx
XII. Hypoglossal nerve	M	Motor for the tongue

spot check 4 Which cranial nerve could be assessed by each of the following: asking the patient to smile, asking the patient to stick out her tongue, and asking the patient to move her head from side to side?

spot check 5 Which cranial nerves would contain bipolar neurons?

6.12 learning outcome

Describe the attachment of nerves to the spinal cord.

Spinal Nerves There are 31 pairs of spinal nerves that connect to the spinal cord. These nerves are numbered according to where they attach to the cord on the vertebral column: C1–8, T1–12, L1–5, S1–5, and Co (coccygeal nerve). See **Figure 6.18**. All spinal nerves carry sensory and motor messages, so they are composed of both unipolar and multipolar neuron axons (bipolar neuron axons are only found in some cranial nerves). Each spinal nerve splits into two nerve roots as it approaches the cord: a dorsal root and a ventral root. Notice in **Figure 6.18b** that the dorsal root has a bulge whereas the ventral root does not. The bulge is a ganglion (group of cell bodies) of the unipolar neurons. If you recall from earlier in the chapter, unipolar neurons have their cell bodies pushed off to the side. So the dorsal root carries afferent (sensory) messages, while the ventral root is composed of multipolar neuron axons that carry efferent (motor) messages.

study hint

How can you keep the direction of messages in the roots straight? Think of naming the spinal cord DAVE ("Dorsal is Afferent; Ventral is Efferent").

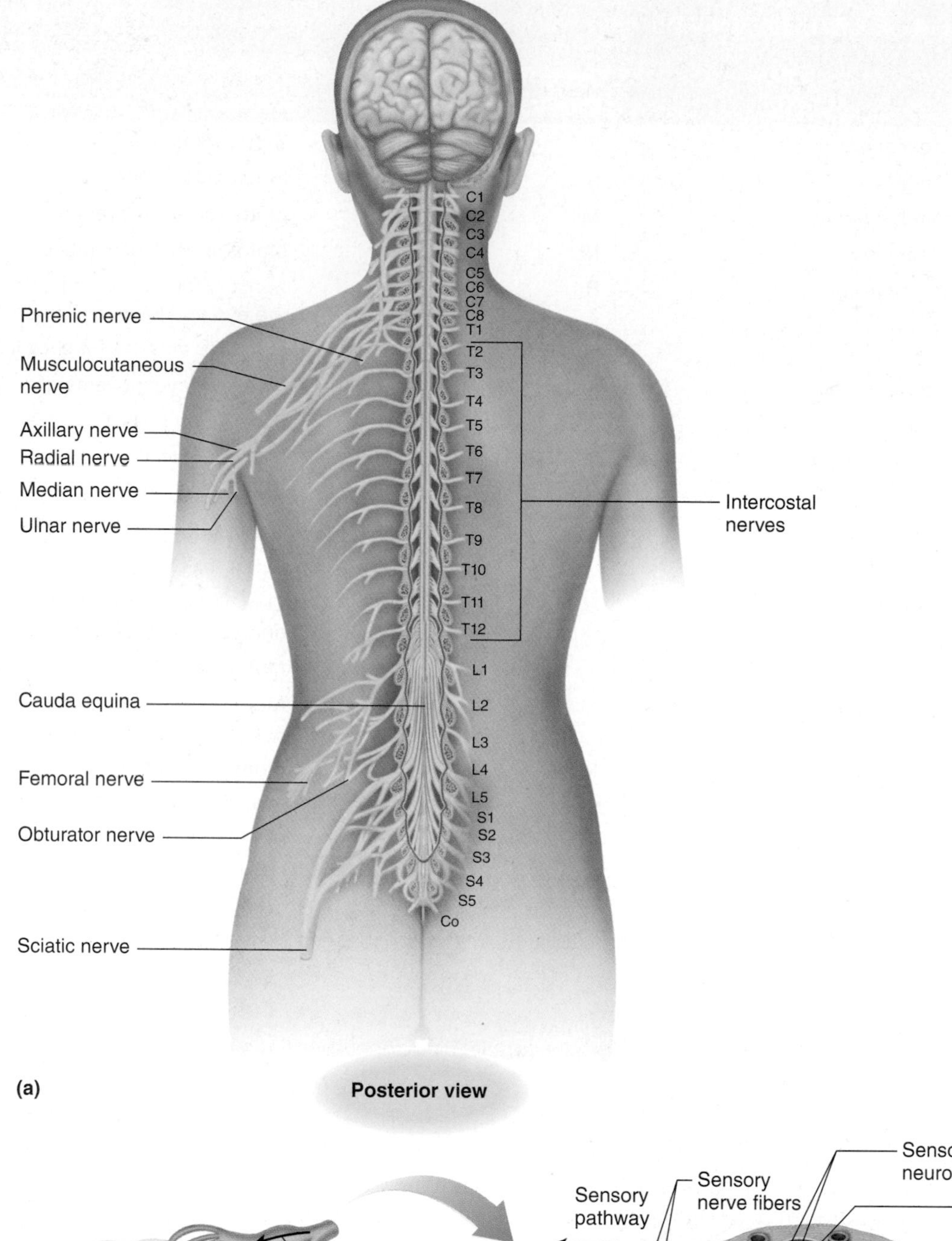

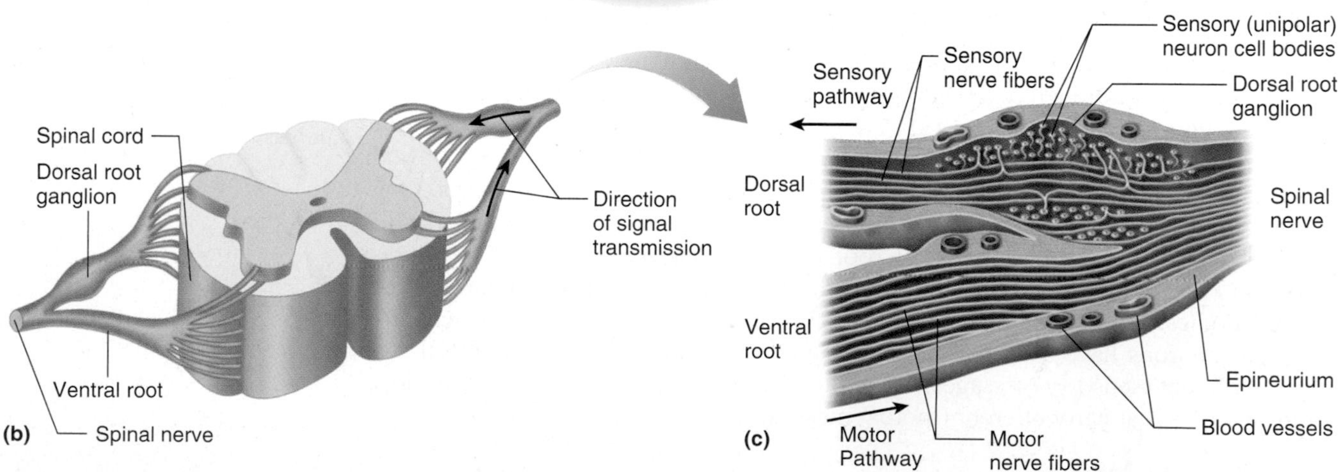

FIGURE 6.18 Spinal nerves: (a) spinal nerves, (b) attachment of spinal nerves to the spinal cord, (c) close-up of dorsal and ventral roots.

Each sensory nerve is responsible in part for carrying messages from specific areas of the skin. These areas are mapped in **dermatomes**. See **Figure 6.19**. Numbness in any given dermatome indicates which spinal nerve is involved.

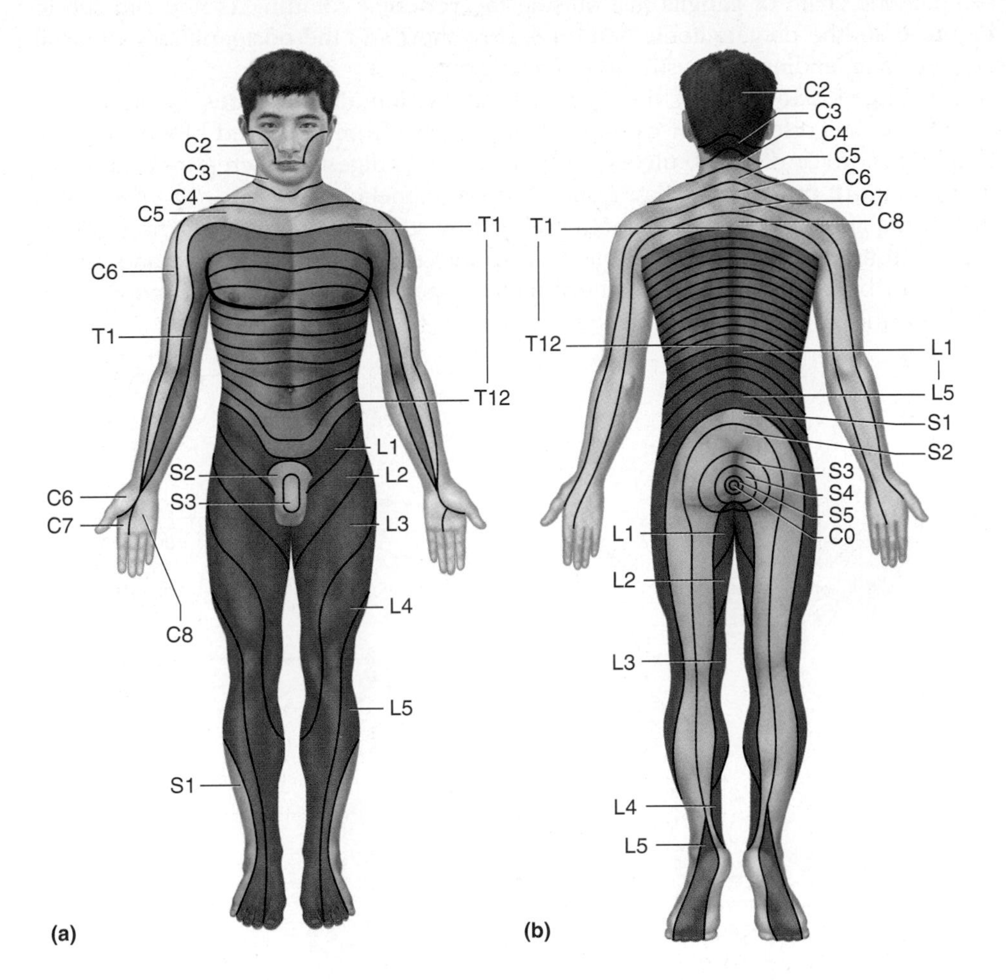

FIGURE 6.19 Dermatomes (a) anterior, (b) posterior.

6.13 learning **outcome**

Compare the parasympathetic and sympathetic divisions of the autonomic nervous system in terms of anatomy and function.

Autonomic Nervous System As you may recall from the organizational chart in the beginning of the chapter (**Figure 6.2**), the autonomic nervous system carries efferent messages. These messages go out primarily to thoracic and abdominal viscera as well as the smooth muscle of blood vessel walls. These messages are involuntary, meaning they are not under conscious control.

The autonomic nervous system is further divided into two parts—the sympathetic and parasympathetic divisions—based on anatomy and function. In both divisions, two neurons are involved in carrying the efferent message from the spinal cord—a preganglionic neuron and a postganglionic neuron. The anatomical differences in these divisions are where they exit the CNS and the location of the ganglia. Each division uses different neurotransmitters to target the same tissue. For example, the parasympathetic division may slow down the heart rate by using acetylcholine as a neurotransmitter, while the sympathetic division may speed up the heart rate by using norepinephrine as a neurotransmitter.

Sympathetic division The sympathetic division carries messages to prepare the body for physical activity, sometimes called *fight or flight*. Efferent messages cause blood vessels to the heart and skeletal muscles to dilate to increase blood flow. The heart is signaled to beat faster. Meanwhile, this division sends other messages to decrease blood flow in blood vessels that deliver blood to the digestive organs. If you are preparing for fight or flight, it is more important to send blood to the heart and skeletal muscles than to the stomach and intestines.

In the sympathetic division, there is a chain of ganglia just outside and lateral to the vertebral column. A short preganglionic neuron leaves the spinal cord from the thoracic and lumbar regions. It synapses with the postganglionic neuron at the sympathetic chain of ganglia just outside the vertebral column. As you can see in **Figure 6.20**, the preganglionic neuron is very short and the postganglionic neuron is quite long, ending at the structure it stimulates.

Prolonged activation of the sympathetic division during times of stress can lead to various diseases. For example, heart disease from increased blood pressure (hypertension) or digestive ulcers due to changes in digestive secretions and motility (amount of movement) may result. The sympathetic division can even weaken the body's immune system (as another effect) by stimulating the production of hormones from the adrenal gland (glucocorticoids) so that more glucose and amino acids can be released for the fight-or-flight response. You will learn more about these hormones in the endocrine system chapter.

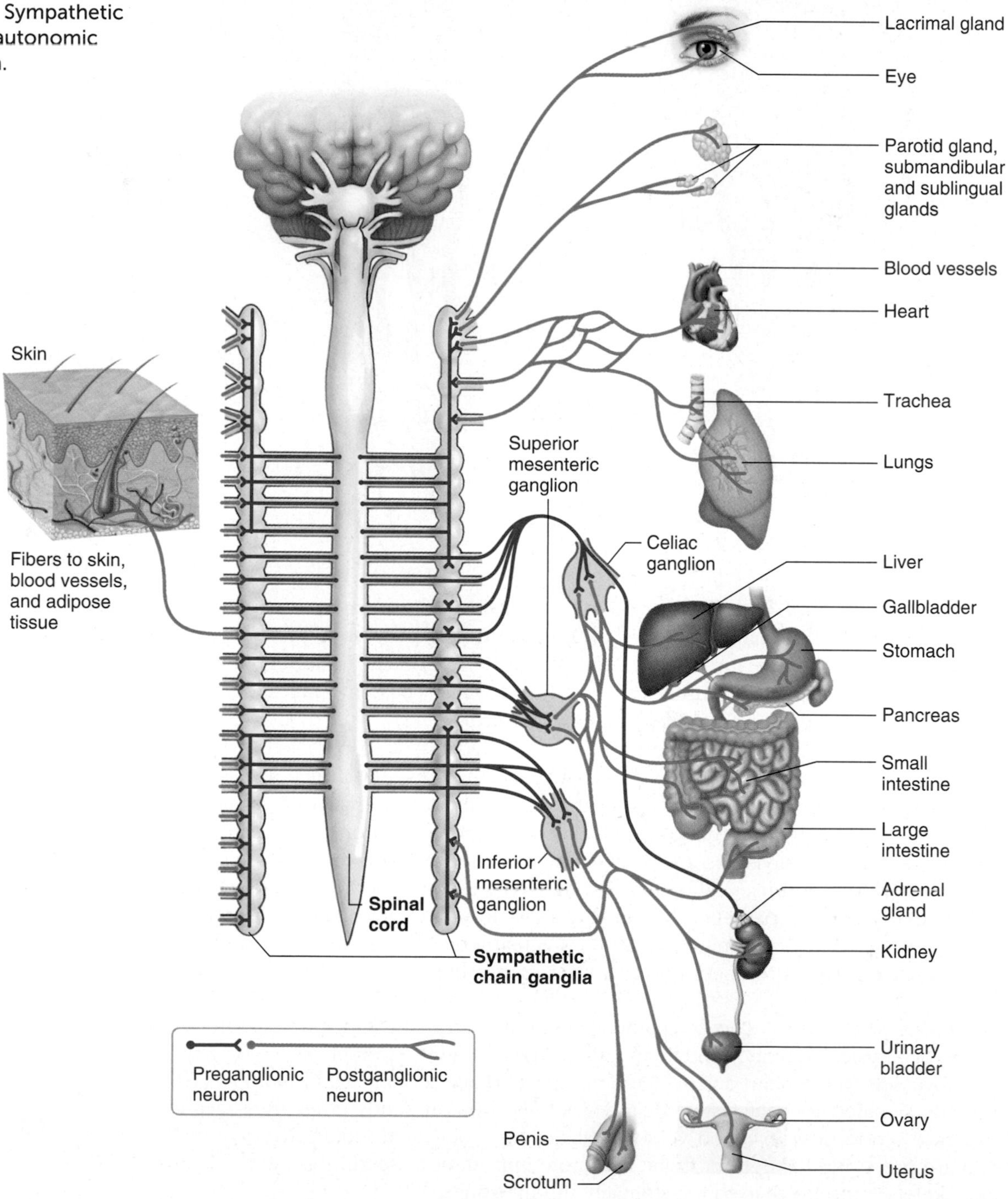

FIGURE 6.20 Sympathetic division of the autonomic nervous system.

spot check ❻ What parts of the preganglionic and postganglionic neurons form the synapse at the ganglia?

Parasympathetic division The parasympathetic division carries messages for everyday body maintenance functions such as digestion and elimination of waste. It has a calming effect on the body. You might think of this as a *rest-and-veg* effect. The parasympathetic division slows down the heart rate and increases blood flow to the digestive organs.

There is no neat chain of ganglia in the parasympathetic division as there is in the sympathetic division. In the parasympathetic division, the preganglionic neurons come off the brain and the sacral region of the spinal cord and synapse with postganglionic neurons in ganglia close to the structure they stimulate. See **Figure 6.21**.

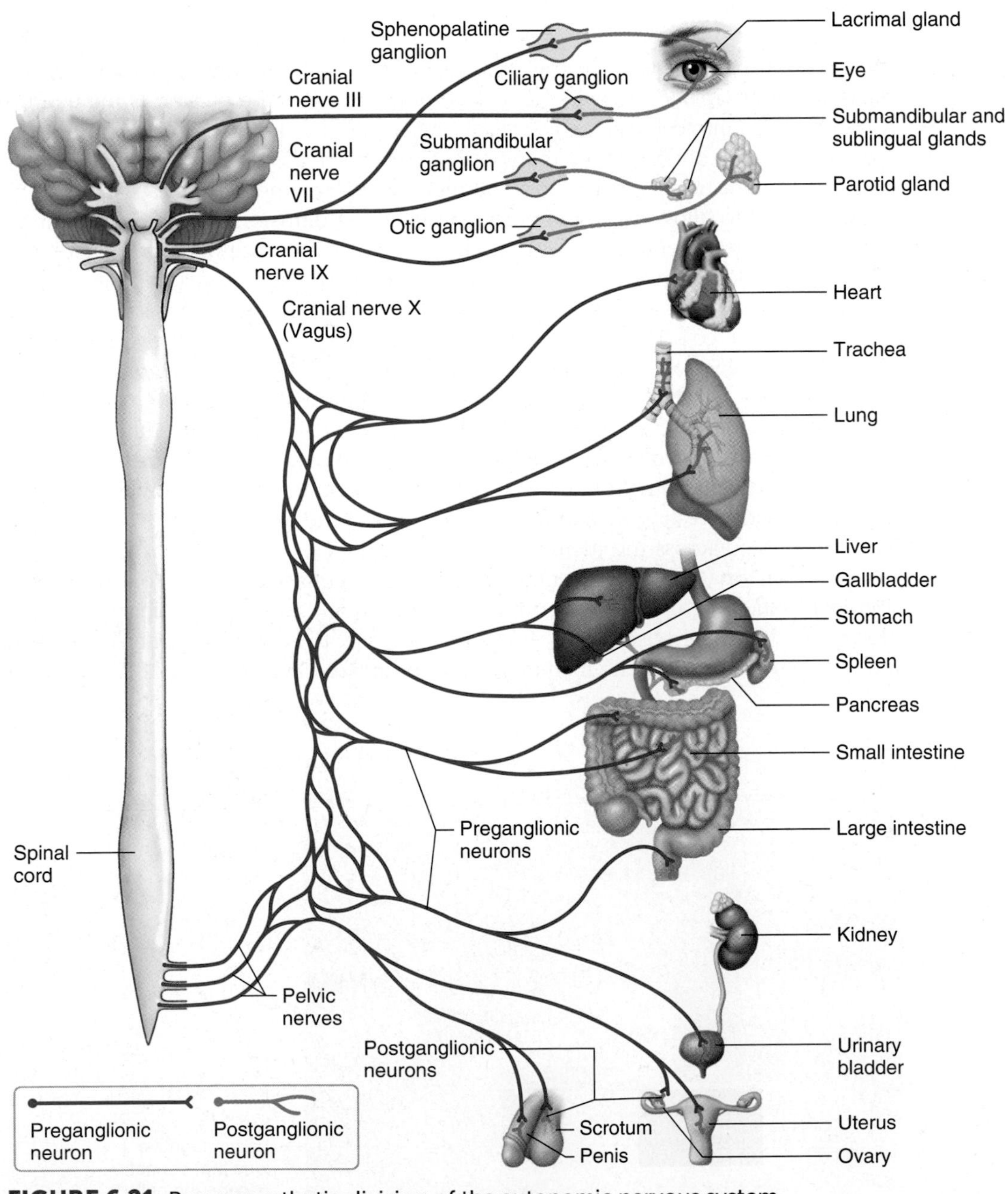

FIGURE 6.21 Parasympathetic division of the autonomic nervous system.

spot check ❼ Compare the length of the preganglionic and postganglionic neurons in the sympathetic and parasympathetic divisions.

You have covered all of the anatomy of the central and peripheral nervous systems. It is time to see how these structures carry out their functions. You will start with a basic nerve impulse.

6.14 learning **outcome**

Describe a resting membrane potential.

Physiology of the Nervous System

Nerve Impulses

As you already know from studying a muscle contraction at the molecular level, a nerve impulse starts as an electrical impulse that travels down a neuron and results in the release of a chemical neurotransmitter at the synaptic knob. To understand how an electrical impulse happens, you need to look at the membrane of a neuron at rest. See **Figure 6.22**.

In the figure, you see the phospholipid bilayer of the neuron membrane. This small section includes two channel proteins: one for sodium ions (Na^+) and one for potassium ions (K^+). *ECF* stands for "extracellular fluid" on the outside of the neuron. *ICF* stands for "intracellular fluid" (cytoplasm) inside the neuron. Notice that most of the Na^+ is on the outside of the membrane and most of the K^+ is on the inside of the cell. Na^+ cannot freely move across the membrane by simple diffusion. It needs an open channel to allow it to cross through facilitated diffusion. Na^+ would naturally diffuse across the membrane from an area of high concentration (outside the cell) to an area of low concentration (inside the cell) if it could, but it cannot because the channel for Na^+ is closed. K^+ can freely cross the membrane, but it is mostly on the inside of the cell. This is because K^+ is attracted to the large negative ions inside the cell that cannot cross the membrane by any means. The presence of many large negative ions inside the cell and many positive sodium ions outside the cell creates a difference in charge across the cell membrane. The outside is positive and the inside is negative, so the membrane is **polarized.** This situation is called a **resting membrane potential.** Electricity is the flow of charges, and there is a potential for a flow of charges with a resting membrane potential. Once the Na^+ channel is opened, the Na^+ can diffuse across the membrane into the cell. This occurs by facilitated diffusion, from an area of high concentration outside the cell to an area of low concentration inside the cell, until the concentrations are equal. This flow of charged ions is electricity. When the positive Na^+ flows into the cell, the difference in charge across the membrane changes. The membrane is **depolarized.** An opening of K^+ channels

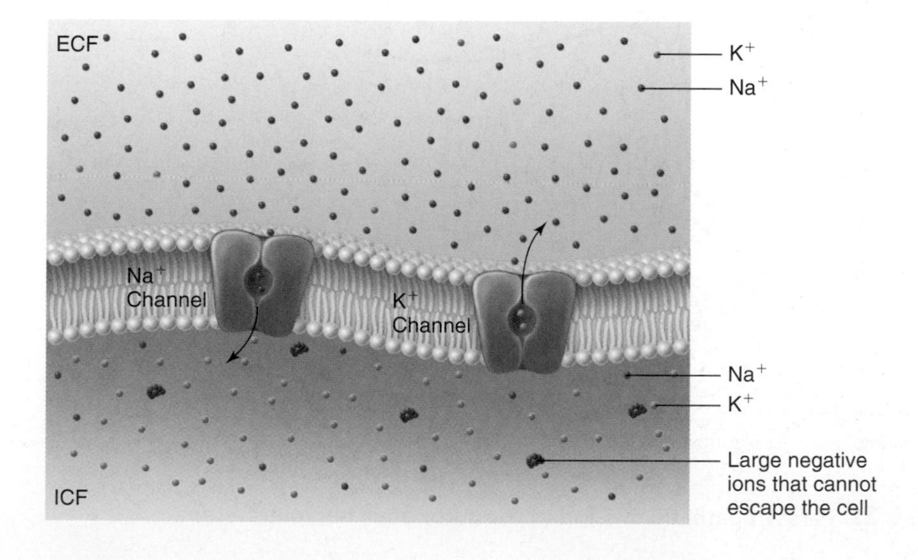

FIGURE 6.22 Resting membrane potential.

follows at a slightly slower rate to allow the flow of K^+ ions to the ECF, which then **repolarizes** the membrane.

The sodium-potassium pump is responsible for reestablishing and maintaining the resting membrane potential by pumping Na^+ out of the cell again through active transport (requiring energy). Running the sodium-potassium pump uses 70 percent of the energy needed by the nervous system. Once the resting membrane potential is restored, the neuron is ready to begin again with another nerve impulse.

What causes the Na^+ channel to open? The process begins at the dendrite of a neuron.

6.15 learning outcome

Compare and contrast a local potential and an action potential.

Local Potential A **local potential** is the flow of electricity begun by stimulating the dendrite of a neuron. It starts with the opening of an Na^+ channel on the membrane of a dendrite. A chemical, heat, light, or mechanical disturbance may cause the Na^+ channel to open. For example, light stimulates sensory neurons in the eye, heat stimulates sensory neurons in the skin, a chemical (like perfume) may stimulate the neurons in the nose, and pressure (caused by touching a table) may stimulate a sensory neuron for touch in the skin of your finger by opening the Na^+ channel on a dendrite.

When that gate opens, Na^+ rushes in and spreads in all directions. Some of it may flow on the inside of the membrane to the next Na^+ channel and open it from the inside. Na^+ rushes into that open channel too. Some of that Na^+ may flow on the inside of the neuron to the next Na^+ channel and open it, and so on, and so on. K^+ channels open following the Na^+ channels for repolarization. Meanwhile, the sodium-potassium pump actively transports the previous Na^+ out of the cell again to restore a resting membrane potential. In effect, this creates a wave of Na^+ moving in and out of the cell, along the dendrite toward the cell body, and on to the trigger zone of the neuron. See **Figure 6.23**.

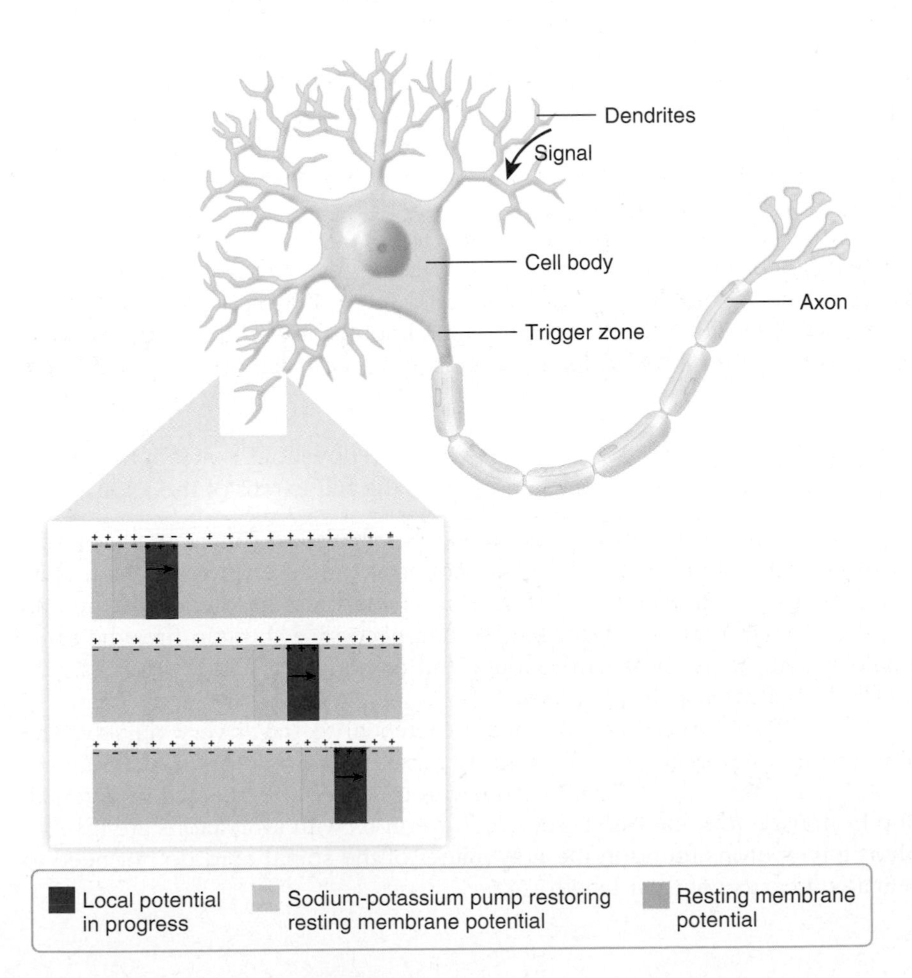

FIGURE 6.23 Local potential: conduction of a nerve impulse along a dendrite.

Local potentials have definite characteristics, which are outlined as follows:

1. The amount of stimulation in a local potential matters. It determines how much Na^+ enters the cell and how many Na^+ channels get opened. Therefore, a local potential is **graded.**

2. The effects of a local potential decrease with distance. Therefore, a local potential is also **decremental.**

3. Local potentials are also **reversible,** meaning that if the stimulation is stopped, the resting membrane potential is quickly restored.

4. In addition, local potentials can be **excitatory,** meaning they cause the neuron to send a signal, or they can be **inhibitory,** meaning they prevent a neuron from sending a signal. For example, a needle used for an injection mechanically disturbs the dendrite of a neuron of a pain receptor. This is excitatory because it results in a message being sent to the brain, and as a result you experience the pain of the needle stick. The chemicals in topical anesthetics are inhibitory. They bind to the dendrite of a pain receptor neuron but prevent a pain signal from being sent.

So far you have only looked at a nerve impulse traveling from a dendrite to the body of a neuron and possibly reaching the trigger zone. If adequate stimulation occurs so that enough channels are opened and the wave of depolarization goes all the way to the trigger zone of the neuron, then an action potential may or may not happen.

Action Potential An **action potential** is the flow of electricity along an axon of a neuron in one direction—from the trigger zone to the synaptic knob. It differs from a local potential in several ways, including:

1. An action potential is not graded. It has a threshold. If that minimal amount of stimulation is present at the trigger zone, the action potential happens. This is an all-or-nothing effect. A stronger stimulus at the trigger zone has no more effect than the threshold amount. If the local potential is subthreshold at the trigger zone, no action potential happens.

2. An action potential is not decremental. It does not decrease with distance. The action potential begun at the trigger zone of a motor neuron located in the spinal cord will be just as strong at the synaptic knob as it was at the trigger zone. This holds true even if the synaptic knob is at a flexor digitorum muscle 2 feet away.

3. An action potential is also not reversible. Once the threshold is met at the trigger zone, the action potential happens and it goes the full extent of the axon.

Trigger zone is a good term because action potentials are similar to shooting a gun. To shoot a gun, you need a minimal amount of pressure (threshold stimulus) on the trigger. It is an all-or-nothing effect. Unless you press hard enough, the gun does not fire. If you press the trigger harder than necessary, the gun fires the same as it would if you pressed just hard enough (all or nothing). If the gun fires, you cannot call the bullet back (irreversible).

Myelination of an axon allows the action potential to travel very quickly. The change in charge (depolarization followed by repolarization) across the membrane needs to happen only at the unmyelinated nodes of Ranvier. So the action potential can jump from node to node. See **Figure 6.24**. Neurons with long axons are myelinated. Short interneurons found in the gray matter of the spinal cord do not need to be myelinated because of their length.

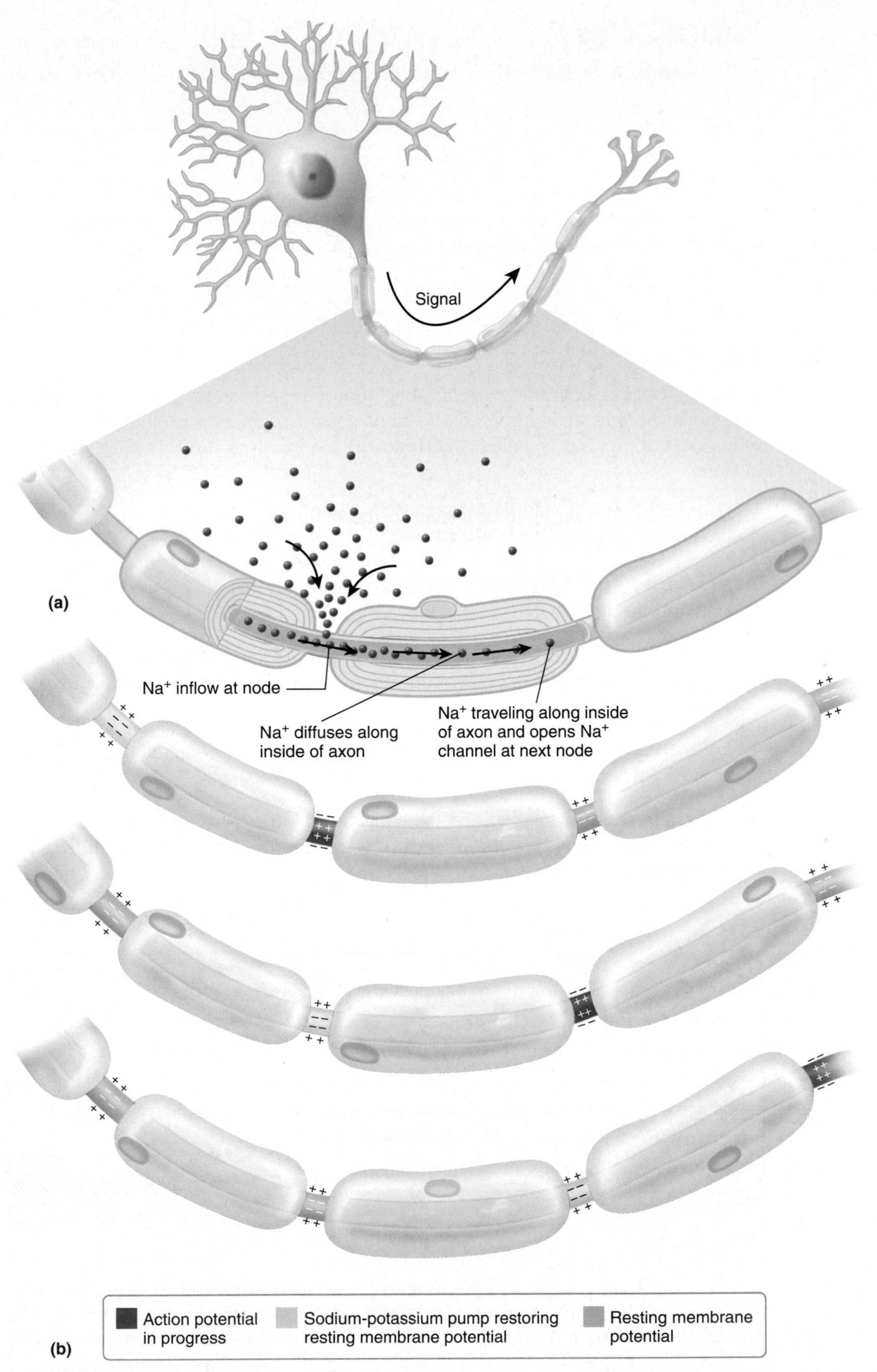

FIGURE 6.24 **Action potential:** (a) movement of Na^+ along an axon, (b) conduction of a nerve impulse along a myelinated axon.

spot check 8 Sound waves cause vibrations in the ear that mechanically disturb bipolar neurons. In terms of local and action potentials, why might you be able to hear a sound that your friend standing next to you cannot?

Now that you have covered the anatomy of the nervous system and the physiology of a nerve impulse, you are ready to see how they are used together in the body in reflexes, memory, and language in carrying out the functions of this system.

6.16 learning outcome

Describe a specific reflex and list the components of its reflex arc.

Reflexes

A **reflex** is an involuntary, predictable, motor response to a stimulus without conscious thought. It is a very fast response, used often as a protective device. A reflex occurs in what is called a **reflex arc** that involves the following anatomy, in order:

1. Receptor: the dendrite of a neuron receiving the stimulus (a chemical, heat, light, or mechanical disturbance).
2. Afferent neuron (sensory): a neuron that has an action potential carrying the signal to the CNS.
3. Integrating center: either the brain or spinal cord, where the signal is received from the afferent neuron and conducted to a motor neuron. This may or may not require an interneuron.
4. Efferent neuron (motor): a neuron that has an action potential carrying a signal away from the CNS.
5. Effector: the structure causing the effect. If this structure is skeletal muscle, it is called a **somatic reflex.** If the effector is a gland or smooth muscle, it is called an **autonomic reflex.**

clinical point

Testing reflexes is an important diagnostic tool for evaluating the condition of the nervous system. Exaggerated, decreased, or absent responses indicate damage to the nervous system from disease or injury. For example, if the spinal cord is damaged, testing reflexes may help determine on what level of the cord the damage occurred. Reflex motor responses above the damage may be normal, while the reflex motor responses below the damage may be decreased or absent.

An example of a somatic reflex is shown in **Figure 6.25**. In this case, the reflex arc begins with the mechanical disturbance of the dendrite of a pain receptor in the skin of the foot. A local potential is generated along the afferent neuron. Notice

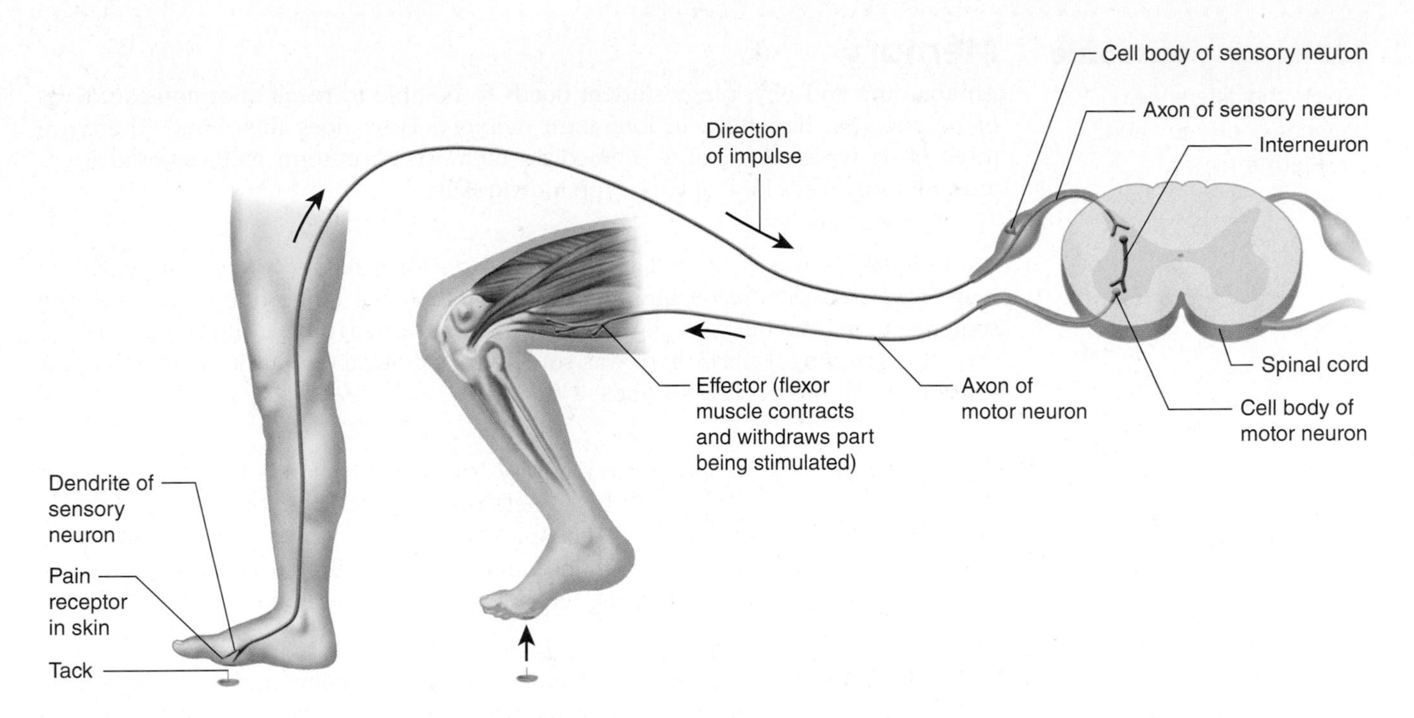

FIGURE 6.25 A reflex arc for a withdrawal reflex.

that this is a unipolar neuron whose body is in the dorsal root ganglion of the spinal cord. An action potential carries the signal to the gray area of the spinal cord (integrating center). Here, the synaptic knob of the afferent neuron synapses with a dendrite of a short, unmyelinated interneuron. A local potential is generated by the neurotransmitter released by the synaptic knob stimulating the dendrite of the interneuron. The local potential travels to the trigger zone of the interneuron and stimulates an action potential to run the short length of the axon. The synaptic knob of the interneuron synapses with the dendrite of an efferent motor neuron and releases a neurotransmitter. A local potential and then an action potential are generated in the motor neuron to carry the electrical impulse out the ventral root to the effector, which, in this case, is flexor muscles in the thigh. The involuntary, predictable, motor response to stepping on a tack is flexion resulting in the withdrawal of the foot and leg.

Not all reflexes use the spinal cord as the integration center. An example of an autonomic reflex is the involuntary, predictable contraction of the iris constricting the size of the pupil within the eye in response to a bright light. In this case, the receptor (dendrite) in the eye is stimulated by light. The afferent neuron is part of a cranial nerve, the integration center is the brain, and the efferent neuron is part of another cranial nerve. This is an autonomic reflex because the effector, the iris, is smooth muscle.

spot check 9 What nerve would contain the afferent neuron in a pupillary light reflex?

You have become familiar with reflexes, which are unconscious responses. Next, you will read about memory, which involves conscious thought.

6.17 learning outcome

Explain the difference between short-term and long-term memory.

Memory

An anatomy and physiology student needs to be able to recall enormous amounts of data stored, hopefully, in long-term memory. How does this work? There are three basic types of memory: immediate memory, short-term memory, and long-term memory. Let's look at each type individually.

Immediate Memory Immediate memory lasts for a few seconds. One example is keeping track of the beginning words in a sentence long enough to finish the sentence to get the full meaning. Another example is what it felt like to put on your shoe this morning. Unless there was something significant about it, memory of this sensation is gone in a few seconds.

Short-Term Memory Short-term memory lasts a few seconds to a few hours. It may be lost if you are distracted by something else, and it is somewhat limited to a few bits of information. Phone companies base the length of phone numbers on this. You may have looked up a phone number, repeated it several times, and still forgotten it by the time you found your phone to enter the number.

Long-Term Memory Long-term memory may last a lifetime and is not limited as to the amount of information it can hold. Each time a dendrite is stimulated, a local potential is generated. But a chemical change takes place inside the neuron as well. If the chemical change takes place often enough, the dendrite is stimulated to grow and make new and broader connections. As mentioned earlier, the more dendrites a neuron has, the more information it can process. The added connections form the memory. So a chemical change at the molecular level causes a physical change in the size and shape of the neuron.

spot check **10** On the basis of how memories are formed, what is the best way to study for an exam on the nervous system?

6.18 learning outcome

Differentiate between Broca's area and Wernicke's area in regard to their location and function in speech.

aphasia: ah-FAY-zee-ah

Language

Just as memory is an important part of human communication, so is language, especially in regard to the ability to retain and recall information that has been communicated verbally and/or in writing. The cerebrum plays a crucial role in both the interpretation of and the ability to produce language. As mentioned earlier in the discussion of the anatomy of the cerebrum, there are two areas in the cerebrum concerned with language: Wernicke's area in the temporal lobe and Broca's area in the frontal lobe. **Aphasia** is any language deficit resulting from damage to either Wernicke's or Broca's area.

Wernicke's Area This area is used to interpret incoming language. People who have had damage to this area may speak clearly but are unable to understand the language directed to them.

Broca's Area This area is used to find the words for outgoing language. If brain damage has occurred in this area, a person is perfectly capable of understanding incoming language but may not be able to *find the words* to respond.

Function of the Nervous System

6.19 learning outcome

Explain the function of the nervous system by writing a pathway for a sensory message sent to the brain to be processed for a motor response.

The function of the nervous system is fast, efficient communication of one body part with another. You have seen how this works for a muscle contraction, and you have seen how this works with a reflex. Now you will put more of the parts of the nervous system together by writing a pathway of nerve impulses that involves a sense and a reaction.

Pathways As you can see in **Figure 6.26**, Miriam is adding an ice cube to her glass of water. You can write a pathway for Miriam's touching the ice cube and saying "cold." You will start with the type of neuron receiving the stimulus and end the pathway with skeletal muscle messages being sent out from the cerebrum:

The pathway begins with a unipolar (sensory) neuron in Miriam's finger

↓

that carries the message to her spinal cord.

From the spinal cord, the message travels to the medulla oblongata,

to the pons,

to the midbrain,

to the thalamus, which acts as a switching station, and

to the parietal lobe for touch (first the general sensory area and then the association area).

Miriam now knows the ice feels cold, but her nerve impulses have to go to Broca's area in the frontal lobe to find the word *cold*.

From Broca's area the message goes to the premotor area in the frontal lobe to plan the skeletal muscle messages to be sent.

Finally, the messages travel from her premotor area to her primary motor area, which actually sends the skeletal muscle messages to say the word *cold*.

FIGURE 6.26 Miriam adds an ice cube to her glass of water.

So the pathway involving multiple neurons is unipolar neuron → spinal cord → medulla oblongata → pons → midbrain → thalamus → parietal lobe (general sensory area, association area) → frontal lobe (Broca's area, premotor area, primary motor area). This demonstrates how complicated, yet how fast and efficient, the nervous system is at communication.

spot check  Why does the pathway not include Wernicke's area in the temporal lobe?

6.20 learning outcome

Explain the nutritional requirements of the nervous system.

Nutritional Requirements of the Nervous System

For the nutritional requirements of the nervous system to be met, it is important throughout life to have adequate sodium and potassium in the diet. This helps to maintain resting membrane potentials. Sodium is found in table salt (NaCl) and most processed foods. The RDA for sodium is less than 2,300 mg for an adult. Having enough sodium in the diet is usually not a problem, as a typical American diet contains much more sodium than is needed. Too much sodium can lead to hypertension, which is discussed in the cardiovascular system chapter. The RDA for potassium is 4,700 mg. Good sources of potassium include red meat, poultry, fish, cereals, spinach, bananas, and apricots. It is also important that children have fat in their diet for proper myelination of developing neurons. See Appendix B for dietary guidelines.

6.21 learning outcome

Explain the effects of aging on the nervous system.

Effects of Aging on the Nervous System

Cognitive function—the ability to think and reason—increases rapidly in the young, remains relatively stable in adulthood, and declines with old age. The definition of old age varies from person to person, due to health, lifestyle, and genetics, but in general the following holds true:

- Short-term memory is affected relatively early.
- Verbal skills and vocabulary usually begin to decline around age 70.
- Intellectual performance may slow but remain high until about age 80.
- Reaction time slows due to a decrease in neuron efficiency.

Overall, the number of neurons in the brain decreases with age. This does not have to have a drastic effect because the brain has more neurons than it needs to function. You have also become aware that existing neurons can grow to make more and broader connections. Exercising the brain through reading and problem-solving activities can help minimize the effects of aging. Having a healthy lifestyle that ensures good cardiovascular health also helps brain function by making sure the brain has an adequate blood supply. This can be accomplished by maintaining a healthy diet, exercising, and not smoking.

We focus next on nervous system disorders that may or may not have a connection to the effects of aging.

6.22 learning outcome

Describe nervous system disorders.

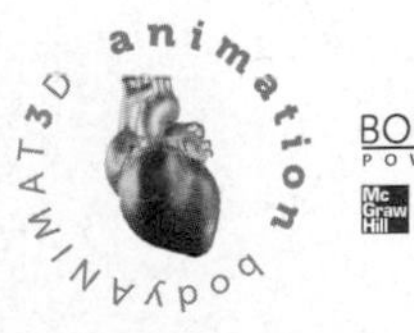

Nervous System Disorders

Cerebrovascular Accident

A cerebrovascular accident (CVA), commonly called a *stroke,* happens when part of the brain dies due to a lack of blood supply. This is commonly the result of a blocked artery, but it can also result from the rupture of an artery feeding part of the brain. The part of the brain past the rupture does not receive blood due to the bleed. The symptoms of a stroke occur suddenly and may include muscle weakness, paralysis, loss in sensation on one side of the body, confusion, vision problems, and difficulty with speech. It is important to determine the cause of the CVA before treating it. Medications to dissolve clots are desirable if the cause is a blockage but can be very harmful if the cause is a ruptured vessel in the brain.

Alzheimer's Disease

Alzheimer's disease is another disorder involving the nervous system. It is a progressive, irreversible disease of the brain that is characterized by dementia. Dementia is the loss of cognitive function, such as thinking, remembering, and reasoning. The symptoms of Alzheimer's disease usually first appear after age 60, but the disease process may have started 10 to 20 years earlier. The exact cause of Alzheimer's disease is unknown. It results in tangles of insoluble proteins within nerve cells and clumps of dead nerve cells, called *plaques*. Parts of the brain degenerate, causing the death of neurons and reducing the responsiveness of other neurons to neurotransmitters. An exact diagnosis is not possible until an autopsy is completed upon death. An alternative explanation for the dementia is always considered first before an Alzheimer's disease diagnosis is made. At this time there is no definitive treatment for Alzheimer's. However, medications to increase neuron sensitivity to neurotransmitters can help alleviate the severity of the symptoms to a point.

putting the pieces **together**

The Nervous System

Integumentary system

Has receptors for the general senses.

Innervates smooth muscle of blood vessels for vasoconstriction and vasodilation to help regulate temperature.

Skeletal system

Provides protection for the brain and spinal cord.

Proprioceptors sense movement of joints and body position.

Muscular system

Muscles carry out movements initiated in the central nervous system.

Stimulates muscle contractions.

Endocrine system

Hormones regulate blood glucose and electrolyte levels needed for neuron function.

Hypothalamus secretes releasing hormones and sends nerve signals to stimulate the pituitary gland.

Cardiovascular system

Provides nutrients and removes wastes.

Innervates smooth muscle of blood vessels for vasoconstriction and vasodilation to regulate blood pressure and flow; medulla oblongata regulates heart rate.

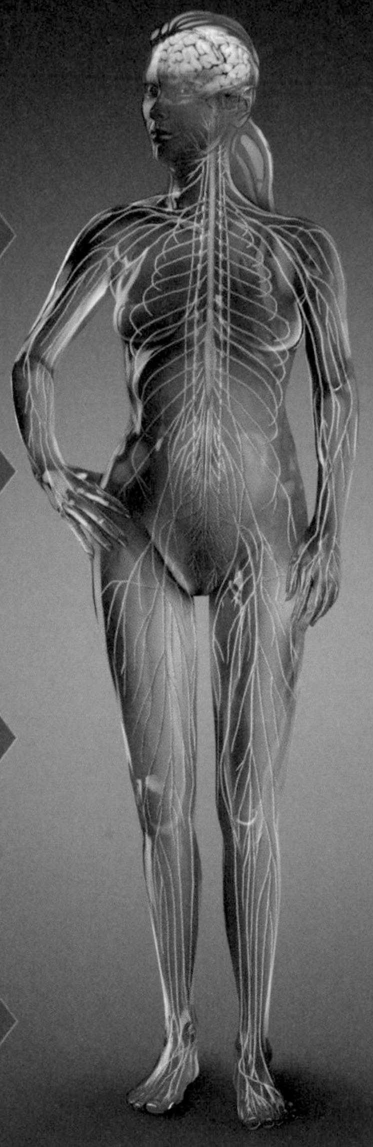

Lymphatic system

Microglia serve as immune system cells to fight pathogens in the CNS.

Respiratory system

Provides O_2 to nervous tissue and removes CO_2.

Medulla oblongata regulates respiratory rate.

Digestive system

Provides nutrients for nervous system tissues.

Parasympathetic division innervates digestive organs.

Excretory/urinary system

Kidney disposes of wastes and maintains electrolyte balance needed by neurons.

Micturition reflex and higher brain centers regulate urine elimination.

Reproductive system

Reproductive hormones affect the production of hypothalamus releasing hormones.

Nerve impulses innervate muscles involved in erection, ejaculation, and childbirth.

FIGURE 6.27 **Putting the Pieces Together—The Nervous System:** connections between the nervous system and the body's other systems.

summary

Overview

- The nervous system is divided into two parts: the central nervous system, consisting of the brain and spinal cord, and the peripheral nervous system, consisting of all the nerves of the body.

Anatomy of the Nervous System

Anatomy of a Neuron

- Neurons have basically three parts: dendrites that receive information, cell bodies that are involved in protein synthesis, and axons that transmit electrical impulses.
- Axons may or may not have a myelin sheath.
- Neurons may be multipolar, bipolar, or unipolar, depending on anatomy.

Neuroglia

There are six types of neuroglial cells that aid in neuron function:

- Oligodendrocytes form myelin in the CNS.
- Ependymal cells produce cerebrospinal fluid in the CNS.
- Astrocytes form the blood-brain barrier in the CNS.
- Microglia seek out and fight pathogens in the CNS.
- Schwann cells form myelin in the PNS.
- Satellite cells control the environment for ganglia in the PNS.

Anatomy of the Central Nervous System

- The brain and spinal cord are covered with three membranes called meninges.
- The brain and spinal cord are bathed in CSF found in the ventricles and subarachnoid space.
- The cerebrum is characterized by gyri and sulci. It is divided into hemispheres that are further divided into four main lobes. Each lobe has a general sensory area and an association area for senses.
- The diencephalon contains the thalamus, which acts as a switching station for incoming sensory messages, and the hypothalamus, which monitors the internal environment and helps regulate homeostasis.
- The brainstem is composed of the medulla oblongata, pons, midbrain, and reticular formation.
- The cerebellum is important for fine-tuning skeletal muscle messages.
- The spinal cord is a solid structure from the foramen magnum to L1.
- The cauda equina extends from the inferior end of the spinal cord.
- The spinal cord has gray matter in the form of an "H" and white matter arranged in columns.

Anatomy of the Peripheral Nervous System

- A nerve is arranged similarly to a muscle, with fascicles and connective tissues.
- Twelve pairs of cranial nerves attach directly to the brain. They can be classified as sensory, motor, or both.
- Thirty-one pairs of spinal nerves attach to the spinal cord by dorsal and ventral roots.
- The autonomic division of the PNS is subdivided into the sympathetic and parasympathetic divisions on the basis of anatomy and function.
- The sympathetic division has a chain of ganglia and prepares the body for physical activity (fight or flight).
- The parasympathetic division has ganglia close to the structure affected and prepares the body for vegetative functions (rest and veg).

Physiology of the Nervous System

Nerve Impulses

- A resting membrane potential is the basis for a nerve impulse and therefore must be maintained.
- Local potentials start at a dendrite and travel toward the trigger zone. They are graded, decremental, and reversible, and may be excitatory or inhibitory.

- Action potentials travel from the trigger zone to the synaptic knob. They require a threshold stimulus and have an all-or-nothing effect. They are not graded, decremental, or reversible.
- Myelination allows for the speed of an action potential.

Reflexes

- A reflex is an involuntary, predictable motor response to a stimulus without conscious thought.
- A reflex occurs in a reflex arc that involves a receptor, an afferent sensory neuron, an integration center in the CNS, an efferent motor neuron, and an effector.

Memory

- There are three types of memory: immediate, short term, and long term.
- Long-term memory results from chemical changes in the neuron. This results in cellular changes, including the growth of dendrites and the formation of new connections.

Language

- Wernicke's area is located in the temporal lobe. It is used to interpret incoming language.
- Broca's area is located in the frontal lobe. It is used to find the words for outgoing language.

Function of the Nervous System

- The function of the nervous system is fast, efficient communication of one part of the body with another part, using action potentials.
- The function can be demonstrated through identifying pathways.

Nutritional Requirements of the Nervous System

- Sodium and potassium are needed throughout life to maintain resting membrane potentials.
- Fat is necessary in the diet, especially for children, to ensure the proper myelination of developing neurons.

Effects of Aging on the Nervous System

- Cognitive ability increases in the young, remains stable in adulthood, and declines in old age.
- The definition of old age differs for each individual.
- Short-term memory is affected early.
- Verbal skills decline around age 70.
- Intellectual performance may remain high until around age 80.
- Reaction times slow as neurons become less efficient.
- The number of neurons in the brain decreases with age.

Nervous System Disorders

Cerebrovascular Accident

- A CVA is commonly called a *stroke* and results in the death of brain tissue.
- The cause may be a blocked artery or a ruptured artery.
- Determining the cause is vital in determining the treatment.

Alzheimer's Disease

- Alzheimer's disease is a progressive irreversible disease characterized by dementia.
- It results in tangles, plaques, and reduced responsiveness of neurons.
- A definitive diagnosis can be determined only upon autopsy.
- Medications to increase neuron sensitivity to neurotransmitters can help alleviate the severity of the symptoms.

key words for review

The following terms are defined in the glossary.

action potential
afferent
autonomic
axon
bipolar
cerebrospinal fluid (CSF)
decremental
dendrite
depolarize
efferent
multipolar
myelin
neuroglia
parasympathetic
reflex
repolarize
resting membrane potential
sympathetic
synapse
unipolar

chapter review questions

Word Building: *Use the word roots listed in the beginning of the chapter and the prefixes and suffixes inside the back cover to build words with the following meanings:*

1. Relating to the coverings of the brain and spinal cord: ______
2. Relating to a group of cells in the PNS: ______
3. Disorder of a neuron: ______
4. Incision of the brain: ______
5. Inflammation of the cerebellum: ______

Multiple Select: *Select the correct choices for each statement. The choices may be all correct, all incorrect, or any combination of correct and incorrect.*

1. Which of the following descriptions of the cerebrum is (are) accurate?
   a. The frontal lobe has gyri and sulci on its surface.
   b. The temporal lobe is for hearing.
   c. Motivation and aggression are located in the parietal lobe.
   d. The occipital lobe is for the sense of smell.
   e. The inability to move the left side of the body indicates a problem with the left hemisphere.
2. Which of the following statements about the spinal cord is (are) accurate?
   a. The spinal cord has enlargements in the thoracic and lumbar regions.
   b. The spinal cord consists of white matter surrounded by gray matter.
   c. The spinal cord has unmyelinated neurons in the horns and columns.
   d. The spinal cord becomes the cauda equina in the lumbar region.
   e. Myelinated axons in the spinal cord travel to and from the brain in the columns.
3. Which of the following statements concerning nerves is (are) true?
   a. A nerve contains fascicles of neurons.
   b. A nerve may contain all afferent, all efferent, or a combination of both types of neurons.
   c. Cranial nerves split to form dorsal and ventral roots.
   d. Unipolar neurons can be found in a ventral nerve root.
   e. Efferent messages enter the spinal cord through the dorsal nerve root.

4. Which of the following statements is (are) true about the peripheral nervous system?
   a. The PNS contains nerves that carry messages in both directions.
   b. The PNS includes all the nerves going out to the body.
   c. The PNS contains an autonomic division that goes to organs, glands, and vessels.
   d. The PNS can be divided into afferent and efferent branches.
   e. The PNS contains a somatomotor division that contains unipolar neurons.
5. Which of the following statements is (are) true about the autonomic nervous system?
   a. The autonomic nervous system includes a sympathetic chain of ganglia.
   b. The autonomic nervous system includes neurons in cranial nerves.
   c. The autonomic nervous system requires a preganglionic neuron synapsing with a postganglionic neuron.
   d. The autonomic nervous system includes ganglia located by organs stimulated by the parasympathetic division.
   e. The autonomic nervous system includes the parasympathetic division that prepares the body for fight or flight.
6. What happens in stimulating a neuron?
   a. Light, heat, mechanical disturbance, or a chemical opens a K^+ channel.
   b. Light, heat, mechanical disturbance, or a chemical closes an Na^+ channel.
   c. Na^+ rushes across a membrane by simple diffusion.
   d. Na^+ enters the cell through a channel, causing the membrane to be polarized.
   e. The membrane becomes depolarized.
7. Which of the following statements is (are) accurate concerning cerebrospinal fluid?
   a. It allows the brain to float somewhat.
   b. It delivers blood to the brain.
   c. It is made in the ventricles.
   d. It rinses away metabolic wastes.
   e. It is constantly being made and constantly being drained through the choroid plexus.
8. Which of the following effects is (are) the result of aging on the nervous system?
   a. Reaction times slow.
   b. Decrease in verbal skills happens early.
   c. Intellectual performance drops at about age 70.
   d. Cognitive ability remains stable throughout life.
   e. Neurons are lost.
9. Which of the following statements is (are) true about meninges?
   a. The dura mater is deep to the pia mater.
   b. CSF can be found in the subpia mater space.
   c. They cover the brain and spinal cord.
   d. They consist of three layers.
   e. They include the arachnoid mater.
10. What should be included in the diet to have a healthy nervous system?
   a. Bananas, as a good source of K^+ for resting membrane potentials.
   b. Table salt, as a good source of Cl^- used to open K^+ channels.
   c. Meat and fish, as good sources of K^+.
   d. Less salt because the typical American gets plenty of Na^+ from processed foods.
   e. Some protein in the diet to ensure proper myelination of developing neurons in children.

Matching: *Match the structures to the description. Some answers may be used more than once.*

_____ 1. Form the blood-brain barrier
_____ 2. Fight pathogens in the CNS
_____ 3. Form myelin in the PNS
_____ 4. Form myelin in the CNS
_____ 5. Produce and circulate CSF

a. Microglia
b. Astrocytes
c. Satellite cells
d. Ependymal cells
e. Oligodendrocytes
f. Schwann cells

Matching: *Match the disorder to the description. Some answers may be used more than once.*

_____ 6. Death of brain tissue due to a blocked artery

_____ 7. Progressive degeneration of the brain that produces tangles, plaques, and loss of cognitive function

_____ 8. Difficulty in speaking or understanding language

_____ 9. Viral infection that travels to the brain by axonal transport

_____ 10. Inflammation of the brain that causes swelling.

a. Aphasia
b. Encephalitis
c. Rabies
d. Alzheimer's disease
e. Cerebrovascular accident

Critical Thinking

1. Explain how encephalitis can result from a bacterium or virus contracted from a mosquito bite.
2. You accidently touch a hot burner on the stove. You pull your hand away and say "ouch." Write two pathways for this event. The first pathway is for the reflex of withdrawing your hand. Include all parts of the reflex arc. Is this a somatic or an autonomic reflex? The second pathway should start with the type of neuron receiving the stimulus and end in the brain, where skeletal muscle messages are being sent to say "ouch."
3. What would happen if there was insufficient ATP in a neuron to run the sodium-potassium pump?

6 chapter mapping

This section of the chapter is designed to help you find where each outcome is covered in this text.

	Outcomes	Readings, figures, and tables	Assessments
6.1	Use medical terminology related to the nervous system.	Word roots: p. 215	Word building: 1–5
6.2	Describe the organization of the nervous system in regard to structure and function.	Overview: pp. 216–217 Figure 6.2	Multiple Select: 4
6.3	Describe the anatomy of a neuron.	Anatomy of a neuron: pp. 217–219 Figure 6.3	Spot Check: 6 Critical Thinking: 1
6.4	Differentiate multipolar, bipolar, and unipolar neurons in terms of anatomy, location, and direction of nerve impulses.	Types of neurons: pp. 219–220 Table 6.1	Spot Check: 1
6.5	Describe neuroglial cells and state their function.	Neuroglia: pp. 220–223 Figures 6.4, 6.5 Table 6.2	Spot Check: 2 Matching: 1–5
6.6	Describe the meninges covering the brain and spinal cord.	Meninges: pp. 223–224 Figure 6.6	Multiple Select: 9

	Outcomes	Readings, figures, and tables	Assessments
6.7	Explain the importance of cerebrospinal fluid, including its production, circulation, and function.	Cerebrospinal fluid: pp. 224–226 Figure 6.7	Spot Check: 3 Multiple Select: 7
6.8	Describe the major landmarks and subdivisions of the brain and state their functions.	Brain: pp. 226–234 Figures: 6.8–6.12 Table 6.3	Multiple Select: 1
6.9	Describe the spinal cord.	Spinal cord: pp. 234–237 Figures 6.13–6.15	Multiple Select: 2
6.10	Describe the anatomy of a nerve and its connective tissues.	Anatomy of a nerve: pp. 237–238 Figure 6.16	Multiple Select: 3
6.11	List the cranial nerves in order, stating their function and whether they are sensory, motor, or both.	Cranial nerves: pp. 238–239 Figure 6.17 Table 6.4	Spot Check: 4, 5, 9
6.12	Describe the attachment of nerves to the spinal cord.	Spinal nerves: pp. 239–241 Figures 6.18, 6.19	Multiple Select: 3
6.13	Compare the parasympathetic and sympathetic divisions of the autonomic nervous system in terms of anatomy and function.	Autonomic nervous system: pp. 241–244. Figures 6.20, 6.21	Spot check: 6, 7 Multiple Select: 5
6.14	Describe a resting membrane potential.	Nerve impulses: pp. 244–245 Figure 6.22	Critical Thinking: 3
6.15	Compare and contrast a local potential and an action potential.	Local potential: pp. 245–246 Figure 6.23 Action potential: pp. 246–248 Figure 6.24	Spot Check: 8 Multiple Select: 6
6.16	Describe a specific reflex and list the components of its reflex arc.	Reflexes: pp. 249-250 Figure 6.25	Spot Check: 9 Critical Thinking: 2
6.17	Explain the difference between short-term and long-term memory.	Memory: p. 250	Spot Check: 10
6.18	Differentiate between Broca's area and Wernicke's area in regard to their location and function in speech.	Language: p. 250	
6.19	Explain the function of the nervous system by writing a pathway for a sensory message sent to the brain to be processed for a motor response.	Function of the nervous system: p. 251 Figure 6.26	Spot Check: 11 Critical Thinking: 2
6.20	Explain the nutritional requirements of the nervous system.	Nutritional requirements of the nervous system: p. 252	Multiple Select: 10
6.21	Explain the effects of aging on the nervous system.	Effects of aging on the nervous system: p. 252	Multiple Select: 8
6.22	Describe nervous system disorders.	Nervous system disorders: pp. 252–253	Matching: 6–10 Critical Thinking: 1

references and resources

Alzheimer's Disease Education & Referral Center. (2010, February 19). *Alzheimer's disease fact sheet*. Retrieved May 20, 2010, from http://www.nia.nih.gov/Alzheimers/Publications/adfact.htm.

Beers, M. H., Porter, R. S., Jones, T. V., Kaplan, J. L., & Berkwits, M. (Eds.). (2006). *Merck manual of diagnosis and therapy* (18th ed.). Whitehouse Station, NJ: Merck Research Laboratories.

Dugdale, D. C. (2009, February 19). *Aging changes in the nervous system*. Retrieved May 16, 2010, from http://www.nlm.nih.gov.medlineplus/ency/article/004023.htm.

Goldman, S. A. (2007, July). Effects of aging: Biology of the nervous system. *Merck manual home edition*. Retrieved May 20, 2010, from http://www.merckmanuals.com/home/sec06/ch076/ch076e.html.

Huang, J. (2008, February). Dementia: Delirium and dementia. *Merck manual home edition*. Retrieved May 20, 2010, from http://www.merckmanuals.com/home/sec06/ch083/ch083c.html.

Porter, R. S., & Kaplan, J. L. (Eds.). (n.d.). Aging and the nervous system. *Merck manual of geriatrics* (chap. 42). Retrieved May 16, 2010, from http://www.merck.com/mkgr/mmg/sec6/ch42/ch42a.jsp.

Porter, R. S., & Kaplan, J. L. (Eds.). (n.d.). Selected physiologic age-related changes. *Merck manual for healthcare professionals*. Retrieved March 19, 2011, from http://www.merckmanuals.com/media/professional/pdf/Table_337-1.pdf?qt=selected physiologic age-related changes&alt=sh.

Portnoy, R. K. (2007, August). Introduction: Pain. *Merck manual home edition*. Retrieved May 25, 2010, from http://www.merckmanuals.com/home/sec06/ch078/ch078a.html.

Saladin, Kenneth S. (2010). *Anatomy & physiology: The unity of form and function* (5th ed.). New York: McGraw-Hill.

Seeley, R. R., Stephens, T. D., & Tate, P. (2006). *Anatomy & physiology* (7th ed.). New York: McGraw-Hill.

Shier, D., Butler, J., & Lewis, R. (2010). *Hole's human anatomy & physiology* (12th ed.). New York: McGraw-Hill.

7 The Nervous System—Senses

It is Saturday morning. Although it is chilly outside, it is warm in the coffee shop you just entered. You can smell the fresh coffee brewing and see the warm cinnamon rolls displayed on the counter. You cannot wait to taste them. Your favorite music even happens to be playing in the background. How sweet it is! This is truly a feast for the senses. At this very moment, sensory neurons are sending all sorts of messages to the brain and spinal cord where they are being processed, contributing to the functions of the nervous system. See Figure 7.1.

7.1 learning outcome

Use medical terminology related to the senses of the nervous system.

learning outcomes

After completing this chapter, you should be able to:

7.1 Use medical terminology related to the senses of the nervous system.

7.2 Classify the senses in terms of what is sensed and where the receptors are located.

7.3 Describe the sensory receptors for the general senses in the skin.

7.4 Explain the types of information transmitted by sensory receptors in the skin.

7.5 Describe the pathway for pain.

7.6 Describe the sensory receptors for taste.

7.7 Describe the different tastes and explain how flavor is perceived.

7.8 Describe the pathway for taste.

7.9 Describe the sensory receptors for smell.

7.10 Explain how odors are perceived.

7.11 Describe the pathway for smell.

7.12 Describe the anatomy of the ear.

7.13 Explain how sound is perceived.

7.14 Describe the pathway for hearing.

7.15 Describe the anatomy of the vestibular apparatus.

7.16 Explain how equilibrium is perceived.

7.17 Describe the pathway for equilibrium.

7.18 Describe the anatomy of the eye.

7.19 Explain how vision is perceived.

7.20 Describe the pathway for vision.

7.21 Describe the effects of aging on the senses.

7.22 Describe disorders of the senses.

word roots & combining forms

audi/o: hearing

aur/o: ear

cochle/o: cochlea

corne/o: cornea

lacrim/o: tears

lith/o: stone

ocul/o: eye

ophthalm/o: eye

opt/o: eye, vision

ot/o: ear

presby/o: old age

propri/o: own

retin/o: retina

scler/o: sclera

tympan/o: eardrum

pronunciation key

chiasm: KYE-asm

choroid: KOR-oid

conjunctiva: kon-junk-TIE-vah

nociceptor: NO-sih-SEP-tor

saccule: SACK-yule

umami: oo-MOM-ee

utricle: YOU-trih-kel

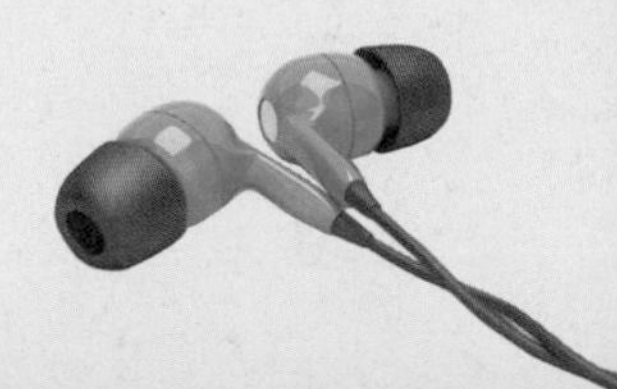

7.2 learning **outcome**

Classify the senses in terms of what is sensed and where the receptors are located.

Overview

The senses can be classified into two categories: general senses and special senses. Various types of simple receptors in the skin, muscles, joints, tendons, and organs are used for general senses. These receptors—located all over the body—detect touch, pressure, stretch, heat, cold, and pain. Special senses involve complex sense organs located only in the head, such as the eye and ear. The special-sense organs are used for taste, smell, hearing, equilibrium, and vision.

This chapter covers the general and special senses. You will study the anatomy of the receptors and how the sense is detected. You will also continue your study of the nervous system by exploring the pathway for each sense to the brain. Your study begins with the general senses detected by receptors in the skin.

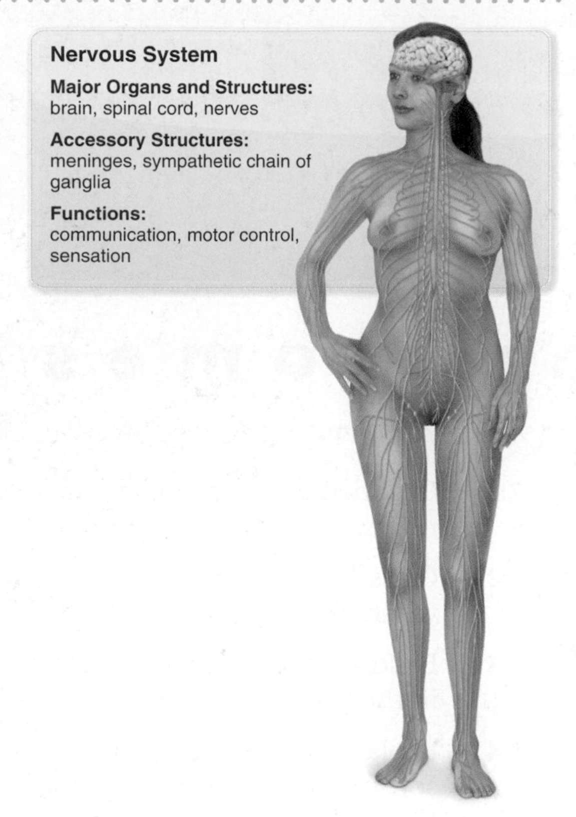

FIGURE 7.1 The nervous system.

7.3 learning **outcome**

Describe the sensory receptors for the general senses in the skin.

nociceptor: NO-sih-SEP-tor

General Senses

There are many ways to classify receptors for the general senses. One way is by the type of stimulus detected:

- **Thermoreceptors** detect heat and cold.
- **Mechanoreceptors** detect touch, vibration, stretch, and pressure. They are stimulated by mechanically disturbing the dendrite of the neuron.
- **Nociceptors** are pain receptors that detect tissue injury or potential tissue injury. These receptors may be stimulated by a chemical, temperature, or mechanical disturbance.

Below, you will explore the anatomy of receptors in the skin and then examine the physiology of how they work.

Anatomy of Receptors in the Skin

The skin contains a variety of receptors that vary by anatomy, location, and function. See **Figure 7.2** and Table 7.1.

spot check 1 In what layer of the skin are most of the general-sense receptors located?

spot check 2 In Chapter 6 you learned about four things that could stimulate a dendrite of a neuron. Which of the four is not used for general senses?

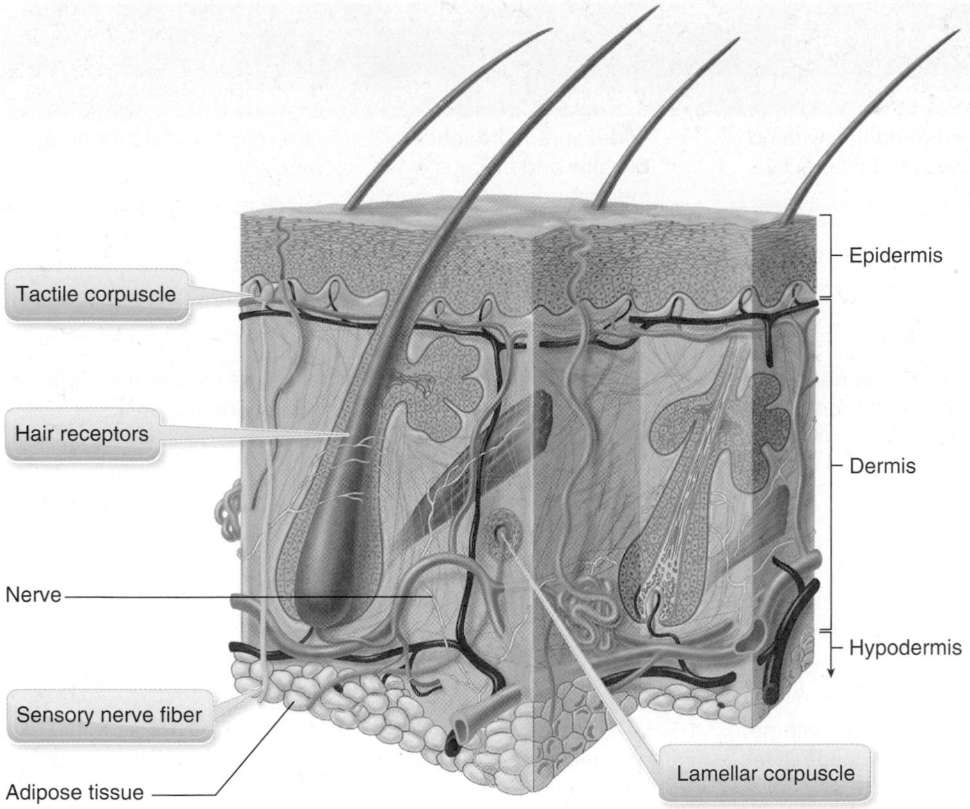

FIGURE 7.2 Receptors in the skin.

Physiology of General Senses in the Skin

7.4 learning outcome

Explain the types of information transmitted by sensory receptors in the skin.

The four kinds of information—type of sensation, location, intensity, and duration—transmitted to the brain from the receptors of the general senses are described below.

1. Type of sensation. The brain knows the function of the neuron sending the signal by the pathway the neuron used. For example, the end point of a pathway for a bulbous corpuscle detecting pressure is different from that of a free nerve ending detecting cold.

2. Location. Each neuron in the skin is responsible for detecting a stimulus in a given area called a **receptive field.** A single sensory neuron on the back has a much larger receptive field than the combined fields of many neurons on the fingertip. Stimulating the neuron anywhere in the receptive field sends the same signal to the brain. The brain does not differentiate where in the field the stimulus happened but knows just that it happened in that particular field. So the brain is better able to distinguish a specific location of a stimulus from the fingertips because of the many neurons with very small fields. See **Figure 7.3**.

3. Intensity. As explained in Chapter 6, an action potential is an all-or-nothing effect. Yet sensory neurons do transmit information on the intensity of the stimulus. They do this in three ways:

   - Some receptors are more sensitive than others. If less sensitive receptors are also sending signals, the stimulus must be intense.

   - The number of signals matters. Therefore, the more receptors stimulated, the more intense the sensation.

TABLE 7.1 Receptors in the skin

Name	Anatomy	Location	Function
Free nerve endings	Bare dendrites with no associated connective tissue	Widespread throughout the skin and mucous membranes	Thermoreceptors for heat and cold Nociceptors for pain
Tactile corpuscles	Two or three nerve fibers among flattened Schwann cells, forming a pearlike structure	Dermal papillae of the skin; highly concentrated in the fingertips and palmar skin	Mechanoreceptor for light touch and texture
Hair receptors	Bare dendrites with no associated connective tissue	Wrapped around the base of a hair follicle	Mechanoreceptors for any light touch that bends a hair
Lamellar corpuscles	Single dendrites surrounded by flattened Schwann cells that in turn are surrounded by fibroblasts, giving a layered appearance like tree rings	Deep in the dermis (especially on the hands, breasts, and genitals)	Mechanoreceptors for deep pressure, stretch, and vibration
Bulbous corpuscles	Long, flattened capsules with a few nerve fibers	Dermis of the skin	Mechanoreceptors for heavy touch, pressure, and stretching of the skin
Tactile disks	Flattened nerve endings	Stratum basale of the epidermis next to specialized tactile cells	Mechanoreceptors for light touch

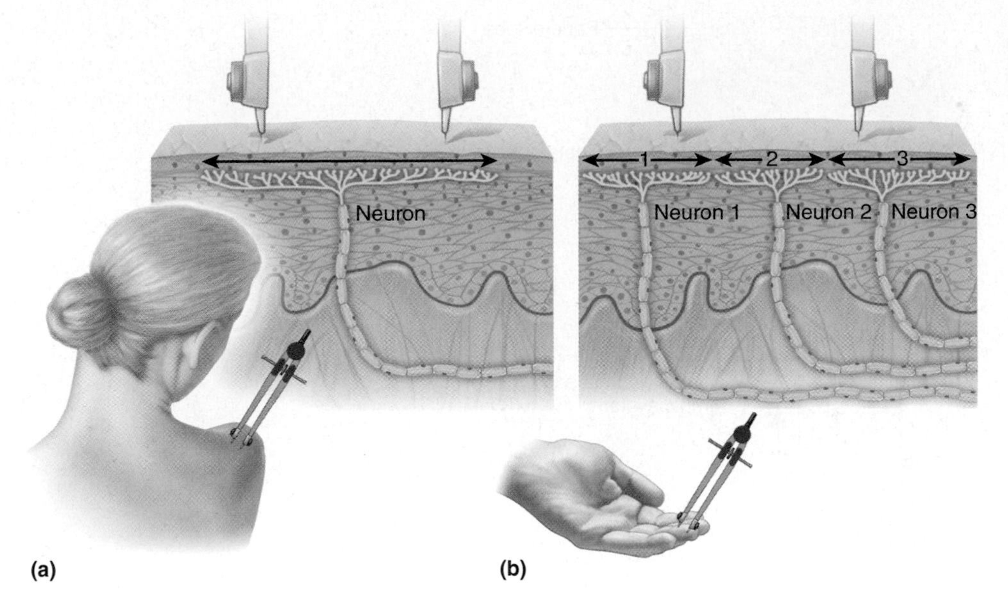

FIGURE 7.3 Receptive fields: a two-point discrimination test. (a) Touching the back with two points in the same receptive field is perceived by the brain as one touch even though the points are relatively far apart. (b) Touching the fingertip with two points the same distance apart is perceived by the brain as two touches because it involves two different receptive fields (field for Neuron 1 and field for Neuron 3).

- The frequency at which the signals are sent makes a difference; the more frequent the nerve impulses, the more intense the sensation. (You may recall from Chapter 5 that this is applicable for motor neurons in the muscular system as well. The more frequent the nerve impulse, the larger the contraction.)

4. **Duration.** Some receptors send a burst of signals when first stimulated but then adapt quickly and slow the frequency of their transmissions. Hair receptors are good examples of this. You may notice the sensation of your shirt on your back when you first put it on, but it quickly becomes insignificant. Other receptors such as lamellar corpuscles are slow to adapt, so you would continue to be aware of a vibration for a longer period of time.

7.5 learning outcome

Describe the pathway for pain.

Pathway for Pain It is time to investigate how a nociceptor for pain works. Pain may be undesirable, but it is necessary to alert us that something wrong in the body needs to be addressed. Damaged tissues secrete chemicals, such as bradykinin, prostaglandins, and histamine. These chemicals stimulate a local potential on the dendrites of nociceptors. Even K^+ and ATP released from ruptured cells can stimulate a nociceptor.

If the nociceptor is in the head, cranial nerve V, VII, IX, or X will carry the sensory message to the brain. If the nociceptor is below the neck, the relevant spinal nerve will carry the message to the spinal cord. Once at the cord, the signal can take either of two pathways to ultimately reach the cortex of the parietal lobe. See **Figure 7.4**. These two pathway options are the following:

1. The pathway can be the same as any other sensory input from the skin:

   Unipolar neuron → spinal cord → medulla oblongata → pons → midbrain → thalamus → parietal lobe

2. The signal may take a different pathway:

   Unipolar neuron → spinal cord → reticular formation → hypothalmus and limbic system

   ↳ parietal lobe

In the second pathway, the signal travels from the spinal cord to the reticular formation in the brainstem. The reticular formation filters the sensory signals it receives

FIGURE 7.4 Pathways for pain.

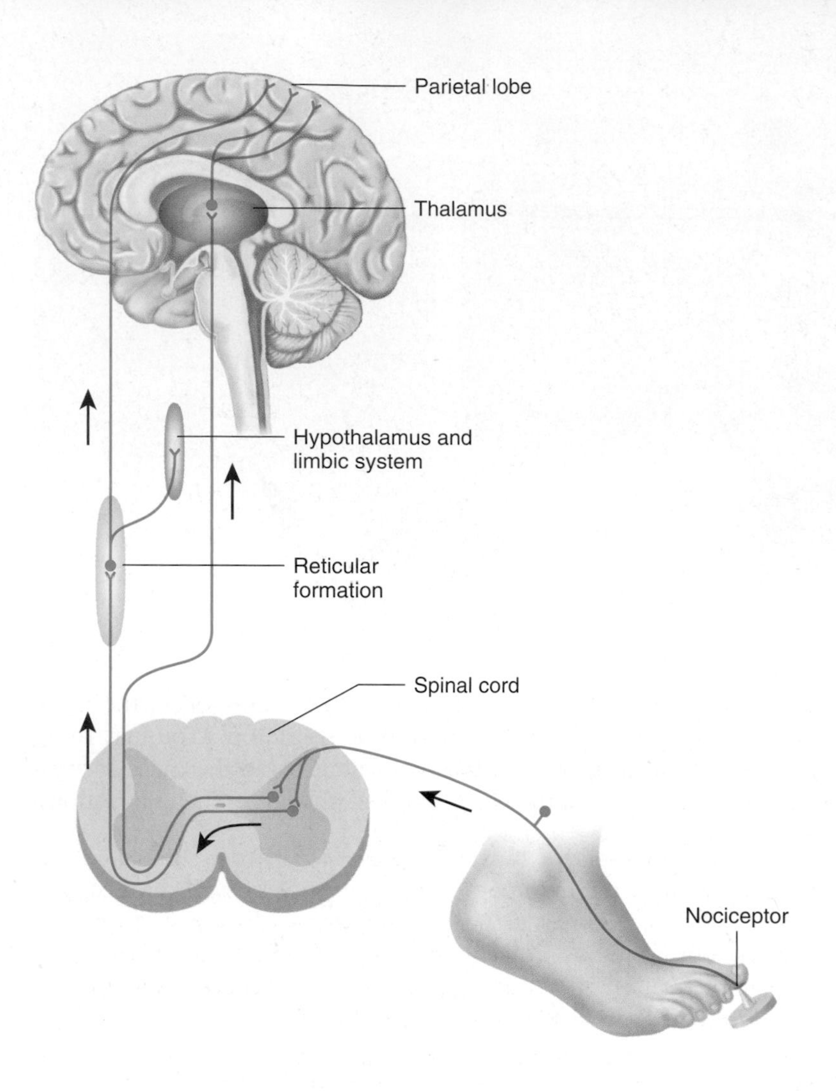

and passes on only important messages—like pain signals—to the cerebrum. The reticular formation sends the pain signal to the parietal lobe directly and to the hypothalamus and limbic system, where the signal may promote physical and emotional responses such as nausea and fear.

Now that you have explored the general senses of the skin, you can move on to the body's special senses, starting with taste.

7.6 learning **outcome**

Describe the sensory receptors for taste.

Taste

Gustation is the term for taste. There are approximately 10,000 **taste buds** in the human mouth. Most of the taste buds are on the tongue, but some are found lining the cheeks and on the soft palate, larynx, and epiglottis. Below, we focus on the anatomy of a taste bud and the receptors for taste and then explore the physiology of how they work.

Anatomy of Receptors for Taste

The surface of the tongue is covered with bumps called **lingual papillae.** Many of them contain multiple taste buds, which are composed of several types of cells. See **Figure 7.5**. These cell types include the following:

- Taste cells are banana-shaped and have hairlike microvilli **(taste hairs)** on their surface. Taste hairs are exposed to molecules taken into the mouth through

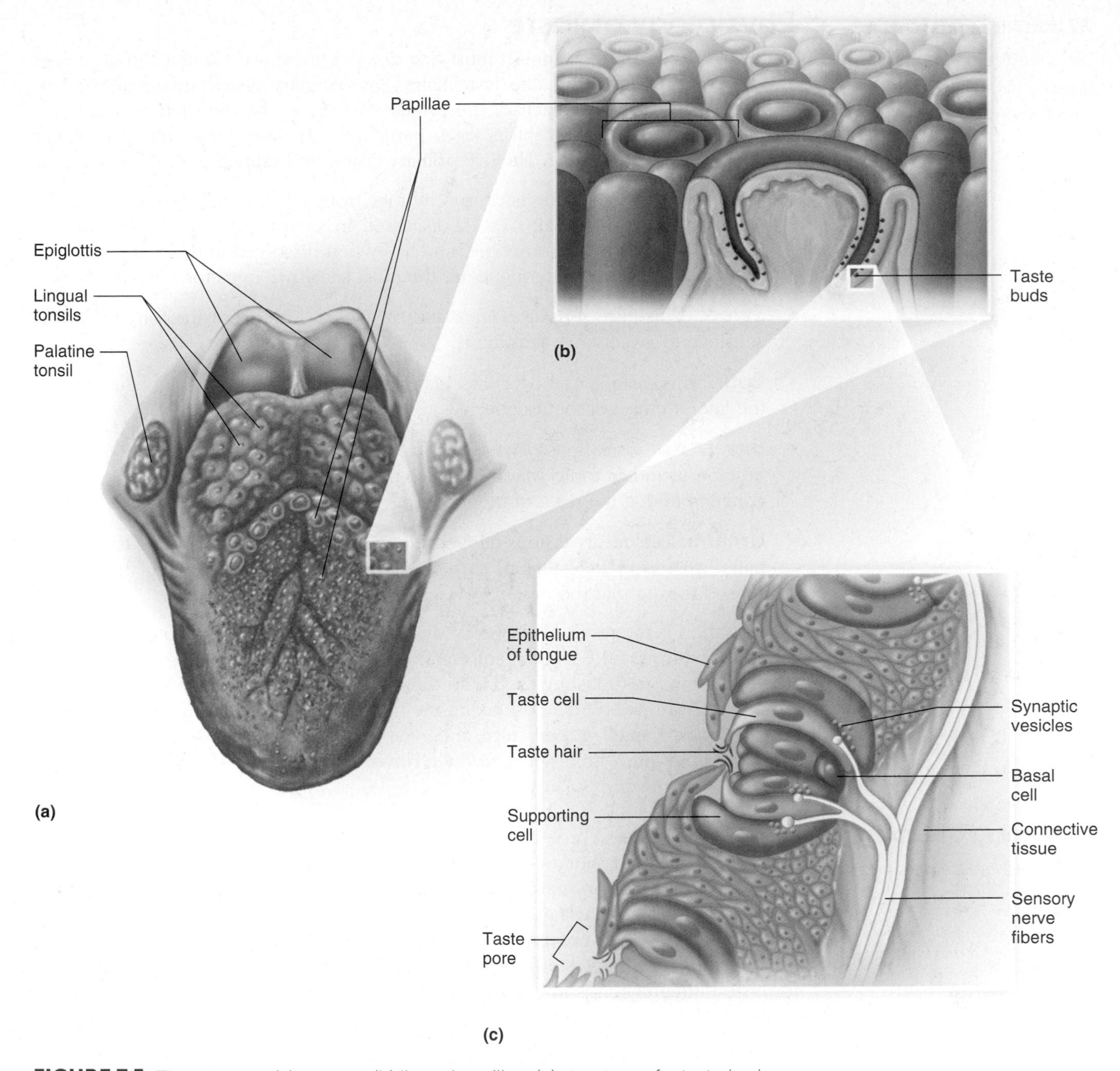

FIGURE 7.5 The tongue: (a) tongue, (b) lingual papillae, (c) structure of a taste bud.

a **taste pore** of the taste bud. The microvilli are the chemoreceptors for taste. Taste cells are not neurons. They are epithelial cells with sensory neurons at their base. When stimulated by a chemical binding to a receptor on the taste hairs, the taste cell secretes a neurotransmitter to stimulate the sensory neuron. Taste cells live for about 10 days.

- **Basal cells** are stem cells in the taste bud that develop to replace taste cells as they die.
- **Support cells** physically support the 50 to 150 taste cells in the taste bud. They do not have a sensory role.

7.7 learning outcome

Describe the different tastes and explain how flavor is perceived.

Physiology of Taste

Molecules taken into the mouth must first dissolve in saliva to enter the taste pore and come in contact with the taste hairs. Five primary tastes are recognized by the taste cells. Although each of the primary tastes can be detected by taste cells throughout the mouth, receptors for a particular taste are concentrated on some specific areas of the tongue. The five primary tastes are outlined in the following list:

- **Salt.** This taste sensation is caused by ions from salts binding to the taste hairs. An example is table salt (NaCl), discussed in Chapter 2. NaCl is an ionically bonded molecule that forms Na^+ and Cl^- when dissolved in the water of saliva. The lateral edges of the tongue are the most sensitive areas for salt sensation.
- **Sweet.** This taste sensation is caused by sugars. The tip of the tongue is most sensitive to sweet taste sensations.
- **Sour.** This taste sensation is associated with acids. It is most concentrated on the lateral edges of the tongue.
- **Bitter.** This taste is associated with alkaloids such as caffeine, nicotine, and quinine found in tonic water. It is also associated with spoiled food. It is most concentrated at the back of the tongue.
- **Umami.** This meaty taste is derived from some amino acids binding to the taste hairs. An example of umami is the taste of beef or chicken broth that can be sensed throughout the mouth.

umami:
oo-MOM-ee

Sensation is the result of a sensory signal sent to the brain. Perception is the way the brain interprets the sensory information. You know from your own experience that there is more to tasting food than these five tastes. The immense variety is not simply a combination effect of mixing the five primary tastes. Other sensory inputs such as texture, smell, temperature, and even pain (for some peppers) are perceived by the brain along with taste to produce **flavors.**

7.8 learning outcome

Describe the pathway for taste.

Pathway for Taste Three cranial nerves—the facial nerve, the glossopharyngeal nerve, and the vagus nerve—carry the sensory messages for taste. The facial nerve (CN VII) carries sensory messages from taste buds in the anterior two-thirds of the tongue. The glossopharyngeal nerve (CN IX) carries sensory messages from taste buds on the posterior one-third of the tongue. And the vagus nerve (CN X) carries sensory messages for taste from the other taste buds in the mouth. The pathway for taste ultimately ends in the association area of the parietal lobe near the lateral sulcus.

Cranial nerve → medulla oblongata → hypothalamus and amygdala
└→ pons → midbrain → thalamus → parietal lobe

The sensory message that takes the path to the hypothalamus and amygdala may trigger salivation, gagging, and emotional responses to taste.

7.9 learning outcome

Describe the sensory receptors for smell.

SMELL

The sense of smell is called **olfaction.** As you have just read, olfaction and taste are often paired to produce the flavors you experience in foods. Even though the human body has about 12 million olfactory receptor cells, a human's olfactory ability is poor compared to that of most mammals.

Receptors for olfaction are located in the mucous membranes of the roof of the nasal cavity, called **olfactory mucosa.** The rest of the mucous membrane lining the nasal cavity, called respiratory mucosa, has no sensory function. Olfactory cells

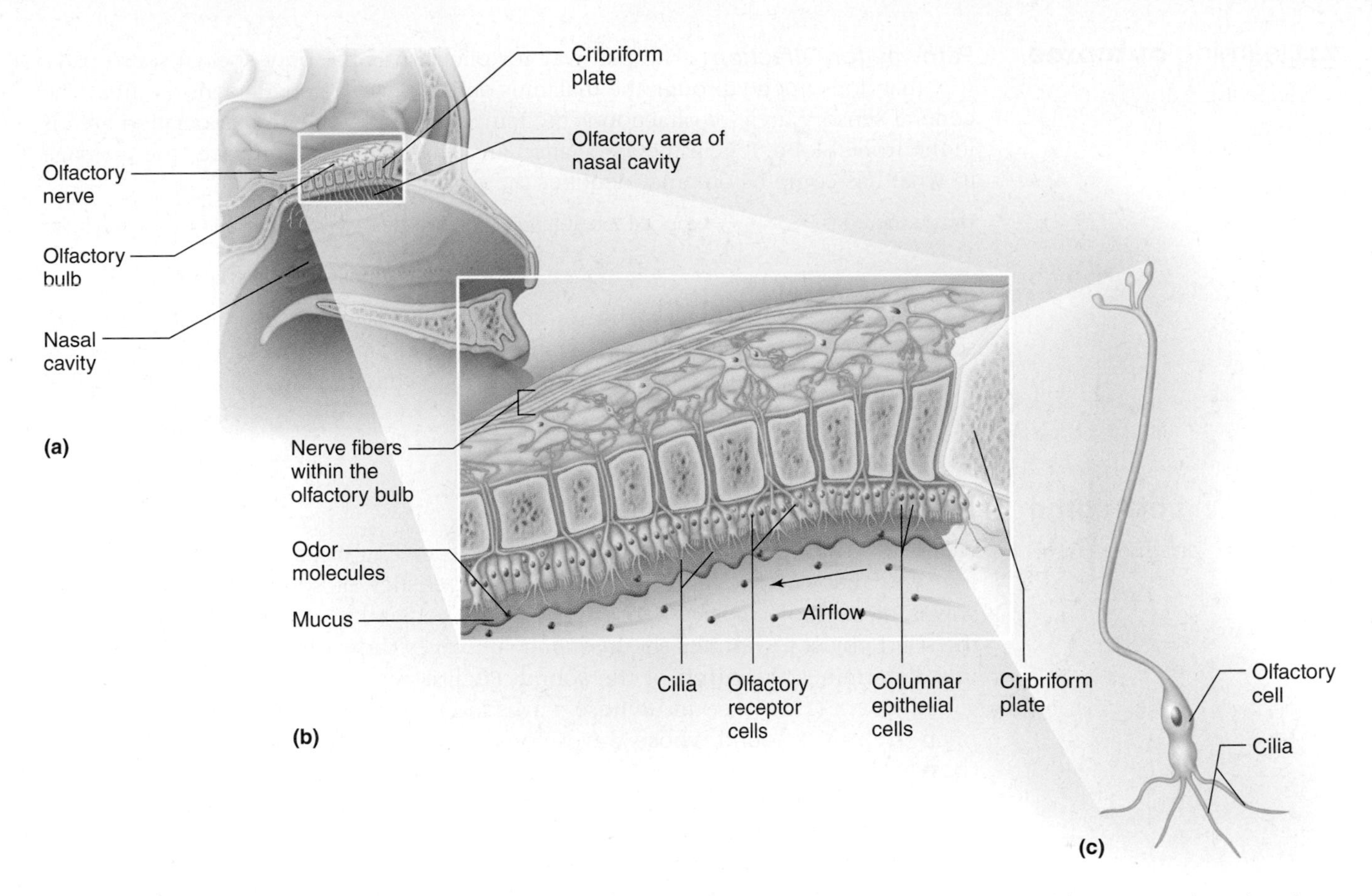

FIGURE 7.6 Olfactory receptors: (a) view of olfactory bulb, cribriform plate; and nasal cavity, (b) close-up view showing the olfactory bulb and cribriform plate; (c) olfactory cell.

access the mucosa through foramen in the cribriform plate of the ethmoid bone (discussed in Chapter 4).

Anatomy of Receptors for Smell

The 12 million **olfactory cells** are bipolar neurons. These clublike cells have cilia called **olfactory hairs** that have binding sites for odor molecules. The hairs are therefore chemoreceptors. The olfactory cells pass from the olfactory bulb of the olfactory nerve (CN I) through the cribriform plate to have their olfactory hairs spread out in the olfactory mucosa. See **Figure 7.6**. Olfactory cells live for approximately 60 days. Nearby basal cells develop to replace olfactory cells as they die.

Physiology of Smell

7.10 learning outcome

Explain how odors are perceived.

Each olfactory cell has one type of receptor to detect one particular odor. The binding of an odor molecule to the receptor initiates a local potential. If there is a threshold stimulus at the trigger zone of the olfactory cell, an action potential is generated that will be passed on to a synapse within the olfactory bulb of CN I.

spot check 3 How might a cold affect your ability to taste your food?

spot check 4 Some people cannot smell garlic. How does their anatomy differ from those who can smell garlic?

7.11 learning **outcome**

Describe the pathway for smell.

Pathway for Olfaction The pathway for olfaction is the only special-sense pathway that does *not* go through the thalamus on its way to a general sensory area. The general sensory area for olfaction is the temporal lobe, while the association area is in the frontal lobe. It is in the association area that the brain compares the message to what has come before and identifies the odor.

Bipolar neuron → CN I → temporal lobe (general sensory area) → frontal lobe (association area)
→ hypothalamus and amygdala

Like other special-sense pathways, the pathway for olfaction splits to carry sensory messages to the hypothalamus and amygdala. Physical reactions initiated by the hypothalamus include coughing or sneezing, while the amygdala is responsible for emotional reactions to odors.

7.12 learning **outcome**

Describe the anatomy of the ear.

Hearing

Hearing is the interpretation of sound waves traveling in air. To understand hearing, you first need to understand the characteristics of sound. Sound travels in waves of air molecules. When a wave of molecules hits a solid object, it causes the solid object to vibrate. The frequency of waves (how often the waves are coming) determines the **pitch** of the sound. Pitch is measured in cycles per second, called **hertz (Hz).** A sound whose waves are coming very frequently has a high pitch (treble). A sound whose waves come at a slower frequency has a low pitch (bass). The most sensitive people can hear frequencies of 20 to 20,000 Hz, or up to 20,000 vibrations per second. The average person can hear frequencies of 1,500 to 5,000 Hz, which includes the normal range for speech.

Volume is determined by the size of the sound wave and therefore the size of the vibration of the solid object. So the bigger the vibration, the louder the volume. Volume is measured in **decibels (dB).** Humans can typically hear from 0 to 120 dB. See **Figure** 7.7.

The ear is an elaborate sensory organ for hearing. It is also the sensory organ for equilibrium. You will investigate that special sense later in the chapter. First, consider the structures of the ear, which are relevant to hearing.

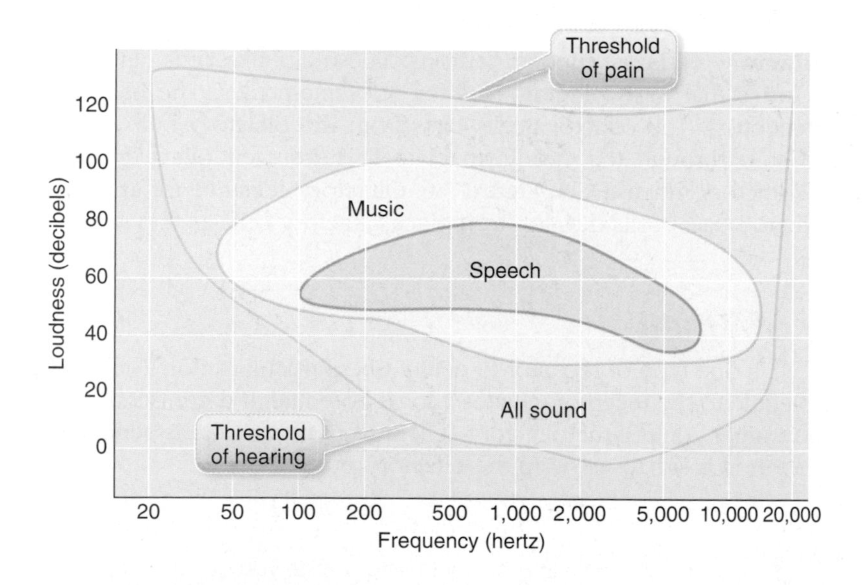

FIGURE 7.7 The range of human hearing.

Anatomy of the Ear

As you can see in **Figure 7.8**, the anatomy of the ear can be divided into three sections: the outer ear, the middle ear, and the inner ear. The outer ear consists of the pinna and the auditory canal. The **pinna** is an external ear flap composed

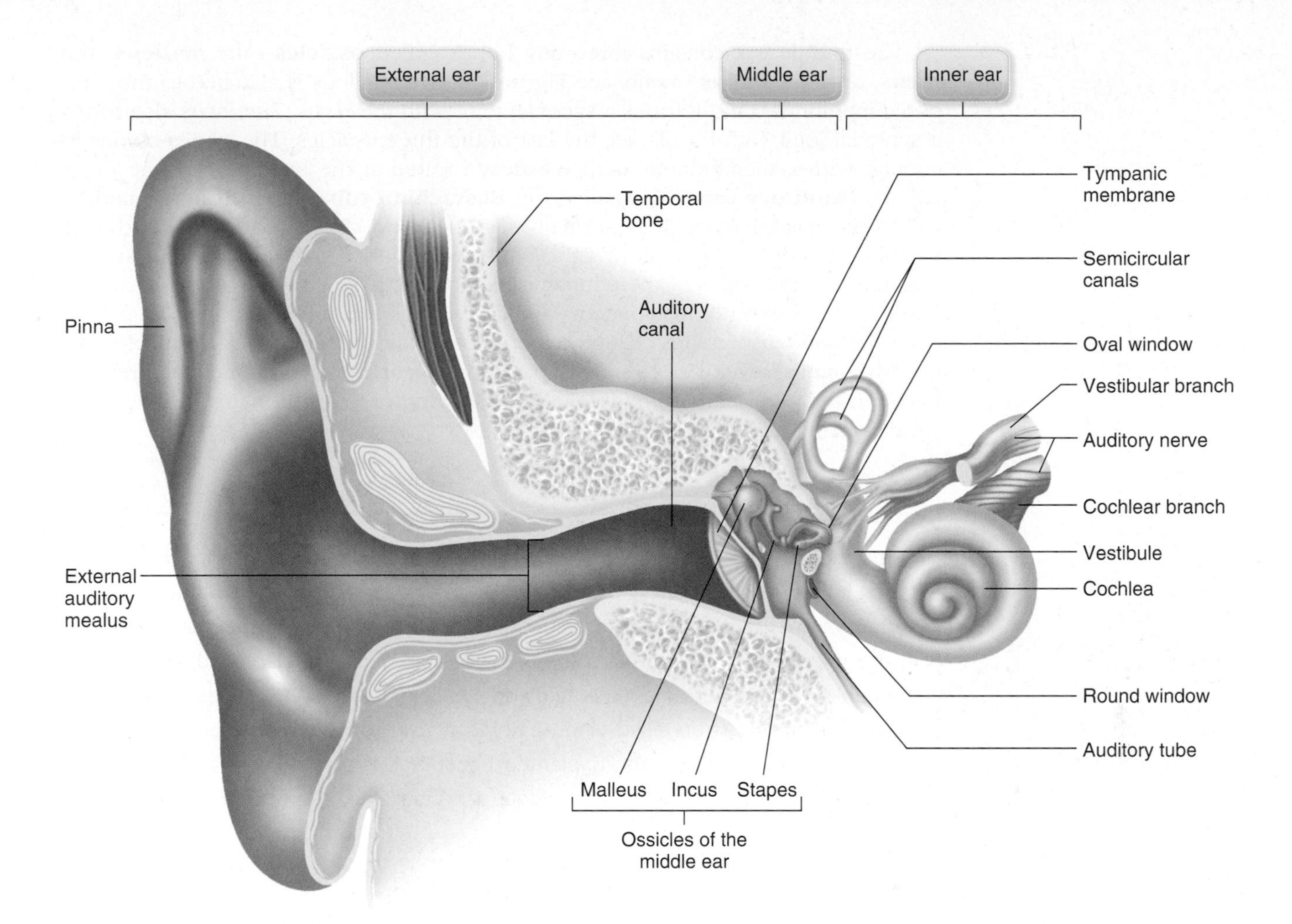

FIGURE 7.8 The internal anatomy of the ear.

of elastic cartilage. It directs sound waves into the ear. The **auditory canal** leads from an opening in the temporal bone, called the **external auditory meatus,** to the **tympanic membrane,** commonly known as the *eardrum*. The auditory canal is lined with skin and has ceruminous glands (specialized sweat glands), which produce cerumen (earwax). The sticky cerumen coats guard hairs in the canal to deter insects from traveling any farther into the ear. Cerumen also has lysozymes and a low pH to deter bacteria. The tympanic membrane separates the outer ear from the middle ear.

clinical point

An infection anywhere in the outer ear is called **otitis externa,** better known as *swimmer's ear*. It is usually associated with repeated exposure of the ear to water, as in swimming. It is typically a bacterial or fungal infection. An infection in the outer ear can usually be distinguished from an infection of the middle ear by pulling on the earlobe. If the pain increases when the ear lobe is pulled, it is typically an outer-ear infection. If the pain is still present but does not increase when the lobe is pulled, it is most likely a middle-ear infection.

The middle ear contains three tiny bones called **ossicles**—the **malleus,** the **incus,** and the **stapes.** Again, see **Figure** 7.8. The malleus is attached to the tympanic membrane, and it forms a synovial joint with the incus. The incus also forms a synovial joint with the stapes, the last of the three ossicles. The stapes comes in contact with a membranous **oval window** located at the beginning of the inner ear. The **auditory tube,** also called the **Eustachian tube,** leads from the middle ear to the nasopharynx. It is normally flattened and closed, but it opens during yawning or swallowing to allow air to enter the middle ear. This allows pressure to equalize on both sides of the tympanic membrane. If you have traveled in an airplane or a fast elevator, you may be familiar with the feeling of your ears "popping" from the sudden pressure change experienced when a plane rises or an elevator quickly climbs several stories. This pain can occur if there is unequal pressure between the outer ear and middle ear. Swallowing or yawning equalizes the pressure and alleviates the pain.

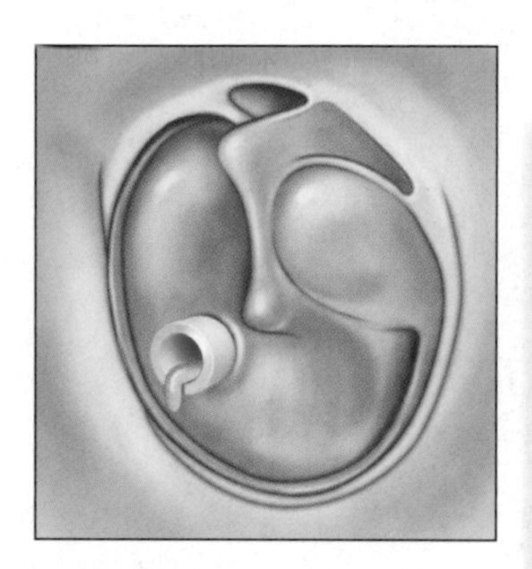

FIGURE 7.9 A tube inserted in the tympanic membrane to drain fluid.

clinical point

Otitis media is an infection of the middle ear. Bacteria can gain entry into this space by way of the auditory tube. The growth of bacteria and the buildup of fluid and pus from the immune system fighting the infection can increase the pressure in the middle ear, causing pain. Tubes may be surgically implanted through the tympanic membrane to drain the fluid and pus from the middle ear. See Figure 7.9. However, care must be taken when bathing or swimming to avoid getting water in the ear if tubes have been implanted because this would open an additional gateway for bacteria to enter the middle ear. Otitis media is usually treated with antibiotics.

The inner ear is a complicated mass of fluid-filled semicircular canals and a spiral tube embedded in a **bony labyrinth** (*labyrinth* means "maze") of the temporal bone. Notice in **Figure 7.8** that the auditory nerve separates into two branches in the inner ear. The cochlear branch is for hearing. The vestibular branch is for equilibrium. As you can also see in **Figure 7.8**, there are semicircular canals in the inner ear. These canals are not involved with hearing. You will read more about them later in regard to the sense of equilibrium.

The stapes of the middle ear comes in contact with the **vestibule** at a membrane called the *oval window.* The oval window separates the stapes from a fluid-filled tube that coils to form a snail-like structure called the **cochlea,** embedded in the bony labyrinth. The cochlea is 2½ twists of the fluid-filled tube, which then ends at the **round window** of the vestibule. **Figure 7.10** shows how this would look if the cochlea were unwound.

The fluid in the tube **(perilymph)** is shown in blue in **Figure 7.10**. This tube begins at the oval window and extends out for some distance before making a U-turn. It ends at the round window. Also notice in **Figure 7.10** that the tube surrounds another tube, called the **cochlear duct,** which is filled with **endolymph** (shown in pink in **Figure 7.10**). The cochlear duct contains **hair cells,** called such because of stiff microvilli on their surface. Hair cells are not neurons, but they have bipolar neurons at their base much like taste cells have neurons at their base. The hair cells are arranged on the **basilar membrane** of the cochlear duct with a stiff **tectorial** membrane suspended over them. This arrangement of structures in the cochlear duct is called the **spiral organ** or **organ of Corti.** See **Figure 7.11**.

Now that you have covered the complex anatomy of the ear for hearing, you are ready to explore how hearing works.

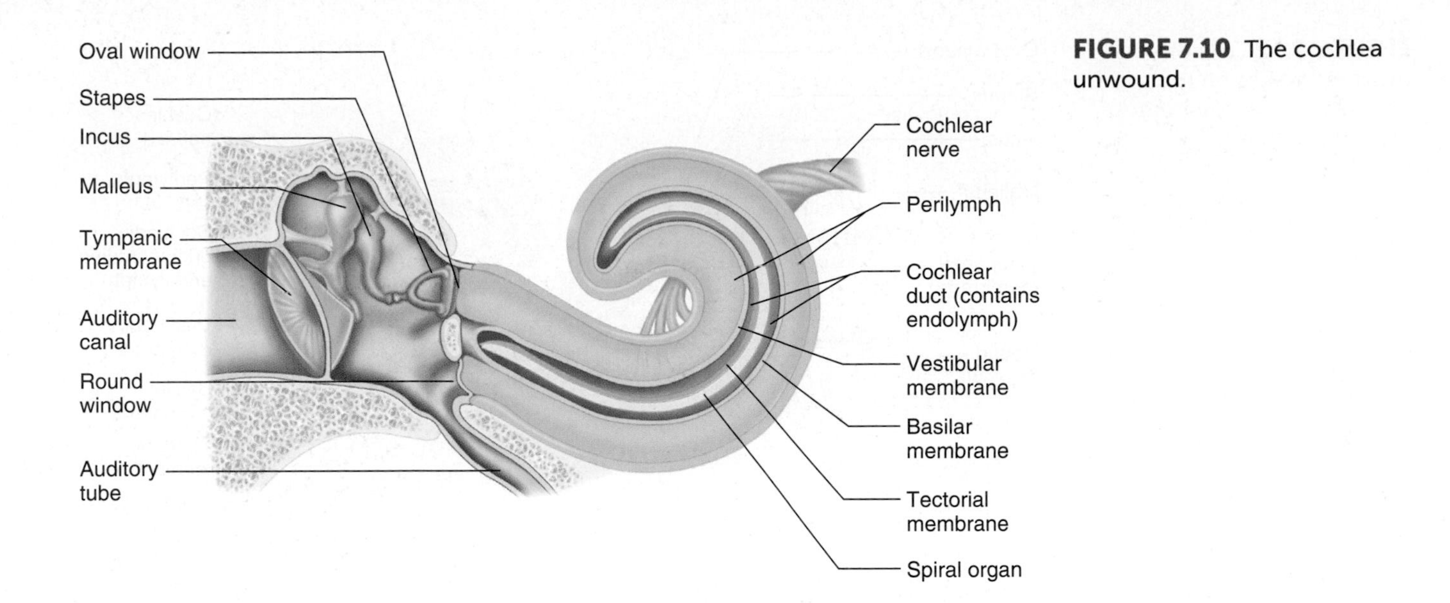

FIGURE 7.10 The cochlea unwound.

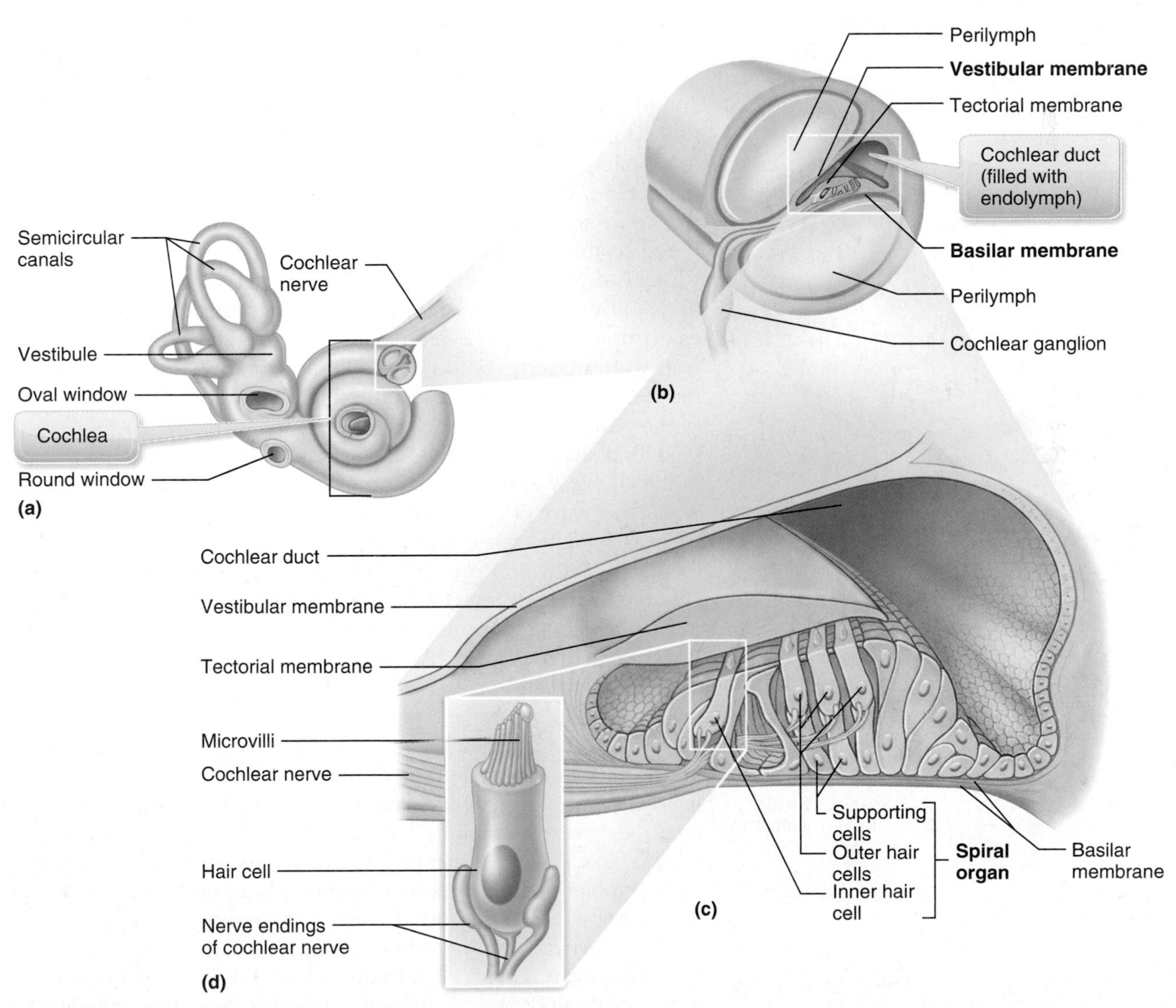

FIGURE 7.11 Anatomy of the cochlea: (a) cross section of the cochlea, with perilymph in blue and endolymph of the cochlear duct in pink, (b) detail of one cross section of the cochlea, (c) detail of the spiral organ, (d) detail of a hair cell and nerve endings of the cochlear branch of the auditory nerve.

FIGURE 7.12 **The effects of sound waves on the cochlea.** Steps 1 to 7 are explained in the text.

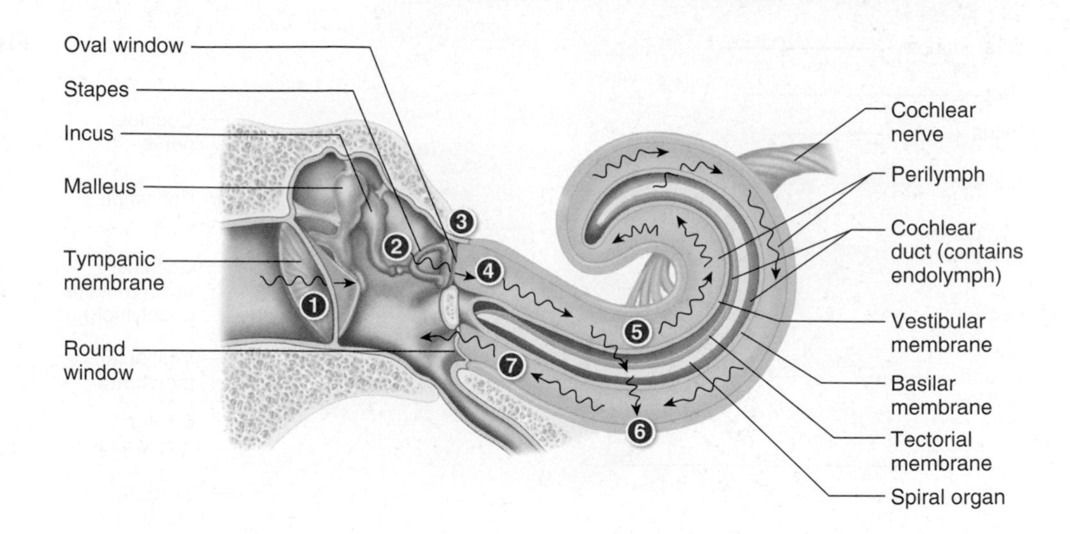

7.13 learning **outcome**

Explain how sound is perceived.

Physiology of Hearing

The physiology of hearing involves converting sound waves outside the head into action potentials in the bipolar neurons of the cochlear nerve. To understand what that means, follow the numbered steps in **Figure 7.12**, which again presents the cochlea unwound, as you read the explanation here. Sound waves are directed into the auditory canal by the pinna. The waves of air molecules hit the tympanic membrane and cause it to vibrate (1). Because the malleus is connected to the tympanic membrane, it too begins to vibrate. That causes the incus and then the stapes to vibrate as well (2). Vibrations of the stapes are transferred to the oval window of the inner ear (3). So far, the sound waves outside the head have been converted to mechanical vibrations in the middle ear and are now being transferred to the inner ear. (But why, then, is a middle ear even needed? Why not have the oval window serve as the tympanic membrane? The middle ear is needed because the bones of the middle ear amplify the vibration of the tympanic membrane 20 times.)

Vibrations of the oval window create waves within the perilymph of the cochlear tube (4). Each time the stapes pushes in on the oval window, a wave of perilymph starts on its way around the tube. This eventually causes the round window to bulge, like a soft water balloon. The perilymph cannot be compressed, so if you push in one side of the tube, it must bulge out somewhere else. The wave can make it all the way around the tube if the frequency is slow enough. If the frequency is faster, the wave pushes on the vestibular membrane, which causes vibrations in the endolymph of the cochlear duct (5). Vibrations in the endolymph cause the basilar membrane to vibrate. The basilar membrane is flexible and can bend and vibrate, unlike the tectorial membrane of the spiral organ, which is stiff. The hair cells are pushed against the tectorial membrane and bend each time the basilar membrane vibrates (6). Hair cells are mechanoreceptors, so each time a hair cell is bent, it releases a neurotransmitter to the bipolar neuron at its base to start a local potential. Eventually, the vibrations reach the round window, where the process ends (7).

The frequency of the vibrations determines which hair cell of the spiral organ is bent. High-frequency sounds bend hair cells close to the oval window. Low-frequency sounds bend hair cells farther away from the oval window. So the pitch of the sound is determined by the hair cell that is bent. How much the hair cells bend determines the volume of the sound. See **Figure 7.13**.

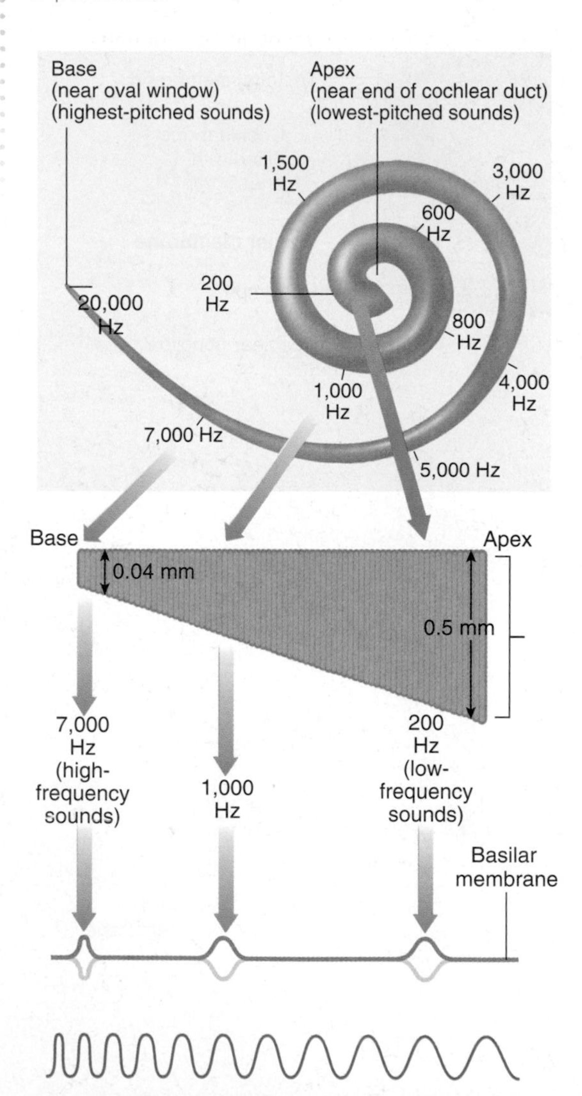

FIGURE 7.13 Frequency response of the basilar membrane in the cochlea.

spot check ❺ It is possible to have arthritis in the joints between ossicles. How might that affect hearing?

7.14 learning outcome

Describe the pathway for hearing.

Pathway for Hearing The pathway for hearing begins with the bipolar neurons at the base of the hair cells. Their axons form the cochlear nerve that joins with the vestibular nerve to form the auditory nerve (CN VIII). The auditory nerve delivers the sensory messages to the pons. From there, the messages go to the inferior colliculi of the midbrain, where the location of the source of the sound is perceived. The sensory messages then travel to the thalamus, which directs them to the temporal-lobe general sensory area and association area for hearing.

Bipolar neurons of CN VIII → pons → midbrain (inferior colliculi) → thalamus → temporal lobe

The ear is also the sensory organ for equilibrium. We focus next on how equilibrium is perceived.

Equilibrium

7.15 learning outcome

Describe the anatomy of the vestibular apparatus.

The sense of equilibrium is based upon the brain knowing where in space the head is located and how it may be moving. There are two kinds of equilibrium: static equilibrium and dynamic equilibrium. **Static equilibrium** is perceived when the head is stationary or moving in a straight line. Holding your head still as you are reading, or while you are accelerating in your car, involves static equilibrium. **Dynamic equilibrium** is perceived when the head is rotating. The rotation would be in the transverse plane if you were spinning in a chair, the coronal plane if you were doing a cartwheel, or the sagittal plane if you were doing a somersault (forward tuck and roll). See **Figure 7.14**. The receptors for both types of equilibrium are located in the **vestibular apparatus.**

FIGURE 7.14 Dynamic equilibrium: rotation in the (a) transverse plane, (b) coronal plane, (c) sagittal plane.

saccule:
SACK-yule

utricle:
YOU-trih-kel

Anatomy of the Vestibular Apparatus

The **saccule** and **utricle** in the vestibule of the inner ear are used to perceive static equilibrium. The saccule is used to perceive vertical movement of the head, as in going up and down in an elevator. The utricle is used for horizontal movement of the head, as in acceleration in a car. Each of these structures contains a patch of hair cells called a **macule** that has a gel-like structure called an **otolithic membrane** over the top of the hair cells. See **Figure 7.15**. Calcium carbonate and protein granules called **otoliths** (*oto* means "ear" and *liths* means "stones") are suspended in the gel. As you can see in **Figure 7.16**, gravity, during a tilt of the head, causes the

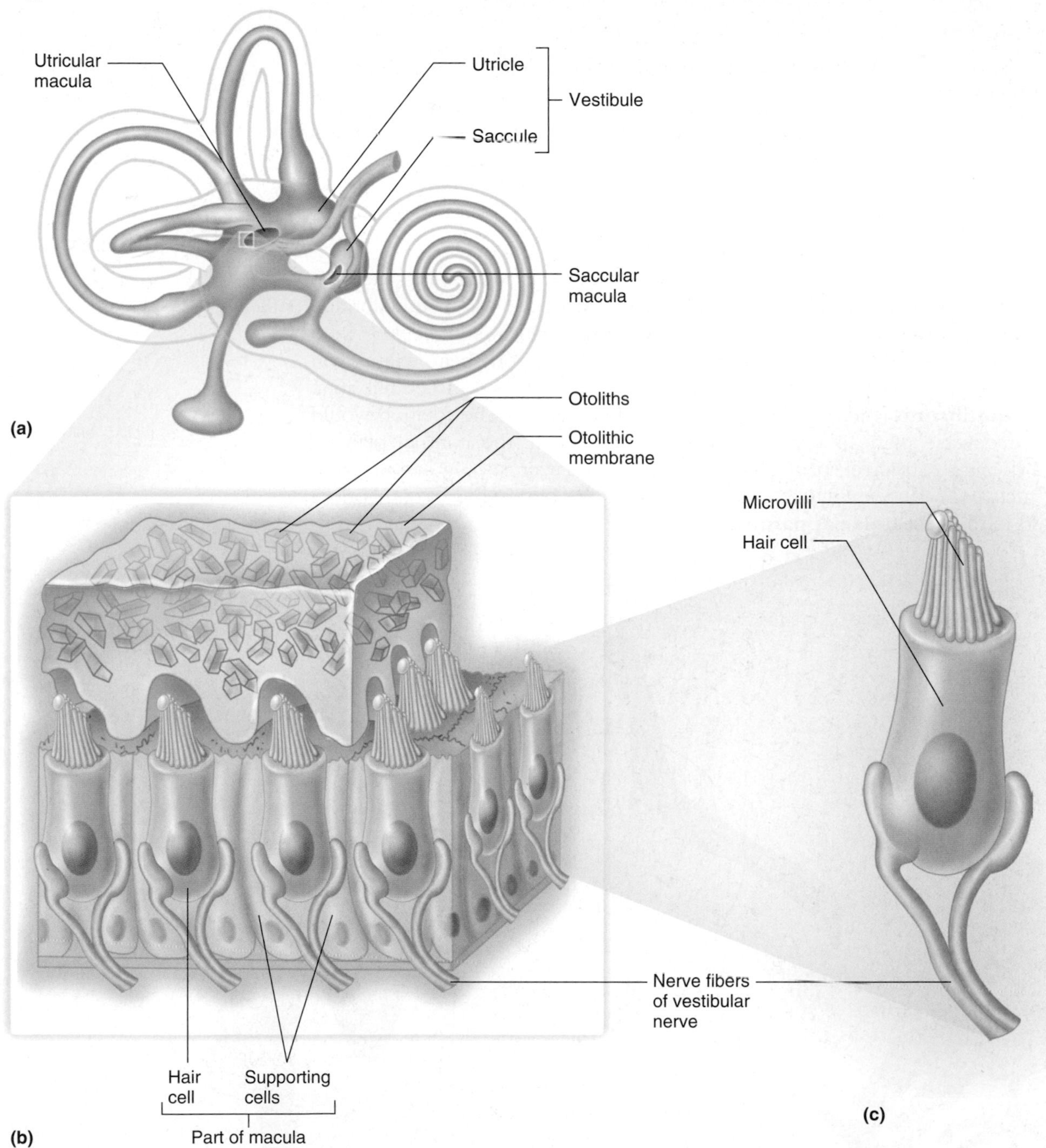

FIGURE 7.15 The saccule and utricle: (a) location of the saccule and utricle, (b) anatomy of the macula, (c) anatomy of a hair cell.

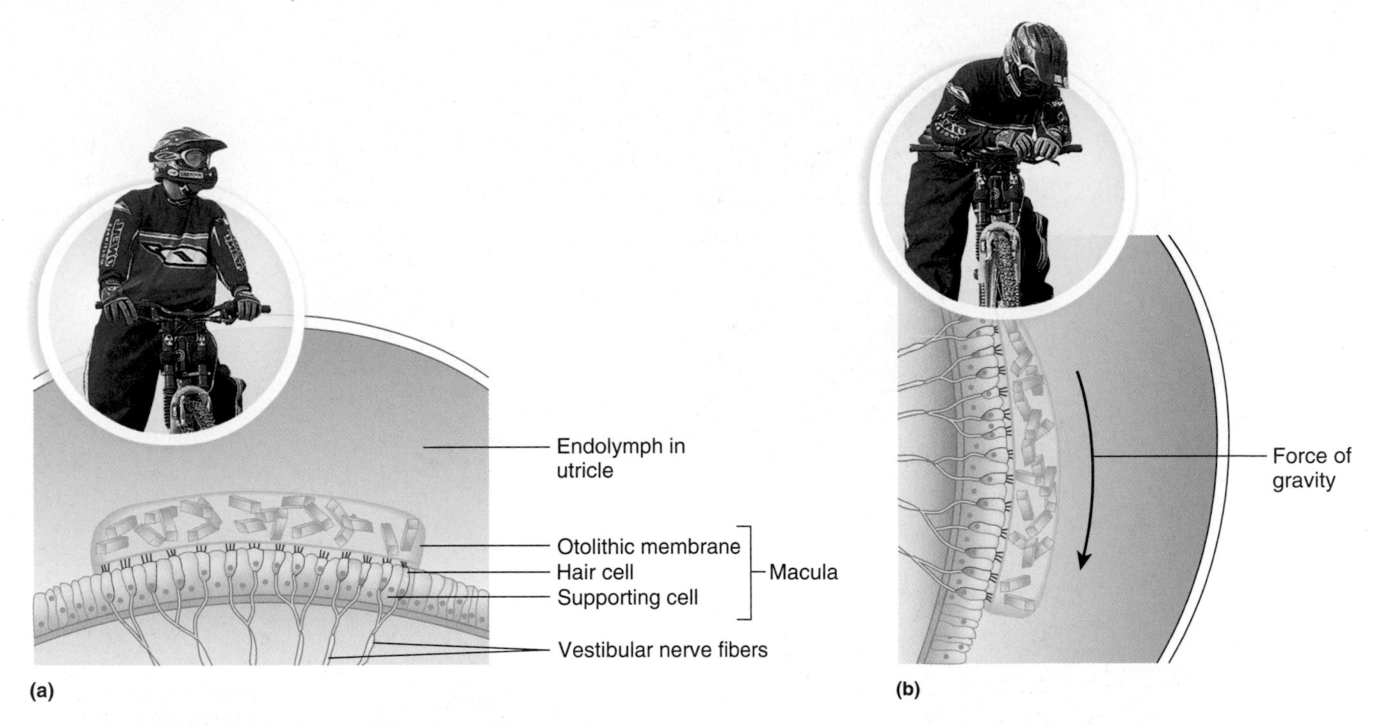

FIGURE 7.16 The effect of gravity on the macula. (a) In an upright position, the otoliths press equally on the hair cells; (b) when the head is tilted, the force of gravity causes the otoliths to bend the hair cells of the macula.

otoliths to move in the direction of the force of gravity, which bends the hair cells. The hair cells are again mechanoreceptors, which, when bent, release a neurotransmitter to initiate a local potential in the bipolar neurons at their base. The axons of these neurons form the vestibular branch of the auditory nerve.

Dynamic equilibrium is perceived from information transmitted by neurons associated with the semicircular canals. See **Figure 7.17**. The three semicircular canals are oriented in different planes to give the brain sensory input concerning rotation of the head in the transverse, coronal, and/or sagittal planes. Each semicircular canal has a patch of hair cells called the **crista ampullaris** located in the bulge at its base, called the **ampulla.** The crista ampullaris has a gel cap called the **cupula.** Each semicircular canal is filled with fluid called **endolymph.** When the head rotates, the endolymph starts flowing in the semicircular canal oriented in that plane. This bends the hair cells of the crista ampullaris. The hair cells respond by releasing a neurotransmitter to stimulate a local potential in the bipolar neurons at their base. The axons of these neurons also form the vestibular nerve.

Physiology of Equilibrium

7.16 learning **outcome**

Explain how equilibrium is perceived.

The sense of equilibrium is somewhat more complicated than these structures in the inner ear can sense alone. To create a sense of equilibrium, the brain uses additional input, such as sight from the eyes and body position from proprioceptors in the muscles and joints, and combines it with input from the inner ear. Many people pay to experience a challenge to the brain's sense of equilibrium by buying a ticket for a roller-coaster or some other amusement park ride.

spot check 6 What type of equilibrium is involved if a roller coaster has a complete vertical loop? What anatomy is used to sense the equilibrium while you are making the loop?

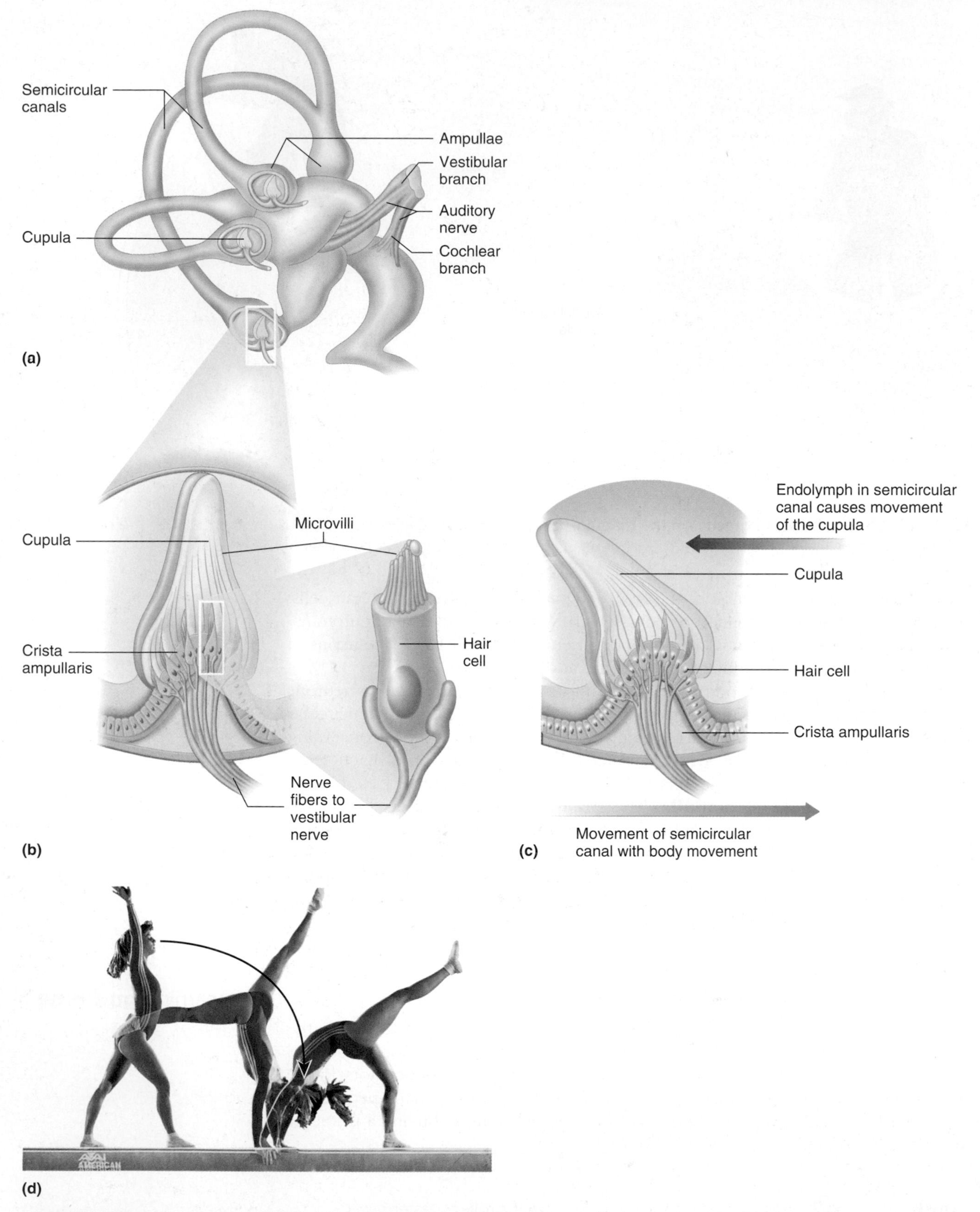

FIGURE 7.17 Semicircular canals: (a) structure of the semicircular canals; (b) detail of the crista ampullaris and hair cell; (c) action of the endolymph on the crista ampullaris and cupula during rotation shown in (d): as the body moves during rotation, the semicircular canals move but the endolymph remains stationary *(red arrow)* in relation to the direction of movement *(blue arrow)*; (d) rotation in the sagittal plane.

clinical point

Motion sickness may result if the brain cannot accurately interpret the information it is receiving. This may happen when you are aboard a boat, in a car, or even in the weightless environment of space. Giving the brain additional input, such as staring at the horizon while in the boat or sitting in the front seat of the car to see where you are going, can help alleviate the symptoms associated with motion sickness, such as nausea.

7.17 learning outcome

Describe the pathway for equilibrium.

Pathway for Equilibrium The pathway for sensory messages for equilibrium has more than one target. It begins with bipolar neurons in the vestibular nerve, which then merges with the neurons of the cochlear nerve to become the auditory nerve, which then synapses with other neurons within the medulla oblongata. The four destinations for these messages are the following:

- **Spinal cord,** which initiates reflexes if the position of the head changes abruptly.
- **Cerebellum,** which uses the information to determine the position of the head for coordination and posture.
- **Neurons in cranial nerves III, IV, and VI,** which coordinate eye movements. They can compensate (balance) the movement of the eyes based upon the position of the head and how it may be moving.
- **Thalamus,** which directs the sensory messages of equilibrium to the frontal and parietal lobes.

Bipolar neurons of CN VIII → medulla oblongata → pons → cerebellum

medulla oblongata → pons → midbrain → thalamus → frontal lobe

thalamus → parietal lobe

medulla oblongata → cranial nerves III, IV, VI

medulla oblongata → spinal cord

Vision

7.18 learning outcome

Describe the anatomy of the eye.

The eye is a complicated sensory organ for vision. Unlike any of the other senses, it uses light to stimulate the receptors for vision, called **photoreceptors.** To understand the anatomy and physiology of vision, you first need to explore some characteristics of light. Light is emitted from a source (e.g., the lightbulb in your room). The light travels in all directions and is reflected (bounced) off objects in straight lines. You can "see" only the straight lines of light that enter your eye. It is important to understand that light travels in straight lines but can be bent as it passes through materials of different densities. This phenomenon is called **refraction.** You can demonstrate this by putting a pencil in a glass half full of water. Light from the source reflects off the portion of the pencil in the water differently than it reflects off the portion of the pencil out of the water, surrounded by air. Water is denser than air, so the light is refracted and the pencil appears bent. As you explore the anatomy of the eye, you will see how all of this works.

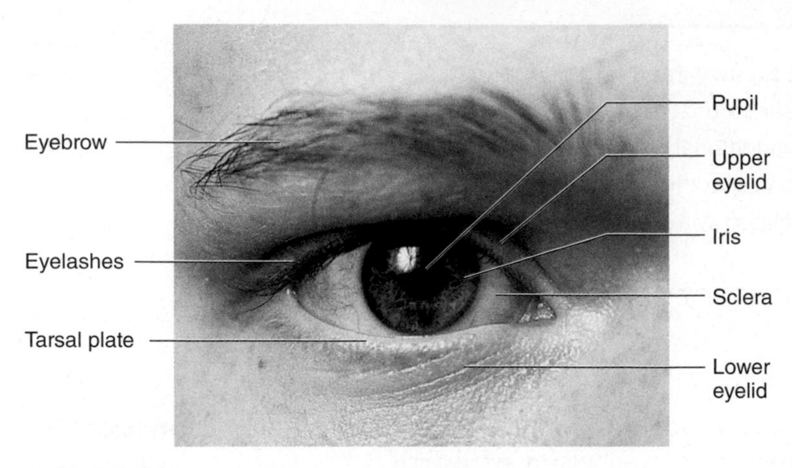

FIGURE 7.18 Orbital region.

Anatomy of the Eye

The Orbital Region The eyes are protected by the **eyebrows, eyelids,** and **eyelashes.** The eyebrows shade the eyes from the glare of the sun and help prevent sweat from entering the eyes. The eyelids blink periodically to distribute moisture across the surface of the eyes. Each eyelid contains **tarsal glands** along its edge **(tarsal plate),** which secrete an oil that helps lubricate the eye. See Figures 7.18 and 7.20. The eyelids are also involved in reflexes that prevent foreign matter, like dust or particles of debris, from entering the eye. The eyelashes, too, help keep debris from entering the eye.

A **lacrimal gland** is located deep to the skin, lateral and superior to each eye. See Figure 7.19. Its function is to produce tears. Tears contain mostly water, to cleanse the eye, and lysozymes to destroy bacteria in order to prevent an eye infection. The tears are secreted through ducts, wash over the eye, and are drained through an opening called the **lacrimal punctum** at the medial, inferior corner of the eye. From there, they are drained through a **lacrimal canal** to the **lacrimal sac** to the **nasolacrimal duct** to the nose. See Figure 7.19.

spot check 7 Which would be the best place to administer eyedrops to a patient to treat an eye infection: the lateral corner of the eye, the middle of the eye, or the medial corner of the eye?

spot check 8 Why does your nose run when you cry?

conjunctiva:
kon-junk-TIE-vah

Figure 7.20 shows another important accessory structure of the eye, the **conjunctiva.** The conjunctiva is a thin, transparent membrane that lines the eyelids and covers the white, exposed surface of the eye. Its purpose is to secrete a mucous film to prevent the eye from drying.

FIGURE 7.19 Lacrimal gland and ducts.

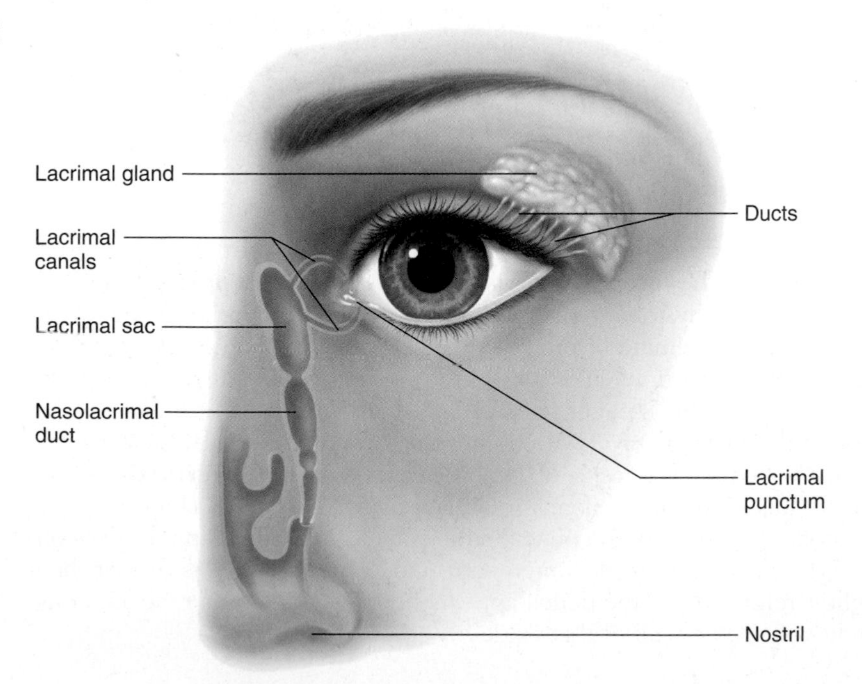

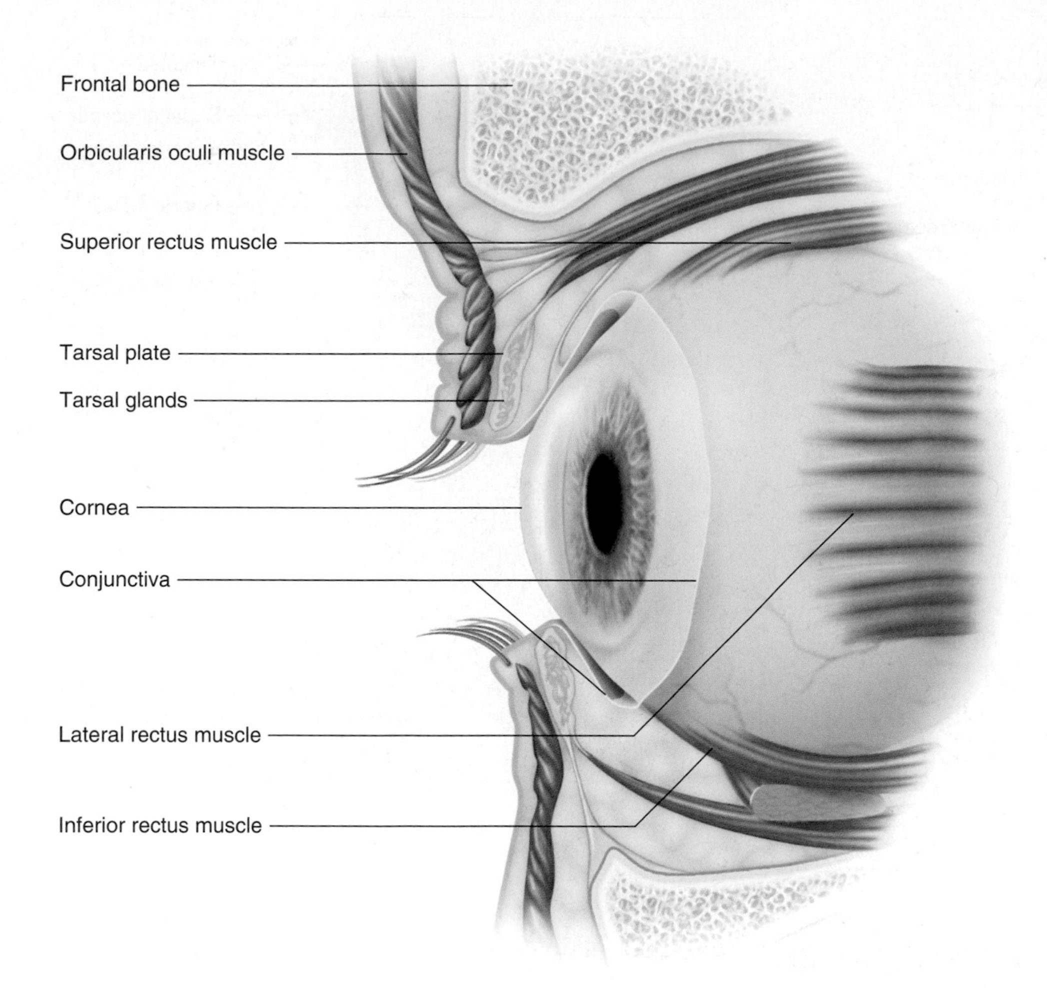

FIGURE 7.20 Accessory structures of the eye.

clinical **point**

Inflammation of the conjunctiva—**conjunctivitis**—results in redness and pain. A common form of conjunctivitis is called *pink eye.* It is a highly contagious, bacterial infection that is treated with antibiotic drops. Two other forms of conjunctivitis can occur with newborns delivered through an infected birth canal. They are *neonatal gonorrheal conjunctivitis* and *neonatal chlamydial conjunctivitis.* Both of these forms of conjunctivitis can cause blindness in the newborn. So all newborns are typically treated with silver nitrate and antibiotic drops in the eyes immediately after birth to prevent either form of conjunctivitis.

Muscles of the Eye Just as there are muscles of the shoulders, neck, and head and all throughout the body, there are also muscles connected to the eye. **Figure 7.21a** shows the muscles that are attached to the eyeball. The muscles are termed *rectus* if they directly approach the front of the eye in a straight line, and they are *oblique* if they approach the eye at an angle to the front of the eye. The muscles control the movement of the eyeball. **Figure 7.21b** shows which cranial nerve controls each muscle.

FIGURE 7.21 **Muscles of the eye:** (a) lateral view, (b) anterior view showing which cranial nerve controls each eye muscle.

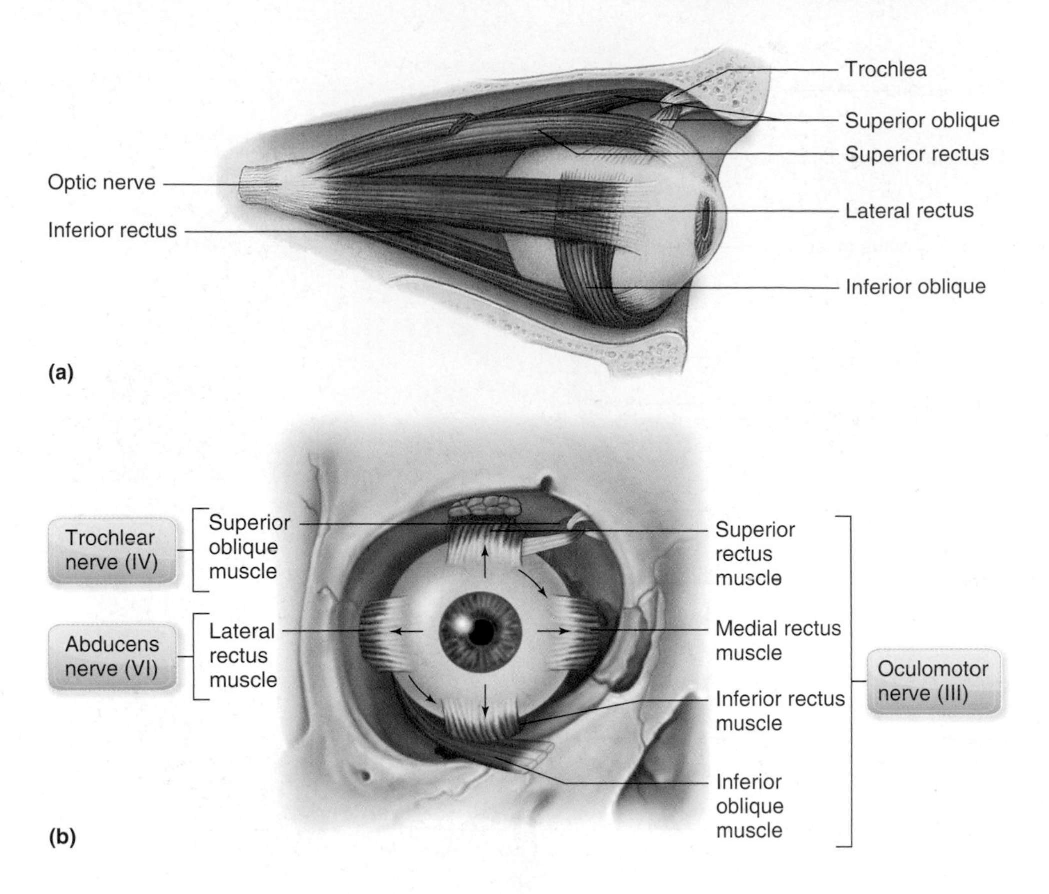

spot check ❾ The muscles of the eye and the cranial nerves that stimulate these muscles can be tested by having the subject move the eye in an "H" pattern. Which eye muscles move the eye in a transverse, side-to-side motion for the bar of the "H"? Which cranial nerves stimulate each of these muscles?

The Eyeball The anatomy of the eyeball is shown in **Figure 7.22**. Refer to the figure as you read the following text. The wall of the eye is composed of three layers:

- **Sclera.** The sclera is the outermost layer. It is a tough, fibrous layer that does not stretch. It can be seen as the white of the eye. The anterior part of the sclera is transparent. This section is called the **cornea.** It needs to be transparent to allow light to enter the eye.
- **Uvea.** The uvea is the middle layer of the eye's wall. It consists of three regions: the **choroid layer,** the **ciliary body,** and the **iris.** The choroid layer is dark and has many blood vessels to feed the neurons located on the inner layer of the eye. It is highly pigmented and dark so that light is absorbed and not reflected inside the eye. The ciliary body is composed of smooth muscle and forms a ring around the lens. **Suspensory ligaments** extend from the ciliary body and suspend the lens in a capsule. The ciliary body, the suspensory ligaments, and the capsule enclosing the lens form a wall within the eye. The ciliary body also produces a thin, watery fluid called **aqueous humor** found in the **anterior** and **posterior chambers** of the eye. Because this fluid is constantly produced, it must constantly be reabsorbed into the blood through a blood vessel called the **canal of Schlemm** (not shown in the figure). The iris of the eye is seen as the colored part of the eye. It is smooth muscle that regulates the size of the pupil, the central opening that allows light to pass to the lens.

choroid: KOR-oid

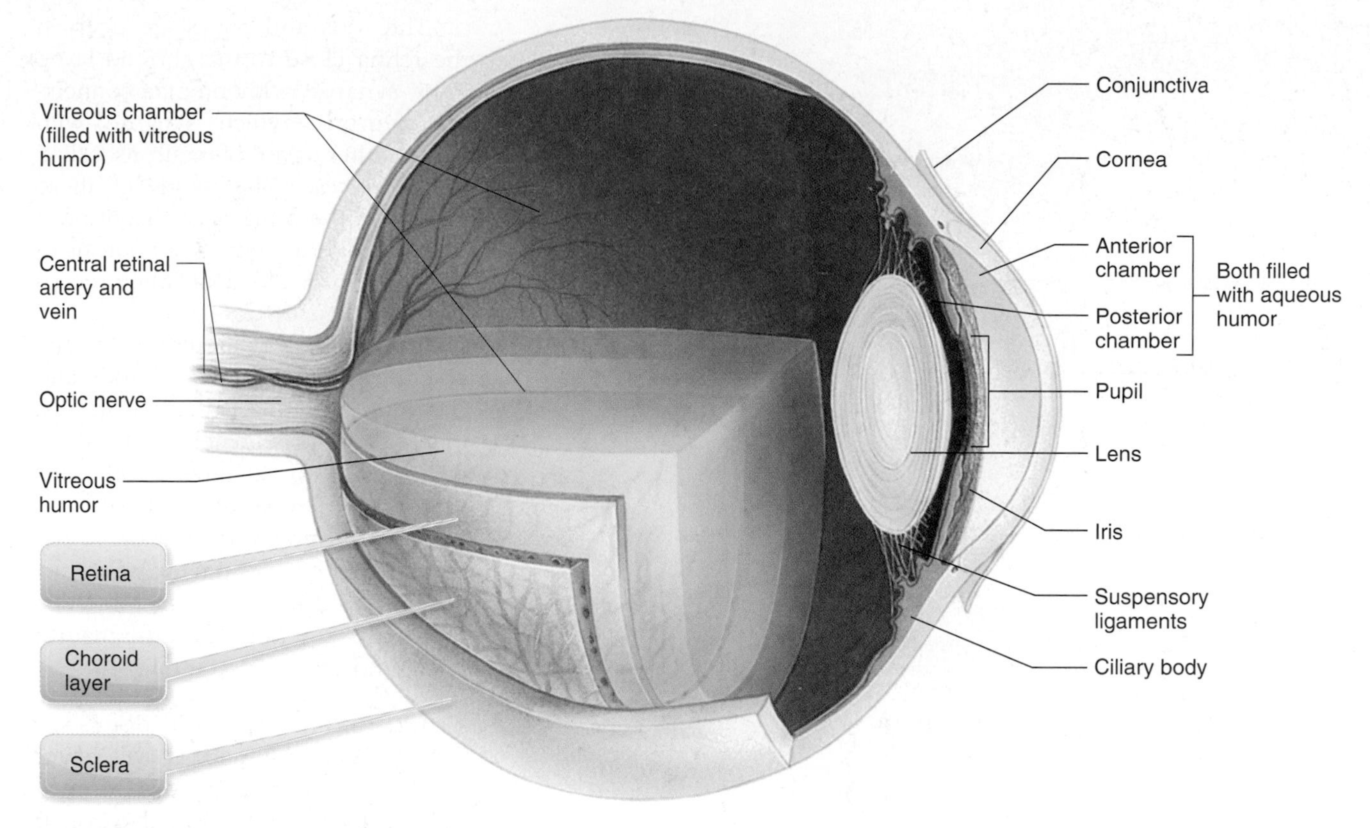

FIGURE 7.22 Sagittal view of the eye.

- Retina. The retina is the innermost layer of the eye's wall. It lines the **vitreous chamber** that is filled with **vitreous humor,** a transparent, gel-like fluid. The retina contains the photoreceptors and associated neurons important for vision. The axons of the bipolar neurons come together to leave the retina of the eye to form the optic nerve.

clinical point

It is vitally important that the drainage of aqueous humor keeps up with its production. If more aqueous humor is produced than is drained, pressure builds up in the anterior portion of the eye. The sclera does not stretch to accommodate the added pressure. The increased pressure instead pushes on the lens, which pushes on the vitreous humor, which pushes on the fragile neurons of the retina and compresses the blood vessels feeding the retina. Without a good blood supply, the neurons on the retina die. This condition is called **glaucoma,** and it is a major cause of blindness. A **tonometer** can be used in a routine eye exam to determine the intraocular pressure (pressure inside the eye). Early detection can stop the progression of glaucoma, but damaged neurons cannot be restored.

There are two types of photoreceptors in the retina of the eye: rods and cones. Rods are photoreceptors used for gray-scale (noncolor vision) and low-light conditions. They contain a chemical called **rhodopsin** that reacts to light to initiate a local potential. Cones are photoreceptors used for color vision and are responsible for the best **visual acuity** (sharpest vision). They contain the chemical **iodopsin** that reacts to light to initiate a local potential. Cones require more light than rods to function. Cones primarily respond to one color, either green, red, or blue. However, each cone will have a mild response to the other two colors.

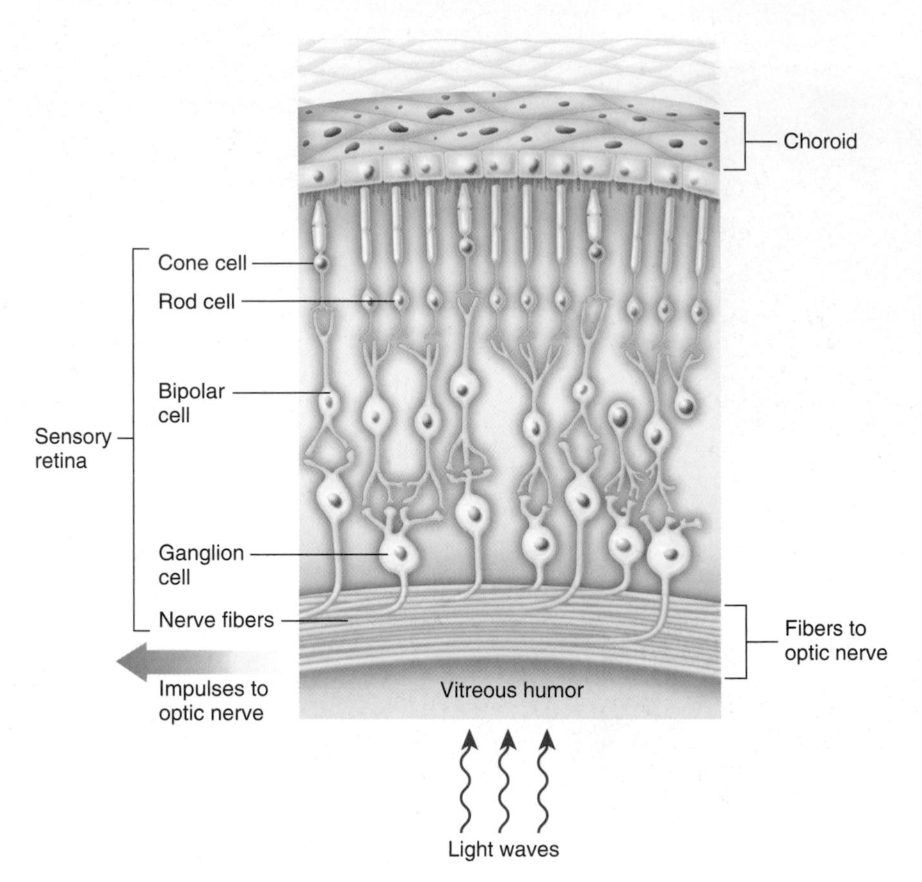

FIGURE 7.23 Neurons of the retina.

The rods and cones are deep in the retina close to the choroid layer. They synapse with bipolar connecting neurons, which synapse with still another layer of neurons called *ganglion cells*. The axons of these cells form the optic nerve. Light must pass through the layers of neurons to reach the rod and cone photoreceptors. See **Figure 7.23**.

There is a difference in the amount and distribution of rods and cones in the retina. More specifically, there are approximately 120 million rods, or 10 to 20 times more rods than cones, in the retina. If you use an **ophthalmoscope** to look at the retina at the back of the eye as shown in **Figure 7.24**, you can see three distinct areas: the optic disc, the macula lutea, and the fovea centralis.

- Optic disc. The optic disc is the area where blood vessels enter the eye and the axons forming the optic nerve leave the eye. It is located medially on the retina. There are no receptor cells at the optic disc, so it is a functional blind spot.
- Macula lutea. The macula lutea is located on the retina directly posterior to the center of the lens. The macula lutea has more cones than rods.
- Fovea centralis. The fovea centralis appears as a small depression in the center of the macula lutea. It contains only cones, so it is the area of sharpest vision. As you read this, you are now moving your eyes across the page to make sure the letters are focused on the fovea centralis in each of your eyes.

The rest of the retina contains a mixture of rods and cones, which allows you to see the color of objects focused anywhere in your field of vision whether you are looking directly at the object or not.

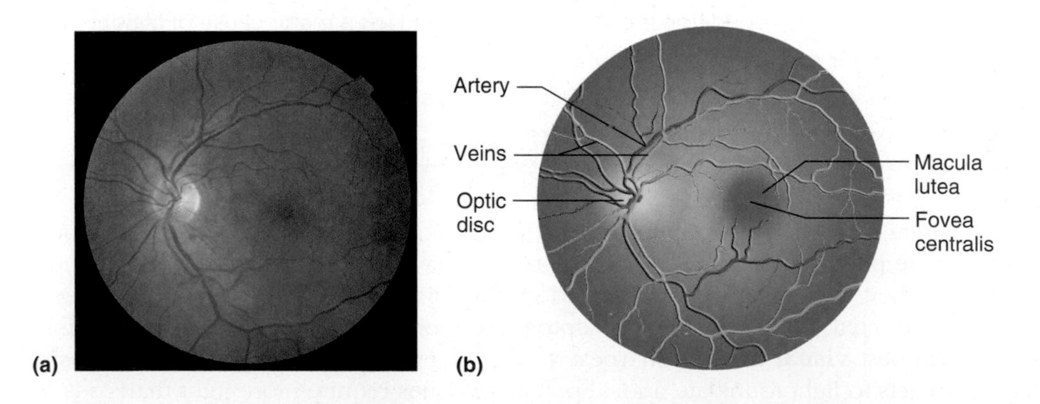

FIGURE 7.24 The retina as seen through an ophthalmoscope. (a) A direct ophthalmoscope shines a fine beam of light into the eye, so the examiner sees a magnified image of the retina where the beam falls; (b) diagram of the retina.

spot check ⑩ You are sitting outside in the evening as it is getting dark. Predict what would be easier: determining the shape of a car driving by or determining whether the color of the car is dark blue or black. Explain in terms of the photoreceptors involved.

You have covered the anatomy of the eye and the photoreceptors involved. Next, you will see how the anatomy and photoreceptors work together to provide a sense of vision.

Physiology of Vision

7.19 learning **outcome**

Explain how vision is perceived.

To understand how the physiology of vision works, look at the two examples in **Figure 7.25**. In **Figure 7.25a**, a tree is shown at a distance (indicated by the broken lines). Light reflected off the top of the tree strikes the cornea and is refracted because the cornea is a different density than the air. The light passes through the aqueous humor and then through the lens, where it is again refracted. The light passes through the vitreous humor and is projected low on the retina. At the same time, light reflected from the bottom of the tree travels in a straight line too. It hits the cornea and is also refracted. This light then passes through the aqueous humor and is refracted by the lens as it passes through the vitreous humor to be projected high on the retina. As you can see in **Figure 7.25a**, the image of the tree on the retina is clear and focused but upside down. All images are projected on the retina upside down, which is okay because the retina is not actually interpreting the information. It is the association area in the occipital lobe that interprets the visual signals.

In this case, the smooth curve of the cornea and the smooth curve of the lens have refracted the light to perfectly focus it on the retina. If the cornea and lens focused the image ahead of the retina, the condition is called **myopia** (nearsightedness). If the cornea and lens focused the image behind the retina, the condition is called **hyperopia** (farsightedness). Either condition can be corrected with artificial lenses or refractive surgery to reshape the cornea to act as a lens. If the cornea or the lens is not a perfectly smooth curve, the light rays will not refract correctly to produce a clearly focused image on the retina. This condition is called an **astigmatism.** It too can be corrected with artificial lenses.

Now compare the near-vision example in **Figure 7.25b** with the example you just reviewed for distant vision. Notice that the image of the "A" in the near-vision example is clear, focused, and upside down on the retina. It is larger for near vision because the

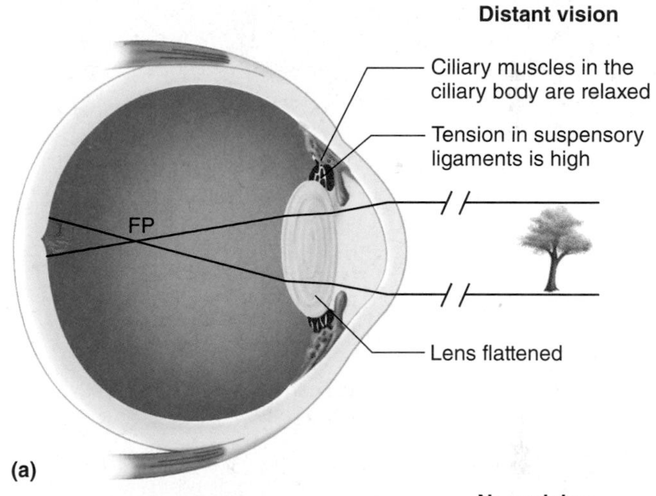

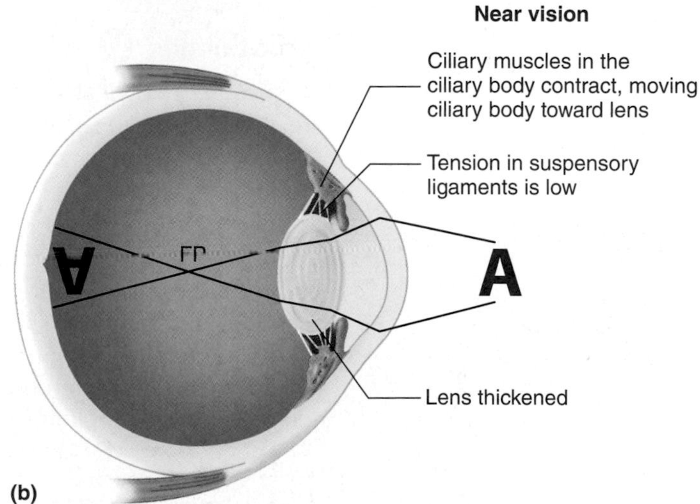

FIGURE 7.25 Focus and accommodation by the eye: (a) distant vision, (b) near vision.

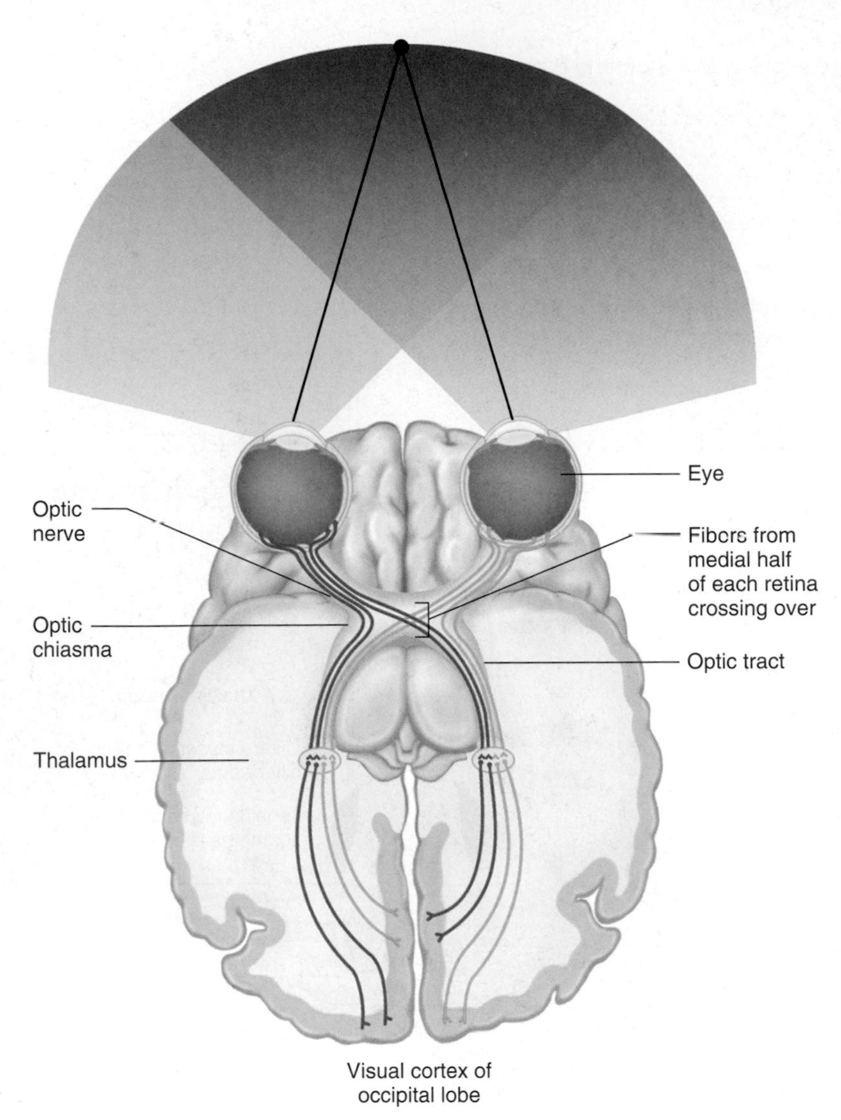

FIGURE 7.26 Pathway for vision.

object is closer. The eye has to accommodate for the change in distance to keep the image in focus on the retina. It does that by changing the shape of the lens. For distant vision, the ring of ciliary muscles around the lens is relaxed (**Figure 7.25a**). This creates tension in the suspensory ligaments that hold the lens in place. The tension pulls on the lens from all sides, causing it to become thinner. But in the near-vision example (**Figure 7.25b**), the ciliary muscles contract to take tension off the suspensory ligaments, allowing the lens to become thicker. Thus, the eye is able to change its focus so that the image of the "A" is perfectly focused on the retina whether it is near or far. This ability is called **accommodation.**

You have now seen the "A", but how do you locate it in space? Humans have two eyes to view objects. This is called **binocular vision.** As you can see in **Figure 7.26**, the occipital lobe has two sets of sensory messages coming in from different angles, one set from the right eye and one set from the left eye. Binocular vision allows the brain to have **depth perception,** knowing where the object is located in space. You can witness this by holding a small tube in front of someone and having that person quickly insert his or her pencil in the tube, first with both eyes open and then with one eye closed. It is much easier to insert the pencil into the small opening if both eyes are open because of binocular vision and depth perception.

spot check **11** You have injured your eye and now need to wear a patch over it while it heals. What activities should you avoid while wearing the patch because of the loss of vision in that eye?

chiasm: KYE-asm

By studying the anatomy of the eye and properties of light, you know that light enters the eye, is refracted by the different mediums it passes through, and is projected on the retina. The light penetrates the retina to activate the rods and cones to start local potentials. The rods work in low-light conditions, and the cones detect color if there is sufficient light. The rods and cones stimulate connecting neurons, which further stimulate other neurons whose axons leave the eye to form the optic nerve. Below, we look at the pathway from this point.

7.20 learning **outcome**

Describe the pathway for vision.

Pathway for Vision Notice in **Figure 7.26** that the two optic nerves (CN II) come together at the **optic chiasm** (or **chiasma**) inferior to the hypothalamus. At the optic chiasm, some of the nerve fibers from each nerve cross to the other side to form optic tracts. Therefore, each optic tract contains neurons carrying messages from both eyes. The optic tracts continue to the thalamus before going on to the occipital lobe's general sensory and association areas for vision. Some of the neurons carry sensory messages to the superior colliculi of the midbrain, where visual reflexes are controlled.

Optic nerve → optic chiasm → optic tracts → thalamus → occipital lobe
↳ superior colliculi of midbrain

Effects of Aging on the Senses

7.21 learning outcome

Describe the effects of aging on the senses.

As you age, the general and special senses are affected, causing a gradual change in the way you experience the various senses. It is not always readily apparent if aging itself or lifestyle choices and disease are the cause of the changes.

Changes in the General Senses

In terms of noticeable differences in the general senses, there may be a changed sensation of pain, vibration, cold, heat, pressure, and touch. Pain sensitivity is reduced usually after age 50. An individual may be aware of the sensation, but it does not bother her as much as it would have at an earlier age. The elderly may not be able to differentiate between cool and cold or warm and hot as easily as when they were younger. Light touch sensitivity, however, may be increased due to the thinning of the skin.

Changes in Taste

The number and size of taste buds decrease beginning at about age 50. A decrease in the sensitivity to taste usually occurs at about age 60. With this decreased sensation, the interest in food may diminish as well. Lifestyle choices such as smoking and diseases such as diabetes may be the cause of the changes in taste perception.

Changes in Smell

The number of nerve endings in the human nose typically decreases by age 70. This may further decrease the perception of flavors of food. Again, environmental, lifestyle, and/or disease factors may be as responsible as aging, or even more responsible, for the decrease in sensitivity of this sense.

Changes in Hearing

Hearing is affected by aging. It is estimated that 30 percent of people over 65 have significant hearing loss.[1] The loss most often involves high-frequency sounds. Repeated exposure to loud noise over a lifetime may account for the loss. Hearing acuity (or sharpness in hearing) typically starts to diminish about age 50. The tympanic membrane thickens with age and becomes less flexible, which diminishes the conduction of vibrations. **Tinnitus,** a persistent abnormal ear ringing or roaring noise, is common in the elderly due to mild hearing loss.

Changes in Equilibrium

The brain uses many sensory clues to sense body position, and an aging effect on any of these sensory clues affects equilibrium. As you age, it becomes more difficult for the brain to adapt to the changing input. For example, age-related changes in the inner ear and the vestibular nerve can affect the sense of equilibrium. This makes walking more difficult, which is why falls are more common among the elderly. It is also why roller coasters affect us more as we age.

Changes in Vision

Vision is affected by aging. Fewer tears are produced, making the eyes drier and more uncomfortable. The lens becomes less flexible and cloudy, making the sun's glare or bright lights more of a problem. The iris is not as fast at changing the size of the pupil in response to changes in the amount of light. Insoluble proteins called *floaters* may form in the fluid of the eye. The ciliary muscles and suspensory ligaments are less able to accommodate the shape of the lens with age, creating a condition called **presbyopia.** The onset of this decreased ability to accommodate is dramatic at about age 40. Distance vision is not affected with presbyopia, but corrective lenses (reading glasses) may be required for near vision.

7.22 learning **outcome**

Describe disorders of the senses.

Disorders of the Senses

Hearing Loss

The two types of hearing loss are dependent upon the location of the problem: conductive hearing loss and sensorineural hearing loss.

Conductive Hearing Loss Conductive hearing loss is caused by a lesion in the outer or middle ear, preventing the proper conduction of vibrations to the inner ear. It could be anything from a thickened or ruptured tympanic membrane to impacted cerumen against the tympanic membrane to arthritis in the joints of the ossicles of the middle ear.

Sensorineural Hearing Loss Sensorineural hearing loss is a problem with the organ of Corti or the auditory nerve.

clinical point

Conductive and sensorineural hearing loss can be determined by testing hearing by air and bone conduction. A tuning fork is used in both tests. First, a vibrating tuning fork is placed close to the ear. Sound waves produced by the fork are received by the ear normally through the air to the outer ear, and then vibrations should be conducted to the middle ear. In the second test, the base of a vibrating tuning fork is placed on the bone of the skull behind the ear, causing the inner ear to vibrate directly. If the person cannot hear the sound produced by the tuning fork through the air but can hear it when the tuning fork is placed on bone, the problem is not with the organ of Corti or the auditory nerve. Instead, this result indicates a conduction problem and conductive hearing loss. If the subject cannot hear in either case, the problem is sensorineural *and* possibly conductive hearing loss.

Cataracts

A cataract is a progressive, painless loss of vision due to the clouding of the lens of the eye. It can occur at any age, but it is far more commonly associated with getting older. Cataracts may be caused by aging, exposure to x-rays, infection, diabetes mellitus, and even some medications. Surgery may be used to remove the clouded lens and replace it with an artificial lens.

summary

Overview

- General senses are touch, pressure, stretch, heat, cold, and pain. Receptors for general senses are located throughout the body.
- Special senses are taste, smell, hearing, equilibrium, and vision. Special sense organs are located in the head.

General Senses

Anatomy of Receptors in the Skin

- General senses are detected by thermoreceptors, mechanoreceptors, and nociceptors.

Physiology of General Senses in the Skin

- The neurons for general senses send messages on the type, location, intensity, and duration of the sensation.
- The pathway includes cranial nerves if the general sense is located in the head. The pathway includes the spinal nerves and spinal cord if the general sense is located below the head.
- Messages sent to the hypothalamus and amygdala may initiate visceral and emotional responses.

Taste

Anatomy of Receptors for Taste

- The special sense organ for taste is the taste bud. Taste buds are located mostly on the tongue.
- Taste cells have taste hairs that are chemoreceptors.

Physiology of Taste

- There are five primary tastes: salt, sweet, sour, bitter, and umami.
- Other sensory inputs are used to produce the sensations of flavors.
- The pathway for taste uses three cranial nerves and ends in the parietal lobe.

Smell

Anatomy of Receptors for Smell

- Olfactory cells are bipolar neurons.
- Olfactory cells access the olfactory mucosa of the roof of the nasal cavity through the foramen of the cribriform plate.
- Olfactory hairs are chemoreceptors.

Physiology of Smell

- The pathway for smell involves the olfactory nerve and does not go through the thalamus on its way to the frontal lobe.

Hearing

Anatomy of the Ear

- Frequency of sound waves, measured in hertz, determines pitch.
- Volume is measured in decibels.
- The ear can be divided into three sections: the outer ear, the middle ear, and the inner ear.

Physiology of Hearing

- Sound waves are directed into the auditory canal and cause the tympanic membrane to vibrate.
- The tympanic membrane causes the ossicles of the middle ear to vibrate.
- The stapes causes the oval window to vibrate.

- Waves of perilymph, caused by the vibrations of the oval window, cause the basilar membrane to vibrate.
- The vibrating basilar membrane pushes hair cells against the tectorial membrane of the organ of Corti, initiating a local potential.
- Pitch is determined by which hair cell is bent.
- Volume is determined by how much hair cells are bent.
- The pathway for hearing starts with the auditory nerve and ends in the temporal lobe.

Equilibrium

Anatomy of the Vestibular Apparatus

- The receptors for equilibrium are housed in the vestibular apparatus of the inner ear.
- Mechanoreceptors in the saccule and utricle detect static equilibrium.
- Mechanoreceptors in the semicircular canals detect dynamic equilibrium.

Physiology of Equilibrium

- The pathway for equilibrium has many destinations.

Vision

Anatomy of the Eye

- The eye uses photoreceptors to detect light.
- Rectus and oblique muscles stimulated by CN III, IV, and VI move the eye.
- The wall of the eye has three layers: the sclera, the uvea, and the retina.
- There are two types of photoreceptors in the eye: rods and cones.
- Rods function well in low light and do not detect color.
- Cones require more light and detect color. They offer the sharpest vision.
- There are more rods than cones on the retina.
- The rods and cones are not evenly distributed. The fovea centralis has only cones.

Physiology of Vision

- Light travels in straight lines and is refracted as it passes through materials of different densities.
- Objects are projected upside down on the retina.
- Binocular vision allows for depth perception.
- The pathway for vision starts with the optic nerve and ends with the occipital lobe.

Effects of Aging on the Senses

In general, the senses are affected by aging, but it is not always clear if it is aging itself or lifestyle choices, environmental exposures, and/or diseases that are actually responsible. Most sensory ability decreases with age, with some senses being more affected than others.

Disorders of the Senses

Hearing Loss

- Conductive hearing loss is caused by a lesion in the outer or middle ear that prevents the proper conduction of vibrations to the inner ear.
- Sensorineural hearing loss is a problem with the organ of Corti or the auditory nerve.

Cataracts

- A cataract is a progressive, painless loss of vision due to the clouding of the lens of the eye.
- Cataracts can be treated by surgically removing the lens and replacing it with an artificial lens.

key words for review

The following terms are defined in the glossary.

accommodation
dynamic equilibrium
gustation
humor
hyperopia
lacrimal
lingual
myopia
nociceptor
olfactory
otoliths
perilymph
presbyopia
receptive field
refraction
sensorineural
static equilibrium
umami
vestibular apparatus
visual acuity

chapter review questions

Word Building: *Use the word roots listed in the beginning of the chapter and the prefixes and suffixes inside the back cover to build words with the following meanings:*

1. Study of the ear: ______________________
2. Person who studies hearing: ______________________
3. Instrument to examine the eye: ______________________
4. Instrument to examine the ear: ______________________
5. Condition of old eyes: ______________________

Multiple Select: *Select the correct choices for each statement. The choices may be all correct, all incorrect, or any combination of correct and incorrect.*

1. What is included in the anatomy of the ear?
   a. A pinna composed of hyaline cartilage that directs sound waves into the ear.
   b. A bone called the *stapes* that transmits vibrations directly to the round window.
   c. Ceruminous glands in the auditory canal that produce perilymph.
   d. An auditory tube that connects the middle ear to the nasopharynx.
   e. A middle-ear portion that is embedded in a bony labyrinth.
2. What is included in the pathway for pain detected in the skin of the arm?
   a. A cranial nerve is involved.
   b. Messages to the hypothalamus may produce an emotional response.
   c. The spinal cord is in both possible pathways.
   d. The thalamus is included in both possible pathways.
   e. The hypothalamus and amygdala are included in one of the pathways.

3. What is included in the senses?
   a. Specialized sense organs for the special senses located all over the body.
   b. Receptors for general senses located only in the head.
   c. Special senses that include thermoreceptors, mechanoreceptors, chemoreceptors, and photoreceptors.
   d. General senses to detect touch, pressure, pain, heat, cold, and stretch.
   e. Special senses used for gustation, olfaction, hearing, equilibrium, and vision.
4. What happens to disrupt function in sensory disorders?
   a. A problem with the organ of Corti or the auditory nerve is responsible for conductive hearing loss.
   b. Presbyopia is caused by a lack of accommodation.
   c. Arthritis between the bones of the middle ear may cause sensorineural hearing loss.
   d. Cataracts can be relieved by opening the canal of Schlemm to drain excess aqueous humor.
   e. Glaucoma is a progressive clouding of the lens.
5. How are odors perceived?
   a. Odor molecules push on mechanoreceptors.
   b. Odor molecules pass the olfactory mucosa located in the floor of the nasal cavity.
   c. Olfactory hairs have receptors for odor molecules.
   d. The sensory messages are interpreted in the frontal lobe.
   e. Each receptor is specific to an individual odor.
6. How is sound perceived?
   a. The frequency of sound waves determines volume.
   b. Bipolar neurons of the vestibular nerve carry sound messages to the temporal lobe.
   c. Volume is measured in hertz.
   d. Volume is perceived in the saccule.
   e. The vibrations produced at the oval window are amplified 20 times by the time they reach the tympanic membrane.
7. How could you describe the vestibular apparatus?
   a. It is located in the middle ear.
   b. It has three semicircular canals, a saccule, and a utricle.
   c. It has hair cells used to detect the position of the head when it is still and when it is moving.
   d. Hair cells in the semicircular canals are in the crista ampularis.
   e. Hair cells in the saccule and utricle are in patches called macules.
8. How is equilibrium perceived?
   a. Static equilibrium is associated with the saccule and utricle.
   b. Dynamic equilibrium is associated with the semicircular canals.
   c. Information from proprioceptors in the tendons is added to sensory messages from the inner ear to determine static and dynamic equilibrium.
   d. Sensory messages from the inner ear are carried by the auditory nerve to multiple destinations to perceive equilibrium.
   e. Going up in an elevator is detected in the saccule and utricle.
9. How is vision perceived?
   a. Light passes first through the cornea, then the vitreous humor, then the lens, and then the aqueous humor and on to the retina.
   b. An image must be focused on the optic disc for the best visual acuity.
   c. All images are interpreted by the frontal lobe, right behind the eyes.
   d. Ciliary muscles contract to put tension on the suspensory ligaments to change the shape of the lens.
   e. Photoreceptors use the chemicals rhodopsin and iodopsin, which react to light.
10. How are images focused on the retina?
    a. Muscles move the eyes so that images can be focused on the fovea centralis.
    b. The cornea and lens are curved to refract light.
    c. Hyperopia is a condition in which images are focused before the retina.
    d. Myopia is a condition in which the image is focused behind the retina.
    e. Astigmatism results if the cornea or lens is misshaped.

Matching: *Match the sense to the cranial nerve involved. Some answers may be used more than once.*

_____ 1. Taste
_____ 2. Equilibrium
_____ 3. Vision
_____ 4. Smell
_____ 5. Hearing

a. CN I
b. CN II
c. CN X
d. CN VIII
e. CN IX

Matching: *Match the sense to the type of receptor involved.*

_____ 6. Taste
_____ 7. Pain
_____ 8. Equilibrium
_____ 9. Vision
_____ 10. Smell

a. Photoreceptor
b. Chemoreceptor
c. Mechanoreceptor
d. Thermoreceptor

Critical Thinking

1. What can you do to minimize the effect of aging on your general senses? Explain.
2. Write the pathway for reading this question aloud, starting at the bipolar neurons in the eye and ending where skeletal muscle messages are being sent. (*Tip:* It may help to review pathways for language in Chapter 6.)
3. There is a faint constellation in the northern hemisphere's night sky called the *seven sisters*. If someone points it out to you, that person is likely to tell you to look just off to the side of where he or she is pointing in order to see it. In terms of photoreceptors and their location on the retina, why is this good advice in this situation?

7 chapter mapping

This section of the chapter is designed to help you find where each outcome is covered in this text.

	Outcomes	Readings, figures, and tables	Assessments
7.1	Use medical terminology related to the senses of the nervous system.	Word roots: p. 263	Word Building: 1–5
7.2	Classify the senses in terms of what is sensed and where the receptors are located.	Overview: p. 264	Multiple Select: 3
7.3	Describe the sensory receptors for the general senses in the skin.	General senses: pp. 264–265 Figure 7.2 Table 7.1	Spot Check: 1, 2 Matching: 7
7.4	Explain the types of information transmitted by sensory receptors in the skin.	Physiology of the general senses of the skin: pp. 265–267 Figure 7.3	
7.5	Describe the pathway for pain.	Pathway for pain: pp. 267–268 Figure 7.4	Multiple Select: 2
7.6	Describe the sensory receptors for taste.	Taste: pp. 268–269 Figure 7.5	Matching: 6

	Outcomes	Readings, figures, and tables	Assessments
7.7	Describe the different tastes and explain how flavor is perceived.	Physiology of taste: p. 270	Spot Check: 3
7.8	Describe the pathway for taste.	Pathway for taste: p. 270	Matching: 1
7.9	Describe the sensory receptors for smell.	Smell: pp. 270–271 Figure 7.6	Matching: 10
7.10	Explain how odors are perceived.	Physiology of smell: p. 271	Spot Check: 4 Multiple Select: 5
7.11	Describe the pathway for smell.	Pathway for olfaction: p. 272	Matching: 4
7.12	Describe the anatomy of the ear.	Hearing: pp. 272–275 Figures 7.7–7.11	Multiple Select: 1
7.13	Explain how sound is perceived.	Physiology of hearing: pp. 276–277 Figures 7.12, 7.13	Spot Check: 5 Multiple Select: 6
7.14	Describe the pathway for hearing.	Pathway for hearing: p. 277	Matching: 5
7.15	Describe the anatomy of the vestibular apparatus.	Equilibrium: pp. 277–279 Figures 7.14–7.17	Matching: 8 Multiple Select: 7
7.16	Explain how equilibrium is perceived.	Physiology of equilibrium: pp. 279–281	Spot Check: 6 Multiple Select: 8
7.17	Describe the pathway for equilibrium.	Pathway for equilibrium: p. 281	Matching: 2
7.18	Describe the anatomy of the eye.	Vision: pp. 281–287 Figures 7.18–7.24	Spot Check: 7–10 Matching: 9 Critical Thinking: 3
7.19	Explain how vision is perceived.	Physiology of vision: pp. 287–288 Figure 7.25	Spot Check: 11 Multiple Select: 9, 10
7.20	Describe the pathway for vision.	Pathway for vision: pp. 288–289 Figure 7.26	Critical Thinking: 2 Matching: 3
7.21	Describe the effects of aging on the senses.	Effects of aging on the senses: pp. 289–290	Critical Thinking: 1
7.22	Describe disorders of the senses.	Disorders of the senses: p. 290	Multiple Select: 4

footnote

1. University of Maryland Medical Center. (2009, February 19). *Aging changes in the senses—Overview.* Retrieved June 3, 2010, from http://www.umm.edu/ency/article/004013.htm.

references and resources

BBC News. (2002, February 4). *Women nose ahead in smell test*. Retrieved February 21, 2011, from http://news.bbc.co.uk/2/hi/health/1796447.stm.

Beers, M. H., Porter, R. S., Jones, T. V., Kaplan, J. L., & Berkwits, M. (Eds.). (2006). *Merck manual of diagnosis and therapy* (18th ed.). Whitehouse Station, NJ: Merck Research Laboratories.

Porter, R. S., & Kaplan, J. L. (Eds.). (n.d.). Selected physiologic age-related changes. *Merck manual for healthcare professionals*. Retrieved March 19, 2011, from http://www.merckmanuals.com/media/professional/pdf/Table_337-1.pdf?qt=selected physiologic age-related changes&alt=sh.

Saladin, Kenneth S. (2010). *Anatomy & physiology: The unity of form and function* (5th ed.). New York: McGraw-Hill.

Seeley, R. R., Stephens, T. D., & Tate, P. (2006). *Anatomy & physiology* (7th ed.). New York: McGraw-Hill.

Shier, D., Butler, J., & Lewis, R. (2010). *Hole's human anatomy & physiology* (12th ed.). New York: McGraw-Hill.

University of Maryland Medical Center. (2009, February 19). *Aging changes in the senses—Overview*. Retrieved June 3, 2010, from http://www.umm.edu/ency/article/004013.htm.

8 The Endocrine System

You may have heard about the "raging hormones" of teenagers or a woman "acting hormonal" after a pregnancy, but hormones produced by the endocrine system affect the homeostasis of the body in many different ways on a daily basis throughout life for both sexes. See Figure 8.1.

8.1 learning outcome

Use medical terminology related to the endocrine system.

learning outcomes

After completing this chapter, you should be able to:

8.1 Use medical terminology related to the endocrine system.

8.2 Compare and contrast the endocrine and nervous systems in terms of type, specificity, speed, and duration of communication.

8.3 Define *gland, hormone,* and *target tissue.*

8.4 List the major hormones, along with their target tissues and functions, of each of the endocrine system glands.

8.5 Locate and identify endocrine system glands.

8.6 Describe the chemical makeup of hormones, using estrogen, insulin, and epinephrine as examples.

8.7 Compare the location of receptors for protein hormones with that of receptors for steroid hormones.

8.8 Differentiate autocrine, paracrine, endocrine, and pheromone chemical signals in terms of the proximity of the target tissue.

8.9 Explain the regulation of hormone secretion and its distribution.

8.10 Explain how the number of receptors can be changed.

8.11 Explain how hormones are eliminated from the body.

8.12 Explain the function of hormones by showing how they interact to maintain homeostasis.

8.13 Explain the effects of aging on the endocrine system.

8.14 Describe endocrine system disorders.

word roots & combining forms

aden/o: gland

adren/o: adrenal glands

adrenal/o: adrenal glands

andr/o: male

cortic/o: cortex

crin/o: secrete

dips/o: thirst

gluc/o: sugar

glyc/o: sugar

gonad/o: sex glands

hormon/o: hormone

pancreat/o: pancreas

ster/o: steroid

thyr/o: thyroid gland

pronunciation key

epinephrine: ep-ih-NEF-rin

gonads: GO-nadz

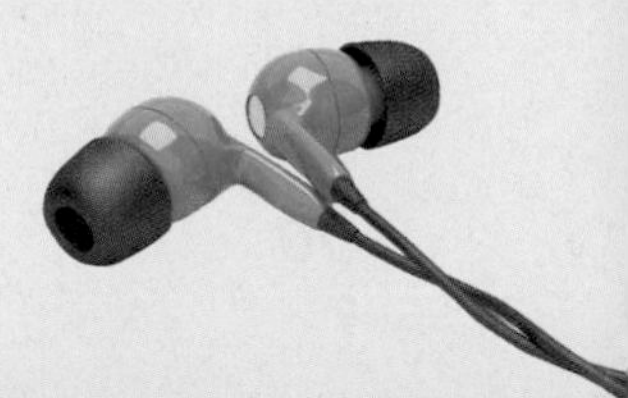

8.2 learning outcome

Compare and contrast the endocrine and nervous systems in terms of type, specificity, speed, and duration of communication.

Overview

Like the nervous system, the endocrine system is all about communication. It involves one part of the body communicating with another part of the body to maintain homeostasis. There are similarities and differences between the two systems. See **Figure 8.2**. Both systems use chemicals as messengers; the nervous system uses chemical neurotransmitters, and the endocrine system uses chemicals called **hormones** to carry messages. As you will remember, the nervous system responds to a stimulus with very fast, electrical action potentials that result in the release of neurotransmitters. Also, the nervous system can be very specific by targeting just a few cells of a muscle. The effects of this communication can be stopped immediately if the nervous system stops sending nerve impulses. In contrast, the endocrine system responds to a stimulus by producing a chemical hormone and secreting it outside the gland that produced it. The blood carries this hormone throughout the body. The hormone will work not just on a few cells but on all of the cells that have a receptor for it. The effects of the hormone will continue until it has been cleared from the target tissue. Communication through the endocrine system is much slower to start, is less specific as to its target, and takes longer to end than communication by the nervous system.

In this chapter, you will investigate the anatomy of the endocrine system including the major glands, hormones, and target tissues. You will study how hormones are regulated, distributed, and eliminated from the body. You will also learn how target tissues regulate their sensitivity to hormones. This chapter explores four scenarios that will help you to understand and see the interaction of glands and their hormones as it relates to their function. Finally, you will cover the effects of aging on the endocrine system and a few endocrine system disorders.

Your study begins with the anatomy of the endocrine system.

Endocrine System

Major Organs and Structures: pineal gland, hypothalamus, thyroid gland, adrenal glands, pancreas, testes, ovaries

Functions: communication, hormone production

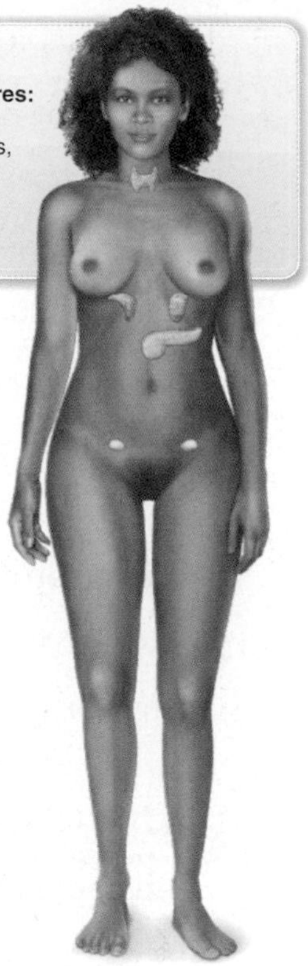

FIGURE 8.1 The endocrine system.

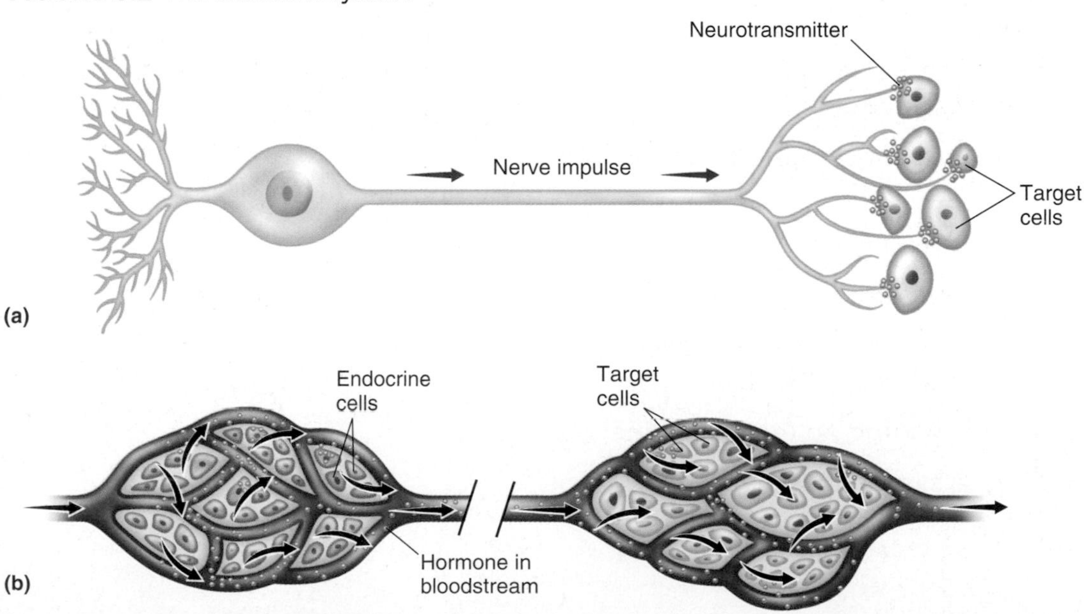

FIGURE 8.2 Communication of the nervous and endocrine systems. (a) A neuron communicates by using action potentials to deliver neurotransmitters to the specific target tissue. (b) Endocrine cells secrete a hormone that is taken up into the bloodstream through capillaries and delivered to a general target tissue, where the hormone leaves the capillaries to fit into receptors.

Anatomy of the Endocrine System

8.3 learning outcome

Define *gland, hormone,* and *target tissue.*

The anatomy of the endocrine system is fairly simple. The system is composed of glands that make chemicals called *hormones* that travel to target tissues to tell them to do something. A **gland** may be a separate structure all on its own, or it may be made up of groups of cells within an organ that function together to produce hormones. The hormones produced by the gland are secreted outside the cells that produce them. There are no special ducts to carry the hormones to their destinations. Instead, the bloodstream is the transportation system. Once secreted by the gland, the hormones are picked up by the blood and travel everywhere the blood travels—to the liver, the eye, even the big toe. Although a hormone travels everywhere, it has an effect only on its **target tissue** because the cells of the target tissue have **receptors** for that specific hormone.

spot check ❶ The target tissue for insulin is most tissues. What do all of the target cells for insulin have in common?

spot check ❷ Can the pancreas target specific cells to respond to insulin?

Glands

8.4 learning outcome

List the major hormones, along with their target tissues and functions, of each of the endocrine system glands.

The glands of the endocrine system are shown in **Figure 8.3**. For a list of the glands and their hormones, target tissues, and functions, see Table 8.1.

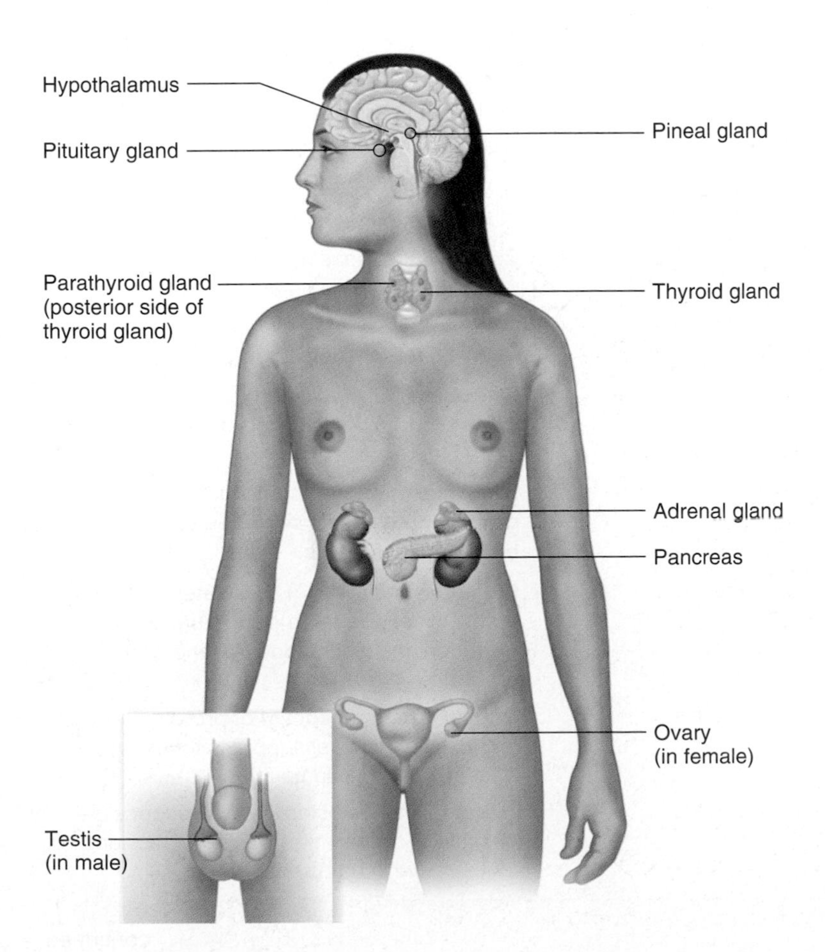

FIGURE 8.3 Endocrine system glands.

TABLE 8.1 Glands, hormones, target tissues, and functions

Gland	Hormone	Target tissue	Function
Pineal	**Melatonin**	Brain Hypothalamus	Helps regulate daily biological rhythms; inhibits GnRH production
Hypothalamus	**GnRH (gonadotropin-releasing hormone)**	Anterior pituitary	Stimulates secretion of FSH and LH
	CRH (corticotropin-releasing hormone)	Anterior pituitary	Stimulates secretion of ACTH
	TRH (thyrotropin-releasing hormone)	Anterior pituitary	Stimulates secretion of TSH
	GHRH (growth hormone–releasing hormone)	Anterior pituitary	Stimulates secretion of GH
Anterior pituitary	**TSH (thyroid-stimulating hormone)**	Thyroid	Stimulates secretion of thyroid hormone and growth of the thyroid
	ACTH (adrenocorticotropic hormone)	Adrenal cortex	Stimulates secretion of glucocorticoids and growth of the adrenal cortex
	FSH (follicle- stimulating hormone)	1. Ovaries 2. Testes	1. Stimulates secretion of estrogen 2. Stimulates sperm production
	LH (luteinizing hormone)	1. Ovaries 2. Testes	1. Stimulates ovulation 2. Stimulates secretion of testosterone
	GH (growth hormone)	Liver, bone, cartilage, muscle, adipose tissue	Stimulates widespread tissue growth
Posterior pituitary	**ADH (antidiuretic hormone)**	Kidneys	Increases water retention
	Oxytocin	1. Uterus 2. Lactating breasts	1. Stimulates uterine contractions 2. Stimulates release of milk
Thyroid	**T_3 and T_4 (thyroid hormone)**	Most tissues	Elevates metabolic rate; increases heart and respiration rates; stimulates appetite
	Calcitonin	Osteoblasts	Stimulates bone deposition
Parathyroids	**PTH (parathyroid hormone)**	1. Osteoclasts 2. Kidneys 3. Small intestine	1. Stimulates bone reabsorption to increase blood calcium levels 2. Stimulates reabsorption of calcium by the kidneys to maintain blood calcium levels 3. Stimulates calcium absorption
Pancreas	**Insulin**	Most tissues, liver	Stimulates cells to take in glucose to lower blood glucose levels; tells liver to store glucose as glycogen
	Glucagon	Liver	Stimulates glycogen conversion to glucose and then its secretion to raise blood glucose levels
Adrenal cortex	**Mineralocorticoids (aldosterone)**	Kidneys	Promote sodium (Na^+) and water reabsorption; promote potassium (K^+) excretion; maintain blood volume and pressure
	Glucocorticoids (cortisol)	Most tissues	Stimulate the breakdown of protein and fat to make glucose; suppress the immune system; reduce inflammation
	Androgens (dehydroepiandrosterone [DHEA])	Most tissues	Precursors to testosterone, are responsible for male secondary sex characteristics and for sex drive in both sexes

continued

TABLE 8.1 concluded

Gland	Hormone	Target tissue	Function
Adrenal medulla	**Epinephrine**	Most tissues	Raises metabolic rate; increases heart and respiration rates; increases blood glucose levels (complements sympathetic nervous system)
Ovaries	**Estrogen**	Most tissues	Stimulates female secondary sex characteristics;* regulates menstrual cycle and pregnancy
Testes	**Testosterone**	Most tissues	Stimulates male secondary sex characteristics,† sex drive, and sperm production
Other tissues	**Prostaglandins**	Many tissues	Have a variety of functions, such as relaxing smooth muscle in respiratory airways and blood vessels and causing contraction of smooth muscle in the uterus

* **Female secondary sex characteristics** include breast development, axillary and pubic hair growth, menstruation, and fat deposition.

† **Male secondary sex characteristics** include muscle and skeletal development, deeper voice, aggression, and facial, axillary, and pubic hair growth.

8.5 learning outcome

Locate and identify endocrine system glands.

Pineal Gland This gland is named *pineal* because it resembles a pine cone. It is located beneath the posterior end of the corpus callosum in the brain. The complete function of the pineal gland is not known, although it may have a function in establishing sleep-wake cycles of daily biological rhythms. What is known is that the pineal gland reaches its maximum size between the ages of 1 and 5 and usually shrinks to one-fourth that size by the end of puberty. The hormone melatonin produced by the pineal gland is believed to suppress gonadotropin-releasing hormone (GnRH) from the hypothalamus. You will learn more about this hormone in one of the scenarios later in the chapter.

Hypothalamus and Pituitary Gland You will study these two glands together because their functions are closely linked. The pituitary gland has two parts—the anterior pituitary and the posterior pituitary—that have very different jobs dependent on the hypothalamus. The hypothalamus plays a major role as a gateway for the brain to control the endocrine system.

The hypothalamus is connected to the pituitary gland's two parts by a stalk called the **infundibulum,** which serves as a passageway. Hormones produced by endocrine cells in the hypothalamus enter the blood through the capillary beds in the hypothalamus. The blood then carries the hormones through the infundibulum directly to capillary beds in the anterior pituitary. Hormones can then leave the bloodstream to bind to their receptors in the anterior pituitary. This system of blood vessels allows for the direct distribution of hormones through the blood from the hypothalamus to the anterior pituitary without traveling to the rest of the body. Hormones from the hypothalamus directly affect the release of hormones from the anterior pituitary. See **Figure 8.4**.

spot check 3 What is common in the naming of hormones produced by the hypothalamus that target the anterior pituitary?

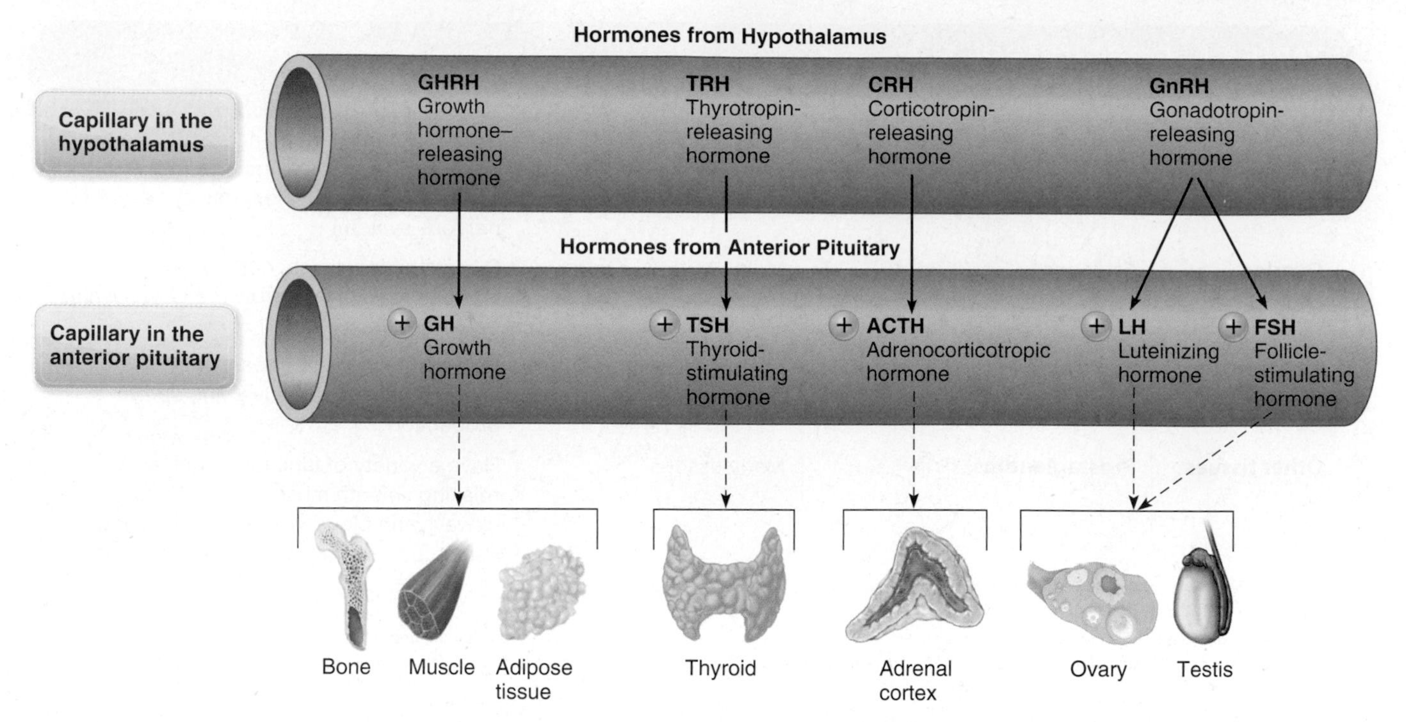

FIGURE 8.4 **Hypothalamus–anterior pituitary target-tissue relationship.** The hormones from the hypothalamus stimulate the release of hormones from the anterior pituitary.

The connection between the hypothalamus and the posterior pituitary is very different from that of the hypothalamus and the anterior pituitary. Groups of neuron cell bodies (nuclei) in the hypothalamus produce antidiuretic hormone (ADH) and oxytocin. These two hormones are delivered by axonal transport through the infundibulum to the posterior pituitary, where they are stored. Nerve signals from the hypothalamus travel through the infundibulum and trigger the release of ADH and oxytocin from the posterior pituitary when they are needed. Although these two hormones are technically made in the hypothalamus, they are referred to as *posterior pituitary* hormones because they are released from there. See **Figure 8.5**.

Thyroid Gland This gland, which resembles a bow tie, is anterior and lateral to the trachea, just inferior to the larynx. T_3 and T_4 are two chemicals produced by the thyroid that collectively are called **thyroid hormone.** Their function is to increase metabolism in most tissues. Calcitonin is also produced in the thyroid gland. It functions to stimulate the deposition of calcium in the bone, making it more relevant for children than adults.

warning

Calcitonin and T_3 and T_4 are hormones made by the thyroid gland, but only T_3 and T_4 together are referred to as *thyroid hormone.*

Thyroid hormone is vital to metabolism regulation in the body. The production of this hormone requires the mineral iodine. The natural nutritional sources of iodine are ocean fish, shellfish, and seaweed. Because these foods are not common

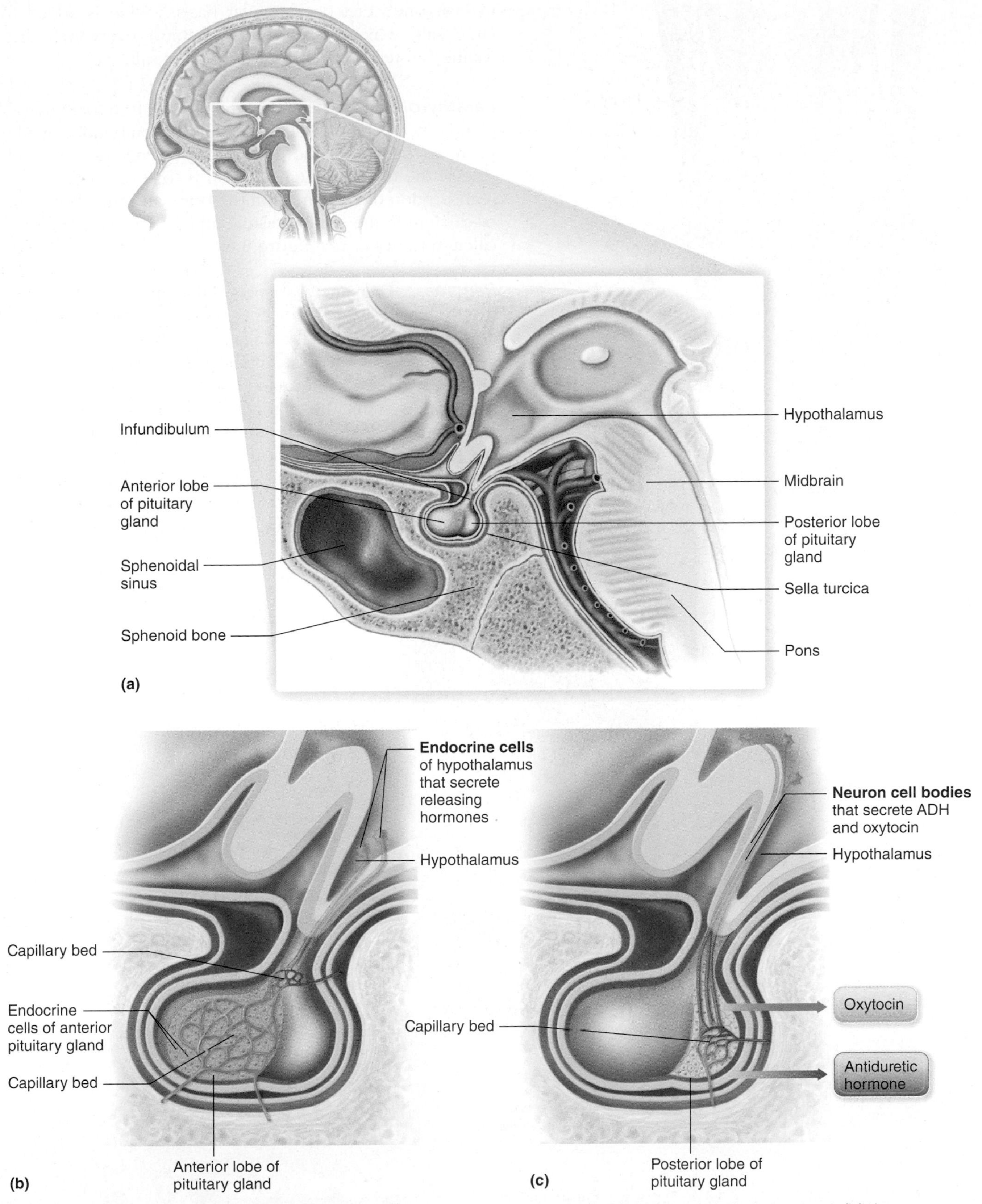

FIGURE 8.5 **Hypothalamus–pituitary relationship:** (a) location of the hypothalamus and pituitary gland; (b) the hypothalmus–anterior pituitary relationship, in which releasing hormones from the hypothalamus travel through the blood to the anterior pituitary; (c) ADH and oxytocin are produced by neuron cell bodies in the hypothalamus and are stored in the posterior pituitary until they are needed, at which time they are released from there.

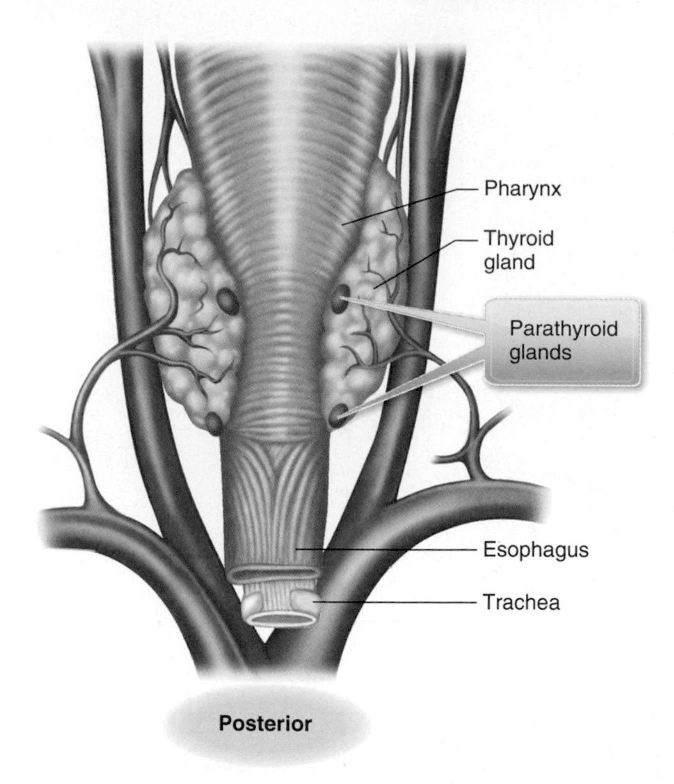

FIGURE 8.6 Parathyroid glands.

in everyone's diet on a regular basis, iodine is added to table salt, which is common in almost everyone's diet. Iodine is listed on packaging as iodized salt.

Parathyroid Glands There are usually four parathyroid glands in the body. They are typically embedded in the posterior surface of the thyroid gland, with two on each side of the trachea. Their function is to stimulate both the reabsorption of calcium from the bone and the absorption of calcium in the small intestine and to prevent the loss of calcium to urine. See **Figure 8.6**.

Pancreas The pancreas is part of the endocrine and digestive systems. This chapter covers only its endocrine function. The pancreas is an elongated gland and has a pebbly appearance. It is inferior and posterior to the stomach. Only about 2 percent of the gland produces hormones for the endocrine system. The endocrine cells are grouped to form 1 to 2 million **pancreatic islets (islets of Langerhans).** See **Figure 8.7**. The two hormones produced by the islets—insulin and glucagon—are important in the regulation of blood glucose levels. You will read about them in detail in one of the scenarios later in the chapter.

FIGURE 8.7 Pancreas. The pancreatic ducts are used for digestive secretions only.

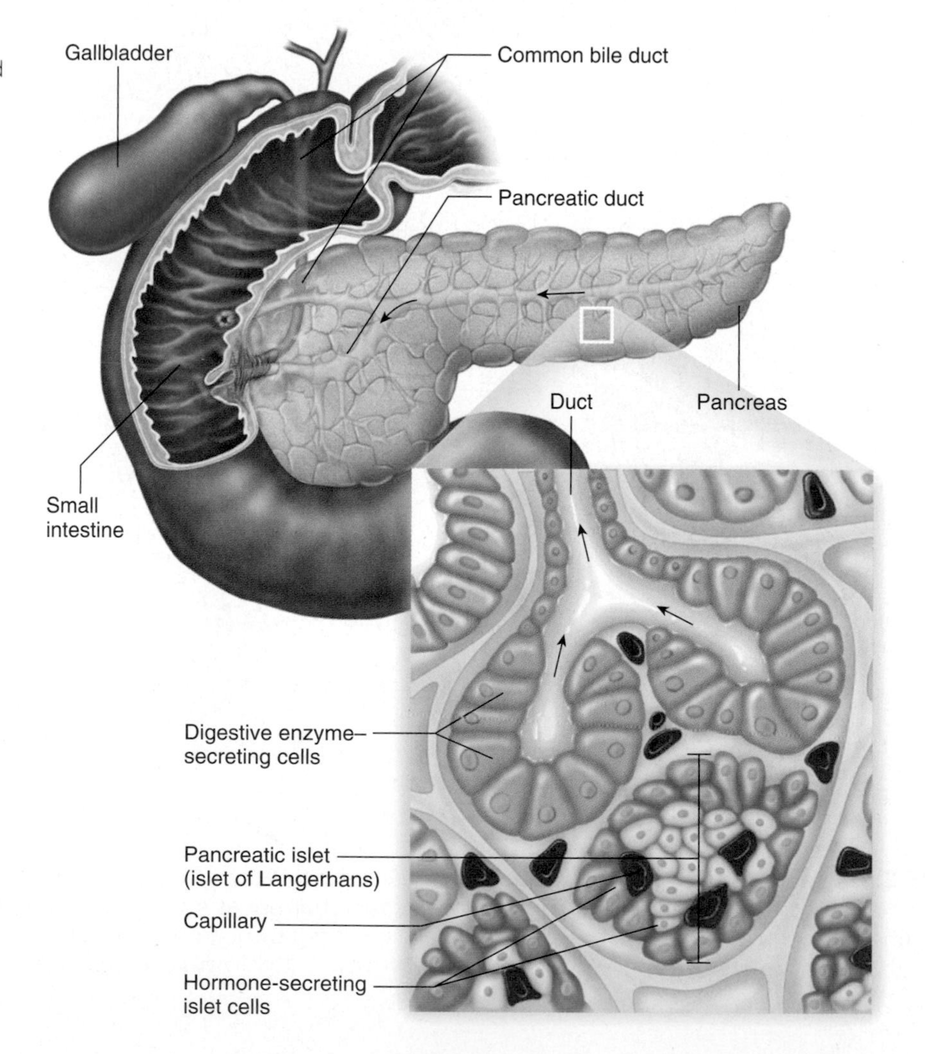

Adrenal Glands The adrenal glands appear to sit as a cap, superior and medial to the kidneys. The two parts of the adrenal gland—the adrenal cortex and the adrenal medulla—function to produce different hormones. See **Figure 8.8**.

Adrenal cortex The adrenal cortex is the outer layer of the adrenal gland. It produces over 25 different hormones classified in three major categories: **mineralocorticoids, glucocorticoids,** and **androgens.** Table 8.1 lists a major example for each class.

Adrenal medulla The adrenal medulla is the middle of the adrenal gland. It is often stimulated by the sympathetic nervous system in situations of fear, pain, and stress. In times of stress, cells from the adrenal medulla can stimulate cells of the adrenal cortex to secrete cortisol.

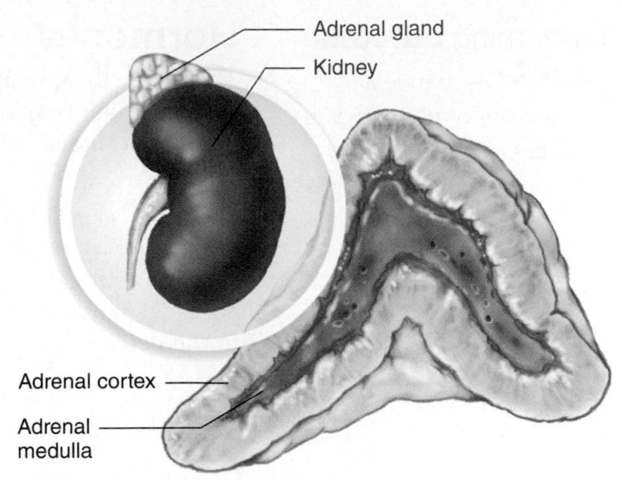

FIGURE 8.8 Adrenal cortex and adrenal medulla.

Gonads The gonads are also endocrine glands. Gonads are the ovaries in a female and the testes in a male, and they function in both the endocrine and reproductive systems. You will examine their endocrine function in this chapter and again with their reproductive function in the reproductive system chapters.

gonads: GO-nadz

Ovaries The ovaries begin to produce their hormones at puberty and continue producing them until menopause. A major hormone produced in the ovaries is estrogen, which is responsible for the development of **secondary sex characteristics** in the female. (See Table 8.1.) You will learn about the cells of the ovaries and their production of hormones in depth in the female reproductive system chapter.

Testes The testes produce the hormone testosterone in the fetus for the development of male anatomy. Testosterone production is dormant from birth to puberty, and then it begins again at puberty to promote the development of male secondary sex characteristics. (See Table 8.1.) Production of testosterone continues after puberty throughout life but significantly decreases after midlife.

spot check ❹ Think back to the integumentary system and the skeletal system. What effects does the increased production of testosterone and estrogen at puberty have on sebaceous glands and osteoblasts?

Other Tissues So far, you have covered the major endocrine system glands, but many other tissues in the body also produce hormones to communicate with other tissues. For example, the heart makes atrial natriuretic hormone (ANH), a hormone that targets the kidney to regulate urine production. The uterus makes prostaglandins, hormones that cause the smooth muscle of the uterus to contract during childbirth. Even endocrine cells of the stomach and small intestine make several hormones to regulate the production and secretion of digestive juices. You will continue to explore the role of these and other hormones in many of the upcoming chapters.

In the next section, you will learn about the chemical composition of hormones produced by endocrine glands.

8.6 learning outcome

Describe the chemical makeup of hormones, using estrogen, insulin, and epinephrine as examples.

Hormones

There are three general categories of hormones based upon chemical structure: steroids, amino acid derivatives, and proteins. The chemical composition matters because it directly affects how the hormone relates to its receptor.

Steroids Steroid hormones are derived from a **cholesterol** molecule. You will read more about cholesterol later in this chapter. Examples of steroid hormones include estrogen, testosterone, progesterone, mineralocorticoids, and glucocorticoids. As lipids, they can pass through cell membranes to reach receptors anywhere in the cell.

clinical point

The ability of steroid hormones to easily pass through cell membranes allows for the clinical delivery of these hormones through the skin. A birth control patch applied to the skin delivers doses of estrogen and progesterone through the skin. Progesterone is covered in detail in the female reproductive system chapter.

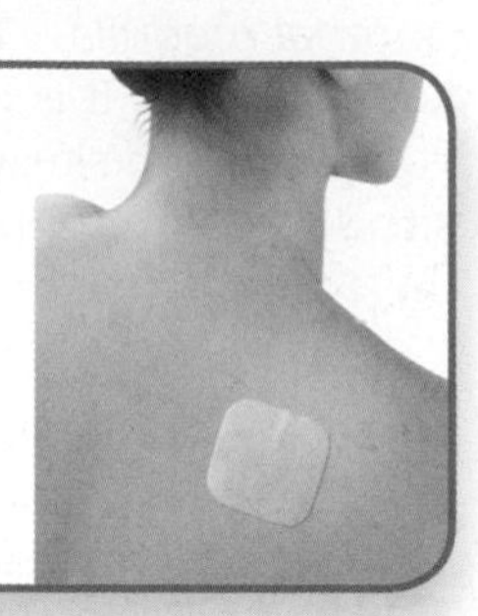

Cholesterol often has a bad reputation as being harmful to the body. However, it is needed to synthesize steroid hormones and other digestive salts that you will explore in the digestive system chapter. Cholesterol can be acquired through the diet and can be produced by the liver. It is a steroid found in high concentration in liver and egg yolks. It is also found in whole milk, butter, cheese, and meats, but cholesterol is not found in plants.

epinephrine:
ep-ih-NEF-rin

Amino Acid Derivatives As the name implies, these hormones are derived from amino acids. Thyroid hormone, **epinephrine,** and melatonin are examples of amino acid derivatives. This type of hormone may or may not be able to cross cell membranes. For example, thyroid hormone *can* easily pass across a cell membrane to reach a receptor inside the cell, while epinephrine *cannot.*

Proteins Protein hormones are made of chains of amino acids. Examples of protein hormones include insulin, thyroid-stimulating hormone (TSH), follicle-stimulating hormone (FSH), luteinizing hormone (LH), growth hormone (GH), parathyroid hormone (PTH), antidiuretic hormone (ADH), adrenocorticotropic hormone (ACTH), glucagon, calcitonin, oxytocin, and hormones from the hypothalamus. Proteins are too large to pass through cell membranes.

spot check 5 What is the chemical composition of the three classes of hormones secreted by the adrenal cortex?

Now that you have reviewed the chemical composition of hormones, you are ready to learn about how it affects their relationship with receptors.

clinical point

As examples of amino acid and protein hormones, epinephrine and insulin cannot pass through a cell membrane, so they cannot be clinically delivered through the skin. The digestive system further complicates their clinical delivery. As you will discover in the digestive system chapter, proteins are partially digested in the stomach and are not absorbed until later, in the small intestine. Therefore, if taken orally, epinephrine and insulin would be broken down before being absorbed into the blood. Medical administration of epinephrine and insulin must be by injection. See Figure 8.9.

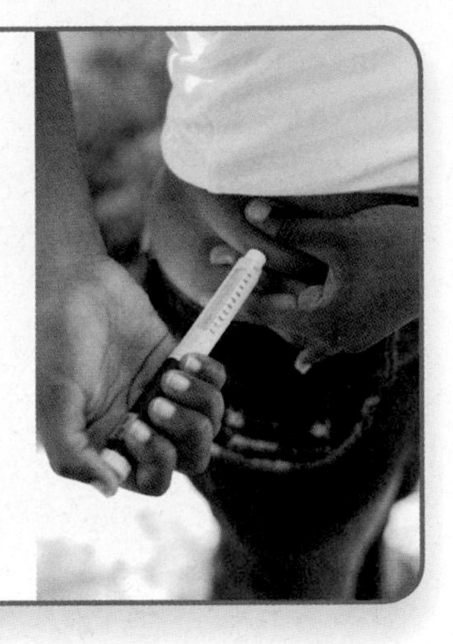

FIGURE 8.9 A teen injecting insulin.

Target Tissues

8.7 learning outcome

Compare the location of receptors for protein hormones with that of receptors for steroid hormones.

There are many target tissues for the various hormones of the body. What makes a tissue a target tissue is the presence of specific receptors for specific hormones based on the hormone's chemical makeup and shape.

Location of Hormone Receptors The two basic classes of receptors for hormones are based on their location—either on the cell membrane or somewhere inside the cell. Protein hormones must bind to receptors on the cell membrane because they cannot enter the cell. They fit into the receptors on the basis of their specific shape—like a key fits into a lock. The hormones may initiate a response directly on the membrane, such as opening a channel, or they may initiate a response inside the cell, such as initiating protein synthesis. If the response is to occur inside the cell, a **second-messenger** system must be used. See **Figure 8.10**. In this system, the binding of a hormone in the receptor on the cell membrane causes a chemical reaction inside the membrane. This reaction creates a second messenger, which then carries the information to where it is needed in the cell to initiate the function of the hormone. This is much like trying to get an important message to a friend at school. You can call the school office and have someone there relay the information you need to get to your friend.

Steroid hormones and thyroid hormone can directly cross the membrane and fit into receptors inside the cell where the function will be carried out. This may be in the cytoplasm or in the nucleus. See **Figure 8.11**. They do not need a second messenger. This process is like calling your friend at school directly on his cell phone.

8.8 learning outcome

Differentiate autocrine, paracrine, endocrine, and pheromone chemical signals in terms of the proximity of the target tissue.

Location of the Target Tissue The location of the target tissue is also relevant to the eventual delivery method of the hormone, as outlined in the list below. Hormone function can be categorized by the location of the cells producing the hormone relative to the location of their target tissues.

- **Autocrine** refers to the secretion of a hormone by cells of the same tissue type that the hormone targets. An example of this is the production of prostaglandins

FIGURE 8.10 Location of receptors—protein hormones. A protein hormone (1) fits into a receptor on the cell membrane (2), causing a chemical reaction that forms a molecule called *cAMP*—a second messenger (3). The second messenger stimulates changes inside the cell (4) as the function of the hormone. In this case, molecule A changes to B, which causes molecule C to change to D (5).

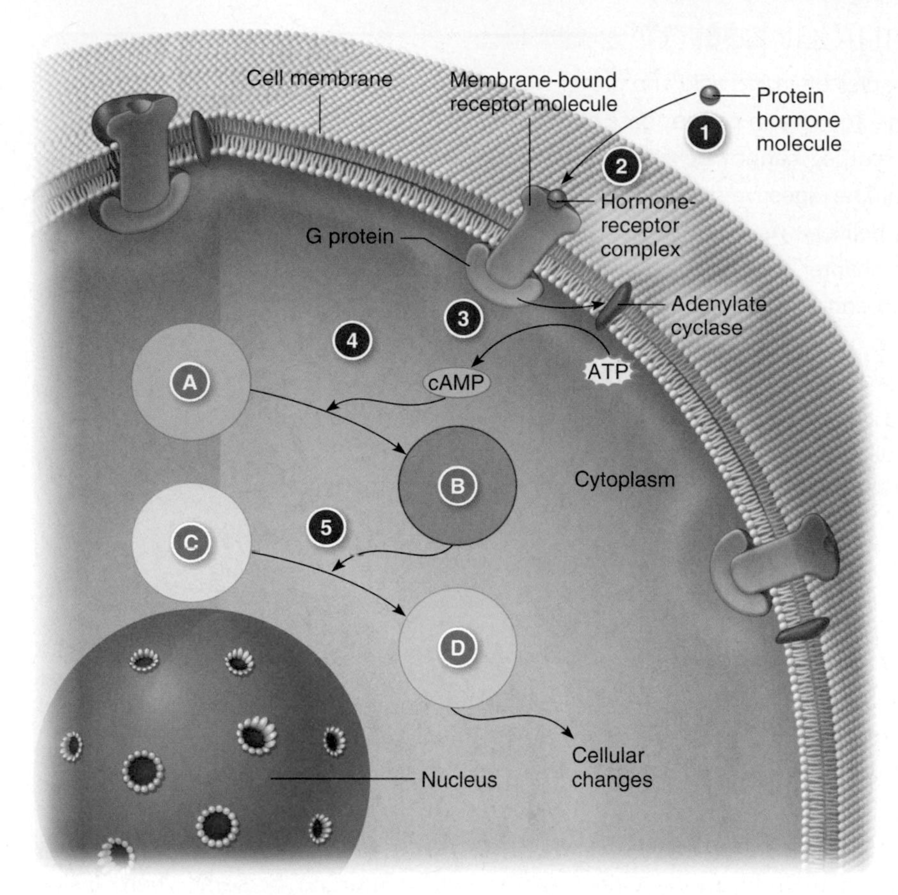

FIGURE 8.11 Location of receptors—steroid hormones. A steroid hormone (1) passes through the cell membrane and fits into a receptor inside the nucleus (2) to stimulate protein synthesis (3 to 5).

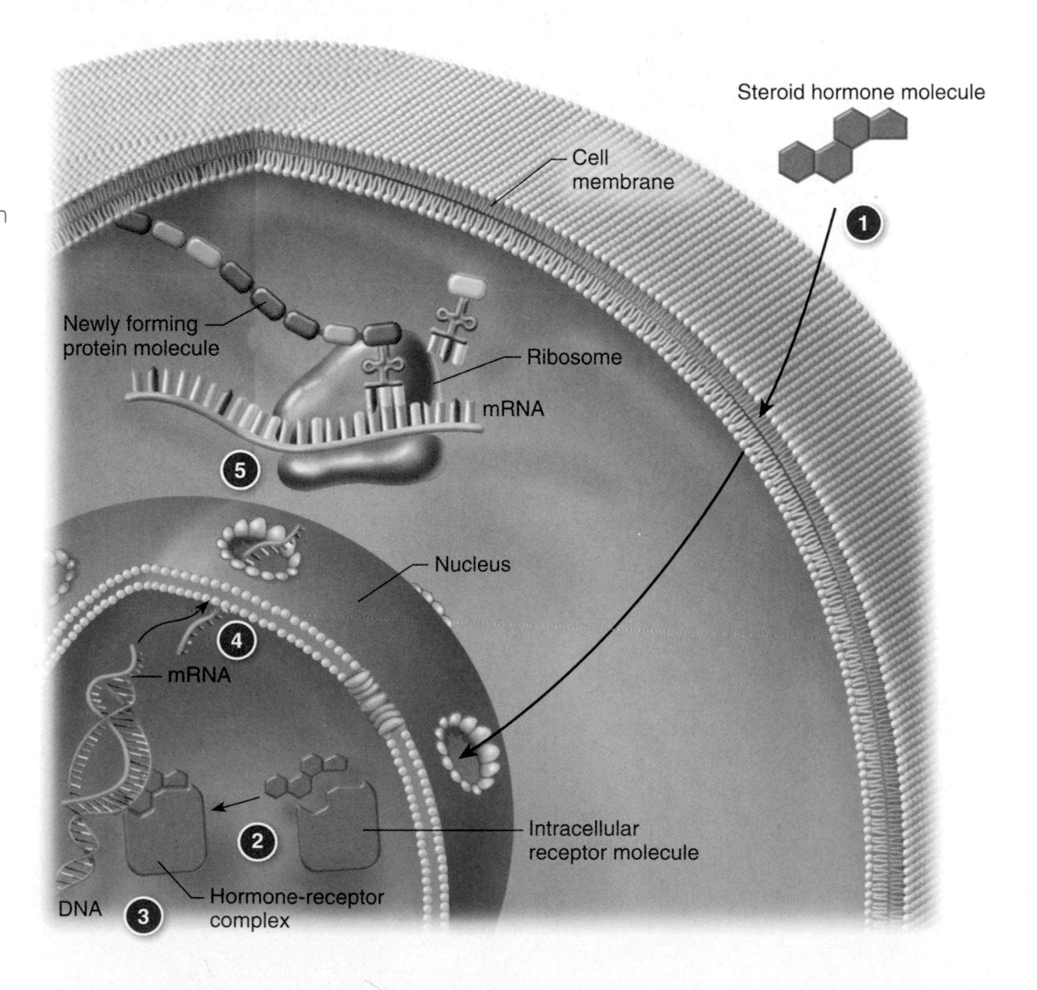

by the smooth muscle of the uterus, which causes the smooth muscle of the uterus to contract during birth.

- **Paracrine** refers to hormones that work on neighboring cells without having to go through the blood to get to the target tissue. An example of this can be seen with endocrine cells in the stomach, which cause neighboring cells to produce hydrochloric acid. You will read more about this in the digestive system chapter.
- **Endocrine** refers to hormones that travel through the blood to get to their target tissue. The hormones discussed in this chapter are endocrine hormones.
- **Pheromone** refers to chemicals that cause a response outside the body, in another individual. In developing fragrances, the perfume industry studies possible pheromones that may attract the opposite sex.

You have now covered the anatomy of this system—the glands, the type of hormones they produce, and their target tissues with their receptors. It is time to examine the physiology of the system.

Physiology of the Endocrine System

8.9 learning outcome

Explain the regulation of hormone secretion and its distribution.

The physiology of this system includes the regulation of hormone secretion and distribution, the regulation of sensitivity of a target tissue to a hormone through the regulation of its receptors, and the elimination of hormones.

Regulation of Hormone Secretion and Distribution

Most hormones are not secreted at a constant rate. They are secreted when there is a need, and their secretion is usually regulated by negative-feedback mechanisms. The secretion of a hormone can be initiated in three ways:

- **Neural stimulation of a gland.** In the nervous system chapter, you studied how neurons can synapse with a gland, a muscle, or the dendrite of another neuron. An example of neural stimulation of a gland occurs when sympathetic neurons stimulate the adrenal medulla to secrete epinephrine in times of pain, fear, and/or stress.
- **Another hormone stimulating a gland.** An example of this can be seen when the hormones released by the hypothalamus stimulate the secretion of hormones from the anterior pituitary. For example, GnRH from the hypothalamus results in the secretion of FSH and LH from the anterior pituitary.
- **A substance other than a hormone stimulating a gland.** An example of this can be seen with the pancreas, which monitors blood glucose levels. Glucose is not a hormone, but if glucose levels are high, the pancreas responds to that stimulation by secreting insulin.

It will be helpful to look at an example of the secretion of a thyroid hormone and the negative-feedback mechanism that regulates it. If you go outside when it is very cool, but you are underdressed, the hypothalamus may try to speed up your metabolism to generate more body heat. See **Figure 8.12**. This figure shows:

1. The hypothalamus releases thyrotropin-releasing hormone (TRH), which goes to its target tissue, the anterior pituitary, and fits into receptors.

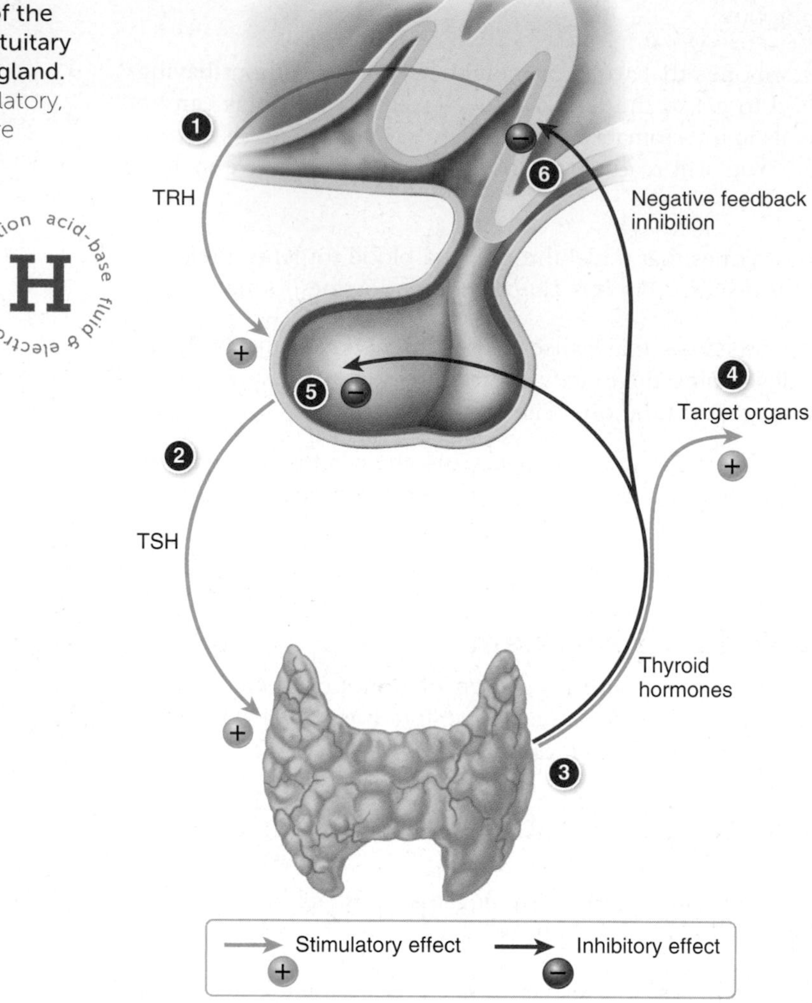

FIGURE 8.12 Negative-feedback inhibition of the hypothalamus and pituitary gland by the thyroid gland. Steps 1 to 4 are stimulatory, while steps 5 and 6 are inhibitory.

2. The anterior pituitary is then stimulated to release TSH, which goes to its target tissue, the thyroid gland, and fits into receptors.
3. The thyroid is then stimulated to release thyroid hormone, which travels to several target tissues.
4. Thyroid hormone travels to most cells, *stimulating* them to increase metabolism (heat is given off as a by-product).
5. Thyroid hormone also travels to the anterior pituitary, where it *inhibits* the further secretion of TSH.
6. Thyroid hormone also travels to the hypothalamus, where it *inhibits* the further secretion of TRH.

This negative-feedback mechanism is the thyroid's method of saying to the anterior pituitary and hypothalamus that it has received the message to secrete thyroid hormone and is following through on the message. It keeps the secretion rate from going too high.

spot check ❻ Which of the three ways of stimulating a gland is used to stimulate the anterior pituitary and thyroid glands?

The chemical composition affects how the hormone is distributed in the blood. Thyroid hormone, in the last example, can easily pass through cell membranes. However, if it were allowed to do that, much of the thyroid hormone would immediately go into the cells, causing a huge spike in metabolism and then a dramatic fall when the hormone was used up. The body has a way to *time-release* thyroid hormone and steroid hormones that can easily pass through cell membranes. Transport proteins, called **plasma proteins,** made by the liver are used to bind to some of the hormone in the blood. The bond is a reversible one, meaning that at any one time, some of the hormone will be free to enter the cell and some of the hormone will be bound to plasma proteins, making it too large to enter the cell. As time goes by, more of the hormone is freed from the plasma proteins so that it can enter the cell. See **Figure 8.13**. The process is similar to the way a time-released cold medicine works: You might take the medicine only once every 12 hours, but its effect is released evenly throughout that time. Hormones that cannot cross a cell membrane do not need to bind to plasma proteins.

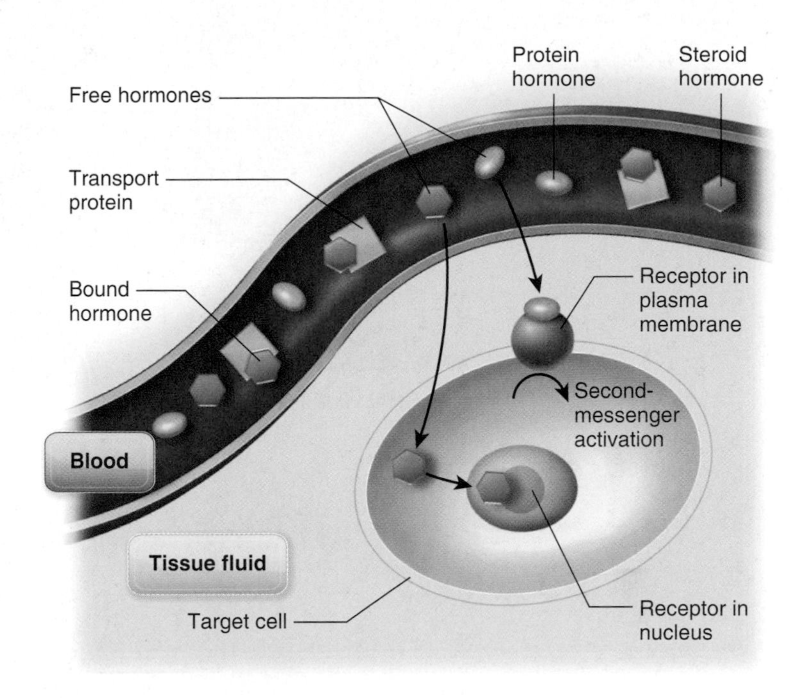

FIGURE 8.13 Transport and action of protein and steroid hormones.

Now that you have reviewed how hormone secretion and distribution are regulated, you are ready to explore how a target tissue can regulate its sensitivity to a hormone.

Receptor Regulation

8.10 learning outcome

Explain how the number of receptors can be changed.

Target tissues can regulate their sensitivity to a hormone by adjusting the number of receptors for that hormone. **Up-regulation** is an increase in the number of receptors for a given hormone. In this case, the cell has more receptors for the hormone to bind, so it has become more sensitive to the hormone and the effects of the hormone are increased. **Down-regulation** occurs when a cell decreases the number of receptors for a hormone. Here, there are fewer opportunities to bind a hormone to a receptor, so the cell is less sensitive to the hormone and the effects of the hormone are reduced. See **Figure 8.14**. Down-regulation is often a response to chronically high levels of a hormone. An example of this can be seen with adipocytes (fat cells) of an obese individual. In this individual, chronically high carbohydrate consumption results in chronically high glucose levels, and the pancreas responds with chronically high insulin levels. The adipocytes respond to the chronically high insulin levels by down-regulating their receptors for insulin. This is similar to the response of individuals in an arena who cover their ears to reduce their sensitivity to a loud sound system.

FIGURE 8.14 **Receptor regulation** (a) up-regulation, (b) down-regulation.

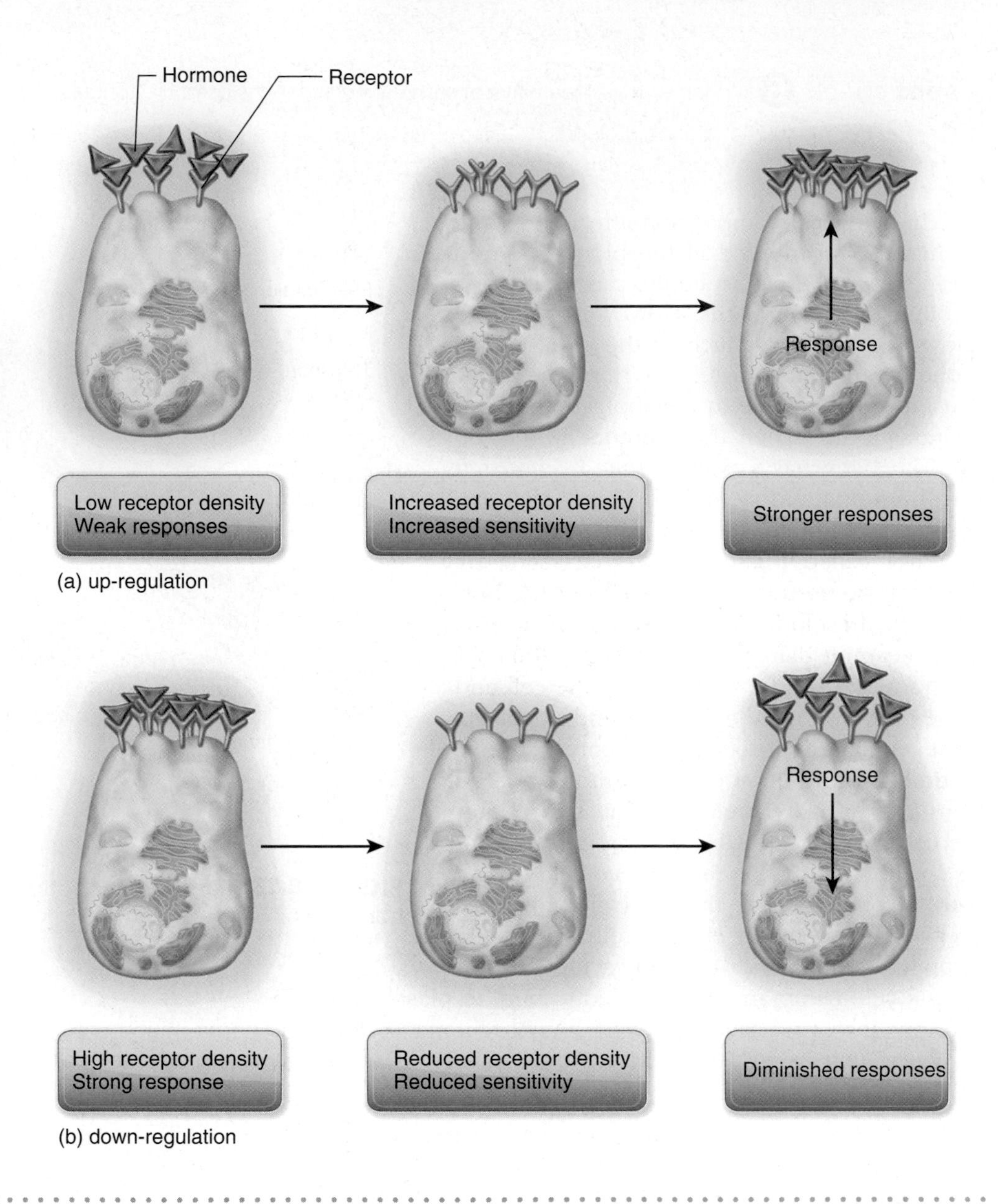

8.11 learning **outcome**

Explain how hormones are eliminated from the body.

Hormone Elimination

The effects of a hormone will continue until the hormone is eliminated from the system. Therefore, the method of hormone elimination is important in regulating the duration of the hormone's effects. Hormones may be eliminated from the body in four ways:

1. **Excretion.** The kidneys can remove a hormone from the blood and excrete it in urine, or the liver can remove a hormone from the blood and excrete it in bile.
2. **Metabolism.** Enzymes in the blood, liver, kidneys, or other target tissues can break down the hormone and excrete it or use it for cellular processes. For example, the breakdown of protein hormones results in amino acids that can be used for protein synthesis.
3. **Active transport.** A hormone such as epinephrine can be taken back up by a cell through active transport so that it can be recycled and released at another time.
4. **Conjugation.** The liver can bind water-soluble molecules to a hormone so that it will be excreted at a faster rate.

Half-life is the length of time it takes for one-half of a substance to be eliminated from the circulatory system. Steroid hormones and thyroid hormone have their half-lives extended by binding to plasma proteins. Their levels tend to be more constant. Protein hormones and epinephrine have relatively short half-lives because they are quickly degraded, recycled, or excreted.

spot check ❼ Why might it be an advantage to have a short half-life for epinephrine?

Functions: Four Scenarios

8.12 learning **outcome**

Explain the function of hormones by showing how they interact to maintain homeostasis.

Four scenarios can help you understand the interaction of the glands and hormones of this system: insulin and glucagon secretion, consequences of reduced melatonin production at puberty, adrenal cortex degeneration, and hormonal regulation of childbirth. It is always helpful to chart the relevant gland, hormone, target tissue, and function when working with endocrine questions. Always remember that if the function of a hormone involves another hormone, you must chart that hormone too.

Insulin and Glucagon Secretion This scenario analyzes the role of insulin and glucagon over a 12-hour period. Paul has a dinner at 6 P.M. He eats pasta and garlic bread with chocolate cake for dessert, a meal rich in carbohydrates. His digestive system will work to break down the carbohydrates into their building blocks, monosaccharides (simple sugars), so that they can be absorbed into the bloodstream. Soon after his meal, his blood sugar level rises above homeostasis. Cells need glucose (blood sugar), but too much sugar in the blood is harmful to the body. Over time, it can lead to coronary artery disease, peripheral nerve damage, and damage to small blood vessels in the kidney and retina. So how does the body maintain normal, homeostatic blood glucose levels?

Look again at the chart of insulin and glucagon. See Table 8.2.

After the meal, Paul's pancreas recognizes the increased blood glucose levels, and it releases the hormone insulin, which travels through the bloodstream to any cell that has an insulin receptor. Once insulin fits into the receptor, that cell takes in glucose from the blood, lowering blood glucose levels. Most cells have an insulin receptor, including the liver. The liver will convert the glucose it takes in to glycogen, a starch that is used as a storage molecule for glucose. Through the secretion of insulin, the cells have taken in the glucose they need, and blood sugar levels are brought back to homeostasis.

H nutrition acid-base fluid & electrolytes

Paul does not snack before going to bed. By 6 A.M., it has been 12 hours since he last ate. His cells have used the available glucose, but they still need more glucose

TABLE 8.2 Insulin and glucagon

Gland	Hormone	Target tissue	Function
Pancreas	**Insulin**	Most tissues, liver	Stimulates cells to take in glucose to lower blood glucose levels; tells liver to store glucose as glycogen
	Glucagon	Liver	Stimulates glycogen conversion to glucose and then its secretion to raise blood glucose levels

to carry out cellular respiration. Blood glucose levels have fallen below homeostasis. In this case, Paul's pancreas recognizes that blood glucose levels are below normal. His pancreas responds by releasing glucagon. Glucagon travels everywhere through the bloodstream, but it affects cells of the liver because they have receptors for glucagon. Glucagon in the liver's receptors causes the liver cells to convert stored glycogen back to glucose for release back into the blood. Homeostatic blood glucose levels are restored.

spot check 8 Which of the three ways to stimulate a gland caused the release of the hormones in this scenario?

spot check 9 What type of feedback mechanism is involved in this case?

spot check 10 Would the pancreas ever release large quantities of insulin and glucagon at the same time? Explain.

Consequences of Reduced Melatonin Production at Puberty Natalie and Nate (sister and brother) are about to enter puberty. Normally, the pineal gland's production of melatonin is reduced at puberty. This creates a chain of events that involves many hormones. What are the consequences for Natalie of this decrease in melatonin?

To answer this question, first look at the chart for melatonin. See Table 8.3. Keep in mind that if the function of a hormone involves another hormone, you should continue charting.

Melatonin normally inhibits the production of GnRH by the hypothalamus. When the melatonin is reduced, so is the inhibitory effect on the hypothalamus. The hypothalamus is free to produce GnRH, which then goes to the anterior pituitary, telling it to produce FSH and LH. FSH targets Natalie's ovaries to stimulate the production of estrogen, which will promote the development of secondary sex characteristics. LH targets Natalie's ovaries to stimulate ovulation.

TABLE 8.3 Effects of melatonin

Gland	Hormone	Target tissue	Function
Pineal	**Melatonin**	Brain Hypothalamus	Helps regulate daily biological rhythms; inhibits *GnRH* production
Hypothalamus	**GnRH** (gonadotropin-releasing hormone)	Anterior pituitary	Releases *FSH and LH*
Anterior pituitary	**FSH** (follicle-stimulating hormone)	Ovaries	Stimulates secretion of *estrogen*
	LH (luteinizing hormone)	Ovaries	Stimulates ovulation
Ovaries	**Estrogen**	Most tissues	Stimulates female secondary sex characteristics; regulates menstrual cycle and pregnancy

spot check 11 What physical changes will we see in Natalie due to these hormone interactions?

spot check 12 How would the charting differ if it had been for Nate instead of Natalie?

spot check 13 What are the physical consequences of reduced melatonin production at puberty for Nate?

Adrenal Cortex Degeneration This scenario looks at how the body deals with an endocrine gland that is not functioning properly. David has just been diagnosed with **Addison's disease.** Addison's disease is the result of adrenal cortex degeneration. His adrenal cortex no longer functions to produce hormones. What are the consequences of this disease for David?

To determine the answer to this question, begin by reviewing the chart for the adrenal cortex. See Table 8.4. Note the three major classes of hormones produced by the adrenal cortex.

TABLE 8.4 Hormones of the adrenal cortex

Gland	Hormone	Target tissue	Function
Adrenal cortex	**Mineralocorticoids (aldosterone)**	Kidneys	Promote sodium (Na^+) and water reabsorption; promote potassium (K^+) excretion; maintain blood volume and pressure
	Glucocorticoids (cortisol)	Most tissues	Stimulate the breakdown of protein and fat to make glucose; suppress the immune system; reduce inflammation
	Androgens (dehydroepiandrosterone [DHEA])	Most tissues	Precursor to testosterone, responsible for male secondary sex characteristics, sex drive in both sexes

Aldosterone is a major mineralocorticoid hormone. It targets the kidney to promote Na^+ and water reabsorption and K^+ excretion. The amount of water reabsorbed affects blood volume and therefore blood pressure. Without production of aldosterone, David will have decreased ability to regulate blood pressure and blood volume. He will also lose more Na^+ in his urine and keep more K^+ in his blood, disrupting his electrolyte balance.

Another hormone, cortisol, is a major glucocorticoid that is used to promote the breakdown of fat and protein to make glucose, thus raising blood glucose levels. It also raises the amount of amino acids in the blood. Cortisol is a major hormone for dealing with stress. It provides the means to acquire the necessary glucose and amino acids in the blood to respond to stress. It also suppresses the immune system, and reduces inflammation. Without cortisol, David's ability to use fat and protein as fuel is reduced, as is his body's ability to deal with stress, and inflammation.

Dehydroepiandrosterone (DHEA) is an androgen produced by the adrenal cortex. It is a precursor to testosterone. The effects for David of decreased DHEA are minimal because the adrenal cortex is not his main source of testosterone. The majority of his testosterone comes from his testes.

Clearly, the body has lost some of its ability to maintain homeostasis if the adrenal cortex is not producing its hormones. That in itself is a move from homeostasis. How does the body try to fix this? The hypothalamus will recognize the need for cortisol, especially in times of stress. To see how it will respond, refer to Table 8.5.

The hypothalamus responds to the need for cortisol by producing corticotropin-releasing hormone (CRH). The CRH goes to the anterior pituitary, telling it to

TABLE 8.5 Hypothalamus response to decreased cortisol

Gland	Hormone	Target tissue	Function
Hypothalamus	**CRH** (corticotropin-releasing hormone)	Anterior pituitary	Releases *ACTH*
Anterior pituitary	**ACTH** (adrenocorticotropic hormone)	Adrenal cortex	Stimulates secretion of glucocorticoids and growth of the adrenal cortex

produce ACTH. The ACTH travels to the adrenal cortex to stimulate its production of glucocorticoids only—not mineralocorticoids or androgens. This negative-feedback response is intended to correct only the low cortisol levels. If the adrenal cortex has degenerated, it will not be able to respond to the ACTH. The hypothalamus will continue to see the need, so it will continue to produce more CRH, which will cause the anterior pituitary to produce more ACTH. The high ACTH levels in the blood, along with low cortisol levels, are one means of diagnosis for Addison's disease.

spot check 14 What other levels in the blood might be examined to diagnose this disease?

spot check 15 Would Addison's disease have a different effect on a female?

Hormonal Regulation of Childbirth This final scenario focuses on one of the hormones involved in childbirth—oxytocin. See Table 8.6.

Dorothy is nine months' pregnant. In **Figure 8.15**, you can see that Dorothy's pregnancy is at full term. The fetal head is pushing on the neck of the uterus (called the *cervix*), which causes the cervix to send nerve impulses to the brain. The hypothalamus in the brain responds by stimulating the posterior pituitary to release oxytocin. Oxytocin goes to its target tissue, the uterus, and causes it to contract. The uterine contractions push on the fetus, causing more pressure on the cervix. The whole cycle keeps repeating itself until the baby is born.

spot check 16 Which of the three ways to stimulate a gland was used to stimulate the posterior pituitary in this scenario?

spot check 17 What type of feedback mechanism was involved in this case?

TABLE 8.6 Hormone involved in childbirth

Gland	Hormone	Target tissue	Function
Posterior pituitary	**Oxytocin**	Uterus	Stimulates uterine contractions

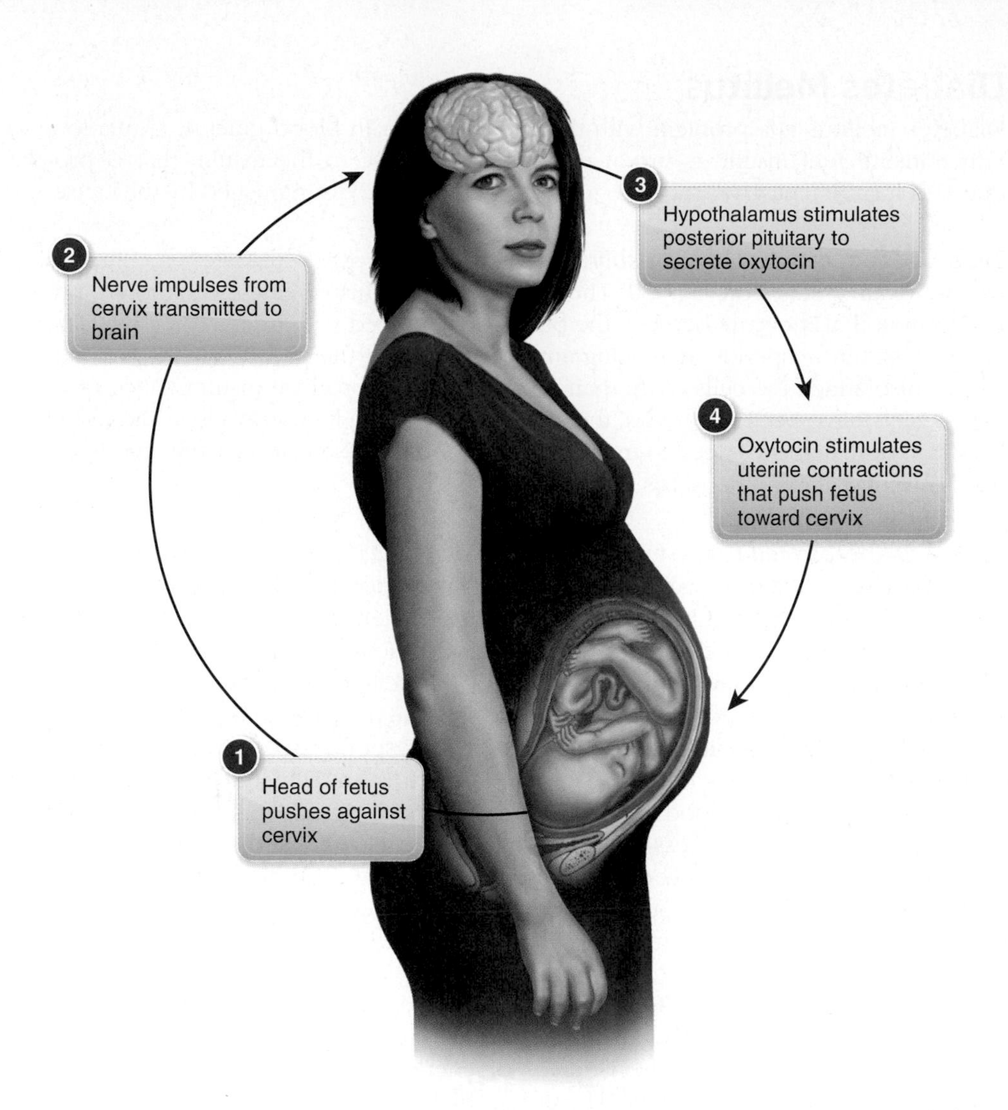

FIGURE 8.15 Oxytocin's effect on childbirth.

Effects of Aging on the Endocrine System

8.13 learning **outcome**

Explain the effects of aging on the endocrine system.

In general, the levels of hormones decline with age. This can be seen most dramatically with the hormones estrogen and testosterone. The production of both hormones is dramatically reduced during midlife. Estrogen production by the ovaries ceases with menopause. Testosterone production continues after midlife, but it gradually declines and at age 80 is about 20 percent of what its peak was at age 20. Both estrogen and testosterone serve as a lock on calcium in the bone. Therefore, the effects of osteoporosis may be seen as the production of these hormones decreases with age.

Even if the levels of some other hormones remain high with age, the sensitivity of their target tissues often decreases with down-regulation of receptors. This is often seen in the development of diabetes mellitus in the elderly. We focus on that disorder next.

Endocrine System Disorders

8.14 learning **outcome**

Describe endocrine system disorders.

By looking at the pathology of a disorder, you can often gain a better understanding of the normal physiology of a system. You will begin this section by examining two disorders termed *diabetes:* diabetes mellitus and diabetes insipidus.

Diabetes Mellitus

Diabetes mellitus is a problem with the use of insulin in blood glucose regulation. Either insufficient insulin is produced, or the response to the insulin that is produced is insufficient. There are two forms of this disease, distinguished by the cause.

Type 1 Diabetes Mellitus (Insulin Dependent) Type 1 diabetes mellitus is usually diagnosed before age 30. The exact cause of this disease is unknown, but many think that it occurs because the body is challenged with a viral infection. The immune system responds by making antibodies to fight the pathogen, but the antibodies also attack the cells of the pancreatic islets that produce insulin. Once most of the islets have been destroyed, the ability to regulate glucose levels in the blood is lost. Type 1 diabetics need to monitor their blood glucose levels closely and inject insulin to maintain homeostasis of blood glucose levels.

Type 2 Diabetes Mellitus Most diabetics have type 2 diabetes mellitus. This form of the disease is characterized by the inability to respond to the insulin produced by the pancreas. The number of receptors is insufficient to adequately respond to the insulin produced in the regulation of blood glucose levels. In the beginning of the disease, insulin levels may be high, but the cells have down-regulated their receptors, so they are not responsive to the insulin produced. As a result, blood glucose levels stay chronically high. The pancreas responds accordingly with more and more insulin that has less and less of an effect. Eventually, the pancreas stops responding to the high blood glucose levels, and insulin levels fall. If the disease is diagnosed early in its progression, type 2 diabetics can be treated with medications to increase the number of receptors, dietary changes that even blood glucose levels, and exercise, which also encourages up-regulation of receptors. If not treated early, however, the type 2 diabetic may have to be treated with insulin as well.

Both types of this disease have similar symptoms. If cells cannot use the glucose in the blood, they must turn to other sources of energy in the body, such as fat and protein. Therefore, diabetes mellitus, left untreated, is a wasting disease characterized by visible weight loss and loss of muscle mass, even though the individual may be eating a high-caloric diet. Normally, the kidney filters out the glucose in the blood while removing wastes. It then reabsorbs all of the glucose so that none is lost in urine. However, in either type of diabetes mellitus, there is too much glucose in the blood. The kidney filters the glucose out but does not have time to completely reabsorb all of it. As a result, some of the glucose is lost in urine **(glucosuria).** Water follows the glucose, so urine output is increased **(polyuria).** The water for the extra urine is taken from the blood, reducing blood volume. The hypothalamus recognizes this and sends the signal for thirst, so the individual drinks to replace the lost water (**polydipsia,** excessive thirst).

Both types of diabetes mellitus have possible life-threatening complications. Therefore, blood glucose levels need to be monitored closely. See **Figure 8.16.** **Hyperglycemia** (too much sugar in the blood) has a devastating effect on the walls of blood vessels. Even blindness and kidney failure can be the direct result of the deterioration of vessels of the retina and kidney. Uncontrolled diabetes mellitus also leads to other degenerative cardiovascular

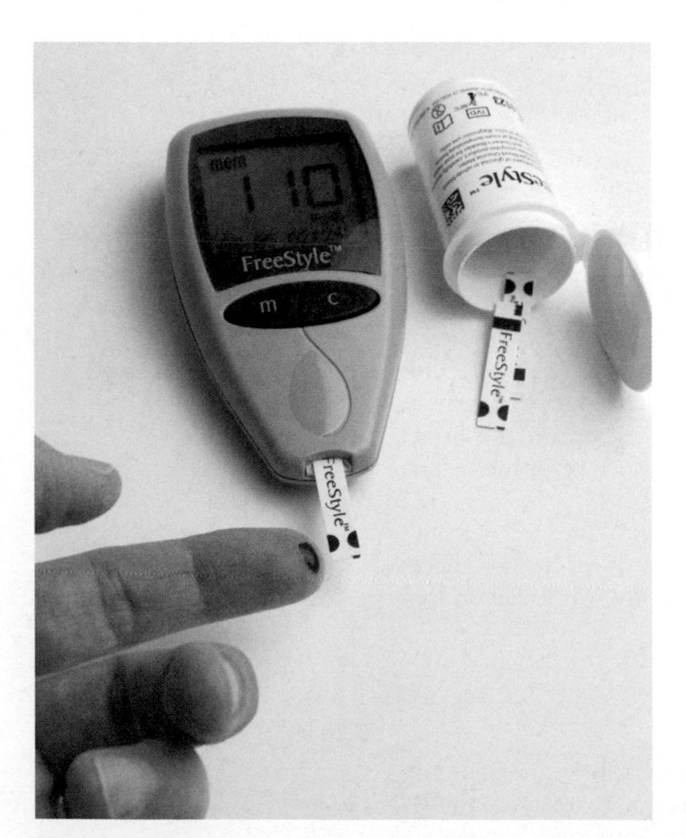

FIGURE 8.16 A glucometer measures the blood glucose level.

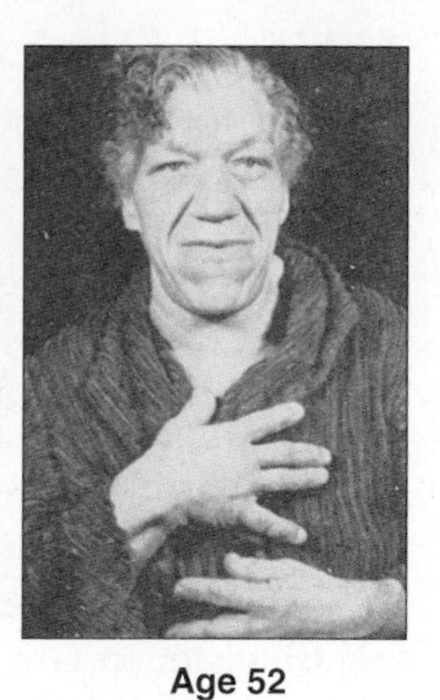

FIGURE 8.17 Acromegaly. The same woman is shown in all photos.

disease complications and neurological diseases. Circulation to the tissues and nerves of the extremities may be compromised, leading to neuropathy (diseases involving the nervous system) and tissue death.

Diabetes Insipidus

Diabetes insipidus is a totally different disease than diabetes mellitus. It has nothing to do with glucose, insulin, or the pancreas. Diabetes insipidus is a problem with the posterior pituitary. In this disease, the posterior pituitary does not release sufficient ADH to cause water reabsorption in the kidneys. Therefore, too much water goes out in urine (polyuria), and the water for the extra urine is taken from the blood, reducing blood volume. The hypothalamus recognizes this and sends the signal for thirst, so the individual drinks to replace the lost water (polydipsia).

Even though diabetes mellitus and diabetes insipidus are totally different diseases that have totally different causes, two symptoms—polyuria and polydipsia—are common to both disorders.

Growth Disorders

In addition to diabetes, other disorders of the endocrine system include growth disorders. Growth disorders can be the result of improper secretion of GH from the anterior pituitary. GH is secreted during childhood to promote the growth of most tissues. Levels of GH normally decrease in adulthood. If there is hypersecretion of GH during childhood, **gigantism** results. If there is hyposecretion of GH during childhood, **pituitary dwarfism** results. If there is normal secretion of GH in childhood but hypersecretion of GH in adulthood, **acromegaly** results. In acromegaly, the epiphyseal plates have closed normally, preventing any further elongation of long bones, but all of the bones become more massive through appositional bone growth (covered in Chapter 4) in adulthood. This can be especially seen in the bones of the face. See **Figure 8.17**. Hypersecretion of GH throughout life can occur and would result in a giant with acromegaly.

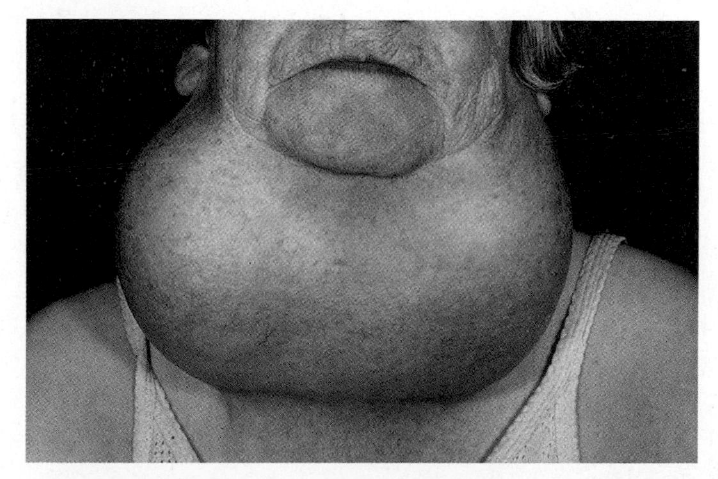

FIGURE 8.18 Endemic goiter resulting from an iodine deficiency.

Goiters

Another endocrine system disorder involves the presence of goiters. A goiter is an enlargement of the thyroid gland. **Figure 8.18** shows an **endemic goiter** that resulted from an iodine deficiency. As stated earlier, iodine is necessary in the diet for the production of thyroid hormone. Normally, the anterior pituitary produces TSH to stimulate the thyroid when metabolism is low. However, if iodine is not available in the diet, hyposecretion of thyroid hormone results. This sets up a positive feedback loop of continued stimulation of the thyroid gland by TSH from the anterior pituitary. Even though the thyroid gland continues to enlarge with all the stimulation, functional thyroid hormone is not produced. Other than the goiter, the effects of hyposecretion of thyroid hormone include weight gain, reduced appetite, constipation, dry skin, and lethargy.

Hypersecretion of thyroid hormone can also result in an enlarged thyroid gland, called a **toxic goiter.** The most common cause of this condition is **Grave's disease.** In Grave's disease, an antibody made by white blood cells to fight a foreign invader mistakenly fits into TSH receptors of the thyroid gland and acts like TSH. The antibody is made for life and has no regard for metabolism levels, so the thyroid gland is constantly stimulated to produce more and more thyroid hormone no matter what the metabolism or anterior pituitary gland would indicate. Other than the goiter, the effects of hypersecretion of thyroid hormone include weight loss, increased appetite, bouts of diarrhea, soft skin, and hyperactivity.

putting the pieces **together**

The Endocrine System

Integumentary system

Provides the precursor molecule for calcitriol (vitamin D).

Reproductive hormones affect hair growth for secondary sex characteristics.

Skeletal system

Sella turcica protects the pituitary gland.

Hormones affect bone deposition and remodeling.

Muscular system

Skeletal muscles protect some glands, such as the adrenal glands.

Hormones regulate blood calcium and glucose levels needed for muscle contractions.

Nervous system

Hypothalamus secretes releasing hormones and sends nerve signals to stimulate the pituitary gland.

Hormones regulate blood glucose and electrolyte levels needed for neuron function.

Cardiovascular system

Blood transports hormones to their target tissues, provides nutrients, and removes wastes.

Hormones regulate blood volume and pressure.

Lymphatic system

Sends white blood cells to fight pathogens in endocrine glands.

Glucocorticoids suppress the immune system and reduce inflammation.

Respiratory system

Provides O_2 for endocrine gland tissue and removes CO_2.

Hormones such as epinephrine increase airflow and respiratory rate.

Digestive system

Provides nutrients for endocrine system glands.

Hormones regulate gastric secretions.

Excretory/urinary system

Kidneys dispose of hormones that are no longer needed.

Hormones regulate urine production.

Reproductive system

Reproductive hormones have a negative-feedback effect on the hypothalamus.

Hormones regulate sexual development, sex drive, menstruation, pregnancy, birth, and lactation.

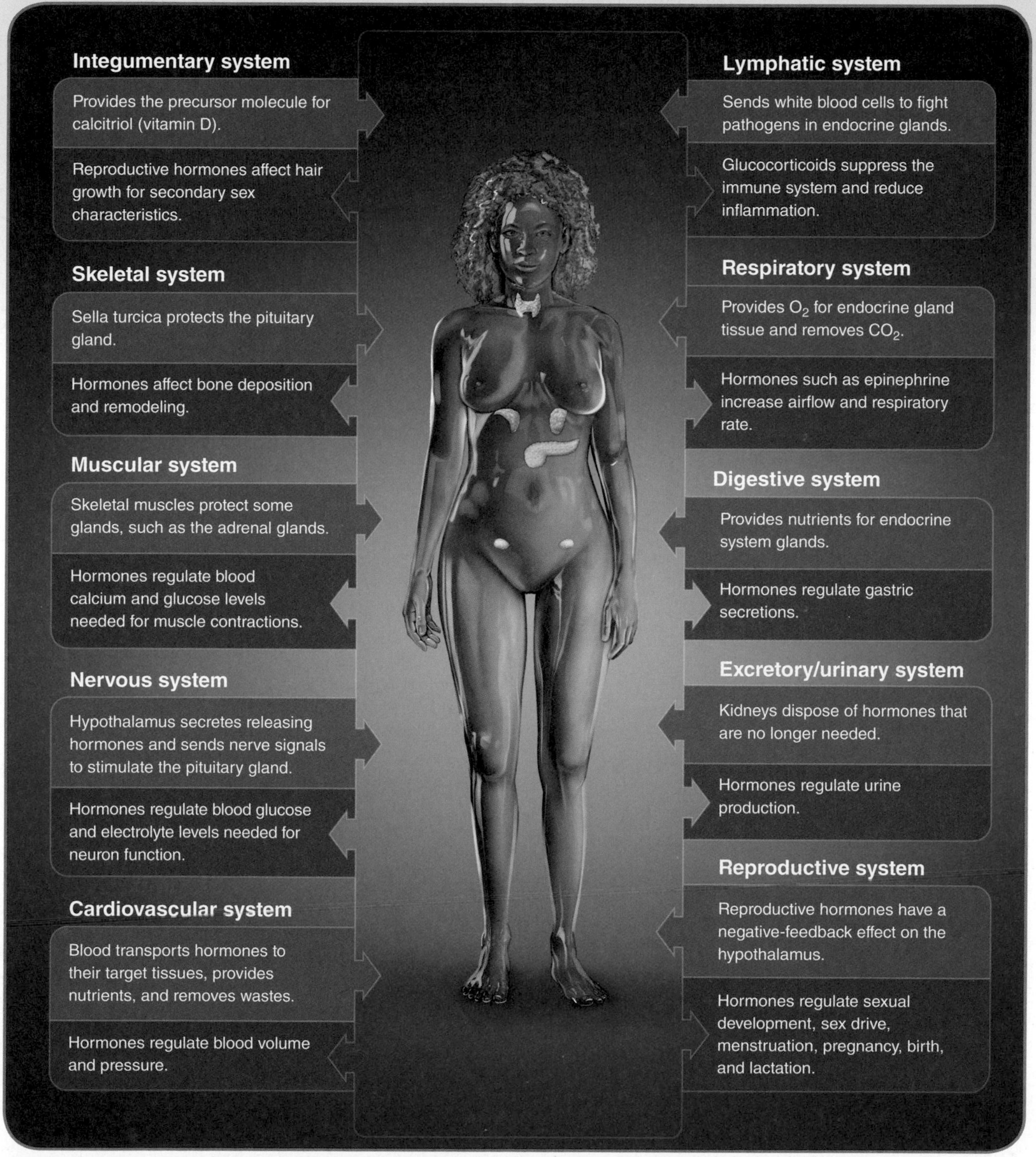

FIGURE 8.19 **Putting the Pieces Together—The Endocrine System:** connections between the endocrine system and the body's other systems.

summary

Overview

- The endocrine system and nervous system are used for communication to maintain homeostasis.
- Both of these systems use chemicals as messengers.
- Communication through the endocrine system is much slower to start, is less specific as to its target, and takes longer to end than communication by the nervous system.

Anatomy of the Endocrine System

- A gland may be a separate structure all on its own, or it may be groups of cells within an organ that function together to produce hormones.
- A target tissue has receptors for a hormone based on shape.

Glands

- Pineal gland function is not completely known.
- The hypothalamus and pituitary gland are connected by the infundibulum and work closely with one another.
- The thyroid gland produces calcitonin and thyroid hormone, which requires iodine.
- The pancreas is part of the endocrine and digestive systems. It produces hormones to regulate blood glucose levels.
- The adrenal gland is divided into two parts: the adrenal cortex and adrenal medulla. Each part produces different hormones.
- The gonads are ovaries and testes.
- Other tissues also make hormones.

Hormones

- There are three categories of hormones based upon chemical composition: steroids, amino acid derivatives, and proteins.
- The composition of a hormone affects how it relates to a receptor.

Target Tissues

- Receptors for hormones can be on the cell membrane or inside the cell.
- The location of the target tissue is relevant to the delivery method of a hormone.

Physiology of the Endocrine System

Regulation of Hormone Secretion and Distribution

- A gland may be stimulated to produce a hormone by a nerve, another hormone, or a substance other than a hormone.
- Thyroid hormone and steroid hormones bind to plasma proteins in the blood, so not all of the hormone enters the cell at once.
- Protein hormones do not bind to plasma proteins because they cannot enter a cell.

Receptor Regulation

- Up-regulation is the increase in the number of receptors for a given hormone. It increases the cell's sensitivity to a hormone and therefore increases the effects of the hormone.
- Down-regulation is the decrease in the number of receptors for a given hormone. It decreases the cell's sensitivity to a hormone and therefore decreases the effects of the hormone.

Hormone Elimination

- Hormones can be eliminated from the system through excretion, metabolism, active transport, or conjugation.
- Half-life is the length of time it takes for one-half of a substance to be eliminated from the circulatory system.

Functions: Four Scenarios

- Hormones such as insulin and glucagon are used to maintain homeostasis.
- Hormones interact to bring about changes in the body.
- The endocrine system uses negative-feedback mechanisms to achieve homeostasis.
- The endocrine system can use positive feedback mechanisms.

Effects of Aging on the Endocrine System

- In general, the levels of hormones decline with age.
- Even if the levels of some hormones remain high with age, the sensitivity of the target tissue often decreases with down-regulation of receptors.

Endocrine System Disorders

Diabetes Mellitus

- Diabetes mellitus is a problem with insulin in blood glucose regulation. It has two forms: type 1 and type 2. Symptoms for both types are excessive thirst, excessive urine production, and glucose in the urine. If uncontrolled, both types may lead to life-threatening complications.

Diabetes Insipidus

- Diabetes insipidus is a problem of insufficient ADH secretion from the posterior pituitary. The symptoms are excessive thirst and excessive urine production.

Growth Disorders

- Growth disorders are a problem of GH secretion from the anterior pituitary.

Goiters

- Goiters can result from hyposecretion or hypersecretion of thyroid hormone.

key words for review

The following terms are defined in the glossary.

androgens
autocrine
down-regulation
endocrine
gland
glucocorticoids
gonads
half-life
hormone
mineralocorticoids
pancreatic islets
paracrine
pheromone
plasma protein
receptor
second messenger
secondary sex characteristics
target tissue
thyroid hormone
up-regulation

chapter review questions

Word Building: *Use the word roots listed in the beginning of the chapter and the prefixes and suffixes inside the back cover to build words with the following meanings:*

1. A substance to stimulate the ovaries and testes: ______________________
2. A lipid hormone made by the outer layer of the adrenal gland: ______________________
3. An effect that works on the same cell that produced the hormone: ______________________
4. Condition of excessive thirst: ______________________
5. Condition of glucose in the urine: ______________________

Multiple Select: *Select the correct choices for each statement. The choices may be all correct, all incorrect, or any combination of correct and incorrect.*

1. Where are the endocrine glands located?
   a. The pituitary gland sits in the cribriform plate.
   b. The parathyroid glands are embedded in the anterior surface of the thyroid gland.
   c. The adrenal cortex is located deep to the adrenal medulla.
   d. The hypothalamus is connected to the pituitary gland by the infundibulum.
   e. The thyroid gland is located in the cervical region.
2. How do the endocrine and nervous systems compare?
   a. Both systems are used for communication.
   b. Both systems use chemicals as messengers.
   c. The nervous system is faster, but it is less specific as to target tissue.
   d. The endocrine system is slower and more general as to target tissue.
   e. Communication from the nervous system can be stopped more quickly than communication from the endocrine system.
3. What is the chemical composition of hormones?
   a. Epinephrine is made from amino acids.
   b. Estrogen is a protein.
   c. Insulin is a protein.
   d. Hormones can be steroids, proteins, or amino acid derivatives.
   e. Protein hormones are modified from a cholesterol molecule.
4. Where are receptors located?
   a. Receptors must be on the cell membrane for steroid hormones.
   b. Receptors may be in the nucleus for protein hormones.
   c. Receptors on the cell membrane may require a second-messenger system.
   d. Receptors may be in the cytoplasm.
   e. The location of the receptor is based on the chemical composition of the hormone.
5. Which of the following statements accurately describe(s) the proximity of a target tissue?
   a. *Pheromone* refers to a hormone that works on neighboring cells.
   b. A sebaceous gland is an example of an endocrine gland.
   c. *Paracrine* refers to a hormone that works outside the body on another organism.
   d. *Autocrine* refers to a hormone that works on the same tissue that produced it.
   e. Endocrine hormones travel to distant target tissues through endocrine ducts.
6. How does a target tissue regulate its sensitivity to a hormone?
   a. The target tissue can down-regulate by decreasing the number of receptors for a hormone.
   b. The target tissue can up-regulate by increasing the number of receptors for a hormone.
   c. Down-regulation decreases the sensitivity of the target tissue to the hormone.
   d. Up-regulation increases the sensitivity of the target tissue to the hormone.
   e. Target tissues tend to up-regulate when exposed to large amounts of a hormone over time.

7. What happens to the endocrine system with aging?
   a. Estrogen levels decrease by age 80 to 20 percent of what they were at their peak.
   b. Some hormones continue to be produced in high levels throughout life.
   c. Target tissues tend to down-regulate with age.
   d. In general, hormone levels diminish with age.
   e. The effects of some hormones are diminished.
8. How are hormones eliminated from the system?
   a. They can be conjugated by the target tissue so that they can be recycled.
   b. They can be excreted by the liver.
   c. They can be excreted by the kidneys.
   d. They can be metabolized by the target tissue.
   e. They can be actively transported to the gland that produced them.
9. What is included in the endocrine system?
   a. Patches of cells in the stomach that make chemicals to communicate with nearby cells.
   b. Nuclei in the hypothalamus.
   c. Pancreatic islets.
   d. Ceruminous glands.
   e. Gonads.
10. What is involved in charting the scenario of the effects of decreased melatonin secretion at puberty?
   a. Melatonin stimulates the production of GnRH.
   b. GnRH targets the ovaries and testes.
   c. LH stimulates sperm production in the male.
   d. ACTH stimulates the anterior pituitary to release FSH and LH.
   e. GnRH is released from the posterior pituitary.

Matching: *Match the gland to the hormone. Some answers may be used more than once.*

_____	1. ACTH	a. Adrenal medulla
_____	2. TSH	b. Thyroid gland
_____	3. ADH	c. Posterior pituitary
_____	4. Cortisol	d. Anterior pituitary
_____	5. GnRH	e. Hypothalamus
		f. Adrenal cortex

Matching: *Match the hormone to its target tissue. Some answers may be used more than once.*

_____	6. ACTH	a. Kidney
_____	7. TSH	b. Anterior pituitary
_____	8. PTH	c. Posterior pituitary
_____	9. TRH	d. Thyroid gland
_____	10. CRH	e. Adrenal cortex
		f. Adrenal medulla

Critical Thinking

1. Grave's disease is a disease of the thyroid gland. In this disease, the immune system mistakenly makes an antibody that fits into receptors on the thyroid gland, stimulating it to produce thyroid hormone. Production of the antibody is continual. Predict the effects of this disease on metabolism, the levels of TRH, and the levels of TSH.
2. The symptoms of diabetes mellitus and diabetes insipidus are very similar. Both include polydipsia and polyuria. What tests might be run to differentiate the two diseases?
3. How would the symptoms of a tumor that prevented the anterior pituitary's production of ACTH be similar to Addison's disease? How would the symptoms of the two conditions differ?

8 chapter mapping

This section of the chapter is designed to help you find where each outcome is covered in this text.

	Outcomes	Readings, figures, and tables	Assessment
8.1	Use medical terminology related to the endocrine system.	Word roots: p. 299	Word Building: 1–5
8.2	Compare and contrast the endocrine and nervous systems in terms of type, specificity, speed, and duration of communication.	Overview: p. 300 Figure 8.2	Spot Check: 2 Multiple Select: 2
8.3	Define *gland, hormone,* and *target tissue.*	Anatomy of the endocrine system: p. 301	Spot Check: 1 Multiple Select: 9
8.4	List the major hormones, along with their target tissues and functions, of each of the endocrine system glands.	Glands: pp. 301–303 Figure 8.3 Table 8.1	Spot Check: 3, 4 Matching: 1–10
8.5	Locate and identify endocrine system glands	Glands: pp. 303–307 Figures 8.3–8.8	Multiple Select: 1
8.6	Describe the chemical makeup of hormones, using estrogen, insulin, and epinephrine as examples.	Hormones: pp. 308–309 Figure 8.9	Spot Check: 5 Multiple Select: 3
8.7	Compare the location of receptors for protein hormones with that of receptors for steroid hormones.	Target tissues: p. 309 Figures 8.10, 8.11	Multiple Select: 4
8.8	Differentiate autocrine, paracrine, endocrine, and pheromone chemical signals in terms of the proximity of the target tissue.	Location of target tissues: pp. 309–311	Multiple Select: 5
8.9	Explain the regulation of hormone secretion and its distribution.	Regulation of hormone secretion and distribution: pp. 311–313 Figures 8.12, 8.13	Spot Check: 6, 8
8.10	Explain how the number of receptors can be changed.	Receptor regulation: pp. 313–314 Figure 8.14	Multiple Select: 6
8.11	Explain how hormones are eliminated from the body.	Hormone elimination: pp. 314–315	Spot Check: 7 Multiple Select: 8
8.12	Explain the function of hormones by showing how they interact to maintain homeostasis.	Function: four scenarios: pp. 315–319 Figure 8.15 Tables 8.2–8.6	Spot Check: 9–17 Multiple Select: 10 Critical Thinking: 1
8.13	Explain the effects of aging on the endocrine system.	Effects of aging on the endocrine system: p. 319	Multiple Select: 7
8.14	Describe endocrine system disorders.	Disorders of the endocrine system: pp. 319–322 Figures 8.16–8.18	Critical Thinking: 2–3

references and resources

Beers, M. H., Porter, R. S., Jones, T. V., Kaplan, J. L., & Berkwits, M. (Eds.). (2006). *Merck manual of diagnosis and therapy* (18th ed.). Whitehouse Station, NJ: Merck Research Laboratories.

Morley, J. E. (2006, May). Effects of aging: Biology of the endocrine system. *Merck manual home edition.* Retrieved June 17, 2010, from http://www.merckmanuals.com/home/sec13/ch161/ch161f.html.

Porter, R. S., & Kaplan, J. L. (Eds.). (n.d.). Selected physiologic age-related changes. *Merck manual for healthcare professionals.* Retrieved March 19, 2011, from http://www.merckmanuals.com/media/professional/pdf/Table_337-1.pdf?qt=selected physiologic age-related changes&alt=sh.

Saladin, Kenneth S. (2010). *Anatomy & physiology: The unity of form and function* (5th ed.). New York: McGraw-Hill.

Seeley, R. R., Stephens, T. D., & Tate, P. (2006). *Anatomy & physiology* (7th ed.). New York: McGraw-Hill.

Shier, D., Butler, J., & Lewis, R. (2010). *Hole's human anatomy & physiology* (12th ed.). New York: McGraw-Hill.

9 The Cardiovascular System—Blood

By this time in your life, you have undoubtedly seen your own blood. You may have seen it oozing from a cut on your finger, being drawn from your arm into a tube at the doctor's office, or forming a bruise under the skin of your knee after a fall. The sight of it may have fascinated you, scared you, or even made you feel faint. Regardless of how the sight of blood makes you feel, you certainly appreciate that blood is vital to your health and the way your cardiovascular system functions. See Figure 9.1

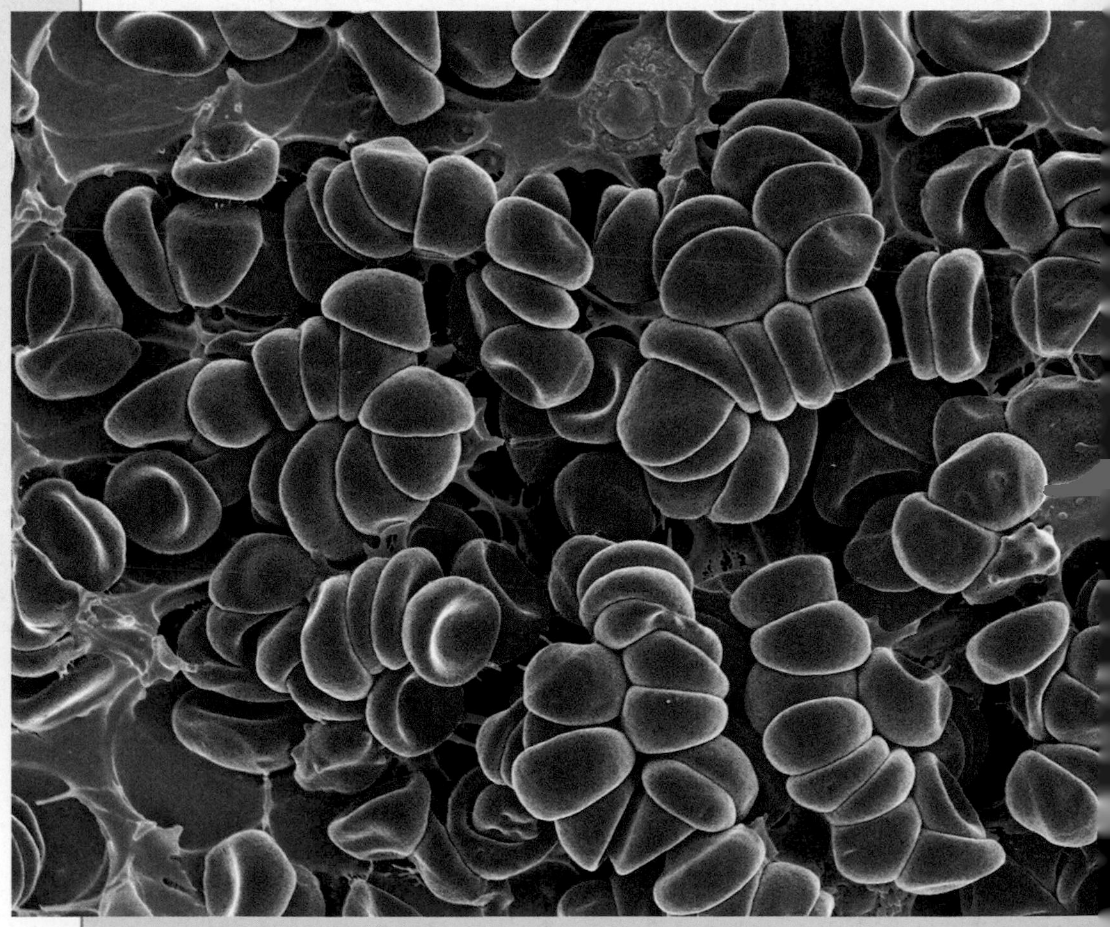

9.1 learning **outcome**

Use medical terminology related to the cardiovascular system.

learning outcomes

After completing this chapter, you should be able to:

9.1 Use medical terminology related to the cardiovascular system.

9.2 Identify the components of blood.

9.3 List the constituents of plasma and their functions.

9.4 Identify the formed elements and list their functions.

9.5 Compare the various forms of hemopoiesis in terms of starting cell, factors influencing production, location, and final product.

9.6 Describe the structure and function of hemoglobin.

9.7 Summarize the nutritional requirements of red blood cell production.

9.8 Describe the life cycle of a red blood cell from its formation to removal.

9.9 Describe the body's mechanisms for controlling bleeding.

9.10 Describe two pathways for blood clotting in terms of what starts each, their relative speed, and the clotting factors involved.

9.11 Describe what happens to blood clots when they are no longer needed.

9.12 Explain what keeps blood from clotting in the absence of injury.

9.13 Explain what determines ABO and Rh blood types.

9.14 Explain how a blood type relates to transfusion compatibility.

9.15 Determine, from a blood type, the antigens and antibodies present and the transfusion compatibility.

9.16 Predict the compatibility between mother and fetus given Rh blood types for both and describe the possible effects.

9.17 Summarize the functions of blood by giving an example or explanation of each.

9.18 Explain what can be learned from common blood tests.

9.19 Describe disorders of the cardiovascular system concerning blood.

word **roots** & combining **forms**

agglutin/o: clumping

blast/o: primitive cell

coagul/o: clot

cyt/o: cell

erythr/o: red

granul/o: granules

hem/o: blood

hemat/o: blood

leuk/o: white

phag/o: eat, swallow

thromb/o: clot

pronunciation **key**

bilirubin: bill-ee-RU-bin

coagulation: koh-ag-you-LAY-shun

eosinophils: ee-oh-SIN-oh-fillz

erythrocyte: eh-RITH-roh-site

hemolytic: HE-moh-LIT-ik

hemopoiesis: HE-moh-poy-EE-sis

hemostasis: HE-moh-STAY-sis

leukopenia: loo-koh-PEE-nee-ah

lymphocyte: LIM-foh-site

macrophages: MAK-roh-fayj-ez

megakaryocyte: MEG-ah-KAIR-ee-oh-site

neutrophils: NEW-troh-fillz

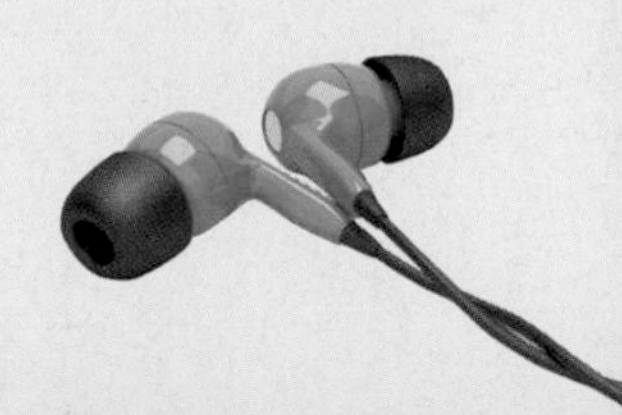

Overview

The cardiovascular system has three components: the heart, blood, and blood vessels. The heart serves as a pump to circulate blood through a closed circuit of blood vessels out to the tissues of the body and back again to the heart. The heart and blood vessels are covered in the next chapter.

This chapter focuses on the anatomy of blood and the physiology of how it works, which includes how blood is produced, serves as a transportation system, fights foreign invaders, and protects the body against its own loss. You will investigate how blood maintains the body's homeostasis by regulating the fluid and electrolyte balance and the acid-base balance. You will also learn about blood typing, common blood tests, and disorders of the blood. Your study begins with the anatomy of blood.

Cardiovascular System

Major Organs and Structures:
heart, aorta, superior and inferior venae cavae

Accessory Structures:
arteries, veins, capillaries

Functions:
transportation, protection by fighting foreign invaders and clotting to prevent its own loss, fluid and electrolyte balance, and temperature regulation

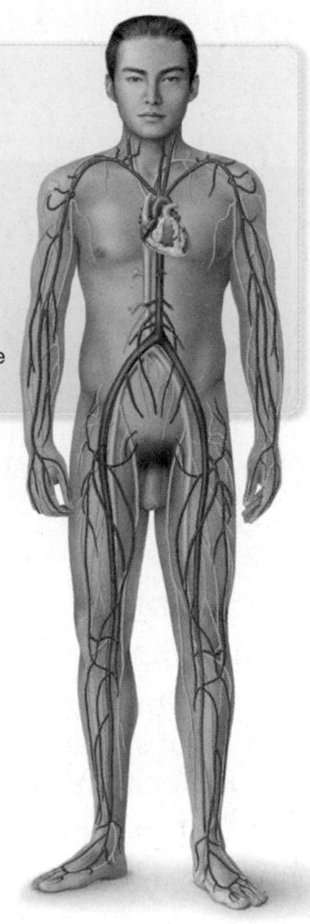

FIGURE 9.1 The cardiovascular system.

9.2 learning outcome

Identify the components of blood.

Anatomy of Blood

As you may recall from Chapter 2, blood is a connective tissue. As such, it is composed of different types of cells and cell parts that exist in a fluid matrix known as **plasma.** The cells and cell parts are called **formed elements.** An average adult has 4 to 6 liters of blood, which is approximately 8 percent of the body's weight. If a tube of whole blood is spun in a centrifuge, the formed elements will separate from the plasma. As you can see in **Figure 9.2**, three layers are formed. Although the formed elements are heavy and settle to the bottom, they are not all the same. Red blood cells (RBCs)—45 percent of whole blood—are the heaviest, so they settle at the very bottom of the tube. Above the red blood cells is a thin, buff-colored layer of white blood cells (WBCs) and platelets—less than 1 percent of whole blood. Above this layer is the clear, straw-colored plasma (55 percent of whole blood). We discuss the components of blood in detail below, beginning with plasma.

spot check  Why is blood classified as a connective tissue?

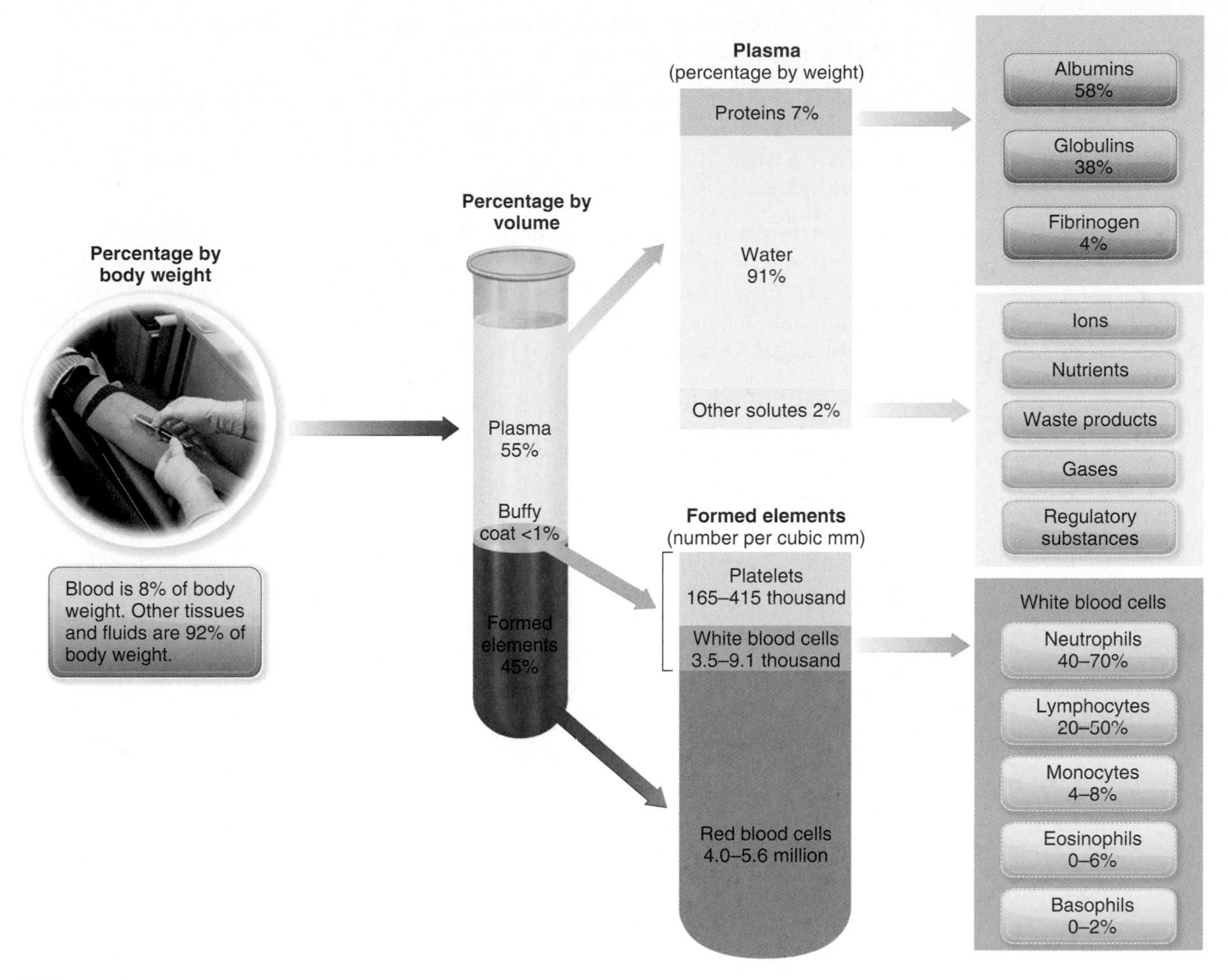

FIGURE 9.2 Composition of blood.

Plasma

9.3 learning outcome

List the constituents of plasma and their functions.

As you can see in **Figure 9.2**, 91 percent of the total volume of plasma is water. The following three types of dissolved proteins make up another 7 percent of plasma's total volume:

- **Albumins.** These are transport proteins dissolved in plasma. An example of an albumin is the plasma protein that binds to thyroid hormone to extend its half-life while traveling in the blood (covered in Chapter 8).
- **Globulins.** The globulins are another class of plasma proteins. Examples of globulins are the antibodies produced by white blood cells to fight foreign pathogens. Antibodies are discussed later in this chapter and again in the lymphatic system chapter.
- **Fibrinogen and clotting factors.** Fibrinogen is a clotting protein dissolved in plasma. Later in this chapter, we explain how fibrinogen comes out of solution to form a clot. Other chemicals, called *clotting factors,* are needed for the clot to happen. **Serum** is plasma with the fibrinogen and clotting factors removed. It is discussed later in this chapter in the section on blood typing.

The last 2 percent of plasma shown in **Figure 9.2** is composed of many items already mentioned in this text:

- **Ions** in solution are electrolytes. You studied several of them in previous chapters. An example of an ion transported in plasma is Ca^{2+}. Osteoblasts deposit excess calcium ions from the blood into bone (Chapter 4). Other ions in plasma include Na^{+}, K^{+}, and Cl^{-}.
- **Nutrients** are necessary chemicals for normal body function. Glucose is a good example. If the blood glucose level falls below normal, the liver converts glycogen to glucose and releases it to the blood (Chapter 8). Other examples of nutrients are amino acids, fatty acids, and vitamins. These and other nutrients are usually absorbed into the blood through the digestive system.
- **Waste products** are by-products of chemical reactions that occur in the cells. They are secreted into the blood for removal. An example is **bilirubin,** a waste produced from the breakdown of worn-out red blood cells. Bilirubin is discussed later in the chapter. Other wastes—such as the nitrogenous wastes removed from the blood by the kidneys—are covered in the excretory system chapter.

bilirubin: bill-ee-RU-bin

- **Gases,** which enter or leave the body through the lungs, may be required by cells for chemical reactions to occur. Gases may also be produced as a waste product of cellular reactions. Examples of gases dissolved in plasma are oxygen and carbon dioxide. However, most of the oxygen and carbon dioxide in the blood is carried by red blood cells—not dissolved in plasma. How these gases are transported is discussed in the respiratory system chapter.
- **Regulatory substances** are chemicals used for communication. Examples of regulatory substances dissolved in plasma are the endocrine hormones, covered in the previous chapter.

It is important to remember that the components of plasma are dissolved. Plasma is a solution with proteins, ions, nutrients, waste products, gases, and regulatory substances as the solutes and water as the solvent. The concentrations of these solutes are important for homeostasis.

spot check ❷ Give an example of one regulatory substance that may be found in plasma. Include where it is produced and its destination.

Now that you have covered the composition of plasma, you are ready to examine the cells and cell parts—also known as *formed elements*—of whole blood.

9.4 learning outcome

Identify the formed elements and list their functions.

Formed Elements

Unlike the components of plasma, the formed elements of blood are not dissolved. They include **erythrocytes (red blood cells), leukocytes (white blood cells),** and **thrombocytes (platelets)**. Erythrocytes and leukocytes are cells, but thrombocytes are cell fragments, not complete cells. It is important to be able to identify each formed element and its function. Each is discussed below and summarized in Table 9.1.

erythrocyte: eh-RITH-roh-site

Erythrocytes These cells are commonly called *red blood cells,* a term that can be used interchangeably with *erythrocytes*. Each erythrocyte is a biconcave disk, thick around its rim and thin at its center. This gives the cell a greater surface area for an exchange of gases in the lungs and at the tissues. Erythrocytes—the most plentiful of the formed elements—function to transport oxygen and carbon dioxide. It is

TABLE 9.1 Formed elements

Formed element	Description	Function
Erythrocytes (RBCs) 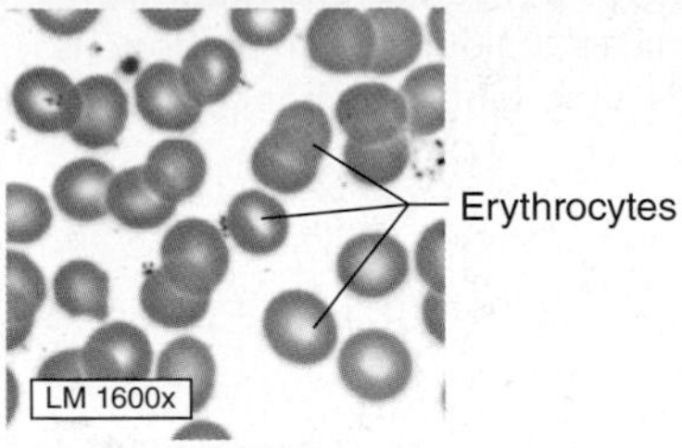 (Erythrocytes, LM 1600x)	Red, biconcave disks with no nucleus; 7.5 μm in diameter	Mainly, transport oxygen and carbon dioxide.
Leukocytes (WBCs):	Spherical cells that must be stained to be seen	Have various functions, depending on type. (See entries below.)
Neutrophil 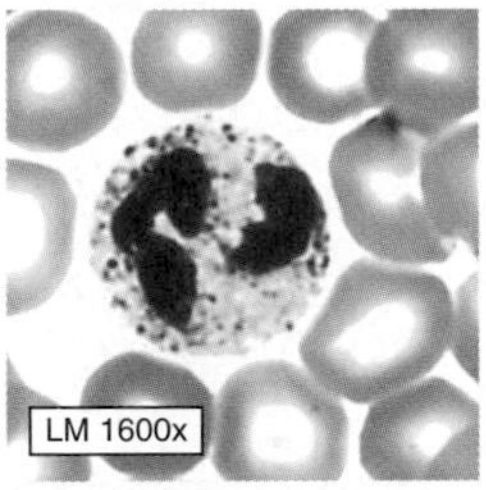 (LM 1600x)	Faint granules; nucleus with multiple lobes connected by a filament; stains pink to purple; 10–12 μm in diameter	Phagocytizes microorganisms. Numbers increase in bacterial infections.
Basophil 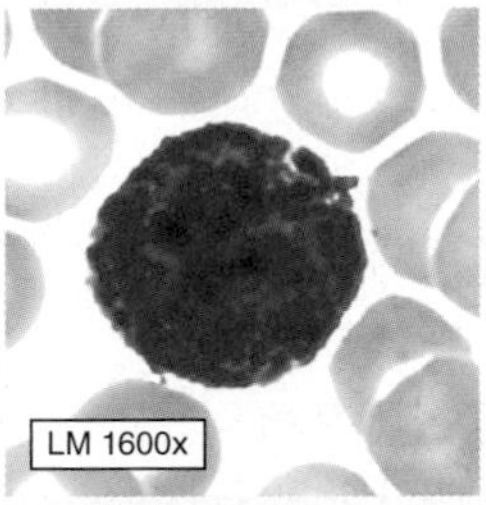 (LM 1600x)	Prominent granules that stain blue-purple; lobed nucleus; 10–12 μm in diameter	Releases histamine to promote inflammation and heparin to prevent unnecessary clot formation. Numbers increase with allergies.
Eosinophil 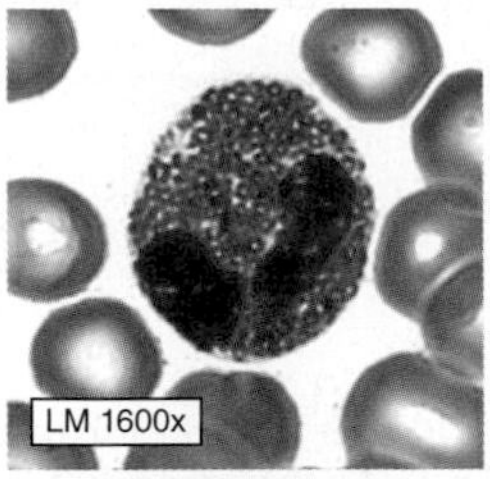 (LM 1600x)	Prominent granules that stain orange to bright red; lobed nucleus; 11–14 μm in diameter	Attacks some worm parasites. Numbers can increase with allergies.
Monocyte 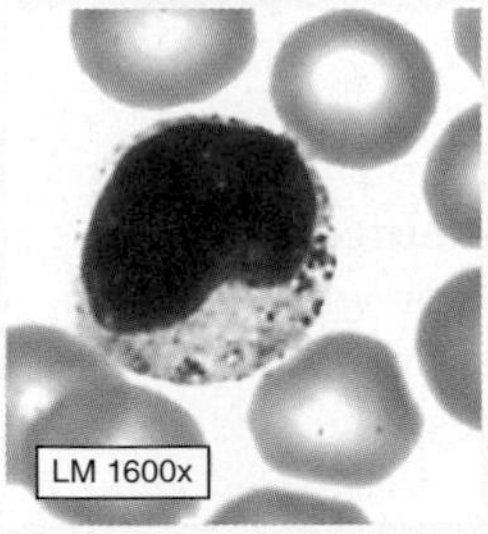 (LM 1600x)	Large cell, 2–3 times the size of an RBC; nucleus that is round, kidney-shaped, or horseshoe-shaped; contains more cytoplasm than a lymphocyte; 12–20 μm in diameter	Leaves the blood to become a macrophage in the tissues. Phagocytizes bacteria, dead cells, and other debris.

continued

TABLE 9.1 concluded

Formed element	Description	Function
Lymphocyte LM 1600x	Round nucleus with little cytoplasm; 6–14 μm in diameter	Is important for the immune system. Produces antibodies and other chemicals to fight foreign pathogens and is important for tumor control.
Thrombocytes (platelets) Platelets LM 1600x	Cell fragments surrounded by a membrane; 2–4 μm in diameter	Forms platelet plugs and releases clotting factors.

Note: μm = micrometer.

important to note the absence of a nucleus. In fact, there are very few organelles in the cytoplasm of a red blood cell. Without mitochondria, erythrocytes cannot carry out aerobic respiration. So red blood cells do not use the oxygen they transport.

clinical point

Without a nucleus, red blood cells do not contain any DNA. Yet you are probably aware of blood DNA testing to establish paternity and forensic testing of blood DNA to determine who committed a crime. The DNA housed in the nuclei of white blood cells is what is actually tested during blood DNA tests. It is amazing how few blood cells are truly needed to supply sufficient DNA for these tests.

spot check 3 What percentage of a single drop of blood is leukocytes?

Red blood cells spend their lives in blood vessels. They do not move out into the tissues unless one of the vessels is broken. You will learn more about the life cycle of a red blood cell later in the chapter.

lymphocyte: LIM-foh-site

eosinophils: ee-oh-SIN-oh-fillz

neutrophils: NEW-troh-fillz

Leukocytes These cells are commonly called *white blood cells.* Each of the five types of leukocytes has a different appearance and function, but all have prominent nuclei that must be stained to be seen. Three of the types—**neutrophils, basophils,** and **eosinophils**—contain small granules that differ in color when stained. These leukocytes are classified as **granulocytes. Monocytes** and **lymphocytes** are **agranulocytes** because they do not contain visible granules. The size and shape of the nucleus, the presence or absence of granules, and the color they stain help to distinguish one type of leukocyte from another.

Unlike erythrocytes, leukocytes can move out of blood vessels into the tissues. They are often in circulation only as a means of getting to the tissues where they perform their functions. These functions provide various defenses against foreign pathogens. How leukocytes perform their functions is discussed in the lymphatic system chapter.

Neutrophils These are the most common type of leukocyte. Neutrophils typically make up 40 to 70 percent of all the white blood cells in circulation. Each neutrophil has a lobed nucleus and faint granules containing lysozymes used to destroy bacteria. The number of neutrophils in circulation rises in response to bacterial infections.

Basophils These are the least common type of leukocyte. Basophils average from 0 percent to less than 2 percent of all the white blood cells in circulation. The large, dark blue-purple granules of a basophil are so prominent that it is difficult to see the S- or U-shaped nucleus. The number of basophils in circulation tends to increase with allergies. A basophil's primary function is to release two chemicals—**histamine** and **heparin**—for defense:

1. Histamine released from basophils causes vessels to **dilate** (expand). This brings more blood to an area and causes blood vessel walls to become more permeable. The increased blood flow and permeability allow more leukocytes to move out of the blood vessels into injured tissues more quickly.
2. Heparin released from basophils is an **anticoagulant,** which means it prevents clotting. This chemical allows other leukocytes to move more freely.

Eosinophils These make up 0 to 6 percent of the total circulating white blood cells. Their granules stain orange to bright red, making them easy to distinguish from other WBCs. Eosinophil numbers increase with parasitic infections and allergies. The chemicals they secrete can be effective against large parasites such as hookworms and tapeworms.

Monocytes Although these cells and lymphocytes stain similarly, there are significant differences between the two types. Monocytes are the largest of the WBCs, measuring 2 to 3 times the size of a red blood cell. They have a large round, kidney-shaped, or horseshoe-shaped nucleus surrounded by abundant cytoplasm. Monocytes migrate to tissues where they become **macrophages** and function to **phagocytize** (eat) dead and dying tissue, microorganisms, and any other foreign matter or debris. The number of monocytes in circulation increases with inflammation and viral infections.

macrophages: MAK-roh-fayj-ez

Lymphocytes These cells are fairly common in circulating blood, making up 20 to 50 percent of the total WBCs. Although lymphocytes and monocytes stain similarly, lymphocytes are smaller and have less cytoplasm. Two of the subclasses of lymphocytes are B cells and T cells. Although they look alike, they have different immune system functions. B cells and T cells are discussed in detail in the lymphatic system chapter.

spot check 4 Why are WBCs not white?

spot check 5 Which leukocyte(s) would you expect to increase in number if you had hay fever?

Thrombocytes These formed elements are commonly called *platelets,* a term that may be used interchangeably with *thrombocytes*. Platelets are not actually cells but cell fragments. Together, platelets and leukocytes make up approximately 1 percent of whole blood, but platelets outnumber leukocytes. Platelets in the human body have many important functions:

- Platelets secrete **vasoconstrictors** (chemicals that reduce the size of broken blood vessels) to slow the flow of blood.
- Platelets secrete clotting factors to promote the formation of blood clots.
- Platelets form platelet plugs, which are discussed later in the chapter.
- Platelets secrete chemicals to attract neutrophils and monocytes to sites of inflammation.
- Platelets destroy bacteria.
- Platelets secrete growth factors to stimulate mitosis to repair vessel walls.

You have covered the anatomy of blood by examining the composition of plasma and studying each of the formed elements. Next, you will focus on the physiology of blood, starting with how blood is produced.

9.5 learning **outcome**

Compare the various forms of hemopoiesis in terms of starting cell, factors influencing production, location, and final product.

Physiology of Blood

Hemopoiesis

Hemopoiesis is blood production. It is a continual process designed to meet the demand of replacing circulating cells that have worn out or been lost through bleeding. There are three types of hemopoiesis: thrombopoiesis, leukopoiesis, and erythropoiesis. It is important to note in each form of production where the production happens, what the beginning cell is, what causes the production to occur, and what the final product is. See Table 9.2.

TABLE 9.2 Hemopoiesis

Type of hemopoiesis	Where	Starting cell	Factor causing production	What produces the factor	Final product
Thrombopoiesis	Red bone marrow	Hemocytoblast	Thrombopoietin	Liver and kidneys	Platelets
Leukopoiesis	Red bone marrow	Hemocytoblast	Colony-stimulating factors	Lymphocytes and macrophages	White blood cells
Erythropoiesis	Red bone marrow	Hemocytoblast	Erythropoietin	Kidneys	Red blood cells

hemopoiesis:
HE-moh-poy-EE-sis

Figure 9.3 will be helpful while you are learning about each form of hemopoiesis. The two most important rows in this figure are the top row (showing the starting cell) and the bottom row (showing the final product).

Notice that the **hemocytoblast** (stem cell) is the starting cell for the production of all the formed elements. This cell is said to be **pluripotent,** which means it can become any one of seven different types of formed elements. Hemocytoblasts are located in red bone marrow. They must be stimulated to grow and divide to become one of the formed elements. It is important for you to know (in each form of hemopoiesis) what chemical **(factor)** causes the hemocytoblast to commit to production of each formed element.

Myeloid hemopoiesis is the production of *all the formed elements* in the red bone marrow. Only lymphocytes can also be produced in lymphoid tissues such as the lymph nodes, thymus, and spleen. This additional production in sites outside the red bone marrow is **lymphoid hemopoiesis**. It is covered in the lymphatic system chapter.

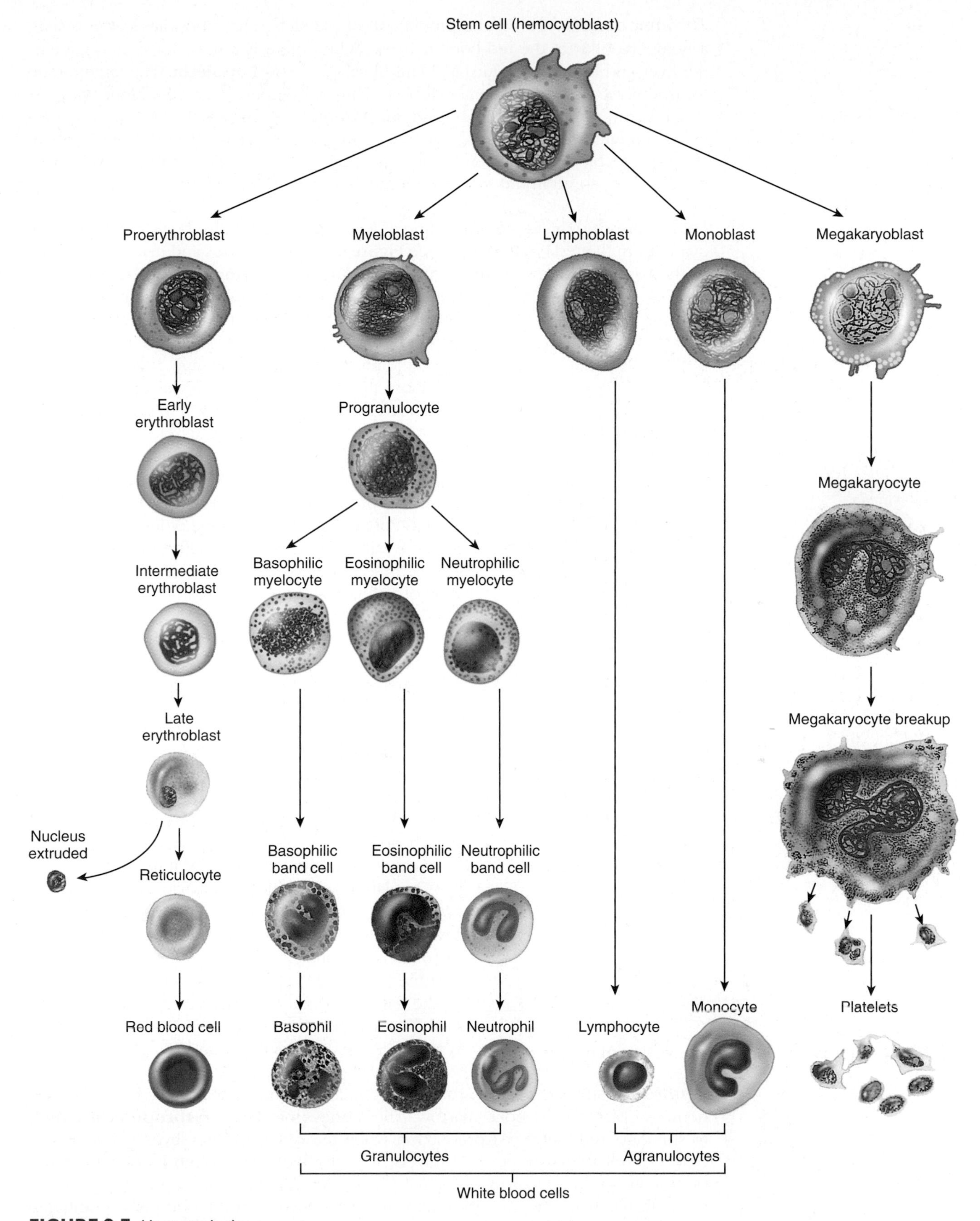

FIGURE 9.3 Hemopoiesis.

Thrombopoiesis This is the production of platelets. Thrombopoiesis begins with a hemocytoblast in the red bone marrow. When there is a need for more platelets, the liver and kidneys produce a chemical called **thrombopoietin.** The target tissue for thrombopoietin is the hemocytoblast. Thrombopoietin causes the hemocytoblast to grow and divide to become a **megakaryocyte.** See **Figure 9.3**. The megakaryocyte eventually breaks apart to become many platelets. Although most of the platelets go into blood circulation immediately, 25 to 40 percent of the newly formed platelets are stored in the spleen until they are needed.

megakaryocyte:
MEG-ah-KAIR-ee-oh-site

Leukopoiesis This is the production of leukocytes, which also begins with a hemocytoblast in the red bone marrow. See **Figure 9.3**. Lymphocytes and macrophages out in the tissues produce chemicals called **colony-stimulating factors (CSFs)** when there is a challenge to the immune system. These factors also target a hemocytoblast. There is a different CSF for each type of leukocyte production. The production of each CSF is dependent upon the immune system challenge. If it is a bacterial infection, the CSF that stimulates neutrophil production is released. The CSF travels to the red bone marrow to fit into a receptor on a hemocytoblast so that the hemocytoblast grows and divides to produce neutrophils. Different CSFs would be produced to promote the production of basophils or eosinophils in the case of allergies. So each type of leukocyte can be produced in abundance, depending upon the need. An increase in any of their numbers can indicate the type of immune system challenge.

Once produced, granulocytes and monocytes are stored in the red bone marrow until they are needed. For example, neutrophils may be stored in the red bone marrow until they are needed to fight an infection in an open wound. There are typically 10 to 20 times more granulocytes and monocytes in red bone marrow than in circulation. Lymphocytes are made in red bone marrow but then migrate to tissues. T lymphocytes migrate to the thymus, while B lymphocytes migrate to other lymphoid tissues such as the lymph nodes and spleen. Lymphocytes are covered in more detail in the lymphatic system chapter.

Circulating white blood cells do not stay in the blood long. They may be in circulation only for a few hours before migrating to the tissues. Monocytes become macrophages once they are in the tissues, and they may last a few years. Lymphocytes migrate to the tissues but may continue to migrate back and forth between the tissues and the blood. Lymphocytes typically last for weeks to decades.

FIGURE 9.4 Negative feedback correction of hypoxemia.

Erythropoiesis This is erythrocyte production. Erythropoiesis, too, begins with a hemocytoblast in red bone marrow. The kidneys produce **erythropoietin (EPO)** to stimulate red blood cell production when the oxygen blood level is low, a condition called **hypoxemia.** EPO travels from the kidneys through blood to the red bone marrow in all cancellous bone. There, EPO fits into receptors on hemocytoblasts to stimulate them to grow and divide to become red blood cells. Additional red blood cells increase the blood's capacity to carry oxygen, thereby increasing the oxygen level in the blood. This is a negative-feedback response to a move away from homeostasis. If blood oxygen levels are too low, more red blood cells are produced to carry more oxygen to restore homeostasis. See **Figure 9.4**.

What causes hypoxemia? Certainly lung disorders may cause low blood oxygen levels, but hypoxemia can also occur without a disorder as the cause. Certain environments and activities can promote hypoxemia, including:

- **High altitudes.** The air is thinner at high altitudes and contains less oxygen. Erythropoiesis is accelerated at high altitudes to provide more red blood cells to carry any available oxygen.
- **Exercise.** Starting an exercise program increases the demand for oxygen in the tissues to perform increased aerobic respiration. Erythropoiesis production increases the number of red blood cells, so more oxygen can be carried to meet the demand.
- **Exposure to carbon monoxide (CO).** Carbon monoxide is a by-product of burning. Exposure to carbon monoxide may happen through car exhaust, improperly vented furnaces, and even cigarette smoke. A smoker inhales carbon monoxide in every puff of a cigarette. Hemoglobin far prefers to carry carbon monoxide than oxygen. Once the hemoglobin picks up carbon monoxide, it never carries oxygen again. This makes the hemoglobin nonfunctional for oxygen transport. Erythropoiesis is accelerated to make up for any nonfunctioning hemoglobin.

warning

It is important that you do not confuse carbon monoxide (CO) with carbon dioxide (CO_2). Carbon dioxide is produced by the body and is a gas normally found in the air we breathe.

- **Blood loss.** There are many reasons for blood loss. The loss may occur from disease, from an injury, or through donation to the local blood drive. The kidneys recognize the diminished blood oxygen level and secrete EPO to stimulate additional erythropoiesis to make up for any loss.

spot check 6 Professional basketball players train to be able to use aerobic respiration continually during a game. How would the red blood count of a Miami Heat player compare to that of a Denver Nuggets player? (*Hint:* Miami is at sea level; Denver is the "mile-high city.") Explain.

spot check 7 How would the red blood count of a Miami Heat player likely compare to that of a University of Miami anatomy and physiology professor? Explain.

Many changes to a hemocytoblast occur during the process of erythropoiesis. For example, one large hemocytoblast becomes many smaller red blood cells. In the process, the nucleus is lost, hemoglobin is formed, and the shape of the cell is changed to a biconcave disk. Again, see **Figure 9.3**. All of this takes place in approximately three to five days in the red bone marrow once EPO has stimulated the hemocytoblast. Below, you will learn more about the hemoglobin that must be produced.

9.6 learning outcome

Describe the structure and function of hemoglobin.

Hemoglobin The cytoplasm of erythrocytes contains **hemoglobin,** a red, complex protein made of four chains of amino acids called **globins.** See **Figure 9.5**. Each chain contains a **heme** group with an iron ion at its center that can bind to one oxygen molecule. So each hemoglobin molecule can carry four oxygen molecules. There are approximately 280 million hemoglobin molecules dissolved in the cytoplasm of each erythrocyte. This makes the cytoplasm of a red blood cell a 33 percent solution of hemoglobin.

In addition to transporting oxygen from the lungs to the tissues, hemoglobin transports hydrogen ions (H^+) and carbon dioxide from the tissues to the lungs. Gas transport is discussed more in the respiratory system chapter. Binding to H^+ means that H^+ will not be free in the blood to lower the blood's pH. So hemoglobin acts as a **buffer** to resist a change in the pH of blood.

Next, you will read about what is needed in a diet to form hemoglobin and carry out erythropoiesis.

applied **genetics**

Some people of African descent have a recessive gene in their DNA that codes for a different form of hemoglobin. This alternative hemoglobin causes red blood cells to change shape and sickle (form a crescent with points) in low blood oxygen situations. See Figure 9.6. Having one copy of the gene results in a condition called **sickle cell trait,** which typically has mild symptoms, if any. Having two copies of the gene results in **sickle cell disease,** a much more serious condition that can cause death if not treated. The sickled cells are sticky and can clump together, blocking blood flow to small vessels. Although the consequences of the gene can be life-threatening, there is an advantage to this form of hemoglobin for people living in areas plagued with malaria. The parasite that causes malaria feeds on hemoglobin, but sickle cell hemoglobin is harmful to the parasite. Thus, people with sickle cell hemoglobin are resistant to malaria.

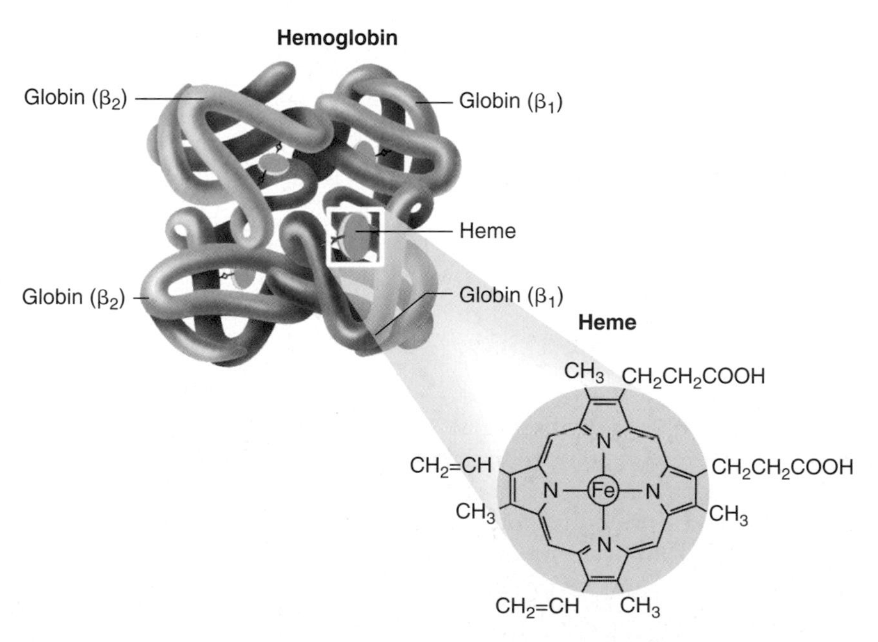

FIGURE 9.5 Hemoglobin molecule composed of four amino acid chains. Each chain has a heme group with an iron ion at its center to carry oxygen.

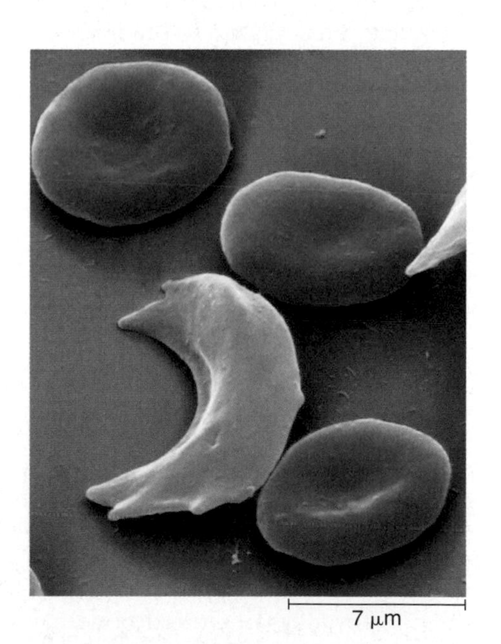

FIGURE 9.6 Sickle cell disease. The crescent-shaped red blood cell has been deformed by sickle cell hemoglobin.

Nutritional requirements for erythropoiesis The key ingredient in hemoglobin is iron because it carries the oxygen for red blood cells to function. Some iron is normally lost each day in urine and feces. The average loss is slightly higher for women than men because of their additional blood loss with menstruation. It is important to note that not all iron from consumed food is absorbed. Acid in the stomach converts only some of the iron consumed to a usable form that can be absorbed. On average, the RDA for iron is 18 mg/day for a female, but a pregnant woman needs 27 mg/day to supply sufficient iron for erythropoiesis for herself and for the developing fetus. Iron can be supplied in a diet through meat, eggs, vegetables, and legumes. (See Appendix B.)

9.7 learning **outcome**

Summarize the nutritional requirements of red blood cell production.

Other nutrients necessary for erythropoiesis include:

- Folic acid and vitamin B_{12} for cell division. Folic acid is supplied in orange juice and vegetables. Vitamin B_{12} is found in meat and dairy products.
- Copper and vitamin C for the enzymes necessary to form hemoglobin. Copper is supplied in the diet through seafood, organ meats, and legumes. Vitamin C is found in fruits and green vegetables.

You have reviewed the production of all the formed elements. Now it is time to examine the life cycle of a red blood cell after it is produced from a hemocytoblast in red bone marrow.

Life Cycle of a Red Blood Cell

9.8 learning **outcome**

Describe the life cycle of a red blood cell from its formation to removal.

The primary function of red blood cells is to transport oxygen and carbon dioxide between the lungs and the tissues. To do this, red blood cells must be in constant circulation, traveling under pressure through smaller and smaller blood vessels and getting squeezed into single file in the capillaries. Because it is not an easy journey, red blood cells survive only 110 to 120 days. After 110 to 120 days, the worn-out red blood cells must be broken down. As these exhausted red blood cells are destroyed, new red blood cells are produced to replace them. Erythropoiesis is an impressive process. The body produces, on average, 2.5 million new red blood cells every single second.

spot check **8** Why can lymphocytes last for decades if erythrocytes last only 110 to 120 days?

The liver and spleen function to remove old, worn-out red blood cells. These two organs break down the cells and process the hemoglobin. In both organs, hemoglobin is broken down to **heme** and **globin.** Globin is the four chains of amino acids. The chains are broken down to free amino acids, which are then recycled to the red bone marrow. These are the exact amino acids that will be needed later to form new hemoglobin during erythropoiesis. The heme is further broken down to iron (recycled to the red bone marrow) and **bilirubin** (a waste that must be removed from the body). The liver secretes bilirubin in **bile,** an important digestive juice that is discussed in the digestive system chapter. The bilirubin is then removed from the body with feces. The spleen secretes bilirubin into the blood for the kidneys to remove and put in urine. As mentioned earlier in the chapter, bilirubin is one of the wastes found in plasma. It is yellow in color, which accounts for the straw color of plasma. The bilirubin that is put in urine by the kidneys gives urine its yellow color. That is why urine is yellow. See **Figure 9.7**.

FIGURE 9.7 The breakdown of hemoglobin by the liver and spleen.

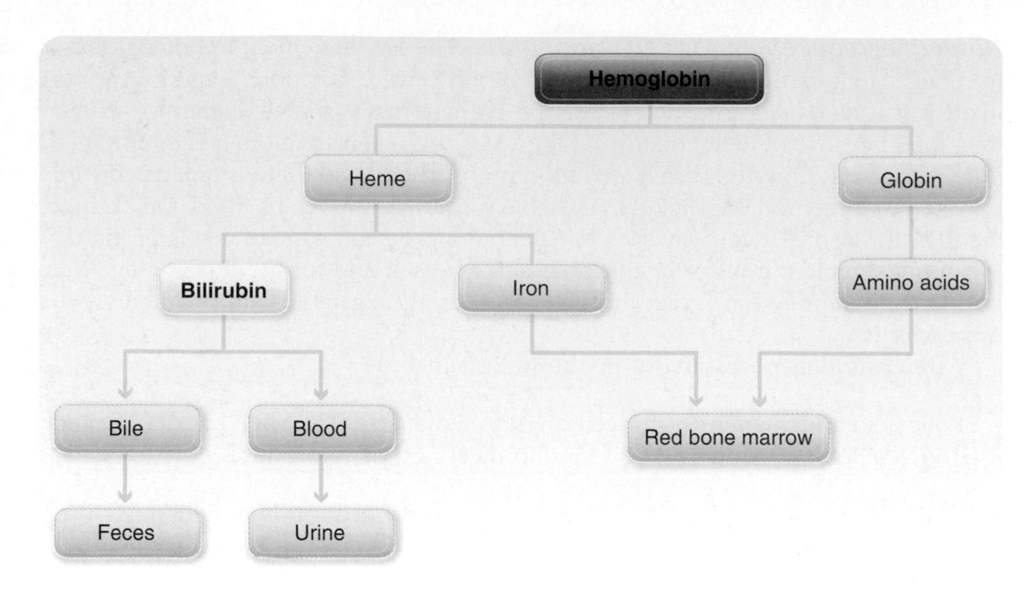

clinical point

If the liver is not functioning, the spleen alone must process all the hemoglobin from worn-out red blood cells. That means all of the bilirubin will be secreted into the blood instead of into bile. The added bilirubin in the blood gives a yellow appearance to the skin and eyes, a condition called **jaundice.** Jaundice is a good indicator of liver malfunction.

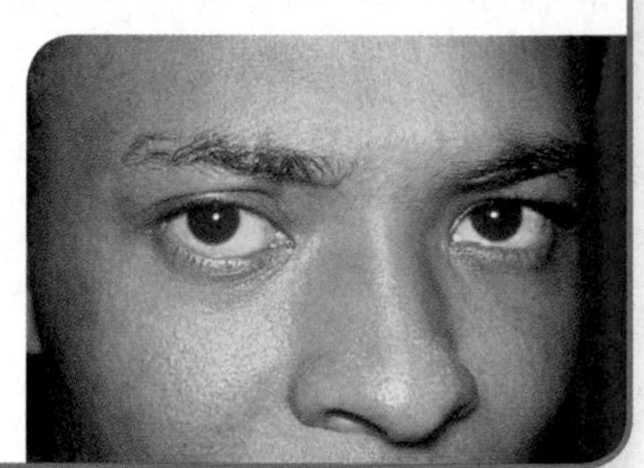

Bacteria, normally found in the intestines, change the color of bilirubin (secreted by the liver in bile) to brown. This is why feces are brown no matter what food was consumed. So light-clay-colored stools are also an indicator of liver trouble.

In the next section, you will see what happens if a blood vessel is broken.

9.9 learning **outcome**

Describe the body's mechanisms for controlling bleeding.

Hemostasis

Hemostasis means "the stopping of bleeding," and it is more than just blood clotting. Hemostasis is a three-step process that usually happens in order: first vascular spasm, then platelet plug formation, and finally blood clotting **(coagulation)**. Platelets are important in all three steps of hemostasis. See **Figure 9.8**.

hemostasis: he-moh-STAY-sis

coagulation: koh-ag-you-LAY-shun

Vascular Spasm This is the immediate constriction of a broken vessel that reduces blood flow. Vascular spasm can be started in several ways:

- Pain receptors in injured tissue can directly stimulate a vessel to constrict.
- Platelets in a broken vessel can release vasoconstrictors.
- Injury to the smooth muscle of a vessel wall can cause the vessel to constrict.

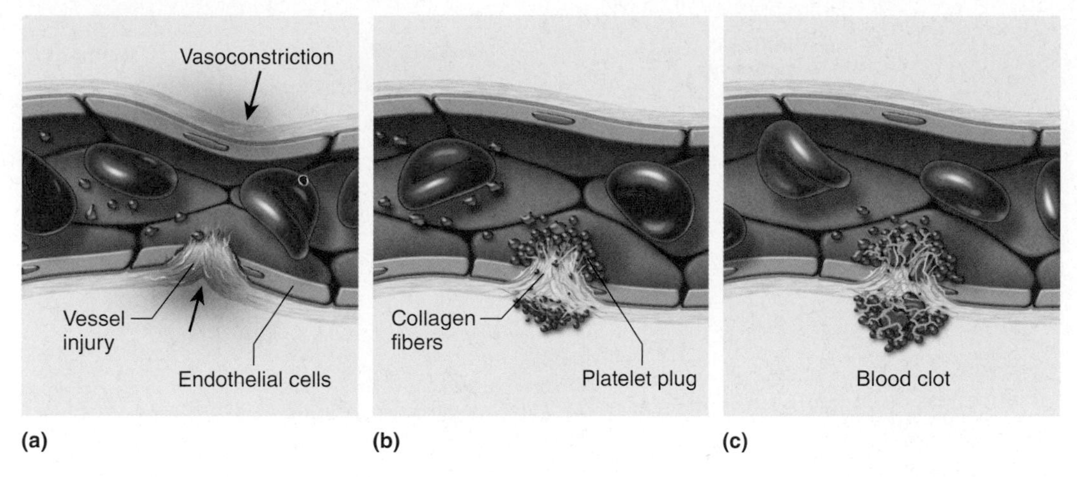

FIGURE 9.8 Hemostasis: (a) vascular spasm, (b) platelet plug formation, and (c) coagulation.

Vascular spasm lasts only a few minutes, but this allows time for the next two hemostatic mechanisms to take effect.

Platelet Plug Formation Normally, platelets are prevented from sticking to vessel walls because the walls are coated with a platelet repellent (like a nonstick, Teflon coating on a frying pan) to ensure that blood flows through the vessels easily. However, if a vessel is broken, collagen fibers of the vessel wall are exposed on the broken edges. Platelets stick to the exposed collagen fibers, forming a platelet plug—another mechanism of hemostasis. It is important to understand that this is different from a blood clot. Once in the plug, platelets begin to secrete factors that initiate the *last* mechanism of hemostasis—coagulation.

Blood Clotting (Coagulation) Blood clotting is the last mechanism of hemostasis to occur, but it has the most lasting effects. The process of coagulation involves a series of events that result in the dissolved protein **fibrinogen** coming out of solution to form a solid fiber called **fibrin.** Fibrin then acts as a net to trap blood cells and platelets to form a solid clot attached to the vessel walls. There are two different pathways to get the process started. See **Figure 9.9**.

9.10 learning **outcome**

Describe two pathways for blood clotting in terms of what starts each, their relative speed, and the clotting factors involved.

Pathways of blood clotting As shown in **Figure 9.9**, the two pathways at the top of the figure merge to form a common pathway at the bottom of the figure. As you study this figure, keep in mind the following important details:

1. The **extrinsic pathway,** shown in the orange box, is begun by *damaged tissues*.
2. The **intrinsic pathway,** shown in the yellow box, is started by *platelets*.
3. Both the extrinsic and intrinsic pathways require the addition of *calcium*.
4. Both the extrinsic and intrinsic pathways require **clotting factors,** but it is a different set of factors for each pathway.
5. There are fewer steps in the extrinsic pathway, so it is faster (15 seconds). The intrinsic pathway involves more steps, so it is slower (3 to 6 minutes).
6. Both the extrinsic and intrinsic pathways lead to a **common pathway,** shown in blue.

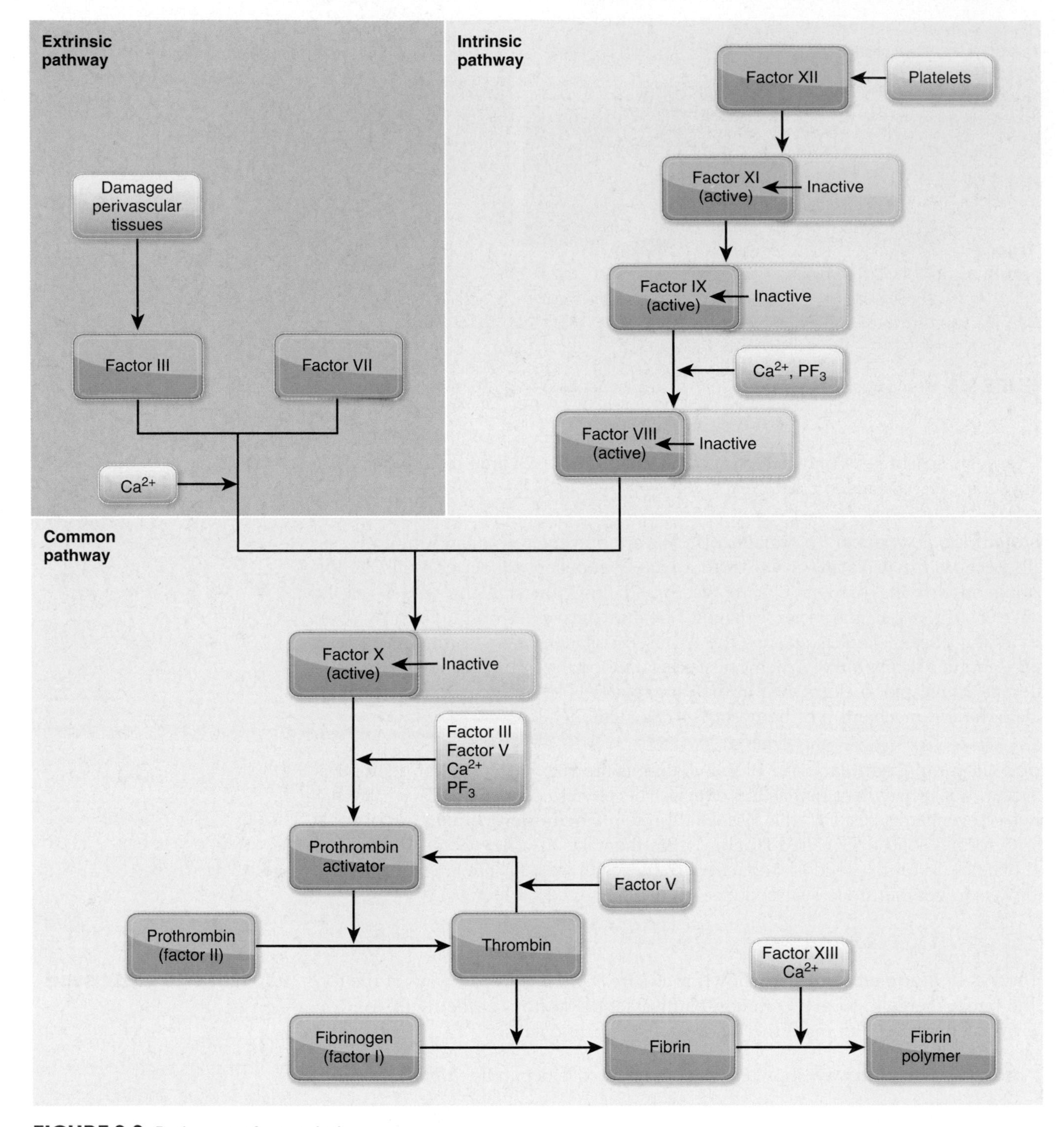

FIGURE 9.9 Pathways of coagulation.

7. Additional, different clotting factors are required in the common pathway.
8. All of the clotting factors are present in the blood at all times but in an inactive form. Each step activates the next step, and this results in what is called a **reaction cascade,** much like falling dominoes.
9. The end result is the formation of a fibrin polymer (solid fibrin strands). See **Figure 9.10**.

It is important to have two pathways to accomplish the same end. For example, if you were missing factor VIII, you would not be able to use the intrinsic pathway but could still form a clot using the extrinsic pathway because it does not require factor VIII. However, if you were missing clotting factors from both the intrinsic and extrinsic pathways, or from just the common pathway, you would not be able to form a clot.

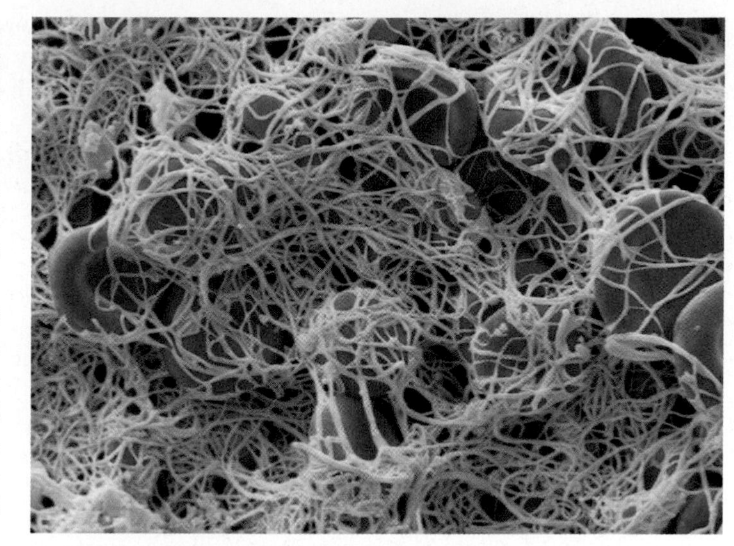

FIGURE 9.10 **Blood clot.** Red blood cells and platelets are caught up in a web of fibrin polymer strands forming a clot.

Most of the clotting factors are produced in the liver, which requires vitamin K for their production. Half of the required vitamin K comes from bacteria in the intestines, while the other half is supplied through diet. Vitamin K is found in liver, cabbage, spinach, and vegetable oils. The adult RDA for vitamin K is 90 micrograms (μg) for adult females and 120 μg for adult males. (See Appendix B.)

Once a clot is formed, platelets shrink to draw the edges of the broken vessel together. This is called **clot retraction.** It is similar to wound retraction, in which the drying scab brings the edges of a wound closer together (discussed in Chapter 3). Platelets then secrete growth factors to encourage cells in the vessel walls to undergo mitosis to repair the break.

spot check  How are platelets involved in each of the three mechanisms of hemostasis?

9.11 learning outcome

Describe what happens to blood clots when they are no longer needed.

Elimination of blood clots A clot is vital if a vessel is broken, but once the vessel is healed, the clot is no longer necessary. As soon as the vessel is repaired, another mini-reaction cascade of events happens to change an inactive enzyme called *plasminogen* to **plasmin.** Plasmin dissolves the now unnecessary blood clot in a process called **fibrinolysis.**

9.12 learning outcome

Explain what keeps blood from clotting in the absence of injury.

Preventing inappropriate clotting Inappropriate clotting—clotting that occurs when vessels are not broken—can have disastrous consequences. Such clots may block vessels and disrupt the flow of blood to the point of causing death. A stationary unwanted clot is called a **thrombus.** A moving unnecessary clot is called an **embolus.** There are several control mechanisms to prevent inappropriate clotting:

- **Platelet repulsion.** This has already been mentioned. The lining of a vessel is very smooth and coated with a platelet repellent to make sure platelets do not stick to vessel walls.
- **Dilution.** Some small amounts of thrombin may always be in circulation, but the circulation of the blood keeps the thrombin diluted and does not allow it to accumulate enough in any one place to change fibrinogen to fibrin. See **Figure 9.9**.
- **Anticoagulants.** These chemicals interfere with the pathways of clotting. For example, the liver produces **antithrombin,** which deactivates circulating thrombin; and basophils release heparin, which blocks **prothrombin activator,** a necessary step in the reaction cascade. See **Figure 9.9**.

You have explored the three mechanisms of hemostasis and the way beneficial clots are removed when they are no longer needed. You have also seen how inappropriate clots are prevented. But what if bleeding cannot be stopped? Then medical intervention and possibly a transfusion may be necessary. In the next section, you will learn about another important topic concerning blood—blood typing.

9.13 learning **outcome**

Explain what determines ABO and Rh blood types.

Blood Typing

A blood type might be A+ or AB−. But what do the letters and positive or negative sign mean? The surface of blood cells may or may not have molecules called **antigens.** These antigens have a unique shape, which marks cells as being *self* (part of an individual, not foreign). The presence or absence of specific antigens on the cells is genetically determined. It is important to remember that *blood type is determined by the antigens on the surface of the cells.*

There are two groups of antigens: ABO and Rh. In the ABO group, there are A antigens, B antigens, but no O antigens— O means *no* antigens of that group. So if cells have just A antigens on their surface, the blood is type A. If cells have just B antigens on their surface, the blood is type B. If cells have both A and B antigens, the blood is type AB. If cells have neither A nor B antigens, the blood is type O.

The Rh group contains several antigens named for the *rh*esus monkeys in which they were first found. If an individual has any antigen from this group, the individual is said to be Rh+ (positive). Most people are Rh+. If the Rh antigen is not present on cells, the blood is Rh− (negative). Thus, if a blood type is AB−, it has A antigens, B antigens, and no Rh antigens. If blood is type O+, it has no A or B antigens but does have Rh antigens on the surface of its cells.

Antigens, however, are only part of the story. **Antibodies** are dissolved proteins in plasma. Some of the globulins mentioned in the "Plasma" section at the beginning of the chapter are antibodies. The role of an antibody is to seek out foreign antigens and mark them for removal. Each antibody is specific to an antigen on the basis of its shape. Anti-A antibodies react to A antigens; anti-B antibodies react to B antigens; and anti-Rh antibodies react to Rh antigens. See **Figure 9.11**.

Dealing with ABO first, consider a young boy named Andre as an example. If Andre is type B, he has B antigens on the surface of his cells and has anti-A antibodies dissolved in his plasma. Anti-A antibodies travel through Andre's blood

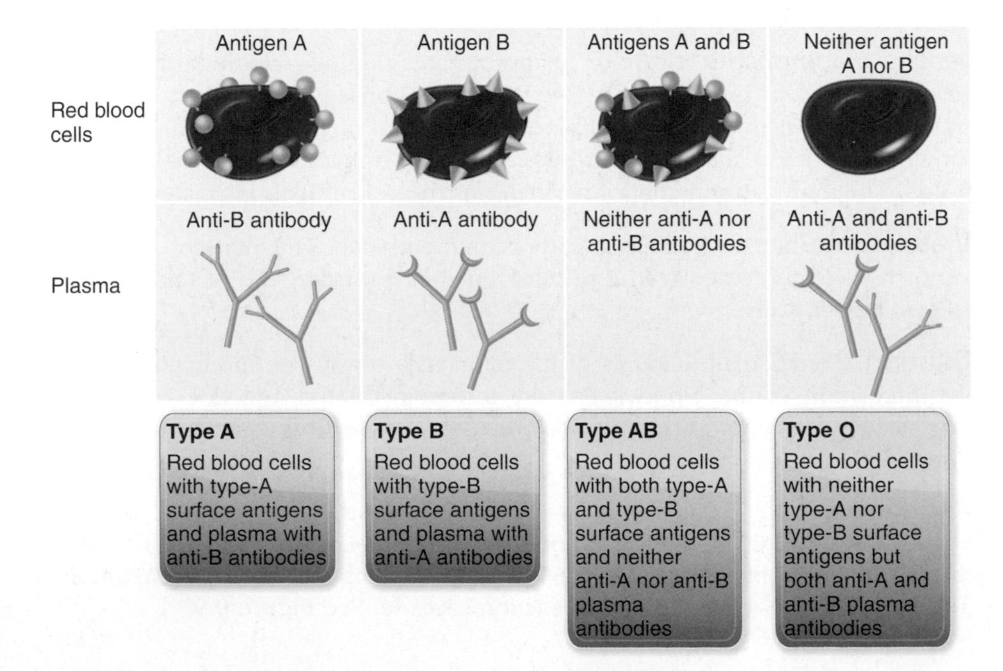

FIGURE 9.11 Antigens and antibodies for each ABO blood type.

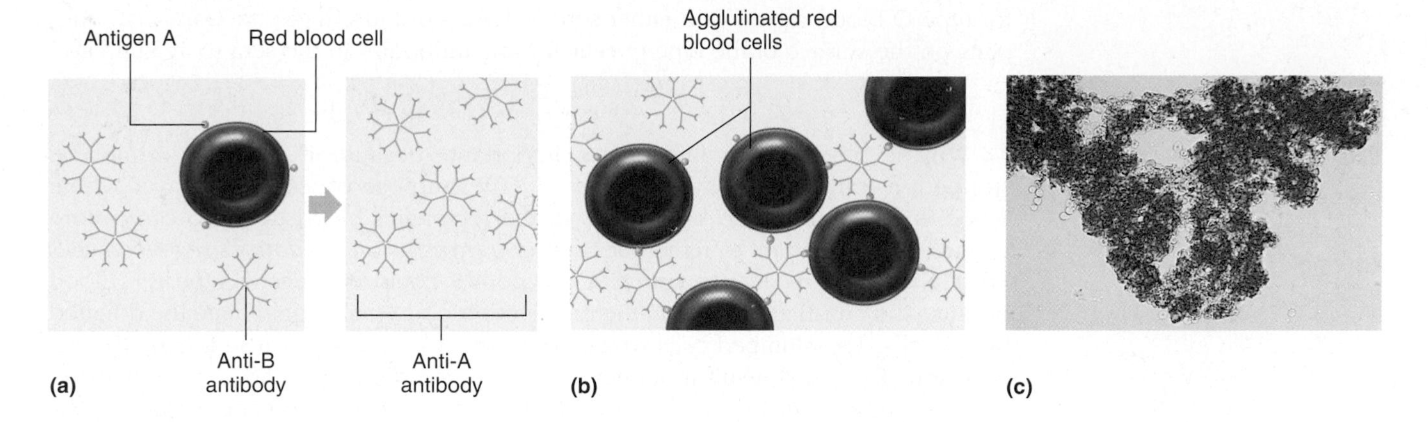

FIGURE 9.12 Agglutination. (a) Foreign type-A blood is introduced to Andre's anti-A antibodies; (b) Andre's anti-A antibodies bind to A antigens on the foreign cells to agglutinate them for removal; (c) micrograph of agglutinated blood.

seeking out any A antigens that would be foreign to him. If the anti-A antibodies find any cells with A antigens, they bind to the antigens and clump up several cells at a time. This clumping is called **agglutination.** See **Figure 9.12**. The agglutination of cells marks them for removal by macrophages. So, if foreign cells with A antigens enter Andre's bloodstream, his anti-A antibodies see them as foreign and mark them for removal through agglutination. Macrophages then come along and get rid of them. This is a safety system to protect the body from foreign pathogens.

Type A blood would never have anti-A antibodies because these antibodies would agglutinate the blood's cells and mark them for removal. Type AB blood would never have anti-A or anti-B antibodies because they would result in the destruction of blood cells. Type O blood would have both anti-A and anti-B antibodies because both A and B antigens would be foreign. Appropriate ABO antibodies for each type are believed to be acquired as a result of a childhood immune system response to bacteria normally present in the intestines or other antigenic stimuli.

spot check **10** What antigens and antibodies would be present in a person with type-A blood? Where would the antigens and antibodies be located?

Anti-Rh antibodies are a little different. An Rh− individual *only* acquires the anti-Rh antibodies if exposed to the Rh antigen. The immune system first recognizes the antigen as foreign and then starts to produce anti-Rh antibodies to fight it. Once started, anti-Rh antibody production will continue for life. This is very important for pregnant women who are Rh−, as you will see shortly.

spot check **11** Would an Rh+ person produce anti-Rh antibodies?

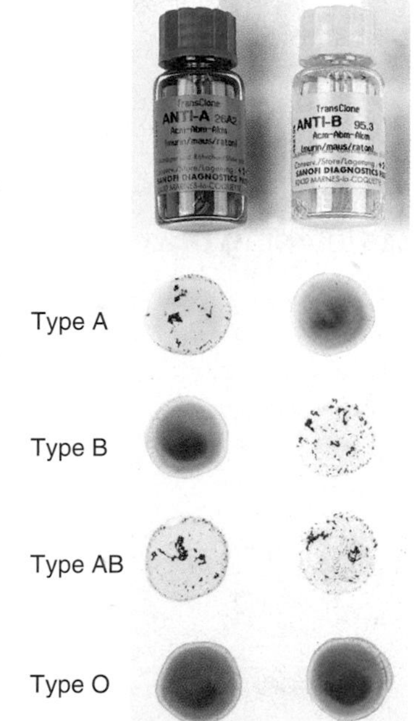

FIGURE 9.13 ABO blood typing. Each row shows the reaction of that type blood with each of the sera. Agglutination appears as clumping of the cells. Type-A blood shows agglutination when mixed with anti-A serum.

9.14 learning outcome

Explain how a blood type relates to transfusion compatibility.

Determining a Blood Type and Transfusion Compatibility Blood type can be determined by mixing a drop of blood with drops of different sera (plural of *serum*). As you may recall from the section on plasma earlier in the chapter, serum is plasma with the clotting proteins removed. Antibodies are present in serum. **Figure 9.13** shows two bottles of serum: The blue bottle contains serum with anti-A antibodies, and the yellow bottle contains serum with anti-B antibodies. In the rows of test slides below the bottles, agglutination appears as clumping of the cells. The clumping can be seen with the naked eye—no microscope is needed. Notice there is no clumping for type-O blood mixed with either serum. This is because there are no A or B antigens on the surface of the type-O cells for the antibodies in the sera to agglutinate.

9.15 learning **outcome**

Determine, from a blood type, the antigens and antibodies present and the transfusion compatibility.

Why is blood-typing information important? Because if bleeding cannot be stopped, a transfusion may be necessary. The purpose of a transfusion of whole blood is to get more red blood cells to the recipient's bloodstream to carry more oxygen. It is important to remember that *in a transfusion the donor's red blood cells must survive the recipient's antibodies.* Type A+ could not receive type B+ blood because the anti-B antibodies in the recipient's blood would agglutinate the donated blood cells. The clumped cells would block small vessels and cause a **transfusion reaction** that could result in kidney failure and death. A type-AB person, with no anti-A or anti-B antibodies, is considered to be a universal recipient. A type-O person, who has neither A nor B antigens, is considered to be a universal donor. The anti-A and anti-B antibodies present in type-O blood limit this type to being able to receive blood only from type-O donors.

spot check 12 To which blood types could a person with type B− donate blood? From which blood types could a person with type B− receive blood?

9.16 learning **outcome**

Predict the compatibility between mother and fetus given Rh blood types for both and describe the possible effects.

Mother-Fetus Blood-Type Compatibility As mentioned before, anti-Rh antibodies can have severe consequences for Rh-negative mothers. For example, if a woman is Rh−, she does not have Rh antigens on her cells and she does not have anti-Rh antibodies in her plasma unless she has been previously exposed to the Rh antigen. Suppose she has not been exposed. If all of her children are also Rh−, there will never be any incompatibility. But assume her first child is Rh+. While she is carrying the fetus, her placenta acts as a barrier to the mixing of blood. The fetal blood cells are kept from the mother's circulation and all is well. At delivery, the placenta tears away from the wall of the uterus so that it can be delivered after the baby. Mixing of some of the fetal blood cells with the mother's blood is likely during the process. As a result, the mother has now been exposed to the Rh antigen and begins to make anti-Rh antibodies. She will continue to produce the anti-Rh antibodies for the rest of her life. Anti-Rh antibodies can cross the placenta to fetal circulation in future pregnancies. If the next fetus is Rh−, there will be no problem because the fetus does not have any Rh antigens for the anti-Rh antibodies to attack. However, if the next fetus is Rh+, the now present anti-Rh antibodies in the mother's blood will cross the placenta and attack all of the fetal blood cells. The agglutinated cells are then cleared away. This condition—called **hemolytic disease of the newborn (HDN)** or **erythroblastosis fetalis**—may result in anemia (the condition of too few red blood cells) so severe as to cause the death of the fetus. See **Figure 9.14**.

clinical **point**

An Rh− mother can be given anti-Rh antibodies by injection during her pregnancy and immediately after delivery. The anti-Rh antibodies quickly target and bind to any Rh antigen that may have entered her system from the fetus. If the Rh antigens are removed before her immune system becomes aware, her production of anti-Rh antibodies is prevented.

This is similar to a child cleaning up milk spilled on the carpet while his mother was at work. If all evidence of the spill is removed before his mother returns, she may never know the spill ever happened.

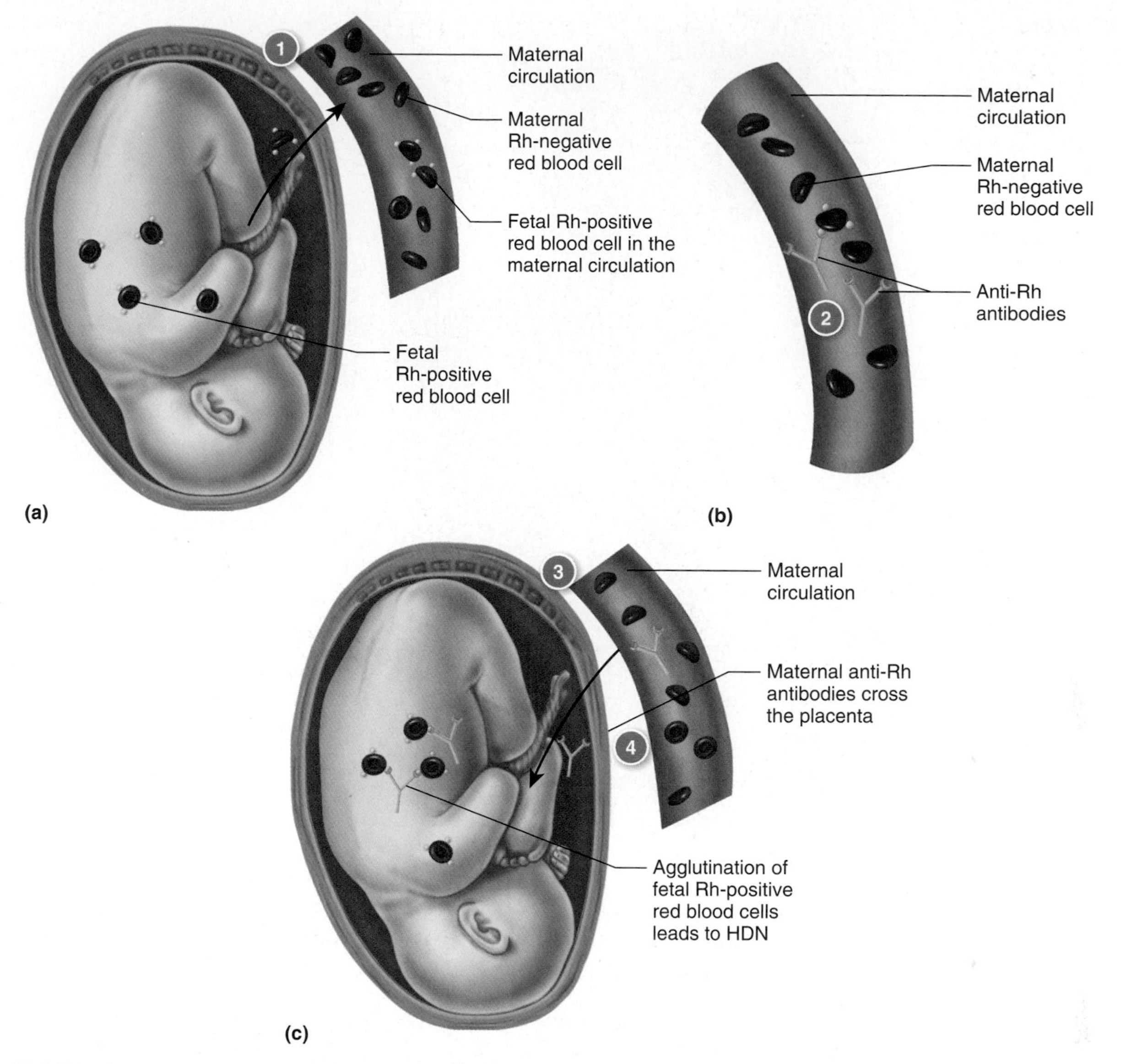

FIGURE 9.14 **Hemolytic disease of the newborn.** (a) First pregnancy: Fetal Rh+ red blood cells enter mother's circulation through a tear in the placenta, usually when the placenta is delivered. (b) Between pregnancies: Mother produces anti-Rh antibodies. (c) Second pregnancy: Mother's anti-Rh antibodies agglutinate fetal Rh+ blood.

Functions of Blood

9.17 learning outcome

Summarize the functions of blood by giving an example or explanation of each.

As you can see in **Figure 9.15**, Andre has severely scraped his knee. His wound is continuing to bleed through its dressing. In this section, you will explore what the blood in his body is doing for Andre under normal circumstances and specifically with his injury.

The functions of blood include:

1. Transportation:
   - Andre's blood is transporting important nutrients throughout his body while carrying away wastes and debris from the wound.
   - Andre's blood is carrying oxygen to his tissues and carbon dioxide away.

FIGURE 9.15 Andre.

- Notice that Andre is wearing shorts. Andre's blood is transporting heat from his core to his skin, where it can radiate off on a warm, sunny day, thereby regulating his temperature.
- Andre's blood is transporting hormones to regulate his body and maintain homeostasis.

2. Protection:
    - Andre's white blood cells are fighting off any foreign pathogens that may have entered his wound. Neutrophils are moving in large numbers to the site to fight off any bacteria, and his macrophages are getting rid of the damaged tissue and debris at the wound site.
    - Andre's blood is actively involved in hemostasis to prevent its own loss. By the time the dressing was applied, Andre had moved past vascular spasm and platelet plug formation to coagulation. By the dressing, you can see coagulation has not been completed yet.
3. Regulation:

- Blood must be maintained in a very narrow range of pH: 7.35 to 7.45. Free hydrogen ions lower the blood pH. The hemoglobin in Andre's blood binds to hydrogen ions at the tissues, thereby acting as a buffer resisting a change in pH. You will learn more about this in the respiratory system chapter.

- A fluid balance is maintained in the tissues by adjusting the volume of plasma. More water can be taken into Andre's blood from the tissues or can be given out from his blood to the tissues, depending upon need.

You have examined the anatomy and physiology of the blood. Next, you will learn about the various common blood tests used in medicine to assess the blood.

Blood Tests

9.18 learning **outcome**

Explain what can be learned from common blood tests.

There are many tests that can be done to assess the health and composition of blood. Table 9.3 shows normal values for some of the more common tests. Instrumentation and laboratory methods may vary from one laboratory to the next, so it is always best to rely on the normal (reference) values from the laboratory performing the test. Notice that there is a difference in values for men and women in the tests pertaining to red blood cells. Men tend to have more red blood cells because testosterone encourages erythropoiesis and because women have more fat deposition. There is an inverse proportion between the amount of fat and erythropoiesis.

Hematocrit The **hematocrit** test measures the percentage of erythrocytes to whole blood. The amount of red blood cells accounts for 38.8 to 46.6 percent of the total blood volume in males and 35.4 to 44.4 percent in females. See **Figure 9.16**.

Hemoglobin A hemoglobin measurement determines the amount of hemoglobin in a given amount of blood. The normal hemoglobin measurement is 13.3 to 16.2 grams per deciliter (g/dL) of blood for males and 12.0 to 15.8 g/dL of blood for females.

TABLE 9.3 Blood test values.

Test	Normal values
Hematocrit	38.8%–46.4% of total blood volume in males
	35.4%–44.4% in females
Hemoglobin	13.3–16.2 g/dL of blood for males
	12.0–15.8 g/dL of blood for females
Red blood cell count	4.30–5.60 million/mm^3 of blood for males
	4.00–5.20 million/mm^3 of blood for females
White blood cell count	3,540–9,060/mm^3 of blood
White blood cell differential:	
Neutrophils	40%–70%
Basophils	0%–2%
Eosinophils	0%–6%
Monocytes	4%–8%
Lymphocytes	20%–50%
Platelet count	165,000–415,000/mm^3 of blood

Source: A. Kratz, M. A. Pesce, D. J. Fink, "Appendix: Laboratory Values of Clinical Importance," in A. S. Fauci, E. Braunwald, D. L. Kaspar, S. L. Hauser, D. L. Longo, J. L. Jameson, et al. (eds.), *Harrison's Principles of Internal Medicine,* (17th ed./New York: McGraw-Hill, 2008).

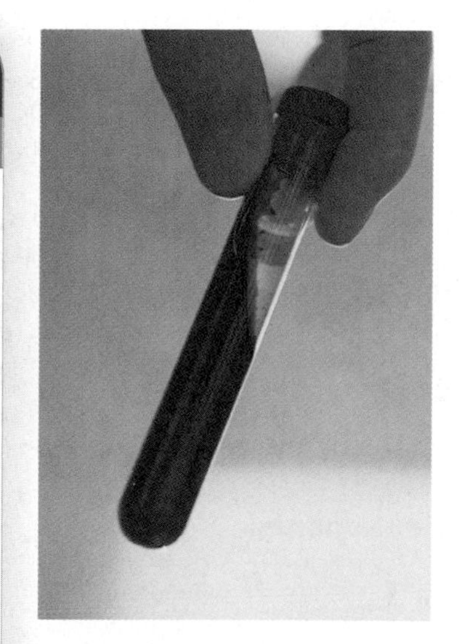

FIGURE 9.16 Hematocrit
Blood is drawn and then placed in a centrifuge to separate the formed elements from the rest of the blood.

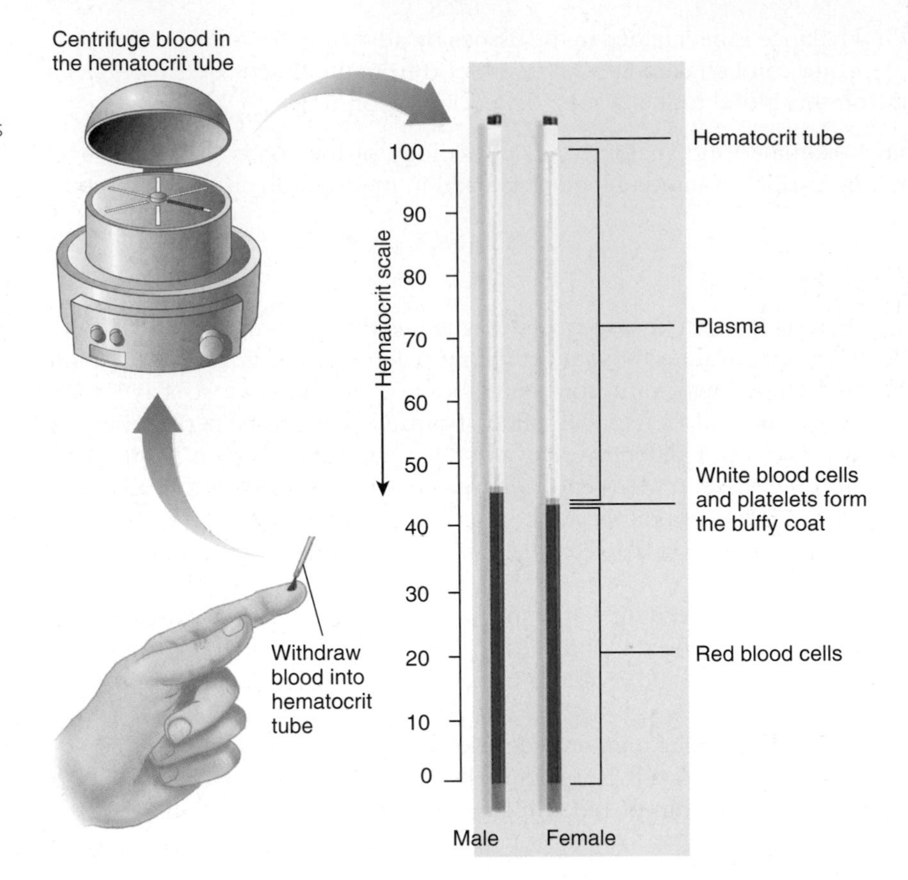

Blood Counts The amount of each formed element in the blood can also be measured.

RBC count The normal number of red blood cells is 4.30 to 5.60 million per cubic millimeter (mm^3) of blood for males and 4.00 to 5.20 million/mm^3 of blood for females.

WBC count The normal number of all the leukocytes is 3,540 to 9,060/mm^3 of blood. **Leukocytosis** is a high white blood cell count. This is a normal response to a challenge to the immune system. Here, more leukocytes are made to fight a particular pathogen. **Leukemia** also involves a high white blood cell count, but the white blood cells in leukemia are immature and incapable of fighting off pathogens. **Leukopenia** is a low white blood cell count.

leukopenia:
loo-koh-PEE-nee-ah

connect.mcgraw-hill.com audio

WBC differential A white blood cell differential gives the percentage of each type of leukocyte in the total number of leukocytes. Normal ranges are:

- Neutrophils: 40 to 70 percent
- Basophils: 0 to 2 percent
- Eosinophils: 0 to 6 percent
- Lymphocytes: 20 to 50 percent
- Monocytes: 4 to 8 percent

Platelet count A normal platelet count is 165,000 to 415,000/mm^3 of blood.

spot check 13 What should the total of all the values of a white blood cell differential equal?

Blood Disorders

9.19 learning outcome

Describe disorders of the cardiovascular system concerning blood.

Each of the above blood tests can be important in assessing the health of an individual and in helping to determine a diagnosis. Below, we focus on some disorders involving the blood.

Polycythemia

Polycythemia is a condition of too many cells in the blood. Since most of the formed elements in blood are erythrocytes, polycythemia is often caused by overproduction of red blood cells. There are two forms of this disorder. **Primary polycythemia** is cancer of the blood. Too many immature cells are produced that do not function. Red blood cell counts can go as high as 11 million RBCs/mm^3 in polycythemia and hematocrits may get as high as 80 percent. **Secondary polycythemia** is far more common. With this form of the disorder, the RBC count typically ranges from 6 to 8 million RBCs/mm^3 of blood. Secondary polycythemia can result from dehydration, lowering the ratio of plasma to formed elements, or accelerated erythropoiesis due to hypoxemia caused by elevation, smoking, or increased oxygen demand due to exercise, pollution, or disease. An elevated hematocrit means there are more formed elements than normal to the amount of plasma. This makes the blood thicker. It is harder for the heart to pump thicker blood in the vessels, so this increases the heart's workload and elevates blood pressure.

Anemias

Anemias are disorders that result from insufficient red blood cells or hemoglobin to carry enough oxygen to maintain homeostasis. There are three categories of anemias:

1. Inadequate erythropoiesis or hemoglobin production. This may be caused by inadequate iron in the diet **(iron deficiency anemia)** or lack of intrinsic factor from the stomach that allows vitamin B_{12} to be absorbed **(pernicious anemia).** Other causes of this type of anemia include kidney failure, resulting in reduced EPO, and red bone marrow destruction by some poisons, drugs, viruses, and radiation **(hypoplastic** or **aplastic anemia).**
2. Excessive bleeding (hemorrhagic anemia). This type of anemia may be caused by trauma, failure to clot, or ulcers.
3. Red blood cell destruction (hemolytic anemia). This type of anemia may result from a drug reaction (penicillin allergy) or a blood incompatibility (hemolytic disease of the newborn). Other causes include inherited factors (sickle cell disease) and parasitic infections (malaria).

hemolytic: he-moh-LIT-ik

Clotting Disorders

Hemophilia is a group of inherited disorders caused by the inability to make one or more clotting factors. This prevents the reaction cascade necessary for clot formation. Hemophilia is a sex-linked disorder, meaning it occurs mostly in males. Purified clotting factors from blood donations are given to patients to treat this disease.

warning

Hemophilia has nothing to do with the number of platelets.

applied genetics

Human DNA consists of 23 pairs of chromosomes. Twenty-two pairs are called **autosomes.** The last pair are the **sex chromosomes.** They are designated as XX for females and XY for males. The mother can contribute only one of her X chromosomes in her egg. The father can contribute an X or a Y chromosome in his sperm. If the resulting child is a girl, she will have inherited an X from her mother and an X from her father. If the resulting child is a boy, he will have inherited an X chromosome from his mother and a Y chromosome from his father. The genes for sex-linked disorders are located on the X chromosome. They are passed from mother to son. See Figure 9.17.

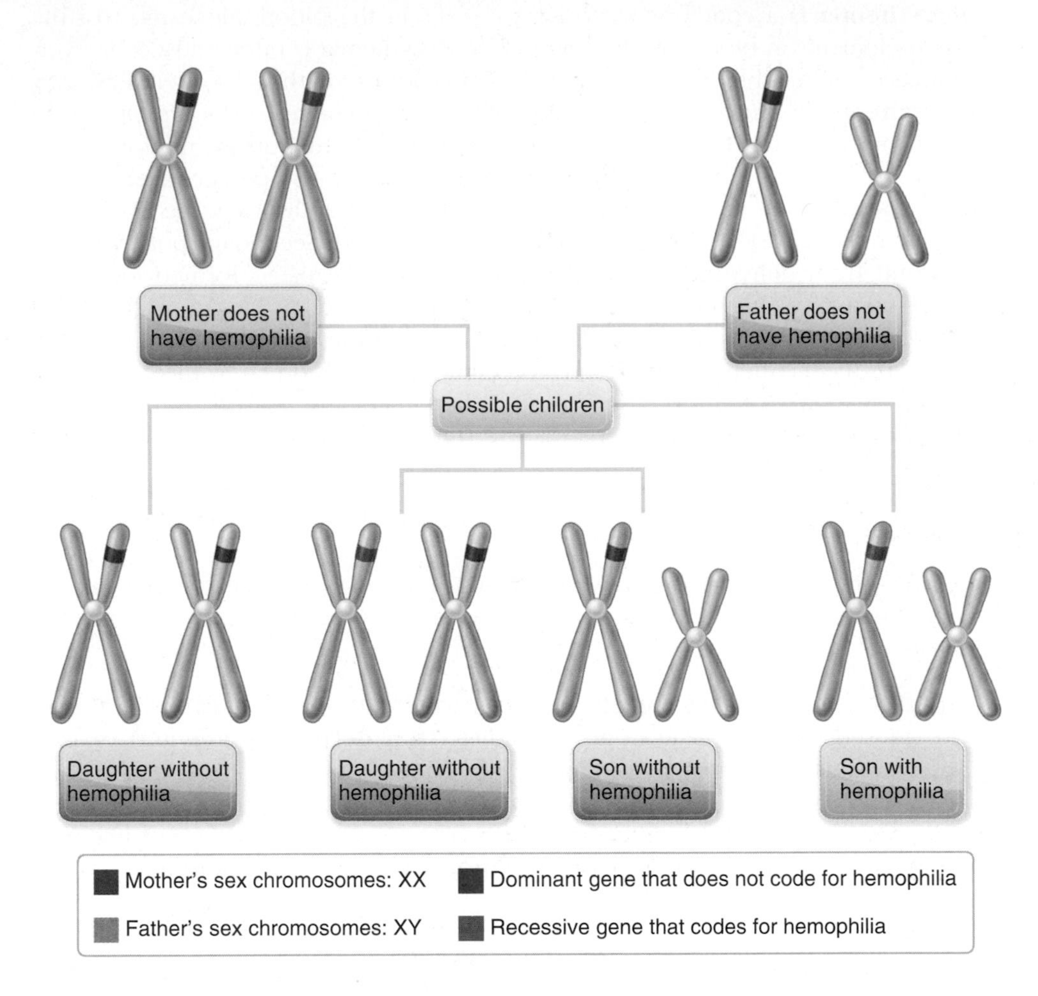

FIGURE 9.17 Inheritance of a sex-linked disease: the sex chromosomes of the possible children when the parents do not have hemophilia but the mother has one X chromosome with the recessive gene for hemophilia.

Thrombocytopenia is a low platelet count (less than 150,000/mm^3 of blood) that may be caused by bone marrow destruction. One of the signs is excessive bruising (hematomas) after minor trauma.

Disseminated intravascular coagulation (DIC) is the widespread coagulation of blood in unbroken vessels. This can be caused by a systemic infection of the blood called **septicemia.** It can also occur if blood flow slows significantly as in a cardiac arrest. Circulating thrombin may then accumulate enough to form clots. In this case, the clots may be more localized to an organ.

summary

Overview

- The cardiovascular system is composed of the heart, blood vessels, and blood.

Anatomy of Blood

- Blood is a connective tissue of formed elements in a matrix of plasma.

Plasma

- Plasma is 91 percent water, 7 percent protein, and 2 percent ions, nutrients, waste products, gases, and regulatory substances.
- Plasma is a solution, and its concentration is important for homeostasis.

Formed Elements

- The formed elements of the blood are erythrocytes, leukocytes, and thrombocytes.
- Erythrocytes are biconcave disks with no nuclei. They contain hemoglobin that has iron to carry oxygen. In addition, red blood cells carry carbon dioxide and hydrogen ions.
- There are five types of leukocytes that must be stained in order to be seen. Each type has its own function. These leukocytes are neutrophils, basophils, eosinophils, monocytes, and lymphocytes.
- Thrombocytes, also called *platelets,* are fragments of cells that have many functions.

Physiology of Blood

Hemopoiesis

- Hemopoiesis is blood production. There are three forms: erythropoiesis, leukopoiesis, and thrombopoiesis.
- Thrombopoiesis starts from a hemocytoblast in the red bone marrow. The liver and kidneys start the process by producing thrombopoietin when there is a need for more platelets.
- Leukopoiesis starts from a hemocytoblast in the red bone marrow. Lymphocytes and macrophages produce CSFs when there is a challenge to the immune system. There is a different CSF for each type of leukocyte production.
- Erythropoiesis starts from a hemocytoblast in the red bone marrow. The kidneys start the process by releasing EPO in situations of hypoxemia. Hypoxemia can result from disease, high elevation, increased exercise, blood loss, and carbon monoxide.
- Iron, folic acid, vitamin B_{12}, copper, and vitamin C are needed for erythropoiesis.

Life Cycle of a Red Blood Cell

- Red blood cells are produced in the red bone marrow.
- Red blood cells carry oxygen and carbon dioxide through the bloodstream for 110 to 120 days before wearing out.
- The liver and spleen remove old, worn-out blood cells.
- Hemoglobin is broken down to heme and globin.
- Heme is further broken down to iron, which is recycled, and bilirubin, a waste product. The liver puts bilirubin in bile, which eventually leaves the body in feces. The spleen secretes bilirubin into the blood, where it is removed by the kidneys and excreted with urine.
- Globin is broken down by the liver and spleen to free amino acids, which are recycled.

Hemostasis

- Hemostasis is the stopping of bleeding.
- Hemostasis is a process with three stages—vascular spasm, platelet plug formation, and coagulation.
- Vascular spasm constricts the broken vessel to slow blood flow.
- Platelet plug formation occurs when platelets stick to exposed collagen fibers of broken vessel walls.
- Coagulation is the last stage to occur, but it is the most effective. It involves two pathways that result in a reaction cascade of one clotting factor activating the next until a clot is formed.
- When blood clots are no longer needed, they are dissolved by a process called *fibrinolysis*.
- Inappropriate clotting is prevented by platelet repulsion, dilution of thrombin, and anticoagulants.

Blood Typing

- Blood typing is based on the presence of ABO and Rh antigens on the surface of cells.
- Antibodies are dissolved proteins in plasma that react to foreign antigens in a process called *agglutination*.
- Antibodies for the ABO group are acquired as a child.
- Antibodies for Rh antigens are acquired only through an exposure to the antigen.

Determining a blood type and transfusion compatibility:

- Blood types can be determined by mixing a drop of blood with sera containing known antibodies.
- In a transfusion, the donor's cells must survive the recipient's antibodies.

Mother-fetus blood-type compatibility:

- Rh− mothers need to be concerned about blood incompatibility with Rh+ babies.

Functions of Blood

- The functions of blood fall into three categories: transportation, protection, and regulation.
- Blood transports nutrients, waste products, gases, regulatory chemicals, and heat.
- Blood protects the body from its own loss through hemostasis. Leukocytes in the blood protect the body from foreign pathogens.
- Blood regulates the fluid and electrolyte balance as well as the body's acid-base balance.

Blood Tests

- A hematocrit test measures the percentage of erythrocytes to whole blood.
- A hemoglobin measurement determines the amount of hemoglobin in a given amount of blood.
- Blood counts can be measured for RBCs, WBCs, and platelets.
- A WBC differential measures the percentage of each type of leukocyte in the total WBC count.

Blood Disorders

Polycythemia

- Polycythemia is too many red blood cells in the blood. Primary polycythemia is cancer of the blood. Secondary polcythemia is due to dehydration or hypoxemia.

Anemias

- Anemias are a group of disorders that provide insufficient red blood cells or hemoglobin to carry enough oxygen to maintain homeostasis. There are three categories of anemia: inadequate erythropoiesis, hemorrhagic anemia, and hemolytic anemia.

Clotting Disorders

- Clotting disorders include hemophilia, thrombocytopenia, and disseminated intravascular coagulation.

key words for review

The following terms are defined in the glossary.

agglutination
clotting factor
coagulation
erythrocyte
erythropoietin
fibrin
fibrinogen
formed elements
granulocyte
hematocrit
hemocytoblast
hemoglobin
hemopoiesis
hemostasis
leukocyte
lymphoid
myeloid
pluripotent
serum
thrombocyte

chapter review questions

Word Building: *Use the word roots listed in the beginning of the chapter and the prefixes and suffixes inside the back cover to build words with the following meanings:*

1. Condition of too many blood cells: ______________________________
2. Higher-than-normal number of white blood cells: ______________________________
3. Lower-than-normal number of white blood cells: ______________________________
4. Production of platelets: ______________________________
5. Blood stem cell: ______________________________

Multiple Select: *Select the correct choices for each statement. The choices may be all correct, all incorrect, or a combination of correct and incorrect.*

1. Which of the following are solutes found in plasma?
   a. Glucose
   b. Platelets
   c. Nitrogen wastes
   d. Fibrinogen
   e. Proteins
2. What does the liver do for the cardiovascular system?
   a. The liver makes antithrombin to inactivate thrombin.
   b. The liver works with the kidneys to remove old, worn-out RBCs.
   c. The liver contributes to blood clotting by making clotting factors.
   d. The liver breaks down heme to bilirubin and iron.
   e. The liver makes clotting factors for fibrinolysis.
3. What is blood clotting?
   a. Blood clotting is the formation of a platelet plug.
   b. Blood clotting is agglutination.
   c. Blood clotting is coagulation.
   d. Blood clotting is fibrinolysis.
   e. Blood clotting requires a reaction cascade.
4. Which of the following statements is (are) true concerning the pathways for blood clotting?
   a. The intrinsic pathway takes less time than the extrinsic pathway.
   b. The intrinsic and extrinsic pathways require sodium.

c. The intrinsic pathway is started by factors from damaged tissues.
d. The intrinsic pathway requires the same clotting factors as the extrinsic pathway.
e. The common pathway results in fibrin changing to fibrinogen.

5. Which mothers need to be concerned about erythroblastosis fetalis?
   a. Rh+ mothers having Rh− babies.
   b. All mothers regardless of blood type.
   c. Rh− mothers having Rh+ babies.
   d. Rh− mothers having Rh− babies.
   e. Rh+ mothers having Rh+ babies.
6. Which of the following statements is (are) true about polycythemia?
   a. Polycythemia increases blood pressure.
   b. Polycythemia can result from cancer, hypoxemia, or dehydration.
   c. Polycythemia is diagnosed from a white blood cell differential test.
   d. Polycythemia is often a problem for smokers.
   e. Polycythemia can occur when the body has chronic exposure to carbon dioxide.
7. What is involved in hemopoiesis?
   a. Hemocytoblasts are the starting cells for the formation of all the formed elements except platelets.
   b. Colony-stimulating factors are made by macrophages.
   c. Red blood cells are the only formed elements produced in the red bone marrow.
   d. Leukopoietin from the kidneys stimulates white blood cell production.
   e. Lymphocytes are produced from megakaryocytes.
8. Which of the following functions pertain(s) to blood?
   a. Transport oxygen.
   b. Transport amino acids.
   c. Form clots to prevent loss.
   d. Transport heat.
   e. Deliver water to and take in water from the tissues.
9. Which of the following describe(s) a characteristic or function of hemoglobin?
   a. Hemoglobin consists of four chains of amino acids.
   b. A hemoglobin molecule contains four iron ions to carry oxygen.
   c. In addition to transporting oxygen, hemoglobin molecules carry carbon dioxide and hydrogen ions.
   d. Hemoglobin is a protein found in all formed elements.
   e. There are four hemoglobin molecules in each red blood cell.
10. How does a person's blood type relate to transfusion compatibility?
    a. A person's blood type is determined by the antibodies on his or her cells.
    b. The antigens in the plasma for a specific blood type may fight the antibodies on the cells of donor blood.
    c. The donor's cells must survive the recipient's antibodies.
    d. The recipient's antibodies must survive the donor's antigens.
    e. A type AB+ individual can receive blood transfusions from several different types.

Matching: *Match the formed element to its function. Some answers may be used more than once. Some functions may have more than one answer.*

_____ 1. Carries oxygen and carbon monoxide
_____ 2. Secretes vasoconstrictors
_____ 3. Stimulates mitosis in vessel walls
_____ 4. Fights bacteria
_____ 5. Becomes a macrophage in the tissues

a. Erythrocyte
b. Leukocyte
c. Platelet

Matching: *Match the formed element to its function. Some answers may be used more than once. Some functions may have more than one answer.*

_____ 6. Becomes a macrophage in the tissues
_____ 7. Fights bacteria
_____ 8. Secretes chemicals to fight worms
_____ 9. Secretes heparin
_____ 10. Secretes histamine

a. Basophil
b. Lymphocyte
c. Monocyte
d. Eosinophil
e. Neutrophil

Critical Thinking

1. Long-term use of antibiotics may kill off the pathogen for which they were prescribed and the normal bacteria of the intestines that produce vitamin K. What might be the visible consequences of a lack of vitamin K? Explain.
2. Answer the following questions concerning a 24-year-old female with type AB− blood: What antigens and antibodies are present in her blood? To which types can she donate blood? From which types can she receive blood?
3. There is an old adage that says: "Good food makes for good blood." Plan a menu for a day that would supply all the necessary nutrients to carry out blood formation and blood physiology. Explain the need in the blood for the nutrients you chose to include in your menu.

9 chapter mapping

This section of the chapter is designed to help you find where each outcome is covered in this text.

	Outcomes	Readings, figures, and tables	Assessments
9.1	Use medical terminology related to the cardiovascular system.	Word roots: p. 331	Word Building: 1–5
9.2	Identify the components of blood.	Anatomy of blood: pp. 332–333 Figure 9.2	Spot Check: 1, 3
9.3	List the constituents of plasma and their functions.	Plasma: pp. 333–334 Figure 9.2	Spot Check: 2 Multiple Select: 1
9.4	Identify the formed elements and list their functions.	Formed elements: pp. 334–338 Table 9.1	Spot Check: 4, 5 Matching: 1–10
9.5	Compare the various forms of hemopoiesis in terms of starting cell, factors influencing production, location, and final product.	Hemopoiesis: pp. 338–341 Figures 9.3, 9.4 Table 9.2	Spot Check: 6, 7 Multiple Select: 7

	Outcomes	Readings, figures, and tables	Assessments
9.6	Describe the structure and function of hemoglobin.	Hemoglobin: pp. 341–342 Figures 9.5, 9.6	Multiple Select: 9
9.7	Summarize the nutritional requirements of red blood cell production.	Nutritional requirements for erythropoiesis: p. 343	Critical Thinking: 3
9.8	Describe the life cycle of a red blood cell from its formation to removal.	Life cycle of a red blood cell: pp. 343–344 Figure 9.7	Spot Check: 8 Multiple Select: 2
9.9	Describe the body's mechanisms for controlling bleeding.	Hemostasis: pp. 344–345 Figure 9.8	Spot Check: 9 Multiple Select: 3 Critical Thinking: 1
9.10	Describe two pathways for blood clotting in terms of what starts each, their relative speed, and the clotting factors involved.	Pathways of coagulation: pp. 345–347 Figures 9.9, 9.10	Multiple Select: 2, 4
9.11	Describe what happens to blood clots when they are no longer needed.	Elimination of blood clots: p. 347	Multiple Select: 2
9.12	Explain what keeps blood from clotting in the absence of injury.	Preventing inappropriate clotting: pp. 347–348 Figure 9.9	Multiple Select: 2
9.13	Explain what determines an ABO and Rh blood types.	Blood typing: pp. 348–349 Figures 9.11, 9.12	Spot Check: 10, 11
9.14	Explain how a blood type relates to transfusion compatibility.	Determining a blood type and transfusion compatibility: pp. 349–350 Figure 9.13	Multiple Select: 10 Critical Thinking: 2
9.15	Determine, from a blood type, the antigens and antibodies present and the transfusion compatibility.	Determining a blood type and transfusion compatibility: p. 350	Spot Check: 12 Critical Thinking: 2
9.16	Predict the compatibility between mother and fetus given Rh blood types for both and describe the possible effects.	Mother-fetus blood type compatibility: pp. 350–351 Figure 9.14	Multiple Select: 5
9.17	Summarize the functions of blood by giving an example or explanation of each.	Functions of blood: pp. 351–353 Figure 9.15	Multiple Select: 8
9.18	Explain what can be learned from common blood tests.	Blood tests: pp. 353–355 Figure 9.16 Table 9.3	Spot Check: 13
9.19	Describe disorders of the cardiovascular system concerning blood.	Blood disorders: pp. 355–356 Figure 9.17	Multiple Select: 6

references and resources

Beers, M. H., Porter, R. S., Jones, T. V., Kaplan, J. L., & Berkwits, M. (Eds.). (2006). *Merck manual of diagnosis and therapy* (18th ed.). Whitehouse Station, NJ: Merck Research Laboratories.

Reprinted with permission from Kratz, A., Pesce, M. A., Fink, D. J. (2008). Appendix: Laboratory values of clinical importance. *Harrison's principles of internal medicine* (17th ed.), editors, Fauci, A. S., Braunwald, E., Kaspar, D. L., Hauser, S. L., Longo, D. L., Jameson, J. L., et. al. Copyright (c) 2008 The McGraw-Hill Companies, Inc.

National Institutes of Health Clinical Center. (2008, November). *Understanding your complete blood count.* Retrieved March 19, 2011, from http://www.cc.nih.gov/ccc/patient_education/pepubs/cbc97.pdf.

Saladin, Kenneth S. (2010). *Anatomy & physiology: The unity of form and function* (5th ed.). New York: McGraw-Hill.

Seeley, R. R., Stephens, T. D., & Tate, P. (2006). *Anatomy & physiology* (7th ed.). New York: McGraw-Hill.

Shier, D., Butler, J., & Lewis, R. (2010). *Hole's human anatomy & physiology* (12th ed.). New York: McGraw-Hill.

10 The Cardiovascular System—Heart and Vessels

You often encounter mechanical fluid pumps in everyday life. For example, your car has an oil pump, a fuel pump, and a water pump to circulate their respective fluids through the appropriate hoses. Your body has a similar pump—the heart—that circulates blood through your body's vessels to carry out the functions of the circulatory system. See Figure 10.1. Unlike the mechanical pumps that run only when occasionally needed, the heart pumps constantly and is far more durable and long-lasting. Your heart began beating around day 22 while you were in your mother's womb, and it will likely continue to beat and pump blood every minute of every day until the day you die. How efficiently your heart works may determine the quality and the length of your life.

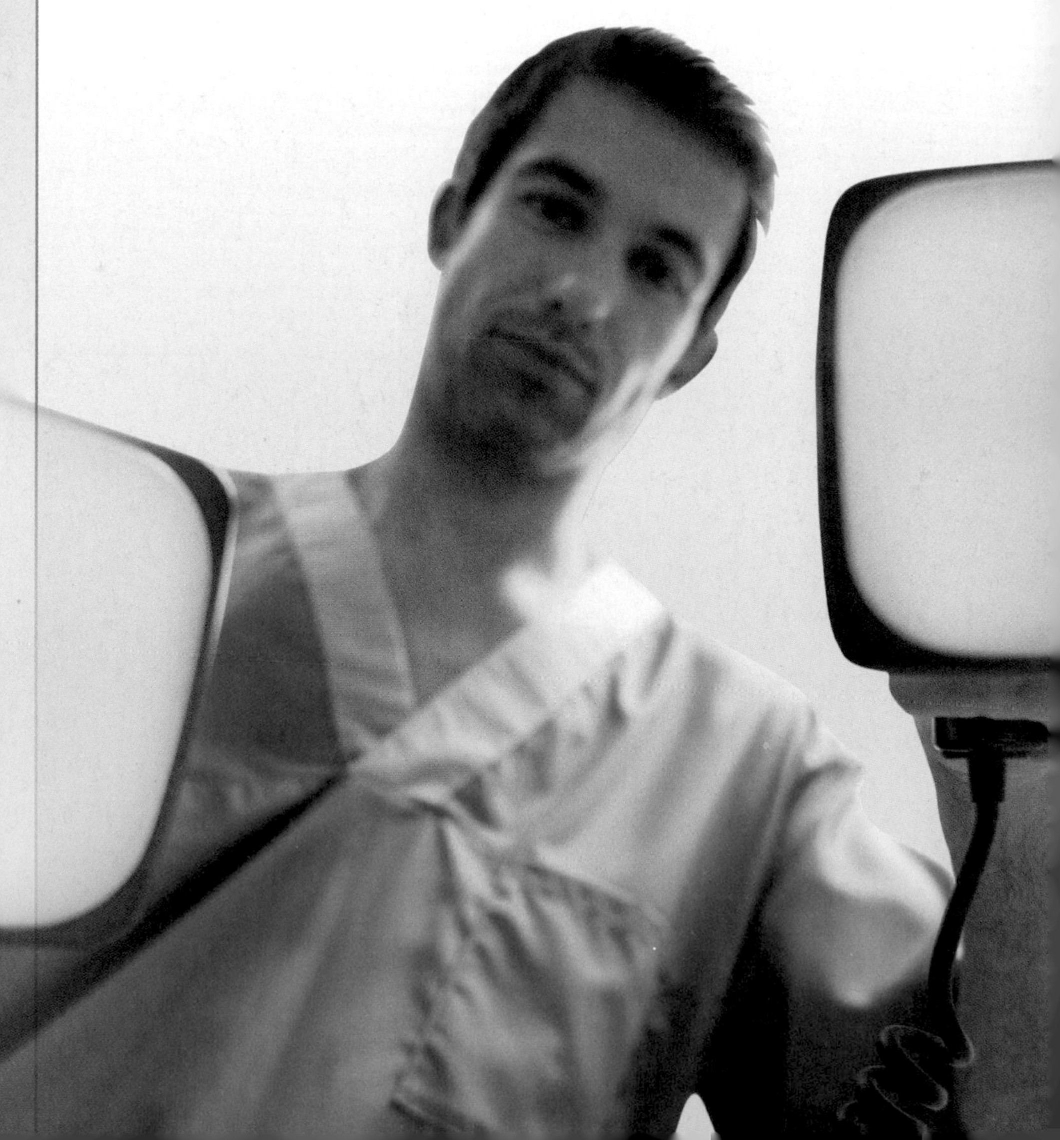

10.1 learning **outcome**

Use medical terminology related to the cardiovascular system.

learning outcomes

After completing this chapter, you should be able to:

10.1 Use medical terminology related to the cardiovascular system.

10.2 Identify the chambers, valves, and features of the heart.

10.3 Relate the structure of cardiac muscle to its function.

10.4 Explain why the heart does not fatigue.

10.5 Trace blood flow through the heart.

10.6 Describe the heart's electrical conduction system.

10.7 Describe the events that produce the heart's cycle of contraction and relaxation.

10.8 Interpret a normal EKG, explaining what is happening electrically in the heart.

10.9 Calculate cardiac output given heart rate and stroke volume.

10.10 Explain the factors that govern cardiac output.

10.11 Summarize nervous and chemical factors that alter heart rate, stroke volume, and cardiac output.

10.12 Locate and identify the major arteries and veins of the body.

10.13 Compare the anatomy of the three types of blood vessels.

10.14 Describe coronary and systemic circulatory routes.

10.15 Explain how blood in veins is returned to the heart.

10.16 Explain the relationship between blood pressure, resistance, and flow.

10.17 Describe how blood pressure is expressed and how mean arterial pressure and pulse pressure are calculated.

10.18 Explain how blood pressure and flow are regulated.

10.19 Explain the effect of exercise on cardiac output.

10.20 Summarize the effects of aging on the cardiovascular system.

10.21 Describe cardiovascular disorders.

word roots & combining forms

arter/o, arteri/o: artery

ather/o: fatty substance

atri/o: atrium

brady/: slow

cardi/o: heart

coron/o: heart

pericardi/o: pericardium

rhythm/o: rhythm

sphygm/o: pulse

steth/o: chest

tachy/: rapid

vas/o: vessel

vascul/o: vessel

ven/o, ven/i: vein

ventricul/o: ventricle

pronunciation key

anastomoses: ah-NAS-tah-MO-seez

arrhythmia: a-RITH-mee-ah

atria: A-tree-uh

auricles: AW-ri-kulz

diastole: die-AS-toe-lee

intercalated: in-TER-kah-lay-ted

ischemia: is-KEE-mee-ah

mitral: MY-tral

proprioceptors: pro-PREE-oh-sep-torz

Purkinje: per-KIN-jee

sphygmomanometer: SFIG-moh-mah-NOM-ih-ter

systole: SIS-toe-lee

tachycardia: tak-ih-KAR-dee-ah

tamponade: tam-po-NAID

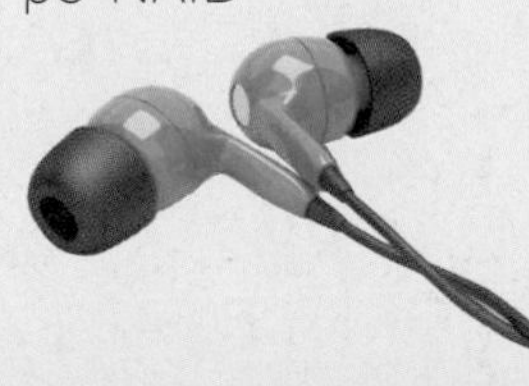

Overview

In Chapter 9, you explored part of the cardiovascular system—the blood. In this chapter, you will study this system further to include the heart and vessels, which work together to distribute blood to the body. You will see the anatomy of the heart and revisit cardiac muscle tissue, which enables the heart to function a lifetime without fatigue. You will explore the physiology of the heart in terms of its blood flow, electrical conduction system, cycle of contraction and relaxation, and regulation. You will then move on to the vessels, covering first their anatomy and then the physiology of how the heart and vessels function together to maintain blood pressure and flow.

Your study begins with heart anatomy.

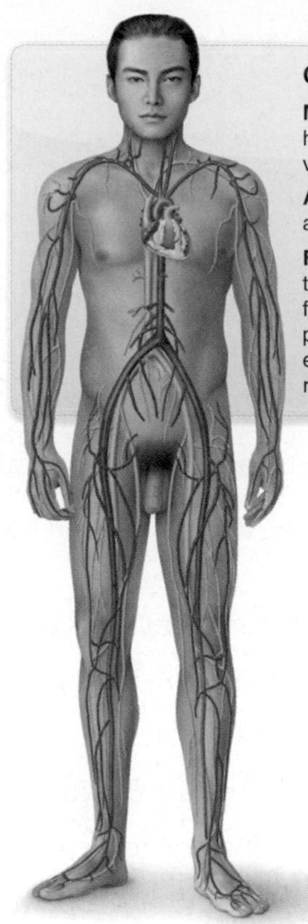

Circulatory System

Major Organs and Structures:
heart, aorta, superior and inferior venae cavae

Accessory Structures:
arteries, veins, capillaries

Functions:
transportation, protection by fighting foreign invaders and clotting to prevent its own loss, fluid and electrolyte balance, temperature regulation

FIGURE 10.1 The cardiovascular system.

10.2 learning outcome

Identify the chambers, valves, and features of the heart.

Heart Anatomy

The heart is a hollow organ about the size of an adult fist and weighs approximately 300 grams (g), or 10 ounces (oz). See **Figure 10.2**. It has a broad, superior **base** attached to the great vessels and a pointed **apex** (end) immediately superior to the diaphragm. The heart is located between the chest's pleural cavities in the mediastinum, deep to the sternum. Normally tilted, approximately two-thirds of the heart rests left of the midsagittal plane. Depending on body size and type, the heart may be positioned more upright (tall bodies) or more tilted (barrel-chested bodies).

spot check  Which is more inferior, the base of the heart or the apex of the heart?

Pericardium

The pericardial sac that you see in **Figure 10.2** is part of the **pericardium**—a serous membrane. As you may recall from Chapter 1, the pericardium is a fluid-filled, double-walled membrane. The **pericardial sac** (parietal pericardium) anchors the heart to the great vessels (**aorta** and **venae cavae**), the posterior wall of the thorax, the sternum, and the diaphragm. It has a tough outer fibrous layer that does not allow for expansion. It also has a thin serous lining of epithelial tissue. The

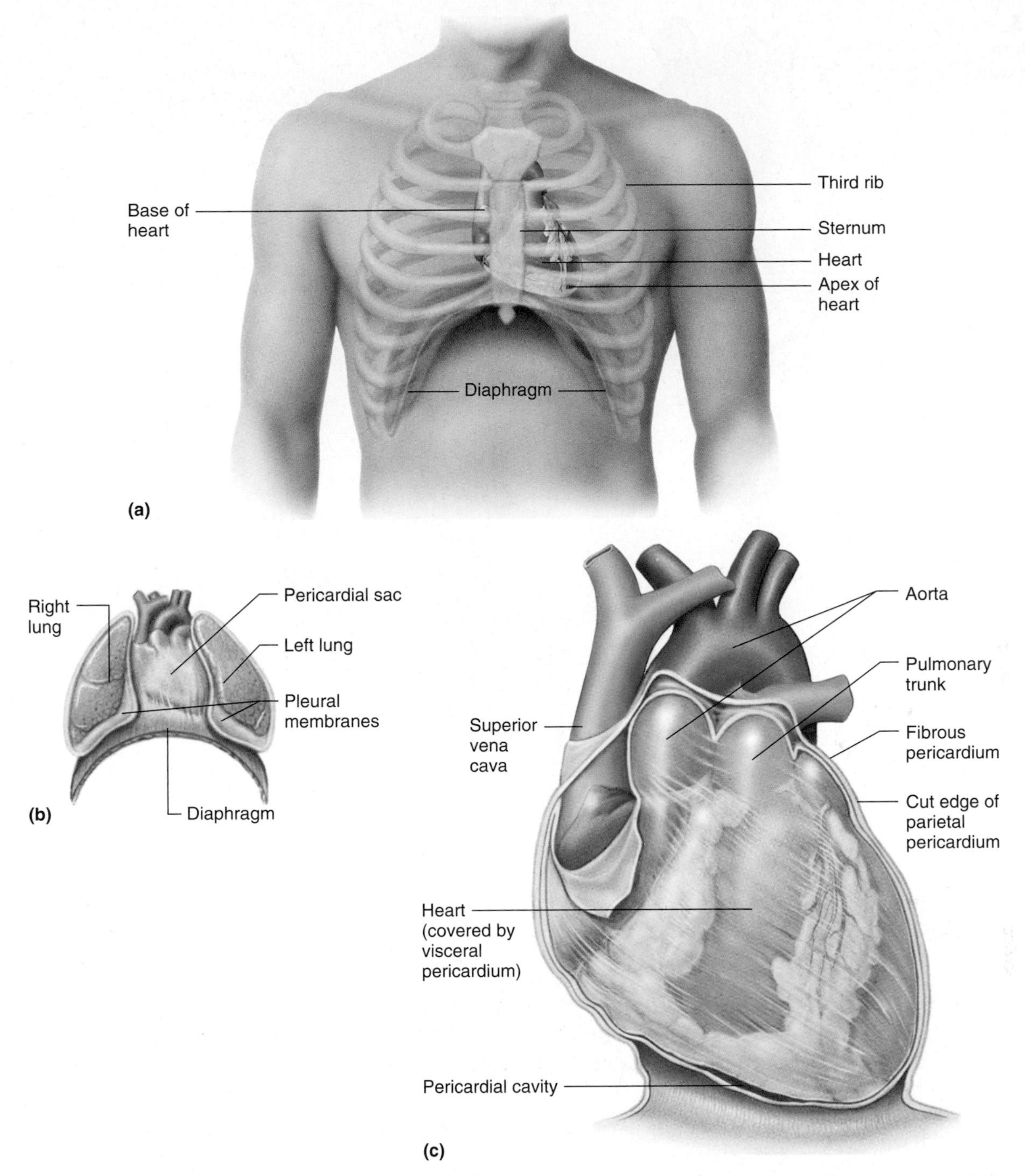

FIGURE 10.2 The position of the heart in the thorax: (a) relationship of the heart to the sternum and ribs, (b) relationship of the heart to the pleural cavities of the lungs and the diaphragm, (c) frontal view with the lungs retracted and the pericardial sac cut.

epicardium (visceral pericardium) is a more delicate layer composed of simple squamous epithelial tissue over loose areolar connective tissue. It is in direct contact with the surface of the heart. Normally, the two layers of the pericardium are close together with pericardial fluid between the layers to reduce the friction caused by the heart's pumping action. However, disease or inflammation may cause the amount of pericardial fluid to increase, restricting room for the heart to expand. **Figure 10.3** shows the pericardium in relation to the heart wall.

FIGURE 10.3
The pericardium and the heart wall.

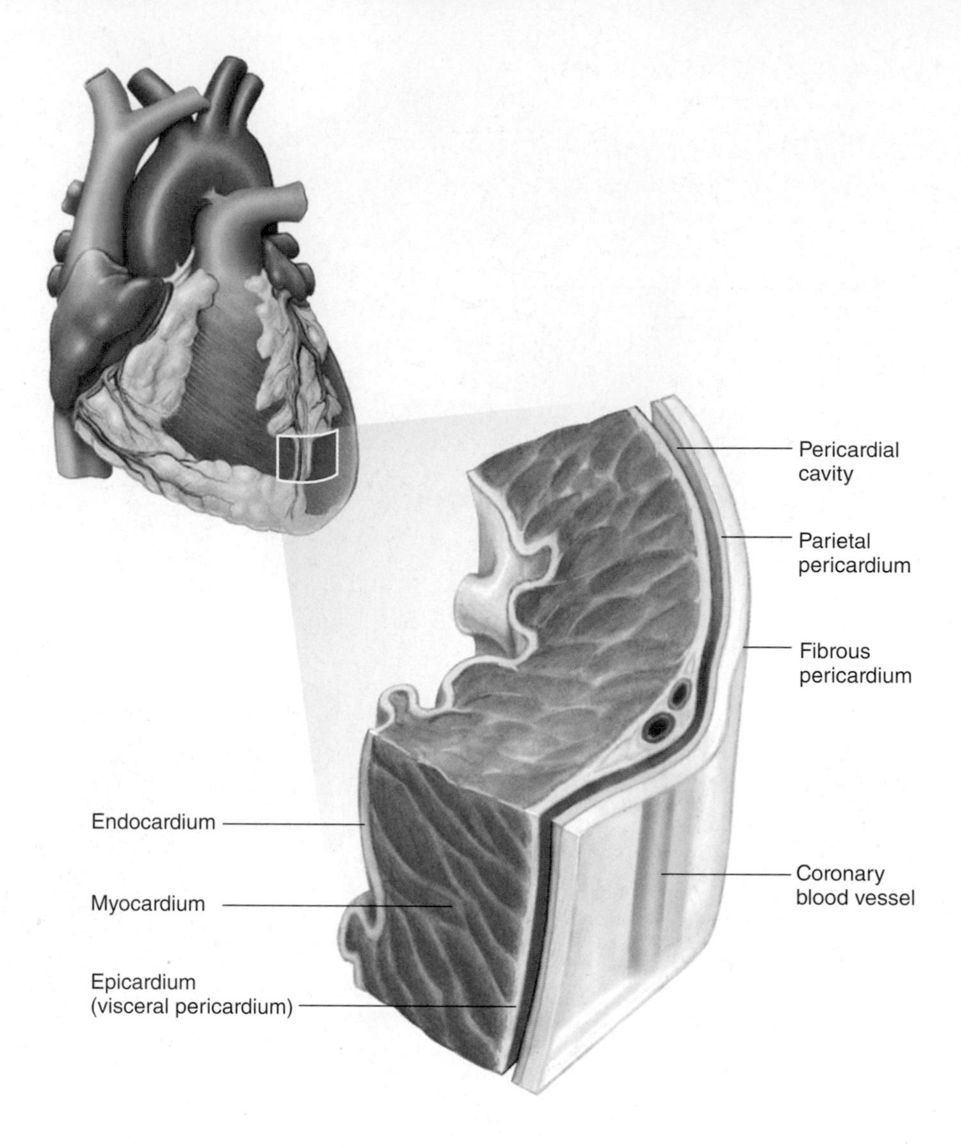

tamponade:
tam-po-NAID

clinical point

Cardiac tamponade is the buildup of excessive fluid within the pericardium. Disease or inflammation can cause this fluid buildup, resulting in more pericardial fluid produced. An injury, causing internal bleeding into the pericardial space, can also result in cardiac tamponade. Because the pericardial sac does not allow for expansion, the excess serous fluid or blood compresses the heart and may lead to heart failure.

Heart Wall

In addition to showing the pericardium, **Figure 10.3** shows the three layers of the heart wall—the epicardium, the myocardium, and the endocardium. These layers are described in the following list:

- The epicardium is the most superficial layer, composed of simple squamous epithelial tissue over loose areolar connective tissue.
- The **myocardium** is composed of cardiac muscle tissue and a fibrous skeleton of collagen and elastic fibers. The cardiac muscle of the myocardium does the heart's work by contracting to pump blood out of the heart and then relaxing

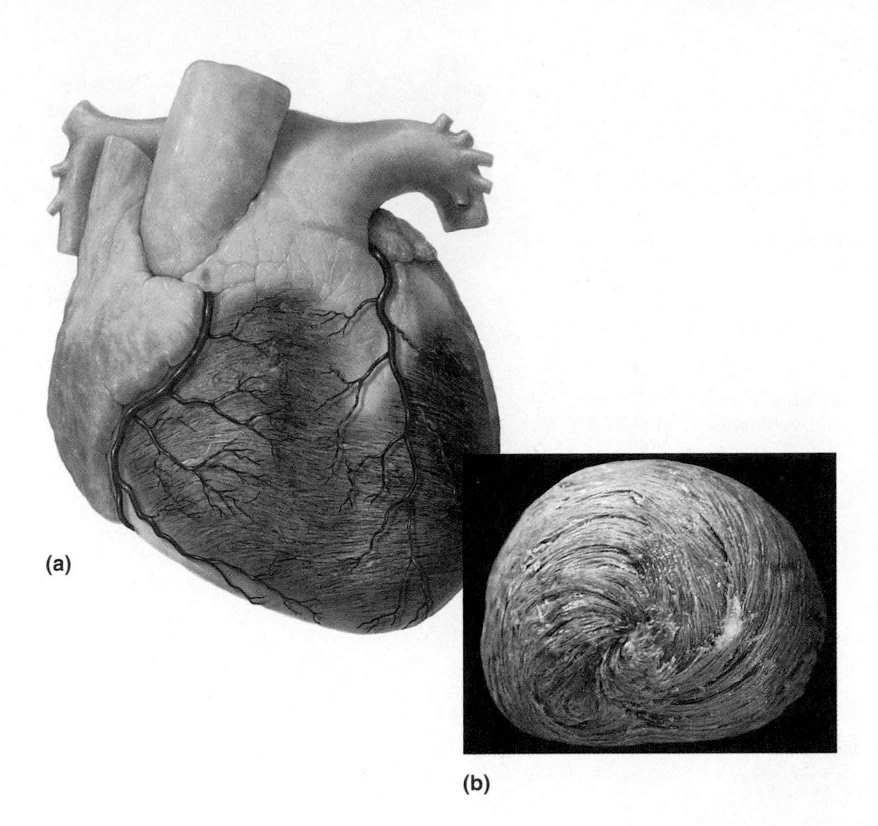

FIGURE 10.4 Spiral arrangement of cardiac muscle in the heart: (a) anterior view, (b) view from the apex of the heart, showing the spiral pattern.

so that the heart can refill. The muscle is arranged in a spiral pattern, which creates a twisting or wringing motion when the heart contracts. See **Figure 10.4**. The collagen fibers in the myocardium are nonconductive, so they act as insulators for cells carrying electrical signals within the heart wall. These fibers also give structural support to the heart around the valves and the vessel openings. The elastic fibers in this layer enable the heart to return to shape after contractions.

- The **endocardium** lines the four chambers of the heart. Like the epicardium, it is also composed of simple squamous epithelial tissue over loose areolar connective tissue. Extensions of the epicardium cover the surface of the heart's valves and continue on as the lining of vessels.

Chambers and Valves

The human heart is divided into four chambers, which can be seen externally in **Figure 10.5**. The two superior chambers—the right and left **atria**—receive blood for the heart, while the two inferior chambers—the right and left **ventricles**—pump blood out of the heart. **Figure 10.5** also shows the atria's small, hollow, earlike flaps called **auricles,** which slightly increase the atria's volume. Sulci are depressions on the surface of the heart and mark the division of the chambers externally. The **coronary sulcus** marks the separation of the atria from the ventricles. The **anterior interventricular sulcus** and **posterior interventricular sulcus** mark the separation of the right and left ventricles.

atria: A-tree-uh

auricles: AW-ri-kulz

If you were to look at the internal workings of the heart, you would see a wall called the **interatrial septum,** which separates the two atria from each other, and another wall, the **interventricular septum,** which separates the two ventricles. These two myocardial walls effectively divide the heart into right and left sides. Each side of the heart serves as a separate pump for blood. The right side of the heart pumps blood to the lungs (short distance), while the left side of the heart pumps blood to the rest of the body (long distance). Because the left side's

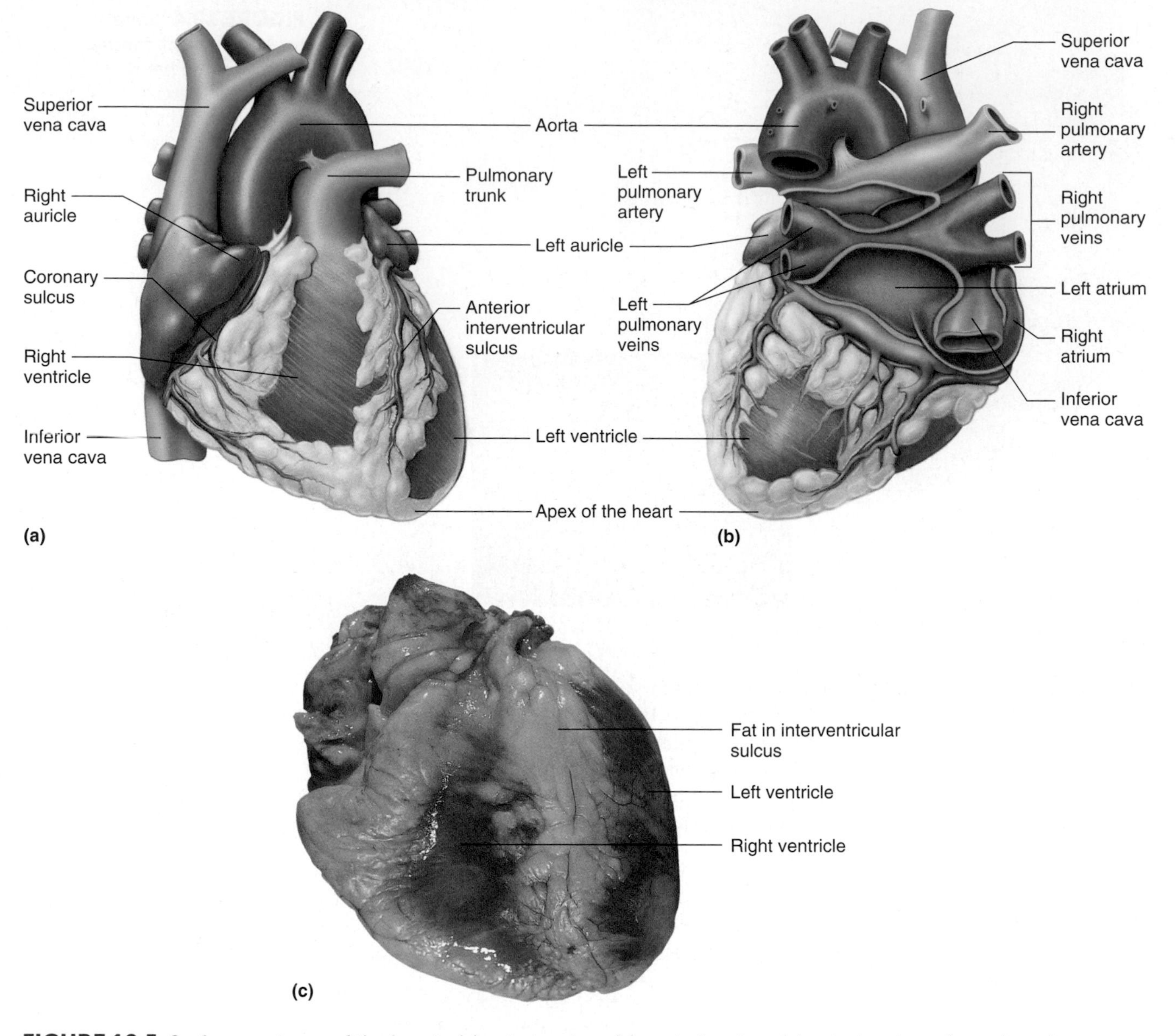

FIGURE 10.5 Surface anatomy of the heart: (a) anterior view, (b) posterior view, (c) anterior view of a cadaver heart.

workload is greater than the right side's, the left ventricle has a thicker myocardium. Always look at the outer walls of the ventricles on a dissected heart or illustration to determine right from left. See **Figure 10.6** for anterior and posterior views of the internal anatomy of the heart.

clinical point

In a fetal heart, an opening called the **foramen ovale** is located in the interatrial septum. This opening allows blood in fetal circulation to bypass the right ventricle and the lungs. A fetus does not need to send blood to its own lungs because it receives oxygenated blood from its mother. This opening is normally closed at birth, becoming the **fossa ovalis.** See Figure 10.6.

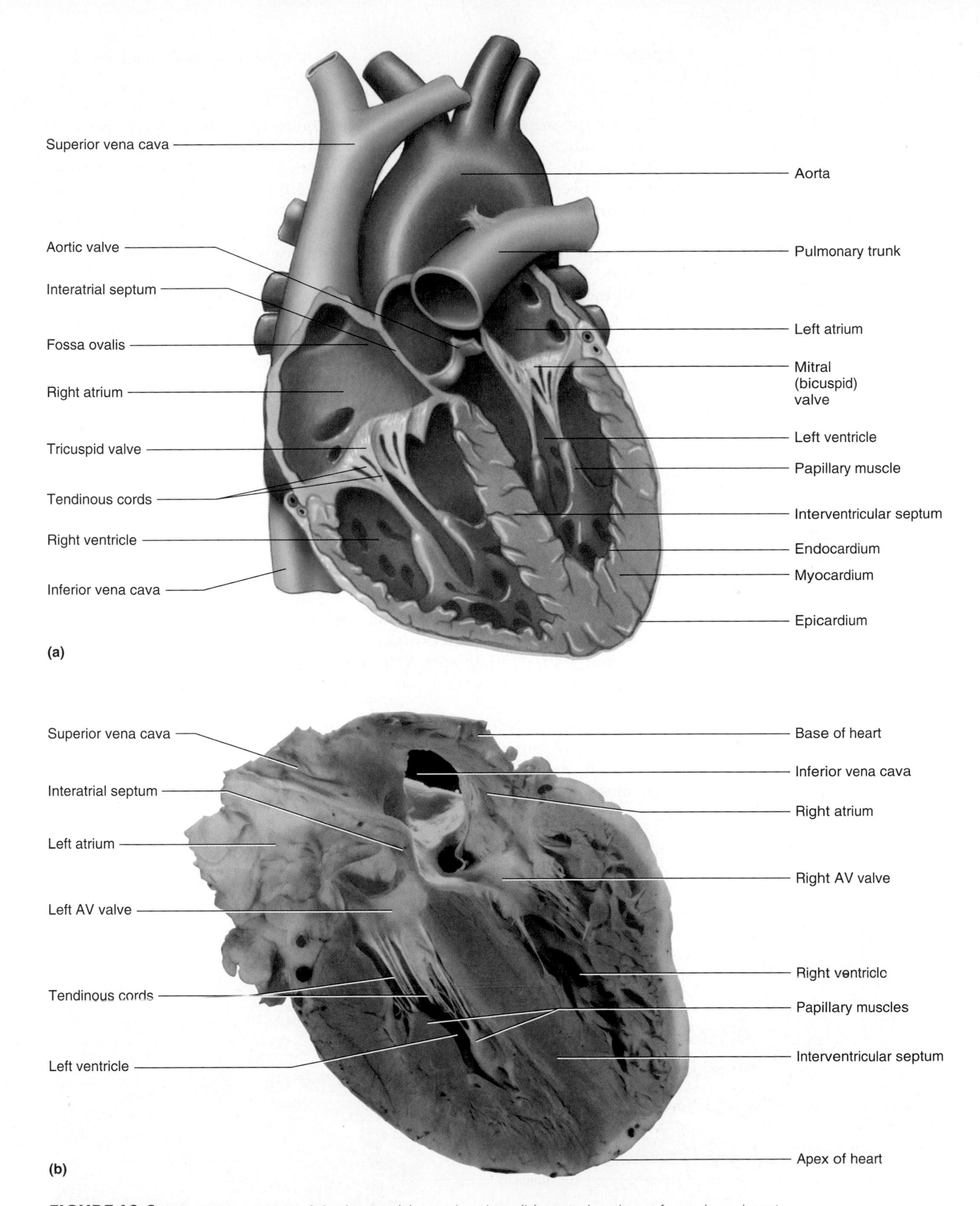

FIGURE 10.6 **Internal anatomy of the heart:** (a) anterior view, (b) posterior view of a cadaver heart.

spot check ❷ How can you determine the right ventricle from the left ventricle?

mitral: MY-tral

You can also see the presence of valves in **Figure 10.6**. The atria are separated from the ventricles by the **atrioventricular (AV) valves**—the **tricuspid valve** lies between the right atrium and the right ventricle, and the **bicuspid (mitral) valve** lies between the left atrium and the left ventricle. Notice that there are **tendinous cords** attaching the cusps of the AV valves to mounds of **papillary muscle** in the ventricles. These cords resemble parachute cords anchoring the valve (parachute) to the muscle. You will learn how these structures work together to direct blood flow when the cardiac cycle is discussed later in the chapter.

Semilunar valves—the **pulmonary valve** and the **aortic valve**—are located between each ventricle and the vessel that carries the pumped blood away from the heart. The pulmonary valve (not shown in **Figure 10.6**) lies between the right ventricle and the pulmonary trunk. The aortic valve lies between the left ventricle and the aorta. See **Figure 10.6**. The purpose of all the heart's valves is to ensure the flow of blood in one direction by preventing backflow. You will learn more about how valves work later in the chapter.

So far, you have become familiar with the anatomy of the heart. Now, it is time to revisit cardiac muscle tissue to expand on what you learned about histology in Chapter 2.

10.3 learning outcome

Relate the structure of cardiac muscle to its function.

Cardiac Muscle Tissue

As you recall from Chapter 2 and can see in **Figure 10.7**, cardiac muscle is striated (striped) and branching and has one nucleus per cell. The specialized junctions between cells **(intercalated disks)** enable the fast transmission of electrical impulses from one cell to another. With this feature, contractions of both atria are stimulated simultaneously, so they contract together as one. The same holds true for the ventricles.

intercalated: in-TER-kah-lay-ted

Unlike skeletal muscle cells that can perform tetany if nerve impulses are frequent enough, all cardiac muscle cells have an absolute refractory period (recall a muscle twitch, discussed in Chapter 5). The absolute refractory period prevents the cardiac muscle from going into tetany and gives the heart chambers a chance to fill between contractions.

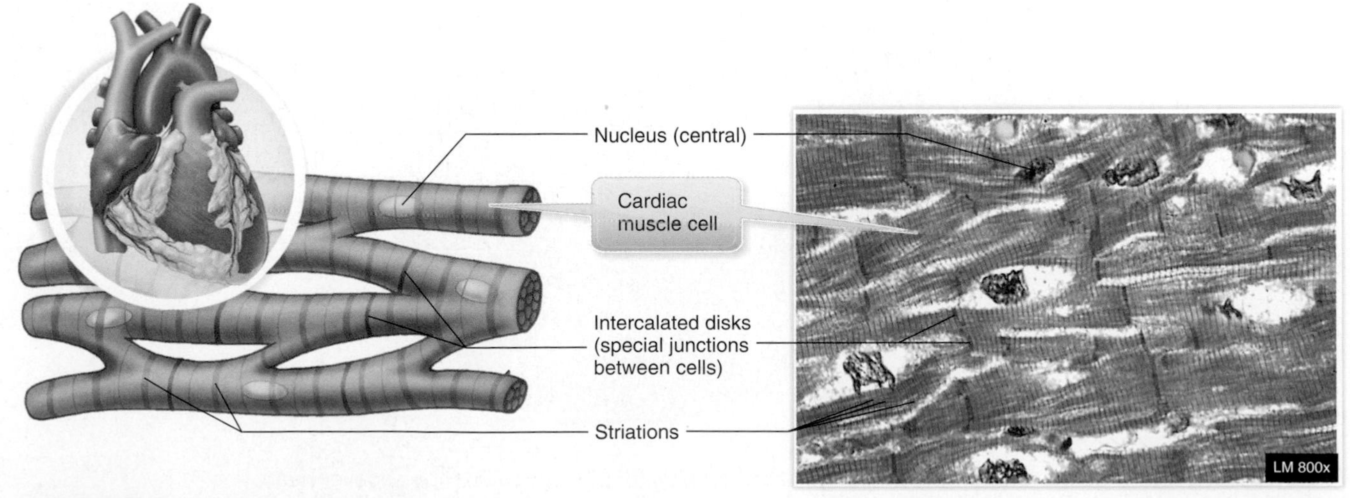

FIGURE 10.7 Cardiac muscle tissue.

By staying aerobic, cardiac muscle cells avoid fatigue, oxygen debt, and the buildup of lactic acid. Cardiac muscle cells also have the following special adaptations that enable them to use aerobic respiration almost exclusively:

- Cardiac muscle cells have many very large mitochondria to perform aerobic respiration.
- Cardiac muscle cells are rich in myoglobin (protein for storing oxygen).
- Cardiac muscle cells are rich in glycogen (a starch that can be converted to glucose to be used as fuel).
- Cardiac muscle cells can use a variety of fuels as energy sources (glucose, fatty acids, amino acids, and ketones).

Some cardiac muscle cells are further adapted to generate and carry electrical impulses. You will learn more about them during the discussion of the heart's electrical conduction system.

10.4 learning outcome

Explain why the heart does not fatigue.

spot check 3 How does myoglobin help to prevent the heart from fatiguing?

Heart Physiology

The function of the heart is to pump blood to meet the needs of the body. It is a sophisticated pump that can be regulated to meet the level of need, whether it is to pump faster or slower. Now that you are familiar with the heart's anatomy, you are ready to explore how the heart works, starting with blood flow through the heart.

Blood Flow through the Heart

10.5 learning outcome

Trace blood flow through the heart.

warning

All blood is red. In Figure 10.8, the colors orange and violet are used to signify the level of oxygen in the blood. Other illustrations or models you may have seen use red and blue. Thus, many people believe oxygen-rich blood is red and oxygen-poor blood is blue. This common misconception is very strong and is unfortunately reinforced by the models, charts, and book illustrations used in education. Some students will insist that oxygen-poor blood in their veins is blue and will try to prove it by pointing to the vessels on the anterior surface of their wrists. However, these vessels are deep to the collagen in the skin, and collagen makes the vessels appear blue. When asked if they have ever had blood drawn and, if so, what the color was, these same students will answer, "Yes, but as soon as the blood hit the air it was exposed to oxygen and then turned red." The reason blood enters a tube when drawn from a vein in the doctor's office or lab is because there is a vacuum in the tube. No oxygen or air is present. All blood is red.

Look at **Figure 10.8** and the following description of its steps closely. It shows the pathway of blood through the heart. You may find it helpful to review this figure several times before moving on.

1. Blood enters the right atrium of the heart from the superior and inferior venae cavae.

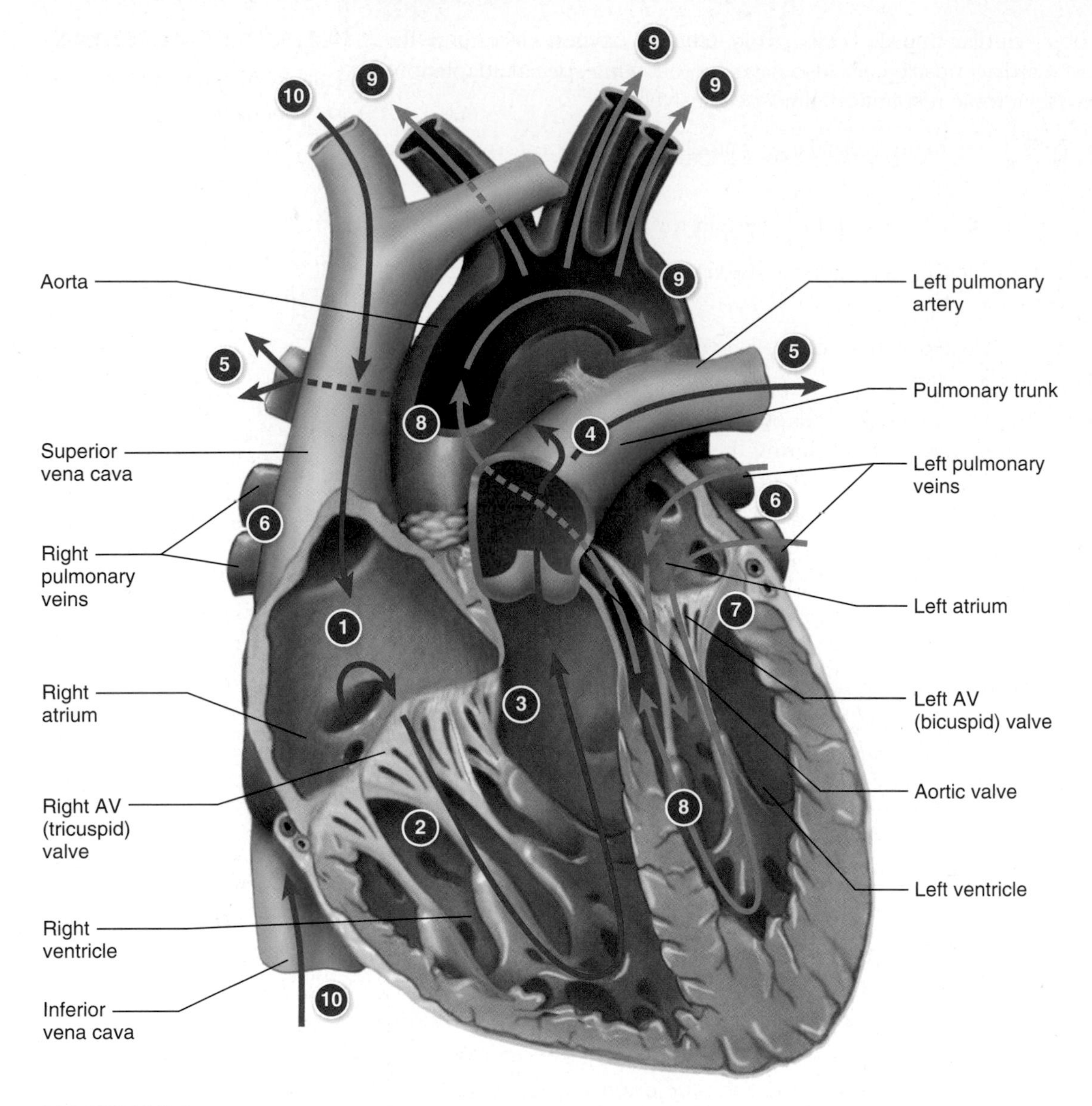

FIGURE 10.8 The pathway of blood flow through the heart. Steps 4 to 6 illustrate the circulation of blood through the lungs. Steps 8 to 10 illustrate the circulation of blood through the body. *Violet* arrows indicate *oxygen-poor* blood, while *orange* arrows indicate *oxygen-rich* blood.

2. Blood passes through the tricuspid valve to the right ventricle.
3. Blood is forced through the pulmonary valve to the pulmonary trunk.
4. Blood travels through the pulmonary trunk to the pulmonary arteries.
5. Blood travels to the lungs, where CO_2 is unloaded and O_2 is loaded.
6. Blood returns to the heart from the lungs through the four pulmonary veins to enter the left atrium.
7. Blood passes through the bicuspid (mitral) valve to the left ventricle.
8. Blood is forced through the aortic valve to the aorta.
9. Blood travels from the aorta to the rest of the body.
10. Blood returns to the heart through the superior and inferior venae cavae.

As you can see in **Figure 10.8**, the right side of the heart pumps blood to the lungs and back **(pulmonary circuit).** At the lungs, CO_2 is unloaded from the blood and O_2 is loaded into the blood. Once the blood is returned from the lungs, the left side of the heart pumps blood out to all parts of the body to be returned once again to the right side of the heart **(systemic circuit).** At the tissues of the body, O_2 is unloaded and CO_2 is loaded. See **Figure 10.9**.

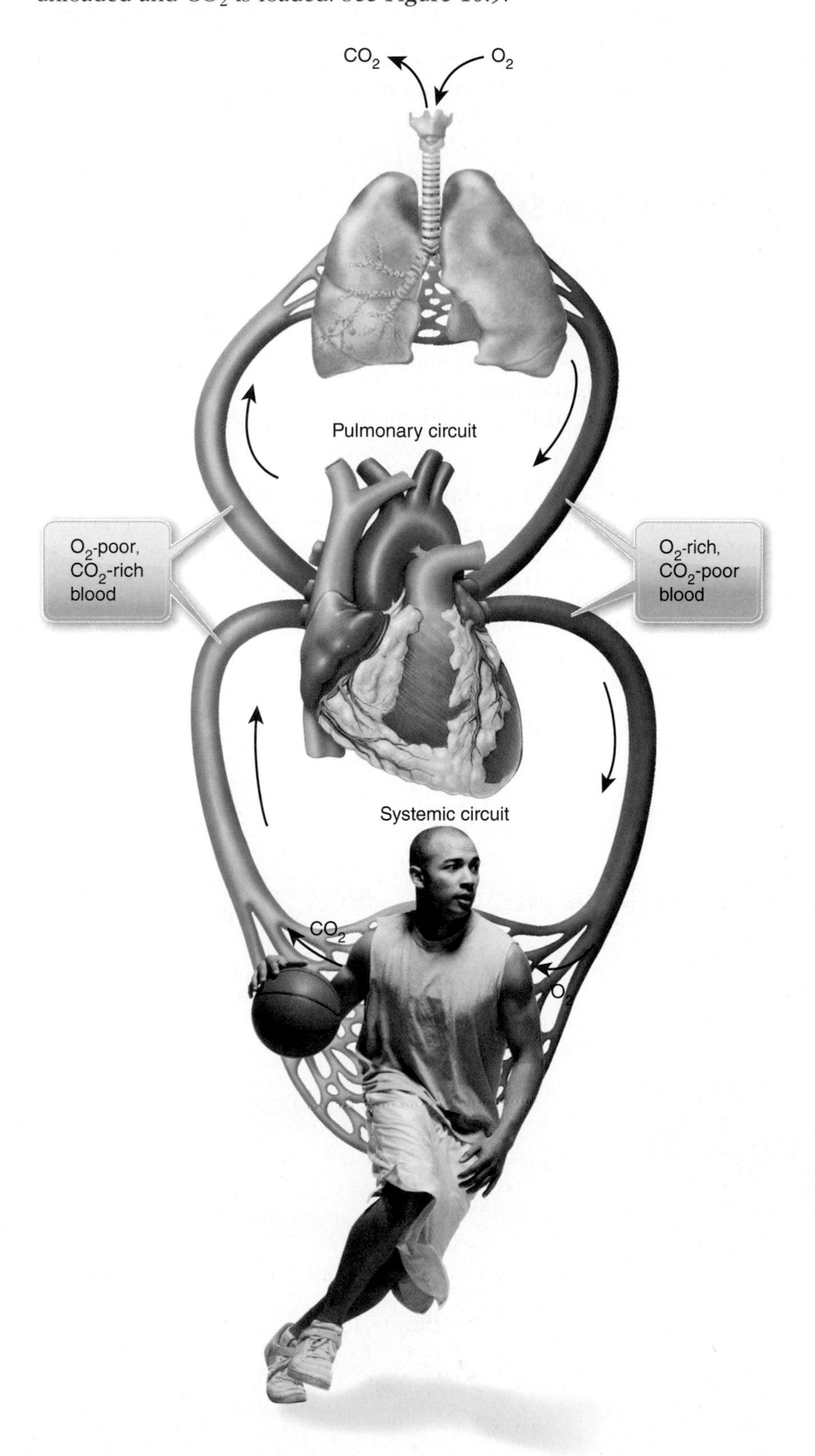

FIGURE 10.9 General diagram of the pulmonary and systemic circuits.

spot check ❹ Where will blood be immediately after passing through the tricuspid valve?

spot check ❺ Through what valve will blood pass next after the left ventricle?

You have now seen the direction of flow as blood moves through the heart, but how is blood pumped? To understand this, you must first understand how the cardiac muscle of the heart is stimulated to contract.

10.6 learning outcome

Describe the heart's electrical conduction system.

Cardiac Conduction System

As you may recall from Chapter 5, cardiac muscle is autorhythmic, meaning it does not need to be stimulated by the brain to contract. Modified cardiac muscle cells initiate and carry the electrical impulses as part of a conduction system that stimulates heart contractions. The nervous system may further modify the amount and frequency of the contractions (as you will see later in the chapter), but nerve impulses are not needed to initiate a heart contraction.

Follow along with **Figure 10.10** to understand the pathway of signals in the heart's conduction system.

1. A **heartbeat** (heart contraction) is started by the **sinoatrial (SA) node,** a patch of specialized cardiac muscle cells located in the wall of the right atrium near the opening for the superior vena cava. The SA node is the heart's **pacemaker.**

2. From the SA node, modified cardiac muscle cells carry electrical impulses across the myocardium of both atria, causing them to depolarize and contract together as one (shown by black arrows).

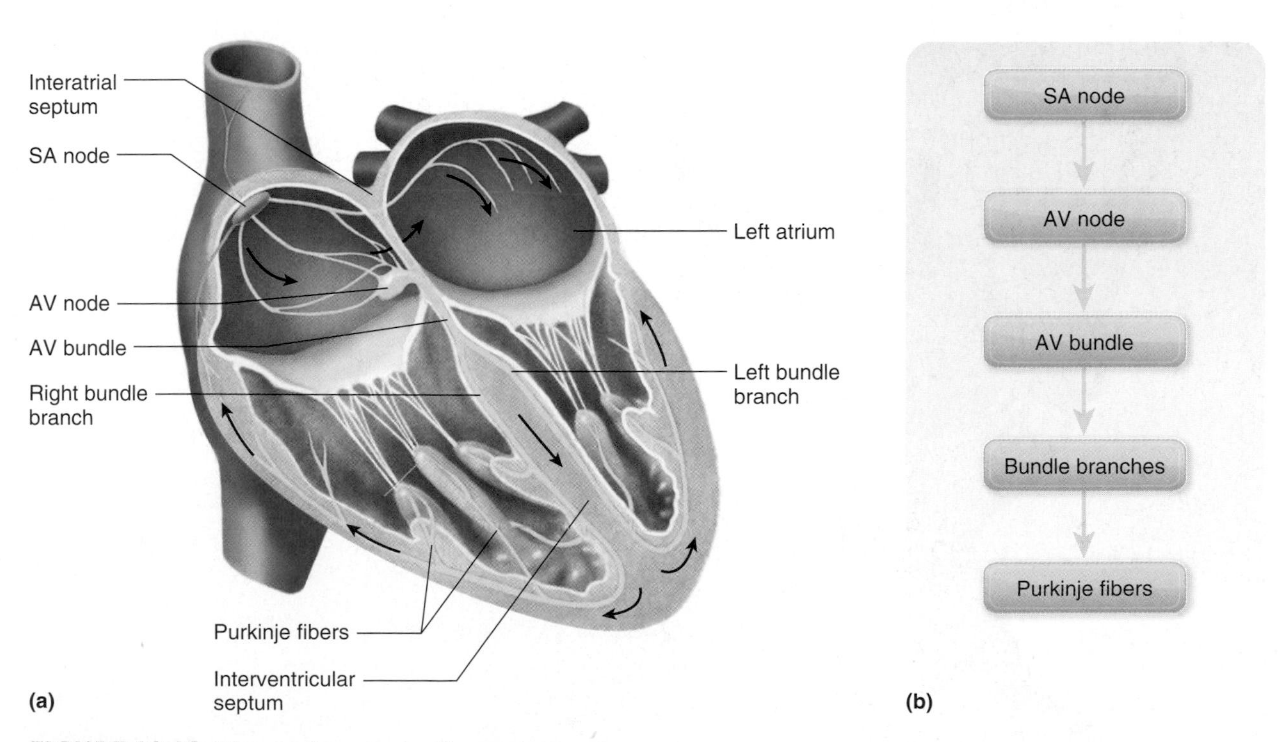

FIGURE 10.10 The cardiac conduction system.

3. While step 2 is happening, other modified cardiac muscle cells carry electrical impulses from the SA node to the **atrioventricular (AV) node,** located on the right atrium's interatrial septum. It is from this node that the ventricles will be stimulated to contract.

4. The **atrioventricular (AV) bundle** divides into branches that continue to carry the electrical impulses from the AV node down the interventricular septum toward the heart's apex. Collagen fibers of the heart's fibrous skeleton insulate the specialized cardiac muscle cells, so the electrical impulses go only to their proper destination.

5. **Purkinje fibers** fan out from the ends of the AV bundle to the walls of the ventricles, stimulating the cardiac muscle cells of the ventricular myocardium to depolarize and contract.

Purkinje: per-KIN-jee

spot check 6 List in order the parts of the electrical conduction system used to stimulate ventricular contractions.

Now that you have studied the blood flow through the heart and the conduction system that stimulates the heart to contract, you are ready to put these two concepts together to begin to understand a cardiac cycle.

Cardiac Cycle

10.7 learning outcome

Describe the events that produce the heart's cycle of contraction and relaxation.

A cardiac cycle is one complete contraction and relaxation of the heart. This cycle may be repeated 70 to 80 times every minute. Before you examine the events of a cardiac cycle, you need to become familiar with the terms *systole* and *diastole*. **Systole** is contraction, and **diastole** is relaxation. These terms can be used for individual chambers, but if no chamber is specified, they usually refer to the action of the ventricles.

systole: SIS-toe-lee

diastole: die-AS-toe-lee

You also need to understand the concepts of volume, pressure, and flow. Think of a heart chamber as a syringe. See **Figure 10.11a**. As you pull back on the plunger of a syringe, the volume inside the syringe increases and the pressure decreases. The air pressure outside the syringe is greater than the air pressure inside the syringe, so air rushes into the syringe until the pressures are equal. Now see **Figure 10.11b**. As you push the plunger into the syringe, the volume inside the syringe decreases and the pressure increases. The air pressure inside the syringe is greater than the air pressure outside the syringe, so air rushes out of the syringe until the pressures are equal. Keep this analogy in mind as you go through the phases of the cardiac cycle.

Phases of the Cardiac Cycle It is important to understand that the right and left atria go through systole and diastole together at the same time and that the right and left ventricles also work together with each other. Similarly, the AV valves function together, as do the semilunar valves. The four phases of a cardiac cycle happen in the following order:

1. **Atrial systole.** After the SA node fires, the atria depolarize and contract together, creating decreased volume and higher pressure in the atria than in the ventricles. The increased pressure pushes blood through the AV valves (tricuspid and bicuspid) into the ventricles. It is important to note that the pressure of the blood due to the atria's contracting is what forces the AV valves to open.

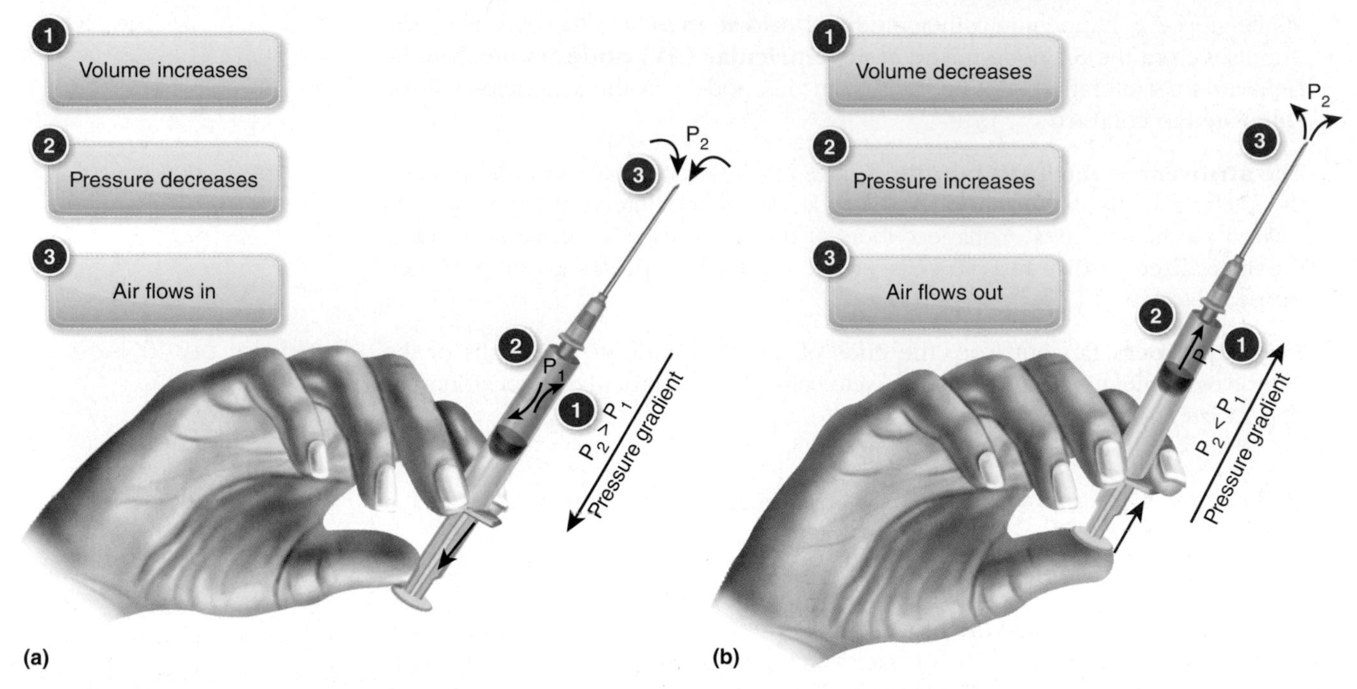

FIGURE 10.11 Concepts of volume, pressure, and flow. (a) As the volume increases, the pressure decreases within the syringe, so air rushes in to equalize the pressure inside and outside the syringe. (b) As the volume decreases, the pressure increases within the syringe, so air rushes out to equalize the pressure inside and outside the syringe.

2. **Atrial diastole.** The atria then repolarize and relax. The elastic fibers in the walls of the atria return the atria to shape, increasing the volume and decreasing the pressure in the atria. The blood pressure in the superior and inferior venae cavae (on the right side of the heart) and the pulmonary veins (on the left side of the heart) is then greater than the pressure inside the atria, so blood rushes in from these vessels to fill the right and left atria.

3. **Ventricular systole.** Once the conduction system has carried the electrical impulses from the AV node to the Purkinje fibers, the ventricles depolarize and contract together. Papillary muscles also contract, pulling on the tendinous cords **(chordae tendineae)** to ensure that the AV valves stay closed and do not swing open toward the atria. This prevents the backflow of blood to the atria—as you can see in **Figure 10.12a**. Contraction of the ventricles decreases the volume and increases the pressure inside the ventricles. As a result, blood is pushed through the pulmonary and aortic valves into the pulmonary trunk and aorta, respectively.

4. **Ventricular diastole.** The ventricles then repolarize and relax. The elastic fibers in the ventricle walls return the ventricles to shape, increasing the volume and decreasing the pressure in the ventricles. The pressure inside the now full atria is greater than that inside the relaxed ventricles, so blood moves through the AV valves from the atria to the ventricles. (*Note:* The atria are not contracting in this phase. Blood is passively moving from the atria to the ventricles due to a difference in pressure.) The pressure is also greater in the pulmonary trunk and aorta than inside the ventricles, but blood is caught in the cup-shaped semilunar pulmonary and aortic valves as it tries to return from these vessels to the ventricles. Once filled with blood, these valves are tightly closed to prevent backflow to the ventricles from the pulmonary trunk and aorta. See **Figure 10.12b**. Blood traveling passively from the atria to the ventricles in this phase reduces the pressure in the atria, so blood also moves passively from the venae cavae and pulmonary veins to the atria. *All four chambers fill with blood during ventricular diastole.*

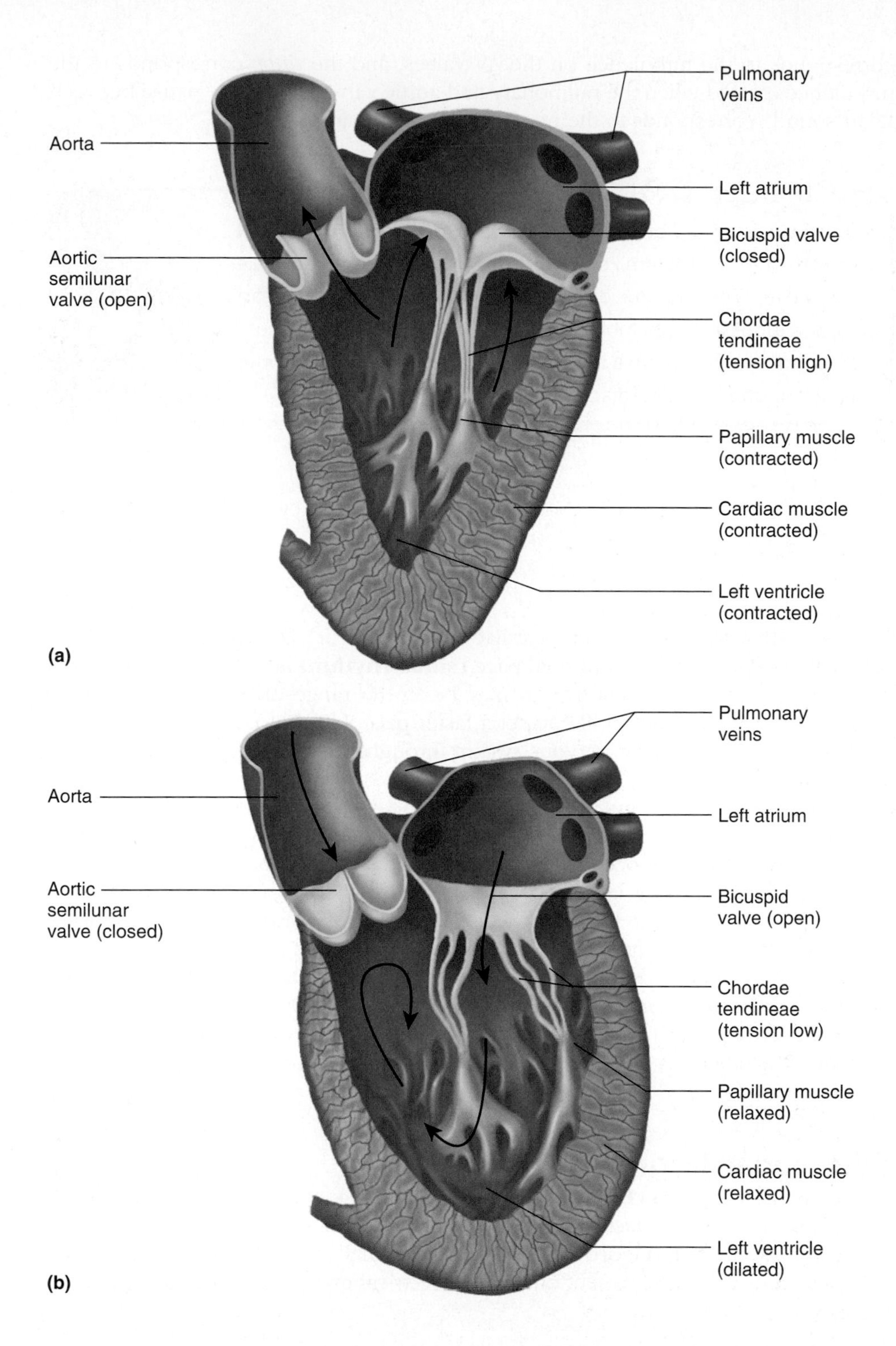

FIGURE 10.12 The action of heart valves on the left side of the heart. (a) *Ventricular systole:* The bicuspid valve is closed because the papillary muscle contracts to keep the cusps of the valve from opening toward the atria, and blood flowing toward the left atrium causes the cusps of the valve to overlap. The aortic semilunar valve is open because the cusps of the valve are pushed open by the blood flowing toward the aorta. (b) *Ventricular diastole:* The bicuspid valve is opened by the blood flowing into the left ventricle. The semilunar valve is closed because the cusps of the valve overlap as they are pushed by the blood in the aorta toward the dilated left ventricle. Similar action is taking place on the right side of the heart at the same time.

The heart rests after the four phases have been completed to ensure enough time for all of the chambers to fill. The contraction of the atria in the next cycle pushes more blood into the already full ventricles, which expands the ventricular walls, making the ventricles more likely to contract.

Heart Sounds during the Cardiac Cycle You may have had an opportunity to listen to a heart through a **stethoscope** (instrument for hearing heart sounds). The sounds you hear—lubb-dupp, lubb-dupp, lubb-dupp—are caused not by the valves themselves but by the turbulence of the blood when the valves close. The *lubb* corresponds to the turbulence on the AV valves, and the *dupp* corresponds to the turbulence created when the pulmonary and aortic valves close. The pause between heart sounds corresponds to the rest between heartbeats.

clinical **point**

A **murmur** is an abnormal heart sound. It may be a **functional murmur** (not a problem) or a **pathological murmur** (possible leaky valve). The murmur often makes a *ssh* sound. If the heart sound is lubbssh-dupp, lubbssh-dupp, the AV valves may be suspected of leaking because the abnormal sound is occurring with the first sound in the cardiac cycle.

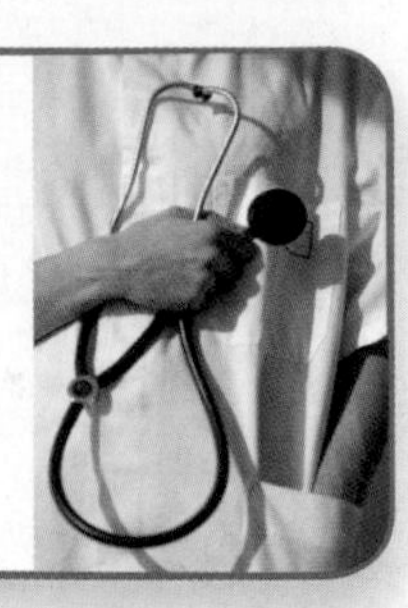

spot check 7 What heart sounds would be produced by a leaky aortic or pulmonary semilunar valve?

Cardiac Rhythm How often cardiac cycles occur is determined by the pacemaker—the SA node. A normal pace **(sinus rhythm)** is usually 70 to 80 beats per minute, although common rates may be in the range of 60 to 100 beats per minute. The SA node would like to set a faster pace, but the pace is normally kept in check by the autonomic nervous system through the vagus nerve. This is called **vagal tone.**

An **ectopic focus** occurs when any part of the conduction system other than the SA node is setting the pace. Any cardiac muscle cell is capable of becoming a pacemaker. A **nodal rhythm** occurs if the AV node is the ectopic focus. Hypoxemia, caffeine, nicotine, electrolyte imbalance, and some drugs may cause an ectopic focus.

arrhythmia: a-RITH-mee-ah

An **arrhythmia** is an abnormal heart rhythm. One cause for an arrhythmia is **heart block,** in which one part of the heart's conduction system fails to send its signals. If the ventricles do not receive the signals from the SA node via the AV node and AV bundle, they will beat at their own pace of 20 to 40 beats per minute. Their contractions, however slow, will not be coordinated with the contractions of the atria, and the efficiency of the heart will be compromised.

10.8 learning **outcome**

Interpret a normal EKG, explaining what is happening electrically in the heart.

Electrocardiogram

An **electrocardiogram (ECG** or **EKG)** is a means of looking at cardiac rhythms. It is a graph showing the electrical activity in the heart, not the amount of contraction of the cardiac muscle. **Figure 10.13** shows a normal ECG. The five waves on an ECG per cardiac cycle represent only three electrical events, outlined in the following list:

- The **P wave** shows the depolarization of the atria.
- The **Q, R,** and **S waves** together represent the ventricles depolarizing.
- The **T wave** represents the ventricles repolarizing.

You may ask, Where does a normal ECG show the atria repolarizing? It does not show that event because repolarization of the atria occurs at the same time that the ventricles are depolarizing (a much larger event).

Figure 10.14 shows the relationship of an ECG and the contraction of the myocardium.

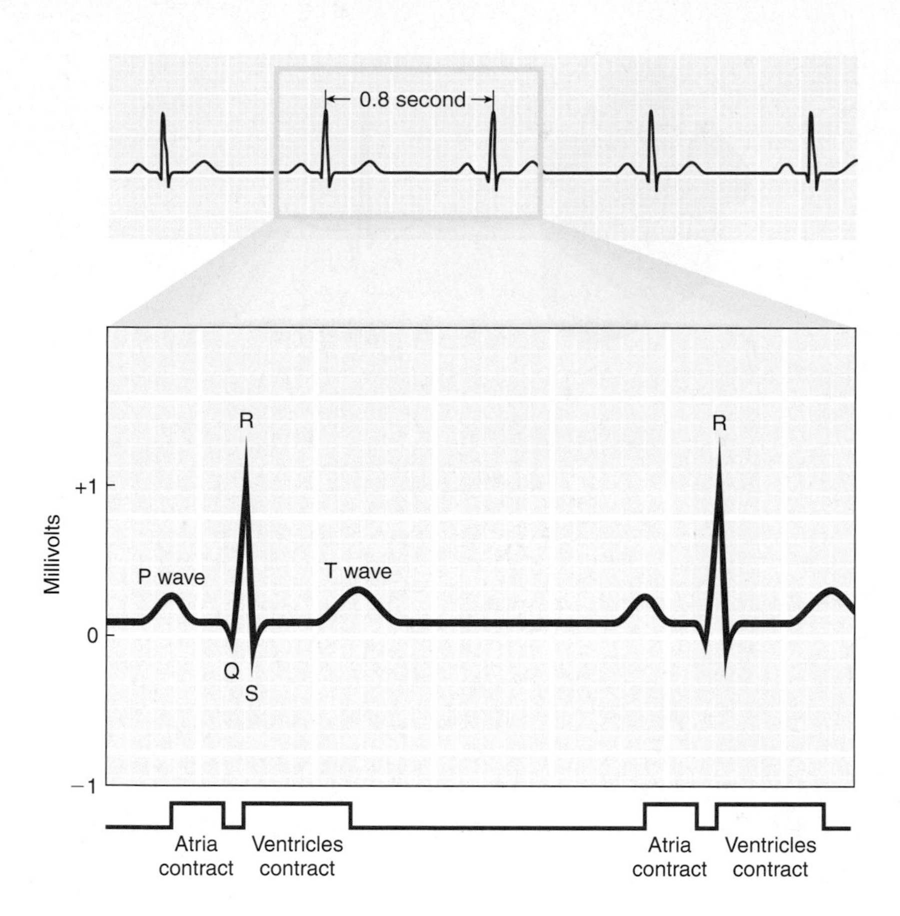

FIGURE 10.13 A normal electrocardiogram.

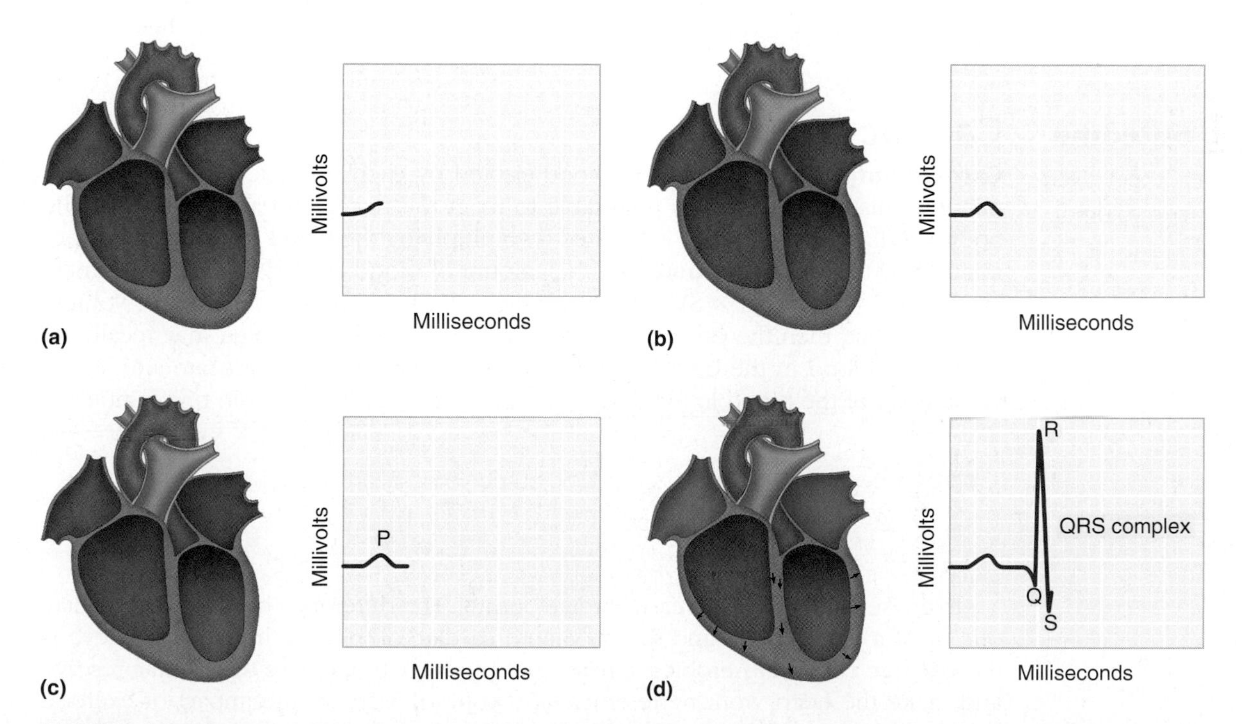

FIGURE 10.14 The relationship of an ECG and contraction of the myocardium. The purple areas in each drawing (a-g) indicate where tissues are depolarizing, and the green areas indicate where tissues are repolarizing. The graphs next to each drawing show the corresponding portion of the ECG pattern.

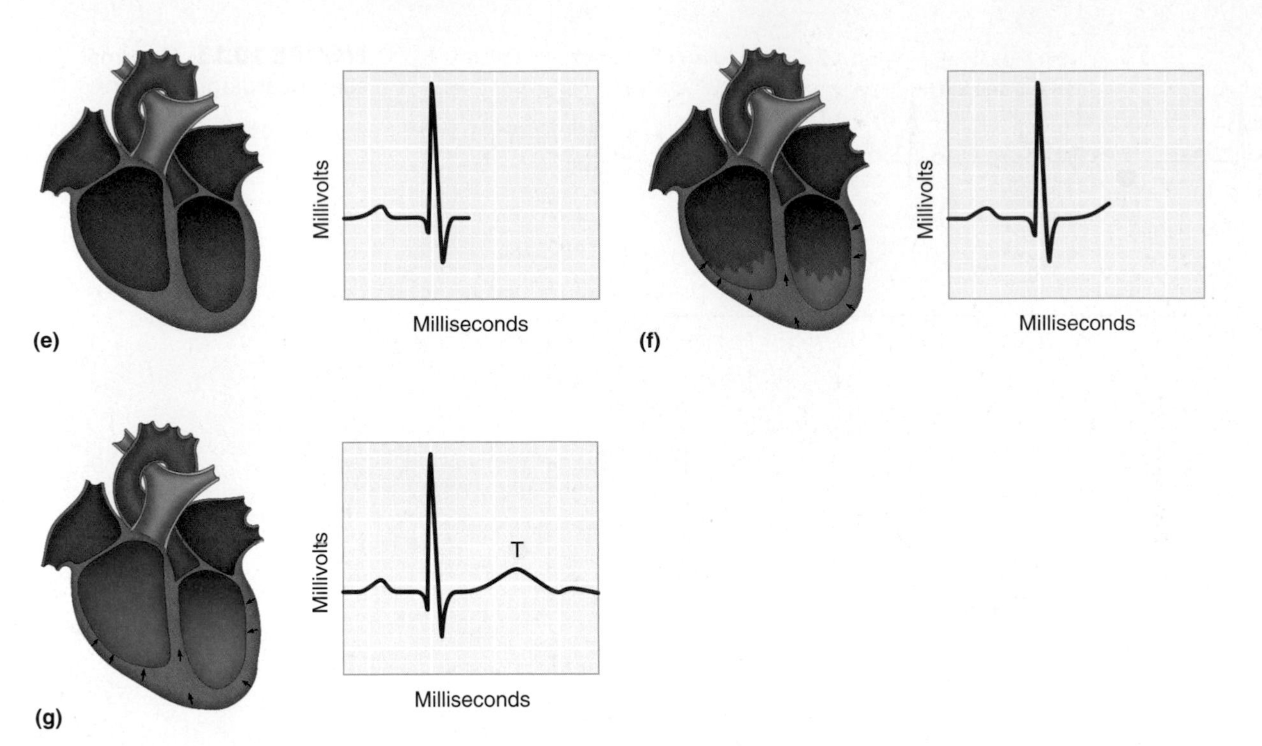

FIGURE 10.14 concluded

spot check 8 What is happening in the atria and ventricles in the interval between the T wave and the next P wave?

The sole purpose of all the physiology you have covered so far in this chapter is to efficiently pump blood out of the heart. But questions remain: How much blood? Does the amount ever change, and if so, how does it change? We answer these questions below, focusing first on the amount of blood pumped by the heart.

10.9 learning **outcome**

Calculate cardiac output given heart rate and stroke volume.

Cardiac Output

Cardiac output is the amount of blood ejected by each ventricle of the heart each minute. This output is highly adjustable to meet the needs and activity level of the body. Cardiac output (CO) is calculated by multiplying the **heart rate** (HR; beats per minute) by the **stroke volume** (SV; the amount of blood ejected from each ventricle per beat). So CO = HR × SV. If the heart rate is 75 beats/min and the stroke volume is 70 mL/beat, then the cardiac output equals 5,250 mL/min. As you may recall, the volume of blood in the body is typically 4 to 6 liters. At this heart rate and stroke volume, all of the blood in the body is pumped through the heart in one minute.

spot check 9 What is the cardiac output if the heart rate is 100 beats/min and the stroke volume is 80 mL/beat?

Exercise can greatly increase cardiac output. The difference between the cardiac output of a heart at rest and the maximum cardiac output the heart can achieve is the **cardiac reserve.** Aerobic exercise and training can increase the cardiac reserve and make the heart work more efficiently. You will explore the impact of exercise on the heart later in the chapter.

Next, you will learn more about cardiac output by studying heart rate and stroke volume.

Heart Rate This measurement is made by feeling a **pulse** in an artery. Each heartbeat creates a surge in pressure that can be felt in the arteries carrying blood away from the heart. Commonly, the radial artery is felt on the anterior surface of the wrist, but arteries in the neck and legs can also be used. See **Figure 10.15**.

As you will learn later in the chapter (under "Heart Regulation"), heart rate varies on a minute-by-minute basis. However, at rest, the normal heart rate for adult males is 64 to 72 beats/min and 72 to 80 beats/min for adult females. The human heart may at times beat too fast or too slow. Tachycardia and bradycardia are two of the conditions in which this may be the case.

10.10 learning outcome

Explain the factors that govern cardiac output.

clinical point

Tachycardia is a persistent resting adult heart rate greater than 100 beats/min. Stress, anxiety, drugs, heart disease, and fever can all cause tachycardia. On the opposite end of the spectrum is **bradycardia,** which is a persistent resting adult heart rate that is less than 60 beats/min. Causes for this condition can be sleep, endurance training for athletes, and hypothermia.

tachycardia: tak-ih-KAR-dee-ah

Stroke Volume This volume also plays a key role in determining cardiac output. Stroke volume is the amount of blood ejected by each ventricle with each heartbeat, but it is *not* the amount of blood each ventricle can hold. Ventricles cannot completely empty themselves with each contraction. There are three factors that affect stroke volume:

- **Preload** is the amount of tension in the myocardium of the ventricular walls. If the walls have been stretched by the amount of blood entering, the ventricles will contract more forcibly. This follows the **Frank-Starling law of the heart,** which basically states that the heart must pump out the amount of blood it receives. If more blood comes in, more blood must go out.
- **Contractility** refers to the responsiveness of the *cardiac muscle* to contract. You may recall that stretching smooth muscle makes it more likely to contract, but this works differently. In this case, calcium makes *cardiac muscle* more excitable, resulting in greater contractility.
- **Afterload** concerns the pressure in the pulmonary trunk and aorta during diastole. As you may remember from the cardiac-cycle discussion, the ventricles return to shape after systole. Blood pressure is then higher in the arteries than in the ventricles, so blood tries to flow back to the ventricles. The cuplike valves in the pulmonary trunk and aorta catch the blood, and this closes the valves. The increased pressure of the blood in these vessels is afterload. This pressure on the valves must first be overcome before the ventricles can eject additional blood. This is like someone going through a swinging door and then standing on the other side. You must apply enough pressure on the door to move the door and the person ahead of you before you can go through the door yourself.

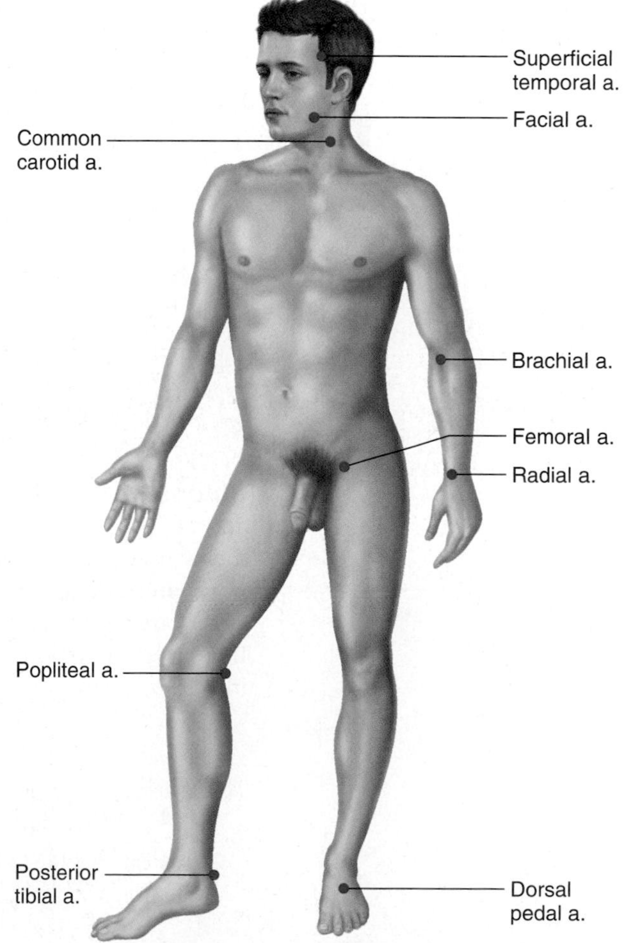

FIGURE 10.15 Arterial pressure points.

10.11 learning **outcome**

Summarize nervous and chemical factors that alter heart rate, stroke volume, and cardiac output.

Heart Regulation

Adjusting either the heart rate or the stroke volume can change the cardiac output. Anything that changes the heart rate is called a **chronotropic factor.** Positive factors increase the heart rate, while negative factors decrease the rate. Consider the examples below.

Chronotropic Factors of the Autonomic Nervous System As mentioned before, the nervous system does not initiate heart contractions, but it can modify their frequency. Two centers located in the medulla oblongata are the **cardiac accelerator center** and the **cardiac inhibitory center:**

- The cardiac accelerator center uses sympathetic neurons to stimulate the SA and AV nodes to speed up the heart rate.
- The cardiac inhibitory center uses parasympathetic neurons of the vagus nerve to keep the SA node at 70 to 80 beats/min (vagal tone). If the vagus nerve is severed, the SA node will typically set the pace at 100 beats/min.

The three types of sensors that feed information to the centers in the medulla oblongata are the proprioceptors, baroceptors, and chemoceptors. They are explained in the following list:

proprioceptors:
pro-PREE-oh-sep-torz

- **Proprioceptors.** As you will remember from the nervous system chapters, these nerve endings are located in the body's muscles, joints, and tendons. The information they send alerts the centers to any change in the body's activity level.
- **Baroreceptors.** These sensors are located in the aorta and carotid arteries. See **Figure 10.16**. They alert the centers to any changes in blood pressure. If blood pressure falls, the cardiac accelerator center stimulates the SA and AV nodes to increase the heart rate in an effort to restore blood pressure to homeostasis.
- **Chemoreceptors.** These sensors monitor pH, carbon dioxide, and oxygen in the blood. They are located at the aortic arch, on carotid arteries, and in the medulla oblongata. See **Figure 10.16**. Although these sensors do have an effect on heart rate, they are much more important for setting the respiratory rate. You will learn more about how they work in the respiratory system chapter.

Chronotropic Effects of Chemicals As you may suspect, epinephrine from the adrenal medulla has a positive chronotropic effect on heart rate, as it complements the sympathetic nervous system. Other positive chronotropic chemicals are caffeine, norepinephrine, nicotine, and thyroid hormone. Potassium ions have a negative chronotropic effect because they interfere with repolarization of the myocardium. Increased potassium can slow the heart. Potassium ions are used as part of a lethal injection to stop a heart.

spot check ⑩ What would be the effect on heart rate of two cups of coffee and a cigarette for breakfast? Explain.

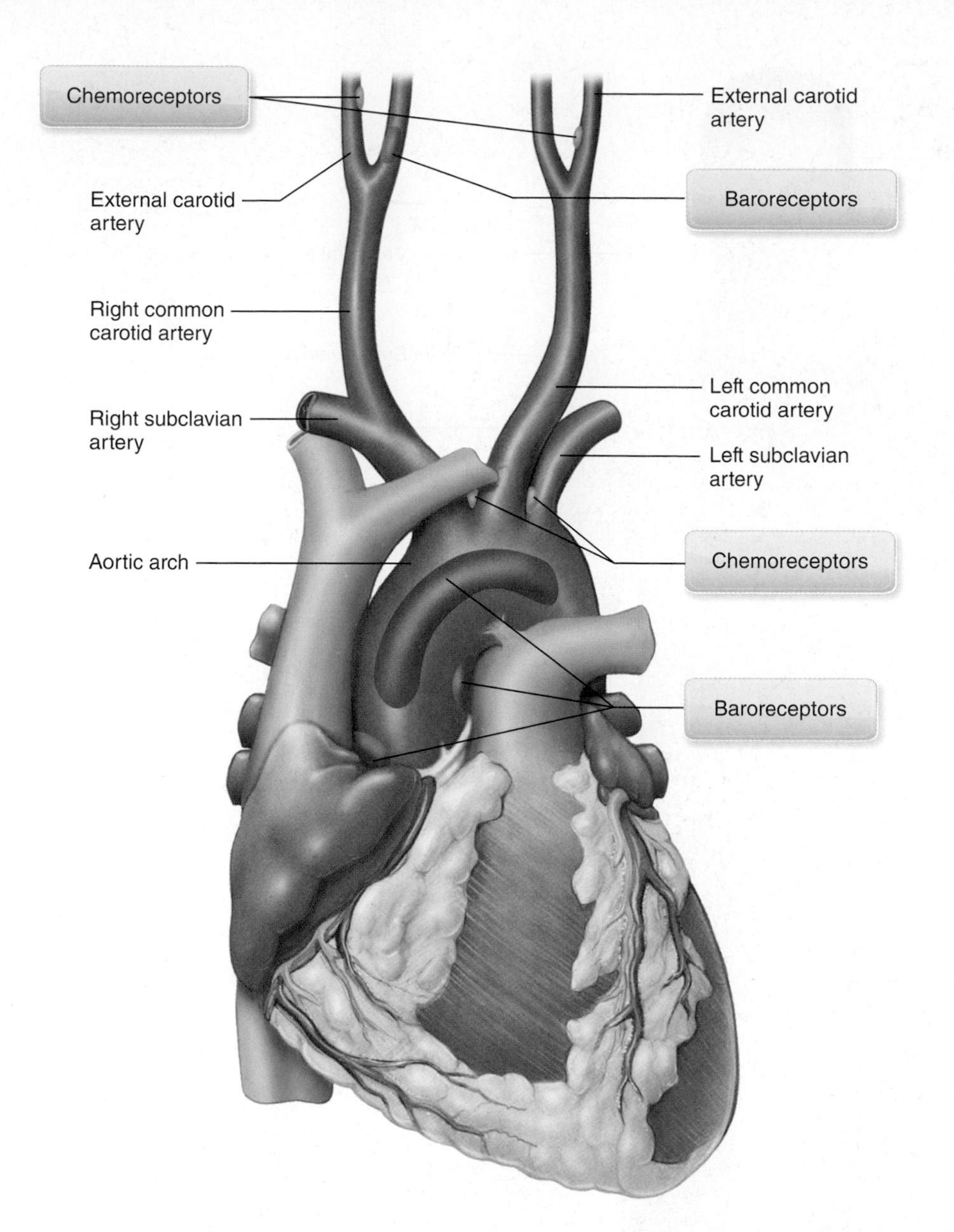

FIGURE 10.16 Baroreceptors and chemoreceptors in arteries superior to the heart.

You will be exploring more about heart physiology once you have studied blood vessel anatomy. Then you will be able to begin to understand how the heart and vessels work together to maintain blood pressure and flow.

Vessel Anatomy

10.12 learning outcome

Locate and identify the major arteries and veins of the body.

The three types of blood vessels in the body differ by location and direction of flow, as explained in the following list:

- **Arteries** carry blood away from the heart to capillaries.
- **Capillaries** allow for the exchange of materials between the blood and tissues.
- **Veins** deliver blood from the capillaries back to the heart.

See **Figures 10.17** and **10.18** for the major arteries and veins in the body.

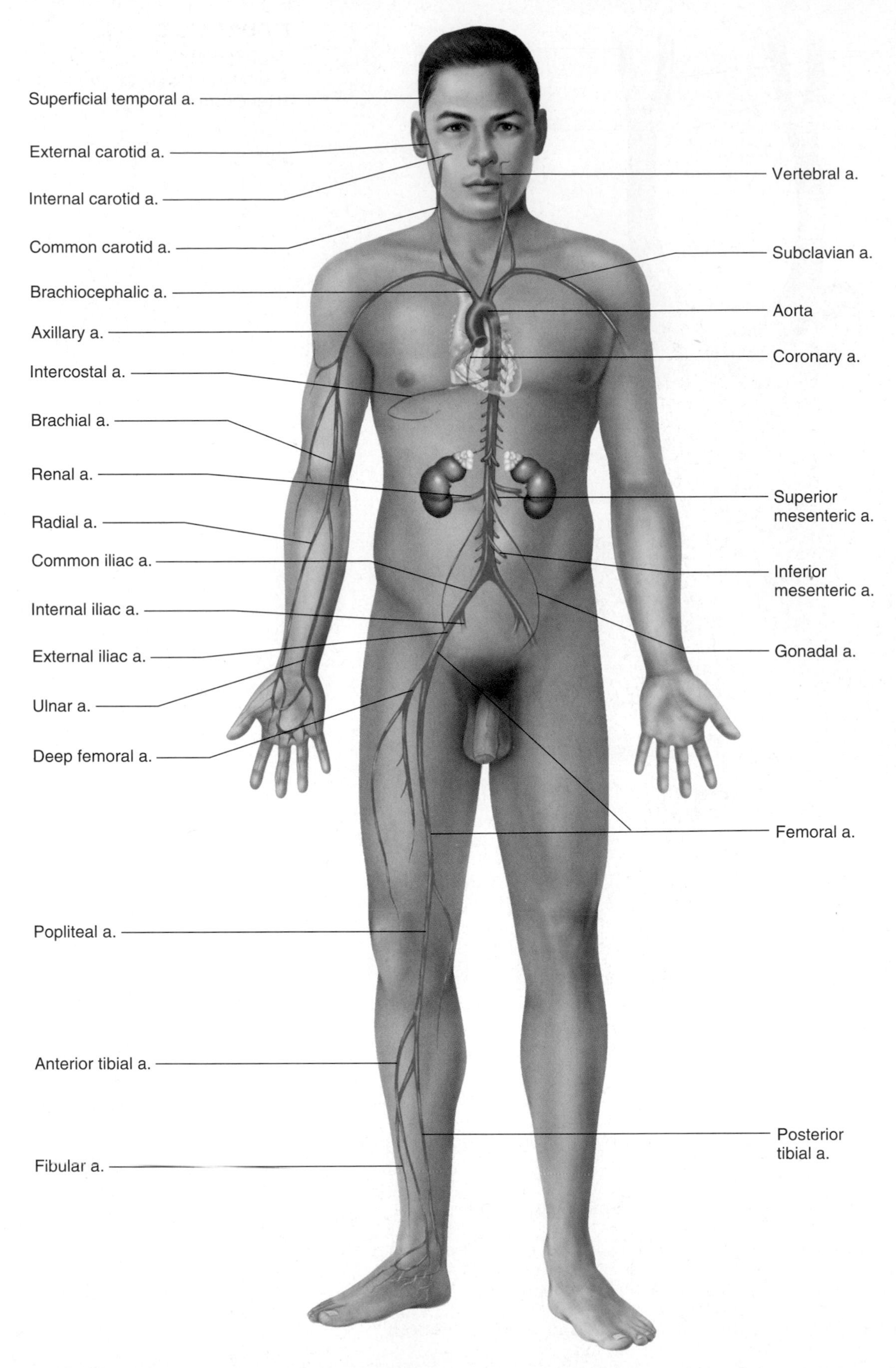

FIGURE 10.17 The major systemic arteries. Many vessels are shown on only one side for clarity but occur on both sides. (a. = artery.)

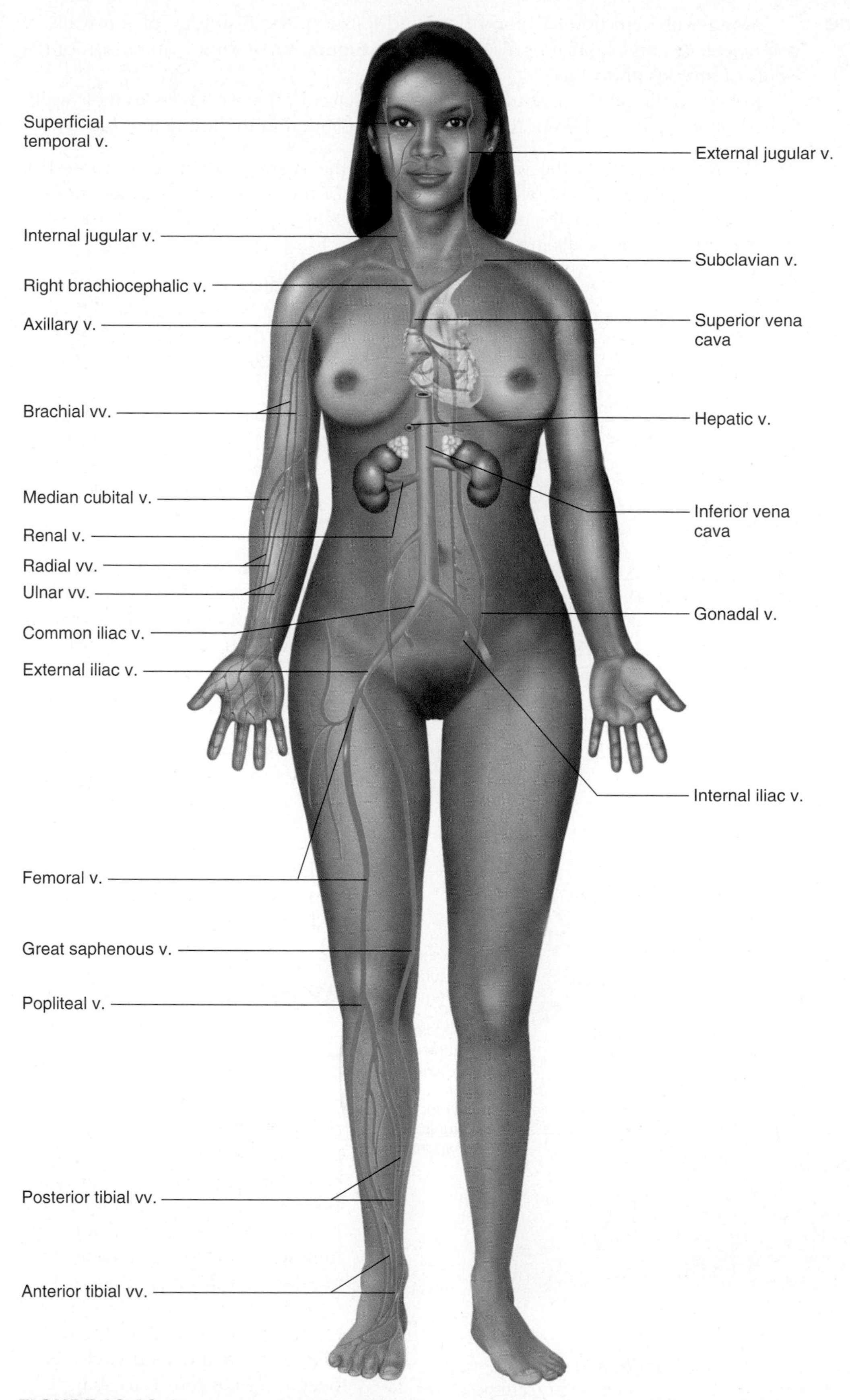

FIGURE 10.18 **The major systemic veins.** Many vessels are shown on only one side for clarity but occur on both sides. (v. = vein, vv. = veins.)

10.13 learning **outcome**

Compare the anatomy of the three types of blood vessels.

Along with direction of blood flow and location, the histology of the walls of arteries, veins, and capillaries also differs. See **Figure 10.19** for a comparison of the walls of arteries and veins.

Notice, in the figure, that arteries and veins have three basic layers to their walls, called tunics. These different tunic layers are explained in the following list:

- **Tunica externa** is the outermost layer of the vessel wall. It is composed of loose connective tissue, which serves to anchor the vessel to surrounding tissue and provide passage for nerves and small vessels supplying blood to the external wall. Internal walls are fed by diffusion.
- **Tunica media** is the middle layer of the vessel wall. It is the thickest layer, composed mostly of smooth muscle. This tunic is more muscular in arteries than veins of comparable size. There may be elastic fibers in this layer, depending upon the vessel. You will cover this shortly.
- **Tunica interna** is the lining of the vessel wall. It is composed of **endothelium** (simple squamous epithelial tissue and fibrous tissue). It is vital that this layer is very smooth and secretes a chemical to repel platelets so that blood can easily flow through the vessel.

Arteries and veins have subtypes based on size. Below, you will explore examples of each type of vessel and the subtypes in the order of blood flow.

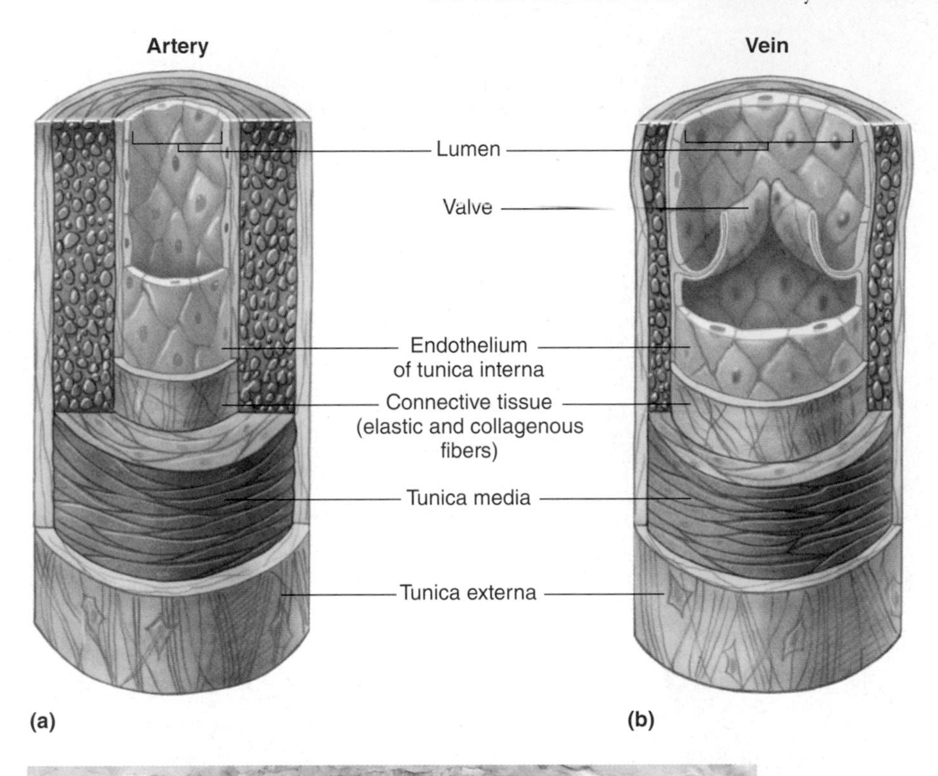

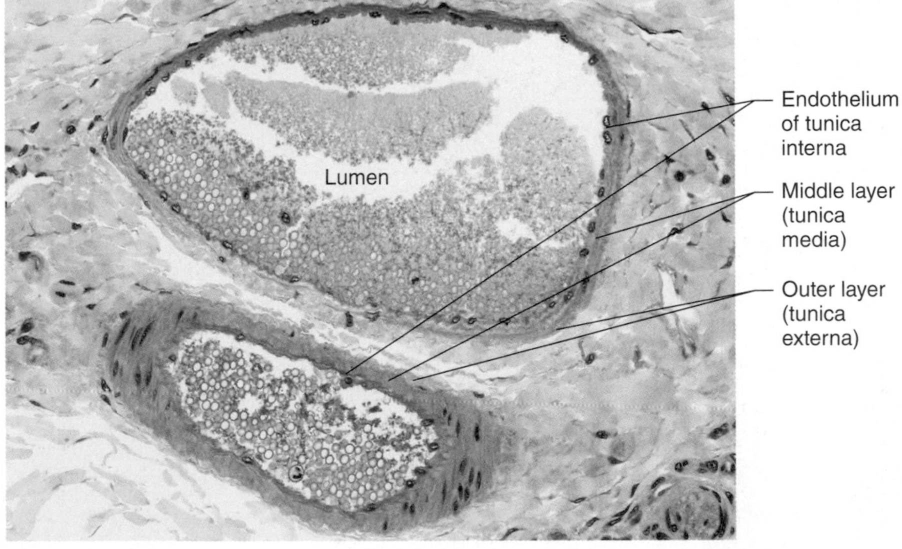

FIGURE 10.19 The structural differences of arteries and veins: (a) artery, (b) vein, (c) micrograph showing cross sections of a small artery (arteriole) at the bottom and a small vein (venule) at the top.

Arteries

Conducting Arteries These are the largest of the arteries. Examples of conducting arteries include the pulmonary arteries, the aorta, and the common carotid arteries. They carry blood away from the heart. Because of their proximity to the heart, they need to withstand the high pressure generated by ventricular systole. For this reason, they have the most muscle and elastic fibers in their walls so that they can expand with each heartbeat and then return to shape.

Distributing Arteries These arteries are medium-size. They distribute blood from the conducting arteries to organs. Examples include the hepatic artery, which carries blood to the liver, and the renal arteries, which carry blood to the kidneys. They have some elastic fibers in their walls to hold their shape, but they do not need to expand as much as conducting arteries with every heartbeat.

Resistance Arteries These are the smallest of the arteries. Examples are the small **arterioles** that deliver blood to the capillaries. Arterioles have little, if any, elastic fibers. Each arteriole can feed a bed of approximately 100 capillaries. **Precapillary sphincters** (circular muscles) in the arterioles open or close to regulate blood flow to the capillaries. See **Figure 10.20**.

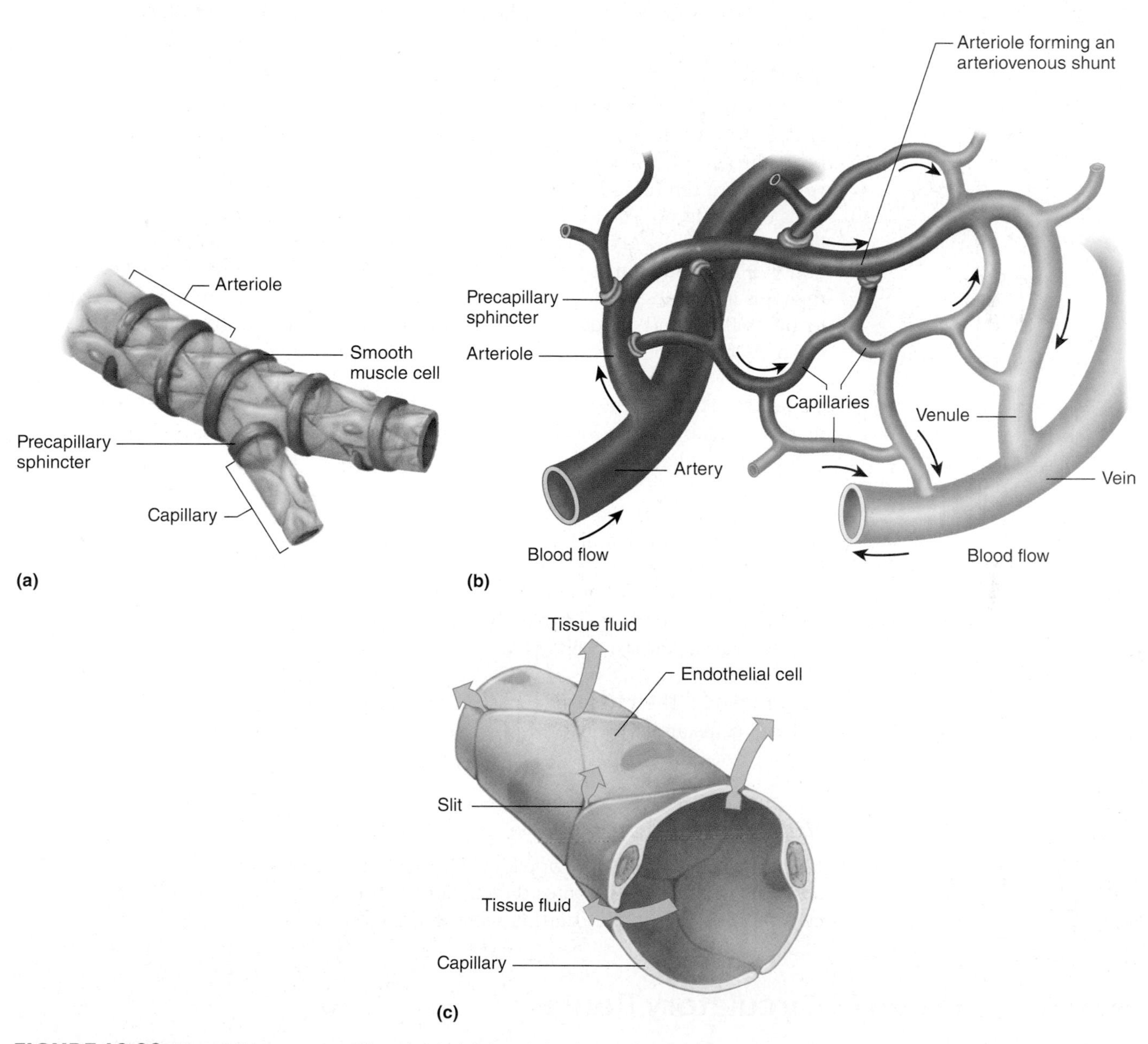

FIGURE 10.20 Blood flow to a capillary bed. (a) Arterioles have circular, smooth muscle cells that act as precapillary sphincters to control blood flow to capillaries. (b) Some arterioles connect directly to small veins (venules) forming an arteriovenous shunt. (c) Capillary walls are a single endothelial cell thick, allowing for the exchange of materials between blood and tissue fluid.

Capillaries

Microscopic capillaries are the smallest of all the vessels. Their one wall is composed of endothelium—only one cell thick with its basement membrane (this is similar to epithelial tissue, discussed in Chapter 2). See **Figure 10.20c**. This thin-walled anatomy is necessary so that fluids and other materials can be exchanged between the tissues and the capillary blood. For example, oxygen is loaded to the red blood cells in the capillaries in the lungs, while carbon dioxide is being unloaded from the capillary blood to the lungs so that it can be exhaled. Blood cells move through the capillaries in single file to ensure the maximum transfer of materials.

The body has approximately a billion capillaries arranged in capillary beds, yet not every cell is bordered by a capillary. How materials are transferred between the capillary blood and these cells is explained in the lymphatic system chapter. From the capillaries, blood moves to the smallest of the veins—the **venules.**

Veins

The volume of the vessels continually decreases as blood moves from larger to smaller arteries to the capillaries. The opposite is true as blood continues on its path to the heart through the veins. Small veins called *venules* lead to medium and then large veins before returning blood to the heart. As the diameter of the veins increases, so does their volume. The total volume of all the veins is greater than that of the arteries. If you remember the discussion earlier about the concepts of volume and pressure, you will understand that the increased volume means blood pressure is less in veins than it is in arteries. The decreased pressure means the walls of the veins do not need to be as thick and sturdy as those of the arteries. Arterial walls will hold an artery open even if empty. The thinner walls of veins collapse when a vein is empty.

Venules These are the smallest of the veins, and they receive blood from the capillaries. Unlike the case with larger veins, there is no smooth muscle in the tunica media of venules.

Medium Veins Unlike venules, medium veins do have smooth muscle in the tunica media of their walls. Examples of medium veins are the radial and ulnar veins in the forearms. Many of these veins, especially in the limbs, have valves formed by folds of the tunica interna. See **Figure 10.19b**. The valves help direct the flow of blood to the heart by preventing backflow. You will see how this works later in the chapter, in the section "Venous Return."

Large Veins These veins have some smooth muscle in all three tunics. Examples of large veins are the venae cavae, pulmonary veins, internal jugular vein, and renal veins.

Vessel Physiology

Blood vessels serve as the pipelines in the heart's distribution system for blood. The anatomy of the various routes directly affects the system physiology. Below, we discuss the possible routes, starting with coronary circulation.

10.14 learning **outcome**

Describe coronary and systemic circulatory routes.

Circulatory Routes

Coronary Route The heart makes up only 0.5 percent of the body's weight, yet receives 5 percent of the circulating blood. This rich blood supply is another reason the heart can stay aerobic. The heart's myocardium is not fed by the blood passing through the chambers. It has its own circulation route composed of coronary arteries and veins. See **Figure 10.21**.

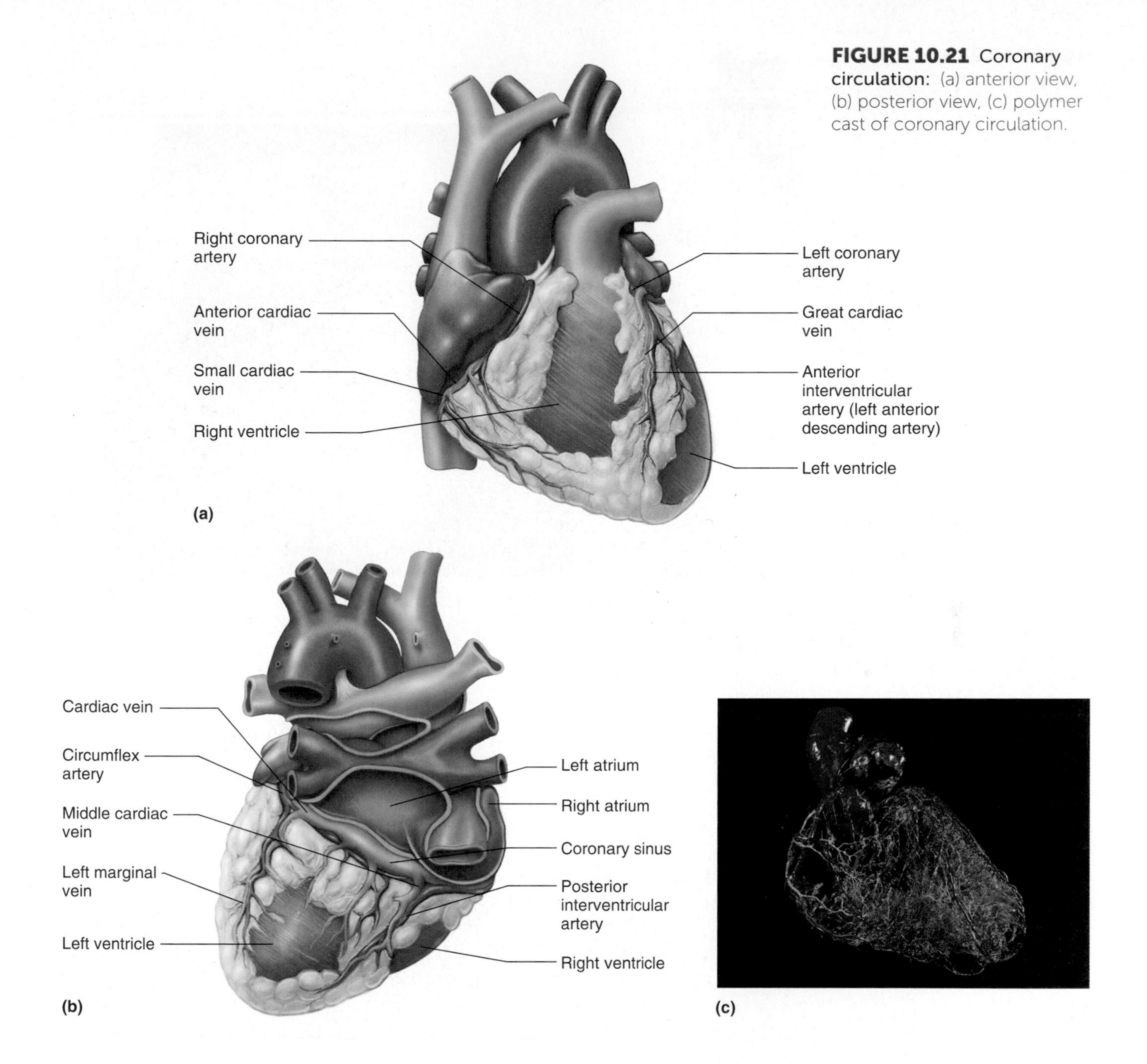

FIGURE 10.21 Coronary circulation: (a) anterior view, (b) posterior view, (c) polymer cast of coronary circulation.

If you take a close look at **Figure 10.22**, you will see two openings within the aortic valve. The pressure of blood filling the cups of the valve as it tries to return to the heart during ventricular diastole forces oxygen-rich blood into the **right** and **left coronary arteries.** These two arteries then branch to form the coronary arteries that lead to capillary beds in the heart's tissues. See **Figure 10.21c.**

Twenty percent of the blood from the capillaries is directly returned to the right atrium of the heart from small veins. The rest of the blood in coronary circulation is collected by the great cardiac vein and the middle cardiac vein and then emptied into the coronary sinus before entering the right atrium. See **Figure 10.21**.

spot check 11 What is the term for the pressure of blood filling the cups of the valve as it is trying to return to the heart during ventricular diastole?

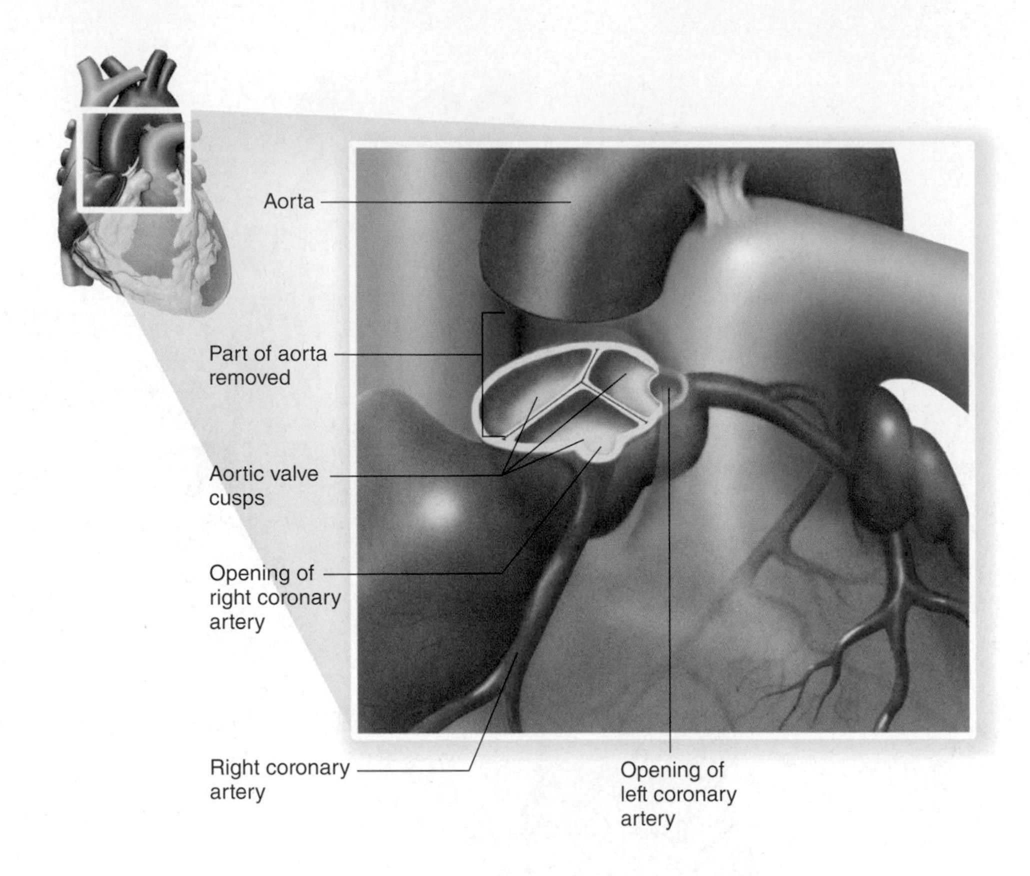

FIGURE 10.22 Superior view of the aortic valve with a section of the aorta removed.

Systemic Routes These routes carry blood from the heart to tissues in the body (other than the heart) and back again. The simplest route goes like this:

Heart → arteries → capillaries → veins → heart

Notice that capillaries are mentioned only once in this route. Consider the following example of this circulatory route.

Blood leaves the heart's right ventricle to travel through the pulmonary trunk to pulmonary arteries, to arterioles, to capillaries covering air sacs (alveoli) in the lungs. Oxygen is loaded to the blood within the capillaries, and carbon dioxide is unloaded. From these capillaries, blood travels through venules to pulmonary veins to the left atrium of the heart. See **Figure 10.23**.

warning

The red and blue colors for the vessels in Figure 10.23 indicate the amount of oxygen in the blood. Red is oxygen-rich, while blue is oxygen-poor. Some students make the mistake of thinking arteries are always shown in red and veins are always shown in blue. Here is the prime example of the error of that misconception: The pulmonary arteries are shown in blue because they are carrying oxygen-poor blood *away* from the heart to the lungs. Pulmonary veins are shown in red because they are *returning* oxygen-rich blood to the heart. It is better to know arteries and veins by direction of flow than by color in an illustration. Arteries carry blood away from the heart, and veins carry blood to the heart.

Alternative Routes There are two types of alternative systemic routes in the body: portal routes and anastomoses.

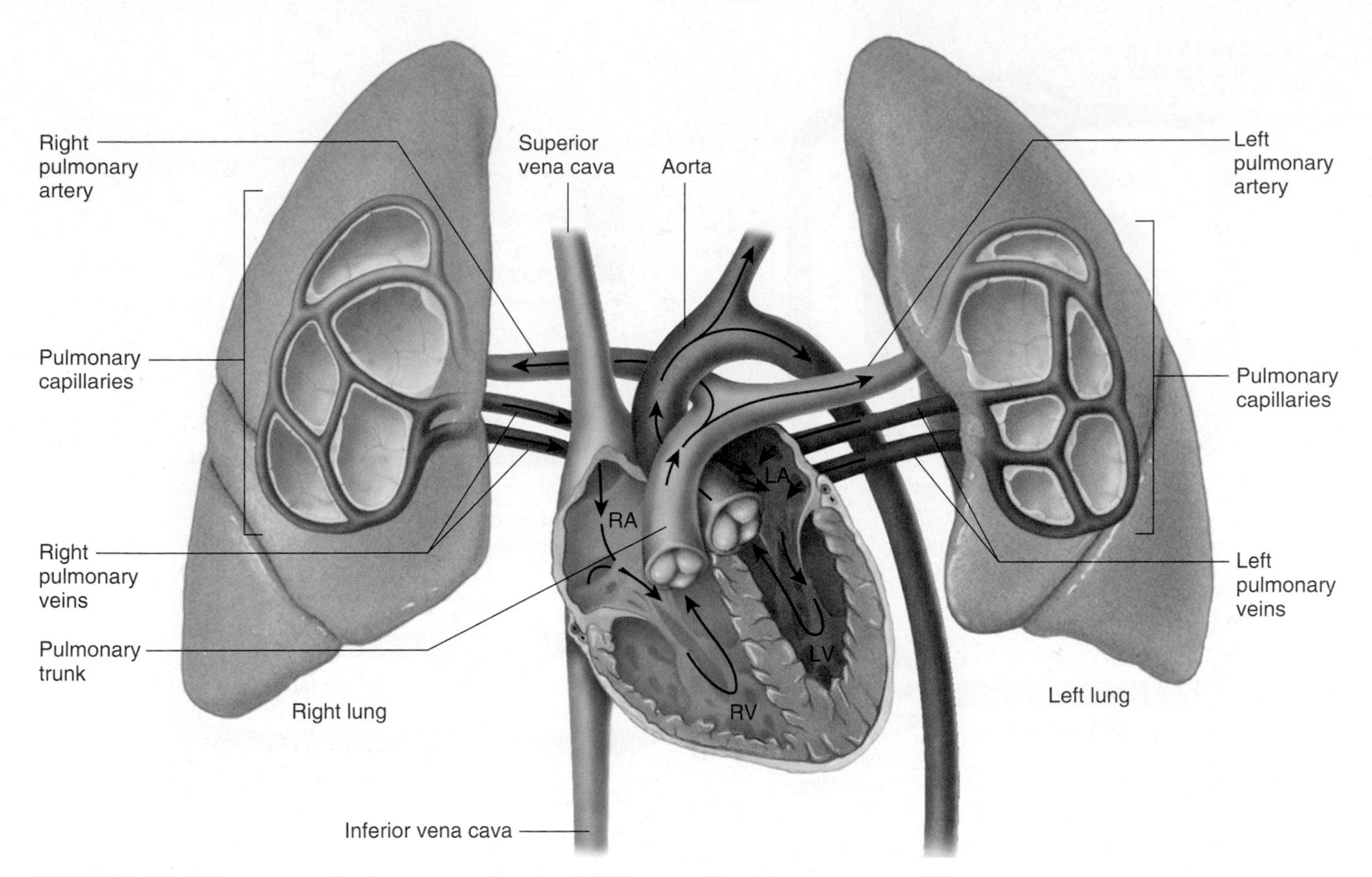

FIGURE 10.23 **Pulmonary circulation (pulmonary circuit).** The microscopic capillaries and the microscopic air sacs in the lungs are greatly enlarged for clarity.

Portal routes A portal route contains two capillary beds before blood is returned to the heart. In this type of circulation, blood travels in the following route:

Heart → arteries → capillaries → intervening vessels → capillaries → veins → heart

Because there are two capillary beds, this type of route allows materials to be exchanged twice between the blood and tissues before returning to the heart. Examples of portal routes can be found between the hypothalamus and pituitary gland (covered in the endocrine system chapter), in the kidney (covered in the excretory system chapter), and between the intestines and the liver.

To better understand portal routes, take a close look at the example of the **hepatic portal route** between the intestines and the liver. See **Figure 10.24**. In this route, blood travels from the heart to arteries to capillary beds in the small intestine and other digestive organs. Here, digested nutrients are absorbed into the blood through capillaries. Blood then travels through small veins leading to the hepatic portal vein (intervening vessels) to capillary beds in the liver, where the nutrients are processed. Blood then exits the liver via the hepatic vein on its way back to the heart. Again, blood travels through two capillary beds in this route before returning to the heart. Understanding this route will be important when you study the digestive system in Chapter 13.

Anastomoses This type of alternative route involves vessels merging together. The three types of anastomoses are explained in the following list:

anastomoses:
ah-NAS-tah-MO-seez

1. **Arteriovenous anastomoses.** This type of route is often called a **shunt.** It merges an artery with a vein, skipping a capillary bed. You may ask why this would ever be done. After all, the point of circulation is to deliver materials

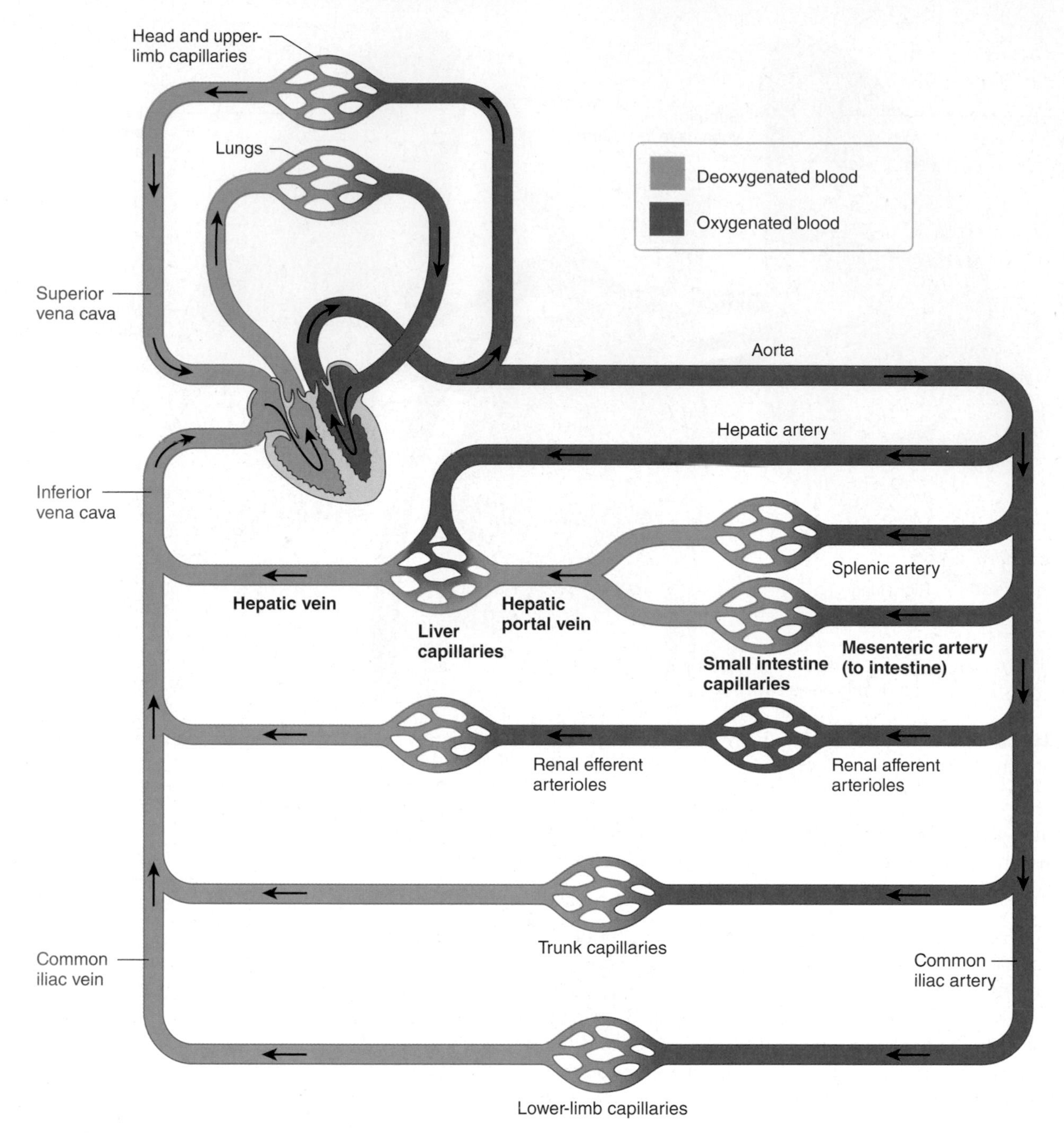

FIGURE 10.24 Hepatic portal route. A diagram of circulatory system routes. Note that blood travels from the mesenteric artery to capillaries in the small intestine, where nutrients are absorbed, and on to the hepatic portal vein that delivers blood to a second bed of capillaries in the liver, where the absorbed materials are processed.

to the tissues and take other wastes away, and that happens in the capillaries. This type of route is used in the fingers, palms, toes, and ears in conditions of extreme cold. The capillary beds are then temporarily bypassed so that heat is not lost. This is a protective mechanism. See **Figure 10.20b**.

2. **Arterial anastomoses.** This type of circulatory route merges two arteries to provide collateral routes to the same area. These routes can be found in the heart, to make sure all parts of the heart are adequately fed, and at joints, where movement may block one of the routes.
3. **Venous anastomoses.** This type of route is the most common of the anastomoses. It merges veins to drain an organ.

Venous Return

10.15 learning outcome

Explain how blood in veins is returned to the heart.

You now understand that the pressure created by ventricular systole propels blood through the arteries and capillaries, but what causes the blood to move through the veins on its way back to the heart?

Five mechanisms aid in **venous return**:

- **Pressure gradient.** Even though there is less pressure in the veins than in the arteries, the pressure in veins due to the action of the heart does propel blood toward the heart.
- **Gravity.** Blood moves through veins above the heart due to gravity, and it flows downhill.
- **Thoracic pump.** The chest expands every time a breath is inhaled. This increases the volume and decreases the pressure within the chest. As air rushes in to equalize the pressure, blood in the veins of the abdominal cavity is sucked into the inferior vena cava of the thoracic cavity by the same principle.
- **Cardiac suction.** You were introduced to this mechanism during the cardiac-cycle discussion. Atria return to shape during atrial diastole. This creates less pressure in the atria than in the superior and inferior venae cavae and the pulmonary veins, so blood is sucked into the atria from the veins.
- **Skeletal muscle pump.** This mechanism is especially effective in the limbs. Skeletal muscle action massages blood through the veins, while the valves in the veins prevent backflow. See **Figure 10.25**.

As you are aware, ventricular contractions force blood out of the heart under pressure. That same pressure causes blood to continue on its journey through the arteries, capillaries, and veins (to a lesser extent) as blood is returned to the heart. We focus next on the dynamics of blood pressure resistance and flow.

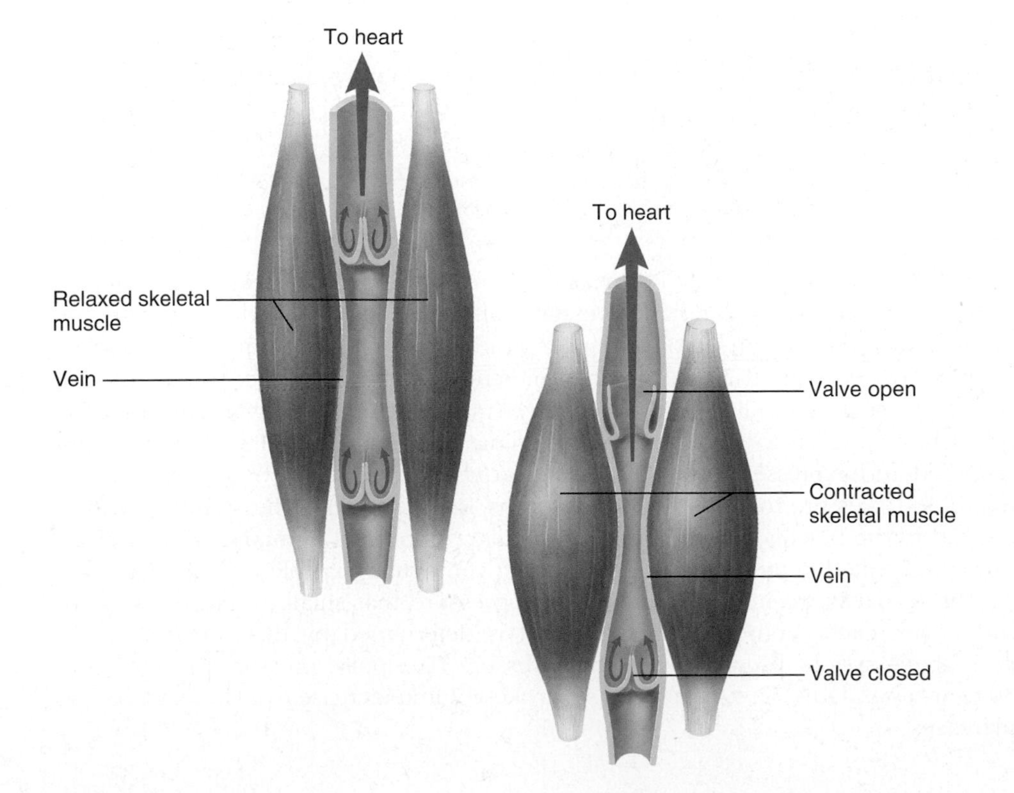

FIGURE 10.25 The skeletal muscle pump.

10.16 learning **outcome**

Explain the relationship between blood pressure, resistance, and flow.

Blood Pressure, Resistance, and Flow

Blood flow is the amount of blood flowing to an area in a given amount of time, and it is usually expressed in milliliters per minute (mL/min). Blood pressure is the force of the blood against the vessel walls, and it is dependent on three things: cardiac output, blood volume, and resistance. You can use a garden hose as an analogy for the vessel. As you read the following list, think of the pressure and flow of water as it moves through a hose:

- Cardiac output is a factor because it takes into account the force of ventricular contractions. (How strong is the pump for the hose?)
- Blood volume makes a difference because more blood exerts a greater force on the vessel walls. (How far is the hose's water faucet turned on?)
- Resistance makes a difference in these three ways:
  1. **Blood viscosity (thickness).** The amount of albumins and red blood cells determines the thickness of blood. Thicker blood offers more resistance to flow and requires more pressure to get it to move. (What if the hose contained honey instead of water?)
  2. **Vessel length.** The greater the vessel length, the more friction occurs between the blood and the vessel walls. Friction slows the blood. (What would be the force of the same amount of water coming out of a 100-foot hose versus a 10-foot hose?) Vessel length becomes a factor in people who are obese. More pressure is needed to propel blood through their longer system of vessels to feed their increased tissue mass.
  3. **Vessel radius.** Vessel radius becomes a factor because the smaller the radius, the more blood comes in contact with the walls of the vessel. (What happens to the amount of flow if the hose diameter is reduced?) As you will see shortly, vessel radius (through vasoconstriction and vasodilation) can be controlled in several different ways to regulate blood pressure.

spot check **12** Would blood flow be faster in the aorta or the femoral artery? Explain in terms of resistance.

10.17 learning **outcome**

Describe how blood pressure is expressed and how mean arterial pressure and pulse pressure are calculated.

sphygmomanometer: SFIG-moh-mah-NOM-ih-ter

Blood pressure is usually measured in the brachial artery by using a **sphygmomanometer** that has a pressure cuff and a device to inflate the cuff. The original instruments (sphygmomanometers) measured blood pressure as the amount of pressure necessary to lift a column of mercury a certain distance. So the units for blood pressure are millimeters of mercury (mmHg). Newer sphygmomanometers typically use a dial to indicate pressure using the same units. Two pressures are recorded and expressed as systolic pressure/diastolic pressure. For example, a normal blood pressure for a 20- to 30-year-old is 120/72 mmHg. The systolic pressure created in the brachial artery during ventricular systole is 120 mmHg. The diastolic pressure created in the brachial artery during ventricular diastole is 72 mmHg.

Pulse pressure indicates the surge of pressure that small arteries must withstand with each ventricular contraction. It is determined by this equation: Pulse pressure = systolic pressure – diastolic pressure. The pulse pressure for our 20- to 30-year-old is 120 – 72 = 48 mmHg. As stroke volume increases, pulse pressure also increases.

clinical point

Prehypertension is diagnosed when the resting systolic pressure is 120 to 139 mmHg and/or diastolic pressure is 80 to 89 mmHg. Hypertension results when resting pressures are greater than 140/90 mmHg. It weakens small arteries and can cause aneurisms, which are a weakness in arterial vessel walls that can balloon out and even rupture. Hypotension is chronic low pressure. It can be caused by dehydration, blood loss, and anemia.

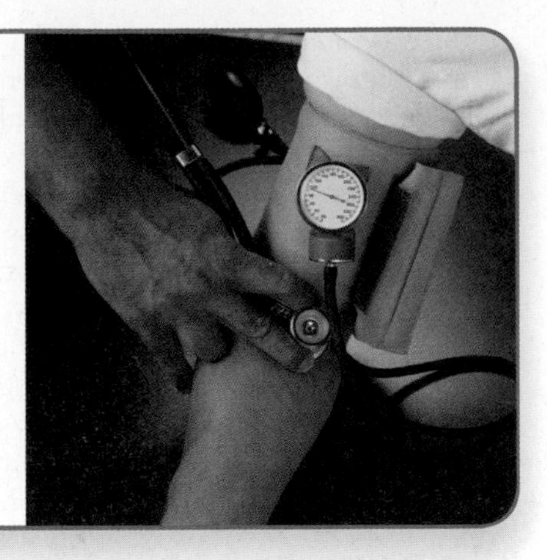

Mean arterial pressure (MAP) is the average pressure arteries must be able to withstand. It is determined by the following equation: MAP = diastolic pressure + 1/3 pulse pressure. But if this is an average, why isn't the equation simply systolic pressure + diastolic pressure divided by 2? This would be the case if the amount of time of ventricular systole in the cardiac cycle was the same as the amount of time of ventricular diastole, but the cardiac cycle includes a rest between heartbeats. So there is more time in a relaxed state than in a contracted state. Blood pressure is further altered by arteries expanding and then recoiling, so pressure does not reach a peak and then drop to zero. The MAP for the 20- to 30-year-old in our example is $72 + \frac{1}{3}(48) = 88$ mmHg.

spot check 13 What are the pulse pressure and mean arterial pressure for a 25-year-old individual whose blood pressure is 150/96 mmHg?

Regulation of Blood Pressure and Flow

10.18 learning outcome

Explain how blood pressure and flow are regulated.

Blood pressure and flow can be regulated locally (by the tissues), hormonally (by the endocrine system), or neurally (by the nervous system). How each of these works is discussed below.

Local Control Tissues can autoregulate their own blood supply in four different ways:

1. **Opening of precapillary sphincters.** If there is inadequate blood flow, wastes such as carbon dioxide, lactic acid, and hydrogen ions build up in the tissues. The presence of wastes stimulates vasodilation and the opening of precapillary sphincters, creating more blood flow. Increased blood flow to the capillaries brings more oxygen to the tissues (to remove the lactic acid) and removes the wastes. The waste removal ends the vasodilation and closes the precapillary sphincters. Think about this: There are about a billion capillaries in the human body arranged in beds. Sphincters control the blood flow to these capillary beds, and three-fourths of the capillary beds are empty at any given time because the sphincters are closed to them. You do not have enough blood to fill all of the capillaries, nor do you have a heart strong enough to pump all the blood that would be necessary to fill all of the capillaries at any one time. This is a lot like a sprinkler system. Imagine a house on a lake. A lake pump is used to water the yard. All of the grass needs to be watered, but Nadine, the owner, does not have a lake pump big enough to run the 70 sprinklers necessary to water all of the grass at one time, so her sprinklers are arranged in zones. At 5 A.M., zone 1

turns on its seven sprinklers, and zone 1 grass is watered for 30 minutes. When that is finished, the lake pump sends water to the seven sprinklers covering zone 2's grass; and so on. By 10 A.M., all of the grass has been watered one zone at a time. All the grass is satisfied, and Nadine did not have to buy a bigger pump. The body's system is even better! Nadine's grass gets watered on a schedule whether it needs it or not. Grass in shady areas is probably overwatered, and grass in sunny areas is underwatered. Local control related to the body's capillaries, however, is based solely on need. When waste builds up in an area, precapillary sphincters open and blood flow increases to take the waste away—immediately—when it is needed, not on some arbitrary schedule. When the wastes are gone, the precapillary sphincters close, so blood is diverted to other areas where it is needed. All the tissues are fed and wastes are taken away efficiently without any more strain on the pump (heart) than is absolutely necessary.

2. **Inflammation.** In response to an injury or the presence of a pathogen, damaged tissues or basophils release vasodilators in the local area to stimulate inflammation. This inflammatory response dilates vessels and makes them more permeable, thereby increasing blood flow to the area. The increased blood flow delivers more white blood cells to the area to fight pathogens and accounts for the redness, heat, and swelling associated with inflammation.

3. **Reactive hyperemia.** If circulation to an area is cut off for a time and then restored, vessels overdilate, flushing an area with blood. An example of this can be seen when you cross your legs at the knee. When you remove the top knee, you may notice a red area on the knee where the leg rested. Here is another example: If you are out in the cold so long that vessels have constricted in the skin to save heat for the core (the skin appears white) and then you come inside where it is warm, the vessels in the skin overdilate and the skin appears red.

4. **Angiogenesis.** Persistent buildup of metabolic waste causes new vessel growth **(angiogenesis)** to increase the blood supply to the area. Heart patients, depending on their condition, may be encouraged to exercise. The buildup of wastes in cardiac muscle tissue promotes the growth of new vessels to provide collateral (additional) circulation in the heart. Fast-growing cancer tumors can also stimulate angiogenesis to feed a tumor and remove its wastes. This is an important area of cancer research. If angiogenesis can be prevented in this case, the tumor can be starved and the tumor's growth may be slowed.

Hormonal Control The endocrine system can also be used to regulate blood pressure and flow by the use of the following four hormones:

- **ADH (also called vasopressin).** As you may recall from Chapter 8, ADH targets the kidney to cause water retention. By preventing water loss in urine, blood volume and therefore pressure are maintained.

- **Aldosterone.** This mineralocorticoid (studied in Chapter 8) targets the kidney to cause sodium to be retained in the blood. Water follows the sodium, so again, by preventing water loss in urine, blood volume and pressure are maintained.
- **Angiotensin II.** Angiotensin II is a vasoconstrictor produced by the liver when blood pressure falls below homeostasis. Widespread vasoconstriction increases blood pressure.

clinical point

An enzyme, ACE, is needed to produce angiotensin II. People who have high blood pressure may be prescribed ACE inhibitors to lower their blood pressure.

- **Epinephrine.** Epinephrine complements the sympathetic nervous system by causing vasoconstriction, which limits blood flow to most vessels except those vessels going to cardiac and skeletal muscle.

Neural Control The brain also regulates blood pressure and flow. The **vasomotor center** in the medulla oblongata uses sympathetic fibers to constrict most vessels except those going to cardiac and skeletal muscles. This center and the cardiac centers of the medulla oblongata are influenced by three reflexes—the baroreflex, chemoreflex, and medullary ischemic reflex—which are explained in the following list:

- **Baroreflex.** Baroreceptors in the aortic arch and carotid arteries (covered earlier in the chapter) constantly send signals. If the pressure is too high, messages from these receptors cause the medulla oblongata's cardiac inhibitory center to increase vagal stimulation of the heart, thereby decreasing the heart rate and blood pressure.
- **Chemoreflex.** Chemoreceptors monitor oxygen, carbon dioxide, and pH. If oxygen levels fall, carbon dioxide levels rise, and pH falls, the vasomotor center will initiate widespread vessel constriction. This increases blood pressure, increasing blood flow to the lungs for more gas exchange. Oxygen levels then rise, carbon dioxide levels fall, and pH also rises.
- **Medullary ischemic reflex.** If blood flow to the brain decreases, the cardiac accelerator center and the vasomotor center send sympathetic signals to increase heart rate and increase vasoconstriction. This increases pressure and blood flow to the brain. The medullary ischemic reflex is effective for even simple actions, like standing suddenly after being in a supine position, such as lying on your back. In this case, change of position reduces blood flow to the brain temporarily.

You have covered a great deal of anatomy and physiology in this system. Now you can put it all together by revisiting Nick and Kate (from Chapter 3) on their morning run to see the effect of exercise on their cardiac output (**Figure 10.26**).

FIGURE 10.26 Nick and Kate on a morning run.

Effects of Exercise on Cardiac Output

10.19 learning **outcome**

Explain the effect of exercise on cardiac output.

Nick and Kate begin their morning run with a resting heart rate and resting cardiac output. Once the run has started, proprioceptors in their joints, muscles, and tendons alert the cardiac accelerator center in the medulla oblongata of the increased activity level. The center sends messages along sympathetic neurons to increase their heart rates.

Meanwhile, Nick's and Kate's muscles are doing increased aerobic respiration, which causes more metabolic wastes to build up in the tissues. Locally, vasodilation and the opening of precapillary sphincters in the muscles increase blood flow, so more of the accumulating wastes are carried away in the blood. This stimulates a chemoreflex to increase blood flow to the lungs. If Nick and Kate make a habit of regular exercise, the chronic buildup of wastes stimulates angiogenesis to improve the collateral circulation in their coronary arteries.

As Nick and Kate continue to run, their muscles are massaging their veins, which increases the venous return to the heart. This added preload results in a greater stroke volume because the heart must pump out what it receives. So Nick's and Kate's heart rates and stroke volumes have increased during the run, and these increases, together, greatly increase their cardiac output. Their cardiac reserve may easily be four to five times larger than their resting cardiac output.

10.20 learning outcome

Summarize the effects of aging on the cardiovascular system.

Effects of Aging on the Cardiovascular System

You have seen how the cardiovascular system works in the previous scenario, but Nick and Kate are relatively young adults. What would be the effects of aging on their cardiovascular systems?

If their cardiovascular systems maintain homeostasis within a normal blood pressure range, there may be little to no effect of aging on their hearts. If Nick or Kate is hypertensive, however, there may be dramatic changes. These changes include the following:

- Vascular resistance increases with age in individuals with hypertension. This increases afterload, which then requires higher diastolic and mean arterial pressure to move the same amount of blood through the system.
- Decreased resting stroke volume decreases cardiac output, which means the heart becomes less efficient.
- The vessels thicken, and become less elastic. This makes the vasculature more susceptible to developing atherosclerosis (discussed in the next section of this chapter).

The effects of lifestyle can also make a difference in how this system ages. Here are some of the effects:

- Physical conditioning can improve aerobic capacity in the elderly by increasing their cardiac output and ability to use oxygen effectively. It is possible for well-conditioned older people to exceed the aerobic capacity of people who are much younger and not as well conditioned.
- Physical conditioning slows or reduces the vascular stiffening that may occur with aging.
- Exercise over a lifetime can significantly increase the collateral circulation in the heart. This provides multiple avenues for blood to feed areas of myocardium that may have otherwise died if a coronary artery becomes blocked.
- A diet low in sodium helps keep blood pressure under control.
- Nonsmokers avoid the associated polycythemia that develops in smokers, whose blood cells carry carbon monoxide instead of oxygen. Not smoking helps to keep the blood from becoming too thick and adding to the workload of the heart.

10.21 learning outcome

Describe cardiovascular disorders.

Disorders of the Heart and Vessels

Atherosclerosis

Atherosclerosis results in the buildup of fatty deposits within arterial walls, which causes the walls to roughen and project to the lumen (open space) within the vessel. See **Figure 10.27**. Atherosclerosis often begins as a result of hypertension or viral infection that weakens the arterial wall. Monocytes stick to the endothelium at the weakened area and then proceed to the tunica media, where they become macrophages. As macrophages, they consume fats and cholesterol from the blood and develop a foamy appearance. The buildup of the fatty deposits **(plaque)** thickens the arterial wall and makes the lining of the artery rough. This obstructs blood flow and provides a surface for platelets to stick. Platelets further complicate the condition by secreting growth factors to stimulate mitosis in the vessel walls, further reducing the size of the lumen. The narrowed, rough interior of the artery is a prime location for developing blood clots to form. If the atheroma (fatty deposit) becomes calcified, the condition is called **arteriosclerosis.**

Myocardial Infarction

Atherosclerosis may block the blood flow in a coronary artery. This lack of blood flow **(ischemia)** can lead to a **myocardial infarction (MI)**—the death of myocardial tissue fed by that artery—commonly called a *heart attack*. Symptoms can vary for men and women, but most people experience:

- A crushing pain in the chest that may radiate to the neck and jaw and down the left arm.
- Shortness of breath or difficult breathing.
- Sweating.
- Nausea.
- A feeling of impending doom.

In addition to the symptoms mentioned above, women may experience lightheadedness, sleep disturbances, indigestion, anxiety, and unusual fatigue lasting for several days.

Angina pectoris is a heaviness or pain in the chest caused by a temporary or reversible myocardial ischemia. The hypoxemia from the reduced blood flow causes the heart to use anaerobic respiration to produce the energy it needs. The buildup of lactic acid produces the associated pain.

Blocked coronary arteries can be treated through **angioplasty** or **coronary bypass surgery.** Angioplasty involves threading a balloon-tipped catheter through a vessel in the leg to the blocked coronary artery. The balloon is inflated within the blockage to open the vessel lumen. A **stent** may be used as part of the balloon system to hold the vessel open after the catheter is withdrawn.

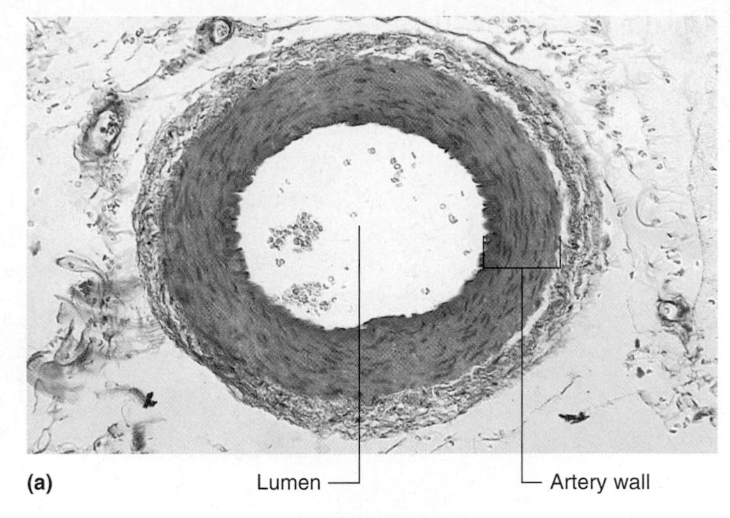

FIGURE 10.27 Atherosclerosis: (a) cross section of a healthy artery, (b) cross section of an artery with atherosclerosis, showing plaque and reduced lumen.

ischemia: is-KEE-mee-ah

Coronary bypass surgery involves harvesting a vessel and inserting its ends before and after the obstruction to effectively bypass the blockage. Several vessels can be used, like the great saphenous vein of the leg or a collateral branch of a mammary artery. But, as you will remember, there is a difference in the anatomy of the walls of arteries and veins in regard to how much pressure they can withstand.

Congestive Heart Failure

Congestive heart failure occurs if one of the ventricles is not working as efficiently as the other. As you can see in **Figure 10.28**, if the right ventricle's output exceeds the left ventricular output, more blood is going to the lungs than can return to the left side of the heart. Blood pressure then builds in the lungs, forcing more fluid out into the pulmonary tissue (pulmonary edema). If the left ventricle's output exceeds the output of the right ventricle, the pressure builds out in the systemic circuit, and systemic edema results (swelling of the hands, fingers, and feet). See **Figure 10.28**.

FIGURE 10.28 Congestive heart failure: (a) pulmonary edema, (b) systemic edema.

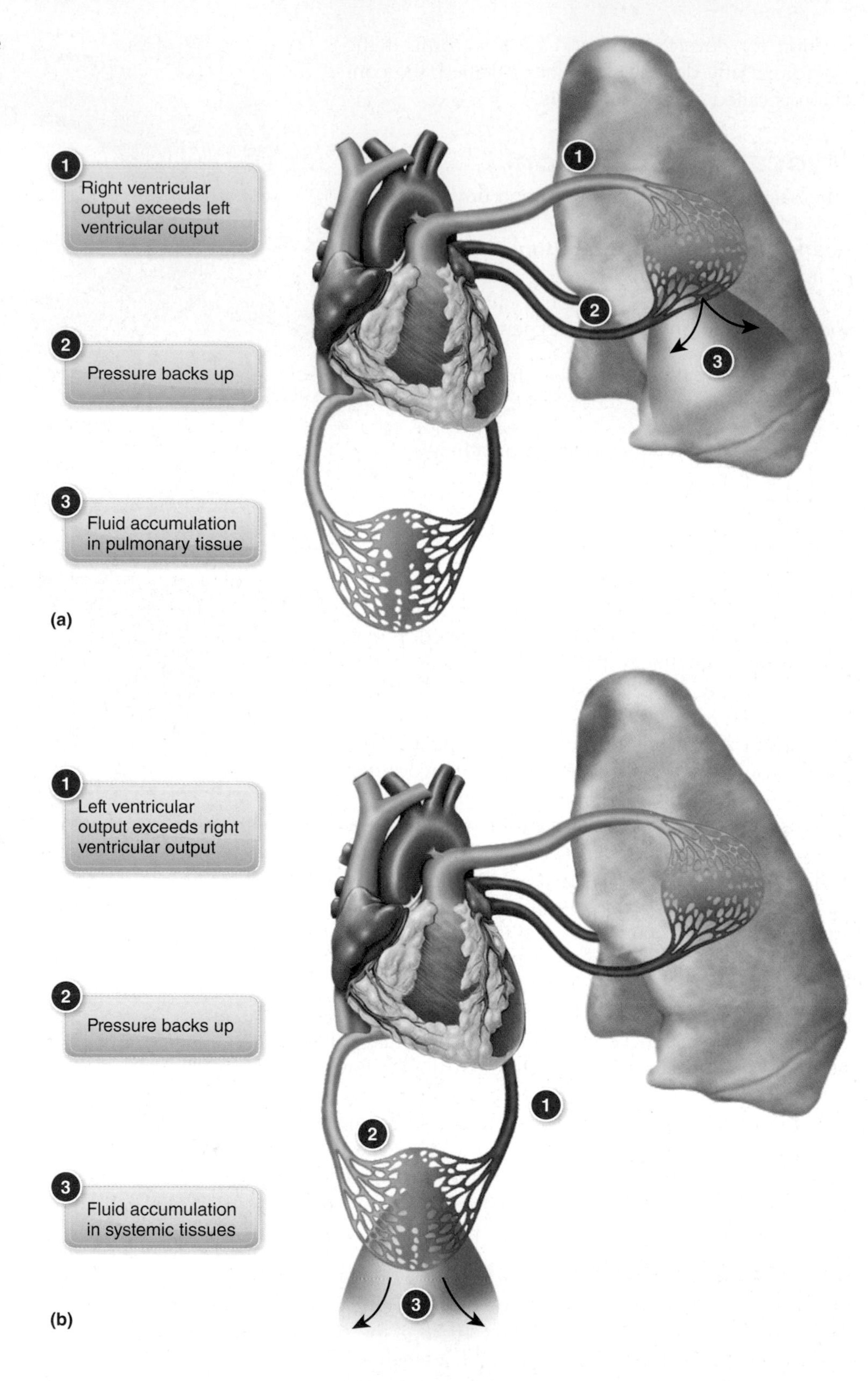

putting the pieces **together**

The Cardiovascular System

Integumentary system

Helps radiate heat of the blood for temperature regulation.

Delivers fluids for sweat production; provides nutrients and removes wastes.

Skeletal system

Red bone marrow produces blood cells and platelets.

Provides nutrients and removes wastes.

Muscular system

Skeletal muscle pump moves blood through veins so that it can return to the heart.

Provides nutrients and removes wastes.

Nervous system

Innervates smooth muscle of blood vessels for vasoconstriction and vasodilation to regulate blood pressure and flow; medulla oblongata regulates heart rate.

Provides nutrients and removes wastes.

Endocrine system

Hormones regulate blood volume and pressure.

Blood transports hormones to their target tissues; provides nutrients and removes wastes.

Lymphatic system

Sends white blood cells to fight pathogens in the cardiovascular system.

Provides fluid for lymph.

Respiratory system

Provides O_2 for cardiovascular tissues and removes CO_2; regulates blood pH.

Blood transports O_2 and CO_2.

Digestive system

Provides nutrients to cardiovascular tissues and to the red bone marrow for hemopoiesis.

Blood transports absorbed nutrients.

Excretory/urinary system

Kidneys dispose of wastes in the blood and regulate blood volume, composition, pressure, and pH.

Blood pressure provides the force for filtration in the kidney.

Reproductive system

Testosterone promotes erythropoiesis.

Blood transports reproductive hormones; vasodilation enables erection.

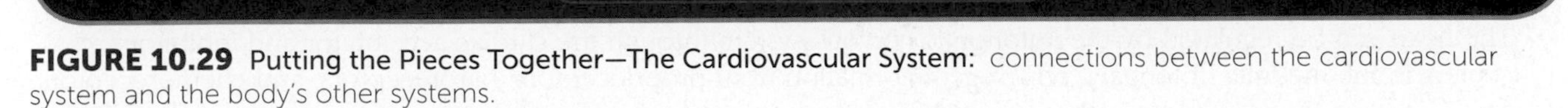

FIGURE 10.29 Putting the Pieces Together—The Cardiovascular System: connections between the cardiovascular system and the body's other systems.

summary

Overview

- The heart serves as a pump to circulate blood through a system of vessels.

Heart Anatomy

- The heart is located in the mediastinum and is tilted, with two-thirds resting left of the midsagittal plane.
- The heart is the size of an adult fist and weighs approximately 300 g, or 10 oz.

Pericardium

- The heart is surrounded by the pericardium.

Heart Wall

- The heart is composed of three walls: the epicardium, the myocardium, and the endocardium.

Chambers and Valves

- The heart has four chambers: two atria, which receive blood; and two ventricles, which pump blood out of the heart.
- There are two AV valves located between the atria and the ventricles, and there are two semilunar valves between the ventricles and the arteries taking blood away from the heart.

Cardiac Muscle Tissue

- Cardiac muscle is striated, is branching, has one nucleus per cell, and has intercalated disks.
- Cardiac muscle tissue is specially adapted to stay aerobic.

Heart Physiology

Blood Flow through the Heart

- Blood flows from the venae cavae to the right atrium, through the tricuspid valve to the right ventricle, through the pulmonary valve to the pulmonary trunk, to pulmonary arteries, and to the lungs. Blood then returns from the lungs through the pulmonary veins, to the left atrium through the bicuspid valve, to the left ventricle through the aortic valve, to the rest of the body.

Cardiac Conduction System

- Cardiac muscle is autorhythmic.
- Electrical impulses start at the SA node and proceed to the right atria and to the AV node, which then send the signals to the AV bundle and bundle branches to the Purkinje fibers.

Cardiac Cycle

- A cardiac cycle is one complete contraction and relaxation of the heart.
- Systole is contraction, and diastole is relaxation.
- The cardiac cycle includes atrial systole, atrial diastole, ventricular systole, ventricular diastole, and a rest.
- The SA node normally sets the cardiac rhythm.

Electrocardiogram

- An electrocardiogram shows the electrical activity of the heart during a cardiac cycle.
- It includes P, Q, R, S, and T waves.

Cardiac Output

- Cardiac output is the amount of blood ejected by each ventricle of the heart each minute.
- Cardiac output is dependent on heart rate and stroke volume.
- $CO = HR \times SV$.
- Stroke volume is dependent on preload, contractility, and afterload.

Heart Regulation

- The heart can be regulated by the autonomic nervous system through the cardiac accelerator and inhibitory centers in the medulla oblongata, which get information from proprioceptors, baroreceptors, and chemoreceptors.
- The heart can be regulated by chemicals.

Vessel Anatomy

Arteries

- The three types of blood vessels are arteries, capillaries, and veins.
- Arteries and veins have three layers (tunics) to their walls, while capillary walls have only one layer.
- Arteries carry blood away from the heart.

Capillaries

- Capillaries are the site of exchange of materials between the blood and tissues.

Veins

- Veins return blood back to the heart.

Vessel Physiology

Circulatory Routes

- The coronary route supplies blood to the heart.
- The typical systemic route includes one capillary bed.
- Alternative routes vary in the number of capillary beds or involve the merging of vessels.

Venous Return

- Blood is returned to the heart through veins by five mechanisms: a pressure gradient, gravity, the thoracic pump, cardiac suction, and the skeletal muscle pump.

Blood Pressure, Resistance, and Flow

- Blood flow is the amount of blood flowing to an area in a given amount of time.
- Blood pressure is the force of blood against the vessel walls.
- Blood pressure is dependent on cardiac output, blood volume, and resistance.
- Blood pressure is measured as systolic pressure/diastolic pressure.
- Pulse pressure = systolic pressure – diastolic pressure.
- MAP = diastolic pressure + $^1/_3$ pulse pressure.

Regulation of Blood Pressure and Flow

- Blood pressure and flow can be regulated locally, hormonally, and neurally.

Effects of Exercise on Cardiac Output

- Exercise increases cardiac output by raising the heart rate and the stroke volume.

Effects of Aging on the Cardiovascular System

- If blood pressure remains normal throughout life, age-related changes to the cardiovascular system may be minimal.
- If an individual is hypertensive, age-related changes may include an increase in vascular resistance, decreased stroke volume, and thicker, less elastic vessels that are prone to atherosclerosis.
- Lifestyle choices like exercising, dieting, and not smoking can make a difference.

Disorders of the Heart and Vessels

Atherosclerosis

- Atherosclerosis is the buildup of fatty deposits within the walls of arteries that reduces the size of the lumen and restricts blood flow.

Myocardial Infarction

- Myocardial infarction is the death of myocardium due to ischemia.

Congestive Heart Failure

- Congestive heart failure occurs when one ventricle is not as efficient as the other ventricle. Blood pressure builds in the circuit before the ventricle, resulting in edema in the preceding tissues.

key words for review

The following terms are defined in the glossary.

afterload
anastomoses
angiogenesis
arrhythmia
atherosclerosis
baroreceptors
cardiac cycle
cardiac output
chronotropic factor
diastole
intercalated disks
ischemia
mean arterial pressure (MAP)
portal route
preload
pulse pressure
stroke volume
systole
tunics
venous return

chapter review questions

Word Building: *Use the word roots listed in the beginning of the chapter and the prefixes and suffixes inside the back cover to build words with the following meanings:*

1. Sudden death of heart muscle: ______________________
2. Inflammation of the heart lining: ______________________
3. Contraction of the pumping chamber of the heart: ______________________
4. Pertaining to veins: ______________________
5. Study of the heart: ______________________

Multiple Select: *Select the correct choices for each statement. The choices may be all correct, all incorrect, or a combination of correct and incorrect.*

1. How is cardiac muscle specialized to avoid fatigue?
   a. It has many large ribosomes to process energy.
   b. It has myoglobin to store glucose.
   c. It has glycogen to store oxygen.
   d. It can use a variety of fuels.
   e. It has adapted to anaerobic respiration almost exclusively.
2. How does the structure of the myocardium relate to its function?
   a. It has myocytes with intercalated disks to insulate electrical messages.
   b. It has collagen fibers to insulate myocytes carrying electrical messages.
   c. It has patches of myocytes that act like neurons sending electrical messages.
   d. It has cells with intercalated disks for fast transmission of electrical signals.
   e. It has elastic fibers that allow the myocardium to come back to shape.
3. Which of the following statements accurately match(es) the cardiac structure to its description or function?
   a. The mitral valve is on the left side of the heart.
   b. The right ventricle has thicker walls.
   c. Oxygen-poor blood travels through the pulmonary valve on its way to the heart.
   d. The AV bundle branches are located in the interatrial septum.
   e. The AV node is located on the interventricular septum.

4. Which of the following statements accurately describe(s) the location or blood flow for the vessel?
   a. The pulmonary veins bring oxygen-poor blood to the lungs.
   b. The aorta carries blood leaving the right ventricle.
   c. The pulmonary arteries carry oxygen-rich blood.
   d. The hepatic portal vein lies between two capillary beds.
   e. The superior and inferior venae cavae deliver blood to the left atrium.
5. What is the anatomy of blood vessels?
   a. Veins and arteries have three tunics.
   b. Capillaries have two tunics.
   c. All veins have valves to prevent backflow.
   d. Conducting arteries have elastic fibers in their walls.
   e. Veins have thinner walls than do arteries of similar size.
6. If Courtney has a blood pressure of 130/88 mmHg, what else is true?
   a. Her MAP is 42 mmHg.
   b. Her pulse pressure is 102 mmHg.
   c. She has hypertension.
   d. She is prehypertensive.
   e. Her ventricular diastolic pressure is 130 mmHg.
7. Given that Norm has a pulse of 136 beats/min and a stroke volume of 80 mL/beat when he works out, what else is true?
   a. Norm's cardiac output is over 10 L/min.
   b. Norm's cardiac output is under 10 L/min.
   c. Norm's cardiac output would decrease at rest.
   d. Norm's cardiac output would increase at rest.
   e. Exercise increases Norm's heart rate but not his stroke volume.
8. How is blood returned to the heart?
   a. Skeletal muscles massage the veins.
   b. Atrial diastole and ventricular diastole suck blood from veins.
   c. Breathing activates the thoracic pump.
   d. Pressure generated during ventricular systole propels blood through the veins.
   e. Valves in veins prevent backflow of blood.
9. Which of the following statements accurately describe(s) cardiovascular disorders?
   a. Pulmonary edema is the result of right ventricular inefficiency.
   b. Systemic edema is the result of left ventricular inefficiency.
   c. Angina pectoris results from the buildup of lactic acid.
   d. Myocardial infarction is death of heart tissue.
   e. Atherosclerosis results in the thickening of the tunica media.
10. What happens during the cardiac cycle?
    a. Papillary muscles open the AV valves during atrial systole.
    b. The pulmonary and aortic valves open during ventricular diastole.
    c. Atria fill during ventricular diastole.
    d. Ventricles fill during ventricular diastole.
    e. The heart rests before atrial systole.

Matching: *Match the hormone to its function. Some answers may be used more than once. Some questions may have more than one correct response.*

_____	1. Maintains blood volume by preventing water loss to urine	a. ADH
_____	2. Causes widespread vasoconstriction	b. Angiotensin II
_____	3. Causes sodium reabsorption in the kidney	c. Epinephrine
_____	4. Causes increased blood flow to cardiac and skeletal muscles	d. Aldosterone
_____	5. Causes the kidneys to reabsorb water	

Matching: *Match the route to the description. Some answers may be used more than once.*

_____ 6. Delivers blood to the myocardium
_____ 7. Has one capillary bed
_____ 8. Has two capillary beds
_____ 9. Multiple vessels to drain an organ
_____ 10. Shunt

a. Venous anastamosis
b. Arterial anastamosis
c. Portal route
d. Typical route
e. Coronary route
f. Arteriovenous anastamosis

Critical Thinking

1. What can you do now to ensure a more efficient cardiovascular system in old age? Explain in terms of physiology.
2. Explain the effects of polycythemia in terms of blood pressure and resistance.
3. In an angioplasty, a balloon is threaded through vessels beginning in the leg to coronary vessels in the heart. Should the balloon be inserted in the femoral artery or the femoral vein? Explain your choice by tracing the balloon through the vessels from the leg to the heart.

10 chapter mapping

This section of the chapter is designed to help you find where each outcome is covered in this text.

	Outcomes	Readings, figures, and tables	Assessment
10.1	Use medical terminology related to the cardiovascular system.	Word roots: p. 365	Word Building: 1–5
10.2	Identify the chambers, valves, and features of the heart.	Anatomy of the heart: pp. 366–372 Figures 10.2–10.6	Spot Check: 1, 2 Multiple Select: 3
10.3	Relate the structure of cardiac muscle to its function.	Cardiac muscle tissue: p. 372 Figure 10.7	Multiple Select: 2
10.4	Explain why the heart does not fatigue.	Cardiac muscle tissue: p. 373	Spot Check: 3 Multiple Select: 1
10.5	Trace blood flow through the heart.	Blood flow through the heart: pp. 373–376 Figures 10.8, 10.9	Spot Check: 4, 5 Multiple Select: 4
10.6	Describe the heart's electrical conduction system.	Cardiac conduction system: pp. 376–377 Figure 10.10	Spot Check: 6
10.7	Describe the events that produce the heart's cycle of contraction and relaxation.	Cardiac cycle: pp. 377–380 Figures 10.11, 10.12	Spot Check: 7 Multiple Select: 10
10.8	Interpret a normal EKG, explaining what is happening electrically in the heart.	Electrocardiogram: pp. 380–381 Figures 10.13, 10.14	Spot Check: 8
10.9	Calculate cardiac output given heart rate and stroke volume.	Cardiac output: p. 382	Spot Check: 9 Multiple Select: 7
10.10	Explain the factors that govern cardiac output.	Cardiac output: p. 383 Figure 10.15	Spot Check: 11
10.11	Summarize nervous and chemical factors that alter heart rate, stroke volume, and cardiac output.	Heart regulation: pp. 384–385 Figure 10.16	Spot Check: 10

	Outcomes	Readings, figures, and tables	Assessment
10.12	Locate and identify the major arteries and veins of the body.	Vessel anatomy: p. 385 Figures 10.17, 10.18	Multiple Select: 4
10.13	Compare the anatomy of the three types of blood vessels.	Vessel anatomy: pp. 388–390 Figures 10.19, 10.20	Multiple Select: 5
10.14	Describe coronary and systemic circulatory routes.	Circulatory routes: pp. 390–394 Figures 10.21–10.24	Matching: 6–10 Critical Thinking: 3
10.15	Explain how blood in veins is returned to the heart.	Venous return: p. 395 Figure 10.25	Multiple Select: 8
10.16	Explain the relationship between blood pressure, resistance, and flow.	Blood pressure, resistance, and flow: p. 396	Spot Check: 12 Critical Thinking: 2
10.17	Describe how blood pressure is expressed and how mean arterial pressure and pulse pressure are calculated.	Blood pressure, resistance, and flow: pp. 396–397	Spot Check: 13 Multiple Select: 6
10.18	Explain how blood pressure and flow are regulated.	Regulation of blood pressure and flow: pp. 397–399	Matching: 1–5
10.19	Explain the effect of exercise on cardiac output.	Effects of exercise on cardiac output: p. 399 Figure 10.26	Multiple Select: 7
10.20	Summarize the effects of aging on the cardiovascular system.	Effects of aging on the cardiovascular system: p. 400	Critical Thinking: 1
10.21	Describe cardiovascular disorders.	Disorders of the heart and vessels: pp. 400–402 Figures 10.27, 10.28	Multiple Select: 9

references and resources

American Heart Association supports lower sodium limits for most Americans. (2009, March 26). Retrieved February 24, 2011, from http://www.newsroom.heart.org/index/php?s=43&item=700.

Beers, M. H., Porter, R. S., Jones, T. V., Kaplan, J. L., & Berkwits, M. (Eds.). (2006). *Merck manual of diagnosis and therapy* (18th ed.). Whitehouse Station, NJ: Merck Research Laboratories.

Mayo Clinic. (2011, March 4). Prehypertension. Retrieved March 8, 2011, from http://www.bing.com/health/article/mayo-126465/Prehypertension?q=pre-hypertension.

Porter, R. S., & Kaplan, J. L. (Eds.). (n.d.). Aging and the cardiovascular system: Cardiovascular function. *Merck manual of geriatrics* (chap. 83). Retrieved July 23, 2010, from http://www.merck.com/mkgr/mmg/sec11/ch83/ch83c.jsp.

Porter, R. S., & Kaplan, J. L. (Eds.). (n.d.). Aging and the cardiovascular system: Cardiovascular structure. *Merck manual of geriatrics* (chap. 83). Retrieved July 23, 2010, from http://www.merck.com/mkgr/mmg/sec11/ch83/ch83b.jsp.

Porter, R. S., & Kaplan, J. L. (Eds.). (n.d.). Aging and the cardiovascular system: Effects of lifestyle. *Merck manual of geriatrics* (chap. 83). Retrieved July 23, 2010, from http://www.merck.com/mkgr/mmg/sec11/ch83/ch83d.jsp.

Porter, R. S., & Kaplan, J. L. (Eds.). (n.d.). Selected physiologic age-related changes. *Merck manual for healthcare professionals*. Retrieved March 19, 2011, from http://www.merckmanuals.com/media/professional/pdf/Table_337-1.pdf?qt=selected physiologic age-related changes&alt=sh.

Rimmerman, C. (2009, September). What are the symptoms of a heart attack? Retrieved November 8, 2010, from http://my.clevelandclinic.org/heart/disorders/cad/mi_symptoms.aspx.

Saladin, Kenneth S. (2010). *Anatomy & physiology: The unity of form and function* (5th ed.). New York: McGraw-Hill.

Seeley, R. R., Stephens, T. D., & Tate, P. (2006). *Anatomy & physiology* (7th ed.). New York: McGraw-Hill.

Shier, D., Butler, J., & Lewis, R. (2010). *hole's human anatomy & physiology* (12th ed.). New York: McGraw-Hill.

11 The Lymphatic System

Deborah is 58, so you might imagine she has had some mixed experiences with vaccinations. As an infant, she had polio before the polio vaccine became available. When she was five, she received a smallpox vaccination, which left a scar on her arm that remains to this day. She also had chickenpox as a child and, as a result, developed some immunity to the virus. However, Deborah is still susceptible to developing shingles later in her life from the same virus that caused her chickenpox. Now her grandchildren receive multiple vaccinations, even one for chickenpox.

All the vaccinations you have received up to this point in your life have activated cells of your lymphatic system to better prepare you to defend yourself against specific pathogens. Along with providing immunity, your lymphatic system does much more. See Figure 11.1.

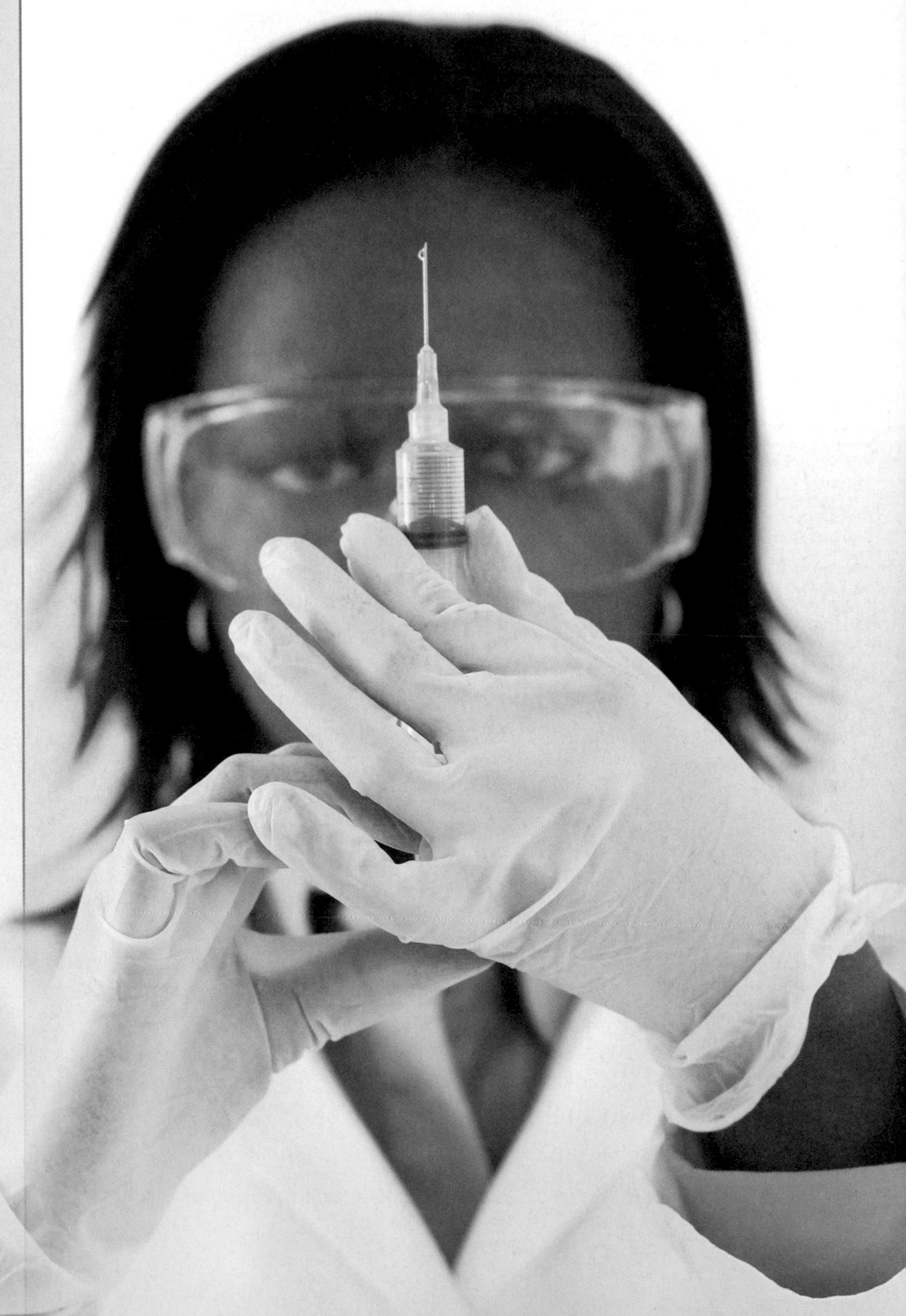

11.1 learning outcome

Use medical terminology related to the lymphatic system.

learning outcomes

After completing this chapter, you should be able to:

11.1 Use medical terminology related to the lymphatic system.

11.2 Explain the origin and composition of lymph.

11.3 Describe lymph vessels.

11.4 Explain the route of lymph from the blood and back again.

11.5 Describe cells of the lymphatic system and list their functions.

11.6 Identify lymphoid tissues and organs and explain their functions.

11.7 Summarize three lines of defense against pathogens.

11.8 Contrast nonspecific resistance and specific immunity.

11.9 Describe the body's nonspecific defenses.

11.10 Explain the role of an APC in specific immunity.

11.11 Explain the process of humoral immunity.

11.12 Explain the process of cellular immunity.

11.13 Compare the different forms of acquired immunity.

11.14 Explain the importance of T_{helper} cells to specific and nonspecific defense.

11.15 Explain the functions of the lymphatic system.

11.16 Summarize the effects of aging on the lymphatic system.

11.17 Describe lymphatic system disorders.

word **roots** & combining **forms**

immun/o: protection

lymph/o: lymph

lymphaden/o: lymph node

splen/o: spleen

thym/o: thymus gland

pronunciation **key**

anaphylaxis: AN-ah-fih-LAK-sis

lymphadenitis: lim-FAD-eh-neye-tis

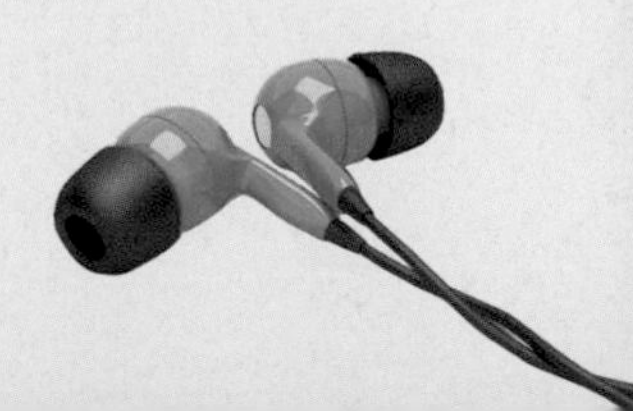

Overview

The human body contains two systems of circulating fluids—the cardiovascular system and the lymphatic system. In Chapters 9 and 10, you studied the cardiovascular system, which deals with the circulation of blood. In this chapter, you will explore the lymphatic system, which involves the circulation of **lymph,** a fluid derived from blood.

Your exploration of the lymphatic system will begin by examining the anatomy—the lymph, vessels, and lymphoid tissues—of lymph circulation. You will learn about the body's three lines of defense against pathogens and see how specific immunity differs from nonspecific defenses. Then, as you did in all of the system chapters, you will explore the effects of aging and disorders of the system. Your first topic is the fluid of the system—lymph.

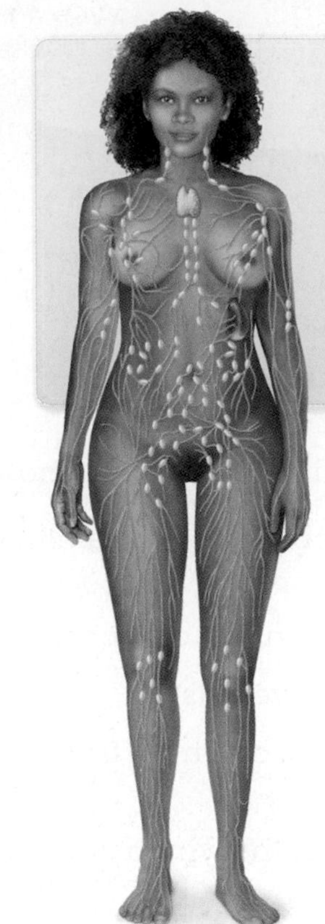

FIGURE 11.1 The lymphatic system.

11.2 learning **outcome**

Explain the origin and composition of lymph.

Anatomy of the Lymphatic System

Lymph and Lymph Vessels

Lymph is a fluid derived from plasma, but it has fewer dissolved proteins. It washes over tissues in the body to deliver nutrients to these tissues and to wash away wastes, cellular debris, viruses, bacteria, and loose (possibly cancerous) cells. Lymph reaches cells that are not immediately adjacent to blood capillaries, so it helps these cells to meet their nutrient and waste-removal needs.

11.3 learning **outcome**

Describe lymph vessels.

Unlike the circulation of blood through a closed system of blood vessels, lymph leaves the system of blood vessels through the capillaries due to blood pressure. Lymph must then return to the cardiovascular system through a network of open lymph capillaries and vessels that drain the tissues of lymph. As you can see in **Figure 11.2**, gaps between the endothelial cells of the lymph capillaries allow lymph, bacteria, and even loose cells to enter the vessel. Valves inside the lymph vessels (shown in **Figure 11.3**) direct the flow of lymph to larger and larger lymph vessels. These vessels eventually drain into one of two collecting ducts—the **thoracic duct** and the **right lymphatic duct.**

11.4 learning **outcome**

Explain the route of lymph from the blood and back again.

The thoracic and right lymphatic ducts deliver lymph to the subclavian veins, where it rejoins the circulating blood. As you can see in **Figure 11.4**, the right lymphatic duct drains lymph from the right side of the head, the right arm, and the right side of the thorax to the right subclavian vein. The thoracic duct delivers lymph from the rest of the body to the left subclavian vein.

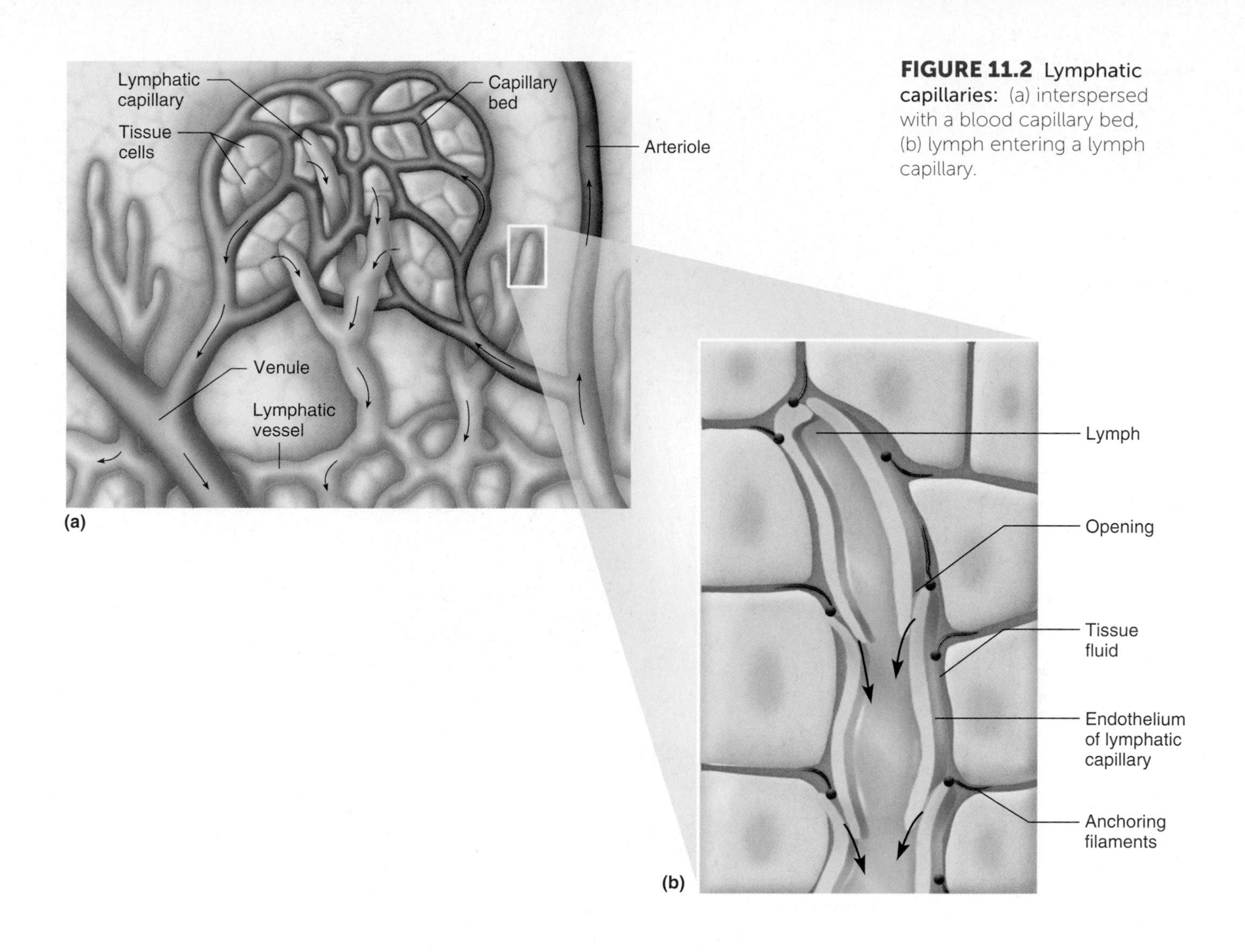

FIGURE 11.2 **Lymphatic capillaries:** (a) interspersed with a blood capillary bed, (b) lymph entering a lymph capillary.

FIGURE 11.3 **Valves in lymphatic vessels:** (a) micrograph, (b) diagram showing one-way direction of flow.

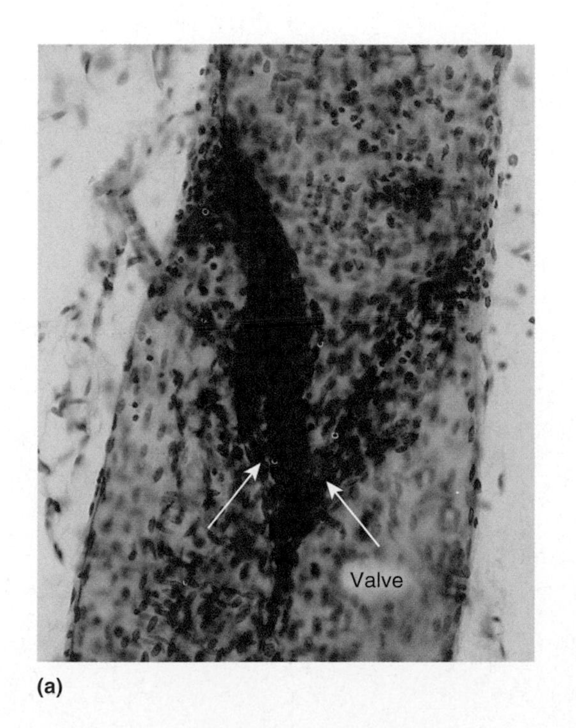

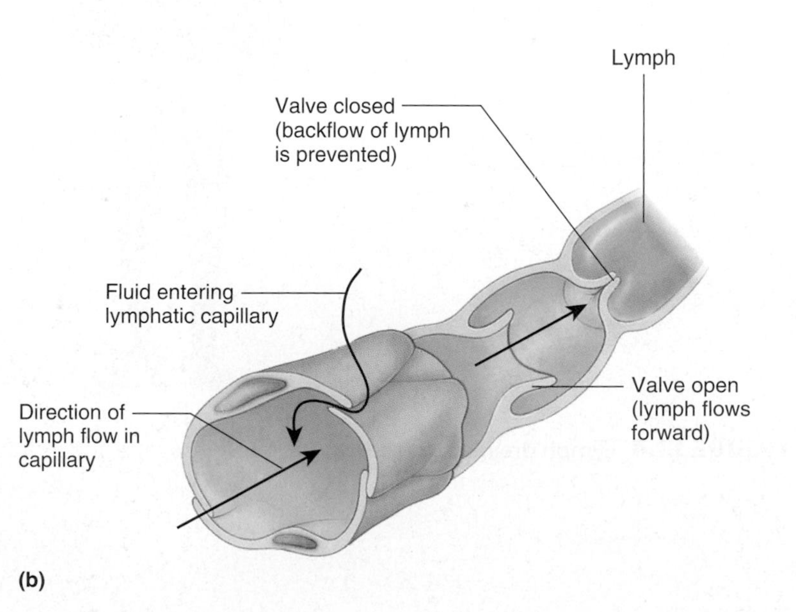

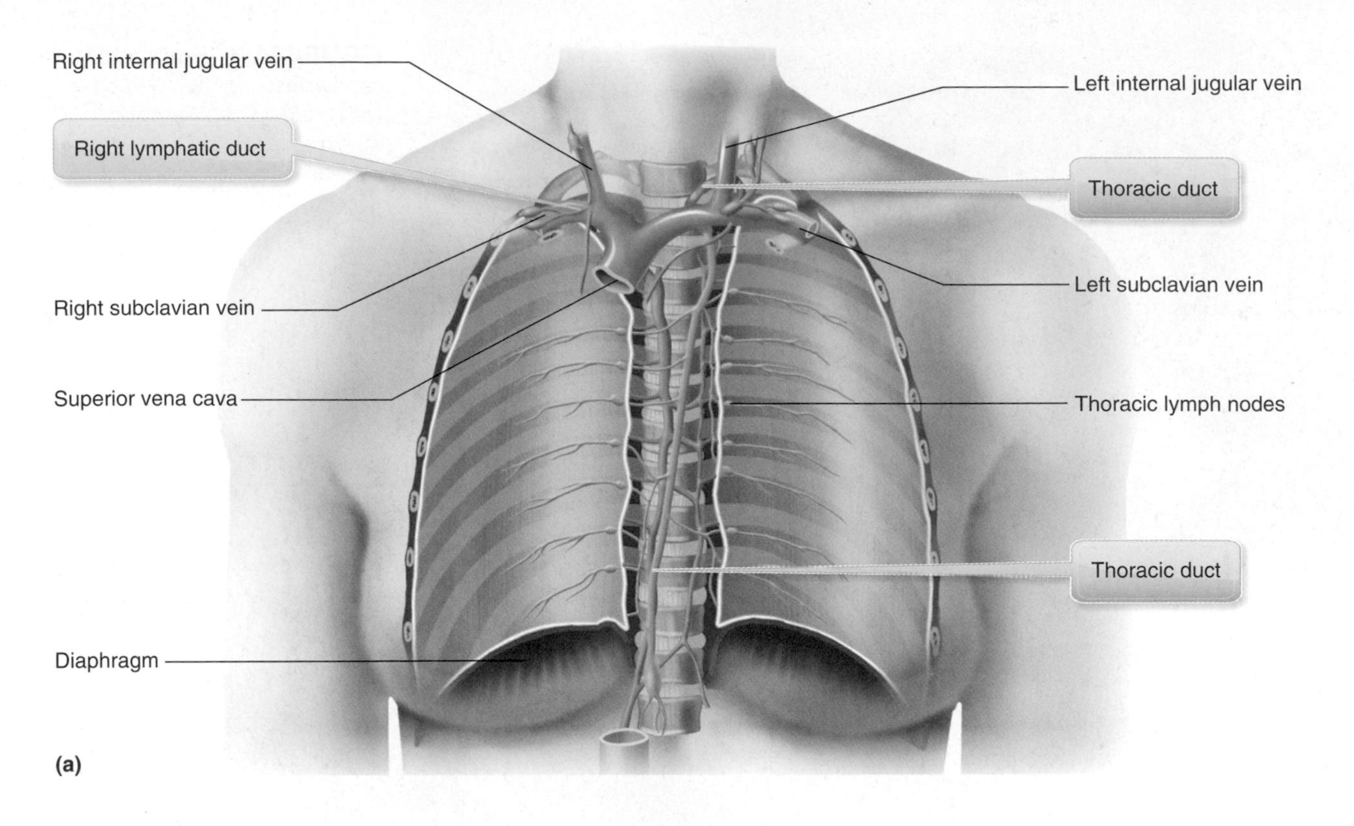

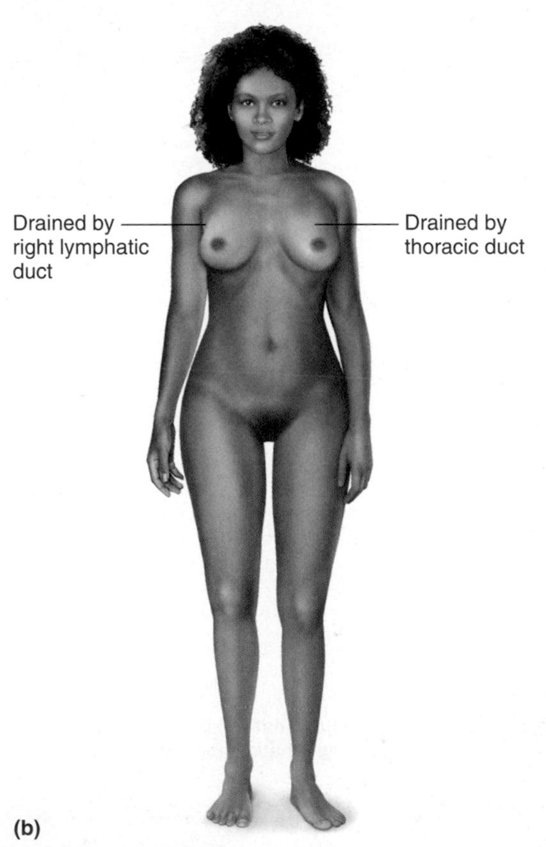

FIGURE 11.4 Lymph drainage to the subclavian veins: (a) in the thoracic region, (b) of the whole body.

The pressure from the buildup of lymph in the tissues causes lymph to enter lymph capillaries. Once lymph is in the capillaries, the skeletal muscle pump moves lymph through the lymph vessels to the collecting ducts and back into blood circulation at the subclavian veins. This is the same skeletal muscle pump that is partially responsible for venous return in the cardiovascular system (Chapter 10).

In summary, the cardiovascular and lymphatic systems are interconnected by this shared fluid—called *plasma* while in the blood and *lymph* as it leaves the blood capillaries with fewer proteins to go to the tissues. It must be returned to the bloodstream to make up for the lost volume at the blood capillaries. See **Figure 11.5**.

spot check 1 Where specifically will lymph in the tissues of the left leg reenter the bloodstream?

spot check 2 Which collecting duct delivers the lymph from the left leg to that location?

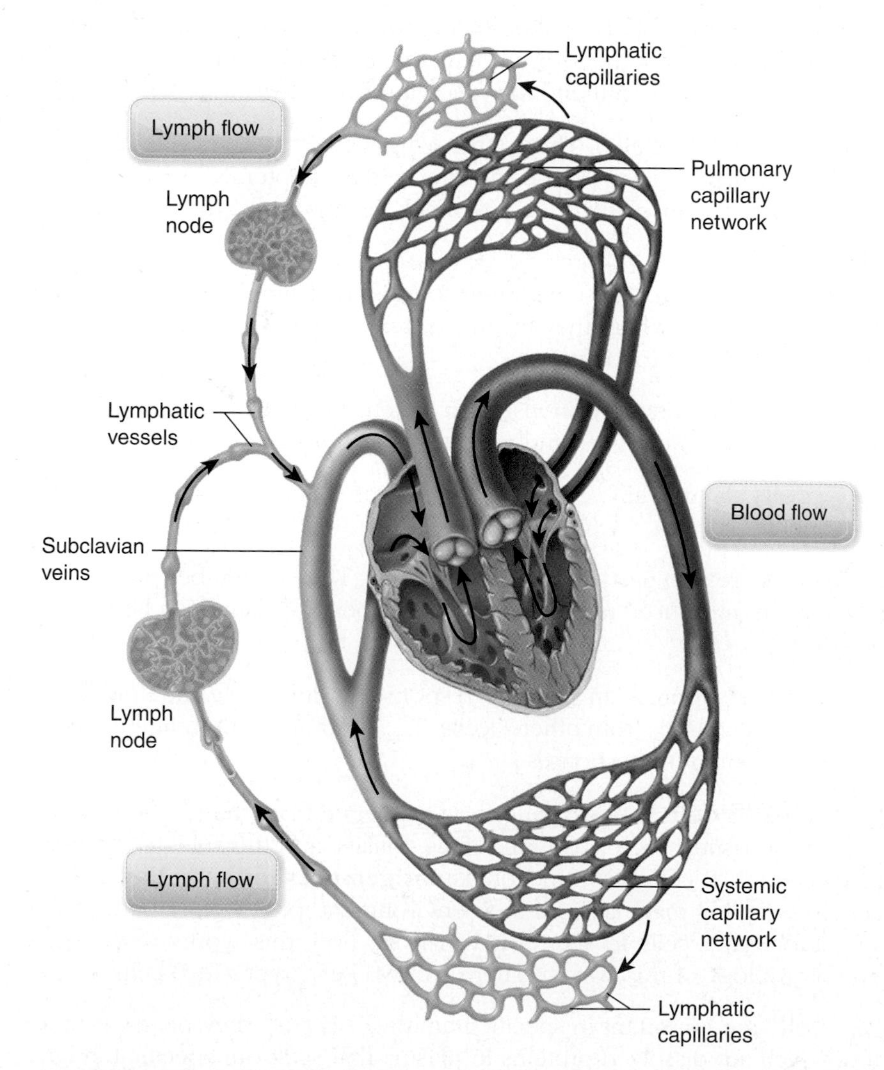

FIGURE 11.5 Fluid exchange between the cardiovascular and lymphatic systems. Arrows within the blood vessels show the path of blood. Arrows outside the blood vessels show the path of lymph leaving the blood capillaries and rejoining the circulatory system at the subclavian veins.

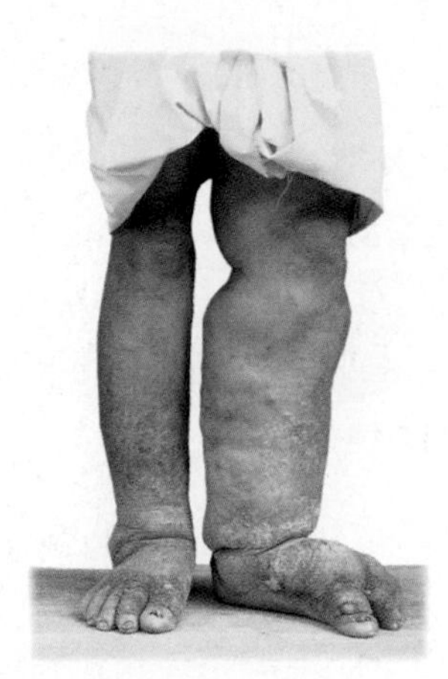

FIGURE 11.6 Elephantiasis of the left leg caused by blocked lymphatic drainage.

clinical point

The lymphatic drainage process is an important part of a healthy, functioning lymphatic system. However, certain conditions may block drainage. For example, elephantiasis is a tropical disease caused by a roundworm that blocks lymphatic drainage. The roundworm gets into the system through a mosquito bite and infects a lymph node (to be discussed shortly), which blocks the flow of lymph, causing edema in the area before the blockage. Typical areas include the legs, arms, breasts, or scrotum. See Figure 11.6.

You now understand how a common fluid links the cardiovascular and lymphatic systems and you are aware of the blood vessels they share for this fluid's delivery to the tissues. Next, you will examine the lymphatic system's cells.

11.5 learning outcome

Describe cells of the lymphatic system and list their functions.

Cells of the Lymphatic System

The primary cells of this system are leukocytes—particularly, lymphocytes. As you may recall from Chapter 9, all of the formed elements are produced in the red bone marrow. The cells of this system fall into the following five categories:

- **Natural killer cells (NK cells).** These large lymphocytes are important in nonspecific defense, which is covered in detail later in this chapter. NK cells destroy bacteria, fight against transplanted tissues, attack cells infected by viruses, and destroy cancer cells.
- **T lymphocytes (T cells).** These lymphocytes migrate from the red bone marrow to the thymus gland, where they mature. Several classes of T cells, based on their function, include the following:
    1. **T_{helper} cells** are important for nonspecific defense and specific immunity by recognizing foreign pathogens and activating the cells to fight them.
    2. **$T_{cytotoxic}$ cells** directly kill cells infected by viruses and cancer cells in specific immunity.
    3. **T_{memory} cells** are also used in specific immunity. They remember pathogens that have been introduced to the body so that repeat exposure can be fought more swiftly.
    4. **$T_{regulatory}$ cells** suppress an immune response by inhibiting multiplication and chemical secretions from other T cells. T_{regs} are important in limiting and preventing autoimmune responses.
- **B lymphocytes (B cells).** These lymphocytes migrate from the red bone marrow to lymphoid tissues, such as lymph nodes, tonsils, and the spleen (covered in the next section). B cells also function as **antigen-presenting cells (APCs)** by constantly sampling material from their environment, processing it, and then displaying it for other cells to see. You will learn how this works when you review the physiology of this system. There are two basic types of B cells:
    1. **B_{plasma} cells** are important in specific immunity because they produce antibodies, which are dissolved proteins in plasma that seek out specific foreign antigens for their destruction (Chapter 9).
    2. **B_{memory} cells,** like T_{memory} cells, remember pathogens that have been introduced to the body so that repeat exposure can be fought more swiftly. You will learn more about this when we discuss humoral immunity later in this chapter.

- **Macrophages.** These cells are not lymphocytes. They are monocytes that were produced in the red bone marrow and have migrated to the tissues to become macrophages. Their purpose, in the nonspecific defense of the body, is to phagocytize bacteria, debris, and dead neutrophils. Like B cells, macrophages are APCs.
- **Dendritic cells.** These are immune system cells of the epidermis that stand guard to alert the body of pathogens entering through the skin. They also function as APCs. You were introduced to these cells in Chapter 3.

Lymphoid tissues were mentioned as the location for several of the preceding cells. These tissues and organs are discussed in the next section.

Lymphoid Tissues and Organs

11.6 learning outcome

Identify lymphoid tissues and organs and explain their functions.

Lymphoid tissues and organs may be as small as a scattering of lymphocytes in mucous membranes or may be full-size organs, such as the spleen. Many of the tissues are shown in **Figure 11.7**.

spot check 3 Red bone marrow is shown in Figure 11.7. Why is red bone marrow relevant to this system?

Mucosa Associated Lymphatic Tissue Mucosa associated lymphatic tissue (MALT) is a scattering of lymphocytes located throughout the mucous membranes lining tracts to the outside environment, such as the tracts for the digestive, respiratory, urinary, and reproductive systems. The purpose of MALT is to stand guard against and fight any pathogens trying to enter the body.

Peyer's Patches These patches of lymphatic tissue are located at the distal end of the small intestine, just before the opening to the large intestine. Peyer's patches are an example of more densely packed pockets of lymphocytes called **nodules.** These particular nodules are meant to fight any bacteria moving into the small intestine from the colon, where they naturally reside.

Lymph Nodes These lymphatic structures act as filters along lymph vessels. Lymph nodes remove anything that may be potentially harmful in lymph much like a water purification filter may remove impurities in the drinking water arriving at your kitchen faucet. Each lymph node has many nodules packed with lymphocytes and macrophages. As you can see in **Figure 11.8**, several lymph vessels direct lymph flow into the lymph node. There, fibers trap debris, cells, and bacteria picked up by the lymph in the tissues. Macrophages phagocytize the debris, while lymphocytes mount an attack on the pathogens. If an infection is present, **germinal centers** (sites for cloning lymphocytes) in the lymph nodes produce more B lymphocytes. Meanwhile, the lymph circulates through the lymph node on its way to larger lymph vessels.

Lymph nodes are located in specific areas, as outlined in the following list, to filter lymph as it is drained from different regions. See **Figure 11.7**.

- Cervical lymph nodes are located in groups in the neck. They filter lymph from the head and neck.
- Axillary lymph nodes are located in the axillary region and the lateral margin of the breast. They filter lymph from the breasts and arms.
- Thoracic lymph nodes are located in the mediastinum surrounding the trachea and bronchi. They filter lymph from organs in the thoracic cavity.
- Abdominal lymph nodes are located in the posterior wall of the abdominopelvic cavity. They filter lymph from the urinary and reproductive systems.

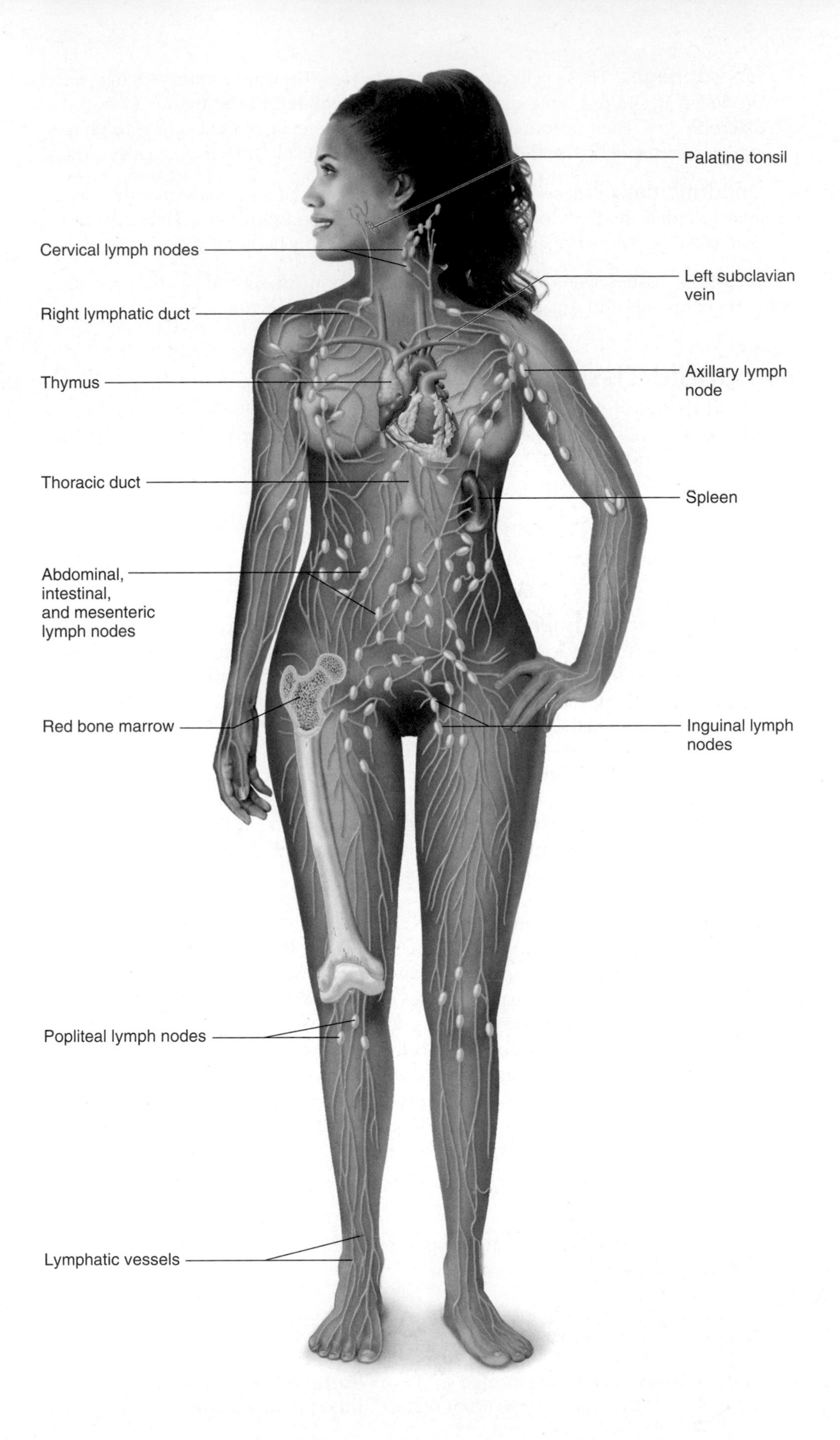

FIGURE 11.7 Lymphoid tissues and organs.

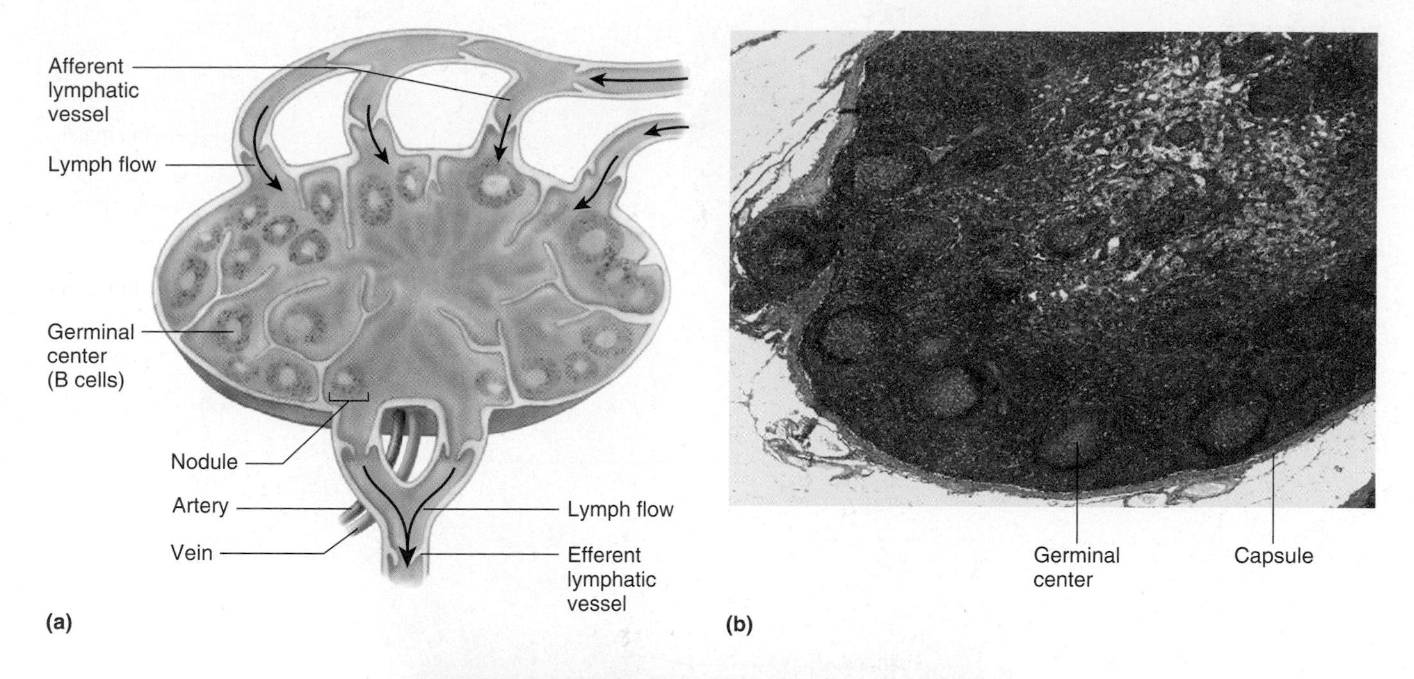

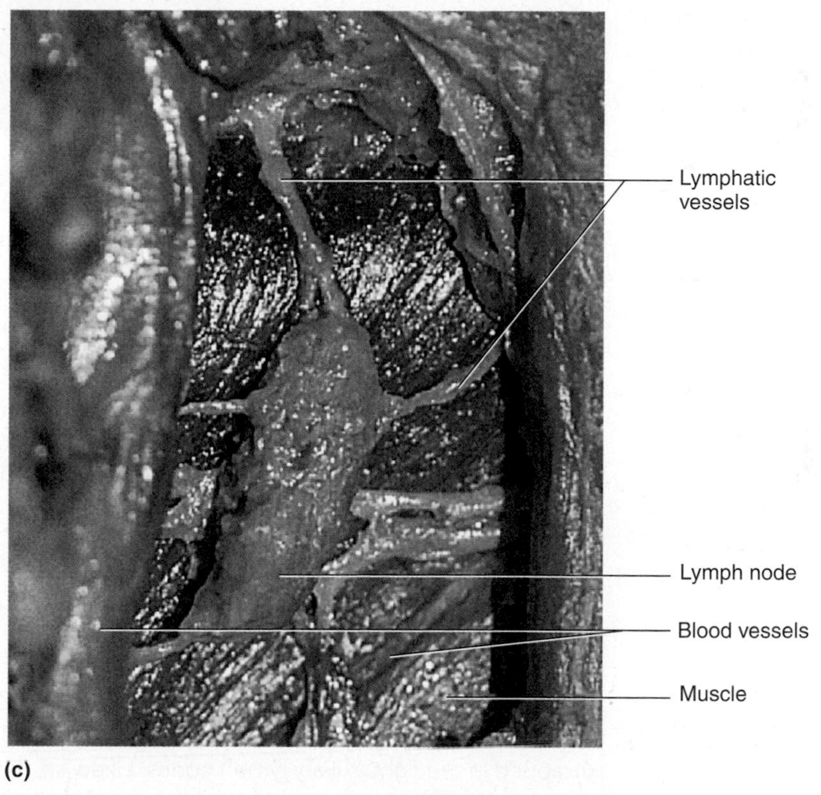

FIGURE 11.8 Lymph node: (a) lymph node showing direction of lymph flow, (b) micrograph of a lymph node, (c) dissection of lymph node and lymph vessels.

- Pelvic lymph nodes are deep in the pelvic region and surround the iliac arteries and veins. They also filter lymph from the urinary and reproductive systems.
- Intestinal and mesenteric lymph nodes are located in the mesenteries and surround the mesenteric arteries and veins. They filter lymph from the digestive organs. Lymph from the digestive system appears milky because it carries the products of lipid digestion.

FIGURE 11.9 Inguinal lymph nodes of a cadaver.

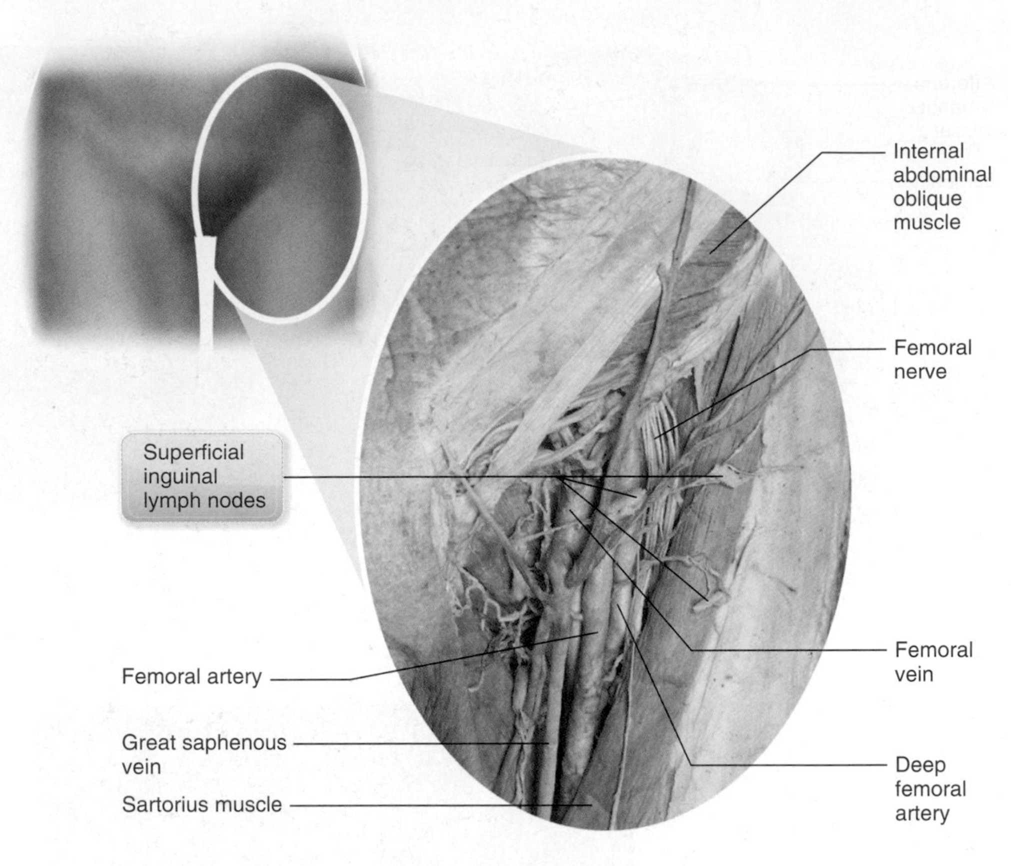

- Popliteal lymph nodes are located in the popliteal region (behind the knee). They filter lymph from the leg.
- Inguinal lymph nodes are superficial in the groin. They filter lymph from the lower limbs. See **Figure 11.9**.

When a pathogen is under attack by the lymph node's lymphocytes, the lymph node may become swollen and painful to the touch. This condition is called **lymphadenitis.** Understanding lymph drainage is helpful in locating the primary site of the cause of the attack.

lymphadenitis: lim-FAD-eh-neye-tis

clinical point

For example, look at Figure 11.10, which shows the lymph drainage for the right breast. Swollen, painful lymph nodes in the right axillary region may indicate the presence of cancer cells that have metastasized from the right breast and have been trapped in the right axillary lymph nodes. Likewise, if breast cancer in the right breast has been found through other means (self-exam, mammography, or needle biopsy), a biopsy of the right axillary lymph nodes may give a good indication of whether the tumor has metastasized. If it has, the lymph should have carried any loose, metastatic cancer cells to the right axillary lymph nodes.

spot check **4** Which lymph nodes may be tender to the touch if Megan has a strep throat infection?

Tonsils Like lymph nodes, tonsils are lymphoid tissue with high concentrations of lymphocytes. As you can see in **Figure 11.11**, three types of tonsils—one pharyngeal, two palatine, and numerous lingual tonsils—ring the **pharynx** (throat) to guard against pathogens entering the body through the nose or mouth. Each tonsil has pits **(crypts)** to give the lymph nodules more exposure to whatever may be passing by. The locations of the tonsils are as follows:

- The **pharyngeal tonsil (adenoids)** is located on the roof of the **nasopharynx** (section of throat at the back of the nasal cavity).
- The **palatine tonsils,** located laterally in the **oropharynx** (section of the throat at the back of the mouth) are commonly recognized as the tonsils. These tonsils often swell and become inflamed during a throat infection and can be seen by looking in the mouth.

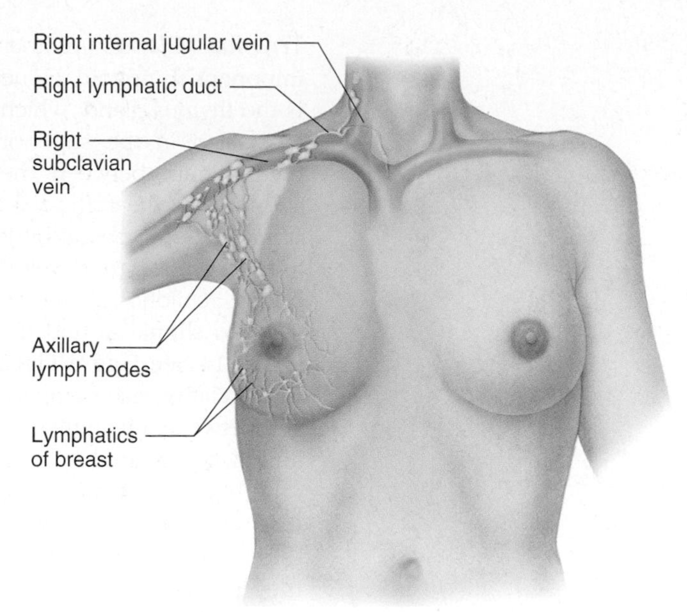

FIGURE 11.10 Lymph drainage of the right breast.

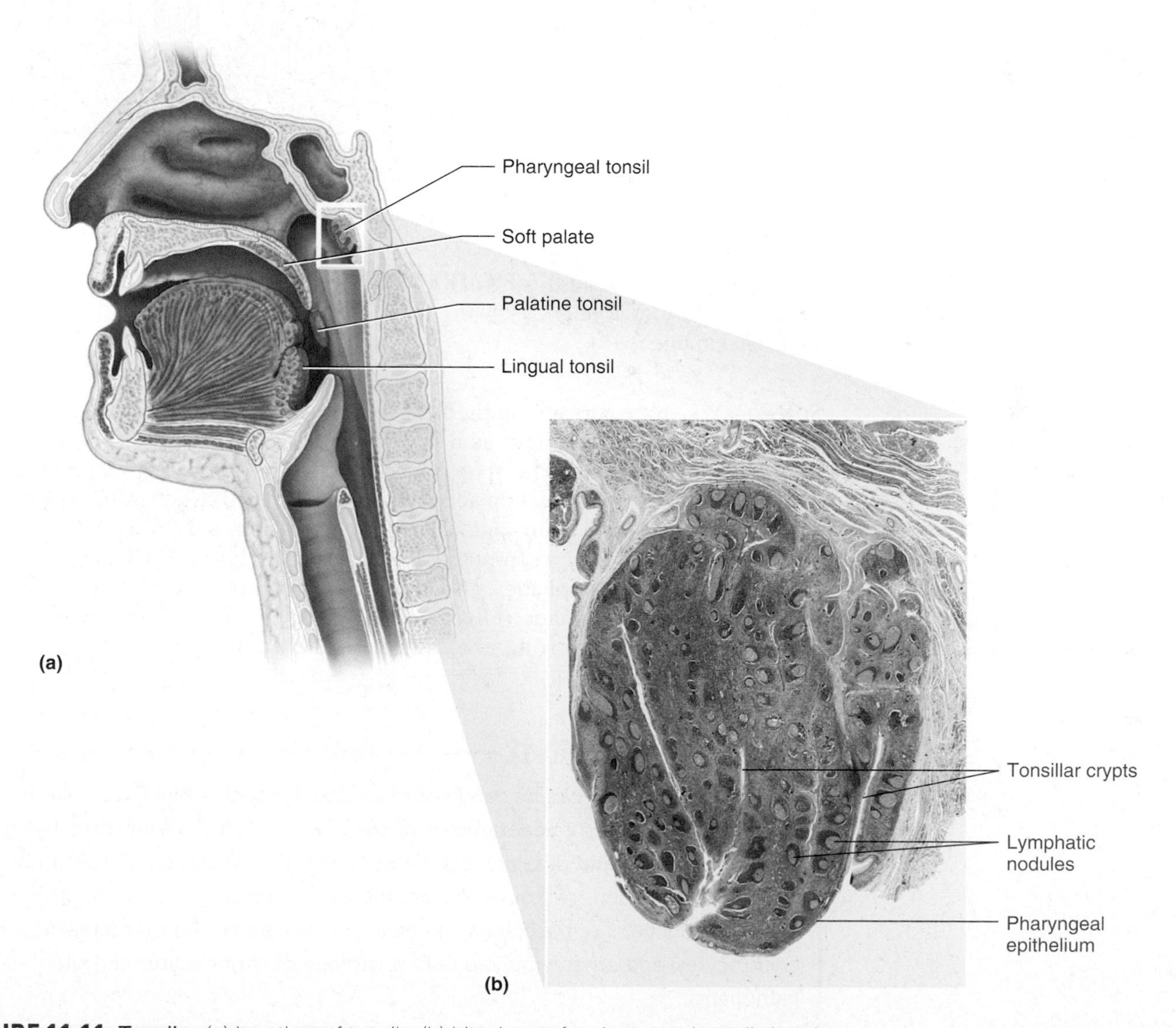

FIGURE 11.11 Tonsils: (a) location of tonsils, (b) histology of a pharyngeal tonsil showing crypts.

Thymus Gland Another important lymphoid tissue is the thymus gland, which is located in the superior mediastinum between the sternum and the aortic arch. It is well developed at birth and continues to develop during childhood, but it starts to shrink around the age of 14. See **Figure 11.12** for a relative size comparison between a fetal thymus and that of an adult.

T cells migrate from the red bone marrow to the thymus gland, where they mature. In the maturation process, the thymus introduces self-antigens to the developing T cells. If the T cell reacts to the self-antigen, the T cell is destroyed. Only those T cells that do not react to self-antigens are stimulated to further develop by chemicals secreted by the thymus. These T cells are an important part of the immune system because they react only to foreign (not self-) antigens.

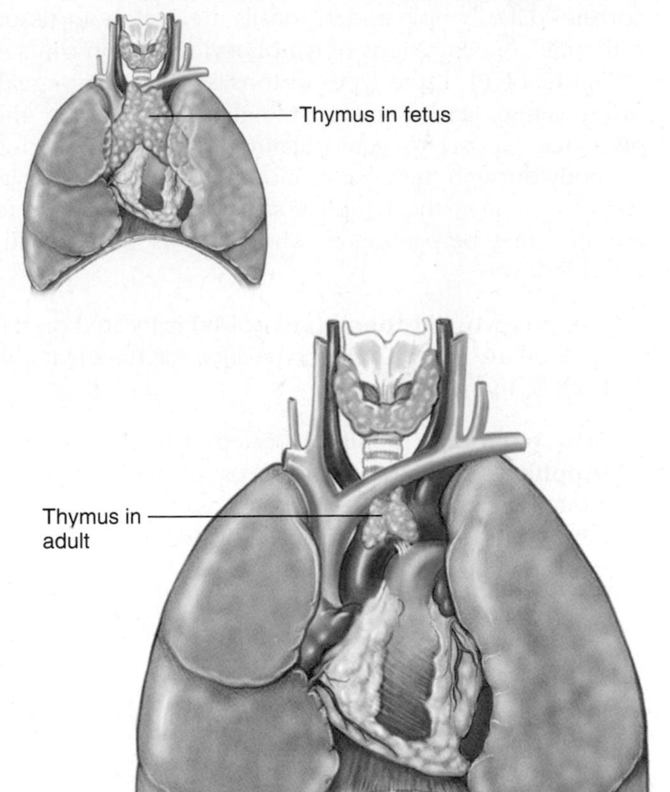

FIGURE 11.12 Thymus gland: (a) of a fetus, (b) of an adult.

Spleen You have already studied the spleen in relation to its role in the life cycle of a red blood cell and as a reservoir for blood, but it also has functions in the lymphatic system. The spleen is located in the upper left quadrant (ULQ), posterior and lateral to the stomach. See **Figure 11.13**. Tissues in the spleen consist of two types—red pulp and white pulp. Red pulp serves as a reservoir for RBCs and destroys old, worn-out red blood cells. White pulp is a reservoir for lymphocytes and macrophages, and it functions similarly to lymph nodes as a site of battle between lymphatic cells and pathogens. The spleen also helps regulate blood volume by transferring excess fluid in the blood to the lymphatic system as lymph.

clinical point

Trauma to the spleen can be dangerous because it is such a highly vascular organ. In cases of trauma, surgical removal of the spleen is often easier than trying to deal with a repair and possible fatal hemorrhaging. It is possible to live a normal life without a spleen. Without a spleen, there is no reservoir for blood. However, the other functions of the spleen can be accomplished by the liver (breakdown of erythrocytes) and other lymphoid tissues (storage of lymphocytes and site to fight pathogens).

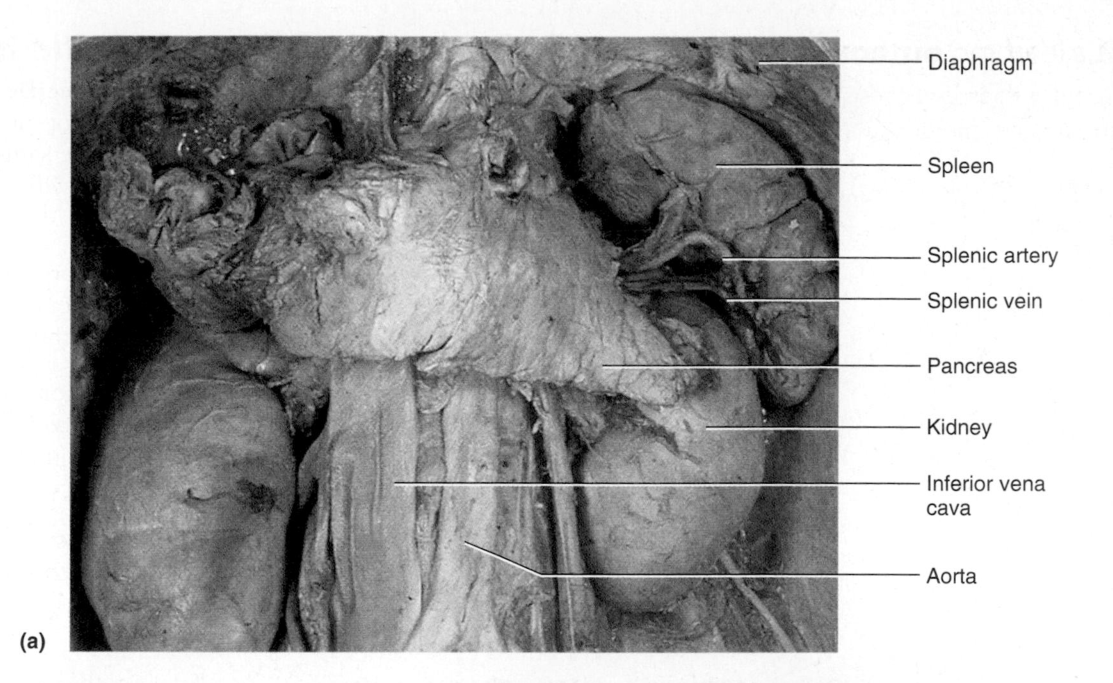

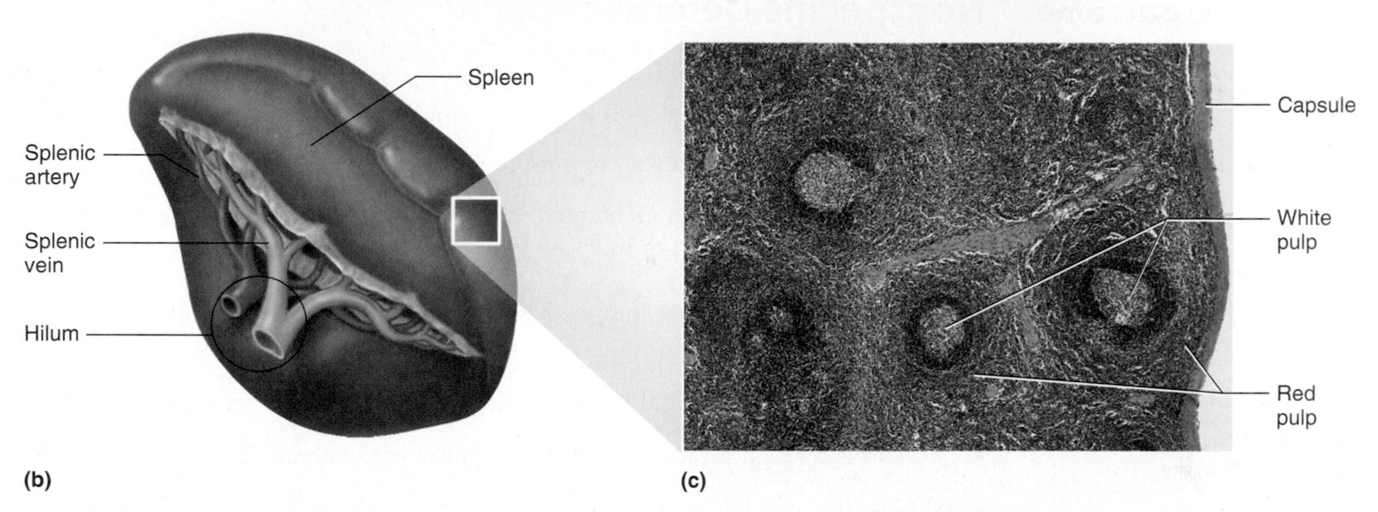

FIGURE 11.13 The spleen: (a) spleen in a cadaver with the stomach removed, (b) medial surface of the spleen, (c) white and red pulp.

Physiology of the Lymphatic System

You are already familiar with the circulation of lymph. It is forced out of capillaries due to blood pressure and washes over tissues to deliver nutrients and remove wastes. The pressure of the lymph in the tissues forces lymph back into lymph capillaries so that the lymph vessels can return it to the bloodstream at the subclavian veins. But how does this system fight pathogens? To understand the answer to this question, you need to study the physiology of defense.

Three Lines of Defense

11.7 learning outcome

Summarize three lines of defense against pathogens.

The three basic lines of defense against pathogens are the following:

1. External barriers.
2. Inflammation, antimicrobial proteins, fever, and other active attacks.
3. Specific immunity.

11.8 learning **outcome**

Contrast nonspecific resistance and specific immunity.

Nonspecific Resistance versus Specific Immunity

The first two lines of defense are considered to be **nonspecific resistance,** while the third is **specific immunity.** The lines of defense are not mutually exclusive, as more than one line of defense is likely to be at work at the same time to eliminate the same pathogen. What is the difference between nonspecific resistance and specific immunity?

- Nonspecific defenses are widespread, meaning they work to fight many pathogens without prior exposure. These defenses work to fend off any pathogen in the same way every time the pathogen comes along in the body.
- Specific immunity is just that—specific. It requires a prior exposure to a pathogen so that the system can *recognize* the pathogen, *react* to the pathogen to fight it off, and then *remember* the specific pathogen so that it can be fought off faster and stronger if it ever occurs in the body again.

Now that you are familiar with the differences of the two types of defenses, you are ready to explore the three lines of defense, starting with the two lines of nonspecific defenses.

11.9 learning **outcome**

Describe the body's nonspecific defenses.

Nonspecific Defenses

The two lines of nonspecific defense are:

1. External barriers.
2. Inflammation, antimicrobial proteins, fever, and other active attacks.

Most of these have been mentioned in chapters you have already covered, but it will be helpful to refresh your memory in regard to how they relate to this system.

External Barriers This first line of defense protects body tissues from pathogens in the outside environment.

Skin Skin acts as an external barrier to pathogens for several reasons:

- Keratin is a tough protein that bacteria cannot easily break through.
- Skin is dry, with few nutrients for bacteria and other pathogens.
- The skin has an acid mantle, which makes it inhospitable for bacteria and other pathogens.

Mucous membranes Mucous membranes (lining all the tracts through the body for the respiratory, digestive, urinary, and reproductive systems) also serve as an external barrier for the following reasons:

- Mucus traps microbes.
- Mucus, tears, and saliva contain lysozymes to destroy pathogens.
- Deep to the mucous membranes is loose areolar connective tissue with fibers to hamper the progress of pathogens.

We turn now to the body's second line of nonspecific defense.

Inflammation, Antimicrobial Proteins, Fever, and Other Active Attacks Like external barriers, each of these nonspecific defenses works against a variety of pathogens in the same way regardless of the number of exposures to the pathogen. You will investigate each of them individually.

Inflammation The functions of inflammation are threefold. See if you can picture how the steps in the inflammatory process meet the following functions of inflammation:

- To limit the spread of pathogens.
- To remove debris and damaged tissue.
- To initiate tissue repair.

As you will recall from Chapter 3, the signs of inflammation are redness, heat, pain, and swelling. (Follow along with **Figure 11.14** as you read more about inflammation.) The steps in the inflammatory process are as follows (in order of occurrence):

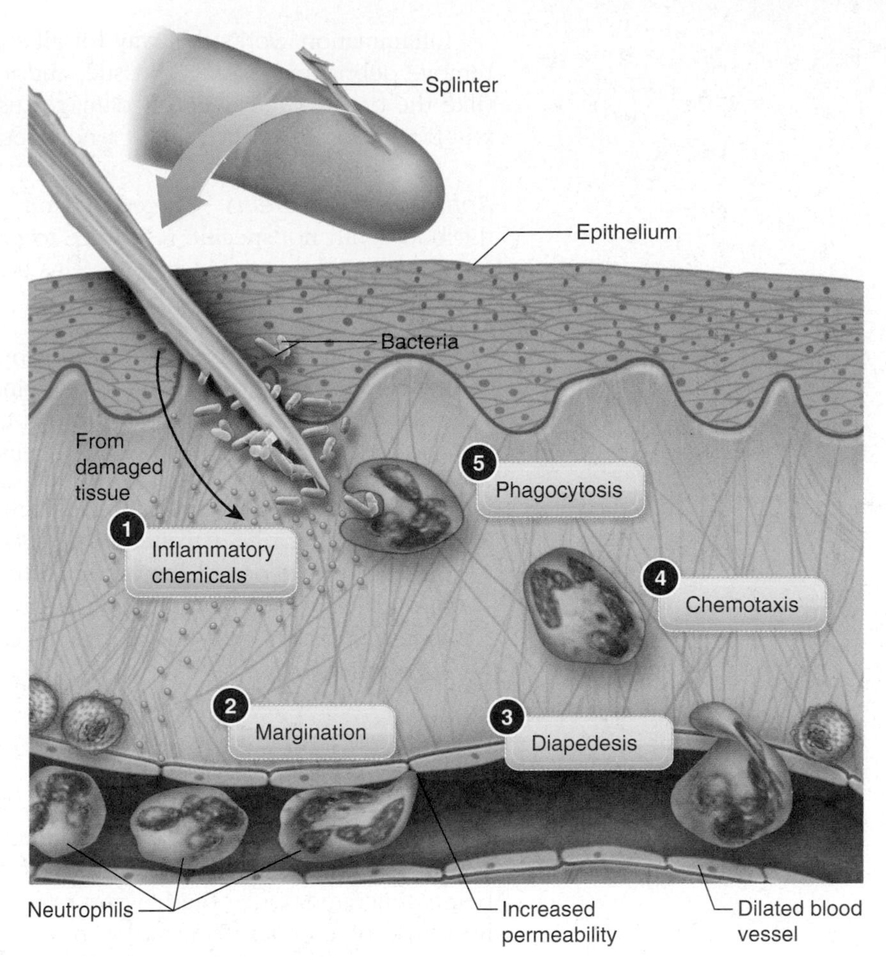

FIGURE 11.14 The inflammatory response, illustrating margination, diapedesis, chemotaxis, and phagocytosis.

1. Chemicals (vasodilators) are released by damaged tissues and basophils. The chemicals diffuse across the surrounding tissues and affect any blood vessels in the area. The dilation of these vessels causes increased blood flow to the area and increased vessel permeability. The increased blood flow accounts for the signs of redness and heat (blood from the core transports heat), while the increased permeability accounts for the swelling and pain (more lymph to tissues causes the swelling, which puts pressure on nerve endings, causing the pain). The heat increases the local metabolic rate to increase cell division and healing. The increased blood flow dilutes possible toxins produced by pathogens, provides cells with more oxygen and nutrients, and removes more wastes. The increased permeability facilitates the movement of leukocytes to the tissues.
2. WBCs stick to the walls of the dilated vessels in the inflamed area **(margination).** See **Figure 11.14**. Neutrophils will be the first on the scene.
3. WBCs crawl through the vessel walls **(diapedesis).**
4. WBCs move to where the concentration of chemicals from damaged tissues is the greatest **(chemotaxis).** Since the chemicals move by diffusion, the greatest concentration will be at the source of the damage.
5. WBCs phagocytize foreign material, debris, and pathogens along the way **(phagocytosis).** The accumulation of WBCs, debris, bacteria, and lymph is called **pus.**

spot check ❺ How does inflammation help with tissue repair?

spot check ❻ How does inflammation help to remove debris and damaged tissue?

spot check ❼ How does inflammation limit the spread of pathogens?

Inflammation works this way for all sorts of pathogens to limit their spread, to remove debris and damaged tissue, and to initiate tissue repair. If another splinter (like the one in **Figure 11.14**) damages tissue a week from Tuesday, the response will be the same because this is a nonspecific line of defense.

Antimicrobial proteins There are two types of antimicrobial proteins that provide the body with nonspecific resistance to pathogens. These antimicrobial proteins—interferons and the 20 inactive proteins that make up the complement system—are explained as follows:

- **Interferons** are chemicals released by virally infected cells. They do not help the cell that produced them. Instead, interferons encourage surrounding healthy cells to make antiviral proteins so that the virus will not invade them. Interferons also activate macrophages and NK cells to fight cancer cells.
- The **complement system** includes 20 inactive proteins (always present in the blood) that may be activated by the presence of a pathogen. Once activated, the proteins initiate one of several different pathways to ensure pathogen destruction through increased inflammation, breaking apart of the pathogen **(cytolysis),** or coating a pathogenic cell to make it easier for a macrophage to phagocytize it **(opsonization).**

Fever Most people view a fever as a bad thing, but it is another method of nonspecific defense. A fever is initiated by the production of chemicals **(pyrogens)** from activated macrophages. These pyrogens travel to the hypothalamus, which then raises the set point for body temperature. The body responds by shivering to produce more heat, while the blood vessels in the skin constrict to preserve the heat being generated. Once the new set point is reached (a stage called **stadium**), the liver and spleen hoard zinc and iron, which are necessary for bacteria growth. This gives time for other defenses to work to defeat the pathogen. Once the pathogen is defeated, the hypothalamus resets the temperature to normal and the brain may initiate sweating to cool the body to homeostasis **(defervescence).** See the graph of body temperature in **Figure 11.15**.

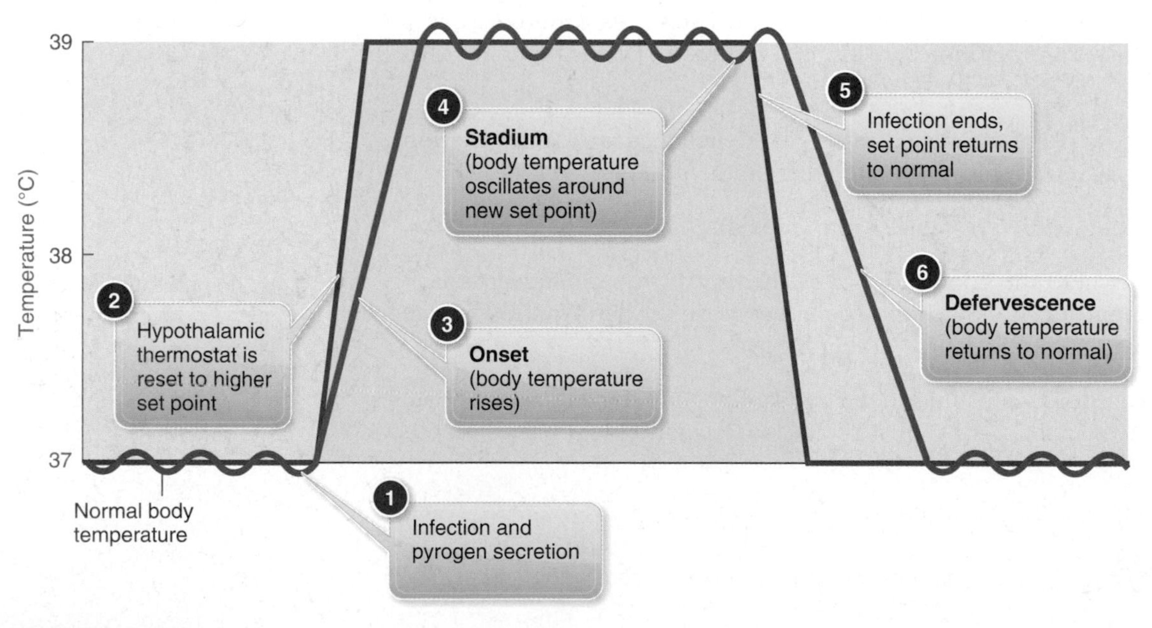

FIGURE 11.15 Graph of a fever.

Other active attacks This line of defense refers to the functions of leukocytes other than lymphocytes. You have already studied them in Chapter 9. As a means of nonspecific resistance, these cells make their attacks with the same speed and strength each time any pathogen enters the body. Here is how each of these cell types works:

- Neutrophils fight bacteria.
- Basophils release histamine to promote inflammation.
- Eosinophils attack worm parasites.
- Monocytes become macrophages to phagocytize bacteria.

Specific Immunity

11.10 learning **outcome**

Explain the role of an APC in specific immunity.

As mentioned earlier, specific immunity differs from nonspecific resistance because it requires a prior exposure to a pathogen in order to work. During the first exposure, the immune system recognizes the specific pathogen as being foreign, reacts to it, and then remembers it. The process starts with an antigen-presenting cell. How does an APC work?

Figure 11.16 shows an APC in the process of presenting an antigen. In this case, the APC is a macrophage, but the process is the same for B cells and other APCs. Imagine that the APC is in an axillary lymph node. Its job is to sample antigens in the surrounding environment by phagocytosis (1). It may sample a foreign antigen or a self-antigen. Next, a lysosome fuses with the vesicle carrying the phagocytized antigen (2). The antigens and the enzymes of the lysosome mix (3). The antigen is broken down to fragments, or degraded (4). Most of the antigen residue is expelled

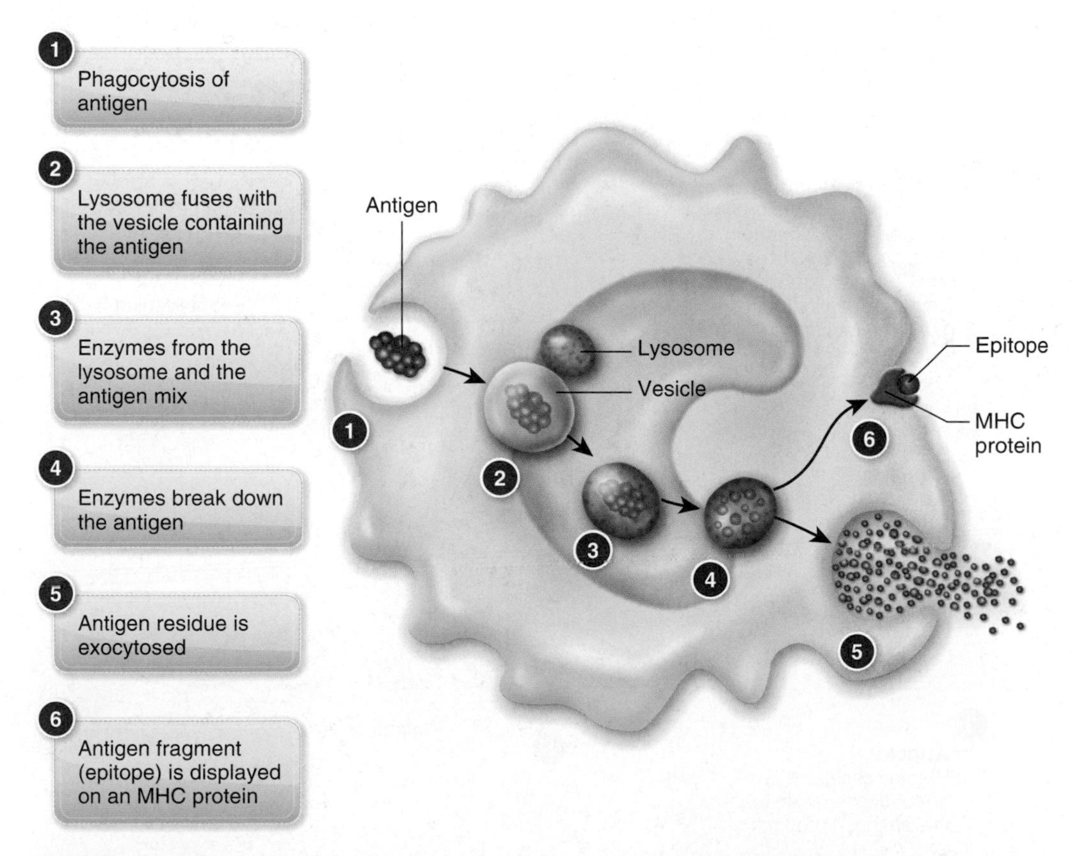

FIGURE 11.16 Antigen-presenting cell in the process of antigen presentation.

from the cell by exocytosis (5). Some of the antigen fragments **(epitopes)** are displayed on an **MHC protein** on the surface of the APC (6).

An MHC protein is like a billboard posting what the APC has sampled. *MHC* stands for **major histocompatibility complex.** Other cells in the body have MHC proteins too. Unlike APC cells that sample their surrounding environment and post what they find, these other body cells present what is inside themselves on their MHC protein. In that case, the MHC protein is posting, "This is me." Therefore, an MHC protein displays what is self and what is foreign. Specific immunity hinges on being able to tell the difference. If the MHC protein displays self-antigens, nothing happens. However, if the epitope in the MHC is foreign, a specific immune response is initiated. Below, we explain how that is accomplished in **humoral immunity,** one form of specific immunity.

11.11 learning outcome

Explain the process of humoral immunity.

Humoral (Antibody-Mediated) Immunity

This form of specific immunity involves B cells making antibodies to attack a foreign antigen. It begins when a B cell (APC) in lymphoid tissue displays an epitope from its environment on an MHC protein. A T_{helper} cell passing by either does nothing because the epitope is self or reacts to it by binding to the B cell because it recognizes the epitope as foreign. (Remember all T cells that react to self are destroyed in the thymus). The T_{helper} cell then communicates to the B cell by releasing a chemical **(interleukin-2)** that tells the B cell that the epitope is foreign and that the B cell should clone itself. The steps up to this point have been the *recognize* stage of specific immunity. See **Figure 11.17.**

So under the direction of the T_{helper} cell, the B cell (still in the lymphoid tissue) begins to clone itself in the germinal centers in the lymphatic nodules. This clone develops **(differentiates)** into two types of B cells—plasma B cells that start to produce specific antibodies (to attack the specific antigen that was displayed previously) and memory B cells (that do nothing now). The antibodies produced by

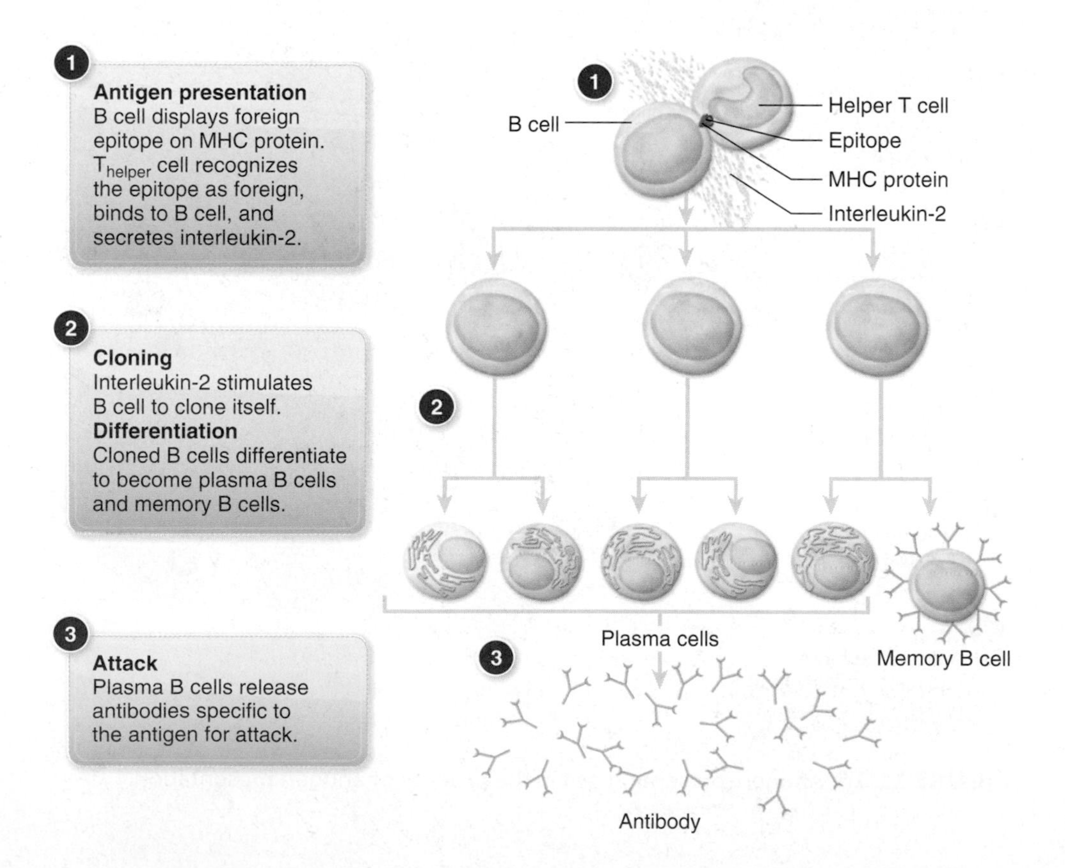

FIGURE 11.17 Humoral immunity: (1) antigen presentation and T_{helper}-cell recognition, (2) cloning and differentiation, (3) antibody production for the attack.

the plasma B cells leave the lymphoid tissue with the lymph and enter the blood at the subclavian veins. From there they travel throughout the body seeking out the specific antigen from the specific pathogen wherever it may be. The antibodies may cover up the binding sites of the foreign invader (rendering the invader harmless), activate the complement system, or agglutinate the antigen so that macrophages can phagocytize it. Note that the B cell does not need to be present at the attack site because antibodies, which act as guided missiles, are sent from bunkers in the lymphoid tissues where the B cells reside. This paragraph describes the *react* stage of humoral immunity. Again, see **Figure 11.17**.

spot check 8 What method of attack do antibodies to blood-typing antigens (A, B, Rh) use? (*Hint:* See Chapter 9.)

It takes 3 to 6 days for humoral immunity to accomplish antibody production in the first exposure to the pathogen. It will take another 10 days before the amount of antibody production reaches its peak. Once the pathogen is defeated, the amount of antibody in the system decreases, but it never totally drops to zero. If the pathogen enters the body again, B_{memory} cells will recognize it immediately. The B_{memory} cells will then increase antibody production to reach a peak in approximately 2 to 5 days, instead of 13 to 16. See **Figure 11.18**. In this way, the pathogen will likely be defeated before any signs of its presence are even noticed. With a repeated exposure like this, antibody production will stay high because the immune system has learned that this pathogen reoccurs. Specific immunity does not prevent a pathogen from entering the body. Instead, it fights it so much faster and stronger with repeated exposure that the pathogen is defeated before it can make you sick. This paragraph describes the *remember* stage of specific immunity.

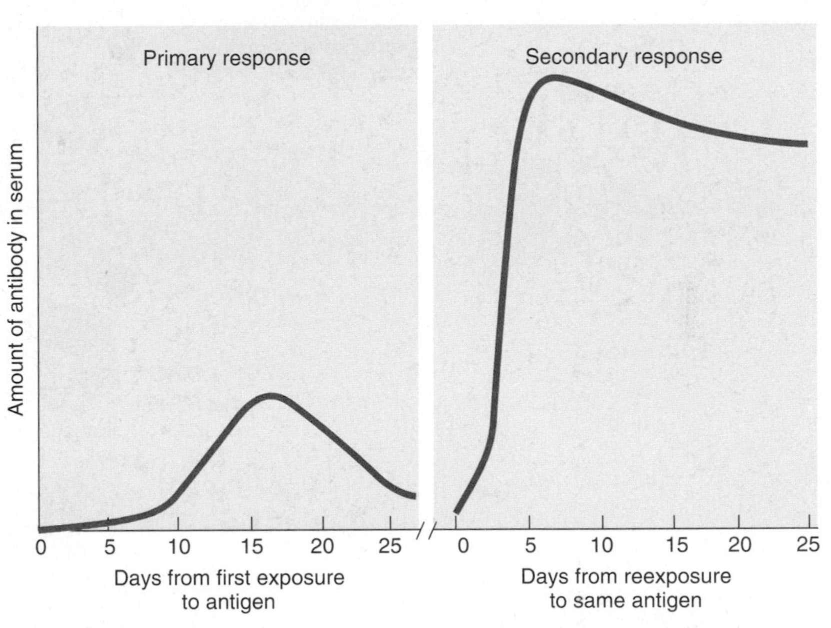

FIGURE 11.18 Graph of primary and secondary response in humoral immunity.

11.12 learning outcome

Explain the process of cellular immunity.

Cellular (Cell-Mediated) Immunity

This is another form of specific immunity, and, as such, it works on the principles of recognize, react, and remember. However, **cellular immunity** is a little more complicated.

Like humoral immunity, cellular immunity starts with an antigen-presenting cell, or any other cell presenting something on an MHC molecule. The epitope can be an antigen the cell sampled from its external environment (such as an APC) or a fragment of something in the cytoplasm of the cell itself (such as other body cells). The epitope may even be a part of an unusual (foreign) protein formed inside a cancer cell. In this form of immunity, a T_{helper} or $T_{cytotoxic}$ cell reacts by binding to the APC because it recognizes the epitope as being foreign. It verifies that the epitope is foreign by binding to a costimulation protein on the APC, if there is one. Then either T cell releases **interleukin-1** to cause the T cell to clone itself to become many $T_{cytotoxic}$ cells (T_C), T_{helper} cells (T_H), and T_{memory} cells (T_M) that all recognize this antigen as foreign (*recognize* stage). See **Figure 11.19**.

FIGURE 11.19 Cellular immunity: (1) antigen recognition, (2) costimulation to verify the epitope is foreign, (3) cloning and differentiation, (4) lethal hit or interleukin secretion to initiate other outcomes.

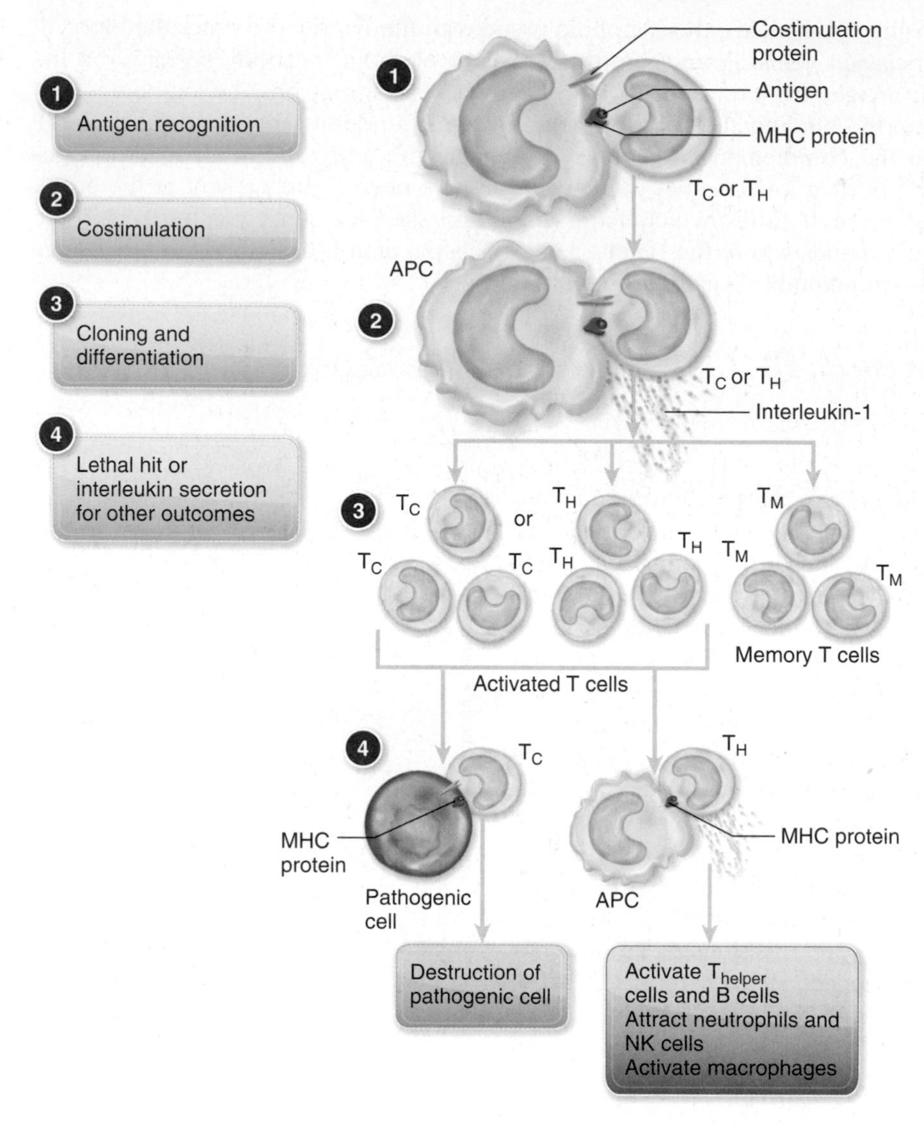

An activated $T_{cytotoxic}$ cell then travels throughout the body seeking cells with this specific foreign antigen. If it finds the foreign antigen, it docks to the cell and delivers a lethal hit to the cell. Unlike the antibodies released from B cells safe and secure in lymphoid tissue, $T_{cytotoxic}$ cells mount a direct cell-to-cell attack. The T_{helper} cells of the clone secrete interleukins to attract neutrophils and NK cells to the area, attract and activate macrophages to clean up any debris, and further activate more $T_{cytotoxic}$ and B cells. Of the T cells, only $T_{cytotoxic}$ cells directly attack a pathogen or cancer cell (*react* stage) in this form of specific immunity. See **Figure 11.20**.

Cellular immunity is effective against virally infected cells. Viruses are basically pieces of nucleic acids surrounded by a protein coat. They penetrate a cell of choice (specific to each virus) and insert their viral (foreign) nucleic acid into the DNA of the cell. The cell then drops its normal function to become a viral factory, producing more and more virus until the cell bursts with all of the virus it has produced. The free virus (enclosed in its protein coat) then seeks out other cells to invade. $T_{cytotoxic}$ cells destroy the self cell that has been turned into a viral factory.

T_{memory} cells stand by until the pathogen reoccurs in the body. If it does reoccur, these cells mount a cellular immunity response that is faster and stronger than the initial response (*remember* stage).

spot check 9 How do the locations of the lymphocytes involved in humoral and cellular immunity differ during the attack on the pathogen?

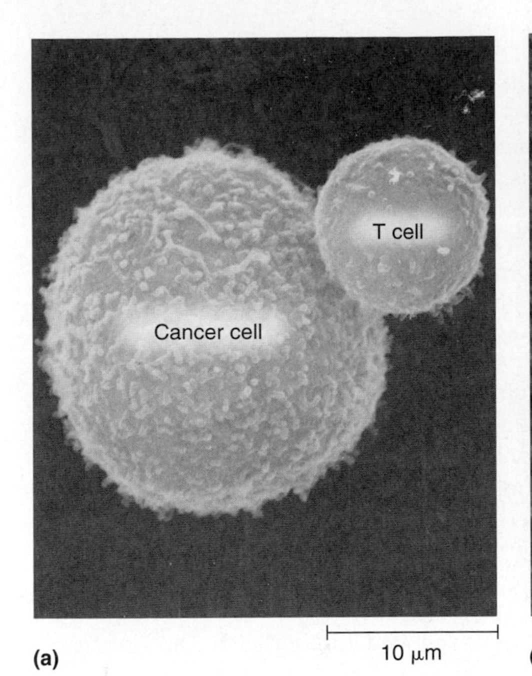

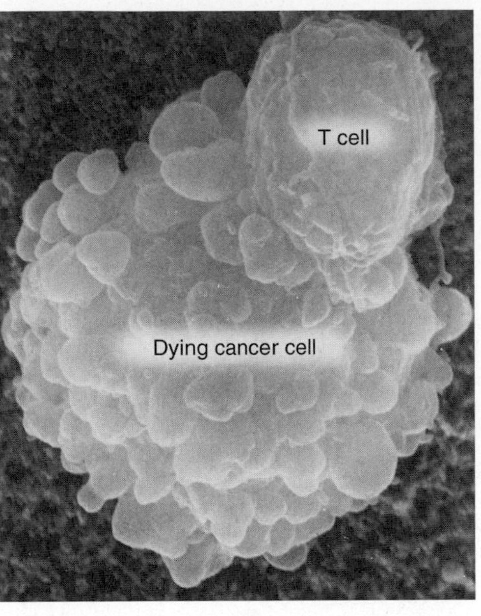

FIGURE 11.20 $T_{cytotoxic}$ **cell delivering a lethal hit to a cancer cell:** (a) $T_{cytotoxic}$ cell docks to a cancer cell, (b) $T_{cytotoxic}$ cell delivers a lethal hit to the cancer cell.

11.13 learning **outcome**

Compare the different forms of acquired immunity.

Forms of Acquired Immunity

Another way of looking at specific immunity is to consider how it is acquired. The four terms that become relevant in this discussion are explained in the following list:

- **Passive** is used to indicate that the immunity was acquired through someone or something else (an animal like a horse or a pig).
- **Active** is used to indicate that the body actively created its own immunity.
- **Natural** is used to indicate that the immunity was accomplished through naturally occurring means.
- **Artificial** is used to indicate that the immunity was not acquired through naturally occurring means.

These terms are used to classify the four types of acquired immunity:

- **Natural active immunity** is what has been described in the previous explanations of humoral and cellular immunity. A pathogen invades the body through everyday activities; the body responds by recognizing the specific pathogen as foreign and reacting to it by producing antibodies or activating $T_{cytotoxic}$ cells to destroy the pathogen. The body then remembers the specific pathogen so that it can fight the pathogen faster and stronger if it reappears. An example of this is catching a cold from someone in your class. Your body recognizes the cold virus as foreign, activates $T_{cytotoxic}$ cells to destroy your virally infected cells, and then remembers that specific cold virus so you do not get that cold again. The fact that you may get a cold every year does not mean your immune system is faulty. It just means that there are many different cold viruses. Specific immunity is specific to each cold virus.
- **Natural passive immunity** means that the body has acquired specific immunity through natural means from someone else. Some antibodies can pass from mother to child through breast milk. The child has specific immunity to some pathogens because the child has acquired its mother's antibodies. This is one of many reasons some mothers choose to nurse their infants.

- **Artificial active immunity** occurs when the body acquires a pathogen in an artificial way and then develops its own humoral or cellular immunity. An example of this is the smallpox, polio, and chickenpox vaccinations mentioned at the beginning of this chapter. Deborah was vaccinated on her arm with weakened antigens from the smallpox virus. So she did not come in contact with the antigens through normal, everyday activities—a health care worker scratched the skin of Deborah's arm and purposely exposed her to the antigen. Deborah's body recognized the antigen as foreign, reacted to it by making antibodies, and now remembers that antigen, so it will attack it faster and stronger if it should ever enter Deborah's body again. The scar on her arm is the result of the attack.
- **Artificial passive immunity** is acquired when an individual receives an injection of serum containing antibodies from another person or an animal such as a horse or pig. The effect is temporary because the body has not actively developed the mechanism to replace the injected antibodies that will eventually be used up. This type of immunity is used for the emergency treatment of tetanus, rabies, and snakebites.

spot check (10) Which of the four types of acquired immunity listed above is an injection of Rhogam for Rh– mothers? (*Hint:* See Chapter 9.)

You have now become familiar with the three lines of defense in the body. Note that T cells are important in more than one line of defense. We summarize their role below.

11.14 learning outcome

Explain the importance of T_{helper} cells to specific and nonspecific defense.

Importance of T_{helper} Cells in Nonspecific Resistance and Specific Immunity

As you can see in **Figure 11.21**, T_{helper} cells provide a vital role in nonspecific defense and specific immunity. T_{helper} cells activate macrophages for nonspecific defenses such as inflammation and fever. T_{helper} cells are also important for both forms of specific immunity. In humoral immunity, these cells first recognize what is foreign and then release interleukin-2 to have B cells react by cloning themselves and producing antibodies. In cellular immunity, T_{helper} cells recognize what is foreign and release interleukin-1 to get $T_{cytotoxic}$ cells to clone themselves and attack. Keep in mind the importance of these cells when you read about disorders and the HIV virus later in the chapter.

You can now combine all of this information concerning anatomy and physiology to clearly understand the functions of this system.

11.15 learning outcome

Explain the functions of the lymphatic system.

Functions of the Lymphatic System

Remember Andre from Chapter 9 (**Figure 11.22**)? In the following list, we explain the different functions of Andre's lymphatic system as a general example of how a healthy lymphatic system works.

- **Fluid balance.** Every minute of every day, Andre loses fluid (lymph) from his cardiovascular system. The lymph washes over his tissues, delivering nutrients and removing wastes. It is collected by open-ended lymph vessels, which return it back to his bloodstream at his subclavian veins.
- **Lipid absorption.** The lymph drained from Andre's digestive system organs carries the products of lipid digestion from the glass of milk he had for breakfast this morning. You will learn more about this in the digestive system chapter.

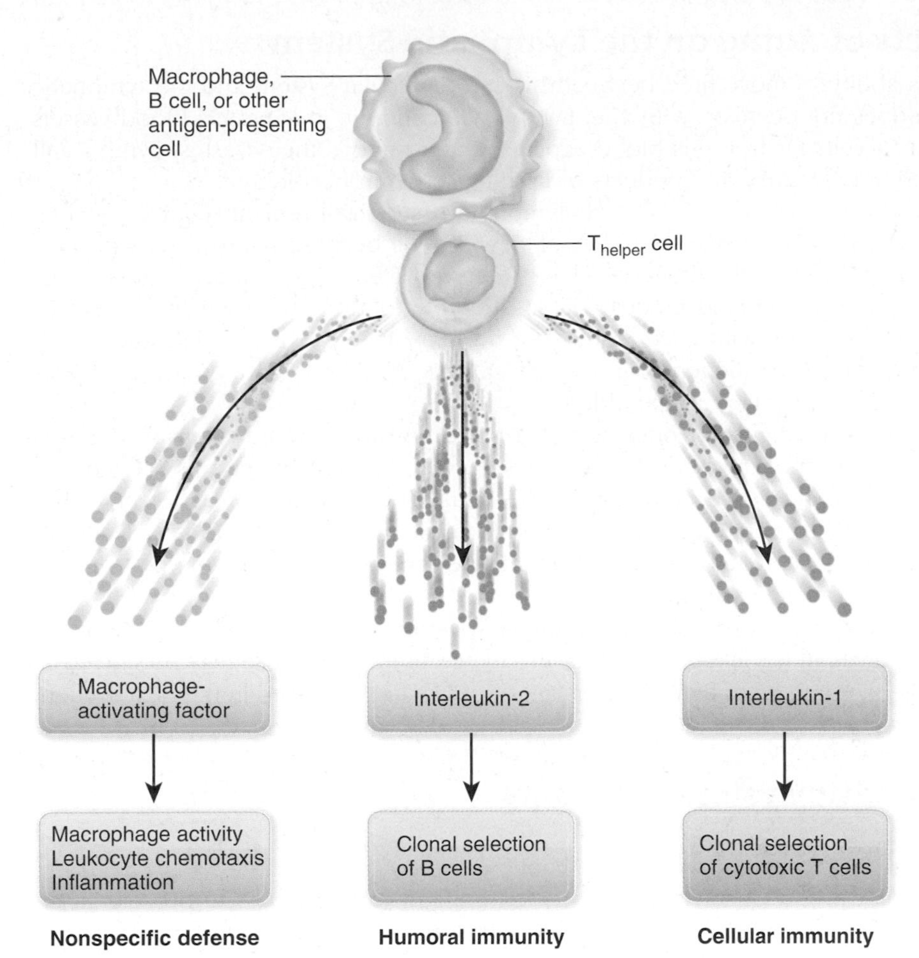

FIGURE 11.21 Importance of T_{helper} cells.

- **Defense against disease.** Although the skin as an external barrier has been broken with the scrape on his knee, Andre's other nonspecific defenses are at work to destroy any pathogens that may have entered the damaged tissue. Dendritic cells in the skin are serving as APCs to present foreign antigens to T_{helper} cells so that macrophages can be activated. These T_{helper} cells may also activate the complement system. The inflammatory process has started, too, and will bring neutrophils to the area. The neutrophils will crawl out of the dilated blood vessels and move to the damaged tissue, consuming bacteria along the way.
- **Immunity.** Andre is still very young. His thymus gland is still growing and maturing the T cells that detect foreign antigens and destroying those that react to his own cells. His T cells will be vital for his lymphatic system to accomplish the third line of defense, whether it be humoral or cellular immunity. Andre's immune system is capable of making at least 10 billion different antibodies, each specific to a particular pathogen. Hopefully, Andre will never be exposed to that many different pathogens, but his capacity for developing specific immunity is there as he needs it.

What will happen to this system when Andre becomes an old man? We focus next on the effects of aging on the lymphatic system.

FIGURE 11.22 Andre.

11.16 learning **outcome**

Summarize the effects of aging on the lymphatic system.

Effects of Aging on the Lymphatic System

Andre's ability to move fluid between his cardiovascular system and the lymphatic system does not decrease with age. Lymph will continue to leave his blood vessels to nourish cells far from his blood capillaries and remove their wastes. Lymph will also continue to carry the products of lipid digestion in his old age.

Andre's number of B cells in his lymphoid tissues will remain relatively stable. What will be affected is his number of new T cells because his thymus will have shrunk and much of the tissue will have been replaced with connective tissue. His T cells in other lymphoid tissues will still be able to clone themselves, but not as many will be made with each clone. The decrease in T_{helper} cells could mean that his recognition of pathogens will be slower. This slowdown may be a reason that cancer is more prevalent in the elderly.

Vaccinations might not offer as much protection as they did when Andre was younger. A good example of this is Deborah, who had chickenpox as a child and developed natural active immunity to the virus. The virus remained latent in her nerves while her immune system kept it from becoming active. But as Deborah ages, her immune system slows down, and the same virus may reemerge to cause painful lesions called **shingles.**

The age-related changes to the immune system may have a positive effect if the elderly individual has allergies because this hyperimmune response may be slowed as well. Allergies and other lymphatic system disorders are discussed in the next section.

11.17 learning **outcome**

Describe lymphatic system disorders.

Lymphatic System Disorders

Allergies

Allergies are hypersensitivities to a foreign antigen **(allergen).** The process in an allergy is the same as that in an immune response—the immune system recognizes the foreign antigen, reacts to it, and then remembers it so that the system can mount a faster and stronger attack if the antigen should ever occur in the body again. The difference between an allergic reaction and a normal immune response is that the allergic response produces undesirable side effects such as increased inflammation. The effect may even be lethal.

anaphylaxis: AN-ah-fih-LAK-sis

clinical **point**

Anaphylaxis is an example of an immediate allergic reaction that can be life-threatening. An immune response to penicillin or bee stings is the most common cause. In anaphylaxis, systemic vasodilation (systemwide dilation of blood vessels) within a few minutes of exposure to the allergen can cause a drop in blood pressure and even cardiac failure.

Another example of an immediate hypersensitivity is asthma, in which the allergen is inhaled. The allergen triggers the release of histamines in the bronchioles of the lungs, and this causes the bronchioles to constrict, making breathing difficult. You will learn more about asthma in the next chapter.

Other allergies, called *delayed hypersensitivities,* may take hours or even days to develop side effects. Examples of these allergies include contact hypersensitivities to poison ivy, poison oak, soaps, or cosmetics. The allergen in this type of allergy comes in contact with epithelial cells (skin, mucous membranes).

Next, T cells initiate inflammation, which causes excessive itching. Scratching the affected area further damages tissue.

Allergies are hypersensitivities to foreign antigens, but some disorders of this system do not involve foreign antigens at all. These disorders are autoimmune disorders, in which the body's immune system attacks the body's own tissues.

Autoimmune Disorders

Rheumatoid arthritis, Graves' disease, and myasthenia gravis are examples of some of the autoimmune disorders you have already studied in previous chapters. So, why does the immune system attack its own tissues? One explanation is **molecular mimicry** (where one molecule is so similar in structure to another molecule that it is mistaken for the other molecule). To understand this, picture an action movie with a highly paid star. The movie studio protects its investment in the star by replacing the star with a stunt double during dangerous action scenes. Since the stunt double's appearance is so similar to the star, the viewer never notices the difference. The same mistaken identity happens in molecular mimicry. An APC presents an epitope for a newly acquired foreign pathogen on its MHC protein. This epitope is unique but is very similar in shape to a self-antigen in the body. A T_{helper} cell recognizes the epitope as foreign and continues the process for humoral immunity, cellular immunity, or both forms of specific immunity. After the pathogen is defeated, the immune system continues to act against the self-antigen that is so similar. Here, the immune system is mistaking self-tissue as foreign tissue. The immune system is designed to fight the pathogen for life, which makes this mistake a lifelong problem that is very difficult to treat. Immunosuppressant drugs may help to manage the progression of the disease.

You have seen how the immune system can overreact, in the case of allergies, and mistakenly react, in autoimmune disorders. The next disorder you will study results when the immune system fails to react whatsoever.

AIDS

AIDS is an acronym for **acquired immunodeficiency syndrome.** *Acquired* means that it is not inherited. *Immunodeficiency* means that the immune system fails to provide protection against pathogens. *Syndrome* means that the disorder is a collection of symptoms and signs of disease. Acquired immunodeficiency syndrome is the final stage of a human immunodeficiency virus (HIV) infection.

How does this virus render the immune system deficient? The answer lies in the type of cell that HIV chooses to invade—T_{helper} cells. By invading these T cells and turning them into viral factories, HIV prevents the processes of specific immunity and nonspecific resistance from even starting (explained in the section on the importance of T_{helper} cells, earlier in the chapter). As soon as the infected T cell has produced its fill of new HIV particles, it bursts and is destroyed. With the loss of T_{helper} cells, the immune system loses its ability to recognize what is foreign. See **Figure 11.23**.

An HIV test to determine whether a person is infected with the virus looks for the presence of HIV antibodies. Even this (confirming an actual HIV infection) is complicated by HIV's invasion of T_{helper} cells. Antibodies should normally peak in 13 to 16 days; but in the case of HIV infections, antibodies may take two to eight weeks to reach detectable levels, and in some rare cases it may take six months before there are sufficient antibodies to produce an HIV-positive result.

A T-cell count is a good indicator of an HIV infection's progression. Note that a normal T cell count is 600 to 1,200 cells per cubic millimeter (mm^3). AIDS is indicated if the T-cell count is less than 200 cells/mm^3. Without enough T_{helper} cells to recognize a foreign pathogen, the body loses its ability to fight **opportunistic infections** (infections normally fought off by healthy immune systems).

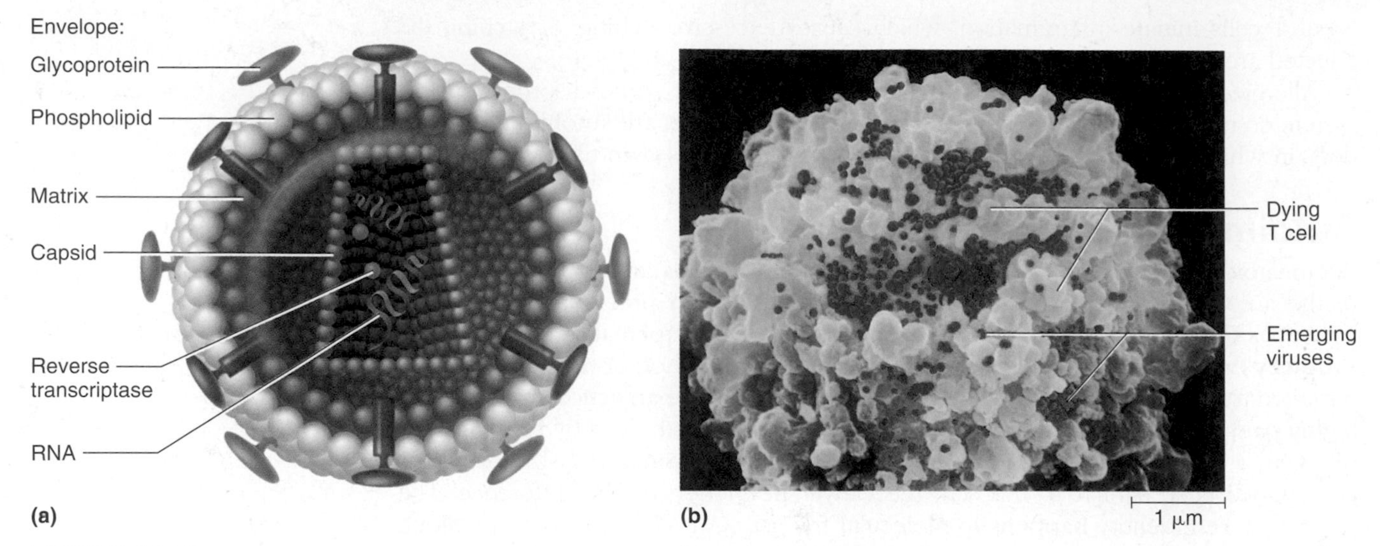

FIGURE 11.23 Infected T cell: (a) HIV virus, (b) dying T cell with emerging HIV.

One such opportunistic infection is Kaposi sarcoma—a common cancer of the epithelial cells lining blood vessels, characterized by bruiselike lesions on the skin. See **Figure 11.24**. Other opportunistic infections include pathogens like *Pneumocystis* (a group of respiratory fungi), herpes simplex virus, tuberculosis bacteria, and cytomegalovirus. Death from infection is inevitable once AIDS has been diagnosed. However, it may take a few months to several years from the time of an AIDS diagnosis.

Currently, there is no cure for AIDS. However, many people have been able to live decades with an HIV infection before it develops into AIDS because of the availability of effective combinations of drugs introduced in the 1990s.

FIGURE 11.24 Kaposi sarcoma.

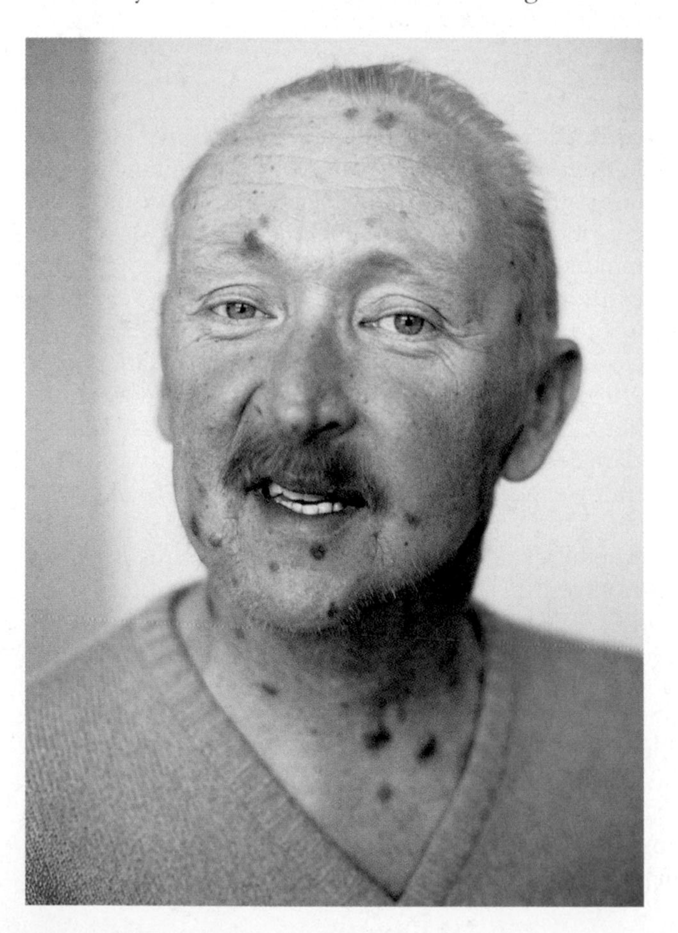

putting the pieces together

The Lymphatic System

Integumentary system

Has dendritic cells to guard against pathogens; protects against fluid loss.

Sends white blood cells to fight pathogens in the integumentary system.

Skeletal system

Red bone marrow produces white blood cells

Sends white blood cells to fight pathogens in the skeletal system.

Muscular system

Moves lymph through lymph vessels so it can be returned to the cardiovascular system.

Sends white blood cells to fight pathogens in the muscular system.

Nervous system

Microglia serve as immune system cells to fight pathogens in the CNS.

Endocrine system

Glucocorticoids suppress the immune system and reduce inflammation.

Sends white blood cells to fight pathogens in endocrine system glands.

Cardiovascular system

Provides fluid for lymph.

Sends white blood cells to fight pathogens in the cardiovascular system.

Respiratory system

Provides O_2 for lymphatic tissues and removes CO_2; thoracic pump helps return lymph to the cardiovascular system.

Sends white blood cells to fight pathogens in the respiratory system.

Digestive system

Provides nutrients to lymphatic tissues.

Transports products of lipid digestion; sends white blood cells to fight pathogens in the digestive system.

Excretory/urinary system

Kidney maintains fluid and electrolyte balance.

Sends white blood cells to fight pathogens in the excretory/urinary system.

Reproductive system

Sustenacular cells form a blood-testis barrier to isolate developing sperm from the immune system.

Sends white blood cells to fight pathogens in the reproductive system.

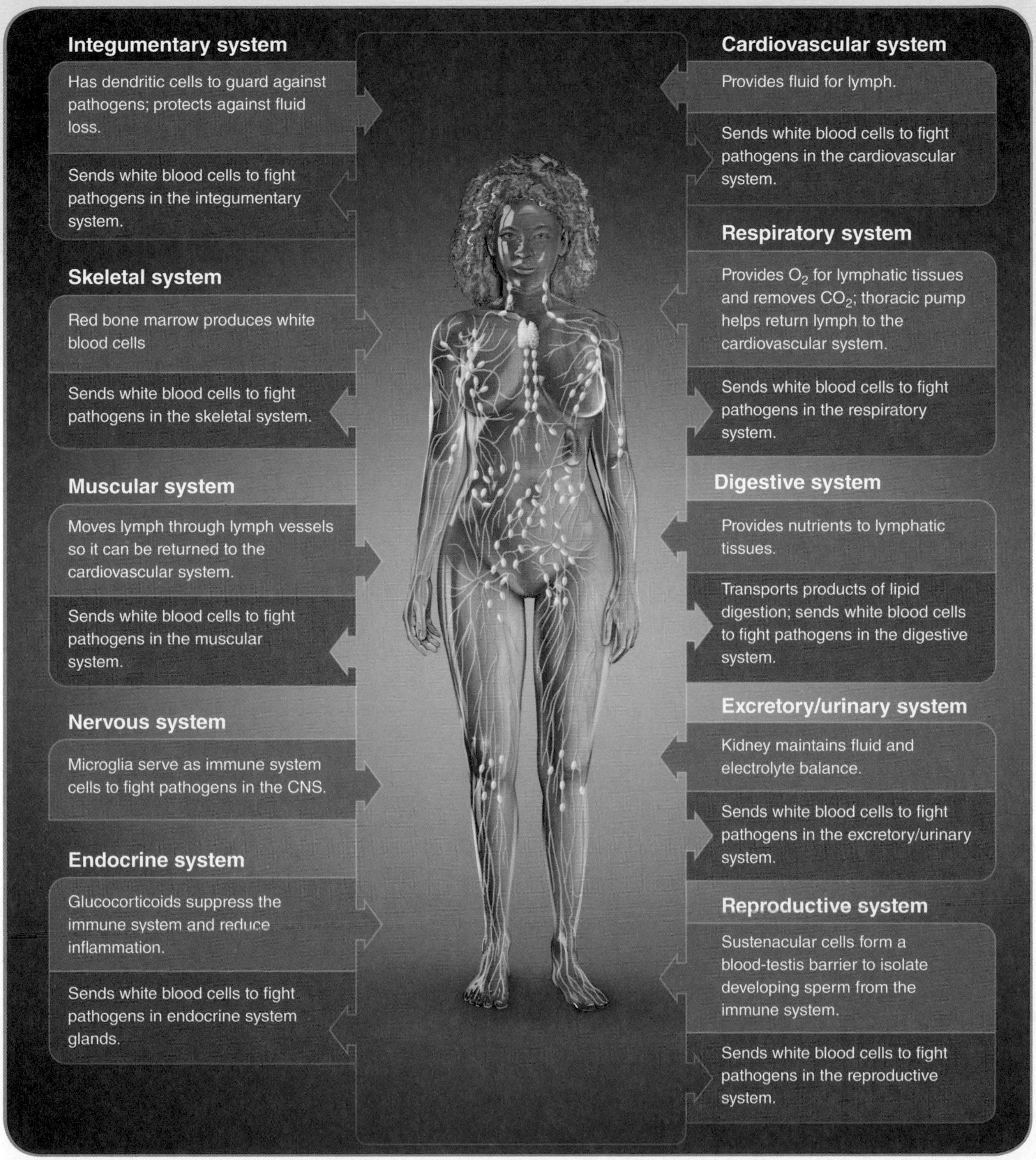

FIGURE 11.25 Putting the Pieces Together—The Lymphatic System: connections between the lymphatic system and the body's other systems.

summary

Overview

- The lymphatic system involves the circulation of lymph, a fluid derived from blood.

Anatomy of the Lymphatic System

Lymph and Lymph Vessels

- Lymph is derived from plasma but has fewer dissolved proteins.
- Lymph washes over tissues to deliver nutrients and remove wastes, cell debris, bacteria, viruses, and loose (possible cancerous) cells.
- Lymph leaves blood vessels due to blood pressure.
- Lymph is returned to the bloodstream by open-ended lymph vessels through the skeletal muscle action.
- Lymphatic collecting ducts return lymph to the bloodstream at the subclavian veins.

Cells of the Lymphatic System

- NK cells are lymphocytes that destroy bacteria, fight against transplanted tissues, attack virally infected cells, and destroy cancer cells.
- T cells are lymphocytes that are important in nonspecific defense and specific immunity.
- There are four types of T cells: T_{helper} cells, $T_{cytotoxic}$ cells, T_{memory} cells, and $T_{regulatory}$ cells.
- B cells are lymphocytes that serve as APCs and are important in humoral immunity because they produce antibodies.
- Macrophages are monocytes that have migrated to the tissues, where they phagocytize bacteria, debris, and dead neutrophils.
- Dendritic cells are located in the epidermis and serve as APCs.

Lymphoid Tissues and Organs

- MALT is a scattering of lymphocytes in mucous membranes lining tracks to the outside environment.
- Peyer's patches are nodules of lymphocytes at the distal end of the small intestine.
- Lymph nodes filter lymph from different regions of the body on its way back to the bloodstream.
- Tonsils ring the pharynx to guard against pathogens entering the body through the nose or mouth.
- The thymus gland matures T cells that recognize foreign antigens and destroys T cells that react to self-antigens.
- The spleen has red pulp to store red blood cells and white pulp to store lymphocytes and macrophages.
- The spleen is a battle site for lymphocytes to attack pathogens. It regulates the amount of fluid in the blood by transferring excess fluid to the lymphatic system as lymph.

Physiology of the Lymphatic System

Three Lines of Defense

- The three lines of defense against pathogens are (1) external barriers; (2) inflammation, antimicrobial proteins, fever, and other active attacks; and (3) specific immunity.
- The first two lines are nonspecific defenses, while the third is specific immunity.

Nonspecific Resistance versus Specific Immunity

- Nonspecific defenses are widespread and function the same way every time.
- Specific immunity requires a prior exposure to a pathogen so that it can recognize, react, and remember the pathogen.
- Specific immunity reacts faster and stronger to repeated exposures to a pathogen.

Nonspecific Defenses

- External barriers include the skin and mucous membranes.
- Inflammation functions to limit the spread of pathogens, to remove debris and damaged tissue, and to initiate tissue repair.
- Inflammation involves the release of vasodilators from damaged tissue and basophils and the margination, diapedesis, and chemotaxis of leukocytes that phagocytize pathogens along the way.

- Antimicrobial proteins include interferon, produced by virally infected cells so that other healthy cells will make antiviral proteins, and the complement system of 20 inactive proteins that, when activated, can destroy pathogens in several ways.
- Fever is a defense initiated by pyrogens from macrophages that cause the hypothalamus to reset the body's temperature.
- Other attacks from leukocytes complete the list of nonspecific defenses.

Specific Immunity

- APCs present epitopes of what they have sampled from their external environment on MHC proteins.
- Other cells present internal self-antigens on MHC proteins.
- Humoral immunity involves B cells producing antibodies.
- Cellular immunity involves $T_{cytotoxic}$ cells directly killing cells with foreign antigens.
- Both types of specific immunity require T_{helper} cells to recognize what is foreign.
- Immunity can be acquired naturally or artificially through passive or active processes.

Importance of T_{helper} Cells in Nonspecific Resistance and Specific Immunity

- T_{helper} cells activate macrophages for nonspecific defense such as inflammation and fever.
- T_{helper} cells recognize what is foreign and release interleukin-2 to activate B cells in humoral immunity.
- T_{helper} cells recognize what is foreign and release interleukin-1 to activate $T_{cytotoxic}$ cells in cellular immunity.

Functions of the Lymphatic System

- The lymphatic system helps maintain the fluid balance in the blood.
- The lymphatic system distributes lymph to wash over tissues to deliver nutrients and remove wastes.
- The lymphatic system carries absorbed products of lipid digestion.
- The lymphatic system provides nonspecific defenses.
- The lymphatic system provides specific immunity against specific pathogens.

Effects of Aging on the Lymphatic System

- The ability to move fluid between the cardiovascular and lymphatic systems does not decrease with age.
- The number of B cells in the lymphoid tissues will remain relatively stable.
- The thymus gland shrinks with age.
- The number of new T cells decreases with age.
- The immune response may slow with age.

Lymphatic System Disorders

- Allergies are hypersensitivities to a pathogen that may have immediate or delayed side effects.
- Autoimmune disorders are the result of the immune system attacking self-antigens.
- AIDS is the final stage of an HIV infection in which the immune system fails to recognize foreign antigens.

key words for review

The following terms are defined in the glossary.

acquired immunity
acquired immunodeficiency syndrome (AIDS)
anaphylaxis
antigen-presenting cell
cellular immunity
chemotaxis
complement system
diapedesis
epitope
humoral immunity
interferons
interleukins
lymph
lymphadenitis
major histocompatibility complex (MHC)
margination
molecular mimicry
nonspecific resistance
pyrogen
specific immunity

chapter review questions

Word Building: *Use the word roots listed in the beginning of the chapter and the prefixes and suffixes inside the back cover to build words with the following meanings:*

1. Swelling due to excess lymph in the tissues: ______________________
2. Tumor of lymphoid tissue: ______________________
3. Removal of the spleen: ______________________
4. Disease of lymph nodes: ______________________
5. Medication to reduce the effectiveness of the immune system: ______________________

Multiple Select: *Select the correct choices for each statement. The choices may be all correct, all incorrect, or a combination of correct and incorrect.*

1. What is lymph, and what does it do?
   a. Lymph is a fluid derived from blood.
   b. Lymph drained from digestive organs carries the products of protein digestion.
   c. Lymph brings nutrients to cells that are not adjacent to blood capillaries.
   d. Lymph may carry cancer cells.
   e. Lymph carries wastes.
2. What are the anatomy and function of lymph vessels?
   a. Lymph capillaries are closed vessels.
   b. The thoracic duct delivers blood to the left subclavian vein.
   c. Valves in lymph vessels direct the flow of lymph to the lymph capillaries.
   d. Lymph vessels are permeable to allow lymph to leave the vessels.
   e. The right lymphatic duct collects lymph from the right side of the head, the right arm, and the right side of the thorax.
3. What is included in the three lines of defense against pathogens?
   a. Neutrophils phagocytizing bacteria.
   b. Inflammation from a mosquito bite.
   c. A fever from the flu.
   d. Natural active immunity.
   e. Skin.
4. How do nonspecific resistance and specific immunity compare?
   a. Nonspecific resistance and specific immunity are mutually exclusive.
   b. Both forms of defense require $T_{cytotoxic}$ cells.
   c. Both forms of defense respond faster and stronger to repeated exposures to a pathogen.
   d. Both forms of defense require an antigen to be presented on an MHC protein in the first exposure.
   e. Both forms of defense go through stages to recognize, react, and remember a specific pathogen.
5. How does an APC function in specific immunity?
   a. An APC produces antibodies.
   b. An APC produces interferons.
   c. An APC presents epitopes.
   d. An APC delivers a lethal hit to virally infected cells.
   e. An APC samples its external environment.
6. Why are T_{helper} cells so important?
   a. T_{helper} cells activate macrophages.
   b. T_{helper} cells tell B cells to make antibodies.
   c. T_{helper} cells tell other T cells what to kill.
   d. T_{helper} cells release interleukins.
   e. T_{helper} cells destroy self-antigens.

7. Which of the following statements describe(s) a function of the lymphatic system?
   a. The lymphatic system helps maintain the fluid balance in the blood.
   b. The lymphatic system distributes lymph to wash over tissues to deliver nutrients and take away wastes.
   c. The lymphatic system carries absorbed products of lipid digestion.
   d. The lymphatic system provides nonspecific defenses.
   e. The lymphatic system provides specific immunity against specific pathogens.
8. What can Nancy expect to happen as her lymphatic system ages?
   a. Lymph will build up in her tissues.
   b. Lymph will not move out of her blood vessels as easily.
   c. The number of her B cells will decrease.
   d. The number of her T cells will remain stable.
   e. Her allergies may diminish as she ages.
9. Mike is an intern at the local hospital. He was accidentally stuck with a contaminated needle while working in the ER. He is concerned about contracting an HIV infection. Which of the following statements is (are) accurate concerning HIV?
   a. HIV's cell of choice to invade is a B cell.
   b. An HIV test checks for the presence of antibodies to HIV.
   c. An HIV infection affects nonspecific defenses.
   d. An HIV infection affects specific immunity.
   e. It may take as long as six months before a conclusive outcome can be determined from an HIV test.
10. Which of the following statements is (are) accurate regarding the different forms of acquired immunity?
   a. Antibodies in breast milk are an example of natural active immunity.
   b. A vaccination of weakened antigens is an example of artificial active immunity.
   c. A shot of rabies antibodies is an example of artificial passive immunity.
   d. Fighting the flu is an example of natural passive immunity.
   e. Fighting a cold is an example of natural artificial immunity.

Matching: *Match the lymphoid tissue to its function. Some answers may be used more than once. Some questions may have more than one correct response.*

_____	1. Transfers excess fluid in the blood to the lymphatic system	a. Thymus gland
_____	2. Matures T cells	b. Lymph nodes
_____	3. Defends against pathogens entering the mouth or nose	c. MALT
_____	4. Protects against pathogens in the urinary tract	d. Spleen
_____	5. Filters lymph	e. Tonsils

Matching: *Match the cell to its function. Some answers may be used more than once. Some questions may have more than one correct response.*

_____	6. Destroys bacteria	a. NK cell
_____	7. Involved in humoral immunity	b. B cell
_____	8. Makes antibodies	c. T_{helper} cell
_____	9. Involved in cellular immunity	d. $T_{cytotoxic}$ cell
_____	10. Involved in nonspecific defense	e. Macrophage

Critical Thinking

1. Andrea found a lump in her right breast. After several rounds of chemotherapy, she consented to have a radical mastectomy on the advice of her oncologist. The surgery included the removal of her right breast and her right axillary lymph nodes, which were found to contain cancerous cells. What effect, if any, might this surgery have on the lymphatic drainage of her right arm? Explain. (*Hint:* Closely examine Figures 11.8 and 11.9.)

2. A mantoux test is administered to health care workers to check for exposure to the pathogen that causes tuberculosis. A small amount of weakened tuberculin antigens is injected into the skin of the anterior forearm, and 48 hours later the test is "read" by observing the location of the injection. Inflammation at the injection site indicates that the individual has likely been exposed to the tuberculosis bacteria. A follow-up chest x-ray is then given to look for the presence of the disease. Explain why there would be inflammation at the site 48 hours after the test was administered if the individual had been previously exposed to the pathogen. Why is the timing so important?
3. Hemophilia patients are treated with injections of the clotting factors they are missing. Anemia patients are sometimes treated with blood transfusions. Would giving transfusions of lymphocytes to HIV-infected patients with low T-cell counts be effective? Explain your answer.

11 chapter mapping

This section of the chapter is designed to help you find where each outcome is covered in this text.

	Outcomes	Readings, figures, and tables	Assessments
11.1	Use medical terminology related to the lymphatic system.	Word roots: p. 411	Word Building: 1–5
11.2	Explain the origin and composition of lymph.	Lymph and lymph vessels: p. 412	Multiple Select: 1
11.3	Describe lymph vessels.	Lymph and lymph vessels: p. 412 Figures 11.2, 11.3	Multiple Select: 2
11.4	Explain the route of lymph from the blood and back again.	Lymph and lymph vessels: pp. 412–416 Figures 11.4–11.6	Spot Check: 1, 2 Critical Thinking: 1
11.5	Describe cells of the lymphatic system and list their functions.	Cells of the lymphatic system: pp. 416–417	Matching: 6–10
11.6	Identify lymphoid tissues and organs and explain their functions.	Lymphoid tissues and organs: pp. 417–423 Figures 11.7–11.13	Spot Check: 3, 4 Matching: 1–5 Critical Thinking: 3
11.7	Summarize three lines of defense against pathogens.	Three lines of defense: p. 423	Multiple Select: 3
11.8	Contrast nonspecific resistance and specific immunity.	Nonspecific resistance versus specific immunity: p. 424	Multiple Select: 4
11.9	Describe the body's nonspecific defenses.	Nonspecific defenses: pp. 424–427 Figures 11.14, 11.15	Spot Check: 5–7
11.10	Explain the role of an APC in specific immunity.	Specific immunity: pp. 427–428 Figure 11.16	Multiple Select: 5
11.11	Explain the process of humoral immunity.	Humoral immunity: pp. 428–429 Figures 11.17, 11.18	Spot Check: 8, 9 Multiple Select: 4 Critical Thinking: 2
11.12	Explain the process of cellular immunity.	Cellular immunity: pp. 429–431 Figures 11.19, 11.20	Spot Check: 9 Multiple Select: 4

	Outcomes	Readings, figures, and tables	Assessments
11.13	Compare the different forms of acquired immunity.	Forms of acquired immunity: pp. 431–432	Spot Check: 10 Multiple Select: 10
11.14	Explain the importance of T_{helper} cells to specific and nonspecific defense.	Importance of T_{helper} cells in nonspecific resistance and specific immunity: p. 432 Figure 11.21	Multiple Select: 6
11.15	Explain the functions of the lymphatic system.	Functions of the lymphatic system: pp. 432–433 Figure 11.22	Multiple Select: 1, 7
11.16	Summarize the effects of aging on the lymphatic system.	Effects of aging on the lymphatic system: p. 434	Multiple Select: 8
11.17	Describe lymphatic system disorders.	Lymphatic system disorders: pp. 434–436 Figures 11.23, 11.24	Multiple Select: 9

references and resources

Beers, M. H., Porter, R. S., Jones, T. V., Kaplan, J. L., & Berkwits, M. (Eds.). (2006). *Merck manual of diagnosis and therapy* (18th ed.). Whitehouse Station, NJ: Merck Research Laboratories.

Centers for Disease Control and Prevention. (2010, August 11). *Basic information about HIV and AIDS*. Retrieved August 24, 2010, from http://www.cdc.gov/hiv/topics/basic/index.htm.

Chikly, B. (1996). *Dr. Chikly's lymph drainage therapy*. Fairview, TN: Arthritis Trust of America/Rheumatiod Disease Foundation.

Porter, R. S., & Kaplan, J. L. (Eds.). (n.d.). Immune system: How the body ages. *Merck manual of health and aging* (chap. 2). Retrieved August 23, 2010, from http://www.merck.com/pubs/mmanual_ha/sec1/ch02/ch02o.html.

Porter, R. S., & Kaplan, J. L. (Eds.). (n.d.). Selected physiologic age-related changes. *Merck manual for healthcare professionals*. Retrieved March 19, 2011, from http://www.merckmanuals.com/media/professional/pdf/Table_337-1 .pdf?qt=selected physiologic age-related changes&alt=sh.

Saladin, Kenneth S. (2010). *Anatomy & physiology: The unity of form and function* (5th ed.). New York: McGraw-Hill.

Seeley, R. R., Stephens, T. D., & Tate, P. (2006). *Anatomy & physiology* (7th ed.). New York: McGraw-Hill.

Shier, D., Butler, J., & Lewis, R. (2010). *Hole's human anatomy & physiology* (12th ed.). New York: McGraw-Hill.

12 The Respiratory System

Everyone is anxiously waiting to hear that first breath as a baby is born. The parents may breathe a sigh of relief when they hear her first cry. It is the result of their baby's first intake of air from outside her body, but it certainly will not be her last. Her body will continue the process of breathing until death—24 hours a day, 365 days a year, for possibly 75 years or more. Rarely will she give her breathing conscious thought, yet day after day her respiratory anatomy will continue to perform the functions of the system. See Figure 12.1.

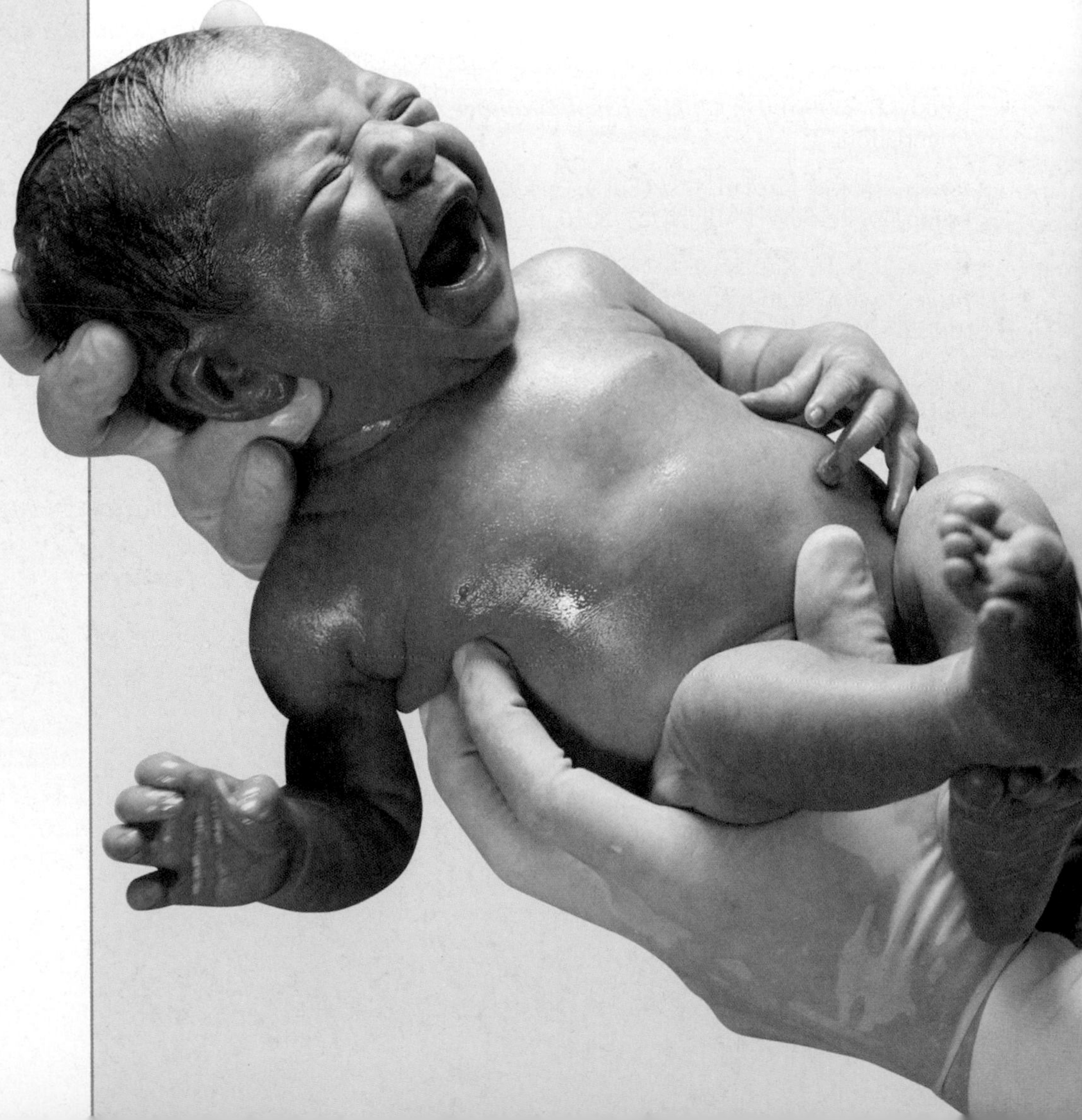

12.1 learning outcome

Use medical terminology related to the respiratory system.

learning outcomes

After completing this chapter, you should be able to:

12.1 Use medical terminology related to the respiratory system.

12.2 Trace the flow of air from the nose to the pulmonary alveoli and relate the function of each part of the respiratory tract to its gross and microscopic anatomy.

12.3 Explain the role of surfactant.

12.4 Describe the respiratory membrane.

12.5 Explain the mechanics of breathing in terms of anatomy and pressure gradients.

12.6 Define the measurements of pulmonary function.

12.7 Define partial pressure and explain its relationship to a gas mixture such as air.

12.8 Explain gas exchange in terms of the partial pressures of gases at the capillaries and the alveoli and at the capillaries and the tissues.

12.9 Compare the composition of inspired and expired air.

12.10 Explain the factors that influence the efficiency of alveolar gas exchange.

12.11 Describe the mechanisms for transporting O_2 and CO_2 in the blood.

12.12 Explain how respiration is regulated to homeostatically control blood gases and pH

12.13 Explain the functions of the respiratory system.

12.14 Summarize the effects of aging on the respiratory system.

12.15 Describe respiratory system disorders.

word roots & combining forms

alveol/o: alveolus, air sac

bronch/o: bronchial tube

bronchi/o: bronchus

bronchiol/o: bronchiole

capn/o: carbon dioxide

cyan/o: blue

laryng/o: larynx

lob/o: lobe

nas/o: nose

pharyng/o: pharynx

phren/o: diaphragm

pneum/o, pneumon/o: air

pulmon/o: lung

rhin/o: nose

sinus/o: sinus

spir/o: breathing

thorac/o: chest

trache/o: trachea

pronunciation key

alveoli: al-VEE-oh-lye

arytenoid: ah-RIT-en-oyd

bronchi: BRONG-kye

bronchus: BRONG-kuss

conchae: KON-kee

corniculate: kor-NIK-you-late

laryngopharynx: lah-RING-oh-FAIR-inks

larynx: LAIR-inks

nares: NAH-reez

pharynx: FAIR-inks

trachea: TRAY-kee-ah

Overview

The word *respiration* has several usages. In Chapter 2, you studied *cellular respiration* as a cellular process performed by mitochondria to release energy from the bonds in a glucose molecule. In Chapter 5, you studied *aerobic* and *anaerobic respiration* as variations of cellular respiration. In this chapter, you will study **respiration** first as the movement of air into **(inspiration)** and out of the lungs **(expiration),** commonly called breathing. Then you will explore respiration as the exchange of gases in two areas—between the air in the lungs and the blood in capillaries and between the blood in the capillaries and the tissues out in the body. Once you understand how the exchange of gases takes place, you will be prepared to investigate how gases are transported in the blood.

As with all of the other human body systems you have covered so far, it is important to understand the anatomy of the system before tackling the physiology. So you will begin below by studying the anatomy of this system.

Respiratory System

Major Organs and Structures:
nose, pharynx, larynx, trachea, bronchi, lungs

Accessory Structures:
diaphragm, sinuses, nasal cavity

Functions
gas exchange, acid-base balance, speech, sense of smell, creation of pressure gradients necessary to circulate blood and lymph

FIGURE 12.1 The respiratory system.

12.2 learning **outcome**

Trace the flow of air from the nose to the pulmonary alveoli and relate the function of each part of the respiratory tract to its gross and microscopic anatomy.

pharynx: FAIR-inks

larynx: LAIR-inks

trachea: TRAY-kee-ah

bronchi: BRONG-kye

alveoli: al-VEE-oh-lye

Anatomy of the Respiratory System

As you can see in **Figure 12.2**, the entire respiratory system's anatomy is housed in the head, neck, and thorax. In general, the anatomy in the head and neck is the **upper respiratory tract,** while the anatomy from the trachea through the lungs is the **lower respiratory tract.**

You have already studied some of this anatomy, such as the pleurae (serous membrane), in Chapter 1. To refresh your memory, a serous membrane is a double-walled, fluid-filled membrane. In the case of the pleurae, the visceral pleura is in contact with the lung's surface, while the parietal pleura is not. The parietal pleura lines the thoracic cavity and covers the diaphragm's superior surface. Fluid exists between the visceral and parietal pleurae. This anatomy will be important when you study the mechanics of breathing, later in the chapter.

Before you get started on the rest of the anatomy, consider the way air enters and moves through the body. Take a deep breath now with your mouth closed, and trace the air in that breath as it travels on its route (follow along with **Figure 12.2**). The air enters the **nasal cavity** through the **nose.** From there it goes to the **pharynx,** to the **larynx,** to the **trachea,** to the **bronchi** (where it enters the lungs), to the **bronchial tree,** and finally to the tiny air sacs called **alveoli** (not shown in the figure). At the alveoli, the second part of respiration—the exchange of gases—takes place.

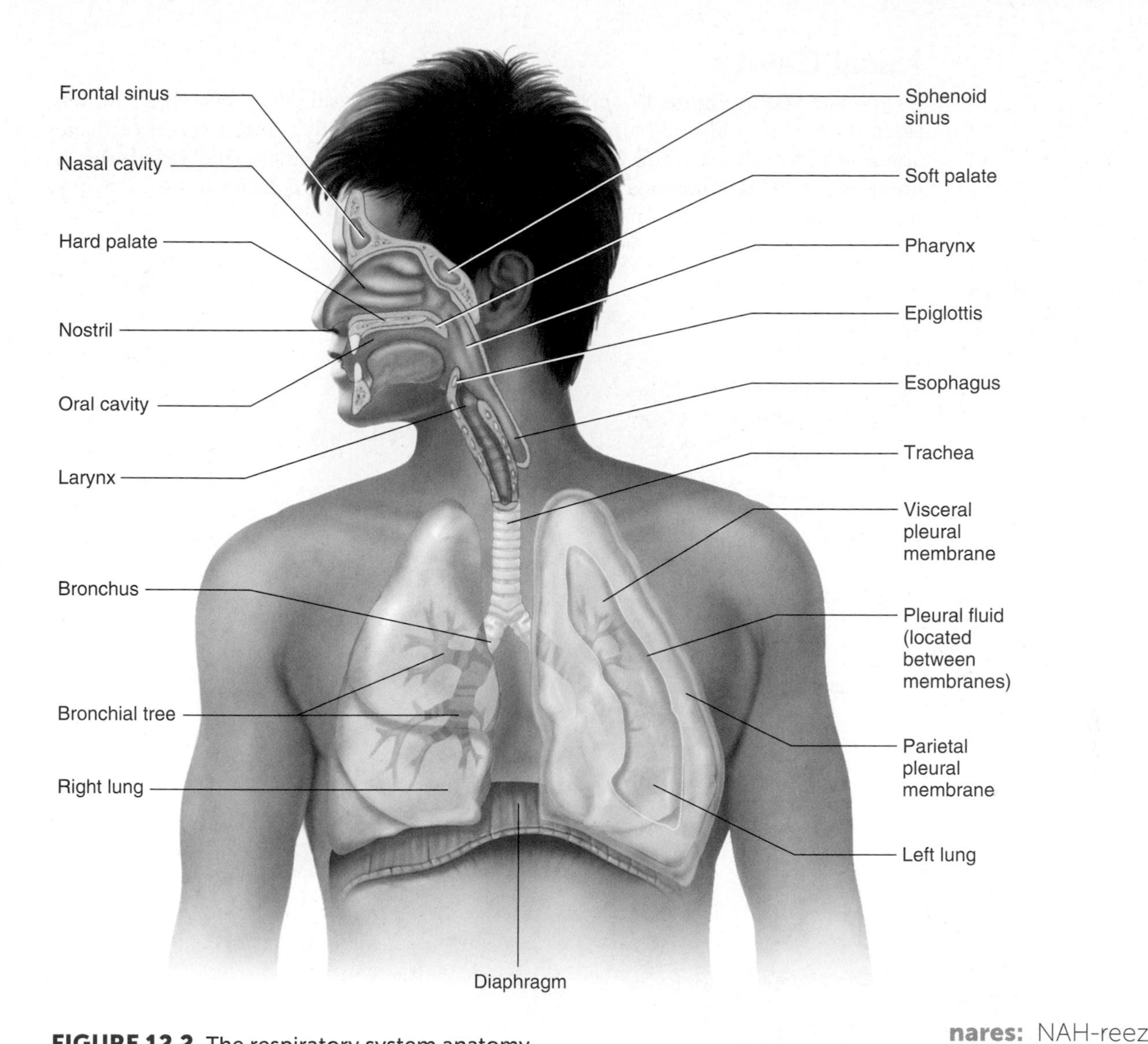

FIGURE 12.2 The respiratory system anatomy.

nares: NAH-reez

audio connect.mcgraw-hill.com

Now you are ready to zoom in on all of the respiratory system's specific anatomy (mentioned above) in the order that the air traveled through it in your deep breath. You will need to become familiar with the gross and microscopic anatomy along the way because this is important in understanding precisely how the anatomy functions.

Nose

Air enters the nasal cavity through the nose's two **nares** (nostrils). The nasal bones superiorly and the plates of hyaline cartilage at the end of the nose are responsible for the nose's shape. You can feel where the nasal bone ends and cartilage begins at the bridge of the nose. See Figure 12.3.

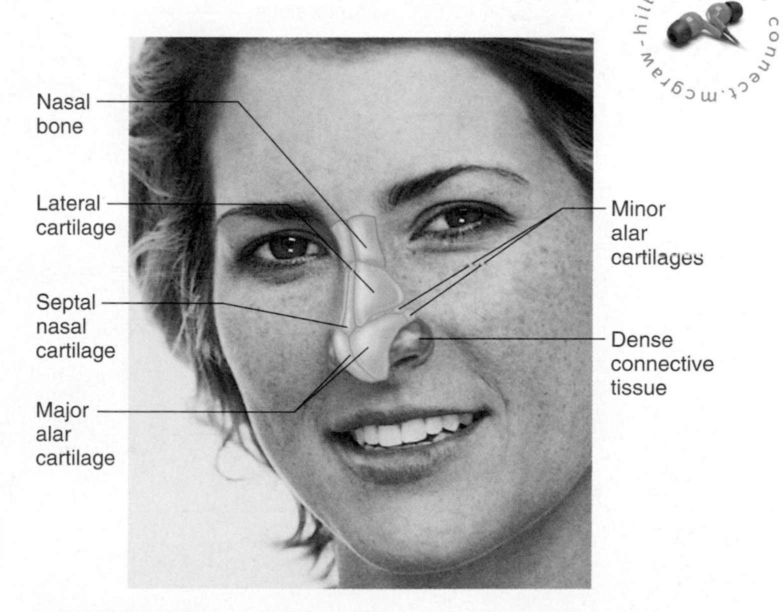

FIGURE 12.3 The nose.

Nasal Cavity

As you can see in **Figure 12.4c**, a septum divides the nasal cavity into right and left sides. The ethmoid bone (superiorly), the vomer (inferiorly), and a septal cartilage anteriorly form the septum. The anterior part of the nasal cavity (the **vestibule**) is lined by stratified squamous epithelial tissue with stiff **guard hairs** to block debris from entering the respiratory tract.

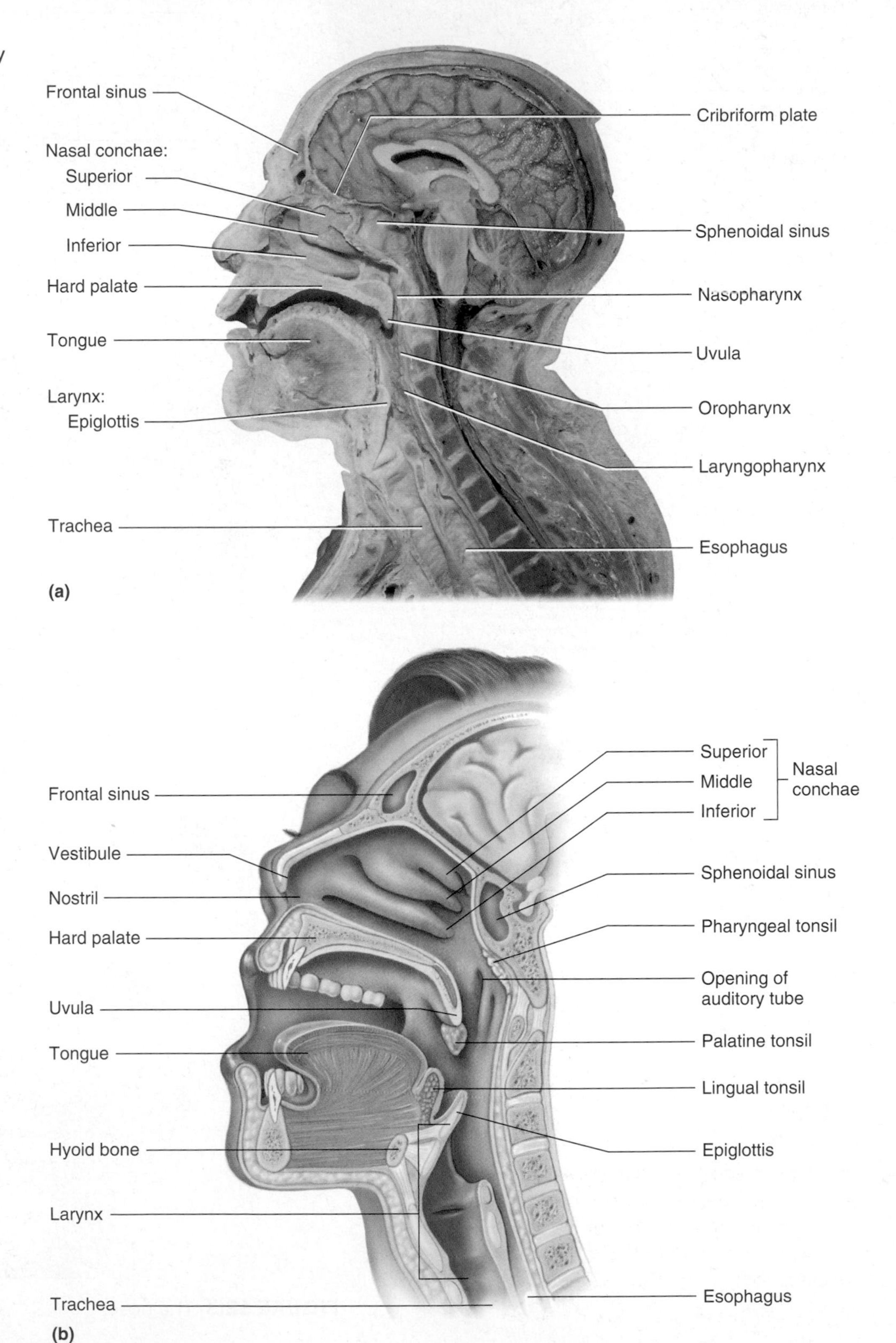

FIGURE 12.4 The anatomy of the upper respiratory tract: (a) sagittal view of cadaver, (b) sagittal section showing internal anatomy (nasal septum has been removed), (c) nasal septum and regions of the pharynx.

FIGURE 12.4 concluded

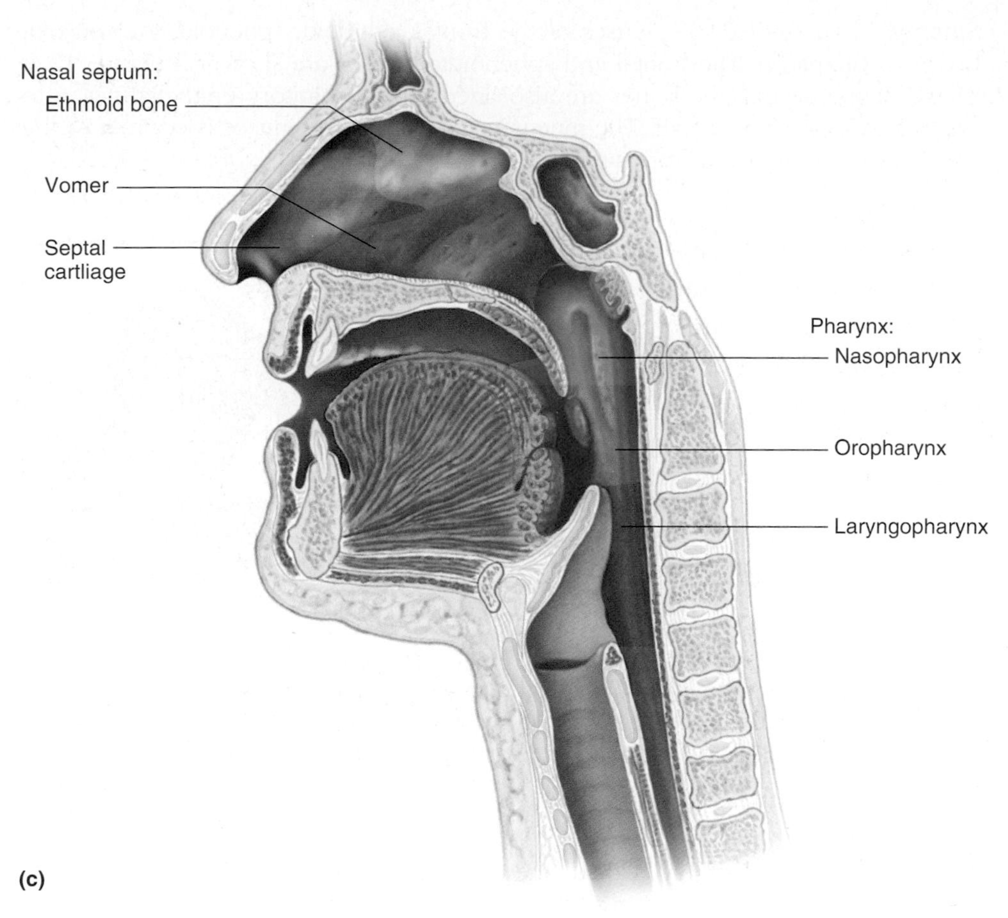

The nasal cavity widens posterior to the vestibule to make room for three bony, lateral ridges called the **nasal conchae.** See **Figure 12.4b**. The ethmoid bone forms the superior and middle nasal conchae, while the inferior nasal concha is a separate bone. This portion of the nasal cavity is lined by mucous membranes that trap debris and warm and moisturize the incoming air. The nasal conchae provide extra surface area for the mucous membranes to function. The mucous membranes are composed of ciliated pseudostratified epithelial tissue. The cilia move mucus and any trapped debris posteriorly so that it can be swallowed. Olfactory neurons located in the roof of the posterior nasal cavity detect odors and provide the sense of smell.

conchae: KON-kee

study **hint**

Take a look up your own nose by using a flashlight and a mirror. You can see the hairs in the vestibule and notice that the posterior nasal cavity appears very red and moist. These are the moist, mucous membranes, and their rich blood supply moisturizes and warms the air. You should also see that there is limited space for air to pass because of the protruding nasal conchae. This causes more air to come in contact with the mucous membranes, so they are better able to function.

spot check ❶ Why do you think the vestibule is stratified epithelial tissue instead of mucous membranes?

Sinuses You studied the sinuses of the frontal, ethmoid, sphenoid, and maxilla bones in Chapter 4. The frontal and sphenoidal sinuses are shown in **Figure 12.4**. These cavities within the bones are also lined with respiratory epithelial tissue to warm and moisturize the air. The mucus produced in the sinuses is drained to the nasal cavity through small openings.

clinical point

Inflammation of the epithelium in the sinuses **(sinusitis)** causes increased mucus production, and the accompanying swelling may block its drainage to the nasal cavity. The pressure within the sinuses created by the buildup of mucus causes a *sinus headache*. Decongestants (vasoconstrictors) help reduce the swelling, thereby improving mucus drainage, which reduces the increased pressure.

At this point in your deep breath, the inspired air leaving the nasal cavity has been partially warmed and moistened, and some of its debris has been trapped. The structure the air encounters next—the pharynx—is explained below.

Pharynx

laryngopharynx: lah-RING-oh-FAIR-inks

The **pharynx,** commonly called the *throat,* is divided into three regions based on location and anatomy—the **nasopharynx,** the **oropharynx,** and the **laryngopharynx.** You will explore these in the paragraphs that follow.

Nasopharynx As you can see in **Figure 12.4c**, the nasopharynx is located posterior to the nasal cavity and the soft palate. This passageway is also lined by ciliated pseudostratified columnar epithelial tissue whose cilia move mucus and trapped debris to the next region of the pharynx so that it can be swallowed. The pharyngeal tonsils and the opening to the auditory tube (eustachian tube) are located in this region.

Oropharynx This region of the pharynx (shown in **Figure 12.4c**) is inferior to the nasopharynx. The oropharynx is common to the respiratory and digestive systems as a passageway for air, food, and drink. For the oropharynx to withstand the possible abrasions caused by the passage of solid food, it must be lined with a more durable tissue—stratified squamous epithelial tissue. In addition, the palatine tonsils are located in this region to deal with any incoming pathogens.

Laryngopharynx This region of the pharynx extends from the level of the epiglottis to the beginning of the esophagus. Like the oropharynx, the laryngopharynx is lined by stratified squamous epithelial tissue to handle the passage of air, food, and drink. See **Figure 12.4c**. Solids and liquids continue on from the laryngopharynx to the esophagus, but inspired air moves through an opening **(glottis)** to the **larynx,** the next structure in the respiratory pathway.

Larynx

The **larynx** is a cartilage box (voice box) of nine separate cartilages, eight of which are composed of hyaline cartilage connective tissue. The **epiglottis** (the ninth cartilage of the larynx) is composed of elastic cartilage connective tissue. As you can see in **Figure 12.4b**, the epiglottis stands almost vertically over the glottis. Its function is to fold over the glottis during swallowing to prevent solids and liquids from entering the larynx. You will learn more about how this works in the next chapter. The epiglottis remains in its vertical position at all other times to ensure the easy passage of air from the laryngopharynx through the glottis to the larynx.

Figure 12.5 gives you a closer look at the larynx. Here you can see the **laryngeal prominence** ("Adam's apple") of the **thyroid cartilage.** It enlarges to be more visible in men than women due to the presence of testosterone. You can also see (in this figure) the two **arytenoid cartilages** and the two **corniculate cartilages** that operate the vocal cords.

arytenoid: ah-RIT-en-oyd

corniculate: kor-NIK-you-late

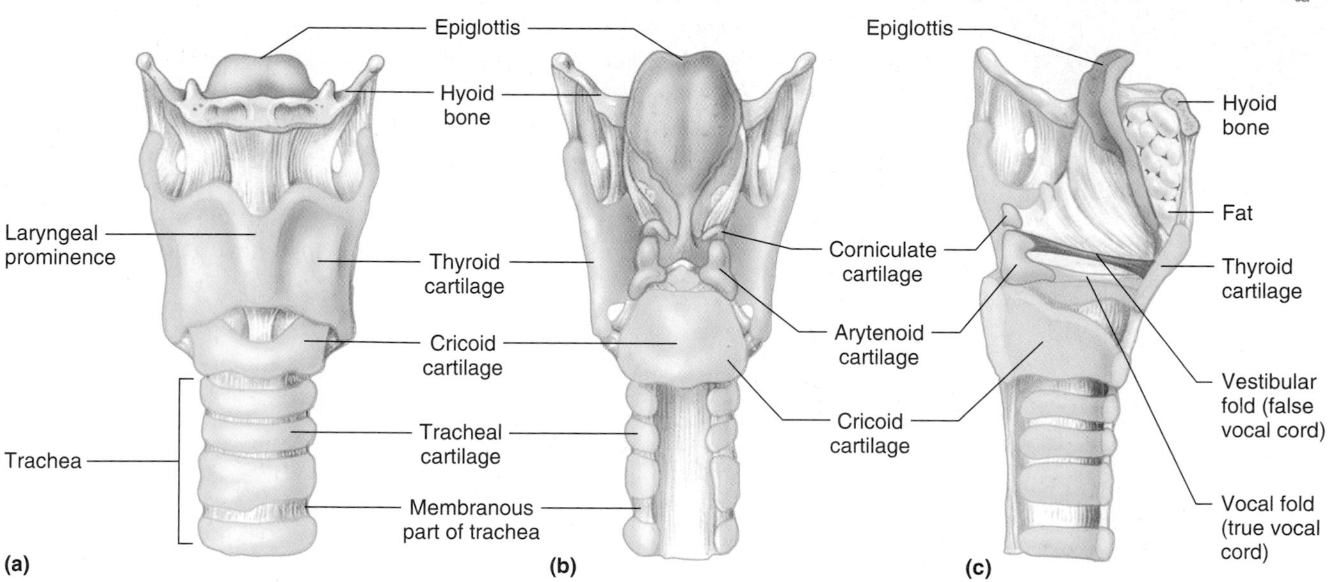

FIGURE 12.5 The larynx: (a) anterior view, (b) posterior view, (c) sagittal section.

Vocal Cords The walls of the larynx are muscular to operate the **vocal cords** shown in **Figures 12.5c, 12.6, and 12.7.** There are two sets of folds in the inner wall of the larynx—the **vestibular folds** and the vocal cords. The vestibular folds have no function in speech. They are important in closing the larynx during swallowing.

Figure 12.6 shows how the vocal cords are *abducted* (spread apart) and *adducted* (brought closer together) by muscles pulling on the arytenoid and corniculate cartilages. The opening formed by *abducting* the vocal cords is the glottis. Air passing through *adducted* vocal cords causes them to vibrate to make sounds of varying pitch depending on the tautness of the cords. So speech is a very active process, which involves muscles pulling on cartilages of the larynx to operate the vocal cords. The larynx at the vocal cords is lined with stratified squamous epithelial tissue to withstand the vibrations.

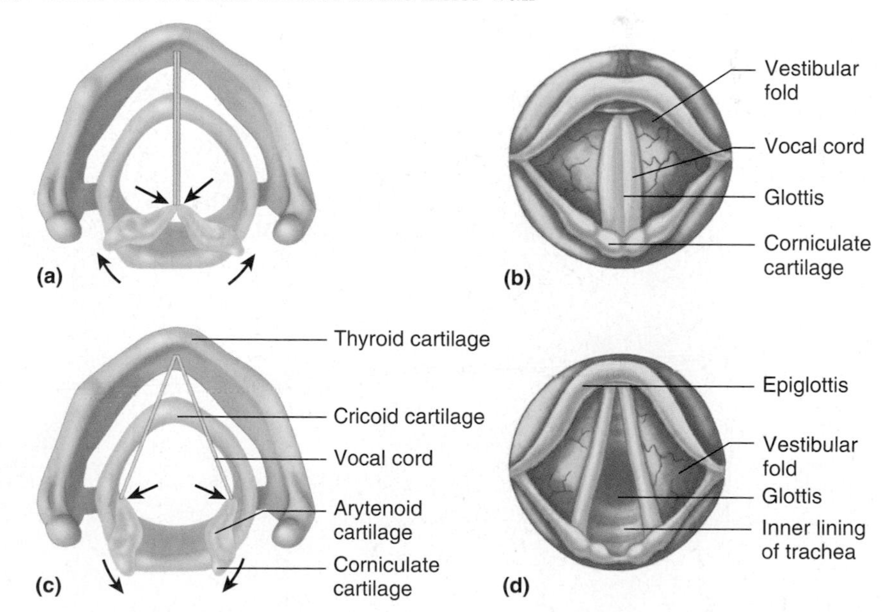

FIGURE 12.6 Action of laryngeal muscles on the vocal cords: (a) adduction showing just the cartilages and the vocal cords, (b) adduction as seen with all tissues present, (c) abduction showing just the cartilages and the vocal cords, (d) abduction as seen with all tissues present.

Trachea

From the larynx, inspired air travels to the **trachea,** a rigid tube with 18 to 20 C-shaped cartilages composed of hyaline cartilage connective tissue. See **Figure 12.8**. These cartilages hold the trachea open for the easy flow of air. The C-shaped cartilages are open posteriorly with smooth muscle bridging the gap. This feature allows the esophagus (directly posterior to the trachea) room to expand into the tracheal space when swallowed food passes on its way to the stomach. If the cartilages were circular instead of C-shaped, a swallowed piece of meat could get hung up on each cartilage as it passed down the esophagus.

Like the nasal cavity and the nasopharynx, the trachea is lined with ciliated pseudostratified columnar epithelial tissue with goblet cells that secrete mucus. See **Figures 12.8** and **12.9**. The air you breathe is full of particles, such as dust, pollen, and smoke particles. You may have seen the dust in

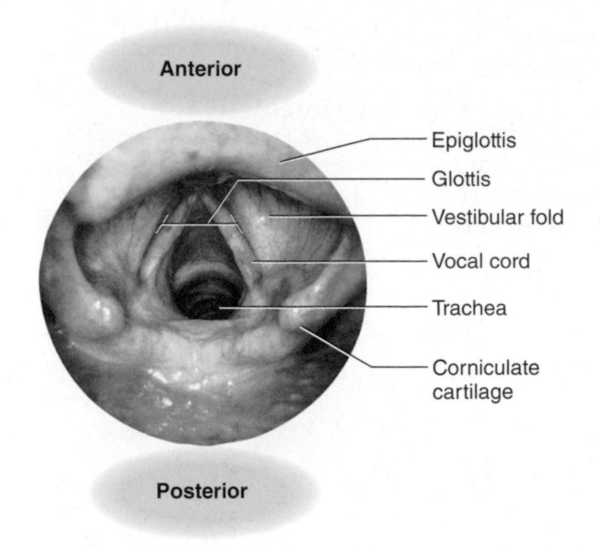

FIGURE 12.7 Endoscopic view of the vocal cords as seen with a laryngoscope.

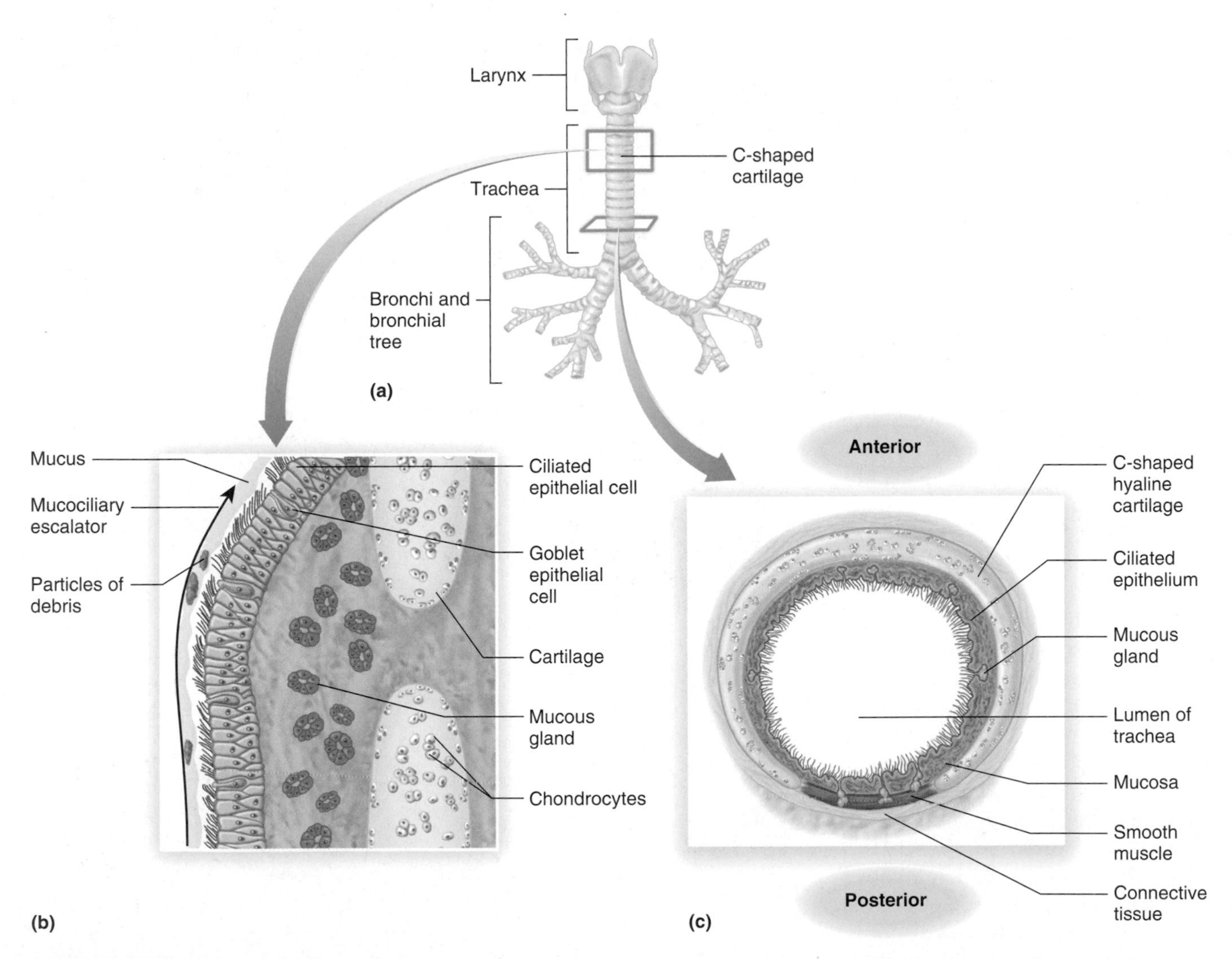

FIGURE 12.8 The trachea and bronchi: (a) anterior view, (b) longitudinal view of the trachea showing cilia moving mucus and debris, (c) transverse section of the trachea showing C-shaped cartilage.

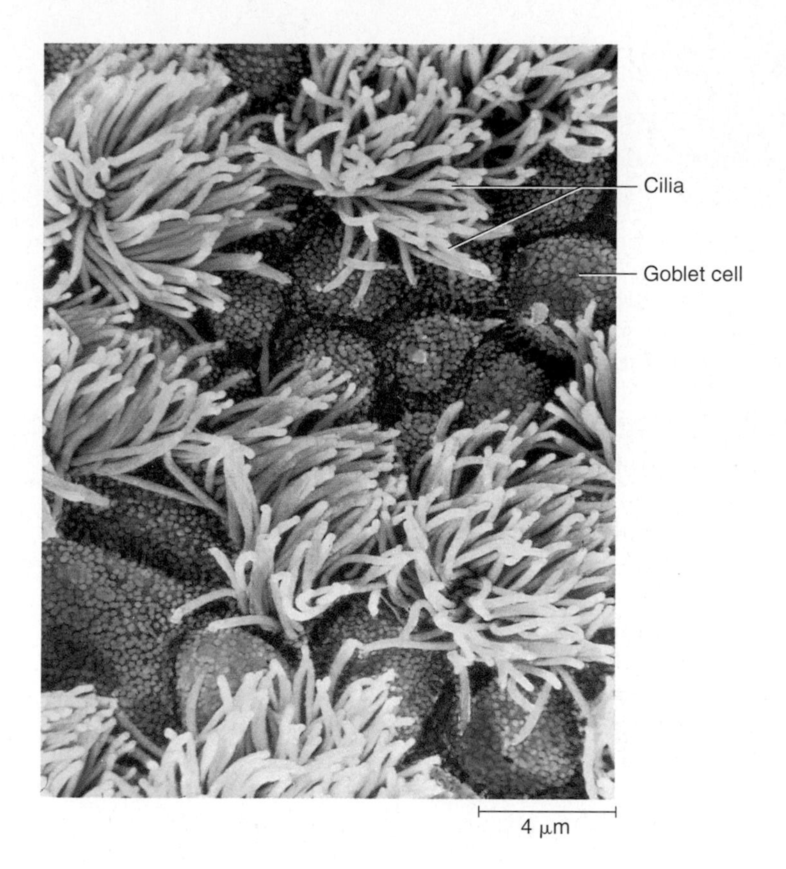

FIGURE 12.9 Lining of trachea. Epithelial tissue of the trachea, showing ciliated cells and goblet cells.

the air as the sun shines through a window. Even during sleep, the cilia of the trachea move mucus and any trapped debris up (like an escalator) toward the pharynx to be swallowed. See **Figure 12.8b**. This prevents the accumulation of debris in the lungs.

clinical point

The smoke inhaled with each drag on a cigarette contains a lot of particles, but the respiratory anatomy is designed to prevent this debris from accumulating in the lungs. However, the increased amount of debris may, over time, cause the lining of a habitual smoker's trachea to go through metaplasia, changing from ciliated epithelial tissue to a more durable, nonciliated tissue. Without the ciliated escalator, the respiratory system resorts to coughing up the debris. As a result, the long-term smoker develops the *smoker's hack* each morning to move the debris inspired each night.

spot check ❷ Compare the direction the cilia move debris in the nasopharynx to the direction they move debris in the trachea. How do they differ?

spot check ❸ A patient on a ventilator has a tube inserted into the trachea through a procedure called a **tracheostomy.** What should be done to the air delivered through a ventilator considering the respiratory anatomy leading to the trachea has been bypassed?

The trachea splits to become the right and left main bronchi, each of which enters its respective lung. You will explore the lungs and bronchial tree together, looking first at their gross anatomy, as shown in **Figure 12.10**.

Lungs and the Bronchial Tree

As you can see in **Figure 12.10b**, the right and left main bronchi each enters its respective lung at an area on the medial surface of the lung called the **hilum.** This is the same location used by pulmonary arteries and veins to enter and leave the lung. The left **bronchus** is slightly more horizontal than the right bronchus due to the location of the heart. The main bronchi and all of their further branches make up the **bronchial tree.** See **Figure 12.10c**.

bronchus: BRONG-kuss

Upon entering the lung, each main bronchus branches to become the **lobar bronchi,** each going to a separate lobe of the lung. The left lung has fewer lobes (two) than the right, again because of the position of the heart. The right lung has three lobes, and therefore three lobar bronchi.

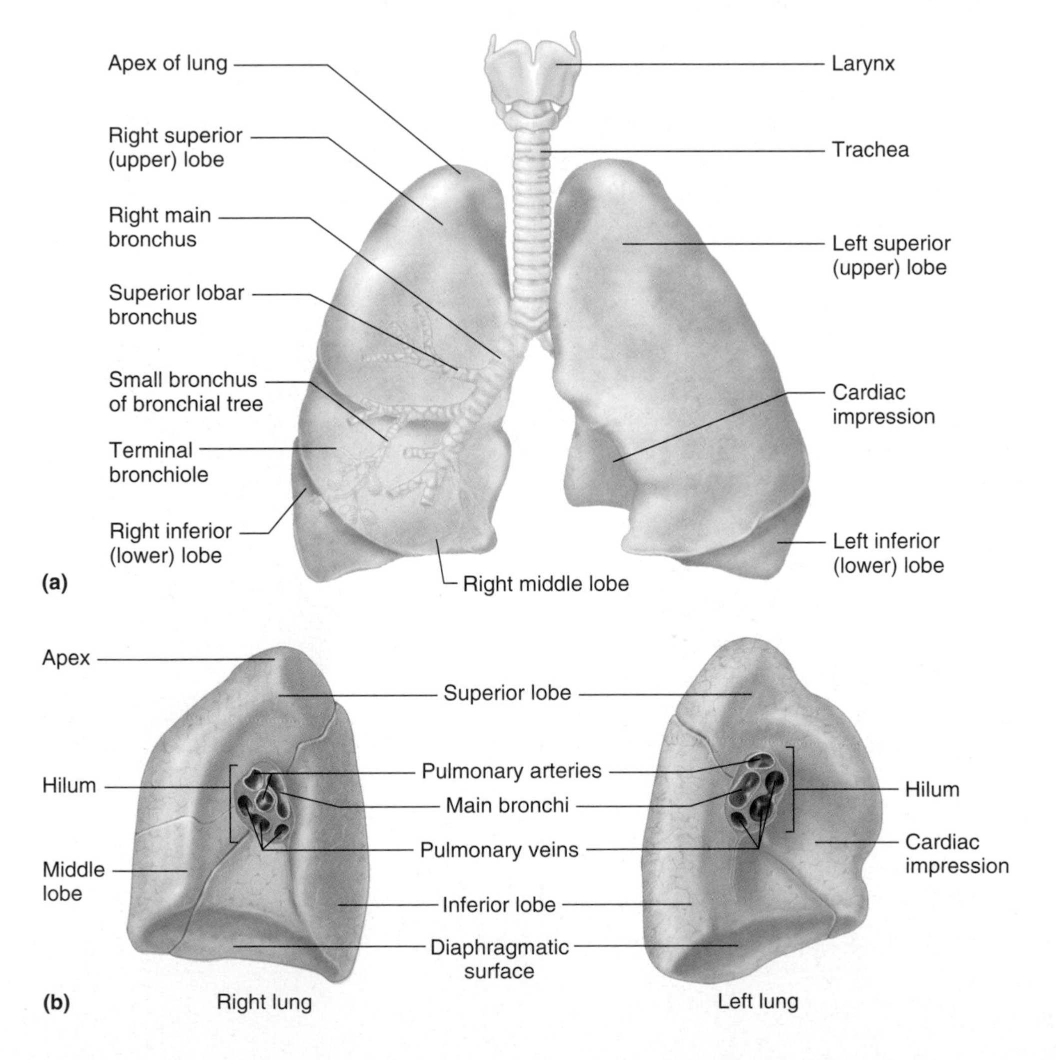

FIGURE 12.10
Gross anatomy of the lungs and bronchial tree:
(a) anterior view, (b) medial views of the right and left lungs, (c) bronchogram (radiograph of the bronchial tree), anterior view.

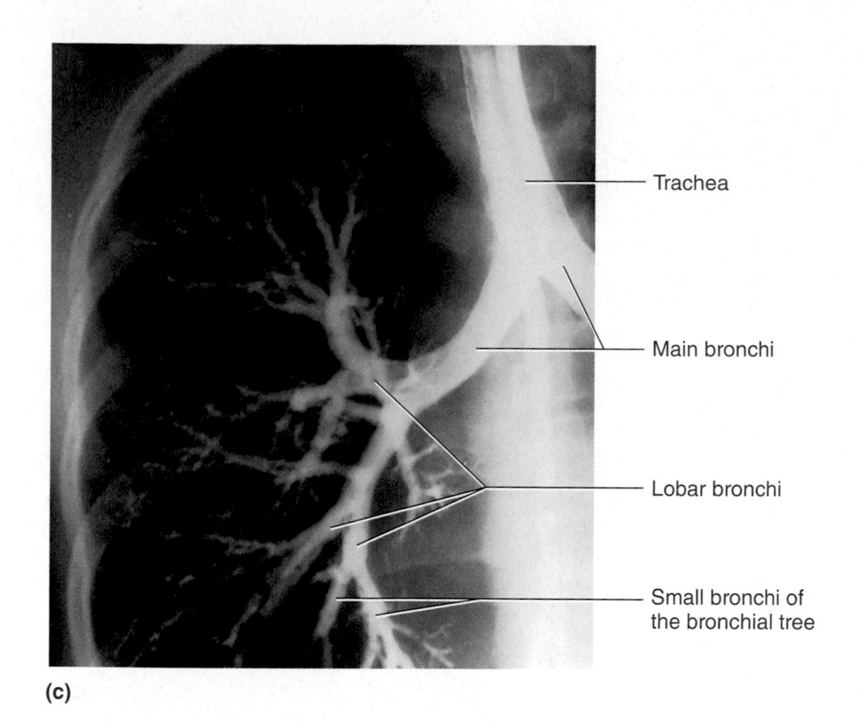

(c)

FIGURE 12.10 concluded

warning

The lungs fill with air, but they are not hollow like a balloon. A cross section of a lung appears solid—more like Styrofoam composed of tiny beads. Each of the tiny beads is a tiny, hollow air sac that can fill with inspired air. See Figure 12.11.

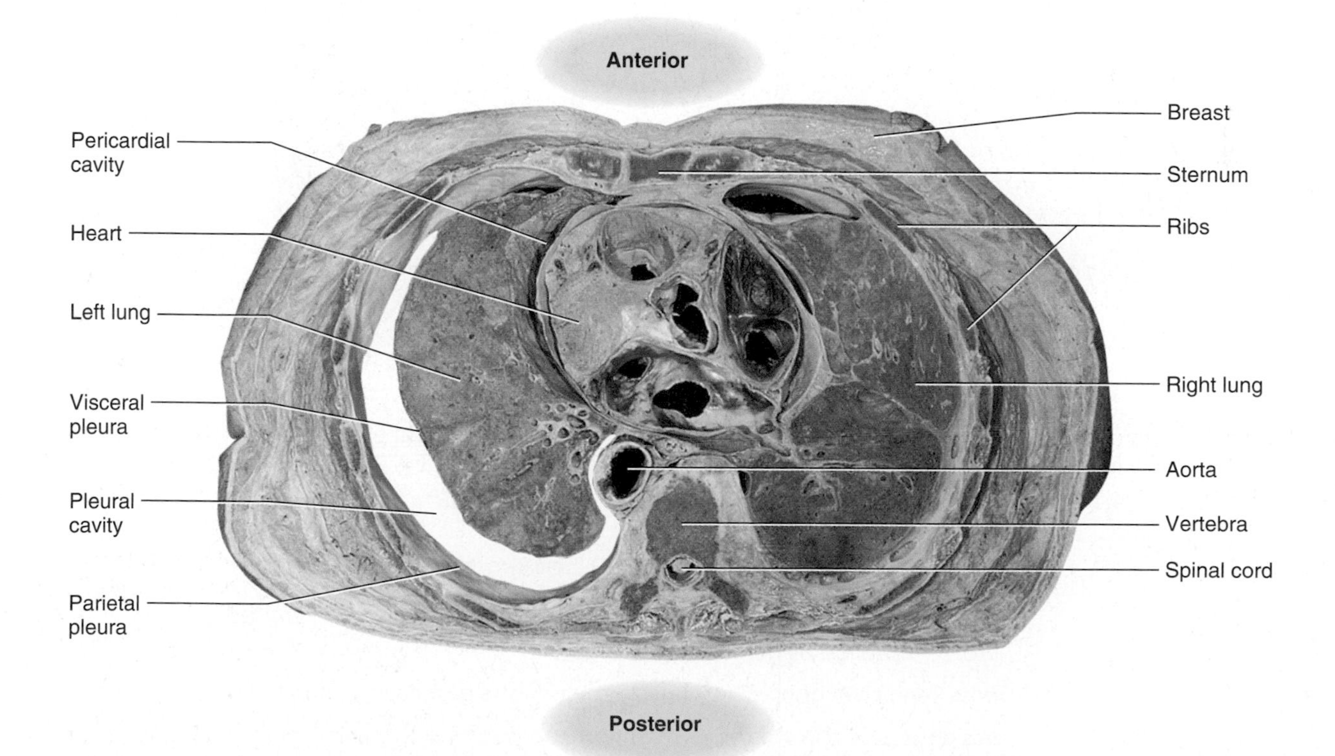

FIGURE 12.11 Cross section of a cadaver through the thoracic cavity.

Lobar bronchi further divide to smaller and smaller bronchi that branch to form the bronchial tree. See **Figure 12.10c**. All of the bronchi are supported by cartilage plates, which hold them open for the easy passage of air. The smallest bronchi further branch to form bronchioles. These small tubes do not have cartilage in their walls. Instead, their walls have smooth muscle that allows them to dilate or constrict to adjust airflow. You will learn more about this later in the chapter. Each bronchiole supplies air to a **lobule** (subsection of a lobe) of the lung composed of tiny air sacs called *alveoli*. See **Figure 12.12a**.

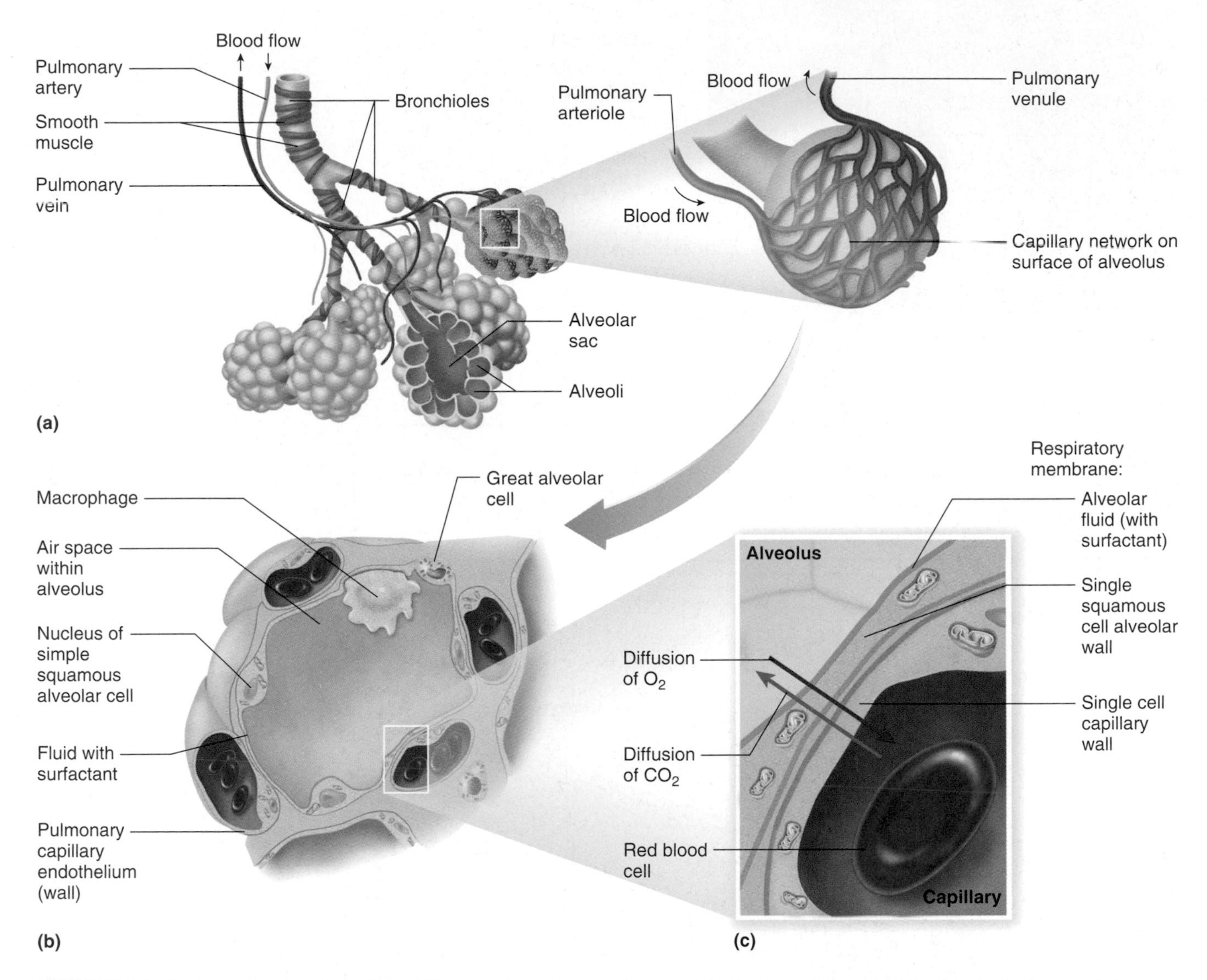

FIGURE 12.12 Bronchiole, alveoli, and the respiratory membrane: (a) clusters of alveoli at the end of a bronchiole and the network of capillaries covering them, (b) cells of the alveoli, (c) respiratory membrane.

spot check 4 Penny is an inquisitive 18-month-old girl who likes to see what fits into what. One morning, she put a small, metal washer that she found on the floor into her nose just as her mother entered the room. Her mother gasped when she saw what Penny had done. This scared Penny, so she gasped, too, and the metal washer was gone. She had inhaled it. What route do you think the metal washer will take (trace the pathway)?

spot check ❺ In which lung will the doctor at the clinic find the metal washer? Explain.

Alveoli The alveoli are clustered like grapes at the end of the bronchiole. As you can see in **Figure 12.12**, a network of capillaries covers the alveoli. This is vital for gas exchange, as you will read shortly. **Figure 12.13** shows the histology of the alveoli with respect to the bronchioles and blood supply to the capillaries. There are approximately 150 million alveoli in each human lung. Each alveolus is a tiny air sac with two types of cells in its walls—simple squamous cells and **great alveolar cells.** Most of the alveolar wall is composed of one layer of thin squamous cells that allow for rapid gas exchange across their surface. The great alveolar cells (shown in **Figure 12.12b**) are important because they secrete a fluid called **surfactant.** Below, you will find out why this fluid is so important.

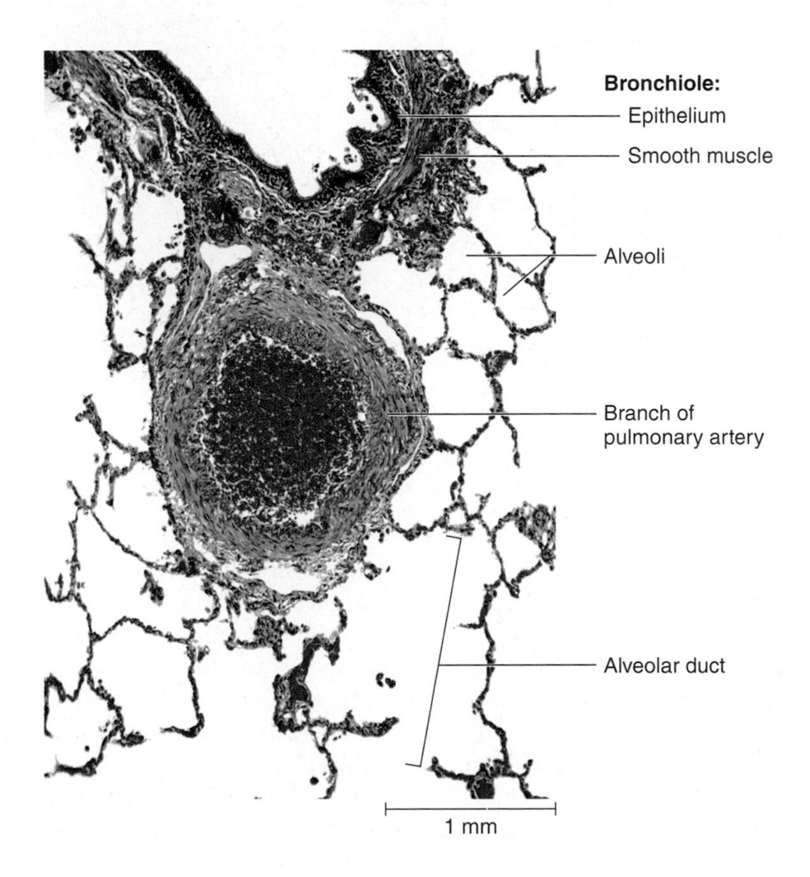

FIGURE 12.13 Histology of the lung: micrograph of alveoli, a bronchiole, and a branch of a pulmonary artery.

spot check ❻ Why must the vessel represented in this figure be an artery and not a capillary, and why must the tube be a bronchiole and not a bronchus of the bronchial tree? (*Hint:* Look at the histology.)

12.3 learning **outcome**

Explain the role of surfactant.

Surfactant To understand the importance of surfactant, you must first understand a property of water: high surface tension. This basically means that water will always try to have the smallest surface area to volume ratio possible. In other words, water forms beads or drops because a sphere has a smaller surface area to volume ratio than a flat sheet. This is why water forms beads or drops on smooth surfaces like glassware in your dishwasher.

Surfactant reduces the surface tension of water much like the rinse agent you may add to your dishwasher to avoid water spots. The rinse agent reduces the surface tension of water (sheeting action), so water sheets off your glassware instead of forming beads that leave water spots as the glasses dry. Surfactant also causes water to form a thin sheet instead of a bead. Why is this important? By the time air has entered the alveoli, it has been thoroughly moisturized by all the mucous membranes it has passed along the respiratory route. If a bead of water were to form inside the tiny alveoli, the plump bead might touch the wall on the opposite side of the air sac and cause the thin, delicate walls of the alveoli to stick together, and this would cause the alveoli to collapse. A thin sheet of water in the alveoli (instead of a plump bead) reduces the chance of the alveoli walls collapsing on each other. Collapsed alveoli do not easily fill with air.

clinical point

A fetal respiratory system does not mature until late in pregnancy. The alveoli in infants born before the lungs are mature often collapse because of the lack of sufficient surfactant. This condition, called **respiratory distress syndrome (hyaline membrane disease),** is a common cause of neonatal death. Oxygen under positive pressure can be administered along with surfactant to keep the lungs (alveoli) inflated between breaths.

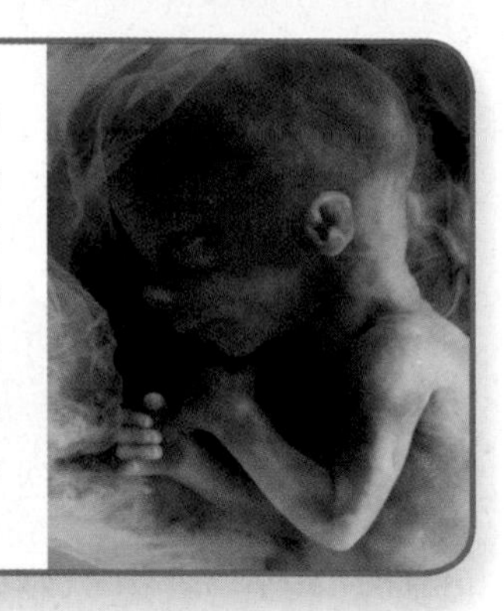

12.4 learning outcome
Describe the respiratory membrane.

Respiratory Membrane

So far, you have seen in **Figure 12.12a** and **b** the relationship of the alveoli to the bronchioles and the cells that make up the alveoli. In **Figure 12.12c**, you can see the structure formed by the capillary network adjacent to the alveoli—the respiratory membrane. This is a very important structure because it is the location of gas exchange in the lung. Take a closer look at this figure. The respiratory membrane is composed of the thin layer of water with surfactant in the alveoli, the single squamous cell alveolar wall, and the single cell capillary wall. If all of the respiratory membrane in one lung were laid out in a single layer, it would cover approximately 70 square meters (m^2), equivalent to the floor of a room 25 feet by 30 feet.

You have now covered all of the anatomy that the air of your deep breath encountered along its way to the respiratory membrane. It is time to explore the way this respiratory anatomy functions, starting with how you took the deep breath in the first place.

12.5 learning outcome
Explain the mechanics of breathing in terms of anatomy and pressure gradients.

Physiology of the Respiratory System

Mechanics of Taking a Breath

Air moves (but is not pushed) along the respiratory passageways on its way to the lungs because of pressure differences within the chest. This is much like the syringe example you became familiar with while studying blood flow through the heart (see **Figure 10.11** in Chapter 10). The syringe example explained the relationship between volume, pressure, and flow. If the volume of space in the syringe is increased, the pressure inside the syringe is decreased, so air flows into the syringe to equalize the pressures inside and outside the syringe. Likewise, if the volume of space in the syringe is decreased, the pressure inside the syringe is increased, so air flows out of the syringe to equalize the pressures. As a result, pushing or pulling on the plunger changes the volume of the syringe.

How does the body change the volume of the chest? See **Figure 12.14**. Concentrate on the major muscles for breathing shown in bold in this figure. As you can see in **Figure 12.14** (a and b), during inspiration the external intercostal, pectoralis minor, and sternocleidomastoid muscles contract to expand the rib cage, and the diaphragm contracts to flatten its dome shape. The combined effect of these contractions is an increase in the size (volume) of the chest cavity. All that needs to be done for normal expiration is to have the same muscles relax. Then the rib cage returns to its normal position and the diaphragm becomes dome-shaped again due to the recoil of abdominal organs. See **Figure 12.14c**. The volume of the chest is decreased, and air flows from the body. Expiration during normal breathing is a passive process (no energy required) involving the relaxation of muscles. But you can also see in **Figure 12.14d** that forced expiration involves the contraction of muscles too. The internal intercostals and abdominal wall muscles do contract in *forced expiration;* however, these muscles are not used for expiration during normal breathing. Forced expiration is intentionally forcing air out of the lungs, which happens when blowing out a candle or inflating a balloon. In this case, energy for muscle contraction is required.

So far in this explanation of the mechanics of breathing, you have seen how the muscles of the chest can increase the volume of the chest, but what about the volume of each lung? How is the volume of the lungs increased? This involves the pleural membranes and the pleural fluid. The parietal pleura is attached to the thoracic

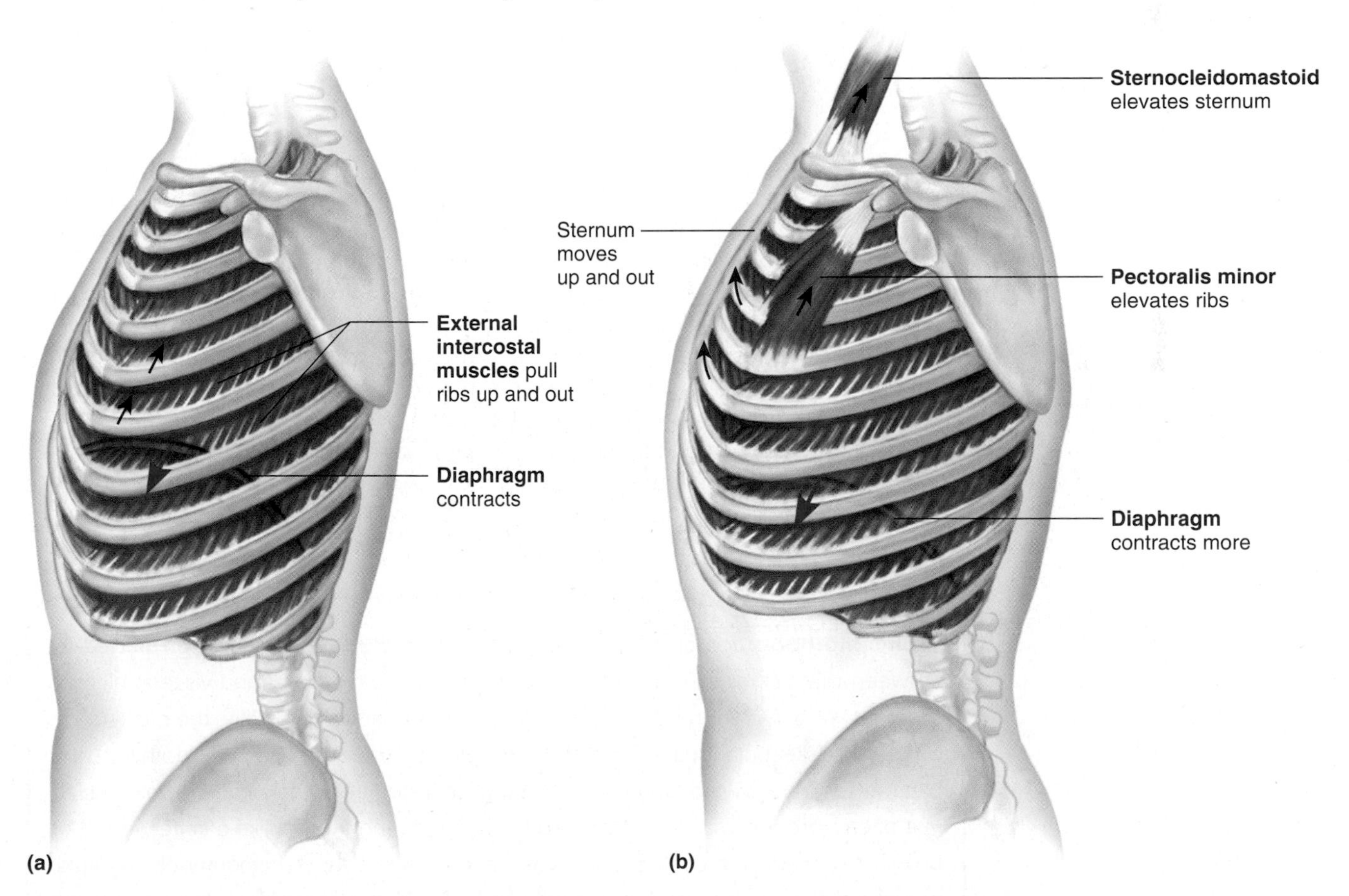

FIGURE 12.14 Respiratory muscles: (a) external intercostal muscles and diaphragm at the beginning of inspiration, (b) additional muscle action to continue inspiration, (c) recoil of abdominal organs, causing diaphragm to dome when it relaxes, (d) muscle actions during forced expiration.

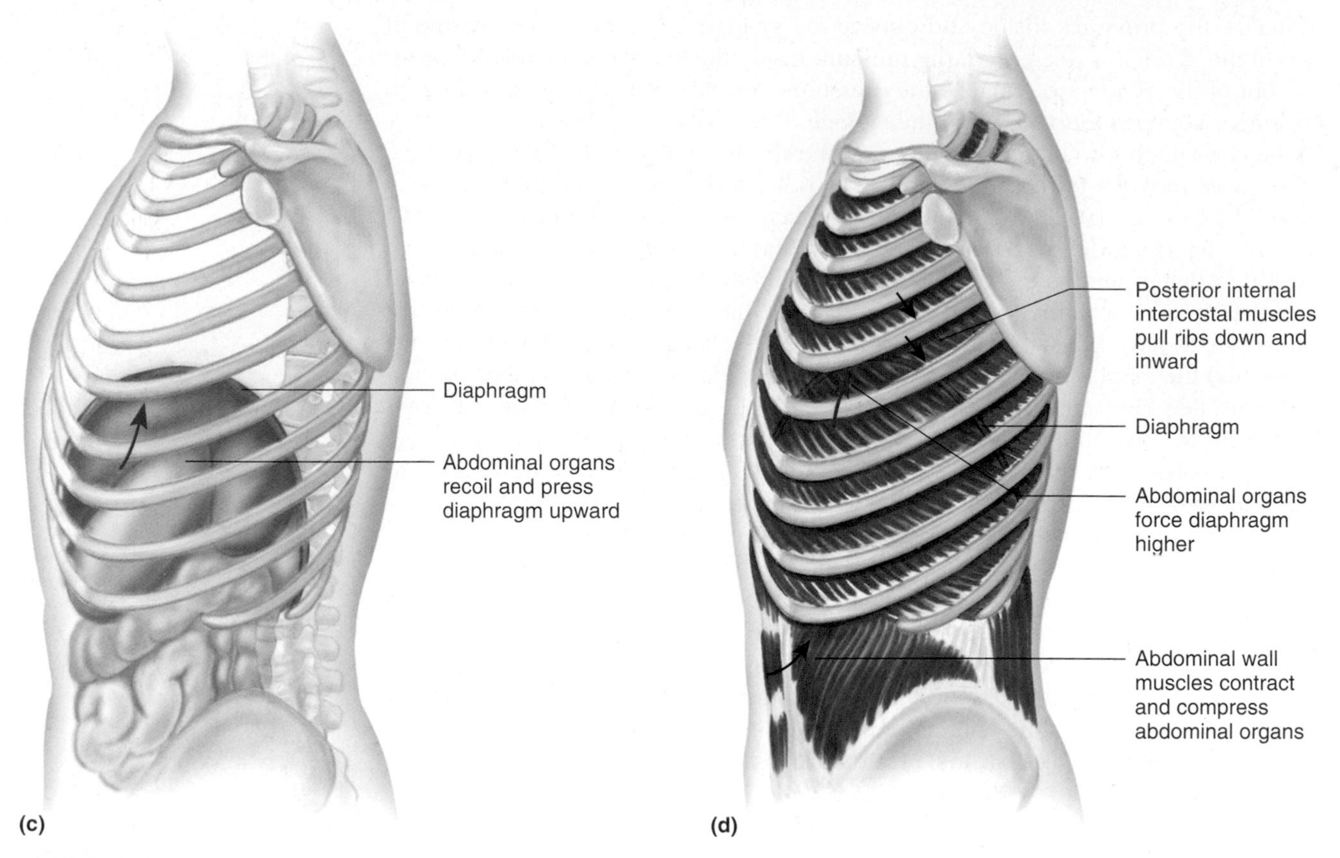

FIGURE 12.14 concluded

wall and diaphragm, while the visceral pleura is attached to the lung. The pleural fluid between the parietal and visceral pleurae cause the two pleurae to stick together and move as one. As the respiratory muscles expand the thoracic wall and flatten the diaphragm, the parietal pleura moves with the wall and diaphragm. As the parietal pleura moves with the thoracic wall and diaphragm, the visceral pleura and the lung move with it—expanding the lung along with the thoracic cavity. As the lung expands, the pressure within the lung (intrapulmonary pressure) decreases, so air moves in until the pressure inside the lung is equal to the pressure outside the body. The intrapulmonary and atmospheric pressures are then equal. When inspiration ends, the thoracic wall returns to its original position and its volume is diminished. The pressure is now greater in the lung than outside the body, so air flows from the body until the intrapulmonary and atmospheric pressures are again equal. **Figure 12.15** shows the muscle action and the pressure changes during inspiration and expiration.

clinical point

A **pneumothorax** (collapsed lung) occurs if air is introduced in the pleural cavity between the pleural membranes. Just as fluid holds the parietal and visceral pleurae together, air between the pleurae allows them to separate. Normally, there is tension on the lung, keeping it partially inflated at all times. However, in a pneumothorax, the pleurae separate, so the lung may recoil and separate from the thoracic wall. The air in a pneumothorax may be introduced by a penetrating trauma like a knife wound or broken rib, medical procedures such as inserting a needle to withdraw pleural fluid, or even a disease like emphysema (covered later in this chapter). In mild cases, the pneumothorax may correct itself without medical intervention. In more severe cases, a chest tube may need to be introduced into the pleural space to remove the air to inflate the lung, and surgery may be required to repair the opening into the pleural space.

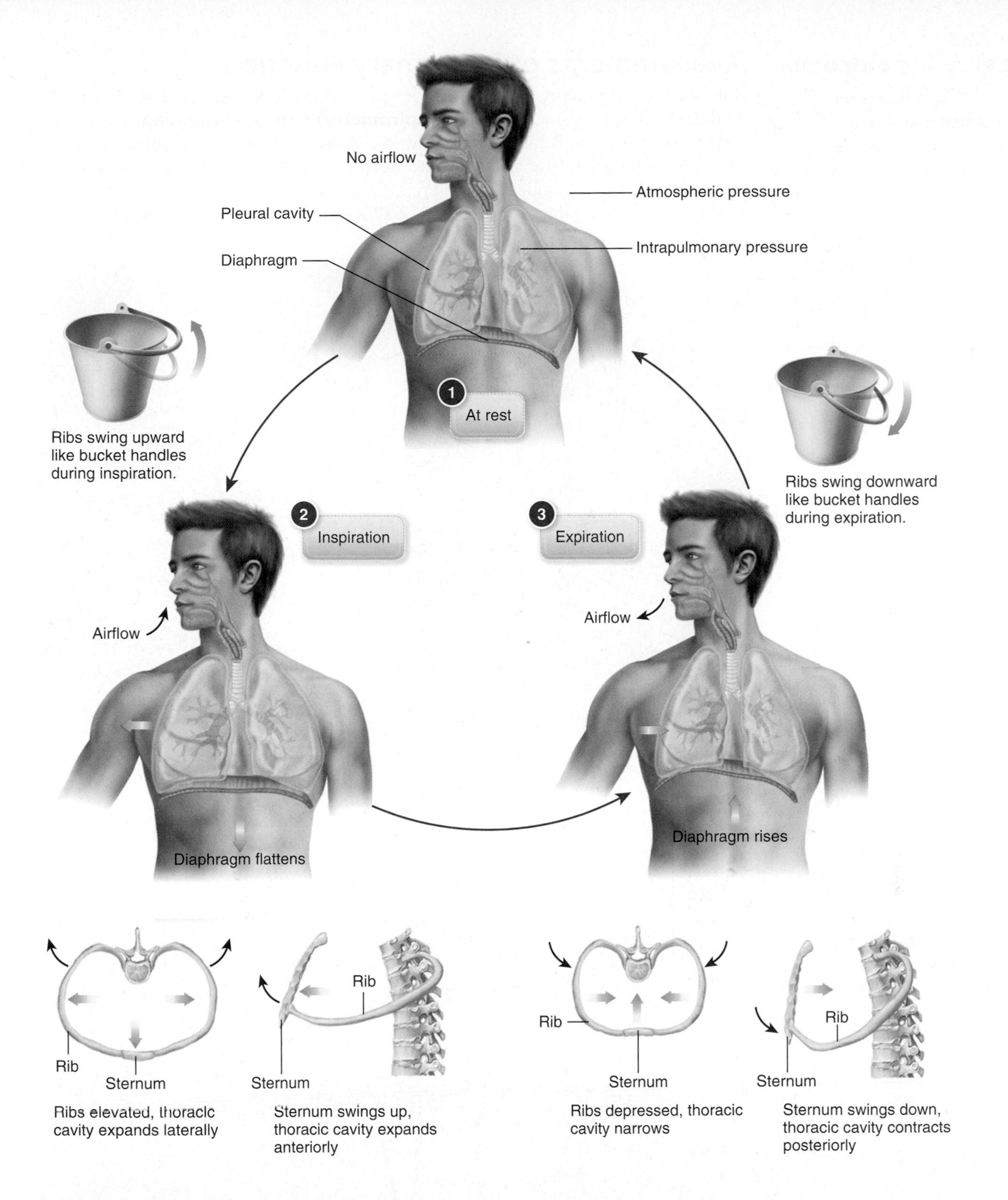

FIGURE 12.15 **A respiratory cycle of inspiration, expiration, and rest:** 1. At rest, atmospheric and intrapulmonary pressures are equal, and there is no airflow. 2. In inspiration, the thoracic cavity expands laterally, vertically, and anteriorly; intrapulmonary pressure falls below atmospheric pressure; and air flows into the lungs. 3. In expiration, the thoracic cavity contracts in all three directions, intrapulmonary pressure rises above atmospheric pressure, and air flows out of the lungs. There is a rest between breaths. The handles of the pails represent ribs.

12.6 learning outcome

Define the measurements of pulmonary function.

Measurements of Pulmonary Function

How well the respiratory system functions to move air into and out of the lungs can be measured in pulmonary function **(spirometry)** tests. A **spirometer** is a device used to measure the volume of air moved. **Figure 12.16** shows a photo of Gabe, who is breathing into the spirometer to determine his various **lung volumes** and **lung capacities** (capacities are determined by adding two volumes). Table 12.1 defines the various values, and **Figure 12.17** shows a graph of Gabe's values.

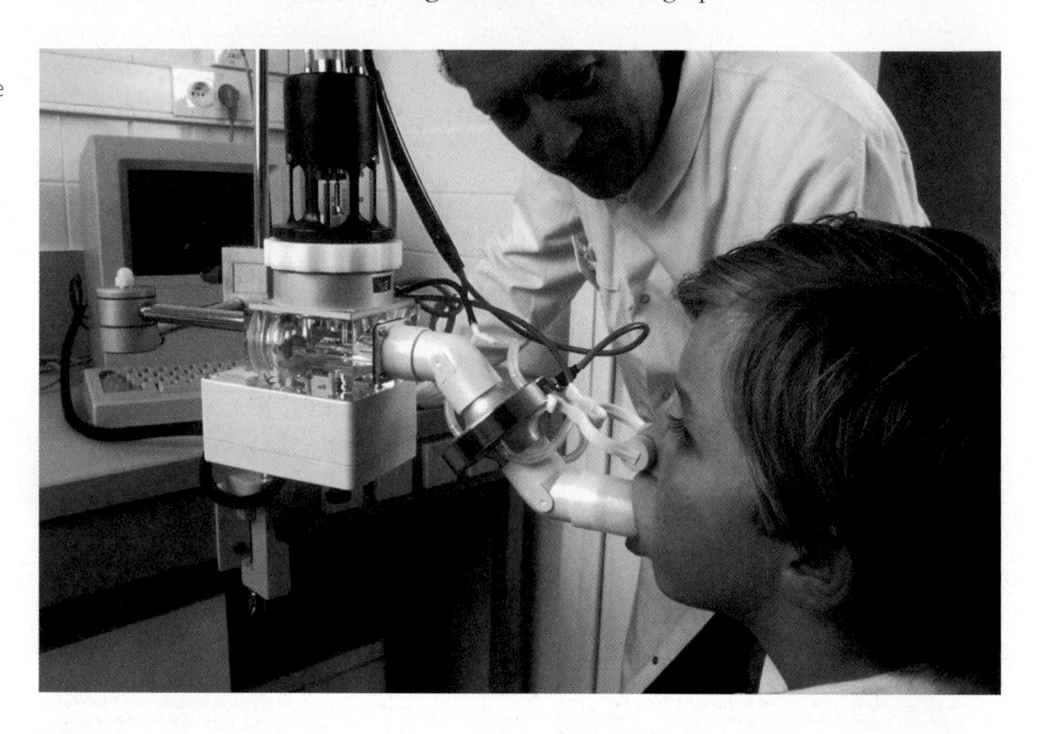

FIGURE 12.16 Spirometry. A spirometer is used to measure lung volumes and capacities.

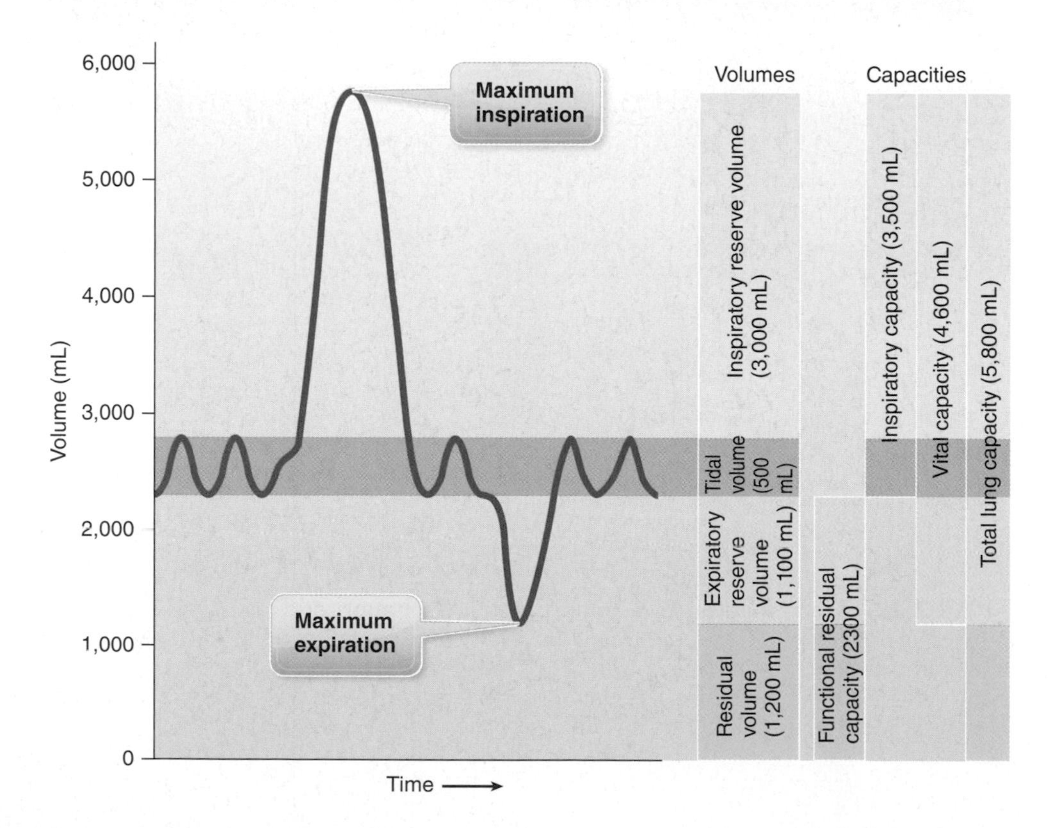

FIGURE 12.17 Graph of pulmonary volumes and capacities.

TABLE 12.1 Lung volumes and capacities

Volume or capacity	Definition	Typical value
Tidal volume (TV)	The tidal volume is the amount of air moved in a normal breath (inspired or expired) at rest.	500 mL
Inspiratory reserve volume (IRV)	The inspiratory reserve volume is the amount of air that can be forcefully inspired beyond the amount inspired in a normal breath at rest.	3,000 mL
Expiratory reserve volume (ERV)	The expiratory reserve volume is the amount of air that can be forcefully expired beyond the amount expired in a normal breath at rest.	1,100 mL
Residual volume (RV)	The residual volume is the amount of air in the lungs that cannot be moved.	1,200 mL
Functional residual capacity (FRC)	The functional residual capacity is the amount of air remaining in the lungs after the expiration of a normal breath at rest. FRC = ERV + RV.	2,300 mL
Inspiratory capacity (IC)	The inspiratory capacity is the maximum amount of air that can be inspired after the expiration of a normal breath at rest. IC = TV + IRV.	3,500 mL
Vital capacity (VC)	Vital capacity is the maximum amount of air that can be moved. VC = IC + FRC.	4,600 mL
Total lung capacity (TLC)	The total lung capacity is the maximum amount of air the lung can hold. TLC = VC + RV.	5,800 mL

Exercise may temporarily increase the tidal volume for an individual, but this does not mean that all of the other values will increase. The maximum amount of air the respiratory system can move (vital capacity) does not change on a temporary (minute-by-minute) basis. So if there is an increase in tidal volume during a workout, there must be a decrease in the inspiratory and expiratory reserve volumes.

Lung volumes and capacities vary from one individual to another due to gender, size, age, and physical condition. In general, a woman's vital capacity is less than a man's; a tall, thin person has a greater vital capacity than someone short and obese; and a trained athlete has a greater vital capacity than someone who has a sedentary lifestyle.

Compliance is another measurement of pulmonary function. It measures how well the lung can expand and return to shape (elasticity). It is harder to expand the lungs and the thorax if there is decreased compliance. This may be due to the buildup of scar tissue in the lung (pulmonary fibrosis), collapse of the alveoli (respiratory distress syndrome), skeletal disorders (scoliosis or kyphosis), or **chronic obstructive pulmonary disorders (COPDs),** such as asthma, chronic bronchitis, emphysema, and lung cancer (discussed later in the chapter).

At this point, you have become familiar with the anatomy of the respiratory system and how it works to deliver air into and out of the lungs. You can now begin to explore the second part of respiration—the exchange of gases—by looking at the gases present in the air you breathe.

12.7 learning **outcome**

Define partial pressure and explain its relationship to a gas mixture such as air.

Composition of Air

Gases diffuse across membranes from high concentration to low concentration until the concentrations are equal. So it is important to be able to talk about quantities of gases. The air you breathe is a mixture of gases—78.6 percent nitrogen, 20.9 percent oxygen, 0.04 percent carbon dioxide, and variable amounts of water vapor depending on humidity levels. Gases fill whatever space is available to them and can be compressed, so volume is not a good measure of the amount of a gas. For example, an open scuba tank (of a given volume) will fill with air, but more air can be pumped under pressure into the same tank before it is sealed (compressed air). Therefore, the amount of a gas is expressed not as volume but in terms of the pressure a gas exerts. In the case of a mixture of gases, like air, the amount of each gas is expressed as a **partial pressure**—the amount of pressure an individual gas contributes to the total pressure of the mixture. So, if the total pressure of the air (atmospheric pressure) is 760 mmHg, then the partial pressure of nitrogen (P_{N_2}) is 78.6 percent of 760, or 597 mmHg; the partial pressure of oxygen (P_{O_2}) is 20.9 percent of 760, or 159 mmHg; the partial pressure of carbon dioxide (P_{CO_2}) is 0.04 percent of 760, or 0.3 mmHg; and the remainder, 3.7 mmHg, is the partial pressure of water vapor. All of the partial pressures of the gases added together equal the total pressure of the air (760 mmHg).

You will need to understand partial pressures as a measurement of the amount of a gas when you study gas exchange in the lung and out at the tissues in the next section of this chapter.

spot check **7** The atmospheric pressure in Miami on Wednesday was 760 mmHg. However, the atmospheric pressure in Denver on the same day was 640 mmHg. What was the partial pressure of CO_2 in Denver that day? What was the partial pressure of O_2?

12.8 learning **outcome**

Explain gas exchange in terms of the partial pressures of gases at the capillaries and the alveoli and at the capillaries and the tissues.

Gas Exchange

Before studying **gas exchange,** it will be helpful for you to keep these two facts in mind: (1) Carbon dioxide is a waste product produced in the tissues through cellular respiration and (2) blood travels to the lungs to be oxygenated. With that stated, we begin by explaining gas exchange at the **respiratory membrane** between an alveolus and a capillary in the lung. In this discussion, we use general symbols—greater than (>), less than (<), and equal (=)—instead of worrying about specific values for the moment. Follow along with the numbered steps in **Figure 12.18** as you read this section on gas exchange.

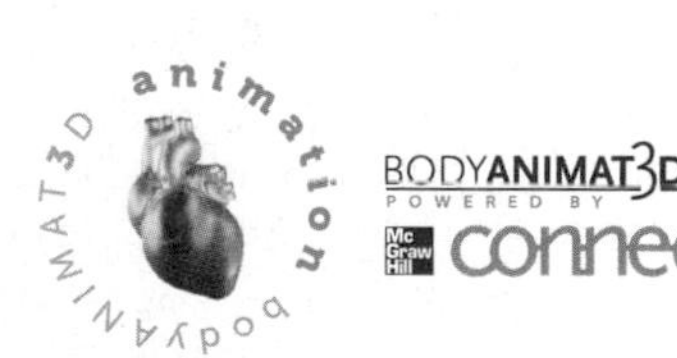

warning

Oxygen and carbon dioxide diffuse across a membrane because of a difference in concentration (concentration gradient) of the same gas. You must compare apples to apples and oranges to oranges, never apples to oranges. In other words, always compare the (P_{O_2}) on one side of the respiratory membrane to the (P_{O_2}) on the other side of the membrane, and then compare the (P_{CO_2}) on one side of the respiratory membrane to the (P_{CO_2}) on the other side of the membrane. Never compare the (P_{O_2}) to the (P_{CO_2}). Oxygen diffuses only if there is a concentration gradient for oxygen across the respiratory membrane. Carbon dioxide diffuses only if there is a concentration gradient for carbon dioxide across the respiratory membrane.

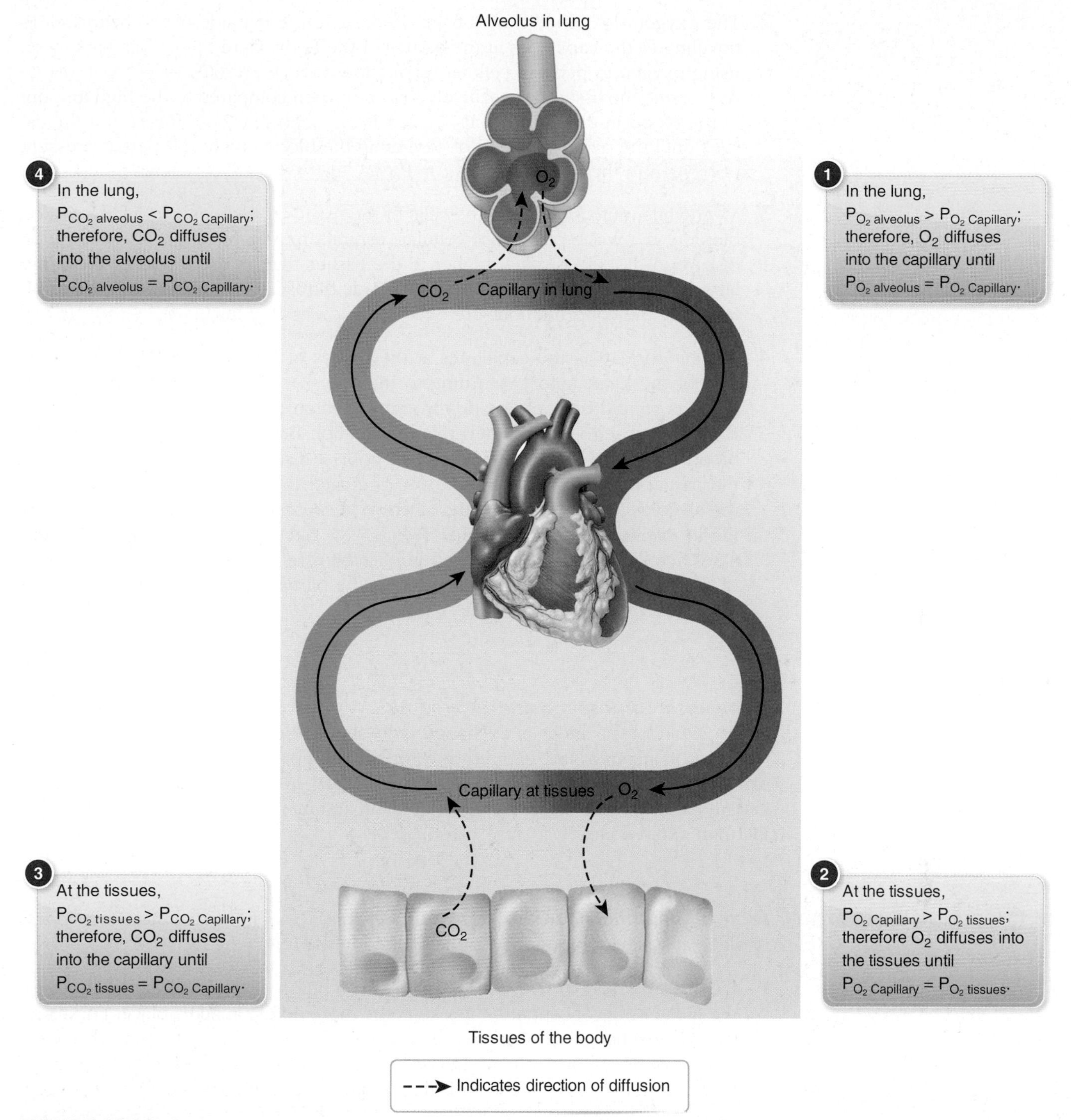

FIGURE 12.18 Gas exchange.

1. Blood coming from the right side of the heart to the lung is low in oxygen. In comparison, the air inspired to the alveolus in the lung is high in oxygen. $P_{O_2\ alveolus} > P_{O_2\ capillary}$, so oxygen diffuses across the respiratory membrane into the blood of the capillary until the partial pressures on both sides of the respiratory membrane are equal: $P_{O_2\ alveolus} = P_{O_2\ capillary}$. Will there still be some oxygen left in the alveolus after the gas exchange has taken place? Yes, because not all of it diffused into the blood; only the amount of oxygen necessary to make the partial pressures equal on both sides diffused. Some oxygen will be expired from the alveolus.

2. The oxygen-rich blood travels from the lung to the left side of the heart before traveling to the capillaries at the tissues of the body. Here the tissues have been using oxygen to perform cellular respiration: $C_6H_{12}O_6 + 6O_2 \rightarrow 6CO_2 + 6H_2O$. As a result, the tissues are relatively low in oxygen compared to the high amount in the blood in the capillary: $P_{O_2\ capillary} > P_{O_2\ tissues}$. So oxygen diffuses into the tissues until the partial pressure of oxygen in the blood equals the partial pressure of oxygen in the tissues: $P_{O_2\ capillary} = P_{O_2\ tissues}$.

3. Meanwhile, mitochondria in the cells of the tissues have been producing carbon dioxide as a waste product of cellular respiration. As a result, the concentration of carbon dioxide is much higher in the tissues than in the blood of the capillary: $P_{CO_2\ tissues} > P_{CO_2\ capillary}$. So carbon dioxide diffuses into the blood of the capillary until the concentrations are equal: $P_{CO_2\ tissues} = P_{CO_2\ capillary}$.

4. The blood leaving the capillaries at the tissues of the body travels to the right side of the heart before returning to the lungs. It has lost some of its oxygen and has gained carbon dioxide through diffusion at the tissues of the body. So it makes sense to have started this explanation of gas exchange by saying the blood coming to the lungs was oxygen-poor. It makes just as much sense to say the partial pressure of carbon dioxide is greater in the blood of the capillary at the alveolus than in the air of the alveolus because there is so little carbon dioxide in inspired air (0.04 percent): $P_{CO_2\ capillary} > P_{CO_2\ alveolus}$. So carbon dioxide diffuses across the respiratory membrane to the alveolus until the partial pressure of carbon dioxide in the capillary equals the partial pressure of carbon dioxide in the alveolus. $P_{CO_2\ capillary} = P_{CO_2\ alveolus}$.

12.9 learning outcome

Compare the composition of inspired and expired air.

Comparison of Inspired and Expired Air Given what you have read about the composition of air and gas exchange, you should be able to compare the composition of inspired and expired air. For this comparison, you will examine gas exchange using specific values. See **Figure 12.19**. As you can see in this figure, inspired air has more oxygen than expired air, and inspired air has less carbon dioxide than expired air.

12.10 learning outcome

Explain the factors that influence the efficiency of alveolar gas exchange.

Factors That Influence Gas Exchange Several factors influence the effectiveness of alveolar gas exchange. They are explained in the following list:

- **Concentration of the Gases.** The concentration of the gases matters because the greater the concentration gradient, the more diffusion takes place. For example, gas exchange of oxygen will increase if a patient is administered oxygen instead of breathing room air. In contrast, gas exchange of oxygen will be less at higher altitudes because the air is thinner and does not contain as much oxygen.

- **Membrane Area.** Membrane area matters because the greater the area of the respiratory membrane, the greater the opportunity for gas exchange. For example, **Figure 12.20** shows alveolar tissue for a healthy individual, a pneumonia patient, and a person with emphysema. You will notice the lack of respiratory membrane for the emphysema patient. This is because emphysema breaks down the alveolar walls. The reduced membrane area means less gas will be exchanged.

- **Membrane Thickness.** The thickness of the respiratory membrane matters because the thicker the membrane, the harder it is for gases to diffuse across it. Look again at **Figure 12.20**. Pneumonia may cause excess fluid in the alveoli and swelling of the alveolar walls, which make gas exchange much more difficult.

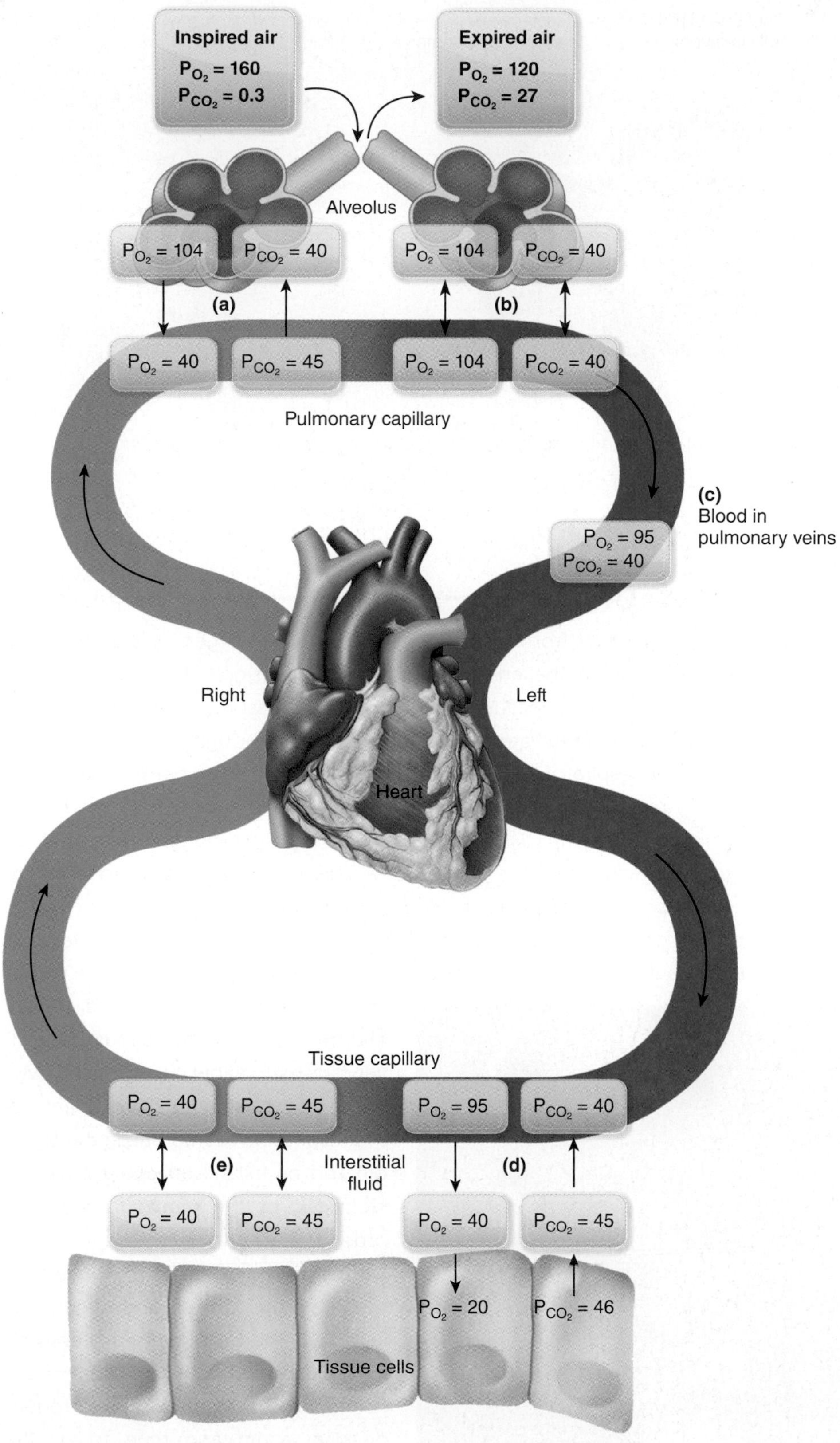

FIGURE 12.19 Changes in P_{O_2} and P_{CO_2} along the respiratory route. Values are expressed in mmHg. (a) Oxygen diffuses into the arterial ends of pulmonary capillaries, and CO_2 diffuses into the alveoli because of differences in partial pressures. (b) As a result of diffusion (at the venous ends of the pulmonary capillaries), the concentrations of O_2 are equal on both sides of the respiratory membrane, as are the concentrations of CO_2 on both sides of the respiratory membrane. (c) The partial pressure of O_2 is reduced in the pulmonary veins due to the mixing of blood drained from the bronchi and bronchial tree. (d) Oxygen diffuses out of the arterial end of capillaries to the tissues, and CO_2 diffuses out of the tissues to the capillaries due to the differences in partial pressures. (e) As a result of diffusion (at the venous ends of tissue capillaries), the concentrations of O_2 are equal in the capillaries and the tissues, as are the concentrations of CO_2 in the capillaries and the tissues.

- **Solubility of the gas.** Gases must be able to dissolve in water if they are to diffuse across a membrane into the blood. For example, nitrogen is 78.6 percent of the air you breathe, but it does not diffuse across the respiratory membrane because it is not soluble at normal atmospheric pressure. Oxygen and carbon dioxide are soluble at normal atmospheric pressure.

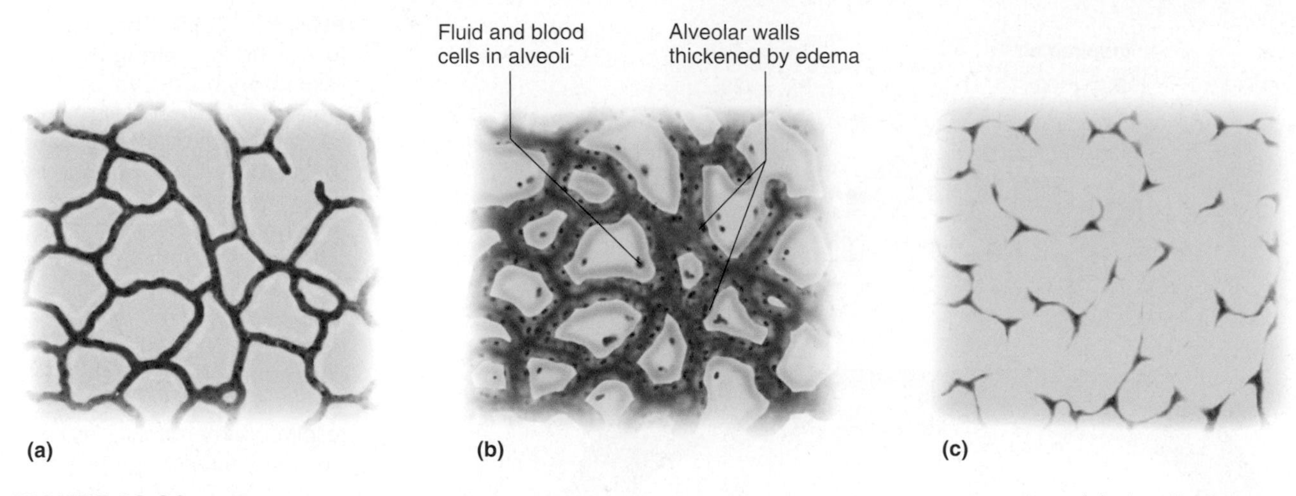

FIGURE 12.20 Influences on gas exchange: (a) normal alveoli, (b) alveoli of pneumonia patient, (c) alveoli of emphysema patient.

clinical point

Scuba divers breathe air from their tanks. It is not pure oxygen; it is air that has been compressed so that the tank can hold more. If divers go to significant depths, nitrogen becomes soluble because of the increased pressure—every 10 meters of water is equal to another full atmosphere of pressure. Although nitrogen can then diffuse across the respiratory membrane, this is alright because nitrogen does not react with anything in the blood. It becomes very relevant, however, during the ascent from the dive. If the diver comes up too quickly, nitrogen comes out of solution as a gas wherever it is in the body. This is similar to club soda being uncapped and poured. See Figure 12.21. Removing the cap relieves the pressure within the can or bottle. The carbon dioxide in the club soda quickly comes out of solution as bubbles with the reduced pressure. The nitrogen bubbles can cause severe damage to the diver's nerves and other tissues. Divers must ascend very slowly to allow nitrogen to slowly come out of solution, diffuse across the respiratory membrane in the alveoli, and be exhaled. **Decompression sickness,** or the **bends,** is the disorder that results if a diver ascends from depths too quickly. The treatment is to put the diver immediately in a hyperbaric (increased-pressure) chamber and put the body under sufficient pressure to have the nitrogen dissolve again and then to slowly decrease the pressure so that the nitrogen can be exhaled.

FIGURE 12.21 Glass of club soda.

- **Ventilation-perfusion coupling.** This basically means that the airflow to the lung must match the blood flow to the lung. Ideally, the maximum amount of air should go to where there is the maximum amount of blood in the lung. This is accomplished through local control in two ways:

    1. **Lung perfusion** (blood flow to alveoli). As blood flows toward alveolar capillaries, it is directed to lobules in the lung where the partial pressure of oxygen is high. How? Alveolar capillaries constrict where the partial pressure of oxygen is low, so blood is diverted to where the partial pressure of oxygen is high.

    2. **Alveolar ventilation** (airflow to alveoli). Smooth muscles in the walls of bronchioles are sensitive to the partial pressure of carbon dioxide. If the partial pressure of carbon dioxide increases, the bronchioles dilate. If the partial pressure of carbon dioxide decreases, the bronchioles constrict. Airflow is therefore directed to lobules where partial pressure of carbon dioxide is high.

  Ventilation-perfusion coupling is very important because it allows the respiratory system to compensate for damaged lung tissue. If an area of the lung is damaged, less air and less blood are directed to that area. See **Figure 12.22**.

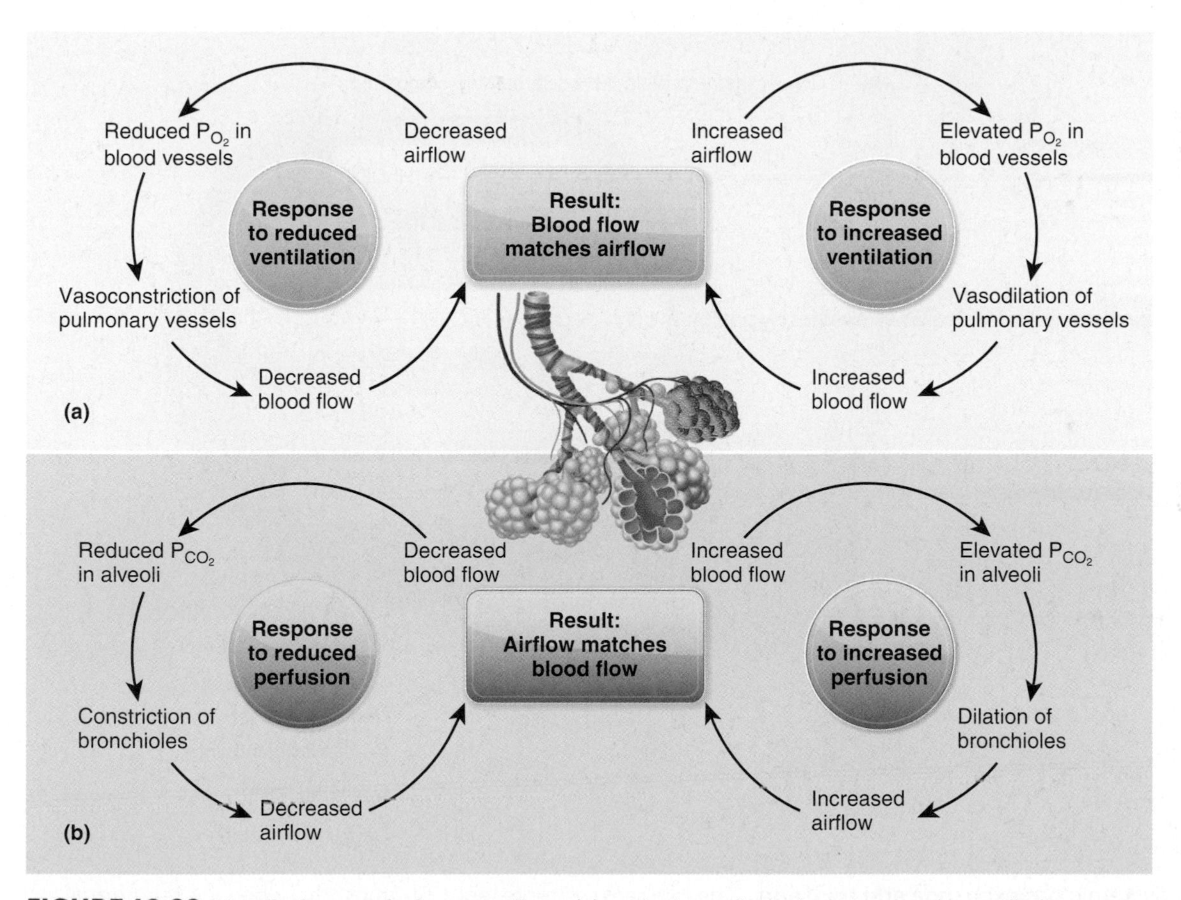

FIGURE 12.22 Ventilation-perfusion coupling: (a) perfusion adjusted to changes in ventilation, (b) ventilation adjusted to changes in perfusion.

spot check 8 The atmospheric pressure is 760 mmHg in New York City and 630 mmHg in Breckenridge, Colorado. In which city should more gas exchange take place? Explain.

12.11 learning outcome

Describe the mechanisms for transporting O_2 and CO_2 in the blood.

Gas Transport

You have now studied how carbon dioxide and oxygen are exchanged across membranes into and out of the blood, but how are these gases carried in the blood from one place to another? To understand gas transport, first look at **Figures 12.23** (systemic gas exchange and transport) **and 12.24** (alveolar gas exchange and transport). Notice the red and blue arrows in these figures. The blue arrows represent CO_2, while the red arrows represent O_2. The thickness of the arrows represents the relative amounts of the gases being exchanged. Although there are three blue arrows and two red arrows in each figure, concentrate on the largest blue arrow (representing 70 percent of the CO_2) and the largest red arrow (representing 98.5 percent of the O_2). These two arrows explain the majority of gas transport in the blood happening at the body's tissues (**Figure 12.23**) and at the alveoli (**Figure 12.24**).

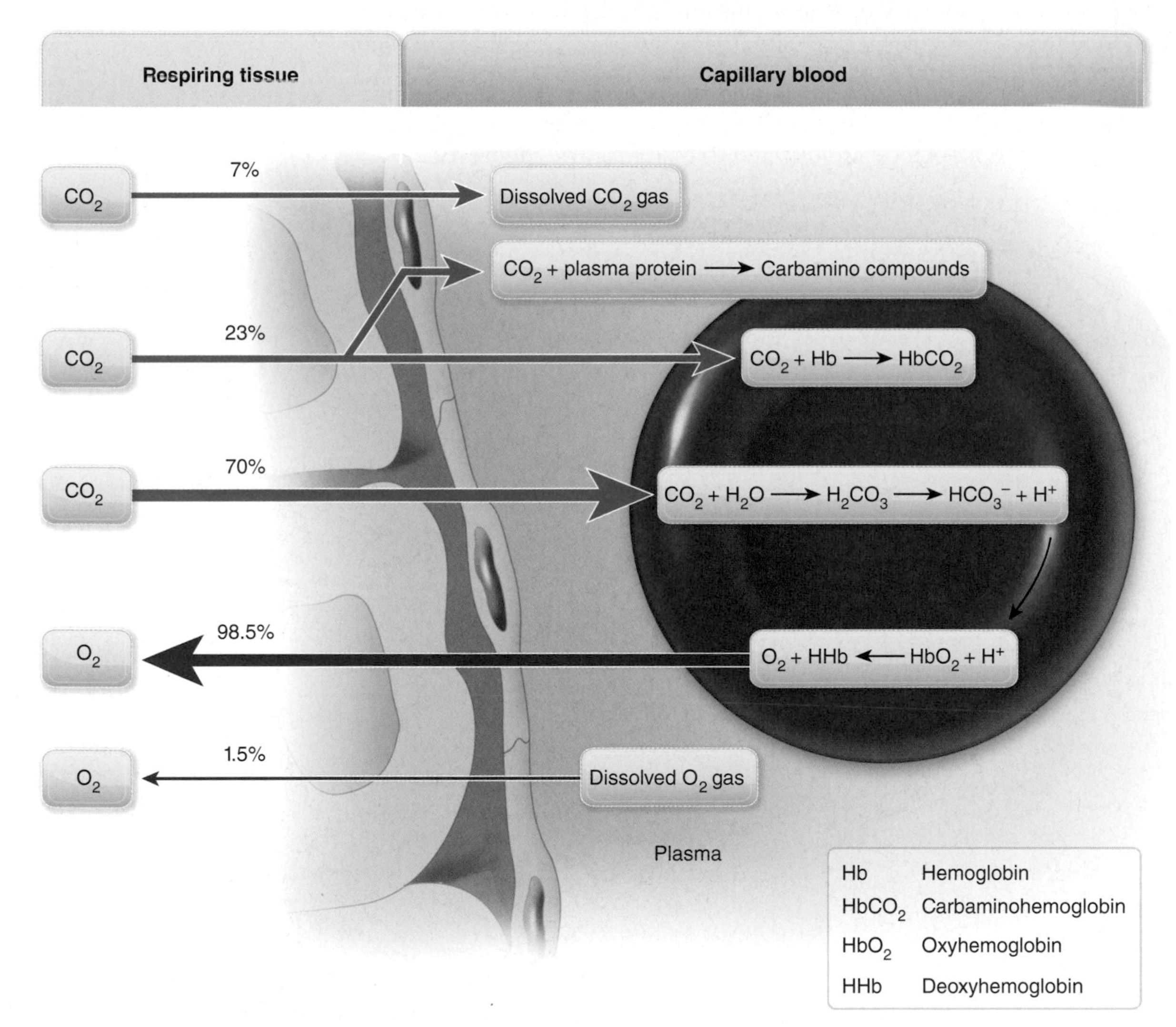

FIGURE 12.23 Systemic gas exchange and transport. The blue arrows represent CO_2 transport, while the red arrows represent O_2 transport. The thickness of the arrows represents the relative amounts of the gases being transported.

Systemic Gas Exchange and Transport You should already be aware that the tissues of the body produce CO_2 as a waste product of cellular respiration and that, because the $P_{CO_2\ tissues} > P_{CO_2\ capillary}$, CO_2 diffuses from the tissues into the capillaries. Now you need to understand that the diffused carbon dioxide mixes with water in the blood to form **carbonic acid** (H_2CO_3). Carbonic acid, because it is in water, separates into its two ions: a **bicarbonate ion** HCO_3^- and a hydrogen ion (H^+)—

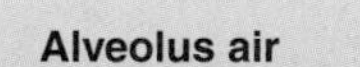

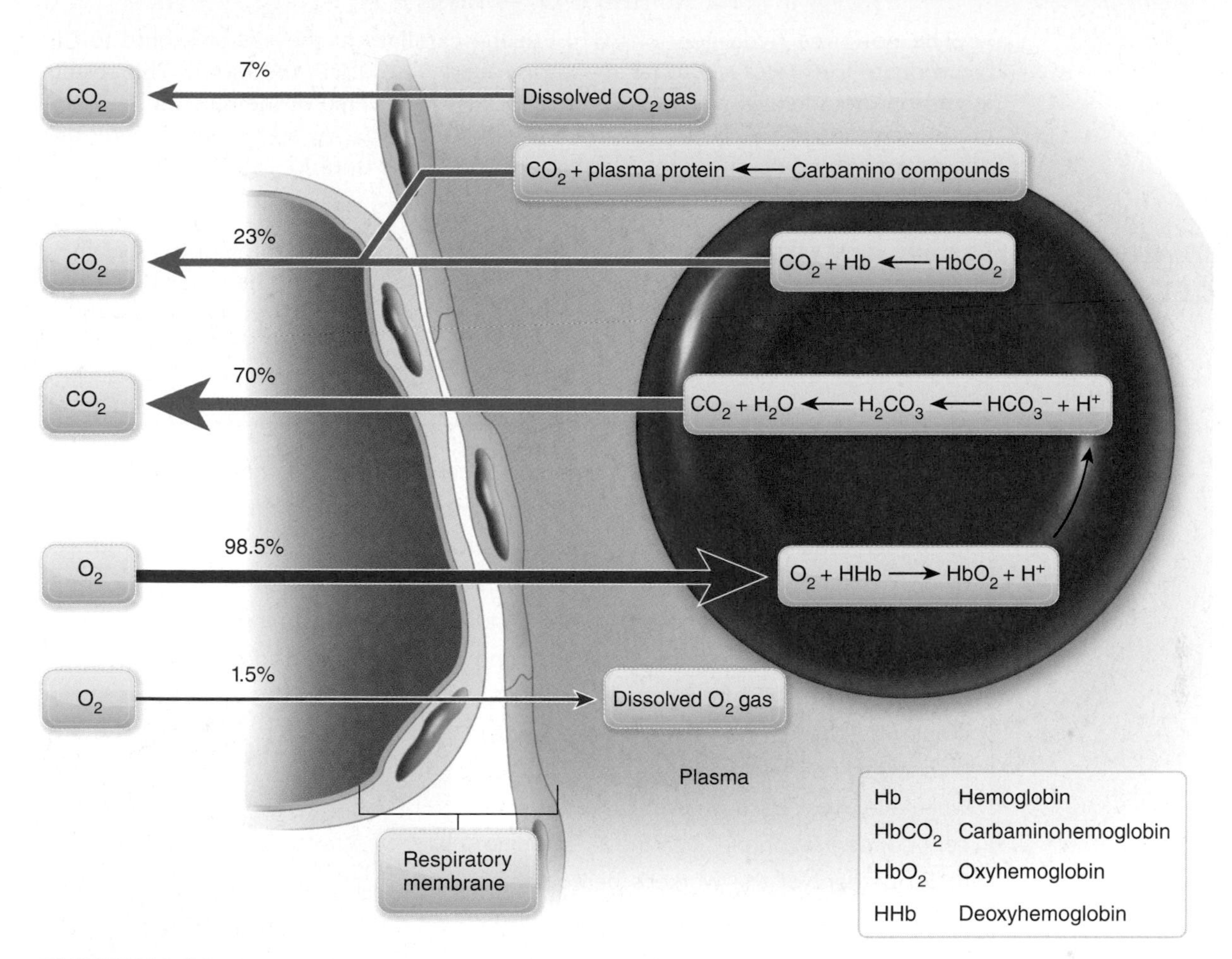

FIGURE 12.24 **Alveolar gas exchange and transport.** The blue arrows represent CO_2 transport, while the red arrows represent O_2 transport. The thickness of the arrows represents the relative amounts of the gases being transported.

remember from Chapter 2 that water allows for ions in solution. This reaction is shown in **Figure 12.23** where the largest blue arrow enters the blood:

$$CO_2 + H_2O \rightarrow HCO_3^- + H^+$$

Free hydrogen ions in the blood would lower the pH of the blood, but notice in **Figure 12.23** that the free hydrogen ion (H^+) reacts with **oxyhemoglobin** (HbO_2) to become **deoxyhemoglobin** (HHb) and oxygen (O_2):

$$H^+ + HbO_2 \rightarrow HHb + O_2$$

Hemoglobin releases oxygen in the presence of a hydrogen ion and then binds to it (H^+). By binding to the free hydrogen ions, hemoglobin acts as a buffer, resisting a change of pH in the blood. $P_{O_2\text{ capillary}} > P_{O_2\text{ tissues}}$, so oxygen diffuses to the tissues until $P_{O_2\text{ tissues}} = P_{O_2\text{ capillary}}$.

The blood containing deoxyhemoglobin and bicarbonate ions continues to the right side of the heart and on to the alveoli of the lung. Below, you will learn what happens in alveolar gas exchange and transport, as shown in **Figure 12.24**.

Alveolar Gas Exchange and Transport Again, you should focus on the largest red and blue arrows representing oxygen and carbon dioxide in **Figure 12.24**. In the alveolus, the $P_{O_2\text{ alveolus}} > P_{O_2\text{ capillary}}$, so oxygen diffuses into the capillaries. When it does, deoxyhemoglobin reacts with oxygen to release hydrogen ions and form oxyhemoglobin:

$$HHb + O_2 \rightarrow HbO_2 + H^+$$

The now free hydrogen ions (H+) in the capillary at the alveolus bind to the bicarbonate ions (HCO_3^-) to form carbonic acid (H_2CO_3) in the blood. This results in carbon dioxide and water. Notice that this is the reverse of the reaction happening for carbon dioxide at the tissues. $P_{CO_2\ capillary} > P_{CO_2\ alveolus}$, so carbon dioxide diffuses across the respiratory membrane to the alveolus until $P_{CO_2\ capillary} = P_{CO_2\ alveolus}$.

$$H^+ + HCO_3^- \rightarrow H_2CO_3 \rightarrow CO_2 + H_2O$$

Basically, most of the oxygen is transported in the blood by hemoglobin as oxyhemoglobin, and most of the carbon dioxide is transported in the blood as bicarbonate ions. Hemoglobin functions to carry oxygen from the lungs to the tissues and hydrogen ions from the tissues to the lungs.

clinical point

You have already read about hemoglobin carrying carbon monoxide (CO) in Chapter 9. Hemoglobin binds to carbon monoxide 210 times more tightly than to oxygen, and it does not carry oxygen as long as it is bound to carbon monoxide. CO is produced during combustion, so it can be emitted from improperly vented furnaces, cars (exhaust), and even cigarettes as they are smoked. Typically, less than 1.5 percent of hemoglobin is bound to carbon monoxide in nonsmokers, while 10 percent of a heavy smoker's hemoglobin may be bound to carbon monoxide. Mechanics need to ventilate their garages while they work because even just a 0.1 percent concentration of CO in the air can bind to 50 percent of the worker's hemoglobin, and a 0.2 percent atmospheric concentration can be lethal.

12.12 learning outcome

Explain how respiration is regulated to homeostatically control blood gases and pH.

Regulation of Respiration

Now that you have become familiar with how oxygen and carbon dioxide are transported in the blood, you are ready to examine how the respiratory system is regulated to homeostatically control blood gases and pH.

The main control centers for respiration are located in the medulla oblongata. See **Figure 12.25**. From there, messages to stimulate inspiration travel through inspiratory (I) neurons that go to the spinal cord and then out to the diaphragm and intercostal muscles (by way of the phrenic and intercostal nerves). Expiratory (E) neurons in the medulla oblongata send signals only for forced expiration.

As you have previously learned, increasing the frequency of nerve impulses causes longer muscle contractions and, therefore, deeper inspirations. If the length of time (duration) is increased for each stimulus, the inspiration is prolonged and the breathing is slower. When nerve impulses from the inspiratory neurons end, muscles relax and expiration takes place.

Information concerning the need for regulation comes to the respiratory centers in the medulla oblongata from several sources. These sources are explained in the following list:

- Stretch receptors in the thoracic wall send signals to the medulla oblongata as to the degree of the chest's expansion. When maximum expansion has been reached, the medulla oblongata stops sending inspiratory messages. This prevents overinflation of the lungs, and it is most important in infants. This action is called the **Hering-Breuer** reflex.

- Proprioceptors in the muscles and joints send signals to respiratory centers during exercise so that ventilation is increased. The respiratory centers in the medulla oblongata can increase the depth and rate of respiration.
- The **pontine respiratory group** in the pons receives input from the hypothalamus, limbic system, and cerebral cortex. It then sends signals to the medulla oblongata to adjust the transitions from inspiration to expiration. In that way, the breaths become shorter and shallower or longer and deeper. This center helps adjust respirations to special circumstances, such as sleep, exercise, or emotional responses like crying.
- The cerebral cortex can exert voluntary control over the respiratory system, but this is limited control. For example, a stubborn child may threaten to hold his breath to get his way. However, if he is allowed to do so, he will eventually pass out and will start breathing again.
- **Peripheral chemoreceptors** in the aortic arch and carotid arteries (discussed in Chapter 10) and **central chemoreceptors** in the medulla oblongata send information to the respiratory centers in the medulla oblongata concerning pH, CO_2, and O_2. The peripheral chemoreceptors (shown in **Figure 12.26**) monitor the blood, while the central chemoreceptors monitor the cerebral spinal fluid (CSF). Why monitor both fluids? Hydrogen ions in the blood cannot pass the blood-brain barrier, but carbon dioxide does cross the barrier. When it does, it reacts with the water in the CSF to form H^+, just as it does at the tissues when mixing with the water in the blood. An increased concentration of hydrogen ions in either fluid means reduced pH. The most important driver of respiration is pH, the next is CO_2, and the driver of minor importance is O_2. Respiration is adjusted by the medulla oblongata to maintain pH in the homeostasis range for blood of 7.35 to 7.45. **Acidosis** occurs if the pH of the blood is less than 7.35. The medulla oblongata then stimulates **hyperventilation** (increased respiratory rate) to blow off CO_2 through expiration to raise the pH. **Alkalosis** results if the pH of the blood is greater than 7.45. The medulla oblongata then stimulates **hypoventilation** (decreased respiratory rate) to keep CO_2 in the blood to lower the pH. **Hypercapnia,** increased carbon dioxide in the blood, causes the pH to fall in both fluids. Oxygen is of minor importance as a driver of respiration because the blood's hemoglobin is usually 97 percent saturated with oxygen during normal breathing. Oxygen drives respiration only during extreme conditions, such as mountain climbing at high altitudes, and when it does, this is called **hypoxic drive.**

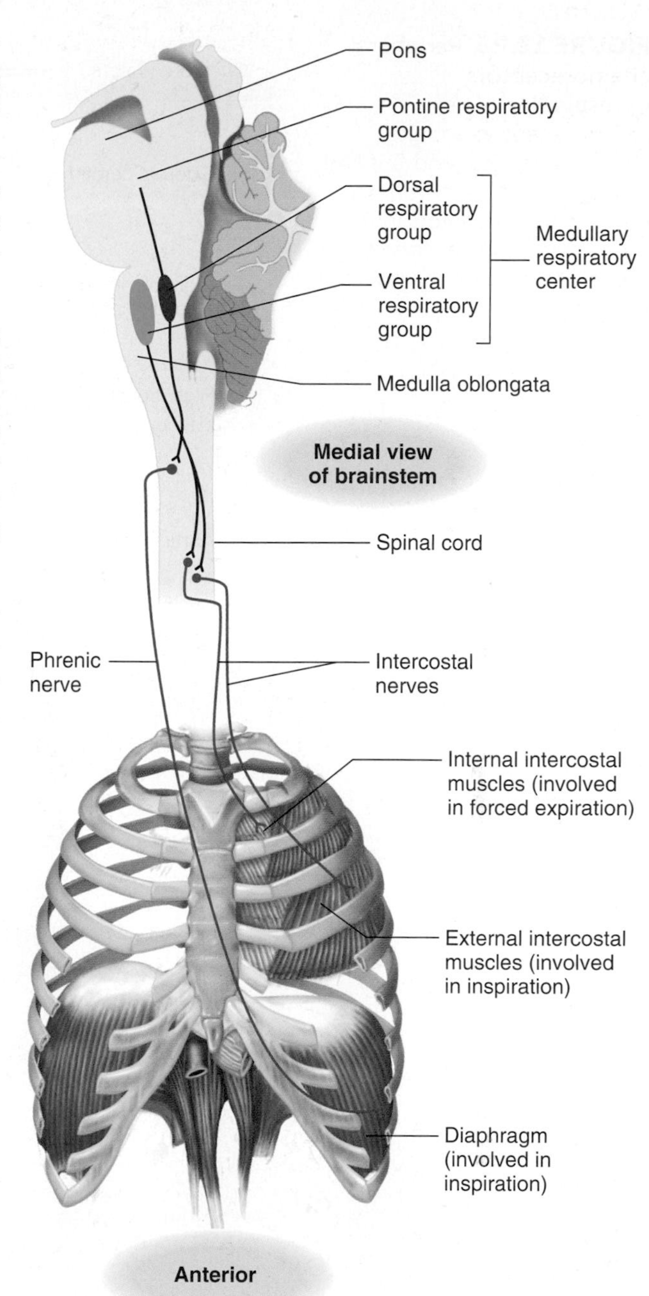

FIGURE 12.25 Control centers for respiration.

You have now explored all of the respiratory anatomy and the physiology of this system. It is time to put that information together to see how the functions of this system are carried out for Carol, who is on a break from her job in the business office at the hospital. See **Figure 12.27**.

FIGURE 12.26 **Peripheral chemoreceptors of respiration.** These chemoreceptors monitor blood gases (CO_2 and O_2) and blood pH. They send signals to the respiratory centers in the medulla oblongata along the vagus and glossopharyngeal nerves.

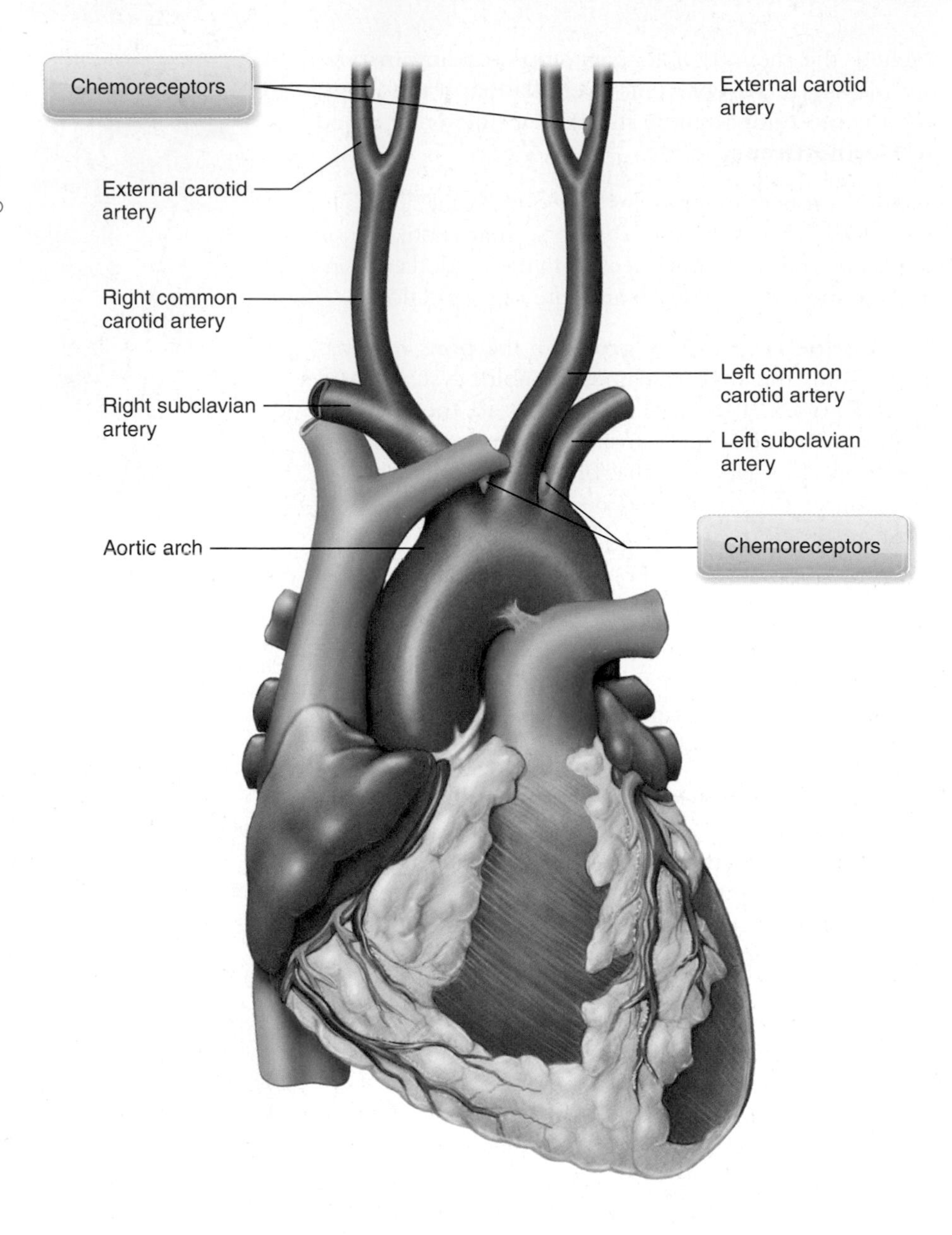

12.13 learning outcome

Explain the functions of the respiratory system.

Functions of the Respiratory System

Carol appears to be a relatively young, healthy woman who is a smoker. Although her respiratory system functions normally at present, her lifestyle choice to continue to smoke may have harmful, long-term effects on her respiratory system. In the list below, we explain the effects on each of the functions of this system:

- **Gas exchange.** Carol's respiratory system functions to exchange carbon dioxide and oxygen across the respiratory membranes of her lungs and out at the tissues of her body. However, each time she inspires through a lit cigarette, she also exchanges carbon monoxide across her respiratory membrane, which then binds to the hemoglobin in her blood. Hemoglobin bound to CO no longer functions to carry oxygen, so her levels of O_2 in the blood will fall (hypoxia). Her kidneys notice the decreased O_2 levels and secrete erythropoietin (EPO, discussed in Chapter 9) to stimulate erythropoiesis to increase the RBC count and the available hemoglobin to carry O_2.

- **Acid-base balance.** Carol's respiratory system regulates her acid-base balance by increasing respirations whenever the pH of her blood begins to fall below homeostasis. Hyperventilation gets rid of more CO_2, so the pH of the blood increases. If Carol's blood pH rises above homeostasis, her respiratory rate will slow (hypoventilation), so any CO_2 in her system remains in the blood longer, thus lowering her pH to normal levels.

- **Speech.** The muscles of Carol's larynx contract to move the arytenoid and corniculate cartilages that operate her vocal cords and cause them to vibrate to produce sound. Even the sinuses connected to her nasal cavity will aid in her speech by giving resonance to her voice. However, her smoking tends to irritate and dry the lining of her larynx. This may lead to laryngitis and a scratchy voice.

- **Sense of smell.** Olfactory neurons in the epithelium of the roof of Carol's nasal cavity detect odors for her sense of smell (covered in detail in Chapter 7). Carol's smoking may cause these receptors to become less sensitive, limiting her ability to sense odors and appreciate flavors.

- **Creation of pressure gradients necessary to circulate blood and lymph.** The muscles used during respiration increase the volume of the thoracic cavity, causing the pressure inside the cavity to fall. As you have learned in this chapter, this pressure decrease will cause air to flow into Carol's lungs. However, you have also studied this mechanism in the circulatory and lymphatic systems as the thoracic pump. This pressure decrease in the thorax also helps Carol's blood return (through the inferior vena cava) to her heart and helps her lymph return (through the thoracic duct) to her left subclavian vein. The thoracic pump created by this system will be less effective if Carol's blood becomes thicker due to smoking-related polycythemia (discussed in Chapter 9).

FIGURE 12.27 Carol.

Carol's respiratory system is functioning now, but what can she expect to be the effects of growing older even if she decides to stop smoking?

Effects of Aging on the Respiratory System

12.14 learning outcome

Summarize the effects of aging on the respiratory system.

Aging has many effects on the respiratory system, which basically lead to a decline in maximum function. These effects are as follows:

- With age, more mucus accumulates in the respiratory tract because the ciliated escalator becomes less efficient. The inability to clear debris efficiently leaves the elderly open to more respiratory infections. So vaccines to prevent infections, such as flu and pneumonia, are highly recommended for the elderly.

- Thoracic wall compliance decreases due to the diminished ability to expand the chest that comes with age. This can be due to weakened respiratory muscles, stiffening of the cartilages in the rib cage, decreased height of the vertebrae from age-related osteoporosis, and kyphosis. The net effect of the reduced compliance is reduced vital capacity because the ability to fill the lungs (inspiratory reserve volume) and the ability to empty the lungs (expiratory reserve volume) are both decreased.

- Some of the alveoli walls may break down, and this reduces the area of the respiratory membrane. The remaining membrane thickens with age, reducing alveolar gas exchange. Tidal volume gradually increases with age to compensate for the reduced area and thickening of the respiratory membrane.
- Obstructive sleep **apnea** (breathing repeatedly stops and starts during sleep) may develop in the elderly as the pharyngeal muscles intermittently relax and block the airway during sleep. This form of apnea may or may not be accompanied by snoring, and it is more prevalent in people who are overweight.

The effects of aging may not be readily apparent in a healthy individual, but they may diminish even the healthy individual's ability to perform vigorous exercise. You will complete your study of this system by examining what can go wrong—respiratory disorders.

12.15 learning outcome

Describe respiratory system disorders.

Respiratory System Disorders

The respiratory disorders discussed below fall into three categories: infections, chronic obstructive pulmonary diseases (COPDs), and lung cancer.

Respiratory Infections

Cold The most common respiratory infection is the common cold, which is commonly caused by a rhinovirus. Its symptoms include congestion, sneezing, and increased mucus production. This infection can easily spread to the sinuses, throat, and middle ear. Typically, a cold runs its course in about a week.

applied genetics

Cystic fibrosis is the most common fatal genetic disease in the United States. As you read in Chapter 2, cystic fibrosis is caused by a single gene in the human DNA that codes for a faulty chloride channel on cell membranes. People with a faulty cystic fibrosis transmembrane regulator gene (CFTR) produce a sticky mucus that cannot be easily moved by the respiratory epithelium's ciliated escalator. As a result, the sticky mucus accumulates in the lungs and airways, and this then leads to infection. Gene therapy for cystic fibrosis began in 1990 when scientists were successful in introducing correct copies of the gene to cells in laboratory cultures. In 1993, common rhinoviruses were tried as a delivery mechanism (vector) to deliver the correct gene. These viruses were tried as vectors because rhinoviruses specifically invade respiratory cells and deliver a piece of nucleic acid to the invaded cell. If the rhinovirus could be modified to carry and insert the correct copy of the CFTR gene, it would deliver it to the appropriate type of cell in a cystic fibrosis patient. Since then, other vectors have been tried in an effort to find the most efficient way of introducing the correct gene to the affected cells. Life span of the respiratory cells also needs to be considered to determine the correct vector and treatment schedule. The research into gene therapy for this disease continues.[1]

Influenza Flu is a respiratory—not digestive—illness caused by a virus. In addition to its cold symptoms, flu is characterized by fever, chills, and muscle aches. The mortality rate for influenza is approximately 1 percent, with most of the deaths occurring in the very young and the elderly. Influenza viruses mutate and change often, so vaccines are created each year to protect against the expected viral flu strains for that year.

Tuberculosis This infection is caused by a bacterium that enters the lungs by way of air, blood, or lymph. The lungs react to the infection by walling off bacterial lesions with scar tissue that diminishes lung compliance. Health care workers are tested for exposure to the bacteria with a Mantoux test.

Pertussis This highly contagious bacterial infection causes the paralysis of cilia in the respiratory epithelium. The accumulation of mucus and debris results in a *whooping cough,* which gives this disorder its common name. Pertussis vaccine is one part of the DPT shot routinely given in the United States to children. (D stands for *diphtheria,* P stands for *pertussis,* and T stands for *tetanus*).

Pneumonia This infection can be caused by bacteria, a virus, a fungus, or even a protozoan. Symptoms include fever, difficulty breathing, and chest pain. In this type of infection, fluid accumulates in the alveoli (pulmonary edema), and inflammation causes the respiratory membrane to thicken, thereby reducing gas exchange.

COPDs

Chronic obstructive pulmonary disorders cause the long-term decrease in ventilation of the lungs. Many COPDs are the result of cigarette smoking.

Chronic Bronchitis This disorder often results from long-term irritation of the epithelium of the bronchial tree. With the subsequent inflammation, cilia are lost and mucus is overproduced. Without the cilia escalator, mucus and debris accumulate, leading to further chronic inflammation and infections. The long-term effect is a decrease in the diameter of the bronchioles, which reduces ventilation of the alveoli. Chronic bronchitis often leads to emphysema.

Emphysema In this disorder, constant inflammation from irritants narrows bronchioles, reducing the airflow to the lungs. The respiratory system tries to clear built-up mucus and debris—often from chronic bronchitis—by coughing. The coughing causes increased pressure in the alveoli that results in rupturing of the alveolar walls. This loss of respiratory membrane reduces gas exchange and reduces the recoil of the lung (compliance). Symptoms of emphysema include shortness of breath and an enlargement of the thoracic cavity (barrel chest).

Asthma This disorder involves increased constriction of the lower respiratory tract due to a variety of stimuli. The symptoms include wheezing, coughing, and shortness of breath. Although no definitive cause has been found, asthma and allergies often go together. Whatever the cause, asthma is characterized by chronic airway inflammation, airway obstruction, and airway hyperreactivity, in which the smooth muscle overreacts to a stimulus by constricting the bronchioles. Treatment involves avoiding the stimulus—if it can be determined—and drug therapy.

Lung Cancer

More people die every year from lung cancer than from any other form of cancer. The most common cause of lung cancer is cigarette smoking. See **Figure 12.28.** According to the Centers for Disease Control (CDC), at least 50 carcinogens can be found in cigarette smoke.[2] There are three forms of lung cancer:

- **Squamous cell carcinoma** originates in the bronchial epithelium. In this form of lung cancer, the ciliated epithelium undergoes metaplasia first, changing to stratified epithelial tissue. Cancerous cells further divide, invading tissues of the bronchial walls and forming tumors that can block airways.

FIGURE 12.28 Effects of Smoking: (a) healthy lung, medial surface, (b) smoker's lung with carcinoma.

- **Adenocarcinoma** originates in the mucous glands of the bronchial tree in the lung. Like squamous cell carcinoma, it also invades other tissues of the bronchial tree and lung.
- **Oat cell carcinoma** is the least common form of lung cancer, but it is the most deadly because it easily metastasizes to other tissues. It usually begins in a main bronchus and then invades the mediastinum and travels to other organs.

putting the pieces **together**

The Respiratory System

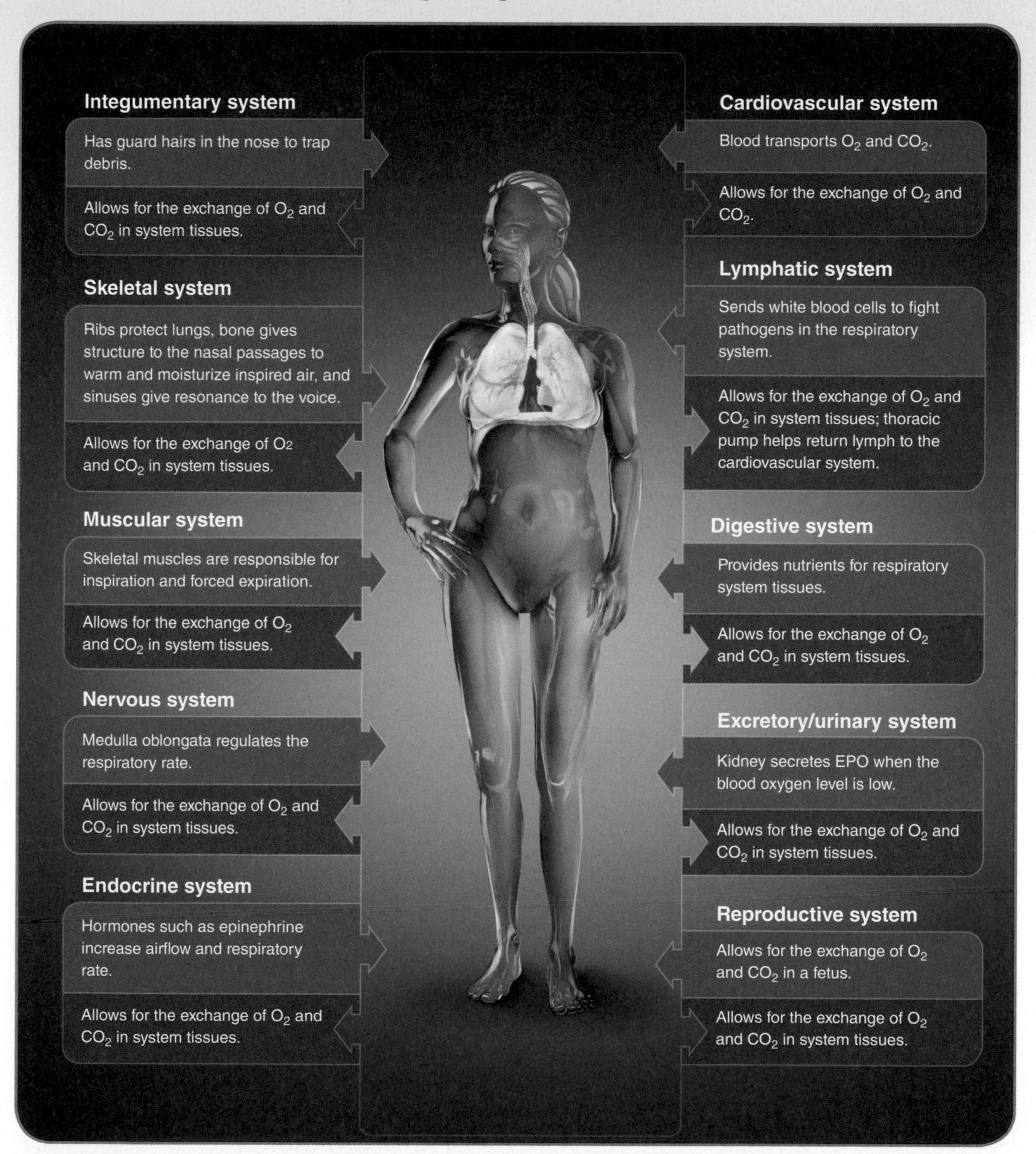

FIGURE 12.29 Putting the Pieces Together—The Respiratory System: connections between the respiratory system and the body's other systems.

summary

Overview

- Cellular respiration is performed by mitochondria in cells to process energy.
- Respiration as a system refers to the movement of gases into and out of the lungs and the exchange of gases between the alveoli and capillaries in the lung and capillaries and tissues in the body.

Anatomy of the Respiratory System

- The air in an inspiration enters the nasal cavity through the nose and continues on to the nasopharynx, oropharynx, and laryngopharynx. It then travels through the glottis to the larynx, to the trachea, to the main bronchi, to the bronchial tree, and finally to the alveoli.

Nose

- The nasal bones and nasal cartilages shape the nose.

Nasal Cavity

- The mucous membranes of the nasal cavity warm and moisturize the air and remove debris
- Nasal conchae provide extra surface area.

Pharynx

- The pharynx is composed of the nasopharynx, oropharynx, and laryngopharynx.
- The epithelial tissue varies in each part of the pharynx based on the materials that travel through each area.

Larynx

- The larynx is a cartilaginous box that contains the vocal cords.
- Muscles in the larynx move cartilages that allow the vocal cords to vibrate to produce sound.

Trachea

- The trachea has 18 to 20 C-shaped cartilages that hold it open for the easy passage of air.
- The trachea splits to form the main bronchi.

Lungs and the Bronchial Tree

- Each main bronchus enters a lung and then further divides to form the bronchial tree.
- Bronchioles have smooth muscle in their walls and lead to small air sacs in the lung called *alveoli*.
- Alveoli have walls of simple squamous cells and great alveolar cells that produce surfactant.
- Surfactant reduces the surface tension of water so that alveoli do not collapse.
- The respiratory membrane is composed of the thin layer of water with surfactant in the alveoli, the single squamous cell alveolar wall, and the single cell capillary wall.

Physiology of the Respiratory System

Mechanics of Taking a Breath

- Inspiration results from intercostal muscles and the diaphragm's contracting to increase the volume of the thoracic cavity, thereby decreasing its pressure.
- Air flows due to pressure gradients.
- Pleural membranes cause the lung to expand with the thoracic cavity.
- Normal expiration is caused by the relaxation of the intercostal muscles and diaphragm.
- Forced expiration is caused by muscle contraction.

Measurements of Pulmonary Function

- A spirometer can be used to measure lung volumes and capacities.
- Compliance measures how well the lung can expand and return to shape.

Composition of Air

- Air is a mixture of gases including nitrogen, oxygen, carbon dioxide, and water vapor.
- Partial pressure is the amount of pressure an individual gas contributes to the total pressure of the mixture.

Gas Exchange

- Gas exchange happens between the alveoli and the capillaries in the lung and between the capillaries and the tissues of the body.
- Gases diffuse across membranes because of a concentration gradient until the concentrations on both sides of the membrane are equal.
- Inspired air has more oxygen and less carbon dioxide than expired air.
- Gas exchange is influenced by concentration of the gases, membrane area, membrane thickness, solubility of the gas, and ventilation-perfusion coupling.

Gas Transport

- Most of the oxygen is transported in the blood by hemoglobin as oxyhemoglobin, and most of the carbon dioxide is transported in the blood as bicarbonate ions.
- Hemoglobin functions to carry oxygen from the lungs to the tissues and to carry hydrogen ions from the tissues to the lungs.

Regulation of Respiration

- Respiration is controlled by respiratory centers in the medulla oblongata.
- The medulla oblongata receives information concerning the need to control respiration from stretch receptors, the pons, the cerebral cortex, and chemoreceptors.
- The drivers of respiration are pH, CO_2, and O_2 (in that order).

Functions of the Respiratory System

- The functions of the respiratory system include gas exchange, acid-base balance, speech, sense of smell, and creation of pressure gradients necessary to circulate blood and lymph.

Effects of Aging on the Respiratory System

- The ciliated escalator becomes less efficient, so more mucus and debris accumulate in the respiratory tract, and this can lead to infection.
- Thoracic wall compliance decreases, causing reduced vital capacity.
- Some alveolar walls break down with age and thicken, thereby reducing gas exchange.
- Obstructive sleep apnea may occur if the pharyngeal muscles block the airway.

Respiratory System Disorders

Respiratory Infections

- Respiratory infections include colds, flu, tuberculosis, pertussis, and pneumonia.

COPDs

- COPDs are often the result of smoking and include chronic bronchitis, emphysema, and asthma.

Lung Cancer

- Lung cancer causes more deaths than any other form of cancer.

keywords for review

The following terms are defined in the glossary.

alveoli
bronchial tree
chronic obstructive pulmonary disorders (COPDs)
compliance
expiration
functional residual capacity (FRC)
gas exchange
gas transport
inspiration
inspiratory reserve volume (IRV)
partial pressure
pharynx
pneumothorax
respiratory membrane
spirometry
surfactant
tidal volume (TV)
ventilation
ventilation-perfusion coupling
vital capacity (VC)

chapter review questions

Word Building: *Use the word roots listed in the beginning of the chapter and the prefixes and suffixes inside the back cover to build words with the following meanings:*

1. Incision into the chest wall: ______
2. Pertaining to the lung: ______
3. Removal of a lobe of a lung: ______
4. Inflammation of the larynx: ______
5. Instrument for examining the air passageway entering the lung: ______

Multiple Select: *Select the correct choices for each statement. The choices may be all correct, all incorrect, or a combination of correct and incorrect.*

1. Which of the following statements is (are) true concerning gas exchange between the capillaries and the tissues?
   a. Gas exchange occurs between the capillaries and the tissues until partial pressures are equal.
   b. Carbon dioxide moves to the capillaries.
   c. The partial pressure of carbon dioxide is greater in the capillaries than in the tissues.
   d. The partial pressure of oxygen is greater in the tissues than in the capillaries.
   e. Oxygen moves from the capillaries to the tissues.
2. What happens at the alveoli?
   a. Oxygen diffuses to the capillaries and combines with hemoglobin to form oxyhemoglobin.
   b. Hemoglobin binds to hydrogen ions.
   c. Gases move across the respiratory membrane by active transport, requiring energy.
   d. Oxygen diffuses across the respiratory membrane until $P_{O_2} = P_{CO_2}$.
   e. Carbon dioxide diffuses across the respiratory membrane until $P_{CO_2\ capillary} = P_{CO_2\ alveoli}$.
3. How does inspired air compare to expired air?
   a. Inspired air is higher in oxygen and lower in carbon dioxide than expired air.
   b. The partial pressure of oxygen in inspired air is 0.03 percent of the total pressure of the air.
   c. Nitrogen is the gas in highest concentration in inspired air.
   d. There is no oxygen in expired air because it all diffuses across the respiratory membrane into the blood.
   e. Expired air contains more carbon dioxide than inspired air.
4. Which of the following factors affect(s) alveolar gas exchange?
   a. Alveolar gas exchange of nitrogen is decreased when a diver dives deep because nitrogen is less soluble under pressure.
   b. Alveolar gas exchange is decreased in pneumonia because the respiratory membrane has become thinner.
   c. Alveolar gas exchange is decreased in emphysema because the respiratory membrane has become more extensive.
   d. Alveolar gas exchange is reduced in asthma because ventilation does not keep up with blood perfusion.
   e. Alveolar gas exchange is increased if the concentration of the gas is increased.
5. What happens when you breathe?
   a. The visceral pleura lining the chest pulls on the parietal pleura attached to the lung due to surfactant.
   b. The diaphragm and intercostal muscles expand the chest.
   c. The Hering-Breuer reflex limits how much is exhaled.
   d. The pons sends inhibitory messages.
   e. The medulla oblongata sends a message by way of inspiratory neurons to cause skeletal muscle contractions.
6. What pathway does air travel through the respiratory system?
   a. Inspired air moves from the bronchial tree to bronchioles to alveolar ducts.
   b. Inspired air goes through the trachea to the larynx.
   c. Inspired air travels from the nasopharynx to the laryngopharynx to the oropharynx.
   d. Air moves from areas of higher pressure to areas of lower pressure.
   e. Inspired air moves through the glottis into the larynx.

7. Which of the following statements is (are) true about hemoglobin?
   a. Hemoglobin functions to carry oxygen from the lungs to the tissues and hydrogen ions from the tissues to the lungs.
   b. Hemoglobin binds to bicarbonate ions.
   c. Hemoglobin is used to carry oxygen to the tissues.
   d. Hemoglobin acts as a buffer to reduce the change in pH of blood when it combines with hydrogen ions to form HHb.
   e. Hemoglobin combines with hydrogen ions to form bicarbonate ions in the capillaries near the tissues.

8. $CO_2 + H_2O \rightarrow H_2CO_3 \rightarrow H^+ + HCO_3^-$. What is true about this equation?
   a. The carbon dioxide is produced in the tissues through cellular respiration.
   b. This reaction happens in the direction indicated at the tissues and capillaries.
   c. The products of this reaction indicate an acid.
   d. The water in this reaction is found in the blood.
   e. This reaction is reversed at the capillaries and alveoli.

9. What would happen if a runner does a sprint at the end of a race and undergoes anaerobic respiration, resulting in a buildup of lactic acid, which enters the blood?
   a. Her respiratory rate would increase.
   b. Her respiratory rate would decrease.
   c. Anaerobic respiration is at the cellular level and would have no effect on the rate of breathing.
   d. Chemoreceptors in the medulla oblongata would detect increased hydrogen ions, and the medulla oblongata would therefore decrease respirations.
   e. Chemoreceptors in the aortic arch and the carotid arteries would send a message to the medulla oblongata to increase respirations.

10. Which of the following statements is (are) true concerning respiratory disorders?
    a. Diphtheria, pertussis, and tuberculosis can be prevented with a DPT shot.
    b. Asthma, chronic bronchitis, and emphysema are examples of COPDs.
    c. Emphysema involves bacterial lesions in the lung walled off by scar tissue.
    d. Flu and tuberculosis are viral infections.
    e. A cold is caused by a rhinovirus.

Matching: *Match volume or capacity to the definition. Some answers may be used more than once. Some questions may have more than one correct response.*

_____ 1. The maximum amount of air that can be moved

_____ 2. The amount of air that is left in the lung after a maximum expiration

_____ 3. The amount of air that can still be inhaled after a normal inspiration at rest

_____ 4. The maximum amount of air that can be inspired after a normal expiration at rest

_____ 5. The maximum amount of air that can be expired after a normal expiration at rest

a. Total lung volume
b. Residual volume
c. Inspiratory reserve volume
d. Expiratory reserve volume
e. Functional residual capacity
f. Vital capacity
g. Inspiratory capacity

Matching: *Match the respiratory disorder with the description. Some answers may be used more than once. Some questions may have more than one correct response.*

_____ 6. A viral infection

_____ 7. A bacterial infection

_____ 8. Degeneration of alveolar walls

_____ 9. COPD

_____ 10. Disorder that results in whooping cough

a. Tuberculosis
b. Emphysema
c. Cold
d. Influenza
e. Asthma
f. Pertussis

Critical Thinking

1. The effects of smoking on Carol's respiratory system function were discussed in this chapter. What other environmental factors and lifestyle choices would adversely affect the respiratory system? Explain.
2. What can be done to minimize the effects of aging on the respiratory system? Explain.
3. Bob participated in an A&P lab during the respiratory system unit. He recorded his respiratory rate and his tidal volume at rest and then again after running around the parking lot at his school. His results were as shown to the right:

	At rest	Immediately after the run
Respiratory rate:	12 breaths/min	20 breaths/min
Tidal volume:	400 mL	650 mL

How much more air per minute was Bob's respiratory system able to move after the run than before it? What caused his values to change? Explain your response in terms of the physiology of the respiratory system.

12 chapter mapping

This section of the chapter is designed to help you find where each outcome is covered in this text.

	Outcomes	Readings, figures, and tables	Assessments
12.1	Use medical terminology related to the respiratory system.	Word roots: p. 445	Word Building: 1–5
12.2	Trace the flow of air from the nose to the pulmonary alveoli and relate the function of each part of the respiratory tract to its gross and microscopic anatomy.	Take a breath: pp. 446–457 Figures 12.2, 12.13	Spot Check: 1–6 Multiple Select: 6
12.3	Explain the role of surfactant.	Surfactant: pp. 457–458	
12.4	Describe the respiratory membrane.	Respiratory membrane: p. 458 Figure 12.12c	
12.5	Explain the mechanics of breathing in terms of anatomy and pressure gradients.	Mechanics of breathing: pp. 458–461 Figures: 12.14, 12.15	Multiple Select: 5
12.6	Define the measurements of pulmonary function.	Measurements of pulmonary function: pp. 462–463 Figures: 12.16, 12.17 Table 12.1	Matching: 1–5
12.7	Define partial pressure and its relationship to a gas mixture such as air.	Composition of air: p. 464	Spot Check: 7
12.8	Explain gas exchange in terms of the partial pressures of gases at the capillaries and the alveoli and at the capillaries and the tissues.	Gas exchange: pp. 464–466 Figure 12.18, 12.19	Multiple Select: 1–3
12.9	Compare the composition of inspired and expired air.	Comparison of inspired and expired air: p. 466 Figure 12.19	Multiple Select: 3
12.10	Explain the factors that influence the efficiency of alveolar gas exchange.	Factors that influence gas exchange: pp. 466–469 Figures 12.20–12.22	Spot Check: 8 Multiple Select: 4

	Outcomes	Readings, figures, and tables	Assessments
12.11	Describe the mechanisms for transporting O_2 and CO_2 in the blood.	Gas transport: pp. 470–472 Figures: 12.23, 12.24	Multiple Select: 2, 7, 8
12.12	Explain how respiration is regulated to homeostatically control blood gases and pH.	Regulation of respiration: pp. 472–474 Figures 12.25–12.27	Multiple Select: 5, 9
12.13	Explain the functions of the respiratory system.	Functions of the respiratory system: pp. 474–475 Figure 12.27	Critical Thinking: 1
12.14	Summarize the effects of aging on the respiratory system.	Effects of Aging on the respiratory system: pp. 475–476	Critical Thinking: 2
12.15	Describe respiratory system disorders.	Respiratory system disorders: pp. 476–478 Figure 12.28	Multiple Select: 10 Matching: 6–10

footnotes

1. National Human Genome Research Institute. (2010, July 2). *Learning about cystic fibrosis*. Retrieved September 22, 2010, from http://www.genome.gov/10001213.
2. Office of the Surgeon General. (2004). *Surgeon general's report: The health consequences of smoking* (chap. 2). Retrieved September 16, 2010, from http://www.cdc.gov/tobacco/data_statistics/sgr/2004/pdfs/chapter2.pdf.

references and resources

Beers, M. H., Porter, R. S., Jones, T. V., Kaplan, J. L., & Berkwits, M. (Eds.). (2006). *Merck manual of diagnosis and therapy* (18th ed.). Whitehouse Station, NJ: Merck Research Laboratories.

Brain, J. D. (2006, August). Effects of aging: Biology of the lungs and airways. *Merck manual home edition*. Retrieved September 9, 2010, from http://www.merckmanuals.com/home/sec04/ch038/ch038g.html.

National Cancer Institute. (n.d.). *Lung cancer*. Retrieved March 17, 2011, from http://www.cancer.gov/cancertopics/types/lung.

National Human Genome Research Institute. (2010, July 2). *Learning about cystic fibrosis*. Retrieved September 22, 2010, from http://www.genome.gov/10001213.

Office of the Surgeon General. (2004). *Surgeon general's report: The health consequences of smoking* (chap. 2). Retrieved September 16, 2010, from http://www.cdc.gov/tobacco/data_statistics/sgr/2004/pdfs/chapter2.pdf.

Porter, R. S., & Kaplan, J. L. (Eds.). (n.d.). Selected physiologic age-related changes. *Merck manual for healthcare professionals*. Retrieved March 19, 2011, from http://www.merckmanuals.com/media/professional/pdf/Table_337-1.pdf?qt=selected physiologic age-related changes&alt=sh.

Saladin, Kenneth S. (2010). *Anatomy & physiology: The unity of form and function* (5th ed.). New York: McGraw-Hill.

Seeley, R. R., Stephens, T. D., & Tate, P. (2006). *Anatomy & physiology* (7th ed.). New York: McGraw-Hill.

Sharma, G., & Goodwin, J. (2006, September). Effect of aging on respiratory system physiology and immunology. *Clinical Interventions in Aging, 1*(3), 253–260.

Shier, D., Butler, J., & Lewis, R. (2010). *Hole's human anatomy & physiology* (12th ed.). New York: McGraw-Hill.

13 The Digestive System

Hungry? Perhaps you would like a cheeseburger? Your burger could be made of beef, turkey, soy, or even tofu. All of these burger choices are good sources of protein. The bun is made of carbohydrates, and the cheese is composed of lipids (fat). So a cheeseburger is a perfect food to use in studying how the digestive system works to digest proteins, carbohydrates, and lipids. Although the digestion of a cheeseburger starts with the first bite, this is only the beginning of the functions of this system. See Figure 13.1.

13.1 learning outcome

Use medical terminology related to the digestive system.

learning outcomes

After completing this chapter, you should be able to:

13.1 Use medical terminology related to the digestive system.

13.2 Differentiate between mechanical digestion and chemical digestion.

13.3 Describe the digestive anatomy of the oral cavity.

13.4 Explain the physiology of mechanical and chemical digestion in the mouth.

13.5 Describe the digestive anatomy from the mouth to the stomach.

13.6 Explain how materials move from the mouth to the stomach.

13.7 Describe the digestive anatomy of the stomach.

13.8 Explain the physiology of mechanical and chemical digestion in the stomach.

13.9 Explain the feedback mechanism of how food moves from the stomach to the small intestine.

13.10 Describe the anatomy of the digestive accessory organs connected to the duodenum by ducts.

13.11 Describe the digestive anatomy of the small intestine.

13.12 Explain the physiology of chemical digestion in the duodenum, including the hormones and digestive secretions involved.

13.13 Explain how nutrients are absorbed in the small intestine.

13.14 Describe the anatomy of the large intestine.

13.15 Explain the physiology of the large intestine in terms of absorption, preparation of feces, and defecation.

13.16 Summarize the types of nutrients absorbed by the digestive system from the diet.

13.17 Trace the circulation of the nutrients once they have been absorbed.

13.18 Explain the control of digestion.

13.19 Summarize the functions of digestion.

13.20 Summarize the effects of aging on the digestive system.

13.21 Describe digestive system disorders, including vomiting, food poisoning, parasites, and peptic ulcers.

word roots & combining forms

chol/e: gall, bile

col/o: colon

cyst/o: bladder, sac

duoden/o: duodenum

emet/o: vomit

enter/o: intestine

esophag/o: esophagus

gastr/o: stomach

gingiv/o: gums

gloss/o: tongue

hepat/o: liver

peps/o: digestion

rect/o: rectum

sigmoid/o: sigmoid colon

pronunciation key

cholecystokinin: KOH-leh-sis-toe-KIE-nin

chyme: KYME

deglutition: dee-glue-TISH-un

duodenum: du-oh-DEE-num

gingiva: JIN-jih-vah

haustra: HAW-stra

ileocecal: ILL-ee-oh-SEE-cal

jejunum: je-JEW-num

lacteal: LAK-tee-al

peristalsis: per-ih-STAL-sis

rugae: ROO-guy

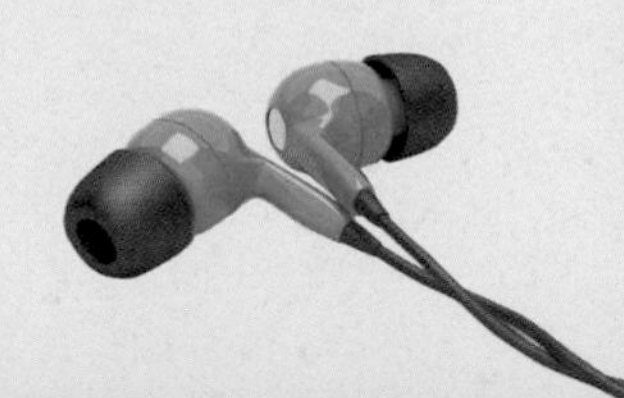

Overview

The anatomy of the digestive system is like a complicated tube—one end of the tube is the mouth and the other end is the anus. This tube—called the **alimentary canal** or **gastrointestinal (GI) tract**—may be as long as 8 meters. See **Figure 13.2**. It twists and turns, enlarges and narrows as it uses muscle contractions to push its contents toward its end. You insert food in the mouth but see a very different product emerge at the anus. In this chapter, you will be exploring the anatomy of the digestive system and what happens to food (a cheeseburger) along its journey through each section of the alimentary canal. You will also look at the way digestion is regulated, the effects of aging on the system, and digestive system disorders.

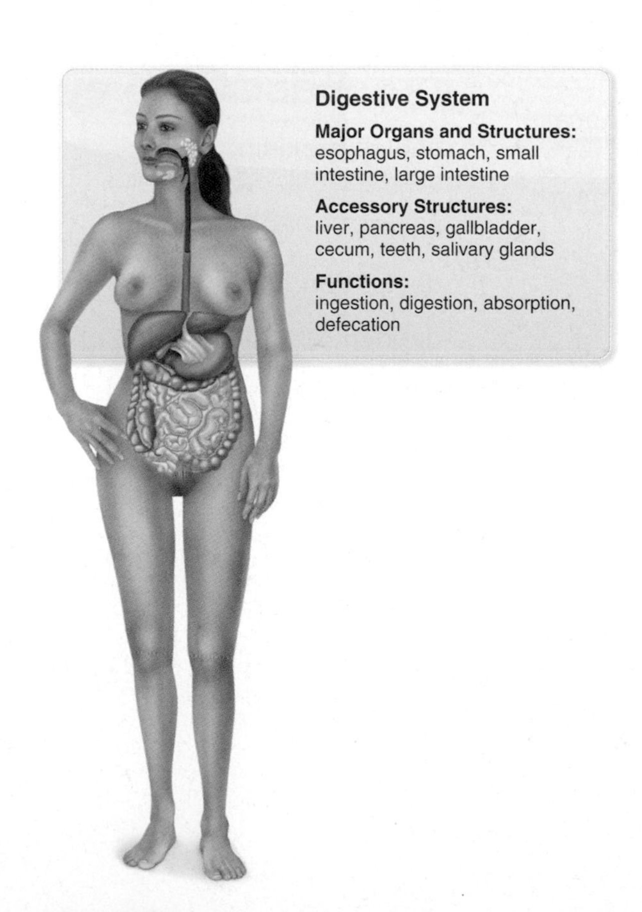

FIGURE 13.1 The digestive system.

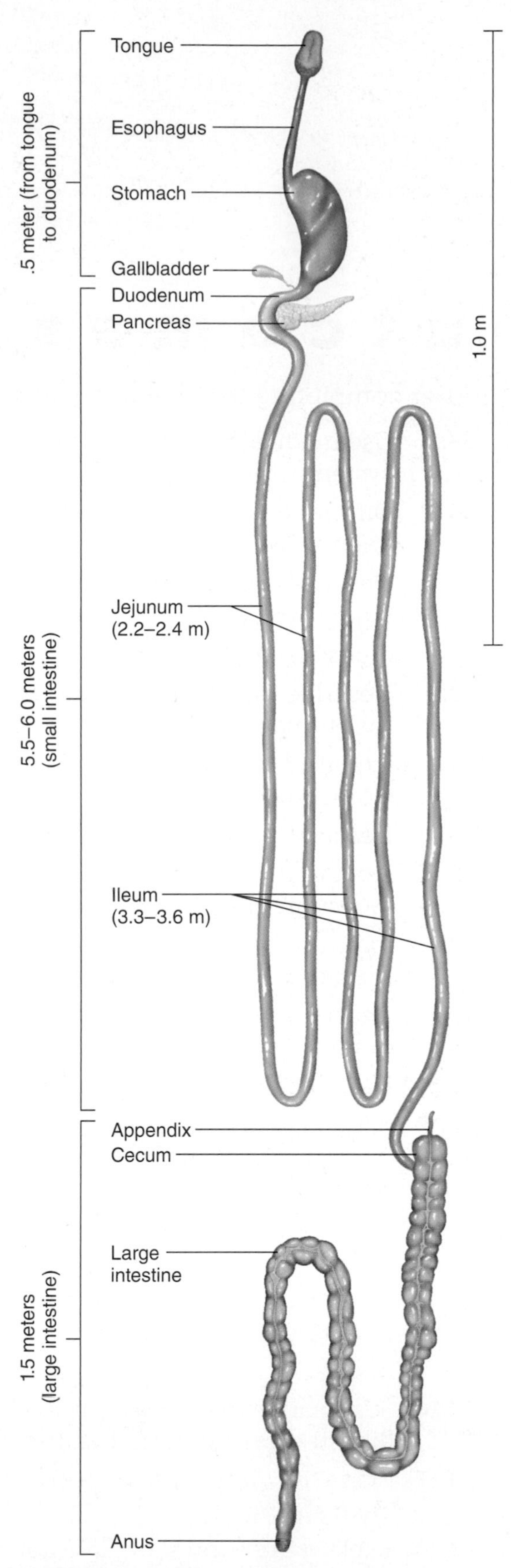

FIGURE 13.2 The alimentary canal, showing relative lengths of each section.

The cheeseburger's journey begins in the mouth and continues to the pharynx, esophagus, stomach, small intestine, large intestine, and anus, where the journey ends. All along the pathway, mucous membranes and mucosa-associated lymphatic tissue (MALT) line the alimentary canal to fight any foreign invaders—such as bacteria—that may enter with the food (discussed in the lymphatic system chapter). Along with studying the structures of the alimentary canal, you will need to study the accessory structures in each section of the tract. These accessory structures—including the teeth, tongue, salivary glands, liver, gallbladder, bile ducts, and pancreas (shown in **Figure 13.3**)—are necessary for the system to carry out its functions.

13.2 learning outcome

Differentiate between mechanical digestion and chemical digestion.

Before you get started, it is important to understand the two types of digestion—mechanical and chemical. **Mechanical digestion** is the breakdown of large pieces of complex molecules to smaller pieces of complex molecules. On the other hand, **chemical digestion** is the breakdown of complex molecules to their building blocks so that they can be absorbed. As a result of chemical digestion, proteins are broken down to their amino acids, carbohydrates to their monosaccharides, and lipids to their fatty acids and glycerol. Mechanical digestion must happen first for chemical digestion to take place.

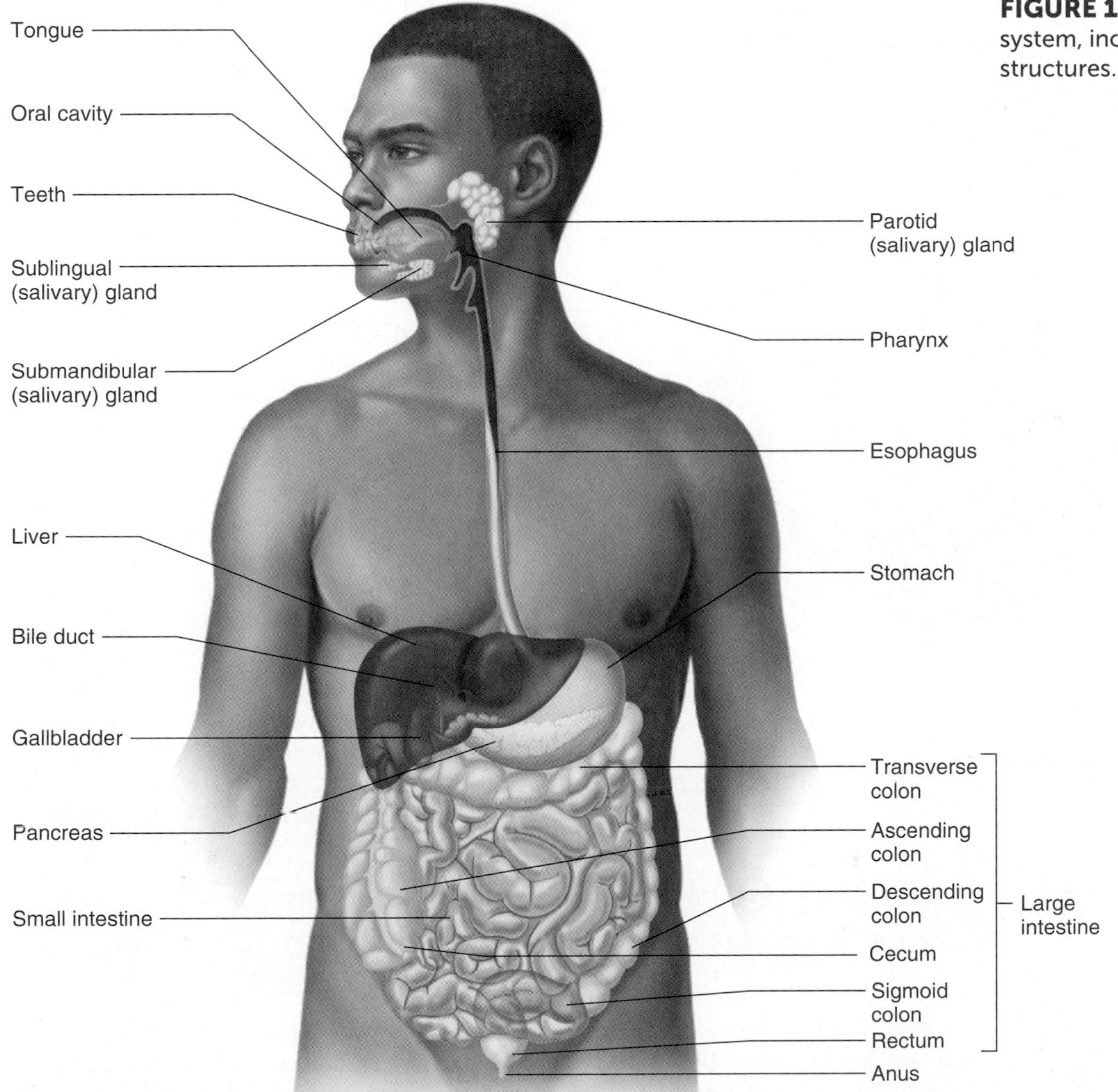

FIGURE 13.3 The digestive system, including accessory structures.

spot check ❶ What would be the end result of mechanical digestion of a complex carbohydrate compared to the end result of chemical digestion of a complex carbohydrate?

Below, you will learn about the structures of the mouth and how they help to mechanically and chemically digest a cheeseburger.

13.3 learning outcome

Describe the digestive anatomy of the oral cavity.

Anatomy and Physiology of the Digestive System

Anatomy in the Mouth

Oral Cavity The mouth can also be called the *oral cavity.* The roof of the oral cavity consists of the **hard palate,** formed by the maxilla and palatine bones, and the **soft palate,** composed of soft tissue. As you can see in **Figure 13.4**, the soft palate ends with the **uvula,** a posterior projection that directs materials downward to the pharynx so that they do not travel to the nose. The sidewalls of the oral cavity are the cheeks, and the floor of the cavity is where the tongue is attached. The oral cavity is lined by stratified squamous epithelial tissue, which is a very durable epithelial tissue that can withstand the abrasions of manipulating solid food in the mouth.

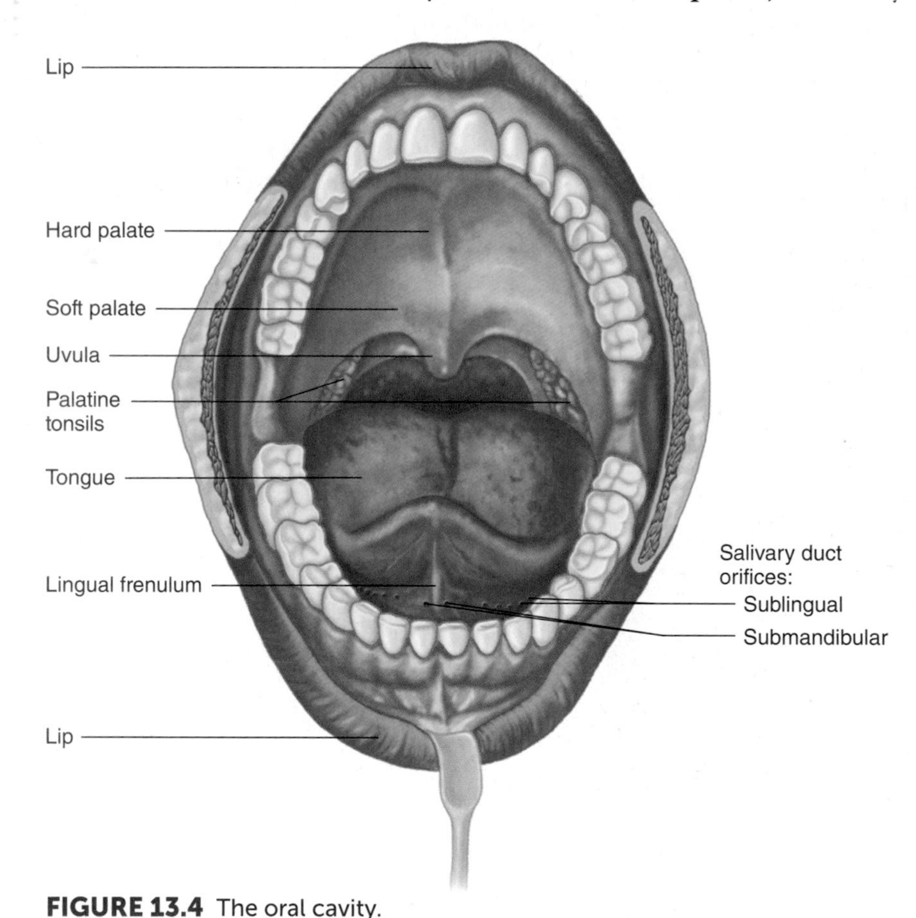

FIGURE 13.4 The oral cavity.

Teeth A baby is not born with teeth but will develop two sets of teeth over its lifetime—a **deciduous** (primary) set and a **permanent** (secondary) set. See **Figure 13.5**. The baby's first set of teeth—the deciduous teeth—begin to erupt, or grow in, at approximately six months and will be complete by the age of two. This primary set consists of 10 teeth in each jaw. Later, as the permanent teeth erupt, they push out the deciduous teeth. See **Figure 13.6**. This secondary set begins to erupt at 6 years of age and will not be fully complete, with 16 teeth in each jaw, until the individual reaches 17 to 25 years of age.

spot check ❷ How many teeth do you have in your mouth? If you do not have 32 teeth, which teeth are missing?

gingiva: JIN-jih-vah

Figure 13.7 shows that a tooth is held in its bony socket **(alveolus)** in the jaw by **periodontal ligaments.** The tissue surrounding a tooth is the **gingiva,** commonly called the *gum.* The portion of the tooth emerging from the gingiva is called the **crown.** The crown is covered by a very hard, smooth, white layer called

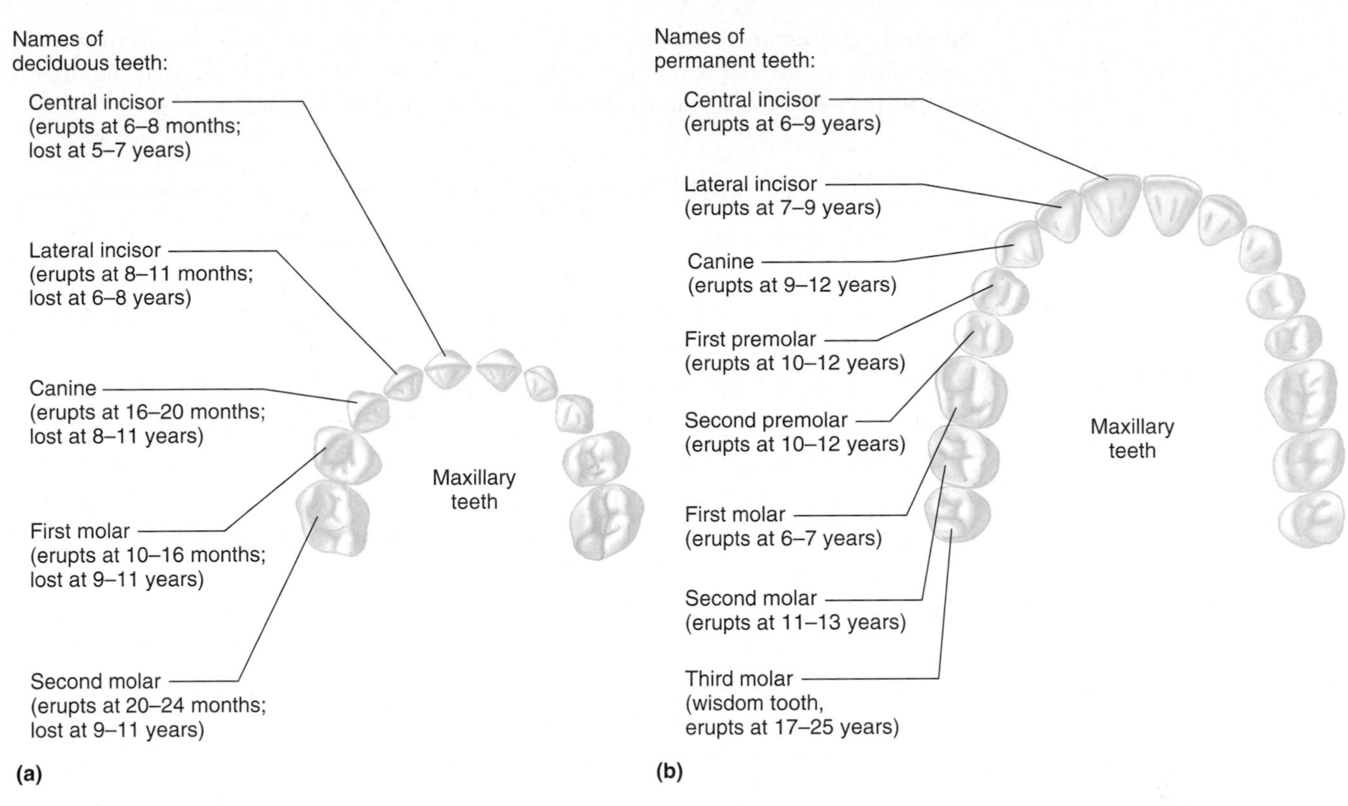

FIGURE 13.5 Teeth of the upper jaw: (a) deciduous teeth, (b) permanent teeth.

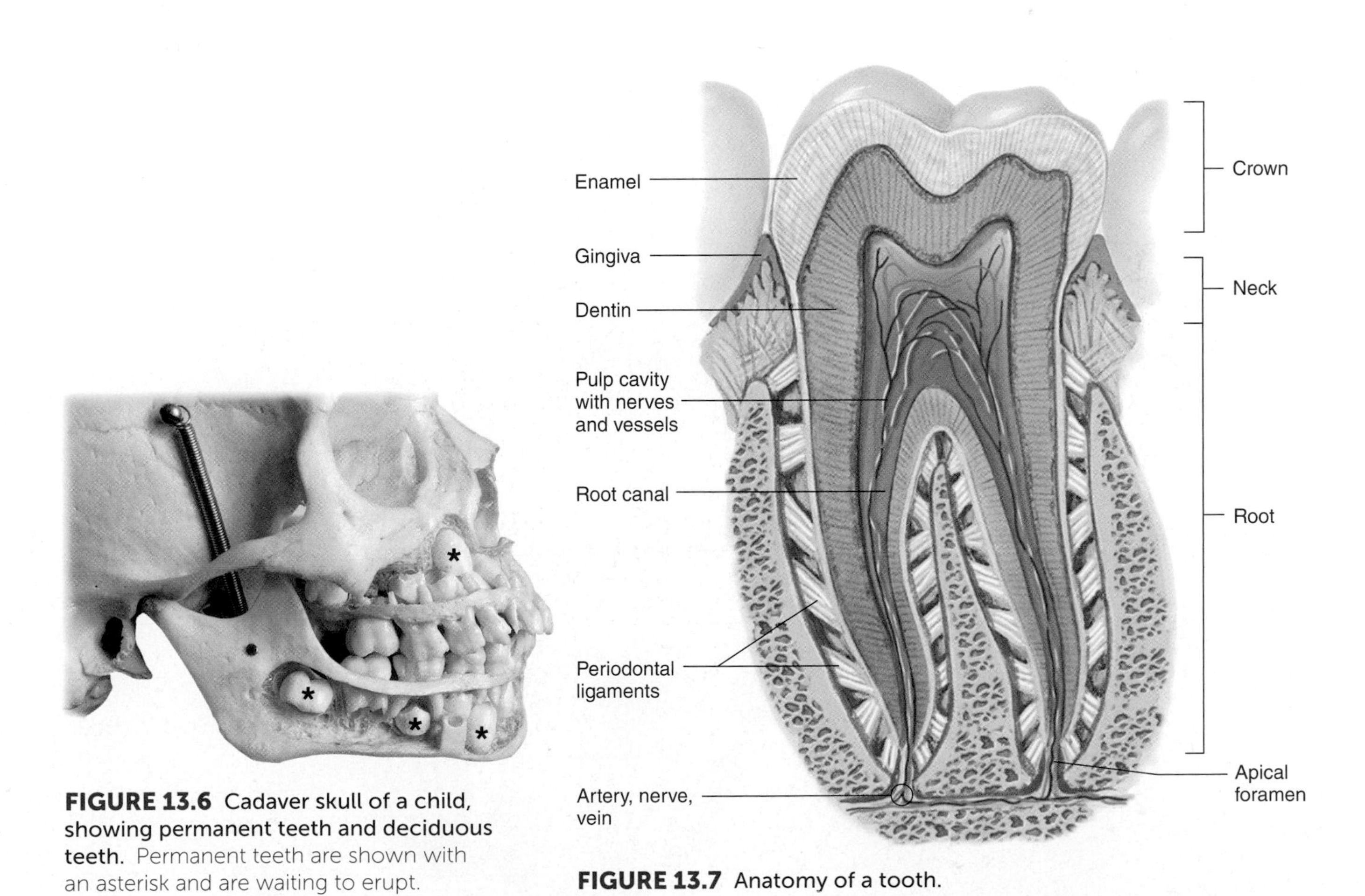

FIGURE 13.6 **Cadaver skull of a child, showing permanent teeth and deciduous teeth.** Permanent teeth are shown with an asterisk and are waiting to erupt.

FIGURE 13.7 Anatomy of a tooth.

enamel. The enamel's function is to protect the underlying layer, the **dentin.** The **root** of the tooth, below the gum line, is not covered by enamel. Deep to the dentin is a **pulp cavity** that contains the blood vessels and nerve for the tooth.

clinical point

A tooth's enamel wears down and thins with age. A **dental caries,** commonly called a *cavity,* is an erosion through the enamel into the dentin. See Figure 13.8. If the erosion continues to the pulp cavity, bacteria may gain access and travel beyond the tooth's root. This infection is called an **abscess.**

The mouth is full of bacteria. Every time you eat, bits of food are wedged between the teeth and between each tooth and the gingiva (see Figure 13.8). Bacteria feed on this buffet left out for them, digest the food, and then excrete their acidic waste in the same location. This waste erodes the enamel to form a caries and irritates the gingiva, causing it to become inflamed **(gingivitis).** As gingivitis progresses, the gingiva pulls away from the tooth and recedes. This allows more food to become wedged between the tooth and the gingiva and more unprotected dentin to be exposed. The daily bacteria buildup that forms on the tooth is **plaque.** Plaque can be flossed and brushed away, but if it is allowed to remain, it hardens to form **tartar,** which must be removed by a dental professional.

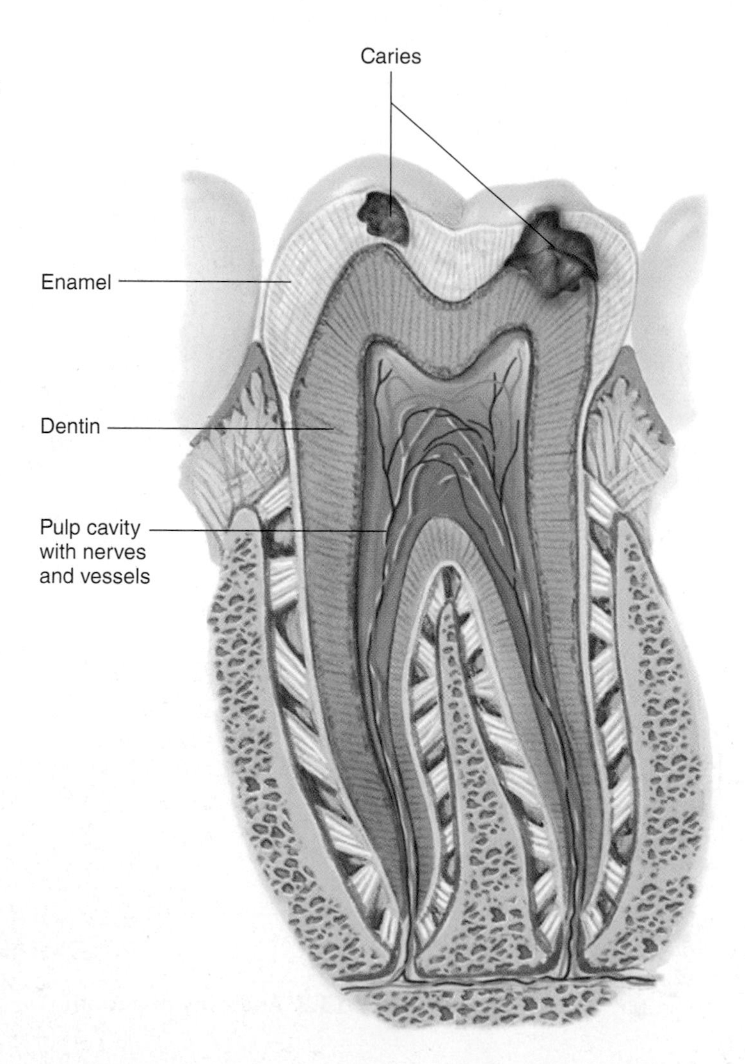

FIGURE 13.8 Dental caries. One carie involves just the enamel, while the other caries shown extends to the dentin.

Tongue The tongue is composed of skeletal muscle tissue anchored to the floor of the oral cavity by a medial fold called the **lingual frenulum** (shown in **Figures 13.4** and **13.9**). On the tongue's superior surface, stratified squamous epithelial tissue covers the lingual papillae, which house the taste buds. As you will recall from Chapter 7, taste buds contain nerve endings that sense sweet, sour, salt, bitter, and umami. The purpose of the tongue is to manipulate what is **ingested** (eaten) and to provide the sense of taste.

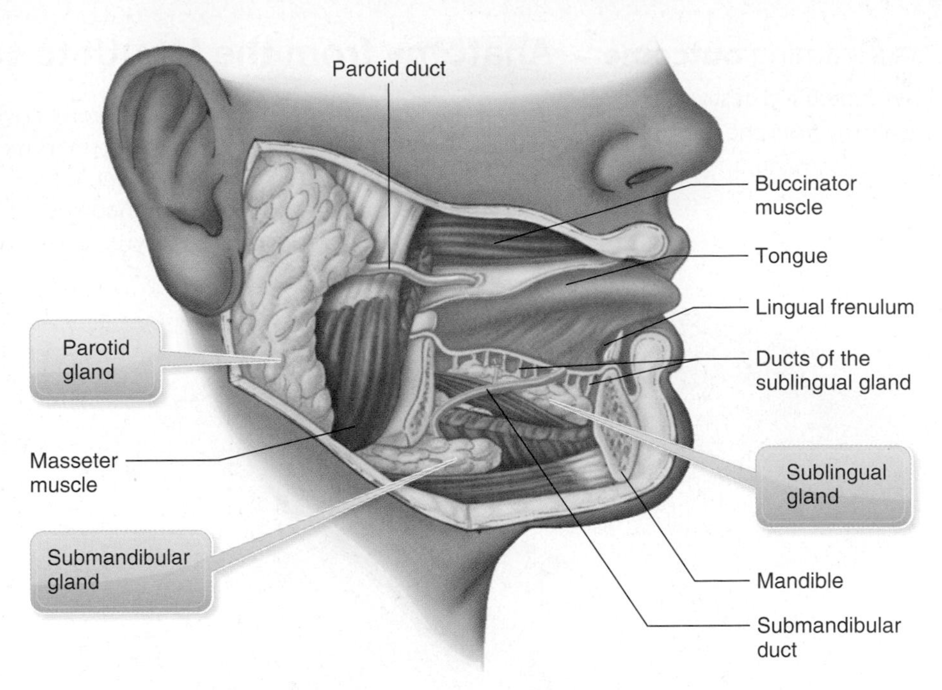

FIGURE 13.9 Salivary glands.

Salivary Glands The salivary glands, which produce about 1.0 to 1.5 liters (L) of saliva a day, consist of the **parotid glands** (anterior to the ears), the **submandibular glands** (inferior to the angle of the mandible on each side), and the **sublingual glands** (below the tongue). See **Figures 13.3** and **13.9**. Each gland produces saliva that travels to the oral cavity through ducts. The saliva, which is mostly water, also contains the enzymes **amylase** and **lingual lipase,** along with mucus, lysozymes, and antibodies. Saliva secretion is initiated by taste receptors that send signals by way of the facial and glossopharyngeal nerves to centers in the medulla oblongata and pons. These control centers also receive other stimuli so that odors, sight, or even the thought of food may stimulate saliva secretion.

Physiology of Digestion in the Mouth

13.4 learning **outcome**

Explain the physiology of mechanical and chemical digestion in the mouth.

Think again of the example of the cheeseburger. You bite into it with your teeth. The process of chewing, called **mastication,** uses the masseter and temporalis muscles to move the jaw in a crushing motion, while the tongue, orbicularis oris, and buccinator muscles work to keep the food between the teeth. This begins mechanical digestion—breaking the bite of cheeseburger into smaller pieces.

Saliva from the salivary glands mixes with the bite of cheeseburger in the mouth. The saliva's pH is 6.8 to 7.0. At this pH, amylase *partially* breaks down the carbohydrates from the bun. This is the beginning of chemical digestion. Lingual lipase does nothing at this pH, but it is activated later by the lower pH of the stomach. Amylase, however, will no longer function at the lower pH in the stomach. Thus, it is important to masticate the bite of cheeseburger thoroughly because doing so allows for mechanical digestion and also gives amylase sufficient time to partially break down the carbohydrates in chemical digestion before the food is swallowed. The mucus in the saliva moistens the bite of food (now called a **bolus**), making it easier to swallow.

The lysozymes and antibodies in the saliva are not used for digestion. They destroy and inhibit the growth of bacteria that may have entered with the bite. Digestion is finished in the mouth when the tongue pushes the bolus to the pharynx.

spot check 3 Saliva is secreted when you are chewing gum. How might chewing sugarless gum prevent the formation of dental caries?

13.5 learning outcome

Describe the digestive anatomy from the mouth to the stomach.

Anatomy from the Mouth to the Stomach

Pharynx The parts of the pharynx were covered in Chapter 12, but it will be helpful to review them here. The nasopharynx leads from the nasal cavity to the oropharynx. The oropharynx is a funnel leading from the oral cavity to the laryngopharynx. This funnel is lined by stratified squamous epithelial tissue and has smooth muscle in its walls. The laryngopharynx leads to the trachea and the esophagus. The respiratory and digestive pathways intersect in the pharynx.

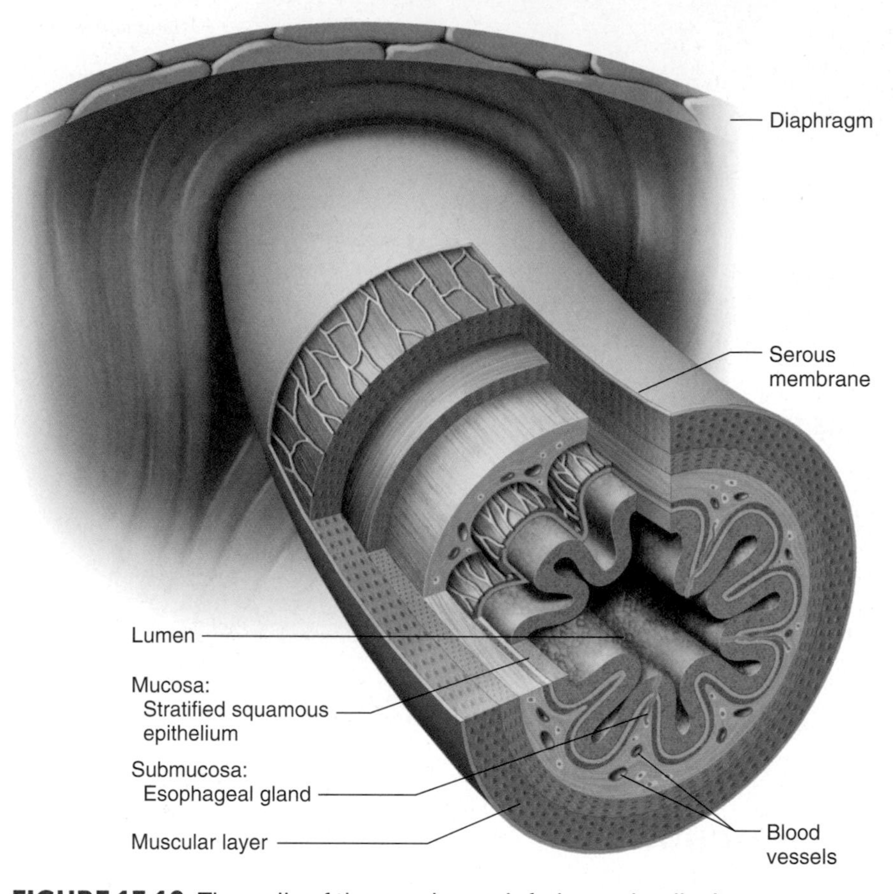

FIGURE 13.10 The walls of the esophagus inferior to the diaphragm.

Epiglottis The epiglottis is made of elastic cartilage connective tissue. It is one of the cartilages of the larynx. It stands guard over the glottis, which is the opening of the larynx.

Esophagus The esophagus is a straight, muscular tube that extends from the laryngopharynx, travels through the mediastinum, penetrates the diaphragm, and connects to the stomach. It is lined by stratified squamous epithelial tissue. Deep to the epithelial lining is a submucosa of connective tissue containing esophageal glands that secrete protective mucus for the esophagus. The upper one-third of the esophagus has skeletal muscle in its walls, while the middle one-third has a mixture of skeletal and smooth muscle and the lower one-third has just smooth muscle in the walls of the esophagus. See **Figure 13.10**. Unlike the trachea that is held open by C-shaped cartilages, the esophagus is normally collapsed.

spot check 4 Trace the bite of cheeseburger from the oral cavity to the stomach. What is the swallowed bite called?

13.6 learning outcome

Explain how materials move from the mouth to the stomach.

deglutition: dee-glue-TISH-un

Physiology of Digestion from the Mouth to the Stomach

Once the bolus has been sufficiently masticated in the mouth, it is time to swallow. Swallowing, called **deglutition,** is a very complex process controlled by the medulla oblongata. It requires four cranial nerves (V, VII, IX, and XII) to stimulate the muscle contractions necessary to move the bolus from the pharynx to the esophagus. Follow along in **Figure 13.11a** to see the steps of the process. Swallowing begins with the tongue pushing the bolus back to the pharynx (1). The larynx pushes up, causing the epiglottis to close over the glottis (2). This ensures

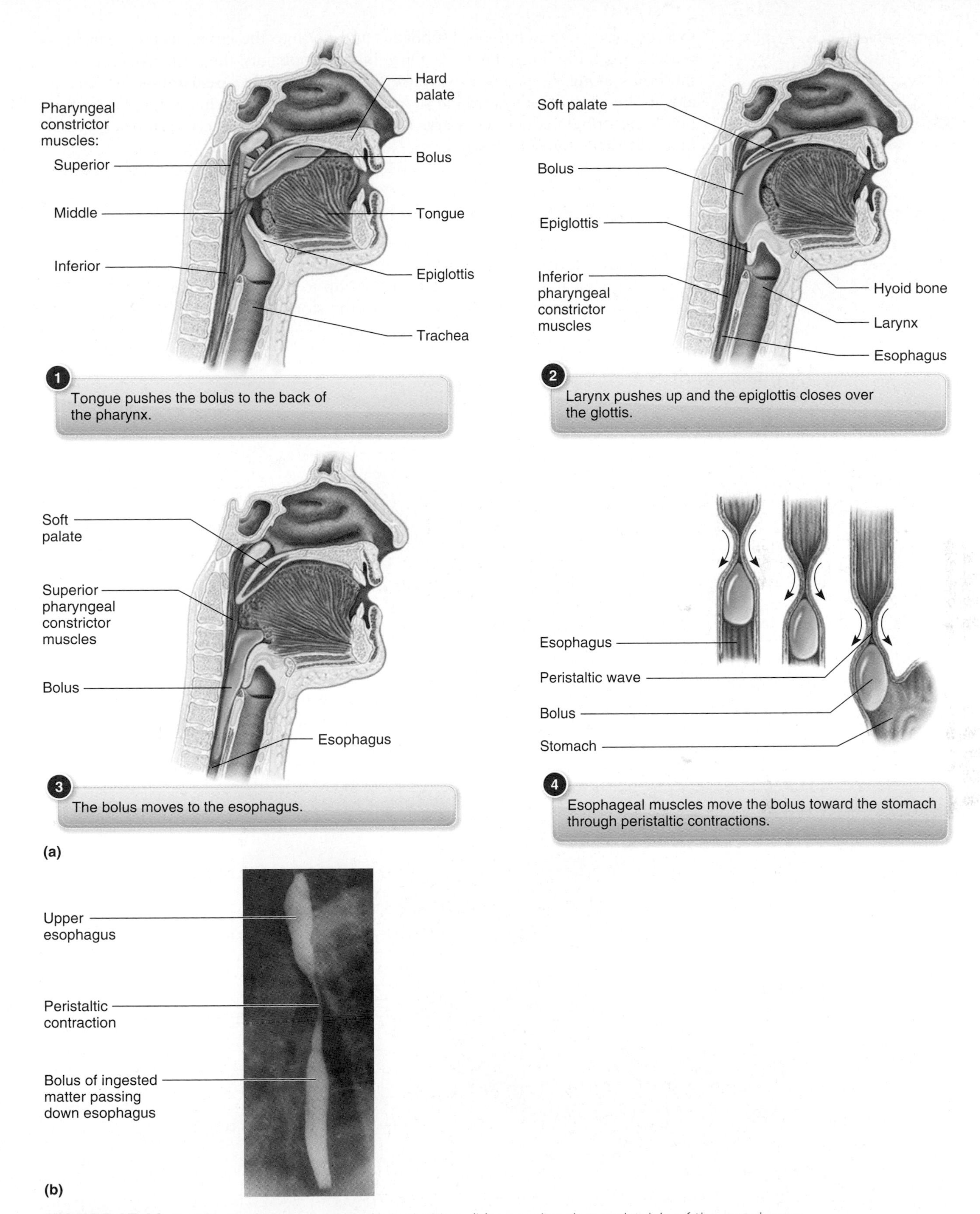

FIGURE 13.11 Swallowing: (a) the steps of deglutition, (b) x-ray showing peristalsis of the esophagus.

peristalsis: per-ih-STAL-sis

connect.mcgraw-hill.com audio

that the bolus moves into the esophagus and *not* into the larynx as the pharyngeal muscles push the bolus down (3). Once in the esophagus, the muscular walls move the bolus along its length in wavelike contractions called **peristalsis** (4). Gravity aids in the movement toward the stomach if the individual is in an upright position, but being upright is not necessary. The bolus can still move to the stomach even if the individual is upside down.

13.7 learning **outcome**

Describe the digestive anatomy of the stomach.

Anatomy of the Stomach

The stomach is a J-shaped organ found in the upper left quadrant of the abdomen, immediately inferior to the diaphragm. It is a muscular sac capable of holding 1.0 to 1.5 L after a meal, but it can stretch to hold up to 4 L when extremely full. Follow along with **Figure 13.12** as the anatomy of the stomach is described.

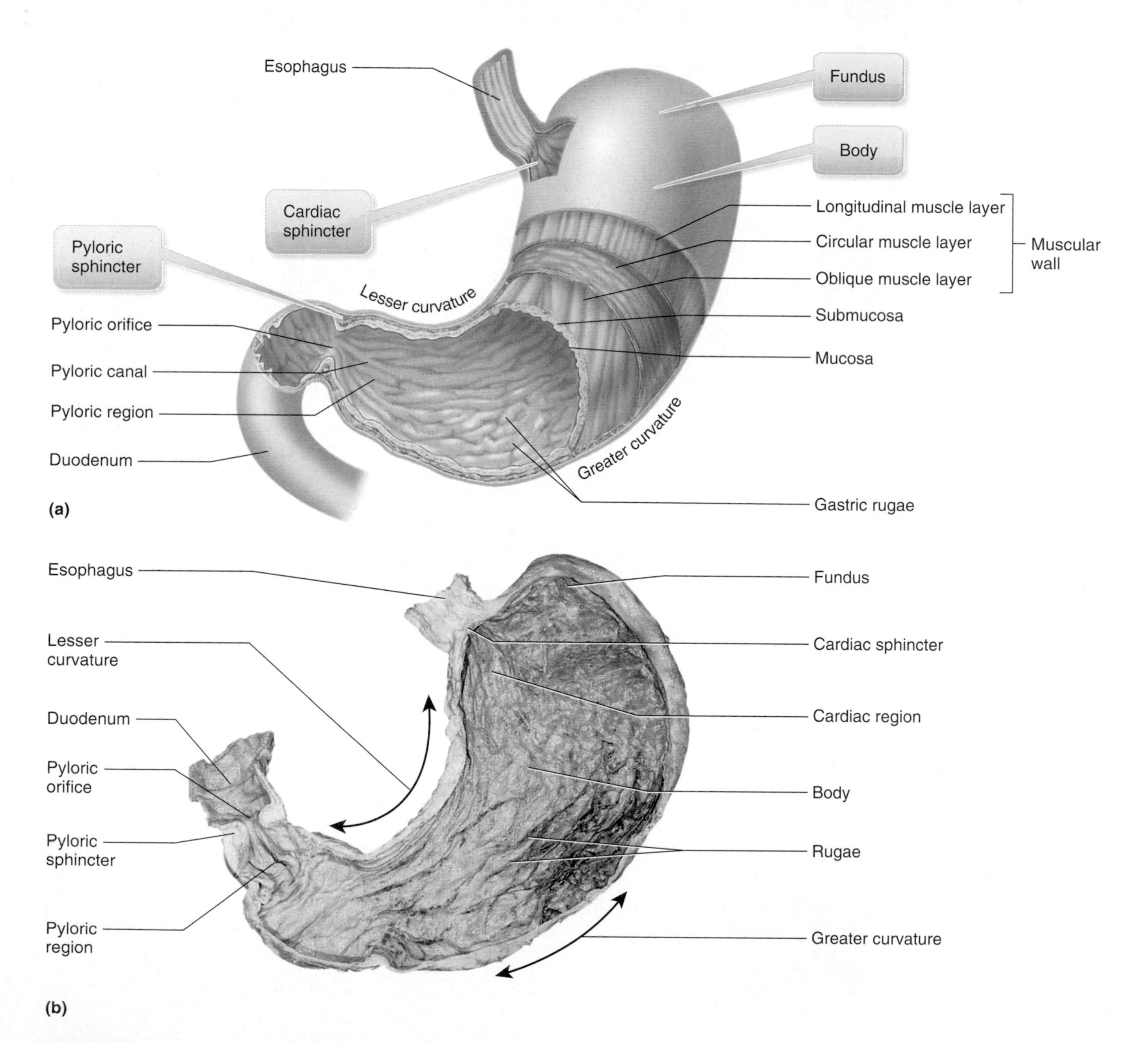

FIGURE 13.12 Anatomy of the stomach: (a) gross anatomy, (b) internal surface of a cadaver's stomach, (c) radiograph of a stomach.

FIGURE 13.12 concluded

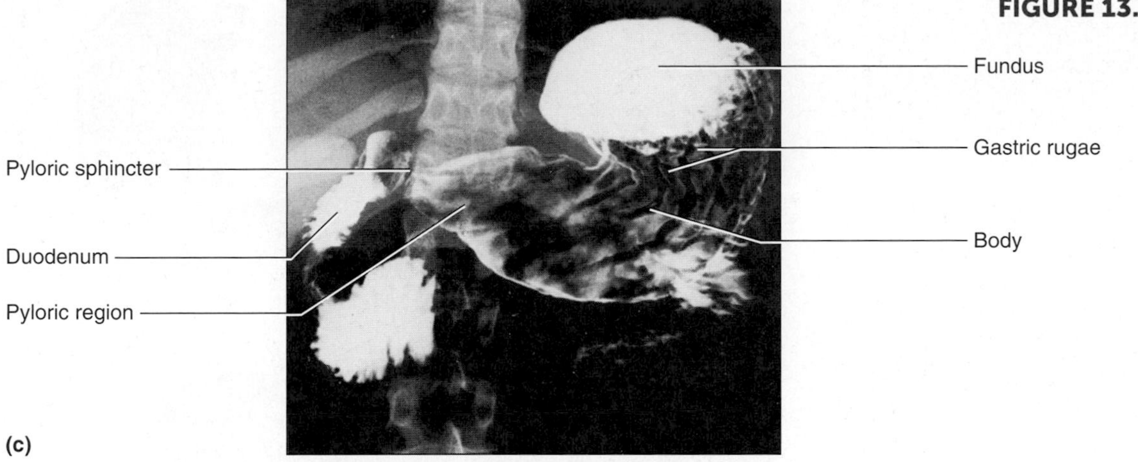

The **cardiac sphincter (lower esophageal sphincter)** controls the opening to the stomach from the esophagus. This circular muscle's purpose is to allow food to enter the stomach and make sure it does not return to the esophagus. The stomach can be described in terms of the following areas: the **lesser curvature** on the inside of the J; the **greater curvature** on the outer side of the J; the **fundus,** superior to the cardiac sphincter; the **body,** making up the majority of the stomach; and the **pyloric region** leading to the smooth muscle **pyloric sphincter,** which regulates the passage of materials to the duodenum. There are three layers of smooth muscle in the walls of the stomach: outer longitudinal muscles, middle circular muscles, and inner oblique muscles. Having muscles oriented in different directions allows for maximum churning of the stomach's contents.

clinical point

It is crucial that the cardiac sphincter closes tightly after the bolus has entered the stomach, as the mucosa lining the esophagus provides insufficient protection from the gastric juices produced in the stomach. Irritation, creating a burning sensation, results if gastric juices leak back to the esophagus. This is commonly called **heartburn** because of the close proximity of the end of the esophagus to the heart. Chronic leakage of gastric juices back to the esophagus is called **gastroesophageal reflux disease (GERD).**

Longitudinal wrinkles called **gastric rugae** can be seen inside the stomach when the stomach is empty. See **Figure 13.12**. These wrinkles become less apparent as the stomach stretches. They also allow for more surface area to accommodate microscopic depressions in the lining, called **gastric pits,** that extend to form **gastric glands.** Five different types of cells line the gastric pits and gastric glands. These cells are shown in **Figure 13.13**, and their functions are described in the list below. The cells and their products are summarized in Table 13.1.

rugae: ROO-guy

- **Mucous cells** secrete a highly alkaline mucus to protect the stomach walls from the hostile environment caused by the acid and digestive enzymes produced in the stomach.
- **Endocrine cells** secrete many hormones, but we will focus on the hormone gastrin. Its function is explained in Table 13.1.

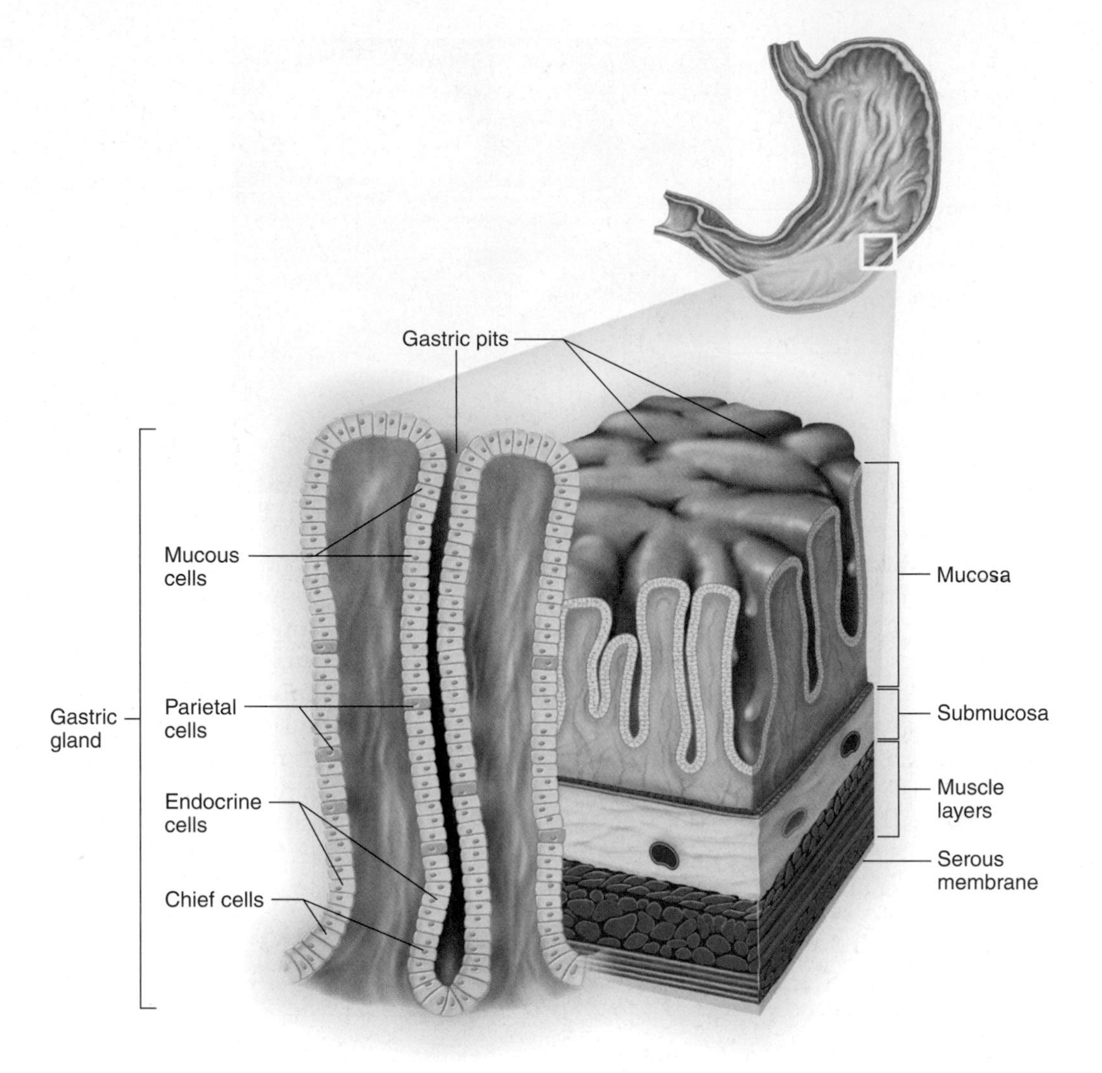

FIGURE 13.13 Gastric pits and gastric glands.
Regenerative cells are not shown.

- **Parietal cells** produce and secrete hydrochloric acid and intrinsic factor. Their functions are explained in Table 13.1.
- **Chief cells** secrete pepsinogen and gastric lipase. Again, their functions are explained in Table 13.1.
- **Regenerative cells** are stem cells that divide and differentiate to replace any of the other cells of the gastric pits and gastric glands. Regenerative cells are very necessary because the cells lining the stomach are short-lived, lasting only three to six days due to the stomach's harsh, acidic environment. The gastric pits' cells must be continually replaced.

The stomach has several mechanisms it uses to protect itself from the harsh environment created by the cells of the gastric pits and gastric glands. These mechanisms include the following:

1. The lining has the highly alkaline mucous coat that resists the hydrochloric acid and digestive enzymes.
2. There is epithelial cell replacement of the lining by the regenerative cells.
3. There are tight junctions between epithelial cells, so acid and enzymes cannot get to the submucosa and smooth muscle walls made of mostly protein.

TABLE 13.1 Gastric juices: chemicals produced in the gastric pits and gastric glands of the stomach

Chemical	Produced by	Function
Gastrin	Endocrine cells	Tells chief and parietal cells to produce their products
Hydrochloric acid	Parietal cells	• Converts pepsinogen to **pepsin,** which *partially* breaks down proteins through chemical digestion • Activates lingual lipase, which, along with gastric lipase, *partially* breaks down lipids through chemical digestion • Converts iron in the diet to a usable form that can be absorbed • Destroys some bacteria
Intrinsic factor	Parietal cells	Allows vitamin B_{12} to be absorbed
Pepsinogen	Chief cells	Changes to pepsin to *partially* break down proteins
Gastric lipase	Chief cells	*Partially* breaks down lipids
Mucus	Mucous cells	Protects the stomach walls

Physiology of Digestion in the Stomach

13.8 learning **outcome**

Explain the physiology of mechanical and chemical digestion in the stomach.

In this section, you will continue to trace the cheeseburger on its journey through the digestive system. During swallowing, the medulla oblongata sends signals to the stomach telling it to relax. As the bolus is moved down the esophagus by peristalsis to the stomach, the stomach's cardiac sphincter opens to allow the bolus to enter. This relaxation of the stomach and the opening of the cardiac sphincter allow the stomach to fill. As the stomach fills, the three layers of smooth muscle in the stomach's walls stretch, causing the muscular walls to contract. These contractions result in peristaltic waves in the direction of the pyloric canal. The pyloric sphincter remains closed, however, making sure the contents stay in the stomach.

As the bolus enters the stomach, the endocrine cells of the gastric pits produce the hormone gastrin. Gastrin targets chief cells and parietal cells, telling the chief cells to produce pepsinogen and gastric lipase and telling the parietal cells to produce hydrochloric acid and intrinsic factor. The hydrochloric acid (HCl) produced by the parietal cells has a pH of 0.8. It converts pepsinogen (produced by the chief cells) to pepsin, which *partially* breaks down proteins (the burger) in the bolus. This is the start of the chemical digestion of proteins in the bolus. The hydrochloric acid also activates lingual lipase from the saliva that mixed with the bolus in the mouth. The activated lingual lipase works together with the gastric lipase produced by the chief cells to *partially* break down the lipids in the bolus (the cheese). This is the start of chemical digestion of the lipids in the bolus. The intrinsic factor produced by parietal cells allows vitamin B_{12} to be absorbed later in the small intestine.

spot check ❺ Consider the composition of muscle tissue. What would happen to the walls of the stomach if the stomach did not have the protective mechanisms mentioned earlier?

13.9 learning **outcome**

Explain the feedback mechanism of how food moves from the stomach to the small intestine.

chyme: KYME

duodenum: du-oh-DEE-num

The churning of the stomach continues mechanical digestion by mixing all the gastric juices with the bolus. This liquefies the contents of the stomach, now called **chyme.** At this point, the carbohydrates have been partially digested in the mouth by amylase, the lipids have been partially digested in the stomach by lingual lipase and gastric lipase, and the proteins have been partially digested in the stomach by pepsin. As the mixing continues, the pH of the chyme falls due to hydrochloric acid's low pH. As the pH of the stomach's contents approaches 2, the endocrine cells of the gastric pits are prevented from producing any more gastrin. With less and less gastrin, the chief and parietal cells are also prevented from producing their products. This is a good example of a negative-feedback mechanism. This low pH also causes the pyloric sphincter to begin to open, allowing approximately 3 milliliters (mL) of chyme to leave the stomach at a time. Digestion in the stomach is complete when the chyme exits the pyloric sphincter.

Chyme travels within the first part of the small intestine—the **duodenum.** However, accessory structures play a large role in the digestion occurring in the duodenum. Therefore, we focus next on the anatomy of these accessory structures—the liver, the gallbladder, the pancreas, and the relevant ducts connecting these structures to the duodenum.

13.10 learning **outcome**

Describe the anatomy of the digestive accessory organs connected to the duodenum by ducts.

Anatomy of Digestive Accessory Structures

Liver The liver is a large, reddish-brown organ immediately inferior to the diaphragm on the right side of the abdominal cavity. See **Figure 13.14**. It has four lobes: the **right and left lobes,** separated by the **falciform ligament;** the **quadrate lobe,** next to the gallbladder; and the **caudate lobe,** which is the most posterior lobe. The falciform ligament is a sheet of mesentery that suspends the liver from the diaphragm and anterior abdominal wall. The **round ligament** is a remnant, or leftover piece, of the umbilical vein, which had delivered blood from the mother's placenta to the liver in the fetus. The liver is a highly vascular organ that is arranged in **hepatic lobules.**

As you can see in **Figure 13.15**, each hepatic lobule has a central vein as a hub and sheets of liver cells **(hepatocytes)** radiating out like spokes on a wheel. The liver receives oxygenated blood from the hepatic artery and nutrient-rich blood from the hepatic portal vein. The hepatic vein drains blood from the liver. The

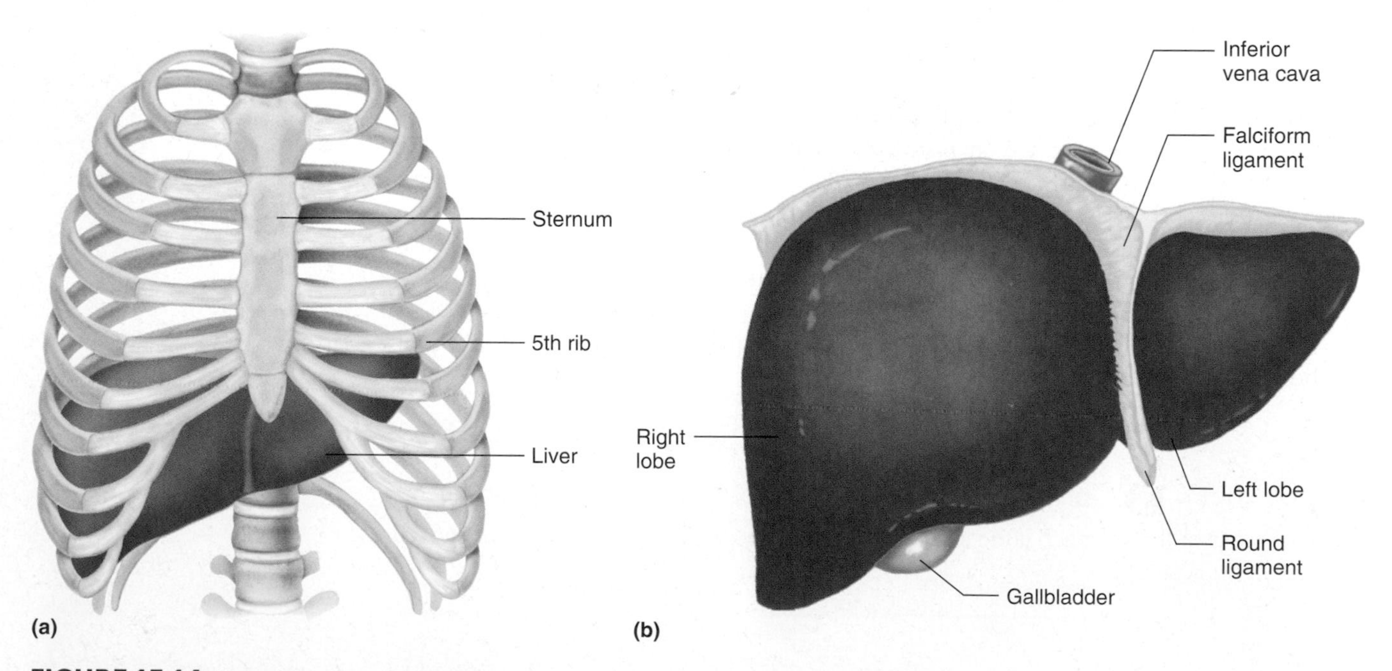

FIGURE 13.14 Gross anatomy of the liver: (a) location, (b) anterior view, (c) inferior view.

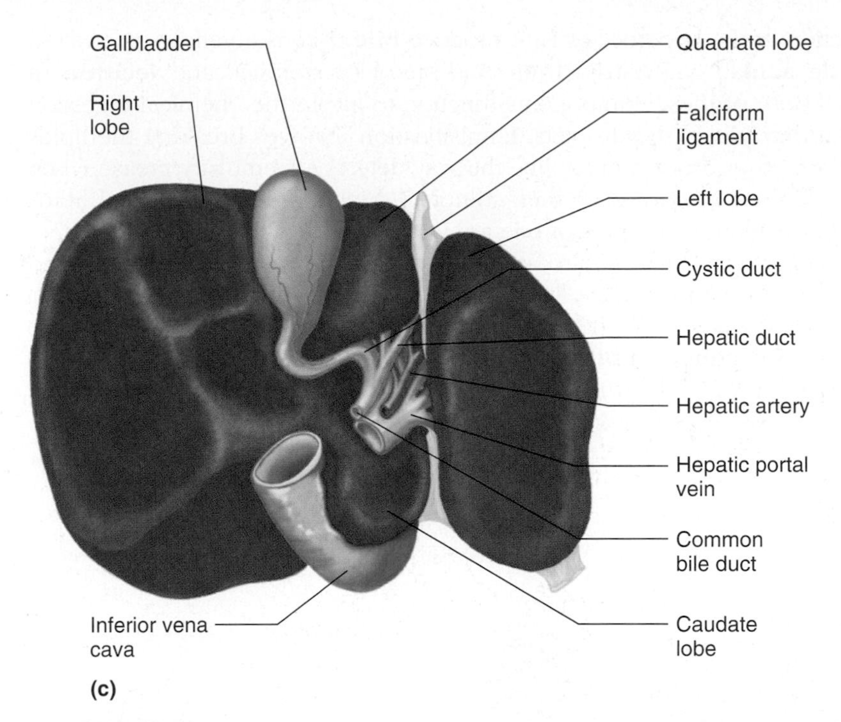

FIGURE 13.14 concluded

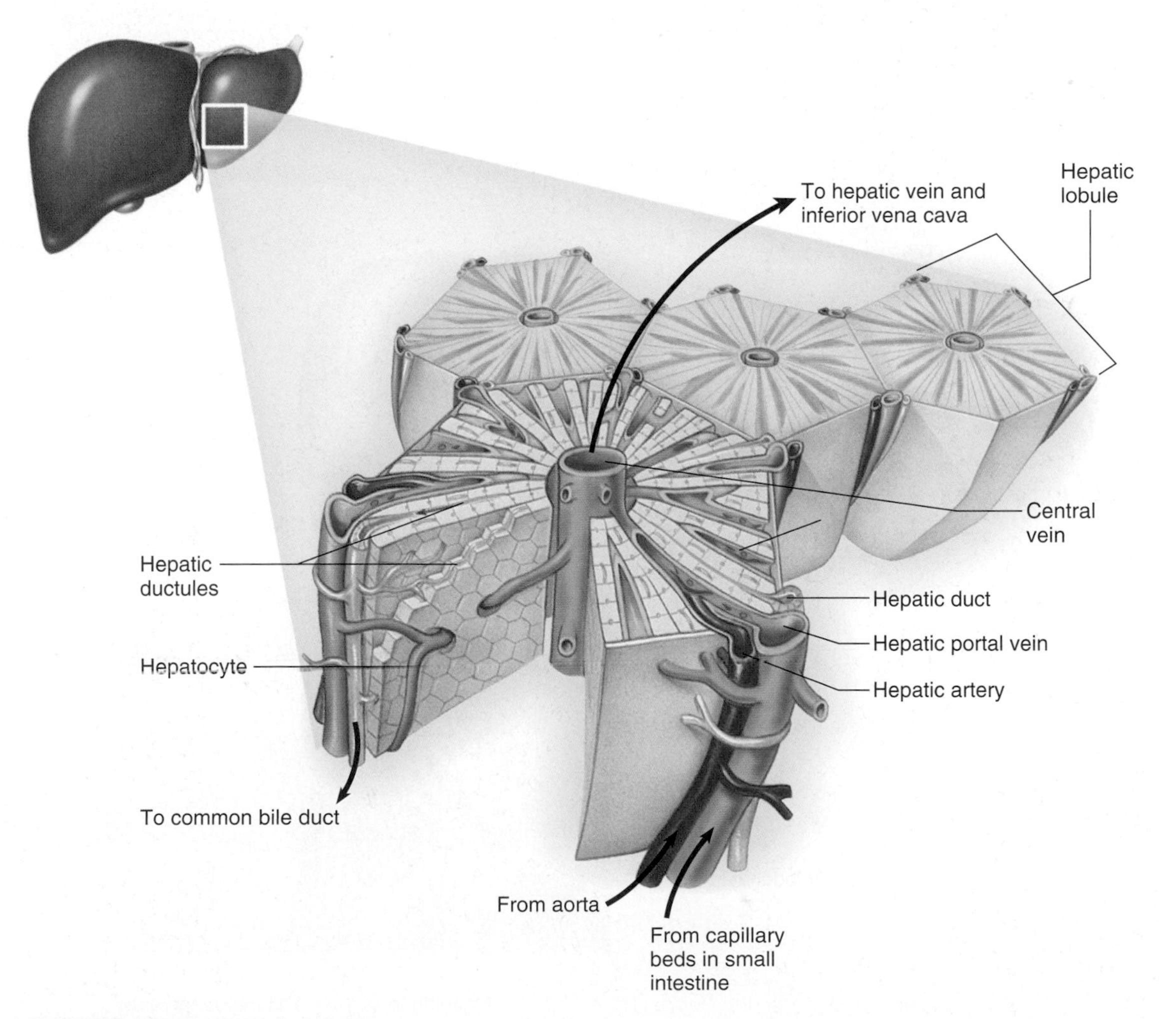

FIGURE 13.15 Hepatic lobule.

digestive function of the hepatocytes is to produce **bile.** Bile is a yellow-green fluid containing **bile acids,** synthesized from cholesterol (a steroid), and **lecithin** (a phospholipid). Both of these components function to aid in the chemical digestion of lipids by **emulsifying** lipid droplets. Emulsification involves breaking the lipids into smaller droplets, a process much like the way detergents emulsify grease when you wash your dishes. Enzymes can then complete the chemical digestion of lipids more efficiently. The other contents of bile are waste products that include bilirubin (from the breakdown of hemoglobin), cholesterol, neutral fats, bile pigments, and minerals. Bile travels within the liver from the hepatocytes, to hepatic ductules, to the right and left hepatic ducts, and to the **common hepatic duct,** which exits the liver and leads to the **common bile duct.** The liver produces approximately 500 to 1,000 mL of bile per day, which is equivalent to one-quarter to one-half of a large 2-L bottle of soda. See **Figures 13.15** and **13.16**.

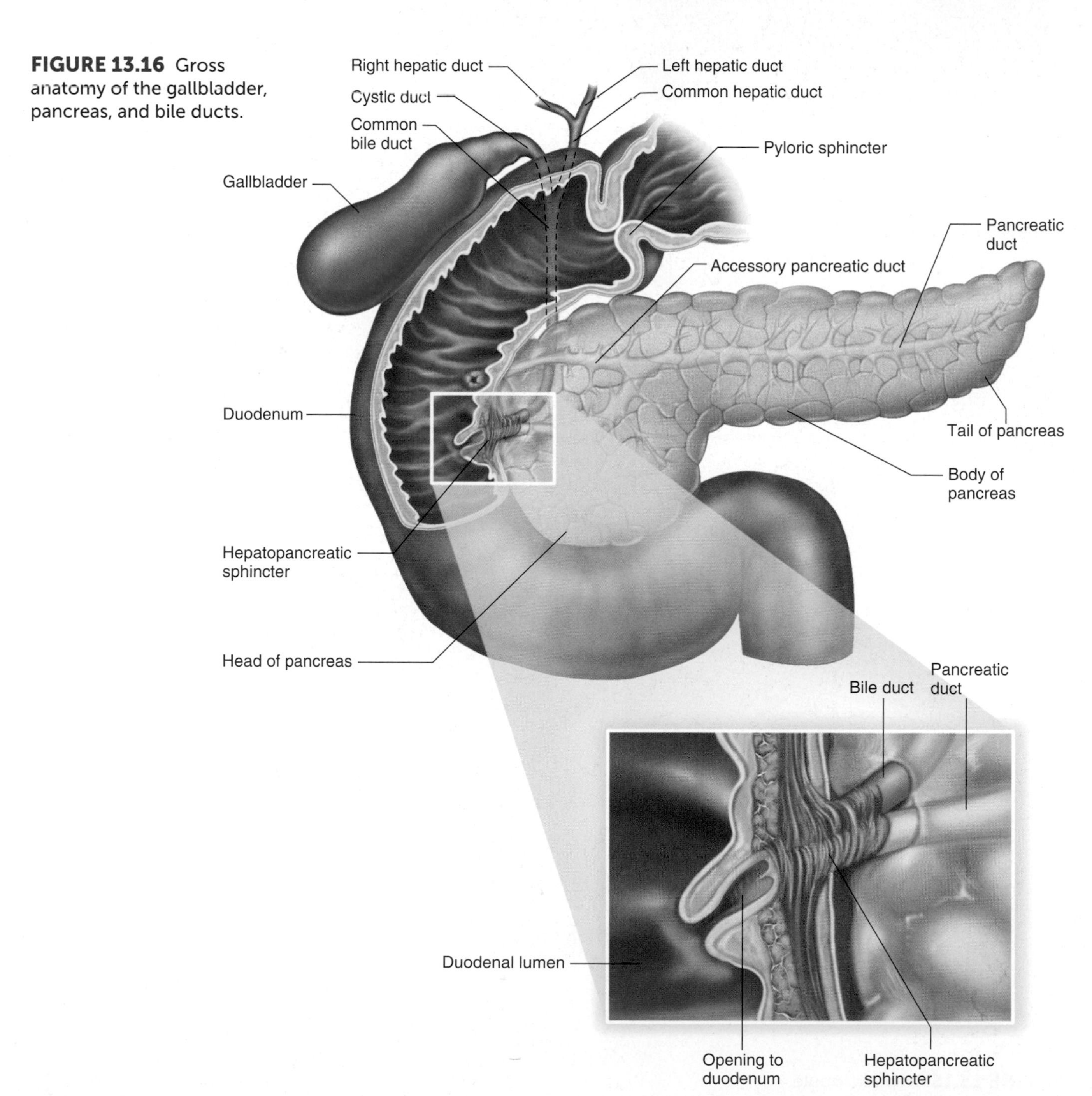

FIGURE 13.16 Gross anatomy of the gallbladder, pancreas, and bile ducts.

Common Bile Duct The common bile duct is a tube running from the common hepatic duct to the duodenum. The **cystic duct** also feeds into the common bile duct. The **hepatopancreatic sphincter** at the opening to the duodenum regulates the passage of materials from the common bile duct and **pancreatic duct** into the duodenum. See **Figure 13.16** to understand the location of these ducts with respect to the liver, gallbladder, pancreas, and duodenum.

Gallbladder As you can see in **Figure 13.16**, the gallbladder is a pear-shaped sac on the inferior side of the liver. It stores and concentrates the bile produced by the liver. As the liver continually produces bile, it fills the common bile duct. Between meals, any overflow of bile in the common bile duct accumulates in the gallbladder through the **cystic duct** because the hepatopancreatic sphincter remains closed. The gallbladder then concentrates the bile by absorbing some of the water and electrolytes. When needed for digestion, the smooth muscle in the walls of the gallbladder contracts, squeezing the bile through the cystic duct to the common bile duct through the relaxed hepatopancreatic sphincter to the duodenum.

clinical point

If the gallbladder concentrates the bile too much, the cholesterol in bile may precipitate (settle out as a solid), forming gallstones. When the gallbladder is directed to release its bile, the stones may block the cystic duct, causing pain. Surgery—a **cholecystectomy**—may be necessary to remove the gallbladder and the gallstones within. See Figure 13.17.

Pancreas The ribbonlike pancreas has a pebbly appearance (again, see **Figure 13.16**), and it is retroperitoneal, meaning it is posterior to the parietal peritoneum. The pancreas functions as two glands: (1) as an endocrine gland, because it produces the hormones insulin and glucagon secreted into the blood, and (2) as an exocrine gland, because it produces the bicarbonate ions and enzymes for protein,

(a)

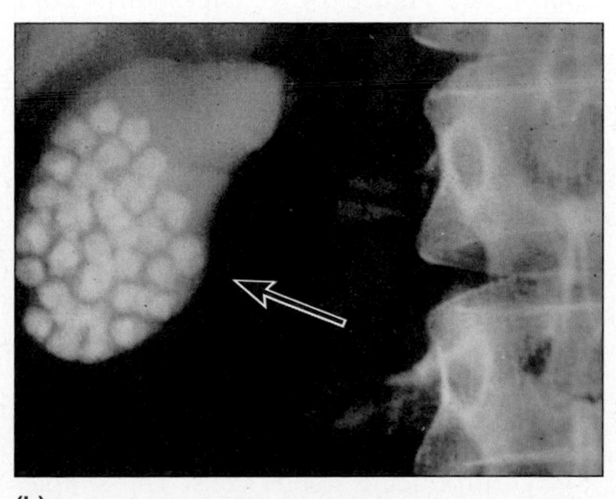

(b)

FIGURE 13.17
Gallstones: (a) gallstones, (b) radiograph of gallbladder containing gallstones (arrow).

lipid, and carbohydrate digestion that are secreted into the **pancreatic duct.** The bicarbonate ions work to neutralize the low pH of the chyme entering the duodenum from the stomach. The pancreatic duct runs the length of the pancreas and joins with the common bile duct as it opens to the duodenum.

It is important to keep in mind that these organs—the liver, gallbladder, and pancreas—are not part of the cheeseburger's path. These organs secrete digestive chemicals that are delivered to the small intestine through ducts. Their digestive juices go to the duodenum; the cheeseburger does not go through these ducts to these accessory organs. See Table 13.2.

TABLE 13.2 Digestive juices from the liver, gallbladder, and pancreas

Accessory structure	Chemical secreted	Route taken to the duodenum	Function
Liver	Bile	Hepatic ductules to the hepatic ducts and then to the common bile duct; possibly overflows through the cystic duct into the gallbladder between meals.	Emulsifies lipids
Gallbladder	Bile	Cystic duct to the common bile duct	Emulsifies lipids
Pancreas	Bicarbonate ions	Pancreatic duct to the duodenum	Neutralizes the acids of chyme
	Enzymes for carbohydrates	Pancreatic duct to the duodenum	Chemically digests carbohydrates
	Enzymes for proteins	Pancreatic duct to the duodenum	Chemically digests proteins
	Enzymes for lipids	Pancreatic duct to the duodenum	Chemically digests lipids

Now that you are familiar with the accessory organs, their secretions, and the ducts associated with the digestive system, you are ready to learn about the small intestine.

spot check 6 What fluid(s) flow(s) through the cystic duct? What fluid(s) flow(s) through the common bile duct? What fluid(s) flow(s) through the pancreatic duct?

13.11 learning outcome

Describe the digestive anatomy of the small intestine.

Anatomy of the Small Intestine

The small intestine is composed of the duodenum, the **jejunum,** and the ileum. See **Figure 13.18**. Digestion is completed in the duodenum, and absorption takes place throughout the small intestine, as you will read shortly. Although it may look as though the small intestine is very unorganized, the mesentery membranes neatly arrange the blood vessels and nerves traveling to and from each section of the small intestine. See **Figure 13.18**.

jejunum: je-JEW-num

The duodenum—the first section of the small intestine—is the next leg of the cheeseburger's journey.

Duodenum The duodenum is the first 25 cm (10 inches) of the small intestine; it is located immediately after the stomach's pyloric sphincter. See **Figures 13.16** and **13.18**. As with the entire small intestinal tract, there is smooth muscle in the duodenal walls, and the duodenal lining has many tiny projections called **villi.** See **Figure 13.19**. The villi are covered with simple columnar epithelial cells and mucus-producing goblet cells. The simple columnar epithelial cells have a brush border of microvilli to give these cells extra surface area for absorbing nutrients. Inside

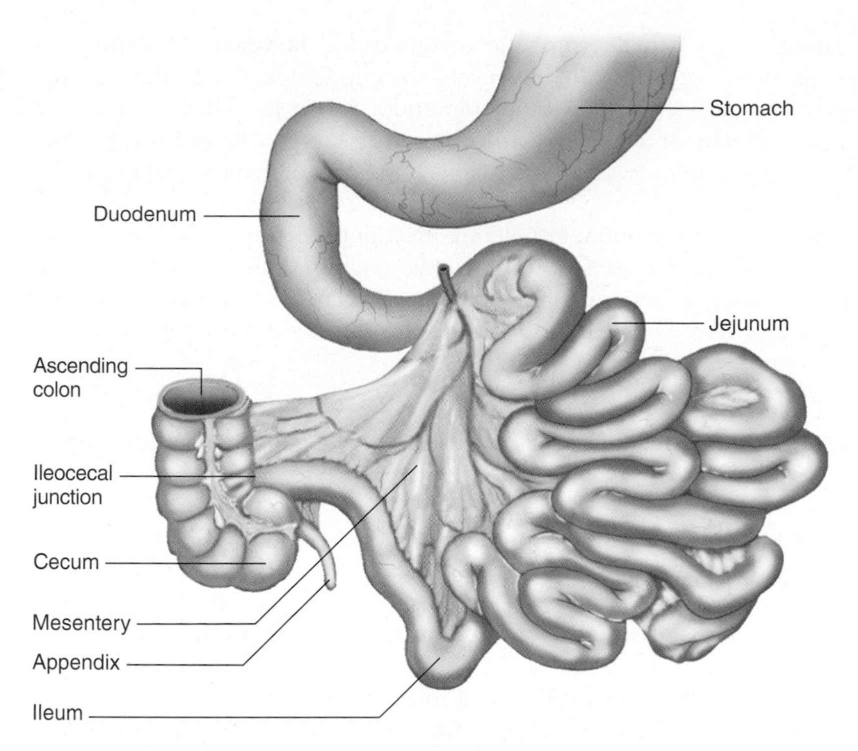

FIGURE 13.18 Gross anatomy of the small intestine.

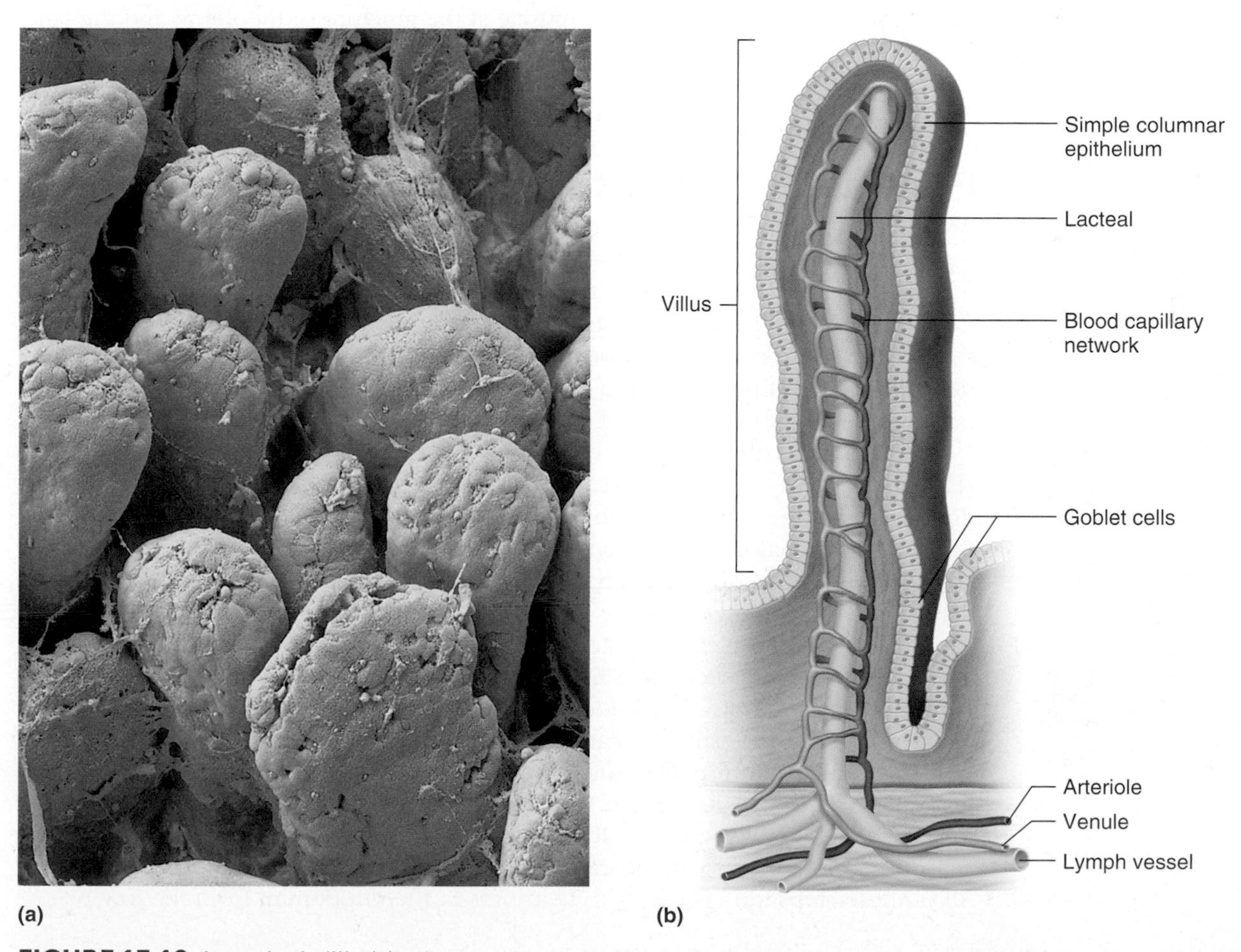

FIGURE 13.19 Intestinal villi: (a) micrograph—each villus in the image is about 1 mm high; (b) structure of a villus, showing simple columnar epithelial cells, goblet cells, capillaries, and a lacteal.

lacteal: LAK-tee-al

cholecystokinin: KOH-leh-sis-toe-KIE-nin

the villi are capillaries and small lymphatic vessels called **lacteals.** Absorption of nutrients takes place through the villi, either into the capillaries or into the lacteals.

The lining of the duodenum also contains endocrine cells. These cells make two hormones—**secretin** and **cholecystokinin**—that target the gallbladder and pancreas, telling them to release bicarbonate ions, digestive enzymes, and bile to be delivered to the duodenum.

The duodenum, like the jejunum and ileum, has tight junctions between cells of the epithelial lining to protect itself from the acidic chyme. Only small amounts of chyme should enter the duodenum at any one time. This helps to keep the mucous lining from becoming overwhelmed, and it gives the duodenum time to neutralize the chyme.

spot check  How much chyme is allowed to enter the duodenum at one time?

Jejunum The jejunum—the second part of the small intestine (see **Figure 13.18**)—has a very rich blood supply that gives it a pink appearance. The jejunum measures approximately 2.2 to 2.4 m in length and its villi are slightly smaller than those in the duodenum. Most of the absorption of nutrients takes place in the jejunum.

ileocecal: ILL-ee-oh-SEE-cal

Ileum The ileum is the last part of the small intestine, measuring 3.3 to 3.6 m in length. Its walls are less muscular and thinner than the jejunum's. The ileum's lining is characterized by nodules of lymphocytes called *Peyer's patches.* These nodules increase in size as they approach the large intestine, and they function to destroy any bacteria or other pathogens entering the small intestine from the large intestine. The **ileocecal valve** is a sphincter muscle at the juncture of the ileum and the large intestine; it regulates the passage of materials from the ileum to the large intestine. See **Figures 13.18** and **13.21**.

13.12 learning **outcome**

Explain the physiology of chemical digestion in the duodenum, including the hormones and digestive secretions involved.

Physiology of Digestion in the Small Intestine

When the acidic chyme enters the duodenum, the endocrine cells of the duodenum begin to secrete their hormones—secretin and cholecystokinin. One minor role of these two hormones is to target the stomach's parietal and chief cells, telling them to stop producing hydrochloric acid and pepsinogen. If chyme is now entering the duodenum, there is no further need for digestion in the stomach. This is a second negative-feedback mechanism to stop digestion in the stomach. It complements the negative-feedback mechanism of low pH in the stomach mentioned earlier.

Another role of these two hormones is to target the pancreas, telling it to release enzymes to complete carbohydrate, lipid, and protein digestion. These digestive enzymes travel through the pancreatic duct to the common bile duct and through the hepatopancreatic sphincter. Then they move into the duodenum to complete lipid, carbohydrate, and protein digestion. You will now investigate how that works.

The low pH of the chyme entering the duodenum stimulates the duodenal endocrine cells to secrete secretin. This hormone mainly targets the pancreas, telling the pancreas to release bicarbonate ions to neutralize the acidic chyme. This bicarbonate solution from the pancreas carries pancreatic enzymes for lipid, protein, and carbohydrate digestion from the pancreas, through the pancreatic duct, to the common bile duct, and to the duodenum. The bicarbonate ions combine with the hydrogen ions of the hydrochloric acid to form carbon dioxide and water. The carbon dioxide is absorbed into the blood and carried to the lungs, where it is eventually expelled. All of these steps are necessary to help protect the duodenum from the low pH.

When partially digested lipids enter the duodenum, the duodenum's endocrine cells release **cholecystokinin,** which travels through the blood to its main target tissues—the gallbladder and the hepatopancreatic sphincter. Cholecystokinin tells the gallbladder to squeeze (contract) and release its bile through the cystic duct to the common bile duct. Cholecystokinin also tells the hepatopancreatic sphincter to relax so that the bile in the common bile duct can enter the duodenum.

spot check 8 How does cholecystokinin help the bicarbonate ions and enzymes from the pancreas reach the duodenum?

Bile helps to complete lipid digestion by **emulsifying** (breaking up) the lipids to tiny droplets so that the lipases (enzymes) from the pancreas can break the lipids down to their building blocks—*fatty acids* and *glycerol.* Bile also helps to activate some of the other digestive enzymes from the pancreas.

The pancreatic enzymes complete protein digestion by breaking the protein molecules down to their building blocks, *amino acids,* while carbohydrate-digesting enzymes from the pancreas break down the chyme's carbohydrates to their building blocks, *monosaccharides* (simple sugars).

In the small intestine, the mechanical and chemical digestion of the cheeseburger introduced in the beginning of the chapter is complete. The fats in the cheese are broken down to *fatty acids* and *glycerol.* The proteins of the burger are broken down to *amino acids.* And the carbohydrates of the bun are broken down to *monosaccharides.* The nutrients, waste products from the bile, and the indigestible materials continue on to the jejunum and ileum, where nutrient absorption occurs through the villi.

The chyme moves through the sections of the small intestine via two types of contractions. See **Figure 13.20. Segmentation** is a stationary constriction of the smooth muscle in ringlike patterns. This type of contraction further churns the chyme, mixing in the bile and digestive enzymes to finish chemical digestion. It also allows for maximum contact between chyme and the villi, facilitating maximum absorption of nutrients. Once the chyme has churned and mixed with the bile and digestive enzymes, it continues to move through the jejunum and ileum through peristalsis (wave-like contractions mentioned earlier during swallowing).

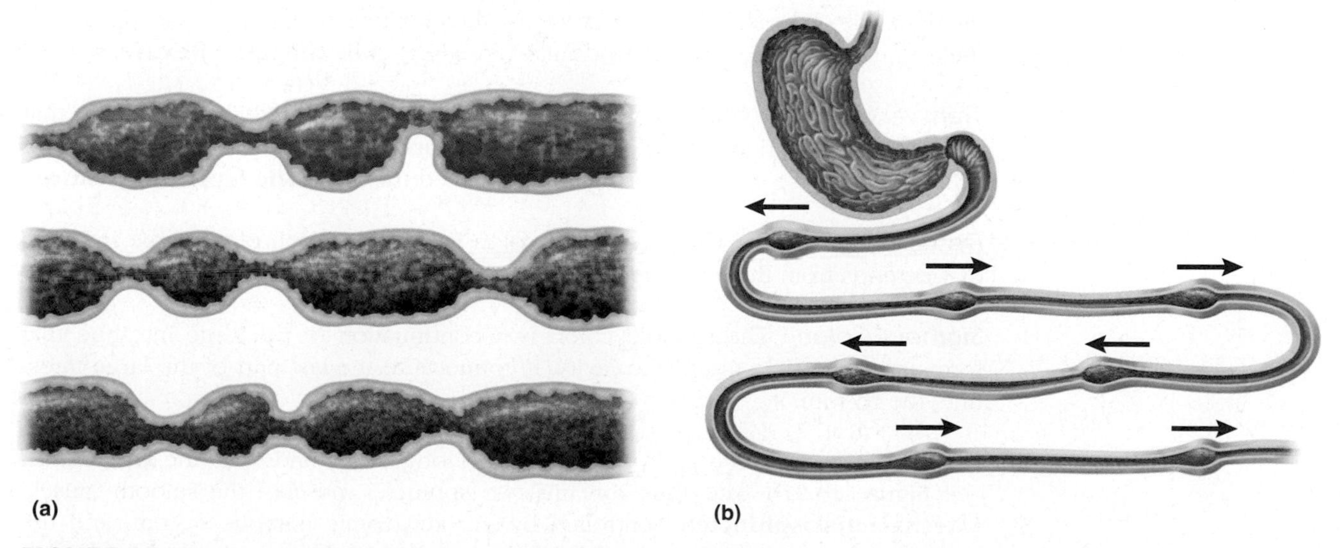

FIGURE 13.20 Segmentation and peristalsis of the small intestine: (a) segmentation, (b) peristalsis.

13.13 learning outcome

Explain how nutrients are absorbed in the small intestine.

Absorption of Nutrients in the Small Intestine

Monosaccharides and amino acids are absorbed into capillaries through the epithelium of the villi by facilitated diffusion. Fatty acids and glycerol are absorbed across the epithelial membranes of the villi by diffusion. They are then coated with proteins and exocytosed to the lacteals, the small lymph vessels located in the villi. They will continue to travel through lymph vessels to the thoracic duct to the left subclavian vein, where they will enter the bloodstream.

The ileum reabsorbs 80 percent of the bile acids in the chyme, while the other 20 percent will leave the body during defecation. This is the body's way of removing cholesterol. The liver will make new bile acids from cholesterol to replace the lost 20 percent of bile acids. What was not absorbed moves through the ileocecal sphincter into the large intestine.

13.14 learning outcome

Describe the anatomy of the large intestine.

Anatomy of the Large Intestine

The large intestine **(colon)** is made up of six regions: the **cecum,** the **ascending colon,** the **transverse colon,** the **descending colon,** the **sigmoid colon,** and the **rectum.** See **Figure 13.21**. All together, these regions measure about 1.5 m in length and 6.5 cm in diameter. Although the large intestine is shorter than the small intestine, it is termed *large* because its diameter is greater.

Cecum The cecum is a blind pouch (does not lead anywhere) inferior to the juncture of the ileocecal valve in the lower right quadrant of the abdomen. The appendix is a dead-end tube extending approximately 7 cm from the inferior portion of the cecum. It contains many lymphocytes.

clinical point

Inflammation of the appendix—**appendicitis**—can be extremely serious because the appendix can rupture and spill its contents into the abdominopelvic cavity. These contents are filled with bacteria, which may infect the entire abdominopelvic cavity if released.

Ascending Colon The ascending colon begins at the ileocecal valve and passes up the right side of the abdominal cavity toward the right lobe of the liver. As it approaches the liver, it forms a right-angle bend called the **right colic (hepatic) flexure.**

Transverse Colon The transverse colon is a continuation of the large intestine that extends from the right colic flexure across the abdomen to the area of the spleen. There, the colon forms another right angle called the **left colic (splenic) flexure.**

Descending Colon The descending colon is a continuation of the large intestine that extends from the left colic flexure down the left side of the abdominal cavity.

Sigmoid Colon The sigmoid colon is a continuation of the large intestine that forms an S shape in the pelvic cavity. It connects to the last part of the large intestine, the rectum.

Rectum The rectum is approximately 15 cm long, and it ends with the **anal canal.** See **Figure 13.21b**. The anus contains two sphincter muscles: the smooth muscle **internal anal sphincter,** controlled by the autonomic nervous system, and the skeletal muscle **external anal sphincter,** controlled by the somatic nervous system.

All of the large intestine's regions contain smooth muscle in the walls, but the ascending, transverse, descending, and sigmoid colons also contain longitudinal bands of smooth muscle called **taenia coli.** The taenia coli cause the large

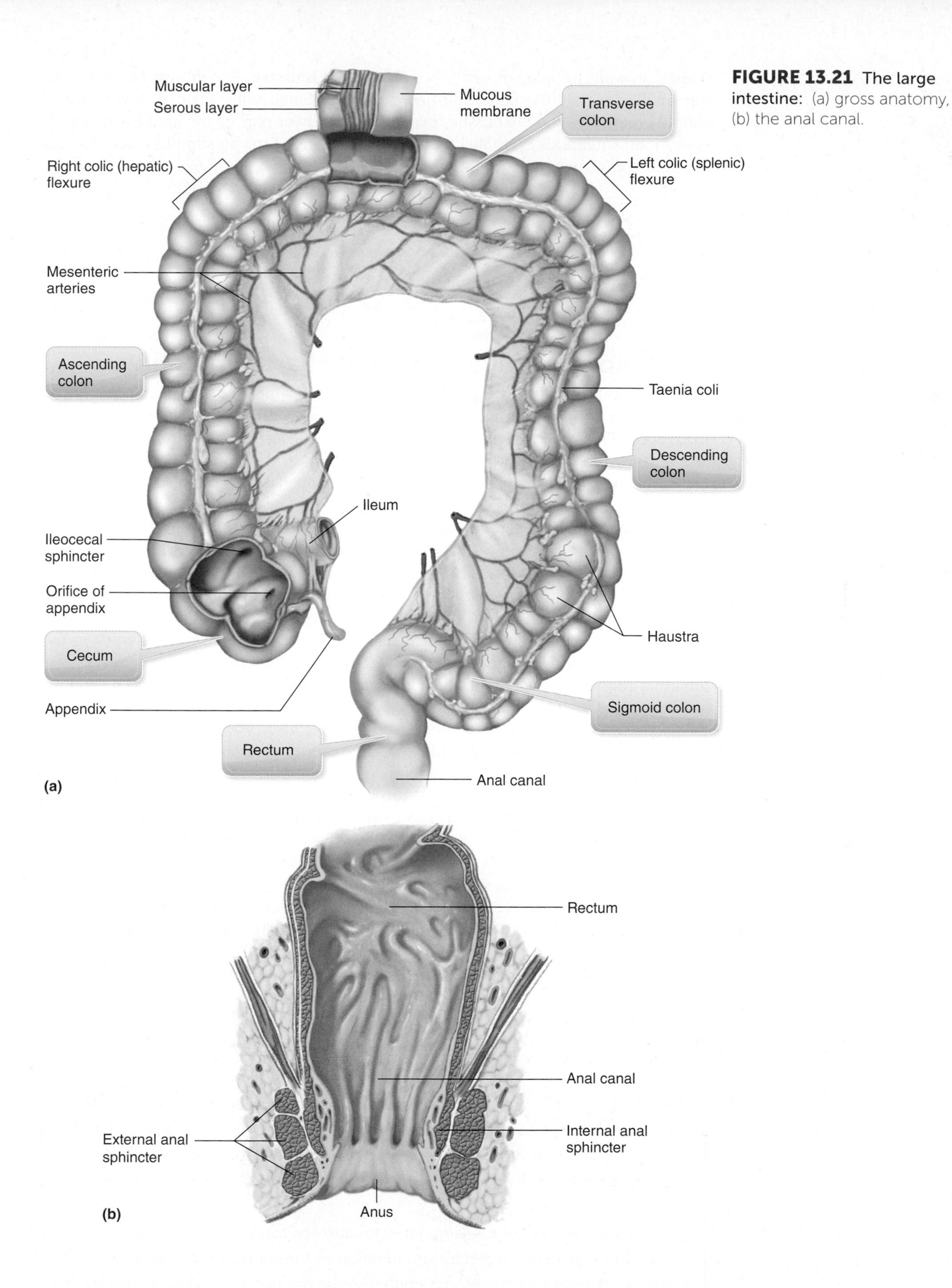

FIGURE 13.21 The large intestine: (a) gross anatomy, (b) the anal canal.

haustra: HAW-stra

intestine's walls to bulge, forming pouches called **haustra.** See **Figure 13.21**. Unlike the small intestine, the large intestine does not contain villi. Instead, it is lined by simple columnar epithelial tissue, except for the lower part of the anal canal, which is stratified squamous epithelial tissue. This tissue needs to be stratified to withstand the abrasion of materials leaving the body.

spot check 9 A major function of the small intestine is to absorb nutrients. Given what you have just read about the anatomy of the large intestine, do you think the large intestine will have a similar function? Explain.

13.15 learning outcome

Explain the physiology of the large intestine in terms of absorption, preparation of feces, and defecation.

Physiology of Digestion in the Large Intestine

Chyme (minus the absorbed nutrients) enters the large intestine in a very liquid state. The large intestine functions to absorb water, and this compacts its contents into **feces.** This process can take 12 to 24 hours. During that time, the large intestine also absorbs some electrolytes (especially sodium and chloride ions) and vitamin K produced by bacteria living in the large intestine. The large intestine then stores fecal matter until it is removed **(defecation).**

Even after the water is absorbed and the feces have been compacted, feces are still typically composed of 75 percent water and 25 percent solid matter. The solid matter consists of bacteria that normally live in the colon, indigestible carbohydrates (dietary fiber), lipids, and a mixture of sloughed-off epithelial cells, digestive juices, mucus, and a small amount of protein. The lipids and proteins are not from the cheeseburger. They are from broken-down epithelial cells and bacteria, which normally live in the colon and have died. Indigestible carbohydrates from the cheeseburger feed the bacteria that reside in the large intestine. In return, the bacteria produce some B vitamins and vitamin K, a necessary vitamin for the production of clotting factors. Although these bacteria provide a very beneficial service, they also produce a gas called **flatus,** which is not so desirable, as it can cause a bloated feeling and an unpleasant odor. The amount of flatus produced depends on the amount of bacteria present in the colon and the type of food ingested. The large intestine normally contains 7 to 10 L of gas. A typical human expels approximately 500 mL of flatus per day.

How do materials move through the large intestine? Upon entering the large intestine, materials pass up the ascending colon by peristalsis to the transverse colon, where the materials stop. Distension (expansion) of the stomach and duodenum causes a **mass movement,** which moves the feces from the transverse colon to the descending colon, to the sigmoid colon, and on to the rectum. Distension of the walls of the rectum triggers the **defecation reflex.** This reflex drives the feces downward and relaxes the internal anal sphincter. Even though this is an involuntary reflex, defecation occurs only if the external anal sphincter is voluntarily relaxed. See **Figure 13.22** for the defecation reflex's reflex arc.

clinical point

If the large intestine absorbs too much water, the feces will become harder to move, leading to **constipation.** Increased fluid intake, increased dietary fiber, and exercise can help move feces along. The increased pressure to push with constipation can cause **hemorrhoids,** which are bulging anal veins. They may be internal to the rectum or external to the anus.

On the other hand, if the large intestine absorbs too little water, **diarrhea** may occur. A runny stool can result from irritation of the intestine caused by bacteria. In the case of diarrhea, the ileum's contents pass through the colon too quickly for adequate water absorption and compaction of feces to take place.

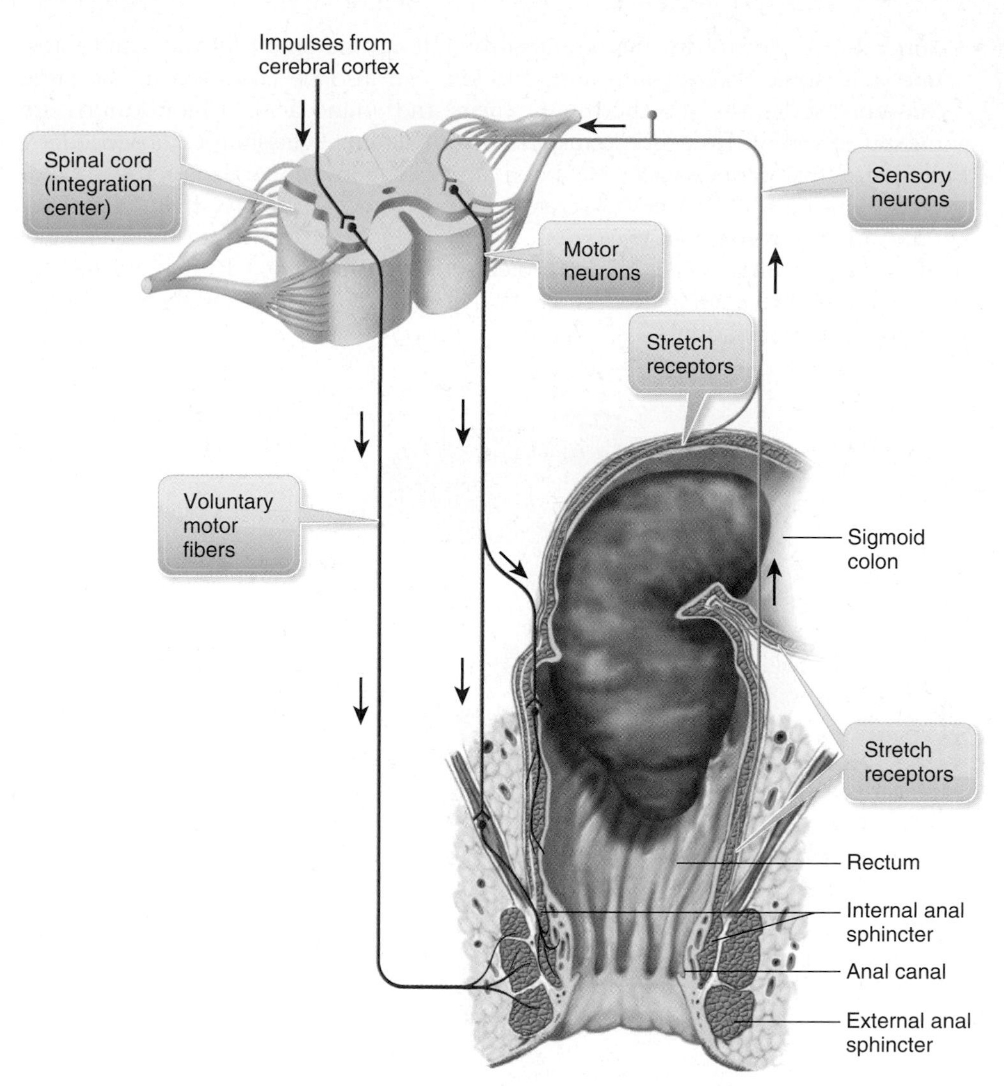

FIGURE 13.22 **Reflex arc for the defecation reflex.** (1) Stretch receptors in the rectal wall send messages (afferent sensory neuron) to the spinal cord (integration center). (2) The spinal cord sends motor messages (efferent motor neuron) to the rectal wall, telling it to contract (effector). (3) The spinal cord also sends messages along motor neurons to the internal anal sphincter, telling it to relax (effector). (4) The cerebrum determines whether the external sphincter will relax so that defecation can occur (this step is not part of the reflex).

Types of Absorbed Nutrients

13.16 learning **outcome**

Summarize the types of nutrients absorbed by the digestive system from the diet.

Until now, you have concentrated on the digestion and absorption of the major nutrients of a cheesesburger: proteins, carbohydrates, and lipids.

- The proteins in the burger were chemically digested to amino acids and were absorbed into the blood of capillaries in the small intestine.
- The carbohydrates in the bun were chemically digested to monosaccharides and were also absorbed into the blood of capillaries in the small intestine.
- The lipids of the cheese were chemically digested to fatty acids and glycerol and were absorbed into the lacteals in the small intestinal villi.

Other nutrients, like vitamins and minerals, are also absorbed by the digestive system.

- Vitamins can be categorized as **fat soluble** or **water soluble.** Fat-soluble vitamins (A, D, E, and K) are absorbed along with the products of lipid digestion, so they must be ingested with fats to be absorbed. On the other hand, water-soluble vitamins (the B complex and C) are absorbed by simple diffusion. Vitamin B_{12} is an exception. It must first bind to intrinsic factor in the stomach and then be endocytosed by cells of the ileum for absorption. A list of vitamins and their RDAs can be found in Appendix B.

- Minerals are electrolytes that are absorbed along the length of the small intestine, and some, like sodium and chloride, can also be absorbed in the large intestine. Sodium is absorbed with sugars and amino acids. Chloride ions are mostly absorbed by active transport in the ileum. Potassium is absorbed by simple diffusion once water has been absorbed. Most minerals are absorbed at a constant rate. The kidneys excrete whatever excess may have been absorbed. Calcium and iron are an exception, as the body absorbs them to meet its level of need. You may recall from Chapter 4 that the hormone PTH regulates the absorption of calcium in the small intestine. A list of minerals and their RDAs can be found in Appendix B.

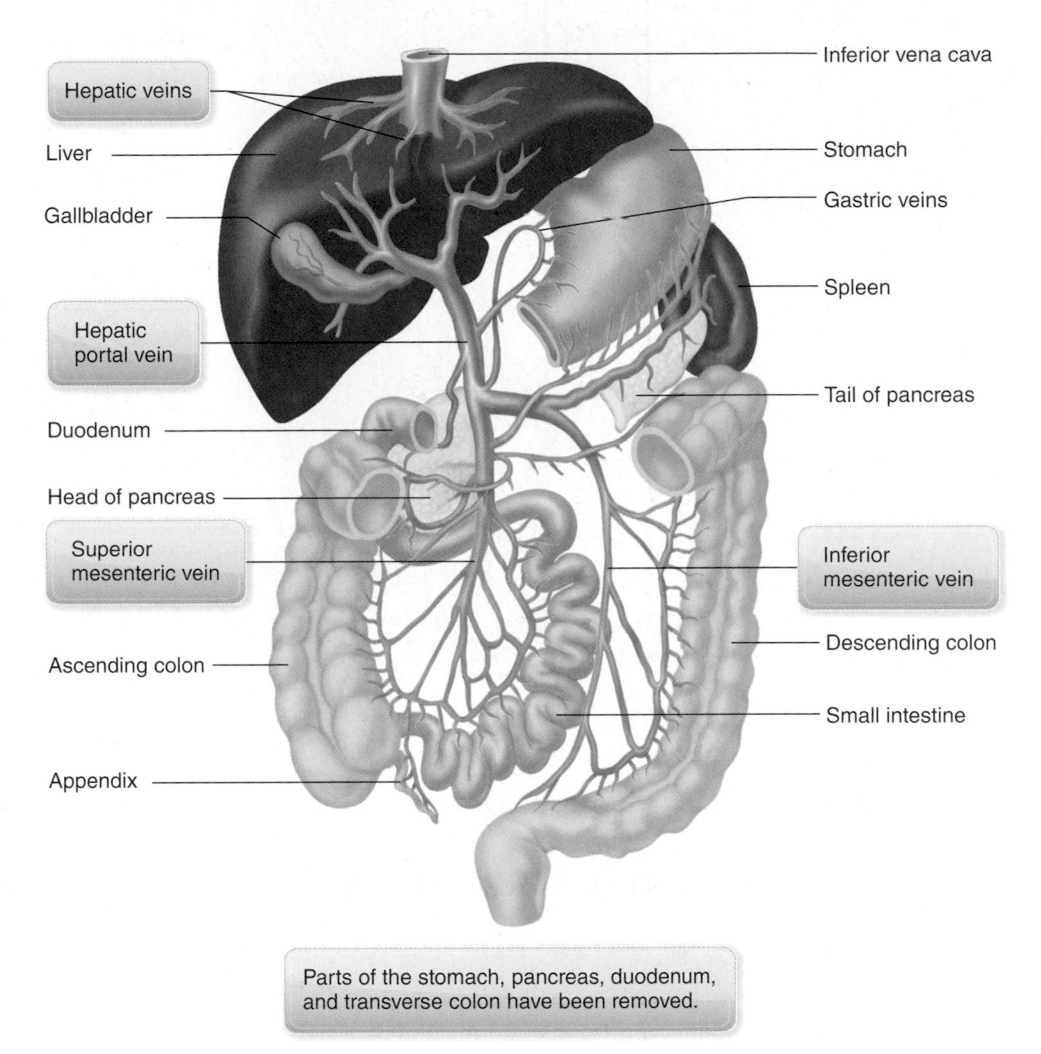

FIGURE 13.23 Veins of the hepatic portal system. Veins drain nutrient-rich blood from capillary beds in the digestive organs and deliver it to capillary beds in the liver through the hepatic portal vein. Once processed, the blood leaves the liver through the hepatic vein on its way to the inferior vena cava and then the heart.

13.17 learning **outcome**

Trace the circulation of the nutrients once they have been absorbed.

Circulation of Absorbed Nutrients

All blood from the capillaries in the stomach and intestines is circulated directly to the hepatic portal system so that it can be processed in the liver. See **Figure 13.23**. The hepatic portal vein drains the nutrient-rich blood from the capillaries in the villi and carries it to the capillary beds in the liver. There, the liver removes excess glucose, amino acids, iron, vitamins, and other nutrients for storage. It also recycles the 80 percent of bile acids reabsorbed from the ileum to form bile for lipid digestion in the future. The fatty acids and glycerol absorbed into lacteals in the villi will join the bloodstream at the subclavian veins and eventually reach the liver through the hepatic artery.

Control of Digestion

13.18 learning **outcome**

Explain the control of digestion.

Control of digestion is through the autonomic nervous system. Parasympathetic fibers of the vagus nerve stimulate digestion, while sympathetic neurons from the celiac ganglion suppress digestion in part by diverting blood to skeletal muscles and the heart.

Functions of the Digestive System

13.19 learning **outcome**

Summarize the functions of digestion.

Now that you have completed the anatomy and physiology involved in the digestive system, it is time to summarize the functions of this system while Lisa enjoys the cheeseburger her mother has prepared. See **Figure 13.24**. These functions include:

- **Ingestion.** This function involves the intake of food into the mouth. By her smile, you can see that Lisa is seeing an orthodontist to straighten her teeth. Healthy teeth and a properly aligned bite will help Lisa to begin the next function of this system.
- **Digestion.** Lisa begins mechanical digestion of her cheeseburger as soon as she starts to chew. Mechanical digestion breaks down large pieces of complex molecules to smaller pieces of complex molecules. This type of digestion continues in the stomach. Chemical digestion breaks complex molecules into their building blocks. It, too, begins in the mouth, where carbohydrates are partially broken down, and it continues in the stomach, where proteins and lipids are partially broken down. Chemical digestion is completed in the small intestine, where the carbohydrates are completely broken down to monosaccharides, the lipids are completely broken down to fatty acids and glycerol, and the proteins are completely broken down to amino acids.
- **Absorption.** The products of digestion are absorbed through the villi in the small intestine. Monosaccharides and amino acids are absorbed through facilitated diffusion into the villi's epithelial cells and then into capillaries to travel to the liver through the hepatic portal vein. Using simple diffusion, the epithelial cells

FIGURE 13.24 Lisa.

of the villi absorb fatty acids and glycerol. They are then coated with protein and endocytosed to lacteals in the villi to travel with lymph to subclavian veins.

- **Defecation.** The stretching of Lisa's stomach and duodenum during the consumption of this meal will initiate a mass movement in her colon. When the previous undigested materials and bacteria stretch the walls of Lisa's rectum, a defecation reflex will result, but it will be Lisa's decision as to when to defecate. It is important to understand that most of the feces removed from the body are not metabolic wastes, produced by chemical processes in cells. They are simply what was inserted in the mouth and never absorbed by the body's cells.

Lisa is a healthy teenager whose digestive system is serving her well. However, what can she expect to happen as she ages?

13.20 learning outcome

Summarize the effects of aging on the digestive system.

Effects of Aging on the Digestive System

The effects of aging on the digestive system can be seen in many of the structures along the alimentary canal, starting with the mouth.

- The effects of aging can be seen in the mouth. The enamel on the teeth thins and the gingiva recedes, allowing for increased tooth decay and loosening of the teeth. This interferes with proper mastication. Proper dental hygiene can minimize these effects. The receptors in the taste buds and nose become less sensitive, leading to a decreased appetite. This may compromise the nutritional status. Although the person may be eating less because of a diminished appetite, there may be weight gain due to a slower metabolism.
- The lining of the stomach begins to atrophy with age. This can result in less intrinsic factor produced. With less intrinsic factor, less vitamin B_{12} is absorbed from the diet, possibly leading to pernicious anemia.
- The liver may metabolize drugs differently as it ages. Geriatric patients may need to have dosages adjusted for drugs they had been taking long-term.
- Movement through the large intestine slows with age. The longer materials stay in the large intestine, the more water is absorbed. This can lead to constipation.

clinical point

With age, the cumulative effects of ingested carcinogens on the walls of the large intestine may lead to **polyps,** which are precancerous growths. A **colonoscopy** is recommended to check for polyps on a routine basis for individuals with a family history of colon cancer. The incidence of **colon cancer** increases with age.

13.21 learning outcome

Describe digestive system disorders, including vomiting, food poisoning, parasites, and peptic ulcers.

Digestive System Disorders

You have already become familiar, through the course of this chapter, with many digestive disorders: dental caries, abscesses, gingivitis, gastroesophageal reflux disease, gallstones, constipation, diarrhea, hemorrhoids, polyps, and colon cancer. To gain a thorough overview of digestive disorders, you will next learn about vomiting, food poisoning, parasites, and ulcers.

Vomiting

Vomiting can result from irritation anywhere along the digestive tract. It is controlled by an **emetic center** in the medulla oblongata. It begins with a deep breath. The hyoid bone and larynx are elevated, closing off the glottis, while the soft palate is elevated, closing off the nasopharynx. The diaphragm and abdominal muscles forcefully contract, putting pressure on the stomach and its contents. The cardiac sphincter opens, and the contents of the stomach are forcefully expelled.

Food Poisoning

What is important to understand about food poisoning are its possible causes. Is it caused by bacteria? Or the toxins the bacteria produce? Heat from cooking may destroy the bacteria but not the toxins they produce. The various types of bacteria and/or their toxins, which can cause food poisoning, are explained in the following list:

- **Staphylococcus** is usually contracted from a food handler. Bacteria contaminating the food make toxins, which cause nausea, diarrhea, and vomiting. The symptoms occur one to six hours after eating the contaminated food.
- **Salmonellosis** is caused by the bacteria in contaminated food (meat, poultry, milk). These bacteria are destroyed by heat. The symptoms of nausea, diarrhea, and vomiting can occur up to 36 hours after eating.
- **Botulism** is caused by a toxin made by a common bacterium found in the soil. This toxin is a powerful neurotoxin that prevents muscle contractions. You may eat a raw green bean directly from the garden and ingest the bacteria. This is not harmful. However, if the green beans are improperly canned and not all of the bacteria are destroyed in the process, the toxin they produce may be fatal.

Parasites

A **parasite** is an organism that lives on or in another organism (the host) and obtains its nourishment there. Parasites may or may not be harmful to the host. With digestive parasites, it is important to understand what they eat. For example, do they eat what is passing by, or do they eat you? The list below contains specific information about the types of parasites that might be found in the digestive system.

- **Pinworms.** These small, white worms commonly live in the digestive tract of humans and feed on the partially digested food going by. They crawl out the anus to lay their eggs, which causes an itching sensation. Contaminated fingers then spread the eggs to surfaces on which they are able to survive. Consider this example of how these worms may be spread: Jimmy is a kindergarten student who has pinworms. As he is coloring, he begins to squirm in his chair because of the itching caused by the worms. It becomes more and more uncomfortable, prompting Jimmy to ask his teacher to be excused to go to the restroom. There, he solves his problem by scratching his itch directly. Having been well trained by his parents, Jimmy washes his hands when he is done. But Jimmy is only five years old. Not all of his fingers necessarily get wet if he is in a hurry, as he is in this case. So Jimmy returns to color the sky blue in his drawing. Sally, who is worm-free, asks to borrow Jimmy's blue crayon. Having learned to share, Jimmy gives her the now pinworm-egg-contaminated crayon.

FIGURE 13.25 Adult tapeworm.

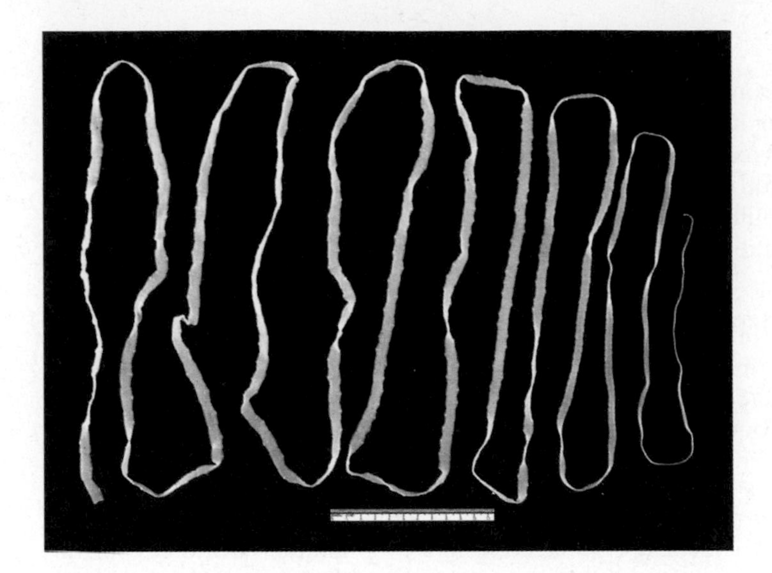

Sally puts the end of the crayon in her mouth as she contemplates what she wants to color blue. Sally has now ingested pinworm eggs. The ingested eggs will hatch in her intestine. These worms are easily spread to other individuals within families and schools.

- **Tapeworms.** The larvae of these worms—from undercooked beef, pork, or fish—infect the digestive tract. They attach to the intestinal wall by suckers and feed off the partially digested materials passing by. Tapeworms are segmented worms, and their segments may break off and appear in the feces. Tapeworms may live in the digestive tract for years, and can grow up to 6 meters in length. See **Figure 13.25**.
- **Roundworms.** The ingested eggs of this parasite hatch into larvae in the upper intestine, enter the bloodstream, and travel to the lungs. There, they cause respiratory symptoms. When coughed to the pharynx, the larvae are then swallowed, returning the worms to the intestine. The adult worms may stay in the intestine, or they may migrate, cutting through intestinal walls. See **Figure 13.26**.
- **Giardia.** These protozoa are prevalent in streams, lakes, and rivers, especially where beavers are present. This infection results from ingesting untreated, contaminated water. The symptoms of nausea, abdominal cramps, and weight loss may last for weeks.

FIGURE 13.26 Roundworms. A CDC technician holds a mass of roundworms passed by a child.

Peptic Ulcers

Peptic ulcers are erosions of the digestive tract lining due to an imbalance of gastric juices (hydrochloric acid and pepsin) and the protection provided by the mucosa. Once through the mucosa, the gastric juices may continue to erode the protein-rich muscular walls. Peptic ulcers are named for where they occur, such as esophageal ulcers, gastric ulcers, or duodenal ulcers. See **Figure 13.27**.

- **Esophageal ulcers** may happen in the lower esophagus if there is reflux of gastric juices through the cardiac sphincter.
- **Gastric ulcers** of the stomach are often the result of a bacterium, *Heliobacter pylori (H. pylori)*. Continued use of nonsteroidal anti-inflammatory drugs (NSAIDs), such as aspirin, may also cause these ulcers.
- **Duodenal ulcers**—the most common—result when the acidic chyme entering the duodenum through the pyloric sphincter is not sufficiently neutralized.

Contrary to popular belief, stress does not cause ulcers. Chronic stress may, however, increase the chances of an ulcer forming and can slow its healing. During chronic stress, the sympathetic nervous system reduces the production of mucus in the digestive tract, lowering the protection from the gastric juices.

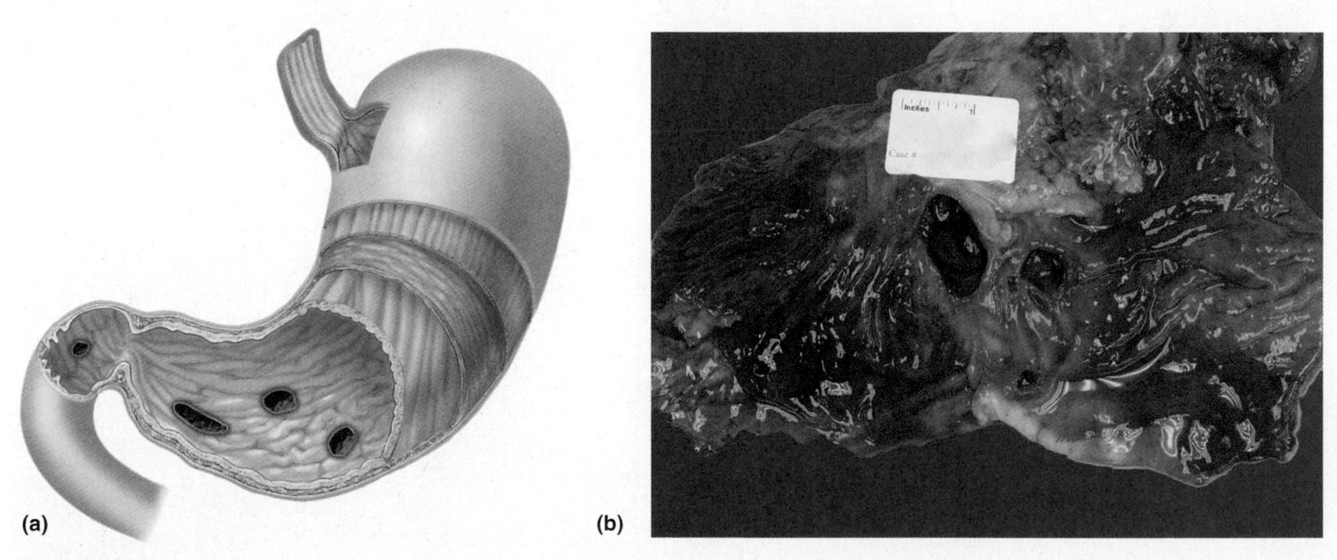

FIGURE 13.27 Peptic ulcers: (a) peptic ulcers, (b) stomach that has been opened and laid flat to show several ulcers.

putting the pieces **together**

The Digestive System

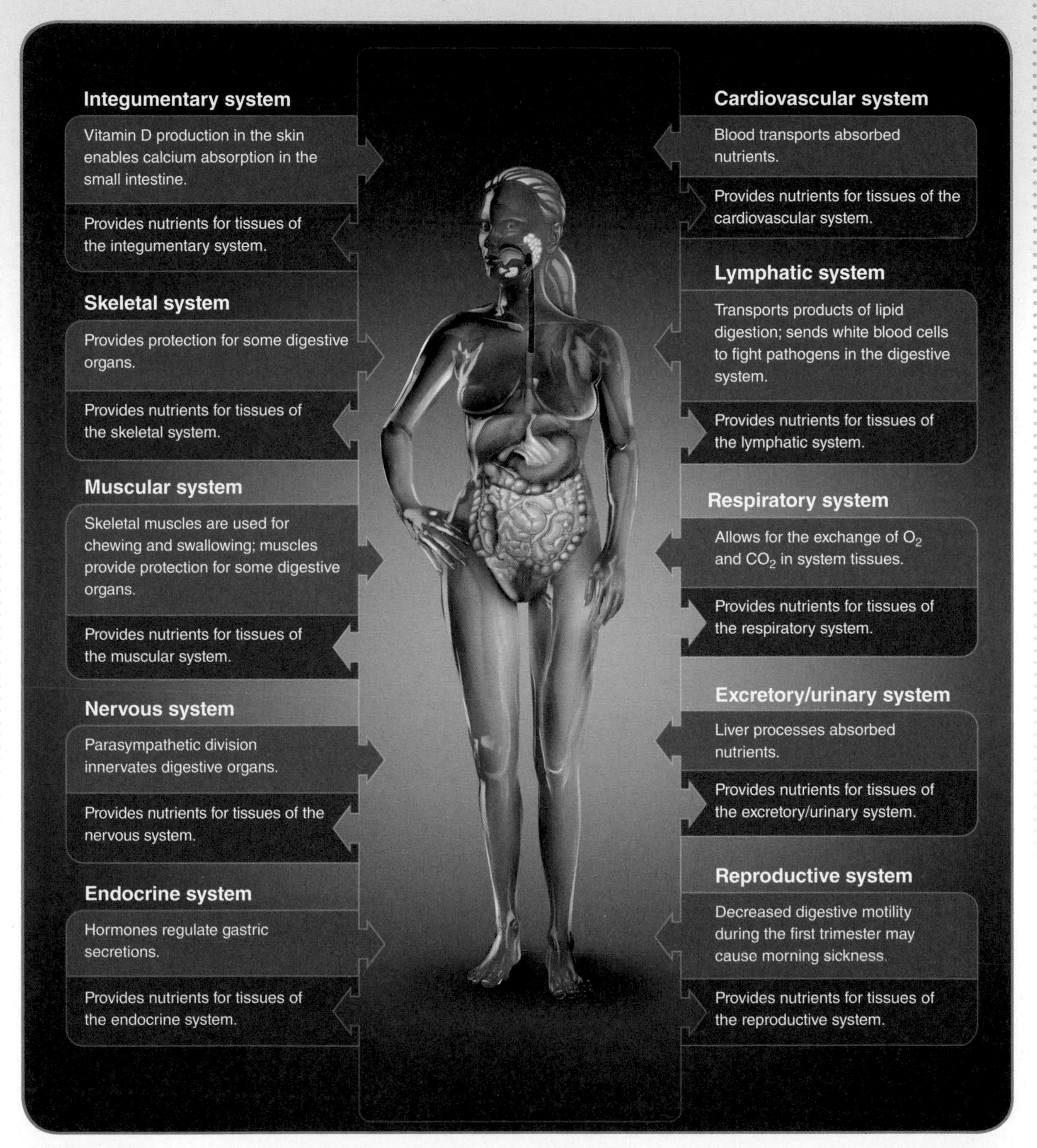

FIGURE 13.28 Putting the Pieces Together—The Digestive System: connections between the digestive system and the body's other systems.

summary

Overview

- Structures of the digestive system form the alimentary canal.
- Mechanical digestion is the breakdown of large pieces of complex molecules to smaller pieces of complex molecules.
- Chemical digestion is the breakdown of complex molecules to their building blocks so that they can be absorbed.

Anatomy and Physiology of the Digestive System

Anatomy in the Mouth

- Deciduous teeth are replaced by permanent teeth.
- The purpose of the tongue is to manipulate what is ingested and to provide the sense of taste.
- The parotid glands, the submandibular glands, and the sublingual glands produce saliva.

Physiology of Digestion in the Mouth

- Mechanical and chemical digestion takes place in the mouth.
- The masseter and temporalis muscles move the jaw for mastication.
- Amylase in saliva partially digests carbohydrates.
- Mucus mixes with the food to make it easier to swallow.
- The bite of food is called a *bolus* before it is swallowed.

Anatomy from the Mouth to the Stomach

- The pharynx is composed of the nasopharynx, the oropharynx, and the laryngopharynx.
- The esophagus is a tube that goes through the diaphragm to connect with the stomach.

Physiology of Digestion from the Mouth to the Stomach

- Deglutition (swallowing) involves four cranial nerves.
- The epiglottis closes off the glottis so that the bolus moves to the esophagus.
- Peristalsis moves the bolus through the esophagus.

Anatomy of the Stomach

- The stomach has three layers of smooth muscle in its walls, each oriented in a different direction.
- The lining of the stomach has rugae for more surface area to accommodate gastric pits that lead to gastric glands.
- Gastric pits and gastric glands are composed of five types of cells: mucus-producing cells, chief cells, parietal cells, endocrine cells, and regenerative cells.
- Parietal cells produce hydrochloric acid and intrinsic factor.
- Chief cells produce pepsinogen and gastric lipase.
- Endocrine cells produce gastrin.

Physiology of Digestion in the Stomach

- During swallowing, the medulla oblongata sends signals to the stomach, telling it to relax.
- The cardiac sphincter opens to allow the bolus to enter.
- Stretching of the stomach walls starts peristaltic contractions.
- The pyloric sphincter remains closed until the pH of the stomach contents reaches 2.
- Hydrochloric acid changes pepsinogen to pepsin so that proteins are partially digested.
- Hydrochloric acid activates lingual lipase, which partially digests lipids along with gastric lipase.
- Intrinsic factor binds to vitamin B_{12} so that it can be absorbed later.
- Once gastric secretions are mixed with the bolus, it is called *chyme*.

Anatomy of Digestive Accessory Structures

- The liver's four lobes are arranged in hepatic lobules.
- Hepatocytes produce bile that contains bile acids and lecithin, both of which aid in the chemical digestion by emulsifying lipids.
- Bile is released into hepatic ductules leading to the hepatic duct.
- The common bile duct is a tube common to the hepatic duct, the cystic duct, and the pancreatic duct.
- The hepatopancreatic sphincter controls the opening of the common bile duct to the duodenum.
- The gallbladder collects the overflow of bile from the common bile duct and concentrates it.
- The pancreas secretes bicarbonate ions and enzymes for carbohydrate, lipid, and protein digestion.

Anatomy of the Small Intestine

- The small intestine is composed of the duodenum, the jejunum, and the ileum.
- All parts of the small intestine have smooth muscle in their walls and are lined by villi.
- Endocrine cells of the duodenum secrete secretin and cholecystokinin.
- The ileocecal valve controls the movement of materials from the small intestine to the colon.

Physiology of Digestion in the Small Intestine

- Secretin is released from endocrine cells of the duodenum in response to the acidic chyme.
- Secretin tells the pancreas to release bicarbonate ions to neutralize the chyme in the duodenum.
- Cholecystokinin is secreted by endocrine cells in the duodenum in response to the presence of lipids.
- Cholecystokinin targets the gallbladder (telling it to release bile) and the hepatopancreatic duct (telling it to relax).
- The release of bicarbonate ions from the pancreas carries the digestive enzymes through the pancreatic duct to the duodenum, where all further chemical digestion is completed.
- Segmentation ensures that all the contents of the small intestine come in contact with villi for absorption.
- Peristalsis further moves the contents through the jejunum and ileum to the ileocecal valve.

Absorption of Nutrients in the Small Intestine

- Monosaccharides and amino acids are absorbed through the epithelium of the villi into capillaries by facilitated diffusion.
- Fatty acids and glycerol are absorbed across the epithelial membranes of the villi by diffusion, coated with proteins, and exocytosed to lacteals.

Anatomy of the Large Intestine

- The colon is composed of the cecum, the ascending colon, the transverse colon, the descending colon, the sigmoid colon, and the rectum.
- The anus contains two sphincter muscles: the smooth muscle internal anal sphincter, controlled by the autonomic nervous system, and the skeletal muscle external anal sphincter, controlled by the somatic nervous system.

Physiology of Digestion in the Large Intestine

- The large intestine absorbs water, compacts materials to form feces, and then stores the feces until they are removed through defecation.
- Bacteria living in the large intestine produce vitamin K and flatus.
- Stretching of the stomach and duodenum causes a mass movement of fecal material from the transverse colon to the rectum.
- Stretch receptors in the rectal walls initiate the defecation reflex.
- Defecation happens voluntarily when the external anal sphincter is relaxed.

Types of Absorbed Nutrients

- Proteins, carbohydrates, lipids, vitamins, and minerals are absorbed in the small intestine.

Circulation of Absorbed Nutrients

- The hepatic portal vein drains nutrient-rich blood from the capillaries in the villi and carries it to the capillary beds in the liver.
- The fatty acids and glycerol absorbed into lacteals in the villi join the bloodstream at the subclavian veins and eventually reach the liver through the hepatic artery.

Control of Digestion

- The autonomic nervous system controls digestion.
- Parasympathetic fibers of the vagus nerve stimulate digestion.
- Sympathetic neurons from the celiac ganglion suppress digestion in part by diverting blood to skeletal muscles and the heart.

Functions of the Digestive System

- The functions of the digestive system include ingestion, digestion, absorption, and defecation.

Effects of Aging on the Digestive System

- Tooth enamel thins, and the gingiva recede.
- The lining of the stomach atrophies.
- The liver may metabolize drugs differently.
- Movement of material through the large intestine slows with age.

Digestive System Disorders

Vomiting

- Vomiting can result from irritation anywhere along the digestive tract, and it is controlled in the medulla oblongata's emetic center.

Food Poisoning

- Food poisoning—such as staphylococcus, salmonella, and botulism—is caused by bacteria or by toxins produced by bacteria.

Parasites

- Digestive parasites—such as pinworms, tapeworms, roundworms, and giardia—live off the food passing by or eat the host.

Peptic Ulcers

- Peptic ulcers are erosions of the digestive tract lining due to an imbalance of gastric juices and the protection provided by the mucosa.

key words for review

The following terms are defined in the glossary.

alimentary canal
bolus
caries
chemical digestion
chyme
constipation
defecation
deglutition
diarrhea
emulsify
feces
flatus
ingestion
lacteals
mass movement
mastication
mechanical digestion
parasite
peristalsis
segmentation

chapter review questions

Word Building: *Use the word roots listed in the beginning of the chapter and the prefixes and suffixes inside the back cover to build words with the following meanings:*

1. Removal of the gallbladder: ______________________
2. Study of the digestive system specific to the stomach and intestine: ______________________
3. Inflammation of the liver: ______________________
4. Diagnostic tool for examining the large intestine: ______________________
5. Procedure for examining the sigmoid colon: ______________________

Multiple Select: *Select the correct choices for each statement. The choices may be all correct, all incorrect, or a combination of correct and incorrect.*

1. What happens in the stomach?
   a. Pepsin changes to pepsinogen.
   b. Lingual lipase becomes activated to digest proteins.
   c. Amylase partially digests carbohydrates.
   d. The parietal cells produce intrinsic factor to digest fats.
   e. Chemical and mechanical digestion take place to form chyme.
2. What happens in the colon?
   a. Constipation results if the colon does its job too well.
   b. The colon absorbs all nutrients of chemical digestion.
   c. Materials move through the cecum, ascending colon, descending colon, transverse colon, sigmoid colon, and rectum, in that order.
   d. The colon absorbs vitamin K.
   e. The colon absorbs water to form feces.
3. How are nutrients circulated?
   a. The liver receives blood rich in monosaccharides and amino acids from the capillaries in the small intestine through the hepatic vein.
   b. The liver receives blood rich in monosaccharides and amino acids from the capillaries in the small intestine through the hepatic portal vein.
   c. Fatty acids and glycerol travel to the heart before going to the liver.
   d. Amino acids and monosaccharides are carried in lymph before blood.
   e. Fatty acids and glycerol are carried in lymph before blood.
4. How is digestion regulated?
   a. Digestion is regulated by the autonomic nervous system.
   b. Digestion is slowed by the parasympathetic nervous system.
   c. Digestion is increased by the sympathetic nervous system.
   d. Digestion is stimulated by the parasympathetic fibers of the vagus nerve.
   e. Sympathetic neurons cause blood to be diverted from digestive organs.
5. What happens in the mouth?
   a. Enzymes from the pancreas finish protein digestion.
   b. Mucus mixes with the food to make it easier to swallow.
   c. Lingual lipase partially breaks down lipids.
   d. Amylase breaks down protein.
   e. Food is masticated using the masseter and temporalis muscles.
6. What happens during swallowing?
   a. Pharyngeal muscles push the bolus down so that it enters the esophagus.
   b. The bolus passes from the mouth to the oropharynx, to the laryngopharynx, and to the esophagus.
   c. The larynx moves up, and the glottis closes off the epiglottis.
   d. The uvula directs the bolus downward.
   e. Four cranial nerves send messages for deglutition.

7. Where does the food you eat go?
   a. It goes through the cystic duct, to the common bile duct, and then to the duodenum.
   b. It goes through the pyloric valve before the cardiac sphincter.
   c. It goes through the ileocecal valve into the large intestine.
   d. It goes through the duodenum, to the ileum, and then to the jejunum.
   e. It goes through the pancreatic duct.
8. How are nutrients absorbed?
   a. Monosaccharides are absorbed into the epithelial cells of villi through facilitated diffusion.
   b. Fatty acids and glycerol are absorbed by diffusion into the villi and are then exocytosed to lacteals.
   c. Amino acids are absorbed into capillaries in the villi.
   d. Fat-soluble vitamins are absorbed with fatty acids and glycerol.
   e. Water-soluble vitamins are absorbed through simple diffusion except for vitamin B_{12}, which must be endocytosed.
9. What can you expect to happen to your digestive system as you age?
   a. Tooth decay may increase due to thinning enamel and receding gingiva.
   b. Even though appetite is likely to be reduced, weight gain can be anticipated.
   c. Constipation is more likely.
   d. Diarrhea is more likely.
   e. Pernicious anemia is more likely due to the increased production of intrinsic factor.
10. How is chyme moved from the stomach to the duodenum?
   a. Each peristaltic wave in the stomach results in some movement of chyme to the duodenum.
   b. It moves in response to a negative-feedback mechanism.
   c. Movement is possible when the pyloric valve opens.
   d. Movement is possible when the cardiac sphincter opens.
   e. Chyme moves to the duodenum as soon as it reaches a pH greater than 2.

Matching: *Match each structure to the chemical it produces. There may be more than one correct answer.*

_____ 1. Chief cells
_____ 2. Parietal cells
_____ 3. Pancreas
_____ 4. Endocrine cells of the duodenum
_____ 5. Endocrine cells of the stomach

a. Gastrin
b. Intrinsic factor
c. Hydrochloric acid
d. Enzymes for carbohydrate digestion
e. Amylase
f. Cholecystokinin
g. Pepsin
h. Gastric lipase

Matching: *Match each parasite to its description. There may be more than one correct answer. Some answers may be used more than once.*

_____ 6. It lays eggs outside the anus.
_____ 7. It is segmented and eats what you eat.
_____ 8. Adults may cut through intestinal walls.
_____ 9. Larva travel to the lung and cause a cough.
_____ 10. It is common in lakes and streams.

a. Protozoan
b. Tapeworm
c. Roundworm
d. Pinworm

Critical Thinking

1. How would an individual who has had a cholecystectomy need to adjust his or her diet? Explain in terms of the physiology involved.
2. What would be the effects of a fat-free diet on the liver and on the absorption of nutrients?
3. The treatment for colon cancer may involve an ileostomy, which creates an opening in the abdominal wall so that the contents of the ileum may be discharged directly from the body, bypassing the colon. What would the discharge look like? How does this surgery affect the functions of this system?

13 chapter mapping

This section of the chapter is designed to help you find where each outcome is covered in this text.

	Outcomes	Readings, figures, and tables	Assessments
13.1	Use medical terminology related to the digestive system.	Word roots: p. 487	Word Building: 1–5
13.2	Differentiate between mechanical digestion and chemical digestion.	Overview: pp. 489–490	Spot Check: 1
13.3	Describe the digestive anatomy of the oral cavity.	Anatomy in the mouth: pp. 490–493 Figures 13.3–13.9	Spot Check: 2
13.4	Explain the physiology of mechanical and chemical digestion in the mouth.	Physiology of digestion in the mouth: p. 493	Spot Check: 3 Multiple Select: 5
13.5	Describe the digestive anatomy from the mouth to the stomach.	Anatomy from the mouth to the stomach: p. 494 Figure 13.10	Spot Check: 4
13.6	Explain how materials move from the mouth to the stomach.	Physiology of digestion from the mouth to the stomach: pp. 494–496 Figure 13.11	Multiple Select: 6
13.7	Describe the digestive anatomy of the stomach.	Anatomy of the stomach: pp. 496–499 Figures 13.12, 13.13 Table 13.1	Multiple Select: 7 Matching: 1, 2, 5
13.8	Explain the physiology of mechanical and chemical digestion in the stomach.	Physiology of digestion in the stomach: p. 499	Spot Check: 5 Multiple Select: 1
13.9	Explain the feedback mechanism of how food moves from the stomach to the small intestine.	Physiology of digestion in the stomach: p. 500	Spot Check: 7 Multiple Select: 10
13.10	Describe the anatomy of the digestive accessory organs connected to the duodenum by ducts.	Anatomy of the digestive accessory structures: pp. 500–504 Figures 13.14–13.17 Table 13.2	Spot Check: 6 Multiple Select: 7 Matching: 3
13.11	Describe the digestive anatomy of the small intestine.	Anatomy of the small intestine: pp. 504–506 Figures 13.18, 13.19	Multiple Select: 7 Matching: 4
13.12	Explain the physiology of chemical digestion in the duodenum, including the hormones and digestive secretions involved.	Physiology of digestion in the small intestine: pp. 506–507 Figure 13.20	Spot Check: 8 Critical Thinking: 1
13.13	Explain how nutrients are absorbed in the small intestine.	Absorption of nutrients in the small intestine: p. 508	Multiple Select: 8
13.14	Describe the anatomy of the large intestine.	Anatomy of the large intestine: pp. 508–510 Figure 13.21	Spot Check: 9
13.15	Explain the physiology of the large intestine in terms of absorption, preparation of feces, and defecation.	Physiology of the digestive system in the large intestine: pp. 510–511 Figure 13.22	Multiple Select: 2 Critical Thinking: 3

	Outcomes	Readings, figures, and tables	Assessments
13.16	Summarize the types of nutrients absorbed by the digestive system from the diet.	Types of absorbed nutrients: pp. 511–512	Multiple Select: 8 Critical Thinking: 2
13.17	Trace the circulation of the nutrients once they have been absorbed.	Circulation of absorbed nutrients: p. 512 Figure 13.23	Multiple Select: 3
13.18	Explain the control of digestion.	Control of digestion: p. 513	Multiple Select: 4
13.19	Summarize the functions of digestion.	Functions of the digestive system: pp. 513–514 Figure 13.24	Critical Thinking: 3
13.20	Summarize the effects of aging on the digestive system.	Effects of aging on the digestive system: p. 514	Multiple Select: 9
13.21	Describe digestive system disorders, including vomiting, food poisoning, parasites, and peptic ulcers.	Digestive system disorders: pp. 514–517 Figures 13.25–13.27	Matching: 6–10

references and resources

Beers, M. H., Porter, R. S., Jones, T. V., Kaplan, J. L., & Berkwits, M. (Eds.). (2006). *Merck manual of diagnosis and therapy* (18th ed.). Whitehouse Station, NJ: Merck Research Laboratories.

Centers for Disease Control and Prevention. (2008, September 3). *Pinworm infection*. Retrieved October 4, 2010, from http://www.cdc.gov/ncidod/dpd/parasites/pinworm/factsht_pinworm.htm.

Porter, R. S., & Kaplan, J. L. (Eds.). (n.d.). Selected physiologic age-related changes. *Merck manual for healthcare professionals*. Retrieved March 19, 2011, from http://www.merckmanuals.com/media/professional/pdf/Table_337-1.pdf?qt=selected physiologic age-related changes&alt=sh.

Saladin, Kenneth S. (2010). *Anatomy & physiology: The unity of form and function* (5th ed.). New York: McGraw-Hill.

Seeley, R. R., Stephens, T. D., & Tate, P. (2006). *Anatomy & physiology* (7th ed.). New York: McGraw-Hill.

Shaheen, N. J. (2006, March). Effects of aging: Biology of the digestive system. *Merck manual home edition*. Retrieved March 17, 2011, from http://www.merckmanuals.com/home/sec09/ch118/ch118j.html.

Shier, D., Butler, J., & Lewis, R. (2010). *Hole's human anatomy & physiology* (12th ed.). New York: McGraw-Hill.

14 The Excretory/Urinary System

As you well know, no matter how much you reduce, reuse, and recycle, you still generate waste in your home. You also know that you must have a system for removing this waste on a regular basis if your household is going to function properly. While the waste hauler in your area removes the waste you place in trash bags, your septic system or city sewer system handles the waste you flush down the drain. Your body handles waste in a similar fashion. For example, the liver and spleen recycle the iron and amino acids when they break down a worn-out red blood cell's hemoglobin, but the bilirubin produced in the process is a metabolic waste that must be completely removed from the body. Bilirubin is carried away in feces during defecation and is flushed out of the body with urine. The excretory system's job is to remove bilirubin and the rest of the body's metabolic wastes. See Figure 14.1.

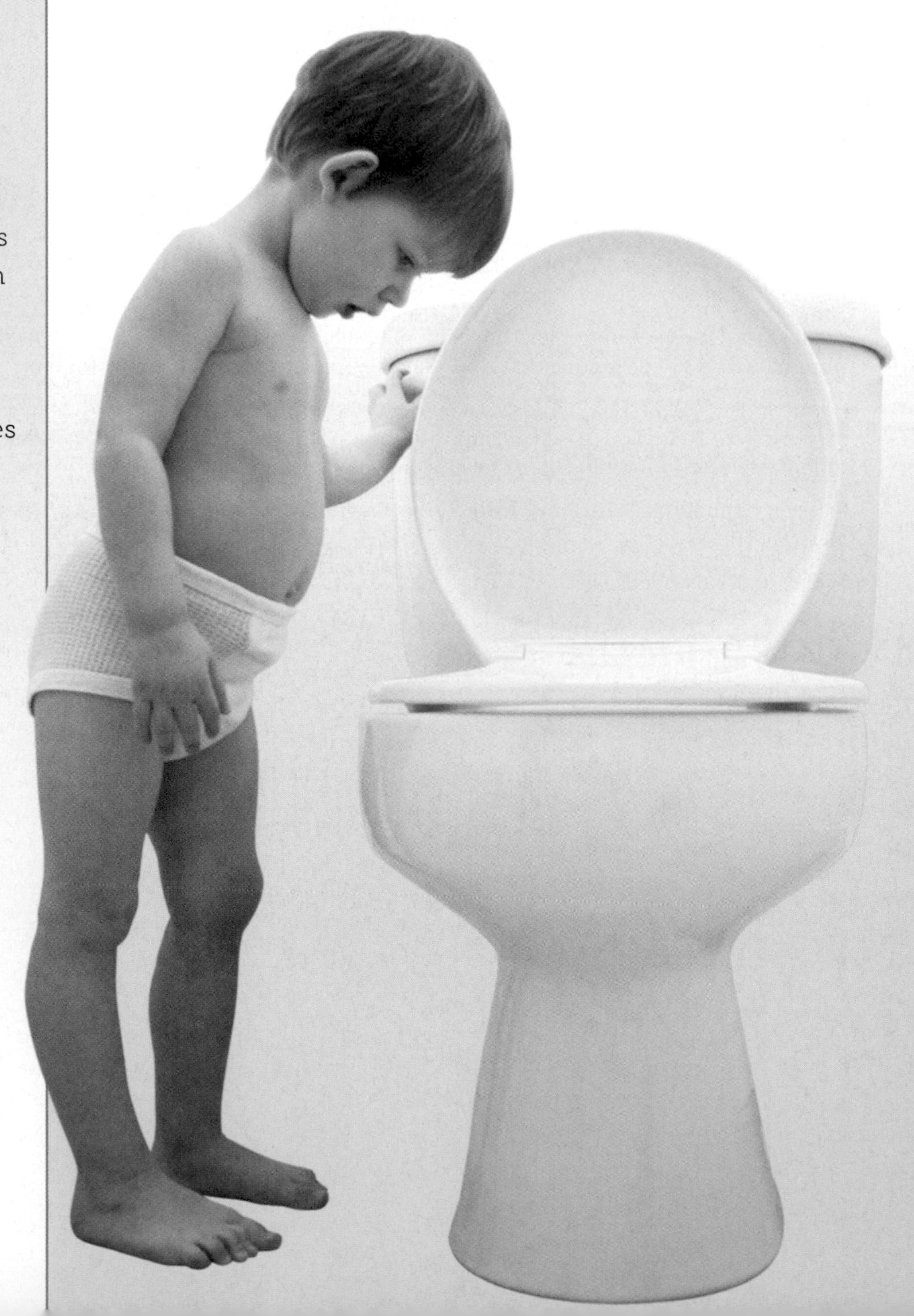

14.1 learning outcome

Use medical terminology related to the excretory system.

learning outcomes

After completing this chapter, you should be able to:

14.1 Use medical terminology related to the excretory system.

14.2 Define *excretion* and identify the organs that excrete waste.

14.3 List the body's major nitrogenous wastes and their sources.

14.4 List the functions of the kidneys in addition to urine production.

14.5 Describe the external and internal anatomy of the kidneys.

14.6 Describe the anatomy of a nephron.

14.7 Trace the components of urine through a nephron.

14.8 Trace the flow of blood through a nephron.

14.9 Describe filtration, reabsorption, and secretion in the kidneys with regard to the products moving in each process, the direction of movement, and the method of movement.

14.10 Describe the fluid compartments of the body and how water moves between them.

14.11 Explain how urine volume and concentration are regulated.

14.12 Explain how diuretics, such as medications, caffeine, and alcohol, affect urine production.

14.13 Describe the anatomy of the ureters, urinary bladder, and male and female urethras.

14.14 Describe the micturition reflex and explain how the nervous system and urinary sphincters control the voiding of urine.

14.15 Summarize the functions of the excretory system.

14.16 Summarize the effects of aging on the excretory system.

14.17 Describe excretory system disorders.

word roots & combining forms

azot/o: nitrogen

cyst/o: urinary bladder

glomerul/o: glomerulus

nephr/o: kidney

pyel/o: renal pelvis

ren/o: kidney

ur/o: urinary tract, urine

ureter/o: ureter

urethr/o: urethra

pronunciation key

calyces: KAY-lih-seez

calyx: KAY-licks

detrusor: de-TRUE-sor

glomerulus: glo-MER-you-lus

juxtaglomerular: JUX-ta-glo-MER-you-lar

micturition: mik-choo-RISH-un

nephron: NEF-ron

renin: REE-nin

trigone: TRY-gon

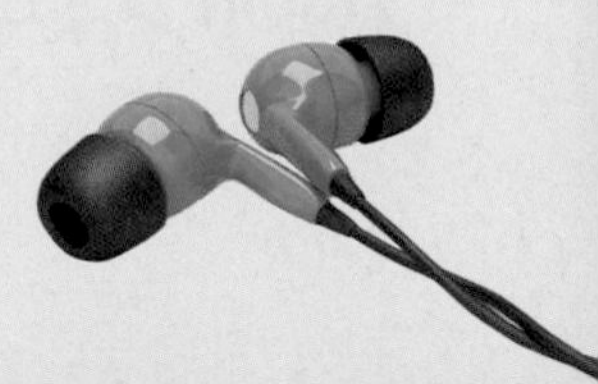

14.2 learning **outcome**

Define *excretion* and identify the organs that excrete waste.

Overview

The excretory system removes the body's metabolic wastes through a process called **excretion.** Excretion of wastes is different from defecation, which you studied in the digestive system chapter. Although you may think of defecation as removing waste, the feces removed from the body through defecation are mostly undigested materials and bacteria that were not previously present in your body's cells. An exception to this is bilirubin, the waste produced from the breakdown of hemoglobin in the liver and spleen. **Metabolic wastes** are wastes produced by the cells—bilirubin is a good example.

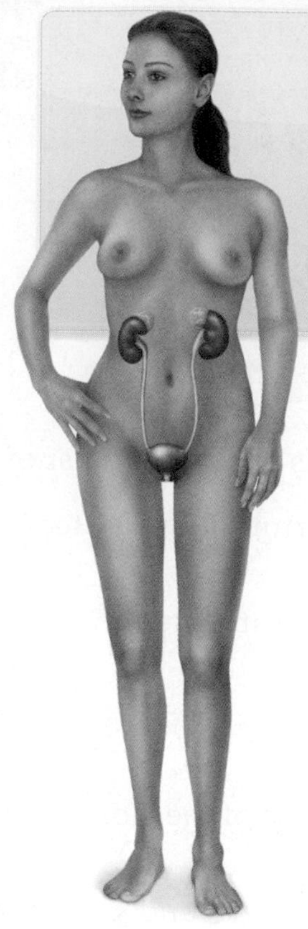

Excretory/Urinary System

Major Organs and Structures:
kidney, ureters, urinary bladder, urethra

Accessory Structures:
lungs, skin, liver

Functions:
removal of metabolic wastes, fluid and electrolyte balance, acid-base balance, blood pressure regulation

FIGURE 14.1 The excretory/urinary system.

Several organs of your body—the skin, lungs, liver, and kidneys—excrete metabolic wastes. See **Figure 14.2**. These organs and the excretions they are responsible for carrying out are explained in the following list:

- The skin removes some salts, lactic acid, and urea with sweat.
- The lungs remove carbon dioxide with every humidified, expired breath.
- The liver removes bilirubin by putting it in bile.
- The kidneys remove **nitrogenous wastes** (wastes containing nitrogen), excess minerals, bilirubin, and excess hydrogen ions by producing urine.

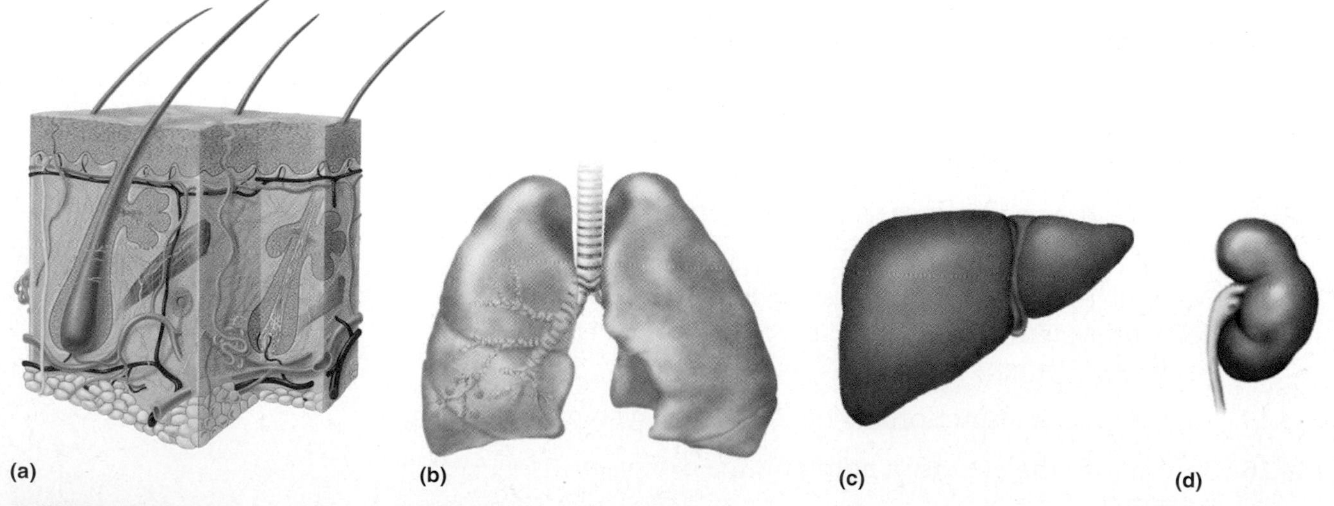

FIGURE 14.2 Organs of the excretory system: (a) skin, (b) lungs, (c) liver, (d) kidney.

As you can tell from the list on page 528, water is often used to eliminate metabolic wastes. Sweat, humidified air, bile, and urine all contain water. Water conservation and its use in excretion are addressed later in the chapter.

14.3 learning **outcome**

List the body's major nitrogenous wastes and their sources.

Kidneys are the primary organs of this system. The excretory system is sometimes referred to as the *urinary system* when the focus is on only the kidneys and their urine production. The nitrogenous wastes removed by the kidneys can be lethal to the body if they are allowed to accumulate in the blood in excessive amounts. These wastes are described in the list below, and their chemical structures are shown in **Figure 14.3**.

- **Ammonia** is produced from the breakdown of amino acids. It is extremely toxic, but it is quickly converted by the liver to urea, a less toxic waste.
- **Urea** is the most common nitrogenous waste produced in the body, accounting for 50 percent of that waste. It is ultimately formed from the breakdown of proteins.
- **Uric acid** is formed from the breakdown of nucleic acids.
- **Creatinine** is formed from the breakdown of creatine phosphate, a stable energy storage molecule (discussed in Chapter 5 under "Muscle Metabolism").

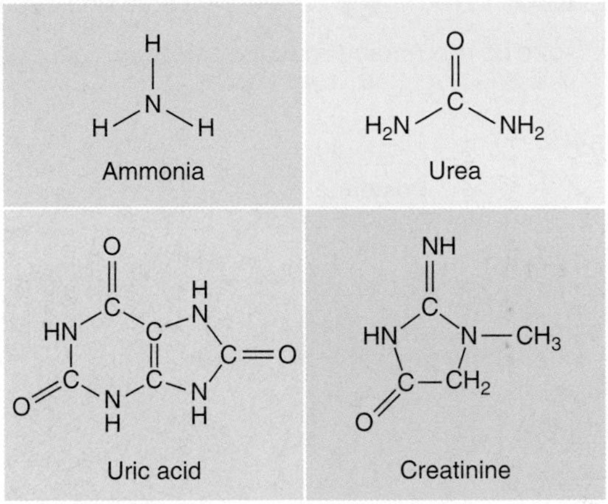

FIGURE 14.3 Nitrogenous wastes. Notice that nitrogen is part of the chemical structure for each waste.

14.4 learning **outcome**

List the functions of the kidneys in addition to urine production.

You have read about the kidneys in earlier chapters because they function in multiple systems. You may recall that the kidneys are important in the body's homeostasis of calcium through their role in vitamin D synthesis (Chapters 3 and 4). They are also important to the cardiovascular system in several ways. For example, the kidneys produce erythropoietin to stimulate red blood cell production when blood oxygen levels are low (Chapter 9). The kidneys also help to regulate blood volume, blood pressure, and the blood's concentration of solutes by adjusting the amount of water they use to produce urine (Chapter 10 and later in this chapter).

Throughout the rest of this chapter, you will explore the following anatomy and physiology of kidneys and the urinary system:

- The kidneys' excretion of wastes in urine.
- The kidneys' role in the regulation of blood volume and pressure by the formation of urine.
- The delivery of urine from the body.
- The way urine production is controlled.

In addition to studying the anatomy and physiology of this system, you will also learn about how aging affects the excretory system and what can go wrong when the anatomy and physiology do not work. You will begin by exploring the anatomy of the kidney.

14.5 learning outcome

Describe the external and internal anatomy of the kidneys.

The Anatomy of the Kidney

The kidneys are dark red, bean-shaped organs about the size of a tightly clenched fist. Like the pancreas, the kidneys are retroperitoneal (posterior to the parietal peritoneum). See **Figure 14.4**. The kidneys extend from T11 to L3 (vertebrae) on each side of the vertebral column and are somewhat protected superiorly by the ribs. The right kidney is slightly lower than the left due to the position of the liver. As **Figure 14.4** shows, a fibrous **renal capsule,** surrounded by adipose tissue **(perirenal fat capsule),** also protects the kidney. This fatty pad absorbs the mechanical shock to the kidney that may occur with a fall. **Renal fascia** (a connective tissue covering) anchors the kidney to the posterior muscle wall of the body's abdomen.

spot check ❶ Once the abdominal organs have been pushed aside, what specific membrane must be pierced to access the kidney?

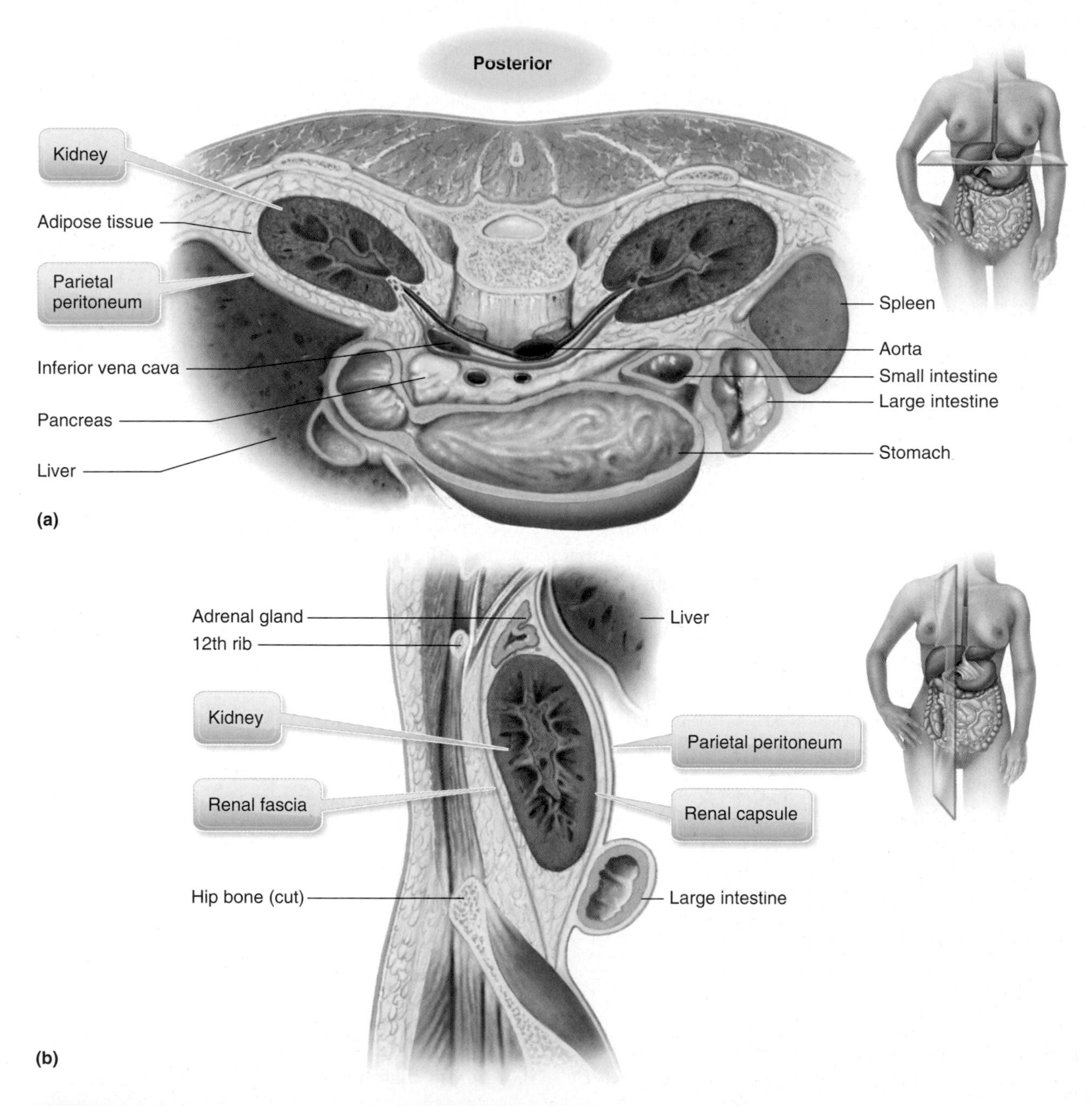

FIGURE 14.4 Retroperitoneal position of the kidney: (a) transverse section through the torso at the level of the kidney, (b) sagittal cut through the torso at the right kidney.

Figure 14.5 shows the kidney's external and internal anatomy. In the figure, you can see a notch on the medial surface of the kidneys. This notch in a kidney is called the *renal hilum*. As with the hilum in the lung, all structures entering or leaving the kidney do so at the hilum. In this figure, you can see that the **renal artery** enters the kidney, while the **renal vein** and the **renal pelvis** leading to the **ureter** exit the kidney. If you were to grab hold and remove the renal pelvis, renal artery, and renal vein from the kidney, you would be left with the space they occupied. This space is the **renal sinus.** Adipose tissue fills whatever space is available in the renal sinus.

Look at **Figure 14.5** (a and b) closely as you explore the internal anatomy of a kidney. Each kidney has three layers: the thin, outer fibrous renal capsule; a layer deep to the capsule called the **renal cortex;** and the inner **renal medulla.** In the diagram, the renal medulla appears to be composed of triangles called *pyramids*. The **renal pyramids** are actually three-dimensional cones, each leading to a

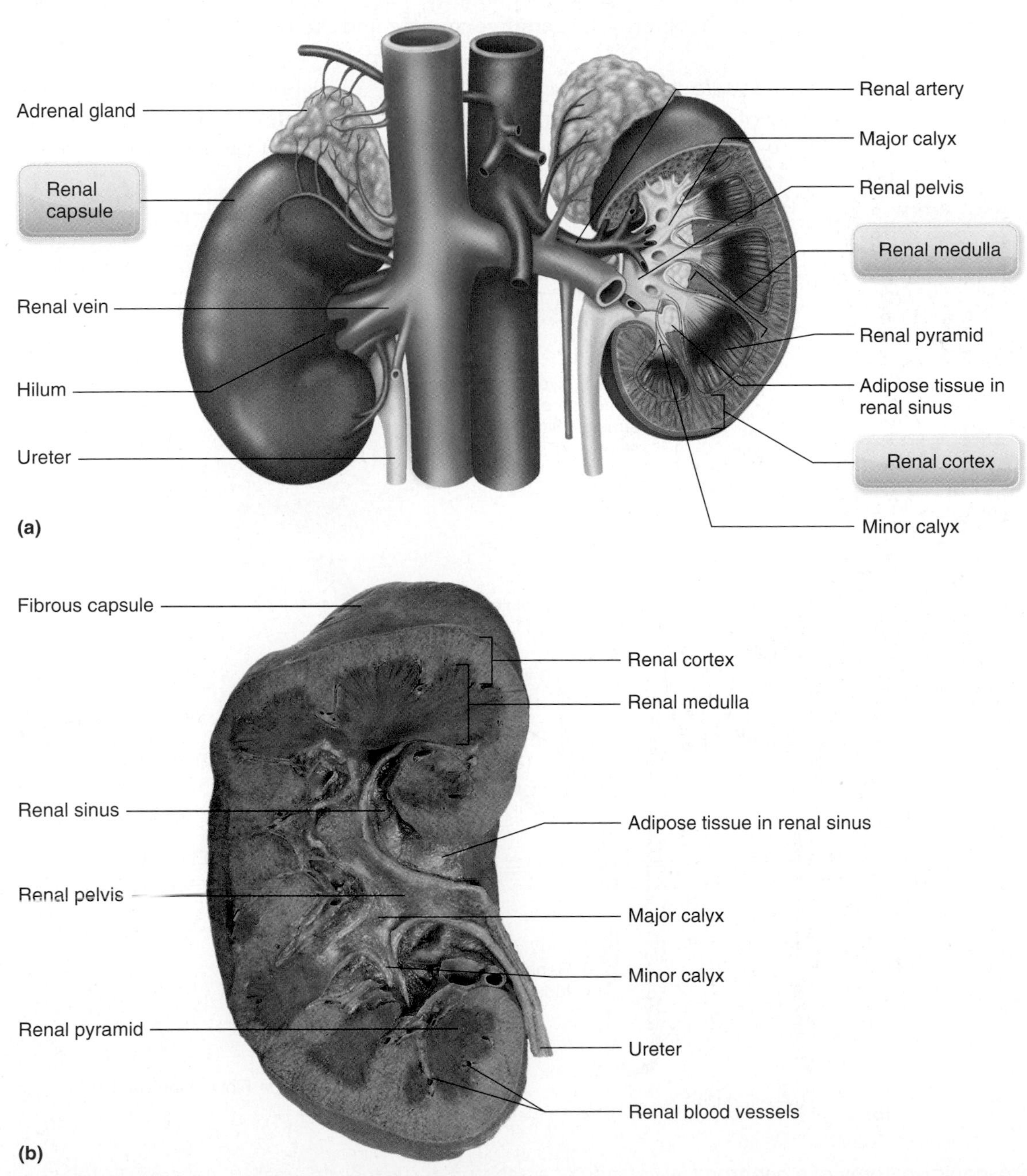

FIGURE 14.5 **Gross anatomy of the kidney:** (a) major anatomical features of the kidneys, (b) coronal section of a cadaver kidney.

calyx: KAY-licks
calyces: KAY-lih-seez
nephron: NEF-ron

connect.mcgraw-hill.com audio

funnel-like structure called a **minor calyx.** Two or more minor **calyces** may merge to form a **major calyx,** which empties into the renal pelvis. The calyces look and act like funnels that collect urine and deliver it to the renal pelvis. You should also note the many blood vessels that are present in this frontal section of the kidney. The kidney has and needs a very rich blood supply to function properly.

All of the structures you have read about so far in this chapter can be seen with a naked eye. However, the functional unit of the kidney—a **nephron**—is microscopic. We explain the anatomy of a nephron next.

14.6 learning **outcome**

Describe the anatomy of a nephron.

Anatomy of a Nephron

Each kidney contains over 1 million nephrons. These structures produce urine. The anatomy of a nephron appears fairly complicated when seen in total. However, **Figure 14.6**, which shows the location of a nephron's anatomy in a kidney, simplifies its arrangement so that you can understand its components. There are two principal parts to a nephron—the **renal corpuscle** and the **renal tubule.**

Renal Corpuscle The renal corpuscle is like an elaborate filter in a cup. It is composed of a **glomerulus** and a **glomerular capsule (Bowman's capsule).** See **Figure 14.6c**. An **afferent arteriole** delivers blood to a capillary bed called the

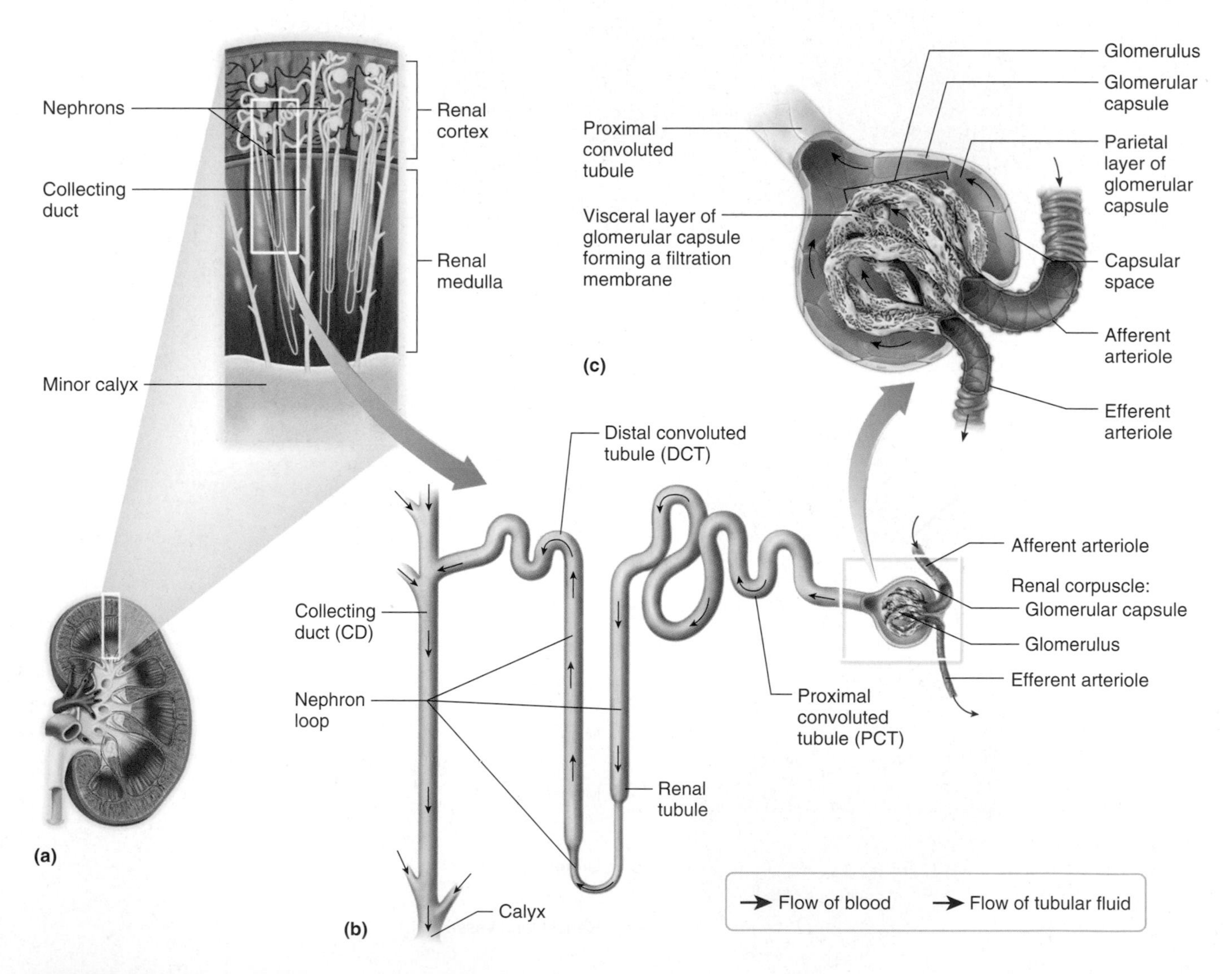

FIGURE 14.6 Microscopic anatomy of a nephron: (a) location of a nephron in a kidney, (b) anatomy of a nephron stretched out so that the parts can be more easily seen, (c) the renal corpuscle.

glomerulus (the filter) inside the glomerular capsule (the walls of the cup). Cells of the capsule extend over each of the capillaries in the glomerulus, forming a filtration membrane. Whatever is filtered out of the blood through this membrane is caught in the glomerular capsule space and delivered to the next part of the nephron—the renal tubule. Meanwhile, the blood in the glomerular capillaries continues on its journey out of the renal corpuscle through the **efferent arteriole.**

glomerulus: glo-MER-you-lus

study **hint**

You have seen the words *afferent* and *efferent* before, when they were used in describing the direction of nerve impulses. The direction is important in the nervous system and here again in the excretory system. Just remember *E* stands for *exit. Efferent* nerve impulses *exit* the CNS, and *efferent* arterioles *exit* the glomerulus.

Renal Tubule A tubule is simply a hollow tube. The renal tubule can be divided into three sections on the basis of anatomy and location—the **proximal convoluted tubule (PCT),** the **nephron loop (loop of Henle),** and the **distal convoluted tubule (DCT).** The walls of the tubule are simple epithelia that allow for the exchange of materials. What flows through the tubules eventually becomes urine. **Figure 14.6b** shows a nephron's renal tubule stretched out so that you can easily see each of the sections and the direction of flow for the components of urine. The proximal convoluted tubule is directly connected to the glomerular capsule. It twists and turns (convolutes) before descending to form the nephron loop in the renal pyramid. The renal tubule then ascends out of the renal pyramid to form the distal convoluted tubule in the renal cortex. Several distal convoluted tubules connect to a shared **collecting duct** that empties at the very end of the renal pyramid into a minor calyx. When you examine the nephron in **Figure 14.6a**, you can see that the renal corpuscle, the proximal convoluted tubule, and the distal convoluted tubule are located in the renal cortex, while the nephron loop and the collecting duct are located in the renal pyramid of the renal medulla.

study **hint**

The word *proximal* refers to the position of the tubule relative to the glomerular capsule. The proximal convoluted tubule is *directly* attached to the glomerular capsule. Although part of the distal convoluted tubule rests right next to the glomerular capsule (as you will see shortly), it is termed *distal* because it is *not* directly attached to the glomerular capsule. In fact, it is the third section of tubule away (*dist*ant) from the glomerular capsule.

Unlike **Figure 14.6**, which shows a nephron stretched out, **Figure 14.7** shows a nephron as it occurs in the kidney. You can see in this close-up of the renal corpuscle that the distal convoluted tubule is very close to the afferent and efferent arterioles as they enter and leave the glomerular capsule. Here, the epithelial cells of the distal convoluted tubule are very close together, forming a structure called the **macula densa.** You can also see specialized smooth muscle cells **(juxtaglomerular cells)** surrounding the afferent arteriole. Together, the juxtaglomerular cells and the macula densa make up a structure called the **juxtaglomerular apparatus.** This structure is discussed later in the chapter in regard to the control of urine production, but now it is time to get back to the structures involved in producing urine.

juxtaglomerular: JUX-ta-glo-MER-you-lar

spot check  List the parts of a nephron in order.

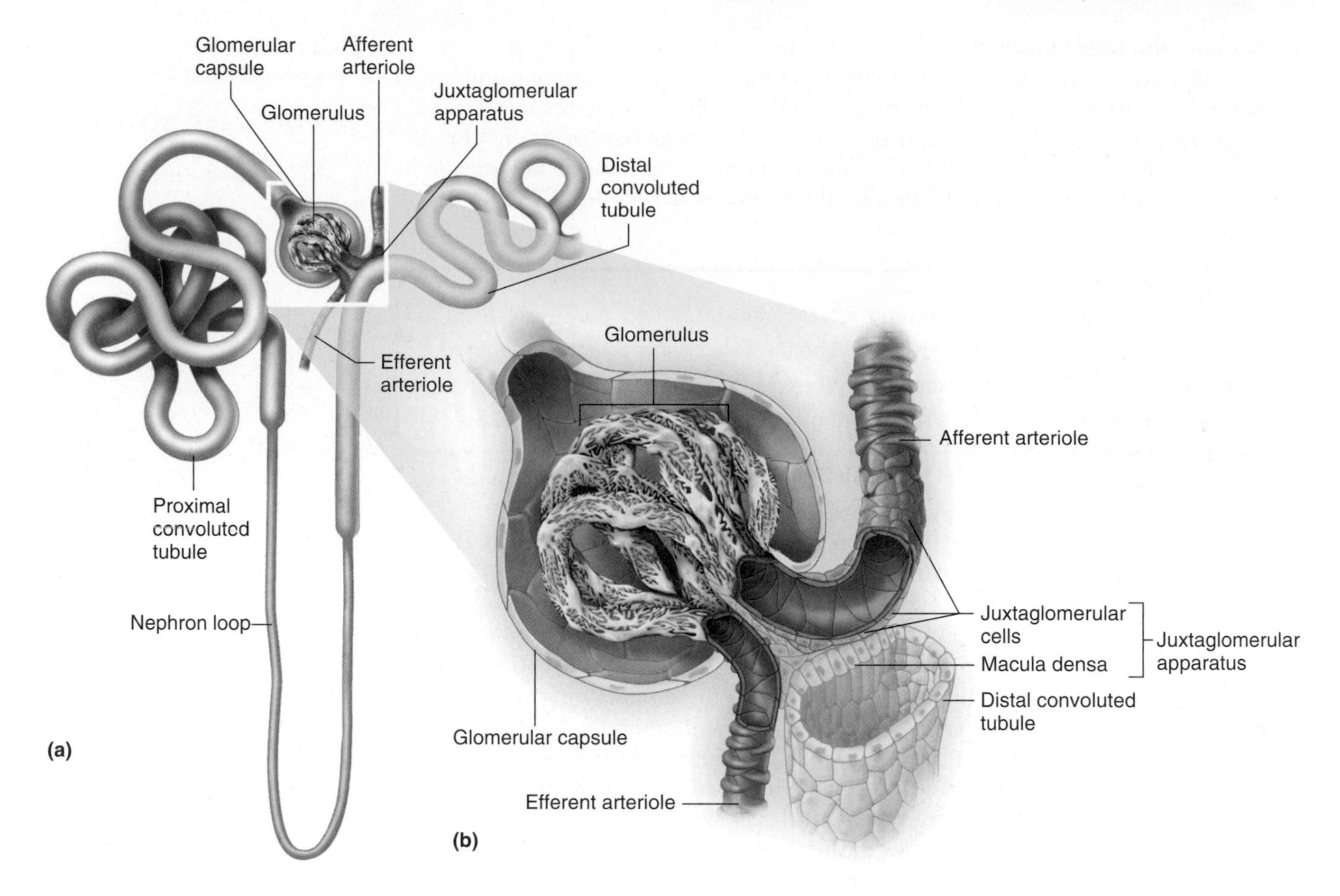

FIGURE 14.7 Nephron: (a) as it occurs in a kidney, (b) juxtaglomerular apparatus.

14.7 learning **outcome**

Trace the components of urine through a nephron.

Flow of Urine Components through a Nephron

The glomerular capsule catches whatever is removed from the blood in the glomerulus. This material collected by the glomerular capsule is called **filtrate.** Later in the chapter, you will explore how filtrate is refined along the way to become the urine that eventually leaves the body. For now, it is important for you to understand the direction of flow through the nephron. This flow is clearly shown in **Figure 14.6b**. Materials flow in this direction:

Glomerular capsule → PCT → nephron loop → DCT → collecting duct → minor calyx

It is time to look at a nephron with the blood flow surrounding it. **Figure 14.8** shows a nephron with two networks of capillaries. This complete figure appears complicated because it does not have the nephron and its tubules stretched out, but much of what it depicts is the same as the explanation you have just read. For example, notice that the renal corpuscle and the convoluted tubules are in the renal cortex, while the nephron loop and collecting duct extend down into the renal medulla. This figure also shows the afferent arteriole leading to the glomerulus and the efferent arteriole exiting out of the glomerular capsule. What is new in this figure is that the efferent arteriole leads to a complex capillary bed surrounding the renal tubule—the **peritubular capillaries.**

The two capillary beds—the glomerulus and the peritubular capillaries—form a portal system so that materials can be exchanged twice between the nephron and the blood in the capillaries before the blood exits the kidney. In the next section, you will trace the blood flow associated with a nephron.

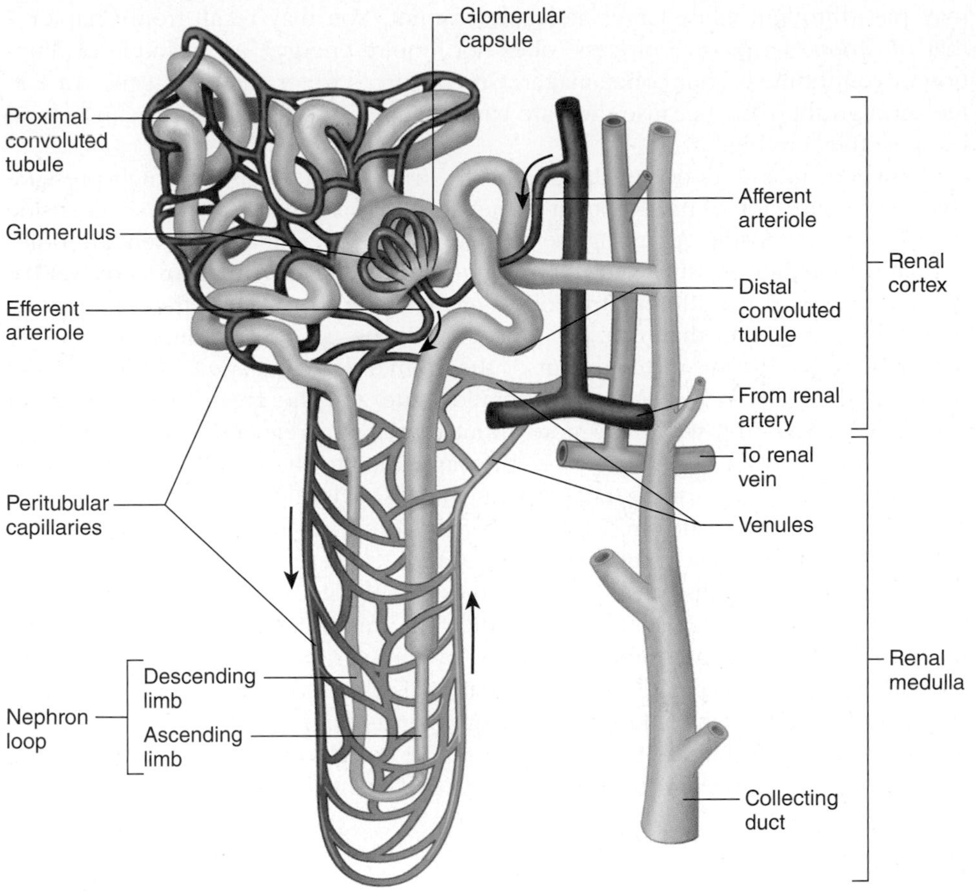

FIGURE 14.8 The renal nephron and the associated blood vessels.

Blood Flow to a Nephron

14.8 learning outcome

Trace the flow of blood through a nephron.

Follow along with **Figure 14.8** as you read this section. Blood enters the kidney through the renal artery. It travels through smaller and smaller arteries leading to the afferent arteriole, which feeds the glomerulus. From the glomerulus, blood flows out the efferent arteriole to the peritubular capillaries, which then feed into venules, to larger and larger veins, and, finally, to the renal vein that exits the kidney.

Through a series of processes, materials are exchanged between capillaries and nephrons in both directions. Now that you have studied the nephron's anatomy and its associated blood vessels, you are ready to explore the processes of urine formation.

spot check  Where does blood go after leaving the efferent arteriole?

Physiology of Urine Production

14.9 learning outcome

Describe filtration, reabsorption, and secretion in the kidneys with regard to the products moving in each process, the direction of movement, and the method of movement.

Urine production involves three processes: filtration, reabsorption, and secretion. For each process, it is important to note what materials are moving and the direction of the movement. Are the materials moving from the capillaries to the tubules or from the tubules to the capillaries? Below, you will start with the first process—filtration.

Filtration

Filtration occurs between the glomerulus and the glomerular capsule of the renal corpuscle. The thin capillary walls of the glomerulus and the cells that cover them act as a filter that allows materials to cross, depending on size—small molecules may pass through, while larger molecules cannot. You may recall from Chapter 2 that *filtration* is a passive process (does not require energy). An example of filtration in your home is your coffee maker: Gravity forces water and the essence of coffee through the filter because they are small, while coffee grounds remain behind because they are big.

However, gravity does not drive filtration in the kidney. Instead, high pressure (blood pressure) forces materials out of the glomerular capillaries to the space inside the glomerular capsule. Adjusting the diameter of the afferent and efferent arterioles regulates this high pressure and ultimately the **glomerular filtration rate (GFR).** If the diameter of the afferent arteriole is greater than that of the efferent arteriole, more blood can enter than can leave the glomerulus. This causes high pressure in the glomerulus, forcing materials out of the glomerular capillaries. The higher the pressure, the greater the glomerular filtration rate and the greater the amount of materials filtered. You will learn more about how this is regulated later. The direction of movement in filtration is from the capillaries to the tubules. The materials moved are water, some nitrogenous wastes, amino acids, glucose, and mineral salts, such as sodium and calcium. These materials comprise the beginnings of urine. Blood cells and proteins do not filter out because they are too big.

If you look at the list of filtered materials closely, you may wonder why these materials are filtered out of the blood. After all, amino acids and glucose are the products of digestion, and they are needed by the cells to build proteins and carry out cellular respiration. The point of filtration is that it is simply based on size. Filtration does filter out some small nitrogenous wastes, but at the same time it also filters out materials that the body should keep in the blood. So a second process—reabsorption—is needed to recapture the materials that should stay in the blood.

Reabsorption

Reabsorption begins between the proximal convoluted tubule and the peritubular capillaries and continues along the renal tubule. In this process, 100 percent of the glucose, 100 percent of the amino acids, and variable amounts of the mineral salts that were filtered out of the blood are *actively transported* (requiring energy) from the tubules to the capillaries. In addition, 99 percent of the water that was filtered into the glomerular capsule is reabsorbed by *osmosis* into the bloodstream.

clinical point

Blood glucose levels are higher than normal in uncontrolled diabetes mellitus. Therefore, more glucose is filtered out of the blood in the glomerulus. However, reabsorption is time-limited: It can occur only while the filtrate is flowing through the renal tubule. If there is more glucose in the filtrate than can be reabsorbed in that amount of time, not all of the glucose will be reabsorbed. So some of the excess glucose will be found in the urine excreted from the body. Although this does bring down the abnormally high blood glucose levels, the levels will rise again with the next meal. This is not a homeostasis mechanism; it is a sign that the body is not using glucose properly.

spot check 4 Could reabsorption happen in the renal corpuscle before filtration? Explain.

At this point in urine production (after reabsorption), the tubules contain some nitrogenous wastes, some mineral salts, and 1 percent of the water that was filtered. The rest of the filtrate has been returned to the blood. The third process of urine production—secretion—completes the process of removing wastes from the blood.

Secretion

In **secretion,** the nephron removes the rest of the wastes that remain in the blood. In this process, materials move from the peritubular capillaries to the tubules. These materials include the rest of the nitrogenous wastes (those that could not be filtered because of their size), excess hydrogen ions, and excess potassium.

Removing excess hydrogen ions is crucial if the blood's pH is to remain in the homeostatic range of 7.35 to 7.45. Acidosis results if the blood's pH falls below this range, and alkalosis results if blood pH rises above this range. Both conditions are potentially lethal. Acidosis may first be seen as disorientation that may lead to coma. Alkalosis may start with hyperexcitability of the PNS along with spontaneous stimulation of muscle contractions and continue to spasms, convulsions, and possible death. What causes these acid-base imbalances, and how are they fixed?

The two forms of acidosis based on cause are described in the following list:

- **Respiratory acidosis** happens if the respiratory system cannot eliminate sufficient CO_2. For example, a patient with emphysema may not have sufficient ventilation of the alveoli in his lungs due to the breakdown of his alveolar walls. So his respiratory system cannot eliminate enough CO_2 to keep up with the CO_2 produced from cellular respiration in his tissues.
- **Metabolic acidosis** happens if there is decreased kidney elimination of hydrogen ions or increased production of acidic substances through metabolism. For example, anaerobic respiration results in lactic acid buildup in the muscles, which makes its way into the bloodstream. Metabolic acidosis can also occur in diabetics who have poor control of their blood sugars. You have already learned that diabetes mellitus is a wasting disease in which the body may need to break down fats and proteins for energy because it cannot use the sugar in the blood. This type of metabolism produces acidic ketones, and their presence in the blood lowers the pH.

Whatever the cause, whenever the body goes into acidosis, both the respiratory and the excretory systems will try to fix the imbalance. The respiratory system increases the respiratory rate (hyperventilation), and the excretory system increases secretion of the excess hydrogen ions to bring the pH of the blood back to homeostasis. See **Figure 14.9**.

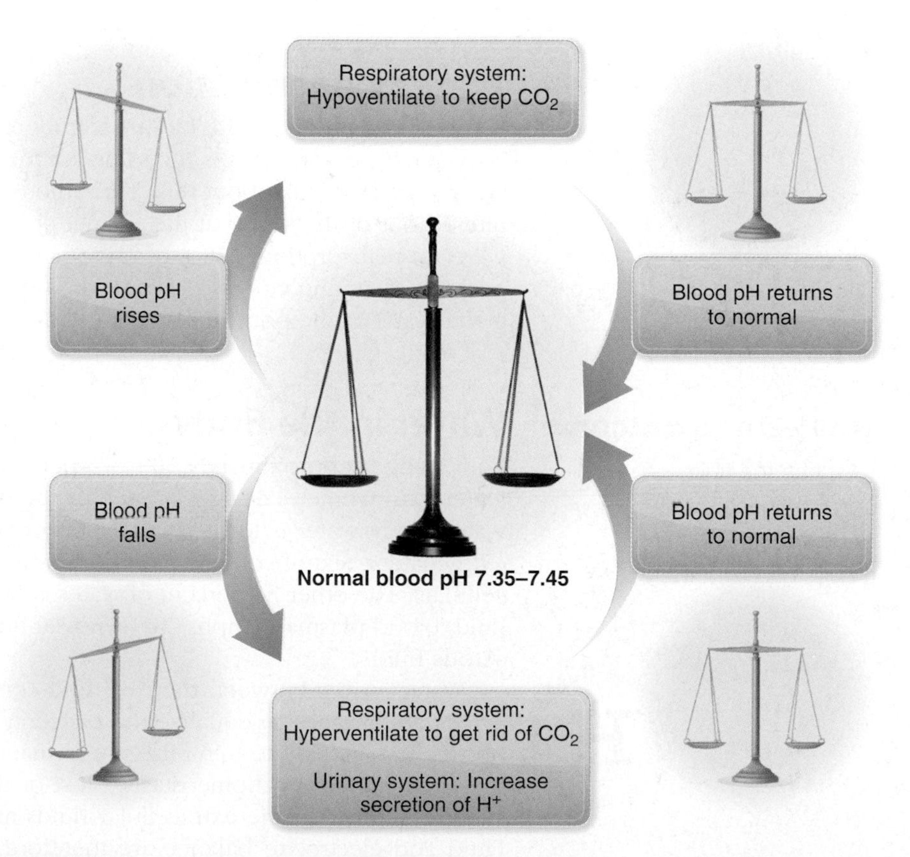

FIGURE 14.9 Homeostasis of blood pH.

Just as there are two forms of acidosis, there are also two forms of alkalosis based on cause:

- **Respiratory alkalosis** occurs during hyperventilation. In this case, too much CO_2 is being blown off.
- **Metabolic alkalosis** is relatively rare, but it can occur if there is prolonged vomiting, which results in the repeated loss of stomach acids.

Whatever the cause, whenever the body goes into alkalosis, the respiratory system will try to fix it by reducing the respiratory rate. This keeps more CO_2 in the blood. The kidneys can raise the pH of the blood by secreting excess H^+ during secretion, but they cannot add more H^+ to the blood to lower the pH. The kidneys can only manage whatever amount of H^+ is in the blood in the first place. They cannot add to it. See **Figure 14.9**.

clinical **point**

Many drugs are also cleared from the blood by secretion in the kidneys. Drug dosages are often determined to keep up with this clearance in the kidneys.

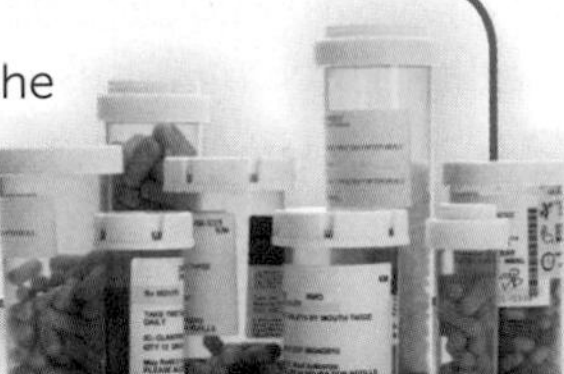

Figure 14.10 summarizes the urine production processes. Through these three processes—filtration, reabsorption, and secretion—urine production has cleared wastes, excess hydrogen ions, and excess potassium from the blood and can now be considered complete. Yet further reabsorption of water may take place in the distal convoluted tubule and the collecting duct for water conservation. We explain this further next.

Water Conservation

Water conservation by the kidney is truly remarkable. Approximately 1.5 L of urine is excreted per day. This is the same amount as three-fourths of a 2-L bottle of your favorite soda. Think about this: This amount represents only 1 percent of the water filtered out of the blood at the glomeruli because the rest was reabsorbed. If there were no reabsorption, 150 L of urine (seventy-five 2-L bottles) would be produced per day, assuming you would have the time to take in that much water. Just where is the water in the body, and where does it come from?

14.10 learning **outcome**

Describe the fluid compartments of the body and how water moves between them.

Water in the Body

Your body is approximately 50 to 75 percent water. Men tend to have slightly more water than women because women deposit more fat, which does not contain much water. Body water is located in two major **fluid compartments**—intracellular and extracellular. Sixty-five percent of body water is in the cytoplasm of your cells (intracellular). The other 35 percent of water is outside your cells (extracellular), as tissue fluid, blood plasma, lymph, CSF, synovial fluid, fluids of the eye (humors), bile, and serous fluid.

Water moves between the two fluid compartments by osmosis, traveling easily across membranes to equalize the concentration of solutes on both sides. Osmosis occurs very quickly to minimize the formation of concentration gradients of solutes in order to maintain homeostasis. Most of the solutes in the fluids are electrolytes, such as sodium in the extracellular fluids and potassium in the intracellular fluids. Fluid and electrolyte balance are therefore tied together. **Figure 14.11** shows the movement of water between the major fluid compartments.

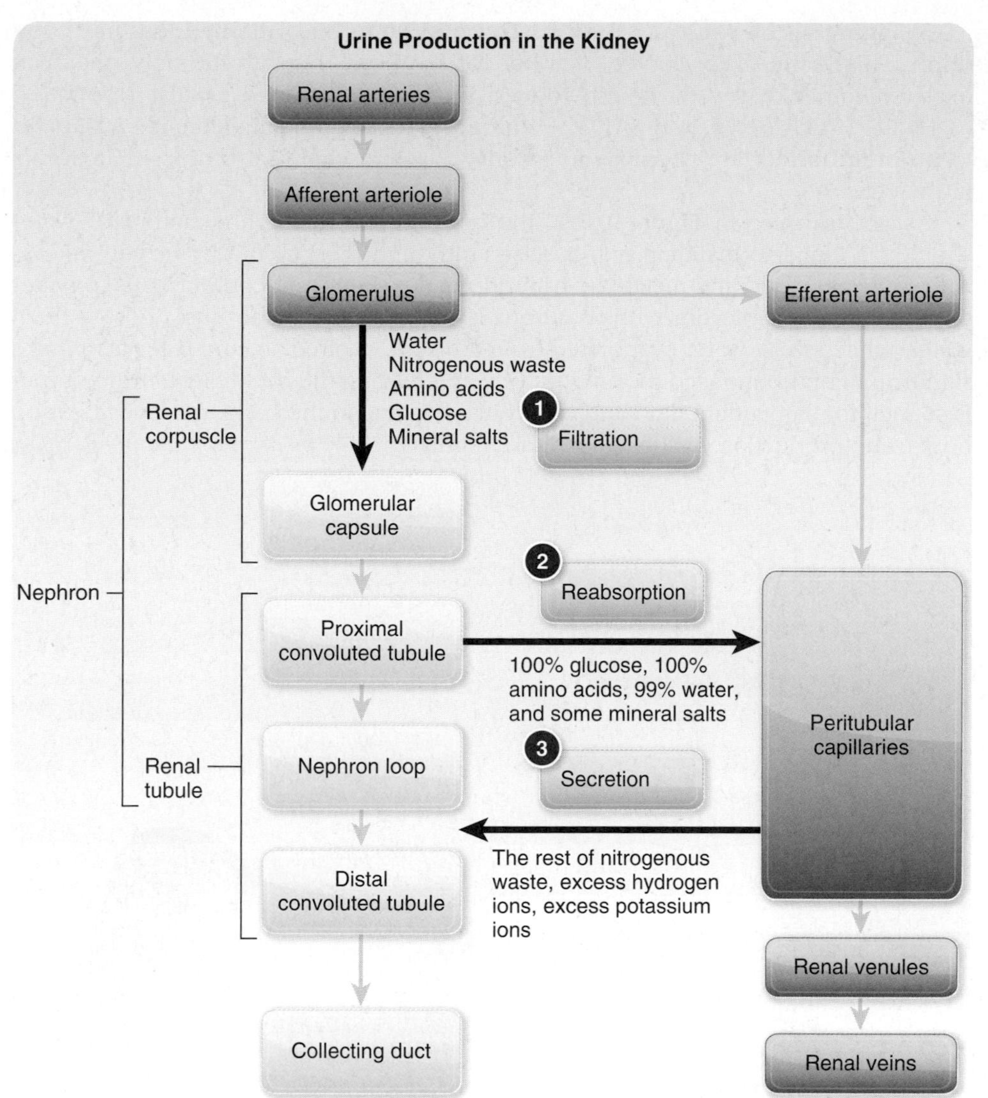

FIGURE 14.10 **The processes of urine production:** (1) filtration, (2) reabsorption, (3) secretion. This figure also shows the pathway for the components of urine (yellow), as well as blood flow associated with a nephron (red). Bold black arrows show the direction materials are moving during each process.

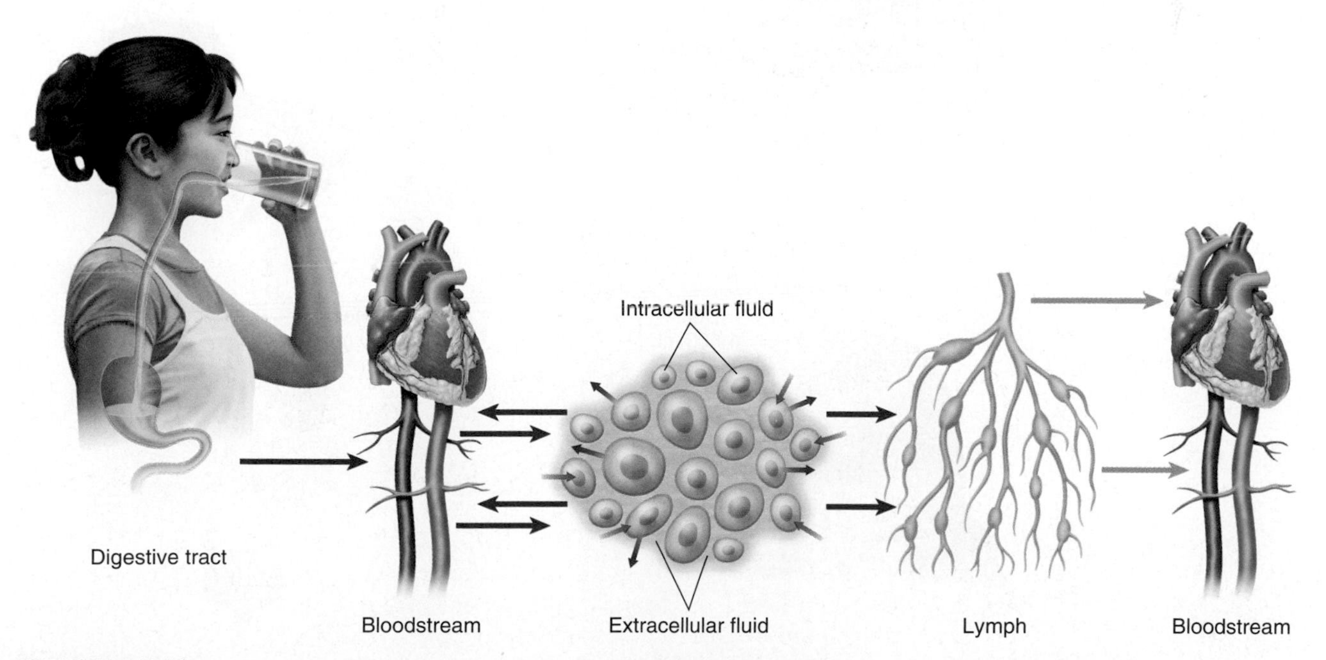

FIGURE 14.11 The movement of water between major fluid compartments.

In **Figure 14.11** you can see that water enters the body through the fluids you drink. This is the major source of water for the body, but not the only one. You may remember that water is also formed in the cells through cellular respiration ($C_6H_{12}O_6 + 6O_2 \rightarrow 6CO_2 + 6H_2O$ + energy). This additional source is considered to be **metabolic water** because it is derived from a chemical process that occurs in the cells.

As you can see in **Figure 14.12**, the body's daily intake and output of water should be equal to maintain homeostasis. Although most of the water you take in is from drinks, food and metabolic water do make significant contributions to water balance. At the same time, urine output is the major way the body rids itself of water, while sweat, water evaporated from the skin, expired air, and feces also make significant contributions to the amount of water leaving the body. Since urine output is so vital to maintaining fluid and electrolyte balance in the body, we focus next on how urine production and its volume are regulated.

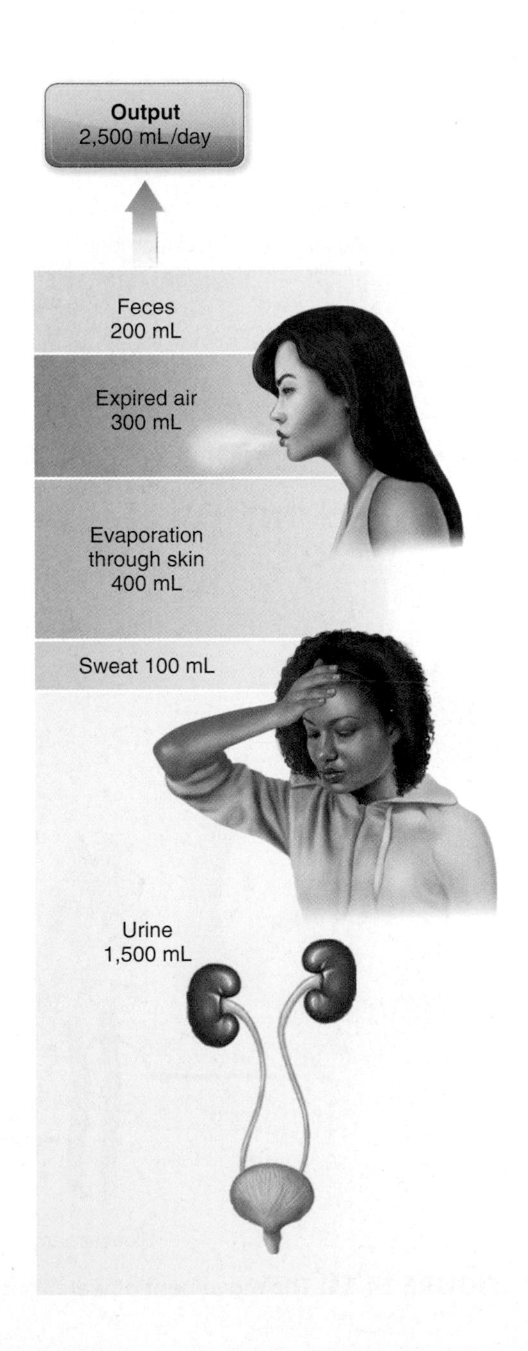

FIGURE 14.12 Daily water intake and output.

Regulation of Urine Volume and Concentration

14.11 learning **outcome**

Explain how urine volume and concentration are regulated.

You can significantly adjust the intake of water and electrolytes by what you consume. However, the kidney's urine production is the only way to significantly adjust the amount of water and electrolytes in the blood through what leaves the body. The kidney cannot increase the amount of water or electrolytes in the blood, but it can prevent their loss by adjusting the amount of water and electrolytes that may exit during urine production. The principal electrolyte in this process is sodium. Where sodium goes, water usually follows. The volume of urine is determined by the amount of water in it, while the concentration of urine is determined by the relative amount of solutes it contains—sodium, in this case. Another important electrolyte in urine production is potassium. As you will see in the explanations that follow, potassium usually moves in the opposite direction of sodium.

Consider the two statements listed below, which you may intuitively agree make sense. They sum up the kidneys' role in the homeostasis of fluids and electrolytes in the body. Once you grasp the goal, you can proceed to investigate how the goal is reached.

- If the blood's concentration of solutes is *higher* than normal, the kidneys will put out *small* volumes of *concentrated* (many solutes) urine. In this way, the kidneys conserve water in the blood and eliminate excess solutes from the blood.
- If the blood's concentration of solutes is *lower* than normal, the kidneys will put out *large* volumes of *dilute* (few solutes) urine. In this case, water is removed from the blood, and the solutes in the blood are conserved.

You may want to keep coming back to these statements as you read about the hormonal, autonomic, and diuretic mechanisms that control urine production.

Hormonal Mechanisms of Control There are three main hormones—antidiuretic hormone (ADH), **aldosterone,** and **atrial natriuretic hormone (ANH)**—that regulate urine production in the kidneys. Both ADH and aldosterone result in less urine produced, but they do so in very different ways. On the other hand, ANH increases urine production. How each of these hormones work is explained below.

ADH As you may recall from Chapter 8, ADH is produced by the hypothalamus, but it is stored and then released from the posterior pituitary when commanded by the hypothalamus. The hypothalamus monitors blood sodium concentration and blood pressure. If blood sodium concentration increases or blood pressure falls, the hypothalamus sends nerve signals to the posterior pituitary telling it to release ADH. ADH targets distal convoluted tubules and collecting ducts in the kidneys. The effect of ADH on these structures is that it makes them more permeable, so more water is reabsorbed. This decreases water loss to urine and therefore helps maintain blood volume and blood pressure. It is important to note that ADH has an effect on only water reabsorption, not sodium reabsorption. Under the influence of ADH, the kidneys conserve water, but not sodium. Sodium is allowed to exit with urine, reducing the sodium concentration in the blood. See **Figure 14.13**.

Aldosterone You studied this hormone in Chapter 8. As you may recall, aldosterone is a mineralocorticoid produced in the adrenal cortex that targets the kidneys. It regulates the amount of active transport in the nephron. If aldosterone levels are up, more sodium ions are actively transported from the tubule to the peritubular capillaries and more potassium ions are secreted. Water follows the sodium by osmosis, so more water moves to the peritubular capillaries too. The result of aldosterone secretion is reduced urine output, with both water and sodium conserved in the blood, and increased potassium in urine.

In the case of ADH, the hypothalamus monitors the blood and sends the signal for ADH release. How does the adrenal cortex know when to secrete aldosterone? The juxtaglomerular apparatus, mentioned earlier (**Figure 14.7**), regulates

FIGURE 14.13 Action of antidiuretic hormone. The pathway shown in red represents the negative-feedback mechanism to restore homeostasis.

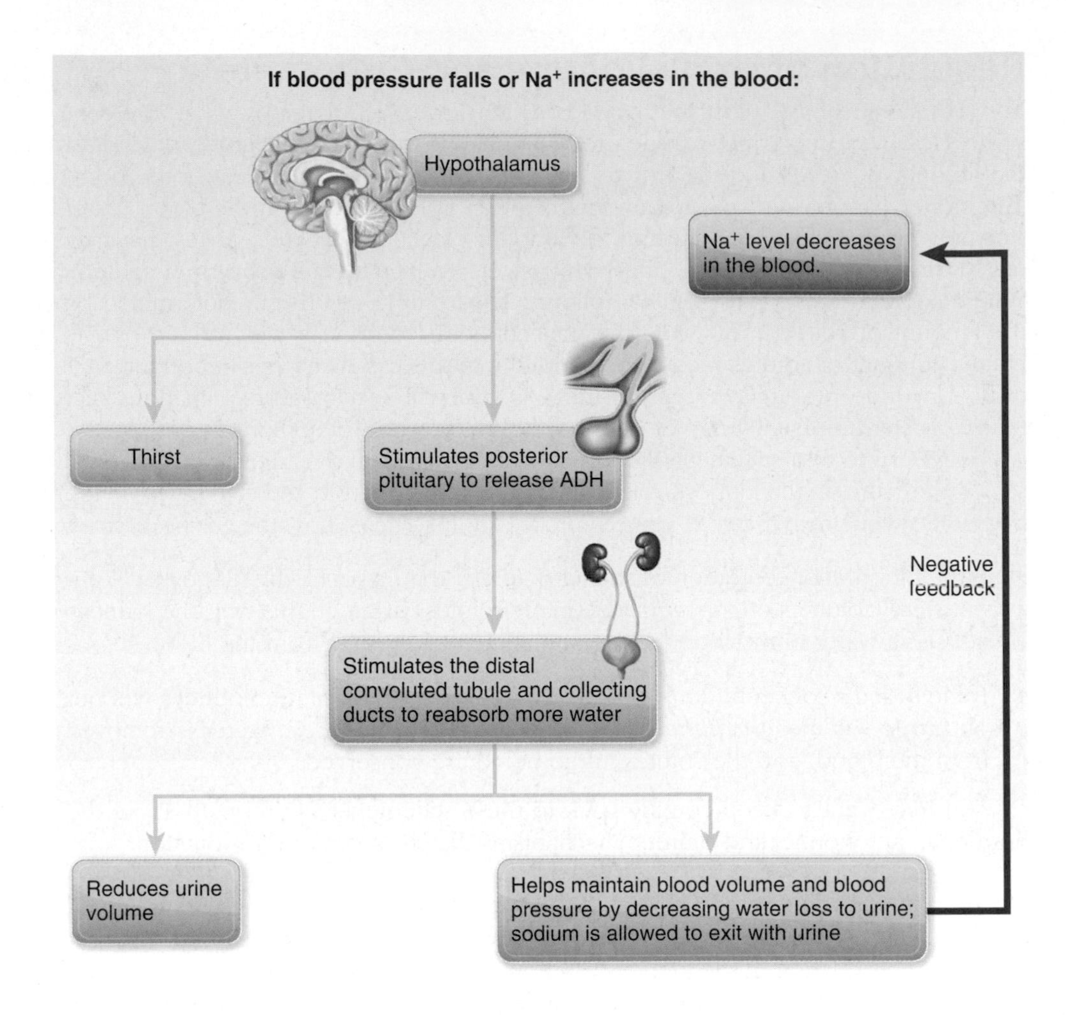

renin: REE-nin

aldosterone secretion. The juxtaglomerular apparatus monitors blood traveling through the afferent arteriole. It also secretes **renin** under any of the conditions listed below and shown in **Figure 14.14**.

- Blood pressure falls (hypotension).
- The level of sodium in the blood is too low **(hyponatremia).**
- The level of potassium in the blood is too high **(hyperkalemia).**

Renin is a chemical that converts a protein from the liver to angiotensin I. An enzyme called **angiotensin-converting enzyme (ACE),** produced in the lungs and kidneys, then converts angiotensin I to angiotensin II. **Angiotensin II** targets the adrenal cortex, telling it to secrete aldosterone. Aldosterone's effect on urine production prevents a further drop in blood pressure from the loss of water to urine, maintains the level of sodium in the blood, and lowers the level of potassium in the blood. See **Figure 14.15**.

Patients with high blood pressure (hypertension) may be prescribed ACE inhibitors. This medication interferes with the enzyme that converts angiotensin I to angiotensin II. Aldosterone is produced in the adrenal cortex only when angiotensin II fits into receptors. So ACE inhibitors inhibit aldosterone secretion. Without aldosterone, urine output is increased and blood volume (and therefore blood pressure) is reduced.

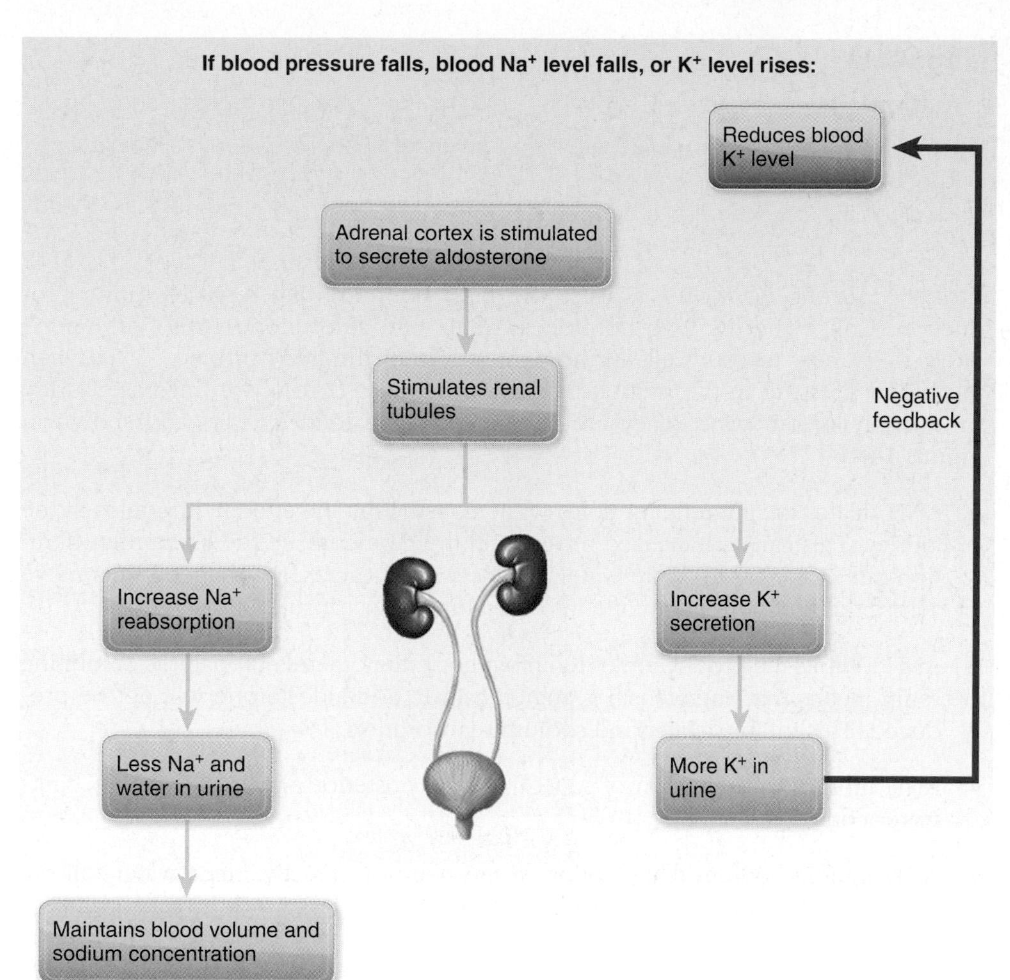

FIGURE 14.14
Action of aldosterone.
The pathways shown in red represent negative-feedback mechanisms to restore homeostasis.

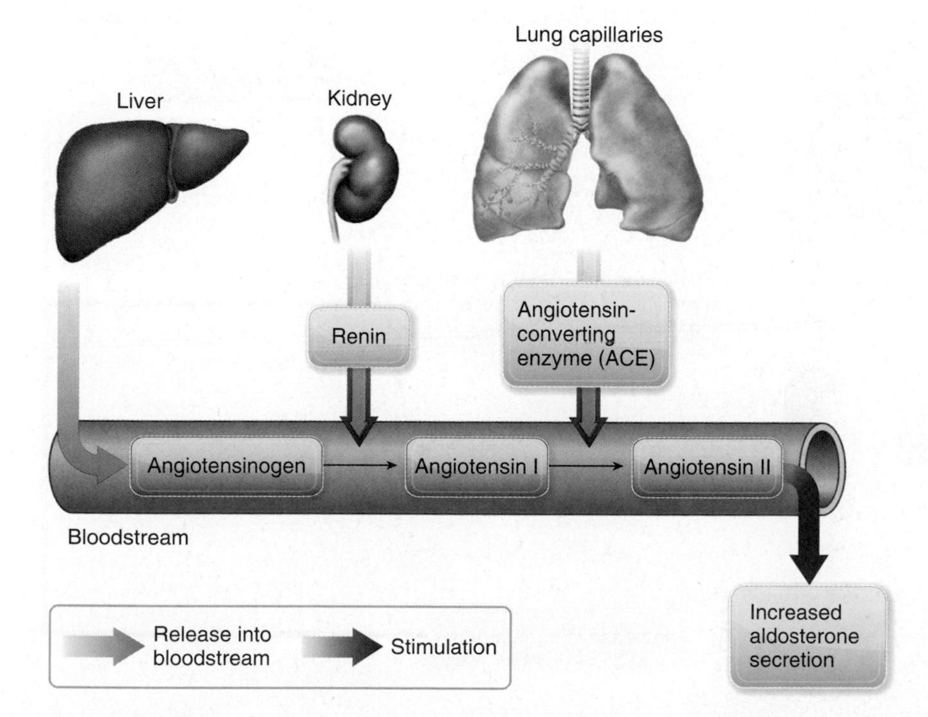

FIGURE 14.15
The renin-angiotensin-aldosterone connection.

spot check 5 Cameron is in a car accident and has internal bleeding. As a result, his blood pressure suddenly falls. What has to happen for the kidneys to reabsorb more sodium, and therefore more water, in an attempt to maintain blood pressure?

Atrial natriuretic hormone This is the third hormone that regulates urine production. Unlike the first two hormones—ADH and aldosterone—this hormone is likely to be new to you. Cells in the right atrium of the heart produce ANH when the blood pressure in the right atrium is too high. ANH results in increased urine production in four ways. They are explained in the following list and shown in **Figure 14.16**.

- ANH dilates the afferent arterioles while constricting the efferent arterioles in the kidney. This causes increased pressure in the glomeruli, so the glomerular filtration rate is increased. More water, glucose, amino acids, and mineral salts move across the filtration membrane into the glomerular capsule.
- ANH inhibits the production of renin by the juxtaglomerular apparatus. Inhibiting renin means that angiotensin I, angiotensin II, and aldosterone will not be produced. This inhibits water and sodium reabsorption.
- ANH inhibits the secretion of ADH from the posterior pituitary. This also limits water conservation.
- ANH inhibits sodium reabsorption in the nephron directly. Since water follows sodium, if less sodium is reabsorbed, it follows that less water is reabsorbed.

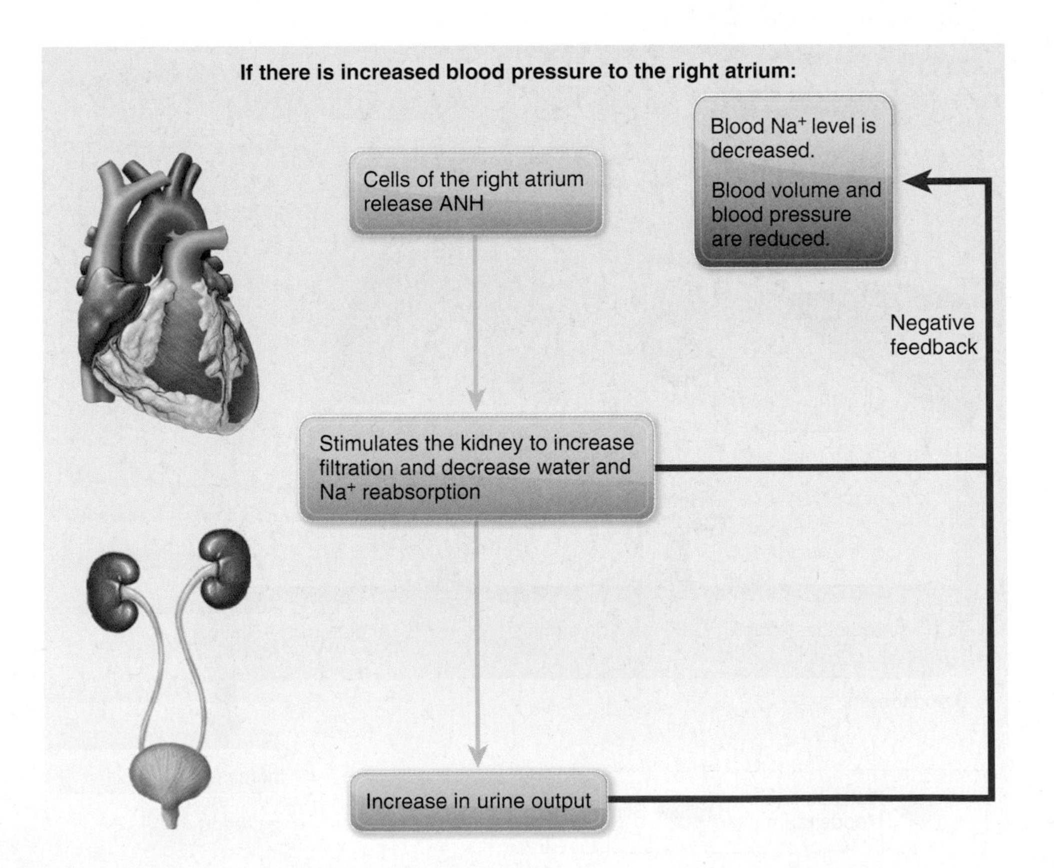

FIGURE 14.16 Action of atrial natriuretic hormone. The pathways shown in red represent negative-feedback mechanisms to restore homeostasis.

Now that you have become familiar with the three hormonal mechanisms that control urine production, you are ready to investigate the autonomic nervous system's control of urine production.

Autonomic Mechanisms of Control The sympathetic nervous system exerts its control of urine production during heavy exercise or acute conditions like a traumatic drop in blood pressure that might occur with sudden blood loss. In these cases, sympathetic neurons cause constriction in the kidney's afferent arterioles. Because this reduces the amount of blood entering the glomeruli, the glomerular filtration rate is also decreased. Blood is diverted from the kidney and sent to the brain, heart, and skeletal muscles instead.

14.12 learning outcome

Explain how diuretics, such as medications, caffeine, and alcohol, affect urine production.

Diuretics In addition to hormonal and nervous mechanisms of control, diuretics, such as alcohol, caffeine, and diuretic drugs, can also affect urine production. A **diuretic** is anything that increases urine volume. The following list contains more specific information about the various types of diuretics and their impact on the body:

- Alcohol inhibits the secretion of ADH. If large quantities of alcohol are consumed, this effect may be so great that the large quantity of dilute urine that is excreted may actually exceed the amount of fluids consumed. This can lead to dehydration and excessive thirst.
- Caffeine increases the blood flow to the kidney, and this increases the glomerular filtration rate. It also decreases the amount of sodium reabsorbed, so the net effect is large volumes of urine containing sodium.
- Diuretic drugs are often prescribed for hypertensive patients to reduce their blood pressure. Many of these drugs work by inhibiting the active transport of sodium. These drugs result in large volumes of urine containing sodium. Although reducing blood volume through increased urine production does decrease blood pressure, care must be given in administering these medications so that the body's electrolytes remain balanced.

You have studied the anatomy of the kidney, the processes of urine production, and the mechanisms that control urine production. Next, you will learn what happens to urine once it is produced. You will begin by exploring the anatomy involved with urine excretion.

14.13 learning outcome

Describe the anatomy of the ureters, urinary bladder, and male and female urethras.

Anatomy of Ureters, Urinary Bladder, and Male and Female Urethras

Figure 14.17 shows the structures of the urinary system. In it, you can see the kidneys, the ureters, the urinary bladder, and the urethra. You have already covered the anatomy of the kidney, but it will be helpful to review the calyces and renal pelvis that lead to the ureters. See **Figure 14.18**.

Ureters

The minor and major calyces of the kidney deliver urine to the renal pelvis. The ureters are an extension of the renal pelvis. Like the kidneys, these muscular tubes are also retroperitoneal. The ureters carry urine from the renal pelvis to the urinary bladder. Each ureter travels posterior to the urinary bladder and enters the bladder at its base. A small flap at the ureter's opening to the bladder prevents backflow of urine to the ureter when the urinary bladder contracts.

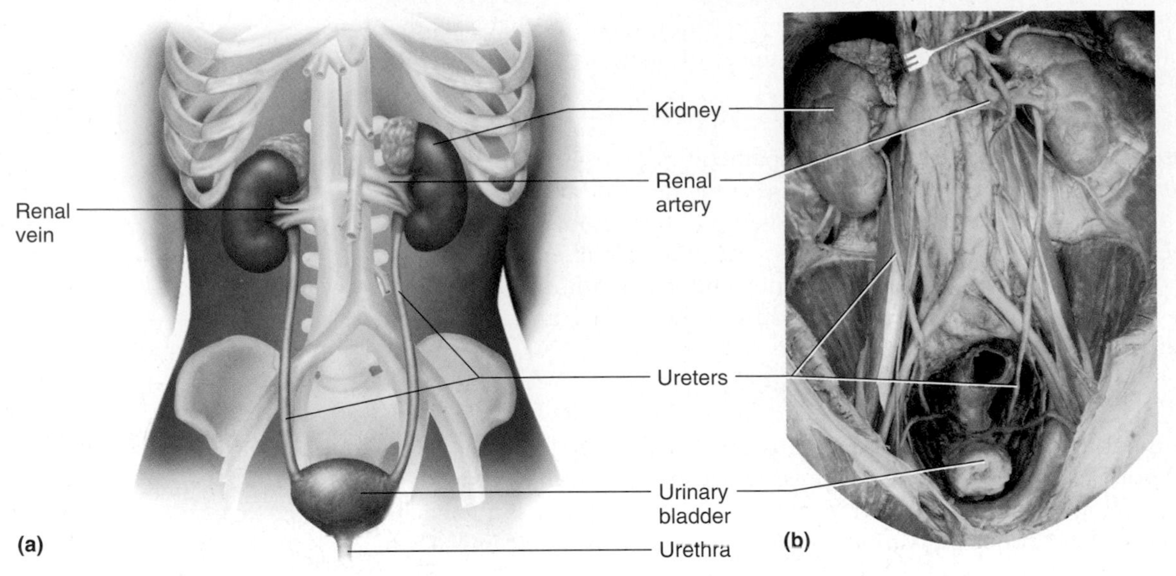

FIGURE 14.17 The urinary system: (a) anterior view, (b) view of a cadaver showing the kidneys, ureters, and urinary bladder (the parietal peritoneum has been removed).

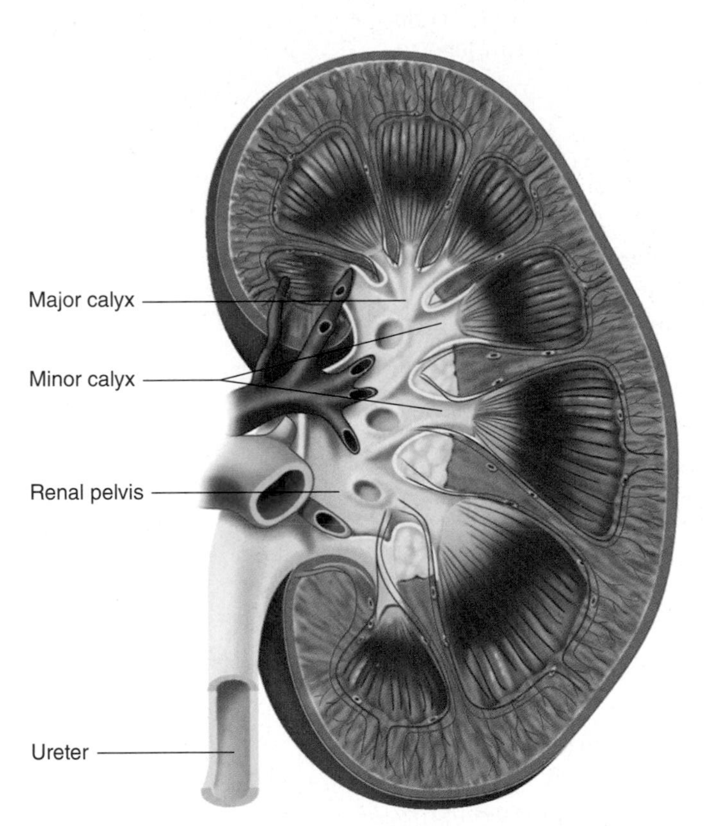

FIGURE 14.18 The calyces and renal pelvis.

trigone: TRY-gon

detrusor: de-TRUE-sor

Urinary Bladder

Figure 14.19 shows the urinary bladder in a female and in a male. Although there are obvious differences in the urethras, the anatomy of the urinary bladder is the same for both sexes. The urinary bladder is covered by the parietal peritoneum superiorly, and it sits posterior to the pubic symphysis. The urinary bladder functions to store urine until its release. So the bladder's anatomy is designed to stretch. The mucosa lining the bladder is transitional epithelial tissue, and the lining has many folds (rugae) that are less conspicuous when the bladder is stretched. Urine fills the bladder from the bottom. As a result, the urinary bladder stretches and expands upward as it fills. The rugae flatten, and the transitional epithelium gets thinner. The maximum amount the bladder can hold is 700 to 800 mL. A feeling of fullness is typically felt at 500 mL. Three openings at the base of the bladder— two ureters and a urethral opening— define a triangular area called the **trigone.** This area is often the site of infection in the urinary bladder. Three layers of smooth muscle make up the **detrusor muscle** of the bladder's walls. This muscle is very important in the physiology of passing urine. At the base of the bladder, the detrusor muscle thickens to form the **internal urethral sphincter.** This muscle compresses the tube leading from the bladder (the **urethra**), so urine remains in the bladder. You will read about passing urine as soon as you have finished studying the anatomy leading out of the body.

warning

Figure 14.19 may lead to a misconception: Although it appears in this figure that the ureters enter the urinary bladder at the top, they actually travel behind the bladder and enter at the bladder's base. The openings where they enter the bladder are labeled *ureteral openings* in this figure.

Urethra

The urethra is a tube that delivers urine from the urinary bladder to the outside. It begins at the internal urinary sphincter at the urinary bladder's base. The urethra passes through the pelvic floor, where it is encircled by a skeletal muscle called the **external urethral sphincter.** Since it is skeletal muscle, the external urinary sphincter is under voluntary control. The biggest difference in male and female urinary anatomy is the length of the urethra. A female urethra is approximately 3 to 4 cm and opens to the outside **(external urethral orifice)** between the clitoris and the vaginal opening. In comparison, the male urethra is approximately 11-18 cm long and can be divided into three distinct areas: The **prostatic urethra** is surrounded by the prostate; the **membranous urethra** penetrates the pelvic floor; and, finally, the longest section, the **penile urethra,** passes through the length of the penis to the external urethral orifice. Male reproductive structures, such as the prostate gland and bulbourethral glands, also secrete fluids into the male urethra. These structures are discussed in the next chapter, on the male reproductive system.

Now that you have become familiar with the anatomy that delivers urine to the outside of the body, it is time to look at how the passing of urine **(micturition)** is initiated.

micturition:
mik-choo-RISH-un

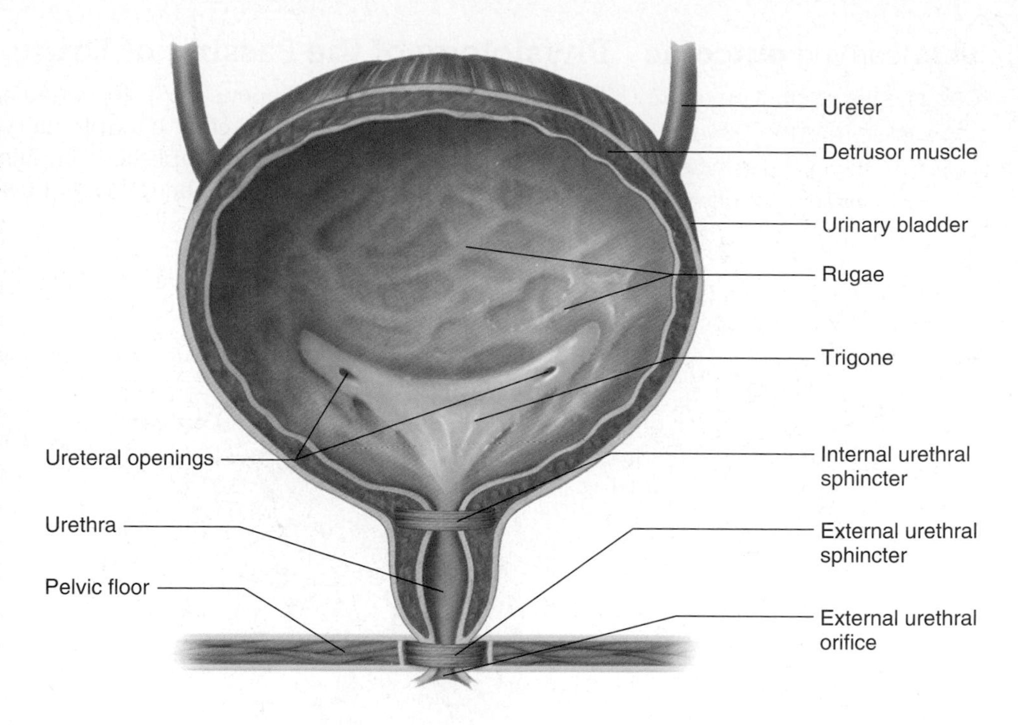

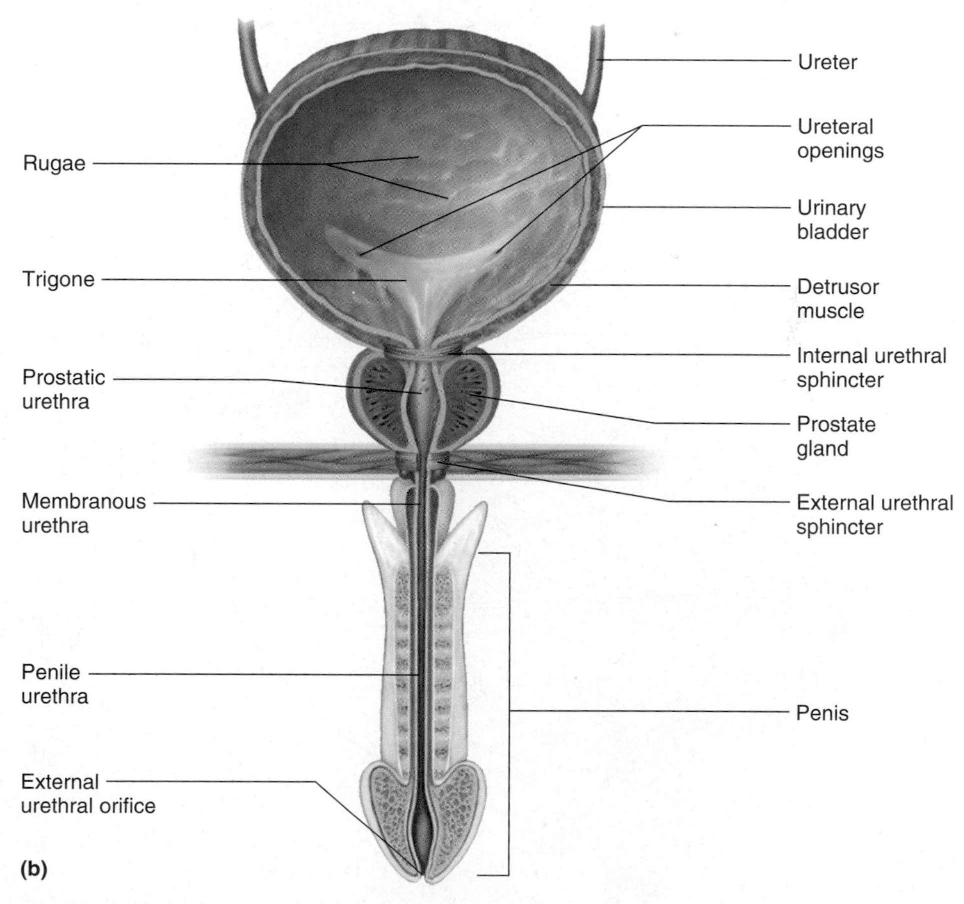

FIGURE 14.19 The urinary bladder and urethra: (a) female, (b) male.

14.14 learning **outcome**

Describe the micturition reflex and explain how the nervous system and urinary sphincters control the voiding of urine.

Physiology of the Passing of Urine

The production of urine happens 24/7. In contrast, micturition is the passing or voiding of urine. A micturition reflex controls the voiding of urine in infants, but once toilet training has been accomplished, impulses from higher centers in the brain can, and likely will, influence the reflex. Follow along with the steps shown in **Figure 14.20** as this reflex is explained.

1. The reflex arc for micturition begins with stretch receptors in the urinary bladder's walls (afferent neurons).
2. As the bladder fills, these receptors send signals to the sacral region of the spinal cord (integration center).
3. Parasympathetic neurons go from the spinal cord to the detrusor muscle (effector), telling it to contract, and to the internal urethral sphincter (effector), telling it to relax. Urine is then voided.

Toilet training affects the outcome of the reflex. Continue to follow along with **Figure 14.20** to see the additional steps for micturition to occur.

4. At the same time as step 3, the afferent signals are also sent to the pons and ultimately to the cerebrum.
5. If it is timely to pass urine, the pons sends a signal to the sacral region of the spinal cord.
6. Motor neurons send signals to the external urethral sphincter, telling it to relax so that urine can be passed. However, if it is not timely to pass urine, the pons will respond to the afferent signals from the stretch receptors by sending inhibitory signals to the external urethral sphincter. These signals will prevent the external sphincter from relaxing in order to retain urine in the bladder.

You have now covered the organs of the excretory system, the anatomy of the urinary system, the processes and control of urine production, and the way urine is passed from the body. You are ready to review how all of this relates to the functions of the excretory system.

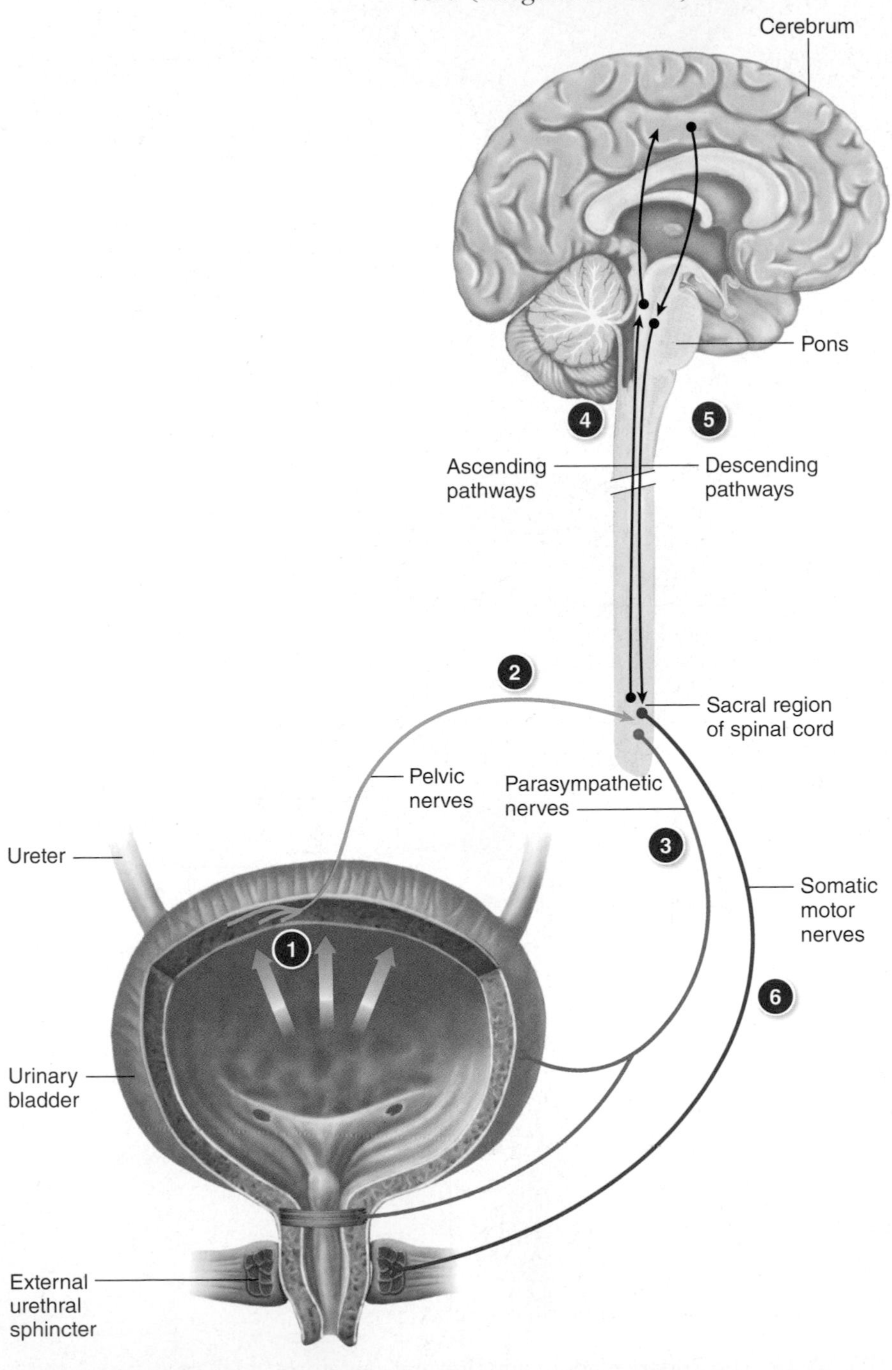

FIGURE 14.20 Neural control of micturition. Steps 1 to 3 involve a reflex, and steps 4 to 6 involve higher brain control.

FIGURE 14.21 Nikki and Chris at the theater.

Functions of the Excretory System

14.15 learning outcome

Summarize the functions of the excretory system.

As you can see in **Figure 14.21**, Nikki and Chris are healthy kids out for a movie. They had enough money for the theater tickets and popcorn but not enough to purchase any beverages. Consider the functions of the excretory system—listed below—as they relate to Chris while he watches his movie.

- **Removal of metabolic wastes.** All the while that Chris is watching the movie, his skin removes small amounts of urea with sweat, his lungs remove CO_2 from his blood, his liver removes bilirubin from the breakdown of his hemoglobin and puts it in bile, his spleen puts bilirubin in plasma for the kidneys to remove, and his kidneys remove nitrogenous wastes.

- **Maintenance of the body's fluid and electrolyte balance.** Chris is eating salty popcorn while he is watching the movie, but he is not drinking fluids. The net result is increased sodium in his blood. His kidneys will conserve water to maintain blood volume while allowing the excess sodium to exit with urine. His net urine output will be a small volume of concentrated urine.

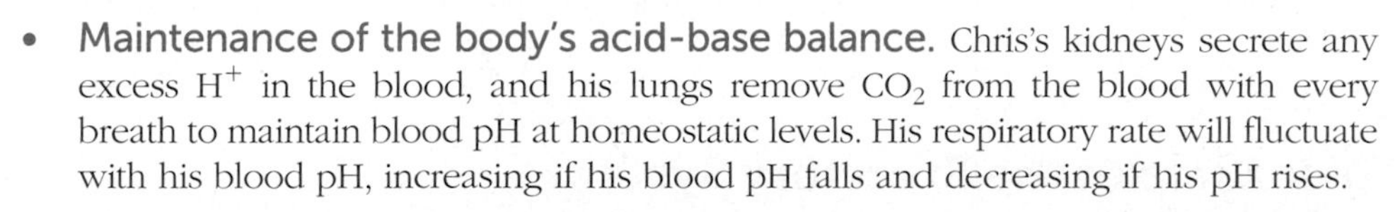

- **Maintenance of the body's acid-base balance.** Chris's kidneys secrete any excess H^+ in the blood, and his lungs remove CO_2 from the blood with every breath to maintain blood pH at homeostatic levels. His respiratory rate will fluctuate with his blood pH, increasing if his blood pH falls and decreasing if his pH rises.

- **Regulation of blood pressure.** All the salty popcorn is likely to increase Chris's thirst. His hypothalamus monitors the sodium level of his blood. Because it is high from all the salty popcorn and lack of additional fluids, his hypothalamus increases his thirst. Once the movie is over and Chris begins to drink fluids again, his blood volume will rise. If the increased fluid intake increases his blood volume and therefore his blood pressure too much, his kidneys will produce an increased volume of urine to reduce his blood volume and reduce his blood pressure.

spot check ❻ What hormone is likely to be released to regulate urine production while Chris is eating all the salty popcorn but not drinking fluids? Explain.

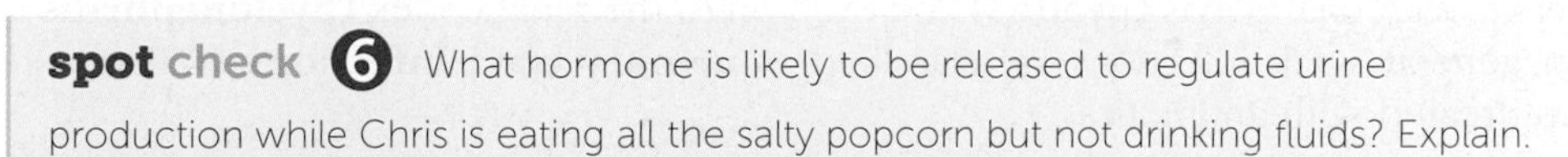

spot check ❼ What hormone is likely to be released to regulate urine production if Chris drinks large amounts of fluids after the movie? Explain.

Now that you have reviewed the functions of the excretory system, you can explore the effects of aging on the system.

14.16 learning **outcome**

Summarize the effects of aging on the excretory system.

Effects of Aging on the Excretory System

Aging of the excretory system primarily affects the urinary system. Some of the effects involve the kidneys' production of urine, but the voiding of urine can be affected as well. The effects on urine production are as follows:

- Typically, the size of the kidneys and the number of functioning nephrons decrease by one-third by the age of 80. This is partly due to the narrowing and hardening of the arteries supplying the kidneys and glomeruli.
- With the reduced number of functioning nephrons, the glomerular filtration rate decreases along with the reserve capacity. Even so, the waste removal by the kidneys is normally sufficient in the elderly. However, other diseases may put pressure on the urinary system and cause it to fail more quickly in the elderly.
- Drugs are cleared less efficiently with age, so drugs remain in circulation longer. Drug dosages may need to be adjusted in the elderly to compensate for the poor clearance.
- Responsiveness to ADH also decreases in the elderly, making water balance a problem. The sense of thirst may also be diminished, which means the elderly may become dehydrated.

The effects of aging on the passing or voiding of urine affect both men and women:

- By the time they are 50 years old, 50 percent of men experience **benign prostatic hyperplasia (BPH).** This will increase to 80 percent of men over 80.[1] In this condition, the prostate enlarges toward its center, compressing the urethra. This makes emptying the bladder more difficult.
- Elderly women are prone to **incontinence** (urine leakage), especially if vaginal childbirths have weakened the pelvic-floor muscles and external urethral sphincter.

Now you are aware of what may occur naturally with aging. As you will learn next, disorders of the excretory system may or may not have anything to do with aging.

14.17 learning **outcome**

Describe excretory system disorders.

Disorders of the Excretory System

Cystitis

Cystitis is an inflammation of the urinary bladder, usually caused by a bacterial infection. It is far more common in women than men because the pathway for the bacteria to the bladder (the urethra) is far shorter. The symptoms include the frequent passing of small amounts of urine accompanied by a burning sensation. The prevalence of cystitis is increased when women become sexually active because of the introduction of more bacteria to the genital area. The infection can travel up the ureters to the renal pelvis **(pyelitis)** and even on to the renal cortex **(pyelonephritis).** In general, such infections are classified as **urinary tract infections (UTIs)** and are treated with antibiotics.

Kidney Stones

Calcium or uric acid can precipitate out of urine to form solid stones in the renal pelvis. Small stones often pass without notice, but larger stones can block the renal pelvis or ureters. The blocked flow of urine increases the pressure within the kidney and can result in damage to nephrons. The kidneys continue to produce urine whether the flow is blocked or not. This increases pressure on the stone. The continued pressure on the sharp-edged stone may cause it to move toward the bladder and may cause intense pain as it passes along the ureter. Possible treatments include medication to dissolve the stone, a procedure using sound waves to break up the stone **(lithotripsy),** and surgery to remove the stone.

Glomerulonephritis

From the name "glomerulonephritis," you know that this disorder is an inflammation of the filtration membrane in the glomerulus of the nephron. There are two forms of this disorder—acute and chronic. **Acute glomerulonephritis** usually occurs one to three weeks after a severe bacterial infection in the body. Antibodies are produced to fight the infection, and in doing so, they attach to antigens. The antibody-antigen complexes batter the walls of the glomeruli in the kidneys due to the increased pressure. As a result of the irritation, the filtration membrane in the renal corpuscle becomes inflamed and more permeable, allowing plasma proteins and leukocytes in the filtrate. Water follows the plasma proteins, resulting in higher-than-normal volumes of urine containing protein and blood cells. Acute glomerulonephritis is usually time-limited. Once the antibody-antigen complexes are cleared from the blood by macrophages, the inflammation is resolved. On the other hand, **chronic glomerulonephritis** is just that—chronic. In this form of glomerulonephritis, the constant irritation to the filtration membrane causes it to thicken and be replaced by connective tissue. This may decrease the amount of filtration to the point of renal failure.

Renal Failure

You really need only part of one kidney to carry out the necessary functions for a normal life. Yet the body has two kidneys, and this provides tremendous reserve capacity. As you have already read, the effects of aging take a toll on kidney function. The ability to clear nitrogenous waste from the blood can be measured by assessing the **blood urea nitrogen (BUN).** This blood test expresses the amount of one of the nitrogenous wastes—urea—in the blood. Slightly higher levels indicate renal insufficiency **(azotemia).** Seriously elevated levels indicate **uremia,** characterized by the vomiting, diarrhea, and arrhythmias associated with increased nitrogenous waste in the blood. Complete kidney failure usually results in convulsions, coma, and death within a few days. Treatment for kidney failure is kidney transplantation or **dialysis,** in which a machine filters the excess fluid, salt, and nitrogenous wastes in the blood.

spot check 8 Dialysis is usually performed at a dialysis center three times a week for three to five hours at a time. Why might it be more difficult to manage blood pressure for a patient on dialysis?

putting the pieces **together**

The Excretory/Urinary System

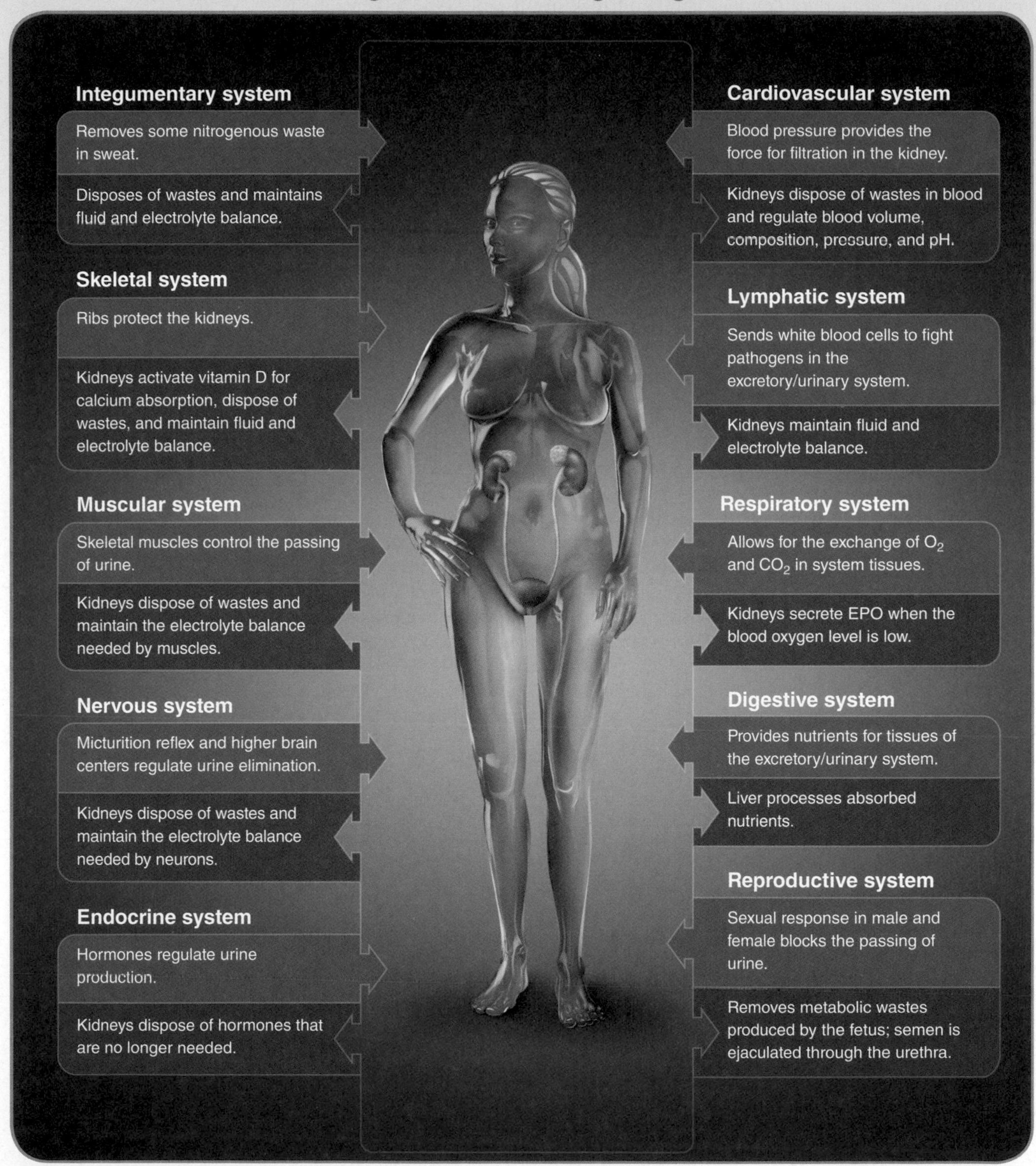

FIGURE 14.22 Putting the Pieces Together—The Excretory/Urinary System: connections between the excretory/urinary system and the body's other systems.

summary

Overview

- Excretion is the removal of metabolic wastes from the body.
- The skin, lungs, liver, and kidneys remove metabolic wastes.
- Nitrogenous wastes include ammonia, urea, uric acid, and creatinine.
- Kidneys function in multiple systems.

The Anatomy of the Kidney

- The kidneys are retroperitoneal.
- Each kidney has three layers: the renal capsule, the renal cortex, and the renal medulla.
- Pyramids are located in the renal medulla.
- Calyces collect urine from the pyramids and deliver it to the renal pelvis.
- The renal sinus is the space occupied by the renal artery, renal vein, and renal pelvis.

Anatomy of a Nephron

- A nephron has two basic parts: the renal corpuscle and the renal tubule.
- The renal corpuscle is composed of the glomerulus and the glomerular capsule.
- The renal tubule is composed of the proximal convoluted tubule, the nephron loop, and the distal convoluted tubule.
- The renal corpuscle and convoluted tubules are located in the renal cortex.
- The nephron loop and collecting duct are located in the pyramid.
- The juxtaglomerular apparatus is located between the afferent arteriole, efferent arteriole, and distal convoluted tubule.

Flow of Urine Components through a Nephron

- The components of urine flow from the glomerular capsule, to the proximal convoluted tubule, to the nephron loop, to the distal convoluted tubule, to the collecting duct, and to a minor calyx.

Blood Flow to a Nephron

- Blood flows from the renal artery, to smaller arteries, to the afferent arteriole, to the glomerulus, to the efferent arteriole, to the peritubular capillaries, to venules, to larger veins, and to the renal vein.

Physiology of Urine Production

- Urine production involves three processes: filtration, reabsorption, and secretion.

Filtration

- Filtration happens in the renal corpuscle.
- Materials move from the glomerulus to the glomerular capsule by filtration.
- The materials moved are water, glucose, amino acids, some nitrogenous wastes, and mineral salts.

Reabsorption

- Reabsorption happens along the renal tubule.
- Materials move from the tubules to the peritubular capillaries.
- The materials moved are 100 percent of the glucose, 100 percent of the amino acids, variable amounts of mineral salts, and water.

Secretion

- Secretion happens along the renal tubule.
- Materials move from the peritubular capillaries to the tubules.
- Materials moved include nitrogenous wastes, excess hydrogen ions, excess potassium ions, and some drugs.
- The secretion of hydrogen ions helps to maintain the blood's acid-base balance.

Water in the Body

- The human body is 50 to 75 percent water, which is held in two fluid compartments.
- Of the body's water, 65 percent is intracellular, while the other 35 percent is extracellular.
- Water moves easily between compartments by osmosis to minimize concentration gradients.
- Daily water intake should equal daily water output.

Regulation of Urine Volume and Concentration

- Urine volume and concentration are regulated through hormones, the nervous system, and diuretics.
- The hormones that control urine production are ADH, aldosterone, and ANH.
- The sympathetic nervous system reduces urine production.
- Diuretics increase urine production.

Anatomy of Ureters, Urinary Bladder, and Male and Female Urethras

Ureters

- The ureters are retroperitoneal.
- The ureters deliver urine from the renal pelvis to the urinary bladder.
- The ureters go posterior to the bladder and enter the bladder at its base.

Urinary Bladder

- The urinary bladder is a storage sac with smooth muscle in its walls.
- The trigone is a triangular area of the bladder floor and is defined by the openings to the ureters and the urethra.
- At the base of the urinary bladder, the detrusor muscle thickens to form the internal urinary sphincter.

Urethra

- The urethra delivers urine from the urinary bladder to the outside.
- The urethra of a male is longer than that of a female.
- The urethra is surrounded by the external urinary sphincter (made of skeletal muscle) as it passes through the pelvic floor.
- The male urethra can be divided into three sections: the prostatic urethra, the membranous urethra, and the penile urethra.

Physiology of the Passing of Urine

- The passing of urine is called *micturition.*
- The micturition reflex controls the passing of urine in infants, but higher brain centers influence the reflex after toilet training.

Functions of the Excretory System

- The functions of the excretory system are removal of metabolic wastes, maintenance of the body's fluid and electrolyte balance, maintenance of the body's acid-base balance, and regulation of blood pressure.

Effects of Aging on the Excretory System

- The size of the kidneys and the number of nephrons decrease.
- The glomerular filtration rate and reserve capacity decrease.
- Drugs are cleared less efficiently.
- Responsiveness to ADH is decreased.
- Eighty percent of elderly men experience benign prostatic hyperplasia, which makes micturition difficult.
- Elderly women often experience incontinence due to weakened pelvic floor muscles.

Disorders of the Excretory System

Cystitis

- Cystitis is an inflammation of the urinary bladder that is more common in women than men.

Kidney Stones

- Kidney stones are calcium or uric acid precipitates that may block the flow of urine in the ureters.

Glomerulonephritis

- Glomerulonephritis can be acute or chronic.
- Acute glomerulonephritis usually occurs after a severe infection and results in protein and white blood cells in the urine.
- Chronic glomerulonephritis results in thickening of the filtration membrane in the nephron and may lead to kidney failure.

Renal Failure

- Renal failure usually begins as azotemia, a renal insufficiency characterized by nitrogen wastes in the blood.
- Treatment of renal failure is dialysis or transplantation.

key words for review

The following terms are defined in the glossary.

cystitis
dialysis
diuretic
excretion
fluid compartments
glomerulonephritis
hyperkalemia
hyponatremia
metabolic acidosis
metabolic alkalosis
metabolic waste
metabolic water
micturition
nephron
nitrogenous wastes
renal corpuscle
renal tubule
respiratory acidosis
respiratory alkalosis
secretion

chapter review questions

Word Building: *Use the word roots listed in the beginning of the chapter and the prefixes and suffixes inside the back cover to build words with the following meanings:*

1. Inflammation of the renal pelvis: ______________________________
2. Removal of a kidney: ______________________________
3. Pertaining to the glomerulus: ______________________________
4. Condition of nitrogen in the blood: ______________________________
5. Bladder infection: ______________________________

Multiple Select: *Select the correct choices for each statement. The choices may be all correct, all incorrect, or a combination of correct and incorrect.*

1. What does the excretory system include?
   a. The heart
   b. The skin
   c. The lungs
   d. The kidneys
   e. The liver

2. What would happen if Chris ate a bag of salty popcorn at the movies but did not drink any liquids?
   a. Atrial natriuretic hormone would be used to restore homeostasis.
   b. His urine would become more concentrated.
   c. His urine would become more dilute.
   d. His ADH levels would rise to increase water reabsorption.
   e. His aldosterone levels would rise, so he could conserve water by actively transporting sodium.
3. What happens in a nephron?
   a. Reabsorption moves materials by active transport and osmosis from the tubules to the peritubular capillaries.
   b. Secretion moves materials from the peritubular capillaries to the tubules.
   c. High blood pressure in the glomerular capillaries allows filtration to take place.
   d. Filtration removes glucose, amino acids, mineral salts, nitrogenous wastes, and water from the blood.
   e. Excess hydrogen ions and sodium move from the tubules to the peritubular capillaries during secretion.
4. What does the kidney do other than produce urine?
   a. The kidney is involved in activating vitamin D.
   b. The kidney monitors blood oxygen levels.
   c. The kidney produces colony-stimulating factors for leukocyte production.
   d. The kidney produces erythropoietin.
   e. The kidney produces intrinsic factor.
5. Where do the components of urine go?
   a. The components of urine are first collected in the renal pelvis.
   b. The components of urine are first collected in the glomerular capsule.
   c. The components of urine travel from the distal convoluted tubule to the nephron loop.
   d. The components of urine travel from the glomerular capsule to the nephron loop to the proximal convoluted tubule.
   e. The components of urine travel from the distal convoluted tubule to the collecting duct.
6. What is the anatomy of the urinary system?
   a. The trigone is defined by the two openings to the urethra and the opening to the ureter.
   b. The detrusor is composed of skeletal muscle.
   c. The ureters are retroperitoneal.
   d. The urinary bladder fills from the top.
   e. The urinary bladder has rugae.
7. How is urine voided?
   a. Voluntary control of micturition is controlled in the medulla oblongata.
   b. Voluntary control of micturition is controlled in the hypothalamus.
   c. Voluntary control of micturition is controlled in the pons.
   d. The reflex arc for micturition controls the external urinary sphincter.
   e. The reflex arc for micturition involves parasympathetic neurons.
8. How does aging affect the urinary system?
   a. Drug clearance by the kidney is reduced.
   b. The glomerular filtration rate slows.
   c. The number of nephrons decreases.
   d. Glomeruli harden and narrow.
   e. Benign prostatic hyperplasia causes incontinence.
9. How would you describe excretory system disorders?
   a. Cystitis is more prevalent in women than men because their ureters are longer.
   b. Acute glomerulonephritis usually happens following a surgery.
   c. Kidney stones are usually precipitates of potassium or uric acid.
   d. Kidney stones may be treated with sound waves.
   e. Azotemia may progress to uremia.

10. How do diuretics work?
    a. Caffeine reduces the glomerular filtration rate.
    b. Alcohol increases the glomerular filtration rate.
    c. Diuretic drugs inhibit the active transport of sodium.
    d. Alcohol inhibits aldosterone.
    e. Caffeine decreases sodium reabsorption.

Matching: *Match the organs of the excretory system with the wastes they remove. Some answers may be used more than once. Some questions may have more than one correct response.*

_____	1. Bilirubin	a. Liver
_____	2. Carbon dioxide	b. Skin
_____	3. Creatinine	c. Kidneys
_____	4. Urea	d. Lungs
_____	5. Lactic acid	

Matching: *Match the waste to its source. Some answers may be used more than once. Some questions may have more than one correct response.*

_____	6. Carbon dioxide	a. Breakdown of nucleic acids
_____	7. Ammonia	b. Breakdown of amino acids
_____	8. Urea	c. Breakdown of proteins
_____	9. Creatinine	d. Breakdown of creatine phosphate
_____	10. Uric acid	e. Cellular respiration

Critical Thinking

1. If renal failure occurs, a patient may be put on dialysis. What functions of the kidney may be sufficiently taken care of by the dialysis? What functions may not be taken care of by dialysis?
2. In what ways could the glomerular filtration rate be increased?
3. Jake is a roofer who is applying asphalt to a roof in Miami in August. His strenuous work in the hot environment results in profuse sweating, producing 3 L of sweat over two hours. His sweat is less concentrated than his extracellular fluid. What effect will this have on his urine volume and concentration? Explain the mechanisms involved.

14 chapter mapping

This section of the chapter is designed to help you find where each outcome is covered in this text.

	Outcomes	Readings, figures, and tables	Assessments
14.1	Use medical terminology related to the excretory system.	Word roots: p. 527	Word Building: 1–5
14.2	Define *excretion* and identify the organs that excrete waste.	Overview: pp. 528–529 igure 14.2	Multiple Select: 1 Matching: 1–5
14.3	List the body's major nitrogenous wastes and their sources.	Overview: p. 529 Figure 14.3	Matching: 6–10
14.4	List the functions of the kidney in addition to urine production.	Overview: p. 529	Multiple Select: 4 Critical Thinking: 1
14.5	Describe the external and internal anatomy of the kidney.	Anatomy of the kidney: pp. 530–532 Figures 14.4, 14.5	Spot Check: 1
14.6	Describe the anatomy of a nephron.	Anatomy of a nephron: pp. 532–534 Figures 14.6, 14.7	Spot Check: 2
14.7	Trace the components of urine through a nephron.	Flow of urine components through a nephron: pp. 534–535 Figure 14.6, 14.8	Multiple Select: 5
14.8	Trace the flow of blood through a nephron.	Blood flow to a nephron: p. 535 Figure 14.8	Spot Check: 3
14.9	Describe filtration, reabsorption, and secretion in the kidney with regard to the products moving in each process, the direction of movement, and the method of movement.	Physiology of urine production: pp. 535–538 Figures 14.9, 14.10	Spot Check: 4 Multiple Select: 3 Critical Thinking: 3
14.10	Describe the fluid compartments of the body and how water moves between them.	Water in the body: pp. 538–540 Figures 14.11, 14.12	Critical Thinking: 3
14.11	Explain how urine volume and concentration are regulated.	Regulation of urine volume and concentration: pp. 541–545 Figures 14.13–14.16	Spot Check: 5–7 Multiple Select: 2 Critical Thinking: 2
14.12	Explain how diuretics, such as medications, caffeine, and alcohol, affect urine production.	Diuretics: p. 545	Multiple Select: 10 Critical Thinking: 2
14.13	Describe the anatomy of the ureters, urinary bladder, and male and female urethras.	Anatomy of ureters, urinary bladder, and male and female urethras: pp. 545–547 Figures 14.17–14.19	Multiple Select: 6
14.14	Describe the micturition reflex and explain how the nervous system and urinary sphincters control the voiding of urine.	Physiology of the passing of urine: p. 548 Figure 14.20	Multiple Select: 7
14.15	Summarize the functions of the excretory system.	Functions of the excretory system: p. 549 Figure 14.21	Critical Thinking: 1
14.16	Summarize the effects of aging on the excretory system.	The effects of aging on the excretory system: p. 550	Multiple Select: 8
14.17	Describe excretory system disorders.	Excretory system disorders: pp. 550–551	Spot Check: 8 Multiple Select: 9 Critical Thinking: 1

footnote

1. Beers, M. H., Porter, R. S., Jones, T. V., Kaplan, J. L., & Berkwits, M. (Eds.). (2006). *Merck manual of diagnosis and therapy* (18th ed.). Whitehouse Station, NJ: Merck Research Laboratories.

references and resources

Androile, G. L. (2008, October). Benign prostatic hyperplasia (BPH): Prostate disorders. *Merck manual home edition*. Retrieved March 17, 2011, from http://www.merckmanuals.com/home/sec21/ch239/ch239b.html.

Beers, M. H., Porter, R. S., Jones, T. V., Kaplan, J. L., & Berkwits, M. (Eds.). (2006). *Merck manual of diagnosis and therapy* (18th ed.). Whitehouse Station, NJ: Merck Research Laboratories.

Cutler, R. E. (2006, September). Effects of aging: Biology of the kidneys and urinary tract. *Merck manual home edition*. Retrieved March 17, 2011, from http://www.merckmanuals.com/home/sec11/ch141/ch141f.html.

Epstein, M. (1996). Aging and the kidneys [Electronic version]. *Journal of the American Society of Nephrology, 7,* 1106–1122.

National Cancer Institute. (n.d.). *Prostate cancer.* Retrieved March 17, 2011, from http://www.cancer.gov/cancertopics/types/prostate.

National Institute of Diabetes and Digestive and Kidney Diseases. (2006, June). *Prostate enlargement: Benign prostatic hyperplasia*. Retrieved March 17, 2011, from http://kidney.niddk.nih.gov/Kudiseases/pubs/prostateenlargement.

National Institute of Diabetes and Digestive and Kidney Diseases. (2006, December). *Treatment methods for kidney failure: Hemodialysis.* Retrieved October 25, 2010, from http://kidney.niddk.nih.gov/Kudiseases/pubs/hemodialysis.

Porter, R. S., & Kaplan, J. L. (Eds.). (n.d.). Selected physiologic age-related changes. *Merck manual for healthcare professionals.* Retrieved March 19, 2011, from http://www.merckmanuals.com/media/professional/pdf/Table_337-1.pdf?qt=selected physiologic age-related changes&alt=sh.

Prostate Institute. (2000). *BPH information*. Retrieved March 18, 2011, from http://www.prostateinstitute.org/bph_home/htm.

Rainfray, M., Richard-Harston, S., Salles-Montaudon, N., & Emeriau, J. P. (2000, July 8). Effects of aging on kidney function and implications for medical practice [Electronic version]. *Presse Médicale, 29*(24), 1373–1378.

Saladin, Kenneth S. (2010). *Anatomy & physiology: The unity of form and function* (5th ed.). New York: McGraw-Hill.

Seeley, R. R., Stephens, T. D., & Tate, P. (2006). *Anatomy & physiology* (7th ed.). New York: McGraw-Hill.

Shier, D., Butler, J., & Lewis, R. (2010). *Hole's human anatomy & physiology* (12th ed.). New York: McGraw-Hill.

University of Maryland Medical Center. (2008, October 27). *Aging changes in the kidneys—Overview.* Retrieved March 17, 2011, from http://www.umm.edu/ency/article/004010.htm.

Verhamme, K. M., Dieleman, L. P., Bleumink, G. S., van der Lei, J., Sturkenboom, M. C., Artibani, W., et al. (2002, October). Incidence and prevalence of lower urinary tract symptoms suggestive of benign prostatic hyperplasia in primary care—the Triumph project. *European Urology, 42*(4), 323–328.

15 The Male Reproductive System

Reproduction is necessary for a species to survive. Each new generation replaces the generation that is dying, and so human beings continue to populate the earth. However, it is not necessary for each individual to reproduce to stay alive. You can live a normal, healthy life without ever having a child. Yet, even if you choose not to reproduce, your reproductive system goes a long way in defining who you are, whether you are male or female. See Figure 15.1.

learning outcomes

After completing this chapter, you should be able to:

15.1 Use medical terminology related to the male reproductive system.

15.2 Explain what is needed for male anatomy to develop.

15.3 Describe the anatomy of the testes.

15.4 Describe the male secondary sex organs and structures and their respective functions.

15.5 Describe the anatomy of a sperm.

15.6 Explain the hormonal control of puberty and the resulting changes in the male.

15.7 Explain the stages of meiosis and contrast meiosis to mitosis.

15.8 Explain the processes of sperm production and differentiate between spermatogenesis and spermiogenesis.

15.9 Explain the hormonal control of the adult male reproductive system.

15.10 Trace the path a sperm takes from its formation to its ejaculation.

15.11 Describe the stages of the male sexual response.

15.12 Explain the effects of aging on the male reproductive system.

15.13 Describe male reproductive system disorders.

15.1 learning **outcome**

Use medical terminology related to the male reproductive system.

word **roots** & combining **forms**

andr/o: male

crypt/o: hidden

epididym/o: epididymis

orch/o, orchi/o, orchid/o: testis, testicle

pen/o: penis

prostat/o: prostate

semin/i: semen

sperm/o, spermat/o: sperm

test/o: testis, testicle

vas/o: duct, vas deferens

pronunciation **key**

epididymis: EP-ih-DID-ih-miss

gubernaculum: GOO-ber-NACK-you-lum

hypospadias: high-poh-SPAY-dee-as

perineum: PER-ih-NEE-um

prostate: PROS-tate

raphe: RAY-fee

rete: REE-tee

seminiferous: sem-ih-NIF-er-us

tunica albuginea: TYU-nih-kah AL-byu-JIN-ee-ah

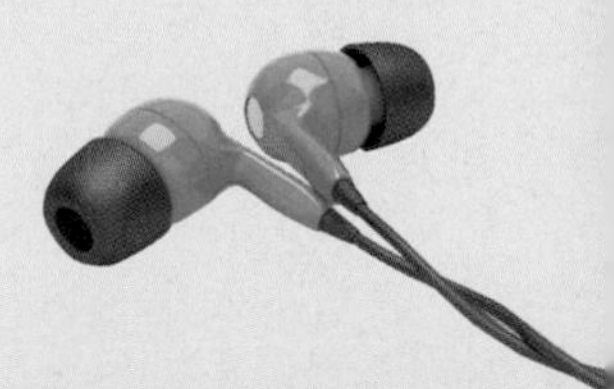

15.2 learning outcome

Explain what is needed for male anatomy to develop.

Overview

Before you explore the specifics of the male reproductive system, you must first understand what is necessary for an individual to develop as a male. All humans start from a **zygote** (fertilized egg). A sperm from the father penetrates an egg from the mother to combine the genetics of both. See **Figure 15.2**. The 23rd pair of chromosomes (sex chromosomes—X, Y) determines the gender of the offspring. All eggs carry an X chromosome. If the sperm carries a Y chromosome, the zygote will be male (XY). If the sperm carries an X chromosome, the resulting zygote will be female (XX). So the father's sperm determines the offspring's gender.

However, sex determination is more complicated than just acquiring the Y chromosome. The Y chromosome contains a gene called the **sex-determining region of the Y (SRY).** This gene codes for the production of a protein that reacts with other genes so that androgen receptors are produced in the developing fetus. As you may recall from Chapter 8, androgens are male hormones—testosterone is a prime example. By eight to nine weeks in the fetus, the developing gonads start producing testosterone if the Y chromosome is present. Testosterone fitting into the androgen receptors causes the reproductive anatomy to develop as male. Without testosterone *and* the androgen receptors, the fetal reproductive anatomy would develop as female under the influence of the abundant estrogen in the mother's circulation.

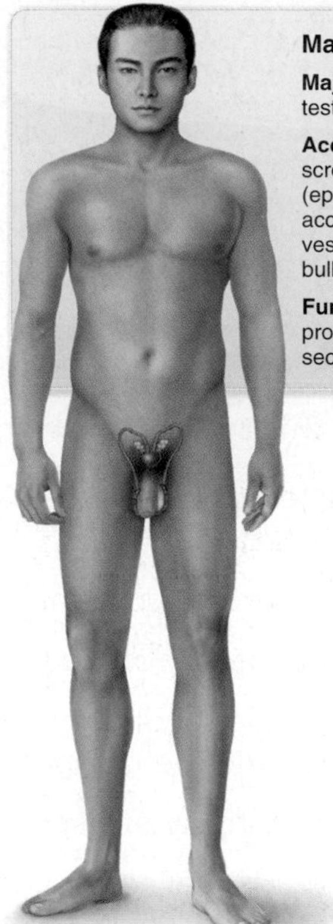

Male Reproductive System

Major Organs and Structures:
testes

Accessory Structures:
scrotum, spermatic ducts (epididymis, ductus deferens), accessory glands (seminal vesicles, prostate gland, bulbourethral glands), penis

Functions:
production and delivery of sperm, secretion of sex hormones

FIGURE 15.1 Male reproductive system.

applied genetics

Androgen-insensitivity syndrome is shown in Figure 15.3. The external reproductive anatomy of these individuals is clearly female, but menstruation at puberty does not occur. These individuals are genetically male (XY). On further examination, internal testes are found in the abdomen, and there are no ovaries, uterus, or vagina. This syndrome is the result of the lack of androgen receptors in the developing fetus. Even though the testes produce testosterone, there are no receptors for it to have an effect.

In this chapter, you will learn about how the male reproductive anatomy develops, how this anatomy functions, and how it is controlled. You will also investigate the effects of aging on this system, as well as male reproductive disorders. Your study begins with the primary reproductive organs in the male—the **testes.**

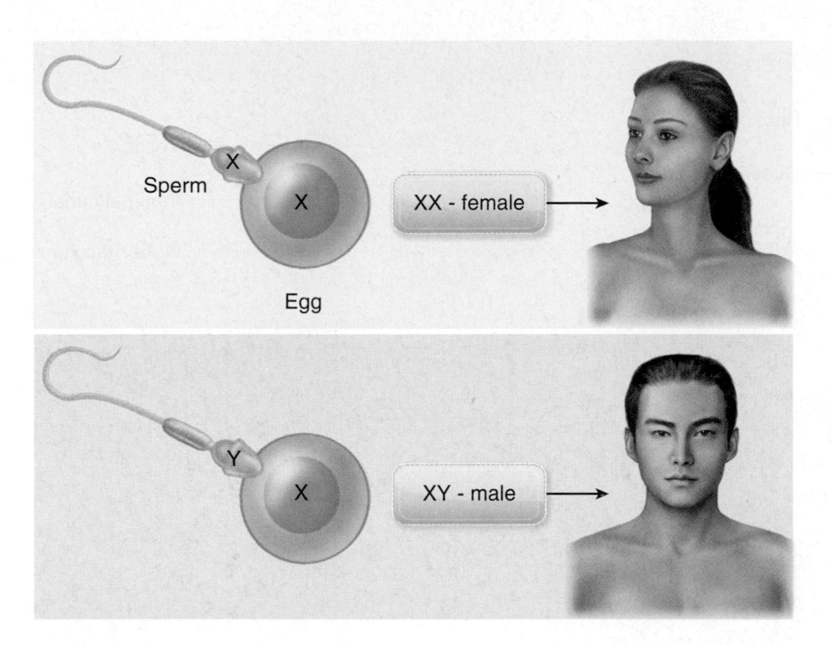

FIGURE 15.2 Chromosomal sex determination.

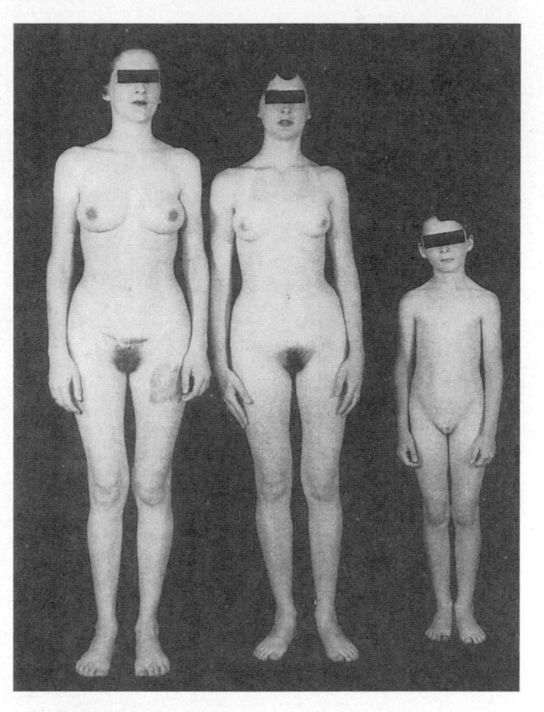

FIGURE 15.3 Androgen-insensitivity syndrome.

Male Reproductive Anatomy

15.3 learning **outcome**

Describe the anatomy of the testes.

The reproductive system is composed of primary organs that produce **gametes** (sex cells, like sperm and eggs) and secondary organs and structures that are necessary for reproduction to occur. In the male, the primary sex organs are the testes.

Testes

The testes **(testicles)** belong to the endocrine and male reproductive systems because they produce the hormone testosterone and they produce sperm. In an adult, each testis (singular of *testes*) is a slightly flattened, oval structure measuring approximately 4 cm long, 3 cm deep (anterior to posterior), and 2.5 cm wide.

Figure 15.4 shows the development of the testes from three months in a fetus to a one-month-old infant. Notice the testes originate retroperitoneal in the abdominal cavity. Each testis is attached to the fetal scrotal swelling by a short cord called the **gubernaculum.** As the fetus grows, all the body structures enlarge except the gubernaculum, which remains firmly attached to the developing scrotum. As you can see in this figure, the net effect of the gubernaculum's stable size is that it pulls the testes down from the abdominal cavity into the developing scrotum, so that by one month the infant's testes have fully descended to the scrotum outside the abdominal cavity. The testes descend through an opening in the abdominal wall called the **inguinal canal.**

gubernaculum: GOO-ber-NACK-you-lum

warning

Although this figure is meant to show the change in *position* of the testes over time in a developing fetus, it does not show the change in *size* of the structures over time. The body size of a one-month-old infant is much larger than a fetus at three months. The only structure in these images that stays the same size over this time period is the gubernaculum. The gubernaculum appears to have shrunk only because all of the other structures relative to it have grown.

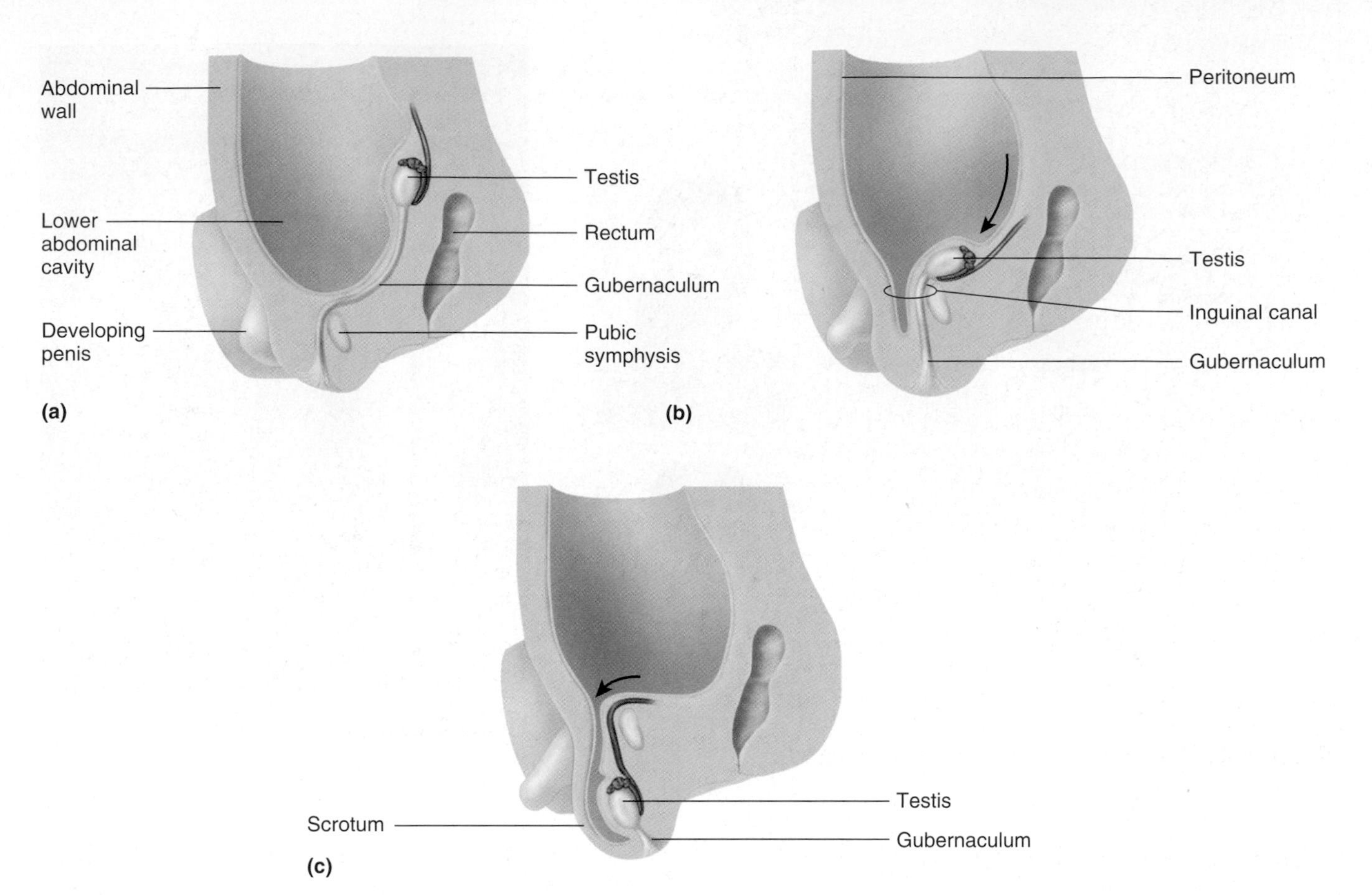

FIGURE 15.4 Descent of the testes: (a) three-month fetus, (b) six-month fetus, (c) one-month-old infant.

Why do testes descend? For the testes to produce viable sperm—those that can survive outside the testes—the temperature of each testis must be approximately 2 degrees Celsius (°C) cooler than the core body temperature (35°C versus 37°C). It would be too warm in the abdominal cavity for viable sperm production if the testes did not descend to the scrotum. How this temperature difference is maintained is discussed later in the chapter.

clinical point

Approximately 3 percent of male infants are born with undescended testes, a condition called **cryptorchidism.** For the majority of infants with this condition, the testes descend sometime during the first year. However, if the testes do not descend within the infant's first year of life, the condition may be corrected by testosterone injections or surgery to guide the testes through the inguinal canal. If not corrected, cryptorchidism can lead to sterility.

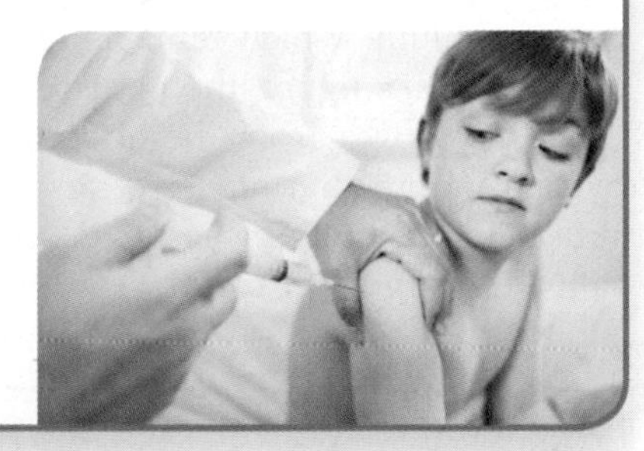

tunica albuginea: TYU-nih-kah AL-byu-JIN-ee-ah

seminiferous: sem-ih-NIF-er-us

spot check 1 Why would cryptorchidism result in sterility?

Each testis has an outer fibrous capsule called the **tunica albuginea.** See Figure 15.5. Inside this capsule, the testis is divided into 250 to 300 wedge-shaped lobules. Each lobule contains one to four **seminiferous tubules.** These tubules

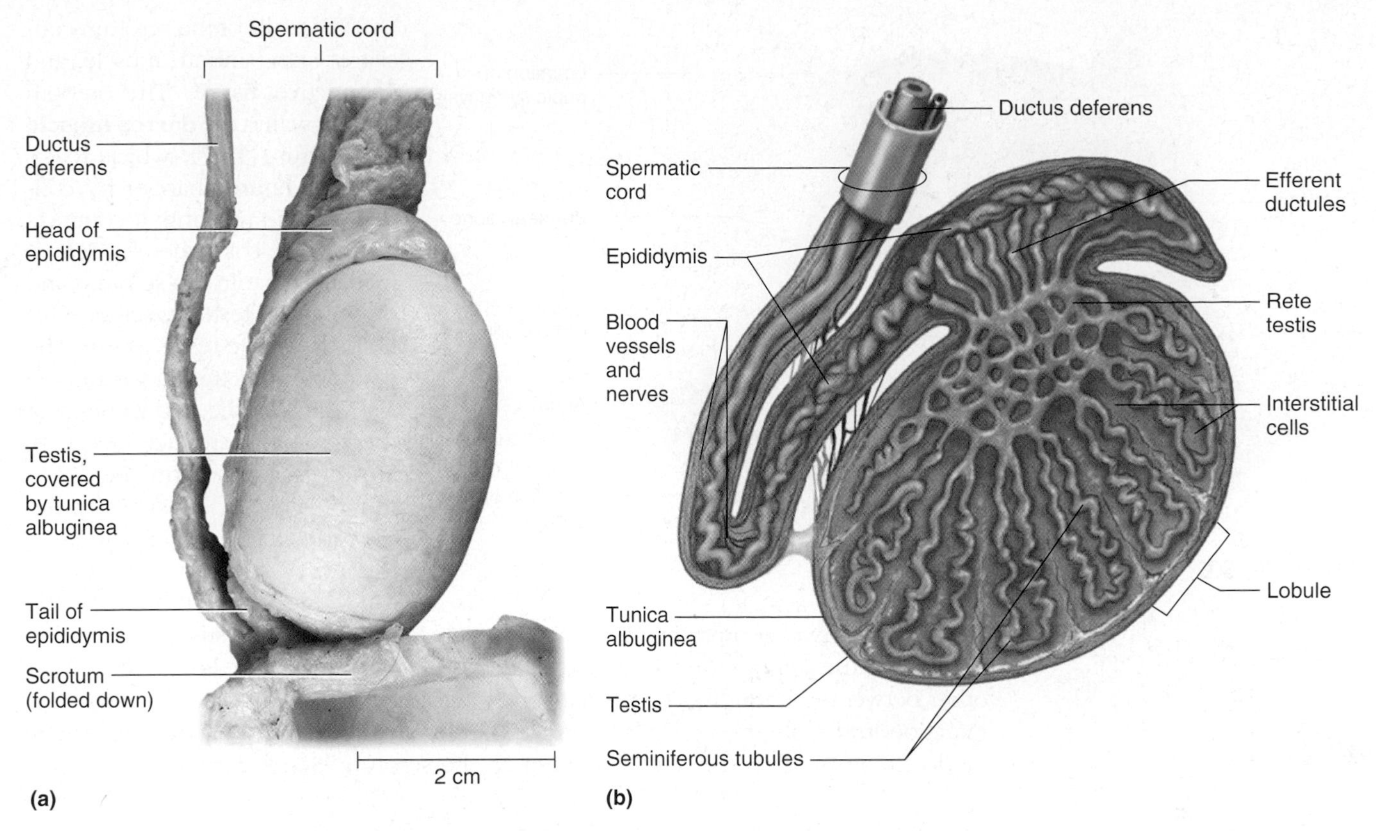

FIGURE 15.5 The testes and associated structures: (a) cadaver testis with the scrotum pulled away and the ductus deferens separated from the spermatic cord, (b) the anatomy of the testis, epididymis, and spermatic cord.

lead to a network of ducts called the **rete testis,** located posteriorly in the testes, still within the fibrous capsule. Sperm are produced in the seminiferous tubules. You will read more about sperm production later in the chapter. **Interstitial cells (cells of Leydig)** lie between seminiferous tubules in the lobules. These are endocrine cells that produce testosterone.

rete: REE-tee

spot check ❷ What is likely the least number of seminiferous tubules per testis? What is likely the greatest number of seminiferous tubules per testis?

Now that you have become familiar with the anatomy of the testes, you can explore the anatomy of the secondary reproductive organs and structures, starting with where the testes are housed—the scrotum.

Secondary Sex Organs and Structures

15.4 learning outcome

Describe the male secondary sex organs and structures and their respective functions.

The secondary reproductive organs and structures in the male are the scrotum, spermatic cord, spermatic ducts, accessory glands, and penis. These structures allow reproduction to occur.

Scrotum The scrotum is a pendulous (hanging from the torso) sac that houses the testes. Along with the penis, the scrotum makes up the external genitalia of the male. Both structures occupy a diamond-shaped space called the **perineum. Figure 15.6** places the perineum between the pubic bone (anteriorly), the ischial tuberosities (laterally), and the coccyx (posteriorly).

perineum: PER-ih-NEE-um

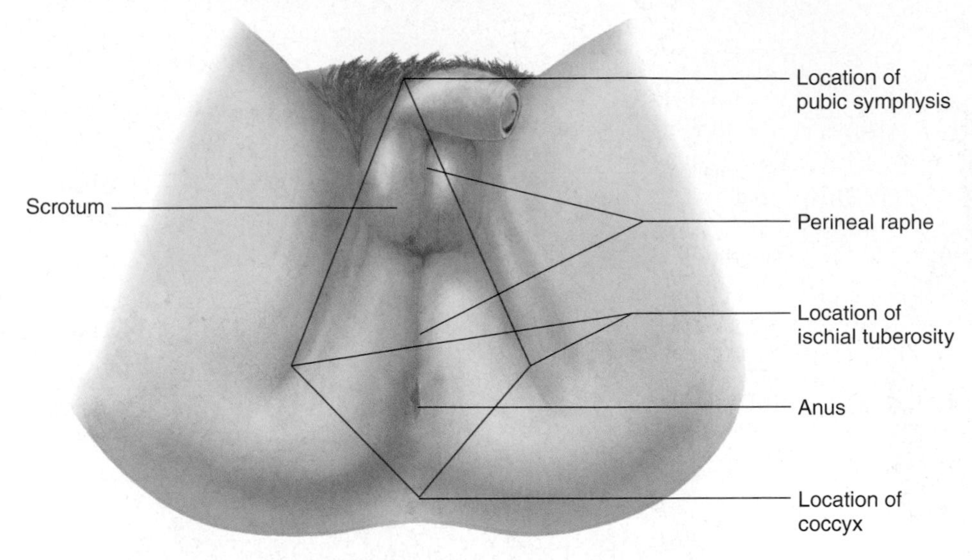

FIGURE 15.6 The male perineum.

The walls of the scrotum consist of skin, smooth muscle, and connective tissue. The smooth muscle wall is the **dartos muscle** (see **Figure 15.7**), which reacts to temperature changes by contracting and relaxing. It contracts when it is cold, thereby reducing the space in the scrotum and drawing the testes together. This helps keep the testes warm. The contraction of the dartos muscle also reduces the surface area of the scrotum to reduce heat loss. However, if it is warm, the dartos muscle relaxes, maximizing the space for the testes to help keep them cool.

raphe: RAY-fee

connect.mcgraw-hill.com audio

A medial wall divides the scrotum into two compartments—one for each testis. The left testis is suspended lower than the right in the scrotal sac, so the testes are not compressed against each other between the thighs. The medial septum also protects each testis from possible infection in the other compartment. An apparent seam **(perineal raphe)** externally marks the location of the medial septum on the scrotum. See **Figure 15.6**.

Spermatic Cord **Figure 15.7** displays the spermatic cord. This structure is a composite of several structures, as the cutaway for the left testis demonstrates. The right spermatic cord (in **Figure 15.7**) shows the **cremaster muscle** as the outer layer of

FIGURE 15.7 The scrotum and spermatic cord.

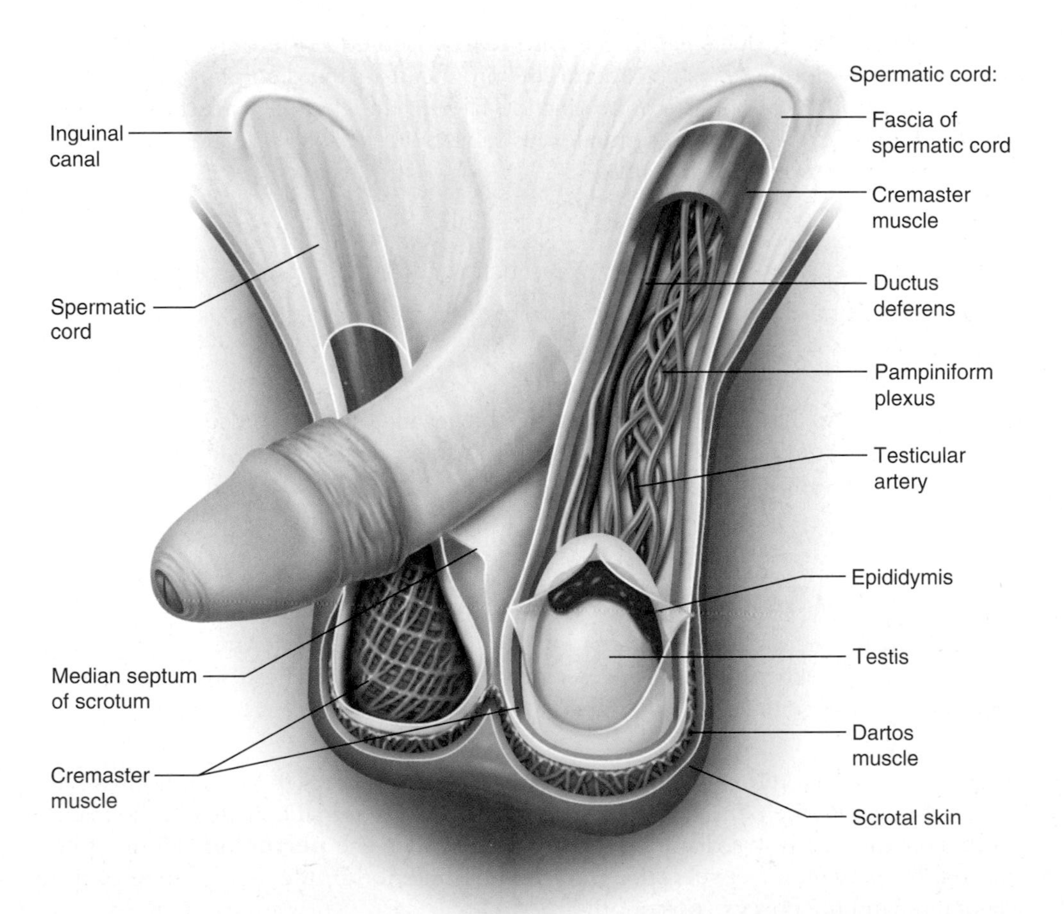

the spermatic cord completely covering the testis. This muscle is derived from the internal abdominal oblique muscle. Like the dartos muscle, the cremaster plays an important role in temperature regulation for the testes. When cold, the cremaster muscle contracts, drawing the testis closer to the body for warmth. When warm, the cremaster muscle relaxes, so the testis hangs farther away from the body to keep cool.

The cremaster muscle has been cut away on the right in **Figure 15.7** so that you can view the other structures of the spermatic cord. You can see the **ductus (vas) deferens,** a red **testicular artery,** and a blue network of veins called the **pampiniform plexus.** All of these structures and the cremaster muscle make up the spermatic cord. The ductus deferens is a tube that carries sperm. It is discussed below as one of the spermatic ducts. Here, we focus on the blood supply to and from the testes.

clinical point

Often, childless couples seek medical help if they are having trouble conceiving. The doctor may first ask what type of underwear the man wears. Briefs and tight jeans hold the testes closer to the body than boxer shorts and loose-fitting pants. The closer the testes are to the body, the warmer they will be. This slight temperature difference may be the cause of the problem.

The testicular artery (which originates from the abdominal aorta) is a very slender artery that has low blood pressure. In fact, the pressure in the testicular artery is so low that it does not produce a pulse. As a result, the testicular artery provides a poor oxygen supply to the testes. Sperm develop large mitochondria to compensate for the diminished oxygen; this adaptation allows them to make use of any available oxygen. Although the pressure of the blood delivered to the testes through the testicular artery is low, the blood is still warm because it comes from the body's core. This warm blood would warm the testes if not for the pampiniform plexus, which works as explained below.

You should be aware that blood cools once it has reached the testes and is away from the body's core. So the blood returning from the testes is cooler than the blood arriving. The pampiniform plexus is not a single vein. It is a network of small veins surrounding the testicular artery. It acts much like the radiator in your car, in that it surrounds the warm artery with cooler veins to cool the arterial blood even more before it arrives at the testes. **Figure 15.8** shows just one vein of the pampiniform plexus network next to the testicular artery so that you can see the transfer of heat.

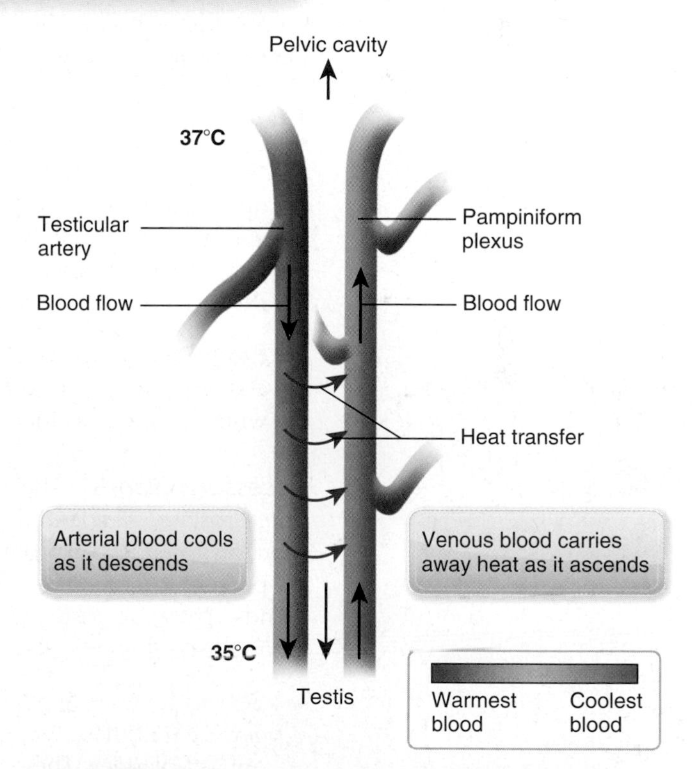

FIGURE 15.8 The transfer of heat from the testicular artery to the pampiniform plexus. Warm blood flowing through the testicular artery is cooled by the blood in the pampiniform plexus network of veins (shown here as a single vein for simplicity). The net effect is a lowering of the temperature of blood arriving at the testes by approximately 2°C.

spot check 3 What structures have an effect on the temperature of the testes?

Spermatic Ducts We continue our discussion of secondary reproductive organs and structures by examining the ducts that carry sperm. They are shown in **Figure 15.9** and described in the following list:

- **Efferent ductules.** As you read earlier, sperm are produced in the seminiferous tubules and move to the rete testis inside each testis. From the testes, sperm travel through efferent ductules to the epididymis. These tiny ducts have ciliated cells to move sperm along.

epididymis: EP-ih-DID-ih-miss

- **Epididymis.** This structure is a single-coiled duct (about 6 m, or 18 ft, long) that forms a ridge adhering to the posterior side of each testis. Sperm mature and are stored here. However, sperm are immature when they first arrive at the head of the epididymis from the efferent ductules. It takes approximately 20 days for sperm to travel from the head to the tail. All along the way, the epididymis reabsorbs excess water secreted by the testes. Once the sperm reach the tail of the epididymis, they are stored. There, sperm remain viable for 40 to 60 days. If not ejaculated, sperm disintegrate and the epididymis reabsorbs them.

clinical point

Just as women are encouraged to perform monthly breast self-exams, men are encouraged to self-examine their testes. The best place to perform this exam is in the shower with a soapy hand to palpate the testes, feeling for lumps. The epididymis can normally be felt as a ridge along the posterior of each testis. It should not be confused with something abnormal.

- **Ductus deferens.** At the tail of the epididymis, the sperm duct makes a 180-degree turn to become the ductus (vas) deferens. This muscular duct travels up the spermatic cord, through the inguinal canal into the abdominal cavity, and goes posterior to the urinary bladder. There, it widens to form an **ampulla** before it merges with a duct from the seminal vesicle to form the **ejaculatory duct.** This pathway is shown in **Figure 15.9**, parts (a) and (b).

prostate: PROS-tate

- **Ejaculatory duct.** The ejaculatory duct carries sperm from the ductus deferens, and fluid from the seminal vesicle, through part of the **prostate** gland to where it opens to the prostatic urethra. Again, see **Figure 15.9**.

Accessory Glands The male has five accessory glands—two **seminal vesicles,** one prostate gland, and two **bulbourethral (Cowper's)** glands. These glands function together to produce **semen,** a fluid ejaculated during orgasm. Semen is 10 percent sperm; the remainder is the different fluids produced by these three glands. These accessory glands are shown in **Figure 15.9** and described in the following list:

- **Seminal vesicles.** The seminal vesicles are approximately the size of a little finger. Each of these two glands is associated with a ductus deferens posterior to the urinary bladder. They secrete a thick, yellowish fluid that makes up 60 percent of semen. It contains sugar and other carbohydrates to nourish sperm and a protein that will help semen adhere to the vaginal walls, where it is deposited during intercourse.

- **Prostate gland.** As you may recall from the previous chapter, the prostate gland surrounds the urethra inferior to the urinary bladder. It produces a thin, whitish fluid that makes up 30 percent of semen. The prostatic fluid is alkaline to help protect sperm from the acidity of the vagina, where the semen is deposited during intercourse.

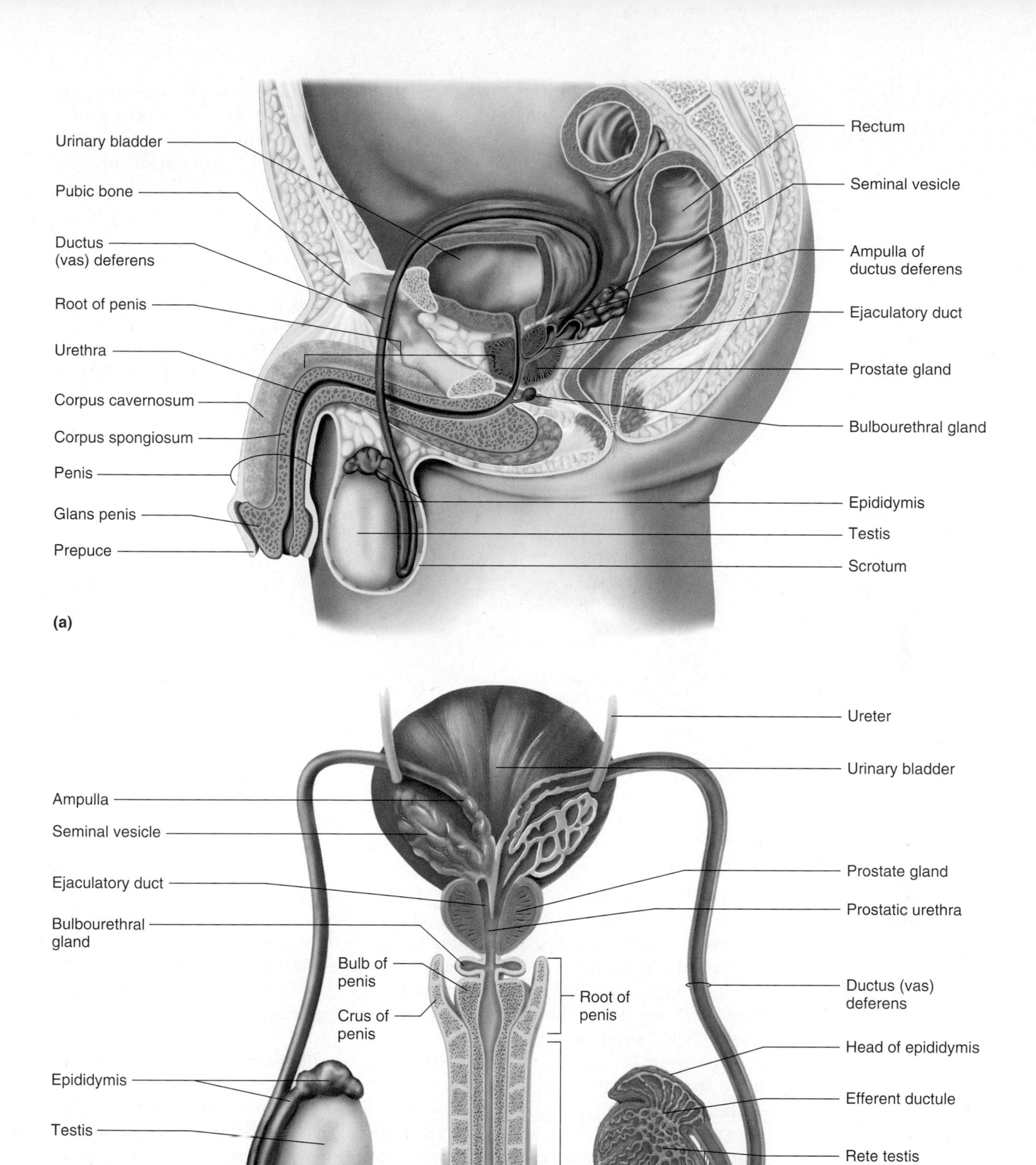

FIGURE 15.9 The male reproductive system: (a) sagittal view, (b) posterior view.

- **Bulbourethral glands.** The bulbourethral glands are named for their position near the bulb of the penis and the urethra. See **Figure 15.9b**. They produce trace amounts of a fluid that lubricates the end of the penis to make intercourse easier and neutralizes the pH of the male urethra, which usually carries acidic urine.

spot check 4 What is the composition of semen? Be specific.

Penis The last secondary reproductive structure you need to understand is the penis. The reproductive function of the penis is to become erect to facilitate the deposition of sperm in the vagina during intercourse. It is the anatomy of the penis that makes this function possible. The anatomy of the penis is shown from different views in **Figures 15.9** and **15.10**. You will find it helpful to compare the views in both figures.

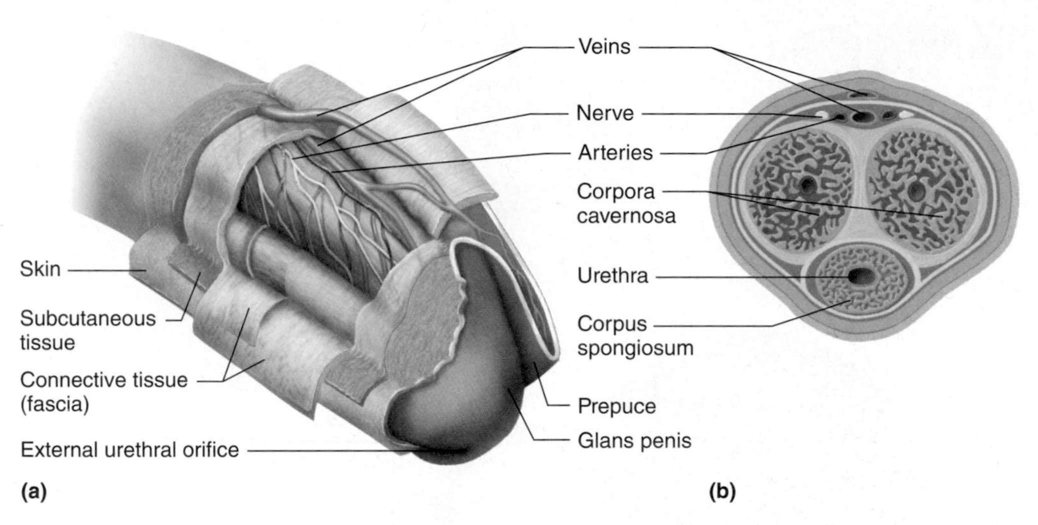

FIGURE 15.10 Anatomy of the penis: (a) superficial dissection, lateral view, (b) cross section of the shaft.

As you can see in **Figure 15.9a**, half of the penis—the **root**—is internal, so it is not seen externally. The other half, which is visible externally, is composed of the **shaft** and **glans.** The shaft and glans of a nonerect penis are approximately 8 to 10 cm (3 to 4 in.) long and 3 cm in diameter, whereas the dimensions of an erect penis range from 13 to 18 cm (5 to 7 in.) in length and 4 cm in diameter.

The glans is the expanded distal tip of the penis. The external urethral orifice (opening) is located here. The glans is highly sensitive because of its many nerve endings for sexual stimulation. The skin covering the shaft of the penis is loose to allow for the enlargement of the penis during erection. This skin extends to form the **prepuce** (foreskin), which covers the skin of the glans. The prepuce is often surgically removed **(circumcision)** from newborns. The skin of the glans and the facing skin of the prepuce have sebaceous glands that produce a waxy substance called **smegma.**

The penis is composed of three columns of erectile tissue. All of these columns have spaces within the tissue that fill with blood to expand the penis and cause it to become erect during sexual arousal. Two of the columns—**corpora cavernosa** (singular: *corpus cavernosum*)—have large spaces, while the third column—**corpus spongiosum**—has smaller spaces. Partitions between the spaces are composed of smooth muscle called **trabecular muscle.** As you can see in **Figure 15.9a**, the penile urethra is located in the corpus spongiosum, and the glans is composed only of corpus spongiosum. The internal end of the corpus spongiosum is called the **bulb.** The **bulbospongiosus muscle,** composed of skeletal muscle tissue, surrounds the bulb. The internal ends of the corpora cavernosa form a Y, and each arm of the Y shape is called a **crus** (plural: *crura*). The crura anchor the penis internally to the pubic arch. The muscles discussed here are new to you, but you will understand their function when you read about the male sexual response, later in the chapter.

You have now become familiar with the reproductive anatomy of the male body. Next, you will continue your study of anatomy by examining the male gamete—a sperm.

Anatomy of a Sperm

15.5 learning outcome

Describe the anatomy of a sperm.

As shown in **Figure 15.11**, a spermatozoon is a single cell with two principal parts—a **head** and a **tail.** The head is basically the nucleus, containing 23 chromosomes and an **acrosome** cap. The acrosome is a lysosome that contains enzymes that are eventually used to penetrate an egg. The tail contains a central flagellum used to propel the sperm on its journey. The section of the tail connected to the head **(midpiece)** is wider than the rest of the tail. This is because of the large mitochondria surrounding the flagellum. These mitochondria make use of any available oxygen to perform cellular respiration, which supplies ATP to energize the tail. As you may notice in this figure, there is an absence of cytoplasm and other organelles. Spermatozoa are developed to be as efficient as possible. Their function is simply to deliver their genetic material to fertilize an egg. Next, you will investigate the physiology of this system in the male that ultimately produces these gametes.

spot check 5 Given the function of a sperm, explain the need for each part of its anatomy.

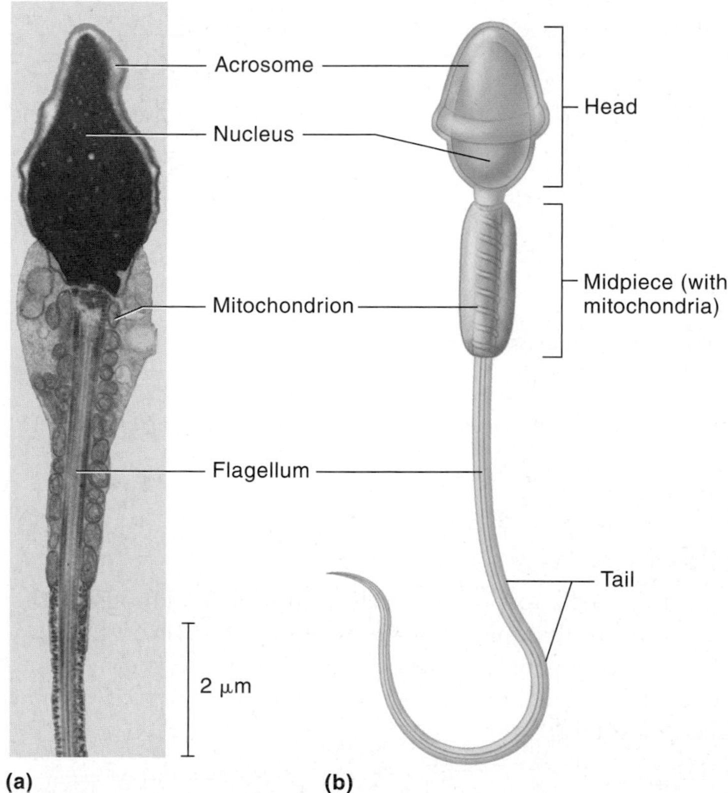

FIGURE 15.11 Mature spermatozoon: (a) head and part of tail, (b) anatomy of a sperm cell.

Physiology of the Male Reproductive System

15.6 learning outcome

Explain the hormonal control of puberty and the resulting changes in the male.

The physiology of the male reproductive system is governed by genetics and the endocrine system. You have already read how the SRY gene results in the formation of androgen receptors in the fetus. The developing gonads in the fetus produce large amounts of testosterone to fit into these receptors so that the male anatomy continues to develop. Testosterone production ceases a few months after birth and remains dormant until **puberty.** Puberty is the first few years of adolescence that begins with the production of FSH and LH at approximately age 10 to 12 in boys. It ends with the first ejaculation of viable sperm at around age 13.

Hormonal Control at Puberty

As you may remember from Chapter 8, the hypothalamus secretes GnRH at puberty, which targets the anterior pituitary. The anterior pituitary responds by producing FSH and LH. FSH targets the testes to stimulate sperm production. It does this by stimulating **sustentacular (Sertoli)** cells in the seminiferous tubules to produce a protein called **androgen-binding protein (ABP).** This protein allows testosterone to accumulate in the tubules to initiate sperm production. Without this protein, testosterone has little effect in the testes.

LH targets the interstitial cells between seminiferous tubules in the testes, telling them to produce testosterone. The testosterone initiates sperm production in the seminiferous tubules and travels throughout the body targeting many tissues to produce the male secondary sex characteristics.

Male Secondary Sex Characteristics These characteristics were originally discussed in Chapter 8, but it will be helpful to review them. The male secondary sex characteristics include:

- Skeletal and muscle development.
- Changes in the larynx that cause a deeper voice.
- Development of axillary and pubic hair with activation of associated apocrine glands.
- Development of facial hair and possible thickening of hair on the torso and limbs.
- Aggression.
- Development of the **libido** (sex drive).

spot check 6 What are the specific target cells for FSH and LH?

It is time to explore how sperm production is accomplished, beginning at puberty.

15.7 learning **outcome**

Explain the stages of meiosis and contrast meiosis to mitosis.

Sperm Production

Sperm production starts at puberty to produce cells with only 23 chromosomes. Until now, you have been aware that mitosis is the type of cell division that produces all of the body's cells except gametes. The body's cells have 46 chromosomes (23 pairs). The type of cell division that produces cells with half the normal number of chromosomes is called **meiosis.**

Mitosis versus Meiosis As you may recall from Chapter 2, mitosis is a one-division process, in which a parent cell of 46 chromosomes replicates and divides to produce two daughter cells. Each daughter cell has a set of 46 chromosomes that is identical to the set of the other daughter cell and to that of the parent cell, which no longer exists after the division takes place. Meiosis works very differently.

As you can see in **Figure 15.12**, meiosis involves two divisions. **Meiosis I** (first division) is shown on the left in this figure. Instead of showing all 46 chromosomes (an amount called **diploid, 2*n***), this figure shows just two pairs of chromosomes so that you can more easily see what is happening. Notice that between mid-to-late prophase and metaphase, the chromosomes no longer look exactly like the original chromosomes. The chromatids have broken and exchanged parts. This is called

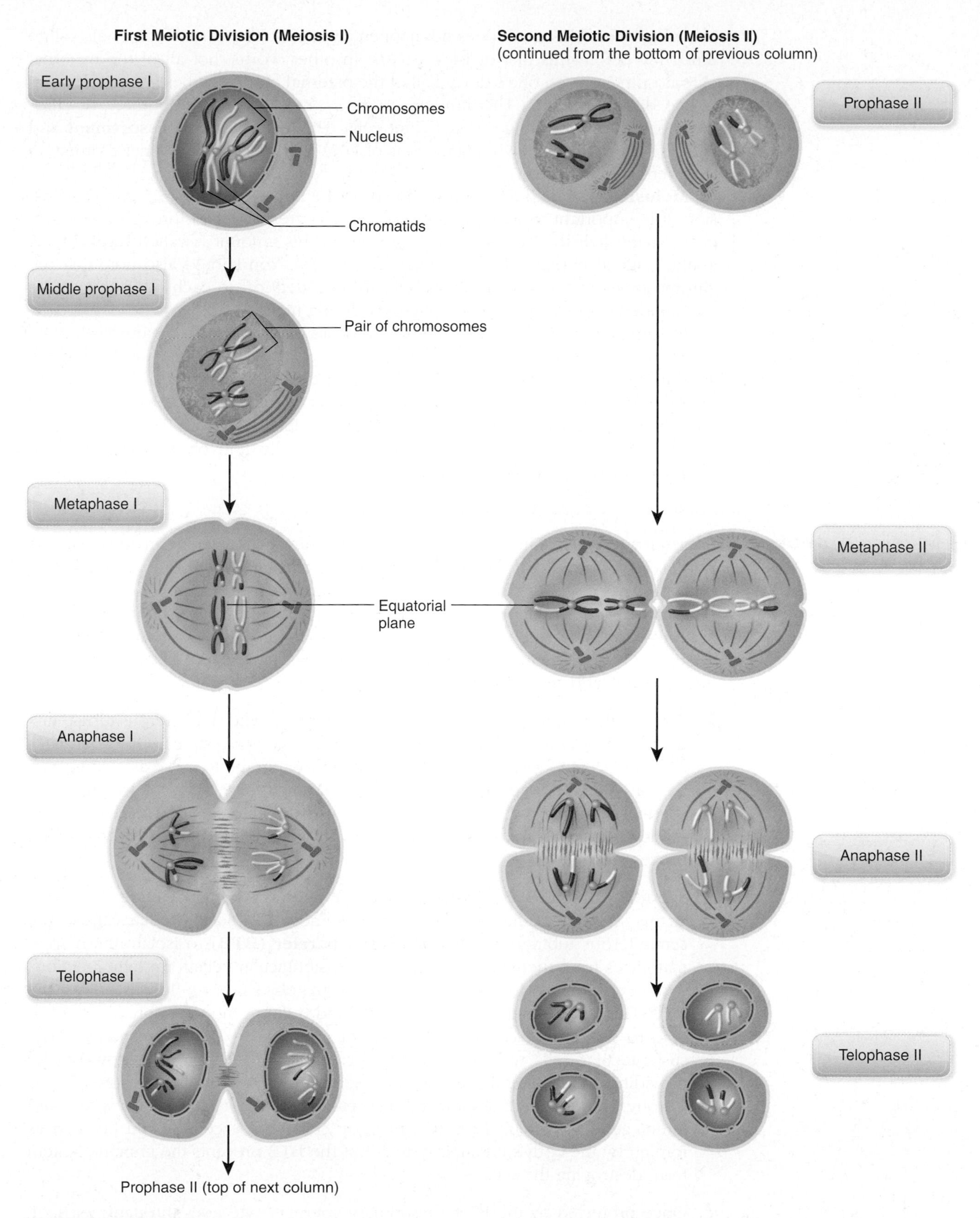

FIGURE 15.12 Meiosis. For simplicity, this diagram shows the stages of meiosis for just two of the 23 pairs of chromosomes. Notice that crossing-over and independent assortment in the two divisions of meiosis result in four gametes that are not identical, but each gamete has half the chromosomes as the starting cell.

crossing-over. This event does not happen in mitosis. The chromosomes also align randomly before this first division occurs. In other words, not all of the maternal chromosomes are on one side and all of the paternal chromosomes are on the other before division occurs. This random arrangement causes an assortment of chromosomes in the resulting two daughter cells. This **independent assortment** and crossing-over creates new combinations of DNA that will provide genetic variety in the eventual sperm.

Meiosis II (second division) is shown on the right in **Figure 15.12**. In this division, it is important to note the result. Four cells are produced through meiosis II, each having half the number of chromosomes (this amount is called **haploid, *n***) as the original parent cell (23 instead of 23 pairs). You should also note that the chromosomes in the four resulting cells are not identical to each other due to the independent assortment and crossing-over during meiosis I.

In summary, meiosis is a two-division process that starts with a parent cell of 46 chromosomes (23 pairs). It results in four daughter cells, each having 23 chromosomes. The set of chromosomes in each daughter cell is different from the sets in the other daughter cells due to independent assortment and crossing-over.

Now that you have seen how meiosis can produce cells with 23 chromosomes, you are ready to examine sperm production specifically. It involves two processes—**spermatogenesis** and **spermiogenesis.**

15.8 learning **outcome**

Explain the processes of sperm production and differentiate between spermatogenesis and spermiogenesis.

Spermatogenesis The first process in sperm production is spermatogenesis. Its purpose is to produce four cells (each with 23 chromosomes) from a specialized stem cell (**germ** cell) with 46 chromosomes. Germ cells are stem cells designed to form gametes. In the male, these germ cells are called **spermatogonia.** These cells are formed in the male fetus, but they remain dormant until testosterone levels rise at puberty, when sperm production begins. Spermatogenesis is explained in the following steps, and shown in **Figure 15.13**.

1. A spermatogonium near the basement membrane of the seminiferous tubule first divides by *mitosis*. This produces two identical spermatogonia. One spermatogonium (type A) remains near the basement membrane to serve as a spermatogonium later. This provides a continual, lifetime supply of germ cells. The other spermatogonium (type B) migrates slightly away from the wall toward the lumen (hollow center) of the seminiferous tubule. It will continue on in the production process.

2. Spermatogonium B enlarges to become a **primary spermatocyte.** Note that at this stage this cell still has 46 chromosomes ($2n$). Sustentacular cells in the seminiferous tubule form a **blood-testis barrier (BTB)** to isolate these spermatocytes from the immune system. The sustentacular cells form tight junctions with each other to isolate the primary spermatocytes on the lumen side of the BTB. This is like shutting a tight door behind them. A blood supply is still available to the sustentacular cells but is not available to the spermatocytes. The sustentacular cells will need to care for the spermatocytes from now on, by providing the developing cells with nutrients and removing their wastes. Meiosis (through independent assortment and crossing-over) produces cells with a different genetic makeup from that of body cells. These cells would be seen as foreign by the body's immune system, but the BTB prevents the immune system from destroying these cells.

3. Once protected by the BTB, the primary spermatocyte goes through meiosis I. This produces two equal-size, genetically unique **secondary spermatocytes,** each having 23 chromosomes (n). The sustentacular cells, which have a blood supply, are responsible for continuing to care for these secondary spermatocytes behind the BTB.

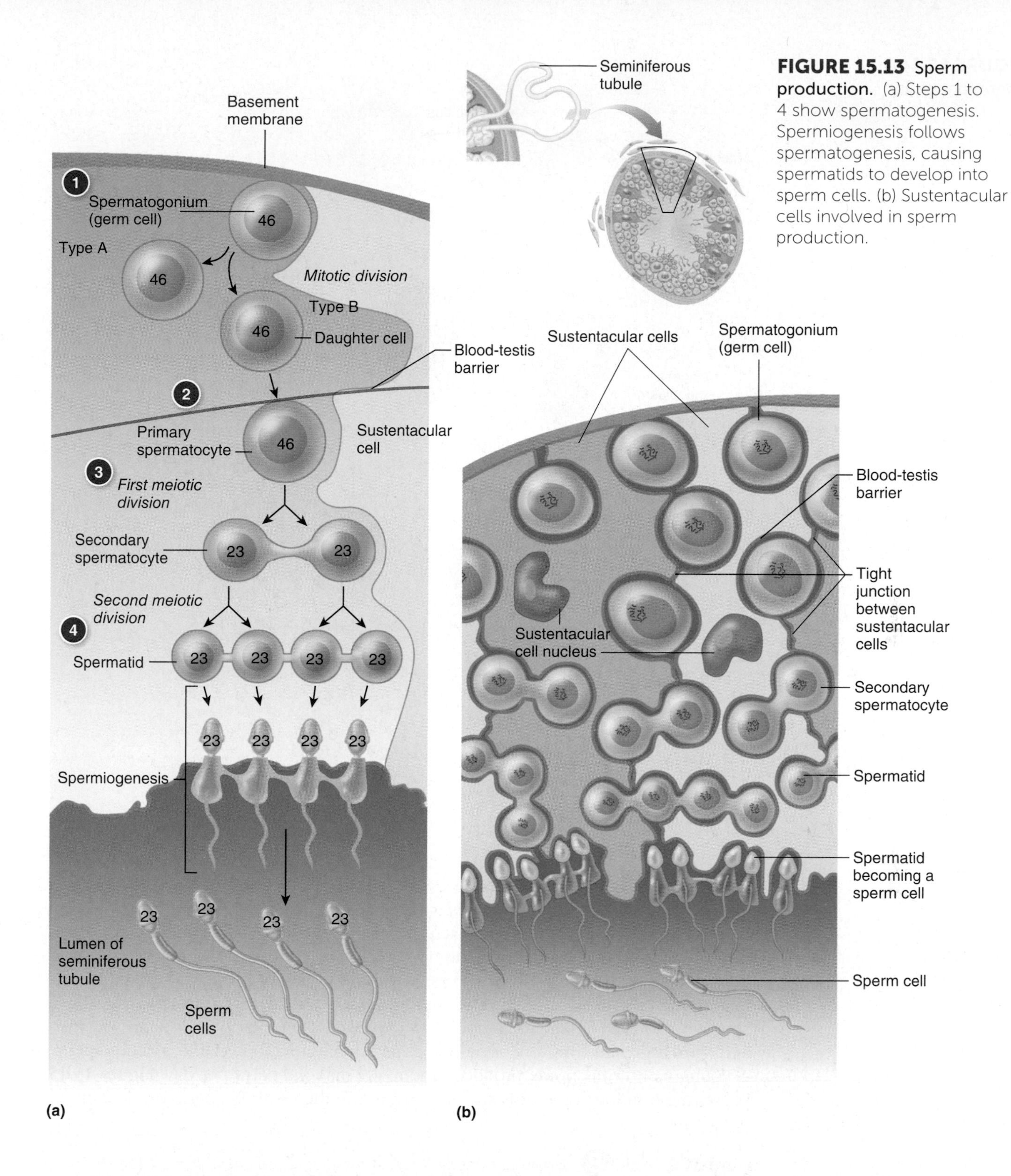

FIGURE 15.13 Sperm production. (a) Steps 1 to 4 show spermatogenesis. Spermiogenesis follows spermatogenesis, causing spermatids to develop into sperm cells. (b) Sustentacular cells involved in sperm production.

4. Each secondary spermatocyte undergoes meiosis II, which produces in total four **spermatids** from the one original spermatogonium type B. Notice in **Figure 15.13** that spermatids look nothing like sperm. They have cytoplasm and no tail. These spermatids are ready to go through the next process—spermiogenesis.

spot check 7 How do the chromosomes compare among the four spermatids?

FIGURE 15.14
Spermiogenesis.

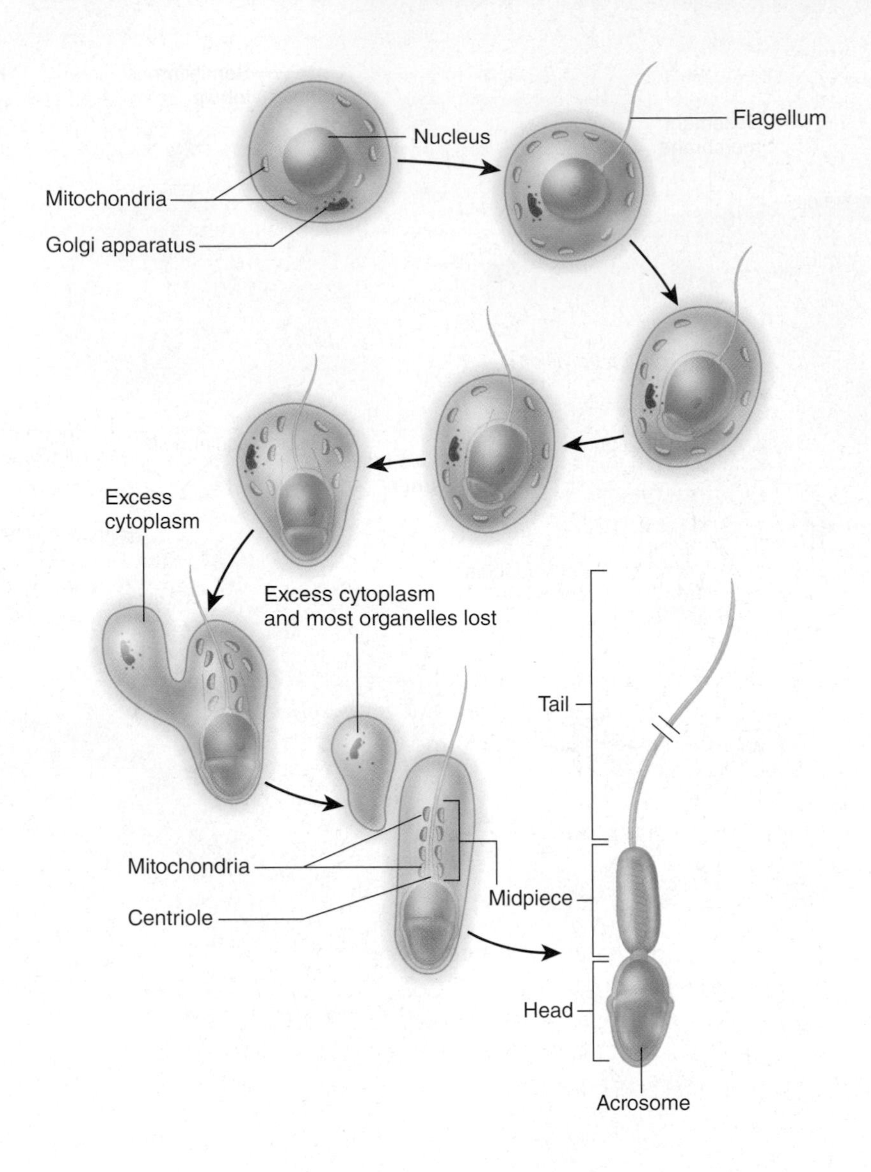

Spermiogenesis This second process in sperm production transforms spermatids to functional sperm. Spermiogenesis is shown in **Figure 15.13** and in more detail in **Figure 15.14**. During this process, each spermatid forms a tail and sheds its cytoplasm to become a sperm having the anatomy you studied earlier. Even though a sperm is fully formed at the end of this process, it will not be able to propel itself until it has matured in the epididymis. Once spermiogenesis is finished, sustentacular cells flush sperm out of the seminiferous tubules to move them on their way. It takes, on average, 74 days to go from one type-B spermatogonium to four viable, mature sperm. A young man in his prime produces about 400 million sperm per day. **Figure 15.15** shows the histology of a testis so that you can see the result of sperm production.

spot check 8 What happens to the sustentacular fluid used to flush sperm out of the seminiferous tubules?

Sperm production, and the male reproductive system as a whole, is under hormonal control by the hypothalamus, pituitary, and testes. In the next section, you will learn how hormones maintain homeostasis for this system.

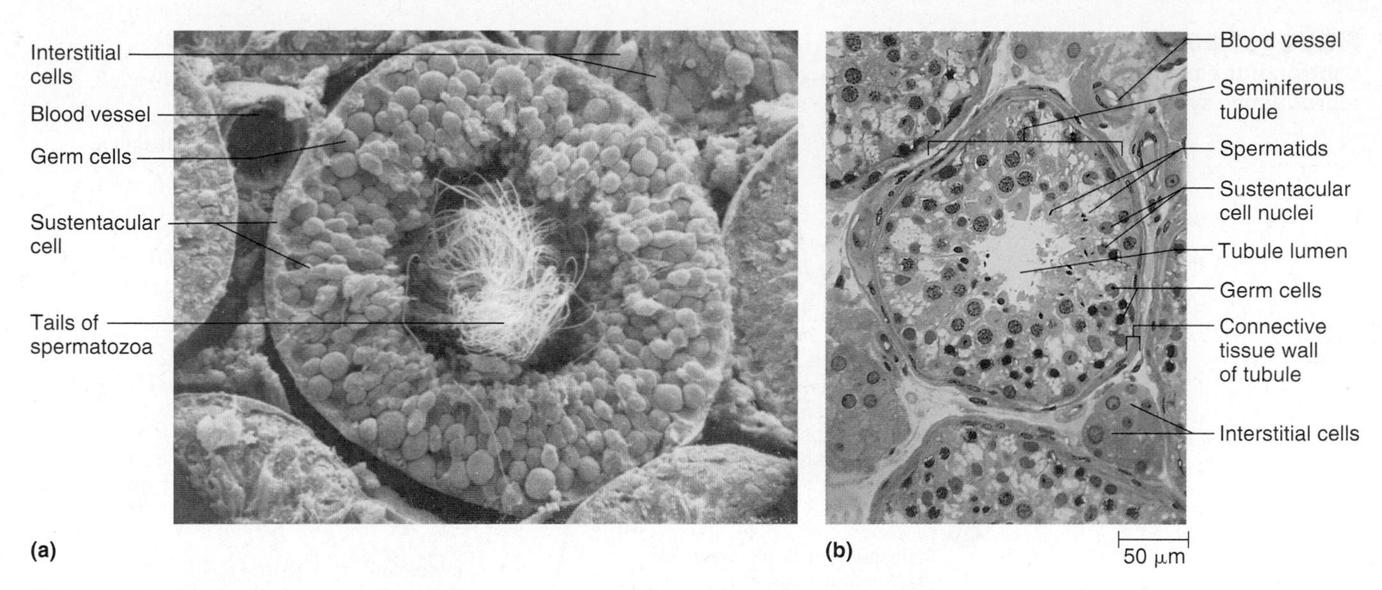

FIGURE 15.15 Histology of a Testis: (a) scanning electron micrograph of a seminiferous tubule with spermatozoa in the lumen, Source: Copyright by R.G. Kessel and R.H. Kardon, *Tissues and Organs: A Text-Atlas of Scanning Electron Microscopy, 1979,* W.H. Freeman, All rights reserved. (b) light micrograph of a seminiferous tubule without spermatozoa at the time.

Hormonal Control in the Adult Male

15.9 learning **outcome**

Explain the hormonal control of the adult male reproductive system.

You are already familiar with the green arrows in **Figure 15.16** because they show the same interactions that begin at puberty (discussed earlier). However, this diagram also shows the negative-feedback mechanisms involved with these hormones. For example, testosterone stimulates spermatogenesis in the testes if there is androgen-binding protein, and it stimulates libido and secondary sex characteristics in many other body tissues. But testosterone also has a negative-feedback effect on the hypothalamus, inhibiting any further production of GnRH, and on the anterior pituitary, reducing its sensitivity to GnRH. Inhibiting GnRH production means that FSH and LH production is also inhibited. This negative-feedback loop prevents the overproduction of testosterone. As a result, testosterone levels are maintained within homeostatic levels.

Figure 15.16 also includes a hormone called **inhibin,** which is produced by sustentacular cells in the seminiferous tubules. Like nurses who care for patients on a ward, sustentacular cells care for developing sperm in the tubules. Sustentacular cells produce ABP when they receive FSH and can handle more sperm production (patients), but they produce inhibin when their ward is full and they cannot handle any more. This is like sending a message to the hospital administration to stop sending any more patients. Inhibin targets the anterior pituitary to inhibit FSH production. As you may recall, FSH stimulates sustentacular cells to produce ABP, and both FSH and ABP are needed in the seminiferous tubules for testosterone to have an effect on sperm production. Without FSH and ABP, sperm production is put on hold. Inhibin has no effect on LH production from the anterior pituitary. It affects only the pituitary's production of FSH.

spot check 

What effect does inhibin have on testosterone production?

FIGURE 15.16 Hormonal control of the male reproductive system.

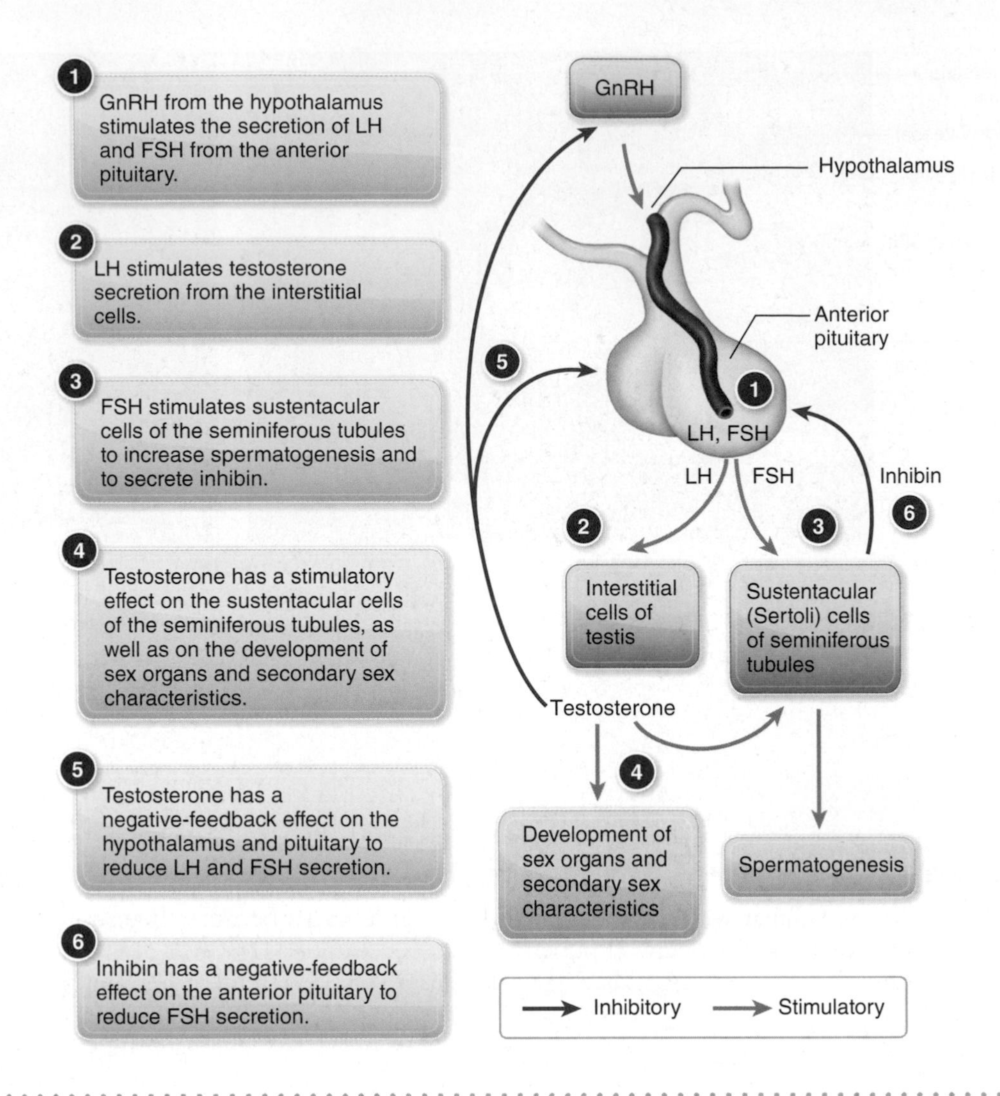

15.10 learning **outcome**

Trace the path a sperm takes from its formation to its ejaculation.

Pathway for Sperm

You have become familiar with the male anatomy, sperm anatomy, sperm production, and the hormonal control of the male reproductive system. But how do sperm get to where they need to go in order to fertilize an egg? First, review the pathway they must travel to leave the male body:

Sperm travel from the seminiferous tubules → to the rete testes → to the efferent ductules → to the epididymis → to the ductus deferens → to the ejaculatory ducts → to the urethra → to outside the body

Now that you have reviewed the pathway, you are ready to study how sperm move from the epididymis to outside the body during the male sexual response.

clinical point

A surgical procedure called a **vasectomy** can be done to make the male sterile as a form of birth control. In a vasectomy, a section of each ductus deferens in the scrotum is excised (removed). The cut ends are then ligated (tied) or cauterized (sealed) to ensure complete blockage. This procedure does not affect sperm or semen production. Rather, it simply prevents sperm from traveling past the excision. As a precaution, the patient's semen should be tested for sperm after the procedure to make sure the vasectomy was successful.

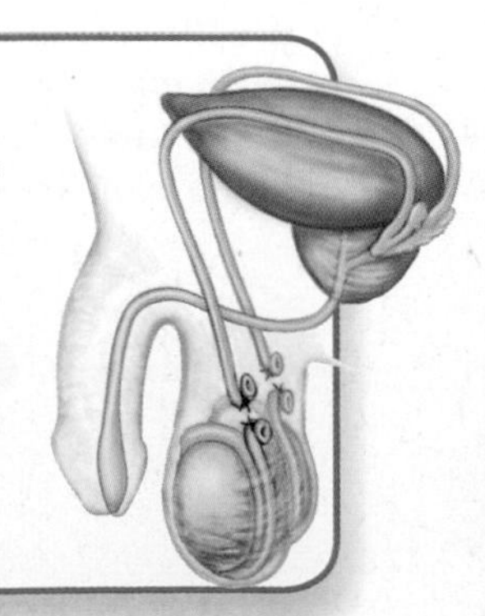

Sexual Response in the Male

15.11 learning outcome

Describe the stages of the male sexual response.

The male sexual response has four stages—**arousal, emission, ejaculation,** and **resolution.** Two of these stages—emission and ejaculation—are orgasm in the male. These stages are described in the following list and summarized in **Figure 15.17**.

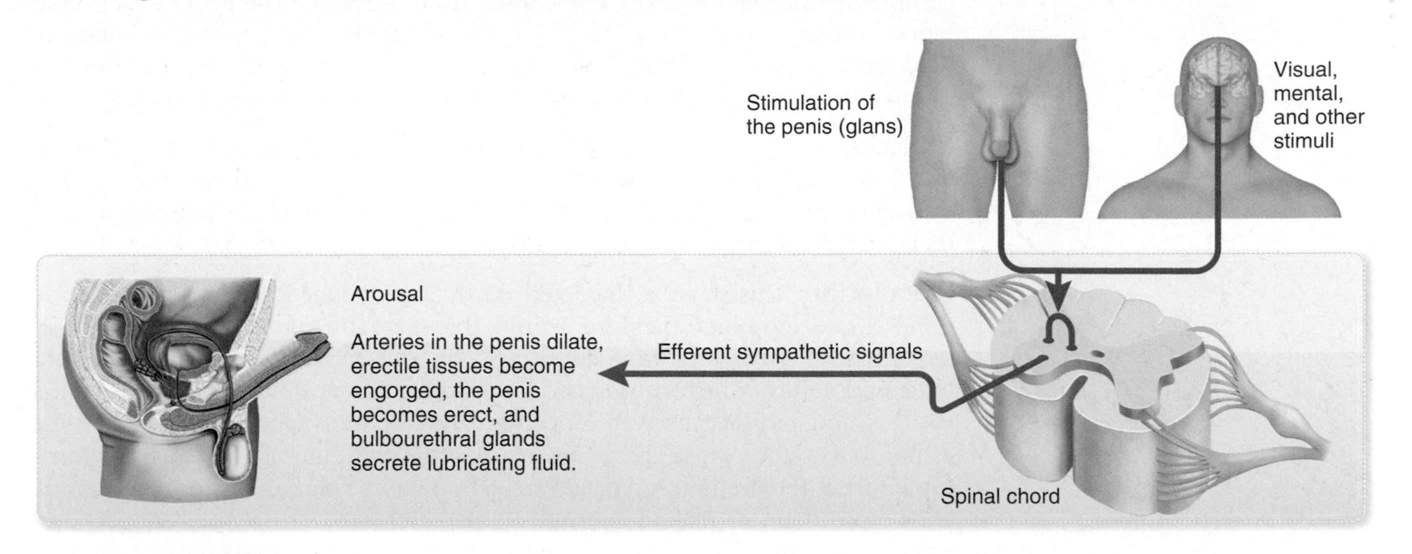

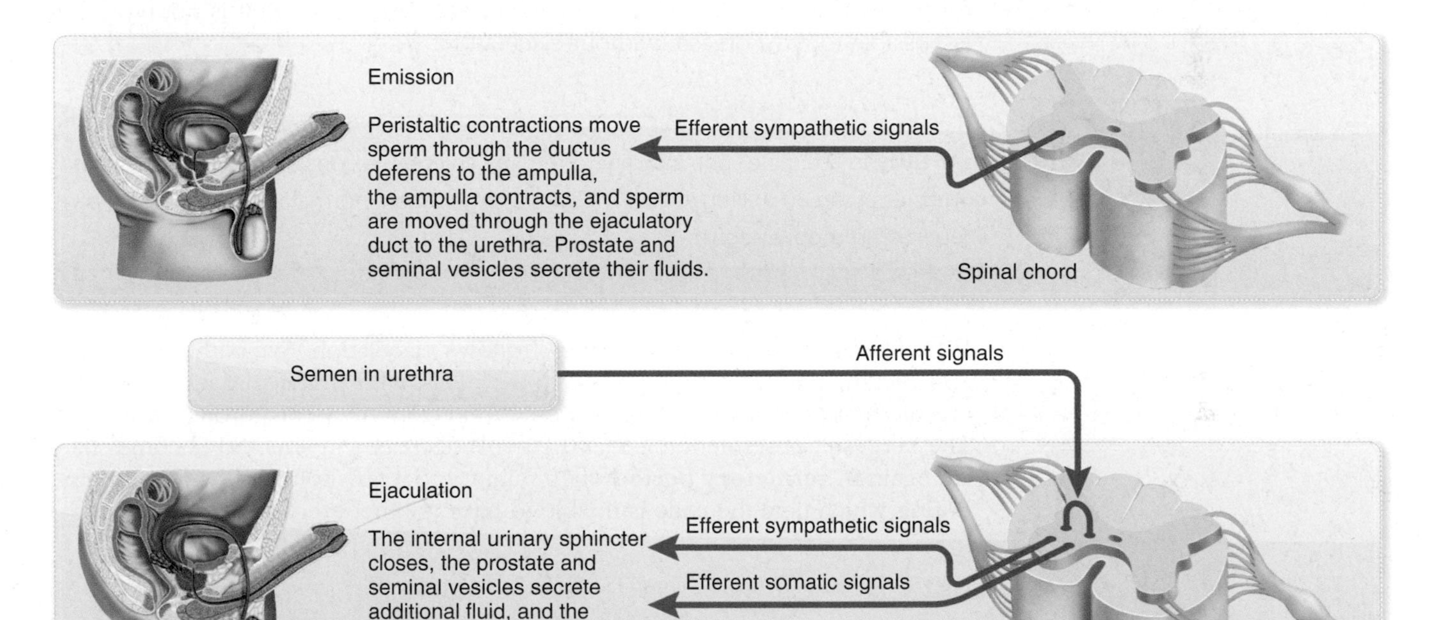

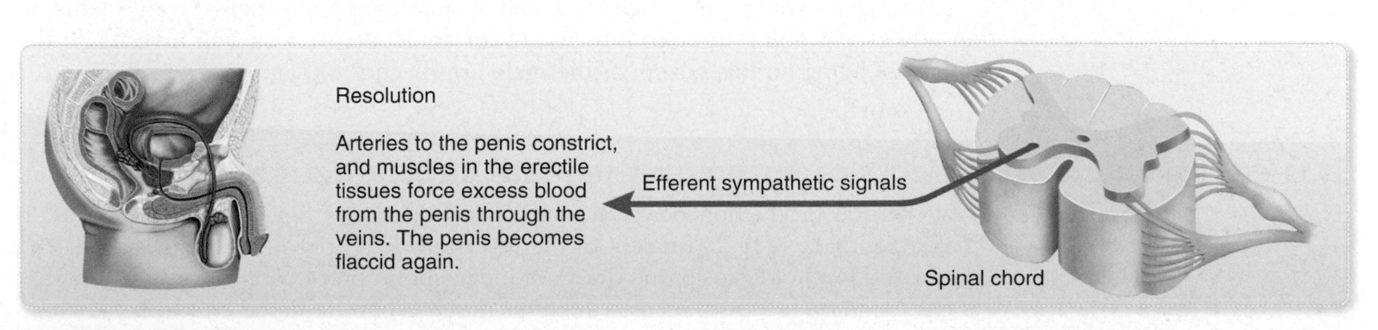

FIGURE 15.17 The male sexual response. This figure summarizes the stages and describes the neural control of each stage.

- **Arousal.** The most obvious sign of this phase is an **erection** of the penis. This can be the result of an autonomic reflex from stimulation of the penis, or it can be initiated by sight, sound, smell, or even thought. Neural signals cause the arteries in the penis to dilate, so the penile erectile tissues engorge with blood. This compresses the veins carrying blood away from the penis. The net effect of more blood coming into the penis than leaving is that the penis becomes enlarged, rigid, and erect. This erection makes penetration of a woman's vagina possible. The bulbourethral glands also secrete their lubricating fluid during this stage.
- **Emission.** During emission, sperm are moved by peristaltic contractions through the ductus deferens to its ampulla. Contractions here move sperm through the ejaculatory duct to the urethra. In addition, the prostate and seminal vesicles add their secretions to semen during this stage.
- **Ejaculation.** This stage is characterized by a pronounced, temporary increase in heart rate and blood pressure. During this stage, the internal urethral sphincter closes, so sperm and urine never mix. The presence of semen in the urethra initiates a reflex, which causes additional secretion of fluids from the seminal vesicles and prostate, as well as contractions of the bulbocavernosus muscle at the root of the penis. The additional fluid and the muscle contractions force semen from the urethra. A typical ejaculation has a volume greater than 2.5 mL with more than 20 million sperm/mL, of which 60% are normal-shaped and motile.[1] The release of tension and accompanying intense, pleasurable sensation for the male during ejaculation is the male **orgasm.** Although it is not typical, it is possible to have orgasm without ejaculation.

clinical point

Infertility in the male—the inability to fertilize an egg—is considered to be a **sperm count** less than 20 million sperm/mL with a larger-than-normal percentage of misshapen or immobile sperm.

- **Resolution.** This stage immediately follows ejaculation. Sympathetic neurons constrict the arteries bringing blood to the penis, while trabecular muscles in the erectile tissues contract to force excess blood from the penis through the veins. The net result of resolution is that the penis decreases in size and becomes flaccid again. A **refractory period** of 10 minutes to a few hours follows resolution, during which time the male is unable to have another erection.

spot check 10 What might stimulate an erection?

spot check 11 What physically happens to cause a penis to become erect?

You have explored the changes in this system over a life span—from a fetus, to an adolescent going through puberty, to an adult. Below, you will investigate the effects of aging on this system as the male approaches old age.

15.12 learning outcome

Explain the effects of aging on the male reproductive system.

Effects of Aging on the Male Reproductive System

Testosterone production peaks at age 20 and declines from there, so at age 80 a male may make only 20 percent of the testosterone he produced at his prime. The decreased level of testosterone does not mean sperm production ceases. Although it is not common to do so, men in their 70s and 80s have fathered children.

The negative-feedback mechanism of high testosterone is also lost by the greatly diminished testosterone levels. So GnRH is not inhibited, and FSH and LH levels increase, rising to significantly higher levels after age 50. The increased levels of these two hormones result in the **male climacteric** or **andropause.** This event is sometimes referred to as the "male menopause," but this term is inaccurate. *Menopause* means the cessation of the menses (menstrual periods in women), an event that does not occur in the male. Most men have few symptoms during andropause, but some may experience hot flashes and mood swings.

Another effect of aging on the male is **erectile dysfunction** (**ED** or **impotence**). An occasional inability to achieve an erection is not considered erectile dysfunction. However, ED is the *frequent* (more than half the time) inability to achieve an erection sufficient to penetrate a vagina. ED occurs in about 20 percent of men in their 60s and increases to 50 percent of men in their 80s. In a normal erection, a nerve stimulus results in the release of a chemical that causes the smooth muscle in the arteries delivering blood to the penis to relax. So the arteries dilate, and the erectile tissues become engorged. The arteries remain dilated until an enzyme degrades the chemical. Drugs, such as sildenafil (Viagra), vardenafil (Levitra), and tadalafil (Cialis), treat erectile dysfunction by inhibiting the enzyme that degrades the chemical so that the arteries remain dilated longer.

Aging in the male also affects the prostate gland. The size of the prostate remains relatively stable in an adult until about age 45. Then it slowly enlarges. Eighty percent of men have **benign prostatic hyperplasia (BPH)** by the age of 80. As the name suggests, this is a noncancerous, nonmetastatic enlargement, which grows toward the middle of the prostate. This enlargement compresses the urethra, making micturition and emptying the bladder more difficult. A man with this condition has frequent micturition because the bladder is not fully emptied and has difficulty maintaining a steady stream of urine.

Male Reproductive System Disorders

15.13 learning outcome

Describe male reproductive system disorders.

During the course of this chapter, you have become familiar with some male reproductive disorders, including androgen-insensitivity syndrome, cryptorchidism, erectile dysfunction, and benign prostatic hyperplasia. Below, you will explore a few more.

Prostate Cancer

Other than skin cancer, prostate cancer is the most common cancer in men over 50. In this condition, the prostatic enlargement occurs to the outside, so there is little effect on the urethra. The lack of the typical symptoms seen with BPH may allow the cancer to proceed undetected. A **digital rectal exam (DRE)** can be performed from the rectum by palpating the prostate to check its size. This cancer can also be detected by a blood test measuring **prostate-specific antigen (PSA)** levels. This antigen is produced in greater amounts when prostate cancer is present. Eighty percent of individuals survive prostate cancer if it is detected and treated early.

Testicular Cancer

Testicular cancer is most common in white males between the ages of 15 and 34. Routine testicular self-exams are recommended for early detection. This form of cancer is highly curable if treated early.

hypospadias: high-poh-SPAY-dee-as

Hypospadias

Hypospadias is a congenital defect (present at birth) in which the urethra opens on the ventral side or base of the penis instead of on the tip of the glans. It can usually be surgically corrected during an infant's first year.

putting the pieces **together**

The Reproductive Systems

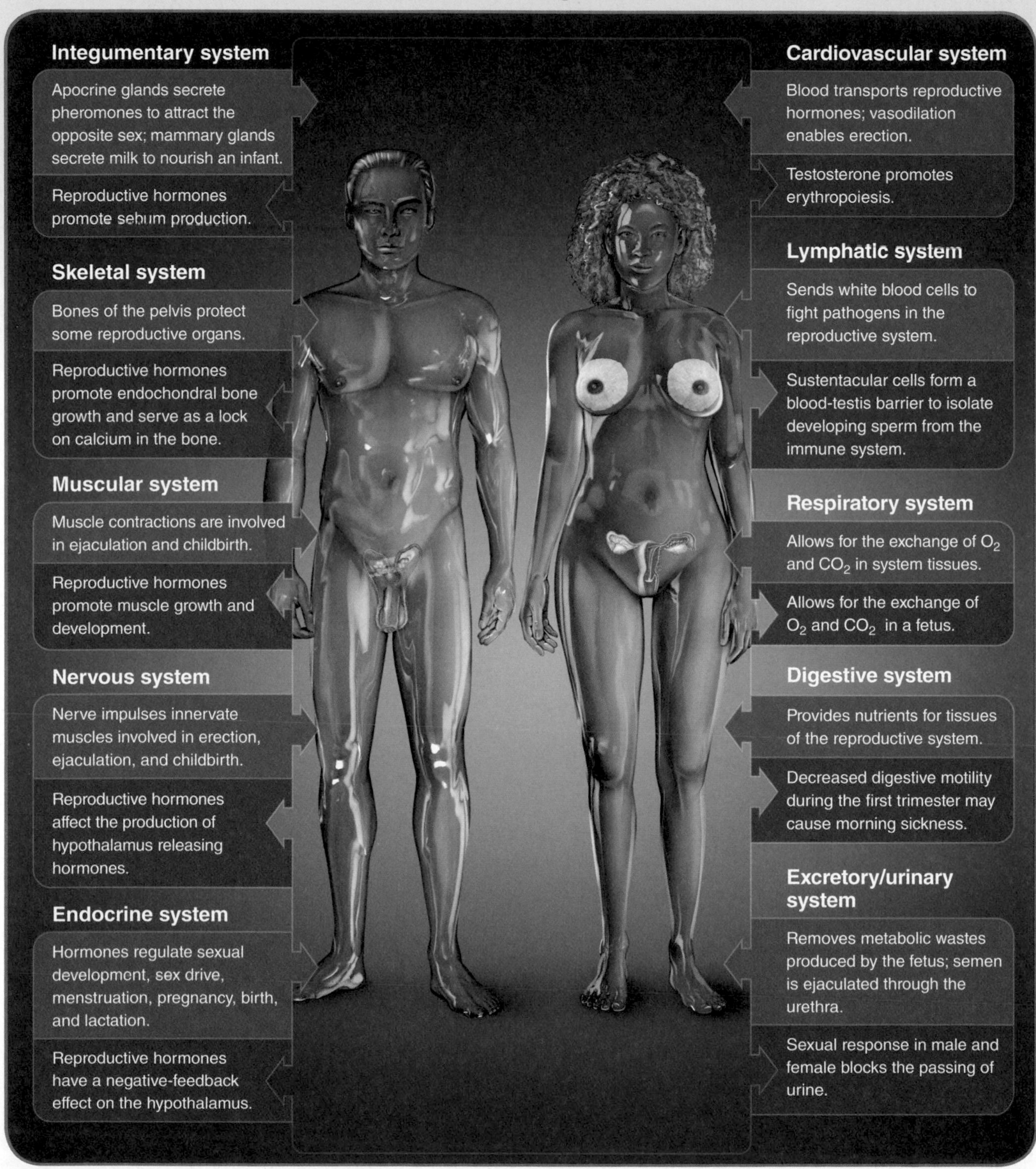

FIGURE 15.18 Putting the Pieces Together—The Reproductive Systems: connections between the male and female reproductive systems and the body's other systems.

summary

Overview

- All humans start from a zygote.
- Gender is determined by sperm from the father.
- The SRY gene on the Y chromosome codes for a protein so that androgen receptors are produced in a male fetus.
- Testosterone and androgen receptors are needed in the fetus for male anatomy to develop.

Male Reproductive Anatomy

Testes

- Testes belong to the endocrine and reproductive systems because they produce testosterone and sperm.
- Testes descend from the abdomen to the scrotum because of the gubernaculum.
- For the testes to produce viable sperm, the temperature of each testis must be approximately 2°C cooler than the core body temperature.
- Each testis is divided into lobules.
- Sperm are produced in the seminiferous tubules.
- Interstitial cells, located between seminiferous tubules, produce testosterone.

Secondary Sex Organs and Structures

- The testes are housed in the scrotum.
- The spermatic cord suspends each testis and is composed of the cremaster muscle, the ductus deferens, the testicular artery, and the pampiniform plexus.
- The pampiniform plexus is a network of veins surrounding the testicular artery.
- The temperature of the testes is maintained by the dartos muscle, the cremaster muscles, and the pampiniform plexus.
- The spermatic ducts are the efferent ductules, the ductus deferens, and the ejaculatory ducts.
- The five accessory glands in the male are two seminal vesicles, one prostate gland, and two bulbourethral glands.
- The accessory glands produce the fluids in semen.
- The penis has an internal root and an external shaft and glans.
- The penis is composed of three columns of erectile tissue.

Anatomy of a Sperm

- A spermatozoon is a single cell with two principal parts—a head and a tail.
- The head contains 23 chromosomes and an enzyme-filled acrosome used to penetrate an egg.
- The tail contains a midpiece with large mitochondria to produce the energy to move the tail's flagellum.

Physiology of the Male Reproductive System

- Testosterone production ceases a few months after birth and does not resume until puberty.
- Puberty begins with the production of FSH and LH, and it ends with the first ejaculation of viable sperm.

Hormonal Control at Puberty

- FSH stimulates sustentacular cells to produce ABP.
- ABP and testosterone are needed for testosterone to have an effect on sperm production.
- LH stimulates interstitial cells to produce testosterone.

Sperm Production

- Mitosis is a one-division process that forms all body cells.
- Meiosis is a two-division process that forms gametes.
- Crossing-over and independent assortment in meiosis create new combinations of DNA that provide genetic variety in sperm.
- Sperm production involves two processes—spermatogenesis and spermiogenesis.
- Spermatogenesis involves meiosis to form four spermatids from one spermatogonium.
- Spermiogenesis is the development of four sperm from four spermatids.

Hormonal Control in the Adult Male

- Testosterone has a positive effect on many tissues, but it also has a negative-feedback effect on the hypothalamus and the anterior pituitary.
- Sustentacular cells produce the hormone inhibin, when they are busy with sperm production, so that FSH from the anterior pituitary is inhibited.

Pathway for Sperm

- Sperm travel from the seminiferous tubules, to the rete testes, to the efferent ductules, to the epididymis, to the ductus deferens, to the ejaculatory ducts, to the prostatic urethra, to the penile urethra, and then to outside the body.

Sexual Response in the Male

- The four stages of the male sexual response during intercourse are arousal, emission, ejaculation, and resolution.
- Arousal results in an erection.
- During emission, sperm move to the urethra and accessory glands secrete fluids for semen.
- During ejaculation, sperm are forcefully expelled from the urethra.
- During resolution, excess blood is forced from the penis, thus causing it to become flaccid.

Effects of Aging on the Male Reproductive System

- Testosterone production peaks at age 20 and declines from there, so at age 80 a male may make only 20 percent of the testosterone he produced at his prime.
- Sperm production continues into old age.
- Increased FSH and LH levels rise significantly after age 50, producing andropause.
- Erectile dysfunction occurs in about 20 percent of men in their 60s and increases to 50 percent of men in their 80s.
- Eighty percent of men have benign prostatic hyperplasia by the age of 80.

Male Reproductive System Disorders

Prostate Cancer

- Other than skin cancer, prostate cancer is the most common cancer in men over the age of 50.
- Prostate cancer can be detected by a digital rectal exam and a blood test that measures PSA levels.

Testicular Cancer

- Testicular cancer is most common in white males between the ages of 15 and 34.
- Routine testicular self-exams are recommended for early detection.

Hypospadias

- Hypospadias is a congenital defect in which the urethra opens on the ventral side or base of the penis instead of on the tip of the glans.

key words for review

The following terms are defined in the glossary.

blood-testis barrier (BTB)
climacteric
crossing-over
cryptorchidism
ejaculation
emission
erection
gamete
infertility
meiosis
pampiniform plexus
puberty
resolution
secondary sex characteristics
semen
smegma
spermatic cord
spermatogenesis
spermiogenesis
zygote

chapter review questions

Word Building: *Use the word roots listed in the beginning of the chapter and the prefixes and suffixes inside the back cover to build words with the following meanings:*

1. Sperm production: ______________________________
2. Inflammation of the epididymis: ______________________________
3. Removal of a testis: ______________________________
4. Enlargement of the prostate gland: ______________________________
5. Pertaining to the penis: ______________________________

Multiple Select: *Select the correct choices for each statement. The choices may be all correct, all incorrect, or a combination of correct and incorrect.*

1. What is the anatomy of a sperm?
   a. A sperm contains 46 chromosomes.
   b. A sperm has an acrosome cap for energy.
   c. A sperm contains abundant cytoplasm.
   d. A sperm has large mitochondria to produce ATP.
   e. A sperm has a flagellum for movement.
2. What happens during sperm production?
   a. Interstitial cells care for sperm.
   b. Sperm are produced in the seminiferous tubules.
   c. Developing sperm undergo meiosis during spermiogenesis.
   d. Interstitial cells protect sperm from the BTB.
   e. Sustentacular cells secrete a fluid to flush sperm from the seminiferous tubules.
3. What happens to sperm after they are produced?
   a. Sperm mature in the rete testes.
   b. Sperm are stored in the epididymis.
   c. Sperm travel from the rete testes to the efferent ductules.
   d. Sperm disintegrate at 40 to 60 days if they are not ejaculated.
   e. Sperm stay in the testes until they are ejaculated.

4. What is the composition of semen?
   a. Semen contains a lubricating fluid from the bulbourethral gland.
   b. Semen is 30 percent sperm.
   c. Sixty percent of semen is an alkaline fluid from the prostate gland.
   d. Semen is composed of sperm and three fluids from the accessory glands.
   e. Semen contains a nourishing fluid from the seminal vesicles.
5. Which of the following is (are) true about the effects of testosterone?
   a. Testosterone is responsible for the development of secondary sex characteristics in the male.
   b. Testosterone has a negative-feedback effect on sustentacular cells.
   c. Testosterone has a negative-feedback effect on most body tissues.
   d. Testosterone stimulates the hypothalamus to produce GnRH.
   e. Testosterone increases the anterior pituitary's sensitivity to GnRH.
6. What is needed for male anatomy to develop?
   a. The mother must supply a Y chromosome in her egg.
   b. The Y chromosome must have an SRY gene.
   c. Testosterone must be produced in the fetus.
   d. Androgen receptors must be formed in the fetus.
   e. ABP must be produced in the fetus.
7. How do meiosis and mitosis differ?
   a. Mitosis is a two-division process, while meiosis involves only one division.
   b. Meiosis results in twice the number of daughter cells than does mitosis.
   c. Meiosis begins with a cell having 46 chromosomes, while mitosis begins with a cell having 23 chromosomes.
   d. Meiosis includes crossing-over, while mitosis does not.
   e. Both mitosis and meiosis result in daughter cells that are genetically identical.
8. Where do sperm travel?
   a. Sperm travel from the ejaculatory duct to the penile urethra.
   b. Sperm travel from the efferent ductules to the rete testes.
   c. Sperm travel from the ductus deferens to the ejaculatory duct.
   d. Sperm leave the body through the internal urethral orifice.
   e. Sperm travel through the prostate gland while in the ejaculatory duct and the prostatic urethra.
9. Which of the following is (are) true concerning the effects of aging on the male reproductive system?
   a. Men can father children in old age.
   b. Testosterone production continues at its peak until age 40, when it begins to decline.
   c. Men may go through andropause due to high levels of inhibin.
   d. The prostate gland enlarges in most men as they age.
   e. Half of 80-year-olds likely have erectile dysfunction.
10. How would you describe male reproductive disorders?
   a. Cryptorchidism occurs when testes fail to descend into the scrotum.
   b. Erectile dysfunction is diagnosed if a male fails to have an erection.
   c. Testicular cancer is most common in middle-aged white males.
   d. Benign prostatic hyperplasia can be detected by a blood test for PSA.
   e. Hypospadias is a low sperm count.

Matching: *Match the stage in the male sexual response to its description. Some answers may be used more than once. Some questions may have more than one correct response.*

_____ 1. Arteries in the penis dilate.
_____ 2. Sperm are moved to the urethra.
_____ 3. Arteries in the penis constrict.
_____ 4. Bulbocavernosus muscles constrict.
_____ 5. Orgasm occurs in the male.

a. Resolution
b. Emission
c. Ejaculation
d. Arousal

Matching: *Match the sperm production process to the description. Some answers may be used more than once.*

_____ 6. Spermatogonia divide by mitosis.
_____ 7. Cells divide by meiosis.
_____ 8. Cells shed cytoplasm.
_____ 9. Cells develop a tail.
_____ 10. Spermatids are formed.

a. Spermiogenesis
b. Spermatogenesis

Critical Thinking

1. Anabolic steroids are drugs like testosterone or drugs that act like testosterone. They are often prescribed by doctors to treat conditions such as delayed puberty or cryptorchidism. As you have already learned about testosterone, anabolic steroids make muscles bigger and bones stronger. Some adults and teens take illegal (not medically prescribed) anabolic steroids to lower body fat and increase muscle size and strength. The dose of illegal steroids is often 10 to 100 times higher than what would normally be prescribed, and more than one steroid may be taken in combination (called *stacking*). Given what you know of the effects of testosterone, predict the effects of illegal anabolic steroid use on (a) body tissues in general, (b) the hypothalamus, (c) the anterior pituitary, (d) levels of FSH and LH, (e) the testes, and (f) sperm production.
2. Physical trauma to the testes can result in sterility because of the immune system's action. Explain how this might happen.
3. What can a male do to protect himself from the effects of reproductive disorders? Be specific.

15 chapter mapping

This section of the chapter is designed to help you find where each outcome is covered in the text.

	Outcomes	Readings, figures, and tables	Assessments
15.1	Use medical terminology related to the male reproductive system.	Word roots: p. 561	Word Building: 1–5
15.2	Explain what is needed for male anatomy to develop.	Overview: pp. 562–563 Figures 15.2, 15.3	Multiple Select: 6
15.3	Describe the anatomy of the testes.	Male reproductive anatomy; Testes: pp. 563–565 Figures 15.4, 15.5	Spot Check: 1, 2 Critical Thinking: 2
15.4	Describe the male secondary sex organs and structures and their respective functions.	Secondary sex organs and structures: pp. 565–570 Figures 15.6–15.10	Spot Check: 3, 4, 8 Multiple Select: 3, 4
15.5	Describe the anatomy of a sperm.	Anatomy of a sperm: p. 571 Figure 15.11	Spot Check: 5 Multiple Select: 1
15.6	Explain the hormonal control of puberty and the resulting changes in the male.	Physiology of the male reproductive system; Hormonal control at puberty: pp. 571–572	Spot Check: 6 Multiple Select: 5
15.7	Explain the stages of meiosis and contrast meiosis to mitosis.	Sperm production; Mitosis versus meiosis: pp. 572–574 Figure 15.12	Multiple Select: 7
15.8	Explain the processes of sperm production and differentiate between spermatogenesis and spermiogenesis.	Spermatogenesis: pp. 574–577 Figures 15.13–15.15	Spot Check: 7 Multiple Select: 2 Matching: 6–10
15.9	Explain the hormonal control of the adult male reproductive system.	Hormonal control in the adult male: pp. 577–578 Figure 15.16	Spot Check: 9 Critical Thinking: 1
15.10	Trace the path a sperm takes from its formation to its ejaculation.	Pathway for sperm: p. 578	Multiple Select: 3, 8
15.11	Describe the stages of the male sexual response.	Sexual response in the male: pp. 579–580 Figure 15.17	Spot Check: 10, 11 Matching: 1–5
15.12	Explain the effects of aging on the male reproductive system.	Effects of aging on the male reproductive system: pp. 580–581	Multiple Select: 9
15.13	Describe male reproductive system disorders.	Male reproductive disorders: p. 581	Multiple Select: 10 Critical Thinking: 3

footnote

1. Beers, M. H., Porter, R. S., Jones, T. V., Kaplan, J. L., & Berkwits, M. (Eds.). (2006). *Merck manual of diagnosis and therapy* (18th ed.). Whitehouse Station, NJ: Merck Research Laboratories.

references and resources

Beers, M. H., Porter, R. S., Jones, T. V., Kaplan, J. L., & Berkwits, M. (Eds.). (2006). *Merck manual of diagnosis and therapy* (18th ed.). Whitehouse Station, NJ: Merck Research Laboratories.

Healthwise. (2009, July 6). *What are anabolic steroids?*. Retrieved November 18, 2010, from http://www.bing.com/health/article/healthwise-1250002102.

Saladin, Kenneth S. (2010). *Anatomy & physiology: The unity of form and function* (5th ed.). New York: McGraw-Hill.

Seeley, R. R., Stephens, T. D., & Tate, P. (2006). *Anatomy & physiology* (7th ed.). New York: McGraw-Hill.

Shier, D., Butler, J., & Lewis, R. (2010). *Hole's human anatomy & physiology* (12th ed.). New York: McGraw-Hill.

Sumfest, Joel M. (2009, January). *Cryptorchidism treatment and management.* Retrieved June 28, 2011, from http://emedicine.medscape.com/article/438378-treatment.

16 The Female Reproductive System

Men and women are similar, but they also very different when it comes to their reproductive systems. For example, a man's reproductive system only has to produce and deliver male gametes (sperm) to contribute to reproduction. On the other hand, a woman's reproductive system must supply a gamete (an egg), receive the male gametes, provide a place for the developing fetus to mature, facilitate the birthing process, and then nourish the infant. See Figure 16.1. The man's reproductive system will produce hormones and gametes every day from puberty into old age, while the woman's will not. Her reproductive system works in monthly cycles, while his does not. However, with all their differences, women are like men in that they do not need to reproduce to survive. And even if they both choose not to reproduce, their reproductive systems define the secondary sex characteristics that contribute greatly to who they are.

16.1 learning **outcome**

Use medical terminology related to the female reproductive system.

learning outcomes

After completing this chapter, you should be able to:

16.1 Use medical terminology related to the female reproductive system.

16.2 Explain what is needed for female anatomy to develop.

16.3 Describe the anatomy of the ovary and its functions.

16.4 Describe the female secondary reproductive organs and structures and their respective functions.

16.5 Explain the hormonal control of puberty and the resulting changes in the female.

16.6 Explain oogenesis in relation to meiosis.

16.7 Explain the hormonal control of the adult female reproductive system and its effect on follicles in the ovary and the uterine lining.

16.8 Describe the stages of the female sexual response.

16.9 Explain the effects of aging on the female reproductive system.

16.10 Describe female reproductive disorders.

16.11 List the four requirements of pregnancy.

16.12 Trace the pathway for a sperm to fertilize an egg.

16.13 Describe the events necessary for fertilization and implantation.

16.14 Explain the hormonal control of pregnancy.

16.15 Explain the adjustments a woman's body makes to accommodate a pregnancy.

16.16 Explain the nutritional requirements for a healthy pregnancy.

16.17 Explain what initiates the birth process.

16.18 Describe the birth process.

16.19 Explain the process of lactation.

16.20 Describe disorders of pregnancy.

word **roots** & combining **forms**

amni/o: amnion

cervic/o: cervix, neck

chorion/o: chorion

episi/o: vulva

gynec/o: female

hyster/o: uterus

lact/o: milk

mamm/o: breast

mast/o: breast

men/o: menses, menstruation

metr/o, metri/o: uterus

o/o: egg

oophor/o: ovary

ov/o: egg

ovari/o: ovary

ovul/o: egg

salping/o: uterine tube

uter/o: uterus

vagin/o: vagina

vulv/o: vulva

pronunciation **key**

atresia: a-TREE-zee-ah

fimbriae: FIM-bree-eye

oogenesis: oh-oh-JEN-eh-sis

oogonia: oh-oh-GO-nee-ah

parturition: PAR-chur-ISH-un

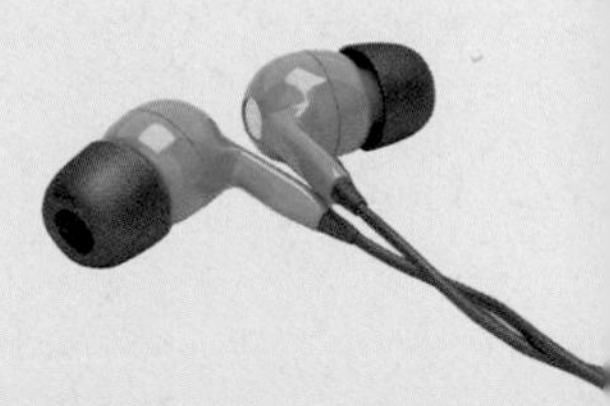

16.2 learning outcome

Explain what is needed for female anatomy to develop.

Overview

Before you start exploring the female reproductive system, you should understand what is necessary for an individual to develop female anatomy. As you may recall from Chapter 15, a female zygote has XX as the sex chromosomes (23rd pair) instead of XY as in the male. Without the SRY gene on the Y chromosome, and therefore the lack of androgen receptors and testosterone, a zygote develops female reproductive anatomy. In this chapter, you will study the female anatomy and how it functions. As with the previous chapter concerning the male, you will explore the hormonal changes at puberty and the hormonal effects of the adult female reproductive system. You will also examine the way gametes develop, the effects of aging on the female reproductive system, and reproductive disorders. However, this chapter does not end there. You will continue your study of this system to include pregnancy and its effects on a woman's body should she choose to reproduce. Your study begins with the female reproductive anatomy.

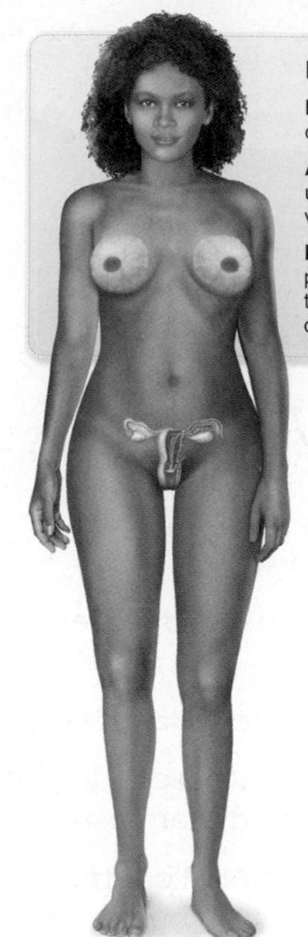

Female Reproductive System

Major Organs and Structures:
ovaries

Accessory Structures:
uterus, uterine tubes, vagina, vulva, breasts

Functions:
production of an egg, housing of the fetus, birth, lactation, secretion of sex hormones

FIGURE 16.1 The female reproductive system.

16.3 learning outcome

Describe the anatomy of the ovary and its functions.

Female Reproductive Anatomy

Female gametes (**ova,** eggs) are produced and developed in the primary sex organ of the female reproductive system—the **ovaries.**

Ovaries

Ovaries are small, almond-shaped organs suspended in the pelvic cavity by ligaments. Each ovary measures 3 cm long, 1.5 cm wide, and 1 cm thick. Like the testes, the ovaries are enclosed in a capsule called the **tunica albuginea.** Inside the capsule, each ovary has two basic layers: an outer cortex containing bubblelike **follicles** that enclose gametes and an inner medulla that contains the ovary's arteries and veins. Follicles secrete hormones each month and go through stages of development along with the ova. In **Figure 16.2**, you can see many of their development stages. These stages are explained more fully later in the chapter, when you will revisit this figure.

spot check ❶ Like the testes, the ovaries belong to the endocrine and reproductive systems. Why do ovaries belong to both systems?

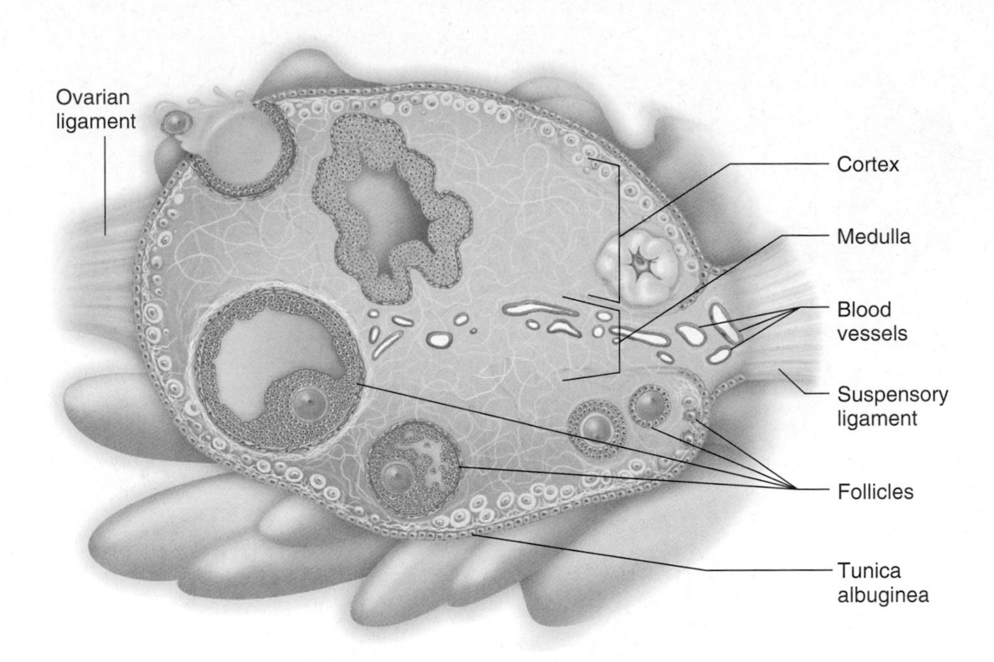

FIGURE 16.2 The anatomy of an ovary.

Figure 16.3 shows the internal female reproductive anatomy. Here you can see the ovaries and their position relative to the uterus. A broad band of connective tissue **(broad ligament)** is an extension of the peritoneum. It holds the ovaries and

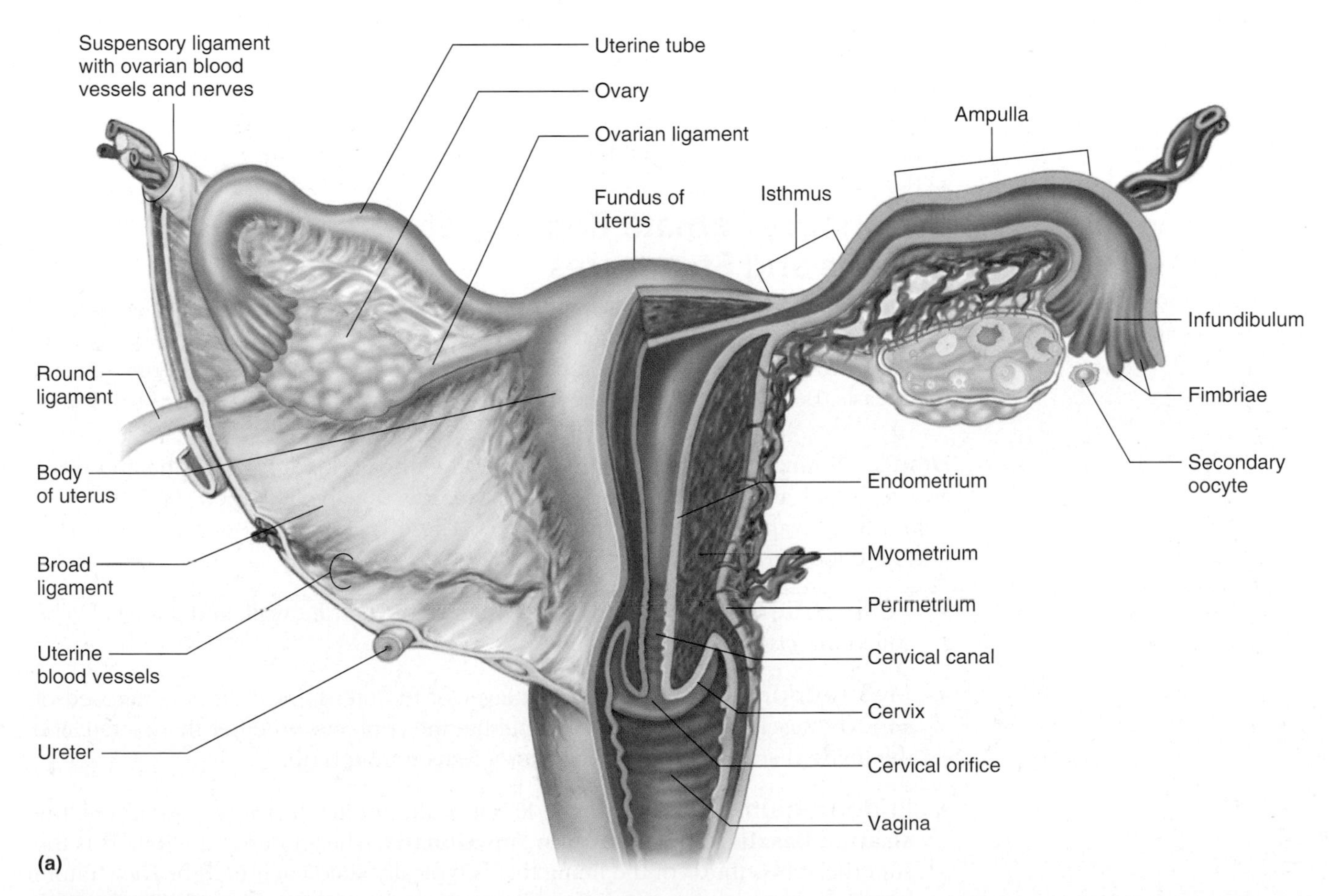

FIGURE 16.3 The internal female reproductive anatomy: (a) posterior view of the reproductive tract, (b) anterior view of the reproductive tract of a cadaver.

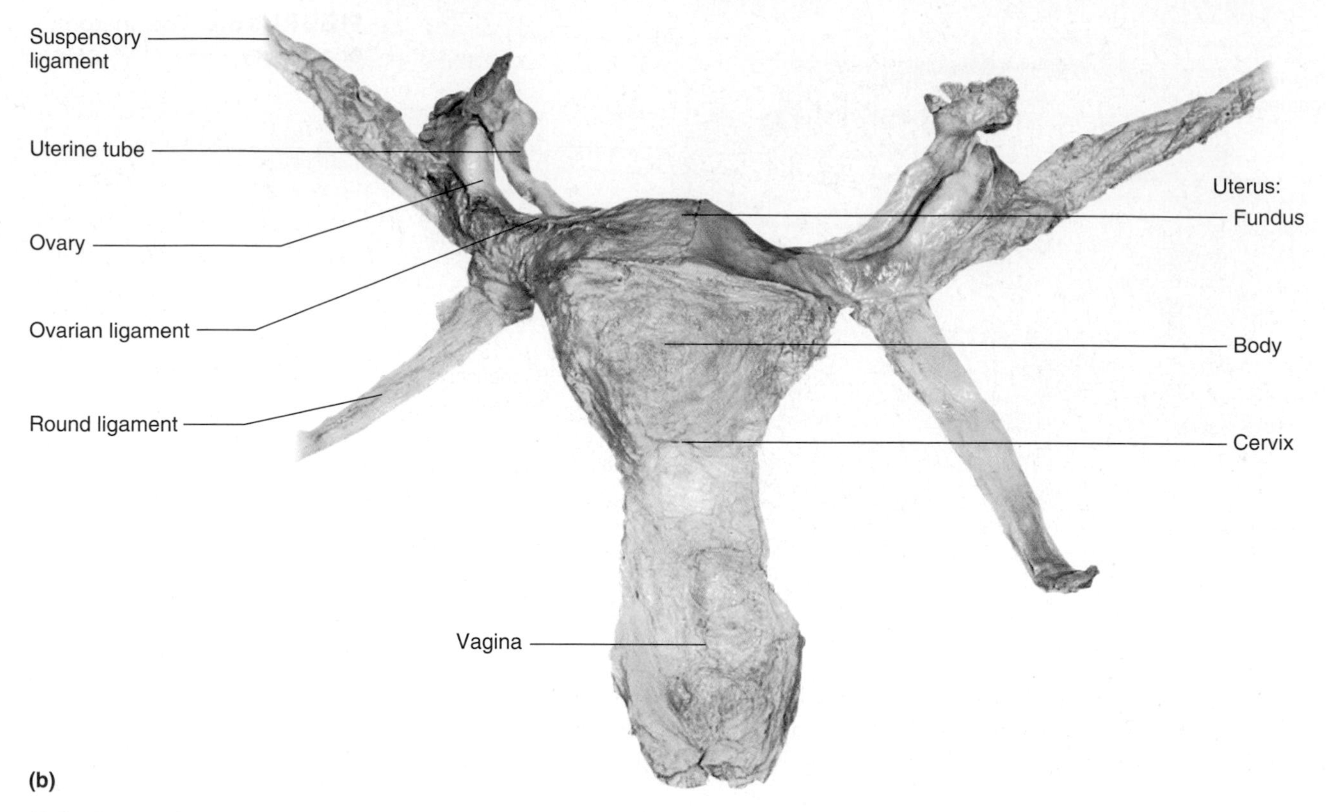

FIGURE 16.3 concluded

uterine tubes in their relative position to the uterus. The **suspensory ligament** attaches the lateral edge of the ovary to the posterior wall of the pelvic cavity and encloses the ovarian artery and vein, while the **ovarian ligament** attaches the medial edge of the ovary to the uterus.

16.4 learning **outcome**

Describe the female secondary reproductive organs and structures and their respective functions.

Secondary Female Reproductive Organs and Structures

Like the ovaries, much of the secondary female reproductive anatomy—the **uterus, uterine tubes,** and **vagina**—is internal. The external secondary structures include the **vulva,** which is composed of the reproductive structures located in the perineum, and the breasts. Below, we focus on the internal secondary reproductive structures first.

Uterus Normally, the uterus is tipped over the urinary bladder and is held in place by the broad and **round ligaments.** The uterus is a pear-shaped, hollow organ with a thick wall composed of three layers. These layers are shown in **Figure 16.3** and described in the following list:

- Perimetrium. This is the outermost layer of the uterine wall, and it may also be called the *visceral peritoneum*.
- Myometrium. This is the thickest layer of the uterine wall. It is composed of smooth muscle that contracts to expel uterine contents, whether the contents be the lining that is shed each month or a fetus during birth.
- Endometrium. This layer is the lining of the uterus. It has two sublayers: the **stratum basalis** and the **stratum functionalis.** The stratum functionalis is the superficial two-thirds of the lining that is typically shed each month. The stratum basalis is the deep one-third of the lining. Its purpose is to generate new stratum functionalis each month. You will learn more about how these layers change later in the chapter.

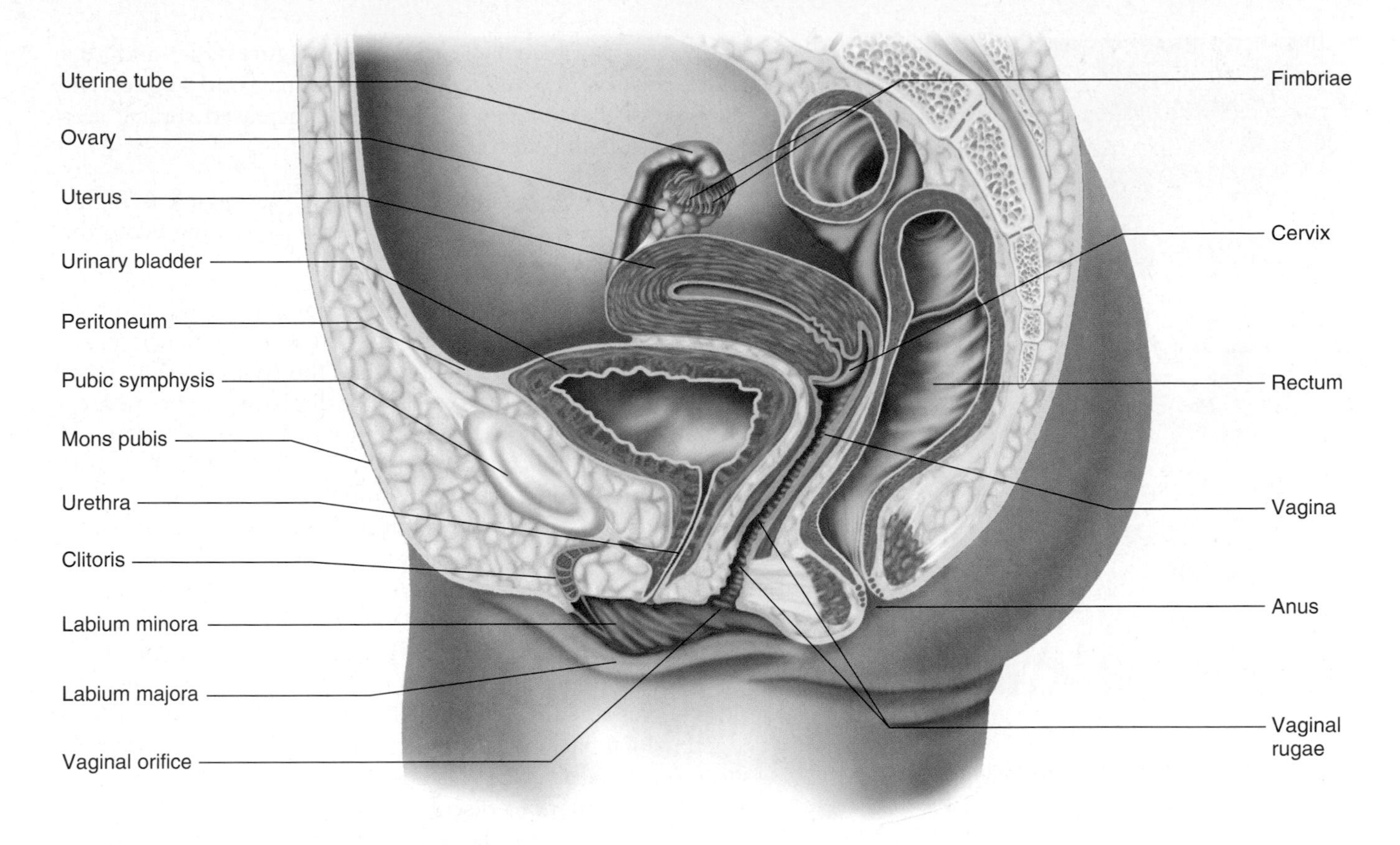

FIGURE 16.4 The female reproductive system.

Three areas describe the uterus: the broad superior curve is the **fundus;** the wide midportion is the **body;** and the narrow, inferior portion is the **cervix** (neck). As you can see in **Figure 16.3**, the uterine tubes open to the uterus between the fundus and the body, and the cervix opens to the vagina. Mucous glands in the lining of the cervix produce a thick mucus to block bacteria from entering the uterus from the vagina. The uterus measures 4 cm wide at the fundus, 7 cm from the cervix to the fundus, and 2.5 cm thick. This organ has a tremendous capacity to expand with a pregnancy, as you will learn later in the chapter. See **Figure 16.4**.

Uterine Tubes Each uterine tube can also be called a **fallopian tube** or **oviduct.** This tube is open-ended and measures approximately 10 cm from the ovary to the uterus. **Figure 16.3** shows that the uterine tube's flared end **(infundibulum)** near the ovary has fingerlike projections called **fimbriae.** It is important to notice that fimbriae are not attached to an ovary, nor do they cover it. Instead, the fimbriae sway to coax an egg released from the ovary into the uterine tube. If that happens, ciliated cells lining the uterine tube move the egg along the wide part of the tube **(ampulla)** to the narrow part of the tube **(isthmus)** before delivering the egg to the uterus. See **Figure 16.5**.

fimbriae: FIM-bree-eye

Vagina The last structure of the female internal reproductive anatomy is the vagina. It is an 8- to 10-cm canal that has three functions:

1. The vagina allows for the flow of the **menses** (shedding of the uterine lining) during a monthly cycle called **menstruation.** Through most of the month, the cervix produces a thick mucus to prevent any microorganisms from entering the uterus from the vagina. However, during menstruation, the uterine lining is shed from the uterus by smooth muscle contractions of the myometrium through the cervical canal, to the vagina, and out of the body.

FIGURE 16.5 Ciliated epithelial lining of the uterine tube.

2. The vagina is a receptacle for semen. **Figures 16.3** and **16.4** show ridges in the vaginal walls called **vaginal rugae.** The purpose of these ridges is to provide increased surface area for extension and stretch.

3. The vagina serves as the birth canal. The vagina is highly extensible (easily stretched) to be able to accommodate the birth of a baby.

The mucous membranes lining the vagina extend and fold inward to form a membrane over the external vaginal opening. This somewhat fragile membrane, called the **hymen,** usually has some holes in it by puberty to allow for discharge of the menses, but it may not be totally ruptured until intercourse. The status of a hymen is not an indication of virginity as it can also be easily ruptured by strenuous exercise or tampon use.

As you may recall from Chapter 2, the vaginal lining in a young girl is composed of simple cuboidal epithelial tissue. However, at puberty, estrogen causes the lining of the vagina to undergo metaplasia. This process changes the lining to a much more durable tissue called *stratified squamous epithelial tissue,* which better accommodates sexual intercourse. This epithelial tissue also serves another purpose. The epithelial cells are rich in glycogen, which bacteria (naturally occurring in the vagina) process to produce lactic acid. This acid lowers the pH of the vagina. A low pH is important to discourage the growth of, or even kill, harmful microorganisms. It is possible for microorganisms to travel from the vagina, through the cervix, up the uterus, and out the open ends of the uterine tubes to the pelvic cavity. So the low pH due to beneficial bacteria in the vagina helps prevent harmful bacteria from infecting the pelvic cavity.

spot check ❷ What else prevents harmful bacteria in the vagina from infecting the pelvic cavity?

The external vaginal opening is located between the urethral orifice and the anus in the female perineum. **Figure 16.6** shows the perineum and the external genitalia, collectively called the *vulva.*

Vulva The vulva is a collection of reproductive structures in the female perineum. They are shown in **Figures 16.4** and **16.6**, and explained in the following list:

- Mons pubis. This is a mound of adipose tissue, covered with pubic hair, and is superficial to the pubic symphysis.
- Labia. The labia are folds of skin and adipose tissue that frame the **vestibule,** which is an area that contains the urethral and vaginal openings. The **labia majora** (lateral folds) are thicker and have hair, while the **labia minora** (medial folds) are thinner and are hairless. The **prepuce** is a fold formed where the labia minora meet anteriorly. It forms a hood over the clitoris.
- Clitoris. This structure is similar to the male penis in that it is composed of two columns of erectile tissue (corpus cavernosa) and includes a **glans** that has many nerve endings for sexual stimulation. See **Figure 16.4**. Unlike the anatomy

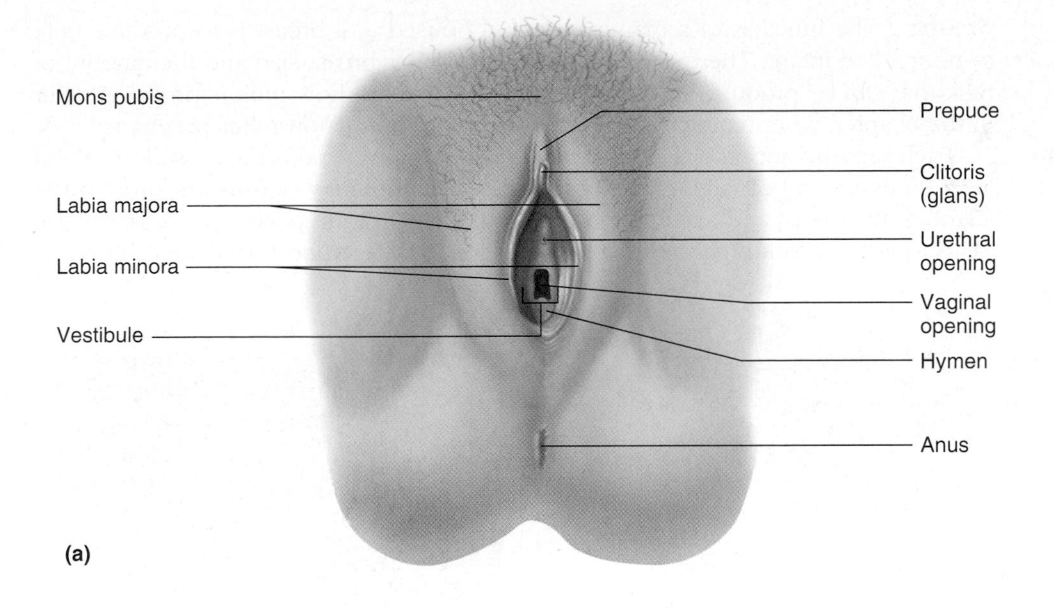

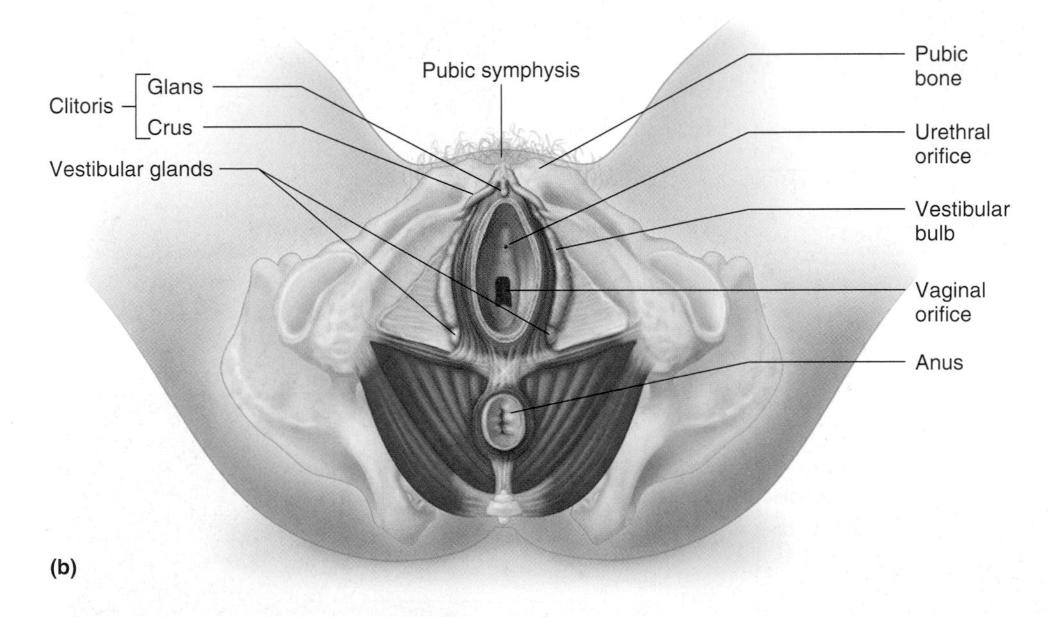

FIGURE 16.6 The female perineum: (a) superficial anatomy, (b) deep anatomy.

of the penis, most of the erectile tissue of the clitoris is internal, with only the glans exposed externally. The clitoris does not enclose the urethra as does the penis, and so it does not have any urinary function.

- **Vestibular bulbs.** The vaginal opening is bracketed on each side by vestibular bulbs of erectile tissue deep to the labia majora. These bulbs become engorged with blood during excitement, so they tighten around the penis. This further increases the sexual stimulation of the penis.
- **Vestibular glands.** Several vestibular glands surround the vaginal opening. They secrete a lubricating fluid during excitement to make sexual intercourse easier. The fluid is released into the lower part of the vagina and the vestibule. These glands are similar to the bulbourethral glands in the male.

Breasts The function of a mammary gland housed in a breast is to produce milk to nourish an infant. There is no correlation between breast size and the amount of milk that can be produced (**Figure 16.7**). You will learn how milk is produced later in the chapter, when you explore **lactation** (milk production) after pregnancy.

Each breast is superficial to a pectoralis major muscle and is composed mostly of adipose tissue and collagen. See **Figure 16.7c**. **Suspensory ligaments** support the shape of the breast and attach the breast to the fascia of the pectoralis major muscle deep to it. Foundation garments (bras) can give added support to these ligaments.

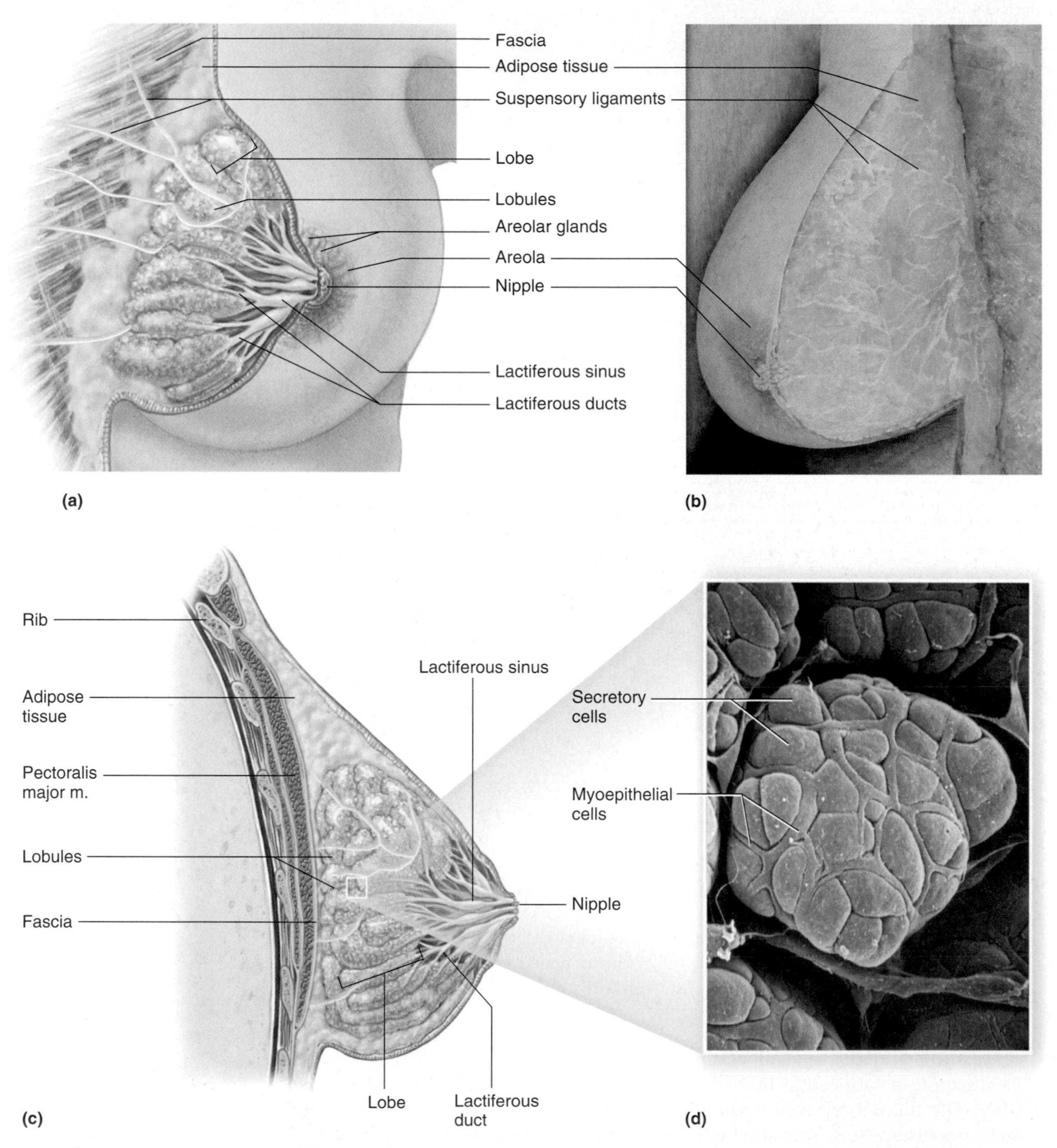

FIGURE 16.7 The female breast: (a) anterior view of a lactating breast, (b) breast of a cadaver, (c) sagittal section of a breast, (d) enlargement of part of a lobe, showing myoepithelial cells important for the release of breast milk.

Externally, each breast contains a nipple surrounded by a darker area called the **areola.** See **Figure 16.7a**. This area has many nerve endings sensitive to cold and touch and responsive to sexual arousal. Smooth muscle deep to the areola is responsible for erecting the nipple. As you can see in this figure, there are small bumps in the areola. These bumps are **areolar glands** that secrete a substance to prevent chafing and cracking of the nipple during nursing.

Although a mammary gland is present in young girls and enlarges during puberty, the mammary gland of each breast does not fully develop until the first pregnancy. You will examine the hormonal controls for the mammary glands' development later in the chapter. For now, you will continue to concentrate on the breast anatomy.

Each mammary gland is divided into 15 to 20 lobes that are further divided into lobules in a lactating breast (one producing milk). See **Figure 16.7** (a, b, and c). Milk is produced by secretory cells in the lobules, released through the action of **myoepithelial cells,** and drained by **lactiferous ducts** that widen to form **lactiferous sinuses** before reaching the nipple. As you may recall from Chapter 3, mammary glands are modified sweat glands.

You have now covered all of the female reproductive anatomy. Next, you will investigate the reproductive physiology, starting with what happens to a young woman at puberty.

Physiology of the Female Reproductive System

16.5 learning outcome

Explain the hormonal control of puberty and the resulting changes in the female.

As with the male, the physiology of the female reproductive system is governed by genetics and the endocrine system. You have already read about the genetics, in which a female zygote has XX as the sex chromosomes. As in a male, a female's hypothalamus and the anterior pituitary of her endocrine system produce the hormones—GnRH, FSH, and LH—to begin puberty.

Hormonal Control at Puberty

Puberty begins between the ages of 8 and 10 for most girls. At that time, the hypothalamus releases GnRH, which triggers the anterior pituitary to release FSH and LH. FSH targets follicles in the ovary, telling them to produce estrogen. Like testosterone, estrogen is a steroid hormone based on a chemically modified cholesterol molecule. There are several forms of estrogen: *estradiol, estriol,* and *estrone*. Together, these forms of estrogen target most tissues to produce the female secondary sex characteristics.

Female Secondary Sex Characteristics It will be helpful to review (from Chapter 8) these sex characteristics, which begin to develop at puberty:

- **Breast development.** This is usually the first sign that puberty has started. The breasts enlarge, and lobules and ducts form, beginning at puberty. As stated earlier, the breasts do not fully develop until the first pregnancy.
- **Development of axillary (armpit) and pubic hair.** The ovaries produce small amounts of androgens, which are responsible for this sex characteristic and the libido (sex drive) in women. As you may recall from Chapter 8, androgens are also produced in a woman's adrenal cortex.
- **Widening of the pelvis.** As you learned in Chapter 4, the female pelvis has a rounded pelvic brim and a large pelvic opening to accommodate the birth of a baby.
- **Fat deposition.** Estrogen causes a girl's body to deposit more adipose tissue. This deposition typically takes place in the breasts, hips, mons pubis, and buttocks. In an adult, a woman's body fat needs to be at a minimum of 22 percent to sustain a pregnancy.

- **Menstruation.** This is the monthly shedding of the uterine lining. It takes longer for this sex characteristic to develop because menstruation cannot begin unless the girl's body fat reaches 17 percent. Menstruation typically starts about age 12. **Ovulation** (the release of an egg from an ovary) is the result of the hormone LH but does not usually occur during the first year of menstrual cycles. You will learn more about the menstrual cycle shortly.

spot check ❸ Female athletes who excessively train and rid their diet of fat often stop menstruating. What might be the reason for this?

Unlike a male, who begins to produce gametes at puberty and continues to produce them every day for the rest of his life, a female produces all of her eggs by the time she is born. At puberty, she merely continues to further develop the eggs that are already present. Below, you will explore how this happens.

16.6 learning outcome

Explain oogenesis in relation to meiosis.

oogenesis: oh-oh-JEN-eh-sis

oogonia: oh-oh-GO-nee-ah

Oogenesis

Egg production is called **oogenesis.** Like spermatogenesis in the male, oogenesis forms gametes (ova, eggs) that have a haploid number of chromosomes (23, n). Unlike spermatogenesis, which produces four viable spermatids from each spermatogonia, oogenesis produces only one viable ovum from each **oogonium** (germ cell). Each ovum is developed along with the **follicle** that surrounds it. A follicle is a group of surrounding cells responsible for caring for the **oocyte**—an immature female reproductive cell—by providing nutrients and removing wastes and also responsible for producing hormones. Like sustentacular cells in the male, the follicle also isolates and protects the egg from the mother's immune system. This protection is necessary because the egg goes through independent assortment and crossing-over during meiosis in oogenesis. This results in genetic variation that would be seen as foreign to the mother's immune system. **Figure 16.8** shows the steps of this development for both the ovum and the follicle. Follow along with this figure as the process is explained in the paragraphs below.

Oogonia undergo mitosis in the ovaries before birth until they number 6 to 7 million. Shortly before birth, oogonia begin to go through meiosis I to become primary oocytes. Not all of the oogonia and primary oocytes survive. In fact, many degenerate by a process called **atresia,** so only approximately 2 million primary oocytes are present at birth. All of the surviving oogonia have become primary oocytes by the time an infant is six months old. Unlike the male, whose germ cells (type A spermatogonia) continue through life, no oogonia remain after the female infant is six months old. The primary oocytes remain dormant in mid-meiosis I until adolescence. Notice in **Figure 16.8** that primary oocytes are diploid (46 chromosomes, $2n$).

atresia: a-TREE-zee-ah

Further atresia reduces the number of primary oocytes to approximately 400,000 by puberty. At adolescence, FSH stimulates some primary oocytes each month to finish meiosis I. In this division, each primary oocyte produces two daughter cells—a secondary oocyte and a first **polar body.** Notice in **Figure 16.8** that the daughter cells are not of equal size. All of the cytoplasm and organelles go to the secondary oocyte. The first polar body simply has 23 chromosomes. Because it lacks cytoplasm and organelles, the first polar body is likely to die before going through meiosis II. This unequal division is very important because the secondary oocyte must provide all the cytoplasm and organelles necessary for a fertilized egg to survive. If fertilization occurs, the sperm will contribute only 23 chromosomes. It had been stripped of its cytoplasm and organelles during spermiogenesis.

The secondary oocyte continues meiosis to mid-meiosis II with a second polar body still enclosed. The secondary oocyte is ovulated (released from the ovary) in this state. If, and only if, a sperm fertilizes the egg, meiosis II is completed and the second polar body is expelled. If fertilization does not occur, the secondary oocyte with its second polar body inside is expelled from the uterus with the menses.

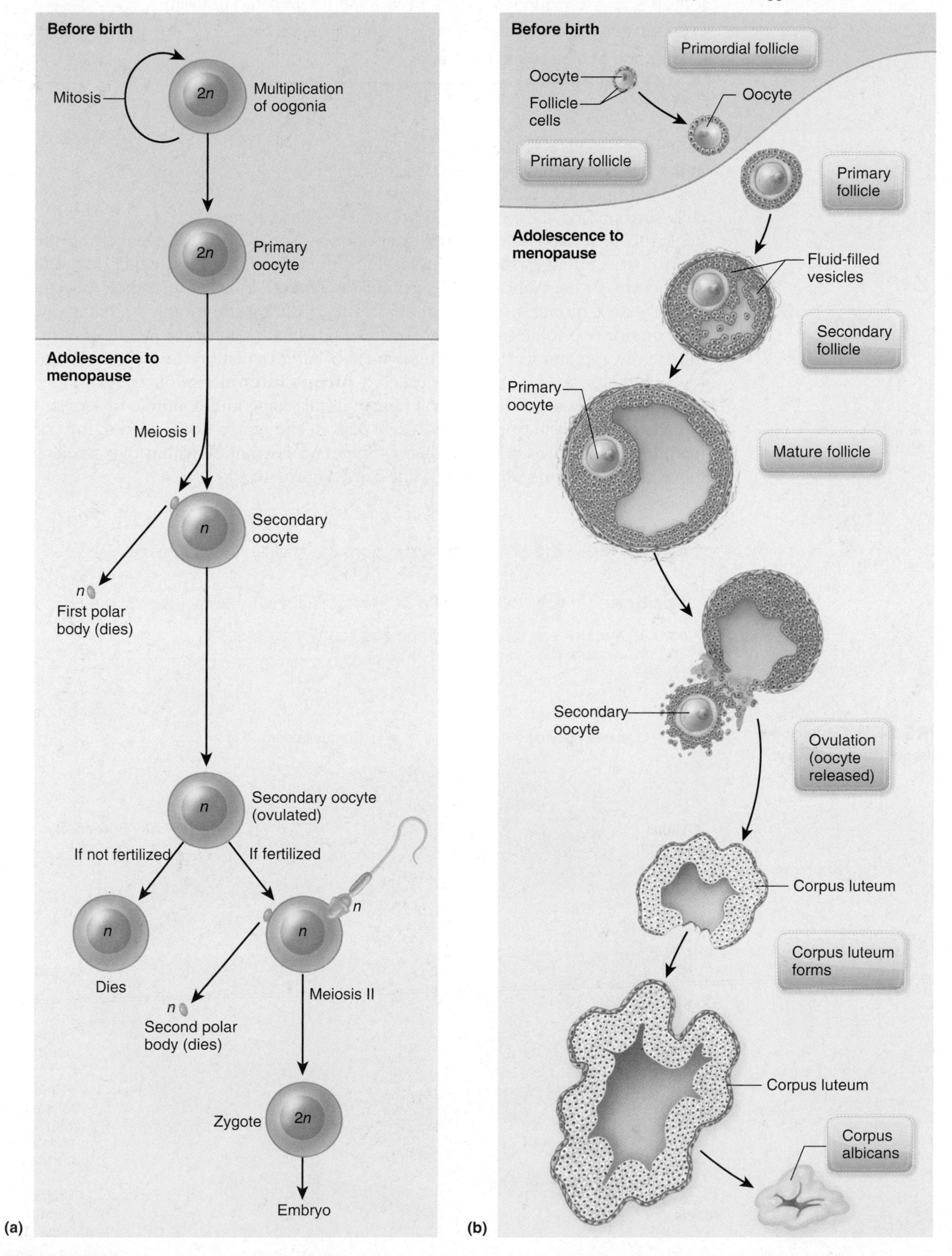

FIGURE 16.8 Oogenesis and folliculogenesis: (a) the production of primary oocytes before birth and their development from adolescence to menopause, (b) the development of follicles from before birth to menopause.

clinical point

If an egg is surrounded by sperm in a lab (in vitro fertilization), fertilization is recognized to have taken place by the expulsion of the second polar body.

Figures 16.8b and **16.9** show the changes in the follicle **(folliculogenesis)** during oogenesis. A primordial follicle is a simple layer of cells surrounding the primary oocyte. If the follicle receives FSH, it begins to grow and starts secreting hormones. You will explore which hormones shortly. Notice that as the follicle develops, fluid is produced in which the secondary oocyte floats. Although FSH stimulates several follicles and their oocytes to develop each month, most will undergo atresia during the month so that, typically, only one or two follicles mature. A mature (graafian) follicle eventually ruptures to release the egg and its fluid during ovulation. After ovulation the follicle changes to a yellow color. At this stage it is called a **corpus luteum** *(yellow body)*. If the egg is fertilized, the corpus luteum will remain in this stage and continue to secrete hormones to support the pregnancy for 90 days. If the egg is not fertilized, the corpus luteum shrinks, turns white, and stops secreting hormones within two weeks. It is then called a **corpus albicans** (*white body,* **Figure 16.9**).

spot check 4 What are the functions of a follicle during oogenesis?

spot check 5 Why would a woman's age affect the condition of her eggs more so than a man's age would affect the condition of his sperm?

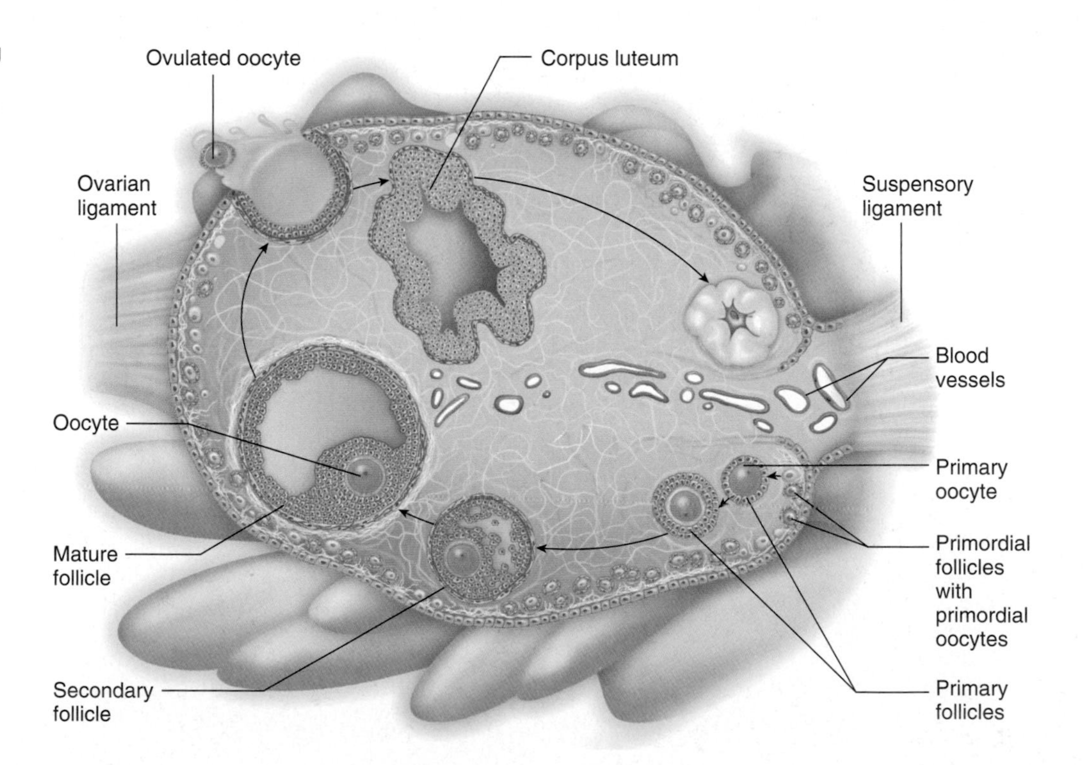

FIGURE 16.9 Developing follicles within the ovary.

By revisiting the ovary, you can see these changes in the follicles. In **Figure 16.9**, each of the follicle stages has been labeled. Notice that the follicles are located in the ovary's cortex, where they undergo their changes.

warning

Figure 16.9 may lead to a misconception. It is meant to show the changes to one follicle over a monthly cycle, and it does a good job of showing the progression of all the stages you have just explored. However, you should understand that the follicle does not move within the ovary. There is no conveyor belt delivering the mature follicle to a specific spot for ovulation, as this diagram may suggest. A follicle goes through all the changes shown in one spot. The first primordial follicle shown on the right in this figure will go through all of its stages, and the ovum will be ovulated from that beginning position near the suspensory ligament. The first primordial follicle shown on the left in this figure will have its ovum ovulated at its beginning position near the ovarian ligament.

You have become familiar with oogenesis and the role of follicles. You have also read about the female reproductive system's cyclic nature. Now it is time to explore the hormones that target the follicles and the hormones that are made by the follicles to produce this system of cycles.

Hormonal Control in the Adult Female

16.7 learning outcome

Explain the hormonal control of the adult female reproductive system and its effect on follicles in the ovary and the uterine lining.

The female **sexual cycle** has two parts: the **ovarian cycle,** which affects the follicles and oocytes in the ovary, and the **menstrual cycle,** which affects the endometrial lining of the uterus. The two cycles are in harmony in the sense that they happen together. **Figure 16.10** separates the two cycles, but you should be aware that the events in the ovarian cycle affect events in the menstrual cycle. This figure charts a 28-day cycle. Although this is the typical length for a sexual cycle, it is not uncommon for a cycle to last anywhere from 20 to 45 days.

Figure 16.10 shows what happens beginning on day 1. Follow along with the figure as you read the explanation below:

- On day 1, the uterine lining (stratum functionalis) begins to shed. This is the first day of the woman's **period.** The menstrual flow will typically continue for four to five days. This is called the **menstrual phase** (days 1 to 5) in the menstrual cycle.

- Also on day 1, the anterior pituitary secretes FSH, which targets the primordial follicle in the ovary. This causes the primordial follicle to develop and to secrete estrogen. As the follicle develops, more and more estrogen is produced. Estrogen levels reach a peak just before day 14. This is called the **follicular phase** (days 1 to 14) in the ovarian cycle.

- The increasing amounts of estrogen secreted by the follicle target the anterior pituitary to inhibit FSH production, so FSH levels go down (day 5). A follicle is already developing, so no further stimulation by FSH is needed. This is a negative-feedback mechanism. Estrogen does not inhibit the anterior pituitary's production of LH, so LH levels continue to rise. A peak of estrogen causes LH levels to reach a peak at day 14.

- The estrogen secreted by the follicle also targets the uterine lining, causing it to thicken. This is the **proliferative phase** (days 5 to 14) of the menstrual cycle.

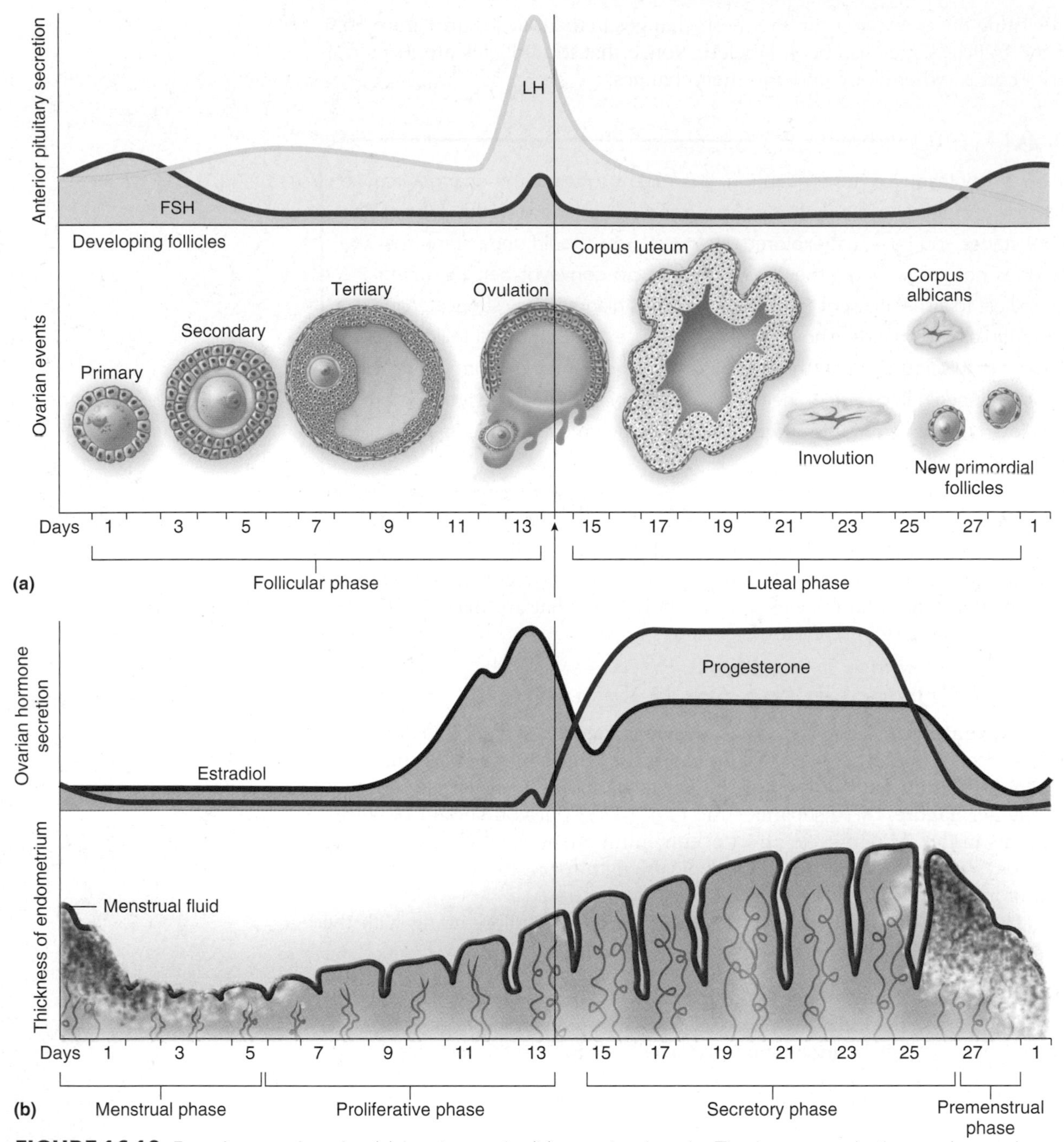

FIGURE 16.10 **Female sexual cycle:** (a) ovarian cycle, (b) menstrual cycle. The hormones in the ovarian cycle are to scale, but the hormones in the menstrual cycle are not. The peak for progesterone production is 17 times greater than the peak for estrogen.

warning

The length of a woman's sexual cycle up to the events on day 14 is highly variable. It can be days or even weeks longer than 14 days. Nutrition, activity levels, stress, illness, and many other variables can cause the length of this portion of the sexual cycle to be variable. This makes it difficult for a woman to predict when she will ovulate. In a typical cycle, ovulation occurs on day 14. However, once ovulation happens, the remaining length of the cycle does not usually vary from another 14 days.

Figure 16.10 also shows what happens during the rest of a woman's sexual cycle, beginning with day 14. Again, follow along with the figure while reading the explanation below:

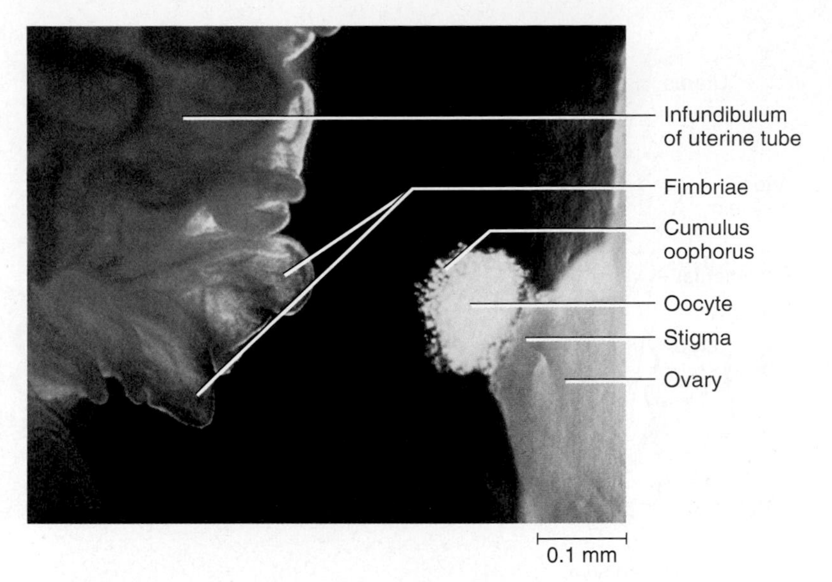

FIGURE 16.11 Ovulation in a human (endoscopic view). The stigma is a bump on the ovary that oozes follicular fluid prior to the rupture of the follicle. Ovulation takes two to three minutes.

- The peak of estrogen combined with a peak of LH causes the mature follicle to rupture and release its egg (ovulation) on day 14. The egg, the cells immediately surrounding it **(cumulus oophorus),** and the follicular fluid are released from the ovary. See **Figure 16.11**. Currents created by the movement of the fimbriae of the uterine tubes draw the egg toward the tube's infundibulum. If the egg reaches the tube, it sticks to it and ciliated cells in the lining of the uterine tube start the egg on its three-day journey to the uterus. If the egg does not reach the tube, it is eventually reabsorbed somewhere in the pelvic cavity. The egg must be fertilized in the uterine tube within 24 hours of ovulation if it is to remain viable.

- Another effect of these ovulation-initiating hormones is the thinning of the mucus created by the cervix. This effect lasts only a short time so that sperm have easier access to the uterus.

- The ruptured follicle remains in the ovary, but it changes to become a corpus luteum. As the corpus luteum, the follicle releases two hormones: estrogen and **progesterone** (another steroid hormone). This is the **luteal phase** in the ovarian cycle. Together these hormones inhibit FSH and LH production in the anterior pituitary. Again, this is a negative-feedback mechanism. If an egg has just been released for a pregnancy to occur, no new eggs need to be developed at this time.

- High levels of estrogen and progesterone cause the uterine lining to continue to thicken and its endometrial glands to secrete glycogen, a nutritional source for a fertilized egg. This **secretory phase** of the menstrual cycle is preparation for a possible pregnancy.

If a sperm fertilizes the egg in the uterine tube, the corpus luteum continues to produce high levels of estrogen and progesterone for 90 days. This hormone production regulates the pregnancy and maintains the rich lining of the uterus until the **placenta** (an organ developed in the uterus to support a pregnancy) is developed to produce pregnancy hormones. You will study the placenta later when you read about pregnancy.

If a sperm does not fertilize the egg in the uterine tube, the corpus luteum will go through **involution** (shrinkage). This happens typically on day 22. So you can continue your investigation beginning with that day:

- Although the involution of the corpus luteum begins on day 22, the estrogen and progesterone output from the corpus luteum are not really affected until the shrinkage is complete on day 26. Then, the levels of these two hormones fall drastically. The corpus luteum has now become a corpus albicans that no longer secretes hormones.

- The drop in estrogen and progesterone on day 26 signals the anterior pituitary that fertilization has not occurred. This signal triggers the production of FSH in the anterior pituitary. It is time to get a new primordial follicle started for the next sexual cycle.

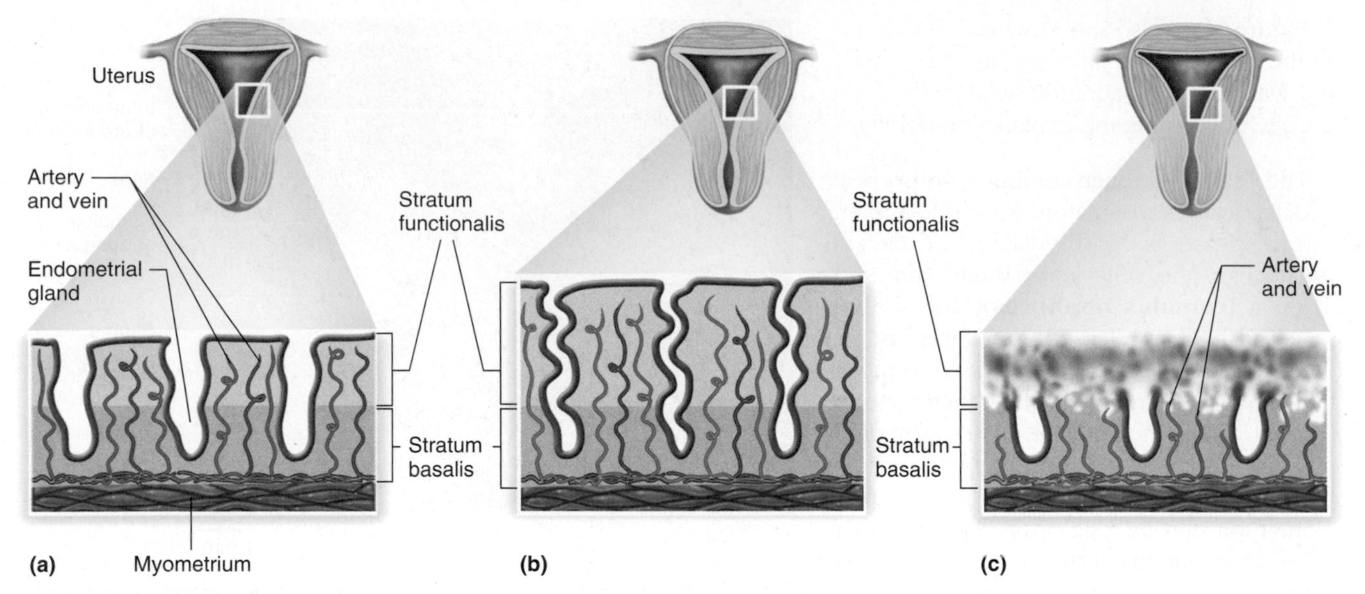

FIGURE 16.12 Changes in the endometrium throughout the menstrual cycle: (a) proliferative phase, (b) secretory phase, (c) menstrual phase.

- The drop in estrogen and progesterone on day 26 also affects the uterine lining. It causes arterioles in the stratum functionalis to spasm, and this disrupts the blood supply and causes menstrual cramps. The disruption to the blood flow causes the stratum functionalis to die and begin to break down. See **Figure 16.12**. This is the **premenstrual phase** of the menstrual cycle. The dead tissue mixes with clotted blood and other fluids to be discharged on day 1 of the next cycle.

Thus, whether this cycle ends and a new cycle begins depends upon whether an egg is fertilized or not. If the egg is fertilized, the corpus luteum continues and a new sexual cycle will not begin again until after the pregnancy ends. However, if the egg is not fertilized, the corpus luteum shrinks to become a corpus albicans and a new cycle begins in a few days. The corpus luteum is in the *ovary*. If an egg is to be fertilized, it will happen in a *uterine tube*. Here is the huge question: How does the corpus luteum know what happened to the egg?

The answer is this: The developing fertilized egg secretes another hormone—**human chorionic gonadotropin (HCG).** HCG targets the corpus luteum so that it continues to produce estrogen and progesterone to support a pregnancy for 90 days, at which time the placenta takes over hormone production. You will learn more about this soon, when the chapter addresses pregnancy.

clinical point

Some HCG is cleared from the system daily by the kidneys, so it can be found in urine. Early home pregnancy tests work by detecting this pregnancy hormone in urine. These tests can be effective as soon as eight to nine days after fertilization (day 22 to day 24 of the sexual cycle). This time frame is before the woman's next cycle would have predictably begun.

spot check ❻ If an early pregnancy test can detect a pregnancy within eight or nine days, why does this correlate to a three-day window in the sexual cycle (day 22, day 23, or day 24)?

Now that you have become familiar with the hormones regulating a woman's sexual cycle, you can move on to investigate her sexual response.

The Female Sexual Response

16.8 learning outcome

Describe the stages of the female sexual response.

The female sexual response goes through four stages—**arousal, plateau, orgasm,** and **resolution.** These stages are shown in **Figure 16.13** and explained in the list below. As with the male, a woman's sexual response may be initiated by

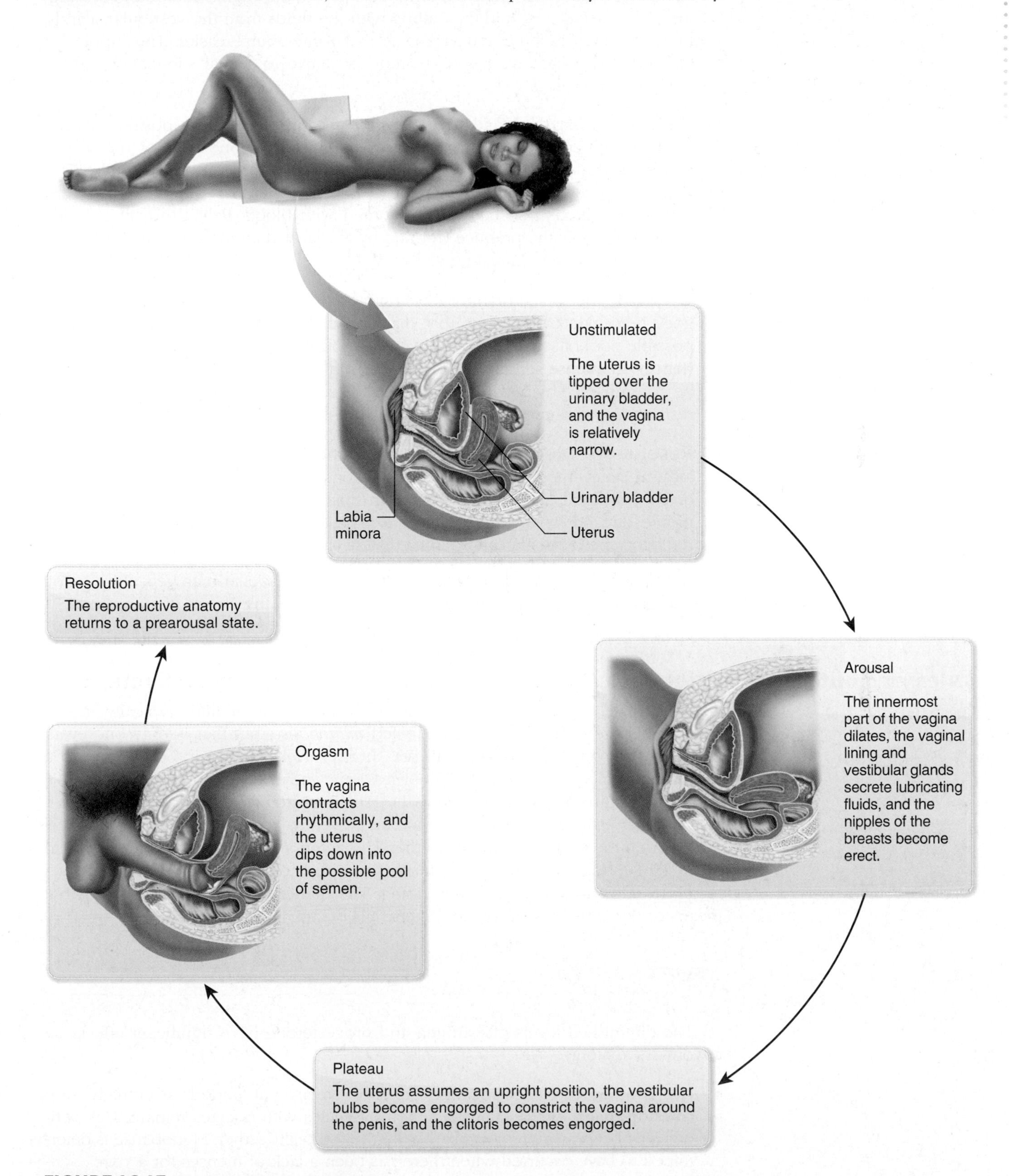

FIGURE 16.13 The female sexual response.

thought, touch, sound, or smell. The neurological control and vascular events of her sexual response are similar to those of the male. Here are some of the features of each stage:

- **Arousal.** During this stage, the innermost part of the vagina dilates and the vaginal lining secretes a fluid that, along with the fluids from the vestibular glands, lubricates the vestibule and vagina to make intercourse easier. The nipples of the breasts become erect and are more sensitive to sexual stimulation, which increases arousal.

- **Plateau.** During this stage, the uterus assumes a more upright position instead of being tipped over the urinary bladder. The lower one-third of the vagina constricts around the penis due to the engorgement of the vestibular bulbs. This constriction, along with the vaginal rugae, stimulates the penis. The erectile tissue of the clitoris also becomes engorged with blood. Pelvic thrusting during intercourse pulls the prepuce (formed by the labia minora) back and forth over the clitoris, further stimulating it.

- **Orgasm.** The vagina rhythmically contracts during this stage. The uterus undergoes peristaltic contractions that may cause the cervix to dip down into the possible pool of semen. This facilitates the journey for sperm. The woman experiences an intense sensation spreading from the clitoris to the pelvis, which may be accompanied by pelvic throbbing and a feeling of warmth. Her heart rate and respiratory rate increase.

- **Resolution.** During this stage, her reproductive anatomy returns to a pre-arousal state—the vagina relaxes, the clitoris and vestibular bulbs are drained of excess blood, the uterus once again tips over the urinary bladder, and the nipples are no longer erect. Unlike a male, a female does not usually have a refractory period—she can be immediately capable of another sexual response.

Although the sexual response stages do not change as a woman gets older, aging does affect her reproductive anatomy. You will explore the effects of aging next.

16.9 learning outcome

Explain the effects of aging on the female reproductive system.

Effects of Aging on the Female Reproductive System

Like the male, a female goes through a climacteric in midlife (typically in her 50s). This stage of her life is often labeled *menopause,* but that is only one event of a broader condition (as you will see shortly). Her climacteric begins due to a decreased number of follicles (1,000 or less) and their decreased sensitivity to FSH and LH. Because of this insensitivity, the remaining follicles in her ovaries release less and less estrogen and progesterone until eventually the secretion of these hormones from the ovaries ceases altogether. Her anterior pituitary initially responds to the lack of estrogen and progesterone by secreting more FSH and LH in a negative-feedback effort to keep the system going. The higher levels of FSH and LH are a means of determining that the climacteric has begun.

spot check  Why has the number of follicles decreased?

The diminished levels of estrogen and progesterone have significant effects on a woman's anatomy:

- Her ovaries stop releasing eggs, and her menstrual periods eventually stop **(menopause).** At this point, few if any follicles with oocytes remain. Her periods may become more irregular before they stop altogether. Menopause is determined to have occurred when there has been a lack of menses for a year.

- **Hot flashes** (intense feelings of warmth accompanied by sweating) are common as blood vessels constrict and then dilate due to the changing hormone levels. They cease once menopause is complete and the hormones have stabilized.
- The tissues of the vagina, labia minora, clitoris, uterus, uterine tubes, and breast atrophy. The thinning of the vagina and decreased secretions from it may cause dryness and make intercourse uncomfortable. If that is the case, the use of lubricants can facilitate more pleasurable intercourse. However, the effects of aging do not have to have an effect on a woman's enjoyment of a sexual relationship.
- Vaginal yeast infections become more common.
- Bone mass declines. As you may recall from Chapter 4, estrogen acts as a lock on calcium in the bone. The risk for osteoporosis increases during and after the climacteric.
- Muscle and connective tissue decrease. This leads to sagging of the breasts and diminished support for the urinary bladder, uterus, vagina, and rectum. As a result, these organs can drop out of position **(prolapse).** This can lead to loss of control of micturition (incontinence) or difficult bowel movements. This effect of aging may be increased if the woman has had children with vaginal deliveries.
- The skin becomes thinner. This is due to decreased fat in the hypodermis.
- Cholesterol levels rise, increasing the risk for cardiovascular disease.

spot check 8 Vaginal yeast infections often occur due to a lack of beneficial bacteria in the vagina. How does aging affect the presence of these bacteria?

The effects of these changes can be treated with **hormone replacement therapy (HRT).** However, estrogen replacement has been associated with increased risks of breast cancer, stroke, and heart disease. A woman and her doctor should weigh the possible risks versus the benefits in each individual situation.

A woman's adrenal glands continue to produce androgens, such as testosterone, throughout and after her climacteric. These hormones are responsible for maintaining her sex drive and slowing her loss of bone and muscle mass. Without high levels of estrogen and progesterone to counteract the masculinizing effects of androgens, a woman's skin may coarsen and some facial hair may develop.

You have read about some of the signs and symptoms associated with the aging of the female reproductive system. It is time to explore some of the disorders associated with this system.

Disorders of the Female Reproductive System

16.10 learning outcome

Describe female reproductive disorders.

Breast Cancer

Breast cancer is the abnormal growth of breast tissue, usually occurring in the lactiferous ducts and lobules of the breast. Other than skin cancer, breast cancer is the most common form of cancer in women. Some of the risk factors are the following:

- Breast cancer risk increases with age. Most women are over 60 years old when they are first diagnosed.
- Family history plays a role. You are more likely to develop breast cancer if a family member has been diagnosed with it.
- Personal history also plays a part. Breast cancer often reoccurs in the other breast.

- Genetic mutations can increase the risk of breast cancer. They are described in the Applied Genetics box below.
- Reproductive history and menstrual history also matter. Breast cancer is more common in women who have not had children, who started menstruating before the age of 12, or who have experienced menopause after the age of 55.
- Lifestyle choices can increase the risk of breast cancer. Obesity, lack of physical activity, and increased alcohol consumption increase the risk of breast cancer.

Detection of breast cancer usually begins with a monthly self-exam and **mammography** (taking an x-ray of a breast). See **Figure 16.14**. Because breast cancer often metastasizes, lymph nodes may be biopsied once the cancer has been detected. Treatment of breast cancer may include chemotherapy, surgery (lumpectomy to remove a tumor, or **mastectomy** to remove the breast and possible underlying tissue), radiation, immunotherapy, and hormone therapy.

clinical point

Mammography and breast self-exams are important tools in breast cancer detection. Women are usually encouraged to have a mammogram before the age of 40 as a baseline. Mammography is then performed every one to two years until the age of 50, after which it is then done annually.

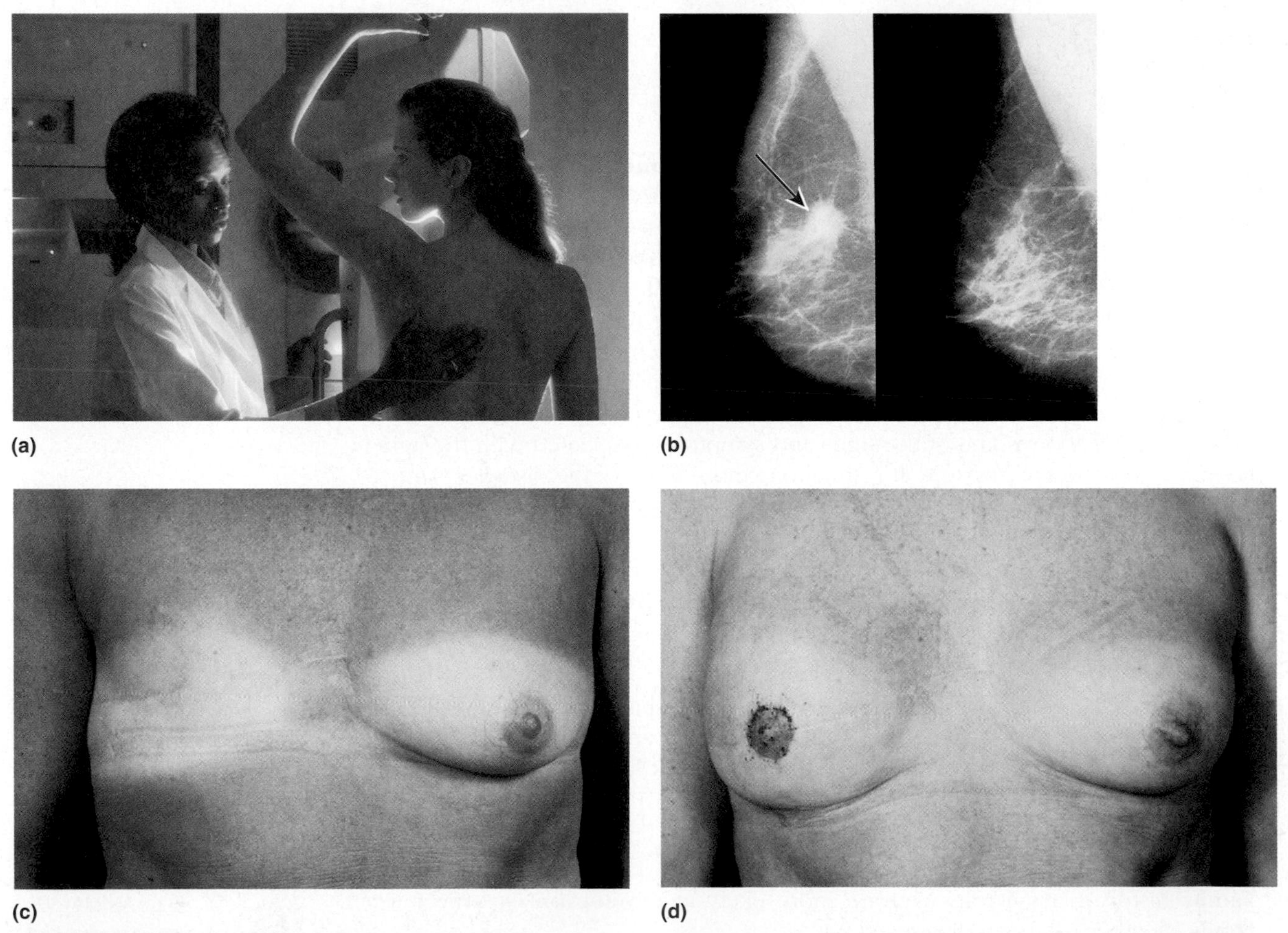

FIGURE 16.14 Breast cancer detection and treatment: (a) mammography, (b) mammogram, with a tumor indicated by the arrow, (c) patient after a mastectomy, (d) same patient after reconstructive surgery.

applied genetics

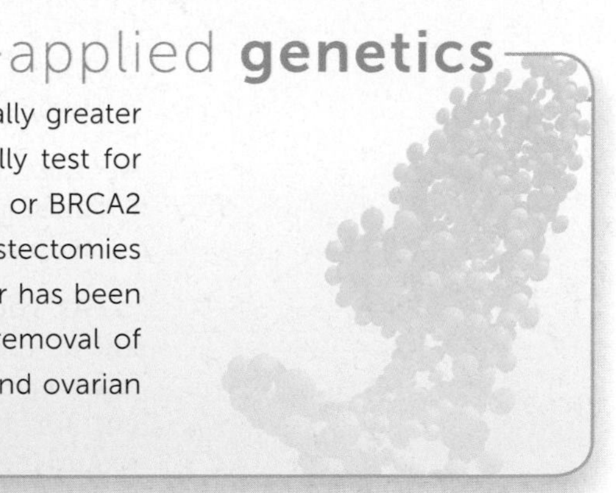

Women with mutations in their BRCA1 or BRCA2 gene have a substantially greater risk of developing breast and ovarian cancer. It is possible to genetically test for these changes. Once the presence of the mutations in either the BRCA1 or BRCA2 gene has been determined, some women choose to have double mastectomies and **oophorectomies** (surgical removal of the ovaries) before any cancer has been detected. The National Cancer Institute reports that this prophylactic removal of the breasts and ovaries is an effective way to reduce the risk of breast and ovarian cancer.[1]

Cervical Cancer

Cervical cancer—a slow-growing cancer that has few symptoms—is often caused by human papillomavirus (HPV) infection. Cervical cancer is typically detected by swabbing the cervix (Pap test) and examining the collected cells for abnormal growth. Treatment options include the removal of the uterus **(hysterectomy).**

Endometriosis

Endometriosis is the growth of endometrium in places other than the uterus. The most likely locations are the uterine tubes or the lining of the pelvic cavity. Because hormones travel throughout the entire system, the endometrium responds to hormones no matter where it is located. Thus, as the endometrium thickens in the uterus in response to the estrogen and progesterone produced in the ovarian cycle, the misplaced endometrium also thickens and becomes rich in the uterine tubes with this condition. Just as the stratum functionalis sheds each month in the uterus, so does the stratum functionalis lining the pelvic cavity in endometriosis. However, in endometriosis, there is nowhere for the shed endometrium to go in the pelvic cavity. Consequently, it irritates pelvic structures until it is eventually reabsorbed. This happens with each sexual cycle, and it may cause painful periods **(dysmenorrhea).** Endometriosis often blocks the uterine tubes, causing infertility because the sperm cannot reach the egg.

spot check 9 What would happen to the signs and symptoms of endometriosis after menopause? Explain.

Pregnancy

16.11 learning outcome

List the four requirements of pregnancy.

Just by looking at Joan and Rick in **Figure 16.15**, you can see that their reproductive systems are functioning to produce their secondary sex characteristics. You can easily determine from these obvious signs who is male and who is female. By their response to the early pregnancy test results that Joan is holding in her hand, you can assume that they have also produced gametes, another function of the reproductive system for both sexes. In the rest of this chapter, you will explore the additional functions Joan's reproductive system performs to house a developing fetus, facilitate the birth, and provide nourishment for the infant.

To begin your study of pregnancy, consider the four basic requirements for a pregnancy to happen:

1. There must be a sperm and an egg.
2. The sperm must meet the egg.

3. The sperm must fertilize the egg.
4. The fertilized egg must implant in the uterus.

As you may already know, the mother and father need not be present for the first three requirements. Eggs and sperm can be collected from the parents and fertilized in a lab. But there is no substitute for a uterus to house the developing fetus.

You are already familiar with sperm and egg production. So you can continue your investigation of pregnancy with the second requirement—how the sperm meets the egg under typical conditions.

FIGURE 16.15 Joan and Rick view an early pregnancy test.

16.12 learning **outcome**

Trace the pathway for a sperm to fertilize an egg.

Pathway for Sperm to Meet an Egg

Joan's egg must be fertilized within 24 hours of ovulation if it is going to survive, but it will have traveled only to the ampulla of her uterine tube in this amount of time. This means Rick's sperm must travel the distance to the ampulla to fertilize the egg. Only one of the millions of sperm deposited in the vagina will be allowed to fertilize the egg. What happens to the rest? Some are likely to be deformed, having two heads or faulty tails. Some will die from the acids of Joan's vagina. Still others will leak out of the vagina and not travel to the uterus at all. Not all of the surviving sperm will accomplish the journey from the vagina along the mucous threads in the cervical canal to the uterus. Some sperm will be destroyed in the uterus by leukocytes before they reach the openings to the uterine tubes. Half the sperm reaching this point will travel into the wrong tube. However, those sperm that do go the distance can survive in the female reproductive tract for up to six days after ejaculation.

The complete journey for a mature sperm is:

Epididymis → ductus deferens → ejaculatory duct → urethra → vagina → uterus (cervix, body) → uterine tube (isthmus, ampulla)

16.13 learning **outcome**

Describe the events necessary for fertilization and implantation.

Fertilization to Implantation

Although sperm can make the journey to the egg within 5 to 10 minutes of ejaculation in the vagina, a sperm cannot fertilize an egg for about 10 hours. During that time, Rick's sperm must first go through a process called **capacitation.** The sperm's cell membrane is reinforced with cholesterol. This helps protect Rick's spermatic ducts from a premature release of the digesting enzymes in the acrosome cap. During capacitation, the acids of the vagina and other fluids of Joan's reproductive tract break down the cholesterol and stimulate the sperm to swim faster, so that after 10 hours, the sperm are able to release the enzymes of the acrosome cap to penetrate an egg.

spot check 10 At what point in a sperm's journey does capacitation begin?

Figure 16.16 shows the process of fertilization. Although only one sperm will be allowed to fertilize the egg, hundreds of sperm may be needed to break through the cells surrounding the egg (granulosa cells). Follow along with the figure as you read the steps below:

1. The capacitated sperm release their acrosomal enzymes to penetrate through the cells surrounding the egg.
2. A sperm fuses with the egg's plasma membrane.

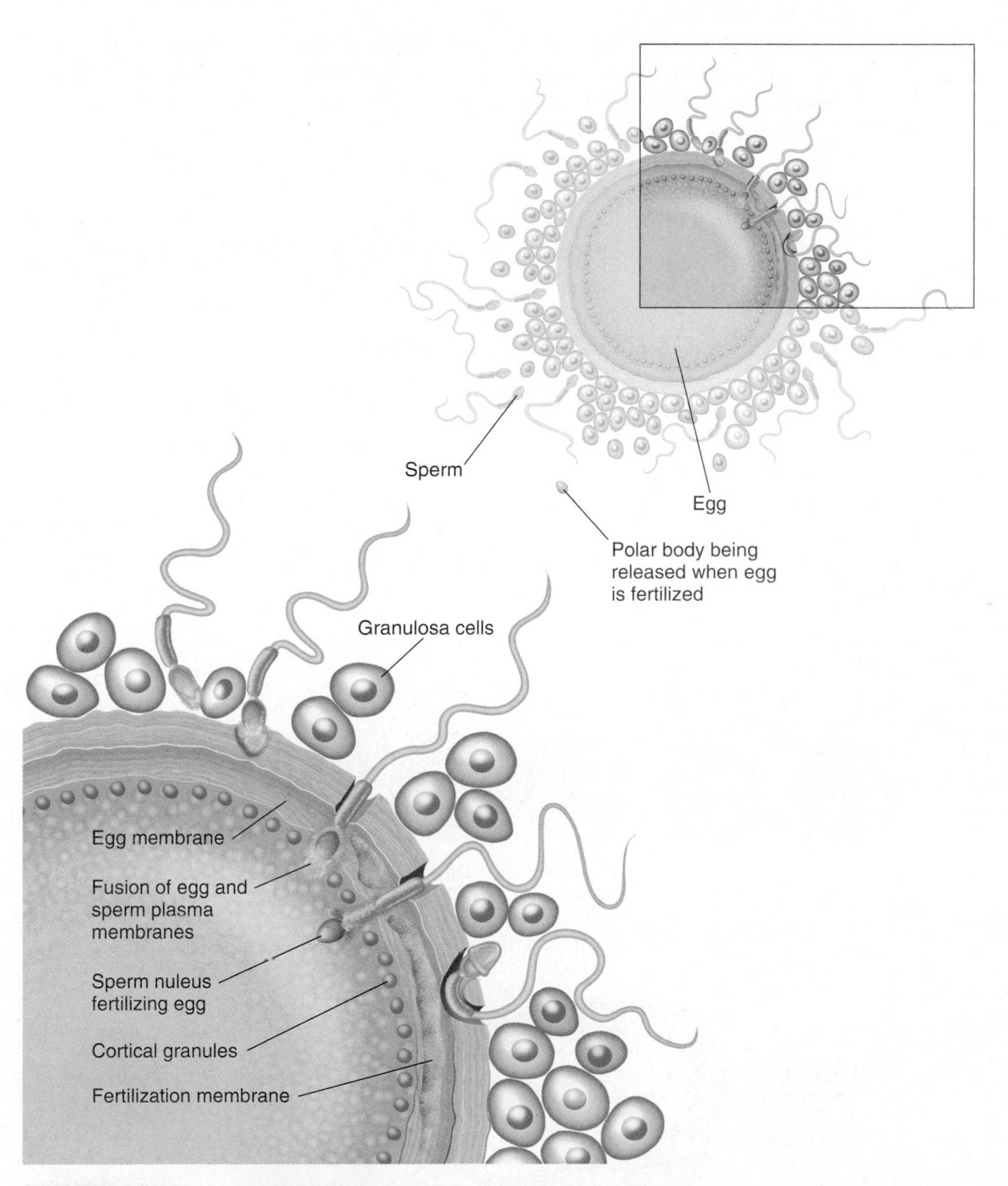

FIGURE 16.16 The process of fertilization. This figure shows just two sperm. The first sperm is shown approaching the egg, making its way through the cells surrounding the egg, fusing with the egg, and fertilizing the egg (steps 1 to 3; step 4 is not shown). The second sperm is shown being rejected by the egg.

3. The sperm nucleus enters the egg, and the egg immediately prevents any further sperm from gaining access by forming a **fertilization membrane** that sperm cannot penetrate. Granules inside the egg form this membrane. The fertilization membrane is important to establish the correct number of chromosomes in the fertilized egg (just 46).

4. The nuclei of the sperm and egg rupture within the fertilized egg, and the chromosomes mix to form a single nucleus. The cell is now called a *zygote*.

The zygote undergoes mitotic division **(cleavage)** as the ciliated cells of the uterine tube lining move it along toward the uterus. **Figure 16.17** shows the stages of the divisions. As you can see in this figure, six to seven days after fertilization, a **blastocyst** implants in the uterine lining. At this point, all four of the pregnancy requirements have been met.

The developing fertilized egg will be called a *blastocyst* from implantation to week 3, an **embryo** from weeks 3 to 9, and a **fetus** from week 9 until birth. **Gestation** (the time from fertilization to birth) is 266 days, or 280 days from the start of Joan's last menstrual period.

You have already become aware of some of the hormones that will control a pregnancy. For example, you have read that a corpus luteum produces hormones for 90 days if an egg is fertilized. Below, you will continue your exploration of the female reproductive system by looking at the many hormones involved in a pregnancy.

16.14 learning **outcome**

Explain the hormonal control of pregnancy.

Hormonal Control of Pregnancy

In addition to hormones secreted by the zygote and corpus luteum, hormones are secreted by the placenta and its associated membranes during pregnancy. The placenta starts to develop soon after the blastocyst implants, and its development will be complete by 90 days. The placenta functions to secrete hormones to regulate Joan's pregnancy, her mammary development, and fetal development. As you may recall from Chapter 9, the placenta also serves as a barrier to the mixing of blood between the mother and the fetus while nutrients are being delivered to the fetus from her blood and wastes are being taken away.

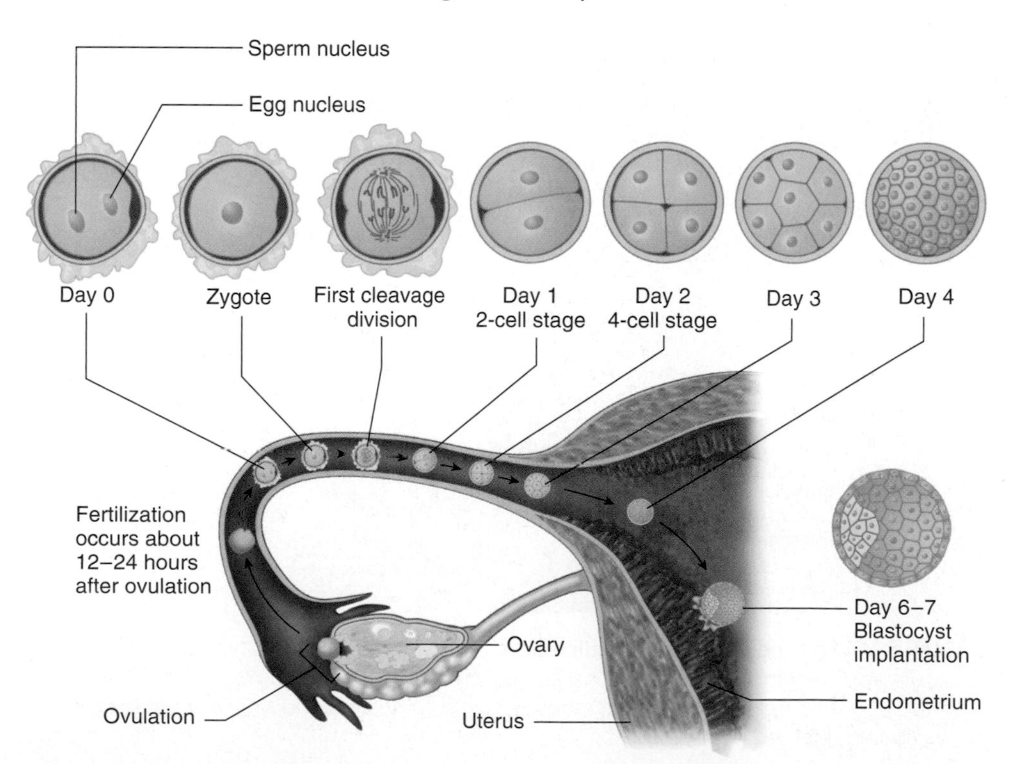

FIGURE 16.17 Migration from fertilization to implantation: the stages of development from fertilization of a secondary oocyte to implantation of a blastocyst.

Each of the hormones that regulate a pregnancy is explained in the following list:

- Human chorionic gonadotropin from the developing blastocyst causes the corpus luteum to continue to secrete estrogen and progesterone.
- Joan's estrogen levels from the corpus luteum and later the placenta rise to 30 times normal during a pregnancy. Over the length of the pregnancy, this hormone causes Joan's breasts to double in size, encourages her external genitalia and uterus to grow, causes her uterus to be more irritable (prone to contract), and causes her pubic symphysis to become more elastic.
- Progesterone is produced first by the corpus luteum and then by the placenta. Along with estrogen, this hormone suppresses FSH and LH secretion so that no additional eggs are developed during the pregnancy. Progesterone suppresses uterine contractions, promotes a rich lining of the uterus for the developing blastocyst and embryo to feed upon, and promotes the development of mammary glands and ducts.
- The placenta produces **human chorionic somatomammotropin (HCS),** which regulates Joan's carbohydrate and protein metabolism so that glucose and amino acids are available in the blood for the developing fetus. However, this is not a self-sacrificing gesture. Although HCS also reduces Joan's sensitivity to insulin so that more glucose stays in her bloodstream for the fetus, it also increases Joan's ability to use fatty acids as a fuel substitute for glucose.

clinical point

The insulin insensitivity effect produced by HCS may be responsible for **gestational diabetes.** This form of diabetes mellitus affects less than 10 percent of women during their pregnancy. Symptoms include increased urine output, excessive thirst, and hyperglycemia. These symptoms and the condition itself usually end with the birth of the baby and the immediate delivery of the placenta. However, the likelihood of developing type 2 diabetes mellitus within 15 years is significantly increased.

- Joan's thyroid hormone levels rise. This hormone increases the metabolism for both Joan and the fetus.
- Joan secretes more parathyroid hormone as her fetus takes more and more calcium from her blood. PTH increases osteoclast activity to keep Joan's calcium levels at homeostasis.

- Joan's adrenocorticotropic hormone levels rise so that more glucocorticoids are produced by her adrenal glands. This results in protein breakdown to produce more glucose for the fetus.
- Joan's aldosterone levels increase for fluid retention to increase her blood volume.

So what are all the effects of these hormones on Joan's body? In the next section, you will explore how her body adjusts to a pregnancy.

A Woman's Adjustment to Pregnancy

16.15 learning outcome

Explain the adjustments a woman's body makes to accommodate a pregnancy.

Pregnancy puts a strain on many of Joan's body systems. Some of the effects are due to hormones, and other effects are simply due to the growth of the fetus putting pressure on Joan's anatomy. These effects are described in the following list and shown in **Figure 16.18.**

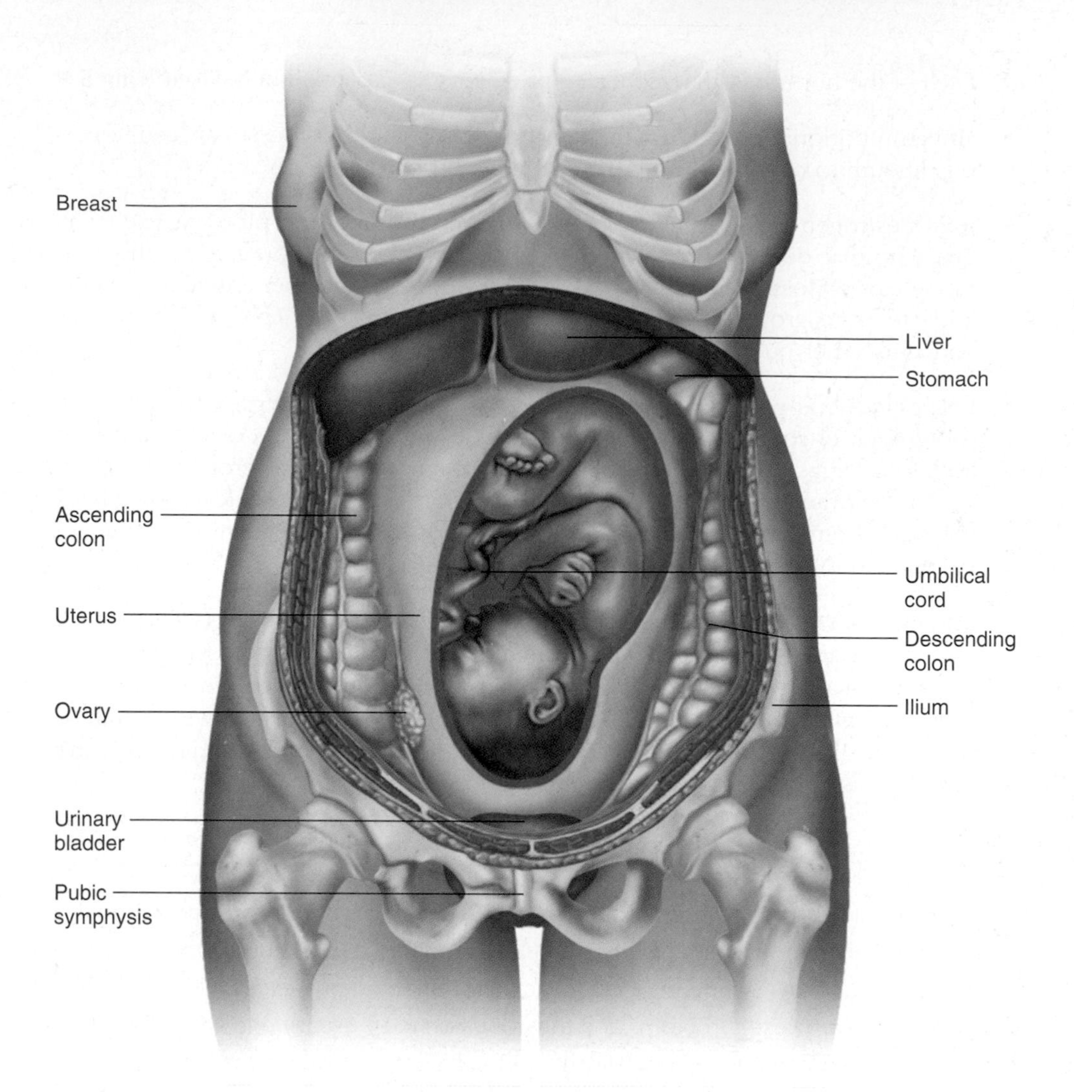

FIGURE 16.18 A full-term fetus in the uterus.

- There is decreased motility (movement) in the digestive system, especially in the first trimester. This may cause nausea **(morning sickness)** for Joan.
- Pressure builds on Joan's stomach from the growing fetus. This causes pressure on the cardiac sphincter of the stomach, which may then leak. The reflux of stomach acid to the esophagus causes heartburn for the mother.
- Joan's cardiac output increases 30 percent due to her increased blood volume. She has more blood volume because the fetus uses oxygen from her blood too. This decreases Joan's blood oxygen levels, so her kidneys respond by secreting EPO to boost her red blood cell production.
- Pressure from the weight of the uterus may cause hemorrhoids or varicose veins.
- Joan's kidneys produce more urine because her blood volume is increased, and they are filtering more waste—waste from the mother *and* the fetus.
- Although more urine is produced, the growth of her uterus limits the amount of urine Joan's urinary bladder can hold. This causes more frequent micturition, which often interrupts Joan's sleep as the pregnancy progresses.
- Joan's respiratory ventilation increases 50 percent. This is because the high level of progesterone makes Joan more sensitive to carbon dioxide. To move the additional air, her body must breathe either faster or deeper. Joan's respiratory rate increases because the depth of her breaths cannot increase due to the pressure on her diaphragm from the growing fetus.

- The integumentary system is affected in several ways: Stretch marks may appear on Joan's rapidly expanding abdomen and breasts, her linea alba may darken and be referred to as the linea nigra, and melanocytes may increase melanin production, producing a *mask of pregnancy* that is visible on her face.
- Increased thyroid secretion raises Joan's basal metabolic rate by 15 percent. She may feel overheated and have an increased appetite.

Nutritional Requirements for a Pregnancy

16.16 learning **outcome**

Explain the nutritional requirements for a healthy pregnancy.

Although Joan's appetite is increased and she may think she is eating for two, she really needs only an additional 300 calories per day to support a pregnancy. It is important not to gain too much weight during a pregnancy. Normal weight gain over the course of a pregnancy is 24 pounds.

A balanced diet is essential. The Mayo Clinic recommends a balanced diet during pregnancy with special emphasis on four nutrients: folic acid, calcium, protein, and iron.[2] These nutrients and their recommended daily allowances are explained in the following list:

- Folic acid is important to prevent birth defects, especially those concerning neural development. The Mayo Clinic recommends 800 μg per day during pregnancy. Sources of folic acid include leafy green vegetables, citrus fruits, cereals, and beans.
- Calcium is necessary for bone development in the fetus and bone maintenance in the mother. The Mayo Clinic recommends 1,000 mg per day during pregnancy. Pregnant teens require 1,300 mg per day. Dairy products are a good source of calcium.
- Protein is very important for fetal growth, especially during the second and third trimesters. The Mayo Clinic recommends 71 g per day during pregnancy. Good sources of protein include lean meat, poultry, fish, eggs, dried beans, peanut butter, and dairy products.
- Iron is important for increased hemoglobin production to maintain the increased blood volume. The Mayo Clinic recommends 27 mg per day during pregnancy. Sources of iron include lean red meat, poultry, fish, nuts, and dried fruit.

Prenatal care is important for the health of both the mother and the developing fetus. A doctor may prescribe prenatal vitamins and nutritional supplements to complement Joan's diet. Proper nutrition is needed to sustain a pregnancy to full term. You will investigate next the events that will naturally end the pregnancy at full term and bring about the birth of Joan's baby.

spot check  What would be a good menu for a day while pregnant? Explain.

Initiating the Birth Process

16.17 learning **outcome**

Explain what initiates the birth process.

parturition: PAR-chur-ISH-un

The developing fetus plays a role in the timing of **parturition** (the process of birth). Follow along with **Figure 16.19** as the steps are explained in the following list:

1. The hypothalamus in the fetus releases corticotropin-releasing hormone (CRH) due to stress (confines of a small space relative to the growing fetus). CRH targets the fetal anterior pituitary to release ACTH.
2. ACTH, in turn, targets the fetal adrenal glands so that glucocorticoids are released. This is similar to what happens to an adult's body under stress (Chapter 8).

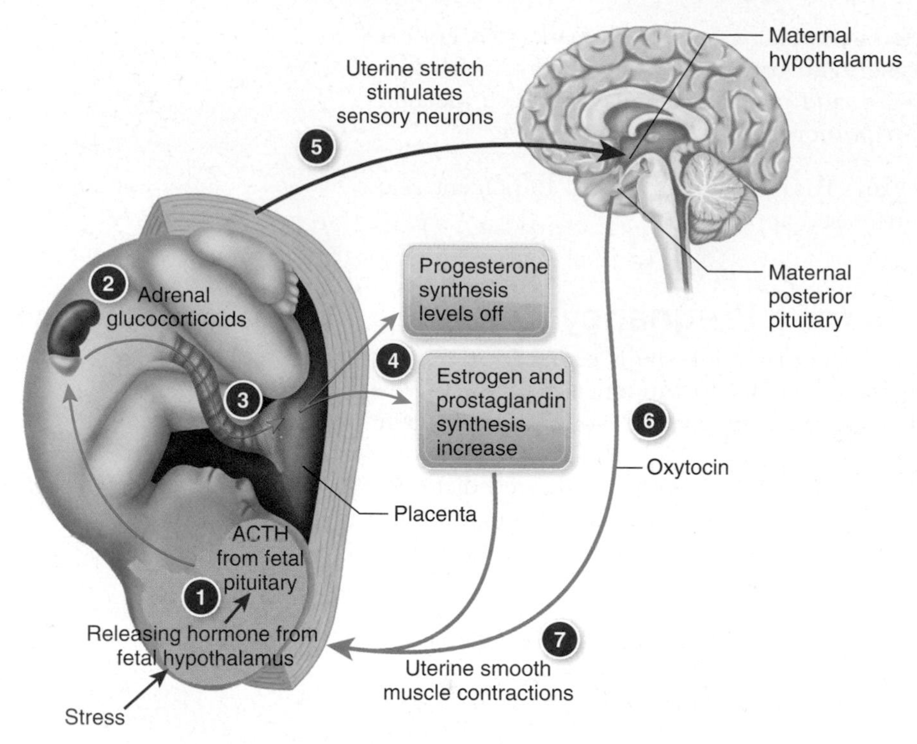

FIGURE 16.19 Initiating the birth process.

3. The fetal glucocorticoids travel to the placenta.

4. The fetal glucocorticoids cause the placenta to adjust the level of its hormones. Progesterone production levels off at around six months, while estrogen levels keep increasing. As stated earlier, estrogen makes the uterus more irritable (likely to contract), while progesterone suppresses contractions. So estrogen begins to have an increased effect, encouraging uterine contractions as the pregnancy reaches full term. The placenta and the membranes associated with it also produce prostaglandins that make the myometrium more likely to contract.

5. Smooth muscle of Joan's myometrium stretches more and more as the pregnancy progresses. As you may recall from Chapter 5, smooth muscle is more likely to contract if stretched. So this makes Joan's uterus more likely to contract. In fact, weak **Braxton Hicks contractions** (false labor) are common during pregnancy. Stretch receptors in the myometrium signal the mother's hypothalamus.

6. Joan's hypothalamus signals her posterior pituitary to secrete oxytocin. This hormone promotes **labor** (uterine contractions to bring about birth) in two ways: Oxytocin directly stimulates the myometrium to contract, and it stimulates the placenta to produce more prostaglandins.

7. Uterine contractions cause the head of the fetus to press harder on the cervix. This causes more signals to be sent to Joan's hypothalamus, so more oxytocin is released. The fetal head pushing on the cervix also stimulates Joan's cervix to release prostaglandins. As you might guess, this becomes a positive-feedback mechanism that will not end until the fetus is expelled from the uterus.

16.18 learning **outcome**

Describe the birth process.

The Birth Process

Now that you are familiar with what starts the birth process, you are ready to explore the three stages of parturition. These stages are shown in **Figures 16.20** and **16.21**.

Stage 1 This stage begins with regular uterine contractions. During this stage, the cervical canal widens **(dilation)** and the cervix thins **(effacement).** Stage 1 ends when the cervical canal reaches 10 cm (the diameter of a fetal head). The amniotic sac, in which the fetus floats, often ruptures during this stage. This event is called the *breaking of the waters.*

Stage 2 During this stage the baby is expelled. The **crowning** of the baby's head is usually seen first because a fetus most often assumes a head-down position in the seventh month. At this point, the doctor may make an incision **(episiotomy)** in Joan's perineum to widen the vaginal opening. The episiotomy is angled away from her anus to prevent the perineum tearing into her anus as the baby is born. The head is the most difficult part to expel. Once it is delivered, the rest of the body is expelled more easily.

Stage 3 The placenta detaches from Joan's uterus and is expelled during this stage. Further contractions of the uterus ensure that all of the placenta and its associated membranes (together, at this time, called **afterbirth**) are expelled. These contractions also help close blood vessels that had led to the placenta. About 350 mL of blood is normally lost when the placenta detaches.

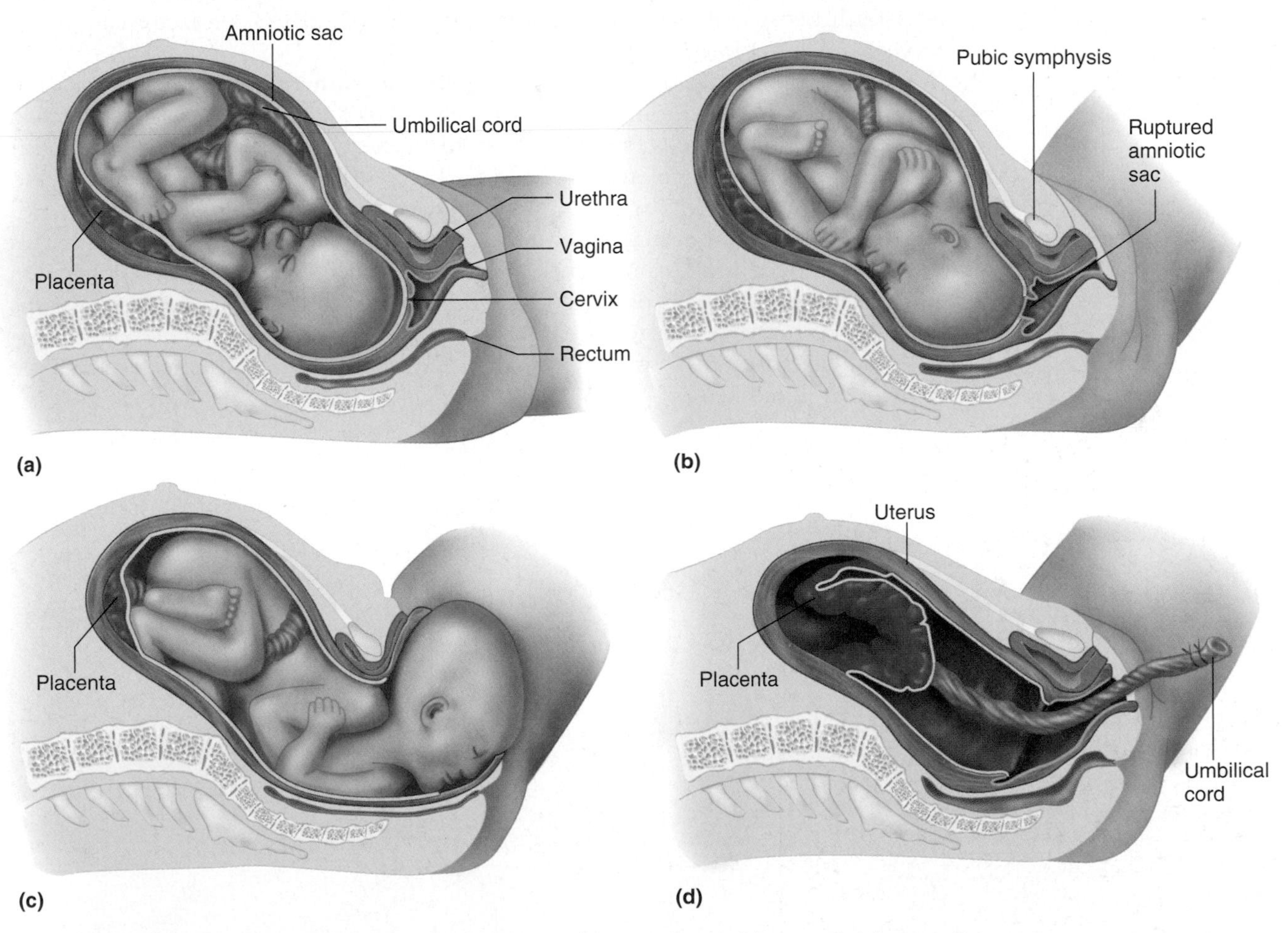

FIGURE 16.20 The birth process: (a) early stage 1, (b) late stage 1, (c) stage 2, (d) stage 3.

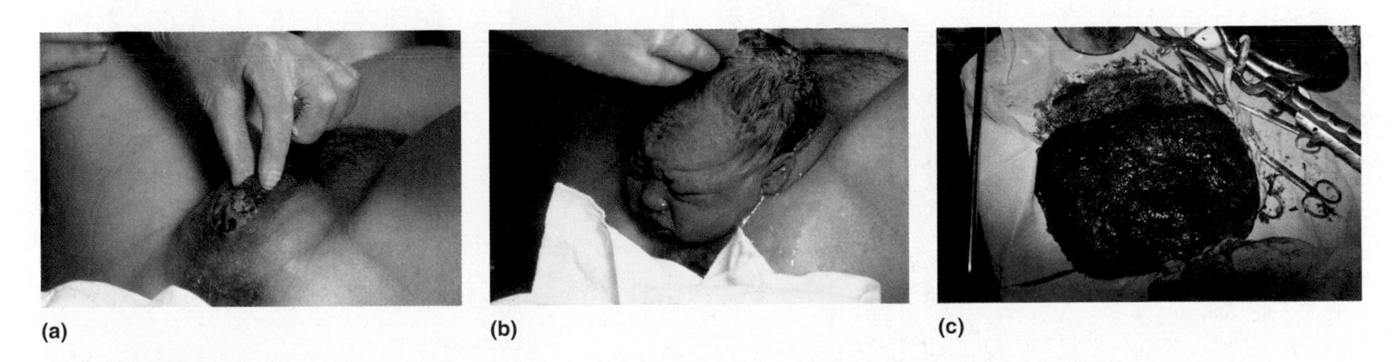

FIGURE 16.21 Childbirth: (a) crowning, (b) expulsion, (c) delivery of the placenta.

16.19 learning **outcome**

Explain the process of lactation.

Lactation

Once the placenta is delivered, the major source of pregnancy hormones is gone. Estrogen and progesterone levels drop drastically. All through the pregnancy, these two hormones together caused breast development but also suppressed the effects of **prolactin** released from the anterior pituitary. Prolactin is a hormone that stimulates milk production (lactation) in the mammary glands. The increased prolactin at birth (without suppression by estrogen and progesterone) causes milk production within a few days.

Late in the pregnancy, Joan's mammary glands produce **colostrum,** a thin, watery fluid containing protein, lactose (milk sugar), and many of her antibodies but one-third less fat than breast milk. This is all right for the first one to three days after birth because the newborn has plenty of fat. By day 3, prolactin causes the mother's milk *to come in*.

Suckling on Joan's breast causes the **milk ejection reflex.** The sensory endings in her nipple send signals to her hypothalamus for oxytocin release. Oxytocin causes myoepithelial cells (see **Figure 16.7d**) in her mammary lobules to contract to release milk to the lactiferous ducts in the breast. This causes the milk to travel to the nipple for the baby. The reflex is repeated every time the infant nurses.

Prolactin levels also surge each time the infant nurses. See **Figure 16.22**. This ensures that there will be ample milk production to meet the needs of the infant. On average, 1.5 L of milk are produced each day. If Joan chooses not to nurse or stops nursing, the surges of prolactin cease and her milk production ends.

There are many reasons why a mother may choose to nurse a newborn. Some of them are listed here:

- Breast milk provides good nutrition.
- Breast milk contains many of the mother's antibodies to provide passive natural immunity to the baby.
- Breast milk has a laxative effect.
- Breast milk helps colonize helpful bacteria in the infant's intestines.

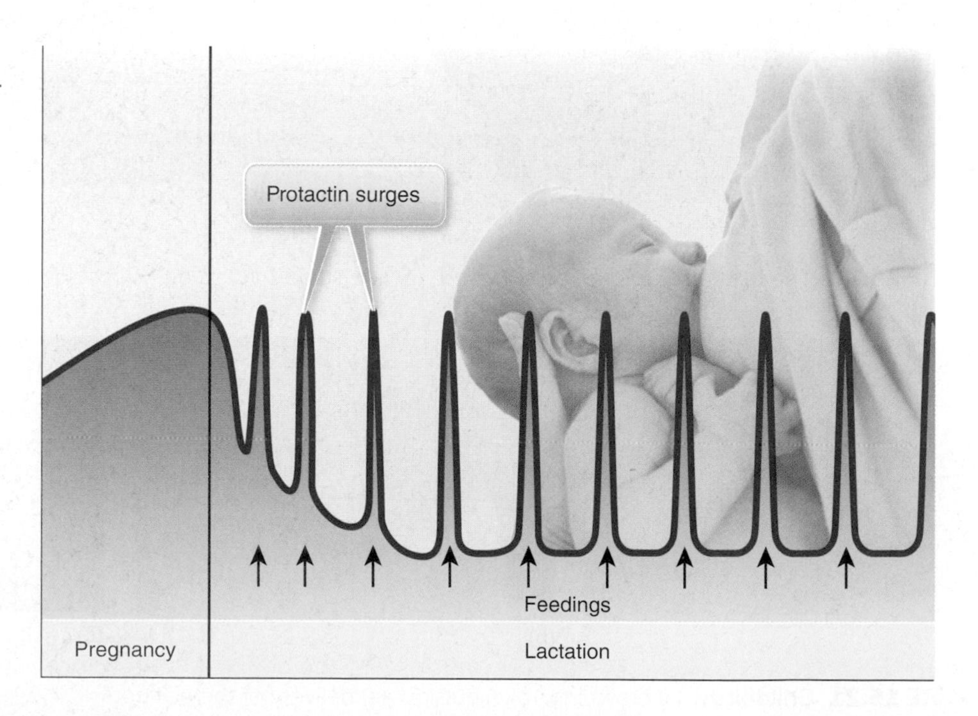

FIGURE 16.22 Prolactin secretion in a nursing mother.

You have covered the physiology of a healthy pregnancy for Joan and Rick. To finish the content of this chapter, you will investigate some of the things that could go wrong in a pregnancy.

Disorders of Pregnancy

16.20 learning outcome

Describe disorders of pregnancy.

Spontaneous Abortion (Miscarriage) Fifty percent of all zygotes are lost before delivery. Most of these will be lost because they never implanted. If that is the case, the woman may never know that her egg had been fertilized.

Of the zygotes that do implant, 10 to 15 percent will end in a spontaneous abortion.[3] The reasons for this include fetal abnormalities, improper implantation, or premature detachment of the placenta (placental abruption).

warning

It is important to differentiate the term *spontaneous abortion* from *abortion*. An abortion is an intentional medical procedure to prematurely end a pregnancy so that the fetus will not survive. On the other hand, a spontaneous abortion is a natural, premature ending to a pregnancy. It can also be called a *miscarriage*.

Ectopic Pregnancy An ectopic pregnancy occurs if the fertilized egg implants anywhere other than the uterus. The most likely location is in a uterine tube. These pregnancies are doomed from the start. A uterine tube cannot support a pregnancy, and the developing embryo cannot be relocated. The growing embryo will eventually cause the uterine tube to rupture, causing a potentially life-threatening situation for the mother.

spot check 12 A uterine tube is usually removed in an ectopic pregnancy. Does this affect the woman's fertility?

Preeclampsia This condition is pregnancy-induced hypertension accompanied by protein in the urine that usually occurs after 20 weeks in the pregnancy. The mother's blood pressure and weight are closely monitored throughout a pregnancy to detect this disorder. The hypertension may be slight (140/90 mmHg or higher). A sudden weight gain (more than 2 pounds in a week) late in pregnancy may indicate edema associated with the hypertension. Although the cause of this condition is not clear, the cure is known—the birth of the baby. Preeclampsia can lead to serious, potentially fatal consequences for the mother and the baby.

putting the pieces **together**

The Reproductive Systems

Integumentary system

Apocrine glands secrete pheromones to attract the opposite sex; mammary glands secrete milk to nourish an infant.

Reproductive hormones promote sebum production.

Skeletal system

Bones of the pelvis protect some reproductive organs.

Reproductive hormones promote endochondral bone growth and serve as a lock on calcium in the bone.

Muscular system

Muscle contractions are involved in ejaculation and childbirth.

Reproductive hormones promote muscle growth and development.

Nervous system

Nerve impulses innervate muscles involved in erection, ejaculation, and childbirth.

Reproductive hormones affect the production of hypothalamus releasing hormones.

Endocrine system

Hormones regulate sexual development, sex drive, menstruation, pregnancy, birth, and lactation.

Reproductive hormones have a negative-feedback effect on the hypothalamus.

Cardiovascular system

Blood transports reproductive hormones; vasodilation enables erection.

Testosterone promotes erythropoiesis.

Lymphatic system

Sends white blood cells to fight pathogens in the reproductive system.

Sustentacular cells form a blood-testis barrier to isolate developing sperm from the immune system.

Respiratory system

Allows for the exchange of O_2 and CO_2 in system tissues.

Allows for the exchange of O_2 and CO_2 in a fetus.

Digestive system

Provides nutrients for tissues of the reproductive system.

Decreased digestive motility during the first trimester may cause morning sickness.

Excretory/urinary system

Removes metabolic wastes produced by the fetus; semen is ejaculated through the urethra.

Sexual response in male and female blocks the passing of urine.

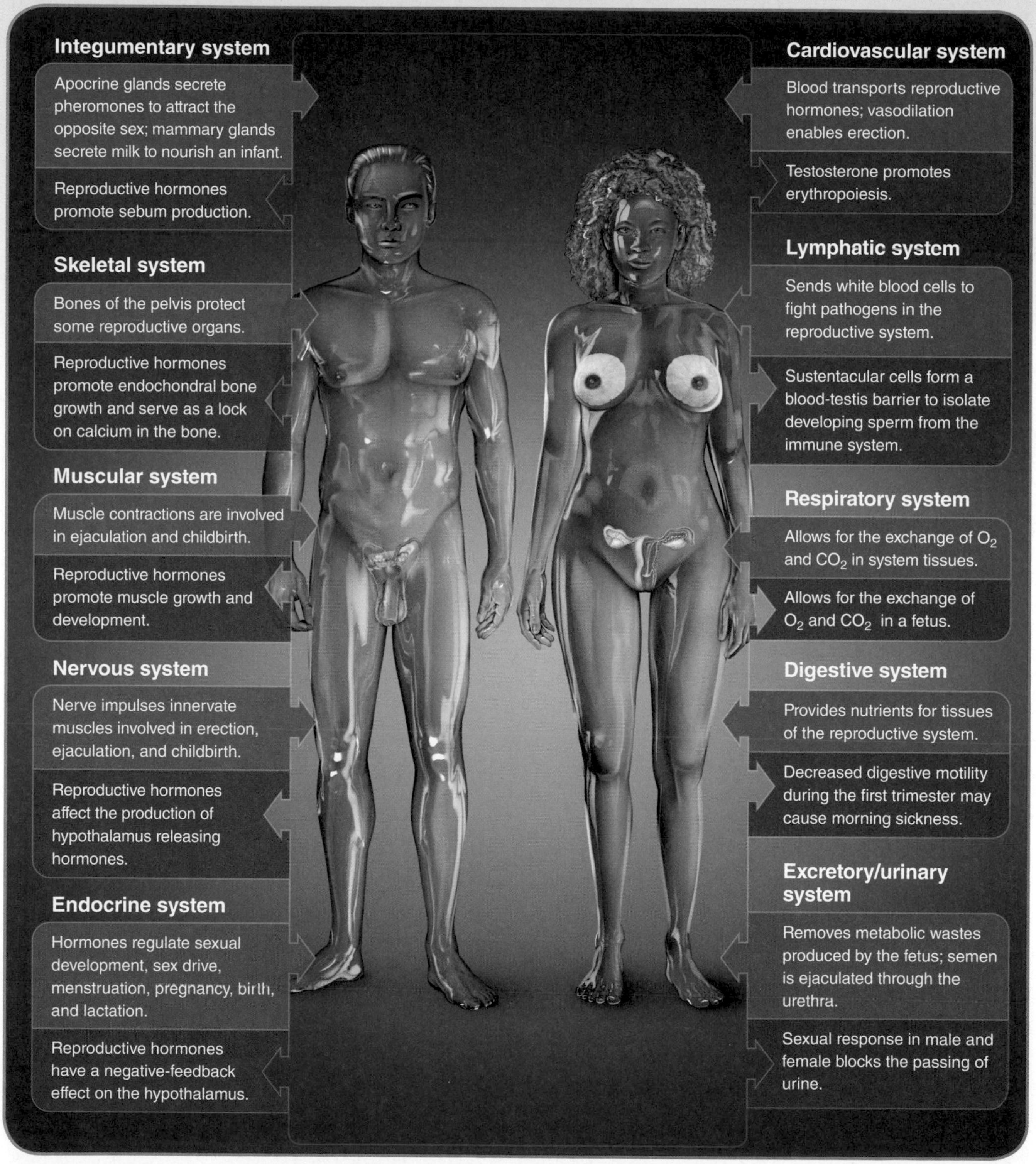

FIGURE 16.23 Putting the Pieces Together—The Reproductive Systems: connections between the female and male reproductive systems and the body's other systems.

summary

Overview

- A female zygote has XX as the sex chromosomes.

Female Reproductive Anatomy

- Ovaries produce ova and are therefore the primary sex organs of the female reproductive system.

Ovaries

- An ovary has two layers: a cortex containing follicles and a medulla containing blood vessels.
- Ligaments suspend the ovaries in the pelvic cavity and anchor them to the uterus.

Secondary Female Reproductive Organs and Structures

- The ovaries, uterus, and vagina are internal.
- The vulva and breasts are external.
- The uterus has three layers: the perimetrium, the myometrium, and the endometrium.
- The uterine tubes transport eggs to the uterus.
- The vagina allows for the flow of the menses, is a receptacle for sperm, and serves as the birth canal.
- The vulva includes the following: the mons pubis, labia, clitoris, vestibular bulbs, and vestibular glands.
- The breasts contain mammary glands that do not fully develop until the first pregnancy.

Physiology of the Female Reproductive System

Hormonal Control at Puberty

- Puberty begins when the GnRH from the hypothalamus stimulates the anterior pituitary to secrete FSH and LH.
- FSH stimulates follicles in the ovaries to produce estrogen.
- Estrogen is responsible for the development of female sex characteristics.

Oogenesis

- One haploid gamete is formed from each oogonium.
- A follicle is responsible for caring for an oocyte and producing hormones.
- All oocytes have been produced from oogonia by birth.
- Many oogonia and oocytes are lost by atresia.
- Oogenesis is halted mid-meiosis I until puberty.
- Secondary oocytes are formed each month.
- Meiosis II is not completed unless an egg is fertilized.
- Follicles develop along with an oocyte.
- Some primordial follicles develop each month after puberty.
- Mature follicles rupture to release an egg during ovulation.
- After ovulation, the follicle becomes a corpus luteum, which secretes hormones.
- If the egg does not become fertilized, the corpus luteum becomes a corpus albicans.

Hormonal Control in the Adult Female

- A sexual cycle is composed of two parts: an ovarian cycle and a menstrual cycle.
- A sexual cycle is typically 28 days and begins on the first day of a woman's period.
- The ovarian cycle has a follicular phase and a luteal phase.
- The menstrual cycle has the following phases: menstrual, proliferative, secretory, and premenstrual.
- Ovulation typically occurs on day 14.

The Female Sexual Response

- The female sexual response includes four stages: arousal, plateau, orgasm, and resolution.
- Unlike the male, a woman does not usually have a refractory period in her sexual response.

Effects of Aging on the Female Reproductive System

- A woman's body goes through a climacteric in midlife due to the decreased production of estrogen and progesterone.
- Menstruation ceases during menopause.
- Hot flashes are common.
- The tissues of the vagina, labia minora, clitoris, uterus, uterine tubes, and breast atrophy.
- Bone mass declines.
- Skin becomes thinner.
- Cholesterol levels rise.

Disorders of the Female Reproductive System

Breast Cancer

- Breast cancer is the abnormal growth of breast tissue, usually occurring in the lactiferous ducts and lobules of the breast.
- Age, family history, genetic mutations, personal history, and lifestyle choices affect breast cancer risk.
- Detection of breast cancer usually begins with a monthly self-exam and mammography.
- Treatment of breast cancer may include chemotherapy, surgery, radiation, immunotherapy, and hormone therapy.

Cervical Cancer

- Cervical cancer is often caused by human papillomavirus (HPV) infection.

Endometriosis

- Endometriosis is the growth of endometrium in places other than the uterus.
- The endometrium goes through the menstrual cycle no matter where it is located.
- Endometriosis can lead to infertility and painful periods.

Pregnancy

- The four requirements of a pregnancy are that there must be a sperm and an egg, the sperm must meet the egg, the sperm must fertilize the egg, and the fertilized egg must implant.

Pathway for Sperm to Meet an Egg

- An egg must be fertilized within 24 hours of ovulation if it is to survive.
- The complete journey for a mature sperm is:

Epididymis → ductus deferens → ejaculatory duct → urethra →
vagina → uterus (cervix, body) → uterine tube (isthmus, ampulla)

Fertilization to Implantation

- Sperm must go through capacitation to fertilize an egg.
- Many sperm may be needed to break through the cells surrounding the egg.
- Only one sperm will be allowed to penetrate the egg—the rest will be rejected.
- The nuclei of the sperm and egg rupture, the chromosomes mix, and a new nucleus forms.
- The zygote undergoes mitotic divisions on its way to the uterus.
- The blastocyst implants in the lining of the uterus six days after fertilization.

Hormonal Control of Pregnancy

- Hormones are made by the fertilized egg, the corpus luteum, and the placenta, along with other endocrine glands during a pregnancy.
- The hormones that control a pregnancy include HCG, estrogen, progesterone, HCS, thyroid hormone, PTH, ACTH, glucocorticoids, and aldosterone.

A Woman's Adjustment to Pregnancy

- Many of the body systems are affected by the pregnancy hormones and the growing fetus's pressure on the female anatomy.

Nutritional Requirements for a Pregnancy

- Only an additional 300 calories are required to sustain a pregnancy.
- A balanced diet rich in folic acid, calcium, protein, and iron is essential for a pregnancy.

Initiating the Birth Process

- The fetus has a role in the timing of parturition.
- The hormones involved in initiating parturition lead to a positive-feedback mechanism that ends with the birth of the baby.

The Birth Process

- There are three stages to the birth process: dilation of the cervix, expulsion of the baby, and delivery of the afterbirth.

Lactation

- Throughout the pregnancy, estrogen and progesterone suppress the effects of prolactin.
- Once the placenta is delivered and the source of estrogen and progesterone is gone, prolactin stimulates milk production.
- Milk production is preceded in the first few days after birth by colostrum.
- Suckling on the breast causes oxytocin release and the milk ejection reflex.
- Prolactin levels surge with each feeding to ensure ample milk production to meet the baby's needs.
- The many reasons for nursing include the following: breast milk provides good nutrition; the antibodies present in breast milk provide the baby some immunity; the milk has a laxative effect on the baby; and breast milk helps colonize helpful bacteria in the baby's intestines.

Disorders of Pregnancy

- Fifty percent of zygotes do not survive. Most are lost before they implant.
- Of the zygotes that do implant, 15 percent are miscarried due to fetal abnormalities, improper implantation, premature detachment of the placenta, and other causes.
- An ectopic pregnancy occurs if the fertilized egg implants anywhere other than in the uterus.
- Ectopic pregnancies are doomed.
- Preeclampsia is pregnancy-induced hypertension accompanied by protein in the urine.
- Preeclampsia can lead to serious, potentially fatal consequences for the mother and the baby.

key words for review

The following terms are defined in the glossary.

afterbirth
atresia
capacitation
colostrum
crowning
effacement
episiotomy
folliculogenesis
gestation
labor
lactation
mammography
menopause
menstrual cycle
milk ejection reflex
oogenesis
ovarian cycle
ovulation
parturition
prolapse

chapter review questions

Word Building: *Use the word roots listed in the beginning of the chapter and the prefixes and suffixes inside the back cover to build words with the following meanings:*

1. Procedure to remove an ovary: ______________________________
2. Inflammation of the lining of the uterus: ______________________________
3. Physician who specializes in women's health: ______________________________
4. Surgical cut to widen the vulva: ______________________________
5. No menstrual flow: ______________________________

Multiple Select: *Select the correct choices for each statement. The choices may be all correct, all incorrect, or a combination of correct and incorrect.*

1. Arun and Felicia want to have a child. What will have to happen for Felicia to become pregnant?
   a. Both of them will have to produce gametes.
   b. Arun's sperm must travel to the ampulla of Felicia's uterus, where fertilization takes place.
   c. After Arun's sperm have gone through capacitation, only one must penetrate Felicia's egg.
   d. Felicia's ovulated ovum must enter a uterine tube.
   e. Even if fertilized in a lab, their embryo must be implanted in Felicia's uterus.
2. What route will Arun's sperm travel to meet Felicia's egg?
   a. They will travel from her vagina to the cervical canal.
   b. They will travel through the lumen of her uterus to the uterine tubes.
   c. Half of Arun's sperm may enter the wrong uterine tube.
   d. They will travel from the ampulla to the isthmus of the uterine tube.
   e. They may reach the infundibulum of the uterine tube.
3. What hormone(s) will control Felicia's body during pregnancy?
   a. Her thyroid gland will secrete more thyroid hormone to raise her metabolism 30 percent.
   b. Her progesterone levels will level off at about six months into her pregnancy.
   c. Her estrogen levels will continue to rise throughout her pregnancy.
   d. Her adrenal cortex will secrete more glucocorticoids.
   e. Her anterior pituitary will continue to secrete FSH and LH.
4. What will happen to Felicia's body if she has a normal, healthy pregnancy?
   a. She will gain approximately 24 pounds.
   b. Her respiratory rate will increase.
   c. Her cardiac output will double.
   d. Hemopoiesis will increase.
   e. Her face may develop dark patches.
5. What will bring about the birth of Felicia's baby?
   a. A positive-feedback mechanism.
   b. A negative-feedback mechanism.
   c. Oxytocin from Felicia's anterior pituitary.
   d. Oxytocin and prostaglandins.
   e. Hormones from the fetus.
6. Felicia took birthing classes during her ninth month of pregnancy. What should she have learned about the birth process?
   a. Stage I involves dilation of the cervix.
   b. Stage II begins with the breaking of the waters.
   c. Stage III begins with the crowning of the baby's head.
   d. An episiotomy may be performed during stage III.
   e. The afterbirth is delivered during stage III.

7. If Felicia chooses to nurse, what can she expect will happen while she is nursing her new baby?
   a. A surge of progesterone will occur each time the baby nurses.
   b. Felicia will provide antibodies in her breast milk for active natural immunity for her baby.
   c. The milk ejection reflex will occur so that milk moves to the nipple.
   d. Milk will move from the lactiferous sinuses, to the lactiferous ducts, and to the nipple.
   e. Oxytocin will be released.
8. What does a zygote need to develop female reproductive anatomy?
   a. An SRY gene.
   b. Testosterone receptors.
   c. XY as the sex chromosomes.
   d. XX as the sex chromosomes.
   e. FSH and LH.
9. Felicia's sister, April, is 11 years old and starting to go through puberty. What is happening in April's body?
   a. Her hypothalamus is producing GnRH.
   b. Her posterior pituitary is secreting FSH and LH.
   c. Some follicles in her ovaries are secreting estrogen.
   d. Her anterior pituitary is secreting progesterone.
   e. The lining of her vagina is undergoing metaplasia due to estrogen.
10. Judy (Felicia's mother) is 52 years old. Which of the following statements describe(s) a normal sign of the aging of her reproductive system?
    a. Sexual intercourse is a little painful due to dryness in her vagina.
    b. She is developing a slight mustache.
    c. Her hips are widening.
    d. She is depositing more fat.
    e. Her periods are becoming more and more irregular.

Matching: *Match the stage in the female sexual response to its description. Some answers may be used more than once. Some questions may have more than one correct response.*

_____	1. The uterus is tipped over the urinary bladder.	a. Plateau
_____	2. The lower one-third of the vagina constricts around the penis.	b. Resolution
_____	3. Vaginal fluids moisten the vagina.	c. Orgasm
_____	4. The vagina rhythmically contracts.	d. Arousal
_____	5. The uterus undergoes peristaltic contractions.	

Matching: *Match the pregnancy hormone to its description. Some choices may be used more than once. Some questions have more than one correct response.*

_____	6. Produced by the placenta	a. Estrogen
_____	7. Produced by the corpus luteum	b. Progesterone
_____	8. Secreted by the posterior pituitary	c. Oxytocin
_____	9. Secreted by a blastocyst	d. HCS
_____	10. Has drastic drop in levels at birth	e. HCG

Critical Thinking

1. Explain ways in which the female reproductive anatomy is designed to accommodate the male reproductive anatomy and bring a male to orgasm during intercourse.
2. Sperm can survive in the female reproductive system for six days, but they need to have 10 hours after ejaculation to undergo capacitation. Ovulation typically occurs on day 14 of a woman's menstrual cycle. An ovulated egg can survive only up to 24 hours if it is not fertilized. Given this information, what days of a woman's sexual cycle would be a window of opportunity for an egg to be fertilized? Explain.
3. Megan began to ovulate when she was 13. She chose not to have children and so never had a pregnancy. She stopped menstruating when she was 53. How many eggs is she likely to have ovulated in her lifetime? How does this number compare to the number of eggs she had at puberty and the number left at menopause?

16 chapter mapping

This section of the chapter is designed to help you find where each outcome is covered in the text.

	Outcomes	Readings, figures, and tables	Assessments
16.1	Use medical terminology related to the female reproductive system.	Word roots: p. 591	Word Building: 1–5
16.2	Explain what is needed for female anatomy to develop.	Overview: p. 592	Multiple Select: 8
16.3	Describe the anatomy of the ovary and its functions.	Ovaries: pp. 592–594 Figures 16.2, 16.3	Spot Check: 1
16.4	Describe the female secondary reproductive organs and structures and their respective functions.	Secondary female reproductive organs and structures: pp. 594–599 Figures 16.3–16.7	Spot Check: 2 Critical Thinking: 1
16.5	Explain the hormonal control of puberty and the resulting changes in the female.	Hormonal control at puberty: pp. 599–600	Spot Check: 3 Multiple Select: 9
16.6	Explain oogenesis in relation to meiosis.	Oogenesis: pp. 600–603 Figures 16.8, 16.9	Spot Check: 4, 5
16.7	Explain the hormonal control of the adult female reproductive system and its effect on follicles in the ovary and the uterine lining.	Hormonal control in the adult female: pp. 603–606 Figures 16.10–16.12	Spot Check: 6
16.8	Describe the stages of the female sexual response.	The female sexual response: pp. 607–608 Figure 16.13	Matching: 1–5
16.9	Explain the effects of aging on the female reproductive system.	Effects of aging on the female reproductive system: pp. 608–609	Spot Check: 7, 8 Multiple Select: 10
16.10	Describe female reproductive disorders.	Disorders of the female reproductive system: pp. 609–611 Figure 16.14	Spot Check: 9
16.11	List the four requirements of pregnancy.	Pregnancy: pp. 611–612 Figure 16.15	Multiple Select: 1
16.12	Trace the pathway for a sperm to fertilize an egg.	Pathway for a sperm to meet an egg: p. 612	Multiple Select: 1, 2
16.13	Describe the events necessary for fertilization and implantation.	Fertilization to implantation: pp. 612–614 Figures 16.16, 16.17	Spot Check: 10 Multiple Select: 1 Critical Thinking: 2
16.14	Explain the hormonal control of pregnancy.	Hormonal control of pregnancy: pp. 614–615	Multiple Select: 3 Matching: 6–10
16.15	Explain the adjustments a woman's body makes to accommodate a pregnancy.	A woman's adjustment to pregnancy: pp. 615–617 Figure 16.18	Multiple Select: 4
16.16	Explain the nutritional requirements for a healthy pregnancy.	Nutritional requirements for a pregnancy: p. 617	Spot Check: 11
16.17	Explain what initiates the birth process.	Initiating the birth process: pp. 617–618 Figure 16.19	Multiple Select: 5
16.18	Describe the birth process.	The birth process: pp. 618–619 Figures 16.20, 16.21	Multiple Select: 6
16.19	Explain the process of lactation.	Lactation: pp. 620–621 Figure 16.22	Multiple Select: 7
16.20	Describe disorders of pregnancy.	Disorders of pregnancy: p. 621	Spot Check: 12

footnotes

1. National Cancer Institute. (2010, September 7). For women with BRCA mutations, prophylactic surgery reduces cancer risk [Electronic version]. *NCI Cancer Bulletin, 7*(17).
2. Mayo Clinic. (2010, December 8). *Pregnancy diet: Essential nutrients when you're eating for two.* Retrieved December 8, 2010, from http://www.mayoclinic.com/health/pregnancy-nutrition/PR00110.
3. Beers, M. H., Porter, R. S., Jones, T. V., Kaplan, J. L., & Berkwits, M. (Eds.). (2006). *Merck manual of diagnosis and therapy* (18th ed.). Whitehouse Station, NJ: Merck Research Laboratories.

references and resources

Beers, M. H., Porter, R. S., Jones, T. V., Kaplan, J. L., & Berkwits, M. (Eds.). (2006). *Merck manual of diagnosis and therapy* (18th ed.). Whitehouse Station, NJ: Merck Research Laboratories.

Hall, D. C. (n.d.). *Nutritional influences on estrogen metabolism.* Retrieved December 5, 2010, from http://www.funimky.com/research_estrogen.htm.

Mayo Clinic. (2009, November 19). *Breast cancer: Causes.* Retrieved December 8, 2010, from http://www.mayoclinic.com/health/breast-cancer/DS00328/DSECTION=causes.

Mayo Clinic. (2009, November 19). *Breast cancer: Risk factors.* Retrieved December 8, 2010, from http://www.mayoclinic.com/health/breast-cancer/DS00328/DSECTION=risk-factors.

Mayo Clinic. (2011, October 10). *Pregnancy diet: Focus on these essential nutrients.* Retrieved October 10, 2011, from http://www.mayoclinic.com/health/pregnancy-nutrition/PR00110.

Mayo Clinic. (2010, March 4). *Endometriosis.* Retrieved December 8, 2010, from http://www.bing.com/health/article/mayo-125699.

Mayo Clinic. (2010, March 4). *Preeclampsia.* Retrieved December 9, 2010, from http://www.bing.com/health/article/mayo-126272.

MedlinePlus. (2008, August 1). *Aging changes in the female reproductive system.* Retrieved December 6, 2010, from http://www.nlm.nih.gov/medlineplus/ency/article/004016.htm.

National Cancer Institute. (2010, September 7). For women with BRCA mutations, prophylactic surgery reduces cancer risk [Electronic version]. *NCI Cancer Bulletin, 7*(17).

National Cancer Institute. (2010, November 23). *Genetics of breast and ovarian cancer(PDQ®).* Retrieved December 8, 2010, from http://www.cancer.gov/cancertopics/pdq/genetics/breast-and-ovarian/HealthProfessional.

National Cancer Institute. (n.d.). *Breast cancer.* Retrieved December 8, 2010, from http://www.cancer.gov/cancertopics/types/breast.

National Cancer Institute. (n.d.). *Cervical cancer.* Retrieved January 24, 2011, from http://www.cancer.gov/cancertopics/types/cervical.

National Institute on Aging. (2009, December 2). *Hormones and menopause.* Retrieved June 17, 2010, from http://www.nia.nih.gov/HealthInformation/Publications/hormones.htm.

Rosenblatt, P. L. (2007, July). Effects of aging: Biology of the reproductive System. *Merck manual home edition.* Retrieved December 6, 2010, from http://www.merckmanuals.com/home/sec22/ch241/ch241f.html.

Saladin, Kenneth S. (2010). *Anatomy & physiology: The unity of form and function* (5th ed.). New York: McGraw-Hill.

Seeley, R. R., Stephens, T. D., & Tate, P. (2006). *Anatomy & physiology* (7th ed.). New York: McGraw-Hill.

Shier, D., Butler, J., & Lewis, R. (2010). *Hole's human anatomy & physiology* (12th ed.). New York: McGraw-Hill.

WebMD. (n.d.). *Braxton hicks or true labor contractions?* Retrieved January 24, 2011, from http://www.webmd.com/baby/guide/true-false-labor.

The Metric System

The metric system is commonly used in science and medicine. This appendix will help you with conversions between the metric and U.S. systems.

Length

TABLE A.1 Units of Length

Metric unit	Metric equivalent	Conversion to U.S. system
Kilometer (km)	1,000 m	0.62 mile; 1.6 km/mile
Meter (m)	100 cm; 1,000 mm	39.4 inches; 1.1 yards
Centimeter (cm)	1/100 m; 0.01 m; 10 mm	0.39 inch; 2.5 cm/inch
Millimeter (mm)	1/1,000 m; 0.001 m	0.039 inch
Micrometer (μm)	1/1,000 mm; 0.001 mm	

Examples of metric lengths:

Measure the length and the width of your thumb in inches and centimeters, using the ruler provided on this page. Also measure the maximum distance you can spread your thumb and index finger apart. Knowing these metric measurements will give you a reference when you read about metric lengths in the text.

Weight

TABLE A.2 Units of Weight

Metric unit	Metric equivalent	Conversion to U.S. system
Kilogram (kg)	1,000 g	2.2 pounds (lb)
Gram (g)	1,000 mg	0.035 ounce (oz); 28.5 g/oz
Milligram (mg)	1/1,000 g; 0.001 g	
Microgram (μg)	1/1,000 mg; 0.001 mg	

Examples of metric weights:

U.S. penny: 2.50 g

U.S. nickel: 5.00 g

5-pound bag of sugar: 2.268 kg

FIGURE A.1
Ruler inch and metric

Volume

TABLE A.3 Units of Volume

Metric unit	Metric equivalent	Conversion to U.S. system
Liter (L)	1,000 ml	1.06 quarts; 0.264 gallon
Deciliter (dL)	1/10 L; 0.1 L	
Milliliter (mL)	1/1,000 L; 0.001 L	0.034 ounce (oz); 29.4 mL/oz
Microliter (μL)	1/1,000 mL; 0.001 mL	

Examples of metric volumes:

Large bottle of soda: 2 L

12-oz can of soda: 355 mL

references and resources

US Mint. (n.d.). *Coin specifications*. Retrieved April 27, 2011, from http://www.usmint.gov/about_the_mint/?action=coin_specifications.

appendix B Nutrition Table

The following table shows the nutritional goals for age-gender groups based on dietary intakes and dietary guidelines recommendations.

TABLE B.1 USDA Dietary Guidelines for Americans, 2010

Nutrient (units)	Source of Goal[a]	Child 1–3	Female 4–8	Male 4–8	Female 9–13	Male 9–13	Female 14–18	Male 14–18	Female 19–30	Male 19–30	Female 31–50	Male 31–50	Female 51≤	Male 51≤
Macronutritents														
Protein (g)	RDA[b]	13	19	19	34	34	46	52	46	56	46	56	46	56
(% of calories)	AMDR[c]	5–20	10–30	10–30	10–30	10–30	10–30	10–30	10–35	10–35	10–35	10–35	10–35	10–35
Carbohydrate (g)	RDA	130	130	130	130	130	130	130	130	130	130	130	130	130
(% of calories)	AMDR	45–65	45–65	45–65	45–65	45–65	45–65	45–65	45–65	45–65	45–65	45–65	45–65	45–65
Total fiber (g)	IOM[d]	14	17	20	22	25	25	31	28	34	25	31	22	28
Total fat (% of calories)	AMDR	30–40	25–35	25–35	25–35	25–35	25–35	25–35	20–35	20–35	20–35	20–35	20–35	20–35
Saturated fat (% of calories)	DG[e]	<10%	<10%	<10%	<10%	<10%	<10%	<10%	<10%	<10%	<10%	<10%	<10%	<10%
Linoleic acid (g)	AI[f]	7	10	10	10	12	11	16	12	17	12	17	11	14
(% of calories)	AMDR	5–10	5–10	5–10	5–10	5–10	5–10	5–10	5–10	5–10	5–10	5–10	5–10	5–10
Alpha=linolenic acid (g)	AI	0.7	0.9	0.9	1.0	1.2	1.1	1.6	1.1	1.6	1.1	1.6	1.1	1.6
(% of calories)	AMDR	0.6–1.2	0.6–1.2	0.6–1.2	0.6–1.2	0.6–1.2	0.6–1.2	0.6–1.2	0.6–1.2	0.6–1.2	0.6–1.2	0.6–1.2	0.6–1.2	0.6–1.2
Cholesterol (mg)	DG	<300	<300	<300	<300	<300	<300	<300	<300	<300	<300	<300	<300	<300
Minerals														
Calcium (mg)	RDA	700	1,000	1,000	1,300	1,300	1,300	1,300	1,000	1,000	1,000	1,000	1,200	1,200
Iron (mg)	RDA	7	10	10	8	8	15	11	18	8	18	8	8	8
Magnesium (mg)	RDA	80	130	130	240	240	360	410	310	400	320	420	320	420
Phosphorus (mg)	RDA	460	500	500	1,250	1,250	1,250	1,250	700	700	700	700	700	700
Potassium (mg)	AI	3,000	3,800	3,800	4,500	4,500	4,700	4,700	4,700	4,700	4,700	4,700	4,700	4,700
Sodium (mg)	UL[g]	<1,500	<1,900	<1,900	<2,200	<2,200	<2,300	<2,300	<2,300	<2,300	<2,300	<2,300	<2,300	<2,300
Zinc (mg)	RDA	3	5	5	8	8	9	11	8	11	8	11	8	11
Copper (μg)	RDA	340	440	440	700	700	890	890	900	900	900	900	900	900
Selenium (μg)	RDA	20	30	30	40	40	55	55	55	55	55	55	55	55
Vitamins														
Vitamin A (μg RAE)	RDA	300	400	400	600	600	700	900	700	900	700	900	700	900
Vitamin D[h] (μg)	RDA	15	15	15	15	15	15	15	15	15	15	15	15	15
Vitamin E (mg AT)	RDA	6	7	7	11	11	15	15	15	15	15	15	15	15
Vitamin C (mg)	RDA	15	25	25	45	45	65	75	75	90	75	90	75	90
Thiamin (mg)	RDA	0.5	0.6	0.6	0.9	0.9	1.0	1.2	1.1	1.2	1.1	1.2	1.1	1.2
Riboflavin (mg)	RDA	0.5	0.6	0.6	0.9	0.9	1.0	1.3	1.1	1.3	1.1	1.3	1.1	1.3

Nutrient (units)	Source of Goal[a]	Child 1–3	Female 4–8	Male 4–8	Female 9–13	Male 9–13	Female 14–18	Male 14–18	Female 19–30	Male 19–30	Female 31–50	Male 31–50	Female 51≤	Male 51≤
Vitamins														
Niacin (mg)	RDA	6	8	8	12	12	14	16	14	16	14	16	14	16
Folate (µg)	RDA	150	200	200	300	300	400	400	400	400	400	400	400	400
Vitamin B_6 (mg)	RDA	0.5	0.6	0.6	1.0	1.0	1.2	1.3	1.3	1.3	1.3	1.3	1.5	1.7
Vitamin B_{12} (µg)	RDA	0.9	1.2	1.2	1.8	1.8	2.4	2.4	2.4	2.4	2.4	2.4	2.4	2.4
Choline (mg)	AI	200	250	250	375	375	400	550	425	550	425	550	425	550
Vitamin K (µg)	AI	30	55	55	60	60	75	75	90	120	90	120	90	120

[a] Dietary guidelines recommendations are used when no quantitative dietary reference intake value is available; apply to ages 2 years and older.
[b] Recommended dietary allowance, IOM.
[c] Acceptable macronutrient distribution range, IOM.
[d] 14 grams per 1,000 calories, IOM.
[e] Dietary guidelines recommendation.
[f] Adequate intake, IOM.
[g] Upper limit, IOM.
[h] 1 µg of vitamin D is equivalent to 40 IU.

Note: AT = alpha-tocopherol; DFE = dietary folate equivalents; RAE = retinol activity equivalents.

Sources: U.S. Department of Agriculture and U.S. Department of Health and Human Services, *Dietary Guidelines for Americans, 2010,* 7th ed. (Washington, DC: GPO, December 2010), appendix 5, available at http://www.health.gov/dietaryguidelines/dga2010/DietaryGuidelines2010.pdf. Based on P. Britten, K. Marcoe, S. Yamini, and C. Davis, "Development of Food Intake Patterns for the MyPyramid Food Guidance System," *Journal of Nutrition Education and Behavior* 38 (6 Suppl), 2006, S78–S92; and Institute of Medicine, *Dietary Reference Intakes: The Essential Guide to Nutrient Requirements* (Washington, DC: National Academies Press, 2006), and *Dietary Reference Intakes for Calcium and Vitamin D* (Washington, DC: National Academies Press, 2010).

references and resources

U.S. Department of Agriculture and U.S. Department of Health and Human Services. (2010, December). *Dietary guidelines for Americans, 2010* (7th ed.) [Electronic version]. Washington, DC: U.S. Government Printing Office.

appendix C spot check answers

Chapter 1: The Basics

1. Answers may vary. Your nose is on the anterior surface of your head, superior to your mouth and medial and inferior to your eyes.
2. The liver is in the right upper quadrant. The liver is in the right hypochondriac and epigastric regions.
3. The parietal peritoneum is attached to the inferior surface of the diaphragm. The parietal pleural membranes and the parietal pericardium are attached to the superior surface of the diaphragm.
4. Negative feedback.

Chapter 2: Levels of Organization of the Human Body

1. Protons: 19; electrons: 19; neutrons: 20.
2. Both liquids are bases. They both release OH^-. Liquid Y is stronger. Liquid Y releases 1,000 times more ions than liquid X.
3. Carbohydrate—the ratio of C:H:O is 1:2:1.
4. Removing heat slows down the chemical reactions of bacteria.
5. A cell in a testicle that produces testosterone would need large amounts of smooth ER and Golgi complexes.
6. The water in the beaker is a hypotonic solution compared to the egg white. The weight of the egg would increase as water is transported across the membrane through osmosis to the inside of the egg.
7. ATP.
8. Codon: CGG; anticodon: GCC.
9. The DNA of a brain cell is identical to the DNA of a bone cell. There is only half of the amount of DNA in a sperm cell as there is in a brain cell.
10. Oncogenes would be found in the DNA of the nucleus. Discussion of other carcinogens will have results that vary.

Chapter 3: The Integumentary System

1. The forearm has thin skin; therefore, its epidermis does not have a stratum lucidum. The lips have thick skin; therefore, their epidermis does have a stratum lucidum.
2. A cut would need to extend into the dermis to bleed.
3. The eyelashes would be extremely long. Growing ½ inch per month for three years could cause eyelashes to be 18 inches in length!
4. The increasing fibers hamper the movement of bacteria.
5. There would be no keratinocytes to generate new waterproofed epidermal cells and no melanocytes to provide melanin for protection from UV light.
6. Hailey has a first-degree burn over approximately 4.5 percent of her body.
7. Discussion question; answers may vary.

Chapter 4: The Skeletal System

1. Any three of the following: parietal, frontal, occipital, temporal, sternum, ribs.
2. Any two of the following: facial bones, ethmoid bone, sphenoid bone, vertebrae, sacrum, coccyx.
3. Ilium, ischium, pubis.
4. Sacroiliac joint.
5. The matrix of compact bone is arranged in osteons, whereas the matrix of cancellous bone is arranged in trabeculae. The matrix of all three types of cartilage is the same—proteoglycans and water with fibers.
6. The matrix of bone is hard and does not allow for diffusion, so bone cells require a blood supply to supply nutrients and remove wastes. Cartilage matrix does allow for diffusion to supply nutrients and remove wastes, so no direct blood supply is necessary.
7. Yes. The shaft (diaphysis) of the humerus is compact bone arranged in osteons. The central canal of each osteon contains blood vessels and a nerve. If the diaphysis is broken, the blood vessels of each osteon would also be broken.
8. Class: fibrous joint; type: syndesmosis.
9. Endochondral ossification.
10. Discussion question; answers may vary. They should include exercise to put stress on the bones and a diet rich in calcium and vitamin D to ensure appositional bone growth.
11. Complete, open, oblique fracture of the humerus.

Chapter 5: The Muscular System

1. Left.
2. Synergists for hip flexion: pectineus, iliacus, psoas major, iliopsoas, sartorius, and rectus femoris. Antagonists to hip flexors: biceps femoris and gluteus maximus.
3. A threshold amount.
4. More motor units are recruited when an egg is crushed.
5. Isometric—muscles tensed but no motion resulted.

6 Jessica's muscles have a higher ratio of fast-twitch fibers adapted for anaerobic respiration. Jennifer's muscles have a higher ratio of slow-twitch fibers adapted for aerobic respiration. Answers may vary for additional events. Jennifer's events should require endurance. Jessica's events should require quick bursts of energy.

7 Vegetarian. A nonvegetarian diet has more complete protein sources.

8 The posture becomes more stooped.

Chapter 6: The Nervous System

1 Multipolar.

2 Astrocytes prevent many medications from crossing the blood-brain barrier. Microglia will seek out and fight the pathogen.

3 It will rise.

4 Smile: CN VII; stick out tongue: CN XII; move head from side to side: CN XI.

5 Olfactory, optic, and auditory.

6 The synaptic knob of the preganglionic neuron synapses with the dendrite of the postganglionic neuron.

7 In the sympathetic division, the preganglionic neuron is short and the postganglionic neuron is long. In the parasympathetic division, the preganglionic neuron is long and the postganglionic neuron is short.

8 The local potential was at threshold when reaching the trigger zone, resulting in an action potential for you, but it must have been subthreshold for the friend, so no action potential resulted.

9 The optic nerve.

10 Answers may vary but should include repetition of input over time to give the dendrites a chance to grow and make new connections for long-term memory.

11 Wernicke's area is for incoming language, of which there is none in this example.

Chapter 7: The Nervous System—Senses

1 Dermis.

2 Light.

3 You would still be able to detect the taste but probably not the flavor of the food due to the impaired sense of smell because of the cold.

4 The people who cannot smell garlic are missing a receptor for garlic.

5 The ossicles are used to amplify the vibration 20 times by the time they reach the oval window. The arthritis may dampen the size of the vibrations reaching the oval window and, therefore, decrease hearing.

6 Dynamic equilibrium; semicircular canals.

7 The lateral corner of the eye so that the drops will wash over the entire surface of the eye before being drained at the lacrimal punctum.

8 Excess tears are being drained to the nose.

9 The medial and lateral rectus muscles move the eye side-to-side. The medial rectus muscle is stimulated by the oculomotor nerve (CN III), and the lateral rectus muscle is stimulated by the abducens nerve (CN VI).

10 It would be easier to determine the shape. Rods can be used in low-light conditions to determine shape. Cones are used to determine color, but they require more light.

11 Answers will vary, but activities mentioned should require depth perception.

Chapter 8: The Endocrine System

1 Receptors for insulin.

2 No, the insulin will travel throughout the body in the bloodstream and will affect all cells that have a receptor for it.

3 They are called *releasing hormones*.

4 Both hormones speed the production of sebum by sebaceous glands and speed the deposition of bone by osteoblasts.

5 Steroids.

6 A hormone stimulating a gland.

7 It is important to have epinephrine when it is needed to complement the sympathetic nervous system. It helps prepare for *fight or flight*. A long half-life of epinephrine during restful times would not be an advantage because it would put a strain on the heart.

8 A substance other than a hormone. In this case, it was glucose.

9 Negative feedback.

10 No. Hormones are secreted when there is a need. Insulin and glucagon have opposite effects, so they would not be used in response to the same need.

11 Breast development, start of menstruation, greater deposition of fat, axillary and pubic hair growth.

12 FSH and LH would target the testes instead of the ovaries, resulting in the production of testosterone and male secondary sex characteristics.

13 Axillary, facial, and pubic hair; skeletal and muscle development; deeper voice; aggression; sperm production.

14 Sodium and potassium.

15 Her sex drive would be reduced. She does not have an alternative source for androgens.

16 Neural stimulation of a gland.

17 Positive feedback.

Chapter 9: The Cardiovascular System—Blood

1 It is cells in a matrix. The cells are the formed elements. The matrix is plasma.

2 Answers will vary. The answer should give a hormone, the endocrine gland that produced it, and its target tissue.

3 Less than 1 percent.

4 They are stained so that they can be seen and differentiated from one another.

5 Basophils and eosinophils.

6 The Denver player would have a higher RBC count because of the increased altitude.

7 The Miami Heat player probably has a higher RBC count because of the increased demand for oxygen due to the level of exercise of a professional athlete.

8 Lymphocytes spend much of their lives in tissues, not in circulation.

9. Platelets secrete vasoconstrictors in vascular spasm, form platelet plugs, and start the intrinsic pathway of coagulation.
10. A type-A person would have A antigens on the surface of the blood cells and anti-B antibodies dissolved in plasma.
11. No, because the Rh antigen is present on the Rh+ person's cells. It is not foreign. Antibodies are produced only to respond to foreign antigens.
12. Type B− could donate to blood types B−, B+, AB−, and AB+. Type B− could receive blood from types B− and O−. You would not want to expose an Rh-negative person to Rh antigens to promote an immune response.
13. 100. They are percentages.

Chapter 10: The Cardiovascular System—Heart and Vessels

1. The apex of the heart.
2. The left ventricle has a thicker outer wall because it has a larger workload.
3. Myoglobin stores oxygen so that cardiac muscle can continue to perform aerobic respiration.
4. The right ventricle.
5. Aortic valve.
6. SA node, AV node, AV bundle, AV bundle branches, Purkinje fibers.
7. Lubb duppssh, lubb duppssh.
8. All the chambers are filling with blood. The heart is at rest.
9. 8 liters per minute.
10. The heart rate would likely increase because of the caffeine and nicotine.
11. Afterload.
12. It would be greater in the aorta because the aorta has a greater diameter than the femoral artery and it is closer to the heart.
13. Pulse pressure = 54 mmHg; MAP

    $= 96 + \frac{1}{3}(54) = 114$ mmHg.

Chapter 11: The Lymphatic System

1. The left subclavian vein.
2. The thoracic duct.
3. It produces the cells used in this system (leukocytes, and specifically lymphocytes).
4. The cervical lymph nodes.
5. Increased heat from increased blood flow raises the local metabolic rate to promote cell division.
6. Increased amount of lymph washes away debris, and increased vessel permeability allows macrophages to have easier access to the area of damage for phagocytosis.
7. Increased vessel permeability allows leukocytes to have access to the area, and chemotaxis draws leukocytes to the site of the damaged tissues and pathogens so that they can be destroyed.
8. The antibodies agglutinate the antigens.
9. The B cell in humoral immunity is safe and secure in lymphoid tissue while antibodies attack the pathogen anywhere in the body. In cellular immunity, the $T_{cytotoxic}$ cell is at the site of the pathogen delivering a direct and lethal cell-to-cell hit.
10. Artificial passive immunity. Rhogam is Rh antibodies.

Chapter 12: The Respiratory System

1. It is a stronger, more durable tissue to withstand probing fingers.
2. The cilia in the nasopharynx move the debris down to the oropharynx, while the cilia of the trachea move the debris up toward the pharynx.
3. The air should be warmed and moistened because it has bypassed much of the mucous membranes of the upper respiratory tract.
4. The washer will travel to the nasal cavity, to the nasopharynx, to the oropharynx, to the laryngopharynx, through the glottis to the larynx, to the trachea, to a main bronchus, and—depending on the size of the washer—to the bronchial tree.
5. The metal washer will most likely go down the right bronchus into the right lung because the right bronchus is more vertical than the left bronchus. The washer is more likely to follow a straight line than make a turn at the left bronchus.
6. The vessel does not have a single layer to its wall, so it cannot be a capillary. The tube does not have any cartilage, so it cannot be a bronchus.
7. $P_{CO_2} = 0.26$ mmHg, $P_{O_2} = 133.7$ mmHg.
8. New York City. The partial pressures of each gas are greater in New York City because the total pressure is greater than that in Breckenridge. So there is a greater concentration of each gas in New York City.

Chapter 13: The Digestive System

1. Mechanical digestion of a complex carbohydrate results in small pieces of complex carbohydrates, while chemical digestion of a complex carbohydrate results in monosaccharides.
2. Answers will vary.
3. Caries form when acids from bacteria erode a tooth's enamel. Lysozymes and antibodies in saliva inhibit bacteria growth. Chewing sugarless gum increases the amount of saliva, and therefore lysozymes and antibodies, without contributing to the feeding of bacteria.
4. The *bolus* travels from the oral cavity, to the oropharynx, to the laryngopharynx, to the esophagus, and to the stomach.
5. The lingual lipase and gastric lipase would digest cell membranes, and the pepsin that forms in the stomach would begin to chemically digest the proteins of the smooth muscle walls.
6. Bile flows through the cystic duct. Bile flows through the common bile duct. Bicarbonate ions and enzymes for carbohydrate, protein, and lipid digestion flow through the pancreatic duct.
7. 3 mL.
8. Cholecystokinin causes the hepatopancreatic sphincter to relax so that the bicarbonate ions and enzymes from the pancreas can enter the duodenum.

9 Answers will vary. The larger diameter and lack of villi would suggest that the large intestine is not designed for nutrient absorption.

Chapter 14: The Excretory/Urinary System

1 The parietal peritoneum.

2 The renal corpuscle (glomerulus and glomerular capsule), the proximal convoluted tubule, the nephron loop, and the distal convoluted tubule, which leads to a collecting duct.

3 The peritubular capillaries.

4 No. High blood pressure is pushing materials out of the capillaries in the renal corpuscle during filtration. This must happen before some of these materials are reabsorbed back into the capillaries.

5 The juxtaglomerular apparatus must first notice the fall in blood pressure. It then secretes renin, which changes a protein from the liver to angiotensin I. ACE converts angiotensin I to angiotensin II. Angiotensin II travels to the adrenal cortex, telling it to secrete aldosterone. Once aldosterone is received by the kidneys, more sodium is actively transported from the tubule to the peritubular capillaries, and water follows sodium through osmosis.

6 ADH will conserve water without conserving sodium.

7 ANH will be released by cells in the right atrium in response to the increased blood pressure as a direct result of the increased blood volume.

8 Fluid intake adds to blood volume. Increased blood volume means increased blood pressure. Fluid intake would have to be monitored closely as excess fluids would be cleared from the system only three times a week.

Chapter 15: The Male Reproductive System

1 The testes are too warm in the abdominal cavity to produce viable sperm.

2 The least number is based on 1 seminiferous tubule per 250 lobules = 250 seminiferous tubules. The greatest number is based on 4 seminiferous tubules per 300 lobules = 1,200 seminiferous tubules.

3 The dartos muscle, the cremaster muscles, and the pampiniform plexus that is in each spermatic cord.

4 Semen is approximately 10 percent sperm, 60 percent nourishing fluid from the seminal vesicles, 30 percent alkaline fluid from the prostate, and trace amounts of a lubricating fluid from the bulbourethral glands.

5 The acrosome cap contains enzymes to penetrate the egg. The 23 chromosomes in the head are the genetic material to fertilize the egg. The large mitochondria are needed to make use of available oxygen to perform cellular respiration. This supplies the necessary ATP to move the tail that propels the sperm to where they have to go—the egg.

6 FSH targets sustentacular cells, and LH targets interstitial cells.

7 Each spermatid has a set of 23 chromosomes, but each set of 23 chromosomes differs from the other sets.

8 It is reabsorbed in the epididymis.

9 None, because it has no effect on LH production.

10 An erection may be stimulated as a reflex by stimulating the penis, or it may be initiated because of sound, smell, sight, or thought.

11 The artery delivering blood to the penis dilates, the erectile tissues fill with blood, and the veins become compressed so that the excess blood cannot leave the penis.

Chapter 16: The Female Reproductive System

1 Ovaries produce gametes (ova) and hormones.

2 Mucous glands in the cervix produce a thick mucus to block the entrance to the uterus.

3 Their body fat is too low, and they have insufficient cholesterol to produce estrogen.

4 A follicle cares for the primary and secondary oocyte by providing nutrients, removing wastes, and protecting it from the woman's immune system.

5 A woman's eggs are as old as she is, whereas a man's sperm are at most 40 to 60 days old once they are mature.

6 The egg is ovulated on day 14. If the egg is fertilized on that day, eight or nine days from then would be day 22 or day 23. However, the egg can remain viable for 24 hours and still be fertilized. So it could also be fertilized on day 15. Then early detection would occur as soon as day 23 or day 24.

7 All of her follicles and oocytes were produced before she was born. Many have died through atresia, and many others have been developed, ovulated, and been discharged from her body with menstruation. Still others began development each month and underwent atresia before being ovulated. She does not continue to develop any more follicles after birth.

8 Beneficial bacteria feed off glycogen, provided by the stratified squamous epithelial lining of the vagina. This results in a low pH in the vagina that discourages the growth of microorganisms, including yeast. Atrophy of the lining decreases the support for the beneficial bacteria.

9 They would cease due to the lack of estrogen and progesterone.

10 The vagina.

11 Answers will vary, but the menu should be balanced and include all of the essential nutrients—folic acid, calcium, protein, and iron—without a high caloric intake.

12 This would affect a woman's fertility only each time an egg is released from the ovary on the side from which the tube was removed. She would still be able to conceive if the egg is released from the ovary on the same side as the remaining tube.

glossary

The following terms are defined in the context they were used in this text.

a

abdominal: pertaining to the belly region.
absorption: the process of putting something into the blood for the first time. For example, the small intestine absorbs calcium from the diet, increasing blood calcium levels.
accommodation: the ability of the eye to change the shape of the lens to keep an image in focus with a change in distance.
acetylcholine (ACh): neurotransmitter released to stimulate a contraction of skeletal muscle tissue.
acid: a molecule that releases a hydrogen ion (H^+) when added to water.
acidosis: condition in which the pH of blood is less than 7.35.
acne: condition in which sebaceous ducts become plugged, bacteria grow in the plugged ducts causing inflammation, hair follicle walls break down due to the increased inflammation, and the pus formed causes pimples.
acquired immunity: term that refers to how the forms of specific immunity are acquired. Examples are natural active immunity, natural passive immunity, artificial active immunity, and artificial passive immunity.
acquired immunodeficiency syndrome (AIDS): the final stage of an HIV infection, in which the immune system fails to provide protection against pathogens.
action potential: the flow of electricity along an axon of a neuron in one direction—from the trigger zone to the synaptic knob.
active transport: movement of materials across a cell membrane from areas of low concentration to areas of high concentration; requires energy.
aerobic respiration: type of cellular respiration requiring oxygen that results in enough energy to generate 36 ATP molecules from every glucose molecule and produces carbon dioxide and water.
afferent: in the direction toward the brain or spinal cord in the nervous system.
afterbirth: the placenta and its associated membranes delivered after the baby.
afterload: the pressure in the pulmonary trunk and aorta during diastole.
agglutination: an immune response in which antibodies clump cells with specific antigens.
alimentary canal: the gastrointestinal tract that begins at the mouth and ends at the anus.
alkalosis: condition in which the pH of blood is greater than 7.45.
alveoli: tiny air sacs in the lung at which gas exchange takes place.
anaerobic respiration: type of cellular respiration in the absence of oxygen that results in enough energy to generate 2 ATP molecules from every glucose molecule and produces lactic acid.
anaphylaxis: systemic vasodilation (systemwide dilation of blood vessels) that occurs within a few minutes of exposure to an allergen and can cause a drop in blood pressure and even cardiac failure.
anastomoses: circulatory routes that involve vessels merging together.
anatomy: the study of body structures.
androgens: hormones produced by the adrenal cortex that are responsible for male secondary sex characteristics and for sex drive in both genders.
angiogenesis: new blood vessel growth.
antagonist: a muscle that has an opposite action.
antibody: a dissolved protein in plasma that carries out an immune response to a specific antigen.
antigen: any molecule whose shape promotes an immune response.
antigen-presenting cell (APC): a cell that samples its environment and posts what it finds. Examples are B cells and macrophages.
aphasia: any language deficit resulting from damage to either Wernicke's or Broca's area.
apoptosis: programmed cell death.
appendicular: pertaining to the body region that includes the arms and legs.
appendicular skeleton: the bones of the arms, legs, and girdles (the bones that attach the arms and legs to the trunk).
appositional bone growth: type of growth that occurs in all bones to make them more massive.
arrhythmia: an abnormal heart rhythm.
arthritis: inflammation of a joint.
articular cartilage: a hyaline cartilage covering over the epiphyses of long bones.
articulation: joint formed where two or more bones meet.
atherosclerosis: the buildup of fatty deposits within arterial walls, which causes the walls to roughen and project to the lumen (open space) within the vessel.
atom: the smallest piece of an element still exhibiting the element's unique set of chemical properties.
atresia: process in which oogonia and primary oocytes degenerate.
atrophy: shrinkage of tissue due to a decrease in cell size or number.
autocrine: term that refers to the secretion of a hormone by the cells of the same tissue that it targets.
autonomic: type of nerve message that goes to glands, the cardiac muscle of the heart, or the smooth muscle of hollow organs and blood vessels.
axial: pertaining to the body region that includes the head, neck, and trunk.
axial skeleton: the bones of the head, neck, and trunk.
axon: portion of a neuron that carries electrical impulses along its length from the cell body to the synaptic knobs at the end of the neuron.

b

baroreceptors: sensors located in the aorta and carotid arteries that detect changes in blood pressure.
base: a molecule that will accept a hydrogen ion, often by releasing a hydroxide ion (OH^-) when added to water.
bilirubin: a waste produced from the breakdown of worn-out red blood cells.
bipolar: type of sensory neuron with one dendrite and one axon; found in the nasal cavity, the retina of the eye, and the inner ear.
blood flow: the amount of blood flowing to an area in a given amount of time, usually expressed in mL/min.
blood-testis barrier: a barrier formed by sustentacular cells to isolate spermatocytes from the immune system in the male.
bolus: a chewed bite of food mixed with saliva.
bronchial tree: series of ever-decreasing-size tubes branching from the bronchi and ending with bronchioles in the lungs.
buffer: something that resists a change in pH.

c

calcitriol: active vitamin D.

callus: collagen and fibrocartilage deposited in a fracture by periosteum stem cells.

cancellous bone: bone connective tissue that has a spongy appearance and is characterized by plates and slivers called *trabeculae*. It is found in the end of long bones and in the middle of flat and irregular bones.

capacitation: process in which the acids of the vagina and other fluids of the female reproductive tract break down the cholesterol of the sperms' cell membrane and stimulate the sperm to swim faster so that, after 10 hours, the sperm are able to release the enzymes of the acrosome cap to penetrate an egg.

carcinogen: chemical or environmental agent that causes cancer.

cardiac cycle: one complete contraction and relaxation of the heart.

cardiac output: the amount of blood ejected by each ventricle of the heart each minute.

cardiac reserve: the difference between the cardiac output of a heart at rest and the maximum cardiac output the heart can achieve.

caries: an erosion through the enamel of a tooth into the dentin.

cellular immunity: form of specific immunity that involves $T_{cytotoxic}$ cells directly attacking cells with a foreign antigen. Also called *cell-mediated immunity.*

cellular respiration: important chemical reaction performed by mitochondria in cells to release energy. (Glucose + oxygen *yields* carbon dioxide + water + energy, $C_6H_{12}O_6 + 6O_2 \longrightarrow 6CO_2 + 6H_2O$ + energy.)

cerebrospinal fluid (CSF): fluid surrounding the brain and spinal cord that is made by ependymal cells lining cavities in the brain called *ventricles.*

chemical digestion: the breakdown of complex molecules to their building blocks so that they can be absorbed.

chemical reaction: interaction between molecules that results in products.

chemotaxis: process in which WBCs move to where the concentration of chemicals from damaged tissues is the greatest.

chondrocyte: cartilage cell that produces a cartilage matrix of proteoglycans and water.

chronic obstructive pulmonary disorders (COPDs): disorders that cause a long-term decrease in ventilation of the lungs.

chronotropic factor: anything that changes the heart rate.

chyme: the liquefied contents of the stomach.

climacteric: a period of life for men and women, usually beginning about age 50, in which reproductive hormone levels change. This period is typically called *andropause* in the male and *menopause* in the female.

closed reduction: procedure that sets the edges of a fracture in proper alignment by manipulating the bone without surgery.

clotting factor: an inactive chemical in the blood that, when activated, promotes a reaction cascade to form a clot.

coagulation: blood clotting; the third step in hemostasis.

colostrum: a thin, watery fluid containing protein, lactose (milk sugar), and many antibodies but one-third less fat than breast milk.

comminuted: a fracture in which the bone is broken into three or more pieces.

compact bone: bone connective tissue arranged in a series of osteons. It is found in the shafts of long bones and the surfaces of flat and irregular bones.

complement system: 20 inactive proteins (always present in the blood) that, when activated by the presence of a pathogen, initiate one of several pathways to ensure pathogen destruction.

complete proteins: proteins in the diet that contain all the amino acids necessary for the human body.

compliance: measurement of how well the lungs can expand and return to shape.

concentration gradient: difference in chemical concentration from one location to another, such as the two sides of a cell membrane.

conjugation: process performed by the liver in which water-soluble molecules are bound to a hormone so that it will be excreted at a faster rate.

connective tissues: tissues that have cells and fibers in a matrix (a background substance).

constipation: condition in which defecation is difficult because too much water has been removed from the feces.

contact inhibition: the end of keratinocyte lateral growth because edges of stratum basale cells are in contact with each other.

cornification: process in which keratinocytes fill with keratin and die as they move toward the surface of the epidermis.

covalent bond: formation of a molecule that occurs when two or more atoms bind by sharing their electrons to fill their outer shells.

cranial: pertaining to the cavity in the head that houses the brain.

crossing-over: an event in meiosis in which chromatids break and exchange parts.

crowning: the appearance of the baby's head at the vaginal opening during birth.

cryptorchidism: a condition in which a male infant is born with undescended testes.

cutaneous: pertaining to the skin.

cystitis: an inflammation of the urinary bladder, usually caused by a bacterial infection.

decremental: decrease with distance.

defecation: the process of removing feces from the body.

deglutition: the process of swallowing.

dendrite: portion of a neuron that receives information.

depolarize: to change the charge across the cell membrane of a neuron by the flow of Na^+ into the cell.

deposition: the process of putting calcium phosphate crystals into the bone.

dialysis: a treatment for renal failure in which a machine filters out excess fluid, salt, and nitrogenous wastes in the blood.

diapedesis: process in which WBCs crawl through vessel walls.

diaphysis: the shaft of a long bone.

diarrhea: condition of a runny stool.

diastole: relaxation of a heart chamber; usually refers to the action of the ventricles.

diuretic: a type of drug often prescribed for hypertensive patients to reduce their blood pressure by increasing urine production.

down-regulation: a decrease in the number of receptors for a given hormone, causing the cell to become less sensitive to the hormone.

dynamic equilibrium: equilibrium perceived when the head is rotating.

effacement: the thinning of the cervix in preparation for birth.

efferent: in the direction away from the brain or spinal cord in the nervous system.

ejaculation: the discharge of semen.

elasticity: the ability to come back to shape after being stretched.

electrolyte: an ion in solution capable of conducting electricity.

embolus: a moving, unnecessary clot.

emission: a stage in the male sexual response in which sperm are moved by peristaltic contractions through the ductus deferens to its ampulla and the prostate and seminal vesicles add their secretions to semen.

emulsify: to break lipids into smaller droplets.

endochondral ossification: process that forms long bones.
endocrine: term that refers to hormones that travel through the blood to get to their target tissue.
endocytosis: movement of materials into the cell in bulk.
epidermis: superficial layer of the skin that is subdivided into four or five general layers called *strata*.
epiphyseal plate: zone of cartilage between the epiphysis and diaphysis in an immature long bone; commonly called the *growth plate*.
epiphyses: clubby ends of long bones.
episiotomy: an incision made in a female's perineum to widen the vaginal opening during birth.
epithelial tissues: tissues that cover and line all body surfaces and have a basement membrane.
epitope: antigen fragment.
erection: a stage in the male sexual response in which the erectile tissues of the penis become engorged with blood, causing the penis to become enlarged, rigid, and erect.
erythrocyte: red blood cell.
erythropoietin (EPO): chemical produced by the kidneys to stimulate red blood cell production when the oxygen blood level is low.
excretion: the removal of metabolic wastes from the body.
exfoliate: to shed dead keratinocytes from the stratum corneum.
exocrine glands: glands, such as sebaceous and sweat glands, that produce and secrete products that are delivered to the appropriate locations through ducts.
exocytosis: movement of materials out of the cell in bulk.
expiration: movement of air out of the lungs.
expiratory reserve volume: the amount of air that can be forcefully expired beyond the amount expired in a normal breath at rest.
extensibility: the ability to be stretched.
extension: action that bends a part of the body posteriorly, such as straightening the arm at the elbow.

f

facilitated diffusion: passive membrane transport method used for molecules that cannot diffuse through the selectively permeable membrane on their own and therefore require help crossing the membrane through a channel protein.
fascicle: group of muscle fibers surrounded by perimysium.
fatigue: the inability of a muscle to fully respond to a nerve impulse.
feces: the contents of the large intestine, composed of 75 percent water and 25 percent solids.
fibrin: a solid protein fiber necessary for blood clot formation.
fibrinogen: protein dissolved in plasma that is the precursor to fibrin.
fibrosis: wound healing with scar tissue; normal function is not returned.
filtration: passive transport method that moves materials across a cell membrane using force but no energy.
first-degree burn: a burn that involves only the epidermis and whose symptoms include redness, pain, and swelling; the most common type of burn.
flatus: gas that is produced by bacteria and causes bloat, discomfort, and an unpleasant odor when released.
flexion: action that bends a part of the body anteriorly, such as flexing the elbow.
fluid compartments: the two locations for water in the body—intracellular and extracellular.
folliculogenesis: the changes in a follicle during oogenesis.
fontanelle: membranous area between flat bones of the skull, eventually replaced by a suture.
foramen magnum: a large opening in the occipital bone that allows the spinal cord to exit the cranial cavity.
formed elements: the cells and cell parts found in blood.
functional residual capacity (FRC): the amount of air remaining in the lungs after the expiration of a normal breath at rest.

gamete: a sex cell; a sperm or egg.
gangrene: tissue necrosis resulting from an insufficient blood supply, often associated with an infection.
gas exchange: the movement of oxygen and carbon dioxide that occurs between capillary blood and the alveoli in the lungs and between capillary blood and the tissues of the body.
gas transport: the movement of gases in the blood to and from the lungs and tissues.
gene: the amount of DNA that must be read for the directions to make one specific protein.
gestation: the time from fertilization to birth.
gland: a structure on its own or groups of cells within an organ that function to produce hormones.
glomerulonephritis: an inflammation of the filtration membrane in the glomerulus of the nephron.
glucocorticoids: hormones produced by the adrenal cortex that stimulate the breakdown of protein and fat to make glucose, suppress the immune system, and reduce inflammation.
gomphosis: fibrous joint holding a tooth in its socket.
gonads: the ovaries in women and the testes in men.
granulation tissue: tissue with many collagen fibers produced by fibroblasts to fill in a wound's clot.
granulocyte: any of three types of leukocytes—neutrophils, basophils, and eosinophils—containing small granules that differ in color when stained.
greater omentum: an extension of the visceral peritoneum that looks like a fatty apron lying over all the abdominal viscera and extends from the inferior margin of the stomach.
gustation: the sense of taste.

h

half-life: the length of time it takes for one-half of a substance to be eliminated from the cardiovascular system.
hematocrit: a test that measures the percentage of erythrocytes to whole blood.
hemocytoblast: starting cell for the production of all the formed elements; stem cell.
hemoglobin: a red, complex protein that is made of four chains of amino acids found in red blood cells and that functions to carry oxygen.
hemopoiesis: blood production.
hemostasis: three-step process—vascular spasm, platelet plug formation, and coagulation—that stops bleeding.
histology: the study of tissues.
homeostasis: an internal environment that the body must maintain for normal functioning.
hormone: chemical used in the endocrine system to carry messages.
humor: fluid found in the eye. Aqueous humor is found in the anterior portion of the eye, and vitreous humor is found in the vitreous chamber of the eye lined with the retina.
humoral immunity: form of specific immunity that involves B cells making antibodies to attack a foreign antigen. Also called *antibody-mediated immunity*.
hydroxyapatite: calcium phosphate mineral salt that makes up the mineral matrix of bone connective tissue.
hypercapnia: condition of increased carbon dioxide in the blood.
hyperkalemia: condition in which the level of potassium in the blood is too high.
hyperopia: farsightedness caused by the cornea and lens focusing an image behind the retina.

hyperplasia: the growth of tissue through the production of more cells.
hypertrophy: the growth of tissue through the growth of existing cells.
hypodermis: *see* **subcutaneous layer.**
hyponatremia: condition in which the level of sodium in the blood is too low.
hypoxemia: condition of low blood oxygen.

i

incomplete proteins: proteins in the diet that are missing one or more of the necessary amino acids for the body.
infarction: the sudden death of tissue, which often results from a loss of blood supply.
infertility: the inability to fertilize an egg.
ingestion: the intake of food into the mouth.
insertion: the attachment of a muscle to a bone or structure that *does move* when the muscle contracts.
inspiration: movement of air into the lungs.
inspiratory capacity: the maximum amount of air that can be inspired after the expiration of a normal breath at rest.
inspiratory reserve volume (IRV): the amount of air that can be forcefully inspired beyond the amount inspired in a normal breath at rest.
integumentary system: body system that includes the skin, hair, nails, and cutaneous glands.
intercalated disks: specialized junctions between cardiac muscle cells that enable the fast transmission of electrical impulses from one cell to another.
interferons: chemicals that are released by virally infected cells and encourage surrounding healthy cells to make antiviral proteins so that the virus will not invade them.
interleukins: chemicals used by lymphocytes to communicate with one another.
intramembranous ossification: process that forms flat bones.
ion: a charged atom.
ionic bond: formation of a molecule that occurs when two or more atoms bind by *giving up or receiving electrons* from each other to fill their outer shells.
ischemia: the lack of blood flow.
isometric: a type of contraction in which the length of the muscle remains constant while the tension in the muscle increases.
isotonic: a type of contraction in which the tension in the muscle remains constant and motion is the result.

j

jaundice: indicator of liver malfunction caused by added bilirubin in the blood that gives a yellow appearance to the skin and eyes.

k

keratin: a hard, waterproof protein found in epidermal cells, hair, and nails.

l

labor: uterine contractions to bring about birth.
lacrimal: pertaining to tears.
lactation: milk production.
lacteals: small lymphatic vessels that are located in the villi of the small intestine and into which the products of lipid digestion are absorbed.
leukocyte: white blood cell.
lever: a rigid object that can be used to lift something. Bones act as levers in lever systems that muscles use to move the body.
lingual: pertaining to the tongue.
local potential: the flow of electricity begun by stimulating the dendrite of a neuron.
lymph: fluid that is derived from blood and is similar to plasma but has fewer proteins.
lymphadenitis: condition in which a pathogen is under attack by a lymph node's lymphocytes, resulting in the lymph node becoming swollen and painful to the touch.
lymphoid: type of hemopoiesis that produces lymphocytes in lymphoid tissues outside the red bone marrow.

m

major histocompatibility complex (MHC): protein molecule used by antigen-presenting cells to display an epitope.
mammography: an x-ray examination of the breast used to detect breast cancer.
margination: process in which WBCs stick to the walls of dilated vessels in an inflamed area.
mass movement: the movement of feces from the transverse colon to the sigmoid colon to the rectum; initiated by distension of the stomach.
mastication: the process of chewing.
mean arterial pressure (MAP): the average pressure arteries must be able to withstand.
mechanical digestion: the breakdown of large pieces of complex molecules to smaller pieces of complex molecules.
mediastinum: the space between the pleural cavities that contains the heart, esophagus, trachea, thymus, and major vessels.
mediators of inflammation: chemicals that are produced by damaged tissues and that diffuse away from the damaged area and cause any blood vessels they meet to dilate.
meiosis: a two-division process, used to create sperm and eggs, that starts with a parent cell of 46 chromosomes (23 pairs). It results in four daughter cells, each having 23 chromosomes. The set of chromosomes in each daughter cell is different from the sets in the other daughter cells due to independent assortment and crossing-over.
melanocytes: cells found in the stratum basale that produce skin pigments called *melanin.*
membrane transport: movement of ions and molecules across the cell membrane.
meninges: three layers of membrane surrounding the brain and spinal cord.
meniscus: fibrocartilage pad found between the femur and the tibia in the knee that serves as a shock absorber.
menopause: cessation of menstrual periods for at least a year.
menstrual cycle: part of the female sexual cycle of events that affects the uterine lining.
mesenteries: sections of the peritoneal membrane where the parietal peritoneum comes back parallel to itself.
metabolic acidosis: condition in which there is decreased elimination of hydrogen ions by the kidneys or increased production of acidic substances through metabolism.
metabolic alkalosis: a relatively rare condition that can occur if there is prolonged vomiting, which results in the repeated loss of stomach acids.
metabolic waste: waste produced by cells.
metabolic water: water formed in the cells through cellular respiration.
metabolism: the sum of all chemical reactions that occur in the body.
metaplasia: the change of a tissue from one type to another.
metastasis: the migration of malignant neoplasm cells.
micturition: the passing of urine out of the body.
milk ejection reflex: a reflex that is initiated by suckling on a breast and results in myoepithelial cells in the mammary lobules contracting to release milk to the lactiferous ducts in the breast.

mineralocorticoids: hormones produced by the adrenal cortex that promote sodium and water reabsorption and potassium excretion in the kidney to maintain blood volume and pressure.

mitosis: the cell division process that creates all body cells other than sperm and eggs.

molecular mimicry: a situation in which one molecule is so similar in structure to another molecule that it is mistaken for the other molecule.

molecule: two or more atoms bonded together.

motor unit: a single nerve cell and all the muscle cells it stimulates.

multipolar: the most common type of neuron in the brain and spinal cord; has multiple dendrites and an axon that may or may not have a collateral branch.

muscle tissue: tissue whose cells have a high concentration of proteins, which allow the cells to contract.

muscle twitch: the contraction of one muscle cell due to one nerve impulse.

mutation: a change in the DNA of a cell that is carried to the next generation of cells. It often results in a change in structure or function.

myelin: a lipid rich intermittent covering over the axons of some neurons. Gaps in the myelin sheath are called *nodes of Ranvier.*

myeloid: type of hemopoiesis that produces all the formed elements in the red bone marrow.

myopia: nearsightedness caused by the cornea and lens focusing an image ahead of the retina.

n

necrosis: the premature death of tissue; caused by disease, infection, toxins, or trauma.

negative feedback: the process the body uses to reverse the direction of movement away from homeostasis.

neoplasia: the uncontrolled growth and proliferation of cells of abnormal or nonfunctional tissue.

nephron: the functional unit of a kidney; composed of a renal corpuscle, proximal convoluted tubule, nephron loop, and distal convoluted tubule that drains into a collecting duct.

nervous tissue: tissue whose cells communicate through electrical and chemical signals. This tissue is composed of nerve cells called *neurons* and many more support cells called *neuroglia* that protect and assist neurons in their function.

neuroglia: cells that aid neurons in their function.

nitrogenous wastes: wastes removed from the blood by the kidneys; include ammonia, urea, uric acid, and creatinine.

nociceptor: pain receptor that detects tissue injury or potential tissue injury.

nonspecific resistance: type of defense that works to fight a variety of pathogens without the need for prior exposure; consists of two lines of defense.

o

olfactory: pertaining to the sense of smell.

oncogenes: potential cancer-causing genes that code for uncontrolled production of cellular growth factors stimulating mitosis or the receptors for the growth factors.

oogenesis: egg production.

open reduction: procedure that sets bones in proper alignment through surgery.

opportunistic infections: infections normally fought off by healthy immune systems.

opsonization: the coating of a pathogenic cell to make it easier for a macrophage to phagocytize it.

organelles: specific structural components of cells (e.g., mitochondria, ribosomes, rough endoplasmic reticulum, and Golgi complexes).

organic molecules: molecules that come from life and contain carbon *and* hydrogen (e.g., carbohydrates, lipids, proteins, and nucleic acids).

origin: the attachment of a muscle to a bone or structure that *does not move* when the muscle contracts.

osmosis: passive movement of water across a cell membrane to equalize the concentrations on both sides of the membrane.

osteocyte: bone cell.

osteomyelitis: a bone infection.

osteon (Haversian system): targetlike arrangement of bone connective tissue found in compact bone.

osteoporosis: condition characterized by a severe lack of bone density.

otoliths: calcium carbonate and protein granules located in the saccule and utricle of the inner ear.

ovarian cycle: part of the female sexual cycle of events that occurs in the ovary.

ovulation: the release of an egg from an ovary.

oxygen debt: the amount of oxygen needed to remove the lactic acid produced through anaerobic respiration.

p

pampiniform plexus: a network of small veins surrounding the testicular artery.

pancreatic islets: 1 to 2 million groups of endocrine cells in the pancreas that produce the hormones insulin and glucagon.

papillae: (1) bumps on the superficial edge of the dermis that are in contact with the epidermis; (2) bumps on the tongue that house the taste buds.

paracrine: term that refers to hormones that work on neighboring cells without having to go through the blood to get to the target tissue.

parasite: an organism that lives on or in another organism (the host) and obtains its nourishment there.

parasympathetic: division of the autonomic nervous system that sends electrical messages to carry out functions for vegetative activities such as digestion, defecation, and urination.

parietal: the part of a serous membrane not in direct contact with an organ.

partial pressure: the amount of pressure an individual gas contributes to the total pressure of a mixture of gases.

parturition: the process of birth.

pathogens: disease-causing foreign invaders.

pectoral girdle: the bones—clavicle and scapula—that connect the arm to the axial skeleton.

pelvic: cavity in the inferior trunk that houses the urinary bladder, rectum, and reproductive organs.

pelvic girdle: the bones—ilium, ischium, and pubis—that connect the leg to the axial skeleton.

perilymph: fluid found in the cochlea of the inner ear.

peristalsis: wavelike contractions used to move materials along the digestive system.

peritoneum: serous membrane located in the abdominopelvic cavity.

pharynx: an area—commonly called the *throat*—that is divided into three sections based on location and anatomy—the nasopharynx, the oropharynx, and the laryngopharynx.

pheromone: term that refers to chemicals that cause a response outside the body, in another individual.

physiology: study of how the body's structures function.

plasma: the fluid matrix of blood connective tissue.

plasma protein: transport protein (made by the liver) that binds to a hormone in the blood to extend its half-life.

pleura: serous membrane surrounding a lung.

pluripotent: the ability to become cells of differing types. For example, a hemocytoblast can become any one of eight different formed elements.

pneumothorax: a condition in which air is introduced in the pleural cavity between the pleural membranes; commonly called a *collapsed lung*.

portal route: a circulatory route that contains two capillary beds before blood is returned to the heart.

positive feedback: the process the body uses to increase movement away from homeostasis.

preload: the amount of tension in the myocardium of the ventricular walls.

presbyopia: a decreased ability of the eye to accommodate.

prolapse: an effect of aging in the female reproductive system in which organs such as the urinary bladder and vagina drop out of position.

protein synthesis: production of proteins in a cell through the processes of transcription and translation.

proximal: closer to the connection to the body

puberty: the first few years of adolescence that begins with the production of FSH and LH.

pulse pressure: the surge of pressure that small arteries must withstand with each ventricular contraction.

pyrogen: chemical that initiates a fever.

r

reabsorption: the process of putting something into the blood again, not the first time. For example, osteoblasts reabsorb calcium from the bone, raising the blood calcium level.

reactants: molecules that come together to interact in a chemical reaction.

receptive field: an area in which a single neuron is responsible for detecting a stimulus.

receptor: shape-specific binding site for a hormone.

recruitment: the process of getting more and more motor units involved in a contraction to create a larger motion. For example, there is rapid recruitment in a boxer's punch.

reflex: an involuntary, predictable, motor response to a stimulus without conscious thought.

refraction: the bending of light as it passes through materials of different densities.

regeneration: wound healing with the same tissue that was damaged; normal function is returned.

renal corpuscle: part of a nephron; composed of a glomerulus and glomerular capsule.

renal tubule: part of a nephron; composed of a proximal convoluted tubule, nephron loop, and distal convoluted tubule.

replication: the process of copying DNA before cell division.

repolarize: to change the charge across the cell membrane of a neuron by the opening of K^+ channels to allow the flow of K^+.

residual volume: the amount of air in the lungs that cannot be moved.

resolution: a stage in the male sexual response in which the penis decreases in size and becomes flaccid again.

respiratory acidosis: condition in which the respiratory system cannot eliminate sufficient CO_2.

respiratory alkalosis: condition in which the respiratory system eliminates too much CO_2.

respiratory membrane: a membrane that is composed of a thin layer of water with surfactant in the alveoli, a single squamous cell alveolar wall, and a single cell capillary wall and across which gas exchange takes place in the lung.

resting membrane potential: a difference in charge across the cell membrane of a neuron created by the presence of many large negative ions inside the cell and many positive sodium ions outside the cell.

s

sarcomere: section of a myofibril extending from one Z line to the next.

sebum: a very oily, lipid-rich substance produced by the sebaceous gland to moisturize the skin and hair.

second-degree burn: a burn involving the epidermis and dermis. Symptoms include redness, pain, swelling, and blisters.

second messenger: chemical created by the binding of a hormone in a receptor on the cell membrane. The second messenger then carries the information to where it is needed in the cell to initiate the function of the hormone.

secondary sex characteristics: gender-specific characteristics developed at puberty due to the action of estrogen (in the female) and testosterone (in the male). Female secondary sex characteristics include breast development, axillary and pubic hair growth, menstruation, and fat deposition. Male secondary sex characteristics include muscle and skeletal development, deeper voice, aggression, and facial, axillary, and pubic hair growth.

secretion: the third process of urine production in the nephron; removes remaining wastes and excess hydrogen ions from the blood.

segmentation: a stationary constriction of the smooth muscle of the small intestine in ringlike patterns to further churn chyme and mix in bile and digestive enzymes to finish chemical digestion.

semen: a fluid—composed of 10 percent sperm, 60 percent nourishing fluid from the seminal vesicles, 30 percent alkaline fluid from the prostate, and trace amounts of a lubricating fluid from the bulbourethral glands—that is ejaculated from the penis during orgasm.

sensorineural: type of hearing loss caused by a problem with the organ of Corti or the auditory nerve.

serous membrane: double-layered membrane that contains fluid between the two layers.

serum: plasma with the fibrinogen and clotting factors removed.

simple diffusion: the passive movement of materials across a selectively permeable membrane from areas of high concentration to areas of low concentration until the concentrations are equal.

sliding filament theory: an explanation of a muscle contraction that involves thick myofilaments grabbing thin myofilaments and pulling them toward the center of the sarcomere.

smegma: a waxy substance produced by the sebaceous glands of the male prepuce.

solution: a substance that is composed of solutes dissolved in a solvent.

somatomotor: type of nerve message used to stimulate skeletal muscles to move the body.

specific immunity: a line of defense that requires a prior exposure to a specific pathogen so that the system can recognize, react, and remember the pathogen to fight it off faster and stronger if the pathogen ever occurs in the body again.

spermatic cord: a composite of several structures, which include the cremaster muscle, pampiniform plexus, testicular artery, and ductus deferens.

spermatogenesis: the first process in sperm production; produces spermatids (each with 23 chromosomes) from a specialized stem cell (germ cell) with 46 chromosomes.

spermiogenesis: the second process in sperm production; transforms spermatids to functional sperm.

spirometry: measurement of lung volumes and capacities.

static equilibrium: equilibrium perceived when the head is stationary or moving in a straight line.

stratum basale: the deepest layer of the epidermis and the only one with cells that actively grow and divide to produce new epidermis.

stroke volume: the amount of blood ejected from each ventricle per beat.

subcutaneous layer: the layer, technically not part of skin, that is deep to the dermis and attaches skin to the rest of the body. Also called *hypodermis*.

surfactant: fluid secreted by great alveolar cells in the lungs that reduces the surface tension of water.

suture: fibrous joint found between cranial bones of the skull.
sweat glands: four types of exocrine glands—merocrine, apocrine, ceruminous, and mammary—that are located in the dermis.
sympathetic: division of the autonomic nervous system that sends electrical messages to prepare the body for physical activity often referred to as *fight or flight*.
symphysis: cartilaginous joint found between the two pubic bones.
synapse: junction formed by the neuron's synaptic knob with another cell—gland cell, muscle cell, or dendrite of another neuron.
synchondrosis: cartilaginous joint found between the epiphysis and diaphysis of long bones in children.
syndesmosis: fibrous joint formed by an interosseus membrane between the radius and the ulna or between the tibia and the fibula.
synergists: muscles that have the same action.
synovial joint: class of joint characterized by a joint capsule lined by a synovial membrane that produces synovial fluid.
synovial membrane: membrane lining the joint space of a synovial joint.
systole: contraction of a heart chamber; usually refers to the action of the ventricles.

t

target tissue: cells of a tissue that have receptors for a specific hormone.
telomeres: sequences of nucleotides that provide a protective cap on the ends of chromosomes.
tetany: a sustained contraction brought about by a high frequency of nerve impulses.
thick skin: epidermis that contains stratum lucidum and lacks hair follicles.
thin skin: epidermis that contains hair follicles and lacks stratum lucidum.
third-degree burn: a burn involving the epidermis, dermis, and hypodermis. Symptoms include charring and no pain due to the destruction of nerve endings.
thoracic: cavity in the superior trunk; commonly called *chest*.
threshold: minimum amount of a neurotransmitter necessary to stimulate a response.
thrombocyte: one type of formed element of the blood; commonly called *platelet*.
thrombus: a stationary, unwanted clot.
thyroid hormone: collective term for both T_3 and T_4.
tidal volume (TV): the amount of air moved in a normal breath (inspired or expired) at rest.
total lung capacity: the maximum amount of air the lungs can hold.
trabeculae: sliverlike or platelike arrangement of bone connective tissue found in cancellous bone.
tunics: layers of vessel walls.

u

umami: a meaty taste derived from some amino acids binding to taste hairs.
unipolar: type of sensory neuron that appears to have one process that serves as dendrite and axon with the cell body pushed off to the side.
up-regulation: an increase in the number of receptors for a given hormone, causing the cell to become more sensitive to the hormone.

venous return: the process of returning blood to the heart through veins.
ventilation: airflow to the lungs.
ventilation-perfusion coupling: the matching of airflow to blood flow in the lungs.
vestibular apparatus: anatomy used to perceive equilibrium; includes the saccule, utricle, and semicircular canals of the inner ear.
visceral: the part of a serous membrane in direct contact with an organ.
visual acuity: ability to discern visual detail. The best visual acuity is the sharpest vision.
vital capacity (VC): the maximum amount of air that can be moved.

wound contracture: scab formation that pulls the edges of the wound closer together as it dries.

z

zygote: a fertilized egg.

photo credits

About the Author
Page v: © 2009, Champa Studios.

Walkthrough

Page viii: © Image Source/Corbis RF; p. x(top right): © The McGraw-Hill Companies, Inc.; p. ix(right): © Footage supplied by Goodshoot/PunchStock RF; p. x: © Stockbyte/Getty Images RF.

Contents

Chapter 1: © Stuart Gregory/Getty Images RF; 2: © Erik Isakson/Getty Images RF; 3: © Image Source/Getty Images RF; 4: © Ingram Publishing/SuperStock RF; 5: © Photodisc/Getty Images RF; 6: © Ingram Publishing RF; 7: © RubberBall Productions RF; 8: © Jeff Von Hoene/Getty Images; 9: © Science Photo Library RF/Getty Images; 10: © IAN HOOTON/SPL/Getty Images RF; 11: © Fuse/Getty Images RF; 12: © Stephen Marks/Getty Images; 13: © Stockbyte/Getty Images RF; 14: © Dennis O'Clair/Getty Images; 15: © Nancy R. Cohen/Getty Images RF; 16: © Alex Mares-Manton/Getty Images.

Chapter 1

Opener: © Stuart Gregory/Getty Images RF; p. 3: © maxuser/iStockphoto; 1.1: © The McGraw-Hill Companies, Inc./Joe De Grandis, photographer; 1.2b: © The McGraw-Hill Companies, Inc./Rebecca Gray, photographer/Don Kincaid, dissections; 1.5: © The McGraw-Hill Companies, Inc./Joe De Grandis, photographer; p. 8: © maxuser/iStockphoto; 1.7a, (bottom left): © McGraw-Hill Higher Education, Inc./Eric Wise, photographer; 1.7b–d: © R. T Hutchings/Visuals Unlimited; p. 11, 13: © maxuser/iStockphoto; 1.12: © The McGraw-Hill Companies, Inc./Rebecca Gray, photographer/Don Kincaid, dissections; p. 14: © maxuser/iStockphoto; 1.14: © Don Farrall/Getty Images RF.

Chapter 2

Opener: © Erik Isakson/Getty Images RF; p. 21: © maxuser/iStockphoto; 2.1: © Mark Andersen/Getty Images RF; p. 29, 37: © maxuser/iStockphoto; 2.15: © Deborah Ann Roiger; p. 42: © maxuser/iStockphoto; 2.18a–c: © Dr. David M. Phillips/Visuals Unlimited; p. 48: © maxuser/iStockphoto; 2.23: © Ed Reschke; 2.25: © Dr. Hesed M. Padilla-Nash, Ph.D., National Cancer Institute, National Institutes of Health, Bethesda, MD USA; p. 51(bottom right): © maxuser/iStockphoto; 2.27: © Ed Reschke; 2.28–2.32b: © Victor Eroschenko; p. 56(top right): © maxuser/iStockphoto; 2.33–2.35: © The McGraw-Hill Companies, Inc./Dennis Strete, photographer; 2.36–2.39: © The McGraw-Hill Companies, Inc./Al Telser, photographer; 2.40: © The McGraw-Hill Companies, Inc./Dennis Strete, photographer; 2.41: © Ed Reschke; 2.42: © Victor Eroschenko; 2.43: © Ed Reschke; 2.44: © Trent Stephens; p. 62: © maxuser/iStockphoto; 2.45b: © The McGraw-Hill Companies, Inc.; Insert: © Image Source/Getty Images RF.

Chapter 3

Opener: © Image Source/Getty Images RF; p. 75: © maxuser/iStockphoto; 3.3: © Photodisc Collection/Getty Images RF; p. 77(center): © Stockbyte/Getty Images RF; pp. 77(right)–79(top): © maxuser/iStockphoto; 3.6a: © The McGraw-Hill Companies, Inc./Dennis Strete, photographer; 3.6a(inset): © Tom & Dee Ann McCarthy/CORBIS; 3.6b: © The McGraw-Hill Companies, Inc./Dennis Strete, photographer; 3.6b(inset): © Creatas/PunchStock RF; p. 81: © maxuser/iStockphoto; 3.8: © BananaStock/PunchStock RF; p. 83(top right): © maxuser/iStockphoto; 3.9b: © CBS/Phototake; 3.10a–d: © The McGraw-Hill Companies, Inc./Joe De Grandis, photographer; p. 85: © maxuser/iStockphoto; 3.12: © Image Source/Corbis RF; p. 92: © Getty Images RF; 3.15a: © SPL/Custom Medical Stock Photo; 3.15b–c: © John Radcliffe/Photo Researchers, Inc.; 3.17a: © NMSB/Custom Medical Stock Photo; 3.17b: © Biophoto Associates/Photo Researchers, Inc.; 3.17c: © James Stevenson/SPL/Photo Researchers, Inc.; 3.18: Centers for Disease Control; 3.19: Centers for Disease Control/Dr. Lucille K. Georg.

Chapter 4

Opener: © Ingram Publishing/SuperStock RF; pp. 105–106: © maxuser/iStockphoto; 4.3–4.8b: © The McGraw-Hill Companies, Inc./Christine Eckel, photographer; p. 111(bottom): © maxuser/iStockphoto; 4.10b–4.12b: © The McGraw-Hill Companies, Inc./Christine Eckel, photographer; pp. 113–114: © maxuser/iStockphoto; 4.14: © Ryan McVay/Getty Images RF; p. 117: © maxuser/iStockphoto; 4.19a–4.22: © The McGraw-Hill Companies, Inc./Christine Eckel, photographer; p. 119(bottom): © Stockbyte Platinum/Alamy RF; p. 119(bottom right): © maxuser/iStockphoto; 4.25a–b: © The McGraw-Hill Companies, Inc./Christine Eckel, photographer; p. 121: © maxuser/iStockphoto; 4.27c–e: © The McGraw-Hill Companies, Inc./Christine Eckel, photographer; pp. 123–125: © maxuser/iStockphoto; 4.30: © RNHRD NHS Trust/Getty Images; 4.31b–d: © The McGraw-Hill Companies, Inc./Christine Eckel, photographer; pp. 126–127: © maxuser/iStockphoto; 4.34c: © The McGraw-Hill Companies, Inc./Christine Eckel, photographer; p. 128: © maxuser/iStockphoto; 4.35b–4.36b: © The McGraw-Hill Companies, Inc./Christine Eckel, photographer; p. 129: © maxuser/iStockphoto; 4.37b: © Walter Reiter/Phototake; 4.38a: © Dr. Don W. Fawcett/Visuals Unlimited; 4.38c: © Dr. Richard Kessel & Dr. Randy Kardon/Visuals Unlimited; pp. 133–139: © maxuser/iStockphoto; 4.43a: © Southern Illinois University/Photo Researchers, Inc.; 4.43b: © CNRI/Science Photo Library/Photo Researchers, Inc.; 4.44a: © SIU/Visuals Unlimited; 4.44b: © Ron Mensching/Phototake; 4.44c: © SIU/Peter Arnold Images/Photolibrary Group; 4.44d: © Mehau Kulyk/SPL/Photo Researchers, Inc.; 4.45: © Brand X Pictures/Jupiterimages RF; 4.46b: © TRBfoto/Getty Images RF; p. 146: © maxuser/iStockphoto; 4.48: © Bananastock/PictureQuest RF; 4.49: © Jim Wehtje/Getty Images RF; 4.50: © Biophoto Associates/Photo Researchers, Inc.; 4.51: © Brand X Pictures/Jupiterimages RF; p. 149: © maxuser/iStockphoto; 4.52: © C Squared Studios/Getty Images RF; 4.54: © Don Farrall/Getty Images RF; 4.55a: © Custom Medical Stock Photo; 4.55b: © Howard Kingsnorth/Getty Images; 4.55c: © Lester V. Bergman/CORBIS; 4.55d: © Custom Medical Stock Photo; 4.56: © SIU/Visuals Unlimited; p. 156: © maxuser/iStockphoto; 4.58a: © Michael Klein/Peter Arnold Images/Photolibrary Group; 4.58b: © Dr. P. Marazzi/Photo Researchers, Inc.; 4.58c: © Yoav Levy/Phototake.

Chapter 5

Opener: © Photodisc/Getty Images RF; pp. 167–168: © maxuser/iStockphoto; 5.2a–5.9b: © The McGraw-Hill Companies, Inc./Timothy L. Vacula, photographer p. 173: © maxuser/iStockphoto; 5.12c: © The McGraw-Hill Companies, Inc./Rebecca Gray, photographer/Don Kincaid, dissections; pp. 176–177: © maxuser/iStockphoto; 5.13c: © Ralph Hutchings/Visuals Unlimited; p. 180: © maxuser/iStockphoto; 5.14b: © The McGraw-Hill Companies, Inc./Rebecca Gray, photographer/Don Kincaid, dissections; p. 181: © maxuser/iStockphoto; 5.15b: © The McGraw-Hill Companies, Inc./Rebecca Gray, photographer/Don Kincaid, dissections; pp. 182–184: © maxuser/iStockphoto; 5.18b: © The McGraw-Hill Companies, Inc./Rebecca Gray, photographer/Don Kincaid, dissections; pp. 185–187: © maxuser/iStockphoto; 5.20c: © The McGraw-Hill Companies, Inc./Photo and dissection by Christine Eckel; p. 188: © maxuser/iStockphoto; 5.23a: © H.E. Huxley; 5.25b: © McGraw-Hill Higher Education, Inc./Carol D. Jacobson Ph.D., Dept. Veterinary Anatomy, Iowa State University; p. 192: © maxuser/iStockphoto; p. 202(left): © The McGraw-Hill Companies, Inc.; p. 202(right): © Ryan McVay/Getty Images RF; 5.31: © Brand X Pictures/Jupiterimages RF.

Chapter 6

Opener: © Ingram Publishing RF; pp. 215–226: © maxuser/iStockphoto; 6.8c–f: © The McGraw-Hill Companies, Inc./Photo and dissection by Christine Eckel; p. 230: © maxuser/iStockphoto; p. 233(center): © Stockbyte/Getty Images RF; p. 233(bottom): © Dynamic Graphics/Jupiterimages RF; 6.17: © The McGraw-Hill Companies, Inc./Photo and dissection by Christine Eckel; p. 248: © Ingram Publishing/SuperStock RF; p. 250: © maxuser/iStockphoto; 6.26: © The McGraw-Hill Companies, Inc./Suzie Ross, photographer.

Chapter 7

Opener: © RubberBall Productions RF; pp. 263–270: © maxuser/iStockphoto; p. 273: © Royalty-Free/CORBIS; p. 278: © maxuser/iStockphoto; 7.16a–b: © Trent Stephens; 7.17d: © Jerry Wachter/Photo Researchers, Inc.; 7.18: © The McGraw-Hill Companies, Inc./Joe De Grandis, photographer; pp. 282–284: © maxuser/iStockphoto; 7.24a: © Mediscan/CORBIS; p. 288: © maxuser/iStockphoto.

Chapter 8

Opener: © Jeff Von Hoene/Getty Images; pp. 299–308: © maxuser/iStockphoto; 8.9: © Ian Hooton/SPL/Getty Images RF; p. 315: © C Squared Studios/Getty Images RF; 8.16: © The McGraw-Hill Companies, Inc./Jill Braaten, photographer; 8.17: Reprinted by permission of publisher from Albert Mendeloff, "Acromegaly, diabetes, hypermetabolism, proteinura and heart failure," *American Journal of Medicine, 20:1, 01-1956*, p. 135 © Elsevier, Inc.; 8.18: © CNR/Phototake.

Chapter 9

Opener: © Science Photo Library RF/Getty Images; p. 331: © maxuser/iStockphoto; 9.2: © liquidlibrary/PictureQuest RF; p. 334: © maxuser/iStockphoto; pp. 335–336: © The McGraw-Hill Companies, Inc./Al Telser, photographer; pp. 336(bottom)–340: © maxuser/iStockphoto; 9.6: © Meckes/Ottawa/Photo Researchers, Inc.; p. 344(center): Centers for Disease Control; p. 344(bottom): © maxuser/iStockphoto; 9.10: © Science Photo Library RF/Getty Images; 9.12c: © George W. Wilder/Visuals Unlimited; 9.13: © Claude Revey/Phototake; 9.15: © Bellurget Jean Louis/Jupiterimages/Getty Images RF; p. 353: © Corbis Images/Jupiterimages RF; pp. 354–355: © maxuser/iStockphoto.

Chapter 10

Opener: © IAN HOOTON/SPL/Getty Images RF; pp. 365–368: © maxuser/iStockphoto; 10.4: Photo and illustration by Roy Schneider, University of Toledo. Plastinated heart model for illustration courtesy of Dr. Carlos Baptista, University of Toledo; p. 369(bottom): © maxuser/iStockphoto; 10.5c–10.6b: © The McGraw-Hill Companies, Inc.; p. 372(left): © maxuser/iStockphoto; 10.7: © Ed Reschke; p. 373: © liquidlibrary/PictureQuest RF; p. 375: © Mike Kemp/Getty Images RF; p. 377: © maxuser/iStockphoto; p. 380(top): © Footage supplied by Goodshoot/PunchStock RF; pp. 380(left)–384: © maxuser/iStockphoto; 10.19c: © The McGraw-Hill Companies, Inc./Al Telser, photographer; 10.21c: © Ralph Hutchings/Visuals Unlimited; pp. 393–396: © maxuser/iStockphoto; p. 397: © Don Farrall/Getty Images RF; 10.26: © Image Source/Corbis RF; 10.27a–b: © Ed Reschke; p. 401(right): © maxuser/iStockphoto.

Chapter 11

Opener: © Fuse/Getty Images RF; p. 411: © maxuser/iStockphoto; 11.3a: © The McGraw-Hill Companies, Inc./Dennis Strete, photographer; 11.6: © SPL/Custom Medical Stock Photo; 11.8b: © The McGraw-Hill Companies, Inc./Al Telser, photographer; 11.8c: © Dr. Kent M. Van De Graaff; 11.9: © The McGraw-Hill Companies, Inc./Rebecca Gray, photographer/Don Kincaid, dissections; p. 420: © maxuser/iStockphoto; 11.11b: © Biophoto Associates/Photo Researchers, Inc.; 11.13a: © The McGraw-Hill Companies, Inc./Dennis Strete, photographer; 11.13c: © The McGraw-Hill Companies, Inc./Al Telser, photographer; 11.20a–b: © Dr. Andrejs Liepins/Photo Researchers, Inc.; 11.22: © Bellurget Jean Louis/Jupiterimages/Getty Images RF; p. 434(right): © Creatas/PunchStock RF; p. 434(left): © maxuser/iStockphoto; 11.23b: © NIBSC/SPL/Photo Researchers, Inc.; 11.24: © Roger Ressmeyer/Corbis.

Chapter 12

Opener: © Stephen Marks/Getty Images; pp. 445–447(right): © maxuser/iStockphoto; 12.3: © Digital Vision/Getty Images RF; 12.4a: © The McGraw-Hill Companies, Inc./Joe De Grandis, photographer p. 449(right): © maxuser/iStockphoto; p. 449(bottom): © Richard Hutchings; p. 450(right): © Ingram Publishing RF; pp. 450(left)–451: © maxuser/iStockphoto; 12.7: © J Siebert/Custom Medical Stock Photo; 12.9: Custom Medical Stock Photo; p. 453(bottom): © Royalty Free/Corbis; p. 454: © maxuser/iStockphoto; 12.10c: © Walter Reiter/Phototake; 12.11: © Ralph Hutchings/Visuals Unlimited; 12.13: © Dr. Gladden Willis/Visuals Unlimited; p. 458: © Brand X Pictures/PunchStock RF; 12.16: © BSIP/Phototake; 12.21: © Brand X Pictures/PunchStock RF; p. 472: © Hisham F. Ibrahim/Getty Images RF; 12.27: © Laurent Delhourme/Getty Images; 12.28a: © The McGraw-Hill Companies, Inc./Dennis Strete, photographer; 12.28b: © Biophoto Associates/Photo Researchers, Inc.

Chapter 13

Opener: © Stockbyte/Getty Images RF; pp. 487–490: © maxuser/iStockphoto; 13.6: © The McGraw-Hill Companies, Inc./Rebecca Gray, photographer/Don Kincaid, dissections; p. 494: © maxuser/iStockphoto; 13.11b: © The McGraw-Hill Companies, Inc./Jim Shaffer, photographer; p. 496: © maxuser/iStockphoto; 13.12b: © The McGraw-Hill Companies, Inc./Rebecca Gray, photographer/Don Kincaid, dissections; 13.12c: © Dr. Kent M. Van De Graaff; p. 497(center): © George Diebold/Getty Images; pp. 497(bottom)–500: © maxuser/iStockphoto; 13.17a(top right): © The McGraw-Hill Companies, Inc./Ken Cavanagh, photographer; 13.17a(left): © Dorling Kindersley/Getty Images; 13.17b: © Mediscan/Visuals Unlimited; p. 504: © maxuser/iStockphoto; 13.19a: © Meckes/Ottawa/Photo Researchers, Inc.; pp. 506–510: © maxuser/iStockphoto; 13.24: © Stockbyte/PictureQuest RF; 13.25: Centers for Disease Control; 13.26: CDC/James Gathany; 13.27b: © Courtesy of Dr. Jill Urban, Dallas County Medical Examiner's Office, Dallas, TX.

Chapter 14

Opener: © Dennis O'Clair/Getty Images; p. 527: © maxuser/iStockphoto; 14.5b: © Ralph Hutchings/Visuals Unlimited; pp. 532–533: © maxuser/iStockphoto; p. 536: © Stockbyte/Getty Images RF; p. 538: © Jeffrey Coolidge/Getty Images RF; p. 542: © maxuser/iStockphoto; 14.17b: © The McGraw-Hill Companies, Inc./Photo and dissection by Christine Eckel; pp. 546–547: © maxuser/iStockphoto; 14.21: © BananaStock/PunchStock RF.

Chapter 15

Opener: © Nancy R. Cohen/Getty Images RF; p. 561: © maxuser/iStockphoto; 15.3: From Bartalos, M., and Baramki, T.A., *Medical Cytogenetics*, © Lippincott Williams & Wilkins, 1967; p. 563(bottom): © maxuser/iStockphoto; p. 564: © LaCoppola-Meier/Getty Images; p. 564(bottom): © maxuser/iStockphoto; 15.5a: © The McGraw-Hill Companies, Inc./Dennis Strete, photographer; pp. 565–568: © maxuser/iStockphoto; 15.11a: © Visuals Unlimited; 15.15a: Copyright by R.G. Kessel and R.H. Kardon, *Tissues and Organs: A Text-Atlas of Scanning Electron Microscopy*, 1979, W.H. Freeman. All rights reserved; 15.15b: © Ed Reschke; p. 581: © maxuser/iStockphoto.

Chapter 16

Opener: © Alex Mares-Manton/Getty Images; p. 591: © maxuser/iStockphoto; 16.3b: © The McGraw-Hill Companies, Inc./Rebecca Gray, photographer/Don Kincaid, dissections; p. 595: © maxuser/iStockphoto; 16.5: © SPL/Photo Researchers, Inc.; 16.7b: From *Anatomy & Physiology Revealed*, © The McGraw-Hill Companies, Inc./The University of Toledo, photography and dissection; 16.7d: © Dr. Donald Fawcett, T. Nagato/Visuals Unlimited; p. 600: © maxuser/iStockphoto; p. 604: © Tom Grill/Getty Images; 16.11: © Landrum B. Shettles, MD; p. 606: © Photodisc Collection/Getty Images RF; 16.14a: © Jeff Kaufman/Getty Images; 16.14b: © UHB Trust/Getty Images; 16.14c–d: © Biophoto Associates/Photo Researchers, Inc.; 16.15: © Digital Vision RF; p. 617: © maxuser/iStockphoto; 16.21a–b: © Petit Format/Photo Researchers, Inc.; 16.21c: © Jason Edwards/Getty Images RF; 16.22: © Bananastock/PictureQuest RF.

index

a

b

C

d

e

f

g

h

i

j

n

O

P

q

r

S

t

U

V

W

X

Y

Z

MEDICAL TERMINOLOGY: PREFIXES, SUFFIXES, AND ADDITIONAL WORD ROOTS

Anatomy and physiology and all health care fields rely on language that includes medical terminology. Word roots with their definitions are listed in each of the system chapters of this text. Here, common prefixes, suffixes, and additional word roots are listed with their definitions. Some of the word roots listed in system chapters are repeated here if they might be used to answer chapter review questions outside the system chapters in which the word roots were introduced.

Term	Definition
a-, an-	no; without; lack of
ab-	away from
-able	capable
acr-	extremity
-ad	toward
-al	pertaining to
-algia	pain
-an	pertaining to
-ant	thing that promotes; thing acted upon
anti-	against; reversed
-ar	pertaining to
arthr/o-	joint
-ary	associated with
-asthenia	lack of strength
auto-	self
bi-	two
bi/o	life
carcin/o	cancerous; cancer
cardi/o	heart
-cele	hernia
-centesis	surgical puncture to remove fluid
-ceptor	taker; receiver
chem/o	drug; chemical
chron/o	time
-cide	killing
-clast	to break
con-	together with
-constriction	narrowing
-crit	separate
cyt/o	cell
-cyte	cell
de-	away from
dia-	through
-dilation	widening; expanding
dist/o	far; distant
-duct-	leading
-dynia	pain
dys-	bad; abnormal
-eal	pertaining to
ec-	out; outside
-ectomy	removal; cut out
-edema	swelling
-emia	blood condition
endo-	in; within
epi-	above; upon; on
erythr/o	red
eti/o	cause
eu-	good; normal
ex-	out; away from
extra-	outside
-ferent	to carry
fibr/o	fiber
-form	resembling
-genesis	origin
ger/o	old age
-graphy	process of recording
hem/o	blood
hemi-	half
home/o	same
hydr/o	water
hyper-	above; excessive
hypo-	under; deficient
-ia	condition
-ic	pertaining to
-ile	pertaining to
-im	not
-in	into; in
inter-	between
intra-	within; inside
-ion	process
-ism	process; condition
-itis	inflammation
-kinesia	movement
lapar/o	abdominal wall; abdomen
-lapse	slide; fall
-logist	person who studies or is an expert in the field
-logy	study of
-lysis	breakdown; destruction
macro-	large
mal-	bad
medi/o	middle
-megaly	enlargement
meta-	change; beyond
-metry	process of measuring
mort/o	death
mut/a	genetic change
my/o	muscle
necr/o	death
neo-	new
-oid	resembling; derived from
-oma	tumor; mass; fluid collection
onc/o	tumor
-opia	vision condition
-or	one who
-ose	full of; pertaining to; sugar
-osis	condition, usually abnormal
-ous	pertaining to
ox/o	oxygen
par-	other than; abnormal
para-	near; beside
-paresis	weakness
-pathy	disease
ped/o	child; foot
-penia	deficiency
-pepsia	digestion
per-	through
peri-	surrounding
-phage	eat; swallow
-pheresis	removal
-plasia	development; formation; growth
-plasm	formation; structure
-poiesis	formation
poly-	many; much
post-	after; behind
pre-	before
pro-	before; forward
pseudo-	false
re-	back; again
retro-	behind; back
-rrhea	flow; discharge
-scope	instrument for visual examination
-scopy	visual examination
semi-	half
-sis	state of; condition
-stenosis	tightening
-stomy	new opening
sub-	under; below
-suppressant	agent that reduces or diminishes activity
sym-	together; with
-tic	pertaining to
-tomy	process of cutting
-tory	pertaining to
trans-	across; through
-trophy	nourishment; development (condition of)
-tropin	stimulate; act on
-um	structure; tissue
uni-	one
-us	structure; thing
-verse	to turn
-y	condition; process